CONTENTS

KIRK-OTHMER

ENCYCLOPEDIA OF
CHEMICAL
TECHNOLOGY

FOURTH EDITION

VOLUME **12**

FUEL RESOURCES
TO
HEAT STABILIZERS

EXECUTIVE EDITOR
Jacqueline I. Kroschwitz

EDITOR
Mary Howe-Grant

KIRK-OTHMER

ENCYCLOPEDIA OF CHEMICAL TECHNOLOGY

FOURTH EDITION

VOLUME 12

FUEL RESOURCES
TO
HEAT STABILIZERS

A Wiley-Interscience Publication
JOHN WILEY & SONS
New York • Chichester • Brisbane • Toronto • Singapore

Library of Congress Cataloging-in-Publication Data

Encyclopedia of chemical technology / executive editor, Jacqueline
    I. Kroschwitz; editor, Mary Howe-Grant.—4th ed.
        p. cm.
    At head of title: Kirk-Othmer.
    "A Wiley-Interscience publication."
    Includes index.
    Contents: v. 12, Fuel resources to heat stabilizers
    ISBN 0-471-52681-9 (v. 12)
    1. Chemistry, Technical—Encyclopedias.   I. Kirk, Raymond E.
    (Raymond Eller), 1890–1957.   II. Othmer, Donald F. (Donald
    Frederick), 1904– .   III. Kroschwitz, Jacqueline I., 1942– .
    IV. Howe-Grant, Mary, 1943– .   V. Title: Kirk-Othmer encyclopedia
    of chemical technology.
    TP9.E685   1992   91-16789
    660′.03—dc20

EDITORIAL STAFF
FOR VOLUME 12

Executive Editor: **Jacqueline I. Kroschwitz**
Editor: **Mary Howe-Grant**
Associate Managing Editor: **Lindy J. Humphreys**
Assistant Editor: **Cathleen A. Treacy**
Copy Editors: **Christine Punzo**
             Lawrence Altieri

CONTRIBUTORS
TO VOLUME 12

Jack H. Adams, *Eagle-Picher Industries, Inc., Quapaw, Oklahoma,* Germanium and germanium compounds

Tejraj M. Aminabhavi, *Southwest Texas State University, San Marcos, Texas,* Heat-resistant polymers

John K. Baird, *Kelco, Division of Merck & Company, Inc., San Diego, California,* Gums

Donald H. Beermann, *Cornell University, Ithaca, New York,* Animal (under Growth regulators)

Walter B. Bienert, *Dynatherm Corporation, Lancaster, Pennsylvania,* Heat pipes (under Heat-exchange technology)

David C. Boyd, *Corning Inc., Corning, New York,* Glass

Carl R. Bozzuto, *ABB Kreisinger Development Laboratory, Windsor, Connecticut,* Furnaces, fuel-fired

A.B. Brennan, *University of Florida, Gainesville,* Glasses, organic–inorganic hybrids

Patrick E. Cassidy, *Southwest Texas State University, San Marcos, Texas,* Heat-resistant polymers

S.M. Cho, *Foster Wheeler Energy Corporation, Clinton, New Jersey,* Heat transfer (under Heat-exchange technology)

Young I. Cho, *Drexel University, Philadelphia, Pennsylvania,* Heat transfer (under Heat-exchange technology)

J.G. Cohn, *Engelhard Corporation, Iselin, New Jersey,* Gold and gold compounds

Robert Collier, *Monsanto Company, St. Louis, Missouri,* Animals (under Genetic engineering)

J. Kevin Cotchen, *Man GHH Corporation, Pittsburgh, Pennsylvania,* Arc furnaces (under Furnaces, electric)

Clarence D. Chang, *Mobil Research & Development Corporation, Princeton, New Jersey,* Liquid fuels (under Fuels, synthetic)

Ronald D. Crooks, *Consultant, Branderis Associates, Inc., Hockessin, Delaware,* Hardness

Daniel A. Crowl, *Michigan Technological University, Houghton,* Hazard analysis and risk assessment

Horace G. Cutler, *USDA/ARS, Athens, Georgia,* Plant (under Growth regulators)

Paul S. Danielson, *Corning Inc., Corning, New York,* Glass

Barry A. Dreikorn, *Dow Elanco, Indianapolis, Indiana,* Fungicides, agricultural

David Duchane, *Los Alamos National Laboratory, New Mexico,* Geothermal energy

G. Yale Eastman, *DTX Corporation, Lancaster, Pennsylvania,* Heat pipes (under Heat-exchange technology)

Janice W. Edwards, *Monsanto Company, Chesterfield, Missouri,* Plants (under Genetic engineering)

William R. Ellis, *Raytheon Engineers and Constructors, New York, New York,* Fusion energy

Donald M. Ernst, *Thermacore Inc., Lancaster, Pennsylvania,* Heat pipes (under Heat-exchange technology)

M.W. Forkner, *Union Carbide Corporation, South Charleston, West Virginia,* Ethylene glycol and oligomers (under Glycols)

Myron Gottlieb, *Gas Research Institute, Chicago, Illinois,* Gas, natural

Scott Han, *Mobil Research & Development Corporation, Princeton, New Jersey,* Liquid fuels (under Fuels, synthetic)

Albert M. Hochhauser, *Exxon Research and Engineering Company, Linden, New Jersey,* Gasoline and other motor fuels

Edward Hohmann, *California State Polytechnic University, Pomona,* Network synthesis (under Heat-exchange technology)

Henry I. Jacoby, *Discovery Research Consultants, Rydal, Pennsylvania,* Gastrointestinal agents

Thomas R. Keenan, *Kind & Knox Gelatine, Inc., Sioux City, Iowa,* Gelatin

Ganesh M. Kishore, *Monsanto Company, Chesterfield, Missouri,* Plants (under Genetic engineering)

Donald L. Klass, *Entech International, Inc., Barryington, Illinois,* Fuels from biomass

Hubert Lauvray, *Rhône-Poulenc, France,* Gallium and gallium compounds

Matthew Lucy, *University of Missouri, Columbia,* Animals (under Genetic engineering)

Alton E. Martin, *The Dow Chemical Company, Freeport, Texas,* Propylene glycols (under Glycols)

Keith Mesch, *Morton International, Inc., Cincinnati, Ohio,* Heat stabilizers

Carol J. Miller, *Wayne State University, Detroit, Michigan,* Groundwater monitoring

T.M. Miller, *University of Florida, Gainesville,* Glasses, organic–inorganic hybrids

Paul E. Minton, *Union Carbide Corporation, South Charleston, West Virginia,* Heat-transfer media other than water (under Heat-exchange technology)

Lowen Morrison, *Procter & Gamble, Cincinnati, Ohio,* Glycerol

Frank H. Murphy, *The Dow Chemical Company, Freeport, Texas,* Propylene glycols (under Glycols)

Vasantha Nagarajan, *E.I. du Pont de Nemours & Co., Inc., Wilmington, Delaware,* Microbes (under Genetic engineering)

Kurt Nassau, *Nassau Consultants, Lebanon, New Jersey,* Gemstone materials; Gemstone treatment (both under Gemstones)

Ralph H. Nielsen, *Teledyne Wah Chang Corporation, Albany, Oregon,* Hafnium and hafnium compounds

W. John Owen, *Dow Elanco Europe, Wantage, Oxon, United Kingdom,* Fungicides, agricultural

T.E. Parsons, *Eastman Chemical Company, Kingsport, Tennessee,* Other glycols (under Glycols)

Linda R. Pinckney, *Corning, Inc., Corning, New York,* Glass-ceramics

Stanley Pohl, *Clairol Inc., Stamford, Connecticut,* Hair preparations

David E. Prinzing, *Enserch, Sacramento, California,* Fuel resources

Philip E. Rakita, *Elf Atochem Japan, Chuo-ku, Tokyo,* Grignard reactions

V. Sreenivasulu Reddy, *Southwest Texas State University, San Marcos, Texas,* Heat-resistant polymers

John H. Robson, *Union Carbide Corporation, South Charleston, West Virginia,* Ethylene glycol and oligomers (under Glycols)

Jean Louis Sabot, *Rhône-Poulenc, Shelton, Connecticut,* Gallium and gallium compounds

Francis J. Schmidt, *University of Missouri, Columbia,* Procedures (under Genetic engineering)

Gary J. Shiflet, *University of Virginia, Charlottesville,* Glassy metals

Gary S. Silverman, *Atochem North America, King of Prussia, Pennsylvania,* Grignard reactions

W.M. Snellings, *Union Carbide Corporation, South Charleston, West Virginia,* Ethylene glycol and oligomers (under Glycols)

Richard A. Sommer, *Consultant, Warren, Ohio,* Introduction; Induction furnaces (both under Furnaces, electric)

James G. Speight, *Laramie, Wyoming,* Gaseous fuels (under Fuels, synthetic)

Eric W. Stern, *Engelhard Corporation, Iselin, New Jersey,* Gold and gold compounds

David M. Stark, *Monsanto Company, Chesterfield, Missouri,* Plants (under Genetic engineering)

L. David Suits, *New York State Department of Transportation, Albany,* Geotextiles

Dennis Thomas, *Eagle-Picher Industries, Inc., Quapaw, Oklahoma,* Germanium and germanium compounds

David A. Thompson, *Corning, Inc., Corning, New York,* Glass

David A. Tillman, *Enserch, Sacramento, California,* Fuel resources; Fuels from waste

Joseph Varco, *Clairol Inc., Stamford, Connecticut,* Hair preparations

Paul Wallace, *Clairol Inc., Stamford, Connecticut,* Hair preparations

Robert R. Walton, *Wellman Furnaces, Inc., Shelbyville, Tennessee,* Resistance furnaces (under Furnaces, electric)

Jeffrey Warshauer, *Enserch, Sacramento, California,* Fuel resources

Leszek J. Wolfram, *Clairol Inc., Stamford, Connecticut,* Hair preparations

Kermit E. Woodcock, *Consultant, Gowanda, New York,* Gas, natural

NOTE ON CHEMICAL ABSTRACTS SERVICE REGISTRY NUMBERS AND NOMENCLATURE

Chemical Abstracts Service (CAS) Registry Numbers are unique numerical identifiers assigned to substances recorded in the CAS Registry System. They appear in brackets in the *Chemical Abstracts* (CA) substance and formula indexes following the names of compounds. A single compound may have synonyms in the chemical literature. A simple compound like phenethylamine can be named β-phenylethylamine or, as in *Chemical Abstracts*, benzeneethanamine. The usefulness of the *Encyclopedia* depends on accessibility through the most common correct name of a substance. Because of this diversity in nomenclature careful attention has been given to the problem in order to assist the reader as much as possible, especially in locating the systematic CA index name by means of the Registry Number. For this purpose, the reader may refer to the CAS Registry Handbook—Number Section which lists in numerical order the Registry Number with the *Chemical Abstracts* index name and the molecular formula; eg, **458-88-8**, Piperidine, 2-propyl-, (*S*)-, $C_8H_{17}N$; in the *Encyclopedia* this compound would be found under its common name, coniine [*458-88-8*]. Alternatively, this information can be retrieved electronically from CAS Online. In many cases molecular formulas have also been provided in the *Encyclopedia* text to facilitate electronic searching. The Registry Number is a valuable link for the reader in retrieving additional published information on substances and also as a point of access for on-line data bases.

In all cases, the CAS Registry Numbers have been given for title compounds in articles and for all compounds in the index. All specific substances indexed in *Chemical Abstracts* since 1965 are included in the CAS Registry System as are a large number of substances derived from a variety of reference works. The CAS Registry System identifies a substance on the basis of an unambiguous computer-language description of its molecular structure including stereochemical detail. The Registry Number is a machine-checkable number (like a Social Security number) assigned in sequential order to each substance as it enters the registry system. The value of the number lies in the fact that it is a concise and unique means of substance identification, which is independent of, and therefore bridges, many systems of chemical nomenclature. For polymers, one Registry Number may be used for the entire family; eg, polyoxyethylene (20) sorbitan monolaurate has the same number as all of its polyoxyethylene homologues.

Cross-references are inserted in the index for many common names and for some systematic names. Trademark names appear in the index. Names that are incorrect, misleading, or ambiguous are avoided. Formulas are given very frequently in the text to help in identifying compounds. The spelling and form used, even for industrial names, follow American chemical usage, but not always the usage of *Chemical Abstracts* (eg, *coniine* is used instead of *(S)-2-propylpiperidine*, *aniline* instead of *benzenamine*, and *acrylic acid* instead of *2-propenoic acid*).

There are variations in representation of rings in different disciplines. The dye industry does not designate aromaticity or double bonds in rings. All double bonds and aromaticity are shown in the *Encyclopedia* as a matter of course. For example, tetralin has an aromatic ring and a saturated ring and its structure

appears in the *Encyclopedia* with its common name, Registry Number enclosed in brackets, and parenthetical CA index name, ie, tetralin [*119-64-2*] (1,2,3,4-tetrahydronaphthalene). With names and structural formulas, and especially with CAS Registry Numbers, the aim is to help the reader have a concise means of substance identification.

CONVERSION FACTORS, ABBREVIATIONS, AND UNIT SYMBOLS

SI Units (Adopted 1960)

The International System of Units (abbreviated SI), is being implemented throughout the world. This measurement system is a modernized version of the MKSA (meter, kilogram, second, ampere) system, and its details are published and controlled by an international treaty organization (The International Bureau of Weights and Measures) (1).

SI units are divided into three classes:

BASE UNITS

length	meter[†] (m)
mass	kilogram (kg)
time	second (s)
electric current	ampere (A)
thermodynamic temperature[‡]	kelvin (K)
amount of substance	mole (mol)
luminous intensity	candela (cd)

SUPPLEMENTARY UNITS

plane angle	radian (rad)
solid angle	steradian (sr)

[†]The spellings "metre" and "litre" are preferred by ASTM; however, "-er" is used in the *Encyclopedia*.

[‡]Wide use is made of Celsius temperature (t) defined by

$$t = T - T_0$$

where T is the thermodynamic temperature, expressed in kelvin, and $T_0 = 273.15$ K by definition. A temperature interval may be expressed in degrees Celsius as well as in kelvin.

DERIVED UNITS AND OTHER ACCEPTABLE UNITS

These units are formed by combining base units, supplementary units, and other derived units (2–4). Those derived units having special names and symbols are marked with an asterisk in the list below.

Quantity	Unit	Symbol	Acceptable equivalent
*absorbed dose	gray	Gy	J/kg
acceleration	meter per second squared	m/s^2	
*activity (of a radionuclide)	becquerel	Bq	1/s
area	square kilometer	km^2	
	square hectometer	hm^2	ha (hectare)
	square meter	m^2	
concentration (of amount of substance)	mole per cubic meter	mol/m^3	
current density	ampere per square meter	$A//m^2$	
density, mass density	kilogram per cubic meter	kg/m^3	g/L; mg/cm^3
dipole moment (quantity)	coulomb meter	C·m	
*dose equivalent	sievert	Sv	J/kg
*electric capacitance	farad	F	C/V
*electric charge, quantity of electricity	coulomb	C	A·s
electric charge density	coulomb per cubic meter	C/m^3	
*electric conductance	siemens	S	A/V
electric field strength	volt per meter	V/m	
electric flux density	coulomb per square meter	C/m^2	
*electric potential, potential difference, electromotive force	volt	V	W/A
*electric resistance	ohm	Ω	V/A
*energy, work, quantity of heat	megajoule	MJ	
	kilojoule	kJ	
	joule	J	N·m
	electronvolt[†]	eV[†]	
	kilowatt-hour[†]	kW·h[†]	
energy density	joule per cubic meter	J/m^3	
*force	kilonewton	kN	
	newton	N	$kg·m/s^2$

[†]This non-SI unit is recognized by the CIPM as having to be retained because of practical importance or use in specialized fields (1).

Quantity	Unit	Symbol	Acceptable equivalent
*frequency	megahertz	MHz	
	hertz	Hz	1/s
heat capacity, entropy	joule per kelvin	J/K	
heat capacity (specific), specific entropy	joule per kilogram kelvin	J/(kg·K)	
heat transfer coefficient	watt per square meter kelvin	W/(m²·K)	
*illuminance	lux	lx	lm/m²
*inductance	henry	H	Wb/A
linear density	kilogram per meter	kg/m	
luminance	candela per square meter	cd/m²	
*luminous flux	lumen	lm	cd·sr
magnetic field strength	ampere per meter	A/m	
*magnetic flux	weber	Wb	V·s
*magnetic flux density	tesla	T	Wb/m²
molar energy	joule per mole	J/mol	
molar entropy, molar heat capacity	joule per mole kelvin	J/(mol·K)	
moment of force, torque	newton meter	N·m	
momentum	kilogram meter per second	kg·m/s	
permeability	henry per meter	H/m	
permittivity	farad per meter	F/m	
*power, heat flow rate, radiant flux	kilowatt	kW	
	watt	W	J/s
power density, heat flux density, irradiance	watt per square meter	W/m²	
*pressure, stress	megapascal	MPa	
	kilopascal	kPa	
	pascal	Pa	N/m²
sound level	decibel	dB	
specific energy	joule per kilogram	J/kg	
specific volume	cubic meter per kilogram	m³/kg	
surface tension	newton per meter	N/m	
thermal conductivity	watt per meter kelvin	W/(m·K)	
velocity	meter per second	m/s	
	kilometer per hour	km/h	
viscosity, dynamic	pascal second	Pa·s	
	millipascal second	mPa·s	
viscosity, kinematic	square meter per second	m²/s	
	square millimeter per second	mm²/s	

Quantity	Unit	Symbol	Acceptable equivalent
volume	cubic meter	m^3	
	cubic decimeter	dm^3	L (liter) (5)
	cubic centimeter	cm^3	mL
wave number	1 per meter	m^{-1}	
	1 per centimeter	cm^{-1}	

In addition, there are 16 prefixes used to indicate order of magnitude, as follows:

Multiplication factor	Prefix	Symbol	Note
10^{18}	exa	E	
10^{15}	peta	P	
10^{12}	tera	T	
10^9	giga	G	
10^6	mega	M	
10^3	kilo	k	
10^2	hecto	h^a	[a]Although hecto, deka, deci, and centi
10	deka	da^a	are SI prefixes, their use should be
10^{-1}	deci	d^a	avoided except for SI unit-multiples
10^{-2}	centi	c^a	for area and volume and nontech-
10^{-3}	milli	m	nical use of centimeter, as for body
10^{-6}	micro	μ	and clothing measurement.
10^{-9}	nano	n	
10^{-12}	pico	p	
10^{-15}	femto	f	
10^{-18}	atto	a	

For a complete description of SI and its use the reader is referred to ASTM E 380 (4) and the article UNITS AND CONVERSION FACTORS which appears in Vol. 24.

A representative list of conversion factors from non-SI to SI units is presented herewith. Factors are given to four significant figures. Exact relationships are followed by a dagger. A more complete list is given in the latest editions of ASTM E 380 (4) and ANSI Z210.1 (6).

Conversion Factors to SI Units

To convert from	To	Multiply by
acre	square meter (m^2)	4.047×10^3
angstrom	meter (m)	1.0×10^{-10}[†]
are	square meter (m^2)	1.0×10^{2}[†]

[†]Exact.

To convert from	To	Multiply by
astronomical unit	meter (m)	1.496×10^{11}
atmosphere, standard	pascal (Pa)	1.013×10^5
bar	pascal (Pa)	$1.0 \times 10^{5\dagger}$
barn	square meter (m^2)	$1.0 \times 10^{-28\dagger}$
barrel (42 U.S. liquid gallons)	cubic meter (m^3)	0.1590
Bohr magneton (μ_B)	J/T	9.274×10^{-24}
Btu (International Table)	joule (J)	1.055×10^3
Btu (mean)	joule (J)	1.056×10^3
Btu (thermochemical)	joule (J)	1.054×10^3
bushel	cubic meter (m^3)	3.524×10^{-2}
calorie (International Table)	joule (J)	4.187
calorie (mean)	joule (J)	4.190
calorie (thermochemical)	joule (J)	$4.184^\dagger$
centipoise	pascal second (Pa·s)	$1.0 \times 10^{-3\dagger}$
centistokes	square millimeter per second (mm^2/s)	$1.0^\dagger$
cfm (cubic foot per minute)	cubic meter per second (m^3/s)	4.72×10^{-4}
cubic inch	cubic meter (m^3)	1.639×10^{-5}
cubic foot	cubic meter (m^3)	2.832×10^{-2}
cubic yard	cubic meter (m^3)	0.7646
curie	becquerel (Bq)	$3.70 \times 10^{10\dagger}$
debye	coulomb meter (C·m)	3.336×10^{-30}
degree (angle)	radian (rad)	1.745×10^{-2}
denier (international)	kilogram per meter (kg/m)	1.111×10^{-7}
	tex‡	0.1111
dram (apothecaries')	kilogram (kg)	3.888×10^{-3}
dram (avoirdupois)	kilogram (kg)	1.772×10^{-3}
dram (U.S. fluid)	cubic meter (m^3)	3.697×10^{-6}
dyne	newton (N)	$1.0 \times 10^{-5\dagger}$
dyne/cm	newton per meter (N/m)	$1.0 \times 10^{-3\dagger}$
electronvolt	joule (J)	1.602×10^{-19}
erg	joule (J)	$1.0 \times 10^{-7\dagger}$
fathom	meter (m)	1.829
fluid ounce (U.S.)	cubic meter (m^3)	2.957×10^{-5}
foot	meter (m)	$0.3048^\dagger$
footcandle	lux (lx)	10.76
furlong	meter (m)	2.012×10^{-2}
gal	meter per second squared (m/s^2)	$1.0 \times 10^{-2\dagger}$
gallon (U.S. dry)	cubic meter (m^3)	4.405×10^{-3}
gallon (U.S. liquid)	cubic meter (m^3)	3.785×10^{-3}
gallon per minute (gpm)	cubic meter per second (m^3/s)	6.309×10^{-5}
	cubic meter per hour (m^3/h)	0.2271

†Exact.

‡See footnote on p. xiii.

To convert from	To	Multiply by
gauss	tesla (T)	1.0×10^{-4}
gilbert	ampere (A)	0.7958
gill (U.S.)	cubic meter (m³)	1.183×10^{-4}
grade	radian	1.571×10^{-2}
grain	kilogram (kg)	6.480×10^{-5}
gram force per denier	newton per tex (N/tex)	8.826×10^{-2}
hectare	square meter (m²)	$1.0 \times 10^{4\dagger}$
horsepower (550 ft·lbf/s)	watt (W)	7.457×10^{2}
horespower (boiler)	watt (W)	9.810×10^{3}
horsepower (electric)	watt (W)	$7.46 \times 10^{2\dagger}$
hundredweight (long)	kilogram (kg)	50.80
hundredweight (short)	kilogram (kg)	45.36
inch	meter (m)	$2.54 \times 10^{-2\dagger}$
inch of mercury (32°F)	pascal (Pa)	3.386×10^{3}
inch of water (39.2°F)	pascal (Pa)	2.491×10^{2}
kilogram-force	newton (N)	9.807
kilowatt hour	megajoule (MJ)	$3.6^{\dagger}$
kip	newton(N)	4.448×10^{3}
knot (international)	meter per second (m/S)	0.5144
lambert	candela per square meter (cd/m³)	3.183×10^{3}
league (British nautical)	meter (m)	5.559×10^{3}
league (statute)	meter (m)	4.828×10^{3}
light year	meter (m)	9.461×10^{15}
liter (for fluids only)	cubic meter (m³)	$1.0 \times 10^{-3\dagger}$
maxwell	weber (Wb)	$1.0 \times 10^{-8\dagger}$
micron	meter (m)	$1.0 \times 10^{-6\dagger}$
mil	meter (m)	$2.54 \times 10^{-5\dagger}$
mile (statute)	meter (m)	1.609×10^{3}
mile (U.S. nautical)	meter (m)	$1.852 \times 10^{3\dagger}$
mile per hour	meter per second (m/s)	0.4470
millibar	pascal (Pa)	1.0×10^{2}
millimeter of mercury (0°C)	pascal (Pa)	$1.333 \times 10^{2\dagger}$
minute (angular)	radian	2.909×10^{-4}
myriagram	kilogram (kg)	10
myriameter	kilometer (km)	10
oersted	ampere per meter (A/m)	79.58
ounce (avoirdupois)	kilogram (kg)	2.835×10^{-2}
ounce (troy)	kilogram (kg)	3.110×10^{-2}
ounce (U.S. fluid)	cubic meter (m³)	2.957×10^{-5}
ounce-force	newton (N)	0.2780
peck (U.S.)	cubic meter (m³)	8.810×10^{-3}
pennyweight	kilogram (kg)	1.555×10^{-3}
pint (U.S. dry)	cubic meter (m³)	5.506×10^{-4}
pint (U.S. liquid)	cubic meter (m³)	4.732×10^{-4}

†Exact.

To convert from	To	Multiply by
poise (absolute viscosity)	pascal second (Pa·s)	$0.10^{\dagger}$
pound (avoirdupois)	kilogram (kg)	0.4536
pound (troy)	kilogram (kg)	0.3732
poundal	newton (N)	0.1383
pound-force	newton (N)	4.448
pound force per square inch (psi)	pascal (Pa)	6.895×10^3
quart (U.S. dry)	cubic meter (m³)	1.101×10^{-3}
quart (U.S. liquid)	cubic meter (m³)	9.464×10^{-4}
quintal	kilogram (kg)	$1.0 \times 10^{2\dagger}$
rad	gray (Gy)	$1.0 \times 10^{-2\dagger}$
rod	meter (m)	5.029
roentgen	coulomb per kilogram (C/kg)	2.58×10^{-4}
second (angle)	radian (rad)	$4.848 \times 10^{-6\dagger}$
section	square meter (m²)	2.590×10^6
slug	kilogram (kg)	14.59
spherical candle power	lumen (lm)	12.57
square inch	square meter (m²)	6.452×10^{-4}
square foot	square meter (m²)	9.290×10^{-2}
square mile	square meter (m²)	2.590×10^6
square yard	square meter (m²)	0.8361
stere	cubic meter (m³)	$1.0^{\dagger}$
stokes (kinematic viscosity)	square meter per second (m²/s)	$1.0 \times 10^{-4\dagger}$
tex	kilogram per meter (kg/m)	$1.0 \times 10^{-6\dagger}$
ton (long, 2240 pounds)	kilogram (kg)	1.016×10^3
ton (metric) (tonne)	kilogram (kg)	$1.0 \times 10^{3\dagger}$
ton (short, 2000 pounds)	kilogram (kg)	9.072×10^2
torr	pascal (Pa)	1.333×10^2
unit pole	weber (Wb)	1.257×10^{-7}
yard	meter (m)	$0.9144^{\dagger}$

†Exact.

Abbreviations and Unit Symbols

Following is a list of common abbreviations and unit symbols used in the *Encyclopedia*. In general they agree with those listed in *American National Standard Abbreviations for Use on Drawings and in Text (ANSI Y1.1)* (6) and *American National Standard Letter Symbols for Units in Science and Technology (ANSI Y10)* (6). Also included is a list of acronyms for a number of private and government organizations as well as common industrial solvents, polymers, and other chemicals.

Rules for Writing Unit Symbols (4):

1. Unit symbols are printed in upright letters (roman) regardless of the type style used in the surrounding text.
2. Unit symbols are unaltered in the plural.
3. Unit symbols are not followed by a period except when used at the end of a sentence.
4. Letter unit symbols are generally printed lower-case (for example, cd for candela) unless the unit name has been derived from a proper name, in which case the first letter of the symbol is capitalized (W, Pa). Prefixes and unit symbols retain their prescribed form regardless of the surrounding typography.
5. In the complete expression for a quantity, a space should be left between the numerical value and the unit symbol. For example, write 2.37 lm, *not* 2.37lm, and 35 mm, *not* 35mm. When the quantity is used in an adjectival sense, a hyphen is often used, for example, 35-mm film. *Exception:* No space is left between the numerical value and the symbols for degree, minute, and second of plane angle, degree Celsius, and the percent sign.
6. No space is used between the prefix and unit symbol (for example, kg).
7. Symbols, not abbreviations, should be used for units. For example, use "A," not "amp," for ampere.
8. When multiplying unit symbols, use a raised dot:

$$N \cdot m \quad \text{for} \quad \text{newton meter}$$

In the case of W·h, the dot may be omitted, thus:

$$Wh$$

An exception to this practice is made for computer printouts, automatic typewriter work, etc, where the raised dot is not possible, and a dot on the line may be used.

9. When dividing unit symbols, use one of the following forms:

$$m/s \quad or \quad m \cdot s^{-1} \quad or \quad \frac{m}{s}$$

In no case should more than one slash be used in the same expression unless parentheses are inserted to avoid ambiguity. For example, write:

$$J/(mol \cdot K) \quad or \quad J \cdot mol^{-1} \cdot K^{-1} \quad or \quad (J/mol)/K$$

but *not*

$$J/mol/K$$

10. Do not mix symbols and unit names in the same expression. Write:

$$\text{joules per kilogram} \quad or \quad \text{J/kg} \quad or \quad \text{J·kg}^{-1}$$

but *not*

$$\text{joules/kilogram} \quad nor \quad \text{joules/kg} \quad nor \quad \text{joules·kg}^{-1}$$

ABBREVIATIONS AND UNITS

A	ampere	AOAC	Association of Official Analytical Chemists
A	anion (eg, HA)		
A	mass number	AOCS	Americal Oil Chemists' Society
a	atto (prefix for 10^{-18})		
AATCC	American Association of Textile Chemists and Colorists	APHA	American Public Health Association
		API	American Petroleum Institute
ABS	acrylonitrile–butadiene–styrene		
		aq	aqueous
abs	absolute	Ar	aryl
ac	alternating current, *n.*	*ar-*	aromatic
a-c	alternating current, *adj.*	*as-*	asymmetric(al)
ac-	alicyclic	ASHRAE	American Society of Heating, Refrigerating, and Air Conditioning Engineers
acac	acetylacetonate		
ACGIH	American Conference of Governmental Industrial Hygienists		
		ASM	American Society for Metals
ACS	American Chemical Society	ASME	American Society of Mechanical Engineers
AGA	American Gas Association	ASTM	American Society for Testing and Materials
Ah	ampere hour		
AIChE	American Institute of Chemical Engineers	at no.	atomic number
		at wt	atomic weight
AIME	American Institute of Mining, Metallurgical, and Petroleum Engineers	av(g)	average
		AWS	American Welding Society
		b	bonding orbital
AIP	American Institute of Physics	bbl	barrel
		bcc	body-centered cubic
AISI	American Iron and Steel Institute	BCT	body-centered tetragonal
		Bé	Baumé
alc	alcohol(ic)	BET	Brunauer-Emmett-Teller (adsorption equation)
Alk	alkyl		
alk	alkaline (not alkali)	bid	twice daily
amt	amount	Boc	*t*-butyloxycarbonyl
amu	atomic mass unit	BOD	biochemical (biological) oxygen demand
ANSI	American National Standards Institute		
		bp	boiling point
AO	atomic orbital	Bq	becquerel

C	coulomb	DIN	Deutsche Industrie Normen
°C	degree Celsius		
C-	denoting attachment to carbon	*dl*-; DL-	racemic
		DMA	dimethylacetamide
c	centi (prefix for 10^{-2})	DMF	dimethylformamide
c	critical	DMG	dimethyl glyoxime
ca	circa (approximately)	DMSO	dimethyl sulfoxide
cd	candela; current density; circular dichroism	DOD	Department of Defense
		DOE	Department of Energy
CFR	Code of Federal Regulations	DOT	Department of Transportation
cgs	centimeter-gram-second	DP	degree of polymerization
CI	Color Index	dp	dew point
cis-	isomer in which substituted groups are on same side of double bond between C atoms	DPH	diamond pyramid hardness
		dstl(d)	distill(ed)
		dta	differential thermal analysis
cl	carload		
cm	centimeter	(*E*)-	entgegen; opposed
cmil	circular mil	ϵ	dielectric constant (unitless number)
cmpd	compound		
CNS	central nervous system	*e*	electron
CoA	coenzyme A	ECU	electrochemical unit
COD	chemical oxygen demand	ed.	edited, edition, editor
coml	commercial(ly)	ED	effective dose
cp	chemically pure	EDTA	ethylenediaminetetra-acetic acid
cph	close-packed hexagonal		
CPSC	Consumer Product Safety Commission	emf	electromotive force
		emu	electromagnetic unit
cryst	crystalline	en	ethylene diamine
cub	cubic	eng	engineering
D	debye	EPA	Environmental Protection Agency
D-	denoting configurational relationship		
		epr	electron paramagnetic resonance
d	differential operator		
d	day; deci (prefix for 10^{-1})	eq.	equation
d-	*dextro*-, dextrorotatory	esca	electron spectroscopy for chemical analysis
da	deka (prefix for 10^1)		
dB	decibel	esp	especially
dc	direct current, *n.*	esr	electron-spin resonance
d-c	direct current, *adj.*	est(d)	estimate(d)
dec	decompose	estn	estimation
detd	determined	esu	electrostatic unit
detn	determination	exp	experiment, experimental
Di	didymium, a mixture of all lanthanons	ext(d)	extract(ed)
		F	farad (capacitance)
dia	diameter	*F*	faraday (96,487 C)
dil	dilute	f	femto (prefix for 10^{-15})

FAO	Food and Agriculture Organization (United Nations)	hyd	hydrated, hydrous
		hyg	hygroscopic
fcc	face-centered cubic	Hz	hertz
FDA	Food and Drug Administration	^{i} (eg, Pr^i)	iso (eg, isopropyl)
		i-	inactive (eg, i-methionine)
FEA	Federal Energy Administration	IACS	International Annealed Copper Standard
FHSA	Federal Hazardous Substances Act	ibp	initial boiling point
		IC	integrated circuit
fob	free on board	ICC	Interstate Commerce Commission
fp	freezing point		
FPC	Federal Power Commission	ICT	International Critical Table
FRB	Federal Reserve Board	ID	inside diameter; infective dose
frz	freezing		
G	giga (prefix for 10^9)	ip	intraperitoneal
G	gravitational constant = 6.67×10^{11} N·m^2/kg^2	IPS	iron pipe size
		ir	infrared
g	gram	IRLG	Interagency Regulatory Liaison Group
(g)	gas, only as in H$_2$O(g)		
g	gravitational acceleration	ISO	International Organization Standardization
gc	gas chromatography		
gem-	geminal	ITS-90	International Temperature Scale (NIST)
glc	gas–liquid chromatography		
g-mol wt; gmw	gram-molecular weight	IU	International Unit
		IUPAC	International Union of Pure and Applied Chemistry
GNP	gross national product		
gpc	gel-permeation chromatography	IV	iodine value
		iv	intravenous
GRAS	Generally Recognized as Safe	J	joule
		K	kelvin
grd	ground	k	kilo (prefix for 10^3)
Gy	gray	kg	kilogram
H	henry	L	denoting configurational relationship
h	hour; hecto (prefix for 10^2)	L	liter (for fluids only) (5)
ha	hectare	l-	$levo$-, levorotatory
HB	Brinell hardness number	(l)	liquid, only as in NH$_3$(l)
Hb	hemoglobin	LC$_{50}$	conc lethal to 50% of the animals tests
hcp	hexagonal close-packed		
hex	hexagonal	LCAO	linear combination of atomic orbitals
HK	Knoop hardness number		
hplc	high performance liquid chromatography	lc	liquid chromatography
		LCD	liquid crystal display
HRC	Rockwell hardness (C scale)	lcl	less than carload lots
		LD$_{50}$	dose lethal to 50% of the animals tested
HV	Vickers hardness number		

LED	light-emitting diode	N-	denoting attachment to
liq	liquid		nitrogen
lm	lumen	n (as n_{D}^{20})	index of refraction (for
ln	logarithm (natural)		20°C and sodium light)
LNG	liquefied natural gas	^{n} (as Bu^{n}),	
log	logarithm (common)	n-	normal (straight-chain
LPG	liquefied petroleum gas		structure)
ltl	less than truckload lots	n	neutron
lx	lux	n	nano (prefix for 10^9)
M	mega (prefix for 10^6);	na	not available
	metal (as in MA)	NAS	National Academy of
M	molar; actual mass		Sciences
$\overline{M}_w$	weight-average mol wt	NASA	National Aeronautics and
$\overline{M}_n$	number-average mol wt		Space Administration
m	meter; milli (prefix for	nat	natural
	10^{-3})	ndt	nondestructive testing
m	molal	neg	negative
m-	meta	NF	*National Formulary*
max	maximum	NIH	National Institutes of
MCA	Chemical Manufacturers'		Health
	Association (was	NIOSH	National Institute of
	Manufacturing Chemists		Occupational Safety and
	Association)		Health
MEK	methyl ethyl ketone	NIST	National Institute of
meq	milliequivalent		Standards and
mfd	manufactured		Technology (formerly
mfg	manufacturing		National Bureau of
mfr	manufacturer		Standards)
MIBC	methyl isobutyl carbinol	nmr	nuclear magnetic
MIBK	methyl isobutyl ketone		resonance
MIC	minimum inhibiting	NND	New and Nonofficial Drugs
	concentration		(AMA)
min	minute; minimum	no.	number
mL	milliliter	NOI-(BN)	not otherwise indexed (by
MLD	minimum lethal dose		name)
MO	molecular orbital	NOS	not otherwise specified
mo	month	nqr	nuclear quadruple
mol	mole		resonance
mol wt	molecular weight	NRC	Nuclear Regulatory
mp	melting point		Commission; National
MR	molar refraction		Research Council
ms	mass spectrometry	NRI	New Ring Index
MSDS	material safety data sheet	NSF	National Science
mxt	mixture		Foundation
μ	micro (prefix for 10^{-6})	NTA	nitrilotriacetic acid
N	newton (force)	NTP	normal temperature and
N	normal (concentration);		pressure (25°C and 101.3
	neutron number		kPa or 1 atm)

NTSB	National Transportation Safety Board	qv	quod vide (which see)
		R	univalent hydrocarbon radical
O-	denoting attachment to oxygen		
o-	ortho	(*R*)-	rectus (clockwise configuration)
OD	outside diameter	*r*	precision of data
OPEC	Organization of Petroleum Exporting Countries	rad	radian; radius
		RCRA	Resource Conservation and Recovery Act
o-phen	*o*-phenanthridine		
OSHA	Occupational Safety and Health Administration	rds	rate-determining step
		ref.	reference
owf	on weight of fiber	rf	radio frequency, *n*.
Ω	ohm	r-f	radio frequency, *adj*.
P	peta (prefix for 10^{15})	rh	relative humidity
p	pico (prefix for 10^{-12})	RI	Ring Index
p-	para	rms	root-mean square
p	proton	rpm	rotations per minute
p.	page	rps	revolutions per second
Pa	pascal (pressure)	RT	room temperature
PEL	personal exposure limit based on an 8-h exposure	RTECS	Registry of Toxic Effects of Chemical Substances
		^{s} (eg, Bu^{s});	
pd	potential difference	*sec*-	secondary (eg, secondary butyl)
pH	negative logarithm of the effective hydrogen ion concentration		
		S	siemens
		(*S*)-	sinister (counterclockwise configuration)
phr	parts per hundred of resin (rubber)		
		S-	denoting attachment to sulfur
p-i-n	positive-intrinsic-negative		
pmr	proton magnetic resonance	*s*-	symmetric(al)
p-n	positive-negative	s	second
po	per os (oral)	(s)	solid, only as in $H_2O(s)$
POP	polyoxypropylene	SAE	Society of Automotive Engineers
pos	positive		
pp.	pages	SAN	styrene-acrylonitrile
ppb	parts per billion (10^9)	sat(d)	saturate(d)
ppm	parts per million (10^6)	satn	saturation
ppmv	parts per million by volume	SBS	styrene–butadiene–styrene
ppmwt	parts per million by weight	sc	subcutaneous
PPO	poly(phenyl oxide)	SCF	self-consistent field; standard cubic feet
ppt(d)	precipitate(d)		
pptn	precipitation	Sch	Schultz number
Pr (no.)	foreign prototype (number)	sem	scanning electron microscope(y)
pt	point; part		
PVC	poly(vinyl chloride)	SFs	Saybolt Furol seconds
pwd	powder	sl sol	slightly soluble
py	pyridine	sol	soluble

soln	solution	*trans-*	isomer in which substituted groups are on opposite sides of double bond between C atoms
soly	solubility		
sp	specific; species		
sp gr	specific gravity		
sr	steradian		
std	standard	TSCA	Toxic Substances Control Act
STP	standard temperature and pressure (0°C and 101.3 kPa)	TWA	time-weighted average
		Twad	Twaddell
sub	sublime(s)	UL	Underwriters' Laboratory
SUs	Saybolt Universal seconds	USDA	United States Department of Agriculture
syn	synthetic		
^{t} (eg, Bu^t), *t-, tert-*	tertiary (eg, tertiary butyl)	USP	*United States Pharmacopeia*
T	tera (prefix for 10^{12}); tesla (magnetic flux density)	uv	ultraviolet
		V	volt (emf)
		var	variable
t	metric ton (tonne)	*vic-*	vicinal
t	temperature	vol	volume (not volatile)
TAPPI	Technical Association of the Pulp and Paper Industry	vs	versus
		v sol	very soluble
		W	watt
TCC	Tagliabue closed cup	Wb	weber
tex	tex (linear density)	Wh	watt hour
T_g	glass-transition temperature	WHO	World Health Organization (United Nations)
tga	thermogravimetric analysis		
		wk	week
THF	tetrahydrofuran	yr	year
tlc	thin layer chromatography	(Z)-	zusammen; together; atomic number
TLV	threshold limit value		

Non-SI (Unacceptable and Obsolete) Units		Use
Å	angstrom	nm
at	atmosphere, technical	Pa
atm	atmosphere, standard	Pa
b	barn	cm^2
bar†	bar	Pa
bbl	barrel	m^3
bhp	brake horsepower	W
Btu	British thermal unit	J
bu	bushel	m^3; L
cal	calorie	J
cfm	cubic foot per minute	m^3/s
Ci	curie	Bq
cSt	centistokes	mm^2/s
c/s	cycle per second	Hz

†Do not use bar (10^5 Pa) or millibar (10^2 Pa) because they are not SI units, and are accepted internationally only for a limited time in special fields because of existing usage.

Non-SI (Unacceptable and Obsolete) Units		Use
cu	cubic	exponential form
D	debye	$C \cdot m$
den	denier	tex
dr	dram	kg
dyn	dyne	N
dyn/cm	dyne per centimeter	mN/m
erg	erg	J
eu	entropy unit	J/K
°F	degree Fahrenheit	°C; K
fc	footcandle	lx
fl	footlambert	lx
fl oz	fluid ounce	m^3; L
ft	foot	m
ft·lbf	foot pound-force	J
gf den	gram-force per denier	N/tex
G	gauss	T
Gal	gal	m/s^2
gal	gallon	m^3; L
Gb	gilbert	A
gpm	gallon per minute	(m^3/s); (m^3/h)
gr	grain	kg
hp	horsepower	W
ihp	indicated horsepower	W
in.	inch	m
in. Hg	inch of mercury	Pa
in. H_2O	inch of water	Pa
in.-lbf	inch pound-force	J
kcal	kilo-calorie	J
kgf	kilogram-force	N
kilo	for kilogram	kg
L	lambert	lx
lb	pound	kg
lbf	pound-force	N
mho	mho	S
mi	mile	m
MM	million	M
mm Hg	millimeter of mercury	Pa
mμ	millimicron	nm
mph	miles per hour	km/h
μ	micron	μm
Oe	oersted	A/m
oz	ounce	kg
ozf	ounce-force	N
η	poise	Pa·s
P	poise	Pa·s
ph	phot	lx
psi	pounds-force per square inch	Pa
psia	pounds-force per square inch absolute	Pa
psig	pounds-force per square inch gage	Pa
qt	quart	m^3; L
°R	degree Rankine	K
rd	rad	Gy
sb	stilb	lx
SCF	standard cubic foot	m^3
sq	square	exponential form
thm	therm	J
yd	yard	m

BIBLIOGRAPHY

1. The International Bureau of Weights and Measures, BIPM (Parc de Saint-Cloud, France) is described in Appendix X2 of Ref. 4. This bureau operates under the exclusive supervision of the International Committee for Weights and Measures (CIPM).
2. *Metric Editorial Guide (ANMC-78-1)*, latest ed., American National Metric Council, 5410 Grosvenor Lane, Bethesda, Md. 20814, 1981.
3. *SI Units and Recommendations for the Use of Their Multiples and of Certain Other Units (ISO 1000-1981)*, American National Standards Institute, 1430 Broadway, New York, N.Y. 10018, 1981.
4. Based on *ASTM E 380-89a (Standard Practice for Use of the International System of Units (SI))*, American Society for Testing and Materials, 1916 Race Street, Philadelphia, Pa. 19103, 1989.
5. *Fed. Regist.*, Dec. 10, 1976 (41 FR 36414).
6. For ANSI address, see Ref. 3.

R. P. LUKENS
ASTM Committee E-43 on SI Practice

F

Continued

FUEL RESOURCES

The wheel is considered to be the greatest invention and fire the greatest discovery of all time. Together, the invention of the wheel and the discovery of fire as a useful force have led to the application of energy. From the invention of the wheel has come such innovations as steam and combustion turbines, rotors and stators used in electricity generation, diesel and Otto-cycle engines for transportation systems, and windmills, water wheels, and hydroelectric turbines. Similarly, the harnessing of fire has led to the use of various materials as fuels: coal (qv), lignite (see LIGNITE AND BROWN COAL), petroleum (qv), natural gas (see GAS, NATURAL), tar sands (qv), oil shale (qv), peat, wood (qv), and the biofuels (see FUELS FROM BIOMASS), organic wastes (see FUELS FROM WASTE), uranium and nuclear power (see NUCLEAR REACTORS), wind, falling water (for hydroelectric power), geothermal steam and hot water (see GEOTHERMAL ENERGY), sunlight (see SOLAR ENERGY), ocean thermal gradients, and the range of conversion products including both electricity and synthesis gas from coal (see FUELS, SYNTHETIC) have been used. These fuels are used both to power the wheel-related inventions and to supply energy for process applications: iron- and steelmaking, nonferrous metal smelting and refining, process heat and steam for pulp (qv) and paper (qv) operations, process energy for chemicals manufacture, etc. Harnessed fuels supply the needs of commercial and residential users as well.

Evaluations of fuel resources or total fuel supply focus on critical economic and environmental issues as well as existence. These issues include availability, utilization patterns, environmental consequences, and related economic considerations.

Historical Patterns in Fuel Utilization

Preindustrial society relied primarily on wood, other biomass, and falling water for energy. These energy sources provided carbon for steelmaking, heat for domestic and commercial purposes, energy for modest shaft power applications, eg, grinding of grain, and fuel for transportation on riverboats and early railroads. These fuels were readily available and could be gathered up or otherwise harnessed with little capital investment and scant attention to technology. U.S. energy consumption by fuel source from 1870 to 1990 is shown in Table 1.

Industrialization in the United States and northern Europe demanded significant sources of carbon for steelmaking, fuels for pumping water from mines, and energy for manufacturing processes. As the process of industrialization gained momentum manufacturing shifted away from optimal sites along rivers and connected regional economies with transcontinental railroads. Industrialization created a national economy, along with strong regional economies, through the use of energy for manufacturing and transportation systems, and coal was the fuel of choice (Table 1). With the advent of industrialization also came the shift in agriculture toward development of mechanized equipment and chemical fertilizers (qv), and petroleum became the dominant fuel. The emergence of pipelines (qv), has enabled natural gas, followed by complemented oil, to be the desired form of energy. Fuel selection factors, in all cases, include availability, energy density (J/kg or J/m^3), energy transportability, fuel cost, and fuel reliability.

Table 1. U.S. Energy Consumption by Source from 1870–1990, EJ[a,b]

Year	Wood and biomass	Coal	Petroleum	Natural gas	Hydroelectric	Nuclear[c]	Other[d]	Total
1870	3.1	1.1						4.1
1880	3.1	2.1	0.1					5.2
1890	2.6	4.3	0.2	0.3				7.5
1900	2.1	7.2	0.2	0.3	0.3			10.1
1910	2.0	13.4	1.1	0.5	0.5			17.5
1920	1.7	16.4	2.7	0.8	0.8			22.4
1930	1.6	14.4	5.7	2.1	0.8			24.5
1940	1.5	13.2	7.9	2.9	1.0			26.3
1950	1.3	13.6	14.2	6.5	1.5			37.1
1960	0.3	10.7	21.2	13.4	1.8			47.4
1970	1.1	13.4	28.9	23.2	2.9	0.2	0.01	69.6
1980	2.53	16.27	36.08	21.51	3.29	2.89	0.08	80.13
1985	2.6	18.44	32.62	18.81	3.54	4.38	0.21	78.01
1986		18.21	33.97	17.63	3.58	4.72	0.23	78.33
1987		19.00	34.68	18.72	3.24	5.18	0.26	81.08[e]
1988		19.89	36.10	19.57	2.79	5.97	0.30	84.61
1989		19.98	36.09	20.45	3.04	5.99	0.26	85.81
1990	3.3	20.17	35.40	20.36	3.11	6.50	0.22	85.76

[a]Refs. 1–6.
[b]To convert EJ to Btu, multiply by 9.48×10^{14}.
[c]Nuclear energy is that generated by electric utilities.
[d]Other includes net imports of coal coke and electricity produced from wood, waste, wind, photovoltaic, and solar thermal sources connected to electric utility distribution systems. It does not include consumption of wood energy other than that consumed by electric utility industry.
[e]An estimated additional 2.5 EJ of wood energy was consumed for residential heating and light industry.

Fuel Production and Consumption Since 1970. In the latter twentieth century, technological and political forces have influenced fuel consumption in the United States and throughout the world. Events such as the oil embargo of 1973, political unrest in the Middle East, and the collapse of the Union of the Soviet Socialist Republics, caused disruptions and shifts in petroleum supply systems. The emergence of the North Sea oil field, construction of the Alyeska Pipeline bringing North Slope, Alaska crude to U.S. refineries, and other technical developments also occurred. Most recently, the selection of fuels has been influenced by environmental concerns such as the potential of the fuel to form air pollutants, eg, particulates, NO_x, SO_2, and most recently air toxics (3, 7–10) (see AIR POLLUTION; AIR POLLUTION CONTROL METHODS; EXHAUST CONTROL, AUTOMOTIVE; EXHAUST CONTROL, INDUSTRIAL). Moreover, there has been the passage of numerous energy and environmental laws within the United States. Legislation has included the Clean Air Act amendments of 1990 and the National Energy Policy Act of 1992. These laws complement the move toward energy conservation, and the emphasis on materials recycling (qv) for resource management. Further, actions by local and state regulatory agencies in the 1990s, including public utility commissions, have further increased the complexity of fuel supply in the United States.

Trends in commercial fuel, eg, fossil fuel, hydroelectric power, nuclear power, production and consumption in the United States and in the Organization of Economic Cooperation and Development (OECD) countries, are shown in Tables 2 and 3. These trends indicate (6,13): (*1*) a significant resurgence in the production and use of coal throughout the U.S. economy; (*2*) a continued decline in the domestic U.S. production of crude oil and natural gas leading to increased imports of these hydrocarbons (qv); and (*3*) a continued trend of energy conservation, expressed in terms of energy consumed per dollar of gross domestic product.

Table 2. U.S. Energy Production and Consumption, 1982–1992, EJ[a,b]

Fuel source	1982	1984	1986	1988	1990	1992
	Consumption					
petroleum	31.90	32.76	33.97	36.11	35.40	35.31
dry natural gas	19.52	19.53	17.63	19.57	20.36	21.44
coal	16.17	18.01	18.21	19.88	20.15	19.96
hydroelectric	3.77	4.01	3.64	2.81	3.11	2.94
nuclear	3.30	3.75	4.72	5.97	6.50	7.02
Total	*74.66*	*78.06*	*78.17*	*84.34*	*85.52*	*86.67*
	Production					
crude oil	19.32	19.89	19.39	18.23	16.43	16.02
natural gas liquids	2.31	2.40	2.27	2.38	2.29	2.49
dry natural gas	19.26	18.92	17.38	18.48	19.37	19.27
coal	19.66	20.8	20.58	21.88	23.70	22.75
hydroelectric	3.45	3.57	3.24	2.46	3.09	2.65
nuclear	3.30	3.75	4.72	5.97	6.50	7.02
Total	*67.30*	*69.33*	*67.58*	*69.40*	*71.38*	*70.20*

[a]Refs. 11 and 12.
[b]To convert EJ to Btu, multiply by 9.48×10^{14}.

Table 3. Total Final Consumption per Gross Domestic Product OECD Countries, 1973–1989[a,b]

Country	1973	1979	1987	1988	1989
North America	0.45	0.41	0.32	0.32	0.31
Canada	0.55	0.52	0.41	0.41	0.41
United States	0.44	0.40	0.31	0.31	0.31
Pacific	0.30	0.26	0.20	0.20	0.20
Australia	0.36	0.35	0.31	0.31	0.31
Japan	0.30	0.25	0.18	0.19	0.18
New Zealand	0.33	0.35	0.39	0.40	0.41
Europe	0.38	0.35	0.30	0.29	0.28
Austria	0.35	0.34	0.30	0.29	0.28
Belgium	0.58	0.51	0.40	0.39	0.38
Denmark	0.36	0.33	0.23	0.23	0.22
Finland	0.50	0.44	0.38	0.36	0.36
France	0.36	0.31	0.26	0.25	0.24
Germany	0.39	0.36	0.30	0.29	0.28
Greece	0.38	0.38	0.40	0.40	0.41
Iceland	0.48	0.35	0.29	0.31	0.33
Ireland	0.45	0.43	0.38	0.36	0.35
Italy	0.34	0.29	0.25	0.24	0.24
Luxembourg	1.43	1.18	0.78	0.76	0.78
Netherlands	0.49	0.47	0.41	0.38	0.37
Norway	0.39	0.36	0.31	0.30	0.29
Portugal	0.39	0.43	0.46	0.49	0.49
Spain	0.31	0.35	0.29	0.30	0.29
Sweden	0.44	0.41	0.32	0.32	0.30
Switzerland	0.21	0.21	0.20	0.19	0.19
Turkey	0.65	0.62	0.61	0.60	0.61
United Kingdom	0.40	0.36	0.30	0.29	0.28
Total OECD	*0.41*	*0.37*	*0.29*	*0.29*	*0.28*

[a]Ref. 13.
[b]Ratio of total final consumption of energy to gross domestic product (GDP). Measured in metric tons of oil equivalent per $1000 of GDP at 1985 prices and exchange rates; changes in ratios over time reflect the combined effects of efficiency improvements, structural changes, and fuel substitution.

U.S. Energy Production, Consumption, and Availability

Production and consumption of commercially available fossil fuel, nuclear power, and hydroelectric power in the United States for the year 1992 is shown in Table 2 (12). Coal production is most significant followed by natural gas and petroleum. Electricity generation and utilization patterns are shown in Table 4. Coal is overwhelmingly the most significant energy source used to generate electricity.

The data presented in Tables 2 and 4 focus on commercially traded sources of energy. During the period 1970–1990, increased emphasis was placed on renewable energy resources (qv), including wood and wood waste; municipal solid waste and refuse-derived fuel; other sources of biomass and waste, eg, agricultural crop wastes, tire-derived fuels, and selected hazardous wastes burned as fuel sub-

Table 4. Electricity Supply and Disposition, 1990[a]

Supply and disposition	Quantity, kW·h × 10⁹	Percent of total
Fuel type for electric utilities generation		
coal	1560	55.6
petroleum	117	4.2
natural gas	264	9.4
nuclear power	577	20.5
pumped storage hydroelectric	− 2	
renewable sources/other[b]	293	10.4
Total	*2808*	*100*
imports	*2*	
Fuel type for nonutilities[c] generation		
coal	33	15
petroleum	5	2.3
natural gas	100	45.9
renewable sources/other[b,d]	80	36.7
Total	*218*	*100*
sales to utilities	*106*	
generation for own use	*111*	
Electricity sales by sector		
residential	924	34.1
commercial/other	843	31.0
industrial	946	34.9
Total	*2713*	*100*

[a]Ref. 14.
[b]Includes hydroelectric, geothermal, wood, wood waste, municipal solid waste, other biomass, and solar and wind power.
[c]Includes cogenerators, small power producers, and all other sources, except electric utilities which produce electricity for self-use or for delivery to the grid. The generation values for nonutilities represent gross generation rather than net generation (net of station use).
[d]Includes waste heat, blast furnace gas, and coke oven gas.

stitutes in cement kilns; wind and solar energy; geothermal steam and hot water; and other unconventional energy sources. Estimates of the contribution of these energy sources vary. As of this writing biofuel utilization in the United States runs about 3.7 EJ/yr (3.5 × 10¹⁵ Btu/yr) in support of process energy needs for industry, cogeneration facilities, and small stand-alone power plants (5), and geothermal energy is about 0.21 EJ/yr (0.2 × 10¹⁵ Btu/yr) (6).

Coal Availability and Utilization. There are vast reserves of coal (qv) and lignite (see LIGNITE AND BROWN COAL) in the United States (Table 5). The total reserve base exceeds 425 billion metric tons equivalent to 11,200 EJ (10.6 × 10¹⁸ Btu) and is distributed throughout 32 states. This reserve base has increased by 8.3% since the 1970s despite the high levels of fuel production (6). Total U.S.

Table 5. U.S. Coal Reserves by State, 1990, EJ[a–d]

State	Reserves	State	Reserves
Alabama	114	Montana	2,848
Alaska	146	New Mexico	106
Arizona	6		
Arkansas	10	North Dakota	229
Colorado	403	Ohio	438
		Oklahoma	38
Illinois	1,857	Pennsylvania	691
Indiana	241	South Dakota	9
Iowa	52	Tennessee	20
Kansas	23	Texas	316
Kentucky	697	Utah	146
Louisiana	12	Virginia	62
Maryland	18	Washington	34
Michigan	3	West Virginia	880
Missouri	143	Wyoming	1,614
Total	*11,155*		

[a]Refs. 6 and 15.
[b]Reserve data are based on demonstrated reserve base. Minable reserves differ from these figures.
[c]Georgia, Idaho, North Carolina, and Oregon also have some reserves.
[d]To convert EJ to Btu, multiply 9.48×10^{14}.

recoverable reserves exceed 240 billion metric tons or 6100 EJ (5.8×10^{18} Btu) and are distributed among three geographic areas: the Appalachian, Interior, and Western coal producing regions. Of these, the western region contains 53.6% of the recoverable reserves, the interior region 25.8%, and the Appalachian region 20.6%. Reserves can also be evaluated in terms of the sulfur (qv) content of the coal. The sulfur is important owing to environmental considerations. Of the recoverable reserves, 34.3% contains <0.6% sulfur, 33.9% contains 0.61–1.67% sulfur, and 31.8% contains >1.68% sulfur.

Coal production and consumption in the 1990s reflects the shift toward the use of western, lower sulfur coal. In 1970, West Virginia, Kentucky, and Pennsylvania ranked 1–3 in coal production, respectively. In 1990, Wyoming, Kentucky, and West Virginia held those ranks, and Texas and Montana entered the top 10 coal producers. Whereas Appalachia remained the most significant energy production region, the western coal producing states surpassed the interior states in solid fossil fuel production (Table 6). The average coal heating value reflected the shift from Appalachia and the interior to the West, declining from 25.8×10^6 J/kg (11.1×10^3 Btu/lb) in 1973 to 24.8×10^6 J/kg (10.7×10^3 Btu/lb) in 1980 and 24.3×10^6 J/kg (10.4×10^3 Btu/lb) in 1990. The shift in coal production toward western coal deposits also reflects the shift in coal utilization patterns (Table 7). Electric utilities are increasing coal consumption on both absolute and percentage bases, whereas coke plants, other industrial operations, and residential and commercial coal users are decreasing use of this solid fossil fuel.

Table 6. Largest Coal-Producing States in 1990, EJ[a,b]

State	1990 production	Rank
Wyoming	4.37	1
Kentucky	4.11	2
West Virginia	4.02	3
Pennsylvania	1.67	4
Illinois	1.43	5
Texas	1.32	6
Virginia	1.11	7
Montana	0.89	8
Indiana	0.85	9
Ohio	0.84	10
Total	*20.63*	

[a]Ref. 6.
[b]To convert EJ to Btu, multiply by 9.48×10^{14}.

Table 7. U.S. Coal Consumption by Sector, 1970–1990, EJ[a,b]

Year	Electric utilities	Industrial Coke plants	Industrial Other	Residential and commercial	Total consumption
1970	7.60	2.29	2.15	0.38	12.42
1975	9.64	1.98	1.51	0.22	13.35
1980	13.51	1.58	1.43	0.15	16.67
1985	16.47	0.98	1.79	0.19	19.43
1990	18.36	0.94	1.81	0.16	21.27

[a]Ref. 6.
[b]To convert EJ to Btu, multiply by 948×10^{14}.

Environmental considerations also were reflected in coal production and consumption statistics, including regional production patterns and economic sector utilization characteristics. Average coal sulfur content, as produced, declined from 2.3% in 1973 to 1.6% in 1980 and 1.3% in 1990. Coal ash content declined similarly, from 13.1% in 1973 to 11.1% in 1980 and 9.9% in 1990. These numbers clearly reflect a trend toward utilization of coal that produces less SO_2 and less flyash to capture. Emissions from coal in the 1990s were 14×10^6 t/yr of SO_2 and 450×10^3 t/yr of particulates generated by coal combustion at electric utilities. The total coal combustion emissions from all sources were only slightly higher than the emissions from electric utility coal utilization (6).

Oil and Natural Gas Availability and Utilization. U.S. resources and reserves of petroleum (qv) and natural gas (see GAS, NATURAL), including natural gas liquids (NGL) are limited. As of January 1, 1992, U.S. reserves of petroleum were some 151 EJ (24.7×10^9 bbl) and U.S. reserves of natural gas were 182 EJ (17.3×10^{16} Btu) (11). Since 1976, the United States has experienced a significant decline in oil reserves. In 1976, proven petroleum reserves totaled 205 EJ ($33.5 \times$

10^9 bbl). Between 1976 and 1993, some 210 EJ (3.4×10^{10} bbl) were added to the reserves, and 263.5 EJ (4.31×10^{10} bbl) were produced, yielding a net reserve loss of 53.8 EJ (8.8×10^9 bbl) (14). Similarly, from 1976 to 1992 there was a net reserve loss of 44.5 EJ (4.22×10^{16} Btu) of dry natural gas (16).

As shown in Table 8, U.S. distribution of oil and natural gas reserves is centered in Alaska, California, Texas, Oklahoma, Louisiana, and the U.S. outercontinental shelf. Alaska reserves include both the Prudhoe Bay deposits and the Cook Inlet fields. California deposits include those in Santa Barbara, the Wilmington Field, the Elk Hills Naval Petroleum Reserve No. 1 at Bakersfield, and other offshore oil deposits. The Yates Field, Austin Chalk formation, and Permian Basin are among the producing sources of petroleum and natural gas in Texas.

The decrease in petroleum and natural gas reserves has encouraged interest in and discovery and development of unconventional sources of these hydrocarbons. Principal alternatives to conventional petroleum reserves include oil shale (qv) and tar sands (qv). Oil shale reserves in the United States are estimated at 20,000 EJ (19.4×10^{18} Btu) and estimates of tar sands and oil sands reserves are on the order of 11 EJ (10×10^{15} Btu) (see TAR SANDS; SHALE OIL). Of particular interest are the McKittrick, Fellows, and Taft quadrangles of California, the Asphalt Ridge area of Utah, the Asphalt, Kentucky, area, and related geographic regions.

The unconventional reserves of natural gas occur principally in the form of recoverable methane from coal beds. As of 1991, reserves of coal bed methane totaled 8.6 EJ (8.2×10^{15} Btu), principally in the states of Alabama, Colorado, and New Mexico (16).

Domestic petroleum, natural gas, and natural gas liquids production has declined at a rate commensurate with the decrease in reserves (see Table 2). Consequently, the reserves/production ratio, expressed in years, remained relatively constant from about 1970 through 1992, at 9–11 years (16). Much of the production in the early 1990s is the result of enhanced oil recovery techniques: water flooding, steam flooding, CO_2 injection, and natural gas reinjection.

Whereas the use of petroleum and natural gas is significant in the electricity generating sector, this usage declined from 1970 to 1990, in part owing to the

Table 8. Crude Oil and Natural Gas Proved Reserves, EJ[a–c]

State	Oil proved reserves	Gas proved reserve[d]
Alaska	37.22	10.22
California	25.80	3.19
Louisiana	4.15	11.79
Oklahoma	4.28	15.83
Texas	41.59	39.87
Wyoming	4.63	11.02
federal offshore	16.03	31.02
Total	*133.72*	*122.94*

[a]Ref. 16.
[b]To convert EJ to Btu, multiply by 9.48×10^{14}.
[c]As of Dec. 31, 1991.
[d]Gas reserves equal dry natural gas plus natural gas liquids.

1977 Fuel Use Act (Table 9). The legislation of the 1990s and the growth of independent power producers (IPP) generating electricity for utilities in combined cycle combustion turbine (CCCT) facilities, may mean a reversal in the trend for oil and natural gas utilization for power generation (qv). In any event, total U.S. oil and gas consumption (Table 1) remains high, and these are the fuels of choice for residential, commercial, industrial, and transportation applications.

Other Fuel Availability and Utilization. As shown in Table 2, nuclear, hydroelectric, and geothermal resources now contribute some 9.8 EJ (9.3×10^{15} Btu) annually to the U.S. economy. Of these energy sources, nuclear power is the dominant force having over 70% of the total. U.S. nuclear power production continued to increase through 1990, but nuclear electricity generation may have peaked at 6.5 EJ for political and social reasons. Hydroelectric power generation remains relatively stable. There are annual variations in supply which depend on local weather, eg, rainfall, snowpack, and regional economic conditions. Geothermal energy (qv) has been developed to only a modest extent.

Biomass and waste fuels contributed some 3.7 EJ to the economy (see FUELS FROM BIOMASS; FUELS FROM WASTE). These fuels include wood and wood waste; spent pulping liquor at pulp and paper mills; agricultural materials such as rice hulls, bagasse, cotton gin trash, coffee grounds, and a variety of manures. When wood waste and numerous other forms of biomass are added to municipal solid waste (MSW), refuse-derived fuel (RDF), methane recovered from landfills and sewage treatment plants, and special industrial and municipal wastes such as tire-derived fuel, these together contribute about 5 EJ (4.7×10^{15} Btu) to the U.S. economy (17). Of these fuels, wood and the biofuels are typically employed in industrial settings either to generate process steam (qv) or to cogenerate electricity and process steam. Some condensing power plants have been built by such utilities as Washington Water Power, Burlington Electric, and several IPP firms. There are some 1500 MW of electricity generating capacity based on wood and the biofuels in existence as of 1993.

MSW incinerators (qv) are typically designed to reduce the volume of solid waste and to generate electricity in condensing power stations. Incineration of unprocessed municipal waste alone recovers energy from about 34,500 t/d or 109 million metric tons of MSW annually in some 74 incinerators throughout the United States. This represents 1.1 EJ (1.05×10^{15} Btu) of energy recovered an-

Table 9. Fuel for Electric Utility Generation of Electricity 1970–1990, kW·h $\times 10^{9a}$

Year	Type of facility			
	Petroleum	Gas-fired	Internal combustion and gas turbine	Total
1970	174	361	22	557
1975	273	288	28	589
1980	238	326	28	592
1985	97	279	16	392
1990	113	246	22	381

[a]Ref. 6.

nually (18). Additionally there are some 20 RDF facilities processing from 200 to 2000 t/d of MSW into a more refined fuel (19). Representative projects are shown in Table 10.

Other sources of energy worth noting are the extensive wind farms, solar projects, and related emerging unconventional technologies. These renewable resources provide only small quantities of energy to the U.S. economy as of this writing.

Trends in Energy Technology and Future Fuel Consumption. Increased economic activity usually means an increase in energy consumption particularly for generation of electricity, manufacturing of products, transportation, and residential and commercial applications. Regulatory and political requirements associated with energy supply and utilization require increasing attention to environmental concerns in order to ensure reliable energy availability without undue environmental degradation. Thus attention is being paid to increasing the efficiency of fuel utilization as well as to reducing the formation of airborne emissions ranging from particulates NO_x and SO_2 to the management of air toxics such as HCl and trace metals.

Coal. Technologies traditionally deployed for coal utilization include using pulverized coal (PC), cyclone, and stoker-fired boilers. For PC boilers, technologies being deployed or developed include the use of micrometer-sized coal, staged fuel–staged air low NO_x burners, limestone injection multistage burners (LIMB), reburning for NO_x control, and advanced techniques for overfire air management. Cyclone-fired boilers also are capitalizing on reburning technologies and air

Table 10. Municipal Waste to Energy Projects[a]

Location	Capacity, t/d
Mass burn	
Hillsborough County, Fla.	1100
Pinnelas County, Fla.	2700
Tampa, Fla.	910
Baltimore, Md.	2000
North Andover, Mass.	1350
Saugus, Mass.	1350
Peekskill, N.Y.	2000
Tulsa, Okla.	1000
Marion County, Oreg.	500
Nashville, Tenn.	1000
Refuse-derived fuel	
Akron, Ohio	1000
Duluth, Minn.	360
Niagara Falls, N.Y.	1800
Dade County, Fla.	2700
Columbus, Ohio	1800
Hartford, Conn.	1800

[a]Ref. 18 and 19.

management techniques. Further, both PC and cyclone-fired boilers are utilizing cofiring techniques, blending nitrogen and sulfur-free biomass fuels, and low sulfur tire-derived fuels with the coal for both cost control and emissions reduction (17,20).

Advanced coal utilization technologies include the development of bubbling, circulating, and pressurized fluidized-bed combustion for electricity generation and process energy production (see COAL CONVERSION PROCESSES; FLUIDIZATION). Since the early 1980s, over 250 fluidized beds have been installed that have capacities ranging from 25 GJ/h to 1 TJ/h. Whereas fluidized beds are fired using every solid fuel available, these beds are predominantly used for coal combustion. Projects include the Shawnee No. 10 boiler of TVA, the Black Dog project of Northern States Power, and the Colorado-Ute circulating fluidized-bed project. Advanced coal utilization technologies also include integrated gasification combined cycle (IGCC) systems where coal is gasified and the low or medium heat-value gas is then utilized in a combustion turbine and the exhaust is ducted to a heat recovery steam generator (HRSG). The initial demonstration of this technology was initiated in 1979 at the Cool Water project near Barstow, California. More recent projects are under development in Polk and Martin counties in Florida, and at the Tracy Station of Sierra Pacific Power Co. in Nevada. Post-combustion technologies including alternative acid gas scrubbing technologies, urea or ammonia injection for NO_x control, and combinations of air pollution control systems are also emerging to support coal utilization.

Petroleum and Natural Gas. The dominant technologies under development for oil and gas include advances in combustion turbine design. These include applying aircraft engine technology to power generation (qv), and placing emphasis on higher temperatures and increased efficiencies in the combustion turbine. Other technologies of significance include dry low NO_x combustion systems employing principles of staged fuel–staged air, and various catalytic and noncatalytic ammonia injection systems in the post-combustion environment.

Other Fuels. The emerging technologies for unconventional and renewable fuels somewhat mirror those associated with coal: cofiring of biofuels and coal in PC and cyclone boilers, fluidized-bed combustion systems fed with biofuel or a blend of various solid fuels, combustion air management, and gasification–combustion systems. Additionally, technologies associated with post-combustion pollution control for the alternative solid fuels are similar to those developed for coal utilization. At the same time, significant advances have been made in such alternative technologies as fuel cells (qv), photovoltaic cells (qv), wind energy, and the broad range of renewable resources. All of these technologies, combined with those for the dominant fossil fuels, are designed to promote fuel supply and utilization within the framework of an environmentally conscious society.

World Fuel Reserves, Production, and Consumption

Energy reserves, production, and consumption in the world economy are given in Tables 11, 12, and 13, respectively. As these tables indicate, the overwhelming sources of petroleum reserves and supply are in the Middle East. Other significant sources of reserves include Russia, the North Sea, North American countries, and

Table 11. World Fossil Fuel Reserves, EJ[a,b]

Region	Coal reserves 1991[c]		Petroleum reserves 1992[d]		Natural gas reserves 1992[d]	
North America	6,570	(6,565)	499	(500)	346	(344)
Central and South America	254	(254)	419	(447)	172	(193)
Western Europe	2,964	(2,578)	90	(136)	187	(222)
Eastern Europe and former USSR	7,818	(8,252)	358	(378)	1,818	(1,929)
Middle East	5	(5)	4,048	(3,651)	1,360	(1,387)
Africa	1,623	(1,623)	370	(462)	320	(344)
Far East and Oceania	7,950	(7,942)	270	(345)	309	(401)
Total	*27,185*	*(27,221)*	*6,054*	*(5,918)*	*4,512*	*(4,820)*

[a]Refs. 11 and 12.
[b]To convert EJ to Btu, multiply by 9.48 × 10^{14}.
[c]Data from the World Energy Council. Values in parentheses are from British Petroleum.
[d]Data from *Oil and Gas Journal*. Data in parentheses are from *World Oil*.

Table 12. World Energy Production 1991, EJ[a,b]

Region	Crude oil	Natural gas liquids	Dry natural gas	Coal	Hydro-electric	Nuclear	Total
North America	26.22	3.75	24.64	24.73	6.64	7.94	102.28
Central and South America	10.56	0.31	2.56	0.82	3.73	0.11	14.90
Western Europe	9.52	0.46	8.21	9.60	5.06	8.10	68.64
Eastern Europe and former USSR	22.86	0.89	29.52	20.97	2.67	2.93	74.12
Middle East	36.83	1.64	4.59	0.03	0.14	0	12.53
Africa	15.04	0.43	2.81	4.64	0.48	0.10	10.49
Far East and Oceania	14.85	0.33	6.29	36.38	4.81	3.21	82.97
Total	*135.88*	*7.81*	*78.61*	*97.18*	*23.51*	*22.40*	*365.40*

[a]Ref. 11.
[b]To convert EJ to Btu, multiply by 9.48 × 10^{14}.

Table 13. World Energy Consumption, 1991, EJ[a,b]

Region	Petroleum	Natural gas	Coal	Hydroelectric	Nuclear	Total
North America	41.95	24.58	21.17	6.64	7.94	102.28
Central and South America	7.88	2.46	0.72	3.73	0.11	14.90
Western Europe	29.67	12.29	13.70	4.98	8.00	68.64
Eastern Europe and former USSR	20.44	27.63	20.27	2.75	3.03	74.12
Middle East	7.78	4.45	0.16	0.14	0.00	12.53
Africa	4.76	1.62	3.53	0.48	0.10	10.49
Far East and Oceania	31.19	6.42	37.34	4.81	3.21	82.97
Total	*143.67*	*79.44*	*96.91*	*23.51*	*22.40*	*365.93*

[a]Ref. 11.
[b]To convert EJ to Btu, multiply by 9.48 × 10^{14}.

Table 14. World Net Electricity Consumption, 1982–1991, kW·h × 10^{9}[a]

Region	1982	1983	1984	1985	1986	1987	1988	1989	1990	1991
North America[b]	390.1	406.3	435.9	458.1	477	499.1	524.2	541.6	543.2	552.5
United States	2,086.4	2,151.0	2,285.8	2,324.0	2,368.8	2,457.3	2,578.1	2,646.8	2,712.6	2,759.3
Central and South America	309.9	333.4	355.6	371.1	391.1	403.7	421.7	427.8	437.0	446.9
Western Europe	1,725.1	1,783.4	1,873.2	1,962.6	2,001.7	2,064.9	2,069.4	2,150.8	2,188.9	2,217.1
Eastern Europe and former USSR	1,555.4	1,610.8	1,694.8	1,748.9	1,727.0	1,787.5	1,825.7	1,853.9	1,838.8	1,751.0
Middle East	111.4	121.7	131.3	143.5	149.4	158.3	169.7	180.2	195.6	185.6
Africa	182.9	202.0	206.5	217.7	248.6	254.7	270.5	280.1	282.0	281.0
Far East and Oceania	1,300.3	1,387.0	1,477.8	1,568.9	1,654.4	1,783.0	1,904.8	2,047.9	2,199.1	2,288.5
World total	*7,661.6*	*7,995.6*	*8,461.0*	*8,794.9*	*9,018.1*	*9,408.5*	*9,764.1*	*10,128.4*	*10,397.2*	*10,481.9*

[a]Ref. 11.
[b]Excluding the United States.

Table 15. Projections of World and U.S. Energy Consumption, 1995–2005, EJ[a,b]

Energy source	1995			2000			2005		
	Base	Low	High	Base	Low	High	Base	Low	High
World projection[c]									
oil	152	141	157	159	147	170	171	152	190
gas	82	80	83	92	89	94	107	100	112
coal	106	102	108	112	107	116	129	118	136
nuclear	24	23	24	26	25	26	31	27	32
other	31	31	32	35	34	36	42	39	46
Total	*395*	*377*	*404*	*424*	*402*	*442*	*480*	*436*	*516*
U.S. projection									
oil	38	37	38	40	39	41	42	40	44
gas	23	23	23	25	24	25	26	25	27
coal	21	21	21	21	21	22	22	22	23
nuclear	7	7	7	7	7	7	7	7	8
other	8	8	8	9	9	10	10	10	10
Total	*97*	*96*	*97*	*102*	*100*	*105*	*107*	*104*	*112*

[a]Ref. 14.
[b]To convert EJ to Btu, multiply by 9.48×10^{14}.
[c]World consumption totals also include the United States.

parts of southeast Asia. There are also significant concentrations of coal reserves in Russia and China. The dominant coal-producing countries include China and the United States, plus Poland, South Africa, Australia, India, Germany, and the United Kingdom. China is the single largest coal producer and consumer, utilizing over 22 EJ/yr (21×10^{15} Btu/yr) of this solid fossil fuel.

In addition to the significant consumption of coal and lignite, petroleum, and natural gas, several countries utilize modest quantities of alternative fossil fuels. Canada obtains some of its energy from the Athabasca tar sands development (the Great Canadian Oil Sands Project). Oil shale is burned at two 1600 MW power plants in Estonia for electricity generation. World reserves of tar sands total some 6400 EJ (6.1×10^{18} Btu), and world reserves of oil shale total some 20,400 EJ (19.3×10^{18} Btu).

Renewable and unconventional energy sources are used more extensively in other parts of the world than in the United States. Tables 12 and 13 document the significance of hydroelectric power throughout industrialized and developing economies. Biofuels are also a significant contributor to certain economies, with proportional contributions as follows: Kenya, 75%; India, 50%; China, 33%; Brazil, 25%; and Scandanavia, 10% (5,21). Peat is a significant source of energy for Russia, Finland, and Ireland.

World electricity generation and consumption, shown in Table 14 increased from 1982 to 1991, and is expected to continue to do so. Further, the industrialized economies are focusing on issues of energy conservation, materials conservation through recycling, and environmental protection. Given the world trends in fuels availability and consumption, projections of energy production and consumption have been made, as shown in Table 15. These projections, to the year 2005, reflect the emphases on fuel availability, energy economics, and environmental awareness of a world community. Further, they reflect the trend toward increased technology development, leading to economically and environmentally sound energy utilization.

BIBLIOGRAPHY

"Fuels, Survey" in *ECT* 1st ed., Vol. 6, pp. 892–902, by R. A. Sherman, Battelle Memorial Institute; in *ECT* 2nd ed., Vol. 10, pp. 179–191, by A. Parker, Consultant; in *ECT* 3rd ed., Vol. 11, pp. 317–333, by H. Perry, Resources for the Future.

1. H. Enzer, W. Dupree, and S. Miller, *Energy Perspectives: A Presentation of Major Energy Related Data*, U.S. Department of the Interior, Washington, D.C., 1975.
2. H. H. Lansberg, L. L. Fishman, and J. L. Fisher, *Resources in America's Future*, Johns Hopkins University Press, Baltimore, Md., 1963.
3. H. C. Hottel, and J. B. Howard, *New Energy Technology: Some Facts and Assessments*, MIT Press, Cambridge, Mass., 1971.
4. *A National Plan for Energy Research, Development, and Demonstration: Creating Energy Choices for the Future*, Energy Research and Development Administration, Washington, D.C., 1976.
5. D. A. Tillman, *The Combustion of Solid Fuels and Wastes*, Academic Press, Inc., San Diego, Calif., 1991.
6. *The U.S. Coal Industry, 1970–1990: Two Decades of Change*, Energy Information Agency, Washington, D.C., 1992.

7. F. W. Brownell, *Clean Air Handbook*, Government Institutes, Inc., Rockville, Md., 1993.
8. S. Bruchey, *Growth of the Modern American Economy*, Dodd, Mead, New York, 1975.
9. N. Rosenberg, *Technology and American Economic Growth*, Harper and Row, New York, 1972.
10. H. M. Jones, *The Age of Energy*, Viking Press, New York, 1970.
11. *International Energy Annual 1991*, Energy Information Agency, Washington, D.C., 1992.
12. *EIA's Annual Energy Review 1992*, Energy Information Agency, Washington, D.C., 1993.
13. *Energy Policies of IEA Countries, 1990 Review*, International Energy Agency, Washington, D.C., 1991.
14. *Annual Energy Outlook, 1993*, Energy Information Agency, Washington, D.C., 1993.
15. *Keystone Coal Industry Manual*, Maclean Hunter Publishing Co., Chicago, 1992.
16. *U.S. Crude Oil, Natural Gas, and Natural Gas Liquids Reserves*, Energy Information Agency, Washington, D.C., 1992.
17. D. Tillman, E. Hughes, and B. Gold, "Cofiring of Biofuels in Coal Fired Boilers: Results of Case Study Analysis," *Proceedings First Biomass Conference of the Americas: Energy, Environment, Agriculture, and Industry*, Burlington, Vt., 1993.
18. D. Tillman, A. Rossi, and K. Vick, *Incineration of Municipal and Hazardous Solid Wastes*, Academic Press, Inc., San Diego, Calif., 1989.
19. J. L. Smith, *Early and Current Systems Utilizing Refuse Derived Fuels*, Combustion Engineering Co., Windsor, Conn., 1986.
20. V. Nast, G. Eirschele, and W. Hutchinson, "TDF Co-firing Experience in a Cyclone Boiler," *Proceedings, Strategic Benefits of Biomass and Waste Fuels Conference*, EPRI, Washington, D.C., 1993.
21. D. O. Hall and R. P. Overend, eds., *Biomass, Regenerable Energy*, John Wiley & Sons, Inc., New York, 1987.

DAVID A. TILLMAN
JEFFREY B. WARSHAUER
DAVID E. PRINZING
ENSERCH

FUELS FROM BIOMASS

The contribution of biomass energy to energy consumption in the late 1970s was over 1.8×10^{15} kJ/yr (850,000 barrels of oil equivalent per day (BOE/d) U.S.) or more than 2% of total energy consumption (1). By 1987, biomass energy had increased to approximately 3.1×10^{15} kJ/yr (1.4×10^6 BOE/d, 3.0×10^{15} Btu/yr), or 3.7% of U.S. primary consumption (2). Projections indicate that by the year 2000, the biomass energy contribution will increase to about 4.2×10^{15} kJ/yr (1.9×10^6 BOE/d), ie, over 4% of total U.S. primary energy consumption (2). Land-

and water-based vegetation, organic wastes, and photosynthetic organisms are categorized as biomass and are nonfossil, renewable carbon resources from which energy, eg, heat, steam, and electric power, and solid, liquid, and gaseous fuels, ie, biofuels, can be produced and utilized as fossil fuel substitutes.

Renewable carbon resources is a misnomer; the earth's carbon is in a perpetual state of flux. Carbon is not consumed such that it is no longer available in any form. Reversible and irreversible chemical reactions occur in such a manner that the carbon cycle makes all forms of carbon, including fossil resources, renewable. It is simply a matter of time that makes one carbon form more renewable than another. If it is presumed that replacement does in fact occur, natural processes eventually will replenish depleted petroleum or natural gas deposits in several million years. Fixed carbon-containing materials that renew themselves often enough to make them continuously available in large quantities are needed to maintain and supplement energy supplies; biomass is a principal source of such carbon.

The capture of solar energy as fixed carbon in biomass via photosynthesis is the initial step in the growth of biomass. It is depicted by the equation

$$CO_2 + H_2O + light \xrightarrow{\text{chlorophyll}} (CH_2O) + O_2$$

Carbohydrate, represented by the building block CH_2O, is the primary organic product. For each gram mole of carbon fixed, about 470 kJ (112 kcal) is absorbed. Oxygen liberated in the process comes exclusively from the water, according to radioactive tracer experiments. There are many unanswered questions regarding the detailed molecular mechanisms of photosynthesis, but the prerequisites for plant biomass production are well established; ie, carbon dioxide, light in the visible region of the electromagnetic spectrum, the sensitizing catalyst chlorophyll, and a living plant are essential. The upper limit of the capture efficiency of incident solar radiation in biomass is estimated to range from about 8% to as high as 15%; in most situations it is generally in the range 1% or less (3).

The primary features of biomass-to-energy technology as a source of synthetic fuels are illustrated in Figure 1. Conventionally, biomass is harvested for feed, food, and materials-of-construction applications, or is left in the growth areas where natural decomposition occurs. Decomposing biomass, and waste products from the harvesting and processing of biomass, disposed of on or in land, in theory can be partially recovered after a long period of time as fossil fuels. This is indicated by the dashed lines in Figure 1. Alternatively, biomass, and any wastes that result from its processing or consumption, can be converted directly into synthetic organic fuels if suitable conversion processes are available. The energy content of biomass also can be diverted to direct heating applications by combustion. Certain species of biomass can be grown, eg, the rubber tree (*Hevea braziliensis*), in which high energy hydrocarbons are formed within the biomass by natural biochemical mechanisms. The biomass serves the dual role of a carbon-fixing mechanism and a continuous source of hydrocarbons without being consumed in the process. Other plants, such as the guayule bush, also produce hydrocarbons but must be harvested to recover them. Thus, conceptually, there are several different pathways by which energy products and synthetic fuels might be manufactured (Fig. 1).

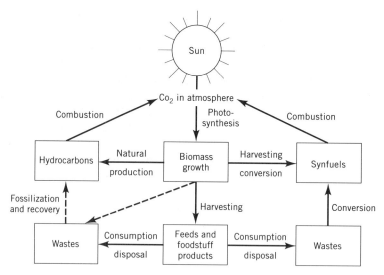

Fig. 1. Biomass-to-energy technology.

Fixed carbon supplies also can be developed from renewable carbon sources by the conversion of carbon dioxide outside the biomass species into synthetic fuels and organic intermediates. The ambient air, which in 1992 contained an average of 350 ppm by volume of carbon dioxide, the dissolved carbon dioxide and carbonates in the oceans, and the earth's large terrestrial carbonate deposits serve as renewable carbon sources. However, because carbon dioxide is the final oxidation state of fixed carbon, it contains no chemical energy and energy must be supplied in a reduction step. A convenient method of supplying the required energy and reducing the oxidation state is to reduce carbon dioxide with elemental hydrogen. The end product can be, eg, methane (CH_4), the dominant component of natural gas.

$$CO_2 + 4\,H_2 \rightarrow CH_4 + 2\,H_2O$$

With all components in the ideal gas state, the standard enthalpy of the process is exothermic by $-165\,kJ$ (-39.4 kcal) per mole of methane formed. Biomass can serve as the original source of hydrogen, which then effectively acts as an energy carrier from the biomass to carbon dioxide, to produce substitute (or synthetic) natural gas (SNG) (see FUELS, SYNTHETIC).

Distribution of Carbon. Estimation of the amount of biomass carbon on the earth's surface is a problem in global statistical analysis. Although reasonable projections have been made using the best available data, maps, surveys, and a host of assumptions, the validity of the results is impossible to support with hard data because of the nature of the problem. Nevertheless, such analyses must be performed to assess the feasibility of biomass energy systems and the gross types of biomass available for energy applications.

The results of one such study are summarized in Table 1 (4). Each ecosystem on the earth is considered in terms of area, mean net carbon production per year,

Table 1. Estimate of Net Photosynthetic Production of Dry Biomass Carbon and Standing Biomass Carbon for World Biosphere[a,b]

Ecosystem	Area, 10^6 km²[c]	Mean net carbon production		Standing biomass carbon	
		t/(hm²·yr)[c]	10^9 t/yr[c]	t/hm²	10^9 t
tropical rain forest	17.0	9.90	16.83	202.5	344
boreal forest	12.0	3.60	4.32	90.0	108
tropical seasonal forest	7.5	7.20	5.40	157.5	118
temperate deciduous forest	7.0	5.40	3.78	135.0	95
temperate evergreen forest	5.0	5.85	2.93	157.5	79
Total	*48.5*		*33.26*		*744*
extreme desert-rock, sand, ice	24.0	0.01	0.02	0.1	0.2
desert and semidesert scrub	18.0	0.41	0.74	3.2	5.8
savanna	15.0	4.05	6.08	18.0	27.0
cultivated land	14.0	2.93	4.10	4.5	6.3
temperate grassland	9.0	2.70	2.43	7.2	6.5
woodland and shrubland	8.5	3.15	2.68	27.0	23.0
tundra and alpine	8.0	0.63	0.50	2.7	2.2
swamp and marsh	2.0	13.50	2.70	67.5	14.0
lake and stream	2.0	1.80	0.36	0.1	0.02
Total	*100.5*		*19.61*		*85*
Total continental	*149.0*		*52.87*		*829*
open ocean	332.0	0.56	18.59	0.1	3.3
continental shelf	36.6	1.62	4.31	0.004	0.1
estuaries excluding marsh	1.4	6.75	0.95	4.5	0.6
algae beds and reefs	0.6	11.25	0.68	9.0	0.5
upwelling zones	0.4	2.25	0.09	0.9	0.04
Total marine	*361.0*		*24.62*		*4.5*
Grand total	*510.0*		*77.49*		*833.5*

[a]Ref. 4.

[b]Dry biomass is assumed to contain 45% carbon.

[c]1 km² = 1 × 10^6 m² (0.3861 sq. mi); to convert t/(hm²·yr) to short ton/(acre·yr), divide by 2.24.

and standing biomass carbon, ie, carbon contained in biomass on the earth's surface and not including carbon stored in biomass underground. Forest biomass, produced on only 9.5% of the earth's surface, contributes more than any other source to the total net carbon fixed on earth. Marine sources of net fixed carbon also are high because of the large area of earth occupied by water. However, the high turnover rates of carbon in the marine environment result in relatively small steady-state quantities of standing carbon. The low turnover rates of forest biomass make it the largest contributor to standing carbon reserves. Forests produce about 43% of the net carbon fixed each year and contain over 89% of the standing biomass carbon of the earth; tropical forests are the largest sources of these carbon reserves. Temperate deciduous and evergreen forests also are large sources of biomass carbon, followed by the savanna and grasslands. Cultivated land is one of the smaller producers of fixed carbon and is only about 9% of the total terrestrial area of the earth.

Human activity, particularly in the developing world, continues to make it more difficult to sustain the world's biomass growth areas. It has been estimated that tropical forests are disappearing at a rate of tens of thousands of hm² per year. Satellite imaging and field surveys show that Brazil alone has a deforestation rate of approximately 8×10^6 hm^2/yr (5). At a mean net carbon yield for tropical rain forests of 9.90 t/hm^2·yr (4) (4.42 short ton/acre·yr), this rate of deforestation corresponds to a loss of 79.2×10^6 t/yr of net biomass carbon productivity.

The remaining carbon transport mechanisms on earth are primarily physical mechanisms, such as the solution of carbonate sediments in the sea and the release of dissolved carbon dioxide to the atmosphere by the hydrosphere (6). The great bulk of carbon, however, is contained in the lithosphere as carbonates in rock. These carbon deposits contain little or no stored chemical energy, although some high temperature deposits could provide considerable thermal energy, and all of the energy for a synfuel system must be supplied by a second raw material, such as elemental hydrogen. These carbon deposits consist of lithospheric sediments and atmospheric and hydrospheric carbon dioxide. Together, these carbon sources comprise 99.9% of the total carbon estimated to exist on the earth. Fossil fuel deposits are only about 0.05% of the total, and the nonfossil energy-containing deposits make up the remainder, about 0.02%.

Biomass carbon is thus a very small, but important, fraction of the total carbon inventory on earth. It helps maintain the delicate balance among the atmosphere, hydrosphere, and biosphere necessary to support all life forms, and serves as a perpetual source of food and materials. Biomass carbon also has served as a primary energy source for the industrialized nations of the world; it continues to do so for developing countries. Biomass carbon may again become a dominant source of energy products throughout the world because of fossil fuel depletion and environmental problems, eg, the effect that large-scale fossil fuel combustion is believed by many to have on atmospheric carbon dioxide build-up (7). The utilization of biomass carbon as a primary energy source does not add any new carbon dioxide to the atmosphere; it is simply recycled between the surface of the earth and the air over a period of time that is extremely short compared to the recycling time of fossil-derived carbon dioxide.

Energy Potential

The percentage of energy demand that could be satisfied by particular nonfossil energy resources can be estimated by examination of the potential amounts of energy and biofuels that can be produced from renewable carbon resources and comparison of these amounts with fossil fuel demands.

The average daily incident solar radiation, or insolation, that strikes the earth's surface worldwide is about 220 W/m^2 (1675 Btu/ft^2). The annual insolation on 0.01% of the earth's surface is approximately equal to all energy consumed (ca 1992) by humans in one year, ie, 321×10^{18}J (305×10^{15} Btu). In the United States, the world's largest energy consumer, annual energy consumption is equivalent (1992) to the insolation on about 0.1 to 0.2% of U.S. total surface.

Based on the state of technology in the early 1990s, the most widespread and practical mechanism for capture of this energy is biomass formation. The energy content of standing biomass carbon, ie, the above-ground biomass reservoir that in theory could be harvested and used as an energy resource (Table 1) is about 110 times the world's annual energy consumption (8). Using a nominal biomass heating value of 16×10^9 J/dry t (13.8×10^6 Btu/short ton), the solar energy trapped in 17.9×10^9 t of biomass, or about 8×10^9 t of biomass carbon, would be equivalent to the world's fossil fuel consumption in 1990 of 286×10^{18} J. It is estimated that 77×10^9 t of carbon, or 171×10^9 t of biomass equivalent, most of it wild and not controlled, is fixed on the earth each year. Biomass should therefore be considered as a raw material for conversion to large supplies of renewable substitute fossil fuels. Under controlled conditions dedicated biomass crops could be grown specifically for energy applications.

A realistic assessment of biomass as an energy resource is made by calculating average surface areas needed to produce sufficient biomass at different annual yields to meet certain percentages of fuel demand for a particular country (Table 2). These required areas are then compared with surface areas available. The conditions of biomass production and conversion used in Table 2 are either within the range of 1993 technology and agricultural practice, or are believed to be attainable in the future.

Figure 2 shows the three yield levels in Table 2 together with the percentage of the U.S. area needed to supply SNG from biomass for any selected gas demand. Although relatively large areas are required, the use of land- or freshwater-based biomass for energy applications is still practical. The area distribution pattern of the United States (Table 3) shows selected areas or combinations of areas that might be utilized for biomass energy applications (9), ie, areas not used for productive purposes. It is possible that biomass for both energy and foodstuffs, or energy and forest products applications, can be grown simultaneously or sequentially in ways that would benefit both. Relatively small portions of the bordering oceans also might supply needed biomass growth areas, ie, marine plants would be grown and harvested. The steady-state carbon supplies in marine ecosystems can conceivably be increased under controlled conditions over 1993 low levels by

Table 2. Potential Substitute Natural Gas in United States from Biomass at Different Crop Yields

Demand, %[a]	Average area required, km^2 [b,c,d]		
	25 t/(hm²·yr)	50 t/(hm²·yr)	100 t/(hm²·yr)
1.58	20,400	10,200	5,100
10	129,000	64,500	32,300
50	645,500	323,000	161,000
100	1,291,000	645,500	323,000

[a]United States demand estimated to be 244×10^8 GJ or 653×10^9 m³ (231×10^{11} standard cubic feet at 15.5°C, 101.5 kPa (60°F, 30.00 in Hg) dry (SCF)). A percentage of 1.58 is equal to a daily production of 28.3×10^6 m³ at normal conditions (1×10^9 SCF) of SNG.

[b]Biomass, whether trees, plants, grasses, algae, or water plants, has a heating value of 15.1×10^9 J/dry t, and is converted in integrated biomass planting, harvesting, and conversion systems to SNG at an overall thermal efficiency of 50%.

[c]1 km² = 0.3861 sq. mi.

[d]Yields expressed as dry t.

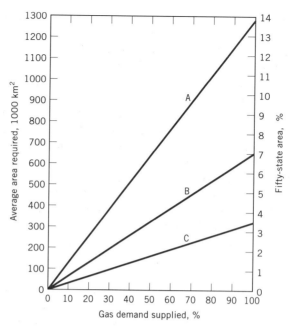

Fig. 2. Average area required and percent of U.S. area vs % gas demand for a total gas demand of 24.4 EJ/yr. A, 25 t/(hm²·yr); B, 50 t/(hm²·yr); C, 100 t/(hm²·yr) (Table 2). To convert t/(hm²·yr) to short ton/(acre·yr), divide by 2.24.

means of marine biomass energy plantations in areas of the ocean dedicated to this objective.

Waste biomass is another large source of renewable carbon supply. It consists of a wide range of materials and includes municipal solid wastes (MSW), municipal sewage, industrial wastes, animal manures, agricultural crop and forestry residues, landscaping and tree clippings and trash, and dead biomass that results from nature's life cycles. Several of these wastes can cause serious health or environmental problems if not disposed of properly. Some wastes, such as MSW, can be considered a source of recyclables such as metals and glass in addition to energy. Waste biomass is a potential energy resource in the same manner as virgin biomass.

To assess the potential availability and impact of energy from wastes on energy demand, the energy contents and availabilities of different types of wastes generated must be considered. For example, in the United States an average of 2.3 kg of MSW/d is discarded per person. From an energy standpoint, one t of MSW has an as-received energy content of about 10.5×10^9 joule (10 million Btu), so about 2.2 EJ/yr of energy potential resides in the MSW discarded in the United States.

The amount of energy that can actually be recovered from a given waste and utilized depends on the waste type. The amount of available MSW is larger than the total amount of available agricultural wastes even though much larger quantities of agricultural wastes are generated. A larger percentage of MSW is collected for centralized disposal than the corresponding amounts of agricultural

Table 3. Land and Water Areas of United States[a]

Area classification	10^6 hm$^{2\,b}$	%
nonfederal land		
forest[c]	179.41	18.8
rangeland[d]	178.66	18.7
other land[e]	279.09	29.2
transition land[f]	14.41	1.5
	651.57	*68.2*
federal land		
forest[c]	102.14	10.7
rangeland[d]	133.10	13.9
other land[e]	25.70	2.7
	260.94	*27.3*
water		
inland water	24.75	2.6
other water	19.28	2.0
	44.03	*4.6*
Total land and water	*956.54*	

[a]Ref. 9. Data for forest, rangeland, and other land as of 1982; data on inland water as of 1980; data on other water as of 1970.
[b]1 hm^2 = 2.471 acre.
[c]At least 10% stocked by trees of any size, or formerly having such tree cover and not currently developed for nonforest use.
[d]Climax vegetation is predominantly grasses, grass-like plants, forbs, and shrubs suitable for grazing and browsing.
[e]Includes crop and pasture land and farmsteads, strip mines, permanent snow and ice, and land that does not fit into any other land cover.
[f]Forest land that carries grasses or forage plants used for grazing as the predominant vegetation.

wastes, most of which are left in the fields where generated; the collection costs would be prohibitive for most agricultural wastes.

Several studies estimate the potential of available virgin and waste biomass as energy resources (Table 4) (10). In Table 4, the projected potential of the recoverable materials is about 25% of the theoretical maximum; woody biomass is about 70% of the total recoverable potential. These estimates of biomass energy potential are based on existing, sustainable biomass production and do not include new, dedicated biomass energy plantations that might be developed.

U.S. Market Penetration. Table 5 shows U.S. consumption of biomass energy in 1990 and projected consumption for 2000 (10,11). The projected consumption for 2000 is about 50% greater than the consumption of biomass energy in 1990.

A projection of biomass energy consumption in the United States for the years 2000, 2010, 2020, and 2030 is shown in Table 6 by end use sector (12). This analysis is based on a National Premiums Scenario which assumes that specific market incentives are applied to all new renewable energy technology deployment. The scenario depends on the enactment of federal legislation equivalent to

Table 4. Potential U.S. Biomass Energy Available in 2000, EJ[a,b]

Energy source	Estimated recoverable	Theoretical maximum
wood and wood wastes	11.0	26.4
municipal solid wastes		
combustion	1.9	2.1
landfill methane	0.2	1.1
herbaceous biomass and agricultural residues	1.1	15.8
aquatic biomass	0.8	8.1
industrial solid wastes	0.2	2.2
sewage methane	0.1	0.2
manure methane	0.05	0.9
miscellaneous wastes	0.05	1.1
Total	*15.4*	*57.9*

[a] Ref. 10. 1 EJ = 0.9488×10^{15} Btu.
[b] Gross heating value of biomass or methane. Conversion of biomass or methane to another biofuel requires that the process conversion efficiency be used to reduce the potential energy available. These figures do not include additional biomass from dedicated energy plantations.

Table 5. U.S. Consumption of Biomass Energy, EJ[a]

Resource	1990,[b] EJ	2000,[c] EJ
wood and wood wastes		
industrial sector	1.646	2.2
residential sector	0.828	1.1
commercial sector	0.023	0.04
utilities	0.013	0.01
subtotal (wood)	*2.510*	*3.35*
municipal solid wastes	0.304	0.63
agricultural and industrial wastes	0.040	0.08
methane		
landfill gas	0.033	0.100
digester gas	0.003	0.004
thermal gasification	0.001	0.002
subtotal (methane)	*0.037*	*0.106*
transportation fuels		
ethanol	0.063[d]	0.1[d]
other biofuels	0	0.1
subtotal (transportation fuels)	*0.063*	*0.2*
Total all resources	*2.954*	*4.37*
primary energy consumption, %	3.3	4.8

[a] 1 EJ = 0.9488×10^{15} Btu. To convert from EJ to barrels of oil equivalent per day (BOE/d), multiply by 448,200.
[b] Refs. 10 and 11. Total energy consumption including biomass energy estimated to be 88.426 EJ in 1990.
[c] Ref. 10. Assumes noncrisis conditions, tax incentives and PURPA in place continued to 2000, no legislative mandates to embark on an off-oil campaign, and total consumption of 91.7 EJ in 2000.
[d] Domestic consumption only.

Table 6. Projected Biomass Energy Consumption in the United States from 2000 to 2030,[a] EJ[b]

End use sector	2000	2010	2020	2030
industry[c]	2.85	3.53	4.00	4.48
electricity[d]	3.18	4.41	4.95	5.48
buildings[e]	1.05	1.53	1.90	2.28
liquid fuels[f]	0.33	1.00	1.58	2.95
Total	7.41	10.47	12.43	15.19

[a]Ref. 12.
[b]1 EJ = 0.9488×10^{15} Btu. Assumes market incentives of 2 ¢/kWh on fossil fuel-based electricity generation, $2.00/$10^6$ Btu on direct coal and petroleum consumption, and $1.00/$10^6$ Btu on direct natural gas consumption.
[c]Combustion of wood and wood wastes.
[d]Electric power derived from present (ca 1992) technology via the combustion of wood and wood wastes, MSW, agricultural wastes, landfill and digester gas, and advanced digestion and turbine technology.
[e]Biomass combustion in wood stoves.
[f]Ethanol from grains, and ethanol, methanol, and gasoline from energy crops.

a fossil fuel consumption tax. Any incentives over and above those in place (ca 1992) for use of renewable energy will have a significant impact on biomass energy consumption.

The market penetration of synthetic fuels from biomass and wastes in the United States depends on several basic factors, eg, demand, price, performance, competitive feedstock uses, government incentives, whether established fuel is replaced by a chemically identical fuel or a different product, and cost and availability of other fuels such as oil and natural gas. Detailed analyses have been performed to predict the market penetration of biomass energy well into the twenty-first century. A range of from 3 to about 21 EJ seems to characterize the results of most of these studies.

U.S. capacity for producing biofuels manufactured by biological or thermal conversion of biomass must be dramatically increased to approach the potential contributions based on biomass availability. For example, an incremental EJ per year of methane requires about 210 times the biological methane production capacity that now exists, and an incremental EJ per year of fuel ethanol requires about 14 times existing ethanol fermentation plant capacity. The long lead times necessary to design and construct large biomass conversion plants makes it unlikely that sufficient capacity can be placed on-line before the year 2000 to satisfy EJ blocks of energy demand. However, plant capacities can be rapidly increased if a concerted effort is made by government and private sectors.

Projections of market penetrations and contributions to primary consumption of energy from biomass are subject to much criticism and contain significant errors. However, even though these projections may be incorrect, they are necessary to assess the future role and impact of renewable energy resources, and to help in deciding whether a potential renewable energy resource should be developed.

Global Market Penetration. The consumption of all energy resources worldwide in 1990, according to the United Nations, is presented in Table 7 (8). Detailed

Table 7. Global Energy Consumption in 1990, EJ[a]

| Region | Fossil fuels[b] | | | Electricity[c] | Biomass[d] | Total[e] |
	Solids	Liquids	Gases			
Africa	2.96	3.36	1.55	0.18	4.68	12.73
N. America	21.55	38.48	22.13	4.69	1.77[f]	88.62
S. America	0.68	4.66	2.09	1.29	2.71	11.43
Asia	35.52	27.58	8.38	2.57	8.89	82.94
Europe[g]	35.18	40.90	37.16	6.25	1.29	120.85
Oceania	1.64	1.70	0.85	0.14	0.19	4.53
Total world	*97.52*	*116.68*	*72.18*	*15.13*	*19.53*	*321.10*

[a]Ref. 8.

[b]Solids are hard coal, lignite, peat, and oil shale. Liquids are crude petroleum and natural gas liquids. Gases are natural gas.

[c]Includes hydro, nuclear, and geothermal sources, but not fossil fuel-based electricity, which is included in fossil fuels.

[d]Includes fuelwood, charcoal bagasse, and animal, crop, pulp, paper, and municipal solid wastes, but does not include derived biofuels.

[e]Sums of individual figures may not equal totals because of rounding.

[f]Less than Table 5 value of 2.954 EJ for the United States because does not include biofuels from biomass.

[g]Includes former Soviet Union.

and time-consuming analysis is necessary to assure validity of the results for an energy resource that is as widespread, dispersed, and disaggregated as biomass, and many nations of the world do not require the archiving of historical energy production and consumption data. Table 7 indicates that biomass energy is a significant source of energy in the developing regions of the world; Africa, 36.8% of total energy consumed; South America, 23.7%; and Asia, 10.9%. It is a small energy resource in the industrialized areas relative to fossil fuels. The markets for biomass energy as replacements and substitutes for fossil fuels are large and have only been developed to a limited extent. As fossil fuels are either phased out because of environmental issues or become less available because of depletion, biomass energy is expected to acquire an increasingly larger share of the organic fuels market.

Chemical Characteristics of Biomass

The chemical characteristics of biomass vary over a broad range because of the many different types of species. Table 8 compares the typical analyses and energy contents of land- and water-based biomass, ie, wood, grass, kelp, and water hyacinth, and waste biomass, ie, manure, urban refuse, and primary sewage sludge, with those of cellulose, peat, and bituminous coal. Pure cellulose, a representative primary photosynthetic product, has a carbon content of 44.4%. Most of the renewable carbon sources listed in Table 8 have carbon contents near this value. When adjusted for moisture and ash contents, it is seen that with the exception of the sludge sample, the carbon contents are slightly higher than that of cellulose, but span a relatively narrow range.

Table 8. Composition and Heating Value of Biomass, Wastes, Peat, and Coal

Analysis	Pure cellulose	Pine wood	Kentucky bluegrass	Giant brown kelp[a]	Feedlot manure	Urban refuse[b]	Primary sewage sludge	Reed sedge peat	Illinois bituminous coal
elemental, wt %									
C	44.44	51.8	45.8	27.65	35.1	41.2	43.75	52.8	69.0
H	6.22	6.3	5.9	3.73	5.3	5.5	6.24	5.45	5.4
O	49.34	41.3	29.6	28.16	33.2	38.7	19.35	31.24	14.3
N		0.1	4.8	1.22	2.5	0.5	3.16	2.54	1.6
S		0.0	0.4	0.34	0.4	0.2	0.97	0.23	1.0
C (MAF)[c]	44.44	52.1	52.9	45.3	45.9	47.9	59.5	57.2	75.6
proximate, wt %									
moisture		5–50	10–70	85–95	20–70	18.4	90–97	84.0	7.3
volatile matter		99.5	86.5	61.1	76.5	86.1	73.47	92.26	91.3
ash		0.5	13.5	38.9	23.5	13.9	26.53	7.74	8.7
high heating value, MJ/kg[d]									
dry	17.51	21.24	18.73	10.01	13.37	12.67[e]	19.86	20.79	28.28
MAF[c]	17.51	21.35	21.65	16.38	17.48		27.03	22.53	30.97
carbon	39.40	41.00	40.90	36.20	38.09		45.39	39.38	40.99

[a]*Macrocystis pyrifera.*
[b]Combustible fraction.
[c]Moisture and ash free.
[d]To convert MJ/kg to Btu/lb, multiply by 430.
[e]As received with metals.

27

The organic components that make up biomass depend on the species. Alphacellulose [9004-34-6], or cellulose as it is more generally known, is the chief structural element and a principal constituent of many biomass types. In trees, eg, the concentration of cellulose is about 40 to 50% of the dry weight; materials such as lignin and compounds related to cellulose, such as hemicelluloses, comprise most of the remaining organic components. However, cellulose is not always the dominant component in the carbohydrate fraction of biomass. For example, it is a minor component in giant brown kelp; mannitol [87-78-5], a hexahydric alcohol that can be formed by reduction of the aldehyde group of D-glucose to a methylol group, and alginic acid [9005-32-7], a polymer of mannuronic and glucuronic acids, are the primary carbohydrates.

Fat and protein content of plant biomass are much less on a percentage basis than the carbohydrate components. Fatty constituents are usually present at the lowest concentration; the protein fraction is much higher in concentration but lower than that of carbohydrate. Crude protein values can be approximated by multiplying the organic nitrogen analyses by 6.25. The average weight percentage of nitrogen in pure dry protein is about 16%, although the protein content of each biomass species can best be determined by amino acid assay. The calculated crude protein values of the biomass species listed in Table 8 range from a low of about 0% for pine wood, to a high of about 30% for Kentucky Blue Grass. For grasses, the protein content is strongly dependent on growing procedures used before harvest, particularly fertilization methods. However, some biomass species, such as legumes, fix nitrogen from the ambient atmosphere and often contain high protein concentrations.

The energy content of biomass is a very important parameter from the standpoint of conversion to energy products and synfuels. The different components in biomass have different heats of combustion because of different chemical structures and carbon content. Table 9 lists heating values for each of the classes of organic compounds. The more reduced the state of carbon in each class, the higher the heating value. As carbon content increases and degree of oxygenation is reduced, the structures become more hydrocarbon-like and heating value increases.

Table 9. Fuel Values of Biomass Components[a,b]

Component	Carbon, %	MJ/kg[c,d]
monosaccharides	40	15.6
disaccharides	42	16.7
polysaccharides	44	17.5
lignin	63	25.1
crude protein	53	24.0
fat	75	39.8
carbohydrate	41–44	16.7–17.7
crude fiber[e]	47–50	18.8–19.8

[a]Ref. 13. Approximate values.
[b]Product water in liquid state.
[c]Dry.
[d]To convert MJ/kg to Btu/lb, multiply by 430.
[e]Contains ca 15–30% lignin.

Fatty components thus have the highest heating values of the components in Table 9. Cellulose, the dominant component in most biomass, has a high heating value of 17.51 MJ/kg (7533 Btu/lb).

Typical low heating values of selected biomass are given in Table 10. The water-based algae, *Chlorella*, has a higher energy content value than woody and fibrous materials because of its higher lipid or protein contents. Oils derived from plant seeds are much higher in energy content and approach the heating value of paraffinic hydrocarbons. High concentrations of inorganic components in a given biomass species greatly affect its heating value because inorganic materials generally do not contribute to heat of combustion, eg, giant brown kelp, which leaves an ash residue equivalent to about 40 wt % of the dry weight, has a high heating value on a dry basis of about 10 MJ/kg, and on a dry, ash-free basis, the heating value is about 16 MJ/kg (Table 8).

Table 10. Low Heating Values[a] of Biomass and Fossil Materials[b]

Material	MJ/kg[c]
wood	
pine	21.03
beech	20.07
birch	20.03
oak	19.20
oak bark	20.36
bamboo	19.23
fiber	
coconut shells	20.21
buckwheat hulls	19.63
bagasse	19.25
green algae	
Chlorella	26.98
seed oils	
cottonseed	39.77
rapeseed	39.77
linseed	39.50
amorphous carbon	33.8
paraffinic hydrocarbon	43.3
crude oil	48.2

[a]Product water in vapor state.
[b]Refs. 14 and 15.
[c]Dry; to convert MJ/kg to Btu/lb, multiply by 430.

Biomass Conversion

Various processes can be used to produce energy or gaseous, liquid, and solid fuels from biomass and wastes. In addition, chemicals can be produced by a wide range of processing techniques. The following list summarizes the principal feed, process, and product variables considered in developing a synfuel-from-biomass process.

Feeds	Conversion processes	Products

Land-based
  trees
  plants
  grasses
Water-based
  single-cell algae
  multicell algae
  water plants
	Separation, combustion,
	  pyrolysis, hydrogena-
	  tion, anaerobic fermen-
	  tation, aerobic fermen-
	  tation, biophotolysis,
	  partial oxidation, steam
	  reforming, chemical hy-
	  drolysis, enzyme hydrol-
	  ysis, other chemical
	  conversions, natural
	  processes
		Energy
		  thermal, steam, electric
		Solid fuels
		  char, combustibles
		Gaseous fuels
		  methane (SNG), hydrogen,
		    low and medium thermal-
		    value gas, light hydro-
		    carbons
		Liquid fuels
		  methanol, ethanol, higher
		    hydrocarbons, oils
		Chemicals

There are many interacting parameters and possible feedstock–process–product combinations, but all are not feasible from a practical standpoint; eg, the separation of small amounts of metals present in biomass and the direct combustion of high moisture content algae are technically possible, but energetically unfavorable.

Moisture content of the biomass chosen is especially important in the selection of a suitable conversion process. The giant brown kelp, *Macrocystis pyrifera*, contains as high as 95% intracellular water, so thermal gasification techniques such as pyrolysis and hydrogasification cannot be used directly without first drying the algae. Anaerobic digestion methods are preferred because the water does not need to be removed. Wood, on the other hand, can often be processed by several different thermal conversion techniques without drying. Figure 3 illustrates the effects of thermal drying on biomass used for synfuel production as SNG. A large portion of a feed's equivalent energy content can be expended for drying, so the properties of the feed must be considered carefully in relation to the conversion process.

Table 11 lists the important feed characteristics to be examined when developing a successful conversion process for a specific biomass feedstock. A particular process also may have specific requirements within a given process type; eg, anaerobic digestion and alcoholic fermentation are both biological conversion processes, but animal manure, which has a relatively high biodegradability, is not equally applicable as a feedstock for both processes. The degree of complexity of the process design also affects practical utility of the conversion process. Some processes, such as combustion, are simple in design, whereas others, such as alcoholic fermentation, consist of several different unit operations and are complex. Capital and operating costs are dictated by the particular process design, the logistics of raw material supply, plant size, and operating conditions. Generally, the more complex the process, the higher the costs.

The need to meet environmental regulations can affect processing costs. Undesirable air emissions may have to be eliminated and liquid effluents and solid residues treated and disposed of by incineration or/and landfilling. It is possible

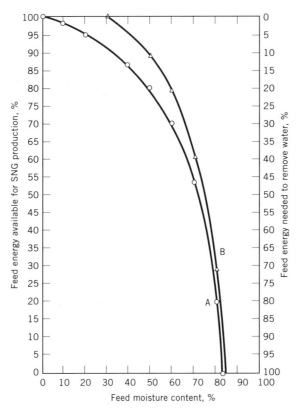

Fig. 3. Effect of feed moisture content on energy available for synfuel production. Assumes feed has a heating value of 11.63 MJ/kg (5000 Btu/lb) dry. A, 0% moisture in dried feed; B, 30% moisture in dried feed. For example, reduction of an initial moisture content of 70 wt % by thermal drying to 30% moisture content requires the equivalent of 37% feed energy content and leaves 63% feed energy available for SNG production.

Table 11. Biomass Feedstock Characteristics that Affect Suitability of Conversion Process

Feedstock characteristic	Process type			
	Physical	Thermal	Biological	Chemical
water content	+	+	+	+
energy content	+	+		+
chemical composition		+	+	+
bulk component analysis	+	+	+	+
size distribution	+	+	+	+
noncombustibles	+	+	+	
biodegradability			+	
carbon reactivity		+		+
organism content/type			+	

for biomass conversion processes that utilize waste feedstocks to combine waste disposal and treatment with energy and/or biofuel production so that credits can be taken for negative feedstock costs and tipping or receiving fees.

The primary types of conversion processes for biomass can be divided into four groups, ie, physical, biological–biochemical, thermal, and chemical.

Physical Processes.   *Particle Size Reduction.*   Changes in the physical characteristics of a biomass feedstock often are required before it can be used as a fuel. Particle size reduction (qv) is performed to prepare the material for direct fuel use, for fabrication into fuel pellets, or for a conversion process. Particle size of the biomass also is reduced to reduce its storage volume, to transport the material as a slurry or pneumatically, or to facilitate separation of the components.

The ultimate particle size required depends on the conversion process used, eg, for thermal gasification processes the particle size of the material converted can influence the rate at which the gasification process occurs. Biological processes, such as anaerobic digestion to produce methane, also are affected by the size of the particle; the smaller the particle, the higher the reaction rate because more surface area is exposed to the organisms. Particle size reduction consists of one or more unit operations that make up the front end of the total processing system. Two basic types of machines are in commercial use (ca 1992) for particle size reduction, ie, wet shredders and dry shredders. Wet shredders utilize a hydropulping mechanism in which a high speed cutting blade pulverizes a water slurry of the feed over a perforated plate. The pulped material passes through the plate and the nonpulping materials are ejected. The two most common types of dry shredders in commercial use are the vertical and horizontal shaft hammermills. Rotating metal hammers on a shaft reduce the particle size of the feed material until the particles are small enough to drop through the grate openings. Particle size reduction units such as agricultural choppers and tree chippers are usually hammermills or are equipped with knife blades that reduce particle size by a cutting or shearing action. Maintenance costs for dry shredders generally are higher than those for wet shredders.

Separation.   It may be desirable to separate the feedstock into two or more components for different applications. Examples include separation of agricultural biomass into foodstuffs and residues that may serve as fuel or as a raw material for synfuel manufacture, separation of forest biomass into the darker bark-containing fractions and the pulpable components, separation of marine biomass to isolate various chemicals, and separation of urban refuse into the combustible fraction, ie, refuse-derived fuel (RDF), and metals and glass for recycling. Common operations such as screening, air classification, magnetic separation, extraction, distillation, filtration, and crystallization often are used as well as industry-specific methods characteristic of farming, forest products, and specialized industries.

Drying.   Drying refers to the vaporization of all or part of the water in the feedstock. In cases where the biomass or waste is thermally processed directly for energy recovery, it may be necessary to partially dry the raw feed before conversion; otherwise, more energy might be consumed to operate the process than that produced in the form of recovered energy or fuels. Open-air solar drying is perhaps the cheapest drying method if it can be used. Raw materials that are not sufficiently stable to be dried by solar methods can be dried more rapidly using spray driers, drum driers, and convection ovens. For large-scale applications, forced-air-type furnaces and driers designed to use stack gases are more efficient. Special driers such as those that use powdered feeds, and hot metal balls separated from the feed for reheating, also have been successful.

Fabrication. Processes for fabricating solid fuel pellets from a variety of feedstocks, particularly RDF, wood, and wood and agricultural residues, have been developed. The pellets are manufactured by extrusion and other techniques and, in some cases, a binding agent such as a thermoplastic resin is incorporated during fabrication. The fabricated products are reported to be more uniform in combustion characteristics than the raw biomass. Depending on the composition of the additives in the pelletized fuel, the heat of combustion can be higher or lower than that of the unpelletized material.

Biological–Biochemical Processes. Fermentation is a biological process in which a water slurry or solution of raw material interacts with microorganisms and is enzymatically converted to other products. Biomass can be subjected to fermentation conditions to form a variety of products. Two of the most common fermentation processes yield methane and ethanol. Biochemical processes include those that occur naturally within the biomass.

Anaerobic Digestion. Methane can be produced from water slurries of biomass by anaerobic digestion in the presence of mixed populations of anaerobes. This process has been used for many years to stabilize municipal sewage sludges for purposes of disposal. Presuming the biomass is all cellulose, the chemistry can be represented in simplified form as follows:

$$(C_6H_{10}H_5)_x + x\ H_2O \xrightarrow{\text{hydrolysis}} x\ C_6H_{12}O_6$$

$$x\ C_6H_{12}O_6 \xrightarrow{\text{acidification}} 3x\ CH_3COOH$$

$$3x\ CH_3COOH \xrightarrow{\text{methanation}} 3x\ CH_4 + 3x\ CO_2$$

Complex organic compounds are first converted in the water slurry to lower molecular weight soluble products, primarily carboxylic acids, by the acidogenic bacteria present in the digester. Methanogenic bacteria then convert these intermediates to a medium heat value (MHV) gas which has heating values ranging from about 19.6 to 29.4 MJ/m^3 at normal conditions (500 to 750 Btu/SCF, dry) at 60°F, 30.00 in Hg (15.5°C, 101.5 kPa). The principal components in the gas are methane and carbon dioxide (Fig. 4). Residual ungasified solids which contain more nitrogen, phosphorus, and potassium than the feed solids also are formed. In some systems, these solids have application as animal feeds and fertilizers. The conventional high rate digestion process is conducted under nonsterile conditions in large, mixed, anaerobic fermentation vessels at near-ambient pressures, temperatures of about 35°C (mesophilic range) or 55°C (thermophilic range), and reactor residence times of 10 to 20 days. The pH is maintained in the range 6.8 to 7.2. The raw digester gas has been used for many years as a fuel to heat the digesters and for steam and electric power production. It also can be upgraded to SNG by removing the carbon dioxide by means of adsorption or acid-gas scrubbing processes.

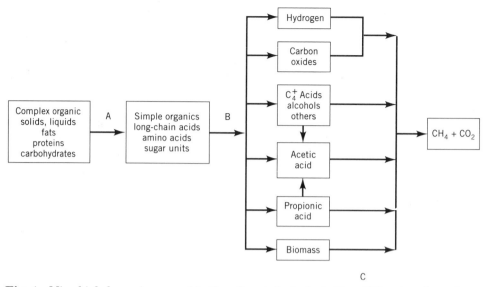

Fig. 4. Microbial phases in anaerobic digestion: A, hydrolysis; B, acidification; C, methane fermentation.

The overall thermal efficiency of the anaerobic digestion of biomass to methane is a function of the process design and the raw material characteristics. Without pretreatment of the feed or the recycled ungasified solids to increase biodegradability, the feed can generally be gasified at overall thermal efficiencies ranging from about 30 to 60%. Yields of methane can range up to 0.30 m^3 at normal conditions/kg (5 SCF/lb) of volatile solids (VS) added to the digesters. A typical biological gasification plant can contain hydrolysis units, anaerobic digesters, gas cleanup and dehydration units, and liquid effluent treatment units (Fig. 5). Advanced digester designs under development include two-phase, plug-flow, packed-bed, fluidized-bed, and sludge blanket digesters (16).

Typical methane yields and volatile solids reductions observed under standard high rate conditions are shown in Table 12. Longer detention times will increase the values of these parameters, eg, a methane yield of 0.284 m^3 at normal conditions/kg VS added (4.79 SCF/lb VS added) and volatile solids reduction of 53.9% for giant brown kelp at a detention time of 18 days instead of the corresponding values of 0.229 and 43.7 at 12 days under standard high rate conditions. However, improvements might be desirable in the reverse direction, ie, at shorter detention times.

Alcoholic Fermentation. Certain types of starchy biomass such as corn and high sugar crops are readily converted to ethanol under anaerobic fermentation conditions in the presence of specific yeasts (*Saccharomyces cerevisiae*) and other organisms (Fig. 6). However, alcoholic fermentation of other types of biomass, such as wood and municipal wastes that contain high concentrations of cellulose, can be performed in high yield only after the cellulosics are converted to sugar concentrates by acid- or enzyme-catalyzed hydrolysis:

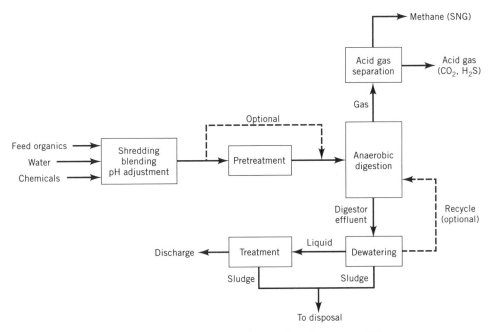

Fig. 5. Methane production by anaerobic digestion of biomass.

$$(C_6H_{10}O_5) + x\,H_2O \xrightarrow{\text{hydrolysis}} x\,C_6H_{12}O_6$$

$$x\,C_6H_{12}O_6 \xrightarrow{\text{fermentation}} 2x\,C_2H_5OH + 2x\,CO_2$$

Advanced processes for conversion of cellulosics are being developed (17). A commercial alcohol fermentation plant for biomass would include units to shred and separate the combustible fermentable organic fraction from the nonfermentable components, hydrolysis units to produce glucose concentrates if the feed were high in low degradability cellulose components, fermenters, distillation towers, and dehydration units (Fig. 7). Under conventional conditions, the degradable organics are converted in the fermenters, at high efficiencies and residence times of 1 to 2 days, to a beer that contains about 10% ethanol. This broth is heated to remove the product alcohol as overhead by distillation, and the resulting 50 to 55% alcohol distillate is distilled again to yield 95% alcohol and by-product aldehydes and fusel oil. Bottoms from the beer still contain low volatility components from the fermentation, called stillage, which is often processed further to yield high protein animal feeds.

The thermal efficiency of ethanol production from fermentable sugars is high, but the overall thermal efficiency of the process is low because of the many energy-consuming steps, the nonfermentable fraction in biomass, and the by-products formed. Alcohol yields are about 40 to 50 wt% of the weight of the fermentable fraction in the feed. Substantial improvements in the overall thermal efficiency of alcoholic fermentation are possible by improving the thermal efficiencies of the auxiliary unit operations. Alcoholic fermentation under reduced

Table 12. Comparison of Methane Fermentation Performance Under High Rate Mesophillic Conditions[a,b]

Component or measure of performance	Primary sewage sludge	Primary activated sludge	RDF–sludge blend	Biomass–waste blend	Coastal Bermuda grass	Kentucky bluegrass	Giant brown kelp	Water hyacinth
carbon, wt% (dry)	43.7	41.8	42.1	43.1	47.1	46.2	26.0	41.0
nitrogen, wt% (dry)	4.02	4.32	1.91	1.64	1.96	4.3	2.55	1.96
phosphorus, wt% (dry)	0.59	1.30	0.81	0.43	0.24		0.48	0.46
ash, wt% of total solids	26.5	23.5	8.4	17.2	5.05	10.5	45.8	22.7
volatile matter, wt% of total solids	73.5	76.5	91.6	82.8	95.0	89.8	54.2	77.3
heating value, MJ/kg (dry)	19.86	18.31	17.20	20.92	19.04	19.19	10.26	16.02
C/N ratio	10.9	9.7	22.0	26.3	24.0	10.7	10.2	20.9
C/P ratio	74.1	32.2	52.0	100	196		54.2	89.1
gas production rate, volume(n)/liquid volume-day	0.74	0.84	0.59	0.52	0.56	0.52	0.62	0.47
methane in gas, mol %	68.5	65.5	60.0	62.0	55.9	60.4	58.4	62.8
methane yield, m³(n)/kg VS added	0.313	0.327	0.210	0.201	0.208	0.150	0.229	0.185
volatile solids reduction, %	41.5	49.0	36.7	33.3	37.5	25.1	43.7	29.8
substrate energy in gas, %	46.2	54.4	39.7	38.3	41.2	27.6	49.1	35.7

[a]Ref. 16.
[b]Daily feeding, continuous mixing, 35°C, pH 6.7 to 7.2, 12-day hydraulic retention time, 1.6 kg volatile solids/m³·d except for kelp, which was 2.1 kg VS/m³·d. All biomass substrates 1.2 mm or less in size.

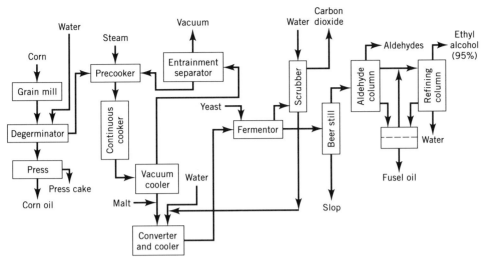

Fig. 6. Flow scheme for manufacture of ethyl alcohol from corn.

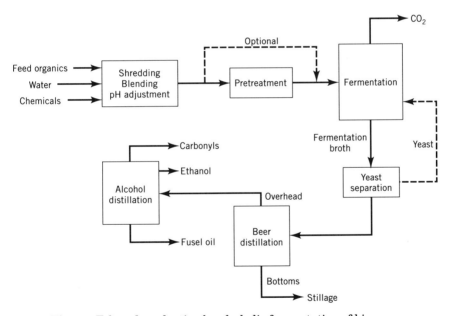

Fig. 7. Ethanol production by alcoholic fermentation of biomass.

pressure, systems in which polysaccharide hydrolysis and fermentation occur together, and improved heat exchange and alcohol-drying processes are under development to provide improved process performance.

Biophotolysis. The decomposition of water (splitting) to hydrogen and oxygen using the radiant energy of visible light and the photosynthetic apparatus of green plants, certain bacteria, and blue-green algae, is called biophotolysis. The concept has been studied in the laboratory, but has not been developed to the point where a practical process exists. Basically, biophotolysis involves the oxidation of water to liberate molecular oxygen and electrons which are raised from the level of the water-oxygen couple, eg, $+0.8$ V, to 0.0 V by Photosystem II.

$$H_2O \xrightarrow{\text{light}} \tfrac{1}{2} O_2 + 2\,H^+ + 2\,e^-$$

The electrons undergo the equivalent of a partial oxidation process in a dark reaction to a positive potential of $+0.4$ V, and Photosystem I then raises the potential of the electrons to as high as -0.7 V. Under normal photosynthesis conditions, these electrons reduce tryphosphopyridine-nucleotide (TPN) to TPNH, which reduces carbon dioxide to organic plant material. In the biophotolysis of water, these electrons are diverted from carbon dioxide to a microbial hydrogenase for reduction of protons to hydrogen:

$$2\,H^+ + 2\,e^- \rightarrow H_2$$

Thus, the overall chemistry is simply the photolysis of water to hydrogen and oxygen.

Synthetic water-splitting membranes that contain the biochemical and other catalysts necessary to form hydrogen also are under development. These membranes mimic natural photosynthesis except that the electrons are directed to form hydrogen. Several sensitizers and catalysts are needed to complete the cycle, but progress is being made. Various single-stage schemes, in which hydrogen and oxygen are produced separately, have been studied, and the thermodynamic feasibility of the chemistry has been experimentally demonstrated.

The upper limit of efficiency of the biophotolysis of water has been projected to be 3% for well-controlled systems. This limits the capital cost of useful systems to low cost materials and designs. But the concept of water biophotolysis to afford a continuous, renewable source of hydrogen is quite attractive and may one day lead to practical hydrogen-generating systems.

Natural Processes.   Hydrocarbon production in land-based biomass by natural chemical mechanisms is a well-known phenomenon. Commercial production of natural rubber, the highly stereospecific polymer *cis*-1,4-polyisoprene [9003-31-0], is an established technology (see ELASTOMERS, SYNTHETIC–POLYISOPRENE; RUBBER, NATURAL). Natural rubber has a mol wt range between about 500,000 and 2,000,000 and is tapped as a latex from the hevea rubber tree (*Hevea braziliensis*). The desert shrub guayule, which grows in the southwestern United States and in northern Mexico, is another biomass species studied as a source of natural rubber almost identical to hevea rubber (18). The idea of growing guayule and extracting the rubber latex from the whole plant was tested in full-scale plantations during the rubber shortage in World War II and found technically feasible. Terpene extraction from pine trees and other biomass species is also established technology.

Many plants native to North America, or that can be grown there, have been tested as sources of oils (triglycerides) and hydrocarbons (19,20). The objectives of this work have been to identify those biomass species that produce hydrocarbons, especially those of lower molecular weight than natural rubber, so that they would be more amenable to standard petroleum refining methods; to characterize hydrocarbon yields and those of other organic compounds; and to learn what controls the structure and molecular weight of hydrocarbons within the plant so that genetic manipulation or other biomass modifications can be applied to control hydrocarbon structures. Some efforts have concentrated on desert plants that might be grown in arid or semiarid areas without competition from biomass grown for foodstuffs. Other work has been aimed at perennial species adapted to wide areas of North America. Several biomass species have been found to contain oils and/or hydrocarbons (Table 13). It is apparent that oil or hydrocarbon formation is not limited to any one family or type of biomass. Interestingly, some species in the Euphorbiaceae family, which includes *Hevea braziliensis*, form hydrocarbons having molecular weights considerably less than that of natural rubber at yields as high as 10 wt% of the plant. This corresponds to hydrocarbon yields of about $3.97 \text{ m}^3/\text{hm}^2 \cdot \text{yr}$ (25 bbl/hm$^2 \cdot$yr).

Figure 8 illustrates one of the processing schemes used for separating various components in a hydrocarbon-containing plant. Acetone extraction removes the polyphenols, glycerides, and sterols, and benzene extraction removes the hydrocarbons. If the biomass species in question contain low concentrations of the nonhydrocarbon components, exclusive of the carbohydrate and protein fractions, direct extraction of the hydrocarbons with benzene or a similar solvent might be preferred.

The principal steps in the mechanism of polyisoprene formation in plants are known and should help to improve the natural production of hydrocarbons. Mevalonic acid, a key intermediate derived from plant carbohydrate via acetyl-coenzyme A, is transformed into isopentenyl pyrophosphate (IPP) via phosphorylation, dehydration, and decarboxylation (see ALKALOIDS). IPP then rearranges to dimethylallyl pyrophosphate (DMAPP). DMAPP and IPP react with each other, releasing pyrophosphate to form another allyl pyrophosphate containing 10 carbon atoms. The chain can successively build up by five-carbon units to yield polyisoprenes by head-to-tail condensations; alternatively, tail-to-tail condensations of two C_{15} units can yield squalene, a precursor of sterols. Similar condensation of two C_{20} units yields phytoene, a precursor of carotenoids. This information is expected to help in the development of genetic methods to control the hydrocarbon structures and yields.

Other sources of natural oils are the oilseed crops, many of which have been used in nonfuel applications, and microalgae (21). Oils from these types of biomass are largely triglycerides and typically contain three long-chain primary fatty acids, each bound to one of the carbon atoms of glycerol via an ester linkage (see FATS AND FATTY OILS). The viscosity and other properties of these oils vary with the degree of saturation of the fatty acid; the more paraffinic oils have higher viscosities and melting points. The oils can be upgraded to diesel fuels or gasoline plus diesel fuels by transesterification or by catalytic cracking or hydrocracking. Transesterification with methanol or ethanol yields monoesters in the C_{15-20} range depending on the oil source (22). The monoesters have viscosity and volatility characteristics similar to conventional diesel fuels. Carbon build-up and crankcase oil contamination in diesel engines vary with the degree of saturation and with the service characteristics of the diesel engine. Catalytic cracking of the

Table 13. Oil- and Hydrocarbon-Producing Biomass Species Potentially Suitable for North America[a]

Family	Genus and species	Common name
Aceraceae	*Acer saccharinum*	silver maple
Anacardiaceae	*Rhus glabra*	smooth sumac
Asclepiadaceae	*Asclepias incarnata*	swamp milkweed
	sublata	desert milkweed
	syriaca	common milkweed
	Cryptostegia grandiflora	Madagascar rubber vine
Buxaceae	*Simmondsia chinensis*	jojoba
Caesalpiniaceae	*Copaifera langsdorfii*	copaiba
	multijuga	
Caprifoliaceae	*Lonicera tartarica*	red tarterium honeysuckle
	Sambucus canadensis	common elder
	Symphoricarpos orbiculatus	corral berry
Companulaceae	*Companula americana*	tall bellflower
Compositae	*Ambrosia trifida*	giant ragweed
	Cacalia atriplicifolia	pale Indian plantain
	Chrysathamnus nauseosus	rabbitbrush
	Circsium discolor	field thistle
	Eupathorium altissimum	tall boneset
	Parthenium argentatum	guayule
	Silphium integrifolium	rosin weed
	laciniatum	compass plant
	terbinthinaceum	prairie dock
	Solidago graminifolia	grass-leaved goldenrod
	leavenworthii	Edison's goldenrod
	rigida	stiff goldenrod
	Sonchus arvensis	sow thistle
	Vernonia fasciculata	ironweed
Curcurbitaceae	*Cucurbita foetidissima*	buffalo gourd
Euphorbiaceae	*Euphorbia denta*	
	lathyris	mole plant, gopher plant
	pulcherima	poinsetta
	tirucalli	African milk bush
Gramineae	*Agropyron repens*	quack grass
	Elymus canadensis	wild rye
	Phalaris canariensis	canary grass
Labiatae	*Pycnanthemum incanum*	western mountain mint
	Teucrium canadensis	American germander
Lauraceae	*Sassafras albidium*	sassafras
Rhamnaceae	*Ceanothus americanus*	New Jersey tea
Rosaceae	*Prunus americanus*	wild plum
Phytolaccaceae	*Phytolacea americana*	pokeweed

[a]Ref. 20.

40

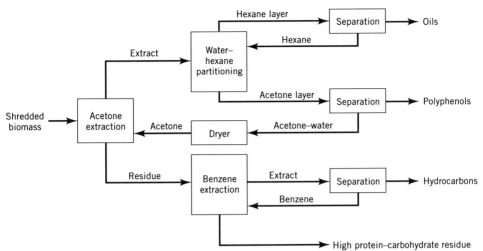

Fig. 8. Operations for processing hydrocarbon-containing biomass.

oils over shape-selective zeolites gives substantial yields of aromatic-rich gasoline-range liquids (23). Catalytic hydrocracking of similar vegetable oils is reported to yield diesel fuel additives (24) or high quality gasolines (25).

Average and potential yields for seed oils are shown in Table 14 (26). If the seed oil is the only product for which revenue is realized, even the high potential yields are insufficient to justify seed oil use as fuel at 1992 petroleum prices, ie, about $20–22 per 159 L (42-gal bbl) of crude oil. Some studies indicate that small-scale transesterification facilities operated as farm cooperatives in the United States can produce biodiesel fuels from seed oils at a profit provided advantage is taken of the Minor Oilseed Provision of the 1990 Farm Bill (27). In this option, the farmer grows rapeseed, for example, on land that is removed from production of a crop such as corn, wheat, cotton, and soybeans. The 1990 Farm Bill permits a farmer to harvest and sell minor oilseed crops grown on this set-aside land without losing his program participation payment. In effect, the farmer is paid land rent by the Government, but can still produce minor oilseed crops.

Work on microalgae has focused on the growth of these organisms under conditions that promote lipid, ie, algal oil, formation. This eliminates the high cost of cell harvest because the lipids often can be separated by simple flotation or extraction (28). Lipid yields greater than 50% of the cell dry weight have been reported when the organisms are grown under nitrogen-limited conditions (29). The oils are high in triglycerides and can be transesterified to form biodiesel fuels in the same manner as seed oils (30). However, the estimated production costs of these fuels still appear to be too high to compete with petroleum-based diesel fuels (28,30).

Thermal Processes. Thermal processes for the production of energy and fuels from biomass and wastes usually involve irreversible chemical reactions, heat, and the transfer of chemical energy from reactants to products. The two largest classes of thermal processes are combustion and pyrolysis. A third class of processes can be described either as a combination of combustion and pyrolysis

Table 14. Commercial Yields of Oilseeds and Seed Oils in the United States[a,b]

| | Seed yield, kg/hm² | | Seed oil yield | | | |
| | | | Average | | Potential | |
Species	Average	Potential	kg/hm²	L/hm²	kg/hm²	L/hm²
castorbean[c] *Ricinus communis*)	950	3,810	428	449	1,504	1,590
Chinese tallow tree (*Sapium sebiferum*)	(12,553)[d]		(5,548)[d]			(6,270)[d]
cotton[c] *Gossypium hirsutum*)	887	1,910	142	150	343	370
crambe[c] (*Crambe abssinica*)	1,121	2,350	392	421	824	940
corn (high oil) (*Zea mays*)		5,940			596	650
flax (*Linum usitatissimum*)	795	1,790	284	309	758	840
peanut (*Arachis hypogaea*)	2,378	5,160	754	814	1,634	1,780
safflower (*Carthamus tinctorius*)	1,676	2,470	553	599	888	940
soybean (*Glycine max*)	1,980	3,360	354	383	591	650
sunflower (*Helianthus annuus*)	1,325	2,470	530	571	986	1,030
winter rape (*Brassica napus*)		2,690			1,074	1,220

[a]Ref. 26.
[b]Growth conditions are dryland unless otherwise noted.
[c]Irrigated.
[d]Not an average yield from several sources; is one reported yield equivalent to 6,270 L/hm² of oil plus tallow. It is believed that yield would be substantially less than this in a managed dense stand, but still higher than that of conventional oilseed crops.

reactions, or as a thermochemical process in which conversion of the feed is facilitated by a reactant such as water or hydrogen. For convenience, these processes are grouped together as miscellaneous processes. Gaseous, liquid, and solid fuels can be produced by pyrolysis processes and several processes in the miscellaneous category.

Combustion.   Complete combustion, eg, incineration, direct firing, burning, is the rapid chemical reaction of the feed and oxygen to form carbon dioxide, water, and heat. The heat released is a function of the enthalpy of combustion of the biomass. Agricultural products, such as bagasse generated in sugarcane plantations, forestry residues, wood chips, RDF, and even raw garbage, have been used as fuels in combustion systems for many years. The recovered heat has been used for steam production, electric power production in a steam-electric plant, and drying.

Many types of combustion equipment are available commercially. The basic differences in various units reside mainly in the design of the combustion chambers, the operating temperature, and the heat transfer mechanism. Refractory-lined furnaces operating at about 1000°C were standard until the introduction in early 1990s of water-wall incinerators. Ash buildup occurs rapidly in refractory-lined furnaces, and excess air must be introduced to limit the wall temperature. The water-wall incinerator has combustion chamber walls containing banks of tubes through which water is circulated, thereby reducing the amount of cooling air needed. Heat is transferred directly to the tubes to produce steam.

Another type of combustion unit operates at about 1600°C to produce a molten slag which forms a granular frit on quenching rather than the usual ash. The higher operating temperature is obtained by preheating the combustion air or by burning auxiliary fuel.

Fluidized-bed combustion represents still another approach. In these systems, air is dispersed through an orifice plate at the bottom of the combustion unit. The dispersed air passes through a bed of sand or residual inorganic particles recovered from combustion causing the effective volume of the bed to increase and the bed to become fluidized. The feed is fed to this rapidly mixed bed, where flameless combustion occurs at about 650°C. This temperature is substantially below flame temperature and because of the lower heat input requirements, many high moisture feeds can be combusted without supplemental fuel. Many other furnace variations, such as stationary and rotating shaft furnaces, suspension firing systems, and stationary and moving grates, are in commercial use or available for biomass and waste combustion applications.

The specific design most appropriate for biomass, waste combustion, and energy recovery depends on the kinds, amounts, and characteristics of the feed; the ultimate energy form desired, eg, heat, steam, electric; the relationship of the system to other units in the plant, independent or integrated; whether recycling or co-combustion is practiced; the disposal method for residues; and environmental factors.

Pyrolysis.   Pyrolysis, eg, retorting, destructive distillation, carbonization, is the thermal decomposition of an organic material in the absence of oxygen. For biomass and wastes, pyrolysis generally starts at temperatures near 300 to 375°C. Chars, organic liquids, gases, and water are formed in varying amounts, depending particularly on the feed composition, heating rate, pyrolysis temperature, and residence time in the pyrolysis reactor. Higher temperatures and longer residence times promote gas production, while higher liquid and char yields result from lower temperatures and shorter residence times. No matter what the pyrolysis conditions are, with the exception of extremely high temperatures, the product mixture has a complex composition and selectivity for specific products is low even with a single feed component.

Depending on the pyrolysis temperature, the char fraction contains inorganic materials ashed to varying degrees, any unconverted organic solids, and fixed carbon residues produced on thermal decomposition of the organics. The liquid fraction contains a complex mixture of organic chemicals having much lower average molecular weights than the feed. For highly cellulosic feeds, the liquid fraction will usually contain acids, alcohols, aldehydes, ketones, esters, heterocyclic derivatives, and phenolic compounds. The pyrolysis gas is a low heat value (LHV) gas having a heating value of 3.9 to 15.7 MJ/m^3 at normal conditions (100 to 400 Btu/SCF). It contains carbon dioxide, carbon monoxide, methane, hydrogen, ethane, ethylene, minor amounts of higher gaseous organics, and water vapor. It is immediately apparent that if pure pyrolysis products are desired, product separation is a significant problem or further processing to refine the products is necessary.

Pyrolysis processes may be endothermic or exothermic, depending on the temperature of the reacting system. For most biomass feeds containing highly oxygenated cellulosic fractions as the principal components, pyrolysis is endothermic at low temperatures and exothermic at high temperatures. Energy to drive the process often is obtained from a portion of the feed or the pyrolysis products such as the char. At low temperatures, pyrolysis generally is reaction-rate controlled; at high temperatures, the process becomes mass-transfer controlled. The experimental data in Tables 15 and 16 show how temperature affects product yields and gas and char compositions with the combustible fraction of municipal solid waste (31). Gas yield increases as the temperature is increased from 500 to 900°C. Although the heating value of the product gas remains about the same, significant increases in hydrogen concentration and energy yield in the gas occur with increasing temperature. Substantial decreases occur in the carbon dioxide concentration over the same temperature range. Also, as the temperature increases from 500 to 900°C, the char yields decrease along with the volatile mat-

Table 15. Product Yields from Pyrolysis of Municipal Solid Waste Organics[a]

Temperature, °C	Products, wt %			Gas yield	
	Gases	Liquids	Char	Combustibles,[b] m^3/kg	Combustibles, MJ/kg
500	12.3	61.1	24.7	0.114	1.39
650	18.6	59.2	21.8	0.166	2.63
800	23.7	59.7	17.2	0.216	3.33
900	24.4	58.7	17.7	0.202	3.05

[a]Ref. 31.
[b]At normal conditions.

Table 16. Char and Gas Composition from Pyrolysis of Municipal Solid Waste Organics[a]

Component	Temperature, °C			
	500	650	800	900
	Gas			
carbon dioxide, mol %	44.8	31.8	20.6	18.3
carbon monoxide, mol %	33.5	30.5	34.1	35.3
methane, mol %	12.4	15.9	13.7	10.5
hydrogen, mol %	5.56	16.6	28.6	32.5
ethane, mol %	3.03	3.06	0.77	1.07
ethylene, mol %	0.45	2.18	2.24	2.43
high heat value (HHV), MJ/m^{3b}	12.3	15.8	15.4	15.1
	Char			
volatile matter, %	21.8	15.1	8.13	8.30
fixed carbon, %	70.5	70.7	79.1	77.2
ash, %	7.71	14.3	12.8	14.5
high heat value (HHV), MJ/kg^c	28.1	28.6	26.7	26.5

[a]Ref. 31.
[b]At normal conditions. To convert MJ/m^3 to Btu/ft^3, multiply by 25.45.
[c]To convert MJ/kg to Btu/lb, multiply by 430.

ter concentration, but the energy value of the char does not undergo similar changes.

Pyrolysis reactor designs are as varied as combustion unit designs. They include fixed beds, moving beds, suspended beds, fluidized beds, entrained feed solids reactors, stationary vertical shaft reactors, inclined rotating kilns, horizontal shaft kilns, high temperature (1000 to 3000°C) electrically heated reactors with gas-blanketed walls, single and multihearth reactors, and a host of other designs. One of the more innovative pyrolysis processes in development for gas production is a fluidized two-bed system (32–34). This system uses two fluidized-bed reactors containing sand as a heat transfer medium. Combustion of char from the pyrolysis reactor takes place within the combustion reactor. The heat released supplies the energy for pyrolysis of the combustible fraction in the pyrolysis reactor. Heat transfer is accomplished by sand flow from the combustion reactor at 950°C to the pyrolysis reactor at 800°C and return of the sand to the combustion reactor (Fig. 9). This configuration separates the combustion and pyrolysis reactions and keeps the nitrogen in air separated from the pyrolysis gas. It yields a pyrolysis gas that can be readily upgraded to SNG by shifting, scrubbing, and methanating without regard to nitrogen separation. The initial pyrolysis gas from high cellulose feeds contains about 37 mol% hydrogen, 35 mol% carbon monoxide, 16 mol% carbon dioxide, and 11 mol% methane. This is an LHV gas having a high heating value of about 13.5 MJ/m^3 at normal conditions (344 Btu/SCF). The projected gas yields are about 667 m^3 at normal conditions (17,000 SCF) of pyrolysis gas and about 196 m^3 at normal conditions (5000 SCF) of methane per dry ton of feed (32).

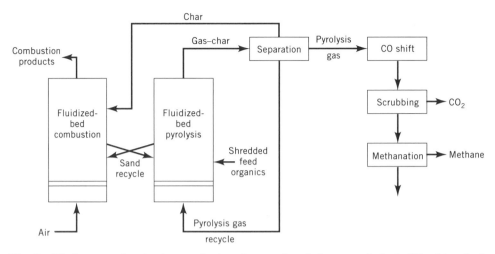

Fig. 9. Methane production by pyrolysis using sand and char recycle in fluidized two-bed system.

An example of a liquid fuel production system under development in a short-residence time pyrolysis process is shown in Figure 10 (35). In this process, high cellulose RDF is dried in a rotary kiln to about 4 wt % moisture content, and finely divided to a particle size of which 80% is smaller than 14 mesh (1200 μm). The feed, about 0.23 kg of recycled char preheated to 760°C per kilogram of this finely divided material, is rapidly passed through the pyrolysis reactor. The raw product mixture, which consists of product gas, the char fed to the reactor, and new char formed on pyrolysis, leaves the reactor at about 510°C. Separation of the gas from the char and rapid quenching to about 80°C yields the liquid fuel. The remaining gas goes through a series of cleanup steps for in-plant use. Part of the gas is used as an oxygen-free solids transport medium and part of it as fuel. The raw product yields are about 10 wt % water, 20 wt % char, 30 wt % gas, and 40 wt % liquid fuel. The product char has a heating value of about 20.9 MJ/kg (9000 Btu/lb), contains about 30 wt % ash, and is produced at an overall yield of about 7.5 wt % of the dry feed. The corresponding values of the liquid fuel are about 24.4 MJ/kg (10,500 Btu/lb), 0.2 to 0.4% ash, and 22.5 wt % of dry feed as received (approximately 1 bbl/short ton of raw refuse). This product has been proposed for use as a heating oil; its properties are compared with a typical No. 6 fuel oil in Table 17. It is apparent that some differences exist, but successful combustion trials in a utility boiler with the liquid fuel have been performed.

A report on the continuous flash pyrolysis of biomass at atmospheric pressure to produce liquids indicates that pyrolysis temperatures must be optimized to maximize liquid yields (36). It has been found that a sharp maximum in the liquid yields vs temperature curves exist and that the yields drop off sharply on both sides of this maximum. Pure cellulose has been found to have an optimum temperature for liquids at 500°C, while the wheat straw and wood species tested have optimum temperatures at 600°C and 500°C, respectively. Organic liquid yields were of the order of 65 wt % of the dry biomass fed, but contained relatively large quantities of organic acids.

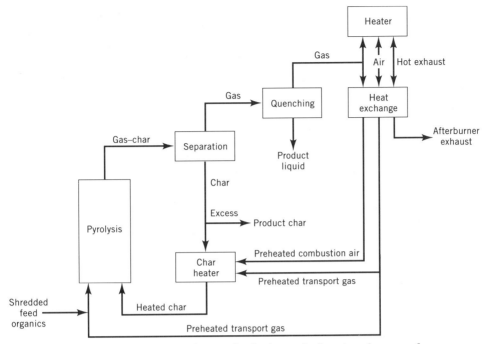

Fig. 10. Liquid-fuel production by flash pyrolysis using char recycle.

Table 17. Properties and Analysis of Liquid Fuel and No. 6 Fuel Oil[a]

Properties	Liquid fuel	No. 6 fuel oil
heating value, MJ/kg[b]	24.6	42.3
density, g/cm^3	1.3	0.98
pour point, °C	32[c]	15–30
flash point, °C	56[c]	65
viscosity at 87.8°C, SUs	1150[c]	90–250
pumping temperature, °C	71	46
atomization temperature, °C	116	104
analysis,[d] wt %		
carbon	57.5	85.7
hydrogen	7.6	10.5
sulfur	0.1–0.3	0.5–3.5
chlorine	0.3	
nitrogen	0.9	2.0
oxygen	33.4	
ash	0.2–0.4	0.5

[a]Ref. 35. Liquid fuel produced by flash pyrolysis using char recycle (Fig. 10).
[b]To convert MJ/kg to Btu/lb, multiply by 430.
[c]Containing 14% water as produced.
[d]Dry basis.

Miscellaneous Thermal Processes.   Many thermal conversion processes can be classified as partial oxidation processes in which the biomass or waste is supplied with less than the stoichiometric amount of oxygen needed for complete combustion. Under these conditions, LHV gases similar to pyrolysis gases are formed that can contain high concentrations of hydrogen and carbon monoxide. Such gaseous mixtures are termed synthesis gases and can be converted to a large number of chemicals and synthetic fuels by established processes (Fig. 11). In some partial oxidation processes, the various chemical reactions may occur simultaneously in the same reactor zone. In others, the reactor may be divided into a combustion zone, which supplies the heat to promote pyrolysis in a second zone, and perhaps a third zone for drying, the overall result of which is partial oxidation. Both air and pure oxygen have been utilized for such systems.

In one system, the three-zoned vertical shaft reactor furnace (Fig. 12), coarsely shredded feed is fed to the top of the furnace. As it descends through the first zone, the charge is dried by the ascending hot gases, which are also partially cleaned by the feed. The gas is reduced in temperature from about 315°C to the range of 40 to 200°C. The dried feed then enters the pyrolysis zone in which the temperature ranges from 315 to 1000°C. The resulting char and ash then descend to the hearth zone, where the char is partially oxidized with pure oxygen. Slagging temperatures near 1650°C occur in this zone and the resulting molten slag of metal oxides forms a liquid pool at the bottom of the hearth. Continuous withdrawal of the pool and quenching forms a sterile granular frit. The product gas is processed to remove fly ash and liquids, which are recycled to the reactor. A typical gas analysis is 40 mol % carbon monoxide, 20 mol % hydrogen, 23 mol % carbon dioxide, 5 mol % methane, and 5 mol % C_2. This gas has an HHV of about 14.5 MJ/m^3 at normal conditions (370 Btu/SCF).

An example of partial oxidation in which air is supplied without zone separation in the gasifier is the molten salt process (39). In this process, the shredded biomass or waste and air are continuously introduced beneath the surface of a sodium carbonate-containing melt which is maintained at about 1000°C. The resulting gas passes through the melt. Acid gases are absorbed by the alkaline salts and the ash is also retained in the melt. The melt is continuously withdrawn for processing to remove the ash and returned to the gasifier. No tars or liquid products are formed in this rather simple process. The heating value of the gases produced depends on the amount of air supplied, and is essentially independent of the type of feed organics. The greater the deficiency of air needed to achieve complete combustion, the higher the fuel value of the product gas. Thus with about

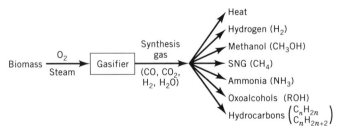

Fig. 11.   Applications of synthesis gas from biomass (37).

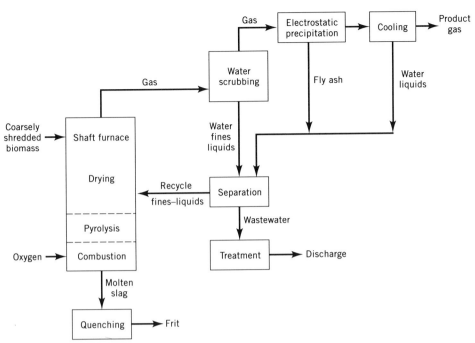

Fig. 12. Production of synthesis gas in three-zone shaft reactor furnace (38).

20, 50, and 75% of the theoretical air, the respective high heating values (HHVs) of the gas are about 9.0, 4.3, and 2.2 MJ/m^3 at normal conditions (230, 110, and 55 Btu/SCF).

Steam also is blended with air in some gasification units to promote the overall process via the endothermic steam–carbon reaction to form carbon monoxide and hydrogen. This was common practice at the turn of the nineteenth century, when so-called producer gasifiers were employed to manufacture LHV gas from different types of biomass and wastes. The producer gas from biomass and wastes had heating values around 5.9 MJ/m^3 at normal conditions (150 Btu/SCF), and the energy yields as gas ranged up to about 70% of the energy contained in the feed. Many gasifier designs were offered for the manufacture of producer gas from biomass and wastes; several types of units are still available for purchase (ca 1992). Thousands of producer gasifiers operating on air and wood were used during World War II, particularly in Sweden, to power automobiles, trucks, and buses. The engines needed only slight modification to operate on LHV producer gas.

Hydrogenation. Another approach to the production of energy products from biomass and wastes is based on hydrogenation. Hydrogen, which can be either generated from the feed or the conversion products, or obtained from an independent source, reacts directly with the feed organics or intermediate process streams at elevated pressures and temperatures to yield substitute fuels. In theory, highly oxygenated feeds should be capable of reduction to liquid and gaseous fuels at any level between the initial oxidation state of the feed and methane.

$$R(OH)_x + y\, H_2 \rightarrow RH_y(OH)_{x-y} + y\, H_2O$$

$$R - R' + H_2 \rightarrow RH + R'H$$

For a cellulosic material containing hydroxyl groups, the reactions might consist of dehydroxylation and depolymerization by hydrogenolysis, during which there is a transition from solid to liquid to gas.

Most of the work on hydrogenation has been concentrated on hydrogasification to produce methane as the final product. One route to methane involve the sequential production of synthesis gas and then methanation of the carbon monoxide with hydrogen to yield methane. The routes shown in Figure 13 involve the direct reaction of the feed with hydrogen (40). In this process, shredded feed is converted with hydrogen-containing gas to a gas containing relatively high methane concentrations in the first-stage reactor. The product char from the first stage is used in a second-stage reactor to generate the hydrogen-rich synthesis gas for the first stage. From experimental results with the first-stage hydrogasifier operated in the free-fall and moving-bed modes at 1.72 MPa and 870°C with pure hydrogen, calculations shown in Table 18 were made to estimate the composition and yield of the high methane gas produced when the first stage is integrated with an entrained char gasifier as the second stage. Although the methane content of the raw product gas is projected to be higher in the moving-bed reactor than in the falling-bed reactor, gas from the first stage must still be adjusted in H_2/CO ratio in a shift converter, scrubbed to remove carbon dioxide, and methanated to obtain SNG.

Another hydrogenation process utilizes internally generated hydrogen for hydroconversion in a single-stage, noncatalytic, fluidized-bed reactor (41). Bio-

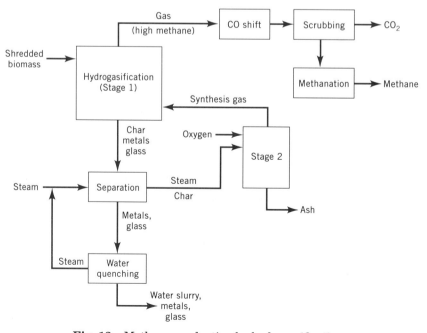

Fig. 13.   Methane production by hydrogasification.

Table 18. Gas Composition and Yield from Integrated Hydrogasification Process at Stage 1[a]

Product	Free fall	Moving bed
composition, mol %		
carbon monoxide	45.9	51.9
hydrogen	31.9	13.3
methane	10.4	17.2
carbon dioxide	10.1	16.1
ethane	1.2	1.1
benzene	0.5	0.4
yield, m³/kg[b] dry feed	1.1	0.95
fraction of total methane produced		
in stage 1 after methanation	0.26	0.52

[a]Fig. 13, estimated.
[b]To convert m³/kg to ft³/lb, multiply by 16.0.

mass is converted in the reactor, which is operated at about 2.1 kPa, 800°C, and residence times of a few minutes with steam-oxygen injection. About 95% carbon conversion is anticipated to produce a medium heat value (MHV) gas which is subjected to the shift reaction, scrubbing, and methanation to form SNG. The cold gas thermal efficiencies are estimated to be about 60%.

Another advancement involves low temperature catalytic gasification of 2 to 10% aqueous biomass slurries or solutions that range from dilute organics in wastewater to waste sludges from food processing (42). The estimated residence time in the metallic catalyst bed is less than 10 min at 360°C and 20,635 kPa (3000 psi) at liquid hourly space velocities of 1.8 to 4.6 L of feedstock/L of catalyst/ hr depending on the feedstock. The product fuel gas contains 45–70 vol % methane, 25–50 vol % carbon dioxide, and less than 5% hydrogen with as much as 2% ethane. The by-product water stream carries residual organics from 40 to 500 ppm COD. The product gas is MHV gas produced directly in contrast to MHV gas- phase processes that require either oxygen in place of air or a two-bed reaction system to keep the nitrogen in air separated from the fuel gas product.

Studies on the gasification of wood in the presence of steam and hydrogen show that steam gasification proceeds at a much higher rate than hydrogasifica- tion (43). Carbon conversions 30 to 40% higher than those achieved with hydrogen can be achieved with steam at comparable residence times. Steam/wood weight ratios up to 0.45 promoted increased carbon conversion, but had little effect on methane concentration. Other experiments show that potassium carbonate- catalyzed steam gasification of wood in combination with commercial methana- tion and cracking catalysts can yield gas mixtures containing essentially equal volumes of methane and carbon dioxide at steam/wood weight ratios below 0.25, with atmospheric pressure and temperatures near 700°C (44). Other catalyst com- binations produced high yields of product gas containing about 2:1 hydrogen/car- bon monoxide and little methane at steam/wood weight ratios of about 0.75 and a temperature of 750°C. Typical results for both of these studies are shown in Table 19. The steam/wood ratios and the catalysts used can have significant ef-

Table 19. Product Gases from Steam Gasification of Wood

Gas or parameter	Value			
gas composition, mol %				
H_2	0	53	29	50
CH_4	52	4	15	17
CO_2	48	12	17	11
CO	0	30	34	17
reactor temperature, °C	740	750	696	762
pressure, kPa[a] (gauge)	0	0	129	159
primary catalyst	K_2CO_3	K_2CO_3	none	wood ash
secondary catalyst	Ni:SiAl	SiAl	none	none
steam/wood weight ratio	0.25	0.75	0.24	0.56
carbon conversion to gas, %	64	77	68	52
feed energy in gas, %	76	78		
heating value of gas,[b] MJ/m^{3c}	20.6	12.1	16.6	17.7
Reference	44[d]	44[d]	45[e]	45[e]

[a]To convert kPa to mm Hg, multiply by 7.5.
[b]At normal conditions.
[c]To convert MJ/m^3 to Btu/ft^3, multiply by 25.45.
[d]Laboratory results with unspecified wood.
[e]Process development unit (PDU) results with unspecified hardwood.

fects on the product gas compositions. The composition of the product gas also can be manipulated depending on whether a synthesis gas or a fuel gas is desired.

Direct hydroliquefaction of biomass or wastes can be achieved by direct hydrogenation of wood chips on treatment at 10,132 kPa and 340 to 350°C with water and Raney nickel catalyst (45). The wood is completely converted to an oily liquid, methane, and other hydrocarbon gases. Batch reaction times of 4 hours give oil yields of about 35 wt % of the feed; the oil contains about 12 wt % oxygen and has a heating value of about 37.2 MJ/kg (16,000 Btu/lb). Distillation yields a significant fraction that boils in the same range as diesel fuel and is completely miscible with it.

A catalytic liquefaction process for heavy liquids production reacts biomass in a water solution of sodium carbonate and carbon monoxide gas at elevated temperature and pressure to form a heavy liquid fuel (46). Biomass and the combustible fraction of wastes have been converted at weight yields of 40 to 60% at temperatures of 250 to 425°C and pressures of 10 to 28 MPa. Lower viscosity products are generally obtained at higher reaction temperatures and solid or semisolid products are obtained when the reaction temperature is below 300°C. However, the high nitrogen and oxygen contents and the boiling characteristics and high viscosity range of the liquid products make it difficult to classify them as synthetic crude oils. Conventional refining methods could not be used to upgrade this kind of material to standard petroleum derivatives. The original process consisted of a sequence of steps: drying and grinding wood chips to a fine powder, mixing the powder with recycled product oil (30% powder to 70% oil), blending the mixture with water containing sodium carbonate, and treatment of the slurry with synthesis gas at about 27,579 kPa (4000 psi) and 370°C. The modified process consists of partially hydrolyzing the wood in slightly acid water and treating the water slurry containing dissolved sugars and about 20% solids with synthesis gas and sodium carbonate at 27,579 kPa and 370°C on a once-through basis. The resulting oil product yield is about 1 bbl/400 kg (158.9 L/400 kg) of chips and is approximately equivalent to No. 6 grade boiler fuel. It contains about 50% phenolics, 18% high boiling alcohols, 18% hydrocarbons, and 10% water.

Study of the mechanism of this complex reduction-liquefaction suggests that part of the mechanism involves formate production from carbonate, dehydration of the vicinal hydroxyl groups in the cellulosic feed to carbonyl compounds via enols, reduction of the carbonyl group to an alcohol by formate and water, and regeneration of formate (46). In view of the complex nature of the reactants and products, it is likely that a complete understanding of all of the chemical reactions that occur will not be developed. However, the liquefaction mechanism probably involves catalytic hydrogenation because carbon monoxide would be expected to form at least some hydrogen by the water-gas shift reaction.

Chemical Processes.   Biological–biochemical and thermal conversion processes are chemical processes, too, but a few specific chemical processes are mentioned separately because they are directed more to conventional chemical processing and production. These processes have been grouped together as chemical processes.

Chemicals have long been manufactured from biomass, especially wood (silvichemicals), by many different fermentation and thermochemical methods. For example, continuous pyrolysis of wood was used by the Ford Motor Co. in 1929 for the manufacture of various chemicals (Table 20) (47). Wood alcohol (methanol) was manufactured on a large scale by destructive distillation of wood for many years until the 1930s and early 1940s, when the economics became more favorable for methanol manufacture from fossil fuel-derived synthesis gas.

In the production of chemicals from biomass, wood is still the raw material of choice for the manufacture of certain chemicals, although many of them cannot compete with fossil-based products. The chemistry of silvichemical production is related directly to the chemical composition of trees, ie, 50% cellulose, 25% hemicelluloses, and 25% lignins. However, specialty chemicals are often manufactured from nonwoody biomass because they occur naturally in certain plant species or can easily be derived from these plants. Examples are the alginic acids from *Macrocystis pyrifera*, ie, giant brown kelp, and physiologically active alkaloids from particular plants. Ethanol has been manufactured for chemical applications from starchy biomass by fermentation for many years. Figure 14 lists some of the more important primary biomass-derived chemicals, the principal intermediates, and the dominant processing methods used. All chemicals listed in Figure 14 are either manufactured commercially in 1992 by the indicated routes or were manufactured in the past. Secondary processing of these primary chemicals would appear to make it possible to manufacture almost all heavy and fine organics produced from fossil raw materials, eg, ethanol (qv) can be converted to ethylene (qv), acetaldehyde (qv), and acetic acid, which can be converted to other organic chemicals by established routes (see ACETIC ACID AND DERIVATIVES).

The availability of C_6 sugars such as glucose is an important factor in the development of a biomass chemicals industry as alluded to in Figure 14. Unfor-

Table 20. Product Yields from Wood Pyrolysis[a,b]

Product	Yield per t dry wood
gas,[c] m^3	156
charcoal, kg	300
ethyl acetate, L	61.1
creosote oil, L	13.6
methanol, L	13.0
ethyl formate, L	5.3
methyl acetate, L	3.9
methyl ethyl ketone, L	2.7
other ketones, L	0.9
allyl alcohol, L	0.2
soluble tar, L	91.8
pitch, kg	33.0

[a]Ref. 47.
[b]Feed: 70% maple, 25% birch, 5% ash, elm, and oak; av temperature, 515°C.
[c]CO_2, 37.9 mol %; CO, 23.4 mol %; CH_4, 16.8 mol %; N_2, 16.0 mol %; O_2, 2.4 mol %; H_2, 2.2 mol %; hydrocarbons, 1.2 mol %. To convert m^3 to ft^3, multiply by 35.3.

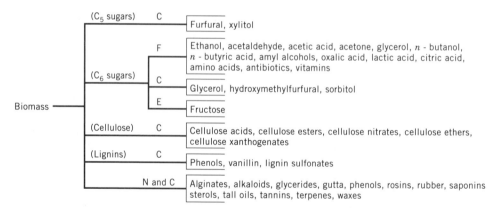

Fig. 14. Primary biomass-derived chemicals. Dominant processing methods are chemical (C), fermentation (F), enzymic (E), and natural (N) processes; products in parentheses represent intermediates.

tunately, most biomass is higher in cellulose than C_6 sugars. Cellulose, because of its relatively low biodegradability, cannot easily be converted to fermentation alcohol and other products without first liberating the monosaccharides. Trees and fibrous biomass, such as plant stalks and reedy plants, contain cellulose in partially crystalline form sometimes complexed with other materials. The low degradability of wood cellulose, which can exist as lignin–cellulose complexes, is attributed to these factors. Such forms of cellulose can be degraded to glucose concentrates by several hydrolytic methods, the most common of which is hydrolysis with dilute sulfuric acid. At the temperatures necessary to form glucose, a large portion of the product is ordinarily converted to by-products such as hydroxymethylfurfural. Glucose yields are usually near 50% of theory because of by-product formation. Enzyme-catalyzed hydrolysis of cellulose has afforded much higher yields of glucose, but the particle size of the cellulosic material subjected to hydrolysis must be reduced to facilitate depolymerization. The cost of particle size reduction tends to outweigh the advantages of higher glucose yields. One approach to improving yield of monosaccharides is first to dissolve the cellulose in a solvent, separate the insolubles from the cellulose solution, and then subject the solution to hydrolysis conditions or precipitate the cellulose before hydrolysis. This method destroys crystallinity and any cellulose complex that may be present, and makes it possible to achieve high yields of glucose even with highly fibrous biomass. Special solvents such as Cadoxen [14874-24-9] and 65% sulfuric acid have been suggested for this application (48). Another approach involves the use of a blocking agent, acetone, which temporarily forms a cyclic ketal (acetonide) with vicinal diols to protect the sugars from degradation under hydrolysis conditions (49). The process has been reported to afford very high yields of sugars and to be equally applicable to high fiber biomass residues.

As mentioned in the biological–biochemical section, another approach to improve alcoholic fermentation combines saccharification and fermentation, ie, simultaneous saccharification and fermentation (SSF). Enzyme-catalyzed cellulose hydrolysis and fermentation to alcohol takes place in the same vessel in the presence of enzyme and yeast (50). Reduced fermenter pressures and enzyme and yeast recycling result in 70 to 80% ethanol yields. These process modifications, coupled with more energy-efficient distillation and heat exchanger improvements, are projected to make fermentation ethanol from low value biomass competitive with industrial ethanol (51).

Multiproduct processes for biomass chemical plants in which the operating conditions can be manipulated to vary the product distribution as a function of demand or other factors, or in which an optimum mix of products is chosen based on feedstock characteristics, appear to have some merit even though no full-scale systems have yet been commercialized. The process depicted in Figure 15 illus-

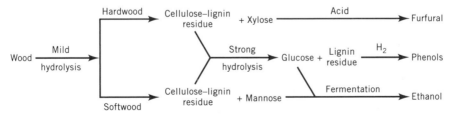

Fig. 15. Furfural, phenols, and ethanol production from wood in a multiproduct process biomass chemical plant (52). Wood (qv) is ca 50% cellulose (qv), 25% lignin (qv), and 25% hemicellulose (qv).

trates how such a plant might function. Mild acid hydrolysis of the hemicelluloses in wood affords either a predominantly xylose or mannose solution, depending on the type of wood feed, and a cellulose–lignin residue. Strong acid treatment of this residue yields a glucose solution, which can be combined with mannose for alcoholic fermentation, and a lignin residue. Phenols can be made from this residue by hydrogenation, and furfural can be made from xylose by strong acid treatment. The products are thus ethanol, furfural, and phenols.

There are many different routes to organic chemicals from biomass because of its high polysaccharide content and reactivity. The practical value of the conversion processes selected for commercial use with biomass will depend strongly on the availability and price of the same chemicals produced from petroleum and natural gas.

Biomass Production

The manufacture of synfuels and energy products from biomass requires that suitable quantities of biomass be grown, harvested, and transported to the conversion plant site. Many variables must be considered when selecting the proper species or mixture of species for operation of a system: growth cycle; fertilization; insolation; temperature; precipitation; propagation and planting procedures; soil and water needs; harvesting methods; disease resistance; growth area competition from biomass for food, feed, and fiber; growth area availability; possibilities for simultaneous or sequential growth of biomass for synfuels and foodstuff or other applications; and nutrient depletion. At least 250,000 botanical species, of which only about 300 are cash crops, are known in the world. A relatively small number are, and will be, used as biomass feedstocks for the manufacture of synfuels and energy products.

In the ideal case, biomass chosen for energy applications should be high yield, low cash-value species that have short growth cycles and grow well in the area and climate chosen for the biomass energy system. Fertilization requirements should be low and possibly nil if the species selected fix ambient nitrogen, thereby minimizing the amount of external nutrients that must be supplied to the growth areas. In areas having low annual rainfall, the species grown should have low water needs and be able to efficiently utilize available precipitation. For land-based biomass, the requirements should be such that the crops can grow well on low grade soils and do not need the best classes of agricultural land. After harvesting, growth should commence again without the need for replanting. Surprisingly, several biomass species meet many of these idealized characteristics and appear to be quite suitable for energy applications. There are a number of important factors that relate to biomass production for energy applications.

Photosynthesis. The basic biochemical pathways in ambient carbon dioxide fixation involve decomposition of water to form oxygen, protons, and electrons; transport of these electrons to a higher energy level via Photosystems I, II, and several electron transfer agents; concomitant generation of reduced nicotinamide adenine dinucleotide [53-57-6] (NADPH) and adenosine triphosphate [56-65-5] (ATP); and reductive assimilation of carbon dioxide to carbohydrate. The initial process is believed to be the absorption of light by chlorophyll [1406-65-1], which promotes decomposition of water. The ejected electrons are accepted by ferredoxin [9080-02-8] (Fd), a nonheme iron protein. The reduced Fd initiates a series of electron transfers to generate ATP from adenosine diphosphate [58-64-0] (ADP), inorganic phosphate, and NADPH. For each of the two light reactions, one photon is required to transfer each electron; a total of eight photons is thus required to fix one molecule of carbon dioxide. Assuming that the carbon dioxide is in the gaseous phase and that the initial product is glucose, the standard Gibbs free energy change at 25°C is $+0.48$ MJ($+114$ kcal) per mole of carbon dioxide assimilated and the corresponding enthalpy change is $+0.47$ MJ ($+112$ kcal).

The maximum efficiency with which photosynthesis can occur has been estimated by several methods. The upper limit has been projected to range from about 8 to 15%, depending on the assumptions made; ie, the maximum amount of solar energy trapped as chemical energy in the biomass is 8 to 15% of the energy of the incident solar radiation. The rationale in support of this efficiency limitation helps to point out some aspects of biomass production as they relate to energy applications.

The relationship of the energy and wavelength of a photon is energy $=$ $\hbar c/\lambda$ where $\hbar$ is Planck's constant, 6.624×10^{-34}; c is velocity of light; and λ is wavelength. Assume that the wavelength of the light absorbed is 575×10^{-9} m and is equivalent to the light absorbed between the blue (400×10^{-9} m) and red (700×10^{-9} m) ends of the visible spectrum. This assumption has been made by several investigators for green plants to calculate the upper limit of photosynthesis efficiency. The energy absorbed in the fixation of one mole of carbon dioxide, which requires 8 photons/molecule, is then given by

$$\text{energy absorbed} = [(6.624 \times 10^{-34})(3.00 \times 10^{8})/(575 \times 10^{-9})] \times 8 \times 6.024 \times 10^{23}$$

$$= 1.67 \text{ MJ}$$

Since 0.47 MJ of solar energy is trapped as chemical energy in this process, the maximum efficiency for total white-light absorption is 28.1%. Further adjustments are usually made to account for the percentages of photosynthetically active radiation in white light, the light that can actually be absorbed, and respiration. The amount of photosynthetically active radiation in solar radiation that reaches the earth is estimated to be about 43%. The fraction of the incident light absorbed is a function of many factors, such as leaf size, canopy shape, and reflectance of the plant; it is estimated to have an upper limit of 80%. This effectively corresponds to the utilization of eight photons out of every 10 in the active incident radiation. The third factor results from biomass respiration. A portion of the stored energy is used by the plant, the amount of which depends on the properties of the biomass species and the environment. For purposes of calculation, assume that about 25% of the trapped solar energy is used by the plant, thereby resulting in an upper limit for retention of the nonrespired energy of 75%. The upper limit for the efficiency of photosynthetic fixation of biomass can now be estimated to be 7.2%, ie, $0.281 \times 0.43 \times 0.80 \times 0.75$. For the case where little or no energy is lost by respiration, the upper limit is estimated to be 9.7%, ie, $0.281 \times 0.43 \times 0.80$. The low efficiency limit might correspond to land-based biomass, while the higher efficiency limit might be closer to water-based biomass such as unicellular algae. These figures can be transformed into dry biomass yields by assuming that all of the fixed carbon dioxide is contained in the biomass as cellulose, $(C_6H_{10}O_5)_x$, from the equation

$$Y = \frac{CIE}{F}$$

where Y is yield of dry biomass, t/hm^2·yr; C is constant, 3.1536; I is average insolation, W/m^2; E is solar energy capture efficiency, %; and F is energy content of dry biomass, MJ/kg. Thus for high cellulose dry biomass, an average insolation of 184 W/m^2 (1404 Btu/ft^2·d), which is the average insolation for the continental United States, a solar energy capture efficiency of 7.2%, and a high heat of combustion of 17.51 MJ/kg for cellulose, the yield of dry biomass is 239 t/hm^2·yr (107 short tons/acre-yr). The corresponding value for an energy capture efficiency of 9.7% is 321 t/hm^2·yr (143 short t/acre·yr). These yields of organic matter can be viewed as an approximation of the theoretical yield limits for land- and water-based biomass. Some estimates of maximum yield are higher and some are lower than these figures, depending on the values used for I, E, and F, but they serve as a guideline to indicate the highest yields that a biomass production system could be expected to achieve under normal environmental conditions. Unfortunately, real biomass yields rarely approach these limits. Sugarcane, for example, which is one of the high yielding species of biomass, typically produces total dry plant matter at yields of about 80 t/hm^2·yr (36 short tons/acre·yr).

Yield is plotted against solar energy capture efficiency in Figure 16 for insolation values of 150 and 250 W/m^2, which spans the range commonly encountered in the United States, and for biomass energy values of 12 and 19 MJ/kg (dry). The higher the efficiency of photosynthesis, the higher the biomass yield. For a given solar energy capture efficiency and incident solar radiation, the yield is projected to be lower at the higher biomass energy values, ie, curves A and C, curves B and D. From an energy production standpoint, this means that a higher energy content biomass could be harvested at lower yield levels and still compete with higher yielding but lower energy content biomass species. It is also apparent that for a given solar energy capture efficiency, yields similar to those obtained with higher energy content species should be possible with a lower energy content species even when it is grown at a lower insolation, eg, curves B and C. Finally, at the solar energy capture efficiency usually encountered in the field, about 1% or less, the spread in yields is much less than at the higher energy capture efficiencies. It is important to emphasize that this interpretation of biomass yield as functions of insolation, energy content, and energy capture, although based on sound principles, is still a theoretical analysis of living systems. Because of the many uncontrollable factors in the field, such as the changes that occur in climate and seasonal changes in biomass composition, departures from the norm can be expected.

The previous discussion of photosynthesis has concentrated on the gross features of ambient carbon dioxide fixation in biomass. However, the biochemical pathways involved in the conversion of carbon dioxide to carbohydrate play an important role in understanding the molecular events of biomass growth. Three

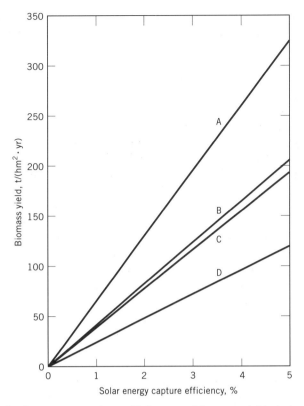

Fig. 16. Effect of solar energy capture efficiency on biomass yield. A, insolation value (I) of 250 W/m^2, and biomass energy value (F) of 12 MJ/kg (dry); B, I of 250 W/m^2, F of 19 MJ/kg; C, I of 150 W/m^2, F of 12 MJ/kg; D, I of 150 W/m^2, F of 19 MJ/kg. To convert W/m^2 to Btu/(ft^2·d), multiply by 7.616; to convert MJ/kg to Btu/lb, multiply by 430.

different biochemical energy transfer pathways occur during carbon dioxide fixation (53–55). One pathway, the Calvin three-carbon cycle, involves an initial three-carbon intermediate of phosphoglyceric acid [*82-11-1*] (PGA). This cycle, often referred to as the reductive pentose phosphate cycle, is used by autotrophic photochemolithotrophic bacteria, algae, and green plants. For every three molecules of carbon dioxide converted to glucose in a dark reaction, nine ATP, six NADPH, and 12 Fd molecules are required. Six molecules of PGA are formed from three molecules each of carbon dioxide and ribulose-1,5-diphosphate (RuDP) in the chloroplasts. After these carboxylation reactions, a reductive phase occurs in which six molecules of PGA are successively transformed into six molecules of diphosphoglyceric acid (DPGA) and then six molecules of 3-phosphoglyceraldehyde, a triose phosphate. Five molecules of the triose phosphate are then used to regenerate three molecules of RuDP, which initiates the cycle again. The other triose phosphate molecule is used to generate glucose-6-phosphate and energy and electron carriers. Plant biomass species that use the Calvin cycle are called C$_3$ plants; it is common in many fruits, legumes, grains, and vegetables. C$_3$ plants usually exhibit low rates of photosynthesis at light saturation, low light saturation points, sensitivity to oxygen concentration, rapid photorespiration, and high carbon dioxide compensation points, ie, about 50 ppm. The carbon dioxide compensation point is the carbon dioxide concentration in the surrounding environment below which more carbon dioxide is respired by the plant than is photosynthetically fixed. Typical C_3 biomass species are peas, sugar beet, spinach, alfalfa, *Chlorella*, *Eucalyptus*, potato, soybean, tobacco, oats, barley, wheat, tall fescue, sunflower, rice, and cotton.

The second pathway is called the C_4 cycle because the carbon dioxide is initially fixed as the four-carbon dicarboxylic acids, malic or aspartic acids. Phosphoenol pyruvate (PEP) reacts with one molecule of carbon dioxide to form oxaloacetate (OAA) in the mesophyll of the biomass, and then malate or aspartate is formed. The C_4 acid is transported to the bundle sheath cells, where decarboxylation occurs to regenerate pyruvate, which is returned to the mesophyll cells to initiate another cycle. The carbon dioxide liberated in the bundle sheath cells enters the C_3 cycle in the usual manner. Thus no net carbon dioxide is fixed in the portion of the C_4 cycle, and it is the combination with the C_3 cycle which ultimately results in carbon dioxide fixation. The subtle differences between the C_4 and C_3 cycles are believed responsible for the wide variations in biomass properties. In contrast to C_3 biomass, C_4 biomass is usually produced at higher yields and has higher rates of photosynthesis, high light saturation points, insensitivity to oxygen concentrations below 21 mol%, low levels of respiration, low carbon dioxide compensation points, and greater efficiency of water usage. C_4 biomass often occurs in areas of high insolation, hot daytime temperatures, and seasonal dry periods. Typical C_4 biomass includes important crops such as corn, sugarcane, and sorghum, and forage species and tropical grasses such as Bermuda Grass. Even crabgrass is a C_4 biomass. At least 100 genera in 10 plant families are known to exhibit the C_4 cycle.

The third pathway is called crassulacean acid metabolism (CAM). CAM refers to the capacity of chloroplast-containing tissues to fix carbon dioxide in the dark via phosphoenolpyruvate carboxylase leading to the synthesis of free malic acid. The mechanism involves the β-carboxylation of PEP by this enzyme and the subsequent reduction of OAA by malate dehydrogenase. CAM has been documented in at least 18 families, including the family Crassulaceae, and 109 genera of the Angiospermae. Biomass species in the CAM category are typically adapted to arid environments, have low photosynthesis rates, and have high water usage efficiencies. Examples are cactus plants and the succulents, such as pineapple. The information developed to date on CAM biomass indicates that CAM has evolved so that initial carbon dioxide fixation can take place in the dark with much less water loss than the fully light-dependent C_3 and C_4 pathways. CAM biomass also conserves carbon by recycling endogenously formed carbon dioxide. Several CAM species show temperature optima in the range 12 to 17°C for carbon dioxide fixation in the dark. The stomates in CAM plants open at night to allow entry of carbon dioxide and then close by day to minimize water loss. The carboxylic acids formed in the dark are converted to carbohydrates when the radiant energy is available during the day. Relatively few CAM plants have been exploited commercially.

Significant differences in net photosynthetic assimilation of carbon dioxide are apparent between C_3, C_4, and CAM biomass species. One of the principal reasons for the generally lower yields of C_3 biomass is its higher rate of photo-respiration; if the photorespiration rate could be reduced, the net yield of biomass would increase. Considerable research is in progress (ca 1992) to achieve this rate reduction by chemical and genetic methods, but as yet, only limited yield improvements have been made. Such an achievement with C_3 biomass would be expected to be very beneficial for foodstuff production and biomass energy applications.

The specific carbon dioxide-fixing mechanism used by a plant will affect the efficiency of photosynthesis, so from an energy utilization standpoint, it is desirable to choose plants that exhibit high photosynthesis rates to maximize the yields of biomass in the shortest possible time. There are numerous factors that affect the efficiency of photosynthesis other than the carbon dioxide-fixing mechanism, eg, insolation; amounts of available water, nutrients, and carbon dioxide; temperature; and transmission, reflection, and biochemical energy losses within or near the plant. For lower plants such as the green algae, many of these parameters are under human control. For conventional biomass growth subjected to the natural elements, it is not feasible to control all of them.

Climate and Environmental Factors. The biomass species selected for energy applications and the climate must be compatible to facilitate operation of fuel farms. The three primary climatic parameters that have the most influence on the productivity of an indigenous or transplanted species are insolation, rainfall, and temperature. Natural fluctuations in these factors remove them from human control, but the information compiled over the years in meteorological records and from agricultural practice supplies a valuable data bank from which to develop biomass energy applications. Ambient carbon dioxide concentration and the availability of nutrients are also important factors in biomass production.

Insolation. The intensity of the incident solar radiation at the earth's surface is a key factor in photosynthesis; natural biomass growth will not take place without solar energy. Insolation varies with location and is high in the tropics and near the equator. The approximate changes with latitude are illustrated in Table 21. At a given latitude, the incident energy is not constant and often exhibits large changes over relatively short distances. A more quantitative summary of insolation values over the continental United States is shown in Table 22. To place the amount of energy that strikes the earth in the proper perspective, the annual insolation on about 0.1 to 0.2% of the surface of the continental United States is

Table 21. Insolation at Various Latitudes for Clear Atmospheres[a]

Location	Latitude	Insolation, W/m²[b]		
		Maximum	Minimum	Average[c]
equator	0°	315	236	263
tropics	23.5°	341	171	263
mid-earth	45°	355	70.9	210
polar circle	66.5°	328	0	158

[a]Ref. 56.
[b]To convert W/m² to Btu/(ft²·d), multiply by 7.616.
[c]Yearly total divided by 365.

Table 22. Daily Solar Radiation in the United States[a]

	Total daily insolation, W/m^2 [b,c]				
Location	January	April	July	October	Annual[c]
Tucson, Arizona	146	289	288	208	229
Fresno, California	93.2	290	338	187	229
Lakeland, Florida	135	260	247	189	210
Indianapolis, Indiana	90.0	188	242	120	157
Lake Charles, Louisiana	109	215	236	175	191
Saint Cloud, Minnesota	75.9	178	275	104	157
Glasgow, Montana	72.0	190	299	118	175
Ely, Nevada	108	257	288	176	210
Oklahoma City, Oklahoma	80.8	212	264	155	183
San Antonio, Texas	113	198	286	182	199
Burlington, Vermont	76.3	182	208	99.7	146
Sterling, Virginia	90.9	173	233	113	159
Seattle–Tacoma, Washington	36.5	179	276	98.1	151

[a]Ref. 57.
[b]To convert W/m^2 to Btu/(ft^2·d), multiply by 7.616.
[c]Average.

equivalent to all the energy consumed by the United States in one year. The production figures shown in Table 23 represent annual yields obtained under good growth conditions (13,14,53,59,60). The estimated solar energy capture efficiencies (SECD) for the biomass listed assumes that all organic matter is cellulose. These are only rough approximations and most are probably too high, but they indicate that C_4 plants are usually better photosynthesizers than C_3 plants, and that high insolation alone does not always correlate with high biomass yield and capture efficiency. Although there are a few exceptions in Table 23, there appears to be a trend in this direction.

Precipitation.   Precipitation as rain, snow, sleet, or hail is governed by movement of air and is generally abundant wherever air currents are predominately upward. The greatest precipitation should therefore occur near the equator. The average annual rainfall in the United States is about 79 cm.

The moisture needs of aquatic biomass presumably are met in full because growth occurs in liquid water, but the growth of land biomass often can be water-limited. Requirements for good growth of many biomass species have been found to be in the range of 50 to 76 cm of annual rainfall (63). Some crops, such as wheat, exhibit good growth with much less water, but they are in the minority. Without irrigation, water is supplied during the growing season by the water in the soil at the beginning of the season and by rainfall. Figure 17 depicts the normal precipitation recorded in the 48-state area during the normal growing season, April to September. This type of information and the established requirements for the growth of land-based biomass can be used to divide the United States into precipitation regions. Regions more productive for biomass generally correlate with precipitation regions. It should be realized, however, that rainfall alone is not quantitatively related to productivity of land biomass because of the

Table 23. Annual Production of Dry Matter and Solar Energy Capture Efficiency (SECD)[a]

Location	Biomass community	Productivity, t/(hm²·yr)[b]	Insolation, W/m²[c]	SECD,[d] %
Sweden	enthrophic lake angiosperm	7.2	106	0.38
Denmark	phytoplankton	8.6	133	0.36
Mississippi	water hyacinth	11.0–33.0	194	0.31–0.94
Minnesota	maize	24.0	169	0.79
New Zealand	temperate grassland	29.1	159	1.02
West Indies	tropical marine angiosperm	30.3	212	0.79
Nova Scotia	sublittoral seaweed	32.1	133	1.34
Georgia	subtropical saltmarsh	32.1	194	0.92
England	coniferous forest, 0–21 yr	34.1	106	1.79
Israel	maize	34.1	239	0.79
New South Wales	rice	35.0	186	1.04
Congo	tree plantation	36.1	212	0.95
Holland	maize, rye, two harvests	37.0	106	1.94
Marshall Islands	green algae	39.0	212	1.02
former West Germany	temperate reedswamp	46.0	133	1.92
Puerto Rico	*Panicum maximum*	48.9	212	1.28
California	algae, sewage pond	49.3–74.2	218	1.26–1.89
Colombia	pangola grass	50.2	186	1.50
West Indies	tropical forest, mixed ages	59.0	212	1.55
Hawaii	sugarcane	74.9	186	2.24
Puerto Rico	*Pennisetum purpurcum*	84.5	212	2.21
Java	sugarcane	86.8	186	2.59
Puerto Rico	napier grass	106	212	2.78
Thailand	green algae	164	186	4.90

[a]Refs. 13, 14, 53, 59, 60.
[b]To convert t/(hm²·yr) to short ton/(acre·yr), divide by 2.24.
[c]Average. To convert W/m² to Btu/(ft²·d), multiply by 7.616.
[d]Approximate estimates of solar energy capture efficiencies; probably too high.

differences in soil characteristics, water evaporation rates, and infiltration. Also, certain areas that have low rainfall can be made productive through irrigation. Finally, some areas of the country that vary widely in precipitation as a function of time, such as many Western states, will produce moderate biomass yields, and often sufficient yields of cash crops without irrigation, to justify commercial production.

Temperature. Most biomass species grow well in the United States at temperatures between 15.6 and 32.3°C. Typical examples are corn, kenaf, and napier grass. Tropical grasses and certain warm-season biomass have optimum growth temperatures in the range 35 to 40°C, but the minimum growth temperature is still near 15°C (64). Cool-weather biomass such as wheat may show favorable growth below 15°C, and certain marine biomass such as the giant brown kelp only survive in water at temperatures below 20 to 22°C (65). The growing season is longer in the southern portion of the United States; in some areas such as Hawaii,

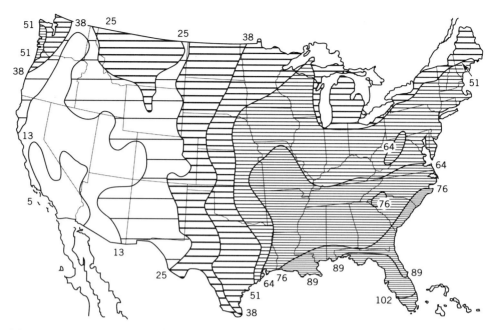

Fig. 17. Numbers indicate approximate precipitation in cm during the growing season, April to September, in the United States (61).

the Gulf states, southern California, and the southeastern Atlantic states, the temperature is usually conducive to biomass growth for most of the year (61,62).

The effect of temperature fluctuations on net carbon dioxide uptake is illustrated by the curves in Figure 18. As the temperature increases, net photosynthesis increases for cotton and sorghum to a maximum value and then rapidly declines. Ideally, the biomass species grown in an area should have a maximum rate of net photosynthesis as close as possible to the average temperature during the growing season in that area.

Ambient Carbon Dioxide Concentration. Many studies have been performed which show that higher concentrations of carbon dioxide than are normally present in air will promote more carbon fixation and increase biomass yields. In confined environmentally controlled enclosures such as hothouses, carbon dioxide-enriched air can be used to stimulate growth. This is not practical in large-scale open systems such as those envisaged for biomass energy farms. For aquatic biomass production, carbon dioxide enrichment of the water phase may be an attractive method of promoting biomass growth if carbon dioxide concentration is a limiting factor; the growth of biomass often occurs by uptake of carbon dioxide from both the air and liquid phase near the surface.

For some high growth-rate biomass species, the carbon dioxide concentration in the air among the leaves of the plant often is considerably less than that in the surrounding atmosphere. Photosynthesis may be limited by the carbon dioxide concentrations under these conditions when wind velocities are low and insolation is high.

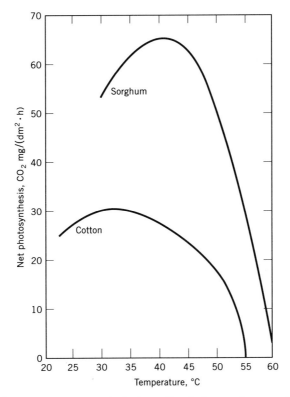

Fig. 18. Effect of temperature on net photosynthesis for sorghum and cotton leaves. To convert mg/(dm^2·h) to lb/(ft^2·h), multiply by 2.373 × 10^{-3}.

Nutrients. All living biomass requires nutrients other than carbon, hydrogen, and oxygen to synthesize cellular material. Principal nutrients are nitrogen, phosphorus, and potassium; other nutrients required in lesser amounts are sulfur, sodium, magnesium, calcium, iron, manganese, cobalt, copper, zinc, and molybdenum. The last five nutrients, as well as a few others not listed, are sometimes referred to as micronutrients because only trace quantities are needed to stimulate growth. For land-based biomass, these elements are usually supplied by the soil, so the nutrients are depleted if they are not replaced through fertilization. Some biomass, such as the legumes, are able to meet all or part of their nitrogen requirements through fixation of ambient nitrogen. Water-based biomass such as marine kelp use the natural nutrients in ocean waters. Freshwater biomass such as water hyacinth is often grown in water enriched with nutrients in the form of wastewater. The growth of the plant is stimulated, and at the same time the influent waste is stabilized because its components are taken up by the plant as nutrients. So-called luxuriant growth of water hyacinth on sewage, in which more than the needed nutrients are removed from the waste, can be used as a substitute wastewater treatment method (28).

Whole plants typically contain 2 wt % N, 1 wt % K, and 0.5 wt % P, so at a yield of 20 t/hm^2·yr (8.9 short ton/acre·yr), harvesting of the whole plant without return of any of the plant parts to the soil corresponds to the annual removal of 400 kg N, 200 kg K, and 100 kg P per hm^2. This illustrates the importance of fertilization, especially of these macronutrients, to maintain fertility of the soil. Biomass growth is often nutrient-limited and yield correlates with fertilizer dose rates. Average nitrogen fertilizer applications for production of wheat, rice, potato, and brussel sprouts are about 73, 134, 148, and 180 kg/hm^2, respectively, in the United States (67). Estimates of balanced fertilizers needed to produce various land biomass species are shown in Table 24. Note that alfalfa does not require added nitrogen because of its nitrogen-fixing ability. It is estimated that this legume can fix from about 130 to 600 kg of elemental nitrogen per hm^2 annually (68).

Normal weathering processes that occur in nutritious soils release nutrients, but they are not available at rates that promote maximum biomass yields. Fertilization is usually necessary to maximize yields. Since nitrogenous fertilizers are largely manufactured from fossil fuels, mainly natural gas, and since fertilizer needs are usually the most energy intensive of all inputs in a biomass production system, careful analysis of the integrated biomass production-conversion system is needed to ensure that net energy production is positive. Trade-offs between synfuel outputs, nonsolar energy inputs, and biomass yields are required to operate a system that produces only energy products.

Land Availability.   The availability of sufficient land suitable for production of land-based biomass can be estimated for the United States by several techniques. One method relies on the land capabilities classification scheme developed by the U.S. Department of Agriculture (69), in which land is divided into eight classes. Classes I to III are suited for cultivation of many kinds of crops; Class IV is suited only for limited production; and Classes V to VIII are useful only for permanent vegetation such as grasses and trees. The U.S. Department of Agriculture surveyed nonfederal land usage for 1987 in terms of these classifications (70). Out of 568 million hm^2, which corresponds to 60% of the 50-state area, 43% of the land (246.3 × 10^6 hm^2) was in Classes I to III, 13% (75.59 × 10^6 hm^2) in Class IV; and 43% (246.3 × 10^6 hm^2) in Classes V to VIII. The actual usage of this land at the time of the survey is shown in Table 25 (70). Table 25 illustrates

Table 24. Fertilizer Requirements of Biomass[a,b], kg

Biomass	N	P_2O_5	K_2O	CaO
alfalfa	0.0	12.3	34.0	20.7
corn	11.8	5.7	10.0	0.0
kenaf	13.9	5.0	10.0	16.1
napier grass	9.6	9.3	15.8	8.5
slash pine[c]	3.8	0.9	1.6	2.3
potato	16.8	5.3	28.3	0.0
sugar beet	18.0	5.4	31.2	6.1
sycamore	7.3	2.8	4.7	0.0
wheat	12.9	5.3	8.4	0.0

[a]Ref. 62.
[b]Estimated per whole dry plant.
[c]Five years.

Table 25. Nonfederal Rural Land Use in United States by Type, 1987,[a] 10^6 hm^2

Land class	Cropland[b]	Pastureland[c]	Rangeland[d]	Forest[e]	Minor	Total
I	11.58	0.82	0.17	0.66	0.23	13.47
II	77.31	12.81	6.58	18.00	2.66	117.36
III	54.28	16.00	18.76	24.26	2.23	115.53
IV	18.60	10.30	21.62	23.56	1.51	75.59
V	1.16	1.86	1.99	7.52	1.03	13.55
VI	6.56	6.84	53.34	37.34	2.63	106.71
VII	1.59	3.91	58.40	46.88	4.08	114.86
VIII	0.035	0.067	1.68	1.40	7.81	11.00
other[f]	0	0	0	0	2.08	2.08
Total	*171.12*	*52.60*	*162.56*	*159.62*	*24.25*	*570.15*
%	30.01	9.22	28.51	28.00	4.25	

[a]Ref. 70. Totals may not be precise summations due to rounding.
[b]Land used for production of crops for harvest alone or in rotation with grasses and legumes.
[c]Land used for the production of adapted, introduced, or native species in a pure stand, grass mixture, or a grass–legume mixture.
[d]Land on which the vegetation is predominantly grasses, grass-like plants, forbs, or shrubs suitable for grazing or browsing.
[e]Land that is at least 10% stocked by forest trees of any size or formerly having had such tree cover and not currently developed for nonforest use.
[f]Land, such as farmsteads, stripmines, quarries, and other lands, that do not fit into any other land class category.

that of all the land judged suitable for cultivation in Classes I to III, only about one-half of it is actually used as cropland (70), and that the combined areas of pasture, range, and forest lands is about 66% of the total nonfederal lands. There is ample opportunity to produce biomass for energy applications on nonfederal land not used for foodstuffs production. Large areas of land in Classes V to VIII not suited for cultivation, and sizable areas in Classes I to IV not being used for crop production, also would appear to be available for biomass energy applications; land used for crop production could be considered for simultaneous or sequential growth of biomass for foodstuffs and energy. Portions of federally owned lands, which are not included in the survey, might also be dedicated to biomass energy applications. Careful design and management of land-based biomass production areas could result in improvement or upgrading of lands to higher land capability classifications.

Water Availability. The production of marine biomass in the ocean, even on the largest scale envisaged for energy applications, would require only a very small fraction of the available ocean areas. The U.S. Navy has estimated that a square area 753 km on each edge off the coast of California may be sufficient to produce enough giant brown kelp for conversion to methane to supply all of the nation's natural gas needs. This large area is very small when compared with the total area of the Pacific Ocean. Also, the benefits to other marine life from a large kelp plantation have been well documented (65). Any conflicts that might arise would be concerned primarily with ocean traffic. With the proper plantation design for marine biomass and precautionary measures to warn approaching ships, it is expected that marine biomass growth could be sustained over long periods.

Freshwater biomass in theory can be grown on the 20 million hm^2 of fresh water in the United States. However, several difficulties mitigate against large-scale freshwater biomass energy systems. About 80% of the fresh water in the United States is located in the northern states, while several of the freshwater biomass species considered for energy applications require a warm climate such as that found in Gulf states. The freshwater areas suitable for biomass production in the southern states, however, are much smaller than those in the North, and the density of usage is higher in southern inland waters. Overall, these characteristics make small-scale aquatic biomass production systems more feasible for energy applications. It may be advisable in the future to examine the possibility of constructing large artificial lakes for this purpose; this does not seem practical in the early 1990s.

Land-Based Biomass. Much effort to evaluate land-based biomass energy applications has been expended. This work aims at selecting high yield biomass species, characterizing physical and chemical properties, defining growth requirements, and rating energy use potential. Several species have been proposed specifically for energy usage, while others have been recommended for multiple uses, one of which is as an energy resource. The latter case is exemplified by sugarcane; bagasse, the fibrous material remaining after sugar extraction, is used in several sugar factories as a boiler fuel. Most land-based biomass plantations operated for energy production or synfuel manufacture also will yield products for nonenergy markets. Land-based biomass for energy production can be divided into forest biomass, grasses, and cultivated plants.

Forest Biomass. About one-third of the world's land area is forestland. Broadleaved evergreen trees are a dominant species in tropical rain forests near the equator (71). In the northern hemisphere, stands of coniferous softwood trees such as spruce, fir, and larch dominate in the boreal forests at the higher latitudes, while both the broadleaved deciduous hardwoods such as oak, beach, and maple, and the conifers such as pine and fir, are found in the middle latitudes. Silviculture, ie, the growth of trees, is practiced by five basic methods: exploitative, conventional extensive, conventional intensive, naturalistic, and short-rotation (71). The exploitative method harvests trees without regard to regeneration. The conventional extensive method harvests mature trees so that natural regeneration is encouraged. Conventional intensive silviculture grows and harvests commercial tree species in essentially pure stands such as Douglas fir and pine on tree farms. The naturalistic method has been defined as the growth of selected mixed tree species, including hardwoods, in which the species are selected to match the ecology of the site. The last method, short-rotation silviculture, ie, short-rotation intensive culture (SRIC) or short-rotation woody crops (SRWC), has been suggested as the most suitable method for energy applications. In this technique, trees that grow quickly are harvested every few years, in contrast to once every 20 or more years. Fast-growing trees such as cottonwood, red alder, and aspen are intensively cultivated and mechanically harvested every 3 to 6 years when they are 3 to 6 m high and only a few centimeters in diameter. The young trees are converted into chips for further processing or direct fuel use and the small remaining stems or stumps form new sprouts by coppice growth and are intensively cultivated again. SRWC production affords dry yields of several tons of biomass per hm^2

annually without large energy inputs for fertilization, irrigation, cultivation, and harvesting.

Historically, trees have been important resources and still serve as significant energy resources in many developing countries. Several studies of temperate forests indicate productivities from about 9 to 28 t/hm^2·yr, while the corresponding yields of tropical forests are higher, ranging from about 20 to 50 t/hm^2·yr (72). These yields are obtained using conventional forestry methods over long periods of time, ie, 20 to 50 years or more. Productivity is initially low in a new forest, slowly increases for about the first 20 years, and then begins to decline. Coniferous forests will grow even in the winter months if the temperatures are not too low; they do not exhibit the yield fluctuations characteristic of deciduous forests.

One of the tree species studied in great detail as a renewable energy source is the *Eucalyptus* (73), an evergreen tree which belongs to the myrtle family, *Myrtaceae*. There are approximately 450 to 700 identifiable species of *Eucalyptus*. The *Eucalyptus* is a rapidly growing tree native to Australia and is a prime candidate for energy use because it reaches a size suitable for harvesting in about seven years. Several species have the ability to coppice, ie, resprout, after harvesting; as many as four harvests can be obtained from a single stump before replanting is necessary. In several South American countries, *Eucalyptus* trees are converted to charcoal and used as fuel. *Eucalyptus* wood has also been used to power integrated sawmill, wood distillation, and charcoal–iron plants in Western Australia. Several large areas of marginal land in the United States may be suitable for establishing *Eucalyptus* energy farms. These areas are in the western and central regions of California and the southeastern United States.

Various species and hybrids of the genus *Populus* are some of the more promising candidates for SRWC growth and harvesting as an energy resource (74). The group has long been cultivated in Europe and more recently in North America. Poplar hybrids are easily developed and the resulting progeny are propagated vegetatively using stem cuttings. Short-rotation growth of poplar hybrids has been reported by several investigators to afford yields of biomass that range as high as 112–202 green t/hm^2·yr (50–90 green short ton/acre·yr) (75). These results were reported with very high density plantings and selected clones; this type of tree growth has been termed woodgrass in which the tree crop is harvested several times each growing season in the same manner as perennial grasses. However, there is some dispute regarding the benefits of woodgrass growth vs SRWC growth (76).

It can be concluded from other studies that deciduous trees are preferred over conifers for the production of biofuels (77). Several species can be started readily from clones, resprout copiously and vigorously from their stumps at least five or six times without loss of vigor, and exhibit rapid initial growth. They also can be grown on sites with slopes as steep as 25%, where precipitation is 50 cm or more per year. It has been estimated that yields between about 18 and 22 dry t/hm^2·yr are possible on a sustained basis almost anywhere in the Eastern and Central time zones in the United States from deciduous trees grown in dense plantings. A representative list of deciduous trees judged to have desirable growth characteristics for methane plantations, and shown to grow satisfactorily at high planting densities on short and repeated harvest cycles, is available (77).

Grasses.   Grasses are very abundant forms of biomass (78). About 400 genera and 6000 species are distributed all over the world and grow in all land habitats capable of supporting higher forms of plant life. Grasslands cover over one-half the continental United States; about two-thirds of this land is privately owned. Grass, as a family Gramineae, includes the great fruit crops, wheat, rice, corn, sugarcane, sorghum, millet, barley, and oats. Grass also includes the many species of sod crops that provide forage or pasturage for all types of farm animals. In the concept of grassland agriculture, grass also includes grass-related species such as the legumes family, ie, the clovers, alfalfas, and many others. Grasses are grown as farm crops, for decorative purposes, for preserving the balance of productive capacity of lands by crop rotation, for controlling errosion on sloping lands, for the protection of water sheds, and for the stabilization of arid areas. Many advances in grassland agriculture have been made since the 1940s through breeding and the use of improved species of grass, alone or in seeding mixtures; cultural practices, including amending the soil to promote herbage growth best suited for specific purposes; and the adoption of better harvesting and storage techniques. Until the mid-1980s, very little of this effort had been directed to energy applications. A few examples of energy applications of grasses can be found, ie, the combustion of bagasse for steam and electric power, but many other opportunities exist that have not been developed.

Perennial grasses have been suggested as candidate raw materials for conversion to synfuels (77). Most perennial grasses can be grown vegetatively, and they reestablish themselves rapidly after harvesting. Also, more than one harvest can usually be obtained per year. Warm-season grasses are preferred over cool-season grasses because their growth increases rather than declines as the temperature rises to its maximum in the summer months. In certain areas, rainfall is adequate to permit harvesting every 3 to 4 weeks from late February into November, and yields between about 18 and 24 $t/hm^2 \cdot yr$ of dry grasses may be obtainable in managed grasslands. Table 26 lists promising warm-season grasses proposed for conversion to synfuels.

Experimental work has shown that cool-season grasses such as Kentucky bluegrass and warm-season grasses such as Coastal Bermuda grass (*Cynodon dactylon*) can be converted to methane by conventional anaerobic digestion techniques (79,80). The compositions of some grasses indicate that fertilization procedures can incorporate certain nutrients into the harvested grass so that they can be converted by biological means without the use of excessive chemical additions to the conversion units (80).

Sugarcane is used commercially as a combination food and fuel crop. A great deal of information has been compiled about sugarcane, and it might well be used as a model for other biomass energy systems. It grows rapidly, produces high yields, the fibrous bagasse is used as boiler fuel, and cane-derived ethanol is used as a motor fuel in gasoline blends, ie, gasohol. Sugarcane plantations and the associated sugar processing and ethanol plants are in reality biomass fuel farms. About one-half of the organic material in sugarcane is sugar and the other half is fiber. Dry cane yields per year have been reported to range as high as 80 to 85 t/hm^2 (36 to 38 short ton/acre). Normal cultivation of sugarcane provides dry annual sugarcane yields of about 50 to 59 t/hm^2 (22 to 26 short ton/acre) (13).

Table 26. Warm-Season Grass Species for U.S. Methane Plantations[a,b]

Species	Localities[c]	Comments
perennial sorghums and their hybrids	Plains, South	Sudan grasses, Johnson grasses, and other warm-season hybrids are promising for localities with alkaline soils; several harvests per year
Bermuda grasses[d]	South and South Central states	most promising of all warm-season grasses, especially for localities with acid soils; can be harvested several times per year
related to sugarcane[e]	Louisiana and Florida	limited suitable sites
related to bamboo[f]	South Central states	
Bahia grasses	Florida and southern coastal plains	competes with Bermuda grasses when fertilized; effect on overall yield is in dispute

[a]Ref. 77.
[b]High annual yields in the range of 18–22 dry t/(hm^2·yr) unless otherwise noted.
[c]Regions in which species grow naturally, have been successfully introduced, or have been tested extensively.
[d]Coastal, midland, and Suwanne grasses.
[e]Very high annual yield up to 45 dry t/(hm^2·yr) in specially suitable sites.
[f]Untested annual yield.

Other productive grasses given serious consideration as raw materials for the production of energy and synfuels include sorghum and their highbreds. Tropical grasses are very productive and normally yield 50 to 60 t of organic matter per hm^2 annually on good sites (13). The tropical fodder grass *Digitaria decumbens* has been grown at yields of organic matter as high as 85 t/hm^2·yr (38 short ton/acre·yr) (13).

There are many grasses and related plants that can be considered for energy applications because they have the desirable characteristics needed for land-based biomass energy systems.

Other Cultivated Crops. Other high yielding land biomass species have been proposed as renewable energy sources (81). Promising species are kenaf, *Hibiscus cannabinus*, an annual plant reproducing by seed only; sunflower, *Helianthus annuus L.*, which is an annual oil seed crop grown in several parts of North America; and a few others, such as the polyisoprene-containing plant species described previously. Kenaf is highly fibrous and exhibits rapid growth, high yields, and high cellulose content. It is a potential pulp crop and is several times more productive than the traditional pulpwood trees. Maximum economic growth usually occurs in less than 6 months, and consequently two croppings may be possible in certain regions of the United States. Without irrigation, heights of 4 to 5 m are average in Florida and Louisiana; 6-m plants have been observed under near-optimum growing conditions. Yields as high as 45 t/hm^2·yr have been observed on experimental test plots in Florida, and it has been suggested that similar yields could be achieved in the Southwest with irrigation.

The sunflower is a prime candidate for biomass energy applications because of its rapid growth, wide adaptability, drought tolerance, short growing season, massive vegetative production, and adaptability to root harvesting. Dry yields have been projected to be as high as 34 t/hm^2 per growing season.

Water-Based Biomass.   The average net annual productivities of dry organic matter on good growth sites for land- and water-based biomass are shown in Table 27. With the exception of phytoplankton, which generally has lower net productivities, aquatic biomass seems to exhibit higher net organic yields than land biomass. Water-based biomass considered to be the most suitable for energy applications include the unicellular and multicellular algae and water plants.

Algae.   Unicellular algae, eg, the species *Chlorella* and *Scenedesmus*, have been produced by continuous processes in outdoor light at high photosynthesis efficiencies. *Chlorella*, for example, has been produced at a rate as high as 1.1 dry t/hm^2·day (82); this corresponds to an annual rate of 401 t/hm^2·yr. These figures are probably in error, but there is no theoretical reason why yields cannot achieve these high values because the process of producing algae can be almost totally controlled.

Algae production is not composed only of surface growth. Algae are produced as slurries in lakes and ponds, so the depth of the biomass-producing area as well

Table 27. Annual Biomass Yields on Fertile Sites[a,b]

Dry organics, t/(hm^2·yr)	Climate	Ecosystem type	Remarks
1	arid	desert	better yield if hot and irrigated
2		ocean phytoplankton	
2	temperate	lake phytoplankton	little influence by humans
3		coastal phytoplankton	probably higher in some polluted estuaries
6	temperate	polluted lake phytoplankton	in agricultural and sewage runoffs
6	temperate	freshwater submerged macrophytes	
12	temperate	deciduous forest	
17	tropical	freshwater submerged macrophytes	
20	temperate	terrestrial herbs	possibly higher yields if grazed
22	temperate	agriculture, annuals	
28	temperate	coniferous forests	
29	temperate	marine submerged macrophytes	
30	temperate	agriculture, perennials	
30		salt marsh	
30	tropical	agriculture, annuals	including perennials in continental climates
35	tropical	marine submerged macrophytes	including coral reefs
38	temperate	reedswamp	
40	subtropical	cultivated algae	better yield if CO$_2$ supplied
50	tropical	rain forest	
75	tropical	agriculture, perennials, reedswamp	

[a]Ref. 13.
[b]Average net values.

as plant yield per unit volume of water are important parameters. The nutrients for algae production can be supplied by sewage and other wastewaters. It should be pointed out that most unicellular algae are grown in fresh water, which limits their energy applications to small-scale algae farms. The high water content of unicellular algae also limits the conversion processes to biological methods.

Macroscopic multicellular algae, or seaweeds, have been considered as renewable energy resources. Candidates include the giant brown kelp *Macrocystis pyrifera* (83–85), the red algae *Rhodophyta*, and the floating algae *Sargassum*. Giant brown kelp has been studied in detail and harvested commercially off the California coast for many years (65). Because of its high potassium content, it was used as a commercial source of potash during World War I; in the 1990s, organic gums and thickening agents and alginic acid derivatives are manufactured from it. Laminaria seaweed is harvested off the East Coast for the manufacture of alginic acid derivatives. In tropical seas not cooled by upwelled water, species of the *Sargassum* variety of algae may be suitable as renewable energy resources. Several species of *Sargassum* grow naturally around reefs surrounding the Hawaiian Islands. However, only a small amount of research has been done on *Sargassum* and little detailed information is available about this algae. A considerable amount of data on yields and growth requirements is available, however, on the *Macrocystis* and *Laminaria* varieties. The high water content of macroscopic algae suggests that biological conversion processes, rather than thermochemical conversion processes, should be used for synfuel manufacture. The manufacture of coproducts from macroscopic algae, such as polysaccharide derivatives, along with synfuel may make it feasible to use thermochemical processing techniques on intermediate process streams.

Water Plants. The productivity of some salt marshes is similar to that of seaweeds. *Spartina alterniflora* has been grown at net annual dry yields of about 33 t/hm^2·yr (14.7 short ton/acre·yr), including underground material, on optimum sites (13). Other emergent communities in brackish water, including mangrove swamps, appear to have annual organic productivities of up to 35 t/hm^2·yr (15.6 short ton/acre·yr) (13); insufficient information is available to judge their value in biomass energy systems. Freshwater swamps may be highly productive and offer opportunities for energy production. Both the reed *Arundo donax* and bulrush *Scirpus lacustris* appear to produce 57 to 59 t/hm^2·yr (25.4–26.3 short ton/acre·yr) yields (13); if these could be sustained, they will be suitable candidates for biomass energy applications.

A strong candidate for energy applications is the water hyacinth, *Eichhornia crassipes* (3,86). This species of aquatic biomass is highly productive, grows in warm climates, and has submerged roots and aerial leaves like reedswamp plants. It is estimated that water hyacinth could be produced at rates up to about 150 t/hm^2·yr (67 short ton/acre·yr) if the plants are grown in a good climate, the young plants always predominated, and the water surface was always completely covered (13). Some evidence has been obtained to support this growth rate (87,88). If it can be sustained on a steady-state basis, water hyacinth may be one of the best candidates as a nonfossil carbon source for synfuels manufacture. It has no competitive uses (ca 1992) and is considered to be an undesirable species on inland waterways. Many attempts have been made to rid navigable streams in Florida of water hyacinth without success; the plant is a very hardy, disease-resistant species (89).

Systems Analysis

The overall design of an integrated biomass-to-synfuel system is very important to its successful operation. The system is large and requires coordination of many different operations, such as planting, growing, harvesting, transporting, and converting biomass to gaseous and liquid synfuels. The detailed design of a biomass-to-synfuel system depends on several parameters, such as the type, size, number, and location of the biomass growth and processing areas. In the ideal case, synfuel production plants are located in or near the biomass growth areas to minimize cost of transporting the harvested biomass to the plant. All nonfuel effluents are recycled to the growth areas as shown in Figure 19. This type of synfuel plantation, if developed, would be equivalent to an isolated system with inputs of solar radiation, air, carbon dioxide, and minimal water, and one output, synfuel. The nutrients are kept within the ideal system so that the addition of external fertilizers and chemicals is not necessary. Also, environmental and disposal problems are minimized.

Various modifications of the idealized design in Figure 19 can be conceived for large-scale usage. One modification in the southern United States might consist of the addition of wastewater influent into the biomass growth area and the growth of water hyacinth for two purposes, ie, the treatment of wastewater by luxuriant uptake of nutrients and the conversion of water hyacinth to synfuels. In this case, inorganic material would build up in the biomass growth area so that the residual material from the conversion plant could be partially removed or bled from the system as the synfuel is produced. This product might be considered to be a coproduct along with the synfuel.

Alternatively, short-rotation hybrid poplar and selected grasses can be multicropped on an energy plantation in the U.S. Northwest and harvested for conversion to liquid transportation fuels and cogenerated power for on-site use in a centrally located conversion plant. The salable products are liquid biofuels and

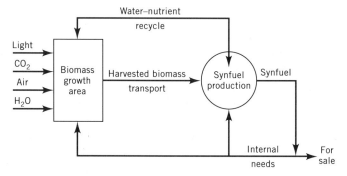

Fig. 19.   Idealized biomass-to-synfuel plantation system.

surplus steam and electric power. This type of design may be especially useful for larger land-based systems.

Another possibility, especially for small-scale farm use, is integration of agricultural crop, farm animal, and biofuel production into one system, eg, a farmer in the Midwest United States might grow corn as feedstock for a farm cooperative fuel ethanol plant. The residual distillers dried grains from the plant is used as hog feed, and the hog manure is used to generate medium heating value (MHV) fuel gas by anaerobic digestion. The fuel gas is used as plant fuel and the residual ungasified solids, which are high in nitrogen, potassium, and phosphorus, are recycled to the fields as fertilizer to grow more corn. The salable products are ethanol and hogs; the residuals are kept within the system.

Still another possibility is a marine biomass plantation such as that envisaged for giant brown kelp grown off the California coast and conversion of the kelp to methane in a system similar to that shown in Figure 19. The location of the SNG plant could be either on a floating platform near the kelp growth area or located on shore, in which case the biomass or fuel transport requirements would be different.

Many different biomass energy system configurations are possible. As the technology is refined and developed to the point where commercialization activities are well under way, optimum designs will evolve. A great deal of attention has been given to the cost factors in the operation of biomass energy systems for the production of energy and biofuels. Of equal importance is the net energy production efficiency of the total system.

Economics. The practical value of biomass energy ultimately depends on the end-user costs of salable energy and biofuels. Consequently, many economic analyses have been performed on biomass production, conversion, and integrated biofuels systems. Conflicts abound when attempts are made to compare results developed by two or more groups for the same biofuels because methodologies are not the same. Technical assumptions made by each group are sometimes so different that valid comparisons cannot be made even when the same economic ground rules are employed. Comparative analyses, especially for hypothetical processes conducted by an individual or group of individuals working together, should be more indicative of the economic performance and ranking of groups of biofuels systems.

Several important generalizations can be made. The first is that fossil fuel prices are primary competition for biomass energy. Table 28 summarizes 1990 U.S. tabulations of average, consumption-weighted, delivered fossil fuel prices by end-use sector (90). The delivered price of a given fossil fuel is not the same to each end user; ie, the residential sector normally pays more for fuels than the other sectors, and large end users pay less.

In the context of biomass energy costs, dry, woody, and fibrous biomass species have an energy content of approximately 20 MJ/kg (8600 Btu/lb) or 20 GJ/t (17.2 MBtu/short ton). If such types of biomass were available at delivered costs of $1.00/GJ ($1.054/MBtu), or $20.00/dry t ($18.14/dry short ton), biomass on a strict energy content basis without conversion to biofuels would cost less than most of the delivered fossil fuels listed in Table 28. The U.S. Department of Energy has set cost goals of delivered biomass energy crops at $1.90–2.13/GJ ($2.00–2.25/MBtu) (91) and $0.18/L ($0.67/gal) for fuel ethanol from biomass without subsidies

Table 28. U.S. Delivered Fossil Fuel Prices to End Users, 1990, $/GJ[a]

Fossil fuel	Residential	Commercial	Industrial	Transportation[b]	Utility electricity[c]
coal	2.87	1.52	1.60		1.38
natural gas	5.34	4.45	2.79		2.20
petroleum	8.43	5.65	5.32	7.90	3.24
LPG	10.38	8.17	5.12	8.03	
kerosene	8.41	6.40	6.25		
distillate fuel	7.60	5.79	5.39	8.03	
motor gasoline		8.68	8.68	8.65	
residual fuel		3.25	2.94	2.83	

[a]Ref. 90. All figures are consumption-weighted averages for all states in nominal dollars and include taxes.
[b]Aviation gasoline is delivered at $8.84/GJ; jet fuel at $5.39/GJ.
[c]Heavy oil, ie, grade nos. 4, 5, and 6, and residual fuel oils; light oils, ie, no. 2 heating oil, kerosene, and jet fuel; and petroleum coke are delivered at $3.13/GJ, $5.33/GJ, and $0.79/GJ, respectively.

in 2000, $0.22 to $0.26/L ($0.85 to $1.00/gal) by the year 2007 for biocrude-derived gasoline, $3.32/GJ ($3.50/MBtu) for methane from the anaerobic digestion of biomass by the year 2000, and 4.5 cents/kWh for electricity from biomass by the late 1990s (92).

An economic analysis of the delivered costs of biomass energy in 1990 dollars has been performed (ca 1992) for herbaceous and woody biomass for different regions of the United States (91). The analysis was done for each decade from 1990 to 2030 for Class I and II lands; results for biomass grown on Class II lands for the years 1990 and 2030 are shown in Table 29. Estimates of the total production costs for biomass were calculated with discounted cash flow models, one for the herbaceous crops switchgrass, napier grass, and sorghum, and one for the short-rotation production of sycamore and hybrid poplar trees. The delivered costs are shown in Table 29 in 1990 $/dry t and 1990 $/MJ and are tabulated by region and biomass species. The yield figures for 1990 were obtained from the literature and the projected yields for 2030 were assumed achievable from continued research. The annual, dry biomass yields per unit area have a great influence on the final estimated costs. This analysis indicates that the lowest cost energy crop of those chosen may be different for different regions of the country. A few of the biomass-region combinations appear to come close to providing delivered biomass energy near the U.S. Department of Energy cost goal. Realizing that there are many differences in the methodologies and assumptions used to compile the 1990 costs for delivered fossil fuels in Table 28 and delivered biomass energy in Table 29, it is evident that many of the biomass energy costs are competitive with those of fossil fuels in several end-use sectors, even without incorporating yield improvements that are expected to evolve from continued research on biomass energy crops.

It is essential to recognize several other factors, in addition to the cost of virgin biomass and its conversion to biofuels, when considering whether the costs of biomass energy are competitive with the costs of other energy resources and fuels. Some potential biomass energy feedstocks have negative values; ie, waste

Table 29. Projected Delivered Costs for Candidate Biomass Energy Crops in 1990 and 2030[a,b]

Region and species	1990			2030		
	Assumed yield[c]	Delivered costs		Assumed yield[c]	Delivered costs	
		$/t	$/GJ		$/t	$/GJ
Great Lakes						
switchgrass	7.6	104.07	5.26	15.5	61.32	3.60
energy sorghum	15.5	62.56	3.17	30.9	36.79	2.16
hybrid poplar	10.1	113.79	5.76	15.9	72.82	4.29
Southeast						
switchgrass	7.6	105.89	5.36	17.3	52.91	3.11
napier grass	13.9	63.72	3.22	30.9	33.31	1.96
sycamore	8.1	88.61	4.49	14.3	53.19	3.13
Great Plains						
switchgrass	5.4	74.32	3.77	10.3	44.05	2.59
energy sorghum	6.3	91.73	4.65	13.7	48.07	2.83
Northeast						
hybrid poplar	8.1	105.26	5.33	11.9	71.69	4.26
Pacific Northwest						
hybrid poplar	15.5	66.69	3.56	23.8	44.73	2.63

[a]Ref. 91. Discounted cash-flow models account for use of capital, working capital, income taxes, time value of money, and operating expenses. Real after-tax return assumed to be 12.0%. Short-rotation model used for sycamore and poplar. Herbaceous model used for other species. Costs in 1990 dollars. Dry tons.

[b]Yields in 1990 obtained from literature on Class II lands. Average total field yields are for entire region on prime to good soil, less harvesting and storage losses. Yields in 2030 assumed to be attained through research and genetic improvements. Short-rotation woody crops (hybrid poplar and sycamore) grown on six-year rotations on six independent plots. Net income is negative for first five years for each plot.

[c]Yield in $t/hm^2 \cdot yr$.

biomass of several types such as municipal biosolids, municipal solid wastes, and certain industrial and commercial wastes must be disposed of at additional cost by environmentally acceptable methods. Many generators of these wastes will pay a service company for removing and disposing of the wastes, and many of the generators will undertake the task on their own. These kinds of feedstocks often provide an additional economic benefit and revenue stream that can help support commercial use of biomass energy.

Another factor is the potential economic benefit that may be realized due to possible future environmental regulations from utilizing both waste and virgin biomass as energy resources. Carbon taxes imposed on the use of fossil fuels in the United States to help reduce undesirable automobile and power plant emissions to the atmosphere would provide additional economic incentives to stimulate development of new biomass energy systems. Certain tax credits and subsidies are already available for commercial use of specific types of biomass energy systems (93).

Energetics. The net energy production efficiency of an integrated biomass energy system is extremely important to its development and practical use. The ultimate goal is to design and operate environmentally acceptable systems to produce new supplies of salable energy whether they be low heat value gas, substitute natural gas, substitute gasolines or diesel fuels, methanol, ethanol, hydrogen, or electric power from biomass at the lowest possible cost and energy consumption. It is necessary to quantify how much energy is expended and how much salable energy is produced in each fully integrated system. An energy budget similar to an economic budget should be prepared because the capital, operating, and salable energy cost projections and the conversion process efficiency are insufficient to choose and design the best systems. These values do not necessarily correlate with net energy production. Also, the capital energy investment consumed during construction of the system should be recovered during its operation. Comparative analyses of similar systems for production of synthetic liquid and gaseous fuels from the same feedstock or of different systems that yield the same fuels from different biomass should be performed by consideration of the economics and the net energetics.

One method of analyzing net energetics of a biomass energy system is to let E_f, E_x, and E_p represent the energy content of the dry biomass feed, E_f, the sum of the external nonsolar energy inputs into the total system, E_x, and the energy content of the salable fuel products, E_p. The ratio $(E_p - E_x)/E_x$, which can be termed the net energy production ratio, indicates how much more, or less, salable fuel energy is produced than that consumed in the integrated system if the external energy consumed is replaced and it is assumed that the biomass feed energy is zero. This is a reasonable assumption because the energy value of biomass is derived essentially 100% from solar radiation. Net energy production ratios greater than zero indicate that an amount of energy equivalent to the sum of the external nonsolar energy inputs and an additional energy increment of salable fuel are produced; the larger the ratio, the larger the increment. The ratio $100E_p/(E_f + E_x)$ is the overall fuel production efficiency of the system or simply the energy output divided by the gross energy input. Finally, another useful ratio is the value of net energy output divided by the gross energy out, $100(E_p - E_x)/E_p$. This ratio expresses the percentage of the energy output that is new energy added to the economy.

A simple model for biomass energy production, excluding conversion to synfuels, is illustrated in Table 30 (94) which presents the results of an analysis of a short-rotation tree plantation that produces dry biomass at a yield of 11.2 t/hm²·yr. The principal sources of energy consumption are fertilization, which includes the energy cost of ammonia production, and biomass transportation. The important result of this study is that the net energy production efficiency is high and that about 14 times more energy is produced in the form of woody biomass than the external nonsolar energy inputs needed to operate the system.

A mixed biomass plantation, including species of short-rotation hardwoods, sunflower, and kenaf, is projected to produce dry biomass at a yield of 67.3 t/hm²·yr (95). Again, fertilization consumes the largest amount of external energy, but the energy cost for irrigation is also high. The net energy production efficiency is high; about 20 times more energy is produced as biomass energy than the external nonsolar energy inputs.

These systems analyses suggest that biomass plantations can be designed to operate at high net energy efficiencies, and that further improvements might

Table 30. Net Energy Analysis of Short-Rotation Wood Biomass Production[a,b]

Operation	Energy consumed per dry wood produced, MJ/t[c]	Total consumption, %
cultivation and planting	9.3	0.8
fertilization	604	52.1
harvesting	87	7.5
transport to trucks	115	9.9
load trucks	55	4.7
80.5-km transport to user	221	19.0
unload	nil	
auxiliary	69	5.9
Total energy consumption	*1160*	*99.9*
biomass energy produced, GJ/(hm²·yr)[d]	196[e]	
energy production efficiency, %	93.4[f]	
net energy production ratio	14.1[g]	

[a]Ref. 94.
[b]Yield is 11.2 t/(hm²·yr) dry; 20-yr planting cycle; fertilization is 224 kg N/(hm²·yr).
[c]To convert MJ/t to Btu/1000 lb, multiply by 430.
[d]To convert GJ to 10^6 Btu, divide by 1.054.
[e]Assumes heating value of biomass is 17.5 MJ/dry kg.
[f](Net energy output ÷ gross energy output) × 100, ie, 100 $(E_p - E_x)/E_p$.
[g]Net energy output ÷ energy consumed, ie, $(E_p - E_x)/E_x$.

best be incorporated in the systems by concentrating on fertilization. The use of nitrogen-fixing biomass and the recycling of nutrients from the conversion facilities may offer additional benefits.

For the total integrated biomass production–conversion system, the arithmetic product of the efficiencies of biomass production and conversion is the efficiency of the overall system. An overall conversion efficiency near 45% would thus be produced by integrating the biomass plantation illustrated in Table 30 with a conversion process that operated at an overall efficiency of 50%. Every operation in the series is thus equally important.

These simplified treatments correspond to net energy analyses using the First Law of Thermodynamics. Some energy analysts feel that only an analysis based on the Second Law can provide the ultimate answers in terms of where more available energy, in the thermodynamic sense, can be found to permit efficiency maximization. Others believe that the conventional energy balance is optimal because it is more realistic and easier to use. Indeed, for integrated synfuel production systems, entropic losses may not always be definable for all segments of the system, and a rigorous Second Law analysis may not be possible.

Location of the system boundaries also is important in the net energy analysis of integrated biomass energy systems. Thus tractors may be used to plant and harvest biomass. The fuel requirements of the tractors are certainly part of E_x, but is the energy expended in manufacturing the tractors also part of E_x? Some analysts believe that a complete study should trace all materials of construction and fossil fuels used back to their original locations in the ground.

Another important factor in net energy analysis concerns energy credits taken for by-products; they have an important effect and can determine whether the net energy production efficiency is positive or negative. An example of this effect is shown in Table 31 (96), which was derived from the integrated alcohol-from-corn production system illustrated in Figure 20. The boundary of the system depicted in Figure 20 circumscribes all the operations necessary to grow and harvest the corn, to collect residual cobs and stalks if they are used as fuel, to operate

Table 31. Net Energy Production Ratios for Ethanol Production from Corn in Integrated System[a]

Salable energy products, E_p	Nonfeed energy inputs, E_x	N^b
alcohol	corn production, fermentation, bottoms drying	−0.65
alcohol	corn production, fermentation	−0.51
alcohol, chemicals	corn production, fermentation	−0.50
alcohol, chemicals, cattle feed	corn production, fermentation, bottoms drying	−0.44
alcohol, chemicals, cattle feed	corn production, fermentation, bottoms drying, 50% residuals[c]	−0.10
alcohol, chemicals, cattle feed	corn production, fermentation, bottoms drying, 75% residuals[c]	0.29
alcohol	corn production, fermentation, 75% residuals[c]	1.43
alcohol, chemicals	corn production, fermentation, 75% residuals[c]	1.47

[a]Ref. 96.
[b]$N = (E_p - E_x)/E_x$.
[c]Percent of cobs and stalks collected and used as fuel within system to replace fossil fuel inputs.

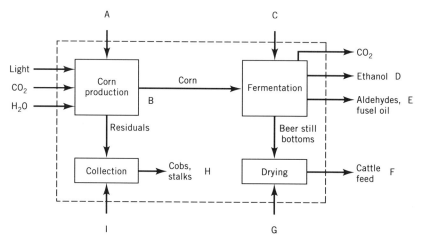

Fig. 20. Energy inputs and outputs to manufacture 3.785 L of anhydrous ethanol from corn. (———) denotes system boundary. All KJ figures are lower heating value (LHV). A, 48,469 KJ; B, 128,314 KJ; C, 113,748 KJ; D, 79,682 KJ; E, 1,138 KJ; F, 47,483 KJ; G, 66,634 KJ; H, 174,768 KJ; I, 1,644 KJ. To convert KJ to Btu, divide by 1.054.

the fermentation plant for the production of anhydrous alcohol and by-product chemicals, and to dry the stillage to produce distillers' dried grains plus soulbles for sale as cattle feed. The capital energy investment in the system is not included within the boundary.

Various net energy production ratios were calculated as shown in Table 31. It is apparent that the ratio can be either positive or negative, depending on whether credit is taken for the by-product chemicals and cattle feed, whether the energy for drying the stillage is included as an input to the system, and whether a portion of the residual corn cobs and stalks is collected and used as fuel within the system to replace fossil fuel inputs. To permit comparisons, net energy analyses must be clearly specified as to all details. On the whole, net energy production in a modern corn-to-ethanol plant would appear to be borderline if petroleum fuels comprise the significant part of the nonfeed energy inputs. However, to improve net energy production, these fuels can be replaced by fuels generated within the system or by renewable fuels, and credit can be taken for the by-products. Another route is to use more of the corn plant or the stillage as feedstock to the fermentation plant. The use of the grain alone as feedstock places a severe limitation on net energy production. Some of the information available on designs of corn wet-milling plants in which corn oil and other products are produced indicate that integration with an alcohol plant may be more efficient than a conventional corn alcohol plant.

Commercial Use of Biomass Energy in the United States

Relatively few biomass energy technologies are in commercial use in 1992. With two possible exceptions, ie, a few small tree plantations, and fuel ethanol from corn, sugarcane, and sugar beet, there is no biomass species grown in the United States specifically for conversion to biofuels. But, conversion processes in commercial use in the early 1990s span the basic technologies of combustion, gasification, and liquefaction. These include combustion of wood, wood wastes, and forestry residues; combustion of agricultural residues such as rice husks and bagasse; combustion of MSW and RDF for simultaneous disposal and energy recovery; biological gasification of animal manures in farm- and feedlot-scale anaerobic digestion systems for simultaneous waste disposal and production of biogas as well as upgraded solids for feed, fertilizer, and animal bedding; biological gasification of municipal wastewater sewage by anaerobic digestion for simultaneous waste disposal and biogas production; biological gasification of MSW in sanitary landfills and recovery of biogas for fuel, which also mitigates environmental and safety problems caused by gas migration in the landfill; thermal gasification of biomass for LHV gas production for on-site use; and alcoholic fermentation of starchy and sugar crops for fuel ethanol for use as a fuel extender and octane enhancer in motor gasoline blends.

Most of these commercial processes have been in use for many years. Some have been greatly improved in the early 1990s, such as alcoholic fermentation that has undergone large process steam requirement reductions, thereby increasing net energy production efficiencies of fuel ethanol. Other commercial processes are relatively new, such as two-phase anaerobic digestion that permits higher plant capacities at low capital costs and the production of higher methane-content biogas at higher rates.

Inventories of commercial usage in the United States are available (97,98) Table 32 offers a summary (2). Although most of the data available does not refer to a specific time or year, wood use as a fuel in the industrial and residential sectors is responsible for the largest portion of biofuels consumption in the United States. Those states that have large forest products industries are principal wood energy users. Similarly, states in the Corn Belt are the largest fuel ethanol producers. With few exceptions, those states having the most populated cities tend to process more municipal solid wastes by simultaneous disposal-energy recovery technologies. The biomass energy industry covers the entire nation; not one state is devoid of commercial biofuels production or utilization. The practical limitations to transport distance of some biomass such as wood, and the requirement for nearby or local processing, correlates with a concentration of biomass energy-processing facilities by state.

Combustion. The Public Utility Regulatory Policies Act of 1978 (PURPA), which provides benefits to cogenerators and small power producers, has stimulated commercialization of biomass combustion for electric power production. To be eligible for benefits under PURPA, small power production systems are limited to 80 MW and must receive 75% or more of their total energy input from renewables. Cogeneration systems do not have these limitations. Emissions limitations must be taken into consideration in developing commercial biofuels combustion projects. The regulations apply in a rather complex fashion for boilers burning either wood or municipal solid wastes alone or in combination with fossil fuels (99); even wood-burning stoves must meet national standards. With few exceptions, all stoves built after July 1988 must comply with a strict set of environ-

Table 32. Biofuels Utilization and Production and Biomass-Fueled Electric Power Plant Capacities in the United States[a]

Utilization	United States, total	Energy equivalents[b]	
		10^{12} kJ/yr	m^3/d
wood fuel utilized			
commercial, 10^6 t/yr	83	1,553	110,594
residential, 10^3 m^3/yr	198,496	1,154	82,151
agricultural wastes utilized as fuels, 10^6 t/yr	9.5	142	10,137
fuel ethanol produced, 10^6 L/yr	2,542	53.5	3,813
biogas produced, 10^6 m^3/yr	9.3	0.2	4,496
MSW converted to energy,[c] 10^6 t/yr	14.0	113	8,103
biomass-fueled electric plant capacity, MW	5,154		

[a]Ref. 2. The indicated biofuels consumption and capacity figures are the estimated values for various time periods since 1985 and do not refer to a specific year.
[b]The energy equivalents were calculated by the author using the following assumptions for Btu values: HHV of commercial wood fuel is 16×10^6 Btu/t, HHV of residential wood fuel is 20×10^6 Btu/cord, HHV of agricultural waste is 12.9×10^6 Btu/t, HHV of ethanol is 75,500 Btu/gal, HHV of biogas is 500 Btu/cf, HHV of MSW is 7×10^6 Btu/t, HHV of 1.00 barrel of oil equivalent (BOE) is 5.8×10^6 Btu.
[c]Ref. 98. These values are from 1987 and 1988. Energy products are steam and/or electric power.

mental regulations established by the U.S. Environmental Protection Agency and some states (100).

Wood Fuels. Wood fuels are the largest contributor to biomass energy usage. Approximately 88% of the total is attributed to wood energy consumption. Distribution of the annual consumption of wood fuel by region and sector from 1980 to 1984 is available (101); the South and the industrial sector are the largest wood energy consumers. Table 33 presents the composition of wood fuel by resource type from 1972 to 1987 (102). The largest concentration of wood fuel usage is in the lumber and wood products industry and the pulpwood and paper industry. It has been reported that the pulp and paper industry meets about 70% of its own fuel needs with wood energy (103). Over the period covered in Table 33, the pulp and paper industry utilized about one-half of total wood energy consumption, primarily as black liquor. This lignin-containing material is a by-product of the pulping process and is not wood as such. It is noteworthy that commercial, utility, and other industrial usage of wood fuels is a very small part of total consumption, and that residential usage more than doubled over the time period 1972–1984, whereas total wood energy usage increased about 61%.

The increase in residential fuelwood consumption over this period parallels the sharp increase in costs of oil, natural gas, and electricity and can be tracked by the number of wood-burning stoves in homes. Between 1950 and 1973, the estimated number of stoves dropped from 7.3 million to 2.6 million, but grew to an estimated 11 to 14 million in 1981 (104). The trend in the wood-burning stove inventory suggests that a four-fold increase in residential fuelwood use may have occurred during the 1970s and 1980s, which can be explained in part by heavy rural wood burning (105). However, since attaining about a one EJ level of consumption in 1984, there has been little gain in total residential wood energy use (106). Steady gains in industrial use have been counterbalanced by a decrease in residential use (106).

The data in Table 33 indicate that there is a small but relatively steady increase in total wood energy usage; that the industries having captive sources of wood or wood-derived products consume the bulk of wood energy, although at a

Table 33. Wood Energy Use,[a] 10^{12} kJ

Sector	1972	1978	1982	1986	1987
lumber and wood products industry	373.1	444.7	362.5	474.3	484.8
pulp and paper industry					
as hog fuel	473.2	104.0	178.1	268.7	271.9
as bark	94.5	99.3	116.9	139.1	142.3
as black liquor	737.8	854.8	811.5	949.6	1001.3
other industry	38.9	47.4	45.3	51.6	51.6
residential	420.5	689.3	985.4	927.5	885.3
commercial	9.5	15.8	23.2	23.1	23.2
utilities	3.2	1.1	3.2	9.4	9.5
Total	*2150.7*	*2256.7*	*2526.7*	*2843.3*	*2869.9*

[a]Refs. 2 and 102.

generally flat rate; that there appears to be much opportunity for growth in wood energy consumption in the utility and commercial sectors; and that residential fuelwood usage has shown the largest incremental growth until the late 1980s.

Municipal Solid Waste. In the early 1990s, the need to dispose of municipal solid waste (MSW) in U.S. cities has created a biofuels industry because there is little or no other recourse (107). Landfills and garbage dumps are being phased out in many communities. Combustion of MSW, ie, mass-burn systems, and RDF, ie, refuse-derived fuel, has become an established waste disposal–energy recovery industry.

In May 1988 the United States had 105 operating MSW-to-energy plants, 29 plants under construction, 61 plants in the advanced planning stage, and 5 plants temporarily shutdown (98). About one-half of the 105 operating plants had been placed in operation since 1985, and about 80% of all operating plants used mass-burn, modular technology. Mass-burn, waterwall designs predominated for plants under construction or in the advanced planning stage. The sum total design capacity of these MSW-to-energy plants was about 30% of the total MSW generated in the United States. A 1988 inventory of 25 MSW-to-energy plants that have been permanently shutdown shows the technologies for these plants consists of both mass-burn and RDF combustion systems, as well as a few pyrolysis plants. They have a combined total design capacity of 13,500 t/d, and most were built in the 1970s (98). The majority of these plants had operating difficulties caused by equipment and environmental concerns as well as cost factors. Problems that can cause permanent shutdown do not appear to be prevalent in more modern designs, presumably because combustion and waste-processing equipment have been improved.

According to the Solid Waste Association of North America, by 1992 there are 137 municipal solid waste-to-energy plants operating in 36 states. They process about 16% of the 185 million tons of solid waste generated in the United States, and produce the equivalent of more than 2300 MW of electricity. Nearly 100 other waste-to-energy projects are in various stages of planning or implementation. The U.S. Environmental Protection Agency has estimated that there will be more than 300 waste-to energy plants in operation in the United States by 2000; these will process about 25% of U.S. municipal solid waste.

Power Production. PURPA has stimulated commercial use of biomass combustion for electric power production. It has prompted many companies to build and operate small power plants fueled with fossil and nonfossil energy resources. The power is used on-site in many cases, and the surplus is injected into the grid for sale to the utility at avoided cost. The number of U.S. filings as of December 31, 1987, submitted to the Federal Energy Regulatory Commission (FERC) for biomass-fueled and all cogeneration and small power production facilities, illustrates the phenomenal growth of this industry from fiscal 1980 to 1988 (Table 34) (108,109). Of the total number of 1730 cogeneration filings that have been qualified by FERC as eligible for PURPA benefits, 138 (8.0% of total) are biomass-fueled and have a capacity of about 2699 MW, ie, 6.0% of total. Similarly, of the 1987 small power production filings that have been qualified by FERC, 468 (23.6% of total) are biomass-fueled and have a capacity of about 6260 MW, ie, 36.8% of total. The number of filings qualified by FERC and the total electric capacity for each biomass type are: 231 and 3703.8 MW of wood wastes, 140 and 654.951 MW

Table 34. Biomass-Fueled Cogeneration and Small Power Production Capacities and Facilities, kW[a,b]

Facility	1980	1985	1988	Biomass total	Qualified total all[c]
Cogeneration					
new	400 (1)	383,003 (23)	13,685 (3)	1,617,390 (122)	41,947,273 (1,705)
existing		115,000 (2)		570,497 (18)	3,341,629 (78)
both[d]	161,000 (1)	88,000 (1)		615,000 (9)	1,784,543 (33)
total	161,400 (2)	586,003 (26)	13,685 (3)	2,802,887 (149)	47,073,445 (1,816)
qualified[c]	161,400 (2)	585,503 (24)	725 (2)	2,698,802 (138)	44,943,616 (1,730)
Small power production					
new	50 (1)	1,078,690 (112)	339,611 (19)	6,219,125 (487)	16,869,289 (1,964)
existing		−800 (2)		116,925 (9)	319,099 (62)
both[d]				11,500 (2)	136,485 (15)
total	50 (1)	1,077,890 (114)	339,611 (19)	6,347,550 (498)	17,324,873 (2,041)
qualified total[c]	50 (1)	1,075,490 (110)	339,611 (19)	6,260,309 (468)	17,006,761 (1,987)

[a]Refs. 108 and 109. Filings under Public Utilities Regulation Policies Act of 1978. Totals are for years 1980–1988.
[b]Number of facilities in parentheses.
[c]Qualified for PURPA benefits, ie, only owners or operators of facilities who claim qualifying status for PURPA benefits would make filings, some filings are submitted after the facilities begin operation. FERC does not review notices of qualifying status, has not completed review of all listed applications for certification, and data provided in the filings are not verified by FERC inspection of the facilities.
[d]Combination of existing and new incremental capacity.

for biogas, 131 and 2546.186 MW for MSW, and 108 and 1719.01 MW for agricultural wastes (109).

Data on the installed electric generating capacity and generation in 1986 by nonutility and utility power producers as of December 31, 1986 are available (110,111). Nonutility generators accounted for 3.45% of total U.S. capacity, and generated 4.31% (112×10^6 MWh) ie, 25.3×10^3 MW of a total 73.3×10^4 MW in 1986, of the power produced in the same year. Biomass-fueled electric capacity and generation was 19.2% (4.9×10^3 MW) and 21.2% (23.7×10^6 MWh) respectively, of total nonutility capacity and generation. Biomass-fueled capacity experienced a 16% increase in 1986 over 1985, the same as natural gas, but it was not possible to determine the percentage of the total power production that was sold to the electric utilities and used on-site. Total production should be substantially more than the excess sold to the electric utilities. Overall, the chemical, paper, and lumber industries accounted for over one-half of the total nonutility capacity in 1986, and three states accounted for 45% of total nonutility generation, ie, Texas, 26% of total; California, 12% of total; and Louisiana, 7% of total. There were 2449 nonutility producers with operating facilities in 1986, a 15.8% increase over 1985; 75% capacity was interconnected to electric utility systems.

Installed nonutility electric generation capacity and generation for each biofuel is presented in Table 35 for 1986. Wood was the largest contributor in 1986 to total capacity and generation, followed in decreasing order by MSW, agricultural wastes, landfill gas, and digester gas. The incremental changes in capacity and generation between 1985 and 1986 were 11.5% and 1.7% for wood, 7.8% and 15.9% for agricultural wastes, 30.4% and 9.2% for MSW, 184% and 207% for landfill gas, and 7.4% and 27.0% for digester gas. All incremental changes were positive.

Utility production of biomass-fueled electric power is much less than nonutility production. In early 1985, there were only 18 facilities having a total capacity of 245 MW, ie, nine fueled with wood (180.7 MW), five fueled with MSW (33.8 MW), two fueled with agricultural residues (22.5 MW), and two fueled with digester gas (8 MW) (112,113). The largest was the 50-MW plant in Burlington, Vermont (114).

Table 35. Installed Nonutility Electricity Generation Capacity and Generation by Biofuel, 1986[a,b]

Biofuel	Cogeneration		Small power producers		Total	
	Capacity, MW	Generation, MWh	Capacity, MW	Generation, MWh	Capacity, MW	Generation, MWh
wood	3,119.8	16,650,778	624.7	2,403,718	3,744.5	19,054,496
agricultural wastes	252.0	1,022,573	68.2	310,387	320.2	1,332,960
municipal solid waste	75.4	217,599	463.9	2,198,941	539.3	2,416,540
landfill gas	0	0	184.6	622,031	184.6	622,031
digester gas	21.2	117,146	49.6	186,750	70.8	303,896
Total	*3,468.4*	*18,008,096*	*1,391.0*	*5,721,827*	*4,859.4*	*23,729,923*

[a]Ref. 110.
[b]Total number of facilities reported in early 1985 was 111 in operation, 50 under construction, and 72 in the planning stages (112,113).

Gasification. Conversion of biomass to gaseous fuels can be accomplished by several methods; only two are used by the biomass energy industry (ca 1992). One is thermal gasification in which LHV gas, ie, producer gas, is produced. The other process is anaerobic digestion, which yields an MHV gas.

Thermal Gasification. A survey of commercial gasifiers in use and under construction in 1983 indicates that the commercialization rate is low, but that several process developers and vendors have installed 30 to 35 operating systems (115). Feed rates of biomass range from 0.1 to 13.6 t/h, LHV gas output ranges from 1.1 to 211 billion J/h (1.0 to 200 million Btu/h) for a wide range of reactor configurations, and the gas is used for several different applications. Several of the large U.S. plants have been shutdown because of operating difficulties, eg, plants in Baltimore, Maryland; Orlando, Florida; and El Cajon, California.

By 1993, many U.S. gasifier vendors had gone out of business or were focusing marketing activities overseas or on other conversion technologies, particularly combustion for power generation, in states where combined federal and state incentives make economic factors attractive. Some existing gasification installations also have been shutdown and placed in a stand-by mode until natural gas prices make biomass gasification competitive again.

A survey of commercial thermal gasification in the United States shows that few gasifiers have been installed since 1984 (115). Most units in use are retrofitted to small boilers, dryers, and kilns. The majority of existing units operate at 0.14 to 1.0 t/h of wood wastes on updraft moving grates. The results of this survey are summarized in Table 36. Assuming all 35 of these units are operated continuously, extremely unlikely, the maximum amount of LHV gas that can be produced is about 0.003 to 0.006 EJ/yr (222–445 m^3/d).

Biological Gasification. Several surveys have been performed on the use of anaerobic digestion for biogas production and waste stabilization. In 1984, 84 farm-scale and industrial anaerobic digestion facilities, exclusive of municipal wastewater treatment plants, were identified in 32 states and were estimated to have a total reactor volume of 90,600 m^3 (117). Individual digesters were found to range in size up to 13,000 m^3, but the majority had volumes in the range 249 to 750 m^3. A comprehensive inventory was conducted in 1985 (118). Exclusive of municipal wastewater treatment, 96 farm-scale and industrial anaerobic digestion systems were found to have been built between 1972 and mid-1985 in 30 states; 2 units in Puerto Rico. Eighty-seven of the systems had digestion volumes of 100 m^3 or more; 60 were operational, 7 were temporarily shutdown, and others were in various stages of design or development. Forty-four of the operational or shutdown systems, ie, 43,900 m^3, were used for digesting animal manures, 35 for dairy or beef cattle manure, and the remainder for swine or poultry manure. Fourteen of the facilities, ie, 107,900 m^3, provided wastewater clean-up services to agricultural product processing plants, breweries, and related food production facilities. The designs that were used extensively for the farm-scale digesters were plug flow and stirred tank configurations and unheated lagoons. Only a few additional commercial digestion systems have been installed since this inventory was completed.

An estimate of the potential methane production possible from existing (ca 1992) municipal wastewater treatment plants that produce and use biogas as a fuel, and from the farm-scale and industrial anaerobic digestion plants identified

Table 36. Commercial Thermochemical Gasifiers, June 1988[a]

Size, 10^6 kJ/h[b]	Number of gasifiers	Feedstock	Gas use
		Updraft reactor	
6.0	2	corn cobs	corn dryer
4.9–25.0	4	wood	space or process heat
0.94–25.9	14	wood	dry kiln, space heat
12.0–69.9	3	wood	brick kilns
		Downdraft reactor	
0.2–6.0	5	wood, peach pits	greenhouse
10.0	2	wood	power boiler
		Fluid-bed reactor	
25.0	2	rice hulls	process heat
82.2	1	wood	power boiler
124.4	2	wood	clay dryers

[a]Ref. 116. All units are LHV gasifiers.
[b]To convert 10^6 kJ/h to 10^6 Btu/h, divide by 1.054.

in the inventory, is presented in Table 37. The maximum amount of methane that could be produced from this commercial anaerobic digestion capacity under conventional operating conditions is about 0.005 EJ/yr (2,400 BOE/d).

Gas production is considerably greater from commercial landfill methane recovery systems. In 1987, 94 plants (50 operational, 44 scheduled) had an estimated design production of 1.2×10^6 m³/d and an estimated actual production of 0.314×10^6 m³/d, or 114×10^6 m³/yr; estimated electric capacity was 231.2 MW (120–123).

The initial biogas recovered is an MHV gas and is often upgraded to high heat value (HHV) gas when used for blending with natural gas supplies. The

Table 37. Potential Methane Production from Commercial U.S. Anaerobic Digestion Systems

Number of plants	Feedstock	Estimated digester volume, 10^6 m³	Estimated methane production potential[a]	
			10^6 m³/d	EJ/yr
209[b]	municipal wastewater	0.213[c]	0.208	0.0028
44	animal manures	0.044	0.043	0.0006
14	industrial wastes	0.108	0.105	0.0015
Total			*0.356*	*0.0049*

[a]Calculated assuming 65 vol % of methane in product gas and 1.5 vol gas/culture vol·d.
[b]Ref. 119. These are treatment plant unit processes, not individual digesters, that produce and use digester gas; the flow capacity is 14.2×10^3 m³/d.
[c]Calculated assuming 15-d hydraulic retention time (HRT).

annual production of HHV gas in 1987, produced by 11 HHV gasification facilities, was 116×10^6 m^3 of pipeline-quality gas, ie, 0.004 EJ (121). This is an increase from the 1980 production of 11.3×10^6 m^3. Another 38 landfill gas recovery plants produced an estimated 218×10^6 m^3 of MHV gas, ie, 0.005 EJ. Additions to production can be expected because of landfill recovery sites that have been identified as suitable for methane recovery. In 1988, there were 51 sites in preliminary evaluation and 42 sites were proposed as potential sites (121).

Liquefaction. Since the 1970s attempts have been made to commercialize biomass pyrolysis for combined waste disposal–liquid fuels production. None of these plants were in use in 1992 because of operating difficulties and economic factors; only one type of biomass liquefaction process, alcoholic fermentation for ethanol, is used commercially for the production of liquid fuels.

Fermentation ethanol, primarily from corn, but also from sugarcane, sugar beet, or derivatives, has shown extraordinarily high production rate increases since 1979 when it was reintroduced in the United States as a blending component in motor fuels. In 1979, 24 operating plants, with a design capacity of 151×10^6 L/yr, produced 75.7×10^6 L/yr; 35 additional plants were planned. By 1988, the number of plants had increased to 55, with a design capacity of 3743×10^6 L/yr and an actual production of 3160×10^6 L/yr; 70 additional plants were shut down, with total unused design capacity of 1400×10^6 L/yr (124). Tax incentives provided by the federal and state governments coupled with generally high gasoline prices, low corn prices, and the phase-out of leaded fuels, have helped establish the fuel ethanol industry.

Because most fuel ethanol manufactured in the United States is made from corn, price plays a crucial role in determining the competitive position of ethanol in an open market. With corn priced at about $2.50/bu, the embedded feedstock cost of product ethanol is about $0.14–0.23/L ($0.52–0.87 gal), depending on overall yield and by-products ignored (125). Fuel ethanol plants may have contingency plans to close if corn prices rise to a certain level, eg, $3.50/bu or above (126).

A listing of fuel ethanol plants in operation, with total anhydrous capacity of 3743.2×10^6 L/yr, is available (124). Leading producers include Archer Daniels Midland, with over one-half of all domestically produced fuel ethanol from four locations; Pekin Energy Co., Pekin, Illinois; South Point Ethanol, South Point, Ohio; and New Energy Company of Indiana, South Bend. Several plants have terminated operations even when the price of corn was in the $2.00/bu range. Continuous operation at a profit is difficult to sustain when the selling price of fuel ethanol must remain competitive with gasoline prices and alternative octane-enhancing methods. Without tax incentives, it is doubtful that fuel ethanol producers, particularly those who operate smaller plants and use older technologies, will be able to survive during times of high corn prices and low crude oil prices.

Biomass Production. In 1992, there was no biomass species grown and harvested in the United States specifically for conversion to biofuels, with the possible exceptions of feedstocks for fuel ethanol and a few tree plantations. This is understandable from an economic standpoint. For example, the average natural gas price in the United States in 1991 at the point of production, not end use cost, was estimated to be $1.51/GJ ($1.59/MBtu) (U.S. Energy Information Agency Washington, D.C.). For biomass to compete on an equivalent basis, it must be grown, harvested, and gasified to produce methane at an average cost of

$1.51/GJ ($1.59/MBtu). Assuming an unrealistic gasification cost of zero, the maximum biomass cost that is acceptable under this condition is $29.73/dry ton. At an optimistic yield of 4.45 dry t/hm^2·yr (10 dry short ton/acre·yr) a biomass energy crop producer for a gasification plant will realize not more than $667.60/hm^2·yr ($270.30/acre·yr), a marginal amount to permit a net return on an energy crop without other incentives. This simplistic calculation emphasizes the effect of depressed fossil fuel prices on biomass energy crops. Negative feedstock costs, ie, wastes, substantial by-product credits, captive uses, and/or tax incentives, are needed to justify energy crop production on strict economic grounds.

Most of the commercial tree plantations that produce wood for captive use as a raw material in manufacturing operations use a portion as fuel. Examples of short-rotation plantations are listed in Table 38 (127). Paper companies in the southeastern United States are reported to have short-rotation plantings also, eg, Weyerhaeuser, James River Corp., Buckeye Cellulose, and Lykes Brothers, but the intensity of maintenance is not known (127).

The advances in biomass growth technologies developed in the United States for agricultural crops, trees, and aquatic species, and that are commercial and being improved further through research, are available for growth of biomass

Table 38. Commercial Tree Production for Energy Use,[a] 1988

Company	Area, hm^2	Species	Rotation, yr	Comments
Simpson Timber Co.[b]	283	eucalyptus		
West Vaco[c]	6475	cottonwood, sycamore	10	primarily for pulp with some to fuel paper mills
Packaging Corp. of America[d]	1214	hybrid poplar		
Hagerstown[e]	202	hybrid poplar		wastewater disposal site, energy use of wood planned
Reynolds Metals Co.[f]	91	hybrid poplar	6	captive energy use of wood planned
Union Corp. of North Carolina	8903	sweetgum	10	captive for pulp with some to fuel paper mills
James River Corp. of Nevada	2975	hybrid poplar	6	captive for fiber and fuel for paper mills, larger plantings are planned

[a]Ref. 127. [b]California. [c]Kentucky.
[d]Michigan [e]Maryland. [f]New York.

energy crops. Multicropping designs and multiple-use crops will be the most likely candidates for biomass energy when conditions warrant commercial plantations.

Economic and Legislative Impacts.

An interminable number of studies have been performed to predict future energy consumption patterns, resources, imports, and prices. If the predictions of higher oil prices had been accurate in the late 1970s, or if the oil price had stabilized at its peak in 1981, the biomass energy industry would have exhibited much greater growth than it has (128).

Biofuels usage has slowly increased since the mid-1980s because of environmental problems, eg, MSW disposal; favorable legislation, eg, tax incentives and PURPA; and combinations of both, eg, oxygenated transportation fuels. Although environmental problems continue to increase, many tax incentives for alternative renewable energy resources have been reduced or eliminated (129). Commercialization of biomass energy is driven by waste disposal, alternative fuels and environmental issues, and the available incentives for PURPA power plants.

Capacity Limitations and Biofuels Markets. Large biofuels markets exist (130–133), eg, production of fermentation ethanol for use as a gasoline extender (see ALCOHOL FUELS). Even with existing (1987) and planned additions to ethanol plant capacities, less than 10% of gasoline sales could be satisfied with ethanol–gasoline blends of 10 vol % ethanol; the maximum volumetric displacement of gasoline possible is about 1%. The same condition applies to methanol and alcohol derivatives, ie, methyl-t-butyl ether [1634-04-4] and ethyl-t-butyl ether.

In 1987, taxable motor gasoline sales were 415.89×10^9 L (109.88×10^9 gal) (131). In the same year, the methanol nameplate capacity was 5.30×10^9 L (1.40×10^9 gal) and actual production was 4.12×10^9 L (1.09×10^9 gal); synthetic ethanol capacity was 0.80×10^9 L (0.21×10^9 gal) and 0.30×10^9 L (0.08×10^9) was actually produced (133,134); fermentation ethanol capacity was 3.62×10^9 L (0.957×10^9 gal) and actual production for blending with gasoline was 2.84×10^9 L (0.750×10^9 gal) (124).

Only a small portion of motor fuel needs could be satisfied if truly large-scale alcohol–gasoline blending or fuel switching occurred via transition to fuel-flexible vehicles and ultimately to neat alcohol-fueled vehicles (132).

Capacities for producing virtually all biofuels manufactured by biological or thermal conversion of biomass must be dramatically increased to approach their potential contribution to primary energy demand (Table 4). An incremental EJ per year of biogas requires about 200 times the existing digestion capacity, including wastewater treatment plants, whereas an incremental EJ per year of ethanol requires about 13 times the existing fermentation plant capacity. Thus, biofuels cannot be expected to satisfy large EJ markets in the short- to mid-term. Since most nonwaste-derived biofuels are not economically competitive with fossil fuels in the early 1990s, large additions to plant capacity will not occur except in those cases where environmental concerns or legislative incentives are governing factors.

Investment Opportunities and Capital Requirements. Despite some of the temporary economic problems that confront the biomass energy industry in the early

1990s, several business opportunities are being developed at rapid rates. These projects are distributed across the nation and include landfill gas recovery plants, MSW-to-energy systems, and nonutility power generation that qualifies under PURPA. Conventional combustion technology is utilized in the majority of plants; gasification seems to have been largely ignored and should offer several advantages (112). A production tax credit equivalent to $0.48/m^3 ($3.00/BOE) indexed to inflation and linked to the price of oil is available; it amounted to about $0.71/GJ ($0.75/MBtu) of product gas in mid-1985 (112,129), and can have a significant beneficial impact on the profitability of a biofuels project. The lower the cost of oil, the greater the credit. Taking the most optimistic view of the language in the law, wastes are included in the definition of biomass, so it appears the production tax credit is applicable to all of the above projects, not just those based on wood and other nonwaste biomass.

The Tax Reform Act of 1986 has resulted in a transition away from capital supplied by individuals for the financing of biofuels projects toward conventional financing and greater use of institutional capital sources (128,134). The capital requirements can be large, eg, $20.7 billion in 10 years to complete 240 early and advanced-planned MSW-to-energy plant projects (98), and $35 million per plant for 300–400 small plants (112) in the United States. However, there are biofuels opportunities that do not involve such large capital needs. Numerous landfill gas recovery plants have been installed for well under $30 million each; most have a capital cost of $5 million or less, depending on scale and end use, although the capital cost of one of the largest landfill gas recovery projects is in the $20 million range (115).

The Energy Policy Act of 1992 (H.R. 776) has liberalized the rules concerning biofuels and provides tax incentives for increased usage. Many states also have gasohol fuel tax exemptions in place, and some have enacted legislation that requires use of oxygenated fuels under certain conditions. Most of these laws impact favorably on biofuels usage.

Many energy analysts believe that it is only a matter of time before petroleum prices, the economic parameter that influences almost all other energy prices, begin to return to market prices of at least $3.97/m^3 ($25/bbl). It is widely believed that the gas bubble, which provided excess gas deliverability in the 1980s, will decline in the 1990s. Thus, energy prices are expected to rise again under any scenario. If petroleum prices stabilize at, or continue to increase to, levels over $3.97/m^3 ($25/bbl) it is expected that this, along with environmental issues, will provide the market forces that will increase biomass energy usage.

Research

A large variety of biomass feedstock developments and advanced conversion processes for the production of energy, fuels, and chemicals are in the research stage in the United States. Many other countries are also developing biomass energy in the laboratory and in the field. The research is aimed at reducing the cost of biomass and increasing the efficiency of production of the final products, eg, new fuels, substitute fossil fuels, and energy, so that biofuels can compete with other energy resources, especially fossil fuels.

Feedstock Development.   Most of the research in process in the United States in the early 1990s on the selection of suitable biomass species for energy applications is limited to laboratory studies and small-scale test plots. Many of the research programs on feedstock development were started in the 1970s or early 1980s.

Herbaceous Biomass.   Considerable research has been conducted to screen and select nonwoody herbaceous plants as candidates for biomass plants that are unexplored in the continental United States; other research has concentrated on cash crops such as sugarcane and sweet sorghum; and still other research has emphasized tropical grasses. In the late 1970s, a comprehensive screening study of the United States generated a list of 280 promising candidates from which up to 20 species were recommended for field experiments in each region of the country (135). The four highest-yielding species recommended for further tests in each region are listed in Table 39 (135). Since many of the plants in the original list of 280 species had not been grown for commercial use, the production costs were estimated as shown in Table 40 for the various classes of herbaceous species and used in conjunction with yield and other data to develop the recommendations in Table 39.

A large number of small-scale field tests on potential herbaceous energy crops have been carried out. The productivity ranges for some of the most important species for the midwestern and southeastern United States are shown in Table 41 (136). The results of this research helped to establish a strategy that these crops should be primarily grasses and legumes, produced using management systems similar to those used for conventional forage crops. It was concluded from this work that the ideal selection of herbaceous energy crops for these areas would consist of at least one annual species, one warm-season perennial species, one cool-season perennial species, and one legume. Production rates, cost estimates, and environmental considerations indicate that perennial species will be preferred to annual species on many sites, but annuals may be more important in crop rotations.

In greenhouse, small-plot, and field-scale tests conducted to screen tropical grasses as energy crops, three categories have emerged, based on the time required to maximize dry-matter yields: short-rotation species (2–3 months), intermediate-rotation species (4–6 months), and long-rotation species (12–18 months) (137). A sorghum–sudan grass hybrid (Sordan 70A), the forage grass napier grass, and sugarcane are outstanding candidates in these categories. Minimum-tillage grasses that produce moderate yields with little attention are wild *Saccharum* clones, and Johnson grass in a fourth category. The maximum yield observed was 61.6 dry t/hm^2·yr for sugarcane propagated at narrow row centers over 12 months. The estimated maximum yield is of the order of 112 dry t/hm^2·yr using new generations of sugarcane and the propagation of ratoon, ie, regrowth, plants for several years after a given crop is planted.

Overall research on the development of herbaceous energy crops shows that a broad range of plant species may ultimately be prime energy crops.

Short-Rotation Woody Crops.   Research to develop trees as energy crops via short-rotation intensive culture (SRIC) made significant progress in the 1980s. Projections indicate that yields of organic matter can be substantially increased by coppicing techniques and genetic improvements. Advanced designs of whole-

Table 39. Reported Maximum Productivities for Recommended Herbaceous Plants[a]

U.S. Region[b]	Species	Yield, dry t/hm^2·yr
southeastern prairie delta and coast	kenaf	29.1
	napier grass	28.5
	Bermuda grass	26.9
	forage sorghum	26.9
general farm and North Atlantic	kenaf	18.6
	sorghum hybrid	18.4
	Bermuda grass	15.9
	smooth bromegrass	13.9
central	forage sorghum	25.6
	hybrid sorghum	19.1
	reed canary grass	17.0
	tall fescue	15.7
Lake states and Northeast	Jerusalem artichoke	32.1
	sunflower	20.0
	reed canary grass	13.7
	common milkweed	12.3
central and southwestern plains and plateaus	kenaf	33.0
	Colorado River hemp	25.1
	switchgrass	22.4
	sunn hemp	21.3
northern and western great plains	Jerusalem artichoke	32.1
	sunchoke	28.5
	sunflower	19.7
	milkvetch	16.1
western range	alfalfa	17.9
	blue panic grass	17.9
	cane bluestem	10.8
	buffalo gourd	10.1
northwestern/Rocky Mountain	milkvetch	12.1
	kochia	11.0
	Russian thistle	10.1
	alfalfa	8.1
California subtropical	Sudan grass	35.9
	Sudan–sorghum hybrid	31.6
	forage sorghum	28.9
	alfalfa	19.1

[a] Ref. 135.
[b] As defined by U.S. Dept. of Agriculture, Agriculture Handbook 296, Mar. 1972; excludes Alaska and Hawaii.

Table 40. Production Costs for Annual Herbaceous Plants[a]

Plant groups	Model crop used	Whole plant yield, dry t/(hm^2·yr)	Cost, $/t
tall grasses	corn	17.3	19.1
short grasses	wheat	9.9	17.2
tall broadleaves	sunflower	15.0	12.7
short broadleaves	sugar beet	13.9	77.1
legumes	alfalfa[b]	13.7	20.9
tubers	potatoes	9.2	136

[a]Ref. 135. Average costs.
[b]Is a perennial.

Table 41. Productivity Rates for Productive Herbaceous Biomass Species in Southeast and Midwest, dry t/hm^2·yr[a]

Biomass type and species	Southeast	Midwest
Annuals		
warm-season		
sorghums[b]	0.2–19.0	1.9–29.1
cool-season		
winter rye[c]	0.0–7.2	2.4–6.1
Perennials[d]		
warm-season		
switchgrass[c]	2.9–14.0	2.5–13.4
weeping lovegrass[c]	5.4–13.7	
napier grass/energycane[b]	20.4–28.3	
cool-season		
reed canary grass[c]		2.7–10.8
legumes		
alfalfa		1.6–17.4
flatpea	2.1–12.9	3.9–10.2
Sericea lespedeza	1.8–11.1	

[a]Ref. 136. Figures are average annual productivities.
[b]Thick-stemmed grass.
[c]Thin-stemmed grass.
[d]Productivity rates after 1–2 yr establishment period.

tree harvesters, logging residue collection and chipping units, and rapid planting machinery have progressed to the point where prototype units are being evaluated in the field. It is expected that several of these devices will be manufactured for commercial use. Some tree species being targeted for research are red alder, black cottonwood, Douglas fir, and ponderosa pine in the Northwest; *Eucalyptus*, mesquite, Chinese tallow, and the leucaena in the West and Southwest; sycamore, eastern cottonwood, black locust, catalpa, sugar maple, poplar, and conifers in the Midwest; sycamore, sweetgum, European black alder, and loblolly pine in the Southwest; and sycamore, poplar, willow, and sugar maple in the East. Generally, tree growth in test plots is studied in terms of soil type and the requirements for site preparation, planting density, irrigation, fertilization, weed control, disease control, and nutrients. Harvesting methods are also important, especially in the case of coppice growth for SRIC hardwoods. Although tree species native to the region are usually included in the experimental design, non-native and hybrid species are often tested too. Advanced biotechnological methods and techniques, such as tissue culture propagation, genetic transformation, and somaclonal variation, are being used in research to clonally propagate individual genotypes and to regenerate genetically modified species.

After an intensive 10-year research effort, short-rotation woody crop yields in the United States, based on data accumulated to 1992, were projected to be 9, 9, 11, 17, and 17 dry $t/hm^2 \cdot yr$ in the Northeast, South/Southeast, Midwest/Lake, Northwest, and Subtropics, respectively (138). The corresponding research goals are 15, 18, 20, 30, and 30 dry $t/hm^2 \cdot yr$. Hybrid poplar, which can grow in many parts of the United States, and *Eucalyptus*, which is limited to Hawaii, Florida, southern Texas, and part of California, have shown the greatest potential thus far for attaining exceptionally fast growth rates (138). Both have achieved yields in the range of 20 to 43 $t/hm^2 \cdot yr$ in experimental trials with selected clones. Research indicates other promising species to be black locust, sycamore, sweetgum, and silver maple.

Hybridizing techniques seem to be leading to super trees that have short growth cycles and yield larger quantities of woody biomass. Fast-growing clones are being developed for energy farms in which the trees are ready for harvest in as little as 10 years and yield up to 30 $m^3/hm^2 \cdot yr$. Genetic and environmental manipulation has also led to valuable techniques for the fast growth of saplings in artificial light and with controlled atmospheres, humidity, and nutrition. The growth of infant trees in a few months is equivalent to what can be obtained in several years by conventional techniques.

Chemical injections into pine trees have been reported to have stimulatory effects on the natural production of resins and terpenes and may result in high yields of these valuable chemicals. Combined oleoresin–timber production in mixed stands of pine and timber trees is under development, and it appears that when short-rotation forestry is used, the yields of energy products and timber can be substantially higher than the yields from separate operations.

One of the largest research projects on SRIC trees in the Western World, the Large European Bioenergy Project (LEBEN), was reported to be scheduled for initiation in the Abruzzo Region of Italy in the mid-1980s and to be established near the end of that decade (139,140). This project integrates SRIC tree production, agricultural energy crops and residues, and biomass conversion to fuels and energy. About 400,000 t/yr of biomass consisting of 260,000 t/yr of woody biomass from 700 hm^2, and 120,000 t/yr of agricultural residues from 700 hm^2 of vineyards and olive and fruit orchards, will be used. Later, 110,000 t/yr of energy crops from 1050 hm^2 will be utilized. The energy products include liquid fuels, ie, biomass-derived oil and charcoal, 200 million kWh/yr of electric power, and waste heat for injection into the regional agro-forestry and industrial sectors. This project is still in the start-up stage.

In the Amazon jungle of Brazil, perhaps the largest SRIC tree plantation in the world is being integrated with energy, pulp, and chemical production facilities (139). Fast-growing Caribbean pine, *Eucalyptus*, and *Gmelina arborea* grown on 51,400 hm^2 are converted to these products. Although gmelina failed on some of the planted sites, it is doing well on about one-third of the planted sites that have the best soil conditions. The other tree species are apparently being grown successfully on the other two-thirds of the sites with sandy and transition soils. It has been reported that the Brazilian Government and industry have taken control of the project from non-Brazilian interests.

Aquatic Biomass. Aquatic biomass, particularly micro- and macroalgae, are more efficient at converting incident solar radiation to chemical energy than are most other biomass species. For this reason, and the fact that most aquatic plants do not have commercial markets, research was performed in the late 1970s and 1980s to evaluate several species as energy crops. The overall goals of the research have generally been directed either to biomass production, often with simultaneous waste treatment, for subsequent conversion to fuels by fermentation, or to species that contain valuable products. The aquatics studied and their main applications are microalgae for liquid fuels, the macrophyte water hyacinth for wastewater treatment and conversion to methane, and marine macroalgae for specialty chemicals or conversion to methane.

Research in the United States on microalgae focuses on the growth of these organisms under conditions that promote lipid formation. This eliminates the high cost of cell harvest because the lipids often can be separated by simple flotation or extraction. The United States Department of Energy research program on microalgae in the 1980s was one of the largest of its kind. It consisted of several projects and emphasized the isolation and characterization of the organisms and the development of microalgae that afford high oil yields. The research included projects on siting studies; collection, screening, and characterization of microalgae; growth of certain species in laboratory and small-scale production systems; exploration of innovative approaches to microalgae production; and innovative methods for increasing oil formation. Some microalgae, such as *Botryococcus braunii*, have been reported to produce lipid yields that are 40–50% of the dry cell weight under nitrogen-limited conditions (28). However, in other research, *B. braunii* has been reported to yield 20–52% of the dry cell weight as liquid hydrocarbons (139).

Conversion. *Combustion.* Biomass combustion accounted for about 4% of total U.S. energy consumption in 1992, primarily in the industrial, residential, and utility sectors. Electric power capacity fueled by biomass grew from 200 MW in the early 1980s to about 6000 MW in 1992. The direct combustion of biomass for heat, steam, and power has been, and is expected to continue to be, the principal end use of biomass energy. Conventional biomass-fired technology uses a variety of combustion equipment designs that are usually capable of burning a wet, nonhomogeneous fuel with large variations in moisture content and particle size (141). Spreader stoker-fired boilers have evolved from the designs of the past to systems which include several designs for controlled fuel distribution and automatic ash removal. Research on biomass combustion has focused on improvements of existing systems with respect to ease of operation, increased efficiency, and lower capital and operating costs; emission controls and abatement; and development of new technologies to permit utilization of solid biomass fuels in a wider range of applications (142). Some of the biomass combustion research developments since the early 1930s include whole-tree burning technologies (143), cyclonic incineration of waste biomass (144), direct wood-fired gas turbines (145), improved combustion cycles for biomass (141), fluid-bed biomass combustion (146), pulverized biomass combustion (147), catalytic wood-burning stoves (148), cofiring of biomass and fossil fuels to reduce emissions (149), and control of biomass combustion to reduce emissions (150). Even though the burning of biomass is one of the oldest energy producing methods used, research continues to make significant advancements in the art and science of biomass combustion. Recent U.S. legislation concerned with air quality and waste biomass disposal has a significant impact on the direction of ongoing research to develop advanced biomass combustion systems (151).

Anaerobic Digestion (Methane Fermentation).   A large amount of research was performed in the 1970s and 1980s on the anaerobic digestion of biomass to develop biological gasification processes capable of producing methane (16). Basic research on methane fermentation provides a better understanding of the kinetics and mechanisms of biomass conversion under anaerobic conditions; improvements in digestion efficiencies in terms of methane yield and volatile solids reduction have been slow to evolve from this knowledge. A large number of agricultural residues, animal wastes, and biomass species have been evaluated as potential feedstocks for methane production in laboratory digesters and small-scale digestion facilities. Considerable laboratory work has also been done to develop pre- and post-digestion treatments that improve biodegradability. A plateau of about 50–60% volatile solids destruction efficiencies and energy recoveries in the product gas seems to exist for most methane fermentation systems.

Research in the early 1990s has addressed several potentially beneficial methods of improving the process, eg, two-phase digestion in which the acetogenic and methanogenic phases are physically separated. Practical implementation of two-phase digestion is achieved by control of the hydraulic retention times in the acid and methane reactors or reaction zones. This process configuration provides several advantages over conventional high rate digestion such as enhanced stability, an optimum environment for acetogenic and methanogenic bacteria, substantial increases in throughput rates for given size reactors, increased gas and methane production rates, and higher methane content in the product gas (16,152). The process has been scaled-up for treatment of municipal biosolids, ie, the Acimet Process (153), and has been applied to industrial wastes (16). From a practical standpoint, two-phase anaerobic digestion of biomass is capable of retrofit to existing digestion systems of any design and is projected to be capable of doubling plant capacity at about 50% of the capital cost of a grassroots plant.

A significant market is anticipated for this advanced technology, particularly for wastewater treatment.

Some of the other research studies have addressed topics such as high solids biomass digestion (154), utilization of superthermophilic organisms (155), advanced reactor designs (156), landfill gas enhancement (157), and microbiology of the mixed cultures involved in methane fermentation (158).

Thermochemical Gasification. Extensive research and pilot studies have been carried out since 1970 to develop thermochemical processes for biomass conversion to energy and fuels. Basic studies on the effects of various operating conditions and reactor configurations have been performed in the laboratory and at the process development unit (PDU) and pilot scales on steam, steam-air, air-blown, and oxygen-blown gasification, and on hydrogasification. Other research has also been done on the rapid pyrolysis of biomass which, in addition to gaseous products, yields coproduct liquids and solids.

Over one million air-blown gasifiers were built during World War II to manufacture LHV gas to power vehicles and to generate steam and electric power. Units are available in a variety of designs, some of which have been retrofitted to gas-fired furnaces. Although some research is in progress to refine air-blown wood gasifiers in North America, particularly portable units, most of the research has been conducted in Europe. The Swedish automobile manufacturers, Volvo and Saab, have ongoing programs to develop a standard gasifier design suitable for mass production.

Research on thermochemical biomass gasification in North America has tended to concentrate on MHV gas production, scale-up of the advanced process concepts that have been evaluated at the PDU scale, and the problems that need to be solved to permit large-scale thermochemical biomass gasifiers to be operated in a reliable fashion for power production, especially advanced power cycles. Many different reactor designs have been evaluated under a wide range of operating conditions. Exemplary advanced gasifiers and gasification systems, some of which are in the scale-up stage, include IGT's single-stage, pressurized, fluid-bed gasifier; the National Renewable Energy Laboratory's (NREL) pressurized, fixed-bed, downdraft gasifier; the fire tube-heated, fluid-bed system of the University of Missouri-Rolla; the indirectly heated, fluid-bed, dual reactor system of Battelle Columbus Laboratory; the pulse-enhanced, indirectly heated, fluid-bed gasifier of M.T.C.I.; and the catalytic, pressurized, gasification system for wet biomass of Pacific Northwest Laboratory.

An example of a scale-up project is the pressurized, fluid-bed Renugas plant in Hawaii (159). The gasifier is designed for 63.5 t/d of sugarcane bagasse. In addition to bagasse, other feedstocks such as wood, waste biomass, and RDF may be evaluated. The demonstration provides process information for both air- and oxygen-blown gasification at low and high pressures. Renugas will be evaluated for both fuel gas and synthesis gas production, and for electric power production with advanced power generation schemes.

Although only limited research has been carried out on small-scale LHV producer gasifiers for biomass, significant design advancements have been made even though they have been used for over 100 years. One development is the open-top, stratified, downdraft gasifier in which air is drawn in through successive reaction strata (160). The unit is simple to operate, inexpensive, and can be close-coupled to an engine–generator set without complex gas-cleaning equipment. The gasifier dimensions are sized to deliver gas to the engine based on its fuel-rate requirements, and no controls are needed.

Emissions from the thermochemical gasification of biomass can affect the operation of advanced power cycles and include particulates, alkalies, oils and tars, and heavy metals. One of the high priority research efforts is to develop hot-gas cleanup methods that will permit biomass gasification to supply suitable fuel gas for these cycles (161). Some of the other research needs that have been identified include versatile feed-handling systems for a wide variety of biomass feed-stocks; biomass feeding systems for high pressure gasifiers; determination of the effects of additives including catalysts for minimizing tar production and capturing contaminants; and suitable ash disposal and wastewater treatment technologies (162).

Liquefaction.   Figure 21 outlines most of the biomass liquefaction methods under development. There are essentially three basic types of biomass liquefaction technologies, ie, fermentation, natural, and thermochemical processes.

Much research has been conducted since the early 1970s to improve the alcohol fermentation process, ie, the technology by which biomass is converted to

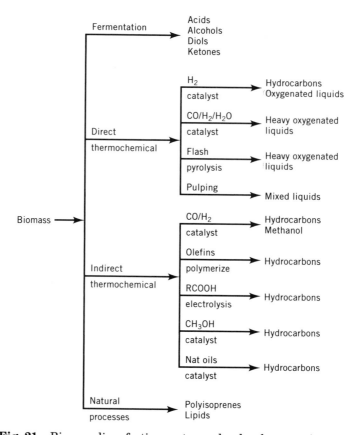

Fig. 21.   Biomass liquefaction routes under development (ca 1992).

fuel ethanol (96,112,125). A large portion of this research has been focused on development of the process so that it will be suitable for conversion of low grade lignocellulosics. This type of biomass generally contains about 50 wt % celluloses, 25 wt % hemicelluloses, and 25 wt % lignins. The hexose sugars obtained on hydrolysis of celluloses are converted by conventional alcohol-forming yeasts to ethanol. The pentose sugars require other specialized organisms for conversion. The lignins are essentially inert.

Research has focused on minimizing the energy inputs for the distillation steps needed to produce 190 and 200 proof ethanol; on the fermentation process itself to increase ethanol yields and reduce fermentation times; on the pretreatment and hydrolysis processes needed to afford high yields of sugars and to make low grade cellulosics suitable feedstocks for fermentation; on development of organisms that ferment pentoses separately, and pentoses and hexoses together; on development of advanced processes that permit simultaneous saccharification and fermentation (SSF); and on development of immobilized organisms and genetically engineered bacteria that afford much shorter fermentation times and avoid yeast recycling.

Several advancements have been developed and incorporated into grain fermentation processes to improve operating efficiencies and reduce energy consumption. Ethanol yields and production rates have improved slightly, but not significantly. In contrast, research on the development of low grade biomass fermentation processes is on the verge of process demonstrations with a variety of feedstocks such as waste paper, agricultural residues, wood biomass, and refuse-derived fuel.

Much of the research to apply ethanol fermentation to low grade biomass is funded by the U.S. Department of Energy. The goal is to produce fuel ethanol at a cost of $0.145/L, or $6.90/GJ, by 1995 (166).

Oilseed and Vegetable Oil Fuels. Limited research has continued on the utilization of seed and vegetable oils as motor fuels, particularly as substitute diesel fuels and diesel fuel extenders (164). Work has focused on studies of the yields and properties of oils from oilseed and vegetable oil crops, the performance of neat oils and oil–diesel fuel blends as fuels for compression ignition engines, improvement of the transesterification process and the fuel characteristics of the resulting esters as diesel fuels, upgrading vegetable oils to gasolines and diesel fuels by hydrocracking processes, and field tests of the liquid fuels made from seed and vegetable oils in trucks and buses. Although several operating problems have been observed, such as lubricating oil deterioration and crystal formation in cooler weather even with some of the lower viscosity seed oil esters, significant advances have been made. Esters of selected vegetable oils are very promising candidates for both indirect and direct fuel-injected engines.

The cost of seed and vegetable oil fuels is still not competitive with petroleum-based diesel fuels. The cost of esterification alone can add up to 50% to the cost of the fuel, depending on the size of the processing operation and the market value of by-product glycerol. In 1992 the cost of oilseed and vegetable oil diesel fuels was about twice that of conventional diesel fuels. One approach to elimination of the cost differential is to use waste vegetable oils from large-scale restaurant operations as the feedstock for transesterification plants (165). This offers the possibility of taking credit for waste oil disposal, ie, the analogue of a tipping fee ·in the solid waste disposal field, and of recycling the oils as fuel. Another approach is to utilize tax credit legislation available in the United States to farmers who produce minor oilseed crops in place of principal commodity crops.

Thermochemical Liquefaction.   Most of the research done since 1970 on the direct thermochemical liquefaction of biomass has been concentrated on the use of various pyrolytic techniques for the production of liquid fuels and fuel components (96,112,125,166,167). Some of the techniques investigated are entrained-flow pyrolysis, vacuum pyrolysis, rapid and flash pyrolysis, ultrafast pyrolysis in vortex reactors, fluid-bed pyrolysis, low temperature pyrolysis at long reaction times, and updraft fixed-bed pyrolysis. Other research has been done to develop low cost, upgrading methods to convert the complex mixtures formed on pyrolysis of biomass to high quality transportation fuels, and to study liquefaction at high pressures via solvolysis, steam–water treatment, catalytic hydrotreatment, and noncatalytic and catalytic treatment in aqueous systems.

Essentially all of these conversion processes are technically feasible and can be used to convert biomass to a wide range of liquid products. Unfortunately, because of the complex composition of biomass and the chemistry of direct thermal cracking of biomass, complex product mixtures are always formed. Selectivities for individual products are low; this is sometimes advantageous. Because of the oxygenated nature of biomass, higher yields of certain oxygenated products can be obtained. This offers the possibility of producing specific organic liquids that have higher intrinsic value as chemicals rather than as fuels. A few research efforts are cited to illustrate the versatility of direct liquefaction processes for biomass.

The principal products in conventional, long-term pyrolysis of biomass at about 400°C or lower are char, gas, organic liquids, and water, as shown in Table 20. Research has shown that fast pyrolysis of biomass at 475–525°C and vapor residence times of a few seconds or less can maximize organic liquid yields. Wood and grasses yield 55–65 wt % and 40–65 wt % of the dry biomass as organic liquids, respectively (168). Products from the fast pyrolysis of wood, for example, contain significant amounts of low molecular weight oxygenated compounds such as hydroxyacetaldehyde, acetaldehyde, formic acid, acetic acid, and glyoxal. Fast pyrolysis of waste lignocellulosics such as newsprint or pulp mill sludges also affords similar liquid products. Pretreatment of the wood before pyrolysis gives dramatic changes in product selectivity.

Ultrafast pyrolysis in the vortex reactor is capable of pyrolyzing biomass at high heat-transfer rates on the reactor wall by ablation and has been found to be useful for a variety of biomass and waste feedstocks. In this reactor, biomass particles are entrained tangentially at high velocities by a carrier gas into the vortex reactor, which causes the biomass particles to be preferentially heated relative to the carrier gas and the pyrolysis vapors (169). Products recovered from this innovative reactor have been demonstrated to be about 55 wt % organic liquids, 14 wt % gases, 13 wt % char, and 12 wt % water (94% closure). The pyrolysis vapors can be condensed to a low viscosity liquid, thought to be suitable for combustion in furnaces and turbines, or can be cracked to form about 15 wt % C_2+ hydrocarbons. Zeolites have been found to catalyze the conversion of the pyrolysis vapors to gasoline range hydrocarbons in yields that approach the theoretical upper limit as determined from stoichiometry (170). The energy conversion efficiency was about 45% for C_2–C_8 range hydrocarbons and 55% for C_2+ hydrocarbons. The ablative fast pyrolysis process has been scaled up to a unit that converts 11,400 t/yr of waste sawdust into 5.3 million L/yr of fuel oil and 1,720 t/yr of charcoal (171). This relatively small plant is expected to have a pretax revenue of $263,000 presuming the feedstock cost is zero, and the fuel oil and char can be sold for $4.74/GJ and $72.60/t, respectively. The capital cost of the plant was $850,000.

Prospects. Despite the slow development of renewable biomass as a primary source of energy, the large research effort in progress on feedstock production and conversion is expected to lead to greater commercialization of advanced energy and organic chemical processes based on biomass. Small-scale systems for the individual farmer are being designed and marketed to make it possible to install and operate complete on-site total energy packages that will supply all of the farm's energy requirements. These systems will be fueled with captive sources of biomass and wastes generated on the farm. It is likely that large building complexes such as schools, apartments, shopping malls, and theme parks in urban areas will be able to incorporate similar systems using captive wastes and delivered biomass. Individual, small-scale, farmers' cooperative and industrial-scale fuel ethanol plants will continue to be built and operated as long as government tax incentives are provided. Tax subsidies for fuel ethanol are expected to become unnecessary as the technologies for use of low grade lignocellulosic feedstocks for fuel ethanol are perfected. New, larger scale, biomass-fueled and waste-to-energy power plants, especially those that incorporate cogeneration, will continue to show modest growth as the technology advances and the disposal of waste biomass in an environmentally acceptable manner is implemented. The development of improved methane fermentation processes for waste biomass such as municipal biosolids, industrial wastes from food-processing and beverage alcohol plants, and refuse-derived fuel, is expected to result in more efficient waste treatment and disposal and increased methane recovery and utilization.

Fossil fuels are still sufficiently low in cost to make the economics of large-scale production of substitute transportation fuels, fuel gases, and fuel oils from biomass borderline or unattractive if the biomass systems are used only to produce energy. Large-scale integrated biomass energy plantations are therefore not expected to be constructed and operated until some time during the first or second quarter of the twenty-first century. Biomass grown strictly as profitable energy crops is expected to occur in that time frame as fossil fuels are phased out or their prices increase because of shortages or additional taxes.

Growing environmental concerns and federal and state environmental regulations are expected to be the driving force behind increased usage of biomass energy. Carbon taxes applied to fossil fuel usage, especially for vehicles and utility power plants, are expected to provide very strong incentives to convert to renewable biomass energy resources for both mobile and stationary applications.

BIBLIOGRAPHY

"Fuels from Biomass" in *ECT* 3rd ed., Vol. 11, pp. 334–392, by D. L. Klass, Institute of Gas Technology.

1. D. L. Klass, in D. L. Klass and J. W. Weatherly III, eds., *Energy from Biomass and Wastes IV*, IGT, Chicago, 1980, pp. 1–41.
2. D. L. Klass, in D. L. Klass, ed., *Energy from Biomass and Wastes XIII*, IGT, Chicago, 1990, pp. 1–46.
3. D. L. Klass, *Chemtech* **4**(3), 161 (1974).
4. R. H. Whittaker and G. E. Likens, in H. Leith and R. H. Whittaker, eds., *Primary Productivity of the Biosphere*, Springer Verlag, New York, 1975.
5. R. Repetto, *Sci. Am.* **262**(4), 36 (1990).
6. B. Bolin, "The Carbon Cycle," in *The Biosphere*, W. H. Freeman and Co., San Francisco, Calif., 1970.
7. J. T. Houghton, G. J. Jenkins, and J. J. Ephraums, eds., *Climate Change: The IPCC Scientific Assessment*, Cambridge University Press, Cambridge, 1990, 365 pp.
8. United Nations, *1990 Energy Statistics Yearbook*, Department of Economic and Social Development, New York, 1992.
9. *RPA Assessment of the Forest and Rangeland Situation in the U.S., 1989*, No. 26, USDA, Forest Service, Washington, D.C., Oct. 1989.
10. D. L. Klass, *Chemtech* **20**(12), 720 (1990).
11. U.S. Department of Energy, *Estimates of U.S. Biofuels Consumption 1990*, DOE/EIA-0548(90), Energy Information Administration, Washington, D.C., Oct. 1991.
12. U.S. Department of Energy, *The Potential of Renewable Energy, An Interlaboratory White Paper*, SERI/TP-260-3674, DE90000322, Office of Policy, Planning and Analysis, Washington, D.C., Mar. 1990.
13. D. F. Westlake, *Biol. Rev.* **38**, 385 (1963).
14. J. S. Burlew, ed., *Algae Culture From Laboratory to Pilot Plant*, Publication 600, Carnegie Institute of Washington, Washington, D.C., 1953, pp. 55–62.
15. C. D. Hodgman, ed., *Handbook of Chemistry and Physics*, 31st ed., Chemical Rubber Publishing Co., Cleveland, Ohio, 1949, p. 1537.
16. D. L. Klass, *Science* **223**, 1021 (1984).
17. L. R. Lynd and co-workers, *Science* **251**, 1318 (1991).
18. *Guayule: An Alternative Source of Natural Rubber*, National Academy of Sciences, Washington, D.C., 1977.
19. R. A. Buchannan and F. O. Otey, "Multi-Use Oil- and Hydrocarbon-Producing Crops in Adaptive Systems for Food, Material, and Energy Production," paper presented at *19th Annual Meeting, Society for Economic Botany*, St. Louis, Mo., June 11–14, 1978.
20. M. Calvin, *Chemtech* **7**(6), 353 (1977); *Bioscience* **29**, 533 (1979); *Die Naturwissenschaften* **67**, 525 (1980); E. K. Nemethy, J. W. Otvos, and M. Calvin, in D. L. Klass and G. H. Emert, eds., *Fuels from Biomass and Wastes*, Ann Arbor Science Publishers, Ann Arbor, Mich., 1981; J. D. Johnson and C. W. Hinman, *Science* **208** 460, (1980).
21. D. K. Schmalzer and co-workers, *Biocrude Suitabilities for Petroleum Refineries*, ANL/CNSV-69, Argonne National Laboratory, Argonne, Ill., June 1988.
22. K. R. Kaufman, in E. D. Shultz and R. P. Morgan, eds., *Fuels and Chemicals from Oil Seeds: Technology and Policy Options*, Westview Press, Boulder, Colo., 1982, pp. 143–174.
23. P. B. Weisz and J. F. Marshall, *Science* **206**, 257 (1979); R. M. Furrer and N. N. Bakshi, in Ref. 2, pp. 897–914.
24. M. Stumborg and co-workers, in D. L. Klass, ed., *Energy from Biomass and Wastes XVI*, IGT, Chicago, 1993, pp. 721–738.
25. E. S. Olson and R. K. Sharma, in Ref. 24, pp. 739–751.
26. E. S. Lipinsky and co-workers, in Ref. 22, pp. 205–223.
27. E. E. Gavett and D. VanDyne, in Ref. 24, pp. 709–719.
28. D. L. Klass, in D. L. Klass, ed., *Energy from Biomass and Wastes IX*, IGT, Chicago, 1985, pp. 1–83.

29. D. M. Tillett and J. R. Benemann, in D. L. Klass, ed., *Energy from Biomass and Wastes XI*, IGT, Chicago, 1988, pp. 771–786.

30. A. M. Hill and D. A. Feinberg, *Fuel Products from Microalgae*, SERI/TP-231-2348, Solar Energy Research Institute, Golden, Colo., 1984.

31. D. A. Hoffman and R. A. Fitz, *Environ. Sci. Technol.* **2**(11), 1023 (1968).

32. S. B. Alpert and co-workers, *Pyrolysis of Solid Wastes: A Technical and Economic Assessment*, NTIS PB 218-231, SRI, Menlo Park, Calif., Sept. 1972.

33. M. A. Paisley, H. F. Feldmann, and H. R. Appelbaum, in D. L. Klass and H. H. Elliott, eds., *Energy from Biomass and Wastes VIII*, IGT, Chicago, 1984, pp. 675–696.

34. R. C. Bailie, in D. L. Klass and J. W. Weatherly III, eds., *Energy from Biomass and Wastes V*, IGT, Chicago, 1981, pp. 549–569.

35. G. T. Preston, in F. Ekman, ed., *Clean Fuels from Biomass, Sewage, Urban Refuse, Agricultural Wastes*, IGT, Chicago, 1976, pp. 89–114.

36. D. S. Scott and J. Piskorz, in D. L. Klass and H. H. Elliott, eds., *Energy from Biomass and Wastes VII*, IGT, Chicago, 1983, pp. 1123–1146.

37. D. L. Klass, in Ref. 35, pp. 21–58.

38. T. F. Fisher, M. L. Kasbohm, and J. R. Rivero, in Ref. 35, pp. 447–459.

39. S. J. Yosim and K. M. Barclay, *Preprints of Papers, 171st National Meeting ACS, Div. of Fuel Chem.* **21**(1), 73 (Apr. 5–9, 1976).

40. H. F. Feldmann and co-workers, *Hydrocarbon Process.* **55**(11), 201 (1976).

41. S. P. Babu, D. Q. Tran, and S. P. Singh, in Ref. 1, pp. 369–385.

42. D. C. Elliott and co-workers, in D. L. Klass, ed., *Energy from Biomass and Wastes XV*, IGT, Chicago, 1991, pp. 1013–1021.

43. H. F. Feldmann and co-workers, in D. L. Klass, ed., *Biomass as a Nonfossil Fuel Source*, ACS Symposium Series 144, American Chemical Society, Washington, D.C., 1980, pp. 351–375.

44. L. K. Mudge and co-workers, "Catalytic Gasification of Biomass," in *3rd Annual Biomass Energy Systems Conference Proceedings: The National Biomass Program*, SERI/TP-33-285, Solar Energy Research Institute, Golden, Colo., 1979, pp. 351–357.

45. D. G. B. Boocock and D. Mackay, in Ref. 1, pp. 765–777.

46. H. R. Appel and co-workers, *Conversion of Cellulosic Wastes to Oil*, U.S. Bur. of Mines, Pittsburgh, Pa., 1975, p. 8013.

47. E. R. Riegel, *Industrial Chemistry*, 2nd ed., The Chemical Catalog Co., New York, 1933, Chapt. 16, p. 257.

48. M. R. Ladisch, C. M. Ladisch, and G. T. Tsao, *Science* **201**, 743 (1978).

49. L. Paszner and co-workers, in Ref. 24, pp. 629–664.

50. U.S. Pat. 4,009,075 (Feb. 22, 1977), W. H. Hoge (BioIndustries, Inc.).

51. L. R. Lynd and co-workers, *Science* **251**, 1318 (1991).

52. R. Katzen Associates, *Chemicals from Wood Wastes*, U.S. Department of Agriculture Forest Products Laboratory, Madison, Wisc., Dec. 14, 1975.

53. R. S. Loomis, W. A. Williams, and A. E. Hall, *Ann. Rev. Plant Physiol.* **22**, 431 (1971).

54. E. I. Rabinowitch, *Photosynthesis*, Vols. 1–2, Interscience Publishers, New York, 1956.

55. C. B. Osmond, *Ann. Rev. Plant Physiol.* **29**, 379 (1978).

56. B. J. Brinkworth, *Solar Energy for Man*, John Wiley & Sons, Inc., New York, 1973.

57. U.S. Dept. of Commerce, *Climatological Data, National Summary*, Vol. 21, Nos. 1–12, U.S. Government Printing Office, Washington, D.C., 1970.

58. H. J. Critchfield, *General Climatology*, 3rd ed., Prentice-Hall, Inc., Englewood Cliffs, N.J., 1974, p. 22.

59. R. S. Loomis and W. A. Williams, *Crop Sci.* **3**, 67 (1963).

60. T. R. Schneider, *Energy Convers.* **13**, 77 (1973).

61. S. S. Visher, *Climatic Atlas of the United States*, Harvard University Press, Cambridge, Mass., 1954.

62. *Statistical Abstracts of the United States*, U.S. Department of Commerce, U.S. Government Printing Office, Washington, D.C., 1976.
63. W. L. Roller and co-workers, *Grown Organic Matter as a Fuel Raw Material Source, NASA Report CR-2608*, Ohio Agricultural Research and Development Center, Washington, D.C., Oct. 1975.
64. M. M. Ludlow and G. L. Wilson, *J. Aust. Inst. Agric. Sci.* **36**, 43 (Mar. 1970).
65. W. J. North, ed., *The Biology of Giant Kelp Beds (Macrocystis) in California*, Cramer, Lehre, Germany, 1971, p. 12.
66. T. A. El-Sharkawy and J. D. Hesketh, *Crop Sci.* **4**, 514 (1964).
67. J. Krummel, in Ref. 35, pp. 359–370.
68. H. J. Evans and L. E. Barber, *Science* **197**, 332 (1977).
69. *Land Capability Classification, Agricultural Handbook 210*, U.S. Department of Agriculture, Soil Conservation Service, Washington, D.C., 1966, 21 pp.
70. *Summary Report 1987 National Resources Inventory*, No. 790, U.S. Department of Agriculture, Soil Conservation Service, Washington, D.C., Dec. 1989, 37 pp.
71. S. H. Spurr, *Sci. Am.* **240**, 76 (1979).
72. A. A. Nichiporovich, *Photosynthesis of Productive Systems*, Israel Program for Scientific Translations, Jerusalem, Israel, 1967.
73. E. O. Mariani, in D. L. Klass and W. W. Waterman, eds., *Energy from Biomass and Wastes*, IGT, Chicago, 1978, pp. 29–38.
74. R. L. Sajdak and co-workers, in Ref. 43, pp. 21–48.
75. J. C. Dula, in Ref. 33, pp. 193–207.
76. L. L. Wright and co-workers, in D. L. Klass, ed., *Energy from Biomass and Wastes XII*, IGT, Chicago, 1989, pp. 261–274.
77. *Solar SNG, Final Report American Gas Association Project IU-114-1*, Prepared by InterTechnology Corp., American Gas Association, Washington, D.C., Oct. 1975.
78. *Grass: The Yearbook of Agriculture 1948*, U.S. Department of Agriculture, U.S. Government Printing Office, Washington, D.C., 1948.
79. D. L. Klass, S. Ghosh, and J. R. Conrad, in Ref. 35, pp. 229–252.
80. D. L. Klass and S. Ghosh, in Ref. 43, pp. 229–249.
81. J. A. Alich, Jr., and R. E. Inman, *Effective Utilization of Solar Energy to Produce Clean Fuel, Grant No. GI 38723*, Final Report for National Science Foundation, Stanford Research Institute, Palo Alto, Calif., June 1974.
82. R. Retovsky, *Continuous Cultivation of Algae, Theoretical and Methodological Bases of Continuous Culture of Microorganisms*, Academic Press, Inc., New York, 1966.
83. D. L. Klass and S. Ghosh, in W. W. Waterman, ed., *Clean Fuels from Biomass and Wastes*, IGT, Chicago, 1977, pp. 323–351.
84. D. L. Klass, S. Ghosh, and D. P. Chynoweth, *Process Biochem.* **14**, 18 (1979).
85. D. P. Chynoweth, D. L. Klass, and S. Ghosh, in Ref. 73, pp. 229–251.
86. D. L. Klass and S. Ghosh, in D. L. Klass and G. H. Emert, eds., *Fuels from Biomass and Wastes*, Ann Arbor Science Publishers, Ann Arbor, Mich., 1981, pp. 129–149.
87. M. G. McGarry, *Process Biochem.* **6**, 50 (1971).
88. J. L. Yount and R. A. Grossman, *J. Water Pollut. Control Fed.* **42**, 173 (1970).
89. E. S. Del Fosse, in Ref. 83, pp. 73–99.
90. *State Energy Price and Expenditure Report 1990*, DOE/EIA-0376(90), U.S. Department of Energy, Energy Information Administration, Washington, D.C., Sept. 1992.
91. M. D. Fraser, in Ref. 24, pp. 295–330.
92. R. F. Moorer, D. K. Walter, and S. Gronich, in Ref. 24, pp. 139–153.
93. S. Lazzari, in Ref. 24, pp. 275–294.
94. N. Smith and T. J. Corcoran, *Preprints of Papers Presented at 171st National Meeting, ACS, Fuel Chemistry Division, Symposium on Net Energetics of Integrated Synfuel Systems* **21**(2), 9 (Apr. 1976).
95. R. E. Inman, in Ref. 94, pp. 21–27.

96. D. L. Klass, *Energy Topics*, 1 (Apr. 14, 1980).
97. National Wood Energy Association, *NWEA State Biomass Statistical Directory*, Arlington, Va., 1988.
98. E. Berenyi and R. Gould, *1988–1989 Resource Recovery Yearbook*, Governmental Advisory Associates, Inc., New York, 1988, 718 pp.
99. R. M. Dykes, in Ref. 76, pp. 379–397.
100. *Stove and Fireplace Catalog IX*, Consolidated Dutchwest, Plymouth, Mass., 1988, 67 pp.
101. *Estimates of U.S. Wood Energy Consumption 1980–1983*, DOE/EIA-0341(83), U.S. Department of Energy, Energy Information Administration, Washington, D.C., Nov. 1984.
102. J. W. Koning, Jr. and K. E. Skog, in D. L. Klass, ed., *Energy from Biomass and Wastes X*, IGT, Chicago, 1986, pp. 1309–1322; J. C. Nicolello, *U.S. Pulp and Paper Industry's Energy Use-Calendar Year 1986*, New York, Apr. 20, 1987; J. C. Nicello, *U.S. Pulp and Paper Industry's Energy Use-Calendar Year 1987*, New York, May 17, 1988; *Annual Energy Review 1987*, DOE/EIA-0384(87). U.S. Department of Energy, Energy Information Administration, Washington, D.C., 1988.
103. National Wood Energy Association, *Wood Energy, America's Renewable Resource*, Arlington, Va., 1988, 2 pp.
104. *Past, Present, and Future Trends in the U.S. Forest Sector: 1952–2040, Review Draft*, U.S. Department of Agriculture, Forest Service, Washington, D.C., June 1988.
105. J. I. Zerbe, *Forum for Applied Research and Public Policy*, 38–47 Winter (1988).
106. Table 1, in Ref. 11.
107. D. L. Klass and C. T. Sen, *Chem. Eng. Prog.* **83**(7), 46 (1987).
108. *The Qualifying Facilities Report*, Federal Energy Regulatory Commission, Washington, D.C., Jan. 11, 1988.
109. J. L. Easterly, personal communication, Meridian Corp., Alexandria, Va., Aug. 16, 1988.
110. D. A. Flint and C. Norris, *1986 Capacity and Generation of Non-Utility Sources of Energy*, Edison Electric Institute, Washington, D.C., July 1988.
111. B. DeCampo, D. A. Flint, and C. Norris, *Electric Perspectives*, 22 (Summer 1988).
112. D. L. Klass, *Resources and Conservation* **15**, 7 (1987).
113. J. L. Easterly, S. Lees, and B. Detwiler, *Electric Power from Biofuels: Planned and Existing Projects in the U.S.*, rev. Jan. 1985, DOE/CE/307841/1, U.S. Department of Energy, Washington, D.C., Aug. 1985.
114. C. Tewksbury, in D. L. Klass, ed., *Energy from Biomass and Wastes X*, IGT, Chicago, 1987, pp. 555–578.
115. D. L. Klass, *Resources and Conservation* **11**, 157 (1985).
116. T. R. Miles and T. R. Miles, Jr., *Biomass* **18**, 163 (1989).
117. R. L. Wentworth, "Anaerobic Digestion in North America," paper presented at *Symposium Anaerobic Digestion and Carbohydrate Hydrolysis of Waste, sponsored by Commission of the European Communities*, Luxembourg, May 8–10, 1984.
118. J. H. Ashworth, Y. M. Bihun, and M. Lazarus, *Universe of U.S. Commercial-Scale Anaerobic Digesters: Results of SERI/ARD Data Collection*, Solar Energy Research Institute, Golden, Colo., May 30, 1985; J. H. Ashworth, *Problems With Installed Commercial Anaerobic Digesters in the United States: Results of Site Visits,* Rev. ed., Solar Energy Research Institute, Golden, Colo., Nov. 6, 1985.
119. *1984 Needs Survey Report to Congress: Planned and Existing Projects in the U.S.*, rev. Jan. 1985, DOE/CE/30784/1, U.S. Department of Energy, Washington, D.C., Aug. 1985.
120. Table 19, in Ref. 28.
121. S. Doelph, *Gas Energy Review* **16**(1), 14 (1988).
122. "Landfill Gas Summary Update," *Waste Age* **19**(3), 167 (1938).

123. "Resource Recovery Activities," *City Currents* **6**(4), 1 (1987).
124. F. L. Potter, personal communication, Information Resources, Inc., Washington, D.C., Aug. 11, 1988.
125. D. L. Klass, *Energy Topics*, 1 (Aug. 1, 1983).
126. *Alc. Update*, Aug. 8, 1988; *Alc. Wk.* **9**(32) (Aug. 8, 1988).
127. L. L. Wright, personal communication, Oak Ridge National Laboratory, Oak Ridge, Tenn., July 1988.
128. S. Fenn, *Institutional Investment in Renewable Energy Technologies*, Renewable Energy Institute, Washington, D.C., Feb. 1987, 50 pp.
129. S. Lazzari, *A History of Federal Tax Policy: Conventional as Compared to Renewable and Nonconventional Energy Resources*, 88-455E, The Library of Congress, Congressional Research Service, Washington, D.C., June 7, 1988.
130. W. A. Rains, paper presented at *1982 Annual Meeting, National Petroleum Refiners Association*, San Antonio, Tex., Mar. 21–23, 1983.
131. W. R. Keene, Lundberg Survey, Inc., N. Hollywood, Calif., Aug. 16, 1988.
132. *Chem. & Eng. News* **66**(25), 40 (June 20, 1988).
133. *Assessment of Costs and Benefits of Flexible and Alternative Fuel Use in the U.S. Transportation Sector, Progress Report One*, DOE/PE-0080, U.S. Department of Energy, Washington, D.C., Jan. 1988.
134. B. Paul, *WSJ LXIX* (217), Sec. 2, 19 (Aug. 19, 1988).
135. K. A. Saterson and M. W. Luppold, *3rd Annual Biomass Energy Systems Conference Proceedings*, SERI/TP-33-285, U.S. Department of Energy, Golden, Colo., June 5–7, 1979, pp. 245–254.
136. J. H. Cushman and A. F. Turhollow, in D. L. Klass, ed., *Energy from Biomass and Wastes XIV*, IGT, Chicago, 1991, pp. 465–480.
137. A. G. Alexander, in Ref. 136, pp. 367–374.
138. L. L. Wright, in L. L. Wright and W. G. Hohenstein, eds., *Biomass Energy Production in the United States: Situation and Outlook*, Oak Ridge National Laboratory, Oak Ridge, Tenn., Aug. 1992, Chapt. 2.
139. D. L. Klass, in Ref. 114, pp. 13–113.
140. G. Grassi, in Ref. 114, pp. 1545–1562.
141. A. Ismail and R. Quick, in Ref. 42, pp. 1063–1100.
142. J. E. Robert and E. N. Hogan, in Ref. 42, pp. 1245–1265.
143. L. D. Ostlie and T. E. Drennen, in Ref. 76, pp. 621–650.
144. A. Rehmat and M. Khinkis, in Ref. 42, pp. 1111–1139.
145. J. T. Hamrick, in Ref. 114, pp. 517–528.
146. M. L. Murphy, in Ref. 29, pp. 371–380; in Ref. 42, pp. 1167–1179; S. C. Bhattacharya and W. Wu, in Ref. 76, pp. 591–601.
147. J. F. L. Lincoln and T. C. Litchney, in Ref. 29, pp. 357–369.
148. S. G. Barnett and S. J. Morgan, in Ref. 136, pp. 191–236.
149. G. A. Norton and A. D. Levine, in Ref. 2, pp. 513–527.
150. C. Tewksbury, in Ref. 42, pp. 95–127.
151. S. M. Turner and D. A. Rowley, in Ref. 42, pp. 43–63; D. R. Patrick, in Ref. 42, pp. 65–72; R. N. Sampson, in Ref. 42, pp. 159–186.
152. U.S. Pat. 4,022,665 (May 10, 1977), S. Ghosh and D. L. Klass (IGT); U.S. Pat. 4,318,993 (Mar. 9, 1982), S. Ghosh and D. L. Klass (IGT); T. L. Miller and S. Ghosh, in Ref. 136, pp. 869–876.
153. *Acimet Technical Briefing*, Illinois Department of Energy and Natural Resources and the DuPage Group, Woodridge, Ill., Oct. 7, 1992.
154. W. J. Jewell and co-workers, in Ref. 28, pp. 669–693; R. Legrand and W. J. Jewell, in Ref. 114, pp. 1077–1095; B. De Wilde and L. De Baere, in Ref. 136, pp. 915–929; C. J. Rivard, in Ref. 24, pp. 1025–1041.
155. J. W. Deming, in Ref. 114, pp. 1097–1111.

156. R. W. Meyer and W. R. Guthrie, in Ref. 28, pp. 857–872; S. R. Harper, G. E. Valentine, and C. C. Ross, in Ref. 29, pp. 637–664; L. M. Safley and P. D. Lusk, in Ref. 136, pp. 955–980; R. R. Dague and S. Sung, in Ref. 24, pp. 1001–1023.
157. J. J. Walsh and co-workers, in Ref. 114, pp. 1115–1125; P. Fletcher, in Ref. 76, pp. 1001–1027.
158. A. J. L. Macario and co-workers, in Ref. 114, pp. 1009–1020; S. J. Schropp and co-workers, in Ref. 114, pp. 1035–1043; D. P. Chynoweth and co-workers, in Ref. 76, pp. 965–981; V. Chitra and K. Ramasamy, in Ref. 24, pp. 1043–1060.
159. A. R. Trenka and co-workers, in Ref. 42, pp. 1051–1061.
160. H. LaFontaine, in Ref. 29, pp. 561–575.
161. *Summary Report for Hot-Gas Cleanup, Compiled by Institute of Gas Technology*, for International Energy Agency, IGT, Chicago, Dec. 1991, 50 pp.
162. *Research Needs for Thermal Gasification of Biomass, Compiled by Studsvik AB Thermal Processes*, for International Energy Agency, IGT, Chicago, Mar. 1992, 4 pp.
163. *Conservation and Renewable Energy Technologies for Transportation*, DOE/CH 10093-84, U.S. Department of Energy, Washington, D.C., Nov. 1990, 20 pp.
164. G. R. Quick, in G. Robbelen, R. K. Downey, and A. Ashri, eds., *Oil Crops of the World*, McGraw-Hill Publishing Co., New York, 1989, pp. 118–131.
165. T. B. Reed, M. S. Graboski, and S. Gaver, in Ref. 42, pp. 907–914.
166. A. V. Bridgwater and G. Grassi, eds., *Biomass Pyrolysis Liquids Upgrading and Utilization*, Elsevier Applied Science, New York, 1991, 377 pp.
167. D. C. Elliott and co-workers, *Energy & Fuels* **5**, 399 (1991).
168. D. S. Scott, J. Piskorz and D. Radlein, in Ref. 24, pp. 797–809.
169. J. Diebold, in Ref. 166, pp. 341–350.
170. J. Diebold and co-workers, in E. Hogan and co-workers, eds., *Biomass Thermal Processing*, The Chameleon Press Ltd., London, 1992, pp. 101–108.
171. D. A. Johnson, G. R. Tomberlin, and W. A. Ayres, in Ref. 42, pp. 915–925.

General References

D. L. Klass, ed., *Energy from Biomass and Wastes*, Vols. 4–16, IGT, Chicago, 1980–1993.
D. L. Klass, ed., *Biomass as a Nonfossil Fuel Source*, ACS Symposium Series 144, American Chemical Society, Washington, D.C., 1981, 564 pp.
D. L. Klass and G. H. Emert, eds., *Fuels from Biomass and Wastes*, Ann Arbor Science Publishers, Inc., Ann Arbor, Mich., 1981, 592 pp.
J. L. Jones and S. B. Radding, eds., *Thermal Conversion of Solid Wastes and Biomass*, ACS Symposium Series 130, American Chemical Society, Washington, D.C., 1980, 747 pp.
S. S. Sofer and O. R. Zaborsky, eds., *Biomass Conversion Processes for Energy and Fuels*, Plenum Press, New York, 1981, 420 pp.
E. Hogan and co-workers, eds., *Biomass Thermal Processing*, The Chameleon Press Limited, London, 1992, 255 pp.
A. V. Bridgwater and G. Grassi, eds., *Biomass Pyrolysis Liquids Upgrading and Utilization*, Elsevier Applied Science, London, 1991, 377 pp.
A. V. Bridgwater, ed., *Thermochemical Processing of Biomass*, Butterworths, London, 1984, 344 pp.
Biomass, International Directory of Companies, Processes & Equipment, Macmillan Publishers, Ltd., New York, 1986, 243 pp.
D. L. Klass, ed., *A Directory of U.S. Renewable Energy Technology Vendors, Biomass, Photovoltaics, Solar Thermal, Wind*, Biomass Energy Research Association, Washington, D.C., 1990, for U.S. Agency for International Development, 74 pp.
P. F. Bente, Jr., ed., *Bio-Energy Directory*, The Bio-Energy Council, Washington, D. C., 1980, 768 pp.
Biofuels Technical Information Guide, SERI/SP-220-3366, Solar Energy Research Institute, Golden, Colo., Apr. 1989, 198 pp.

A Guide to Federal Programs in Biomass Energy, Meridian Corporation and Science Applications International Corp., Washington, D.C., Sept. 1984, 157 pp.

W. H. Smith, ed., *Biomass Energy Development*, Plenum Press, New York, 1986, 668 pp.

First Biomass Conference of the Americas: Energy, Environment, Agriculture, and Industry, NREL/CP-200-5768, DE930/0050, Proceedings Vols. I–III, National Renewable Energy Laboratory, Golden, Colo., 1993, 1942 pp.

DONALD L. KLASS
Entech International, Inc.

FUELS FROM WASTE

A significant number and variety of organic wastes are combusted in energy recovery systems including municipal solid waste (MSW), various forms of refuse-derived fuel (RDF) produced from MSW, and municipal sewage sludge; bark and other wood wastes from sawmills and other forest industry operations; spent pulping liquor from chemical pulp mills such as kraft and sulfite mills; wastewater treatment solids (WTS) or sludges from pulp and paper operations; agribusiness wastes including bagasse from sugar-refining operations, rice hulls, orchard and vineyard prunings, cotton gin trash, and a host of other food and fiber-producing operations; manure from feedlots and dairy cattle, chickens, and other agricultural animals; methane-rich gases generated from municipal-waste landfills; industrial trash and specific wastes such as demolition debris, broken pallets, unrecyclable paper wastes, and related materials; off-gases from pulp mills and chemical manufacturers; incinerable hazardous wastes generated regularly as a function of production processes, eg, spent solvents, or found on Superfund sites targeted for clean-up; and a broad range of other specific specialty wastes. The practice of incinerating these materials has become increasingly prevalent (ca 1990) in order to accomplish disposal in a cost-effective, environmentally sensitive manner. The combustion of such wastes already contributes some 5 EJ (5×10^{15} Btu) to the U.S. economy and over 15 EJ ($>14 \times 10^{15}$ Btu) to the economies of the industrialized world (1). Combustion of such wastes reduces the volume of material which must be disposed of in a landfill, reduces the airborne emissions resulting from plant operations and landfill operations, and permits some economic benefit through energy recovery.

The technologies used to combust wastes depend on the form and location of components to be burned. Typically solid wastes are burned, alone or in combination, and both with and without supplementary fossil fuels. Solid wastes can be burned in mass-burn or pile-burning systems such as hearth furnaces, spreader–stokers, ashing and slagging rotary kilns, or fluidized beds. The choice of combustion technology depends on the degree of waste preparation which is practical; the availability of existing combustion systems, eg, a spreader–stoker for hog fuel utilization, adapted to the cofiring of hog fuel and WTS; and the type of energy recovery contemplated. Energy recovery from the solid wastes can be accomplished in the form of medium or high pressure steam, eg, 4.5–8.6 MPa (44–85 atm) (672–783 K), suitable for cogeneration or condensing power generation purposes; low pressure steam, eg, 314–1030 kPa (3.1–10.2 atm), saturated, suitable for process purposes; or the direct production of process heat in the form of heated air or hot combustion products. Energy recovery from gaseous wastes can be accomplished through electricity generation from gas-fired boilers, combustion turbines, or internal combustion engines. Alternatively, these gaseous fuels can be used to generate process heat in conventional fashion.

The success of waste-to-energy programs using municipal wastes and biowastes reduces the volume of material being interred in the ground in landfills. This action also changes the character of materials being landfilled, reducing the organic content with its associated generation of methane gas, and leachates with their significant concentrations of organic compounds. Waste-to-energy, applied to municipal and biomass wastes, can simultaneously provide renewable energy while addressing environmental issues.

Critical concerns associated with energy generation from wastes include fuel composition characteristics; combustion characteristics; formation and control of airborne emissions including both criteria pollutants and air toxics, eg, trace metals; and the characteristics of bottom and flyash generated from waste combustion. These issues are particularly important given the U.S. Clean Air Act Amendments of 1990, the Resource Recovery and Conservation Act (RRCA), and related state and regional regulations. Further, these issues are of critical importance given the capital intensive nature of organic waste-to-energy systems.

Fuel Characteristics of Organic Wastes

Fuel characteristics of organic wastes include physical characteristics such as state, specific gravity, bulk density, porosity, and void volume, and related thermal properties; traditional chemical analyses such as proximate and ultimate analyses, including chlorine; calorific content; elemental analyses of the ash, including trace metal contents, base–acid, slagging, and fouling ratios of the various ash products; and certain chemical structural analyses such as aromaticity. These characteristics are governed by the sources of waste-based fuels. They determine the performance of materials in fuel preparation systems such as particle size reduction and drying systems, and also govern the combustion characteristics of the various wastes being burned.

Sources of Waste-Based Fuels. The general architecture of waste-based fuels is a function of waste origination. MSW characteristics are governed by the product composition of the waste stream, as shown in Table 1. The composition of RDF is governed by the processing technologies used to generate the fuel. RDF production technologies involve, at a minimum, coarse shredding of the MSW stream, followed by magnetic separation of ferrous metals. Primary separation techniques for concentration of combustibles involve trommels, air classifiers, or eccentric screens. Trommels have become the most popular separation systems; their overall separation efficiency can be as high as 98.5% (Table 2). Process flow sheets using trommel separation of MSW follow the pattern shown in Figure 1. The composition of MSW and RDF ultimately is a function not only of the general

Table 1. Product Composition for Municipal Solid Waste,[a] Wt %

Product	1990[b]	2000[c]
paper and paperboard	38.3	41.0
yard waste	17.0	15.3
food waste	7.7	6.8
plastics	8.3	9.8
wood	3.7	3.8
textiles	2.2	2.2
rubber and leather	2.5	2.4
glass	8.8	7.6
metals	9.4	9.0
miscellaneous	2.1	2.1

[a]Ref. 2.
[b]Approximate.
[c]Estimated.

Table 2. MSW Separation Efficiencies for Trommels as a Function of Waste Component,[a] Wt %

Waste component	Separation efficiency
paper, plastic	61.1–69.4
other combustibles[b]	74.6–86.8
ferrous metal	61.6–80.1
aluminum	76.7–93.6
glass, stones, and other	96.6–100
fines	97.0–98.0
overall efficiency	*81.0–98.5*

[a]Ref. 3.
[b]For example, wood.

composition of the waste stream and the RDF production technology, but also of community and industrial recycling programs. Such programs are accelerating and will influence the amount and relative concentration of paper, plastic, aluminum, and other commodities in the waste stream.

The basic architecture of wood-waste fuels is governed by sawmill or plywood mill configuration, and the consequent blend of bark, trim ends, sawdust, planer shavings, and related residuals. All chippable wastes typically are directed to pulp chips. Planer shavings and some sawdust may be directed to alternative products including oriented strand board (OSB), particleboard, animal bedding, a range of other materials applications, and fuel. The characteristics of pulp-mill wastes, eg, bark, WTS, and spent pulping liquor, also are determined by the production processes. The characteristics of wastes from food processing, eg, bagasse, rice hulls, peach pits, cotton gin trash, etc, are governed by the basic product manufacturing technology and its efficiency of separation.

Physical Properties.   Physical properties of waste as fuels are defined in accordance with the specific materials under consideration. The greatest degree

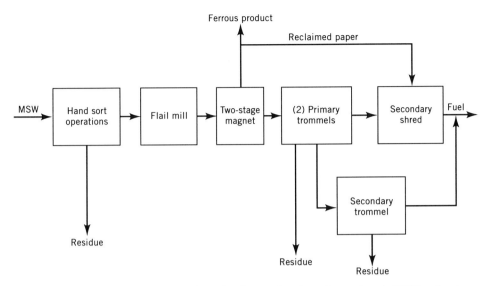

Fig. 1. Simplified schematic flow sheet for the production of a moderate RDF, using trommel separation.

of definition exists for wood and related biofuels. The least degree of definition exists for MSW, related RDF products, and the broad array of hazardous wastes. Table 3 compares the physical property data of some representative combustible wastes with the traditional fossil fuel bituminous coal. The solid organic wastes typically have specific gravities or bulk densities much lower than those associated with coal and lignite.

Specific gravity is the most critical of the characteristics in Table 3. It is governed by ash content of the material, is the primary determinant of bulk density, along with particle size and shape, and is related to specific heat and other

Table 3. Physical Properties of Waste-Based Fuels[a]

Fuel	Specific gravity[b]	Bulk density,[c] kg/m³	Moisture content, wt %
municipal waste		160–320	25–35
waste paper	1.2–1.4	80–160	15–25
waste wood	0.37–0.65	100–320	5–15[d]
			40–65[e]
bagasse			50–55
rice hulls			7–10
orchard and vineyard prunings	0.45–0.55		20–40
bituminous coal	1.12–1.35	672–1393	3.5–5.0

[a]Refs. 1 and 4.
[b]Ovendry.
[c]To convert kg/m³ to lb/ft³, divide by 16.01.
[d]Dry waste.
[e]Wet waste.

thermal properties. Specific gravity governs the porosity or fractional void volume of the waste material, ie,

$$FVV = (1 - SG)/1.5 \tag{1}$$

where FVV is fractional void volume, SG is specific gravity, and the value 1.5 is the approximate specific gravity of the cell wall in wood fiber (5). Specific gravity and moisture content (MC) together determine thermal conductivity characteristics, k, of cellulosic waste-based fuels:

$$k_{MC<30\%} = SG \ (1.39 + 0.028 \times MC) + 0.165 \tag{2}$$

and

$$k_{MC>30\%} = SG \ (1.39 + 0.038 \times MC) + 0.165 \tag{3}$$

Specific gravity is directly related to the bulk density of waste fuels prepared in a variety of ways. Solid oven-dry (OD) wood, for example, has a typical bulk density of 48.1 kg/m^3 (30 lb/ft^3). In coarse hogged form, eg, <1.9-cm minor dimension, this bulk density declines to about 24 kg/m^3 (15 lb/ft^3). In pulverized form, at a particle size <0.16 cm, this bulk density declines to 16–19 kg/m^3 (10–12 lb/ft^3). Similar relationships hold for municipal waste, agricultural wastes, and related fuels.

Chemical Composition.   Chemical compositional data include proximate and ultimate analyses, measures of aromaticity and reactivity, elemental composition of ash, and trace metal compositions of fuel and ash. All of these characteristics impact the combustion processes associated with wastes as fuels. Table 4 presents an analysis of a variety of wood-waste fuels; these energy sources have modest energy contents.

The analysis of solid fuels (Table 4) contains the bases for calculating reactivities, ie, volatile:fixed carbon ratios, volatile carbon:total carbon ratios, hydrogen:carbon and oxygen:carbon ratios, and aromaticity, which is estimated from the chemical components of the waste stream. The typical source of aromatic carbon in waste fuels, such as municipal solid waste and wood waste, is lignin (qv) found in either groundwood-based papers or wood products. Lignin [9005-53-2] has a typical empirical formula (1) of $C_9H_{10}(OCH_3)_{0.9-1.7}$, and a higher heating value (HHV) of 26.7 MJ/kg (11,500 Btu/lb), resulting from its basic building blocks of phenyl propane units. These basic building blocks contain aromatic structures. Other sources of aromatic structures in waste fuels include plastic polymers in the waste stream. The aromaticity of a solid fuel can be estimated as a function of H:C atomic ratio (6):

H:C atomic ratio	Carbons as aromatic carbons, %
1.5	0.0
1.2	3.0
0.7	9.0
0.5	16.0

Table 4. Analysis of Wood-Based Fuels,[a] Wt %

Wood material	Volatile matter	Fixed carbon	Ash	Element C	H	O	N	S	HHV,[b] MJ/kg
big leaf maple	87.9	11.5	0.6	49.9	6.1	43.3	0.14	0.03	16.9
douglas fir	87.3	12.6	0.1	50.6	6.2	43.0	0.06	0.02	18.3
douglas fir bark	73.6	25.9	0.5	54.1	6.1	38.8	0.17	c	19.7
oak									
black[d]	85.6	13.0	1.4	49.0	6.0	43.5	0.15	0.02	16.8
tan[e]	87.1	12.4	0.5	48.3	6.1	45.0	0.03	0.03	17.2
oak bark			5.3	49.7	5.4	39.3	0.2	0.1	17.5
pine bark			2.9	53.4	5.6	37.9	0.1	0.1	18.4
pitch pine			1.1	59.0	7.2	32.7			21.8
poplar			0.7	51.6	6.3	41.5			17.2
red alder	87.1	12.5	0.4	49.6	6.1	43.8	0.13	0.07	17.3
red alder bark	77.3	19.7	3.0	50.9	5.5	40.7	0.39		17.2
western hemlock[f]	87.0	12.7	0.3	50.4	5.8	41.4	0.1	0.1	19.8

[a]Ref. 4.
[b]Higher heating value (OD basis); to convert MJ/kg to Btu/lb, multiply by 430.3.
[c]Trace amounts.
[d]Black oak bark has 81.0 wt % volatile matter, 16.9 wt % fixed carbon, and 2.1 wt % ash.
[e]Tan oak bark has 76.3 wt % volatile matter, 20.8 wt % fixed carbon, and 2.9 wt % ash.
[f]Western hemlock bark has 73.9 wt % volatile matter, 24.3 % fixed carbon, and 0.8 wt % ash.

Typically, 40–50% of the carbon atoms in lignite are in aromatic structures while 60–70% of the carbon atoms in Illinois bituminous coal are in aromatic structures (7,8). By all of these measures, waste fuels are significantly more reactive than coal, peat, and other combustible solids.

The chemical analysis of waste fuels also demonstrates that the wood-based fuels contain virtually no sulfur and little nitrogen. Unless the hog fuel contains bark from logs previously stored in salt-water, the chlorine content is very modest to nonmeasurable.

Municipal waste contains, nominally, about 0.5% nitrogen and 0.5% chlorine, the latter coming largely from plastics. Municipal waste also contains moderate amounts of sulfur. The actual composition of MSW, or RDF generated from MSW, is a function of the relative percentages of various components in the waste stream as shown in Table 5. Use of these wastes provides a means for reducing acid gas emissions from energy generation.

Ash Characteristics. The elemental ash composition of biomass waste and municipal solid waste differs dramatically from that of coal (qv). Wood wastes have ash compositions that are quite alkaline (Table 6) and that have consequent low ash fusion temperatures (Table 7). When firing solid wastes with coal or lignite, the potential exists to have eutectic mixtures formed by the two ash products.

The Clean Air Act of 1990 has made trace metal content in fuels and wastes the final ash-related compositional characteristic of significance. Considerable attention is paid (ca 1993) to emissions of such metals as arsenic, cadmium, chromium, lead, mercury, silver, and zinc. The concentration of these metals in both grate ash and flyash is of significance as a result of federal and state requirements;

Table 5. Components of Municipal Solid Waste[a]

Material	Carbon	Hydrogen	Oxygen	Nitrogen	Chlorine	Sulfur	Moisture	Ash	HHV, MJ/kg[b]
				Components, wt %					
corrugated paper	36.79	5.08	35.41	0.11	0.12	0.23	20.0	2.26	13.0
newsprint	36.62	4.66	31.76	0.11	0.11	0.19	25.0	1.55	13.0
magazine stock	32.93	4.64	32.85	0.11	0.13	0.21	16.0	13.13	11.4
other paper	32.41	4.51	29.91	0.31	0.61	0.19	23.0	9.06	11.5
plastics	56.43	7.79	8.05	0.85	3.00	0.29	15.0	8.59	24.2
rubber and leather	43.09	5.37	11.57	1.34	4.97	1.17	10.0	22.49	17.6
wood	41.20	5.03	34.55	0.24	0.09	0.07	16.0	2.82	14.5
textiles	37.23	5.02	27.11	3.11	0.27	0.28	25.0	1.98	13.8
yard waste	23.29	2.93	17.54	0.89	0.13	0.15	45.0	10.07	8.37
food waste	17.93	2.55	12.85	1.13	0.38	0.06	60.0	5.10	6.82
fines[c]	15.03	1.91	12.15	0.50	0.36	0.15	25.0	44.90	5.41

[a]Ref. 9.
[b]To convert MJ/kg to Btu/lb, multiply by 430.3.
[c]Smaller than 2.54 cm (1 in.).

Table 6. Elemental Analysis of Wood Waste Ash[a]

Compound	CAS Registry Number	Source, wt %		
		Pine bark	Oak bark	Spruce bark
SiO_2	[14808-60-7]	39.0	11.1	32.0
Fe_2O_3	[1309-37-1]	3.0	3.3	6.4
TiO_2	[13463-67-7]	0.2	0.1	0.8
Al_2O_3	[1344-28-1]	14.0	0.1	11.0
MnO_4	[12502-70-4]	b	b	1.5
CaO	[1305-78-8]	25.5	64.5	25.3
MgO	[1309-48-4]	6.5	1.2	4.1
Na_2O	[12401-86-4]	1.3	8.9	8.0
K_2O	[12136-45-7]	6.0	0.2	2.4
SO_3	[7446-11-9]	0.3	2.0	2.1
Cl	[7782-50-5]	b	b	b

[a]Ref. 10.
[b]Trace amounts.

Table 7. Ash Fusion Temperatures for Some Wood Waste Ash,[a] K

Wood species	Initial		Softening		Fluid	
	Oxidizing	Reducing	Oxidizing	Reducing	Oxidizing	Reducing
tan oak	1663	1650	1713	1711	1730	1727
pine bark	1483	1467	1522	1500	1561	1540
oak bark	1744	1750	1772	1766	1783	1777

[a]Refs. 4 and 10.

of particular importance is the mobility of metals. This mobility, and the consequent toxicity of the ash product, is determined by the Toxic Characteristic Leaching Procedure (tclp) test. Tables 8–10 present trace metal contents for wood wastes and agricultural wastes, municipal waste, and refuse-derived fuel, respectively. In Table 8, the specific concentration of various components in the RDF governs the expected average concentration of trace metals.

Biofuels, ie, wood and agricultural waste, are relatively low in metal contents, and typically have a lower metals content when compared to coals being burned for energy generation. However, municipal waste and its derivative fuels (RDF) can be quite high in trace metals. The RDF production process removes approximately 67% of the incoming metals content, but significant quantities of components such as lead and cadmium remain in some compositions of RDF. These metals must be controlled for safe energy generation from such combustible materials. The wood waste in RDF is typically not the same as the wood waste from forest products manufacture. Commonly the wood in RDF is treated with compounds, eg, copper chromium arsenate (CCA), which make it more suitable in outdoor service, such as in deck construction. Such wood treating adds trace metals to the fuel feed (13).

Table 8. Trace Metal Concentrations[a] in Ash from Agricultural Biofuels and Wood-Fired Boilers, mg/kg

Metal	Agricultural biofuel			Wood-fired boilers[b]
	Cotton gin trash	Orchard prunings	Vineyard prunings	
barium	120	220	41	130
silver	<0.08	<0.08	<0.08	<0.08
arsenic	12	5.5	3.4	3.0–6.3
beryllium	0.1	0.1	0.06	0.1
cadmium	1.1	0.36	0.39	1.5–16
cobalt	14	9.0	2.8	8.5–20
chromium	20	12	11	16.8–25
copper	23	14	31	40–76.9
mercury	<0.05	<0.05	<0.05	<0.05–<0.5
molybdenum	16	2	2	3.0–14
nickel	4.6	5.8	4.4	11–50
lead	21	22	55	38–70
antimony	10	10	10	10
selenium	<0.2	<0.2	<0.2	5.0
vanadium	20	12	11	26–27
zinc	87	190	40	130–560
thallium	15	10	2	6.5

[a]Ref. 1.
[b]Range of concentrations from various locations.

Table 9. Trace Metals in Municipal Solid Waste and Solid Waste Ash,[a,b] ppm by wt

Metal	Solid waste[c]	Solid waste ash
arsenic	0.73–12.5	2.9–50
barium	19.8–675	79–2,700
beryllium	ND–0.6	ND–2.4
boron	6–43.5	24–174
cadmium	0.05–25	0.18–100
chromium	3.0–375	12–1,500
cobalt	0.43–22.75	1.7–91
copper	10–1,475	40–5,900
lead	7.75–9.15	31–36,600
magnesium	175–4,000	700–16,000
molybdenum	0.6–72.5	2.4–290
manganese	3.5–782.5	14–3,130
mercury	ND–4.38	0.05–17.5
nickel	3.25–3,228	13–12,910
selenium	0.03–12.5	0.10–50
strontium	3.0–160	12–640
zinc	23–11,500	92–46,000

[a]Range of concentration. Ref. 11.
[b]ND = nondetectable.
[c]Based on ash measurements. Imputed to waste.

Table 10. Analysis of Refuse-Derived Fuel[a]

Parameter	Glossy paper	Nonglossy paper	Cardboard	Film plastics	Rubber, leather, and hard plastics	Wood and textiles	Other organics	Total RDF
Trace metals, mg/kg fuel								
arsenic	3.1	3.3	3.5	2.7	2.5	5.2	4.6	4.0
barium	285.1	78.9	48.7	186.5	724.3	96.7	210.0	173.2
beryllium	1.1	1.3	1.2	0.5	0.4	1.5	1.5	1.3
cadmium	1.1	1.3	3.8	7.7	17.3	3.0	3.1	3.4
chromium	23.8	37.3	23.2	69.4	95.9	34.8	44.5	42.7
copper	74.8	40.3	27.0	2740.7	12.1	202.3	61.4	220.1
lead	88.4	621.2	66.2	836.6	668.1	747.6	475.1	495.5
manganese	61.2	137.6	101.1	311.8	83.1	183.9	367.3	260.4
mercury	0.3	0.7	0.4	1.0	0.4	0.9	1.2	0.9
nickel	10.4	15.5	25.5	45.6	170.4	27.4	17.7	24.4
selenium	3.1	2.9	3.3	2.1	2.0	3.3	3.0	2.9
strontium	62.4	73.2	47.8	88.5	88.6	198.9	474.8	283.2
zinc	164.5	227.6	161.4	482.2	2494.5	449.4	360.0	380.2
Ultimate analysis, wt %								
carbon	43.4	47.3	49.6	59.8	53.8	50.1	34.6	41.1
hydrogen	5.3	6.1	6.4	8.2	8.9	6.0	4.3	5.3
oxygen	27.5	32.0	35.7	13.8	23.3	31.5	41.1	35.2
nitrogen	0.62	1.58	0.72	1.01	0.83	1.07	1.07	1.12
sulfur	0.25	0.25	0.24	0.56	0.57	0.28	0.38	0.34
chlorine	0.04	0.04	0.05	0.10	0.05	0.05	0.01	0.07
ash	23.0	12.7	7.4	16.6	12.5	11.0	18.3	16.5
higher heating value, MJ/kg[b]	14.7	19.7	18.5	31.0	25.4	21.0	16.5	18.7

[a]Fuel produced in Tacoma, Wash. Values on ovendry (OD) fuel basis. Ref. 12.　　[b]To convert MJ/kg to Btu/lb, multiply by 430.3.

Combustion of Solid Waste-Based Fuels

It is useful to examine the combustion process applied to solid wastes as fuels and sources of energy. All solid wastes are quite variable in composition, moisture content, and heating value. Consequently, they typically are burned in systems such as grate-fired furnaces or fluidized-bed boilers where significant fuel variability can be tolerated.

Combustion characteristics of consequence include the overall mechanism of solid waste combustion, factors governing rates of waste fuels combustion, temperatures associated with waste oxidation, and pollution-formation mechanisms.

Mechanisms and Rates of Combustion. All solid fuels and wastes burn according to a general global mechanism (Fig. 2). The solid particle is first heated. Following heating, the particle dries as the moisture bound in the pore structure and on the surface of the particle evaporates. Only after moisture evolution does pyrolysis initiate to any great extent. The pyrolysis process is followed by char oxidation, which completes the process.

The rate of solid waste combustion is controlled by diffusion, rather than by reaction kinetics. In general, the time required for combustion of a single particle of waste (1) can be expressed as:

$$T_b = T_h + T_d + T_p + T_{co} \qquad (4)$$

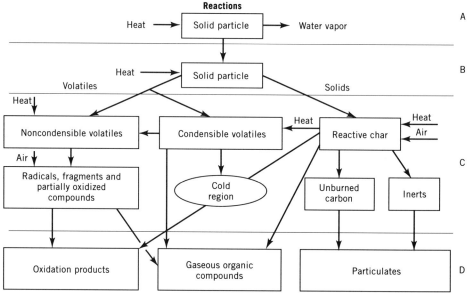

Fig. 2. Overall schematic of solid fuel combustion (1). Reaction sequence is A, heating and drying; B, solid particle pyrolysis; C, oxidation; and D, post-combustion. In the oxidation sequence, left and center comprise the gas-phase region, right is the gas–solids region. Noncondensible volatiles include CO, CO_2, CH_4, NH_3, H_2O; condensible volatiles are C-6–C-20 compounds; oxidation products are CO_2, H_2O, O_2, N_2, NO_1; gaseous organic compounds are CO, hydrocarbons, and polyaromatic hydrocarbons (PAHs); and particulates are inerts, condensation products, and solid carbon products.

where T_b is time for complete particle burnout, T_h is time for initial particle heat-up, T_d is time required for particle drying, T_p is time required to pyrolyze the particle into volatiles and char, and T_{co} is time for char oxidation. The first two terms, initial heating plus drying, $T_h + T_d$, can be taken as the drying step. This time component is governed by the temperature of the environment, the particle size, the moisture content of the particle, and the porosity of the particle. The term T_p is strictly governed by heat transfer through the particle (14,15). The rate of pyrolysis is governed by the heat capacity of the solid waste, its porosity, and its thermal conductivity. The T_{co} term is mass-transfer-limited, with diffusion of oxygen to the surface of the char particle being rate-limiting. Of these steps, either drying or char oxidation may be rate-limiting, as shown in Figure 3, depending on the moisture content of the solid waste.

Temperatures Associated with Combustion. The temperatures achieved by solid waste combustion are typically lower than those associated with fossil fuel oxidation, and are governed by the following general equation (1):

$$T_{f,\ \text{solid waste}} = [695 - 10.1\ \text{MC}_t + 1734(1/\text{SR}) + 0.61(A - 298)]\ \text{K} \quad (5)$$

Where T_f is flame temperature in K; MC_t is moisture content of the waste, expressed on a total weight basis; SR is defined as stoichiometric ratio or moles O_2 available/moles O_2 required for complete oxidation of the carbon, hydrogen, and sulfur in the fuel, ie, $1/\text{SR}$ = equivalence ratio; and A is temperature of the combustion air, expressed in K. In English units, this equation is as follows:

$$T_{f,\ \text{solid waste}} = [3870 - 15.6\ \text{MC}_t - 130.4\ \text{EO}_2 + 0.59(A - 77)]\ °\text{F} \quad (6)$$

where EO_2 is excess oxygen in the stack gas, ie, total, not dry, basis.

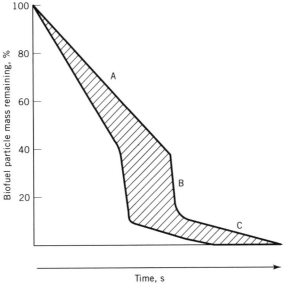

Fig. 3. Schematic of the sequential nature of waste fuel combustion (1). A, particle heating and drying; B, solid particle pyrolysis; and C, char oxidation. A and C may be rate-limiting.

Whereas solid wastes can achieve significant flame temperatures, they are substantially below those associated with fossil fuels. These differences are largely caused by the lower calorific value; the chemical composition, eg, oxygen content of the waste; and the higher moisture and ash contents commonly associated with various solid wastes. Typically for grate-fired systems the use of wastes as fuel requires maintaining temperatures in excess of 1256 K (1800°F) and for residence times exceeding 2 s in order to ensure complete combustion and minimize dioxin and furan formation. As shown from equation 5, such temperatures are readily achieved under most conditions.

Given the mechanisms and temperatures, waste combustion systems typically employ higher percentages of excess air, and typically also have lower cross-sectional and volumetric heat release rates than those associated with fossil fuels. Representative combustion conditions are shown in Table 11 for wet wood waste with 50–60% moisture total basis, municipal solid waste, and RDF.

Formation of Airborne Emissions.   Airborne emissions are formed from combustion of waste fuels as a function of certain physical and chemical reactions and mechanisms. In grate-fired systems, particulate emissions result from particles being swept through the furnace and boiler in the gaseous combustion products, and from incomplete oxidation of the solid particles, with consequent char carryover. If pile burning is used, eg, the mass burn units employed for unprocessed MSW, typically only 20–25% of the unburned solids and inerts exit the combustion system as flyash. If spreader–stoker technologies are employed, between 75 and 90% of the unburned solids and inerts may exit the combustion system in the form of flyash.

Sulfur dioxide [7446-09-5] is formed as a result of sulfur oxidation, and hydrogen chloride is formed when chlorides from plastics compete with oxygen as an oxidant for hydrogen. Typically the sulfur is considered to react completely to form SO_2, and the chlorine is treated as the preferred oxidant for hydrogen. In practice, however, significant fractions of sulfur do not oxidize completely, and at high temperatures some of the chlorine atoms may not form HCl.

Nitrogen oxide, NO_x, formation results from conversion of nitrogen in the fuel to NO, since combustion temperatures are below those typically associated

Table 11. Combustion Conditions for Conventional Waste Fuel Boilers[a]

Parameter	Wood[b]	MSW	RDF
maximum fuel moisture, %	55–60	30–40	20–35
grate fuel feed rate, kg/m²h[c]	1000–1500		
grate heat release, GJ/m²h[d]	8.5–11.35	3.4	5.7–8.5
volumetric heat release, MJ/m³h[e]	480–560	335–410	450–480
stoichiometric ratio	1.25–1.5	1.8–2.0	1.6–1.8
excess air, %	25–50	80–100	60–80
representative overfire air, %	20–40	30–40	30–40

[a]Refs. 1 and 16.
[b]Wet wood waste, 50–60% moisture total basis.
[c]To convert kg/(m²h) to lb/(ft²h), multiply by 0.204.
[d]To convert GJ/(m²h) to Btu/(ft²h), multiply by 8.8 × 10⁴.
[e]To convert MJ/(m³h) to Btu/(ft³h), multiply by 26.9.

with thermal NO_x formation, eg, 1483°C as the threshold for thermal NO_x has been documented (17,18). The conversion of fuel nitrogen to NO_x typically proceeds along the pathways of nitrogen volatilization followed by oxidation of the nitrogen volatiles in the presence of excess oxygen. In the absence of available oxygen, the nitrogen volatiles react with each other to form N_2. Conversion of nitrogen from waste fuels into NO_x is typically 15–25% of the fuel nitrogen converted, depending on combustion technology and firing conditions.

Formation of emissions from fluidized-bed combustion is considerably different from that associated with grate-fired systems. Flyash generation is a design parameter, and typically >90% of all solids are removed from the system as flyash. SO_2 and HCl are controlled by reactions with calcium in the bed, where the limestone fed to the bed first calcines to CaO and CO_2, and then the lime reacts with sulfur dioxide and oxygen, or with hydrogen chloride, to form calcium sulfate and calcium chloride, respectively. SO_2 and HCl capture rates of 70–90% are readily achieved with fluidized beds. The limestone in the bed plus the very low combustion temperatures inhibit conversion of fuel N to NO_x.

Trace metal emissions from waste combustion are a function of metal content in the feed, combustion temperatures, and the percentage of ash exiting the combustion chamber as flyash. They are also a function of the temperatures in the air pollution control system, eg, the precipitator, baghouse, or scrubber, and the consequent degree to which these metals undergo homogeneous nucleation and become a fine fume, eg, submicron particles in the flyash, or undergo heterogeneous condensation on existing flyash particles. For some metals, such as arsenic and lead, emissions are also a function of the combustion system and the presence of lime. In fluidized beds, it has been shown that the arsenic and lead are captured and stabilized by the presence of reactive lime (1,19).

Dioxin [828-00-2] and furan [110-00-9] emissions are the final pollutants of consideration and are of most concern for combustors using MSW, RDF, or hazardous waste. Dioxins are formed, at some concentration level ranging from inconsequential to problematical, whenever aromatic compounds and trace quantities of chlorine are present in the boiler feed. Several mechanisms have been postulated for dioxin and furan emission formation including the passage of such molecules, unreacted, through the furnace; the formation of dioxins and furans from such precursors as lignins and trace concentrations of chlorine; and the formation of dioxins in post-combustion reactions in the economizer section at temperatures of about 550–700 K (2,20,21). Of these, the post-combustion mechanism is shown to dominate. However, the impact of this mechanism can be minimized by maintaining temperatures in excess of 983°C for 2 s, while ensuring complete mixing of fuels and oxidants and ensuring the absence of cool zones in the furnace. A general equation for approximating dioxin and furan emissions is as follows (22):

$$D + F = 0.0376(EA) - 3.305 \tag{7}$$

where $D + F$ is dioxins plus furans (in $\mu g/nm^3$) in the gaseous combustion products, corrected to 12% CO_2, and EA is the percentage of excess air above about 70%. Dioxin emissions are a problem more for MSW burners than other types of waste fuel systems, largely as a result of chlorine in the waste feed. Dioxin emissions typically are minimized in fluidized-bed combustion as a consequence of the solids mixing and solids turbulence in the bed.

Applications of Fuels From Waste

Because fuels from combustible organic wastes have long been economic in specific industries such as pulp mills, sawmills, sugar mills or factories, and other biomass processing operations, and because municipal waste-to-energy is becoming increasingly cost effective, these systems are continuing to be installed. The typical industrial system is used either to generate process steam or to generate both electricity and steam in a cogeneration application. Typically these applications involve power boilers which are the essential source of process energy in pulp mills and food processing operations. Typical larger installations generate some 200–300 t/h of steam used to generate 25–35 MW plus process heat. As stand-alone electricity generating stations, these units are capable of 50—60 MW, with a typical thermal efficiency of 65–75%. Thermal efficiency depends on moisture and ash content of the feed waste, consequent firing conditions employed, and extent of heat-transfer surface available for combustion air heating as well as steam generation.

Since the early 1980s, there have been several stand-alone power plants built to fire biomass wastes including such materials as wood waste, rice hulls, and vineyard prunings; these facilities typically generate 20–50 MW_e for sale to electric utilities. MSW and RDF are typically consumed in condensing power plants generating 15–50 MW_e while reducing the volume of solids to be landfilled. These units have thermal efficiencies comparable to the large power boilers of the pulp and paper industry, depending again on waste fuel condition and firing strategy.

There has been increased interest in firing wood waste as a supplement to coal in either pulverized coal (PC) or cyclone boilers at 1–5% of heat input. This application has been demonstrated by such electric utilities as Santee-Cooper, Tennessee Valley Authority, Georgia Power, Delmarva, and Northern States Power. Cofiring wood waste with coal in higher percentages, eg, 10–15% of heat input, in PC and cyclone boilers is being carefully considered by the Electric Power Research Institute (EPRI) and Tennessee Valley Authority (TVA). This practice may have the potential to maximize the thermal efficiency of waste fuel combustion. If this practice becomes widespread, it will offer another avenue for use of fuels from waste.

BIBLIOGRAPHY

"Fuels From Waste" in *ECT* 3rd ed., Vol. 11, pp. 393–410, by D. A. Tillman, Consultant.

1. D. A. Tillman, *The Combustion of Solid Fuels and Wastes*, Academic Press, San Diego, Calif., 1991.
2. W. R. Seeker, W. S. Lanier, and M. P. Heap, *Municipal Waste Combustion Study: Combustion Control of MSW Combustors to Minimize Emissions of Trace Organics*, EER Corporation, Irvine, Calif., 1987.

3. J. Barton, *Evaluation of Trommels for Waste to Energy Plants, Phase I*, National Center for Resource Recovery, Washington, D.C., 1982.

4. A. J. Rossi, in D. A. Tillman and E. Jahn, eds., *Progress in Biomass Conversion*, Vol. 5, Academic Press, New York, 1984, pp. 69–99.

5. U.S. Forest Service, *Wood Handbook: Wood as an Engineering Material*, U.S. Government Printing Office, Washington, D.C., 1974.

6. F. Shafizadeh and Y. Sekuguchi, *Carbon* **21**(5), 511–516 (1983).

7. K. E. Chung and I. B. Goldberg, in *Proceedings of the 12th Annual EPRI Contractors' Conference on Fuel Science and Conversion*, EPRI, Palo Alto, Calif., 1988.

8. K. E. Chung, I. B. Goldberg, and J. J. Ratto, *Chemical Structure and Liquefaction Reactivity of Coal*, EPRI, Palo Alto, Calif., 1987.

9. R. E. Kaiser, "Physical-Chemical Character of Municipal Refuse," *Proceedings of the 1975 International Symposium on Energy Recovery from Refuse*, University of Louisville, Louisville, Ky., 1975.

10. *Steam: Its Generation and Use*, 40th ed., Babcock & Wilcox, Barberton, Ohio, 1991.

11. A. M. Ujihara and M. Gough, *Managing Ash From Municipal Waste Incinerators*, Resources for the Future, Washington, D.C., 1989.

12. D. A. Tillman and C. Leone, "Control of Trace Metals in Flyash at the Tacoma, Washington Multifuels Incinerator," *Proceedings of the American Flame Research Committee Fall International Symposium*, San Francisco, 1990.

13. D. A. Tillman, "The Fate of Arsenic at the Tacoma Steam Plant #2," paper presented at *1992 Fall International Symposium*, American Flame Research Committee, Boston, Mass., 1992.

14. M. Hertzberg, I. A. Zlochower, and J. Edwards, *Coal Particle Pyrolysis Temperatures and Mechanisms*, RI 9169, U.S. Department of the Interior, Bureau of Mines, Washington, D.C., 1988.

15. M. Hertzberg, I. Zlochower, R. Conti, and K. Cashdollar, *Am. Chem. Soc.* **32**(3), 24–41 (1987).

16. D. A. Tillman, A. J. Rossi, and K. M. Vick, *Incineration of Municipal and Hazardous Solid Wastes*, Academic Press, San Diego, Calif., 1989.

17. D. W. Pershing and J. Wendt, *Proceedings of the 16th International Symposium*, The Combustion Institute, Pittsburgh, Pa., 1976.

18. D. W. Pershing and co-workers, *Proceedings of the 17th International Symposium*, The Combustion Institute, Pittsburgh, Pa., 1978.

19. T. C. Ho and co-workers, "Metal Capture During Fluidized Bed Incineration of Wastes Contaminated with Lead Chloride," presented at the *Second International Congress on Toxic Combustion By-Products: Formation and Control*, Salt Lake City, Utah, Mar. 26–29, 1991.

20. R. G. Barton, W. O. Clark, W. S. Lanier, and W. R. Seeker, "Dioxin Emissions During Waste Incineration," presented at *Spring Meeting, Western States Section of the Combustion Institute*, Salt Lake City, Utah, 1988.

21. *National Incinerator Testing and Evaluation Program: Mass Burning Incinerator Technology*, Vol. II, Lavalin, Inc. Quebec City, Quebec, Canada, 1987.

22. M. Beychok, *Atmos. Envir.* **21**(1), 29–36 (1987).

DAVID A. TILLMAN
Enserch Environmental

FUELS, SURVEY.   See FUEL RESOURCES.

FUELS, SYNTHETIC

GASEOUS FUELS

Substitute or synthetic natural gas (SNG) has been known for several centuries. When SNG was first discovered, natural gas was largely unknown as a fuel and was more a religious phenomenon (see GAS, NATURAL) (1). Coal (qv) was the first significant source of substitute natural gas and in the early stages of SNG production the product was more commonly known under variations of the name coal gas (2,3). Whereas coal continues to be a principal source of substitute natural gas (4) a more recently recognized source is petroleum (qv) (5).

Gas from Coal

Coal can be converted to gas by several routes (2,6–11), but often a particular process is a combination of options chosen on the basis of the product desired, ie, low, medium, or high heat-value gas. In a very general sense, coal gas is the term applied to the mixture of gaseous constituents that are produced during the thermal decomposition of coal at temperatures in excess of 500°C (>930°F), often in the absence of oxygen (air) (see COAL CONVERSION PROCESSES, GASIFICATION) (3). A solid residue (coke, char), tars, and other liquids are also produced in the process:

$$C_{coal} + heat \rightarrow C_{char} + tar/liquid + CO + CO_2 + H_2$$

The tars and other liquids (liquor) are removed by condensation leaving principally hydrogen, carbon monoxide, and carbon dioxide in the gas phase. This gaseous product also contains low boiling hydrocarbons (qv), sulfur-containing gases, and nitrogen-containing gases including ammonia (qv) and hydrogen cyanide. The solid residue is then treated under a variety of conditions to produce other fuels which vary from a purified char to different types of gaseous mixtures (see also COAL CONVERSION PROCESSES, CARBONIZATION). The amounts of gas, coke, tar, and other liquid products vary according to the method used for the carbonization (especially the retort configuration), and process temperature, as well as the nature (rank) of the coal (3,11).

The recorded chronology of the coal-to-gas conversion technology began in 1670 when a clergyman, John Clayton, in Wakefield, Yorkshire, produced in the laboratory a luminous gas by destructive distillation of coal (12). At the same time, experiments were also underway elsewhere to carbonize coal to produce coke, but the process was not practical on any significant scale until 1730 (12). In 1792, coal was distilled in an iron retort by a Scottish engineer, who used the by-product gas to illuminate his home (13).

The conversion of coal to gas on an industrial scale dates to the early nineteenth century (14). The gas, often referred to as manufactured gas, was produced in coke ovens or similar types of retorts by simply heating coal to vaporize the volatile constituents. Estimates based on modern data indicate that the gas mixture probably contained hydrogen (qv) (ca 50%), methane (ca 30%), carbon monoxide (qv) and carbon dioxide (qv) (ca 15%), and some inert material, such as nitrogen (qv), from which a heating value of approximately 20.5 MJ/m^3 (550 Btu/ft^3) can be estimated (6).

Blue gas, or blue-water gas, so-called because of the color of the flame upon burning (10), was discovered in 1780 when steam was passed over incandescent carbon (qv), and the blue-water gas process was developed over the period 1859–1875. Successful commercial application of the process came about in 1875 with the introduction of the carburetted gas jet. The heating value of the gas was low, ca 10.2 MJ/m^3 (275 Btu/ft^3), and on occasion oil was added to the gas to enhance the heating value. The new product was given the name carburetted water gas and the technique satisfied part of the original aim by adding luminosity to gas lights (10).

Coke-oven gas is a by-product fuel gas derived from coking coals by the process of carbonization. The first by-product coke ovens were constructed in France in 1856. Since then they have gradually replaced the old and primitive method of beehive coking for the production of metallurgical coke. Coke-oven gas is produced in an analogous manner to retort coal gas, with operating conditions, mainly temperature, set for maximum carbon yield. The resulting gas is, consequently, poor in illuminants, but excellent as a fuel. Typical analyses and heat content of common fuel gases vary (Table 1) and depend on the source as well as the method of production.

Table 1. Analyses of Fuel Gases

Type of fuel gas	Gas composition, vol %							Heat value,[a] MJ/m^3
	CO	CO$_2$	H$_2$	N$_2$	O$_2$	CH$_4$	Illuminants	
blast-furnace	27.5	10.0	3.0	58.0	1.0	0.5		3.8
producer (bituminous)	27.0	4.5	14.0	50.9	0.6	3.0		5.6
blue-water	42.8	3.0	49.9	3.3	0.5	0.5		11.5
carburetted water	33.4	3.9	34.6	7.9	0.9	10.4	8.9[b]	20.0
retort coal	8.6	1.5	52.5	3.5	0.3	31.4	2.2[c]	21.5
coke-oven	6.3	1.8	53.0	3.4	0.2	1.6	3.7[d]	21.9
natural								
mid-continent		0.8		3.2		96.0		36.1
Pennsylvania				1.1		67.6	31.3[e]	46.0

[a]To convert MJ/m^3 to Btu/ft^3, multiply by 26.86.
[b]6.7 vol % C$_2$H$_4$ plus 2.2 vol % C$_6$H$_6$.
[c]1.1 vol % each of C$_2$H$_4$ and C$_6$H$_6$.
[d]2.7 vol % C$_2$H$_4$ plus 1.0 vol % C$_6$H$_6$.
[e]31.3 vol % C$_2$H$_6$.

In Germany, large-scale production of synthetic fuels from coal began in 1910 and necessitated the conversion of coal to carbon monoxide and hydrogen.

Water gas reaction $C_{coal} + H_2O \rightarrow CO + H_2$

The mixture of carbon monoxide and hydrogen is enriched with hydrogen from the water gas catalytic (Bosch) process, ie, water gas shift reaction, and passed over a cobalt–thoria catalyst to form straight-chain, ie, linear, paraffins, olefins, and alcohols in what is known as the Fisher-Tropsch synthesis.

$$n\,CO + (2n + 1)\,H_2 \xrightarrow[\text{catalyst}]{\text{cobalt}} C_nH_{2n+2} + n\,H_2O$$

$$2n\,CO + (n + 1)\,H_2 \xrightarrow[\text{catalyst}]{\text{iron}} C_nH_{2n+2} + n\,CO_2$$

$$n\,CO + 2n\,H_2 \xrightarrow[\text{catalyst}]{\text{cobalt}} C_nH_{2n} + n\,H_2O$$

$$2n\,CO + n\,H_2 \xrightarrow[\text{catalyst}]{\text{iron}} C_nH_{2n} + n\,CO_2$$

$$n\,CO + 2n\,H_2 \xrightarrow[\text{catalyst}]{\text{cobalt}} C_nH_{2n+1}OH + (n - 1)\,H_2O$$

$$(2n - 1)\,CO + (n + 1)\,H_2 \xrightarrow[\text{catalyst}]{\text{iron}} C_nH_{2n+1}OH + (n - 1)\,CO_2$$

In Sasolburg, South Africa, a commercial plant using the Fischer-Tropsch process was completed in 1950 and began producing a variety of liquid fuels and chemicals. The facility has been expanded to produce a considerable portion of South Africa's energy requirements (15,16).

In 1948, the first demonstration of suspension gasification was successfully completed by Koppers, Inc. The product gas was of 11.2 MJ/m³ (300 Btu/ft³) calorific value and consisted primarily of a mixture of hydrogen and oxides of carbon. In the United States, so-called second-generation coal gasification processes came into being as a result of the recognized need to develop reliable, domestic energy sources to replace the rapidly diminishing supply of conventional fuels (see FUEL RESOURCES) (3,9). More recently, the biological conversion of coal and synthesis gas (carbon monoxide–hydrogen mixtures) into liquid fuels by methanogenic bacteria has received some attention (17–19).

Gas Products. The originally designated names of the gaseous mixtures are used herein, with the understanding that since their introduction there may be differences in means of production and in the make-up of the gaseous products. Properties of fuel gases are available (20). There are standard tests to determine properties and character of gaseous mixtures. These tests are accepted by the American Society for Testing and Materials (ASTM), by the British Standards Institution (BSI), by the Institute of Petroleum (IP), and by the International Standards Organization (ISO) (1,10,20).

Low Heat-Value Gas. Low heat-value (low Btu) gas (7) consists of a mixture of carbon monoxide and hydrogen and has a heating value of less than 11 MJ/m³ (300 Btu/ft³), but more often in the range 3.3–5.6 MJ/m³ (90–150 Btu/ft³). The gas is formed by partial combustion of coal with air, usually in the presence of steam (7).

$$2\,C_{coal} + O_2 \rightarrow 2\,CO$$
$$C_{coal} + H_2O \rightarrow CO + H_2$$
$$CO + H_2O \rightarrow CO_2 + H_2$$

This gas is of interest to industry as a fuel gas or even, on occasion, as a raw material from which ammonia and other compounds may be synthesized.

The first gas producer making low heat-value gas was built in 1832. (The product was a combustible carbon monoxide–hydrogen mixture containing ca 50 vol % nitrogen). The open-hearth or Siemens-Martin process, built in 1861 for pig iron refining, increased low heat-value gas use (see IRON). The use of producer gas as a fuel for heating furnaces continued to increase until the turn of the century when natural gas began to supplant manufactured fuel gas (see FURNACES, FUEL-FIRED).

The combustible components of the gas are carbon monoxide and hydrogen, but combustion (heat) value varies because of dilution with carbon dioxide and with nitrogen. The gas has a low flame temperature unless the combustion air is strongly preheated. Its use has been limited essentially to steel (qv) mills, where it is produced as a by-product of blast furnaces. A common choice of equipment for the smaller gas producers is the Wellman-Galusha unit because of its long history of successful operation (21).

Medium Heat-Value Gas. Medium heat-value (medium Btu) gas (6,7) has a heating value between 9 and 26 MJ/m^3 (250 and 700 Btu/ft^3). At the lower end of this range, the gas is produced like low heat-value gas, with the notable exception that an air separation plant is added and relatively pure oxygen (qv) is used instead of air to partially oxidize the coal. This eliminates the potential for nitrogen in the product and increases the heating value of the product to 10.6 MJ/m^3 (285 Btu/ft^3). Medium heat-value gas consists of a mixture of methane, carbon monoxide, hydrogen, and various other gases and is suitable as a fuel for industrial consumers.

High Heat-Value Gas. High heat-value (high Btu) gas (7) has a heating value usually in excess of 33.5 MJ/m^3 (900 Btu/ft^3). This is the gaseous fuel that is often referred to as substitute or synthetic natural gas (SNG), or pipeline-quality gas. It consists predominantly of methane and is compatible with natural gas insofar as it may be mixed with, or substituted for, natural gas.

Any of the medium heat-value gases that consist of carbon monoxide and hydrogen (often called synthesis gas) can be converted to high heat-value gas by methanation (22), a low temperature catalytic process that combines carbon monoxide and hydrogen to form methane and water.

$$CO + 3\,H_2 \rightarrow CH_4 + H_2O$$

Prior to methanation, the gas product from the gasifier must be thoroughly purified, especially from sulfur compounds the precursors of which are widespread throughout coal (23) (see SULFUR REMOVAL AND RECOVERY). Moreover, the composition of the gas must be adjusted, if required, to contain three parts hydrogen to one part carbon monoxide to fit the stoichiometry of methane production. This is accomplished by application of a catalytic water gas shift reaction.

$$CO + H_2O \rightleftarrows CO_2 + H_2$$

The ratio of hydrogen to carbon monoxide is controlled by shifting only part of the gas stream. After the shift, the carbon dioxide, which is formed in the gasifier and in the water gas reaction, and the sulfur compounds formed during gasification, are removed from the gas.

The processes that have been developed for the production of synthetic natural gas are often configured to produce as much methane in the gasification step as possible thereby minimizing the need for a methanation step. In addition, methane formation is highly exothermic which contributes to process efficiency by the production of heat in the gasifier, where the heat can be used for the endothermic steam–carbon reaction to produce carbon monoxide and hydrogen.

$$C + H_2O \rightarrow CO + H_2$$

Carbonization. Next to combustion, carbonization represents one of the largest uses of coal (2,24–26). Carbonization is essentially a process for the production of a carbonaceous residue by thermal decomposition, accompanied by simultaneous removal of distillate, of organic substances.

$$C_{organic} \rightarrow C_{coke/char/carbon} + liquids + gases$$

This process may also be referred to as destructive distillation. It has been applied to a whole range of organic materials, more particularly to natural products such as wood (qv), sugar (qv), and vegetable matter to produce charcoal (see FUELS FROM BIOMASS). However, in the present context, coal usually yields coke, which is physically dissimilar from charcoal and appears with the more familiar honeycomb-type structure (27).

The original process of heating coal in rounded heaps, the hearth process, remained the principal method of coke production for over a century, although an improved oven in the form of a beehive was developed in the Durham-Newcastle area of England in about 1759 (2,26,28). These processes lacked the capability to collect the volatile products, both liquids and gases. It was not until the midnineteenth century, with the introduction of indirectly heated slot ovens, that it became possible to collect the liquid and gaseous products for further use.

Coal carbonization processes are generally defined according to process operating temperature. Terms are defined in Table 2.

Low Temperature Carbonization. Low temperature carbonization, when the process does not exceed 700°C, was mainly developed as a process to supply town gas for lighting purposes as well as to provide a smokeless (devolatilized) solid fuel for domestic consumption (30). However, the process by-products (tars) were also found to be valuable insofar as they served as feedstocks (qv) for an emerging chemical industry and were also converted to gasolines, heating oils, and

Table 2. Coal Carbonization Methods[a]

Carbonization process	Final temperature, °C	Products	Processes
low temperature	500–700	reactive coke and high tar yield	rexco (700°C) made in cylindrical vertical retorts; coalite (650°C) made in vertical tubes
medium temperature	700–900	reactive coke with high gas yield, or domestic briquettes	town gas and gas coke (obsolete); phurnacite, low volatile steam coal, pitch-bound briquettes carbonized at 800°C
high temperature	900–1050	hard, unreactive coal for metallurgical use	foundry coke (900°C); blast-furnace coke (950–1050°C)

[a]Ref. 29.

lubricants (see GASOLINE AND OTHER MOTOR FUELS; LUBRICATION AND LUBRICANTS) (31).

Coals preferred for the low temperature carbonization were usually lignites or subbituminous, as well as high volatile bituminous, coals (see LIGNITE AND BROWN COAL). These yield porous solid products over the temperature range 600–700°C. Certain of the higher rank (caking) coals were less suitable for the process, unless steps were taken to destroy the caking properties, because of the tendency of these higher rank coals to adhere to the walls of the carbonization chamber.

The options for efficient low temperature carbonization of coal include vertical and horizontal retorts which have been used for batch and continuous processes. In addition, stationary and revolving horizontal retorts have also been operated successfully, and there are also several process options employing fluidized or gas-entrained coal. Coke production from batch-type carbonization of coal has been supplanted by a variety of continuous retorting processes which allow much greater throughput rates than were previously possible. These processes employ rectangular or cylindrical vessels of sufficient height to carbonize the coal while it travels from the top of the vessel to the bottom and usually employ the principle of heating the coal by means of a countercurrent flow of hot combustion gas. Most notable of these types of carbonizers are the Lurgi-Spulgas retort and the Koppers continuous steaming oven (2).

High Temperature Carbonization. When heated at temperatures in excess of 700°C (1290°F), low temperature chars lose their reactivity through devolatilization and also suffer a decrease in porosity. High temperature carbonization, at temperatures >900°C, is, therefore, employed for the production of coke (27). As for the low temperature processes, the tars produced in high temperature ovens are also sources of chemicals and chemical intermediates (32).

A newer concept has been developed that is given the name mild gasification (33). It is not a gasification process in the true sense of the word. The process temperature is some several hundred degrees lower, hence the term mild, than the usual gasification process temperature and the objective is not to produce a gaseous fuel but to produce a high value char (carbon) and liquid products. Gas is produced, but to a lesser extent.

Documented efforts at cokemaking date from 1584 (34), and have seen various adaptations of conventional wood-charring methods to the production of coke including the eventual evolution of the beehive oven, which by the mid-nineteenth century had become the most common vessel for the coking of coal (2). The heat for the process was supplied by burning the volatile matter released from the coal and, consequently, the carbonization would progress from the top of the bed to the base of the bed and the coke was retrieved from the side of the oven at process completion.

Some beehive ovens, having various improvements and additions of waste heat boilers, thereby allowing heat recovery from the combustion products, may still be in operation. Generally, however, the beehive oven has been replaced by wall-heated, horizontal chamber, ie, slot, ovens in which higher temperatures can be achieved as well as a better control over the quality of the coke. Modern slot-type coke ovens are approximately 15 m long, approximately 6 m high, and the width is chosen to suit the carbonization behavior of the coal to be processed. For example, the most common widths are ca 0.5 m, but some ovens may be as narrow as 0.3 m, or as wide as 0.6 m.

Several (usually 20 or more, alternating with similar cells that contain heating ducts) of these chambers are constructed in the form of a battery over a common firing system through which the hot combustion gas is conveyed to the ducts. The flat roof of the battery acts as the surface for a mobile charging car from which the coal (25–40 t) enters each oven through three openings along the top. The coke product is pushed from the rear of the oven through the opened front section onto a quenching platform or into rail cars that then move the coke through water sprays. The gas and tar by-products of the process are collected for further processing or for on-site use as fuel.

Most modern coke ovens operate on a regenerative heating cycle in order to obtain as much surplus gas as possible for use on the works, or for sale. If coke-oven gas is used for heating the ovens, the majority of the gas is surplus to requirements. If producer gas is used for heating, much of the coke-oven gas is surplus.

The main difference between gas works and coke oven practice is that, in a gas works, maximum gas yield is a primary consideration whereas in the coke works the quality of the coke is the first consideration. These effects are obtained by choice of a coal feedstock that is suitable to the task. For example, use of lower volatile coals in coke ovens, compared to coals used in gas works, produces lower yields of gas when operating at the same temperatures. In addition, the choice of heating (carbonizing) conditions and the type of retort also play a principal role (10,35).

Gasification.   The gasification of coal is essentially the conversion of coal by any one of a variety of processes to produce combustible gases (7,8,11,22,36–38). Primary gasification is the thermal decomposition of coal to produce mixtures containing various proportions of carbon monoxide, carbon dioxide, hydrogen, water, methane, hydrogen sulfide, and nitrogen, as well as products such as tar, oils, and phenols. A solid char product may also be produced, and often represents the bulk of the weight of the original coal.

Secondary gasification involves gasification of the char from the primary gasifier, usually by reaction of the hot char and water vapor to produce carbon monoxide and hydrogen.

$$C_{char} + H_2O \rightarrow CO + H_2$$

The gaseous product from a gasifier generally contains large amounts of carbon monoxide and hydrogen, plus lesser amounts of other gases and may be of low, medium, or high heat value depending on the defined use (Table 3) (39,40).

The importance of coal gasification as a means of producing fuel gas(es) for industrial use cannot be underplayed. But coal gasification systems also have undesirable features. A range of undesirable products are also produced which must be removed before the products are used to provide fuel and/or to generate electric power (see POWER GENERATION) (22,41).

Chemistry. Coal gasification involves the thermal decomposition of coal and the reaction of the carbon in the coal, and other pyrolysis products with oxygen, water, and hydrogen to produce fuel gases such as methane by internal hydrogen shifts

$$C_{coal} + H_{coal} \rightarrow CH_4$$

or through the agency of added (external) hydrogen

$$C_{coal} + 2\,H_2 \rightarrow CH_4$$

although the reactions are more numerous and more complex as can be seen in Table 4.

Table 3. Gas Composition Requirements for Substitute Natural Gas and for Power Generation

	Product requirement	
Characteristic	Synthetic natural gas	Power generation
methane content	high, less synthesis required	low, probably means no tars
H_2/CO ratio	high, less shifting required	low, CO more efficient fuel
moisture content	high, steam required for shift	low, lower condensate treatment costs
outlet temperature	low, maximizes methane minimizes sensible heat loss	high, precludes tar formation provides for steam generation
	leads to high cold gas efficiency	reduces cold gas efficiency
gasifier oxidant	O_2 only, cost of N_2 removal excessive	air or O_2, low heating gas value acceptable fuel

Table 4. Gasification Reactions

Reaction	ΔH, kJ[a]
Gasification zone, 595–1205°C	
$C + CO_2 \rightarrow 2\,CO$	9.65[b]
$CO + H_2O \rightarrow CO_2 + H_2$	-1.93[c]
$C + H_2O \rightarrow CO + H_2$	7.76[d]
$C + 2\,H_2 \rightarrow CH_4$	-5.21
$C + 2\,H_2O \rightarrow 2\,H_2 + CO_2$	5.95
Combustion zone, >1205°C	
$2\,C + O_2 \rightarrow 2\,CO$	-22.6
$C + O_2 \rightarrow CO_2$	-6.74
Methanation reactions	
$CO + 3\,H_2 \rightarrow CH_4 + H_2O$	-12.9
$CO_2 + 4\,H_2 \rightarrow CH_4 + 2\,H_2O$	-11.0
$2\,C + 2\,H_2O \rightarrow CH_4 + CO_2$	0.62
Other reactions	
$2\,H_2 + O_2 \rightarrow 2\,H_2O$	
$CO + 2\,H_2 \rightarrow CH_3OH$	
$2\,C + H_2 \rightarrow C_2H_2$	
$CH_4 + 2\,H_2O \rightarrow CO_2 + 4\,H_2$	

[a]To convert kJ to kcal, divide by 4.184.
[b]Boudouard reaction.
[c]Water gas shift reaction.
[d]Water gas reaction.

If air is used as a combustant, the product gas contains nitrogen and, depending on design characteristics, has a heating value of approximately 5.6–11.2 $\times$ 10^3 MJ/m^3 (150–300 Btu/ft^3). The use of pure oxygen, although expensive, results in a product gas having a heating value of 11–15 MJ/m^3 (300–400 Btu/ft^3) and carbon dioxide and hydrogen sulfide as by-products.

If a high heat-value gas (33.5–37.3 MJ/m^3 (900–1000 Btu/ft^3)) ie, SNG, is the desired product, efforts must be made to increase the methane content.

Shift conversion reaction

$$CO + H_2O \rightarrow CO_2 + H_2$$

$$CO + 3\,H_2 \rightarrow CH_4 + H_2O$$

$$2\,CO + 2\,H_2 \rightarrow CH_4 + CO_2$$

$$CO + 4\,H_2 \rightarrow CH_4 + 2\,H_2O$$

The gasification is performed using oxygen and steam (qv), usually at elevated pressures. The steam–oxygen ratio along with reaction temperature and pressure determine the equilibrium gas composition. The reaction rates for these reactions are relatively slow and heats of formation are negative. Catalysts may be necessary for complete reaction (2,3,24,42,43).

Process Parameters. The most notable effects in gasifiers are those of pressure (Fig. 1) and coal character. Some initial processing of the coal feedstock may be required. The type and degree of pretreatment is a function of the process and/or the type of coal (see COAL CONVERSION PROCESSES, CLEANING AND DESULFURIZATION).

Depending on the type of coal being processed and the analysis of the gas product desired, some or all of the following processing steps may be required: (*1*) pretreatment of the coal (if caking is a problem); (*2*) primary gasification of the coal; (*3*) secondary gasification of the carbonaceous residue from the primary gasifier; (*4*) removal of carbon dioxide, hydrogen sulfide, and other acid gases; (*5*) shift conversion for adjustment of the carbon monoxide–hydrogen mole ratio to the desired ratio; and (*6*) catalytic methanation of the carbon monoxide–hydrogen mixture to form methane. If high heat-value gas is desired, all of these processing steps are required because coal gasifiers do not yield methane in the concentrations required.

An example of application of a pretreatment option occurs when the coal displays caking or agglomerating characteristics. Such coals are usually not amenable to gasification processes employing fluidized-bed or moving-bed reactors. The pretreatment involves a mild oxidation treatment, usually consisting of low temperature heating of the coal in the presence of air or oxygen. This destroys the caking characteristics of coals.

Gasification technologies for the production of high heat-value gas do not all depend entirely on catalytic methanation, that is, the direct addition of hydrogen to coal under pressure to form methane.

The hydrogen-rich gas for hydrogasification can be manufactured from steam by using the char that leaves the hydrogasifier. Appreciable quantities of methane are formed directly in the primary gasifier and the heat released by methane formation is at a sufficiently high temperature to be used in the steam–carbon reaction to produce hydrogen so that less oxygen is used to produce heat

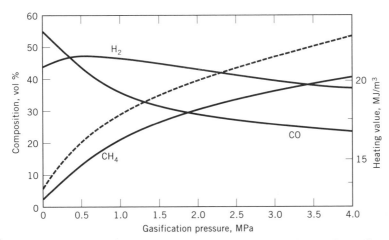

Fig. 1. Variation of (——) gas composition and (- - -) heating value with gasifier pressure. To convert MJ/m^3 to Btu/ft^3, multiply by 26.86. To convert MPa to psi, multiply by 145.

for the steam–carbon reaction. Hence, less heat is lost in the low temperature methanation step, thereby leading to a higher overall process efficiency.

There are three fundamental reactor types for gasification processes: (1) a gasifier reactor, (2) a devolatilizer, and (3) a hydrogasifier (Fig. 2). The choice of a particular design is available for each type, eg, whether or not two stages should be involved depending on the ultimate product gas desired. Reactors may also be designed to operate over a range of pressure from atmospheric to high pressure.

Gasification processes have been classified on the basis of the heat-value of the produced gas. It is also possible to classify gasification processes according to the type of reactor and whether or not the system reacts under pressure. Additionally, gasification processes can be segregated according to the bed types, which differ in the ability to accept and use caking coals. Thus gasification processes can be divided into four categories based on reactor (bed) configuration: (1) fixed bed, (2) moving bed, (3) fluidized bed, and (4) entrained bed.

In a fixed-bed process the coal is supported by a grate and combustion gases, ie, steam, air, oxygen, etc, pass through the supported coal whereupon the hot produced gases exit from the top of the reactor. Heat is supplied internally or from an outside source, but caking coals cannot be used in an unmodified fixed-bed reactor. In the moving-bed system (Fig. 3), coal is fed to the top of the bed and ash leaves the bottom with the product gases being produced in the hot zone just prior to being released from the bed.

The fluidized-bed system (Fig. 3) uses finely sized coal particles and the bed exhibits liquid-like characteristics when a gas flows upward through the bed. Gas flowing through the coal produces turbulent lifting and separation of particles

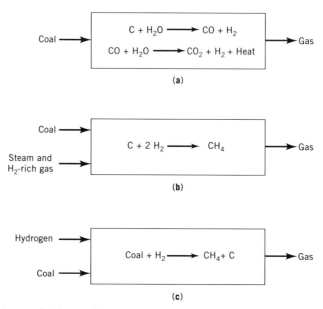

Fig. 2. Chemistry of (a) gasifier; (b) hydrogasifier; and (c) devolatization processes. The gaseous product of (a) is of low heating value; that of (b) and (c) is of intermediate heating value.

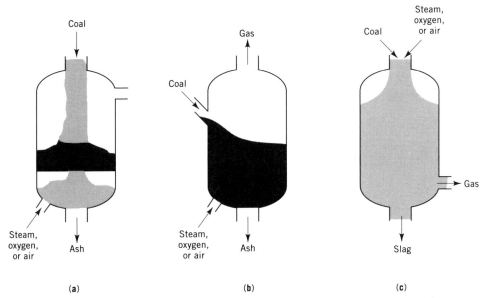

Fig. 3. Gasifier systems: **(a)**, moving bed (dry ash); **(b)**, fluidized bed; and **(c)**, entrained flow.

and the result is an expanded bed having greater coal surface area to promote the chemical reaction. These systems, however, have only a limited ability to handle caking coals (see FLUIDIZATION).

An entrainment system (Fig. 3) uses finely sized coal particles blown into the gas steam prior to entry into the reactor and combustion occurs with the coal particles suspended in the gas phase; the entrained system is suitable for both caking and noncaking coals.

Gasifier Options. The standard Wellman-Galusha unit, used for noncaking coals, is an atmospheric pressure, air-blown gasifier (Fig. 4) fed by a two-compartment lockhopper system. The upper coal-storage compartment feeds coal intermittently into the feeding compartment, which then feeds coal into the gasifier section almost continuously except for brief periods when the feeding compartment is being loaded from storage. In small units up to 1.5 m internal diameter ash is removed by shaker grates. In larger units, ash is removed continuously from the bottom of the gasifier into an ash-hopper section by a revolving grate.

The grate is constructed of flat, circular steel plates set one above the other with edges overlapping. The grate, revolving eccentrically within the gasifier, causes ash to fall from the coal bed as the space between the grate and the shell increases, and then pushes the ash down into the ash-hopper as the space decreases. The smaller size units are brick-lined, although the larger sizes are unlined and water-jacketed. Combustion air, provided by a fan, passes over the warm (82°C) jacket water causing the water to vaporize to provide the necessary steam for gasification. Gas leaves from above the fuel bed at 428–538°C for bituminous coal.

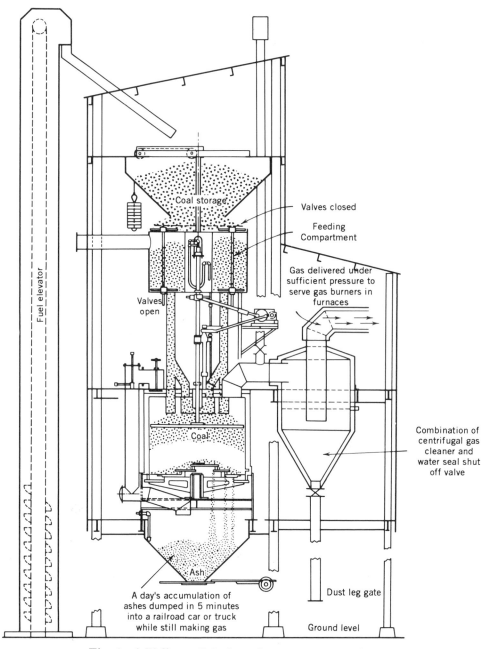

Coal storage

Valves closed

Feeding
Compartment

Gas delivered under
sufficient pressure to
serve gas burners in
furnaces

Valves
open

Fuel elevator

Coal

Combination of
centrifugal gas
cleaner and
water seal shut
off valve

Ash

A day's accumulation of
ashes dumped in 5 minutes
into a railroad car or truck
while still making gas

Dust leg gate

Ground level

Fig. 4. A Wellman-Galusha agitator-type gas producer.

138

On the other hand, the agitator type of Wellman-Galusha unit, used for gasification of any type of coal, uses a slowly revolving horizontal arm which also spirals vertically below the surface of the fuel bed to retard channeling and maintain a uniform bed. Use of the agitator not only allows operation with caking coals but also can increase the capacity of the gasifier by about 25% for use with other coals.

The Winkler gasifier (Fig. 5) is an example of a medium heat-value gas producer which, when oxygen is employed, yields a gas product composed mainly of carbon monoxide and hydrogen (43).

In the process, finely crushed coal is gasified at atmospheric pressure in a fluidized state; oxygen and steam are introduced at the base of the gasifier. The coal is fed by lockhoppers and a screw feeder into the bottom of the fuel bed. Sintered ash particles settle on a grate, where they are cooled by the incoming oxygen and steam; a rotating, cooled rabble moves the ash toward a discharge port. The ash is then conveyed pneumatically to a disposal hopper.

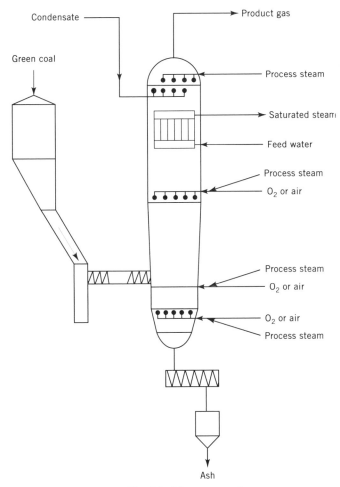

Fig. 5. The Winkler gas producer.

The gas, along with entrained ash and char particles, which are subjected to further gasification in the large space above the fluid bed, exit the gasifier at 954–1010°C. The hot gas is passed through a waste-heat boiler to recover the sensible heat, and then through a dry cyclone. Solid particles are removed in both units. The gas is further cooled and cleaned by wet scrubbing, and if required, an electrostatic precipitator is included in the gas-treatment stream.

The Koppers-Totzek process is a second example of a process for the production of a medium heat-value gas (44,45). Whereas the Winkler process employs a fluidized bed, the Koppers-Totzek process uses an entrained flow system. In the Koppers-Totzek process, dried and pulverized coal is conveyed continuously by a screw into a mixing nozzle. From there a high velocity stream of steam and oxygen entrains the coal into a gasifier. The gasifier (Fig. 6) is a cylindrical vessel with a refractory-lining that is designed to conduct a selected amount of heat to a surrounding water jacket in which low pressure process steam is generated. The lining is thin (about 5 cm) and made of a high alumina cast material. In a two-headed gasifier two burner heads are placed 180° apart at either end of the vessel. Four burner heads, 90° apart, are used in a four-headed gasifier. The largest gasifiers are 3–4 m diameter at the middle, tapering to 2–3 m at the burner ends and are about 19 m long. The reactor volume is about 30 m^3 for the two-headed design, and 64 m^3 for the larger, four-headed models.

The process is carried at moderate (slightly above atmospheric) pressures, but at very high temperatures that reach a maximum of 1900°C. Even though the reaction time is short (0.6–0.8 s) the high temperature prevents the occurrence of any condensable hydrocarbons, phenols, and/or tar in the product gas. The absence of liquid simplifies the subsequent gas clean-up steps.

Normally ca 50% of the coal ash is removed from the bottom of the gasifier as a quenched slag. The balance is carried overhead in the gas as droplets which

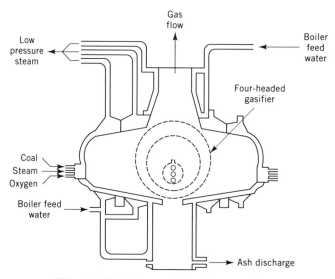

Fig. 6.   The Koppers-Totzek gas producer.

are solidified when the gas is cooled with a water spray. A fluxing agent is added, if required, to the coal to lower the ash fusion temperature and increase the molten slag viscosity.

   Conversion of carbon in the coal to gas is very high. With low rank coal, such as lignite and subbituminous coal, conversion may border on 100%, and for highly volatile A coals, it is on the order of 90–95%. Unconverted carbon appears mainly in the overhead material. Sulfur removal is facilitated in the process because typically 90% of it appears in the gas as hydrogen sulfide, H_2S, and 10% as carbonyl sulfide, COS; carbon disulfide, CS_2, and/or methyl thiol, CH_3SH, are not usually formed.

   The production of synthetic natural gas can be achieved by use of the Lurgi gasifier (Fig. 7), which is similar in principle to the Wellman-Galusha unit and is

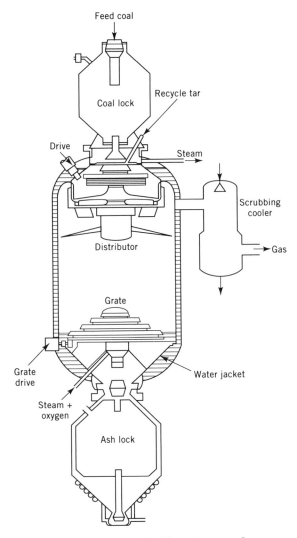

Fig. 7. The pressurized Lurgi gas producer.

designed for operation at pressures up to 3.1 MPa (450 psi) (46). Three distinct reaction zones are identifiable in a pressurized (1.9–2.9 MPa (280–425 psi)) Lurgi reactor: (*1*) the drying/devolatilization and pyrolysis zone, 370–595°C, nearest the coal feed end, commences the process by converting the coal to a reactive char; (*2*) the gasification zone (595–1205°C); and (*3*) the combustion zone (>1205°C), nearest the discharge end, which provides the heat requirements for the endothermic steam–carbon reaction. Equations and reaction enthalpies for the last two zones are given in Table 4.

The operating conditions in the gasifier (temperature and pressure) and the reaction kinetics (residence time, concentration of the constituents, and rate constants) determine the extent of conversion or approach to equilibrium.

The coal is fed through a lockhopper mounted on the top of the gasifier where a rotating distributor provides uniform coal feed across the bed. When processing caking coals, blades attached to the distributor rotate within the bed to break up agglomerates. A revolving grate supports the bed at the bottom and serves as a distributor for steam and oxygen. Solid residue is removed at the bottom of the gasifier through an ash lockhopper. The entire gasifier vessel is surrounded by a water jacket in which high pressure steam is generated.

Crude gas leaves from the top of the gasifier at 288–593°C depending on the type of coal used. The composition of gas also depends on the type of coal and is notable for the relatively high methane content when contrasted to gases produced at lower pressures or higher temperatures. These gas products can be used as produced for electric power production or can be treated to remove carbon dioxide and hydrocarbons to provide synthesis gas for ammonia, methanol, and synthetic oil production. The gas is made suitable for methanation, to produce synthetic natural gas, by a partial shift and carbon dioxide and sulfur removal.

As in most of the high heat-value processes, the raw gas is in the medium heat-value gas range and can be employed directly in that form. Removing the carbon dioxide raises the heating value, but not enough to render the product worthwhile as a high heat-value gas without methanation.

Methanation of the clean desulfurized main gas (less than 1 ppm total sulfur) is accomplished in the presence of a nickel catalyst at temperatures from 260–400°C and pressure range of 2–2.8 MPa (300–400 psi). Equations and reaction enthalpies are given in Table 4.

Hydrogenation of the oxides of carbon to methane according to the above reactions is sometimes referred to as the Sabatier reactions. Because of the high exothermicity of the methanization reactions, adequate and precise cooling is necessary in order to avoid catalyst deactivation, sintering, and carbon deposition by thermal cracking.

Catalytic methanation processes include (*1*) fixed or fluidized catalyst-bed reactors where temperature rise is controlled by heat exchange or by direct cooling using product gas recycle; (*2*) through wall-cooled reactor where temperature is controlled by heat removal through the walls of catalyst-filled tubes; (*3*) tube-wall reactors where a nickel–aluminum alloy is flame-sprayed and treated to form a Raney-nickel catalyst bonded to the reactor tube heat-exchange surface; and (*4*) slurry or liquid-phase (oil) methanation.

To enable interchangeability of the SNG with natural gas, on a calorific, flame, and toxicity basis, the synthetically produced gas consists of a minimum of 89 vol % methane, a maximum of 0.1% carbon monoxide, and up to 10% hydrogen. The specified minimum acceptable gross heating value is approximately 34.6 MJ/m^3 (930 Btu/ft^3).

In a combined power cycle operation, clean (sulfur- and particulate-free) gas is burned with air in the combustor at elevated pressure. The gas is either low or medium heat-value, depending on the method of gasification.

The hot gases from the combustor, temperature controlled to 980°C by excess air, are expanded through the gas turbine, driving the air compressor and generating electricity. Sensible heat in the gas turbine exhaust is recovered in a waste heat boiler by generating steam for additional electrical power production.

The use of hot gas clean-up methods to remove the sulfur and particulates from the gasified fuel increases turbine performance by a few percentage points over the cold clean-up systems. Hot gas clean-up permits use of the sensible heat and enables retention of the carbon dioxide and water vapor in the gasified fuel, thus enhancing turbine performance. Further, additional power may be generated, prior to combustion with air, in an expansion turbine as the hot product fuel gas is expanded to optimum pressure level for the combined cycle. Future advances in gas turbine technology (turbine inlet temperature above 1650°C) are expected to boost the overall combined cycle efficiency substantially.

More recently, advanced generation gasifiers have been under development, and commercialization of some of the systems has become a reality (36,41). In these newer developments, the emphasis has shifted to a greater throughput, relevant to the older gasifiers, and also to high carbon conversion levels and, thus, higher efficiency units.

For example, the Texaco entrained system features coal–water slurry feeding a pressurized oxygen-blown gasifier with a quench zone for slag cooling (Fig. 8). In fact, the coal is partially oxidized to provide the heat for the gasification reactions. The Dow gasifier also utilizes a coal–water slurry fed system whereas the Shell gasifier features a dry-feed entrained gasification system which operates at elevated temperature and pressure. The Kellogg Rust Westinghouse system and the Institute of Gas Technology U-Gas system are representative of ash agglomerating fluidized-bed systems.

In response to the disadvantage that the dry ash Lurgi gasifier requires that temperatures have to be below the ash melting point to prevent clinkering, improvements have been sought in the unit; as a result the British Gas-Lurgi GmbH gasifier came into being. This unit is basically similar to the dry ash Lurgi unit insofar as the top of the unit is identical but the bottom has been modified to include a slag quench vessel (Fig. 9). Thus the ash melts at the high temperatures in the combustion zone (up to 2000°C) and forms a slag that runs into the quench chamber, which is in reality a water bath where the slag forms granules of solid ash. Temperatures and reaction rates are high in the gasification zone so that coal residence time is markedly reduced over that of the dry ash unit.

In summary, these second-generation gasifiers offer promise for the future in terms of increased efficiency as well as for use of other feedstocks, such as biomass. The older, first-generation gasifiers, however, continue to be used.

Combustion. Coal combustion, not being in the strictest sense a process for the generation of gaseous synfuels, is nevertheless an important use of coal as a source of gaseous fuels. Coal combustion, an old art and probably the oldest

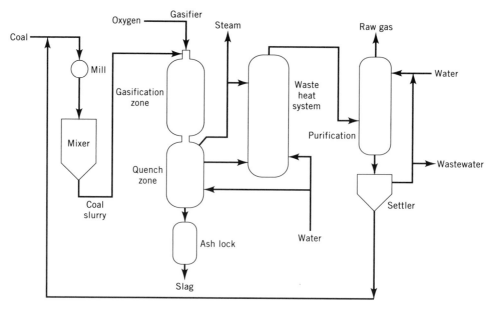

Fig. 8. The Texaco gasification process.

known use of this fossil fuel, is an accumulation of complex chemical and physical phenomena. The complexity of coal itself and the variable process parameters all contribute to the overall process (8,10,47–50) (see also COMBUSTION SCIENCE AND TECHNOLOGY).

There are two principal methods of coal combustion: fixed-bed combustion and combustion in suspension. The first fixed beds, eg, open fires, fireplaces, and domestic stoves, were simple in principle. Suspension burning of coal began in the early 1900s with the development of pulverized coal-fired systems, and by the 1920s these systems were in widespread use. Spreader stokers, which were developed in the 1930s, combined both principles by providing for the smaller particles of coal to be burned in suspension and larger particles to be burned on a grate (10).

A significant issue in combustors in the mid-1990s is the performance of the process in an environmentally acceptable manner through the use of either low sulfur coal or post-combustion clean-up of the flue gases. Thus there is a marked trend to more efficient methods of coal combustion and, in fact, a combustion system that is able to accept coal without the necessity of a post-combustion treatment or without emitting objectionable amounts of sulfur oxides, nitrogen oxides, and particulates is very desirable (51,52).

The parameters of rank and moisture content are regarded as determining factors in combustibility as it relates to both heating value and ease of reaction as well as to the generation of pollutants (48). Thus, whereas the lower rank coals may appear to be more reactive than higher rank coals, though exhibiting a lower heat-value and thereby implying that rank does not affect combustibility, environmental constraints arise through the occurrence of heteroatoms, ie, noncarbon

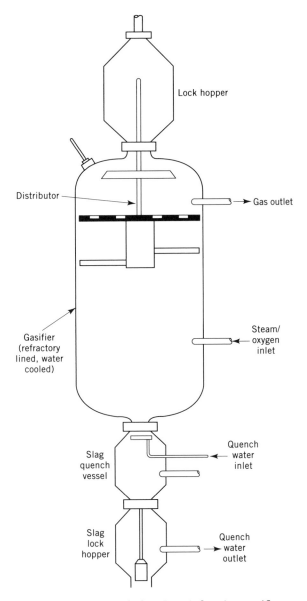

Fig. 9. The British Gas-Lurgi slagging gasifier.

atoms such as nitrogen and sulfur, in the coal. At the same time, anthracites, which have a low volatile matter content, are generally more difficult to burn than bituminous coals.

Chemistry. In direct combustion coal is burned to convert the chemical energy of the coal into thermal energy, ie, the carbon and hydrogen in the coal are oxidized into carbon dioxide and water.

$$2\ H_{coal} + O_2 \rightarrow H_2O$$

After burning, the sensible heat in the products of combustion can then be converted into steam that can be used for external work or can be converted directly into energy to drive a shaft, eg, in a gas turbine. In fact, the combustion process actually represents a means of achieving the complete oxidation of coal.

The combustion of coal may be simply represented as the staged oxidation of coal carbon to carbon dioxide

$$2\ C_{coal} + O_2 \rightarrow 2\ CO$$
$$2\ CO + O_2 \rightarrow 2\ CO_2$$

with any reactions of the hydrogen in the coal being considered to be of secondary importance. Other types of combustion systems may be rate-controlled as a result of the onset of the Boudouard reaction (see Table 4).

The complex nature of coal as a molecular entity (2,3,24,25,35,37,53) has resulted in the chemical explanations of coal combustion being confined to the carbon in the system. The hydrogen and other elements have received much less attention but the system is extremely complex and the heteroatoms, eg, nitrogen, oxygen, and sulfur, exert an influence on the combustion. It is this latter that influences environmental aspects.

For example, the conversion of nitrogen and sulfur, during coal combustion, to the respective oxides during combustion cannot be ignored:

$$S_{coal} + O_2 \rightarrow SO_2$$
$$2\ SO_2 + O_2 \rightarrow 2\ SO_3$$
$$2\ N_{coal} + O_2 \rightarrow 2\ NO$$
$$2\ NO + O_2 \rightarrow 2\ NO_2$$
$$N_{coal} + O_2 \rightarrow NO_2$$

The sulfur and nitrogen oxides that escape into the atmosphere can be converted to acids by reaction with moisture in the atmosphere (see also AIR POLLUTION; ATMOSPHERIC MODELS)

Sulfurous acid	$SO_2 + H_2O \rightarrow H_2SO_3$
Sulfuric acid	$2\ SO_2 + O_2 \rightarrow 2\ SO_3$
	$SO_3 + H_2O \rightarrow H_2SO_4$
Nitrous acid	$NO + H_2O \rightarrow H_2NO_2$
Nitric acid	$2\ NO + O_2 \rightarrow 2\ NO_2$
	$NO_2 + H_2O \rightarrow HNO_3$

Combustion Systems. Combustion systems vary in nature depending on the nature of the feedstock and the air needed for the combustion process (54). However, the two principal types of coal-burning systems are usually referred to as layer and chambered. The former refers to fixed beds; the latter is more specifically for pulverized fuel.

Fixed or Slowly Moving Beds.   For fuel-bed burning on a grate, a distillation effect occurs. The result is that liquid components which are formed volatilize before combustion temperatures are reached; cracking may also occur. The ignition of coal in a bed is almost entirely by radiation from hot refractory arches and from the flame burning of volatiles. In fixed beds, the radiant heat above the bed can only penetrate a short distance into the bed.

Consequently, convective heat transfer determines the intensity of warming up and ignition. In addition, convective heat transfer also plays an important part in the overall flame-to-surface transmission. The reaction of gases is greatly accelerated by contact with hot surfaces and, whereas the reaction away from the walls may proceed slowly, reaction at the surface proceeds much more rapidly.

Fluidized Beds.   Fluidized-bed combustion occurs in expanded beds (Fig. 3) and at relatively lower (925°C) temperatures; high convective transfer rates exist resulting from the bed motion. Fluidized systems can operate under substantial pressures thereby allowing more efficient gas clean-up. Fluidized-bed combustion is a means for providing high heat-transfer rates, controlling sulfur, and reducing nitrogen oxide emissions from the low temperatures in the combustion zone.

There are, however, problems associated with pollution control. Whereas the sulfur may be removed downstream using suitable ancillary controls, the sulfur may also be captured in the bed, thereby adding to the separations and recycle problems. Capture during combustion, however, is recognized as the ideal and is a source of optimism for fluidized combustion.

A fluidized bed is an excellent medium for contacting gases with solids, and this can be exploited in a combustor because sulfur dioxide emissions can be reduced by adding limestone, $CaCO_3$, or dolomite, $CaCO_3 \cdot MgCO_3$, to the bed.

$$2 \, SO_2 + O_2 \rightarrow 2 \, SO_3$$
$$SO_3 + CaCO_3 \rightarrow CaSO_4 + CO_2$$

or

$$2 \, SO_2 + O_2 + 2 \, CaCO_3 \rightarrow 2 \, CaSO_4 + 2 \, CO_2$$

The spent sorbent from fluidized-bed combustion may be taken directly to disposal and is much easier than the disposal of salts produced by wet limestone scrubbing. Alternatively, the spent sorbent may be regenerated using synthesis gas, CO/H_2.

$$CaSO_4 + H_2 \rightarrow CaO + H_2O + SO_2$$
$$CaSO_4 + CO \rightarrow CaO + CO_2 + SO_2$$

The calcium oxide product is supplemented with fresh limestone and returned to the fluidized bed. Two undesirable side reactions can occur in the regeneration of spent lime leading to the production of calcium sulfide:

$$CaSO_4 + 4 \, H_2 \rightarrow CaS + 4 \, H_2O$$
$$CaSO_4 + 4 \, CO \rightarrow CaS + 4 \, CO_2$$

which results in the recirculation of sulfur to the bed.

Entrained Systems. In entrained systems, fine grinding and increased re-tention times intensify combustion but the temperature of the carrier and degree of dispersion are also important. In practice, the coal is introduced at high veloc-ities which may be greater than 30 m/s and involve expansion from a jet to the combustion chamber.

Types of entrained systems include cyclone furnaces, which have been used for various coals. Other systems have been developed and utilized for the injection of coal–oil slurries into blast furnaces or for the burning of coal–water slurries. The cyclone furnace, developed in the 1940s to burn coal having low ash-fusion temperatures, is a horizontally inclined, water-cooled, tubular furnace in which crushed coal is burned with air entering the furnace tangentially. Temperatures may be on the order of 1700°C and the ash in the coal is converted to a molten slag that is removed from the base of the unit. Coal fines burn in suspension; the larger pieces are captured by the molten slag and burn rapidly.

Gas from Other Fossil Fuels

As of this writing natural gas is a plentiful resource, and there has been a marked tendency not to use other fossil fuels as SNG sources. However, petroleum and oil shale (qv) have been the subject of extensive research efforts. These represent other sources of gaseous fuels and are worthy of mention here.

Petroleum. Thermal cracking (pyrolysis) of petroleum or fractions thereof was an important method for producing gas in the years following its use for increasing the heat content of water gas. Many water gas sets operations were converted into oil-gasification units (55). Some of these have been used for base-load city gas supply, but most find use for peak-load situations in the winter.

In the 1940s, the hydrogasification of oil was investigated as a follow-up to the work on the hydrogasification of coal (56,57). In the ensuing years, further work was carried out as a supplement to the work on thermal cracking (58,59), and during the early 1960s it became evident that light distillates having end boiling points <182°C and containing no sulfur could be catalytically reformed by an autothermic process to pure methane (60). This method was extensively used in England until natural (North Sea) gas came into use.

Prior to the discovery of plentiful supplies of natural gas, and depending on the definition of the resources (1), there were plans to accommodate any shortfalls in gas supply from solid fossil fuels and from gaseous resources by the conversion of hydrocarbon (petroleum) liquids to lower molecular weight gaseous products.

$$CH_{petroleum} \rightarrow CH_4$$

Thermal Cracking. In addition to the gases obtained by distillation of crude petroleum, further highly volatile products result from the subsequent processing of naphtha and middle distillate to produce gasoline, as well as from hydrodesul-furization processes involving treatment of naphthas, distillates, and residual fuels (5,61), and from the coking or similar thermal treatment of vacuum gas oils and residual fuel oils (5).

The chemistry of the oil-to-gas conversion has been established for several decades and can be described in general terms although the primary and secondary reactions can be truly complex (5). The composition of the gases produced from a wide variety of feedstocks depends not only on the severity of cracking but often to an equal or lesser extent on the feedstock type (5,62,63). In general terms, gas heating values are on the order of 30–50 MJ/m^3 (950–1350 Btu/ft^3).

Catalytic Processes. A second group of refining operations which contribute to gas production are the catalytic cracking processes, such as fluid-bed catalytic cracking, and other variants, in which heavy gas oils are converted into gas, naphthas, fuel oil, and coke (5).

The catalysts promote steam reforming reactions that lead to a product gas containing more hydrogen and carbon monoxide and fewer unsaturated hydrocarbon products than the gas product from a noncatalytic process (5). Cracking severities are higher than those from thermal cracking, and the resulting gas is more suitable for use as a medium heat-value gas than the rich gas produced by straight thermal cracking. The catalyst also influences the reactions rates in the thermal cracking reactions, which can lead to higher gas yields and lower tar and carbon yields (5).

The basic chemical premise involved in making synthetic natural gas from heavier feedstocks is the addition of hydrogen to the oil:

$$CH_3(CH_2)_nCH_3 + (n + 1)H_2 \rightarrow (n + 2)CH_4$$

In general terms, as the molecular weight of the feedstock is increased, similar operating conditions of hydrogasification lead to decreasing hydrocarbon gas yields, increasing yields of aromatic liquids, with carbon also appearing as a product.

The principal secondary variable that influences yields of gaseous products from petroleum feedstocks of various types is the aromatic content of the feedstock. For example, a feedstock of a given H/C (C/H) ratio that contains a large proportion of aromatic species is more likely to produce a larger proportion of liquid products and elemental carbon than a feedstock that is predominantly paraffinic (5).

Another option for processing crude oils which are too heavy to be hydrogasified directly involves first hydrocracking the crude oil to yield a low boiling product suitable for gas production and a high boiling product suitable for hydrogen production by partial oxidation (57). Alternatively, it may be acceptable for carbon deposition to occur during hydrogasification which can then be used for heat or for hydrogen production (64,65).

Partial Oxidation. It is often desirable to augment the supply of naturally occurring or by-product gaseous fuels or to produce gaseous fuels of well-defined composition and combustion characteristics (5). This is particularly true in areas where the refinery fuel (natural gas) is in poor supply and/or where the manufacture of fuel gases, originally from coal and more recently from petroleum, has become well established.

Almost all petroleum fractions can be converted into gaseous fuels, although conversion processes for the heavier fractions require more elaborate technology to achieve the necessary purity and uniformity of the manufactured gas stream

(5). In addition, the thermal yield from the gasification of heavier feedstocks is invariably lower than that of gasifying light naphtha or liquefied petroleum gas(es) because, in addition to the production of hydrogen, carbon monoxide, and gaseous hydrocarbons, heavy feedstocks also yield some tar and coke.

As in the case of coal, synthetic natural gas can be produced from heavy oil by partially oxidizing the oil to a mixture of carbon monoxide and hydrogen

$$2\,CH_{petroleum} + O_2 \rightarrow 2\,CO + H_2$$

which is methanated catalytically to produce methane of any required purity. The initial partial oxidation step consists of the reaction of the feedstock with a quantity of oxygen insufficient to burn it completely, making a mixture consisting of carbon monoxide, carbon dioxide, hydrogen, and steam. Success in partially oxidizing heavy feedstocks depends mainly on details of the burner design (66). The ratio of hydrogen to carbon monoxide in the product gas is a function of reaction temperature and stoichiometry and can be adjusted, if desired, by varying the ratio of carrier steam to oil fed to the unit.

To make synthetic natural gas by partial oxidation, virtually all of the methane in the product gas must be produced by catalytic methanation of carbon monoxide and hydrogen. The feed is mixed with recycled carbon and fed, together with steam and oxygen, to a reactor in which partial combustion takes place. Heat from the reaction gasifies the rest of the feed, and by-product coke is formed. The heavier feedstocks tend to produce more carbon than can be consumed through recycling; thus some must be withdrawn. After carbon separation by water scrubbing, the synthesis gas that is available can be converted into hydrogen or into synthetic natural gas by methanation.

Steam Reforming. When relatively light feedstocks, eg, naphthas having ca 180°C end boiling point and limited aromatic content, are available, high nickel content catalysts can be used to simultaneously conduct a variety of near-autothermic reactions. This results in the essentially complete conversions of the feedstocks to methane:

$$CH_3(CH_2)_3CH_3 + 2\,H_2O \rightarrow 4\,CH_4 + CO_2$$

Because of limitations on the activity of practical catalysts, this reaction must be carried out in stages, the first of which is carried out at 425–480°C and 1.5–2.9 MPa (200–400 psi) and amounts approximately to the following reaction:

$$C_5H_{12} + 3\,H_2O \rightarrow 3\,CH_4 + CO_2 + 3\,H_2 + CO$$

In ensuing catalytic stages, usually termed hydrogasification and methanation (not to be confused with the noncatalytic, direct hydrogasification processes described above), the remaining carbon monoxide and hydrogen are reacted to produce additional methane.

Oil Shale. Oil shale (qv) is a sedimentary rock that contains organic matter, referred to as kerogen, and another natural resource of some consequence that could be exploited as a source of synthetic natural gas (67–69). However, as of this writing, oil shale has found little use as a source of substitute natural gas.

Biomass. Biomass is simply defined for these purposes as any organic waste material, such as agricultural residues, animal manure, forestry residues, municipal waste, and sewage, which originated from a living organism (70–74).

Biomass is another material that can produce a mixture of carbonaceous solid and liquid products as well as gas:

$$C_{organic} \rightarrow C_{coke/char/carbon} + liquids + gases$$

Whereas biomass has not received the same attention as coal as a source of gaseous fuels, questions about the security of fossil energy supplies related to the availability of natural and substitute gas have led to a search for more reliable and less expensive energy sources (75). Biomass resources are variable, but it has been estimated that substantial amounts (up to 20×10^6 mJ (20×10^{15} Btu)) of energy, representing ca 19% of the annual energy consumption in the United States. In addition, environmental issues associated with the use of coal have led some energy producers to question the use of large central energy generating plants. However, biomass may be a gaseous fuel source whose time is approaching (see FUELS FROM BIOMASS).

The means by which synthetic gaseous fuels could be produced from a variety of biomass sources are variable and many of the known gasification technologies can be applied to the problem (70,71,76–82). For example, the Lurgi circulatory fluidized-bed gasifier is available for the production of gaseous products from biomass feedstocks as well as from coal (83,84).

Gas Treating

The reducing conditions in gasification reactors effect the conversion of the sulfur and nitrogen in the feed coal to hydrogen sulfide, H_2S, and ammonia, NH_3. Some carbonyl sulfide, COS, carbon disulfide, CS_2, mercaptans, RSH, and hydrogen cyanide, HCN, are also formed in the gasifier. These compounds, along with carbon dioxide, are removed simultaneously, either selectively or nonselectively, from the gas stream in the clean-up stages of the process using commercially available physical or chemical solvents and scrubbing agents (1,5,85–88).

Solvents used for hydrogen sulfide absorption include aqueous solutions of ethanolamine (monoethanolamine, MEA), diethanolamine (DEA), and diisopropanolamine (DIPA) among others:

$$2\ RNH_2 + H_2S \rightarrow (RNH_3)_2S$$
$$(RNH_3)_2S + H_2S \rightarrow 2\ RNH_3HS$$
$$2\ RNH_2 + CO_2 + H_2O \rightarrow (RNH_3)_2CO_3$$
$$(RNH_3)_2CO_3 + CO_2 + H_2O \rightarrow 2\ RNH_3HCO_3$$
$$2\ RNH_2 + CO_2 \rightarrow RNHCOONH_3R$$

These solvents differ in volatility and selectivity for the removal of H_2S, mercaptans, and CO_2 from gases of different composition. Other alkaline solvents used for the absorption of acidic components in gases include potassium carbonate, K_2CO_3, solutions combined with a variety of activators and solubilizers to improve gas–liquid contacting.

Whereas most alkaline solvent absorption (qv) processes result in gases of acceptable purity for most purposes, it is often essential to remove the last traces of residual sulfur compounds from gas streams. This is in addition to ensuring product purity such as the removal of water, higher hydrocarbons, and dissolved elemental sulfur from liquefied petroleum gas. Removal can be accomplished by passing the gas over a bed of molecular sieves (qv), synthetic zeolites commercially available in several proprietary forms. Impurities are retained by the packed bed, and when the latter is saturated it can be regenerated by passing hot clean gas or hot nitrogen, generally in a reverse direction (see also CATALYSTS, REGENERATION).

By-product water formed in the methanation reactions is condensed by either refrigeration or compression and cooling. The remaining product gas, principally methane, is compressed to desired pipeline pressures of 3.4–6.9 MPa (500–1000 psi). Final traces of water are absorbed on silica gel or molecular sieves, or removed by a drying agent such as sulfuric acid, H_2SO_4. Other desiccants may be used, such as activated alumina, diethylene glycol, or concentrated solutions of calcium chloride (see DESICCANTS).

BIBLIOGRAPHY

"Manufactured Gas" in *ECT* 1st ed., Vol. 8, pp. 765–800, by W. H. Fulweiler, Consultant; "Gas, Manufactured" in *ECT* 2nd ed., Vol. 10, pp. 353–442, by M. A. Elliot, Illinois Institute of Technology, and H. R. Linden, Institute of Gas Technology; "Carbon Monoxide–Hydrogen Reactions" in *ECT* 2nd ed., Vol. 4, pp. 446–489, by H. Pichler and A. Hector, Carl-Engler-und Hans-Bunte-Institut für Mineralöl und Kohleforschung der Technischen Hochschule Karlsruhe; "Gaseous" under "Fuels, Synthetic" in *ECT* 3rd ed., Vol. 11, pp. 410–446, by J. Huebler and J. C. Janka, Institute of Gas Technology.

1. J. G. Speight, ed., *Fuel Science and Technology Handbook*, Marcel Dekker, Inc., New York, 1990.
2. J. G. Speight, *The Chemistry and Technology of Coal*, Marcel Dekker, Inc., New York, 1983.
3. R. A. Hessley, in Ref. 1, pp. 645–734.
4. E. J. Parente and A. Thumann, eds., *The Emerging Synthetic Fuel Industry*, Fairmont Press, Atlanta, Ga., 1981.
5. J. G. Speight, *The Chemistry and Technology of Petroleum*, 2nd ed., Marcel Dekker, Inc., New York, 1991.
6. A. Kasem, *Three Clean Fuels from Coal: Technology and Economics*, Marcel Dekker, Inc., New York, 1979.
7. L. L. Anderson and D. A. Tillman, *Synthetic Fuels from Coal*, John Wiley & Sons, Inc., New York, 1979.
8. A. D. Dainton, in G. J. Pitt and G. R. Millward, eds., *Coal and Modern Coal Processing: An Introduction*, Academic Press, Inc., New York, 1979.
9. D. M. Considine, *Energy Technology Handbook*, McGraw-Hill Book Co., Inc., New York, 1977.
10. A. Francis and M. C. Peters, *Fuels and Fuel Technology*, Pergamon Press, Inc., New York, 1980.

11. J. L. Johnson, in M. A. Elliott, ed., *Chemistry of Coal Utilization*, 2nd suppl. vol. John Wiley & Sons, Inc., New York, 1981, Chapt. 23.
12. A. Elton, in C. Singer and co-workers, eds., *A History of Technology*, Vol. 4, Oxford University Press, Oxford, U.K., 1958, Chapt. 9.
13. L. Shnidman, in H. H. Lowry, ed., *Chemistry of Coal Utilization*, John Wiley & Sons, Inc., New York, 1945, Chapt. 30.
14. C. M. Jarvis, in Ref. 12, Vol. 5, Chapt. 10.
15. P. F. Mako and W. A. Samuel, in R. A. Meyers, ed., *Handbook of Synfuels Technology*, McGraw-Hill Book Co., Inc., New York, 1984, Chapt. 2–1.
16. *Oil Gas J.* **90**(3), 53 (1992).
17. K. T. Klasson and co-workers, C. Akin and J. Smith, eds., *Gas, Oil, Coal, and Environmental Biotechnology II*, Institute of Gas Technology, Chicago, 1990, p. 408.
18. B. D. Faison, *Crit. Revs. Biotechnol.* **11**, 347 (1991).
19. S. R. Bull, *Energy Sources* **13**, 433 (1991).
20. R. C. Reid, J. M. Prausnitz, and T. K. Sherwood, *The Properties of Gases and Liquids*, McGraw-Hill Book Co., Inc., New York, 1977.
21. *Wellman Galusha Gas Producers*, research bulletin no. 576A, McDowell-Wellman Engineering Co., Cleveland, Ohio, May 1976.
22. R. F. Probstein and R. E. Hicks, *Synthetic Fuels*, pH Press, Cambridge, Mass., 1990.
23. J. S. Sinninghe Damste and J. W. de Leeuw, *Fuel Processing Technol.* **30**, 109 (1992).
24. N. Berkowitz, *Introduction to Coal Technology*, Academic Press, Inc., New York, 1979.
25. R. K. Hessley, J. W. Reasoner, and J. T. Riley, *Coal Science*, John Wiley & Sons, Inc., New York, 1986.
26. M. O. Holowaty and co-workers, in R. A. Meyers, ed., *Coal Handbook*, Marcel Dekker, Inc., New York, 1981, Chapt. 9.
27. W. Eisenhut, in Ref. 11, Chapt. 14.
28. F. W. Gibbs, in Ref. 12, Vol. 3, Chapt. 25.
29. G. J. Pitt and G. R. Millward, eds., in Ref. 8, p. 52.
30. L. Seglin and S. A. Bresler, in Ref. 11, Chapt. 13.
31. E. Aristoff, R. W. Rieve, and H. Shalit, in Ref. 11, Chapt. 16.
32. D. McNeil, in Ref. 11, Chapt. 17.
33. C. Y. Cha and co-workers, *Report No. DOE/MC/24268-2700 (DE89000967)*, United States Department of Energy, Washington, D.C., 1988.
34. F. M. Fess, *History of Coke Oven Technology*, Gluckauf Verlag, Essen, Germany, 1957.
35. M. A. Elliott, ed., in Ref. 11.
36. D. Hebden and H. J. F. Stroud, in Ref. 11, Chapt. 24.
37. R. A. Meyers, ed., in Ref. 15.
38. D. R. Simbeck, R. L. Dickenson, and A. J. Moll, *Energy Prog.* **2**, 42 (1982).
39. W. W. Bodle and J. Huebler, in Ref. 26, Chapt. 10.
40. L. K. Rath and J. R. Longanbach, *Energy Sources* **13**, 443 (1991).
41. S. Alpert and M. J. Gluckman, *Ann. Rev. Energy* **11**, 315 (1986).
42. R. A. Meyers, ed., in Ref. 26.
43. J. H. Martin, I. N. Banchik, and T. K. Suhramaniam, *Report of the Committee on Production of Manufactured Gases*, report no. 1GU/B-76, London, 1976.
44. J. F. Farnsworth, *Ind. Heat.* **41**(11), 38 (1974).
45. J. F. Farnsworth, *Proceedings Coal Gas Fundamentals Symposium*, Institute of Gas Technology, Chicago, 1979.
46. J. C. Hoogendoorn, *Proceedings of the Ninth Pipeline Gas Symposium*, Chicago, Oct. 31–Nov. 2, 1977.
47. R. Essenhigh, in Ref. 11, Chapt. 19.
48. M. A. Field and co-workers, *Combustion of Pulverized Coal*, British Coal Utilization Research Association, Leatherhead, Surrey, U.K., 1967.

49. A. Levy and co-workers, in Ref. 26, Chapt. 8.
50. N. Chigier, *Combustion Measurements*, Hemisphere Publishing Corp., New York, 1991.
51. United States Congress, *Public Law 101-549, An Act to Amend the Clean Air Act to Provide for Attainment and Maintenance of Health Protective National Ambient Air Quality Standards, and for Other Purposes*, Nov. 15, 1990.
52. United States Department of Energy, *Clean Coal Technology Demonstration Program*, DOE/FE-0219P, U.S. Dept. of Energy, Washington, D.C., Feb. 1991.
53. J. E. Funk, in J. A. Kent, ed., *Riegel's Handbook of Industrial Chemistry*, Van Nostrand Reinhold Co., New York, 1983, Chapt. 3.
54. F. J. Ceely and E. L. Daman, in Ref. 26, Chapt. 20.
55. J. M. Reid, *Proceedings SNG Symposium 1*, Institute of Gas Technology, Chicago, May 12–16, 1973.
56. F. J. Dent, *Gas J.* **288**(12), 600, 606, 610 (1956).
57. F. J. Dent, *Gas World*, **144**(11), 1078, 1080 (1956).
58. H. R. Linden and E. S. Pettyjohn, *Research Bulletin No. 12*, Institute of Gas Technology, Chicago, 1952.
59. G. B. Schultz and H. R. Linden, *Research Bulletin No. 29*, Institute of Gas Technology, Chicago, 1960.
60. F. J. Dent, *Proceedings of the Ninth International Gas Conference*, The Hague (Scheveningen), the Netherlands, Sept. 1–4, 1964.
61. J. G. Speight, *The Desulfurization of Heavy Oils and Residua*, Marcel Dekker, Inc., New York, 1981.
62. B. B. Bennett, *J. Inst. Fuel* **35**(8), 338 (1962).
63. H. R. Linden and M. A. Elliot, *Am. Gas J.* **186**(2), 22 (1959).
64. *Oil Gas J.* **71**(7), 36, 37 (1973).
65. *Oil Gas J.* **71**(15), 32, 33 (1973).
66. C. J. Kuhre and C. J. Shearer, *Oil Gas J.* **71**(36), 85 (1971).
67. P. Nowacki, *Oil Shale Technical Data Handbook*, Noyes Data Corp., Park Ridge, N.J., 1981.
68. V. D. Allred, ed., *Oil Shale Processing Technology*, Center for Professional Advancement, East Brunswick, N.J., 1982.
69. C. S. Scouten, in Ref. 1.
70. J. S. Robinson, *Fuels from Biomass: Technology and Feasibility*, Noyes Data Corp., Park Ridge, N.J., 1980.
71. J. L. Jones and S. B. Radding, eds., *Thermal Conversion of Solid Wastes and Biomass, Symposium Series No. 130*, American Chemical Society, Washington, D.C., 1980.
72. M. P. Kannan and G. N. Richard, *Fuel* **69**, 747 (1990).
73. L. Jimenez, J. L. Bonilla, and J. L. Ferrer, *Fuel* **70**, 223 (1991).
74. L. Jimenez and F. Gonzalez, *Fuel* **70**, 947 (1991).
75. *Energy World* **145**(3), 11 (1987).
76. M. P. Sharma and B. Prasad, *Energy Management (New Delhi)* **10**(4), 297 (1986).
77. A. A. C. M. Beenackers and W. P. M. van Swaaij, *Thermochemical Processing of Biomass*, Butterworths, London, 1984, p. 91.
78. G. J. Esplin, D. P. C. Fung, and C. C. Hsu, *Can. J. Chem. Eng.* **64**, 651 (1986).
79. S. Gaur and co-workers, *Fuel Sci. Technol. Int.* **10**, 1461 (1992).
80. M. A. McMahon and co-workers, *Preprints, Div. Fuel Chem.* **36**(4), 1670 (1991).
81. K. Dura-Swamy and co-workers, *Preprints, Div. Fuel Chem.* **36**(4), 1677 (1991).
82. J. T. Hamrick, *Preprints, Div. Fuel Chem.* **36**(4), 1986 (1991).
83. R. Reimert and co-workers, *Bioenergy 84*, Vol. 3, Elsevier Applied Science Publishers, London, p. 102.

84. P. K. Herbert and J. C. Loeffler, *Proceedings, Opportunities in the Synfuels Industry: SynOps '88*, Energy and Environmental Research Center, University of North Dakota, Grand Forks, 1988, p. 141.
85. H. A. Grosick and J. E. Kovacic, in Ref. 11, Chapt. 18.
86. A. V. Slack, in Ref. 11, Chapt. 22.
87. H. E. Benson, in Ref. 11, Chapt. 25.
88. C. H. Taylor in Ref. 15, Chapt. 3–13.

JAMES G. SPEIGHT
Consultant

LIQUID FUELS

The creation of liquids to be used as fuels from sources other than natural crude petroleum (qv) broadly defines synthetic liquid fuels. Hence, fuel liquids prepared from naturally occurring bitumen deposits qualify as synthetics, even though these sources are natural liquids. Synthetic liquid fuels have characteristics approaching those of the liquid fuels in commerce, specifically gasoline, kerosene, jet fuel, and fuel oil (see AVIATION AND OTHER GAS TURBINE FUELS; GASOLINE AND OTHER MOTOR FUELS). For much of the twentieth century, the synthetic fuels emphasis was on liquid products derived from coal (qv) upgrading or by extraction or hydrogenation of organic matter in coke liquids, coal tars, tar sands (qv), or bitumen deposits. More recently, however, much of the direction involving synthetic fuels technology has changed. There are two reasons.

The potential of natural gas, which typically has 85–95% methane, has been recognized as a plentiful and clean alternative feedstock to crude oil (see GAS, NATURAL). Estimates (1–3) place worldwide natural gas reserves at ca 1 × 10^{14} m^3 (3.5 × 10^{15} ft^3) corresponding to the energy equivalent of ca 1 × 10^{11} m^3 (637 × 10^9 bbl) of oil. As of this writing, the rate of discovery of proven natural gas reserves is increasing faster than the rate of natural gas production. Many of the large natural gas deposits are located in areas where abundant crude oil resources lie such as in the Middle East and Russia. However, huge reserves of natural gas are also found in many other regions of the world, providing oil-deficient countries access to a plentiful energy source. The gas is frequently located in remote areas far from centers of consumption, and pipeline costs can account for as much as one-third of the total natural gas cost (1) (see PIPELINES). Thus tremendous strategic and economic incentives exist for on-site gas conversion to liquids.

In general, the proven technology to upgrade methane is via steam reforming to produce synthesis gas, CO + H$_2$. Such a gas mixture is clean and when converted to liquids produces fuels substantially free of heteroatoms such as sulfur and nitrogen. Two commercial units utilizing the synthesis gas from natural gas technology in combination with novel downstream conversion processes have been commercialized.

The direct methane conversion technology, which has received the most research attention, involves the oxidative coupling of methane to produce higher hydrocarbons (qv) such ɪs ethylene (qv). These olefinic products may be upgraded to liquid fuels via catalytic oligomerization processes.

A second trend in synthetic fuels is increased attention to oxygenates as alternative fuels (4) as a result of the growing environmental concern about burning fossil-based fuels. The environmental impact of the oxygenates, such as methanol (qv), ethanol (qv), and methyl *tert*-butyl ether (MTBE) (see OCTANE IMPROVERS) is stɪɪɪ under debate, but these alternative liquid fuels are gaining new prominence. The U.S. Alternative Fuels Act of 1988, and the endorsement of oxygenate fuels that act contains, clearly underscore the idea that economics is no longer the sole consideration with regard to alternative fuels production (5).

Despite reduced prominence, coal technology is well positioned to provide synthetic fuels for the future. World petroleum and natural gas production are expected ultimately to level off and then decline. Coal gasification to synthesis gas is utilized to synthesize liquid fuels in much the same manner as natural gas steam reforming technology (see COAL CONVERSION PROCESSES). Although as of this writing world activity in coal liquefaction technology is minimal, the extensive development and detailed demonstration of processes for converting coal to liquid fuels should serve as solid foundation for the synthetic fuel needs of the future.

Coal, tar, and heavy oil fuel reserves are widely distributed throughout the world. In the Western hemisphere, Canada has large tar sand, bitumen (very heavy crude oil), and coal deposits. The United States has very large reserves of coal and shale. Coal comprises ca 85% of the U.S. recoverable fossil energy reserves (6). Venezuela has an enormous bitumen deposit and Brazil has significant oil shale (qv) reserves. Coal is also found in Brazil, Colombia, Mexico, and Peru. Worldwide, the total resource base of these reserves is immense and may constitute >90% of the hydrocarbon resources in place (see FUEL RESOURCES).

The driving force behind the production of combustible liquids before 1900 was the search for low cost lighting. Gas produced during coal distillation was used to light homes at the end of the eighteenth century (7). Large-scale use of coal, which began in England in the nineteenth century, led to significant reductions in the costs of hydrocarbon liquids. The production of coal tar, and the separation therefrom of various coal liquids concomitant to the production of illuminating gas, probably predates production from the coking operations associated with iron (qv) production. The coal tars produced in gas works may have been the first synthetic liquids turned to fuel use in quantity.

Proof of the existence of benzene in the light oil derived from coal tar (8) first established coal tar and coal as chemical raw materials (see FEEDSTOCKS, COAL CHEMICALS). Soon thereafter the separation of coal-tar light oil into substantially pure fractions produced a number of the aromatic components now known to be present in significant quantities in petroleum-derived liquid fuels. Indeed, these separation procedures were for the recovery of benzene–toluene–xylene (BTX) and related substances, ie, benzol or motor benzol, from coke-oven operations (8) (see BTX PROCESSING).

By the middle of the nineteenth century it was realized, both in England and in the United States, that kerosene, or coal oil, distilled from coal, could produce a luminous combustion flame. Commercialization was rapid. By the time of the U.S. Civil War, ~87 m³/yr (23,000 gal/yr) of lamp oil was produced in the United States from the distillation not only of coal, but also of oil shale and natural bitumen. In 1859, high gravity, low sulfur crude oil was discovered in the United States. This produced high quality kerosene with minimal processing, and the world's first oil boom erupted. Until the end of the nineteenth century, kerosene was the only substance of value extracted from natural crude oil. It cost too much for heating purposes, but was used widely for lighting until replaced by electricity. Refiners slowly learned to use the residues from kerosene production, and as the market for lamp oil collapsed, gasoline increased in value. The widespread use of liquids as fuels dates from that time.

Liquid fuels possess inherent advantageous characteristics in terms of being more readily stored, transported, and metered than gases, solids, or tars. Liquid fuels also are generally easy to process or clean by chemical and catalytic means. The energy densities of clean hydrocarbon liquids may be very high relative to gas, solid, or semisolid fuel substances. Moreover, liquid fuels are the most compatible with the twentieth century world fuel infrastructure because most fuel-powered conveyances are designed to function only with relatively clean, low viscosity liquids. In general, liquid hydrocarbon fuels possess an intermediate hydrogen-to-carbon content. Production of synthetic fuels from alternative feedstocks to natural petroleum crude oil is based on adjusting the hydrogen-to-carbon ratio to the desired intermediate level.

There is an inherent economic penalty associated with producing liquids from either natural gas or solid coal feedstock. Synthetic liquid fuels technologies are generally not economically competitive with crude oil processing in the absence of extraneous influences such as price supports or regulations.

Indirect Liquefaction/Conversion to Liquid Fuels

Indirect liquefaction of coal and conversion of natural gas to synthetic liquid fuels is defined by technology that involves an intermediate step to generate synthesis gas, $CO + H_2$. The main reactions involved in the generation of synthesis gas are the coal gasification reactions:

Combustion

$$C + O_2 \rightleftharpoons CO_2 \qquad \Delta H_{298\,K} = -394 \text{ kJ/mol} \,(-94.2 \text{ kcal/mol}) \qquad (1)$$

Gasification

$$C + 1/2\,O_2 \rightleftharpoons CO \qquad \Delta H_{298\,K} = -111 \text{ kJ/mol} \,(-26.5 \text{ kcal/mol}) \qquad (2)$$

$$C + H_2O\,(g) \rightleftharpoons CO + H_2 \qquad \Delta H_{298\,K} = +131 \text{ kJ/mol} \,(31.3 \text{ kcal/mol}) \qquad (3)$$

$$C + CO_2 \rightleftharpoons 2\,CO \qquad \Delta H_{298\,K} = +172 \text{ kJ/mol} \,(41.1 \text{ kcal/mol}) \qquad (4)$$

Water gas shift

$$CO + H_2O\,(g) \rightleftharpoons CO_2 + H_2 \qquad \Delta H_{298\,K} = -41 \text{ kJ/mol} \,(-9.8 \text{ kcal/mol}) \qquad (5)$$

the methane steam reforming reactions:

Partial oxidation

$$CH_4 + 1/2\,O_2 \rightleftharpoons CO + 2\,H_2 \qquad \Delta H_{298\,K} = -36\ kJ/mol\ (-8.6\ kcal/mol) \qquad (6)$$

Reforming

$$CH_4 + H_2O\,(g) \rightleftharpoons CO + 3\,H_2 \qquad \Delta H_{298\,K} = 206\ kJ/mol\ (49.2\ kcal/mol) \qquad (7)$$

and the water gas shift reaction (eq. 5), used to increase the H_2/CO ratio of the product synthesis gas.

Coal gasification technology dates to the early nineteenth century but has been largely replaced by natural gas and oil. A more hydrogen-rich synthesis gas is produced at a lower capital investment. Steam reforming of natural gas is applied widely on an industrial scale (9,10) and in particular for the production of hydrogen (qv).

The conversion of coal and natural gas to liquid fuels via indirect technology can be achieved by the routes shown in Figure 1. Two pathways from synthesis gas can be taken. Both have been commercialized. One pathway involves coupling with Fischer-Tropsch technology to produce fuel range hydrocarbons directly or upon further processing. Using coal feedstock, this route has been commercialized in South Africa since the 1950s and a process using natural gas was commercialized in Malaysia by Shell Oil Co. in 1993. An alternative route relies on the production of methanol from synthesis gas and subsequent transformation of the methanol to fuels using zeolite catalyst technology introduced by Mobil Oil Corp. (see MOLECULAR SIEVES). This route was commercialized using indigenous natural gas in New Zealand in 1985.

Coal Upgrading via Fischer-Tropsch. The synthesis of methane by the catalytic reduction of carbon monoxide and hydrogen over nickel and cobalt catalysts at atmospheric pressure was reported in 1902 (11).

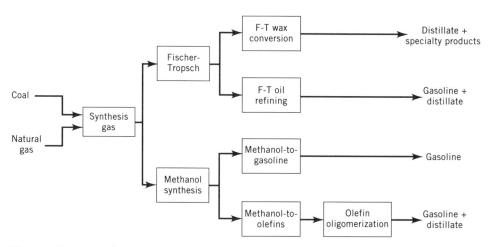

Fig. 1. Routes to liquid fuels from natural gas and coal via synthesis gas. F-T is the Fischer-Tropsch process.

$$CO + 3H_2 \rightarrow CH_4 + H_2O \ (l) \qquad \Delta H_{298\ K} = -250\ kJ/mol\ (-59.8\ kcal/mol) \qquad (8)$$

$$2\ CO + 2\ H_2 \rightarrow CH_4 + CO_2 \qquad \Delta H_{298\ K} = -247\ kJ/mol\ (59.0\ kcal/mol) \qquad (9)$$

In the early 1920s Badische Anilin- und Soda-Fabrik announced the specific catalytic conversion of carbon monoxide and hydrogen at 20–30 MPa (200–300 atm) and 300–400°C to methanol (12,13), a process subsequently widely industrialized. At the same time Fischer and Tropsch announced the Synthine process (14,15), in which an iron catalyst effects the reaction of carbon monoxide and hydrogen to produce a mixture of alcohols, aldehydes (qv), ketones (qv), and fatty acids at atmospheric pressure.

In the classical normal pressure synthesis (16), higher hydrocarbons are produced by net reactions similar to those observed in the early 1900s, but at temperatures below the level at which methane is formed:

$$n\ CO + 2n\ H_2 \rightarrow C_nH_{2n} + n\ H_2O + heat \qquad (10)$$

$$2n\ CO + n\ H_2 \rightarrow C_nH_{2n} + n\ CO_2 + heat \qquad (11)$$

In the mid-1930s improvements in catalysts and techniques (17–19) culminated in the licensing of the process to Ruhrchemie to produce liquid hydrocarbons and paraffin waxes using precipitated cobalt-on-kieselguhr catalysts. Subsequently, a medium pressure synthesis was developed (20) at 0.5–2 MPa (5–20 atm) using dispersed cobalt catalysts which improved hydrocarbon yields by 10–15%. The yields of paraffin wax, in particular, could be increased to 45% of the total liquid product. Hydrotreating of catalyst (required in the normal pressure process) was avoided, and catalyst life was extended (21–23). There is a marked influence of pressure on product yields. Beyond the optimum pressure of about 2 MPa (20 atm), paraffin yield decreases. Little change is found in the gasoline and gas oil yields, however.

Furthermore it was discovered that reasonable yields could be obtained using precipitated iron catalysts at 1–3 MPa (10–30 atm), and that very high melting waxes could be synthesized at 10–100 MPa (100–1000 atm) over ruthenium catalysts. At the same time a related process, the oxo synthesis, was announced (24) (see OXO PROCESS). Early in World War II, the iso-synthesis process was developed for the production of low molecular weight isoparaffins at high temperatures and pressures over thoria and mixtures of alumina and zinc oxide (25–28). In the early 1960s polymethylenes were synthesized using activated ruthenium catalysts at high pressures.

Industrial operation of the Fischer-Tropsch synthesis involved five steps: (1) synthesis gas manufacture; (2) gas purification by removal of water and dust, and hydrogen sulfide and organic sulfur compounds; (3) synthesis of hydrocarbons; (4) condensation of liquid products and recovery of gasoline from product gas; and (5) fractionation of synthetic products. Only the synthesis reactor and its method of operation were unique to the process. For low pressure synthesis the reactor incorporated elaborate bundles of cooling tubes immersed in the catalyst, whereby circulating water removed the heat of reaction, limiting the conversion to methane which produced high temperatures. In the pressure process, bundles of concentric tubes, with catalyst arranged in the annuli, through and around which cooling water flowed, served as conversion units. In both systems, the conversion units each contained about 10 m^3 (ca 350 ft^3) of catalyst, and were rated at a capacity of ca 4.8 m^3 (30 bbl) of liquid product per day.

During World War II, nine commercial plants were operated in Germany, five using the normal pressure synthesis, two the medium pressure process, and two having converters of both types. The largest plants had capacities of ca 400 m^3/d (2500 bbl/d) of liquid products. Cobalt catalysts were used exclusively.

Developments Outside Germany.   In the late 1930s experimental work in England (29–31) led to the erection of large pilot facilities for Fischer-Tropsch studies (32). In France, a commercial facility near Calais produced ca 150 m^3 (940 bbl) of liquid hydrocarbons per day. In Japan, two full-scale plants were also operated under Ruhrchemie license. Combined capacity was ca 400 m^3 (2500 bbl) of liquids per day.

In the mid-1930s Universal Oil Products reported (33,34) that gasoline of improved quality could be produced by cracking the high boiling fractions of Fischer liquids, and a consortium, the Hydrocarbon Synthesis, Inc., entered into an agreement with Ruhrchemie to license the Fischer synthesis outside Germany.

In 1955 the South African Coal, Oil, and Gas Corp. (Sasol) commercialized the production of liquid fuels utilizing Fischer-Tropsch technology (35). This Sasol One complex has evolved into the streaming of second-generation plants, known as Sasol Two and Three. The Sasol One process, shown in Figure 2**a** (36), combines fixed-bed Ruhrchemie-Lurgi Arge reactor units with fluidized-bed Synthol process technology (37). For Sasol One, 16,000 t/d of coal is crushed and gasified with steam and oxygen. After a number of gas purification steps in which by-products and gas impurities are removed, the pure gas is processed in both fixed- and fluidized-bed units simultaneously. Table 1 gives product selectivity comparisons of fixed-bed and Synthol operations. Conversion to hydrocarbons is higher in the Synthol unit and the H$_2$/CO ratio is also higher. Because the fixed-bed Arge reactor favors the formation of straight-chain paraffins, there is greater production of diesel and wax fractions than the Synthol unit. The Arge reactor products have lower gasoline octane number but higher diesel cetane number relative to Synthol. The high wax production using the Arge reactor was disadvantageous at the time owing to market limitations of wax fuels. Sasol One produced a vast array of chemical and fuel products, including gasoline at 1.5 × 10^6 t/yr.

The 1973 oil crisis resulted in the Sasol Two unit which started up in early 1980 followed by the nearly identical Sasol Three unit two years later. Figure 2**b** gives the schematic flow diagrams for the Sasol Two and Three processes. Sasol Two uses 36 Lurgi gasifiers in parallel to process ca 31,000 t/d of sized coal. By-product effluents and gas impurities are removed in Rectisol (sulfur compounds and CO$_2$ removal), Phenosolvan (oxygen compounds and ammonia removal), and tar separation units. Synthol fluid-bed units were employed because of the product distribution and ease of design scale-up. Approximately 80,000 t/d of coal are needed for the two plants. Composition and manufacturing information for Sasol Fischer-Tropsch catalysts are trade secrets, but the catalyst is widely accepted as being an alkali-metal promoted iron-based material.

More recently, Sasol commercialized a new type of fluidized-bed reactor and was also operating a higher pressure commercial fixed-bed reactor (38). In 1989, a commercial scale fixed fluid-bed reactor was commissioned having a capacity

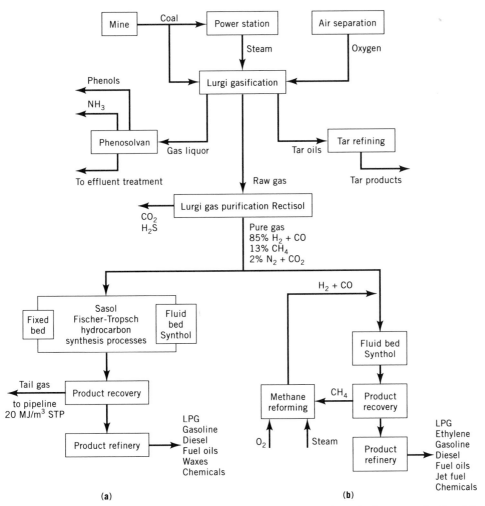

Fig. 2. Process flow sheet of (**a**) Sasol One and (**b**) the Sasol synfuel process for Sasol Two and Three (36). LPG is liquefied petroleum gas; other terms are defined in the text. To convert J to cal, divide by 4.184.

similar to existing commercial reactors at Sasol One (39). This effort is aimed at expanded production of higher value chemicals, in particular waxes (qv) and linear olefins.

Properties. Fischer-Tropsch liquid obtained using cobalt catalysts is roughly equivalent to a very paraffinic natural petroleum oil but is not so complex a mixture. Straight-chain, saturated aliphatic molecules predominate but mono-olefins may be present in an appreciable concentration. Alcohols, fatty acids, and other oxygenated compounds may represent less than 1% of the total liquid product. The normal pressure synthesis yields ca 60% gasoline, 30% gas oil, and 10% paraffin (mp 20–100°C). The medium pressure synthesis yields 35% gasoline, 35% gas oil, and 30% paraffin. The octane rating of the gasoline is too low for direct use as motor fuel (40).

Table 1. Product Selectivities for Commercial Fixed-Bed and Synthol Units[a]

Product	Fixed-bed	Synthol
methane	2.0	10
ethylene	0.1	4
ethane	1.8	4
propylene	2.7	12
propane	1.7	2
butenes	3.1	9
butanes	1.9	2
C_5 and higher	83.5	51
soluble chemicals	3.0	5
water-soluble acids	0.2	1

[a]Ref. 36.

Most of the German gasoline production was blended into motor fuels using benzene derived from coking. The gas oil could be used directly as a superior diesel fuel; some was also used in soap (qv) manufacture. The paraffin (referred to as gatsch) was used primarily for the synthesis of fatty acids and hard soaps. The propane and butane gases were also used as motor fuels. Some propylene and butylenes were polymerized over phosphoric acid to high octane gasoline, and some olefins to lubricating oils. Typical values for the composition of the technical-scale reaction products of the normal and medium pressure synthesis are available (41).

Natural Gas Upgrading via Fischer-Tropsch. In the United States, as in other countries, scarcities from World War II revived interest in the synthesis of fuel substances. A study of the economics of Fischer synthesis led to the conclusion that the large-scale production of gasoline from natural gas offered hope for commercial utility. In the Hydrocol process (Hydrocarbon Research, Inc.) natural gas was treated with high purity oxygen to produce the synthesis gas which was converted in fluidized beds of iron catalysts (42).

Shell Middle Distillate Synthesis. The Shell middle distillate synthesis (SMDS) process developed by Shell Oil Co., uses remote natural gas as the feedstock (43–45). A simplified flow scheme is given in Figure 3. This two-step process involves Fischer-Tropsch synthesis of paraffinic wax called the heavy paraffin synthesis (HPS). The wax is subsequently hydrocracked and hydroisomerized to yield a middle distillate boiling range product in the heavy paraffin conversion (HPC). In the HPS stage, wax is maximized by using a proprietary catalyst having high selectivity toward heavier products and by the use of a tubular, fixed-bed Arge-type reactor. The HPC stage employs a commercial hydrocracking catalyst in a trickle flow reactor. The effect of hydrocracking light paraffins is shown in Figure 4. The HPC step allows for production of narrow range hydrocarbons not possible with conventional Fischer-Tropsch technology.

After years of bench-scale and pilot-plant studies, construction was begun on a gca 1600 m³/d (10,000 bbl/d) unit in Sarawak, Malaysia, by Shell in a joint venture with Mitsubishi and the Malaysian government. Plant commissioning

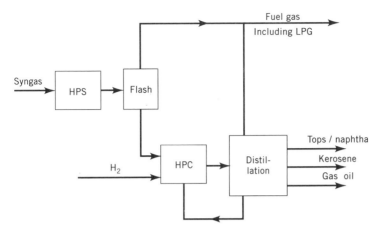

Fig. 3. The Shell middle distillate synthesis (SMDS) process. HPS = heavy paraffin synthesis. HPC = heavy paraffin conversion.

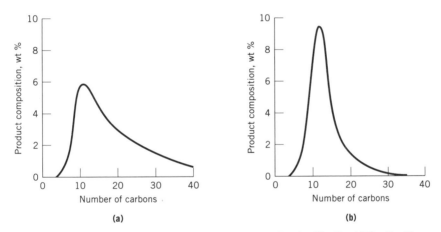

Fig. 4. Product compositions as a function of carbon number for the Shell middle distillate synthesis process: (**a**) the Fischer-Tropsch product following HPS, and (**b**) the final hydrocracking product following HPC. See text (45).

was in early 1993 at a capital investment of ca $600–700 × 10^6. The plant uses natural gas from offshore fields and is located adjacent to the existing Malaysian liquefied natural gas (LNG) plant. The production of liquid transportation fuels via SMDS cannot compete economically with fuels derived from crude oil. However, economics vary greatly with site location, and subsidies from the Malaysian government, eg, reduced natural gas cost, brought this plant to commercialization. In addition, premium selling prices for the high quality products made from SMDS are a primary influence on commercialization potential (44).

A similar process to SMDS using an improved catalyst is under development by Norway's state oil company, den norske state olijeselskap AS (Statoil) (46).

High synthesis gas conversion per pass and high selectivity to wax are claimed. The process has been studied in bubble columns and a demonstration plant is planned.

Properties.   Shell's two-step SMDS technology allows for process flexibility and varied product slates. The liquid product obtained consists of naphtha, kerosene, and gas oil in ratios from 15:25:60 to 25:50:25, depending on process conditions. Of particular note are the high quality gas oil and kerosene. Table 2 gives SMDS product qualities for these fractions.

The products manufactured are predominantly paraffinic, free from sulfur, nitrogen, and other impurities, and have excellent combustion properties. The very high cetane number and smoke point indicate clean-burning hydrocarbon liquids having reduced harmful exhaust emissions. SMDS has also been proposed to produce chemical intermediates, paraffinic solvents, and extra high viscosity index (XHVI) lubeoils (see LUBRICATION AND LUBRICANTS) (44).

Liquid Fuels via Methanol Synthesis and Conversion.   Methanol is produced catalytically from synthesis gas. By-products such as ethers, formates, and higher hydrocarbons are formed in side reactions and are found in the crude methanol product. Whereas for many years methanol was produced from coal, after World War II low cost natural gas and light petroleum fractions replaced coal as the feedstock.

Methanol-to-Gasoline.   The most significant development in synthetic fuels technology since the discovery of the Fischer-Tropsch process is the Mobil methanol-to-gasoline (MTG) process (47–49). Methanol is efficiently transformed into C_2–C_{10} hydrocarbons in a reaction catalyzed by the synthetic zeolite ZSM-5 (50–52). The MTG reaction path is presented in Figure 5 (47). The reaction sequence can be summarized as

$$n/2 \ [2 \ CH_3OH \rightleftharpoons CH_3OCH_3 + H_2O] \rightarrow C_nH_{2n} \rightarrow n[CH_2] \qquad (12)$$

where $[CH_2]$ represents an average paraffin–aromatic mixture.

How the initial C—C bond is formed from the C_1 progenitor is unknown and much debated (48,53–55). Light olefins are key intermediates in the reaction sequence. These undergo further transformation, ultimately forming aromatics and

Table 2. SMDS Product Quality[a,b]

Parameter	SMDS	Specification
	Gas oil	
cetane number	70	40 to 50
cloud point, °C	-10	-10 to $+20$
	Kerosene	
smoke point, mm	45	19
freezing point, °C	-47	-47

[a]Ref. 44.
[b]The tops/naphtha fraction is similar to straight-run material.

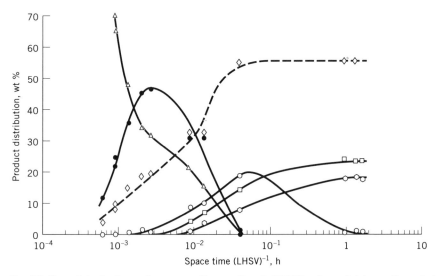

Fig. 5. Methanol-to-hydrocarbons reaction path at 371°C, where (△) is methanol; (●), dimethyl ether;(◇), water; (□), paraffins (and C_6 + olefins); (○), aromatics; and (○), C_2–C_5 olefins. LHSV = Liquid hourly space velocity.

light paraffins. Table 3 lists a typical MTG product distribution. A unique characteristic of these products is an abrupt termination in carbon number at around C_{10}. This is a consequence of molecular shape–selectivity (56–58), a property of ZSM-5. The composition and properties of the C_5 + fraction are those of a typical premium aromatic gasoline. Interestingly, C_{10} also is the end point of conventional gasoline.

The MTG process was developed for synfuel production in response to the 1973 oil crisis and the steep rise in crude prices that followed. Because methanol can be made from any gasifiable carbonaceous source, including coal, natural gas, and biomass, the MTG process provided a new alternative to petroleum for liquid fuels production. New Zealand, heavily dependent on foreign oil imports, utilizes the MTG process to convert vast offshore reserves of natural gas to gasoline (59).

Two versions of the MTG process, one using a fixed bed, the other a fluid bed, have been developed. The fixed-bed process was selected for installation in the New Zealand gas-to-gasoline (GTG) complex, situated on the North Island between the villages of Waitara and Motonui on the Tasman seacoast (60). A simplified block flow diagram of the complex is shown in Figure 6 (61). The plant processes over 3.7×10^6 m³/d (130×10^6 SCF/d) of gas from the offshore Maui field supplemented by gas from the Kapuni field, first to methanol, and thence to 2.3×10^3 m³/d (14,500 bbl/d) of gasoline. Methanol feed to the MTG section is synthesized using the ICI low pressure process (62) in two trains, each with a capacity of 2200 t/d.

A flow diagram of the MTG section is shown in Figure 7. Methanol feed, vaporized by heat exchange with reactor effluent gases, is converted in a first-stage reactor containing an alumina catalyst to an equilibrium mixture of methanol, dimethyl ether (DME), and water. This is combined with recycle light gas,

Table 3. MTG Product Distribution[a,b]

Hydrocarbon	Distribution, wt %
methane	1.0
ethane	0.6
ethylene	0.5
propane	16.2
propylene	1.0
i-butane	18.7
n-butane	5.6
butenes	1.3
i-pentane	7.8
n-pentane	1.3
pentenes	0.5
C_6+ aliphatics	4.3
benzene	1.7
toluene	10.5
ethylbenzene	0.8
xylenes	17.2
C_9 aromatics	7.5
C_{10} aromatics	3.3
$C_{11}+$ aromatics	0.2

[a]Reaction conditions of 371°C and LHSV of 1.0 h^{-1}.
[b]100% conversion.

which serves to remove reaction heat from the highly exothermic MTG reaction, and enters the reactors containing ZSM-5 catalyst. As indicated in Figure 7, five parallel swing reactors are used. Four reactors are on feed while the fifth is under regeneration. The multiple-bed configuration is used to minimize pressure drop as well as to control product selectivity. Reaction conditions are 360–415°C, 2.17×10^3 kPa (315 psia), and 9/1 recycle/fresh feed ratio. The overall thermal efficiency of the plant is ca 53%.

A fluid-bed version of the MTG process has been developed (60,63–65) and demonstrated in semiworks scale of 15.9 m^3/d (100 bbl/d), but has not been commercialized to date (ca 1993). Heat management of the exothermic MTG reaction is greatly facilitated by use of fluid-bed reactors. The turbulent bed, with its excellent heat-transfer characteristics, ensures isothermality through the reaction zone and permits steam generation by direct exchange with steam coils in the bed. A schematic diagram appears in Figure 8. The reactor system consists of three principal parts: the reactor, the catalyst regenerator, and an external catalyst cooler. The reactor is also equipped with internal heat-exchanger tubes. Methanol is converted in a single pass at 380–430°C, 276–414 kPa (40–60 psia). The methanol feed rate is 500–1050 kg/h. The fluid-bed demonstration was carried out in 1982–1983 (66).

Properties. Table 4 contains typical gasoline quality data from the New Zealand plant (67). MTG gasoline typically contains 60 vol % saturates, ie, paraffins and naphthenes; 10 vol % olefins; and 30 vol % aromatics. Sulfur and ni-

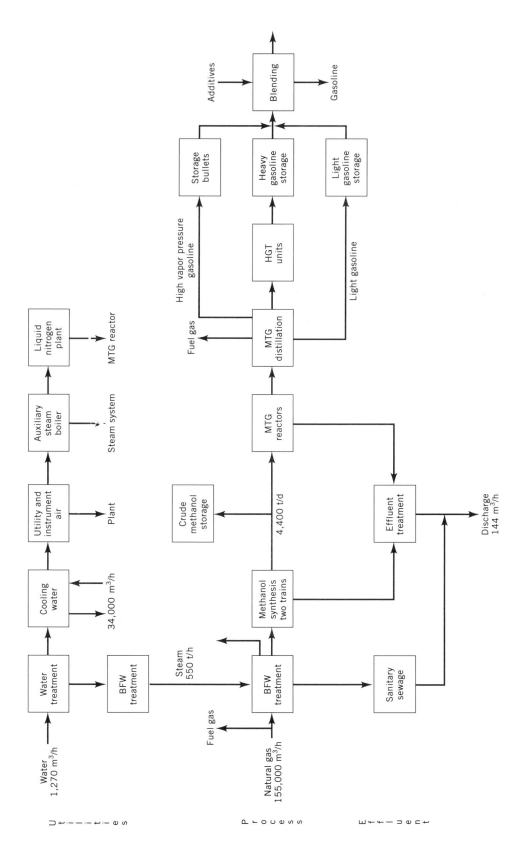

Fig. 6. Simplified block flow diagram for the New Zealand gas-to-gasoline (GTG) plant (61). To convert m³/h to gal/min, multiply by 4.40. HGT = heavy gasoline treatment facility; MTG = methanol-to-gasoline; BFW = boiler feed water.

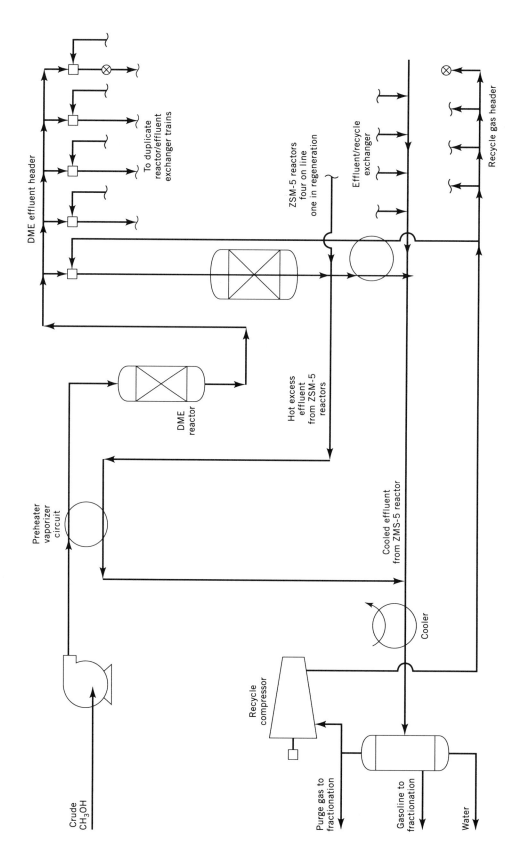

Fig. 7 A methanol to gasoline (MTG) fixed bed process flow diagram. DME ... diagram (1).(1)

DME effluent header

To duplicate reactor/effluent exchanger trains

ZSM-5 reactors four on line one in regeneration

Effluent/recycle exchanger

Recycle gas header

DME reactor

Hot excess effluent from ZSM-5 reactors

Cooled effluent from ZMS-5 reactor

Preheater vaporizer circuit

Cooler

Crude CH₃OH

Recycle compressor

Purge gas to fractionation

Gasoline to fractionation

Water

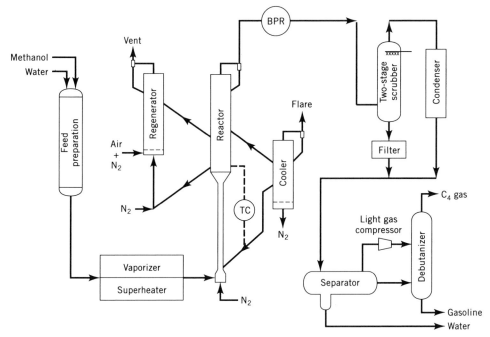

Fig. 8. Fluid-bed MTG demonstration plant schematic diagram. BPR = Back pressure regulator; TC = temperature controller.

Table 4. MTG Gasoline Quality[a]

Parameter	Average	Range
density at 15°C, kg/m^3	730	728–733
Reid vapor pressure, kPa[b]	86.2	83.4–91.0
octane number		
research	92.2	92.0–92.5
motor	82.6	82.2–83.0
durene content, wt %	2.0	1.74–2.29
induction period, min	325	260–370
distillation, % evaporation		
at 70°C	31.5	29.5–34.5
at 100°C	53.2	51.5–55.5
at 180°C	94.9	94.0–96.5
distillation end point, °C	204.5	196–209

[a]Ref. 67.
[b]To convert kPa to psia, multiply by 0.145.

169

trogen levels in the gasoline are virtually nil. The MTG process produces ca 3–7 wt % durene [*95-93-2*] (1,2,4,5-tetra-methylbenzene) but the level is reduced to ca 2 wt % in the finished gasoline product by hydrodealkylation of the durene in a separate catalytic reactor.

Methanol-to-Olefins and Olefins-to-Gasoline-and-Distillate.   Because the MTG process produces primarily gasoline, a variation of that process has been developed which allows for production of gasoline and distillate fuel (68). This process integrates two known technologies, methanol-to-olefins (MTO) and Mobil olefins-to-gasoline-and-distillate (MOGD). The MTO/MOGD process schematic is shown in Figure 9. The combined process produces gasoline and distillate in various proportions and, if needed, olefinic by-products.

In the MTO process, methanol is converted over ZSM-5 giving high (up to ca 80 wt % hydrocarbons) olefin yields and low ethylene and light saturate yields. The low ethylene yields are desirable in achieving high distillate yields using MOGD. Figure 5 shows that the production of olefins rather than gasoline from methanol is governed by the kinetics of methanol conversion over ZSM-5 catalyst (69). Generally, catalyst and process variables which increase methanol conversion decrease olefins yield.

The MTO process employs a turbulent fluid-bed reactor system and typical conversions exceed 99.9%. The coked catalyst is continuously withdrawn from the reactor and burned in a regenerator. Coke yield and catalyst circulation are an order of magnitude lower than in fluid catalytic cracking (FCC). The MTO process was first scaled up in a 0.64 m^3/d (4 bbl/d) pilot plant and a successful 15.9 m^3/d

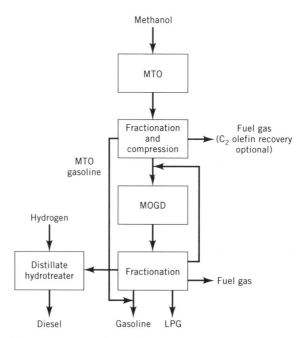

Fig. 9.   Methanol-to-olefins (MTO) and Mobil olefins-to-gasoline (MOGD) and distillate process schematic.

(100 bbl/d) demonstration plant was operated in Germany with U.S. and German government support.

The MOGD process oligomerizes light olefins to gasoline and distillate products over ZSM-5 zeolite catalyst. Gasoline and distillate selectivity is >95% of the feed olefins and gasoline/distillate product ratios can vary, depending on process conditions, from 0.2 to >100. High octane MTO gasoline is separated before the MOGD section and blended with MOGD gasoline. Some MOGD gasoline may be recycled. The distillate product requires hydrofinishing. Generally, the process scheme uses four fixed-bed reactors, three on-line and one in regeneration. A large-scale MOGD refinery test run was conducted by Mobil in 1981.

Properties. The gasoline product from the integrated MTO/MOGD process is predominately olefinic and aromatic. The gasoline quality (ca 89 octane) is comparable to FCC gasoline. Typical distillate product properties are given in Table 5. After hydrofinishing, the distillate product is mostly isoparaffinic and has high cetane index, low pour point, and negligible sulfur content. MOGD diesel fuel has somewhat lower density than typical conventional fuels (0.8 vs 0.86). Low aromatics levels contribute to a stable jet fuel with very little smoke emission during combustion. MOGD diesel and jet fuels meet or exceed all conventional specifications.

Table 5. MTO/MOGD Distillate Properties

Parameter	Total distillate	Jet fuel	Diesel fuel
quantity, vol %	100	30	70
density, g/mL	0.792	0.774	0.800
pour point, °C	−50		−30
freeze point, °C	−60	−60	
flash point, °C	60	50	100
cetane number	50		52
smoke point, mm	25	25	
aromatics, vol %	4	5	
sulfur, ppm	50		

Direct Conversion of Natural Gas to Liquid Fuels

The capital costs associated with indirect natural gas upgrading technology are high, thus research and development has focused on direct conversion of natural gas to liquid fuels. Direct conversion is defined as upgrading methane to the desired liquid fuels products while bypassing the synthesis gas step, ie, direct transformation to oxygenates or higher hydrocarbons. Direct upgrading routes which have been extensively studied include direct partial oxidation to oxygenates, oxidative coupling to higher hydrocarbons, and pyrolysis to higher hydrocarbons. Owing to the inert nature of methane, the technology is limited by the yields of desired products which in turn affects the process economics. Only one direct oxidative methane conversion process has been commercialized. A plant at Copsa Mica (Romania) in the 1940s (70) produced formaldehyde directly from methane and air by partial oxidation. This plant is no longer in operation. Plants to produce acetylene from methane by high temperature pyrolysis routes have been commercialized (see HYDROCARBONS, ACETYLENE).

Generally, the most developed processes involve oxidative coupling of methane to higher hydrocarbons. Oxidative coupling converts methane to ethane and ethylene by

$$2\ CH_4\ +\ 1/2\ O_2 \rightarrow H_3CCH_3\ +\ H_2O \tag{13}$$

$$H_3CCH_3\ +\ 1/2\ O_2 \rightarrow H_2C{=}CH_2\ +\ H_2O \tag{14}$$

The process can be operated in two modes: co-fed and redox. The co-fed mode employs addition of O_2 to the methane/natural gas feed and subsequent conversion over a metal oxide catalyst. The redox mode requires the oxidant to be from the lattice oxygen of a reducible metal oxide in the reactor bed. After methane oxidation has consumed nearly all the lattice oxygen, the reduced metal oxide is reoxidized using an air stream. Both methods have processing advantages and disadvantages. In all cases, however, the process is run to maximize production of the more desired ethylene product.

Direct conversion of natural gas to liquids has been actively researched. Process economics are highly variable and it is unclear whether direct natural gas conversion technologies are competitive with the established indirect processes. Some emerging technologies in this area are presented herein.

ARCO Gas-to-Gasoline Process. A two-step process using oxidative coupling to upgrade natural gas to liquid fuels has been proposed by ARCO (Atlantic Richfield Co.) (71,72). A simplified process scheme is given in Figure 10. Methane is passed through a redox-mode oxidative coupling reactor which generates C_2+ hydrocarbons such as ethylene. The olefinic products are then oligomerized over a zeolite catalyst in a second reactor to produce gasoline and distillate. Unreacted methane is recycled. ARCO claims 25% conversion of methane with 75% C_2+ selectivity (ethylene to ethane ratio up to 10) in the oxidative coupling first stage

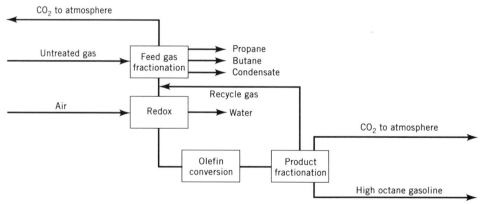

Fig. 10. Simplified flow diagram depicting the ARCO gas-to-gasoline process for a conceptual gasoline production plant (72).

and 95% ethylene conversion with 70% selectivity to gasoline and distillate in the olefin oligomerization second stage. This technology has been developed through bench-scale and pilot-plant stages.

OXCO Process. The OXCO process for upgrading natural gas has been proposed by the Commonwealth for Scientific and Industrial Research Organization (CSIRO) in Australia (73,74). This process involves C_2 + pyrolysis and oxidative coupling of natural gas in a two-stage reactor; the entire concept is shown schematically in Figure 11. The methane in natural gas is separated and oxidatively coupled in a fluidized-bed reactor operating in co-fed mode to produce ethylene and ethane. The higher alkanes from the natural gas as well as the product ethane from the first stage are injected into an oxygen-free pyrolysis stage to make additional ethylene. The heat from the coupling reactor is used in the pyrolysis reaction. The overall carbon conversion to unsaturates plus CO_2 per pass is 30% with an overall carbon selectivity to unsaturates of 86%. The ethylene may be subsequently upgraded by oligomerization to liquid fuels. This process, which produces higher yields of ethylene and has a more favorable heat balance than conventional oxidative coupling technology, has been demonstrated in 30- and 60-mm fluidized-bed reactors.

Properties. Liquid fuels derived from oxidative coupling/olefin oligomerization processes would be expected to have properties similar to those derived from olefin oligomerization pathways such as MTO/MOGD.

Oxygenate Fuels

Alcohols and ethers, especially methanol, ethanol, and methyl *tert*-butyl ether [1634-04-4] (MTBE), have been widely used separately or in blends with gasolines (reformulated gasoline) and other hydrocarbons to fuel internal combustion engines. Fuel properties of key oxygenates are presented in Table 6 (5). These compounds, as a class, may be considered to be partially oxidized, ie, each has a mole of oxidized hydrogen. They differ from the hydrocarbons that make up gasoline principally in lower heating values and in higher vaporization heat requirements. This constitutes a serious disadvantage to the substitution of oxygenates, especially lower alcohols, for motor gasoline. For example, the heating value of methanol is about half that of gasoline on an equivalent volume basis. Other properties which greatly influence the potential of oxygenates as fuels include octane performance, solubility in gasoline, effect on gasoline vapor pressure, sensitivity to water, and evaporative/exhaust emissions. Oxygenate fuels tests are often debated because the tests employed were developed for conventional gasolines.

The addition of small percentages of oxygenates to gasoline can produce large gains in octane. Thus, as blending components in gasoline, oxygenates improve octane quality. As neat fuels for spark-ignition engines, octane values for oxygenates are not useful in determining knock-limited compression ratios for vehicles because of the lean carburetor settings relative to gasoline. Neither do these values represent the octane performance of oxygenates when blended with gasoline.

In part because neat alcohols are insufficiently volatile to enable a cold engine to start, even at moderate temperatures, the use of neat alcohols for auto-

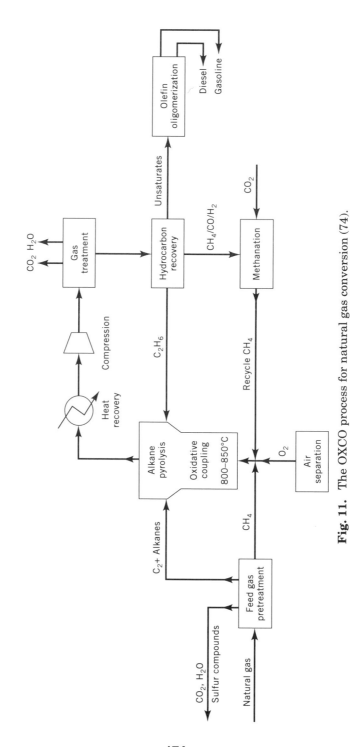

Fig. 11. The OXCO process for natural gas conversion (74).

Table 6. Fuel Oxygenates Properties[a]

Oxygenate	Blending octane, 1/2(RON + MON)[b]	Heat of combustion, MJ/L[c]	Specific gravity	Boiling point, °C
methanol	101	18.0	0.79	64.6
ethanol	101	21.3	0.79	78.5
2-propanol	106	26.4	0.79	82.4
2-butanol	99	28.3	0.80	99.5
tert-butyl alcohol	100	28.1	0.80	82.6
MTBE[d]	108	30.2	0.75	55.4
ETBE[e]	111	32.5	0.74	72.8
TAME[f]	102	31.2	0.77	86.3
gasoline	87	34.8	0.74	

[a]Ref. 5.
[b]RON = research octane number; MON = motor octane number.
[c]To convert MJ/L to Btu/gal, multiply by 3589.
[d]MTBE = methyl tert-butyl ether.
[e]ETBE = ethyl tert-butyl ether.
[f]TAME = tert-amyl methyl ether.

motive motor fuel is problematic (see ALCOHOL FUELS). Manufacturers have, however, reported that alcohol-powered cars, after being started and warmed up, can have the same or better driveability as gasoline cars (75–77). As for gasoline vehicles, port fuel injector fouling has occurred in some methanol vehicles and has affected driveability and emissions. Other problems related to high alcohol content gasoline in conventional engines include vapor lock and corrosion. Flexible-fuel vehicles (FFV), which can operate on either neat methanol or gasoline, or mixtures thereof, are being evaluated.

Gasoline blends containing oxygenates change the emissions characteristics of a motor vehicle designed for gasoline. Oxygenates and oxygenate-blends approved for use by the U.S. government are expected to have desirable emissions features as automotive fuels, and governmental environmental mandates and regulations have necessitated increased examination and implementation of oxygenates as fuels. As of this writing, however, no process can produce alcohols or ethers at equivalent or lower cost per volume than gasolines derived from natural petroleum.

Methanol. Methanol (qv) production in the 1990s is dominated by reaction of synthesis gas produced from natural gas. The economics of producing methanol as fuel are highly variable and site-specific. The natural gas feedstock has a broad range of values depending on location. Delivered costs for methanol are probably double that for gasoline from petroleum. Impurities, including water, vary according to the synthesis process employed. The term methyl fuel describes certain methanol products that may also contain significant quantities of water and higher alcohols as well as other oxygenated compounds. Water removal from such mixtures by distillative processes is generally complicated because of the formation of azeotropes. Methanol is used to produce MTBE, another oxygenate fuel.

Methanol as a fuel has been proposed in various ratios with gasoline. In gasoline formulations having relatively low methanol content, eg, M3 (3% methanol) and M15 (15% methanol), solubilizers are used and stringently dry conditions must be maintained. High methanol content fuels, M85 (85% methanol) and M100 (neat methanol), have special engine requirements. The use of high methanol content fuels is limited by methanol cost plus poor compatibility with the existing gasoline infrastructure.

Methanol is more soluble in aromatic than paraffinic hydrocarbons. Thus varying gasoline compositions can affect fuel blends. At room temperature, the solubility of methanol in gasoline is very limited in the presence of water. Generally, cosolvents are added to methanol–gasoline blends to enhance water tolerance. Methanol is practically insoluble in diesel fuel.

Concerns about using methanol–gasoline blend fuels include problems with vapor lock, cold start, and warmup. Oxygenate–gasoline blends, and in particular those containing methanol, have unusual volatility characteristics and cannot be accurately characterized using test methods developed for gasoline. Vapor pressure is an important volatility parameter that is adversely affected by the addition of methanol. Gasoline blends containing oxygenates form nonideal solutions with varying characteristics. In general, methanol–gasoline blend fuels exhibit increases in Reid vapor pressure (RVP) over that of gasoline itself. This effect contributes to vapor lock and evaporative emissions. The data available for assessing methanol's impact on exhaust emissions, and consequently on air quality, are limited. Formaldehyde (qv) has been reported in the exhausts of cars fueled with straight methanol or methanol–gasoline blends (see also EXHAUST CONTROL, AUTOMOTIVE).

The use of methanol as a motor fuel has been discussed since the 1920s. Straight methanol has long been a preferred fuel for racing engines because of the much higher compression ratios at which methanol may be combusted relative to gasolines. This is translatable for racing purposes at equivalent power outputs to engines of considerably reduced weight. However, fuel consumptions are roughly three times that of gasoline on a km/L (mi/gal) basis, and extremely high emissions of unburned fuel and carbon monoxide can result (78).

In Germany in the early 1950s, a 50:50 mixture of methanol and 2-propanol was blended with gasoline, first at a level of 7.5% and later at 1.5% (79). Complaints about stalling, power loss, and phase separation caused the ratio to be changed to 60:40::methanol:2-propanol but this apparently aggravated the problems. The practice was discontinued in 1970 when a tax was placed on alcohol.

In the United States, the Clean Air Act of 1970 imposed limitations on composition of new fuels, and as such methanol-containing fuels were required to obtain Environmental Protection Agency (EPA) waivers. Upon enactment of the Clean Air Act Amendments of 1977, EPA set for waiver unleaded fuels containing 2 wt % maximum oxygenates excluding methanol (0.3 vol % maximum). Questions regarding methanol's influence on emissions, water separation, and fuel system components were raised (80).

In 1979 Sun Oil Co. was granted a waiver for a gasoline blend containing 2.75 vol % methanol and an equal volume of *tert*-butyl alcohol (TBA) (2 wt % total oxygen). Cosolvents such as TBA were shown to reduce adverse effects of methanol on volatility and water tolerance. ARCO obtained EPA waiver in 1981 for a 3.5 wt % oxygen fuel blend containing Oxinol, also comprising equal parts of methanol and TBA. In 1985 a waiver was granted to Du Pont, Inc. for a gasoline blend containing 5 vol % maximum methanol with at least 2.5 vol % higher alcohol cosolvents. The waiver incorporated a water tolerance or phase separation requirement (81).

The most extensive worldwide program on methanol blend gasoline was in Italy where from 1982 to 1987 a 1.9×10^4 m^3/yr (5×10^6 gal/yr) plant produced a mixture containing 69% methanol. The balance contained higher alcohols. This mixture was blended into gasoline at the 4.3% level and marketed successfully as a premium gasoline known as Super E (82).

Methanol, a clean burning fuel relative to conventional industrial fuels other than natural gas, can be used advantageously in stationary turbines and boilers because of its low flame luminosity and combustion temperature. Low NO$_x$ emissions and virtually no sulfur or particulate emissions have been observed (83). Methanol is also considered for dual fuel (methanol plus oil or natural gas) combustion power boilers (84) as well as to fuel gas turbines in combined methanol/electric power production plants using coal gasification (85) (see POWER GENERATION).

Owing to its properties, methanol is not recommended for aircraft or marine fuel uses. Methanol cannot be used in conventional diesel-powered vehicles without modifications to the fuel system and engine. Simple methanol–diesel blends are not possible because of insolubility. Heavy-duty diesel engines have been adapted to use neat methanol by many U.S. manufacturers, and several are being used in field demonstrations (82) (see ALCOHOL FUELS).

Ethanol. Ethanol (qv) is produced both from ethylene (qv) derived from the cracking of petroleum fractions and by the fermentation of sugars derived from grains or other biomass (see SUGAR). Many of its relevant properties are similar to those of methanol. Although ethanol may be a more desirable fuel or fuel component than methanol, its significantly higher cost (volume basis) may outweigh these advantages. Broad implementation of ethanol-containing fuels would require government action, eg, in the form of subsidies to farmers and fuel waivers.

The term gasohol has come into wide usage to identify, generally, a blend of gasoline and ethanol, with the latter derived from grain. The term may also be applied to blends of methanol or other alcohols in gasolines or other hydrocarbons, without regard to sources of components.

Brazil's Alcohol Program. In Brazil, the enactment of legislation in 1931 made ethanol addition to gasoline compulsory at a level of 5% (86). Excess molasses and sugar were converted to alcohol in distilleries attached to sugar mills as a means to stabilize sugar prices. Production of fuel ethanol in the 1990s is mostly from biomass.

Starting in the city of Sao Paulo in 1977, and extending to the entire state of Sao Paulo in 1978, a gasohol incorporating 20% ethanol was mandated. Brazil's National Alcohol Program (Proalcool) set an initial goal of providing the 20% fuel mixture nationwide by 1980–1981 and a system of special tax, warranty, and price considerations were enacted to advance the aims of Proalcool.

For a considerable period, >90% of the new cars in Brazil operated on E96 fuel, or a mixture of 96% ethanol and 4% water (82). The engines have high compression ratios (ca 12:1) to utilize the high knock resistance of ethanol and deliver optimum fuel economy. In 1989 more than one-third of Brazil's 10 million automobiles operated on 96% ethanol/4% water fuel. The remainder ran on gasoline blends containing up to 20% ethanol (5).

Gasohol in the United States.   Over 90% of the fuel ethanol in the United States is produced from corn. Typically, 0.035 m³ (1 bushel) of corn yields 9.5 L (2.5 gal) of ethanol. Ethanol is produced by either dry or wet milling (87). Selection of the process depends on market demand for the by-products of the two processes. More than two-thirds of the ethanol in the United States is produced by wet milling. Depending on the process used, the full cost of ethanol after by-product credits has been estimated to be between $0.25–0.53/L ($1–2/gal) for new plants (88). Feedstock costs are a significant factor in the production of fuel ethanol. A change in corn price of $0.29/m³ ($1.00/bushel) affects the costs of ethanol by $0.08/L ($0.30/gal).

Ethanol can also be produced from cellulose (qv) or biomass such as wood (qv), corn stover, and municipal solid wastes (see FUELS FROM BIOMASS; FUELS FROM WASTE). Each of these resources has inherent technical or economic problems. The Tennessee Valley Authority (TVA) is operating a 2 t/d pilot plant on converting cellulose to ethanol.

After the oil embargo in 1973, gasohol use was stimulated by tax incentives. An application for EPA waiver of gasohol fuels (up to 10 vol % ethanol) was granted in 1979. From 1981 to 1983 the California Energy Commission field tested alcohol-powered cars equipped with a gasoline-assist starting system, ie, having an onboard auxiliary supply of volatile fuel for cold start tests. In 1989 about 8% of U.S. gasoline contained 10% ethanol plus a corrosion inhibitor (82). As of this writing, government waivers of RVP standards for gasohol fuels are being considered (89).

Methyl *t*-Butyl Ether.   MTBE is produced by reaction of isobutene and methanol on acid ion-exchange resins. The supply of isobutene, obtained from hydrocarbon cracking units or by dehydration of *tert*-butyl alcohol, is limited relative to that of methanol. The cost to produce MTBE from by-product isobutene has been estimated to be between $0.13 to $0.16/L ($0.50–0.60/gal) (90). Direct production of isobutene by dehydrogenation of isobutane or isomerization of mixed butenes are expensive processes that have seen less commercial use in the United States.

More than 95% of MTBE produced worldwide is used to blend with gasoline. In 1987 U.S. production of MTBE exceeded 3.8×10^6 m³/yr (1×10^9 gal/yr) (82). The worldwide capacity for MTBE is increasing, especially in the United States and Europe, and has been projected to exceed production for years to come.

MTBE's gain in prominence as a fuel-blend component is a result of inherent technical advantages over other oxygenates, especially the lower alcohols. MTBE has a high blending octane number (Table 6) although this number varies somewhat with gasoline composition. The low vapor pressure relative to the lower alcohols results in no increase in RVP for MTBE-gasoline blends and consequently better evaporative emission and vapor lock characteristics. No phase separation occurs in blends with other fuels. MTBE, in blends of <20 vol % with gasoline, does not deleteriously affect other fuel or driving characteristics such as cold start, fuel consumption, and engine materials compatibility.

MTBE has been used in motor fuels in Europe since the early 1970s and is undergoing rapid growth, particularly in the United States. MTBE-blended gasoline containing up to 11 vol % MTBE received EPA waiver in 1981. Later legislation increased the MTBE waiver up to 15 vol %. In 1987–1988 Colorado began mandating use of winter oxygenate-based fuels in the Denver region. About 90% of the fuel in this period used a gasoline blend containing 8 vol % MTBE and in 1988–1989 the fuel was required to contain at least 2% oxygen (11 vol % MTBE). Based on the success of this program and EPA assessments that CO reductions of 10–20% over the next decade were possible with oxygenate-blend fuels, numerous state governments enacted legislation requiring the use of these fuels in winter and in cities having high ozone (smog) concentrations. The Clean Air Act Amendments of 1990 have mandated the use of reformulated gasolines, especially in serious ozone problem areas, by 1995.

The effectiveness of MTBE, however, is under discussion (91). Based on Denver, Colorado vehicle emissions data from 1981 to 1991 and theoretical models, Colorado scientists have claimed that the use of MTBE-blended fuels had no statistically significant effect on atmospheric CO levels, but increased pollutants such as formaldehyde. A drop in CO levels in Denver during this time period was attributed to fleet turnover of older, more polluting cars being replaced by newer cars having cleaner burning engines. In addition, health problems associated with direct exposure to MTBE in Fairbanks, Alaska has resulted in EPA exemption of the oxygenated fuel requirement in that area (91).

Direct Liquefaction of Coal

Direct liquefaction, the production of liquids from feed coal in a single processing scheme without a synthesis gas intermediate step, includes two routes for the upgrading of coal: hydrogenation and pyrolysis. In hydrogenation, the conversion of coal to liquids having higher hydrogen-to-carbon ratio involves the addition of hydrogen. Generally, the additional hydrogen required is added either from molecular hydrogen or from a hydrogen-donor solvent such as tetralin. Processes classified under pyrolysis are those which produce liquids by removal of carbon. This occurs when coal is thermally processed under inert or reducing atmospheric conditions. The use of hydrogen in a pyrolytic process to increase yields of distillate products is known as hydropyrolysis. Coal carbonization to produce metallurgical coke involves much the same chemistry as pyrolysis.

Coal and Coal-Tar Hydrogenation. If paraffinic and olefinic liquids are extracted from solid fuel substances, the hydrogen content of the residual material is reduced even further, and the residues become more refractory. The yields of liquids so derivable are generally low, even when a significant fraction of the hydrogen is extractable. Thus production of fuel liquids from nonliquid fuel substances such as coal and coal tars may be enhanced only by the introduction of additional hydrogen in a synthesis process. The principal differences in the processes are from the modes in which hydrogen is introduced and the catalysts used.

Hydrogenation of coal and other carbonaceous matter using high pressure hydrogen has been patented (92), and subsequently the Nobel Prize in chemistry was won for this accomplishment. By 1922, a 1 t/d plant was operating and using hydrogen at 10 MPa (100 atm) and 400°C to treat brown coal tar to give a liquid that comprised 25 wt % gasoline boiling at 75–210°C and 40 wt % middle oil, 210–300°C (see LIGNITE AND BROWN COAL). The pitch residue had a specific gravity of 1.04, and a solidification point of 15°C. The degree of liquefaction was shown to increase with decrease in oil rank (93). Liquid products were of low quality, being high in oxygen, nitrogen, and sulfur content, owing to low hydrogenation rates and polymerization of primary products (94).

In 1935 an ICI coal hydrogenation plant at Billingham, U.K., produced ca 136,000 t/yr motor fuel from bituminous coal and coal tar. By 1936, 272,000 t/yr of motor fuel were produced by improved hydrogenation of brown coal and coal tar at a facility constructed at Leuna and some 363,000 t/yr was being produced in three other German plants (95). Two years later the total German output from these facilities was ca 1.4×10^6 t/yr (96). The number of coal hydrogenation plants in Germany increased during World War II to 12, with total capacity of about 4×10^6 t/yr (100,000 bbl/d) of aviation and motor gasolines.

Experimental plants for hydrogenating coal or coal tar were operated in Japan, France, Canada, and in the United States before or during World War II. Much of that technology has remained proprietary. In general, coal-in-oil slurries containing iodine or stannous oxalate catalyst were subjected to liquid-phase hydrogenation at pressures of 25–70 MPa (250–700 atm). Liquids produced were fractionated, and the middle oils were then subjected to vapor-phase hydrogenation over molybdenum-, cobalt-, or tungsten sulfide-on-alumina catalysts (97). About 1 t of crude motor fuel was recovered from 4.5 t of coal, from which all necessary hydrogen and power requirements for the production were also obtained.

Developments in the United States.   A large number of proprietary coal hydrogenation process variants have been proposed. Much of the technology originally directed to the catalytic hydrogenation of coals and coal tars in Germany has been applied to the hydrorefining of petroleum fractions, but U.S. commercial interest in coal hydrogenation was offset by the relative abundance of domestic petroleum up to World War II.

The huge demand for liquid fuels during World War II prompted the passage of the Synthetic Liquid Fuels Act of 1944. There were various programs relating to demonstration plants to produce liquid fuels from coal, oil shale, and other substances, including agricultural and forestry products. The Bureau of Mines had begun work on coal liquefaction in 1936, at which time a 45 kg/d experimental coal hydrogenation unit was constructed (98). The expanded program, after 1944, culminated in the construction and operation of a 45 t/d coal hydrogenation demonstration plant at Louisiana, Missouri, in 1949 (99), where a variety of problems and processing variations were investigated (100,101). Cost studies (102) showed that production of gasoline from coal hydrogenation could not compete with using natural petroleum as a gasoline source. The demonstration plant operations were terminated in 1953.

Work on coal hydrogenation continued by the Bureau of Mines on a laboratory scale (103–105). In one of these variants (106) coal-oil pastes admixed with catalyst in tubular reactors were hydrogenated at high pressure and low residence times to give improved yields of liquid products. The original thrust of the work was to hydrodesulfurize coal economically to produce environmentally acceptable boiler fuel (107). In the mid-1970s, a process sponsored by the Bureau of Mines named Synthoil (108) was developed, but the efforts were terminated by 1978 owing to limited catalyst lifetimes.

H-Coal Process. The H-coal process (Hydrocarbon Research, Inc., HRI, subsidiary of Dynalectron Corp.), for the conversion of coal to liquid products (109), is an application of HRI's ebullated-bed technology for the conversion of heavy oil residues into lighter fractions. Coal is dried, pulverized, and slurried with coal-derived oil (110). The coal-oil slurry is charged continuously with hydrogen to a reactor of unique design (111) containing a bed of ebullated catalyst, where the coal is hydrogenated and converted to liquid and gaseous products. The liquid product is a synthetic crude oil that can be converted to gasoline or heating oil by conventional refining processes. Alternatively, under milder operating conditions, a clean fuel gas and low sulfur fuel oils may be produced. The relative yields of these products depend on the desired sulfur level in the heavy fuel oil. In general, reaction products are separated by fractionation and absorption (qv). Unreacted coal may be fed into a fluid coker that produces gas, gas oil, and dry char. The coker gas oil, along with gas oils separated from the main reactor effluent, may be subjected to hydrocracking for conversion to lighter products.

In 1976, Ashland Oil (Ashland Synthetic Fuels, Inc.) was awarded the prime contract to construct a 540 t/d H-coal pilot plant adjacent to its refinery at Catlettsburg, Kentucky, by an industry–government underwriting consortium. Construction was completed in 1980 (112). The pilot-plant operation ended in early 1983.

Properties. The properties of naphtha, gas oil, and H-oil products from an H-coal operation are given in Table 7. These analyses are for liquids produced from the syncrude operating mode. Whereas these liquids are very low in sulfur compared with typical petroleum fractions, they are high in oxygen and nitrogen levels. No residual oil products (bp >540°C) are formed.

Solvent-Refined Coal Process. In the 1920s the anthracene oil fraction recovered from pyrolysis, or coking, of coal was utilized to extract 35–40% of bituminous coals at low pressures for the purpose of manufacturing low cost newspaper inks (113). Tetralin was found to have higher solvent power for coals, and the I. G. Farben Pott-Broche process (114) was developed, wherein a mixture of

Table 7. Properties of Syncrude from H-Coal Process[a]

Property	Boiling range			Total
	Initial to 190°C	190–343°C	343–524°C	
specific gravity, (°API)[b]	0.767 (53.0)	0.915 (23.2)	1.05 (3.5)	0.863 (32.4)
vol % of total	40.0	54.2	5.8	100.0
analysis, wt %				
carbon	84.5	88.8	89.4	87.3
hydrogen	13.6	11.0	10.2	11.9
oxygen	1.7			0.6
nitrogen	0.1	0.1	0.1	0.1
sulfur	0.1	0.1	0.3	0.1
Total	*100*	*100*	*100*	*100*

[a]Ref. 111.

[b]°API $= \dfrac{141.5}{\rho} - 131.5$ where ρ is specific gravity.

cresol and tetralin was used to dissolve ca 75% of brown coals at 13.8 MPa (2000 psi) and 427°C. The extract was filtered, and the filtrate vacuum distilled. The overhead was distilled a second time at atmospheric pressure to separate solvent, which was recycled to extraction, and a heavier liquid, which was sent to hydrogenation. The bottoms product from vacuum distillation, or solvent-extracted coal, was carbonized to produce electrode carbon. Filter cake from the filters was coked in rotary kilns for tar and oil recovery. A variety of liquid products were obtained from the solvent extraction-hydrogenation system (113). A similar process was employed in Japan during World War II to produce electrode coke, asphalt (qv), and carbonized fuel briquettes (115).

In the United States there was little interest in solvent processing of coals. A method to reduce the sulfur content of coal extracts by heating with sodium hydroxide and zinc oxide was, however, patented in 1940 (116). In the 1960s the technical feasibility of a coal deashing process was studied (117), and a pilot plant able to process ca 45 t/d was completed in late 1974 (118).

A flow diagram of the solvent-refined coal or SRC process is shown in Figure 12. Coal is pulverized and mixed with a solvent to form a slurry containing 25–35 wt % coal. The slurry is pressurized to ca 7 MPa (1000 psig), mixed with hydrogen, and heated to ca 425°C. The solution reactions are completed in ca 20 min and the reaction product flashed to separate gases. The liquid is filtered to remove the mineral residue (ash and undissolved coal) and fractionated to recover the solvent, which is recycled.

The liquid remaining after the solvent has been recovered is a heavy residual fuel called solvent-refined coal, containing less than 0.8 wt % sulfur and 0.1 wt % ash. It melts at ca 177°C and has a heating value of ca 37 MJ/kg (16,000 Btu/lb), regardless of the quality of the coal feedstock. The activity of the solvent is apparently more important than the action of gaseous hydrogen in this type of uncatalyzed hydrogenation. Research has been directed to the use of petroleum-derived aromatic oils as start-up solvents (118).

In the early 1970s production of low sulfur, ashless (solid) boiler fuel was the preferred commercial application (119). This basic process (SRC-I) yielded small amounts of liquid oil products with additional processing. Liquid output was significantly increased by the coal-oil-gas (COG) refinery concept (120–122) which incorporated high degrees of hydroconversion and hydrotreating. A SCR-II process has been developed, in which hydrocracking occurs in the solution (hydrogenation) vessels (123). A low viscosity fuel oil is the primary distillate product in this case, although naphtha and LPG are also recovered.

Two pilot plants have been built and operated to demonstrate the feasibility of the SRC process. These included a 6 t/d plant at Wilsonville, Alabama (vide infra) and a 50 t/d plant at Ft. Lewis, Washington which was operated from 1974 to 1981.

In an effort to obtain higher value products from SRC processes, a hydrocracking step was added to convert resid to distillate liquids. The addition of a hydrocracker to the SRC-I process was called nonintegrated two-stage liquefaction (NTSL). The NTSL process was essentially two separate processes in series: coal liquefaction and resid upgrading. NTSL processes were inefficient owing to the inherent limitations of the SRC-I process and the high hydrocracker severities required.

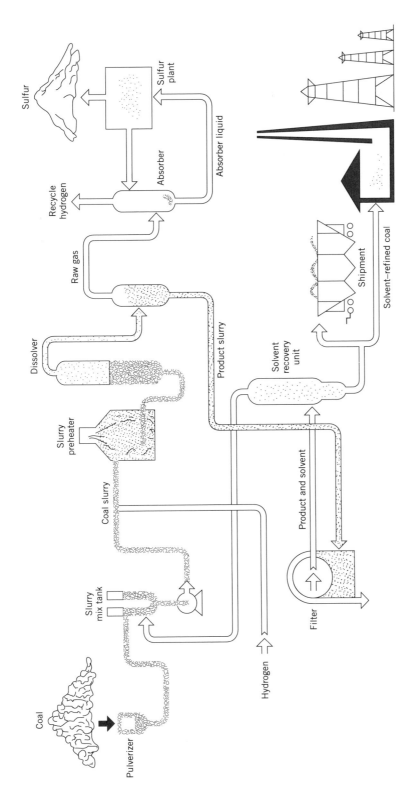

Fig. 12. Solvent-refined coal process (119).

183

Properties.   The properties of the liquid fuel oil produced by the SRC-II process are influenced by the particular processing configuration. However, in general, it is an oil boiling between 177 and 487°C, having a specific gravity of 0.99–1.00, and a viscosity at 38°C of 40 SUs (123). Pipeline gas, propane and butane (LPG), and naphtha are also recovered from an SRC-II complex.

Exxon Donor Solvent-Coal Liquefaction Process.   The EDS process from Exxon is a hydrogenation process using a donor solvent for the direct conversion of a broad range of coals to liquid hydrocarbons (124). In the process sequence, shown in Figure 13, the feed coal is crushed, dried, and slurried with hydrogenated recycle solvent (the donor solvent) and fed to the reactor with hydrogen. The reactor is an upward plug-flow design operating at 430–480°C and at ca 14 MPa (2000 psi) total pressure.

The reactor effluent is separated by conventional distillation into recycle solvent, light gases, C_4 to 537°C bp distillate, and a heavy vacuum bottoms stream containing unconverted coal and ash. The recycle solvent is hydrogenated in a separate reactor and sent back to the liquefaction reactor.

The heavy vacuum bottoms stream is fed to a Flexicoking unit. This is a commercial (125,126) petroleum process that employs circulating fluidized beds at low (0.3 MPa (50 psi)) pressures and intermediate temperatures, ie, 480–650°C in the coker and 815–980°C in the gasifier, to produce high yields of liquids or gases from organic material present in the feed. Residual carbon is rejected with the ash from the gasifier fluidized bed. The total liquid product is a blend of streams from liquefaction and the Flexicoker.

The EDS process was developed starting from 1976 in a 10-year joint undertaking between DOE and private industry (127). Under the direction of Exxon

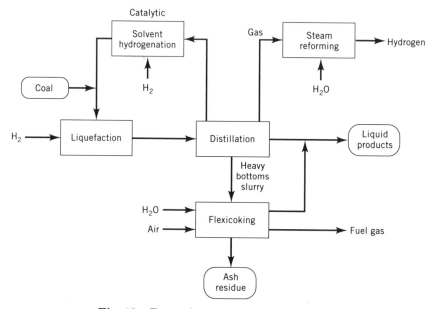

Fig. 13.   Exxon donor solvent process (124).

Co. USA, a 250 t/d pilot plant was operated at Baytown, Texas. Operation of this unit began in 1980 and was completed by late 1982.

Properties. Pilot-unit data indicate the EDS process may accommodate a wide variety of coal types. Overall process yields from bituminous, subbituminous, and lignite coals, which include liquids from both liquefaction and Flexicoking, are shown in Figure 14. The liquids produced have higher nitrogen contents than are found in similar petroleum fractions. Sulfur contents reflect the sulfur levels of the starting coals: ca 4.0 wt % sulfur in the dry bituminous coal; 0.5 wt % in the subbituminous; and 1.2 wt % sulfur in the dry lignite.

Table 8 shows that the naphthas produced by the EDS process have higher concentrations of cycloparaffins and phenols than do petroleum-derived naphthas, whereas the normal paraffins are present in much lower concentrations. The sulfur and nitrogen concentrations in coal naphthas are high compared to those in petroleum naphthas.

Gas oil fractions (204–565°C) from coal liquefaction show even greater differences in composition compared to petroleum-derived counterparts than do the naphtha fractions (128). The coal-gas oils consist mostly of aromatics (60%), polar heteroaromatics (25%), asphaltenes (8–15%), and saturated compounds (<10%).

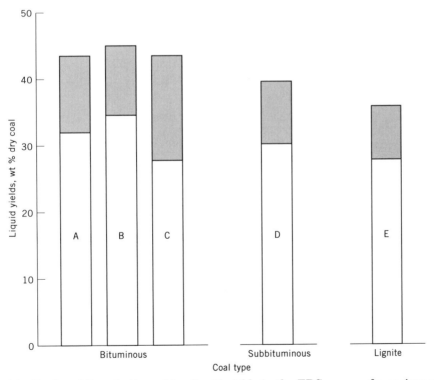

Fig. 14. Preferred liquefaction-coking liquid yields in the EDS process for various coals where ▨ represents Flexicoking liquids and ▢, liquefaction liquids (124). A, Ireland (West Virginia); B, Monterey (Illinois); C, Burning Star (Illinois); D, Wyodak (Wyoming); and E, Big Brown (Texas).

Table 8. Composition of Naphthas[a] from Various Sources[b]

Component, wt %	EDS coal liquefaction		Petroleum naphthas	
	Illinois coal	Wyodak coal	Cycloparaffinic[c]	Paraffinic[d]
saturated compounds	69.9	60.8	77.9	81.3
paraffins	13.4	19.4	36.0	58.2
cycloparaffins	56.5	41.4	41.9	23.1
olefins[e]	0.5	3.7		
aromatics	17.0	28.6	22.1	18.7
benzenes	11.7	25.1	20.8	17.2
indans, tetralins	5.1	3.5	1.0	1.2
indenes	0.2	0.0		
naphthalenes	0.01	0.0	0.3	0.3
phenols	12.6	6.9	traces	traces
Total	*100.0*	*100.0*	*100.0*	*100.0*
sulfur	0.57	0.10	0.035	0.049
nitrogen	0.15	0.18	0.0001	0.0001
oxygen	1.60	2.49		

[a]C_5 to 204°C bp.
[b]Ref. 128.
[c]Prudhoe Bay.
[d]Arab Light.
[e]Values are approximate.

Petroleum-gas oils, on the other hand, contain more than 50% saturated compounds, less than 5% polar heteroaromatics, and no asphaltenes. Furthermore, the aromatics of petroleum-gas oils have longer side chains.

Coal Liquefaction at Wilsonville. Starting in 1974 the Advanced Coal Liquefaction R&D Facility at Wilsonville, Alabama operated a 6 t/d pilot plant and studied various coal liquefaction processing schemes. The facility, cosponsored by the DOE, the Electric Power Research Institute (EPRI) and Amoco Oil Co, was shut down in early 1992.

Initial operation at the Wilsonville pilot plant was in SRC-I mode and later evolved into a two-stage process (129) by operation in NTSL mode. NTSL limitations described previously combined with high hydrogen consumptions resulted in subsequent focus on a staged integrated approach, which was to be the basis for all further studies at Wilsonville.

The integrated two-stage process (ITSL) combined short contact time liquefaction in one reactor with ebullated-bed hydrocracking in a second stage (130). The short contact time conditions permitted better hydrogen transfer from the solvent rather than from the gas phase. The hydrocracking step operated at lower severity resulting in lowered gas make and improved hydrogen efficiency. Recycle solvent was generated from the hydrocracked distillates and coupled the two reaction stages. Results of ITSL processing of Illinois No. 6 coal at Wilsonville are given in Table 9. Distillate yields and coal throughput for ITSL were higher than those obtained by NTSL.

Further developments of the ITSL process resulted in incremental gains in distillate yields (131). Reconfigured integrated two-stage liquefaction (RITSL) in-

Table 9. Wilsonville Plant Operating Conditions and Yields for ITSL and CTSL Modes[a]

Parameter	Mode of operation[b]	
	ITSL	CTSL
Operating conditions		
run number	7242BC; 243JK/244B	253A
catalyst	Shell 32M	Shell 317
First stage[c]		
average reactor temperature, °C	460; 432	432
space velocity	690; 450[d]	4.8[e]
pressure, MPa[f]	17; 10–17	17.9
Second stage		
average reactor temperature, °C	382	404
space velocity, feed/catalyst[e]	1.0	4.3
catalyst age, resid/catalyst	278–441; 380–850	100–250
Yields[g]		
C_1–C_3 gas	4; 6	6
C_4+ distillate	54; 59	70
resid	8; 6	−1
hydrogen consumption	4.9; 5.1	6.8
Other		
hydrogen efficiency, C_4+ distillate/H_2 consumed	11; 11.5	10.3
distillate selectivity, C_1–C_3/C_4+ distillate	0.07; 0.10	0.08
energy content of feed coal reject to ash concentrate, %	24; 20–23	20

[a] Feed is Illinois No. 6 coal.
[b] CTSL = catalytic two-stage liquefaction; ITSL = integrated two-stage liquefaction.
[c] First stage is thermal for ITSL.
[d] Value given is coal space velocity at temp >371°C in kg/m^3.
[e] Value given is in h^{-1}.
[f] To convert MPa to psia, multiply by 145.
[g] Wt % on a moisture- and ash-free (MAF) coal basis.

volved placing the solvent deasher after the hydrocracker thus producing a recycle solvent consisting of deashed resid and distillate. This resulted in reduction of feed to the deasher and reduced organic rejection. Close coupled integrated two-stage liquefaction (CC-ITSL) linked the two reactors and removed several operations between the two stages. A deleterious effect of these two processing modes was increased hydrogen consumption over ITSL.

From 1985 to 1992, development activity at Wilsonville was on catalytic two-stage liquefaction (CTSL). CTSL, initiated by HRI (132), consists of catalytic processing in two ebullated-bed reactors which lower reaction temperatures and increase distillate yields, up to 78% yield. CTSL results from Wilsonville for Illinois No. 6 coal are also given in Table 9. Distillate yields were shown to be significantly higher for CTSL over ITSL; however, hydrogen consumption was somewhat increased.

Properties. CTSL distillates have qualities comparable to or better than No. 2 fuel oil and have good hydrogen content and low heteroatom contents. Distillates having a higher boiling point distribution from Wilsonville CTSL operation (131) showed 26.8°API gravity with heteroatom levels of 0.11 wt % sulfur, <1 wt % oxygen, and 0.16 wt % nitrogen.

Coal Pyrolysis.   Pyrolysis is the destructive distillation of coal in the absence of oxygen typically at temperatures between 400 and 500°C (133). As the temperature of carbonaceous matter is increased, decomposition ultimately occurs. Melting and dehydration may also occur. Coals exhibit more or less definite decomposition temperatures, as indicated by melting and rapid evolution of volatile components, including potential fuel liquids, during destructive distillation (134). Table 10 summarizes an extensive survey of North American coals subjected to laboratory pyrolysis. The yields of light oils so derived average no more than ca 8.3 L/t (2 gal/short ton), and tar yields of ca 125 L/t (30 gal/short ton) are optimum for high volatile bituminous coals (135).

Coal pyrolysis has been studied at both reduced and elevated pressures (136), and in the presence of a variety of agents and atmospheres (137). Although important to the study of coal structure and reactions, coal pyrolysis, as a means to generate liquids, has proved to have limited commercial value.

COED Process.   Sponsored by the Office of Coal Research of the U.S. Department of the Interior, the COED process was developed by FMC Corp. as Project Char-Oil-Energy Development (COED) through 1975 (138–140). Bench-scale experiments led the way to construction in 1965 of a process development unit employing multistage, fluidized-bed pyrolysis to process 45 kg/h (141). Correlated studies included hydrotreating of COED oil (142), high temperature hydrodesulfurization of COED char, and investigations of char-oil and char-water slurry pipelining economics (143). A pilot plant capable of processing up to 33 t/d and hydrotreating 4.7 m³/d (30 bbl/d) was started up in 1970 (144), and was operated successfully for a number of years (145).

The COED concept (139), designed to recover liquid, gaseous, and solid fuel components, consists of four stages. Heat is generated by the reaction of oxygen with a portion of the char in the last pyrolysis stage and is also introduced by the air combustion of gas to dry feed coal. The number of stages in the pyrolysis, and the operating temperatures in each, may be varied to accommodate high volatile bituminous and subbituminous feed coals with widely ranging caking or agglomerating properties.

Oil condensed from the released volatiles from the second stage is filtered and catalytically hydrotreated at high pressure to produce a synthetic crude oil. Medium heat-content gas produced after the removal of H_2S and CO_2 is suitable as clean fuel. The pyrolysis gas produced, however, is insufficient to provide the fuel requirement for the total plant. Residual char, 50–60% of the feed coal, has a heating value and sulfur content about the same as feed coal, and its utilization may thus largely dictate process utility.

Table 10. Average Yields and Range of Yields of Fischer Assay of Various Coals[a,b]

Rank of coal	Coke, %		Tar, L/t[c]		Light oil, L/t[c]		Gas, m³/t[d]		Water, %	
	Average	Range	Average	Range	Average	Range	Average	Range	Average	Range
semianthracite			3.2		0.14					
low volatile bituminous	89.7	85.8–93.3	39.6	29.0–58.4	4.69	3.36–7.41	59.8	54.4–66.6	3.2	1.1–6.6
medium volatile bituminous	83.3	77.4–90.4	86.9	44.6–117.8	7.68	4.92–10.58	66.0	47.3–76.2	4.1	2.8–7.0
high volatile A bituminous	75.5	68.8–81.4	142.1	105.3–187.2	10.53	6.81–15.09	67.0	57.5–80.2	6.0	3.0–9.2
high volatile B bituminous	70.4	66.0–73.2	139.4	111.8–198.3	10.03	7.13–15.82	68.3	56.4–82.3	11.1	10.2–13.1
high volatile C bituminous	67.1	65.4–68.6	124.2	85.1–178.5	8.65	5.93–12.47	61.2	53.0–70.4	15.9	12.0–19.1
high volatile C bituminous or subbituminous A	59.1		94.3	84.6–112.2	7.59	6.26–8.88	90.4		23.4	
subbituminous A					6.21	6.12–6.26				
subbituminous B	57.6	54.8–59.9	81.9	81.0–82.8	6.12	5.24–7.13	90.4	62.2–93.8	27.8	23.3–30.4
lignite	36.5		70.8	60.7–76.8	5.47	2.90–8.69	71.4		44.0	
cannel	58.8	44.1–69.0	338.1	247.0–498.2	23.28	16.84–34.13	61.5	51.0–72.1	3.7	2.0–4.8

[a]Ref. 135.
[b]As-received basis; maximum temperature, 500°C.
[c]To convert L/t to gal/short ton, divide by 4.6.
[d]To convert m³/t to ft³/short ton, multiply by 29.4.

Properties.   The properties of char products from two possible coal feeds, a low sulfur Western coal, and a high sulfur Midwestern coal, are shown in Table 11. The char derived from the low sulfur Western coal may be directly suitable as plant fuel, with only minor addition of clean process gas to stabilize its combustion. Flue gas desulfurization may not be required. Flue gas from the combustion of the char derived from the high sulfur Illinois coal, however, requires desulfurization before it may be discharged into the atmosphere.

Typical COED syncrude properties are shown in Table 12. The properties of the oil products depend heavily on the severity of hydroprocessing. The degree of severity also markedly affects costs associated with hydrogen production and compression. Syncrudes derived from Western coals have much higher paraffin and lower aromatic content than those produced from Illinois coal. In general, properties of COED products have been found compatible with expected industrial requirements.

Occidental Petroleum Coal Conversion Process.   Garrett R&D Co. (now the Occidental Research Co.) developed the Oxy Coal Conversion process based on mathematical simulation for heating coal particles in the pyrolysis unit. It was estimated that coal particles of 100-mm diameter could be heated throughout their volumes to decomposition temperature (450–540°C) within 0.1 s. A large pilot facility was constructed at LaVerne, California, in 1971. This unit was reported to operate successfully at feed rates up to 136 kg/h (3.2 t/d).

Hot product char carries heat into the entrained bed to obtain the high heat-transfer rates required. Feed coal must be dried and pulverized. A portion of the char recovered from the reactor product stream is cooled and discharged as product. The remainder is reheated to 650–870°C in a char heater blown with air. Gases from the reactor are cooled and scrubbed free of product tar. Hydrogen sulfide is removed from the gas, and a portion is recycled to serve as the entrainment medium.

Table 11. Properties of COED Char Product[a]

Property	Utah coal	Illinois No. 6
volatile matter, wt %	6.1	2.7
fixed carbon, wt %	80.2	77.0
ash, wt %	13.7	20.3
higher heating value, MJ/kg,[b] dry	28.6	25.6
elemental analysis, wt %, dry		
carbon	81.5	73.4
hydrogen	1.3	0.8
nitrogen	1.5	1.0
sulfur	0.5	3.4
oxygen	1.5	1.0
chlorine	0.006	0.1
iron[c]	0.28	

[a]Ref. 139.
[b]To convert MJ/kg to Btu/lb, multiply by 430.
[c]Included in ash above.

Table 12. Typical COED Syncrude Properties[a]

Property	Utah A-seam	Illinois No. 6 seam
specific gravity, (° API)[b]	0.934 (20)	0.929 (22)
pour point, °C	16	−18
flash point, closed cup, °C	24	16
viscosity, at 38°C, mm^2/s (= cSt)	8	5
ash, wt %	0.01	0.01
moisture, wt %	0.1	0.1
metals, ppm	10	10
elemental analysis, wt %		
C	87.2	87.1
H	11.0	10.9
N	0.2	0.3
O	1.4	1.6
S	0.1	0.1
ASTM distillation initial bp, °C	138	88
10%	221	134
30%	277	199
50%	349	270
70%	416	316
90%	493	362
end point (95%)	510	397
hydrocarbon type analysis, liquid vol %		
paraffins	23.7	10.4
olefins	0	0
naphthenes	42.2	41.4
aromatics	34.1	48.2

[a]Ref. 139.

[b]°API $= \dfrac{141.5}{\rho} - 131.5$ where ρ is specific gravity.

Properties. A high volatile western Kentucky bituminous coal, the tar yield of which by Fischer assay was ca 16%, gave a tar yield of ca 26% at a pyrolysis temperature of 537°C (146–148). Tar yield peaked at ca 35% at 577°C and dropped off to 22% at 617°C. The char heating value is essentially equal to that of the starting coal, and the tar has a lower hydrogen content than other pyrolysis tars. The product char is not suitable for direct combustion because of its 2.6% sulfur content.

The TOSCOAL Process. The Oil Shale Corp. (TOSCO) piloted the low temperature carbonization of Wyoming subbituminous coals over a two-year period in its 23 t/d pilot plant at Rocky Falls, Colorado (149). The principal objective was the upgrading of the heating value in order to reduce transportation costs on a heating value basis. Hence, the solid char product from the process represented 50 wt % of the starting coal but had 80% of its heating value.

Furthermore, 60–100 L (14–24 gal) oil, having sulfur content below 0.4 wt %, could be recovered per metric ton coal from pyrolysis at 427–517°C. The re-

covered oil was suitable as low sulfur fuel. Figure 15 is a flow sheet of the Rocky
Flats pilot plant. Coal is fed from hoppers to a dilute-phase, fluid-bed preheater
and transported to a pyrolysis drum, where it is contacted by hot ceramic balls.
Pyrolysis drum effluent is passed over a trommel screen that permits char product
to fall through. Product char is thereafter cooled and sent to storage. The ceramic
balls are recycled and pyrolysis vapors are condensed and fractionated.

Properties.   Results for the operation using subbituminous coal from the
Wyodad mine near Gillette, Wyoming, are shown in Table 13. Char yields decreased with increasing temperature, and oil yields increased. The Fischer assay
laboratory method closely approximated the yields and product assays that were
obtained with the TOSCOAL process.

The volatiles contents of product chars decreased from ca 25–16% with temperature. Char (lower) heating values, on the other hand, increased from ca 26.75
MJ/kg (11,500 Btu/lb) to 29.5 MJ/kg (12,700 Btu/lb) with temperature. Chars in
this range of heating values are suitable for boiler fuel application and the low
sulfur content (about equal to that of the starting coal) permits direct combustion.
These char products, however, are pyrophoric and require special handling in
storage and transportation systems.

Properties of the tar oil products are given in Table 14. The oils change only
slightly with change in the retorting temperature; sulfur levels are low. The fraction boiling up to 230°C contains 65 wt % of phenols, cresols, and cresylic acids.

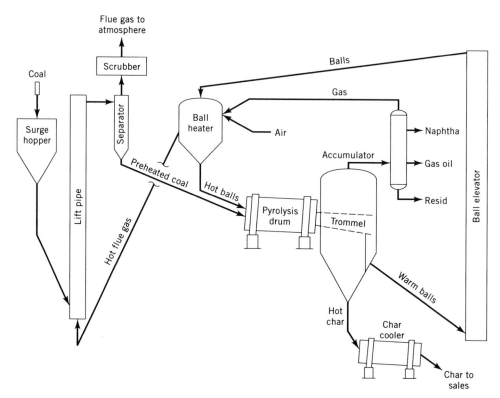

Fig. 15.   TOSCOAL process (149).

Table 13. TOSCOAL Retorting of Wyodak Coal[a,b]

Component	Retort temperature		
	427°C	482°C	521°C
char	524.5	505.8	484.4
gas, C_3 and lighter	59.5	78.4	63.0
oil, C_4 and heavier	57.0	71.5	93.1
water[c]	351.0	351.0	351.0
Total	*992.0*	*1006.7*	*991.5*
Recovery, %	*99.2*	*100.7*	*99.1*

[a]Ref. 150.
[b]Product yields, kg/t, of as-mined coal.
[c]Value assumed from Fischer assay and moisture content. The addition of steam to the process prevented accurate measurement of water produced in retorting.

TOSCO tar oils have high viscosity and may not be transported by conventional pipelines. Heating values of product gas on a dry, acid gas-free basis are in the natural gas range if butanes and heavier components are included.

Coal Carbonization. In the by-product recovery of a modern coke oven, coal tar is removed first by cooling the gases emanating, and light oil is removed last by scrubbing the gas with solvents. Other products, including ammonia, phenols, pyridine, or naphthalene, may be recovered between these operations. The constituents of coal tar, light oil, and gas usually overlap considerably, ie, the fractional condensation does not effectively separate individual components. Assuming the lowest boiling coal tar constituent to be benzene (bp 80.09°C), and the highest boiling to be naphthalene, the overlapping compositions of gas, light oil, and tar may be as shown graphically in Figure 16. Many chemical compounds have been identified (8) in these substances. Included are most of the significant constituents of petroleum-derived fuel liquids, although only a few components are present in sufficient quantity to make commercial recovery feasible.

The precise compositions of the light oil and coal tar recovered from coke-oven gas is a distinct function of the design of the recovery system, as well as of the properties of the starting coal. In general, 12.5–16.7 L/t (3–4 gal/m light oil per short ton) of coal carbonized is recovered from high temperature coke-oven operations. Light oil may contain 55–70% benzene, 12–20% toluene, and 4–7% xylene. Unrecovered light oil appearing in the effluent coal gas may comprise ca 1 vol % and contribute ca 5% of the gas's heating value. Refining of light oil consists mainly of sulfuric acid washing, followed by fractional distillation.

Large-scale recovery of light oil was commercialized in England, Germany, and the United States toward the end of the nineteenth century (151). Industrial coal-tar production dates from the earliest operation of coal-gas facilities. The principal bulk commodities derived from coal tar are wood-preserving oils, road tars, industrial pitches, and coke. Naphthalene is obtained from tar oils by crystallization, tar acids are derived by extraction of tar oils with caustic, and tar bases by extraction with sulfuric acid. Coal tars generally contain less than 1%

Table 14. TOSCOAL Oil Properties[a]

Properties	Retort temperature		
	427°C[b]	482°C	521°C
analysis, wt %			
carbon	81.4	80.7	80.9
hydrogen	9.3	9.1	8.7
oxygen	8.3	9.4	9.3
nitrogen	0.48	0.7	0.7
sulfur	0.43	0.2	0.2
chlorine	0.0	0.0	0.0
ash	0.0	0.2	0.1
Total	*99.91*	*100.3*	*99.9*
heating values			
gross, MJ/kg[c]	38.59	37.72	37.13
net, MJ/kg[c]	36.61	35.75	35.26
specific gravity, (°API)[d]			
primary oil	1.015 (7.9)	1.040 (4.5)	1.061 (1.9)
calculated, with C_4 and heavier components of gas added	0.978 (13.2)	0.985 (12.1)	1.027 (6.2)
pour point, °C	32	38	35
Conradson carbon, wt %	7.6	9.9	11.4
distillation,[e] °C			
2.5 vol %	212	216	199
20 vol %	302	288	235
50 vol %	407	413	385
viscosity, SUs			
at 82°C	122	123	128
at 90°C	63	66	69

[a]Ref. 150.
[b]Feed coal was different from that used for 482 and 521°C.
[c]To convert MJ/kg to Btu/lb, multiply by 430.
[d]°API $= \dfrac{141.5}{\rho} - 131.5$ where ρ is specific gravity.
[e]Combination of true boiling point and D1160 distillations.

benzene and toluene, and may contain up to 1% xylene. The total U.S. production of BTX from coke-oven operations is insignificant compared to petroleum product consumptions.

Other Processes

Shale Oil.   In the United States, shale oil, or oil derivable from oil shale, represents the largest potential source of liquid hydrocarbons that can be readily processed to fuel liquids similar to those derived from natural petroleum. Some countries produce liquid fuels from oil shale. There is no such industry in the

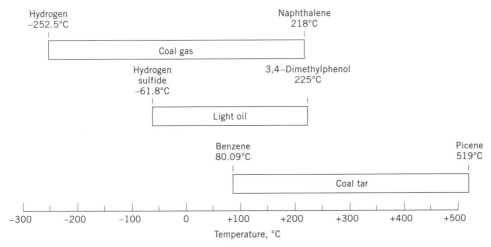

Fig. 16. Boiling ranges of carbonization products (8).

United States although more than 50 companies were producing oil from coal and shale in the United States in 1860 (152,153), and after the oil embargo of 1973 several companies reactivated shale-oil process development programs (154,155). Petroleum supply and price stability has since severely curtailed shale oil development. In addition, complex environmental issues (156) further prohibit demonstration of commercial designs.

Heavy Oil. The definitions used to distinguish among naturally occurring heavy petroleum oils, bitumens, asphalts, and tars, are subject to broad variations. More than 10% of the world's current crude oil production has an API gravity below 20°, or a specific gravity greater than 0.934_{15}^{15}. Oils having sp $gr_{15}^{15} >$ 0.904–0.934 (20–25°API) are considered heavy oils in most classifications. However, Safaniyah crude oil produced in Saudi Arabia having sp gr_{15}^{15} of 0.893 (27°API) carries the designation Arabian heavy, and in petroleum parlance is generally referred to as heavy crude oil. Yet its production method does not differ from that of Arabian medium or Arabian light crude oils.

Energy in the form of injected water or CO_2 may be supplied to increase the rate of production of light crude oils. Application of heat to the reservoirs, eg, using hot water, steam, heated CO_2, fireflood, or *in situ* combustion, however, is generally associated with the production of heavier, viscid crudes.

Heavy crude oil is widely distributed, and it is difficult to estimate reserves separate from normal crude oil reserves or from tar sands deposits. Estimates of petroleum reserves frequently include a large heavy oil component, which can only be produced at significantly higher cost than light oil.

Most heavy oil production is concentrated in California, Canada, and Venezuela. There is significant production of heavy oil in California from the Kern River field near Bakersfield and in Canada from the Cold Lake deposit in Alberta. Production generally involves steam drives, or the injection of steam into reservoirs through special wells in prescribed sequences. Oil–water mixtures are recovered, and often separated water is treated and reinjected.

Heavy oil may be upgraded through two main routes: coking and hydro-processing. Virtually all established upgrading schemes involve some variant of those two routes. The challenges in upgrading and refining are from the low hydrogen content and specific gravity and high sulfur, nitrogen, and metals content of the heavy oil.

Tar Sands. Tar sands (qv) are considered to be sedimentary rocks having natural porosity where the pore volume is occupied by viscous, petroleum-like hydrocarbons. The terms oil sands, rock asphalts, asphaltic sandstones, and malthas or malthites have all been applied to the same resource. The hydrocarbon component of tar sands is properly termed bitumen.

Distinctions between tar sands' bitumens and heavy oils are based largely on differences in viscosities. The bitumen in oil sand has a specific gravity of less than 0.986 g/mL (12°API), and thus oil sands may be regarded as a source of extremely heavy crude oil. Whereas heavy oils might be produced by the same techniques used for the lighter crude oils, the bitumens in tar sands are too viscous for these techniques. Consequently these oil-bearing stones have to be mined and specially processed to recover contained hydrocarbon.

Tar sands have been reported on every continent except Australia and Antarctica. The best known deposits are the Athabasca of Canada, where almost $60,000 \text{ km}^2$ in northeastern Alberta is underlain with an estimated $138 \times 10^9 \text{ m}^3$ $(870 \times 10^9 \text{ bbl})$ of recoverable bitumen (157). The Alberta deposits may contain up to $215 \times 10^9 \text{ m}^3$ $(1350 \times 10^9 \text{ bbl})$ of bitumen reserves. Venezuela may have the largest accumulations in the world; the Orinoco heavy-oil belt has been estimated by some (157) to contain as high as $636 \times 10^9 \text{ m}^3$ $(4000 \times 10^9 \text{ bbl})$. The Olenek reserves in the former USSR may contain ca $95 \times 10^9 \text{ m}^3$ $(600 \times 10^9 \text{ bbl})$. The United States is estimated to have deposits of about $4.5 \times 10^9 \text{ m}^3$ $(28 \times 10^9$ bbl).

The Great Canadian Oil Sands, Ltd. (GCO) (Sun Oil Co.) has been operating a plant at Fort McMurray, Alberta, Canada, since 1967. Initially, some 8050 t/d (55,000 bbl/d) of synthetic crude oil was produced from coking (158) with the project expanding to 9220 t/d (63,000 bbl/d). Since 1978, Syncrude Canada has been producing ca $22,000 \text{ m}^3/\text{d}$ (140,000 bbl) synthetic crude oil by fluid coking from their plant at Cold Lake, Alberta, Canada (159) with expansion planned for ca $35,000 \text{ m}^3/\text{d}$ (225,000 bbl/d).

Economic Aspects of Synthetic Fuels

As of this writing, processes for production of synthetic liquid fuels by upgrading natural gas, coal, or heavy oil are generally not directly competitive with crude oil upgrading (160). The key controlling factors in the economics are crude oil price and availability. Many economic analyses for synthetic liquid fuels give a crude oil price target whereupon the alternative technology becomes attractive, but these studies sometimes neglect the fact that the natural gas, coal, and heavy oil prices often track those of crude oil. In addition, conversion of a refractory gas (methane) or solid (coal) to liquids is a greater technical challenge than that of processing crude oil. Thus there are processing cost penalties which inevitably exist even after considerable technological development. Nevertheless, synthetic fuels technology is projected to play a primary role in providing liquid fuels once crude oil depletion is of concern. Economic competitiveness plays a reduced role in commercialization only when environmental legislation mandates the use of certain fuels such as oxygenates.

The commercialization potential of synthetic fuels technology relies on site-specific economic and political factors. This complex network of factors may include capital costs, crude oil price, product yields and value, government subsidies, strategic impact, alternative uses for the feed, and environmental and geographical constraints. Whereas no direct coal liquefaction process has gone to commercial stage, technologies involving indirect conversion of natural gas or coal have been commercialized. In all cases, special conditions allowed the technology to progress. In the Sasol project at South Africa, coal upgrading was possible due to factors such as no indigenous petroleum, minimal environmental standards, and cheap labor (160). The New Zealand GTG process became economically feasible owing to high oil prices, abundance of indigenous natural gas, and government commitment to energy self-sufficiency (59). Government support and long-term strategic benefits were also keys to Shell's SMDS project in Malaysia.

At 1994 crude oil prices of ca $94–125/m^3 (ca $15–20/bbl), conversion of natural gas to liquid fuels exists only in unique situations. For natural gas upgrading via New Zealand-type technology, economics by Mobil for a 1987 plant start-up on the U.S. Gulf Coast (161) indicated an investment of $895 $\times$ 10^6 was required for a 2.3 $\times$ 10^3 m^3/d (14,600 bbl/d) gasoline production unit. Thus this and other natural gas-to-fuels processes are highly capital-intensive and capital recovery remains the dominant factor even with incremental advances in conventional technology. This is especially the case using indirect upgrading of natural gas because the cost of synthesis gas manufacture may account for more than 50% of the total process capital cost (44). An analysis by Shell of the SMDS process published in 1988 showed capital expense for a 1600 m^3/d (10,000 bbl/d) plant to be $300 $\times$ 10^6 for a developed site in an industrialized country and $600 $\times$ 10^6 for a developing site in a developing country (44). Direct upgrading of natural gas to gasoline and distillate by oxidative coupling plus olefin oligomerization has been evaluated to be ca 10% costlier in capital than upgrading via indirect technologies (162).

Natural gas upgrading economics may be affected by additional factors. The increasing use of compressed natural gas (CNG) directly as fuel in vehicles provides an alternative market which affects both gas price and value (see GASOLINE AND OTHER MOTOR FUELS; GAS, NATURAL). The hostility of the remote site environment where the natural gas is located may contribute to additional costs, eg, offshore sites require platforms and submarine pipelines.

The economic feasibility of coal upgrading, and in particular direct coal liquefaction, are closely tied to crude oil price and capital expense. H-coal technology was evaluated as a base case in 1981 and the results showed economic feasibility was possible only at crude oil price of $630/m^3 ($100/bbl) or greater (160). A more recent analysis by HRI on coal/oil coprocessing technology indicated the required light crude price was $138–182/m^3 ($22–29/bbl) for economic feasibility (163). The cost of capital could add over $60/m^3 ($10/bbl) to the cost of products. A study of EDS, H-coal, ITSL, and two-stage Wilsonville systems showed capital costs for a 30,000 t/d plant processing Illinois No. 6 coal to run between $4100–$4700 $\times$

10^6 (131). Required break-even selling prices for products from these technologies ranged from $226–314/m^3 ($36–50/bbl). The two-stage system was the most economical at $229.94/m^3 ($36.56/bbl). An evaluation of coprocessing Cold Lake vacuum bottoms using Alberta subbituminous coal in a 3200 m^3/d (20,000 bbl/d) synthetic crude oil production unit indicated a selling price of $189–220/m^3 ($30–35/bbl) was necessary for the process to be competitive (164). In general, the economics of direct coal liquefaction depend more on the high cost of liquefaction rather than the cost for upgrading product coal liquids (165).

Factors which may affect the cost of coal upgrading are environmental considerations such as toxicity, hazardous waste disposal, and carcinogenic properties (131). These and other environmental problems from process streams, untreated wastewaters, and raw products would figure significantly into the cost of commercialization.

BIBLIOGRAPHY

"Fuels, Synthetic Liquid" in *ECT* 1st ed., Vol. 6, pp. 960–983, by J. H. Arnold and H. Pichler, Hydrocarbon Research, Inc.; "Carbon Monoxide–Hydrogen Reactions" in *ECT* 2nd ed., Vol. 4, pp. 446–489, by H. Pichler and A. Hector, Carl-Engler und Hans-Bunte-Institut für Mineralöl und Kohleforschung der Technischen Hochschule Karlsruhe; "Fuels, Synthetic (Liquid)" in *ECT* 3rd ed., Vol. 11, pp. 447–489, by C. D. Kalfadelis and H. Shaw, Exxon Research and Engineering Co.

1. H. Mimoun, *New J. Chem.* **11**, 4 (1987).
2. U. Preuss and M. Baerns, *Chem. Eng. Technol.* **10**, 297 (1987).
3. "Liquid Fuels From Natural Gas," *Petrole Informations*, API 34-5250, 96 (May 1987).
4. *Chem. Eng. News*, 25 (Aug. 14, 1989).
5. E. E. Ecklund and G. A. Mills, *Chemtech*, 549 (Sept. 1989).
6. L. Haar, in W. P. Earley and J. W. Weatherly, eds., *Advances in Coal Utilization Technology IV*, Institute of Gas Technology, Chicago, 1981, pp. 787–952.
7. L. Shnidman, in H. H. Lowry, ed., *Chemistry of Coal Utilization*, Vol. 2, John Wiley & Sons, Inc., New York, 1945, pp. 1252–1286.
8. W. L. Glowacki, in Ref. 7, pp. 1136–1231; E. O. Rhodes, in Ref. 7, pp. 1136–1231.
9. J. R. Rostrup-Nielsen, *Steam Reforming Catalysts*, Teknisk Forlag A/S, Copenhagen, 1975.
10. *Catalyst Handbook*, Wolfe Scientific Texts, London, 1970.
11. P. Sabatier, *Catalysis, Then and Now*, Part II, Franklin Publishing Co., Englewood, N.J., 1965.
12. Fr. Pat. 571,356 (May 16, 1924), (to Badische Anilin- und Soda-Fabrik).
13. Fr. Pat. 580,905 (Nov. 19, 1924), (to Badische Anilin- und Soda-Fabrik).
14. F. Fischer and H. Tropsch, *Ber.* **56**, 2428 (1923).
15. F. Fischer, *Die Umwandlung der Kohle in Ole*, Borntraeger, Berlin, 1923, p. 320.
16. F. Fischer and H. Tropsch, *Ber.* **59**, 830, 832, 923 (1926).
17. F. Fischer, *Brennstoff-Chem.* **11**, 492 (1930).
18. F. Fischer and H. Koch, *Brennstoff-Chem.* **13**, 61 (1932).
19. F. Fischer and K. Meyer, *Brennstoff-Chem.* **12**, 225 (1931).
20. F. Fischer and H. Pichler, *Brennstoff-Chem.* **20**, 41, 221 (1939).
21. K. Fischer, *Comparison of I. G. Work on Fischer Synthesis, Technical Oil Mission Report, Reel 13*, Library of Congress, Washington, D.C., July 1941.
22. H. Pichler, *Medium Pressure Synthesis on Iron Catalyst, (Pat. Appl.), Technical Oil Mission Report, Reel 100*, Library of Congress, Washington, D.C., 1937–1943.

23. H. Pichler, *Medium Pressure Synthesis on Iron Catalyst, Technical Oil Mission Report, Reel 101*, Library of Congress, Washington, D.C., June 1940.
24. U.S. Pat. 2,327,066 (Aug. 17, 1943), O. Roelen.
25. F. Fischer, *Ole Kohle* **39**, 517 (1943).
26. H. H. Storch, N. Golumbic, and R. B. Anderson, *The Fischer-Tropsch and Related Synthesis*, John Wiley & Sons, Inc., New York, 1951.
27. W. G. Frankenburg, V. I. Komarewsky, and E. D. Rideal, *Advances in Catalysis*, Vol I., Academic Press, Inc., New York, 1948, pp. 115–156.
28. H. H. Storch, in Ref. 7, p. 1797.
29. O. C. Elvins and A. W. Nash, *Fuel* **5**, 263 (1926).
30. O. C. Elvins, *J. Soc. Chem. Ind. (London)* **56**, 473T (1927).
31. A. Erdeley and A. W. Nash, *J. Soc. Chem. Ind. (London)* **47**, 219T (1928).
32. W. W. Myddleton, *Chim. Ind.* **37**, 863 (1937); *J. Inst. Fuel* **11**, 477 (1938); *Colliery Guardian* **157**, 286 (1938).
33. G. Egloff, *Brennst.-Chem.* **18**, 11 (1937).
34. G. Egloff, E. F. Nelson, and J. C. Morrell, *Ind. Eng. Chem.* **29**, 555 (1937).
35. F. Mako and W. A. Samuel, in R. A. Meyers, ed., *Handbook of Synfuels Technology*, McGraw-Hill, Inc., New York, 1984, pp. 2-5–2-43.
36. J. C. Hoogendoorn, *Phil. Trans. R. Soc. Lond. A* **300**, 99 (1981).
37. W. B. Johnson, *Pet. Ref.* **35** (Dec. 1956).
38. M. E. Dry, "Fischer-Tropsch Synthesis Over Iron Catalysts," paper presented at *1990 Spring AIChE National Meeting*, Orlando, Fla., Mar. 18–22, 1990.
39. B. Jager and co-workers, in Ref. 38.
40. A. E. Sands, H. W. Wainwright, and L. D. Schmidt, *Ind. Eng. Chem.* **40**, 607 (1948); A. E. Sands and L. D. Schmidt, *Ind. Eng. Chem.* **42**, 2277 (1950).
41. H. Pichler, *Technical Oil Mission Report, Reel 259*, Library of Congress, Washington, D.C., 1947, frames 467–654.
42. P. C. Keith, *Oil Gas J.* **45**(6), 102 (1946).
43. *Oil Gas J.*, 74 (Feb. 17, 1986).
44. M. J. v. d. Burgt and co-workers, in D. M. Bibby and co-workers, eds., *Methane Conversion*, Elsevier Science, Inc., New York, 1988, pp. 473–482.
45. I. E. Maxwell and J. E. Naber, *Catal. Lett.* **12**, 105 (1992).
46. P. T. Roterud and co-workers, in Ref. 38.
47. C. D. Chang and A. J. Silvestri, *J. Catal.* **47**, 249 (1977).
48. C. D. Chang, *Catal. Rev. -Sci. Eng.* **25**, 1 (1983).
49. C. D. Chang and A. J. Silvestri, *Chemtech* **17**, 624 (1987).
50. U.S. Pat. 3,702,886 (1972), R. J. Argauer and G. R. Landolt (to Mobil Oil Corp.).
51. G. T. Kokotailo and co-workers, *Nature* **272**, 437 (1978).
52. D. H. Olson, G. T. Kokotailo, and S. L. Lawton, *J. Phys. Chem.* **85**, 2238 (1981).
53. C. D. Chang, in Ref. 44, pp. 127–143.
54. G. J. Hutchings and R. Hunter, *Catal. Today* **6**, 279 (1990).
55. F. Bauer, *ZfI-Mitt.* **156**, 31 (1990).
56. P. B. Weisz and V. J. Frilette, *J. Phys. Chem.* **64**, 382 (1960).
57. S. M. Csicsery, *ACS Monograph* **171**, 680 (1976).
58. N. Y. Chen, W. E. Garwood, and F. G. Dwyer, *Shape Selective Catalysis in Industrial Applications*, Marcel Dekker, Inc., New York, 1989.
59. C. J. Maiden, in Ref. 44, pp. 1–16.
60. J. E. Penick, W. Lee, and J. Maziuk, *ACS Symp. Ser.* **226**, 19 (1983).
61. J. Z. Bem, in Ref. 44, pp. 663–678.
62. P. L. Rogerson, in Ref. 35, pp. 2-45–2-73.
63. A. Y. Kam, M. Schreiner, and S. Yurchak, in Ref. 35, pp. 2-75–2-111.
64. D. Liederman and co-workers, *Ind. Eng. Chem. Proc. Des. Devel.* **17**, 340 (1978).

65. H. R. Grimmer, N. Thiagarajian, and E. Nitschke, in Ref. 44, pp. 273–291.
66. K. H. Keim and co-workers, *Erdol. Erdgas, Kohle* **103**, 82 (1987).
67. K. G. Allum and A. R. Williams, in Ref. 44, pp. 691–711.
68. A. A. Avidan, in Ref. 44, pp. 307–323.
69. C. D. Chang, *Catal. Rev.-Sci. Eng.* **26** (3&4), 323 (1984).
70. M. M. Holm and E. H. Reichl, *Fiat Report No. 1085*, Office of Military Government for Germany (U. S.), Mar. 31, 1947.
71. J. A. Sofranko, "Gas to Gasoline: The ARCO GTG Process," paper presented at *Bicentenary Catalysis Meeting*, Sydney, Australia, Sept. 1988.
72. J. A. Sofranko and J. C. Jubin, "Natural Gas to Gasoline: The ARCO GTG Process," paper presented at *International Chemical Congress of Pacific Basin Societies*, Honolulu, Hawaii, Dec. 1989.
73. J. H. Edwards, K. T. Do, and R. J. Tyler, in Ref. 72.
74. J. H. Edwards, K. T. Do, and R. J. Tyler, in E. E. Wolf, ed., *Methane Conversion by Oxidative Processes*, Van Nostrand Reinhold, New York, 1992, pp. 429–462.
75. R. J. Nichols, "Applications of Alternative Fuels," *SAE Special Publication SP-531*, Society of Automotive Engineers, Warrendale, Pa., Nov. 1982.
76. R. A. Potter, "Neat Methanol Fuel Injection Fleet Alternative Fuels Study," paper presented at *Fourth Washington Conference on Alcohol*, Washington, D.C., Nov. 1984.
77. N. D. Brinkman, *Ener. Res.* **3**, 243 (1979).
78. T. Powell, "Racing Experiences with Methanol and Ethanol Based Motor-Fuel Blends," paper 750124, *Automotive Engineering Congress and Exposition*, Society of Automotive Engineers, Detroit, Mich., Feb. 1975.
79. American Petroleum Institute, Task Force EF-18 of the Committee on Mobile Source Emissions, *Alcohols—A Technical Assessment of Their Application as Fuels, Publication No. 4261*, API, New York, July 1976.
80. U.S. Environmental Protection Agency, *Fed. Reg.* **46**(144) (July 28, 1981).
81. U.S. Environmental Protection Agency, *Fed. Reg.* **50**(12), 2615 (Jan. 17, 1985).
82. G. A. Mills and E. E. Ecklund, *Chemtech*, 626 (Oct. 1989).
83. KVB, Inc., *KVB Report Number 72-804830-1998*, Vol. 2, California Energy Commission, Irvine, Calif., Mar. 1985, pp. 1–2.
84. A. J. Weir and co-workers, "Methanol Dual-Fuel Combustion," paper presented at *1987 Joint Symposium on Stationary Combustion NO_x Control*, New Orleans, La., Mar. 23–26, 1987.
85. S. B. Alpert and D. F. Spencer, *Methanol and Liquid Fuels From Coal - Recent Advances*, Electric Power Research Institute, Palo Alto, Calif., 1987.
86. V. Yand and S. C. Trindade, *Chem. Eng. Prog.*, 11 (Apr. 1979).
87. *Alcohols: Economics and Future U. S. Gasoline Markets*, Information Resources, Inc., Washington, D.C., 1984.
88. H. L. Muller and S. P. Ho, "Economics and Energy Balance of Ethanol as Motor Fuel," paper presented at *1986 Spring AIChE National Meeting*, New Orleans, La., Apr. 1986.
89. *Chem. Eng. News*, 7 (Nov. 2, 1992).
90. ARCO Chemical Co., *Testimony to the Colorado Air Quality Control Commission on Proposed Regulation No. 13 (Oxygenate Mandate Program)*, Denver, Colo., June 4, 1987.
91. *Chem. Eng. News*, 28 (Apr. 12, 1993).
92. U.S. Pat. 1,251,954 (Jan. 1, 1918), F. Bergius and J. Billwiller.
93. F. Fischer and H. Tropsch, *Ges. Abhandl. Kenntis Kohle* **2**, 154 (1918).
94. F. Bergius, *Pet. Z.* **22**, 1275 (1926).
95. *Gas World* **104**, 421 (1936).
96. *Pet. Times* **42**, 641 (1939).
97. H. H. Storch and co-workers, *U. S. Bur. Mines. Tech. Pap.* **622** (1941).

98. A. C. Fieldner and co-workers, *U. S. Bur Mines Tech. Pap.* **666** (1944).
99. M. L. Kastens and co-workers, *Ind. Eng. Chem.* **41**, 870 (1949).
100. J. L. Wiley and H. C. Anderson, *U. S. Bur. Mines Bull.* **485**, I (1950), II (1951), III (1952).
101. C. C. Chaffee and L. L. Hirst, *Ind. Eng. Chem.* **45**, 822 (1953).
102. Bituminous Coal Staff, *U. S. Bur. Mines Rep. Invest.* **5506** (1959).
103. E. L. Clark and co-workers, *Ind. Eng. Chem.* **42**, 861 (1950).
104. L. L. Newman and A. P. Pipilen, *Gas Age* **119**(10), 16 (1957); **119**(11), 18 (1957).
105. U. S. Pat. 2,860,101 (Nov. 11, 1958), M. G. Pelipetz (to the United States of America).
106. S. Akhtar, S. Friedman, and P. M. Yavorsky, *U. S. Bur. Mines Tech. Prog. Rep.* **35** (1971).
107. S. Akhtar and co-workers, "Process for Hydrodesulfurization of Coal," paper presented at *71st National AIChE Meeting*, Dallas, Tex., Feb. 20, 1972.
108. B. Linville and J. D. Spencer, *U. S. Bur. Mines Inf. Cir.* **8612** (1973).
109. U.S. Pat. 3,321,393 (May 23, 1967), S. C. Schuman, R. H. Wolk, and M. C. Chervenak (to Hydrocarbon Research, Inc.).
110. *Coal Age*, 101 (May 1976).
111. G. A. Johnson and co-workers, "Present Status of the H-Coal Process," paper 30, *IGT Coal Symposium*, Chicago, 1973.
112. J. E. Papso, in Ref. 35, pp. 1-47–1-63.
113. "High Pressure Hydrogenation at Ludwigshafen-Heidelberg," Vol. IA, *FIAT Final Report No. 1317, ATI No. 92,762*, Central Air Documents Office, Dayton, Ohio, 1951.
114. H. H. Lowry and H. J. Rose, *U. S. Bur. Mines Inf. Cir.* **7420** (1947).
115. A. Baba and co-workers, *Rep. Resources Res. Inst. Jpn.*, (22) (1955).
116. U.S. Pat. 2,221,866 (Nov. 19, 1940), H. Dreyfus.
117. D. L. Kloepper and co-workers, *Solvent Processing of Coal to Produce a De-ashed Product*, Contract 14-01-0001-275, OCR Report No. 9, U. S. Government Printing Office, Washington, D.C., 1965.
118. V. L. Brant and B. K. Schmid, *Chem. Eng. Prog.* **68**(12), 55 (1969).
119. B. K. Schmid and W. C. Bull, "Production of Ashless, Low-Sulfur Boiler Fuels From Coal," paper presented at *ACS Division of Fuel Chem. Symposium on Pollution Control*, New York, Sept. 12, 1971.
120. *Demonstration Plant, Clean Boiler Fuels From Coal*, OCR R&D report no. 82, Interim report no. 1, Vols. 1–3, Ralph M. Parsons Co., Los Angeles, Calif., 1973–1975.
121. M. E. Frank and B. K. Schmid, "Economic Evaluation and Process Design of a Coal–Oil–Gas (COG) Refinery," paper presented at *Symposium on Conceptual Plants for the Production of Synthetic Fuels From Coal, AIChE 65th Annual Meeting*, New York, Nov. 26, 1972.
122. U.S. Pat. 3,341,447 (Sept. 12, 1967), W. C. Bull and co-workers (to the United States of America and Gulf Oil Corp.).
123. B. K. Schmid and D. M. Jackson, "The SRC-II Process," paper presented at *Third Annual International Conference on Coal Gasification and Liquefaction*, University of Pittsburgh, Aug. 3–5, 1976; D. M. Jackson and B. K. Schmid, "Production of Distillate Fuels by SRC-II," paper presented at *ACS Div. of Ind. and Eng. Chem. Symposium*, Colorado Springs, Col., Feb. 12, 1979.
124. W. R. Epperly and J. W. Taunton, "Exxon Donor Solvent Coal Liquefaction Process Development," paper presented at *Coal Dilemma II ACS Meeting*, Colorado Springs, Colo., Feb. 12, 1979.
125. D. E. Blaser and A. M. Edelman, "Flexicoking for Improved Utilization of Hydrocarbon Resources," paper presented at *API 43rd Mid-Year Meeting*, Toronto, Canada, May 8, 1978.
126. S. F. Massenzio, in Ref. 35, pp. 6-3–6-18.

127. T. A. Cavanaugh, W. R. Epperly, and D. T. Wade, in Ref. 35, pp. 1-3–1-46.

128. L. E. Swabb, Jr., G. K. Vick, and T. Aczel, "The Liquefaction of Solid Carbonaceous Materials," paper presented at *The World Conference on Future Sources of Organic Raw Materials*, Toronto, Can., July 10, 1978.

129. E. L. Huffman, *Proceedings of the Third Annual International Conference on Coal Gasification and Liquefaction*, Pittsburgh, Pa., 1976.

130. H. D. Schindler, J. M. Chen, and J. D. Potts, *Final Technical Report on DOE Contract No. DE-AC22-79ET14804*, Department of Energy, Washington, D.C., 1983.

131. H. D. Schindler, *Final Technical Report on DOE Contract No. D-AC01-87ER30110*, Vol. 2, Department of Energy, Washington, D.C., 1989.

132. A. G. Comolli and co-workers, *Proceedings of the DOE Direct Liquefaction Contractors' Review Meeting*, Pittsburgh, Pa., 1986.

133. M. G. Thomas, in B. R. Cooper and W. A. Ellingson, eds., *The Science and Technology of Coal and Coal Liquefaction*, Plenum Press, New York, 1984, pp. 231–261.

134. M. J. Burges and R. V. Wheeler, *Fuel* **5**, 65 (1926).

135. W. A. Selvig and W. H. Ode, *U.S. Bur. Mines. Bull.* **571**, (1957).

136. H. C. Howard, in Ref. 7, Vol. 1, pp. 761–773.

137. *Ibid.*, Suppl. Vol., pp. 340–394.

138. J. F. Jones and co-workers, *Chem. Eng. Prog.* **62**(2), 73 (1966).

139. J. A. Hamshar, H. D. Terzian, and L. J. Scotti, "Clean Fuels From Coal by the COED Process," paper presented at *EPA Symposium on Environmental Aspects of Fuel Conversion Technology*, St. Louis, Mo., May 1974.

140. C. D. Kalfadelis and E. M. Magee, *Evaluation of Pollution Control in Fossil Fuel Conversion Processes, Liquefaction, Section 1, COED Process*, EPA-650/2-74-009e, Environmental Protection Agency, Washington, D.C., 1975.

141. R. T. Eddinger and co-workers, *Char Oil Energy Development, Office of Coal Research R&D Report No. 11*, Vol. 1 (PB 169,562) and Vol. 2 (PB 169,563), U.S. Government Printing Office, Washington, D.C., Mar. 1966.

142. J. F. Jones and co-workers, *Char Oil Energy Development, Office of Coal Research R&D Report No. 11*, Vol. 1 (PB 173,916) and Vol. 2 (PB 173,917), U.S. Government Printing Office, Washington, D.C., Feb. 1967.

143. M. E. Sacks and co-workers, *Char Oil Energy Development, Office of Coal Research Report 56, Interim Report No. 2*, GPO Cat. No. 163.10:56/Int.2, U. S. Government Printing Office, Washington, D.C., Jan. 1971.

144. J. F. Jones and co-workers, *Char Oil Energy Development, Office of Coal Research R&D Report No. 56, Final Report*, GPO Cat. No. 163.10:56, U. S. Government Printing Office, Washington, D. C., May 1972.

145. J. F. Jones and co-workers, *Char Oil Energy Development, Office of Coal Research R&D Report No. 73, Interim Report No. 1*, GPO Cat. No. 163.10:73/Int 1, U. S. Government Printing Office, Washington, D. C., Dec. 1972.

146. A. Sass, "The Garrett Research and Development Company Process for the Conversion of Coal into Liquid Fuels," paper presented at *65th Annual AIChE Meeting*, New York, Nov. 29, 1972.

147. *Oil Gas J.*, 78 (Aug. 26, 1974).

148. A. Sass, *Chem. Eng. Prog.* **70**(1), 72 (1974).

149. F. B. Carlson and co-workers, *Chem. Eng. Prog.* **69**(3), 50 (1973).

150. F. B. Carlson, L. H. Yardumian, and M. T. Atwood, "The TOSCOAL Process for Low Temperature Pyrolysis of Coal," paper presented at *American Institute of Mining, Metallugical, and Petroleum Engineers*, San Francisco, Calif., Feb. 22, 1972, and to *American Institute of Chemical Engineers*, New York, Nov. 29, 1972.

151. W. Tiddy and M. J. Miller, *Am. Gas J.* **153**(3), 7 (1940).

152. M. J. Gavin, *U.S. Bur. Mines Bull.* **210**, (1922).

153. M. J. Gavin and J. S. Desmond, *U.S. Bur. Mines Bull.* **315** (1930).

154. H. Shaw, C. D. Kalfadelis, and C. E. Jahnig, *Evaluation of Methods to Produce Aviation Turbine Fuels From Synthetic Crude Oils-Phase I, Technical Report AFAPL-TR-75-10*, Vol. I, Air Force Aero Propulsion Laboratory, Wright-Patterson Air Force Base, Dayton, Ohio, Mar. 1975.
155. C. D. Kalfadelis, *Evaluation of Methods to Produce Aviation Turbine Fuels From Synthetic Crude Oils-Phase II, Technical Report AFAPL-TR-75-10*, Vol. II, Air Force Aero Propulsion Laboratory, Wright-Patterson Air Force Base, Dayton, Ohio, May 1976.
156. Colony Development Operation, *An Environmental Impact Analysis for a Shale Oil Complex at Parachute Creek, Colorado*, Vols. 1–3, Denver, Colo., 1974.
157. H. L. Erskine, in Ref. 35, pp. 5-3–5-32.
158. R. D. Hynphreys, F. K. Spragins, and D. R. Craig, "Oil Sands—Canada's First Answer to the Energy Shortage," *Proceedings 9th World Petroleum Congress*, Tokyo, Japan, May 11, 1975, Vol. 5, p. 17.
159. C. W. Bowman, R. S. Phillips, and L. R. Turner, in Ref. 35, pp. 5-33–5-79.
160. M. Crow and co-workers, *Synthetic Fuel Technology Development in the United States—A Retrospective Assessment*, Praeger Publishing, New York, 1988.
161. S. Yurchak and S. S. Wong, "Mobil Methanol Conversion Technology," *Proceedings IGT Asian Natural Gas Seminar*, Singapore, 1992, pp. 593–618.
162. J. L. Matherne and G. L. Culp, in Ref. 73, pp. 463–482.
163. J. E. Duddy, S. B. Panvelker, and G. A. Popper, "Commercial Economics of HRI Coal/Oil Co-Processing Technology," paper presented at *1990 SummerAIChE National Meeting*, San Diego, Ca., 1990.
164. M. Ikura and J. F. Kelly, "A Techno-Economic Evaluation of CANMET Coprocessing Technology," *Proceedings Annual International Pittsburgh Coal Conference*, 1990, pp. 719–728.
165. J. G. Sikonia, B. R. Shah, and M. A. Ulowetz, "Technical and Economic Assessment of Petroleum, Heavy Oil, Shale Oil and Coal Liquid Refining," paper presented at *Synfuels' 3rd Worldwide Symposium*, Washington, D.C., Nov. 1–3, 1983.

General Reference

American Petroleum Institute, *Alcohols and Ethers-A Technical Assessment of Their Application as Fuels and Fuel Components*, API Publication 4261, American Petroleum Institute, Washington, D.C., July 1988.

SCOTT HAN
CLARENCE D. CHANG
Mobil Research and Development Corporation

FULLER'S EARTH. See CLAYS, USES.

FULMINATES. See EXPLOSIVES AND PROPELLANTS, EXPLOSIVES.

FUMARIC ACID. See MALEIC ACID, FUMARIC ACID, AND MALEIC ANHYDRIDE.

FUMIGANTS. See INSECT CONTROL TECHNOLOGY.

FUNCTIONAL FLUIDS. See HYDRAULIC FLUIDS.

FUNGICIDES, AGRICULTURAL

Pathogenic fungi cause a substantial reduction in expected crop yields; further losses can result during storage of harvested crops. Although there are over 100,000 classified fungal species, no more than 200 are known to cause serious plant disease. Most plants are resistant to the majority of potential pathogenic fungi in their environment. However, a limited number of fungal pathogens are able to delay or prevent the onset of defense responses of certain plant species, or have developed mechanisms to counteract specific plant defense reactions. For those fungi that can seriously affect economically important plants (Table 1), means have been sought to control these infections by crop rotation and husbandry, genetic manipulation of the plant species (see GENETIC ENGINEERING, PLANTS), and external treatment of plants using agricultural fungicides.

Agricultural fungicide application accounts for about 20% of all pesticide use. More than 5.6×10^9 was spent worldwide on these fungicides in 1991 (1). Agricultural fungicides can be applied to the soil to control fungi that are resident there, to the seed or foliage of the plant to be protected, or to harvested produce to prevent storage losses. Those applied to the soil are in many instances nonselective, volatile soil sterilants, such as formaldehyde (qv), which kill all soil organisms, including fungi. Soil and crop storage fungicides, which represent only a very small fraction of the fungicides used, are covered elsewhere (2,3) (see SOIL CHEMISTRY OF PESTICIDES). Seed and foliar-applied agricultural fungicides, listed in Table 2, are discussed herein.

The word fungicide might suggest a compound that nonselectively kills all fungi, but even compounds having an unspecified mode of action can exhibit a remarkable degree of selectivity against different fungi. In addition, some fungicides are more properly called fungistats because their action controls the spread of disease without actually killing the pathogen.

Because of the wide diversity of chemical structures encountered, fungicides are classified herein as being nonsystemic or systemic. The nonsystemic fungicides have a protectant mode of action and must be applied to the surface of a plant generally before infection takes place. These do not translocate from the site of application. The systemic fungicides can penetrate the seed or plant and are then redistributed within to unsprayed parts or subsequent new growth, rendering protection from fungal attack or eradicating a fungus already present.

Nonsystemic Fungicides

From 20 to 25 nonsystemic fungicides are utilized in agriculture, although use is declining. These are some of the oldest known fungicides and cover a wide range of chemistry from simple inorganic salts to highly complex organic structures. Selective accumulation by spores plays a dominant role in the toxicity of many of these compounds. The majority are regarded as general cell poisons and can be used only when they are not able to penetrate host plant tissue in appreciable amounts. The fungal pathogen is controlled before it infects the plant so that the

Table 1. Important Diseases of Crop Plants

Fungal class	Pathogen	
	Scientific name	Common name
Phycomycetes	*Phytophthora infestans*	potato late blight
subclass oomycetes	*Plasmopara viticola*	downy mildew of grape
	Pseudoperonospora cubensis	cucumber downy mildew
	Pythium spp.	damping off diseases
Ascomycetes	*Erysiphe graminis*	powdery mildew of wheat/barley
	Gaeumannomyces graminis	take-all of oats and wheat
	Podosphaera leucotricha	apple powdery mildew
	Pyrenophora teres	net blotch of barley
	Pyricularia oryzae	rice blast
	Rhynchosporium secalis	leaf scald of barley
	Sclerotinia spp.	brown rot of pome fruit
		leaf spot of brassicas and legumes
	Sphaerotheca fuliginea	cucurbit powdery mildew
	Uncinula necator	grape powdery mildew
	Venturia inaequalis	scab of apple
	Mycosphaerella fijiensis	sigatoka disease of bananas
Basidiomycetes	*Puccinia* spp.	leaf rusts of wheat and oats
	Rhizoctonia spp.	black scurf of potato
		sheath blight of rice
		sharp eyespot of wheat
	Tilletia spp.	bunts of wheat
	Uromyces spp.	bean rusts
	Ustilago spp.	smuts of wheat, barley, oat, and maize
Deuteromycetes		early blight of potato
	Alternaria spp.	tobacco brown spot
		leaf spot of brassicas
	Botrytis spp.	grey mold of grape and other crops
	Cercospora spp.	leaf spot of sugarbeet
		brown eyespot of coffee
	Fusarium spp.	wilts, broad range of hosts
		ear blight of wheat
		root and foot rot of wheat
	Helminthosporium spp.	leaf spot of maize
	Pseudocercosporella herpotrichoides	eyespot of wheat
	Septoria nodorum	glume blotch of wheat
	Septoria tritici	wheat leaf blotch

Table 2. Alphabetical List of Fungicides

Common name	Trademark	Company	Year of intro-duction	Molecular formula	Structure number
anilazine	Dyrene	Bayer AG	1955	$C_9H_5Cl_3N_4$	(**34**)
benalaxyl	Galben	Agrimont SpA	1981	$C_{20}H_{23}NO_3$	(**77**)
benomyl	Benelate	E. I. du Pont de Nemours	1968	$C_{14}H_{18}N_4O_3$	(**45**)
blasticidin S	Bla-S	Karen, Kumiai, Nihon	1959	$C_{17}H_{26}N_8O_5$	(**91**), (**92**)
bupirimate	Nimrod	ICI Plant Protection	1972	$C_{13}H_{24}N_4O_3S$	(**82**)
buthiobate	Denmert	Sumitomo Chemical Co.	1975	$C_{21}H_{28}N_2S_2$	(**55**)
captafol	Difolatan	Chevron Chemical Co.	1961	$C_{10}H_9Cl_4NO_2S$	(**15**)
captan	Orthocide	Chevron Chemical Co.	1949	$C_9H_8Cl_3NO_2S$	(**13**)
carbendazim	Bavistin	BASF AG	1972	$C_9H_9N_3O_2$	(**44**)
carboxin	Vitavox	Uniroyal Inc.	1966	$C_{12}H_{13}NO_2S$	(**35**)
chinomethionat	Morestan	Bayer AG	1960	$C_{10}H_6N_2OS_2$	(**29**)
chloroneb	Demosan	E. I. du Pont de Nemours	1967	$C_8H_8Cl_2O_2$	(**18**)
chlorothalonil	Bravo	Fermenta Plant Protection	1975	$C_8Cl_4N_2$	(**21**)
chlozolinate	Serinal	Agrimont SpA	1980	$C_{13}H_{11}Cl_2NO_5$	(**24**)
cymoxanil	Curzate	E. I. du Pont de Nemours	1977	$C_7H_{10}N_4O_3$	(**93**)
cyproconazole	Alto	Sandoz AG	1982	$C_{15}H_{18}ClN_3O$	(**65**)
dichlofluanid	Euparen	Bayer AG	1965	$C_9H_{11}Cl_2FN_2O_2S_2$	(**16**)
dichlone	Phygon	Uniroyal Inc.	1943	$C_{10}H_4Cl_2O_2$	(**20**)
dicloran	Allisan	Boots (now Schering AG)	1960	$C_6H_4Cl_2N_2O_2$	(**19**)
diclomezine	Monguard	Sankyo Co. Ltd.	1988	$C_{11}H_8Cl_2N_2O$	(**30**)
dimethirimol	Milcurb	ICI Plant Protection	1968	$C_{11}H_{19}N_3O$	(**81**)
dinocap	Karathane	Rohm and Haas	1946	$C_{18}H_{24}N_2O_6$	(**17**)
dithianon	Delan	E. Merck	1963	$C_{14}H_4N_2O_2S_2$	(**28**)
dodemorph	Milan	BASF AG	1967	$C_{18}H_{35}NO$	(**71**)
dodine	Cyprex	American Cyanamid Co.	1957	$C_{15}H_{33}N_3O_2$	(**27**)
ediphenphos	Hinosan	Bayer AG	1968	$C_{14}H_{15}O_2PS_2$	(**83**)
ethirimol	Milcap	ICI Plant Protection	1969	$C_{11}H_{19}N_3O$	(**80**)
etridazole	Terrazole	Uniroyal Inc.	1969	$C_5H_5Cl_3N_2OS$	(**31**)
fenarimol	Rubigan	Eli Lilly (now DowElanco)	1975	$C_{17}H_{12}Cl_2N_2O$	(**57**)
fenfuram	Pano-ram	Shell Research Ltd.	1974	$C_{12}H_{11}NO_2$	(**38**)
fenpiclonil	Beret	CIBA-GEIGY AG	1988	$C_{11}H_6Cl_2N_2$	(**32**)
fenpropidin	Patrol	Dr. Maag (now CIBA-GEIGY)	1986	$C_{19}H_{31}N$	(**74**)
fenpropimorph	Corbel	Dr. Maag (now CIBA-GEIGY)	1979	$C_{20}H_{33}NO$	(**73**)
fentin acetate[a]	Brestan	Hoechst AG	1954	$C_{20}H_{18}O_2Sn$	(**2**)
fentin hydroxide[b]	Du-ter	N.V. Philips-Duphar	1954	$C_{18}H_{16}OSn$	(**3**)
ferbam	Fermate	E. I. du Pont de Nemours	1931	$C_9H_{18}FeN_3S_6$	(**4**)
flusilazole	Nustar	E. I. du Pont de Nemours	1982	$C_{16}H_{15}F_2N_3Si$	(**67**)
flutriafol	Impact	ICI Plant Protection	1982	$C_{16}H_{13}F_2N_3O$	(**69**)
flutolanl	Moncut	Nihon Nohyaku Co. Ltd.	1976	$C_{17}H_{16}F_3NO_2$	(**37**)
folpet	Phaltan	Chevron	1952	$C_9H_4Cl_3NO_2S$	(**14**)
fosetyl-Al	Aliette	Rhône-Poulenc Agrochimie	1977	$C_9H_{18}AlO_9P_3$	(**94**)
fuberidazole	Voronit	Bayer AG	1966	$C_{11}H_8N_2O$	(**43**)
furalaxyl	Fongarid	CIBA-GEIGY AG	1976	$C_{17}H_{19}NO_4$	(**76**)
imazalil	Fungaflor	Janssen Pharmaceuticals	1973	$C_{14}H_{14}Cl_2N_2O$	(**59**)
imibenconazole	Manage	Hokko Chem. Ind. Ltd.	1988	$C_{17}H_{13}Cl_3N_4S$	(**70**)
iprobenphos	Kitazin P	Kumiai Chemical Ind.	1966	$C_{13}H_{21}O_3PS$	(**84**)
iprodione	Rovral	Rhône-Poulenc Agrochimie	1970	$C_{13}H_{13}Cl_2N_3O_3$	(**25**)
isoprothiolane	Fuji-one	Nihon Nohyaku Co. Ltd.	1975	$C_{12}H_{18}O_4S_2$	(**85**)

Table 2. (*Continued*)

Common name	Trademark	Company	Year of intro- duction	Molecular formula	Structure number
kasugamycin	Kasumin	Hokki Chem Ind. Ltd.	1965	$C_{14}H_{25}N_3O_9$	(**92**)
mancozeb	Dithane M-45	E. I. du Pont de Nemours	1961		(**10**)
maneb	Dithane M-22	E. I. du Pont de Nemours	1950	$C_4H_6MnN_2S_4$	(**9**)
mepronil	Basitac	Kumiai Chem. Ind. Ltd.	1981	$C_{12}H_{19}NO_2$	(**39**)
metalaxyl	Ridomil	CIBA-GEIGY AG	1977	$C_{13}H_{21}NO_4$	(**75**)
methfuroxam	Trivax	Uniroyal Inc.	1976	$C_{14}H_{15}NO_2$	(**40**)
metsulfovax	Provax	Uniroyal Inc.	1986	$C_{12}H_{12}N_2OS$	(**41**)
myclobutanil	Systhane	Rohm and Haas Co.	1984	$C_{15}H_{17}ClN_4$	(**68**)
nabam	Parzate	E. I. du Pont de Nemours	1943	$C_4H_6N_2Na_2S_4$	(**7**)
nuarimol	Trimidal	DowElanco	1976	$C_{17}H_{12}ClFN_2O$	(**58**)
ofurace	Oturanic	Chevron Chem. Co.	1982	$C_{14}H_{16}ClNO_3$	(**78**)
oxadixyl	Sandofan	Sandoz AG	1979	$C_{14}H_{18}N_2O_4$	(**79**)
oxycarboxin	Plantvax	Uniroyal Inc.	1966	$C_{12}H_{13}NO_4S$	(**36**)
polyoxin B	Polyoxin AL	Hokko Chem. Ind. Co.	1968	$C_{17}H_{25}N_5O_{13}$	(**91**)
polyoxin D	Polyoxin Z	Hokko Chem. Ind. Co.	1968	$C_{17}H_{23}N_5O_{14}$	(**92**)
prochloraz	Sportak	Boots (now Shering AG)	1974	$C_{15}H_{16}Cl_3N_3O_2$	(**60**)
procymidone	Sumisclex	Sumitomo Chemical Co.	1969	$C_{13}H_{11}Cl_2NO_2$	(**26**)
propiconazole	Tilt	Janssen Pharmaceuticals	1979	$C_{15}H_{17}Cl_2N_3O_2$	(**63**)
pyroquilon	Funorene	Pfizer Inc.	1980	$C_{11}H_{11}NO$	(**87**)
quintozene	Botrilex	I.C. Farben (now Bayer AG)	1930	$C_6Cl_5NO_2$	(**17**)
tebuconazole	Folicur	Bayer AG	1983	$C_{16}H_{22}ClN_3O$	(**64**)
tetraconazole	Eminent	Agrimont SpA	1986	$C_{13}H_{11}Cl_2F_4N_3O$	(**66**)
thiabendazole	Mertect	Merck and Co.	1986	$C_{10}H_7N_3S$	(**42**)
thiophanate methyl	Topsin M	Nippon Soda Co. Ltd.	1969	$C_{12}H_{14}N_4O_4S_2$	(**46**)
thiram	Tersan	E. I. du Pont de Nemours	1931	$C_6H_{12}N_2S_4$	(**6**)
triadimefon	Bayleton	Bayer AG	1975	$C_{14}H_{16}ClN_3O_2$	(**62**)
triarimol	Trimidal	DowElanco	1969	$C_{17}H_{12}Cl_2N_2O$	(**56**)
tricyclazole	Beam	DowElanco	1972	$C_9H_7N_3S$	(**86**)
tridemorph	Calixin	BASF AG	1969	$C_{19}H_{39}NO$	(**72**)
triforine	Cela W524	Celamerck (now Shell)	1967	$C_{10}H_{14}Cl_6N_4O_2$	(**54**)
vinclozolin	Ronilan	BASF AG	1975	$C_{12}H_9Cl_2NO_3$	(**23**)
zineb	Dithane Z-78	Rohm and Haas	1943	$C_4H_6N_2S_4Zn$	(**8**)
ziram	Milbam, Zerlate	E. I. du Pont de Nemours	1930	$C_6H_{12}N_2S_4Zn$	(**5**)

[a]CAS Registry Number [*900-95-8*].
[b]CAS Registry Number [*76-87-9*].

resulting efficacy is primarily achieved through protecting the plant rather than curing the disease. The mode of action, ie, biochemical basis for activity, of most known nonsystemic fungicides is generally nonspecific, and inhibition at multiple sites results ultimately in interference with energy producing or transferring processes which disrupts fungal respiration and membranes (4).

   Sulfur. Sulfur became firmly established as an agricultural fungicide in the nineteenth century, when the preparation of lime sulfur was reported in 1802 to control mildew on fruit trees (5). Elemental sulfur, in the form of flowers of sulfur, was the first effective nonsystemic protectant fungicide. Although toxicologically one of the safer fungicides, sulfur must be applied frequently and in large quantities to be effective, causing handling difficulties and leading in some instances to phytotoxicity. Sulfur is an effective inhibitor of fungal spore germination and may affect several target sites in fungal cells. It probably exerts fungicidal efficacy *in vivo* by reduction to H_2S, which both reacts with proteins (qv) and chelates to heavy metals to disrupt cellular processes, including respiration.

Copper. Although copper sulfate was used for treating the seed-borne disease wheat bunt (*Tilletia* spp.) as early as 1761, widespread use was limited by its inherent phytotoxicity. In 1882, it was observed (6) that grapevines that had been coated with a mixture of copper sulfate and lime to deter grape pilferage, were not infected with grape downy mildew (*Plasmopara viticola*). This observation resulted in the development of a fungicide called Bordeaux mixture, the exact composition of which is unclear. Many copper fungicides are available for a wide variety of applications, eg, the sulfates (Bordeaux mixture), oxides and oxychlorides, and a variety of organic salts such as copper naphthenates and copper quinolinates. Crops protected using copper compounds include vines, fruit, coffee (qv), cocoa, and vegetables. Most copper fungicides work by inhibiting fungal spore germination. Sensitive fungi are affected by the uptake of copper salts and its subsequent accumulation, which then complexes with amino, sulfhydryl, hydroxyl, or carboxyl groups of enzymes resulting in inactivation of the fungus (7).

Mercury. The first successful use of mercury as a fungicide occurred in 1913 (8). The first seed treatment compound developed was chloro(2-hydroxyphenyl)mercury [90-03-9] (1). Subsequently, a number of organic mercury derivatives having general formula RHgX have been used. These compounds are extremely restricted because of high toxicity and persistence in the environment, and are totally banned in many countries.

(1)

(2) A^- = acetate
(3) A^- = hydroxide

Tin. The fungicidally active tin compounds are organotins. Many of the most fungicidal compounds, eg, the tripropyl and tributyl tins, are too phytotoxic for direct application to plants and the success of tin compounds as fungicides require balancing efficacy against phytotoxicity. The triphenyl tin compounds fentin acetate [900-95-8] (2) and fentin hydroxide [76-87-9] (3), used primarily in controlling diseases of potato and sugar beet, are less phytotoxic. These compounds control disease by inhibition of the mitochondrial adenosine triphosphatase (ATP-ase) involved in oxidative phosphorylation (9). The high cost of tin, the concerns about heavy metals in the environment, and the phytotoxic potential of these compounds continue to be critical factors influencing use.

Thiocarbamate and Thiurame Derivatives. The thiocarbamate family of fungicides was discovered in the 1930s as a result of research in the rubber industry for accelerators in the curing of rubber, and the subsequent broad screening of those compounds (10). These are broad-spectrum fungicides (with the exception of the powdery mildews) that have a multisite action on the fungus and interfere with its metabolism in many ways. The key products to emerge from this group (Fig. 1) are the thiocarbamates: ferbam [14484-64-1] (**4**); ziram [137-30-4] (**5**); and thiram [137-26-8] (**6**), and the ethylene bis-dithiocarbamates: nabam [142-59-6] (**7**); zineb [12212-67-7] (**8**); maneb [12427-38-2] (**9**); and mancozeb [8018-01-7] (**10**). These compounds are still widely used on many crops, especially top fruit (orchard fruits), vines, and field vegetables. In the case of the dimethyl-thiocarbamates, the anion (**11**), generated *in vivo*, acts as an inhibitor of essential copper-containing enzymes, whereas the ethylene bis-dithiocarbamates are converted to the ethylene diisothiocyanate (**12**), the primary toxic agent, which binds preferentially to SH groups of fungal enzymes (11). Because these compounds do not have a specific mode of action on the fungus but interfere with it in a number of ways, there is only a low risk of fungal resistance developing. The principal pressure against continued use has come from regulatory concerns about the effects of residues such as ethylene thiourea on human health.

Phthalimides and Some Trichloromethylthiocarboximides. The fungicidal efficacy of this chemistry was recognized in the 1950s. The phthalimide derivatives are excellent, broad-spectrum fungicides which can be applied to the foliage, roots, or seed of a crop. The most important products in this chemistry are captan [133-06-02] (**13**), folpet [133-07-3] (**14**), captafol [2425-06-1] (**15**), and, more re-

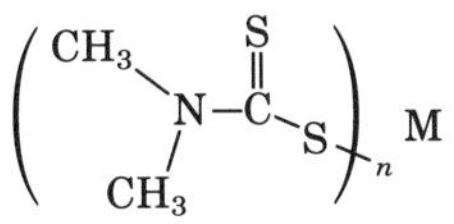

(**4**) M = Fe, $n = 3$

(**5**) M = Zn, $n = 2$

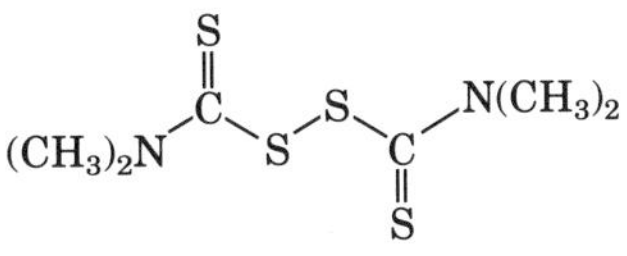

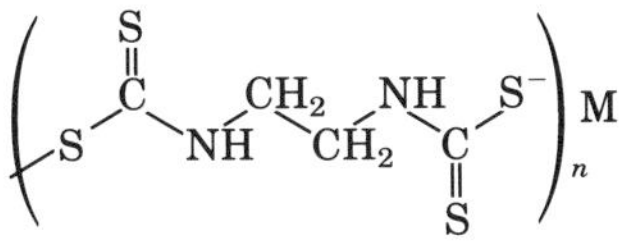

(**7**) M = Na, $n = 2$

(**8**) M = Zn, $n = 1$

(**9**) M = Mn, $n = 1$

(**10**) M = Mn, Zn, $n = 2$

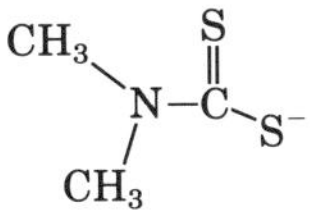

(**6**)

(**11**)

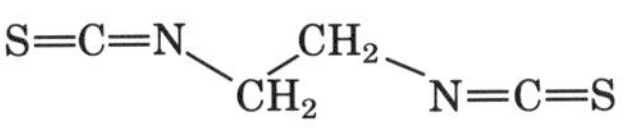

(**12**)

Fig. 1. Thiocarbamate nonsystemic fungicides.

cently, dichlofluanid [*1085-98-9*] (**16**), which is structurally different but has a similar mode of action. As in the case of the dithiocarbamates, these compounds react with thiol groups in fungi, releasing thiophosgene and H_2S (12). The thiophosgene may subsequently react with thiol and amino groups in the enzymes (13). Differences in uptake between fungal species are considered responsible for the differences in fungicidal spectrum between specific compounds. The U.S. Environmental Protection Agency (EPA) and other regulatory bodies have raised questions concerning the safety of the phthalimides, in particular captan and captafol, and some restrictions have been imposed on their usage (14).

Aromatic Hydrocarbons. The aromatic hydrocarbon fungicides were among the earliest compounds to replace sulfur in the treatment of powdery mildews of fruit. This diverse group of substituted aromatic hydrocarbons is shown in Figure 2. Two of the earliest were quintozene [*82-68-8*] (**17**) (also known as

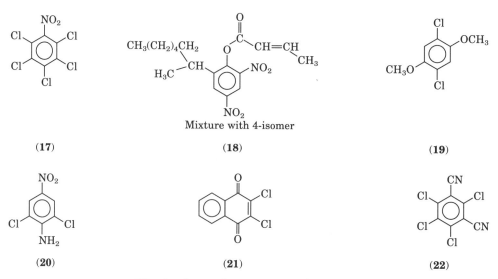

Fig. 2. Aromatic nonsystemic fungicides.

PCNB) and dinocap [*131-72-6*] (**18**). Some of the more important compounds in this class in the 1990s include chloroneb [*2675-77-6*] (**19**) and dichloran [*99-30-9*] (**20**). Most of these compounds were utilized for the control of soil and seed-borne diseases. Whereas quintozene and chloroneb are known to possess some whole plant systemicity, this is not generally the case for the group as a whole. The efficacy of some members of this class of fungicides is also facilitated by marked vapor action. They tend to be inhibitory to mycelial growth rather than the preinfection stages of the various pathogens and control a broad spectrum of Oomycetes (see Table 1) and other pathogenic fungi, including *Rhizoctonia, Botrytis, Ustilago, Alternaria,* and *Helminthosporium* (15). The mode of action is considered to be inhibition of the enzyme NADPH-cytochrome C reductase, which results in the generation of free radicals and/or peroxide derivatives of flavin which oxidize adjacent unsaturated fatty acids to disrupt membrane integrity (16) (see ENZYME INHIBITORS).

Although compounds such as dichlone [*117-80-6*] (**21**) and chlorothalonil [*1897-45-6*] (**22**) are also aromatic hydrocarbons and widely effective against a broad range of pathogens, these apparently have a different mode of action involving binding to SH groups of fungal enzymes. Use of the various aromatic hydrocarbons has declined because of replacement by more efficacious fungicides having broader antifungal spectrum.

Dicarboximides. The dicarboximides, introduced in the early 1970s, are characterized by a cyclic imide group represented by an oxazolidinedione, eg, vinclozolin [*50471-44-8*] (**23**) and chlozolinate [*84332-86-5*] (**24**); a hydantoin, eg, iprodione [*36734-19-7*] (**25**); and a succinimide, eg, procymidone [*32809-16-8*] (**26**). Although the fungicidal spectrum is similar to that of the aromatic hydrocarbons (17), dicarboximides inhibit spore germination more effectively than mycelial growth and cause increased branching and swelling of the germ tubes and hyphal tips. The mode of action, as in the case of the aromatic hydrocarbons, is the inhibition of the enzyme NADPH-cytochrome C reductase (18).

(**23**) R = CH=CH$_2$
(**24**) R = COOCH$_2$CH$_3$

(**25**)

(**26**)

Miscellaneous Nonsystemics. A wide variety of other types of compounds which cannot be easily grouped chemically have been developed and used as protectant fungicides (Fig. 3). These are important fungicides for specialty markets. Some of the more significant examples are a guanidine salt, eg, dodine [*2439-10-*

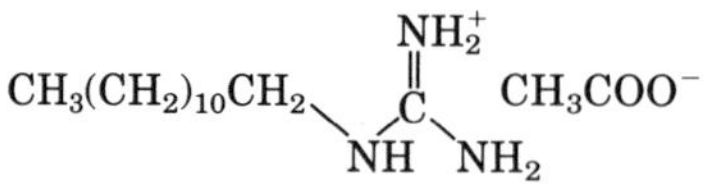

Fig. 3. Miscellaneous nonsystemics.

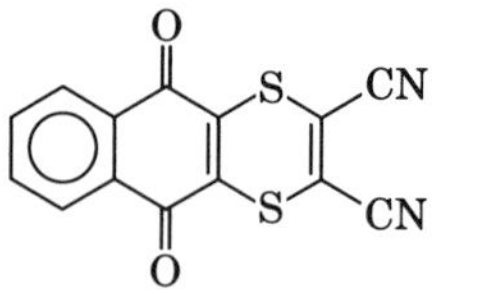

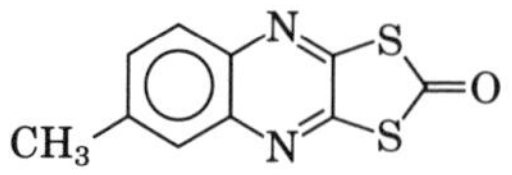

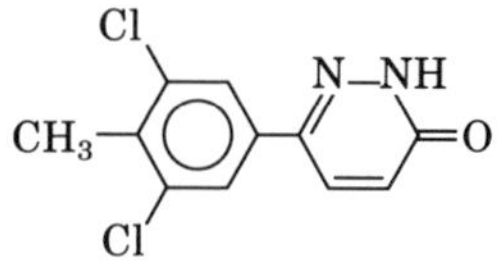

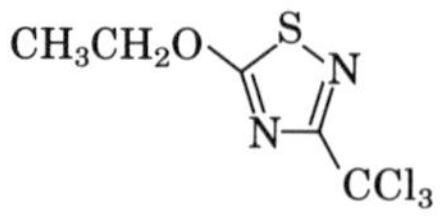

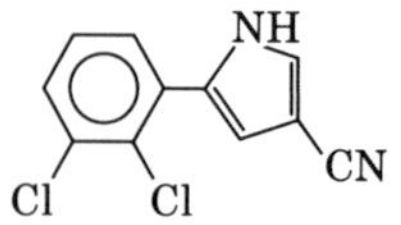

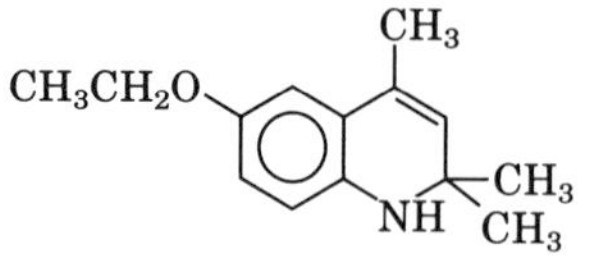

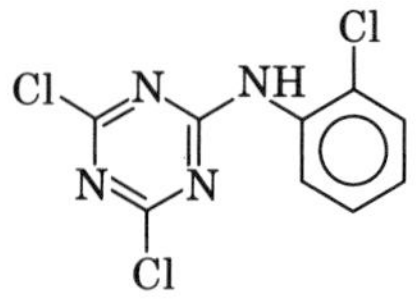

3] (**27**); a quinone, eg, dithianon [*3347-22-6*] (**28**); a quinoxaline, eg, chinome-thionat [*2439-01-2*] (**29**); a pyridazine, eg, diclomezine [*62865-36-5*] (**30**); a thia-diazole, eg, etridiazole [*2593-15-9*] (**31**); a pyrrole, eg, fenpiclonil [*74738-17-3*] (**32**); a quinoline, eg, ethoxyquin [*91-53-2*] (**33**); and a triazine, eg, anilazine [*101-05-3*] (**34**). These compounds are mostly enzyme poisons, binding with -SH or amino groups of fungal enzymes or interfering with fungal membrane structure and function.

Site-Specific Systemic Fungicides

In general, the systemic fungicidal treatment of crop plants is only possible using inhibitors of fungal-specific targets, and there has been considerable progress in developing agricultural fungicides having high levels of fungal specificity. Eluci-dation of the biochemical mechanisms of action of compounds has led to the dis-covery of some novel compounds. Many of the fungicides introduced since the 1970s have been systemic fungicides which inhibit fungal growth at various stages of fungal development. These fungicides are often active at very low levels compared with nonsystemics and tend to exhibit a much narrower activity spectrum as a consequence of their action against a specific biochemical target. Precise biochemical targets have been defined for many of the different classes of fungicide chemistries. Some have a biochemical target site in common. The selectivity of systemic fungicides can be attributed to differences in a number of factors. These include uptake and accumulation in the fungal cell, inherent differences at the target site, differences in metabolism of the fungicide by the plant or fungi, and the degree of importance of the target system to the survival of the fungus.

Mitochondrial Respiration Inhibitors. The carboxanilides, discovered in 1964, were among the first systemic commercial fungicides capable of protecting the unsprayed new growth of plants from fungal attack (19). The principal fungicides in this class of mitochondrial respiration inhibitors are carboxin [5234-68-4] (35); oxycarboxin [5259-88-1] (36); flutolanil [66332-96-5] (37); fenfuram [24691-80-3] (38); mepronil [55814-41-0] (39); methfuroxam [28730-17-8] (40); and metsulfovax [21452-18-6] (41). These compounds, shown in Figure 4, were mainly active against the Basidiomycetes (see Table 1), a class of fungi which includes such important pathogens as the rusts (*Puccinia* spp.), smuts (*Ustilago* spp.), and bunts of cereals. They are used as both seed and foliar fungicides against rusts of coffee beans (*Uromyces* spp.) and ornamentals (plants grown for decorative reasons). The mode of action of the carboxanilides involves interference with succinate metabolism. Studies using whole cells (20) and later isolated mitochondrial preparations (21,22) led to the conclusion that the primary target is the succinate dehydrogenase complex of the mitochondrial respiratory chain, which is inhibited by these compounds. Genetic and molecular biology studies (23–26) have confirmed this conclusion.

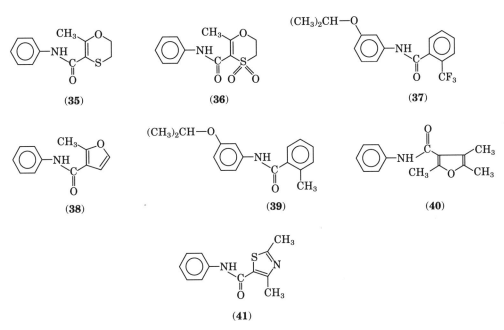

Fig. 4. Mitochondrial respiration inhibitors.

Because of the initial selectivity of carboxanilides to Basidiomycete fungi, it was thought at one time that the molecular site of action of carboxin (**35**) might be unique to Basidiomycetes. Subsequently, however, analogues emerged which demonstrated activity against non-Basidiomycete fungi (27,28). This suggested that selectivity was based not on differing carboxamide affinities for the succinate dehydrogenase complexes of various fungi, but rather on metabolic factors or permeability differences, such as mitochondrial penetration. Resistance to the carboxanilide fungicides has been noted across this class of compounds, and has contributed to a decline in usefulness.

Microtubulin Polymerization Inhibitors. The benzimidazoles were first reported to have systemic fungicidal activity in 1964 (29). Prominent examples include thiabendazole [148-79-8] (**42**); fuberidazole [3878-19-1] (**43**); carbendazim [10605-21-7] (**44**); benomyl [17804-35-2] (**45**); and thiophanate methyl [23564-05-8] (**46**). Benomyl (**45**), the most widely used member of this group is almost certainly inactive as a fungicide until it is converted in plants and soil to carbendazim (**44**). Likewise, thiophanate and thiophanate methyl (**46**) are nonfungitoxic until converted to carbendazin (**44**).

(**42**) R = [thiazole]

(**43**) R = [furan]

(**44**) R = —NHCOOCH$_3$

(**45**)

(**46**)

Whereas most Ascomycetes, Deuteromycetes (see Table 1), and Basidiomycetes are sensitive to the fungicides, the Phycomycetes (Oomycetes and Zygomycetes) are inherently resistant. The primary mode of action has been identified as specific binding to the β-tubulin subunit of fungal tubulin. Because β-tubulin is a principal component of the fungal cytoskeleton, the resulting interference with assembly of the microtubules leads to a disruption of both mitosis and meiosis (30). All organisms except bacteria and blue-green algae possess microtubules. The benzimidazoles are characterized, however, by a remarkable selectivity for fungi that probably depends on differences in molecular structures of the binding sites of the microtubules. There is a high margin of safety of carbendazim and related compounds to plants. This may have a basis in low tubulin binding in plants, as has been demonstrated for mammalian tubulin. Resistance, observed shortly after the introduction of benomyl (**43**), has increased throughout the world (31). Studies have suggested that single-site mutations in the β-tubulin gene are responsible for the resistance (32,33).

Inhibitors of Sterol Biosynthesis. The discovery of compounds that inhibit ergosterol biosynthesis in fungi was one of the most significant advances in the history of fungicide research (34). Sterols are known to be essential for all eukaryotes, either synthesized *de novo* from acetate or taken up from the environment. In fungi, the early steps in the pathway from acetate culminate with the cyclization of squalene epoxide to produce lanosterol [79-63-0] (**47**), $C_{30}H_{50}O$. Figure 5 presents the steps involved in the biosynthesis of the principal sterol in most fungi, ergosterol [57-87-4] (**33**), $C_{28}H_{44}O$, a component in membrane structure.

Fungicides that inhibit sterol biosynthesis have utility only against those fungi that synthesize their sterol complement. Consequently, these compounds are generally not efficacious against grape downy mildew and potato late blight (*Phytophthora infestans*), which satisfy their sterol requirements by mycelial uptake. Because of the importance of ergosterol in fungal membranes, any reduction in its availability to fungi increases the permeability to electrolytes and leads to

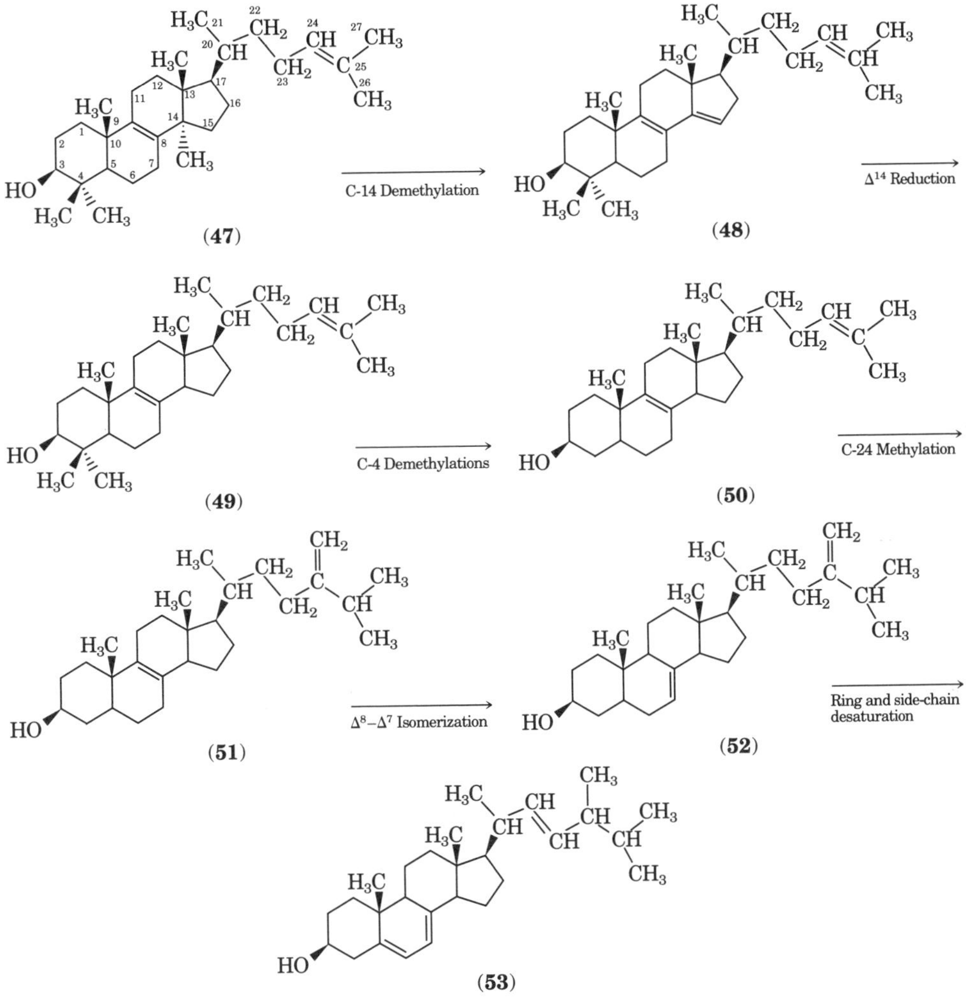

Fig. 5. Ergosterol biosynthesis pathway.

a severe leakiness of membrane-enclosed compartments. This structural role can only be filled by ergosterol and even closely related sterols are apparently inadequate. The primary result of sterol biosynthesis inhibition is the accumulation of sterol precursors and the depletion of the demethylated sterol pool. The precursors are incorporated into the plasma membrane, eventually replacing ergosterol and arresting fungal growth and reproduction. This effect is fungistatic in character so that removal of the inhibitor results in full recovery of cell viability. The accumulation of sterol precursors not only affects the permeability of membranes, but also the active transport of nutrients such as amino acids (qv) (35), and the activity of membrane-located enzymes. Finally, ergosterol and related fungal sterols may also serve as precursors for steroid hormones necessary for sexual reproduction events in fungi and, possibly, other biological processes (36).

Although sterol biosynthesis-inhibiting compounds have been used since the 1950s to control human fungal diseases, the success and impact of agricultural fungicides with target sites in ergosterol biosynthesis has, since the late 1960s, revolutionized the control of plant diseases (37). The impact of these new classes of inhibitors has been particularly marked against Ascomycetes, Deuteromycetes, and Basidomycetes.

C-14 Demethylation Inhibitors. Piperazines, pyridines, pyrimidines, and azoles all inhibit the C-14 demethylation step catalyzed by the cytochrome P450-dependent 14α-demethylase (Fig. 5). The only apparent common feature of the various demethylation inhibitors is the presence of a heterocycle containing at least one nitrogen atom. Basic amines, pyridines, and azoles have long been known to exhibit a strong affinity for various cytochrome-P450 monoxygenases and may act by the hydrophobic substituents of the various compounds binding to the site on the demethylase normally occupied by lanosterol (47). This binding positions the basic nitrogen of the essential heterocycle such that it blocks the binding of oxygen to the cytochrome P-450 cofactor, which is a prerequisite for demethylation (38–40).

The earliest commercial fungicides recognized to inhibit ergosterol formation in fungi were the piperazines, eg, triforine [26644-46-2] (**54**); pyridines, eg, buthiobate [51308-54-4] (**55**); and the pyrimidine carbinols, represented by triarimol [26766-27-8] (**56**), fenarimol [60168-88-9] (**57**), and nuarimol [63284-71-9] (**58**) (Fig. 6). Imidazoles, also very active, were among the earliest azoles developed. Imazalil [35554-44-0] (**59**) is effective against a wide variety of fruit, vegetable, and cereal diseases. It is used primarily as a seed or post-harvest treatment. Prochloraz [67747-09-5] (**60**) has found a niche in the control of eyespot (*Pseudocerosporella herpotrichoides*), glume blotch (*Septoria nodorum*), and leaf blotch (*Septoria tritici*) diseases of cereals, whereas triflumizole [99387-89-0] (**61**), a more recent example, is primarily used in controlling fruit diseases.

Probably the most important fungicides in this group are the triazoles which have in common a 1,2,4-triazole group attached through the 1-nitrogen to a large lipophillic group (Fig. 7). The most important members of this family are triadimefon [43121-43-3] (**62**), introduced in 1973 as a highly active compound against powdery mildews (*Erysiphe graminis*) and rusts of cereals; propiconazole [60207-90-1] (**63**), introduced in 1979 with an extremely broad spectrum of cereal disease activity; tebuconazole [107534-96-3] (**64**); cyproconazole [113096-99-4] (**65**); and

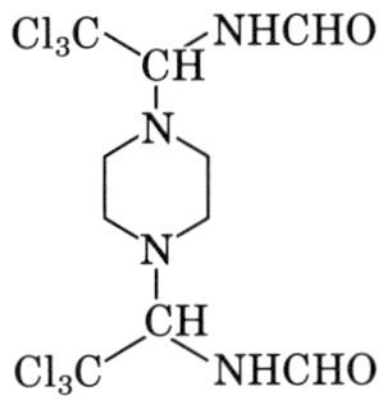

Fig. 6. C-14 demethylation inhibitors.

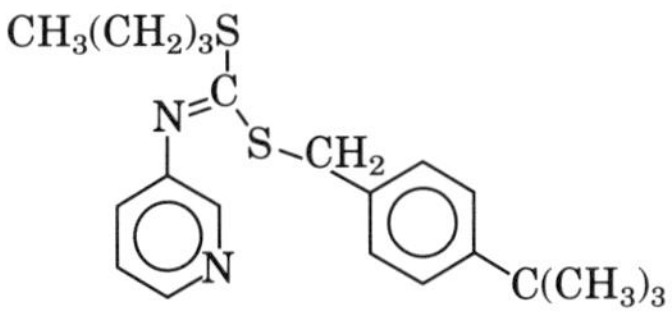

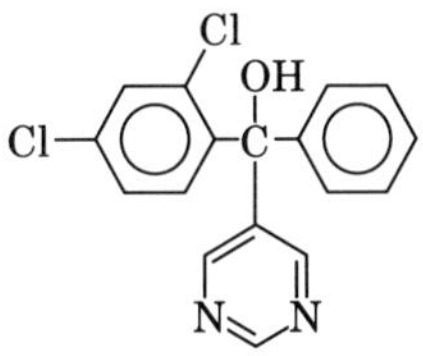

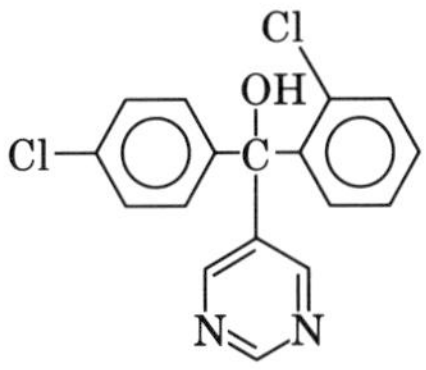

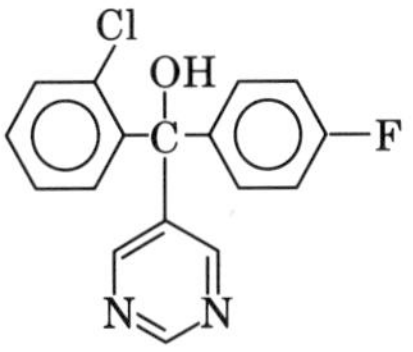

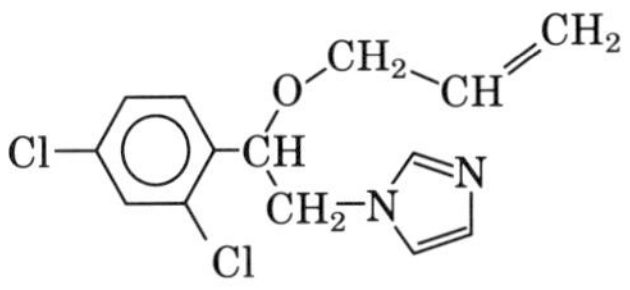

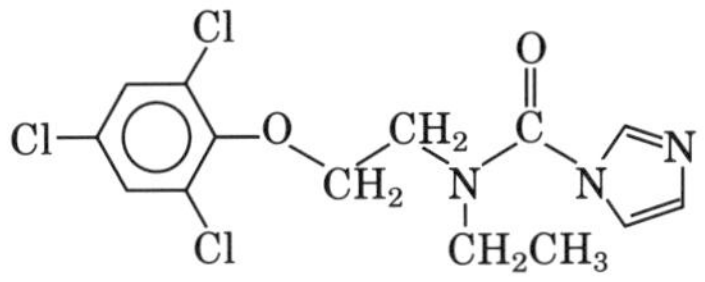

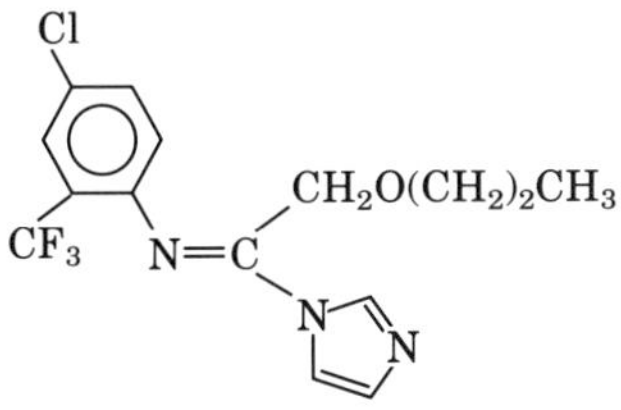

tetraconazole [*112281-77-3*] (**66**), all highly efficacious, broad-spectrum fungicides recently introduced for use on both cereals and top fruit. Some of the more unusual structures in this class are flusilazole [*85509-19-9*] (**67**), which incorporates a silicon atom; myclobutanil [*88671-89-0*] (**68**), with a nitrile; flutriafol [*76674-21-0*] (**69**), which mimics the pyrimidine carbinols; and imibenconazole [*86598-92-7*] (**70**), which incorporates an imine in the alkyl chain. Additional structures of commercialized azoles have been outlined in detail in a number of publications (41,42).

The almost exclusive use of triazoles for cereal powdery mildew control up to the mid-1980s has resulted in a shift in *Erysiphe* populations toward reduced sensitivity or resistance to this class of fungicides. However, use has continued because field performance of most triazoles has remained adequate (43). Decreased azole sensitivity in the *Septoria* population to azoles has also been noted.

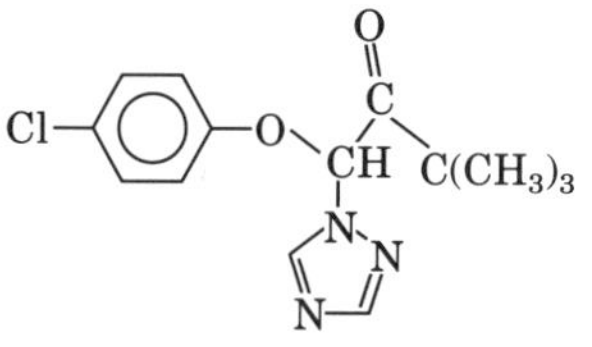

(**62**)

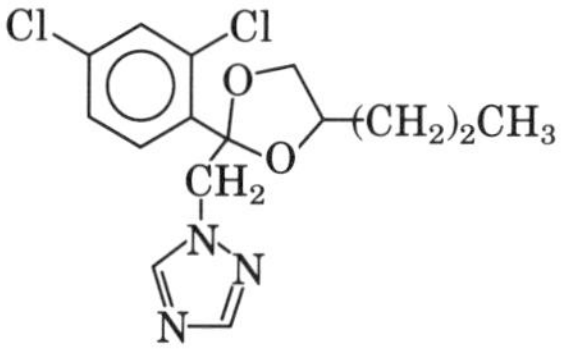

(**63**)

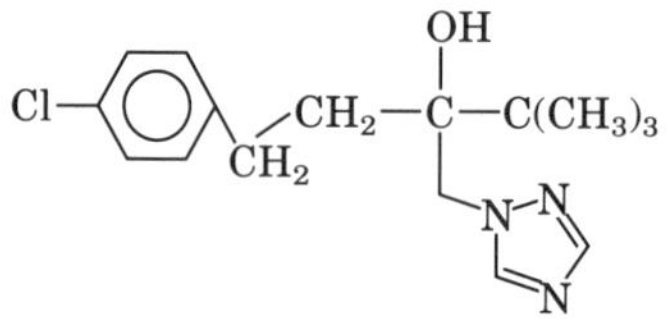

(**64**)

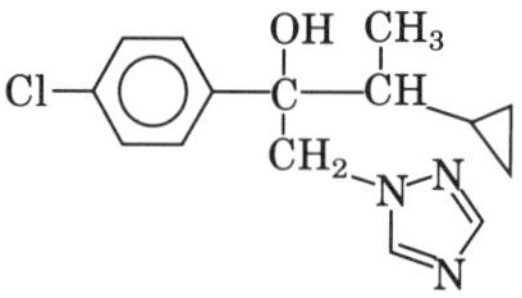

(**65**)

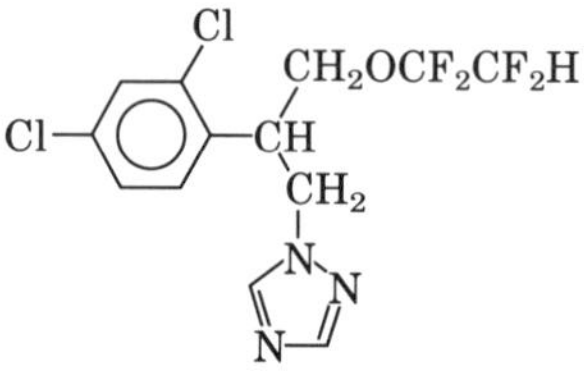

(**66**)

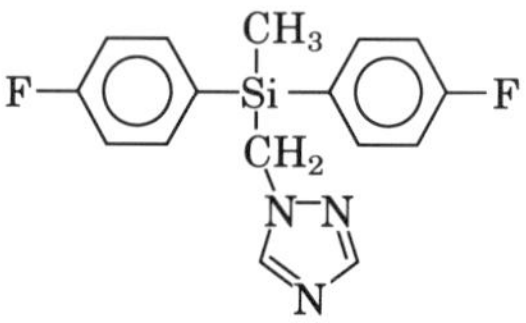

(**67**)

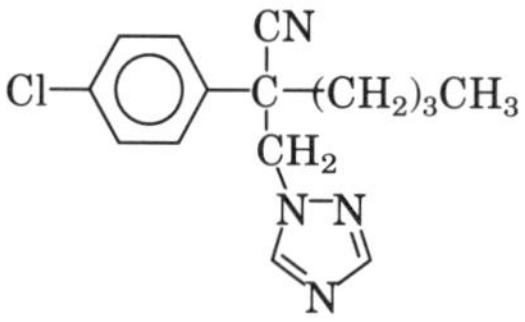

(**68**)

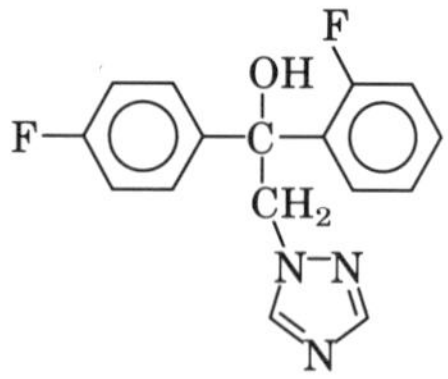

(**69**)

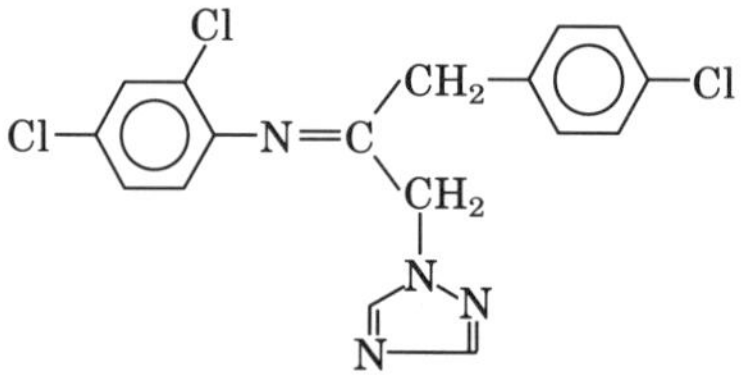

(**70**)

Fig. 7. Triazole C-14 demethylation inhibitors.

Δ^{14}-Reduction and Δ^{8}–Δ^{7}-Isomerization Inhibitors. Only one group of compounds that act as inhibitors for both the Δ^{14}-reduction and Δ^{8}–Δ^{7}-isomerization steps (Fig. 6), the morpholines, have been developed as commercial agricultural fungicides. The earliest compounds of this class, dodemorph [*1593-77-7*] (**71**) and tridemorph [*81412-43-3*] (**72**), were first introduced in the 1960s. These were followed in 1979 by fenpropimorph [*67306-03-0*] (**73**) and fenpropidin [*67306-00-7*]

(**74**), which are especially active, both as eradicants and protectants, against the Ascomycete (particularly powdery mildews) and, to a lesser extent, Basidiomycete (eg, rusts) diseases of cereals and ornamentals. The $\Delta^8-\Delta^7$ isomerase step (**49**) to (**52**) in Figure 5, was initially proposed as the primary target of the morpholines following accumulation, in tridimorph-treated fungi, of such sterols as fecosterol (**51**) (44). However, specific morpholines also inhibit the Δ^{14}-reduction step, (**48**) to (**49**), to a greater or lesser extent (45). Although laboratory resistance to morpholines has been reported in several fungi (46,47), significant resistance problems have not yet surfaced in the field (43,48), and these compounds are widely used particularly for cereals.

RNA Biosynthesis Inhibitors.   Phenylamides and hydroxypyrimidines function as RNA biosynthesis inhibitors. The phenylamide fungicides are comprised of the acylalanines, eg, metalaxyl [137414-52-9] (**75**), furalaxyl [57646-30-7] (**76**), and benalaxyl [71626-11-4] (**77**); the butyrolactones, eg, ofurace [58810-48-3] (**78**); and the oxazolidinones, eg, oxadixyl [77732-09-3] (**79**). These compounds are readily taken up by roots and foliage, and have good activity against the oomycetes including grape downy mildew and potato late blight. Studies of phenylamides on various steps in the infection process indicate little effect on the release, mobility, encystment, and germination of zoospores of these oomycete pathogens or on host penetration and primary haustorium (the specialized fungal structure which absorbs nutrients from the plant host) formation (49). Studies using other *Phytophthora* species firmly established RNA-polymerase I as the primary biochemical target of the phenylamides (50,51). As a consequence of interference with this enzyme target new ribosome formation is inhibited and protein synthesis becomes impaired, leading to fungal growth inhibition. Inhibition of RNA biosynthesis leads to accumulation of nucleoside triphosphate precursors which promote fungal β-(1,3)-glucan synthetase, and thus stimulate the biosynthesis of key cell wall constituents resulting in inhibition of cell growth. After repeated and exclusive use of metalaxyl in the field against late blight of potato, resistance to all the phenylamides rapidly developed (52,53) and strategies involving combination sprays and mixtures with other fungicides have been developed to address this problem.

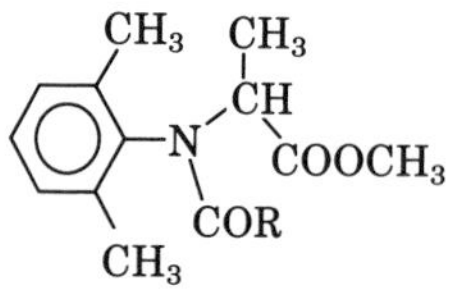

(75) R = —CH$_2$OCH$_3$

(76) R =

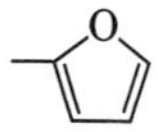

(77) R = —CH$_2$—C$_6$H$_5$

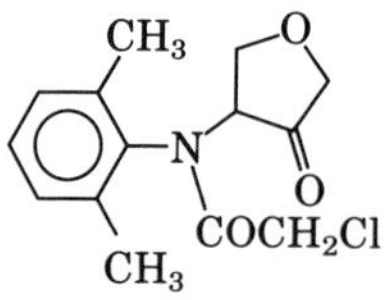

(78)

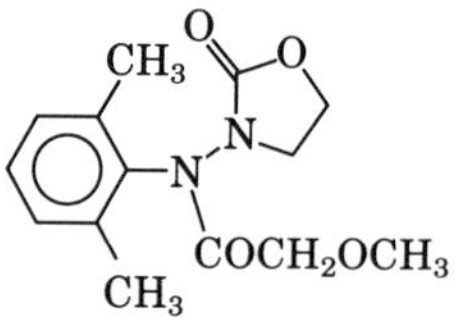

(79)

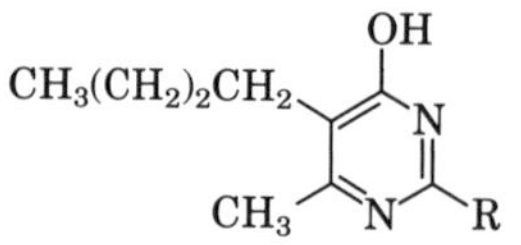

(80) R = —NHCH$_2$CH$_3$

(81) R = —N(CH$_3$)$_2$

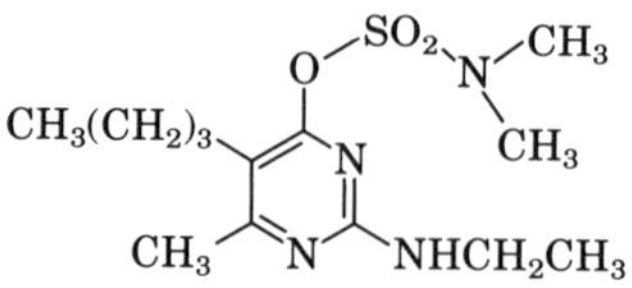

(82)

The 4-hydroxypyrimidine derivatives, eg, ethirimol [23947-60-6] (80), dimethirimol [5221-53-4] (81), and bupirimate [41483-43-6] (82), have a selective and systemic activity against powdery mildews of cereals, field vegetables, and ornamentals, both as foliar and seed treatment compounds. These interfere with several stages of the infection process of powdery mildew, but particularly with appressorium formation (54) and germ tube extension. Biochemically, ethirimol (80) may interfere with purine metabolism. Reversal experiments showed that ethirimol was antagonized by metabolites such as adenine, adenosine, guanine, and folic acid (55). Later studies have shown that the target enzyme is adenosine deaminase, which appears to be inhibited specifically in powdery mildews (55). Powdery mildews do not synthesize purines *de novo* and it is believed that adenosine deaminase is essential to these fungi for utilization of purines acquired from the host during the infection process. Kinetin (6-fufuryladenine) and isopentenyladenine have also been observed to inhibit appressorium formation, and ethirimol-resistant isolates of barley powdery mildew exhibit cross-resistance to these growth regulators (see GROWTH REGULATORS, PLANTS) (56). Resistance problems were encountered in the field shortly after introduction of the hydroxypyrimidines, and use of these compounds has been restricted to cereal seed treatment and greenhouse applications.

Phospholipid Biosynthesis Inhibitors. The organophosphate compounds ediphenphos [17109-49-8] (83) and iprobenphos [26087-47-8] (84) represent a class of fungicides that target a site in the lipid biosynthesis pathway (57). The structurally unrelated compound isoprothiolane [50512-35-1] (85), introduced in 1975, appears to have the same mechanism of action despite the absence of a phosphorus atom. All three are readily taken up by both roots and leaves of rice, subsequently translocated to control rice blast (*Pyricularia oryzae*), and appear to be more toxic to mycelium growth and sporulation of rice blast than to spore germination or appressorium formation (58,59). The biochemical mode of action has been identified as the inhibition of methyl transfer to phosphatidyl ethanol-amine (59,60) in phosphatidyl choline biosynthesis, resulting in membrane disruption. After 10 years of field usage strains of rice blast resistant to iprobenphos (**85**) have emerged.

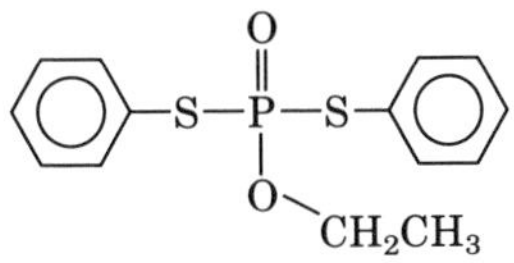

(**83**)

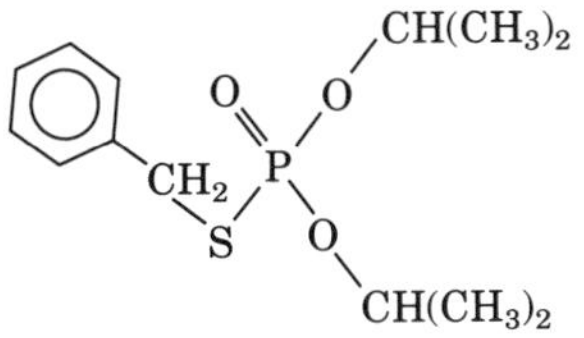

(**84**)

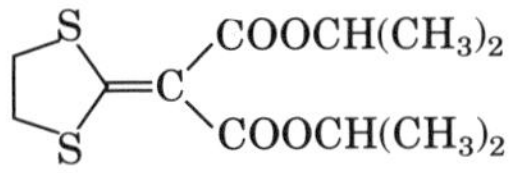

(**85**)

Melanin Biosynthesis Inhibitors. The discovery in 1969 of tricyclazole [*41814-78-2*] (**86**) and its protectant activity against rice blast led to understanding of the importance of melanin inhibition as a means of controlling this pathogen (61). Since then other compounds of differing structures have been shown to similarly inhibit melanin formation, eg, pyroquilon [*57369-32-1*] (**87**) and the experimental compound PP389 [*89342-33-5*] (**88**). Melanization of the appressorial walls of the rice blast fungus is essential for the development of infection hyphae and successful penetration of the leaf (62,63). Appressoria formed in the presence of tricyclazole are devoid of melanin and this reduces the mechanical strength of the infection peg and prevents leaf penetration (64,65). Biochemical studies have indicated that these compounds inhibit the polyketide pathway of melanin biosynthesis at two sites (66,67). No evidence of rice blast resistance to tricyclazole has been observed in the field.

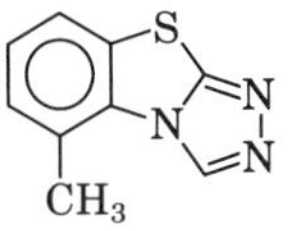

(**86**)

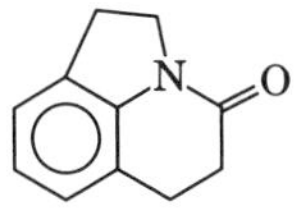

(**87**)

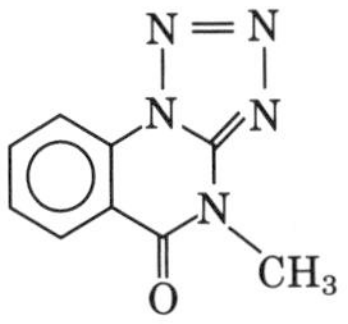

(**88**)

Fungal Protein Biosynthesis Inhibitors. Two fermentation products have been shown to be very effective against rice blast. The first, blasticidin S [*2079-00-7*] (**89**), is produced by *Streptomyces grieschromogens*. The second, kasuga-mycin [*6980-18-3*] (**90**), is a water-soluble base obtained from *Streptomyces kasugaensis*. Both have been used to control rice blast in Japan since 1965 as protectant and curative rice blasticides. They inhibit the growth of the rice blast fungus at levels of 5–10 µg/mL and primarily work by inhibiting protein biosynthesis. These compounds exert this effect by binding respectively to the larger, 60S, and smaller, 30S, subunits of fungal ribosomes (68–71). Resistance has been observed to both compounds in fields where they have been extensively used.

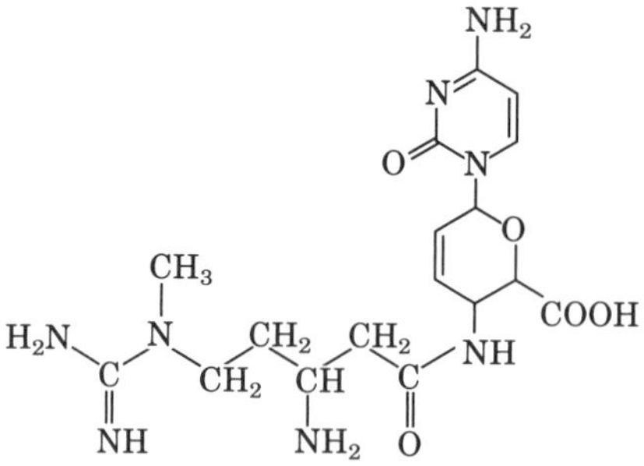

(89)

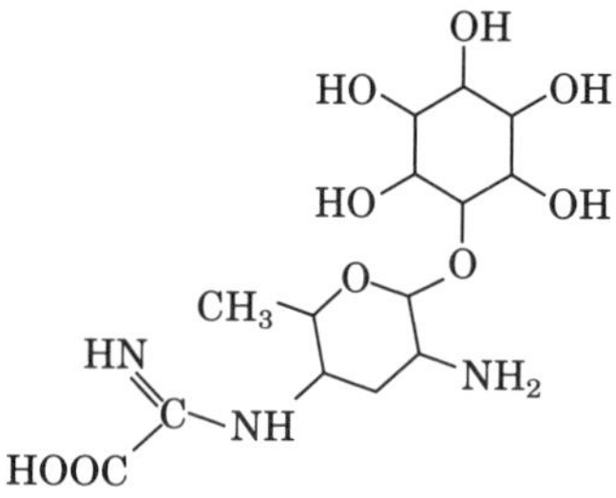

(90)

Cell Wall Biosynthesis Inhibitors. Polyoxin B [19396-06-6] (**91**) and polyoxin D [22976-86-9] (**92**), both from *Streptomyces cacaoi* var. *asoensis*, are closely related antibiotics (qv) that act as highly selective inhibitors of fungi containing chitin in their walls. For this reason the polyoxins are not active against Oomycetes that contain cellulose as the principal cell wall constituent. The mode of action of the polyoxins has been determined to be the inhibition of chitin synthase (chitin UDP-*N*-acetylglucosaminyl transferase), which is localized in the plasma membrane of growing hyphae (72–74). These nucleoside antibiotics show structural similarities to UDP-*N*-acetylglucosamine with which they compete for the chitin synthase active site (75).

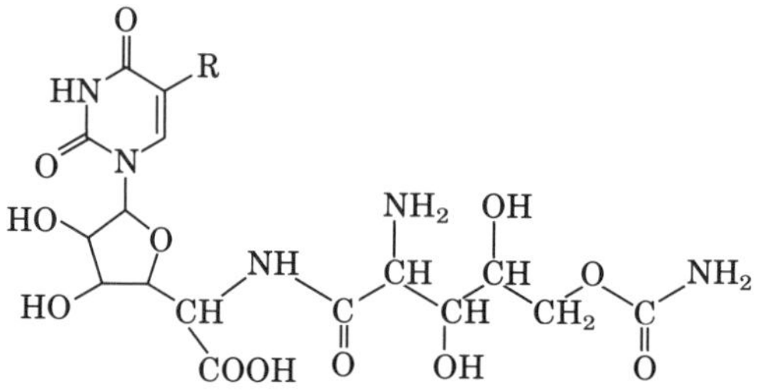

(**91**) R = CH$_2$OH
(**92**) R = COOH

Incompletely Elucidated Modes of Action. A number of compounds have been successfully used as agricultural fungicides but their modes of action have yet to be completely understood. Cymoxanil [57966-95-7] (**93**) is a systemic compound that shows curative and protectant activity against the oomycetes, notably grape downy mildew and potato late blight (76). It has been shown to interfere with RNA and protein synthesis in some fungi but it is not clear whether cymoxanil or some metabolite is the active moiety (77). Fosetyl-Al [39148-24-8] (**94**) is also highly active against the oomycetes, especially grape downy mildew and a variety of other diseases (78). It can be applied as a foliar spray, root drench, or by stem injection, and translocates in both the xylem and phloem systems of the plant. Though several lines of evidence indicate that fosetyl-Al (**94**) has a direct action on target pathogens (79,80), treatment of tobacco plants with fosetyl-Al increased the synthesis of the natural phytoalexin capsidiol. It is possible that accumulation of capsidiol in tobacco and of other phytoalexins such as stilbenes and flavonoids in grape is associated with its activity (79,80).

(**93**)   (**94**)

Resistance to Fungicides

Development of fungicide resistance continues to be one of the primary problems in plant disease control. Resistance can be defined as a stable inheritable adjustment by a fungus to a fungicide that results in less than normal sensitivity to that fungicide (81). In the case of the protectant fungicides, because these are generally multisite inhibitors (with the possible exception of dicarboximides), fungi have little chance, in the short term, of developing resistance. By contrast, systemic fungicides frequently are single-site inhibitors and thus carry a greater potential for resistance to develop. Mutation of a single gene can result in a modified target site with reduced affinity to the fungicide. When under selection pressure by the fungicide, buildup of a residual resistant population occurs and, in the extreme case, may result in the failure of disease control.

Well over 100 plant pathogens have become resistant to various fungicides under field conditions. Failure of the acyl alanines, benzimidazoles, thiophanates, carboxanilides, dicarboximides, hydroxypyrimidines, some organophosphates, and most of the antibiotics has occurred. In other cases, a moderate decrease in sensitivity without a rapid loss of disease control has been observed as in the case of sterol biosynthesis inhibitors (triazoles, pyrimidines, and imidazoles) and organophosphates. The most effective approach is to use fungicides having different modes of action in combination, either as mixtures or in alteration, possibly utilizing both specific site and multisite inhibitors. Because of resistance problems great importance is attached to chemistries that inhibit novel fungal enzyme targets.

Economic Aspects

Growers regard the use of fungicides as part of their broad crop management strategy, in both planning and implementation. The conventional approach to crop production and pesticide use has been via economically justified maximum yield responses, and has led to applications being either made routinely or targeted to specific risks, with a wide range of frequency of applications. Within a particular market segment the pricing of fungicide products from the various manufacturers is extremely uniform and tends to be dictated at least in part by the cost of established products that have stood the test of time balanced against the needs of the grower to demonstrate a clear cost-benefit advantage from their use. As of 1993, fungicide costs for some of the key market segments ranged from $26 per treatment equivalent to $76 per season for European cereals, $25 per treatment or $150 per season for pome fruit to $18–42 per treatment or $110–250 per season for the prevention of grape downy mildew. These three markets together generated sales of $1.7 billion at the manufacturers' level.

Newer fungicides, in order to retain cost-effectiveness, need to be very highly active, which also serves to achieve efficacy in the field at low dose rates thus keeping environmental pollution problems as small as possible. More in-depth knowledge of fungal biochemistry and the molecular events involved in host/pathogen interactions should facilitate the identification of novel fungal targets for use in a biorational approach to fungicide discovery, through the application of computer-aided molecular design (CAMD) approaches to the molecular modeling (qv) of the target to design new fungicides (82). Recombinant DNA technologies are expected to play an escalating role in the validation of such biorational targets (see GENETIC ENGINEERING).

Government regulations for the registration and utilization of all plant protection agents require exhaustive studies on topics such as mammalian toxicology, effects on various forms of wildlife (eg, fish toxicity), soil leachability, and residue levels in crops, soil, and water. It is also required that compounds persist in the environment only as long as is necessary to control crop diseases before being biodegraded. As a consequence, some previously registered fungicides have been withdrawn from the market.

BIBLIOGRAPHY

"Fungicides, Agricultural" in *ECT* 1st ed., Vol. 6, pp. 984–991, by J. C. Horsfall, Connecticut Agricultural Experiment Station; in *ECT* 2nd ed., Vol. 10, pp. 220–228, by D. C. Torgeson, Boyce Thompson Institute for Plant Research; in *ECT* 3rd ed., Vol. 11, pp. 490–498, by E. J. Butterfield and D. C. Torgeson, Boyce Thompson Institute for Plant Research.

1. *Reference Volume of the Agrochemical Service*, County NatWest WoodMac Securities Ltd., London, 1992, pp. 48–69.
2. A. P. Sinha, K. Singh, and A. N. Mukhopadhyay, *Soil Fungicides*, Vols. I & II, CRC Press, Inc., Boca Raton, Fla., 1988.
3. J. W. Eckert, in H. R. Siegel and H. D. Sisler, eds., *Antifungal Compounds*, Vol. 1, Marcel Dekker, Inc., New York, 1977, pp. 290–352.
4. H. Buchenauer, in G. Haug and H. Hoffmann, eds., *Chemistry of Plant Protection*, Vol. 6, Springer-Verlag, Berlin, 1990, pp. 221–226.
5. W. Forsyth, *A Treatise on the Culture and Management of Fruit Trees*, Nichols & Son, London, 1802.
6. A. Millardet, *J. Agr. Prat.* **49**, 801–805 (1885), English trans. F. J. Schneiderhan, *Phytopathol. Classics* **3**, 18–25 (1933).
7. Ref. 4, p. 222.

8. S. E. A. McCallen, in D. C. Torgeson, ed., *Fungicides, an Advanced Treatise*, Vol. 1, Academic Press, Inc., New York, 1967, pp. 1–37.

9. Ref. 4, pp. 240–241.

10. U.S. Pat. 1,972,961 (1934), W. H. Tisdale and I. Williams (to E. I. du Pont de Nemours & Co., Inc.)

11. Ref. 4, p. 223.

12. R. J. Lukens and H. D. Sisler, *Phytopathology* **48**, 235 (1958).

13. M. R. Siegel, *Pestic. Biochem. Physiol.* **1**, 225 (1971).

14. *Chem. Week* **136**(23), 40 (1985).

15. T. Kato, in I. Miyamoto and P. C. Kearney, eds., *Pesticide Chemistry*, Vol. 3, Pergamon Press, Oxford, U.K., 1983, p. 153.

16. H. Lyr, in H. Lyr, ed., *Modern Selective Fungicides—Properties, Application and Mechanism of Action*, Gustav Fischer Verlag, Jena, Germany, 1987, p. 75.

17. E. H. Pommer and G. Lorenz, in Ref. 15, p. 91.

18. H. Sisler, in C. J. Delp, ed., *Fungicide Resistance in North America*, APS Press, St. Paul, Minn., 1988, p. 52.

19. B. von Schmeling and M. Kulka, *Science* **152**, 659 (1966).

20. D. E. Mathre, *Phytopathology* **60**, 671 (1970); N. N. Ragsdale and H. D. Sisler, *Phytopathology* **60**, 1422 (1970).

21. D. E. Mathre, *Pestic. Biochem. Physiol.* **1**, 216 (1971).

22. J. T. Ulrich and D. E. Mathre, *J. Bacteriol* **110**, 628 (1972).

23. S. G. Georgopoulos, E. Alexandri, and M. Chrysayi, *J. Bacteriol* **110**, 809 (1972).

24. S. G. Georgopoulos, M. Chrysayi, and G. A. White, *Pestic. Biochem. Physiol.* **5**, 543 (1975).

25. I. A. U. N. Gunatilleke, H. N. Arst, and C. Scazzochio, *Aspergillus nidulans*, *Genet. Res. Cambridge* **26**, 297 (1975).

26. J. P. R. Keon, G. A. White, and J. A. Hargreaves, *Current Genetics* **19**, 475–481 (1991).

27. L. V. Edington and G. L. Barron, *Phytopathology* **57**, 1256 (1967).

28. G. A. White and S. G. Georgopoulos, *Pestic. Biochem Physiol.* **25**, 188 (1986).

29. T. Staron and C. Allard, *Phytiat–Phytopharm.* **13**, 163 (1964).

30. H. Lyr, *Plant Disease* **1**, 239 (1977).

31. P. Leroux and B. Besselat, *Phytoma–Defense des Cultures* **359**, 25 (1984).

32. H. Ishii, H. Yanase, and J. Dekker, *Meded. Fac. Landbouwwet. Rijksuniv. Gent.* **49**(2a), 163 (1984).

33. M. Fujimura and co-workers, in M. B. Green, H. M. LeBaron, and W. K. Moberg, eds., *Managing Resistance to Agrochemicals*, American Chemical Society, Washington, D.C., 1990, p. 224.

34. D. Berg and M. Plempel, eds., *Sterol Biosynthesis Inhibitors—Pharmaceutical and Agrochemical Aspects*, Ellis Horwood, Chichester, 1988.

35. J. F. Ryley, R. G. Wilson, and K. J. Barrett-Bee, *J. Med. Vet. Mycol.* **22**, 53 (1984).

36. W. Köller in W. Köller, ed., *Target Sites of Fungicide Action*, CRC Press, Inc., Boca Raton, Fla., 1992, pp. 139–141.

37. K. H. Kuck and H. Scheinpflug, in G. Haug and H. Hoffmann, eds., *Chemistry of Plant Protection*, Vol. 1, Springer-Verlag, Berlin, 1986, pp. 65–96.

38. Y. Yoshida, in M. R. McGinnis, ed., *Current Topics in Medical Mycology*, Vol. 2, Springer-Verlag, New York, 1988, p. 388.

39. H. Vanden Bossche, in Ref. 40, Vol. 1, 1985, p. 313.

40. H. Vanden Bossche, in Ref. 36, p. 79.

41. C. R. Worthing and R. J. Hance, eds., *The Pesticide Manual*, 9th ed., British Crop Protection Council, Farnham, U.K., 1991.

42. W. Kramer, in Ref. 39, pp. 25–64.

43. M. A. De Waard, in I. Denholm, A. L. Devonshire, and D. W. Hollomon, eds., *Resistance 91: Achievements and Developments in Combating Pesticide Resistance*, Elsevier Applied Science, London, 1992, pp. 48–60.
44. T. Kabo, M. Shoami, and Y. Kaurase, *J. Pestic. Sci.* **5**, 69 (1980).
45. R. I. Baloch and E. I. Mercer, *Phytochemistry* **26**, 663 (1987).
46. A. Keerkenaar, in R. A. Fromtling, ed., *Recent Trends in the Discovery, Development and Evaluation of Antifungal Agents*, J. R. Prouse, Science Publishers, Barcelona, Spain, 1987, p. 523.
47. E. I. Mercer, in Ref. 36, p. 120.
48. W. Köller, in D. Pimentel, ed., *Handbook of Pest Management in Agriculture*, Vol. 3, 2nd ed., CRC Press, Boca Raton, Fla., 1991, p. 679.
49. T. Staub, H. Dahmen, and F. J. Schwinn, *Z. Pfl. Krankh. Pfl. Schutz* **87**, 83 (1980).
50. L. C. Davidse, A. E. Hoffman, and G. C. M. Velthuis, *Mycology* **7**, 344 (1983).
51. R. Wollgiehn and co-workers, *Z. Allgem. Mikrobiologie* **24**, 269 (1984).
52. F. J. Schwinn, in D. C. Erwin, S. Bartnicki-Garcia, and P. H. Tsao, eds., *Phytophthora: Its Biology, Taxonomy, Ecology*, American Phytopathic Society, St. Paul, Minn., 1983, p. 327.
53. L. C. Davidse, O. C. M. Gerritsma, and A. E. Hoffman, *Phytotiatrie-Phytopharmacie* **30**, 235 (1981).
54. D. W. Hollomon, in N. R. McFarlane, ed., *Crop Protection Agents*, Academic Press, Inc., London, 1977, p. 505.
55. D. W. Hollomon, *Pestic. Biochem. Physiol.* **10**, 181 (1979).
56. J. Dekker, *Nature* **197**, 1027 (1963).
57. Ref. 7, pp. 253–254.
58. S. H. Ou, *Plant Disease* **64**, 439 (1980).
59. O. Kodama, K. Yamashita, and T. Akatsuka, *Agric. Biol. Chem.* **44**, 1015 (1980).
60. M. Yoshida, S. Moriya, and Y. Uesugi, *J. Pestic. Sci.* **9**, 703 (1984).
61. J. D. Froyd and co-workers, *Phytopathology* **66**, 1135 (1976).
62. C. P. Woloshuk, H. D. Sisler, and E. L. Vigil, *Physiol. Plant Pathol.* **22**, 245 (1983).
63. P. M. Wolkow, H. D. Sisler, and E. L. Vigil, *Physiol. Plant Pathol.* **23**, 55 (1983).
64. C. P. Woloshuk and H. D. Sisler, *J. Pestic. Sci.* **7**, 161 (1982).
65. S. Inoue, T. Vematsu, and T. Kato, *J. Pestic. Sci.* **9**, 689 (1984).
66. C. P. Woloshuk and co-workers, *Pestic. Biochem. Physiol.* **14**, 256.
67. M. H. Wheeler, *Exp. Mycol.* **6**, 171 (1982).
68. H. Yamaguchi, C. Yamamoto, and N. Tanaka, *J. Biochem (Tokyo)* **57**, 667 (1965).
69. Ref. 3, p. 399.
70. N. Tanaka, in J. W. Corcoan and F. E. Hahn, Eds., *Antibiotics*, Vol. 3, Springer-Verlag, New York, 1975, p. 340.
71. B. Poldermans, N. Goosen, and P. H. van Knippenberg, *J. Biol. Chem.* **254**, 9085 (1979).
72. A. Endo and T. Misato, *Biochem. Biophys. Res. Commun.* **37**, 718 (1969).
73. S. Bartnicki-Garcia and E. Lippman, *J. Gen. Microbiol.* **71**, 301 (1972).
74. B. Bowers, G. Levin, and E. Cabib, *J. Bacteriol.* **119**, 564 (1974).
75. K. Endo and co-workers, *Nipon Nogei Kakatu Kaishi* **44**, 356 (1970).
76. H. L. Klopping and C. J. Delp, *J. Agric. Food Chem.* **28**, 467 (1980).
77. R. Fritz, D. Despreaux, and P. Leroux, *Tag.-Ber., 222, Akad. Landwirtsch.-Wiss, DDR, Berlin*, 65 (1984).
78. T. Staub and H. Hubele, in R. Wegler, ed., *Chemie der Pflazenschutz und Schädlingsbekämpfungsmittle*, Vol. 6, Springer-Verlag, Berlin, 1981, p. 389.
79. O. Langcake, *Phil. Trans. Royal Soc. London Ser. B* **295**, 83 (1981).
80. D. I. Guest, *Physiol. Plant Pathol.* **25**, 125 (1984).
81. J. Dekker, in Ref. 15, pp. 39–52.

82. A. Dearing, in H. Frehse, ed., *Proceedings 7th International Congress of Pesticide Chemistry (IUPC) Hamburg 1990*, VCH Verlagsgesellschaft, Weinheim, Germany, 1991, pp. 61–72.

General References

D. C. Torgeson, ed., *Fungicides: An Advanced Treatise*, Vols. 1 & 2, Academic Press, Inc., New York, 1967.

N. N. Melnikov, in F. A. Gunther and J. D. Gunther, eds., *Residue Reviews* Vol. 36, Springer-Verlag, Berlin, 1971.

R. W. Marsh, ed., *Systemic Fungicides*, 2nd ed., Longman, London, 1977.

M. R. Siegel and H. D. Sisler, eds., *Antifungal Compounds*, Vols. 1 & 2, Marcel Dekker, Inc., New York, 1977.

J. Dekker and S. G. Georgopoulos, eds., *Fungicide Resistance in Crop Protection*, Centre for Agricultural Publishing and Documentation, Wageningen, the Netherlands, 1982.

T. H. Staub and A. Hubele, in R. Wegler, ed., *Chemie der Pflanzenschutz-und Schädlings-bekämpfungsmittel*, Vol. 6, Springer-Verlag, Berlin, 1981.

G. Haug and H. Hoffman, eds., *Chemistry of Plant Protection, Vol. 1: Sterol Biosynthesis Inhibitors and Anti-Feeding Compounds*, Springer-Verlag, Berlin, 1986.

H. Buchenauer, in G. Haug and H. Hoffmann, eds., *Chemistry of Plant Protection*, Vol. 6, Springer-Verlag, Berlin, 1990.

H. Frehse, ed., *Pesticide Chemistry: Advances in International Research, Development and Legislation, Proceedings of the 7th International Congress of Pesticide Chemistry (IUPAC)*, Hamburg, 1990, VCH Verlagsgesellschaft, Weinheim, Germany, 1991.

C. R. Worthing and R. J. Hance, eds., *The Pesticide Manual*, 9th ed., The British Crop Protection Council, Farnham, U.K., 1991.

W. Köller, ed., *Target Sites of Fungicide Action*, CRC Press, Boca Raton, Fla., 1992.

I. Denholm, A. L. Devonshire, and D. W. Holloman, eds., *Resistance 91: Achievements and Developments in Combating Pesticide Resistance*, Elsevier Applied Science, London, 1992.

BARRY A. DREIKORN
W. JOHN OWEN
DowElanco

FUNGICIDES, INDUSTRIAL. See INDUSTRIAL ANTIMICROBIAL
AGENTS.

FURFURAL AND OTHER FURAN COMPOUNDS. See FURAN
DERIVATIVES.

FURNACES, ELECTRIC

INTRODUCTION

The term electric furnace applies to all furnaces that use electrical energy as their sole source of heat. The definition distinguishes such apparatus from traditional fuel-fired furnaces (see FURNACES, FUEL-FIRED) in which heat is produced directly by combustion of fossil fuels (eg, coal, oil, or gas). Electric furnaces are used mainly for heating solid materials to desired temperatures below their melting points for subsequent processing, or melting materials for subsequent casting into desired shapes, ie, electric heating furnaces or electric melting furnaces. The latter includes so-called holding furnaces which store a molten charge received from separate melting furnaces.

Classification is by the manner in which the electrical energy is converted into heat. Thus three distinct types of widely used industrial furnaces can be distinguished: electric resistance furnaces (qv), electric arc furnaces (qv), and electric induction furnaces (qv). The conversion of electrical energy into heat in each type of furnace is schematically illustrated in Figure 1. Common to all is a charge that is to be heated or melted, A, a refractory furnace lining, B, and an electric power supply, C. The refractory lining separates the hot furnace interior from the work area. It must withstand high temperatures and provide thermal insulation to conserve energy. The lining is often contained in a steel structure. The power supply is normally ac (dc is rarely used) at standard power line frequency (60 Hz in North America) and it usually includes a transformer so that the most suitable furnace voltage is obtained. However, certain induction furnaces require a higher frequency.

Electric Resistance Furnaces

In all electric resistance furnaces, the power supply is connected directly to an electric resistance in the interior of the furnace lining in which all electrical energy is converted to heat. Widely used and best known is the indirect-heat resistance furnace (Fig. 1a), also known as the electric resistor furnace. It resembles an electric toaster or kitchen oven. Resistors, D, are attached to the furnace roof or walls, or both, and are connected electrically to the power supply. Conventional metallic resistors serve to ca 1150°C; special resistor materials are needed for higher temperatures. Electrical energy is converted to heat in the resistors, and then the heat is transmitted to the charge by radiation and convection. This furnace type is suited for heating or melting, and there is no inherent dependence on the electrical conductivity of the charge; however, temperatures above 1700°C are not easily attained.

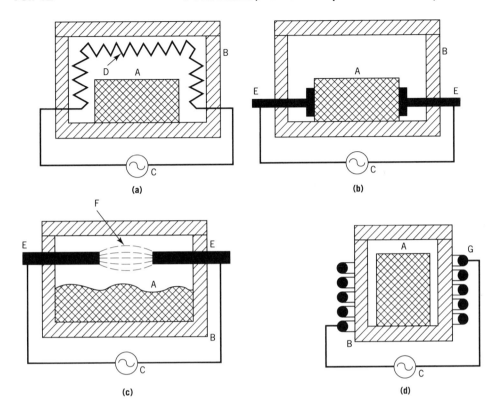

Fig. 1. Main types of electric furnaces: (**a**) resistance furnace, indirect heat (resistor furnace); (**b**) resistance furnace, direct heat; (**c**) arc furnace; (**d**) induction furnace. A, charge to be heated or melted; B, refractory furnace lining; C, electric power supply; D, resistors; E, electrodes; F, electric arc; G, induction coil.

The other form of electric resistance furnace is the direct-heat type (Fig. 1**b**). Here the resistance is the charge, A, which may be a solid mass, an agglomeration of solid particles, or a pool of molten material. The power supply is connected to this resistance through so-called electrodes, E, that are hot on the inside end and cold on the outside end. Use of direct-heat resistance furnaces is limited by the electrical conductivity of the charge and by the electrode system. Such use is well-suited for carbon, molten glass, and molten salts. However, direct-heat resistance furnaces cannot be used for nonconducting charges and generally are not practical for metals because of their high conductivity.

Electric Arc Furnaces

In electric arc furnaces (Fig. 1**c**), the conversion of electrical energy to heat occurs primarily within the electric arc, F, formed between spaced electrodes to which the power supply is connected. The arc has a high temperature (3700°C), and heat is transmitted to the charge primarily by radiation. A widely used electrode material is graphite, which can withstand high temperatures. Figure 1c illustrates the pure arc furnace principle; however, in most industrial furnaces, the arcs are formed between one or more electrodes and an electrically conducting charge. Most arc furnaces are large. They are used primarily for melting steel, ferro alloys, and nonmetallic substances. Operating temperatures are 1000–3000°C.

Electric Induction Furnaces

In electric induction furnaces (Fig. 1d), the conversion of electrical energy to heat takes place in the charge; however, no electrodes are used. Instead, the power supply is connected to a cold induction primary coil, G, on the exterior of the furnace lining. Electrical energy then is transmitted by transformer effect, causing a high secondary current to flow in the charge. Therefore, functioning of this furnace does not require the presence of any material other than the charge in the hot zone enclosed by the lining. This is a rather unique advantage. When necessary, extremely high temperatures can be attained. Induction furnaces require a charge of relatively high electrical conductivity. Thus they are best suited for heating and melting all metals and for heating graphite.

Economic Aspects

Selection of an industrial furnace for a given purpose usually involves deciding between electric and fuel-fired furnaces. Only electric furnaces are practical above 1700°C, but both types generally are considered for lower temperatures. Electric furnaces frequently have a higher capital (amortization) cost. Generally, there is not a decisive difference in energy cost as long as the cost of electricity is based on its generation from fossil fuels. However, in addition to amortization and energy the following potentially significant furnace operating cost factors should always be examined: (1) operating labor cost, (2) charge material loss (oxidation and vaporization), (3) furnace related raw material cost differences, (4) cost of rejects (effect of process control), (5) environmental control costs (investment and operation), and (6) furnace maintenance costs (including relining and electrodes). Electric furnaces often are selected in preference to fuel-fired furnaces because the former are characterized by a significantly lower total operating cost resulting from savings in the six areas just named. The same total operating cost analysis should be used when making the selection between various electric furnace types suitable for the same job.

The increasing importance of environmental considerations tends to favor selection of electric furnaces. Fuel-fired furnaces emit large volumes of hot products of combustion that contain objectionable gases and particles, which may include vapors or fumes from certain molten charges. The cost of treating these products of combustion to meet clean air standards can be very high, and will increase as standards are tightened. In electric furnaces (particularly resistance and induction furnaces), air pollution control is required only where pollutants emanate from the charge, but cost is much lower in the absence of a large hot gas volume.

Electric furnaces have a much higher efficiency and therefore release considerably less heat into their surroundings, thereby minimizing the need to cool the work area. Induction furnaces are best in this respect because the charge is heated internally and there are no elements operating above charge temperature. Noise control is also increasingly important. Resistance furnaces are quiet. Coil vibration noise may be a problem in induction furnaces, but can be minimized by coil design. Arc furnaces and fuel-fired furnaces generate noise which may have to be controlled (see INSULATION, ACOUSTIC; NOISE POLLUTION AND ABATEMENT METHODS).

BIBLIOGRAPHY

"Furnaces, Electric" in *ECT* 1st ed., Vol. 7, pp. 1–23, by V. Paschkis, Columbia University; in *ECT* 2nd ed., Vol. 10, pp. 252–278, by V. Paschkis, Columbia University; "Introduction" under "Furnaces, Electric" in *ECT* 3rd ed., Vol. 11, pp. 528–531, by M. Tama, Ajax Magnethermic Corp.

General References

W. Trinks, *Industrial Furnaces, Fuels, Furnace Types and Furnace Equipment—Their Selection and Influence Upon Furnace Operation*, 4th ed., Vol. II, John Wiley & Sons, Inc., New York, 1967, 358 pp., emphasis placed on heating furnaces (fuel-fired and electric) rather than melting furnaces.
A. G. E., Robiette, *Electric Melting Practice*, John Wiley & Sons, Inc., New York, 1972, 412 pp., arc, induction resistor, and special melting furnaces are discussed.
K. Kegel, ed., *Elektrowaerme, Theorie und Praxis*, Union Internationale d'Electrothermie, Verlag W. Girardet, Essen, Germany, 1974, 902 pp., a complete and up-to-date reference for all electric furnaces.
J. R. Rotella and G. A. Walzer, *Estimating Cupola and Induction Melting Using a Computer Model*, AFS Transactions, American Foundryman's Society, Chicago, Ill., pp. 331–335.
A. Muhlbaur, "Electrical Energy for Heating Processes for the Future," *Electrowarme Int.* **49**(B3), B58 (1991), a review of various electric heating and melting industrial applications.
Foundry Yearbook, 1977, Part III, Foundry Trade Journal, Redhill, Surrey, U.K., p. 254; Table 21 is a detailed application chart for fuel-fired and electric foundry melting furnaces.
E. Doetsch and H. Doliwa, "Economical and Process Technology Aspects of Cast Iron Melting," *Electrowarme Int.* **37**(B3), B157 (1979), contains an economic comparison: fuel-fired and electric iron foundry melting furnaces.

RICHARD A. SOMMER
Consultant

ARC FURNACES

Arc furnaces used in electric melting, smelting, and electrochemical operations are of two basic designs: the indirect and the direct arc. The arc of the indirect-arc furnace is maintained between two electrodes and radiates heat to the charge. The arcs of the direct-arc furnace are maintained between the charge and the electrodes, making the charge a part of the electrical power circuit. Not only is heat radiated to the charge, but the charge is heated directly by the arc and the current passing through the charge.

Indirect-Arc Furnaces

Indirect-arc furnaces have been used primarily in foundries for melting copper, copper alloys, and other nonferrous metals having a low melting point (see COPPER). They have also been used for producing molten iron and, occasionally, molten steel (see IRON; STEEL). The typical indirect-arc furnace is a single-phase furnace utilizing two horizontally mounted graphite electrodes, each of which project into an end of a refractory-lined horizontally mounted cylindrical steel shell. One electrode is set manually and the other electrode is automatically adjusted to maintain the preset voltage and current of the arc. The electrode pair is connected to a multivoltage tap transformer and reactor through flexible copper cables and bus bars.

  Although rocking of the furnace to intermittently cover and hence protect up to 90% of the refractory, as well as improved refractories, has done much to make the indirect-arc furnace more viable, these furnaces are becoming less common, primarily due to high operating costs as a result of erosion of the refractory by the intense arc radiation.

Direct-Arc Furnaces

  Open-Arc Furnaces. Most of the open-arc furnaces are used in melting and refining operations for steel and iron (Fig. 1). Although most furnaces have three electrodes and operate utilizing three-phase a-c power to be compatible with power transmission systems, d-c furnaces are becoming more common. Open-arc furnaces are also used in melting operations for nonferrous metals (particularly copper), slag, refractories, and other less volatile materials.

  A standard melting furnace consists of a refractory-lined steel shell with water-cooled upper sidewalls and roof (the lower portion is refractory to contain the molten metal); graphite electrodes; electrical equipment, bus bars, and flexible conductors to energize the electrode(s) (Fig. 2); equipment to regulate the position of the electrodes and thereby control the energy input; a means to access the inside of the furnace through a door; a method to tilt the furnace to empty it; and a means to allow the furnace to be recharged. Practically all furnace shells are short vertical cylinders made of welded steel plate with reinforcing and water-cooled segments. The bottom usually is comprised of a steel dished head so that the refractory bottom lining can form an inverted arch to ensure the refractory's integrity. The dished bottom also results in a more even heat flux and allows the hearth to expand and contract freely with temperature changes but without overstressing the refractories.

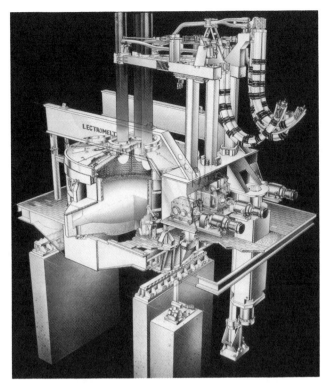

Fig. 1. A cut-away schematic of a typical a-c open-arc, steelmaking, eccentric bottom tapping (EBT) furnace. Courtesy of Lectromelt Corp.

Many shells are horizontally split to facilitate refractory repairs so that the nonproductive furnace time required to replace these refractories is minimized (1). Tapered shells are sometimes used to increase the charge capacity or hot metal capacity of existing furnaces. The conventional furnace shell contains a tapping spout to direct the molten contents when the furnace is emptied. More recently furnaces are being designed for eccentric bottom tapping (EBT) as shown in Figure 1. This design, where the tap hole is contained in the bottom of an extension to the furnace shell, allows the furnace to be completely drained by tilting the furnace only 15° as opposed to 45° for conventional furnaces. This allows a larger portion of the furnace sidewall to be water-cooled, which lowers refractory consumption. It also allows faster tapping and hence lower temperature losses during tapping. A water-cooled door is located diametrically across from the tap hole for the addition of alloys, fluxes, oxygen injection, etc, and allows the removal of slag. Additional openings are used on some furnaces to facilitate gunning, ie, spraying granular refractory material to rebuild the eroded lining; oxygen injection; or the introduction of oxy-fuel burners as a supplementary heat source.

The roof, in the form of a dome, is either comprised of refractory brick held in place by a water-cooled steel roof ring, or it may be composed of water-cooled panels. Sometimes water-cooled rings or glands are placed on the roof around the electrodes to maintain the refractory. On high power furnaces refractory is used around the electrodes to minimize the possibility of electrical short circuits.

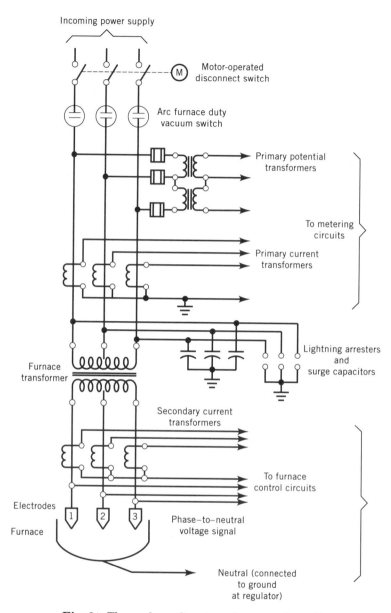

Fig. 2. Three-phase direct-arc furnace schematic.

Refractory Linings.   The refractory linings (2,3) for the hearth and lower walls of furnaces designed for melting ferrous materials may be acidic, basic, or neutral (see REFRACTORIES). Silica has been widely used in the past, and is still being used in a number of iron and steel foundries. Alumina, a neutral refractory, is normally used for furnace roofs and in the walls for iron foundries, but basic brick can also be used in roofs (4).

Magnesite or dolomite, basic refractories, are used primarily in furnaces where the sulfur or phosphorus content of the metal, or both, must be reduced.

Usually the bottom is lined with one or more courses of refractory brick in either an inverted dome or stadium configuration to serve as a safety lining. Compatible granular material is rammed over the safety lining until the desired bottom contour is attained. For fully refractory-lined furnaces, the wall is lined with compatible brick. For basic lined furnaces the slag line refractory often contains 10 to 20% added carbon and other metallic additives to enhance the thermal conductivity and decrease the reactivity of the refractory to increase its service life. Insulating brick is not used because it increases the rate of refractory consumption by shifting the isotherms toward the furnace shell.

Acid linings are the least expensive linings and are used wherever the melting process allows it. However, their use often requires careful selection of the charge materials to minimize residual element concentrations, and acid refractories are subject to spalling and thermal damage. Therefore, a more expensive lining may be more economical in an intermittent operation. In this case fireclay and linings containing a higher percentage of alumina are used, or, as in most cases, basic practice is adopted for an overall lower cost operation.

Water-cooled cast or fabricated panels (5,6) are popular and are used to replace up to 95% of the wall and roof refractories above the sill or slag line for EBT furnaces and 75% for conventionally designed furnaces. Energy consumption per ton of melt with water-cooled panels is generally the same as refractory linings. The typical life of the refractory walls and roofs in ferrous melting furnaces ranges from 200 cycles (heats) and can be greater than 1000 cycles, especially with patching or gunning, or both. In ultrahigh power furnaces (2000 kW/m^2 hearth area), the refractory life may be substantially reduced; thus special operating practices such as using foamy slags (7) are integral to the success of the arc furnace operation. The use of water-cooled panels has extended wall life to 1500 cycles, or more. The refractory hearth usually lasts at least six months and up to five years or more because it can be patched easily between cycles. As of the early 1990s, inert gas stirring through porous refractories or tuyeres in the hearth is being practiced. Energy and alloy savings are being claimed as well as increased yields due to improved slag/metal mixing.

Uses. The standard three-phase arc furnaces are available in sizes from 200 kg to 500 t and shell diameters of 1–12 m. Furnace transformer ratings are available from 200 to >160,000 kVA. The power density of steelmaking furnaces has gradually increased to the point at which extra ultrahigh power furnaces exhibit power densities in excess of 3500 kVA/m^2 of hearth.

Nearly all open-arc furnaces used in foundries and steel mills are three-phase and contain individually controlled jib-type electrode arms, each supporting a vertical column of graphite electrodes. The electrode arms are raised or lowered to maintain the desired arc characteristics, arc voltage, and current. This action takes place within a fraction of a second after the error signals are generated; the speed of the movement depends on the strengths of these signals as does the distance traveled. Each electrode arm is moving almost constantly because its arc characteristics are changing continually as scrap falls away from or against the electrodes, as the electrodes erode, as the atmosphere in the furnace changes, etc. Each electrode arm's electrical conductors are connected through flexible cables to the bus bars or tubes of the delta closure extensions and onto a multivoltage tap transformer. Generally, smaller furnace transformers (<7500 kW) and some larger transformers also contain a multitap reactor to provide sufficient inductive reactance to offset the negative characteristic of the arc so as to provide the desired arc stability.

Electrodes.   Almost all the electrodes (8,9) used in open-arc furnaces are prefabricated and are made of regular or dense graphite. Carbon electrodes seldom are used in melting furnaces, and those that are used are being replaced by graphite electrodes because of the latter's higher conductivity, lower weight, and smaller diameter which results in a smaller diameter electrode circle for a given size transformer. This increases the distance between the refractory wall and the arc which generally improves refractory life. Dense graphite electrodes are used instead of regular density electrodes whenever greater mechanical strength or slightly higher density, and accordingly conductivity, is required. Dense graphite electrodes are available in diameters of 32–762 mm. Regular density electrodes are commercially available in sizes of 178–610 mm diameter. With the advent of higher power d-c furnaces the pressure to increase the range of high density electrode sizes available has increased; thus electrodes up to 914 mm may be commercially available in the near future. Carbon electrodes are usually rated at electric current densities of 4.5–9 A/cm^2 and graphite electrodes at 15.5–46.5 A/cm^2.

The electrode diameter normally is selected on the basis of its current-carrying capability and its mechanical strength. The principal cause of electrode consumption usually is oxidation because of the high furnace temperatures and oxidizing furnace atmosphere. This oxidation rate is further accelerated by the electrode's surface temperature being increased by passage of current. Another factor is the stress that is imposed on the electrode by tilting the furnace, swinging the electrodes aside, scrap falling against the electrode, arc forces, etc.

Electrode consumption for ferrous melting a-c furnaces usually averages 2.5–6 kg/t of molten metal dependent on the particular furnace practices. D-c furnaces have electrode consumptions that are about 30% lower for similar operations. A typical energy consumption for a typical high productivity ministeel mill practice is 400 kW·h/t. In comparison, power consumptions exceeding 600 kW·h/t in foundries is not unusual because of longer furnace cycle times.

Voltage.   The voltage chosen for open-arc furnaces must be high enough to compensate for the voltage drops caused by the resistance and inductance of the primary and secondary electrical circuits and still have the required power input available to sustain the arcs (10). In the smaller furnaces, the voltage must be high enough to penetrate any thin oxide coatings on the scrap. Also, it must provide a sufficient area of meltdown; otherwise, the electrodes bore a small hole through the scrap, melting insufficient metal to cover the hearth resulting in high consumption of the bottom refractories. The highest phase-to-phase no-load voltage for a 200 kVA production furnace usually is 200 V, and 1000 V for a 120,000 kVA furnace is not uncommon. Lower voltages are also available for the operator to use during a furnace refining cycle; the lowest voltage is approximately one-third of the highest voltage. However, high productivity operations generally do not make use of the lower voltage taps.

For a given voltage tap, the operating electrode current can be <20 to >100% of the rated current depending on the quality of the electrode regulator and positioner. The current also reaches zero when there is no arc, because the electrode is too far from the charge, and maximum when the scrap falls against two electrodes and causes a short circuit. The power factor and arc voltage are highest at very low currents, whereas the maximum power input is attained at a power factor slightly higher than 0.707 (the cosine of 45°) depending on the electrical characteristics of the primary and secondary circuits. However, maximum power input does not necessarily equate with maximum efficiency (11). Typical operating power factors range from 0.68 to 0.82.

Vacuum-Arc Furnace.   Another type of open-arc furnace is the vacuum-arc furnace (12) which is used for melting metals that have high temperature melting points, eg, titanium, molybdenum, and tungsten, or for upgrading alloy steels (Fig. 3). An electrode of the material to be melted is cast or formed from metal powder. An arc is formed between the starter button, B, and the electrode, E. As the electrode is melted, an ingot is formed in the water-cooled copper or steel mold, M. The furnace or mold cavity usually is under a vacuum, but it may also be filled with an inert gas. In some instances, a water-cooled tungsten-tip electrode is used and the material to be melted is dropped into the melting chamber and the rate of feed is coordinated with the power input.

Plasma-Arc Furnace.   The plasma-arc furnace, sometimes used in the production of castings of high alloy steels and special alloys or for the smelting of fine materials, usually has a furnace shell similar to that of the three-phase conventional open arc-furnace used in the production of iron, steel, or ferroalloys. However, water-cooled nonconsumable electrodes are used to conduct direct current and argon (to serve as the plasma base) to the sealed furnace interior. The plasma torch can be of either the transferred or nontransferred type. The two types are distinguished by the electric current conduction path. There, most commonly, are one (14) or more (15) fixed electrodes in the furnace roof or sidewall and a water-cooled bottom electrode extending through the refractory hearth so that it is in contact with the molten metal serving as the anode. It is said that

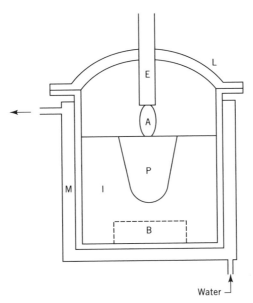

Fig. 3.   Consumable-electrode melting in the vacuum-arc furnace (13). A, arc; B, button; E, electrode; I, ingot; L, lid; M, water-cooled mold; and P, pool of molten metal.

the increased operating costs for argon and refractories are compensated for by the savings in alloy and graphite electrodes, but inherent longer arcs in plasma furnaces, if not properly contained, may reduce efficiency.

As with a plasma-arc furnace, various gases and pneumatically conveyed solids have been added to the various types of a-c open-arc furnaces to decrease alloy loss, to stabilize the arc, and to decrease the noise level, but with mixed commercial success.

D-C Arc Furnace. With the advent of more economical thyristor-controlled d-c power supplies, as well as limitations imposed by power companies on arc furnace-generated flicker, d-c furnaces (16) have become more common, particularly in countries with weak power grids such as Japan. These furnaces are nearly identical to their counterparts, except they typically have a single electrode passing through the roof and a means to collect the current through a furnace bottom electrode. Bottom electrode designs include full conductive bottoms and electrodes, made of various conducting materials, that protrude through the refractory to make contact with the melt. The latter type generally includes some water-cooled parts under the furnace.

Due to their similarity to a-c furnaces, d-c furnaces can be substituted for nearly any a-c furnace including the open-arc, submerged-arc, and arc-resistance furnaces, provided that design criteria, particularly electrical parameters, are properly chosen. Currently, steel and ferrochrome is being made commercially in d-c furnaces and a silicon metal pilot plant is being built.

There are substantiated claims that d-c furnaces exhibit advantages in power consumption, graphite consumption, noise, and power transmission line disturbance. Initially these may be offset by a slightly higher initial cost for these furnaces. In addition, production from a d-c furnace is claimed to be higher than a comparable a-c furnace, however, the required periodic bottom electrode maintenance may negate this gain over longer time periods.

Submerged-Arc Furnace. Furnaces used for smelting and for certain electrochemical operations are similar in general design to the open-arc furnace in that they are usually three-phase, have three vertical electrode columns and a shell to contain the charge, but direct current may also be utilized. They are used in the production of phosphorus, calcium carbide, ferroalloys, silicon, other metals and compounds (17), and numerous types of high temperature refractories.

When a smelting campaign is started, carbonaceous material, eg, coke or a conductive material, is placed on the hearth. The electrodes are lowered and arc on this material is at a low power input. The charge materials (ores, reductants, etc) are added slowly. As the material becomes molten and conductive, the power input and charging rate are correspondingly increased to the desired production rate when the furnace is filled with the charge materials. The electrodes continue to arc on the pool of molten metal on the furnace hearth, and the furnace exhibits electrical characteristics similar to open-arc steelmaking furnaces which have an attendant current voltage phase shift. Submerged describes the operation in that the electrode tip is surrounded by charge material in different stages of melting or reduction. In the submerged-arc case, the space immediately beneath the electrode is filled with ionized gases through which the arcs travel, which is typical of silicon alloy production operations. These gases travel upward through the burden, preheating it, and burn at the top with open flames. Generally the gases are collected to separate toxic materials and/or to recover waste heat.

Arc-Resistance Furnace. The arc-resistance furnace is similar to the submerged-arc furnace except the electrodes of the former are most often in direct contact with material, usually slag or a nonmetallic material, but they may also arc to the slag layer. Even when the electrode is in contact with the melt there are still minute arcs between the bottom and sides of the electrode, because it is not wetted by the slag, and the majority of the heat is developed in the melt in the immediate vicinity of the electrode tip. The furnace interior may be filled with a burden of unmelted charge above the melt, as in a submerged-arc furnace, or may contain a bare molten bath. The primary difference between the arc-resistance and submerged-arc furnace is that the former exhibits ohmic conductance. Often the two types are confused and hence misnamed. Most of the submerged-arc and arc-resistance furnaces do not tilt and sometimes do not have roofs. Where the volatilized materials and gases are toxic or cannot be exposed to air, the shell is covered with a refractory roof or hood so that the gases and vapors can be ducted away from the furnace for subsequent collection.

Most furnace shells are short vertical cylinders but may also be triangular, elliptical, or rectangular in plan view. Single-phase furnaces may have one or two movable electrodes. Three-phase furnaces usually have three movable electrodes, but some have six (three pairs, two electrodes for each phase). This is more common for larger smelting furnaces used to produce ferronickel, ferromanganese, silicon, and copper mattes. A few of the smaller furnaces must tilt to expedite emptying a portion of the molten material since the transfer must be extremely rapid to keep the high melting point material from freezing. In cases of materials having melting points over 2200°C, the vertical portion of the furnace shell may be removable to allow cooling of the materials for a day or two. After cooling, the fused material can be broken into smaller pieces for further processing and the nonfused material can be added to the next furnace charge. Generally, these furnaces have a refractory lining. When there is no appropriate refractory material to withstand the high temperature and the chemical reaction with the material to be processed, then unmelted material is used for lining the furnace. In other instances, the shell is cooled by water sprays to freeze the melt, thus utilizing a self-lining concept.

Electrodes. Because of the numerous different processes, there are many different types of electrodes in use (9), eg, prefabricated graphite, prefabricated carbon, self-baking, and composite electrodes (see CARBON). Graphite electrodes are used primarily in smaller furnaces or in sealed furnaces. Prebaked carbon electrodes, made in diameters of <152 cm or 76 by 61 cm rectangular, are used primarily in smelting furnaces where the process requires them. However, self-baking electrodes are preferred because of their lower cost.

The self-baking electrode (8–18) consists of a cylindrical steel casing that has internal radial fins. These casings are periodically filled with a carbonaceous paste. As the electrode is consumed and automatically lowered, the heat from the furnace and the electrical current passing through the electrode softens and subsequently bakes the carbonaceous paste to form a solid, monolithic mass. New sections of casing are welded on as the electrode is consumed. Self-baking electrodes having diameters of <165 cm are used on numerous types of ferroalloy furnaces when the process can tolerate the iron impurity (see also ALUMINUM AND ALUMINUM ALLOYS). Testing has been performed on iron-free self-baking electrodes, but they have not yet become commercially accepted. Electrodes over

165 cm are seldom used because the skin effect of a-c current prevents proper utilization of such a large conductor.

A typical large three-phase ferroalloy furnace using prebaked carbon electrodes is shown in Figure 4. The hearth and lower walls where molten materials come in contact with refractories are usually composed of carbon blocks backed by safety courses of brick. In the upper section, where the refractories are not

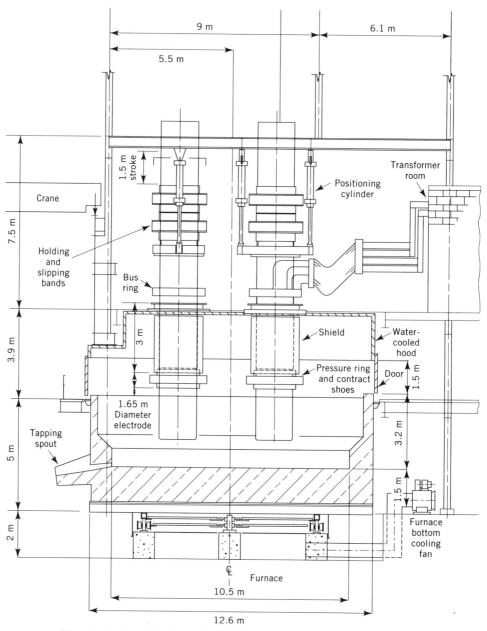

Fig. 4. Design of a ferroalloy furnace. Courtesy of Lectromelt Corp.

exposed to the higher temperatures, superduty or regular firebrick may be used. The walls of the shell also may be water-cooled for extended life. Usually, the furnace shell is elevated and supported on beams or on concrete piers to allow ventilation of the bottom. When normal ventilation is insufficient, blowers are added to remove the heat more rapidly. The shell also may rest on a turntable so that it can be oscillated slightly more than 120° at a speed equivalent to 0.25–1 revolution per day in order to equalize refractory erosion or bottom buildup.

The larger electrodes are usually supported from overhead by chains, cables, or steel bands connected to a regulating winch or hydraulic cylinders. Alternatively, the columns may be supported directly by hydraulic cylinders. The electrodes are often held in alignment by insulated vertical guides. The arcs are quite stable because of the atmosphere in the arcing zone that allows the use of slower yet more sensitive regulators and positioners as compared to steelmaking furnaces.

Table 1 shows some of the typical electrode consumption figures for various submerged-arc furnace operations.

Table 1. Electrode^a Consumption, kg/MW·h

standard ferromanganese	5.0
pig iron	5.0
75% ferrosilicon and silicon metal	6.0
aluminum silicon	15.0
calcium carbide	7.0
phosphorus	1.5
nickel matte	5.0

^{a}Submerged arc furnace.

Energy consumptions for submerged-arc and arc-resistance furnaces are generally higher than for open-arc furnaces, because the latter is used primarily for melting and refining metals. In contrast, the submerged-arc and arc-resistance furnaces are used in melting of compounds, eg, ores, slags, etc, that have much higher specific heats. Furthermore extra energy must also be furnished to allow the desired endothermic reactions to proceed to completion. Therefore, in a reduction operation, energy consumption figures can vary from 700 to over 13,000 kW·h/t of product, depending on the materials to be processed and the product desired. In submerged-arc and arc-resistance furnaces, the powering, furnace size, electrode size, electrode spacing, diameter–height ratio of the lined shell, voltage, and current are critical factors that must be closely coordinated to optimize energy consumption, electrode consumption, and furnace production (19–21).

Health and Safety

Because intense heat is generated in these furnaces it is understandable that the arc volatilizes such metals as tin, zinc, lead, cadmium, and the like. In addition, both melting and smelting furnaces may generate large amounts of carbon monoxide. As a result all new furnace installations require pollution control equipment. This normally consists of off-gas afterburning (sometimes with energy recovery), and dust collection equipment, typically a baghouse. Most dusts collected are considered hazardous wastes because of their heavy-metal content and accordingly must be treated and/or disposed of in a prescribed manner.

For arc furnace worker safety, high power electrical systems require proper design and precautions, and handling of molten materials requires a minimum of fire-retardant clothing and often dust masks. Water must be prevented from coming in contact with the melt. Furthermore, since open-arc furnace noise levels commonly exceed 100 dBA, hearing protection is a necessity. Noise is normally not a problem with smelting furnaces.

BIBLIOGRAPHY

"Furnaces, Electric" in *ECT* 1st ed., Vol. 7, pp. 1–23, by V. Paschkis, Columbia University; in *ECT* 2nd ed., Vol. 10, pp. 252–278, by V. Paschkis, Columbia University; "Arc Furnaces" under "Furnaces, Electric" in *ECT* 3rd ed., Vol. 11, pp. 531–541, by D. Oakland, Whiting Corp.

1. J. W. Wild, *ISS-AIME Electric Furnace Proceedings*, Vol. 34, I&SS, St. Louis, Mo., 1976, pp. 301–303.
2. *AIME Electric Furnace Steelmaking*, Vol. 1, Wiley-Interscience, Inc., New York, 1962, pp. 153–174.
3. C. R. Taylor, ed., *Electric Furnace Steelmaking*, Iron and Steel Society, Warrendale, Pa., 1985, pp. 63–70.
4. C. R. Davis and B. H. Baker, *ISS-AIME Electric Furnace Proceedings*, Vol. 33, Houston, Tex. 1975, pp. 136–144.
5. R. Assenmacher, E. Eisner, and D. Ameling, *Proceedings of the Third International Iron and Steel Congress*, Chicago, Apr. 16–20, 1978, pp. 570–579.
6. D. DiMicco, M. Parish, and D. Arthur, *ISS Electric Furnace Proceedings*, Vol. 47, Orlando, Fla., 1989, pp. 195–200.
7. H. Grippenberg and co-workers, *Iron and Steel Engineer*, 44–53 (July 1990).
8. *Electrode Use in Electric Arc Furnaces*, Iron and Steel Society, Warrendale, Pa., 1986.
9. J. A. Persson, *ISS-AIME Electric Furnace Proceedings*, Vol. 35, Chicago 1977, pp. 115–126.
10. J. A. Ciotti, *TMS-AIME Electric Furnace Proceedings*, Vol. 28, Pittsburgh, Pa. 1970, pp. 130–139.
11. K. H. Klein and co-workers, *Second European Electric Steel Congress*, Florence, Italy, Oct. 1986, pp. R3.3/1–R3.3/32.
12. R. Schlatter, *J. Metals*, (Apr. 1970).
13. V. Paschkis and J. Persson, *Industrial Electric Furnaces and Appliances*, 2nd ed., Interscience Publishers, Inc., New York, 1960.
14. K. Upadhya, J. J. Moore, and K. J. Reid, *J. Metals*, (Feb 1984).
15. C. McCombe, *Foundry Trade J. Int.*, 60 (Dec. 1978).
16. Y. Vigneron and co-workers, *ISS-Electric Furnace Proceedings*, Vol. 44, Dallas, Tex., 1986, pp. 37–41.
17. W. Fettweis, G. Rath, and W. Haiduk, in Ref. 9, pp. 21–26.
18. R. E. Scherrer and M. L. Stott, *AIME Electric Furnace Proceedings*, Vol. 22, Buffalo, N.Y., 1964, pp. 136–141.
19. W. M. Kelley, *Carbon and Graphite News* **5** (1) (Apr./May 1958).
20. J. A. Persson, in Ref. 10, pp. 168–169.
21. J. Westly, in Ref. 4, pp. 37–41.

J. KEVIN COTCHEN
MAN GHH Corporation

INDUCTION FURNACES

Induction furnaces utilize the phenomena of electromagnetic induction to produce an electric current in the load or workpiece. This current is a result of a varying magnetic field created by an alternating current in a coil that typically surrounds the workpiece. Power to heat the load results from the passage of the electric current through the resistance of the load. Physical contact between the electric system and the material to be heated is not essential and is usually avoided. Nonconducting materials cannot be heated directly by induction fields.

Utility power distribution grids normally operate at a fixed frequency of 50 or 60 Hz. These frequencies can be utilized directly for the induction process if the load characteristics are appropriate. If they are not, specific applications can be optimized by the use of variable and higher frequencies produced by solid-state frequency power converters connected between the supply and the load.

The efficiency of an induction furnace installation is determined by the ratio of the load useful power, P_n, to the input power, P_o, drawn from the utility. Losses that must be considered include those in the power converter (transformer, capacitors, frequency converter, etc), transmission lines, coil electrical losses, and thermal loss from the furnace. Figure 1 illustrates the relationships for an induction furnace operating at a constant load temperature with variable input power. Thermal losses are constant, coil losses are a constant percentage of the coil input power, and the useful out power varies linearly once the fixed losses are satisfied.

Induction Heating

Design. The coil of an induction heater typically encircles the load, as shown in Figure 2. The current intensity within the load is greatest at the surface and diminishes to zero at the center (Fig. 2**a**) (1). This crowding of the current close to the surface is known as skin effect. The rate at which the current intensity decreases from its maximum value at the surface is a function of the applied frequency, the resistive and magnetic properties of the load, and the load diameter. A useful term in induction design is reference depth, which is defined as the thickness of a shell that with a constant value of current equal to the current at the surface of the load results in developing the same power as the actual load (Fig. 2**b**). The electrical efficiency of an optimized induction heating coil and load combination as a function of reference depth is shown in Figure 3. The curve suggests that a minimum load diameter of four times the reference depth is desirable for reasonable efficiency.

Power Supplies and Controls. Induction heating furnace loads rarely can be connected directly to the user's electric power distribution system. If the load is to operate at the supply frequency, a transformer is used to provide the proper load voltage as well as isolation from the supply system. Adjustment of the load voltage can be achieved by means of a tapped transformer or by use of a solid-state switch. The low power factor of an induction load can be corrected by installing a capacitor bank in the primary or secondary circuit.

Some induction heating furnaces must operate at frequencies higher than the supply frequency. Formerly, rotating motor alternator frequency converters

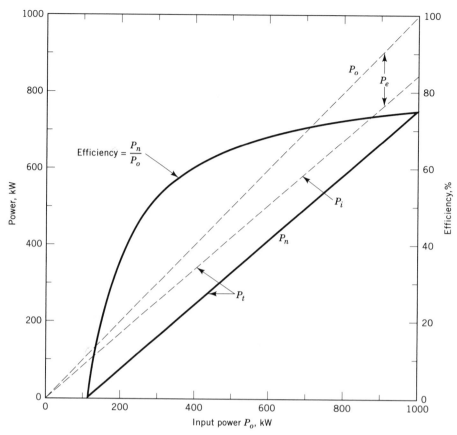

Fig. 1. Induction furnace efficiency. Typical characteristics of a 1000 kW furnace. Example: P_e = 15% of P_o and P_t = 100 kW. P_n = useful power; P_o = power input; P_e = electrical loss; P_i = induced power; and P_t = thermal loss.

were used. Now the availability of high speed, high power silicon controlled rectifiers for use in frequency converters has made rotary converters obsolete. Modern units operate at higher efficiency, cost less, require less factory space, and coordinate readily with process controls (2).

Economics. Induction heating equipment installations can require significant investment in electric power components as well as the work handling equipment made necessary by the process. These costs can be offset by savings in plant space, reduction in metal loss, precise control of product temperature, and reduced in-process inventory. A typical continuous induction heating line consumes about 360 kW·h/t heating carbon steel bars to 1230°C.

Applications. A unique capability of induction heating is apparent in its ability to heat the surface of a part to a high temperature while the interior remains at room temperature. Proper selection of material, high frequency, and high power density can produce a thin surface hardness with a heat unaffected core (3). Figure 4 shows the cross section of a typical automotive shaft heated with 10 kHz at various power densities. The required hardness depth is selected to satisfy

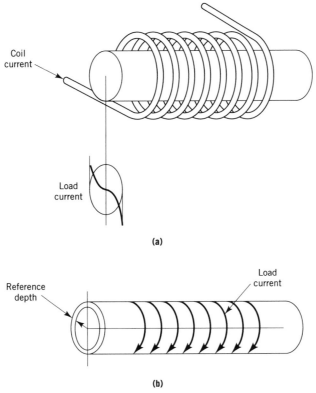

Fig. 2. Induction heating coil and load showing (**a**), current distribution in load, and (**b**), reference depth.

the product requirements. The ability to precisely control the power and length of the induction heating cycle allows it to be integrated into complex work handling equipment.

Induction heating using low frequency and low power density when applied to a stationary or moving bar can produce a uniformly heated part suitable for introduction to a rolling mill (4). A coil line capable of producing 32 t/h of 17.8 cm (7 in.) diameter alloy steel bars heated to 1177°C is shown in Figure 5.

Induction heating is used to heat steel reactor vessels in the chemical process industry (5). The heat produced in the walls is conducted to the material within. Multisectioned coils are used to provide controlled heat input to the process material as it passes through the reactor. Figure 6 illustrates a cross section of such a typical installation.

High process temperatures generally not achievable by other means are possible when induction heating of a graphite susceptor is combined with the use of low conductivity high temperature insulation such as flake carbon interposed between the coil and the susceptor. Temperatures of 3000°C are routine for both batch or continuous production. Processes include purification, graphitization, chemical vapor deposition, or carbon vapor deposition to produce components for

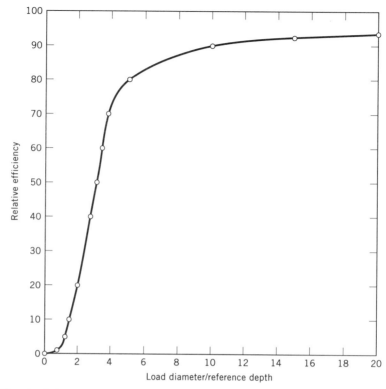

Fig. 3. Relative coil efficiency vs ratio of load diameter to reference depth.

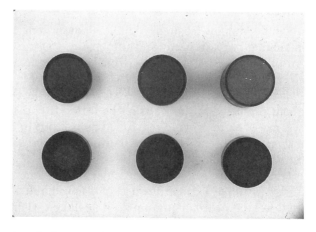

Fig. 4. Automotive shaft cross sections showing the effect of power density vs hardened depth. Courtesy of Ajax Magnethermic Corp.

Fig. 5.   Continuous bar heating line. Courtesy of Ajax Magnethermic Corp.

the aircraft and defense industry. Figure 7 illustrates a furnace suitable for the production of aerospace brake components in a batch operation.

A special coil configuration is used to heat thin strips of metal that cannot be heated efficiently with a coil that encircles the load, as the strip thickness is small compared to the depth of penetration. The transverse flux induction coil is positioned on either side of a strip to produce a uniformly heated strip with good efficiency in a much smaller space than conventional radiant or convective strip heating furnaces (6).

Induction Melting

Induction melting applications almost always contain the liquid metal charge within a hearth formed by a suitable refractory material. It is possible to design the hearth to satisfy a wide variety of application requirements ranging from a few kilograms to hundreds of tons of metal and for operation in normal or hostile environments. As the heat is developed within the charge, the metal and the

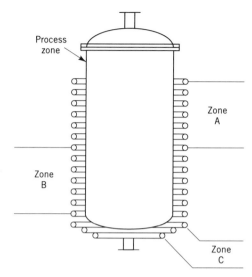

Fig. 6. Process reactor.

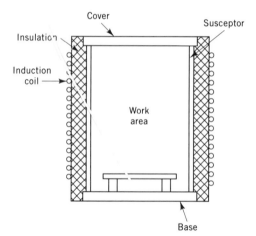

Fig. 7. High temperature processor.

furnace refractory are not exposed to excessive temperatures that may be present in either electric arc or fuel-fired furnaces (7,8). Operation is practical in vacuum or inert atmospheres for the production of critical materials that require protection from oxygen or other gases. The environmental impact of an induction furnace is generally less than that of an equivalent fuel-fired furnace.

Coreless Induction Furnaces

Coreless furnaces derive their name from the fact that the coil encircles the metal charge but, in contrast to the channel inductor described later, the coil does not encircle a magnetic core. Figure 8 shows a cross section of a typical medium sized furnace. The coil provides support for the refractory that contains the metal being heated and, therefore, it must be designed to accept the mechanical loads as well as the conducted thermal power from the load. In small coreless furnaces the coil itself may possess sufficient strength to allow satisfactory operation. Larger furnaces provide support to the coil from surrounding structures.

Frequency Selection.   When establishing the specifications for a coreless induction furnace, the material to be melted, the quantity of metal to be poured for each batch, and the quantity to be produced per hour must be considered simultaneously. Graphs have been developed that combine these factors with practical experience to indicate possible solutions for a specific requirement.

Skin effect is utilized in the design of coreless furnaces. It is particularly evident when the furnace is full of molten metal. Current and power are distributed within the volume of metal just as they are in an induction heating load. In both cases power density at the center of the coil is greater than at the ends of the coil, and in the coreless furnace this results in a circulation of metal. This circulation assists in the melting process by carrying the charge below the surface of the melt and assures a uniform bath temperature and metallurgical homogeneity. The use of lower frequencies produces stronger circulation in the same furnace at the same power level.

Operation.   Small and medium sized coreless induction furnaces powered from high frequency power supplies can be started with a charge of metal pieces at room temperature, usually scrap material of appropriate alloy. The charge material is selected to allow a reasonable power to be drawn from the power

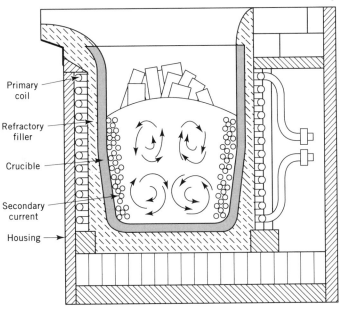

Fig. 8.   Small coreless-induction furnace, 500-kg high frequency furnace with insulating board housing and crucible.

supply. As the metal charge begins to melt, a molten pool is established and the charge compacts, allowing additional charge to be added. Alloy additions and temperature adjustments complete the melting cycle (9). Higher operating efficiency is achieved if the next cycle is initiated promptly after the charge is poured off so that the stored energy in the hearth refractory is not lost to the coil cooling water. Large coreless furnaces operating at line frequency are often started with a molten initial charge, although it is possible to start with a charge of solid material. Typical operation of these furnaces involves dispensing 20 to 30% of the furnace capacity and immediately recharging dry or preheated material into the bath as power is applied. These furnaces are usually held full during off shifts to maximize refractory life. An alternative is to empty the furnace and maintain the refractory continuously warm with supplemental heat. Furnaces with capacities of 4.5 to 13.6 t with input power ratings of 825 to 1100 kW/t produce liquid iron at a consumption rate of 550 to 600 kW·h/t.

Hearth. The induction melting coil is almost always round and in the form of a right cylinder. It is highly desirable that the refractory lining within the coil be uniform in thickness, so most hearths are cylindrical whether they hold a few kg or 59 t. There are a few instances of a smaller coil being attached to the bottom of a larger hearth, so the hearth could be modified to suit a particular requirement (10). Oval coils have been built and operated satisfactorily, but they are rare.

Channel Induction Furnaces

The term channel induction furnace is applied to those in which the energy for the process is produced in a channel of molten metal that forms the secondary circuit of an iron core transformer. The primary circuit consists of a copper coil which also encircles the core. This arrangement is quite similar to that used in a utility transformer. Metal is heated within the loop by the passage of electric current and circulates to the hearth above to overcome the thermal losses of the furnace and provide power to melt additional metal as it is added. Figure 9 illustrates the simplest configuration of a single-channel induction melting furnace. Multiple inductors are also used for applications where additional power is required or increased reliability is necessary for continuous operation (11).

Inductor. The channel inductor assembly consists of a steel box or case that contains the inductor refractory and the inductor core and coil assembly. The channel is formed within the refractory. Inductor power ratings range from 25 kilowatts for low temperature metals to 5000 kilowatts for molten iron. Forced air is used to cool the lower power inductors, and water is generally used to cool inductors rated 500 kilowatts or more.

Metal contained in the channel is subjected to forces that result from the interaction between the electromagnetic field and the electric current in the channel. These inward forces produce a circulation that is generally perpendicular to the length of the channel. It has been found that shaping the channels of a twin coil inductor shown in Figure 10 produces a longitudinal flow within the channel and significantly reduces the temperature difference between the channel and the hearth (12).

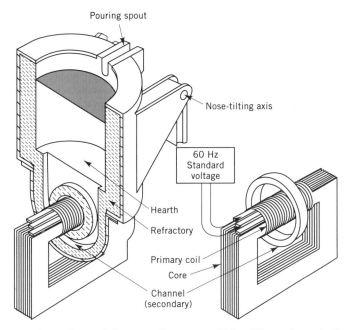

Fig. 9. Basic channel furnace. Courtesy of Ajax Magnethermic Corp.

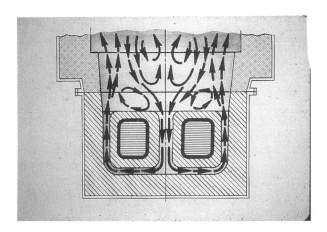

Fig. 10. Twin coil, jet flow inductor. Courtesy of Ajax Magnethermic Corp.

Hearth. The hearth of a channel induction furnace must be designed to satisfy restraints that are imposed by the operating inductor, ie, the inductor channels must be full of metal when power is required, and it is also necessary to provide a sufficient level of metal above the channels to overcome the inward electromagnetic pressure on the metal in the channel when power is applied. Once these requirements are satisfied, the hearth can then be tailored to the specific application (13). Sizes range from stationary furnaces holding a few hundred kilograms of aluminum to rotating drum furnaces with a useful capacity of 1500 t of liquid iron.

The refractory used to construct the hearth can be in the form of bricks, preformed shapes, or monolithic. Often a furnace design utilizes all three. Openings or passageways through the walls are fashioned in the same manner as windows in a brick building.

The steel shell that encloses the refractory is exposed to significant forces from the expansion of the refractory as well as the load from the refractory and the charge within the furnace. Similarly, the structures that support the furnace and the foundations must be designed to assure safe operation. A failure of any component can have serious consequences.

Operation. Channel furnaces can be used for melting or holding metal. In either case, the inductor and the hearth refractory are preheated to avoid thermal shock as the liquid metal is introduced at start up. Once the inductor channel has been flooded, it is rarely emptied until the inductor is taken out of service. Inductor life can vary from six months to a number of years depending on the metal alloy and the size and power rating of the inductor. Channel melting furnaces are often designed so that a large portion of their total capacity can be discharged by tilting or rotating the furnace. Dry or preheated metal is added to the furnace at the melt rate of the furnace.

Holding furnaces usually operate with a relatively constant metal level. Included in this category are furnaces that supply metal to various casting processes and large pots that hold metal for continuous coating lines. Multiple inductor furnaces are designed so that individual inductors can be replaced without emptying the remaining inductors.

Applications. Small and medium sized foundries producing castings for automotive and other similar applications often utilize iron melting channel melting furnaces. They allow melting off shift at lower power demands and make their total working batch available at the start of the pouring shift. Power consumption under these operation conditions ranges from 600 to 880 kW·h/t. More continuous operation can reduce this figure. Furnaces have been designed to superheat liquid iron delivered in 90 t batches prior to its introduction into a basic oxygen furnace (BOF) for conversion to steel. Similar furnaces are utilized for duplexing in conjunction with cupolas in large foundries.

A typical melter installed in a medium sized brass foundry contains 4500 kg of brass and its inductor is rated 500 kilowatts. Brass is an alloy containing copper and zinc. Zinc vaporizes at temperatures well below the melting temperature of the alloy. The channel inductor furnace's low bath temperature and relatively cool melt surface result in low metal loss and reduced environmental concerns. Large drum furnaces have found use in brass and copper continuous casting installations.

A combination of a channel induction holding furnace with an induction heating coil is shown in Figure 11. Steel strip is introduced into a zinc bath in a coating pot. The process is called continuous galvanizing. In this installation further heating of the strip extends the alloying of iron and zinc to produce a "galvannealed" strip for automobile bodies with improved fabrication and corrosion resistance characteristics (14,15). The control provided by the use of induction furnaces results in a superior product compared to fuel-fired alternatives.

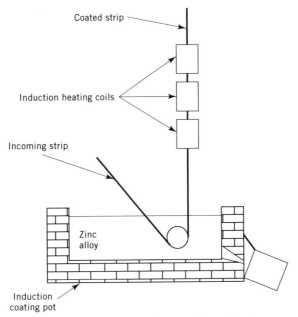

Fig. 11. Galvanize–galvanneal installation.

BIBLIOGRAPHY

"Induction Furnaces" under "Furnaces, Electric" in *ECT* 1st ed., Vol. 7, pp. 1–23, by V. Paschkis, Columbia University; in *ECT* 2nd ed., Vol. 10, pp. 252–278, by V. Paschkis, Columbia University; in *ECT* 3rd ed., Vol. 11, pp. 542–550, by M. Tama, Ajax Magnethermic Corp.

1. J. T. Vaughan and J. W. Williamson, *AIEE Trans.* **65**, 887–892 (1946).
2. G. F. Bobart, *Ind. Heat.*, 22–26 (June 1989).
3. R. F. Kern, *Heat Treat.*, 20–24 (Dec. 1991).
4. S. B. Lasday, *Ind. Heat.*, 42–46 (Feb. 1992).
5. K. G. Webley, "Induction Heating of Steel Reactor Vessels," *Chemical Process Industry Symposium*, AICHE, Philadelphia, Pa., June 5–8, 1978.
6. S. B. Lasday, *Ind. Heat.*, 43–45 (Oct. 1991).
7. S. B. Lasday, *Ind. Heat.*, 31 (Sept. 1991).
8. *Foundry Manage. Technol.*, B3–B13 (Dec. 1990).
9. H. Roth, *ABB Review*, 25–33 (June 1990).
10. H. G. Heine and J. B. Gorss, *Metallurgical Trans. A*, 489–513 (Nov. 1990).
11. M. Tama, *J. Metals*, (Jan. 1974).
12. U.S. Pat. 3,595,979 (July 27, 1971), W. E. Shearman (to Ajax Magnethermic).
13. H. Roth, *ABB Review*, 25–33 (June 1990).
14. U.S. Pat. 4,895,736 (Jan. 23, 1990), R. A. Sommer, G. Havas, and M. Tama.
15. T. J. Logan, *Steel Technol. Int.*, 227 (1992).

General References

American Foundrymen's Society, Inc., *Refractories Manual*, 2nd ed., Des Plaines, Ill., 1989.
American Society for Metals, *Metals Handbook, Heat Treating*, Vol. 4, 9th ed., Metals Park, Ohio, 1991.
Materials Engineering Institute, *Course 60, Induction Heating*, American Society for Metals, Metals Park, Ohio, 1986.
S. L. Semiatin and D. E. Stutz, *Induction Heat Treatment of Steel*, American Society for Metals, Metals Park, Ohio, 1986.
W. Trinks, *Industrial Furnaces*, Vol. 1, 4th ed., John Wiley & Sons, Inc., New York, 1951.
C. A. Tudbury, *Basics of Induction Heating*, John Rider, New York, 1960.
S. Zinn and S. L. Semiatin, *Elements of Induction Heating*, Electric Power Research Institute, Palo Alto, Calif., 1988.

RICHARD A. SOMMER
Consultant

RESISTANCE FURNACES

The most widely used and best known resistance furnaces are indirect-heat resistance furnaces or electric resistor furnaces. They are categorized by a combination of four factors: batch or continuous; protective atmosphere or air atmosphere; method of heat transfer; and operating temperature. The primary method of heat transfer in an electric furnace is usually a function of the operating temperature range. The three methods of heat transfer are radiation, convection, and conduction. Radiation and convection apply to all of the furnaces described. Conductive heat transfer is limited to special types of furnaces.

Operating temperature ranges are classified as low, medium, and high; there is no standard or precise definition of these ranges. Generally, a low temperature furnace operates below 760°C, medium temperature ranges from 760–1150°C, and furnaces operating above 1150°C are high temperature furnaces. There is often indiscriminate use of the words furnace and oven. The term oven should be used when temperatures are below 760°C, and the word furnace applied for higher temperatures. The term furnace is used here regardless of operating temperature.

Batch and Continuous Furnaces

The determination of the need for either a batch or continuous furnace is dependent on production rate and the physical size and weight of the work to be processed.

Batch Furnaces. In batch furnaces the desired time–temperature cycle for the product to be processed is accomplished by subjecting the entire furnace and its contents or charge of work to the particular cycle. Batch furnaces are most often used for very large and/or heavy charges, low production rates, infrequent operation, variable time–temperature cycle, and processing material that must be in batches because of previous or subsequent operations. Larger batch furnaces are often of the elevator or car-bottom type. A typical electric elevator furnace is shown in Figure 1. In this furnace, the charge of one or many pieces is loaded onto the hearth of a car. The car is moved under the furnace and is hoisted into the furnace by way of an elevator mechanism which is part of the furnace. Very large or heavy loads are often processed in a car-bottom furnace similar to the elevator furnace except that the furnace is not elevated and the car carries the work into the furnace through a door at one end.

Medium-sized loads are often processed in a bell furnace, as shown in Figure 2. The operation of this furnace is opposite to that of an elevator furnace: the work load is placed on a stationary hearth and the furnace is lowered over the hearth. Bell furnaces are often arranged with two or more bases (hearths) which permit more efficient use of the furnace because one base can be unloaded/loaded as the furnace carries out a heating cycle on another base.

Small loads are commonly processed in a box furnace. The product is placed on the furnace hearth through a door. Box furnaces may be single-ended or double-ended. A single-ended box furnace is usually used in an air atmosphere application where the product can be removed hot from the furnace for cooling. A double-ended box furnace is usually used in a controlled atmosphere application. In this case a water cooler is attached to one end. The product can be placed on the hearth (in the heat chamber) through the front door, then after the product reaches temperature, it is manually transferred into the water cooler for cooling before it is manually removed out the exit door on the other end of the water cooler.

Other versions include the pit furnace, which is a box furnace with the door on top and which is often installed in a pit with the top of the furnace near floor level.

Fig. 1. Elevator furnace. Courtesy of Wellman Furnaces, Inc.

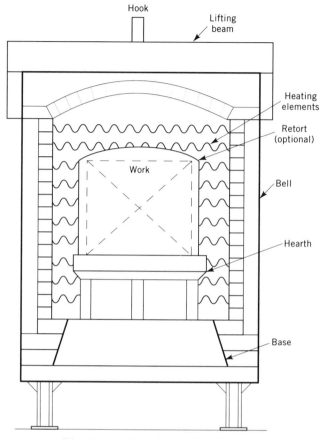

Fig. 2.   Typical electric bell furnace.

Continuous Furnaces.   These furnaces are applicable for uniform charges of work that arrive at the furnace continuously, moderate to high production rates, constant time–temperature cycle, and continuous operation over at least one and preferably two or three shifts per day. The desired time–temperature cycle is designed into the continuous furnace. The charge is subjected to this cycle by moving it through chambers operating at different temperatures. Although temperatures may be varied, a sequence, eg, heat, cold, and cool, is a part of the furnace design, and it may be difficult and expensive to change this cycle once the furnace is built. Although multicycle continuous furnaces are possible, they are expensive and are subject to design limitations. Continuous furnaces are usually named for the method used to convey the material through the furnace.

The roller-hearth furnace shown in Figure 3 is used to process a wide variety of parts. Unless the configuration of the work permits it to roll on a roller conveyor, the charge is carried on a tray or, in the case of small parts, in a basket which is in turn carried by a tray. For any given operating temperature, the weight that can be carried per unit length of the furnace is limited by the strength of the rolls.

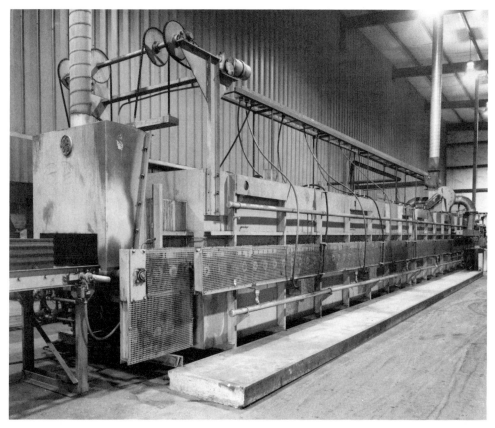

Fig. 3. Roller-hearth furnace. Courtesy of Wellman Furnaces, Inc.

Light loads are often processed in a mesh-belt furnace which usually carries the work load directly on the mesh belt. At a given operating temperature, loading per unit area of the belt is limited by its tensile strength. Cast-link belt furnaces function in the same manner as mesh-belt furnaces except that the former carry heavier loads because the belt is made from suitable alloy castings instead of woven wire. The belt is normally contained in the furnace on both the working and return sides, whereas the mesh-belt usually exits the furnace with the work load and returns outside the furnace. Because of its large weight, it is uneconomical to let the cast-link belt cool on the return and reheat it with the work load.

In pusher furnaces, the product (work load) is pushed through the furnace in steps by a hydraulic or electromechanical mechanism that pushes each load into the furnace, thus pushing all work in the furnace ahead one work space. The walking-beam furnace lifts the work load on a walking beam, advances the load a step, and returns the work to the hearth. The walking beam then returns to its original position (under the hearth) in preparation for the next step.

Furnace Atmospheres

Electric furnaces can operate either with air in the interior of the furnace or with a protective atmosphere; the choice is dictated by the process requirements of the work. The furnace must be designed for the atmosphere to be used, because the combination of temperature and atmosphere are significant factors in selecting internal materials used in the furnace construction; this applies particularly to the selection of heating element (resistor) material. It is feasible and common to design an electric furnace that can operate in both air and protective atmospheres although shortened element life generally results from frequent alternating between reducing atmospheres and oxidizing atmospheres. There are exceptions to this rule as some resistor materials must be periodically oxidized, if used in a reducing atmosphere. Other resistor materials are limited to a particular atmosphere.

Air-Atmosphere Furnaces.   These furnaces are applied to processes where the work load can tolerate the oxidation that occurs at elevated temperatures in air. In some special applications, the oxidation is not only tolerable but is desired. Some furnaces heat the work solely to promote oxidation. Furnaces designed for air operation are not completely gas-tight which results in somewhat lower construction costs. There are no particular problems encountered in selecting the insulation systems because almost all refractory insulations are made up of oxides. Heating element materials are readily available for the common temperature ranges used with air atmospheres.

Protective-Atmosphere Furnaces.   These furnaces are used where the work cannot tolerate oxidation or where the atmosphere must provide a chemical or metallurgical reaction with the work. In some cases, mainly in high temperature applications, the atmosphere is required to protect the electric heating element from oxidation.

Protective-atmosphere furnaces are of two general types. In one type, the work is inside a muffle (retort) and the protective atmosphere is inside the muffle. The outside of the muffle and the interior of the furnace operate in air and are designed accordingly. The other type is gas-tight, and the atmosphere is introduced directly into the furnace, obviating the expensive and expendable retort or muffle. It does require careful selection of the internal furnace parts which must not be adversely affected by the atmosphere. The selection of electric heating elements must be carefully made with respect to operating temperature and atmosphere. The best material to use for a given application is a function of the combination of temperature and atmosphere.

The true operating temperature range and atmosphere must be specified in a description of an electric controlled-atmosphere furnace. Frequently, a higher temperature is specified than required for the contemplated operations, apparently with the thought that the higher temperature construction results in a safety factor. In the case of heating elements it can result in more expensive materials which, at the true operating temperature, are actually inferior to less expensive element materials. In addition to heating elements, there are other materials used in furnace construction that are satisfactory for one temperature range but are not suitable for a lower temperature operation.

Low Temperature Convection Furnaces

Low temperature convection furnaces are designed to transfer the heat from the heating elements by forced convection. Convection is normally used in furnaces operating below 760°C because it is the most effective means of heat transfer that can maintain good uniformity of temperature on various workload configurations. Convection furnaces also are used (in this range of temperatures) where it is important that no part of the work load exceed the controlled temperature. This is accomplished by shielding the work load from any view of the heating elements and by controlling the temperature of the air or atmosphere, which carries the heat from the heating elements to the work, at the desired maximum temperature.

One design for a low temperature convection furnace shown in Figure 4 utilizes an external circulating fan, heating chamber, and duct system. The fan draws air (or a protective atmosphere) from the furnace and passes through the external heating chamber and back into the furnace past the work. This system minimizes the chance that the work receives any direct heat radiation. In theory it is less efficient because the external blower, heating chamber, and ductwork add external surfaces that are subject to heat losses.

Another design, shown in Figure 5, functions similarly but all components are inside the furnace. An internal fan moves air (or a protective atmosphere) down past the heating elements located between the sidewalls and baffle, under the hearth, up past the work and back into the fan suction. Depending on the specific application, the flow direction may be reversed if a propeller-type fan is used. This design eliminates floorspace requirements and eliminates added heat losses of the external system but requires careful design to prevent radiant heat transfer to the work.

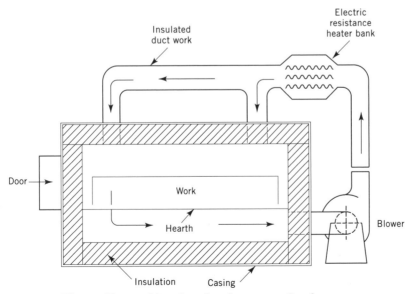

Fig. 4. External heating chamber convection furnace.

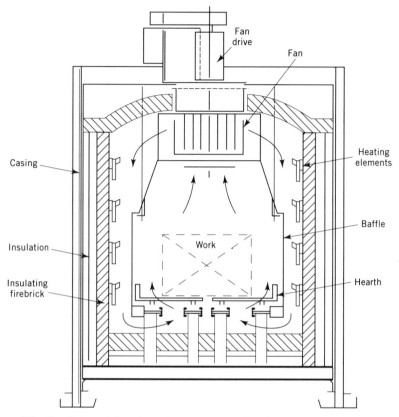

Fig. 5.   Internal heating-element convection furnace cross section.

Heating elements operating <760°C are almost always of a chrome–nickel resistance alloy and are in the form of ribbon, cast alloy, open wire coils, or sheathed construction. Several alloys are suitable in this temperature range and all are satisfactory if properly applied. In general, the more expensive alloys are used when physical space limitations dictate higher watts per area dissipation from the element.

Radiation Furnaces

Low Temperature Radiation Furnaces.   These are of the infrared heater type. Heat transfer is by direct radiation from a high temperature heating element. Control of the heat is obtained by controlling the time of exposure to the heat radiation. This type of furnace is normally used for such applications as drying of paint films. Heating elements are nickel–chrome resistance wire which is wound on ceramic supports or contained in sheaths. Other materials include tungsten resistors in glass or quartz envelopes which exclude oxygen from the resistors.

Medium Temperature Radiation Furnaces. The temperature range is generally 760–1150°C. Most of the heat is transferred directly to the work by radiation from the heating elements and by radiation to the furnace refractory which reradiates the energy to the work. Heating elements may be located in the sidewalls, roof, or floor of the furnace. Location of heating elements must be selected to assure uniform temperatures in the work zone. Elements located below the hearth level require protection from falling work or other contaminants. Figure 6 shows a typical cross section through a radiation furnace.

From 760 to 960°C, circulating fans, normally without baffles, are used to improve temperature uniformity and overall heat transfer by adding some convection heat transfer. They create a directional movement of the air or atmosphere but not the positive flow past the heating elements to the work as in a convection furnace. Heating elements are commonly chrome–nickel alloys in the forms described previously. Sheathed elements are limited to the very low end of the temperature range, whereas at the upper end silicon carbide resistors may be used. In this temperature range the selection of heating element materials, based on the combination of temperature and atmosphere, becomes critical (1).

High Temperature Radiation Furnaces. These furnaces are similar in construction to medium temperature radiation furnaces, but operate above 1150°C.

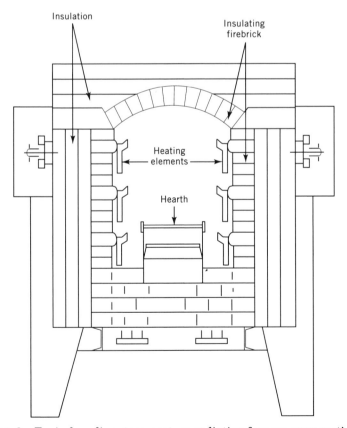

Fig. 6. Typical medium temperature radiation furnace cross section.

The insulation system must be designed to withstand the high temperatures, and internal structural parts become critical.

At temperatures above 1150°C, alloys used for the hearth or material handling systems in low and medium temperature furnaces lose strength rapidly (2) and temperatures are reached where ceramic refractories are required to support the work. This results in less use of roller-hearth and belt-type hearths and greater use of pushers or walking-beam designs for continuous furnaces.

Chrome–nickel alloy heating elements that commonly are used in low temperature furnaces are not suitable above the very low end of the range. Elements commonly used as resistors are either silicon carbide, carbon, or high temperature metals, eg, molybdenum and tungsten. The latter impose stringent limitations on the atmosphere that must be maintained around the heating elements to prevent rapid element failure (3), or the furnace should be designed to allow easy, periodic replacement.

Refractory selection becomes critical with high temperature radiation furnaces that have molybdenum or tungsten heating elements. Although these elements are stable in vacuum or inert atmospheres, many applications require a reducing atmosphere, often high in hydrogen content and very dry. In these cases, there is the possibility of reducing the oxides that make up refractory insulations. Published data are available relating temperature and dew point at which hydrogen reduces the various oxides present in insulations (4). If this point is reached, the reduction process begins to destroy the insulation system of the furnace, thus limiting the maximum practical operating temperature to less than the capability of the heating element material (see REFRACTORIES).

Vacuum Radiation Furnaces. Vacuum furnaces are used where the work can be satisfactorily processed only in a vacuum or in a protective atmosphere. Most vacuum furnaces use molybdenum heating elements. Because all heat transfer is by radiation, metal radiation shields are used to reduce heat transfer to the furnace casing. The casing is water-cooled and a sufficient number of radiation shields between the inner cavity and the casing reduce the heat flow to the casing to a reasonable level. These shields are substitutes for the insulating refractories used in other furnaces.

Conduction Furnaces

Conduction furnaces utilize a liquid at the operating temperature to transfer the heat from the heating elements to the work being processed. Some furnaces have a pot filled with a low melting metal, eg, lead, or a salt mixture, eg, sodium chloride and potassium chloride, with a radiation-type furnace surrounding the pot. Although final heat transfer to the work is by conduction from the hot lead or salt to the work, the initial transfer of heat from the resistors to the pot is by radiation.

Conduction furnaces are of three general types. One has a pot or crucible with suitable exterior insulation. Sheathed resistance elements are inside the pot which contains molten lead or another low melting metal. The molten metal can be the conductive medium that transfers heat to the work immersed in it, or the molten metal may be the work. Such furnaces are often used to supply molten-type metal, lead, zinc, etc. As the molten metal is removed, bars of the metal are added for melting. The initial charge of solid metal does not provide good surface contact with the heating elements and, because of this, the metal around the heating elements is often melted initially with a torch or other auxiliary heat during start up.

The salt-bath furnace is another type of conduction furnace. A molten salt not only provides the medium for conductive heat transfer, the salt is the heating resistor. These furnaces commonly are applied for temperatures ranging from slightly above the melting point of the salt to 1260°C. The salt-bath furnace shown in Figure 7 consists of a metal (or for higher temperatures, ceramic) crucible that is surrounded by a suitable insulating refractory and the outer casing. Metal electrodes immersed in the salt are connected to a low voltage power source. Placement of electrodes is important because the mean path through the salt determines the resistance of the salt which serves as heating resistor. The current should flow from electrode to electrode with a negligible amount of current passing through the work when it is immersed in the bath.

Cold sodium chloride, either granulated or solidified after melting, has a high electrical resistance and low heat conductance (see SODIUM COMPOUNDS). It is, therefore, necessary to make provisions for melting the initial charge of salt. This can be done by melting it in a radiation furnace and pouring the molten salt into the salt-bath furnace or by melting the salt in the salt-bath furnace with a torch. The torch method is necessary when a salt bath solidifies as a result of a planned maintenance shutdown or a prolonged power failure. If the shutdown is planned, heat conducting metal tubes can be placed in the bath to improve heat transfer when remelting the salt.

The third type of conduction furnace is a fluidized bed. In this design the product to be heated is submerged in sand, which is supported by a high porosity plate. Heated air (or atmosphere) is recirculated through the porous plate and

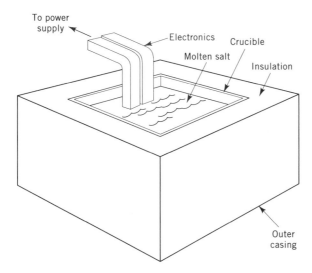

Fig. 7. Salt-bath furnace.

sand, which gives a high heat-transfer efficiency to the product. The disadvantage of this furnace is the product usually has a lot of warpage. To correct this, fixturing usually is required, which means more mass needs to be heated. Fluidized-bed furnaces are usually used at low and medium temperature ranges; the higher the temperature the more maintenance is needed on the recirculating fan.

Direct-Heat Electric-Resistance Furnaces

Direct-heat electric furnaces use the material to be heated as the resistor, and the furnace consists of an insulated enclosure to retain the heat, a power source of suitable voltage, and means of attaching the power leads to the work (Fig. 8). This type of furnace has several limitations that have prevented widespread use. Since the work is the resistor, it must have a uniform cross section between power connection points, and the material must be homogeneous. Varying sections or nonuniformities in the material can produce hot or cold spots in proportion to the change in electrical resistance. Also, a given furnace must be designed for work in which each piece to be heated has about the same resistance and power requirements. Although voltage and power can be controlled, a furnace designed to heat a part with a given cross section and length probably does not have the voltage required to heat a part of twice the length and half the cross section or have the current capacity to heat a part of half the length and twice the cross section.

There are additional problems in making the electrical connections to the work to be heated. The connection must have low electrical resistance to prevent overheating at the point of contact, but such a connection has a low resistance to heat transfer, thus conducting heat away from the work. These problems have limited the use of direct-resistance heating mainly to heating of pipe, tubing, bars, or small identical parts. This heating is a one-piece-at-a-time batch operation and often does not use an insulated housing; instead it is used as a preheater for a forming operation that takes place as soon as the work reaches the desired temperature.

There are large-scale operations using direct-heat resistance furnaces. These are mainly in melting bulk materials where the liquid material serves as a uni-

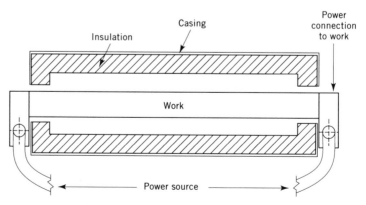

Fig. 8. Direct-heat resistance furnace.

form resistor. The material is contained in a crucible of fixed dimensions which, coupled with a given resistivity of the material, fixes the total resistance within reasonable limits. The most common application for this type of direct-heat electric resistance furnace is the melting of glass (qv) and arc furnaces for the melting of steel (qv).

Applications

Electric furnaces are used for annealing, brazing, carburizing, galvanizing, forging, hardening, melting, sintering, enameling, and tempering metals, most notably aluminum, copper, iron and steel, and magnesium alloys. Protection of metals is done in exothermic (lean and rich), prepared nitrogen (lean and rich), endothermic (lean and rich), charcoal, exothermic–endothermic (lean and rich), dissociated ammonia, and combusted ammonia (lean and rich) atmosphere.

BIBLIOGRAPHY

"Resistance Furnaces" under "Furnaces, Electric" in *ECT* 1st ed., Vol. 7, pp. 16–23, by V. Paschkis, Columbia University; in *ECT* 2nd ed., Vol. 10, pp. 269–278, by V. Paschkis, Columbia University; in *ECT* 3rd ed., Vol. 11, pp. 551–562, by R. M. Miller, Wellman Thermal Systems Corp.

1. C. D. Starr and F. M. Cole, *Ind. Heat Mag.* (Oct. 1963).
2. *High Alloy Data Sheets, Heat Series*, Alloy Casting Institute Division, Steel Founders' Society of America, New York, 1973.
3. *How to Make Out With Moly*, Schwarzkopf Development Corp., Holliston, Mass.
4. *Hydrogen Atmospheres, Their Influence on Serviceability of Refractories*, paper 2, Babcox & Wilcox Co., Refractories Division, Augusta, Ga., 1971.

General References

W. Trinks and M. Mawhinney, *Industrial Furnaces, Principles of Design and Operation*, 5th ed., Vol. 1, John Wiley & Sons, Inc., New York, 1961, 486 pp.
W. Trinks and M. Mawhinney, *Industrial Furnaces, Fuels, Furnace Types and Furnace Equipment—Their Selection and Influence Upon Furnace Operation*, 4th ed., Vol. 2, John Wiley & Sons, Inc., New York, 1967, 358 pp.
V. Paschkis and J. Persson, *Industrial Electric Furnaces and Appliances*, 2nd ed., Interscience Publishers, a division of John Wiley & Sons, Inc., New York, 1960.

ROBERT R. WALTON
Wellman Furnaces, Inc.

FURNACES, FUEL-FIRED

A furnace is a device (enclosure) for generating controlled heat with the objective of performing work. In fossil-fuel furnaces, the work application may be direct (eg, rotary kilns) or indirect (eg, plants for electric power generation). The furnace chamber is either cooled (waterwall enclosure) or not cooled (refractory lining). In this article, furnaces related to metallurgy such as blast furnaces are excluded because they are covered under associated topics (see IRON; COMBUSTION TECHNOLOGY; COAL; FURNACES, ELECTRIC).

Among the technologies in existence by ca 4000 BC, which included the manufacture of synthetic lapis lazuli, the development of the first true pottery kilns must rate as a significant achievement (1). For polychrome pottery to be successfully manufactured, it was essential to separate the fire (fuel) from the work (clay pottery). The excavations performed in the near east (Mesopotamia in antiquity) indicate that these early kilns were probably of beehive construction. Subsequent Egyptian pottery kilns of the period ca 3000 BC were the familiar chimney shape. With the smelting of copper in pit hearths predating by perhaps a millenium the start of the Bronze Age at ca 3000 BC, another important advance was the invention of the bellows at ca 2000 BC. Bellows supply combustion air where it is needed and are used as a means of raising furnace temperature.

Waterwall furnaces were employed by the ancient Greeks and Romans for household services. A water boiler, found in Pompeii, was constructed of cast bronze and incorporated the water-tube principle (2). The earliest recorded instance of boilers performing mechanical work (130 BC) was Hero's engine (3) which is the earliest known reaction turbine. Furnaces, in general, and waterwall furnaces, in particular, were neglected for about the next 1600 years. In part, this may be ascribed to the fact that steam as a working fluid had no application until the invention of the first commercially successful steam engine at the end of the seventeenth century, followed by Newcomen's engine in 1705 with self-regulating steam valves. These machines were generally used for pumping water from coal mines (2). Their sequence furnace-boiler-work foreshadowed the eventual emergence of the utility power plant which, in essence, is the same concept.

In fire-tube furnaces developed in the nineteenth century, such as typified by the Scotch-Marine boiler (Fig. 1), thin currents of water contact a multiplicity of tubes; thus, the hot gases transmit heat simultaneously to all regions of the bulk of the water. Therefore, this boiler–furnace combination steams readily and responds promptly to load changes, and is, for a given amount of heating surface, the least expensive of all furnace–boiler installations (4). Furnaces of this type, such as the steam locomotive furnace–boiler design, had the obvious disadvantage that pressure was limited to ca 1 MPa (150 psi). The development of seamless, thick-wall tubing for stationary power plants (ie, water-tube furnaces) and other engines for motive power, such as diesel–electric, has in many cases eclipsed the fire-tube boiler. For applications calling for moderate amounts of lower pressure steam, however, the modern fire-tube boiler continues to be the indicated choice (5).

A key development in water-tube furnace design was the Babcock and Wilcox boiler of 1877 (Fig. 2) (3). This can be considered the direct evolutionary ancestor of the 1000 MW steam power plants a century later (see STEAM).

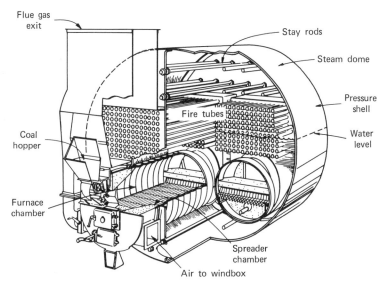

Fig. 1. Scotch-Marine boiler, ca 1930.

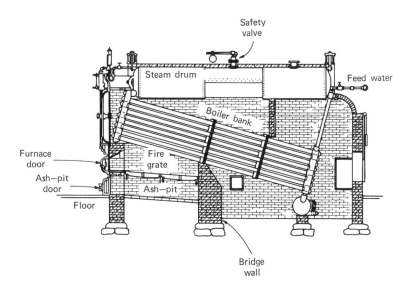

Fig. 2. Early Babcock and Wilcox boiler, 1877.

Classification

Furnaces are either cooled (water- or air-cooled chamber) or not cooled (refractory-lined chamber). In general, the basic structure roughly resembles either a rectangular box or a cylinder with variations for generally good reasons (6). For example, if the material being processed is a liquid, the furnace bottom must provide a bath. Hence, the open-hearth furnace is used for steel melting and refining, the reverbatory furnace for copper, and glass tanks for various materials. If the material tends to be lumpy, as in smelting, cupola melting, or lime burning, the furnace is constructed vertically to make gravity feeding possible.

The roof of the typical furnace collapses unless it is built as a sprung or self-supporting arch, or is flat but suspended, along with the rest of the furnace, from an enclosing structural framework. All modern power-plant waterwall furnaces of any significant megawatt rating are of suspended construction.

If a refractory-wall furnace is very large, the walls must be relatively thick to provide the necessary structural strength required to withstand the weight of the roof and thermal stresses as well as to retain heat. Large furnaces are built with two or even three different types of brick, with the outer brick generally the more highly insulating and the inside brick able to withstand the highest temperatures. The middle bricks are, therefore, often of lower quality and carefully selected since their softening temperature must not be exceeded.

Wall losses through most refractory walls are ca 10% of the heat supplied by the fuel. Losses increase with rising operating temperature. In special cases, eg, in glass tanks, losses can be as high as 30–35%. In these instances, very high values are required to maintain the refractory at a temperature below which it does not melt or collapse.

A furnace may be direct-fired or indirect-fired. The indirect-fired is known as a muffle furnace, and in such furnaces the combustion gases are separated from the stock being heated to prevent contamination.

Fuels

Fuel-fired furnaces primarily utilize carbonaceous or hydrocarbon fuels. Since the purpose of a furnace is to generate heat for some useful application, flame temperature and heat transfer are important aspects of furnace design. Heat transfer is impacted by the flame emissivity. A high emissivity means strong radiation to the walls.

The carbon–hydrogen ratio of fuels is a variable used widely in fuel technology to estimate emissivity as shown in Figure 3. However, its effect on flame temperature is usually misrepresented (7). The adiabatic flame temperature, T_{ad}, is the theoretical maximum flame temperature. It is commonly believed that flame temperature is a significant function of heat of combustion. This is only true at very low values of heat of combustion, below ca 13.9 MJ/kg (6000 Btu/lb), because the largest weight of material moving through most furnaces is atmospheric nitrogen (7). This usually outweighs the contribution from the fuel by 10 to 1 or more (Table 1). Coupled with a statistical requirement of roughly 0.0283 m^3 (1 ft^3) of cold air for every 105 kJ (100 Btu) released, fossil-fuel combustion is roughly equivalent to a release of 3.7 MJ/m^3 (100 Btu/ft^3) of combustion mixture at STP (or 5.4 MJ/m^3 (145 Btu/ft^3) for the combustion of CO and H$_2$). Consequently, most adiabatic flame temperatures are ca 2000 $\pm$ 100°C (6). This holds for the heat of combustion range of ca 14–140 MJ/kg (6,000–60,000 Btu/lb). The range includes CO and H$_2$ although they are not strictly fossil fuels.

Figure 4 illustrates the trend in adiabatic flame temperatures with heat of combustion as described. Also indicated is the consequence of another statistical result, ie, flames extinguish at a roughly common low limit (1200°C). This corre-

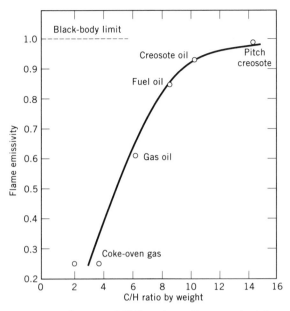

Fig. 3. Influence of C/H ratio on flame emissivity.

Table 1. Theoretical Air Requirement for Stoichiometric Combustion

Fuel	Air (dry), wt/wt fuel
gas	
blast furnace	1.54
average natural	15.86
coke oven	11.36
oil, bunker C	13.83
coal	
coke by-product	10.03
anthracite	10.64
bituminous, Eastern low vol	10.80
bituminous, Eastern high vol	10.38
bituminous, Midwest	8.54
subbituminous, Wyo.	7.03
lignite, N. Dak.	5.33
wood	
30% moisture	4.06
50% moisture	2.90
bagasse, 50% moisture	2.61

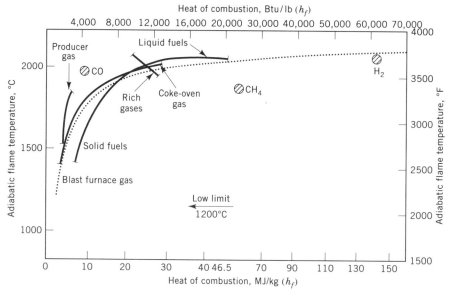

Fig. 4. Variation of adiabatic flame temperature with heat of combustion where $T_{ad} = T_o + \dfrac{2083}{1 + 1.7/h_f}$ (°C). Note change of scale at 46.5 MJ/kg (20,000 Btu/lb).

sponds to heat-release density of ca 1.9 MJ/m³ (50 Btu/ft³) of fuel–air mixtures, or half that for the stoichiometric ratio. It also corresponds to flame temperature, as indicated, of ca 1220°C. Because these are statistical quantities, the same numerical values of flame temperature, low limit excess air, and so forth, can be expected to apply to coal–air mixtures and to fuels derived from coal (see FUELS, SYNTHETIC).

The insensitivity of adiabatic flame temperature to heat of combustion does not necessarily apply to the operational flame temperature, T_o, which is the flame temperature found in an actual furnace (remembering that this refers to some average temperature). The higher excess air requirements at higher C/H ratios coupled with greater thermal loads on longer flames generally results in markedly lower operational temperatures as the C/H ratio increases.

The fuels listed in Table 2 are generally representataive of fuels to be encountered over the range of industrial furnaces and, depending on the type (cooled or refractory wall), exhibit operating temperatures considerably different from adiabatic values. The choice of fuel is dependent upon a number of factors including cost, availability, cleanliness, emissions, reliability, and operations. Small furnaces tend to burn cleaner, easier to use fuels. Large furnaces can more effectively use coal.

The transition from a choice of multiple fossil fuels to various ranks of coal, with the subbituminous varieties a common choice, does in effect entail a fuel-dependent size aspect in furnace design. A controlling factor of furnace design is the ash content and composition of the coal. If wall deposition thereof (slagging) is not properly allowed for or controlled, the furnace may not perform as predicted. Furnace size varies with the ash content and composition of the coals used. The ash composition for various coals of industrial importance is shown in Table 3.

Table 2. Fuel Analyses, Wt %

Fuel	Carbon	Hydro-gen	Sulfur	Nitro-gen	Oxygen	Ash	Mois-ture	HHV,[a] MJ/kg[b]
gas								
blast furnace	15.86	0.20		53.97	28.91		1.06	2.62
natural	69.20	23.09	0.35	5.81	1.55			51.02
coke oven	48.62	20.03		4.91	26.44			39.95
oil (fuel)	85.5	11.5	1.6	0.7	0.7			43.03
coal								
coke by-product	85.0	0.7	1.0	1.3	0.5	10.7	0.8	29.51
anthracite	83.9	2.9	0.7	1.3	0.7	8.0	2.5	31.91
bituminous, low vol	80.7	4.5	1.8	1.1	2.4	6.2	3.3	33.28
bituminous, high vol A	76.6	4.9	1.3	1.6	3.9	9.1	2.6	31.65
bituminous, high vol B	67.6	4.3	1.8	1.4	5.4	13.5	6.0	28.37
bituminous, Midwest	63.4	4.3	2.3	1.3	7.6	8.7	12.4	26.56
bituminous, West	66.7	5.2	0.5	1.2	9.7	8.7	8.0	27.91
subbituminous A	60.6	4.6	0.5	1.3	13.3	7.3	12.4	24.65
subbituminous B	54.6	3.8	0.4	1.0	13.2	3.8	23.2	21.91
subbituminous C	47.3	3.3	0.7	0.6	12.1	9.5	26.5	19.12
lignite, N. Dak.	42.4	2.8	0.7	0.7	12.4	6.2	34.8	16.77
lignite, Tex.	41.0	3.1	0.7	0.7	12.3	9.9	32.2	16.86
wood								
30% moisture	34.3	4.2			30.8	0.7	30.0	15.39
50% moisture	24.5	3.0			22.0	0.5	50.0	11.00
bagasse	22.5	3.0			23.5	1.0	50.0	9.76

[a]High heating value.
[b]To convert MJ/kg to Btu/lb, multiply by 430.

Table 3. Ash Properties of Typical Coals

Property	Eastern bituminous	Midwest bituminous	Western subbituminous	N. Dak. lignite
Ash composition, wt %				
acid SiO_2	44.8	44.7	33.5	13.8
acid Al_2O_3	28.0	18.4	18.0	10.7
base Fe_2O_3	17.8	18.3	7.3	8.4
base CaO	2.0	5.3	18.4	24.2
base MgO	0.6	1.1	4.7	7.6
base Na_2O	0.3	1.8	0.2	9.0
base K_2O	1.7	2.1	0.2	0.4
acid TiO_2	1.4	0.8	0.7	0.4
acid P_2O_5	0.3	0.2	0.2	0.6
acid SO_3	0.8	5.8	15.3	20.3
other	2.3	1.5	1.5	4.6
base–acid ratio	0.30	0.45	0.59	2.0
Fe_2O_3–CaO ratio	8.9	3.5	0.4	0.35
Ash fusibility temperatures, °C				
initial deformation	1210	1120	1130	1130
softening	1350	1150	1200	1200
fluid	1420	1240	1270	1350

Power-Plant Furnaces

In 1991, 22% of the electric power consumed in the United States was generated by nuclear energy (see NUCLEAR REACTORS). It is expected that before the year 2000 this will have increased to 25% with the remainder supplied by fossil-fuel and hydroelectric power (8,9). The bulk of electric power generation will probably be supplied by coal-fired power-plant furnaces supplying steam to turbogenerators. In terms of megawatts supplied, the coal-fired power plant is, therefore, the foremost component of importance in the energy supply system. Power-plant furnaces are of waterwall type and are generally designed for steam pressures in the range of 12.4–24.1 MPa (1800–3500 psi); the latter value is referred to as supercritical, ie, >22.1 MPa abs (3208 psia).

An example of a modern, tangentially fired, supercritical, lignite-fuel furnace is shown in Figure 5. This unit, at maximum continuous ratings, supplies 2450 metric tons per hour superheat steam at 26.6 MPa (3850 psi) and 544°C, and 2160 t/h reheat steam at 5.32 MPa (772 psi) and 541°C. These are the values at the superheater and reheater outlet, respectively. Supercritical fluid-pressure installations are, however, only rarely needed. Most power plants operate at subcritical pressures in the range of 12.4–19.3 MPa (1800–2800 psi).

Although furnace sizes (dimensions for a given MW production) do not vary too widely between principal manufacturers, the type of firing employed by each is generally quite distinctive. This indicates that the furnace size is not strongly controlled by the type of firing system, particularly for pulverized-coal firing (10) (70% through 74-μm (200 mesh) screen for a mean size of ca 40–50 μm). The furnace needs to be sufficiently large to permit the oxygen enough time to penetrate (diffuse through) the blanketing CO_2 layer evolving from the burning coal particle. The residual ash particles are, of course, considerably smaller than the parent coal particle, on the order of a mean size of ca 10 μm before post-combustion agglomeration. Although flame temperatures should be high for combustion efficiency in order to minimize CO formation and combustible carbon loss, it is further required that the combustion products (gases) are sufficiently cooled to enter the convection banks below the temperature at which slagging occurs. These contradictory conditions (aside from pollution control requirements) influence the furnace size and have led to solutions such as tangential firing. Another popular and widely applied solution was the introduction of separate combustion chambers auxiliary to the main furnace. This is generally referred to as cyclone firing or the cyclone furnace (9). This furnace (Fig. 6) is a horizontal, refractory-lined, water-cooled cylinder firing crushed coal, 95% through a 4.76-mm (4-mesh) screen. Fuel and air are mixed by a swirling (centrifugal) motion which promotes turbulence. The fuel is burned at high heat-release rates in the range of 16.7–30.0 GJ/(m·h) ((4.5–8.0) × 10^5 Btu/(ft³·h)) at combustion temperatures in excess of 1649°C. At these temperatures the coal ash content forms a liquid slag film on the water-cooled refractory-lined wall because of the centrifugal force imparted to the coal particles. The gaseous products of combustion are discharged into the main gas-cooling boiler furnace and the film of molten slag on the walls continually drains away from the burner end and discharges through the tap opening into the main boiler furnace from which it drains into a tank for further disposition. In tangential firing systems, a cyclone (ie, vortex flame) is created in the

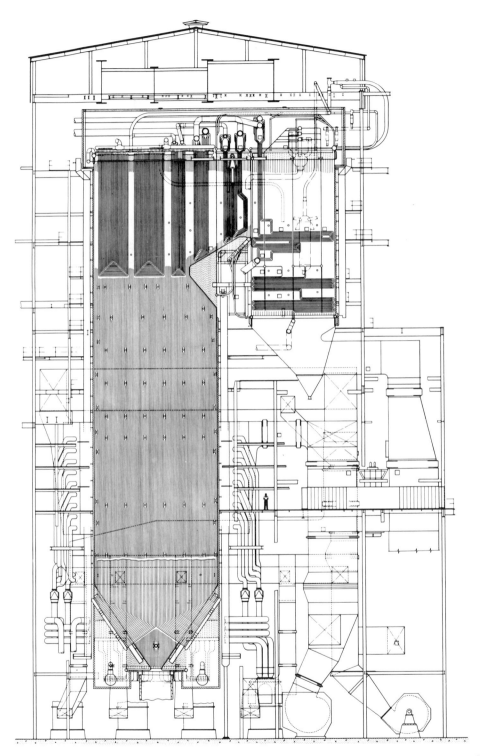

Fig. 5. Martin Lake unit of Texas Utilities Services, Inc., 750 MW; a representative of modern supercritical design. Courtesy of Combustion Engineering, Inc.

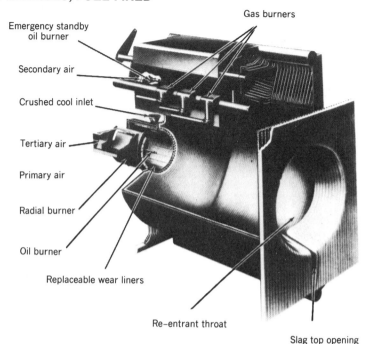

Gas burners

Emergency standby
oil burner

Secondary air

Crushed cool inlet

Tertiary air

Primary air

Radial burner

Oil burner

Replaceable wear liners

Re-entrant throat

Slag top opening

Fig. 6. Cyclone furnace: in effect a high temperature, slag removal combustion chamber auxiliary to the gas-cooling boiler furnace. Courtesy of Babcock & Wilcox, Inc.

main furnace, ie, the fuel and air admission system is corner installed. For opposed firing, the fuel–air system is generally designed to impart an initial swirl to the reactants before furnace entry.

Industrial Furnaces

Generally speaking, industrial furnaces are an order of magnitude smaller than power-plant furnaces since the applications are usually on an individual basis (hospital complex, chemical plant, paper mill, etc) rather than feeding power to a regional electric grid (4). Like the power-plant furnace, the function of the industrial furnace usually is to generate steam, generally for a chemical process, mechanical power, or heating application, rather than electric power generation. There are also many fired heaters that utilize the hot exhaust gases directly for heating, drying, roasting, calcining, etc.

Package Boilers.   These boilers are also known as shop-assembled boilers. Low capacity units are shipped complete with fuel-burning equipment, safety and combustion controls, and boiler trim. Large capacity units, because of tunnel and overpass shipping clearances and railroad flatcar limitations, are designed as multiple integrated-component packages such as the complete furnace–burner assembly, forced-draft fan assembly, and the like. In this category of steam generators, the fire-tube furnace competes with waterwall furnaces, particularly for duty in the capacity range of 205 kPa to 1.8 MPa (15–250 psig) and 1800–9100 kg steam/h. Figure 7 shows a modern fire-tube furnace of four-pass design. Like the fire-tube furnace, the waterwall package furnace, because of the temperature quenching effect of the immediate proximity of the cold sink (steam and water), requires a long flame path (ie, residence time) to permit the combustion reaction to go to completion.

Paper Mill (Recovery) Furnaces.   The conversion of wood chips into cellulose and lignins by chemical pulping requires considerable amounts of steam. Paper is made from the cellulose fibers, which amount to ca 50% of the decomposition products, with the remainder (including bark) being burned to generate the process steam (see PULP). In terms of power generated, the paper industry ranks second (after electric utilities), all of the generated power being consumed in-plant. The pulp is produced by either the kraft or the soda process. The main chemical constituent is sodium sulfate (kraft process, on the order of 80% of the production) with the sodium carbonate process (soda ash) roughly accounting for the remaining production. In the kraft process, wood chips are cooked under pressure (by steam heating) in an aqueous solution of NaOH and Na_2S, known as white liquor. In this operation the lignin binder is dissolved, freeing the cellulose fibers. The spent cooking liquor (containing the lignin), called black liquor, is successively steam and stack-gas dried for concentration; sodium sulfate, Na_2SO_4, is added to make up chemical losses. The main function of the recovery furnace is to reduce the sodium sulfate content of the black liquor to sodium sulfide, Na_2S. The heat obtained from the combustion of the organic black liquor constituent as

Fig. 7.   Four-pass fire-tube boiler. Courtesy of Cleaver Brooks.

well as from the supplementary fuel produce steam and molten smelt, largely composed of Na_2CO_3 and Na_2S. It is tapped from the furnace bottom and dissolved in water to form green liquor for further processing. In the recovery furnace shown in Figure 8, liquor spray nozzles are used to inject the heavy black liquor into the furnace of the recovery unit.

The burning of the ligneous portion of the black liquor produces sufficient heat in the furnace to sustain flash drying of residual moisture, salt-cake reduction, and chemical smelting. The heat in the gas passing through the furnace, boiler, and economizer produces steam for power and process.

Large Industrial Furnaces (Waterwall). In contrast to the shop-assembled waterwall furnace (package boilers), the large industrial furnaces are erected on

Fig. 8. Black liquor recovery furnace. Courtesy of Combustion Engineering, Inc.

site. Steam capacities are in the range of 45–450 t of superheated steam/h and generally serve industries such as steel, oil refining, automotive, sugar, and chemical. In distinction to the power-plant utility furnace, the industrial furnace is of considerably smaller capacity and is likely to be of modular design and operationally flexible. It generally is designed to burn two or more fuels of the industrial range comprised of coal, oil, gas, refuse, wood products, bagasse, sewage sludge, and other solid, liquid, or gaseous waste products. Consequently, the furnace is often equipped with several firing methods in order to accommodate different fuels. For example, it may be arranged for tangential, horizontal, or stoker firing of single or multiple fuels.

Refractory-Wall Furnaces. Many types of refractory-wall furnaces are used for melting or heat treatment of metals, at temperatures ranging from less than 540°C to as high as 1760°C, such as ceramic and glass kilns and furnaces, ovens and dryers operating from 50°C to 600°C, and stills and retorts (7) (see CLAYS). Heaters can be directly fired when the products of combustion do not seriously affect the process stream (see PETROLEUM). Indirect heaters use heat exchangers to transfer heat to the process medium. Incinerators are special furnaces designed to consume waste products that must be disposed of without creating pollution problems. Trash or solid waste incinerators usually employ a grate or rotating kiln to accomplish mixing of the air with the waste for combustion.

These furnaces may operate batchwise or continuous. In the batch, intermittent, or periodic types, the content is heated at the desired temperature for the stipulated time and then removed. In the continuous type, the charge moves at a predetermined rate through one or more heating zones to emerge in most cases at the end opposite the point of entry. Figures 9 and 10 are representative examples of typical, industrial refractory-wall furnaces.

Fig. 9. Car-type annealing furnace.

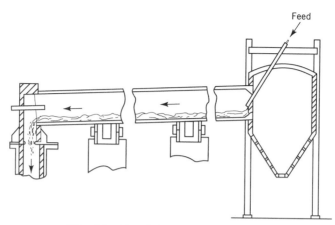

Fig. 10. Revolving-drum cement kiln.

Analysis

The rising demand for fuel efficiency and performance guarantees are imposing increasing requirements for analytical complexity. Furnace orientation in this area has developed only since the early 1920s. Starting with the rosin analysis of the 1920s (11), subsequent development of furnace and combustion analysis proceeded along distinctively dissimilar lines. The Russian approach (12,13) in particular includes considerable theoretical (mathematical) emphasis, whereas the European–British–American approach concentrates on a fuel and flame chemistry orientation (14–16). Analytical emphasis has shifted toward a heat-transfer oriented view of predicting furnace performance: (1) wall absorption rate and gas temperature profiles, and (2) pollutant formation. Among the various models (17), the Hottel zoning method (18) is still popular (19). Another analytical model that lends itself to engineering analysis of furnaces is generally referred to as PSR theory (perfectly stirred reactor theory) (20) and is, in fact, applied on an industrial basis (21) where it is important to obtain a quantitative evaluation of the emissivity of the combustion products (22). Commercial computer codes that calculate complex flow fields with reacting flows are available but not yet validated. A detailed treatment of combustion modeling has been given (23) (see COMBUSTION SCIENCE AND TECHNOLOGY).

The analytical mechanisms for predicting the corresponding pollutant formation associated with fossil-fuel-fired furnaces lag the thermal performance prediction capability by a fair margin. The most firmly established mechanism at this time is the prediction of thermal NO_x formation (24). The chemical kinetics of pollutant formation is, in fact, a subject of research.

Fluidized-Bed Combustion

New furnace concepts in evolutionary stages include fluidized-bed furnaces (25), coal gasification furnaces (26), and MHD furnaces (27,28). Of these technologies, fluidized-bed combustion has reached commercial-scale operations (Fig. 11).

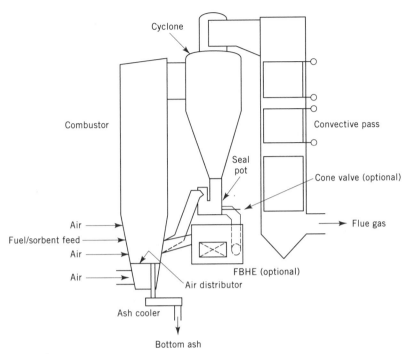

Fig. 11.   Typical circulating fluidized-bed (CFB) steam generator. FBHE = Fluidized bed, high efficiency.

Fluidization refers to the condition in which solid materials are given free-flowing, fluid-like behavior (29). As a gas is passed upward through a bed of solid particles, the flow of gas produces forces which tend to separate the particles from one another. At low gas flows, the particles remain in contact with other solids and tend to resist movement. This condition is referred to as a fixed bed. As the gas flow is increased, a point is reached at which the forces on the particles are just sufficient to cause separation. The bed then becomes fluidized. The gas cushion between the solids allows the particles to move freely, giving the bed a liquid-like characteristic.

For decades, fluidized-bed reactors have been used in noncombustion reactions in which the thorough mixing and intimate contact of the reactants in a fluidized bed result in high product yield with improved economy of time and energy. Although other methods of burning solid fuels also can generate energy with very high efficiency, fluidized-bed combustion can burn coal efficiently at a temperature low enough to avoid ash slagging and NO_x formation from combustion in the other modes.

BIBLIOGRAPHY

"Furnaces, Fuel-Fired" in *ECT* 1st ed., Vol. 7, pp. 23–37, by M. H. Mawhinney, Consulting Engineer; in *ECT* 2nd ed., Vol. 10, pp. 279–294, by M. H. Mawhinney, Consulting Engineer; in *ECT* 3rd ed., Vol. 11, pp. 563–579, by K. A. Bueters, Combustion Engineering, Inc., and R. H. Essenhigh, Ohio State University.

1. H. Hodges, *Technology in the Ancient World*, Alfred A. Knopf Publishing Co., New York, 1970.
2. T. Croft and R. B. Purdy, *Steam Boilers*, 2nd ed., McGraw-Hill Book Co., Inc., New York, 1937.
3. *Steam—Its Generation and Use*, 38th ed., Babcock & Wilcox Publishing, New York, 1972.
4. O. de Lorenzi, ed., *Combustion Engineering—A Reference Book on Fuel Burning and Steam Generation*, Combustion Publishing Co., New York, 1947.
5. J. Griswold, *Fuels, Combustion, and Furnaces*, 1st ed., McGraw-Hill Book Co., Inc., New York, 1946.
6. R. H. Essenhigh, "Energy Use, Efficiency, and Conservation in Industry," *Proceedings of the EPA/ERDA Symposium on Environment and Energy Conservation*, Denver, Colo., Nov. 1975, EPA-600/2-76-212, ERDA-47.
7. R. H. Essenhigh, *Future Fuel Supplies for Industry: Bases for Choice, ASHRAE Transactions*, Vol. 84, American Society of Heating, Refrigerating, and Air-Conditioning Engineers, New York, 1978, Part I.
8. J. C. Fisher, *Energy Crises in Perspective*, 1st ed., Wiley-Interscience, New York, 1974.
9. D. M. Considine, ed., *Energy Technology Handbook*, 1st ed., McGraw-Hill Book Co., Inc., 1977.
10. R. H. Essenhigh, *J. Inst. Fuel*, (Jan. 1961).
11. R. H. Essenhigh, *Ind. Eng. Chem.* **59**(7), 52 (1967).
12. L. N. Khitrin, *Physics of Combustion and Explosion*, University of Moscow Press, CIS 1957, trans. Israel Scientific Translations Program, Office of Technical Services, *OTS 61-31205*, U.S. Dept. of Commerce, Washington, D.C., 1962.
13. L. A. Vulis, *Thermal Regimes of Combustion*, Kazakh State University, CIS, G. C. Williams, trans., ed., McGraw-Hill Book Co., Inc., New York, 1961.
14. M. A. Field, and co-workers, *Combustion of Pulverized Coal*, BCURA, Leatherhead, Surrey, U.K. 1967.
15. M. W. Thring, *The Science of Flames and Furnaces*, John Wiley & Sons, Inc., New York, 1952.
16. B. Lewis and C. von Elbe, *Combustion, Flames and Explosions of Gases*, Academic Press, Inc., New York, 1961.
17. T. J. Tyson, "The Mathematical Modeling of Combustion Devices," paper presented at *Proceedings of the Stationary Source Combustion Symposium*, Vol. I, U.S.E.P.A., Washington, D.C., EPA-600/2-76-1522, June 1976.
18. H. C. Hottel and E. S. Cohen, "Radiant Heat Exchange in a Gas-Filled Enclosure," paper presented at *ASME/AIChE Joint Heat Transfer Conference*, University Park, Pa., Aug. 1957, AIChE Paper No. 57-HT-23.
19. K. K. Boon, "A Flexible Mathematical Model for Analyzing Industrial P. F. Furnaces," M.S. thesis, University of Newcastle, Australia, Sept. 1978.
20. R. H. Essenhigh, "A New Application of Perfectly Stirred Reactor (P.S.R.) Theory to Design of Combustion Chambers," *Technical Report FS67-1(u)*, Pennsylvania State University, Dept. of Fuel Science, University Park, Pa., Mar. 1967.
21. K. A. Bueters, J. G. Cogoli, and W. W. Habelt, "Performance Prediction of Tangentially Fired Utility Furnaces by Computer Model," paper presented at the *Fifteenth Symposium on Combustion*, Tokyo, Japan, Aug. 25–31, 1974, The Combustion Institute, Pittsburgh, Pa., 1975.
22. K. A. Bueters, *Combustion* **45**(9) (1974).
23. L. Douglas Smoot, ed., *Coal Science and Technology, Fundamentals of Combustion*, Vol. 20, Elsevier, New York, 1993.

24. K. A. Bueters and W. W. Habelt, NO_x *Emissions from Tangentially Fired Utility Boilers*, AIChE Symposium Series, Vol. 71, American Institute of Chemical Engineers, New York, 1975, Parts I–II.
25. J. B. Anderson and W. R. Norcross, *Combustion* **50**(8), 9 (1979).
26. J. T. Stewart and T. D. Pay, "Coal Gasification Processes and Equipment Available for Small Industrial Applications," paper presented at the *Fifth Annual International Conference on Coal Gasification Liquefaction and Conversion to Electricity*, University of Pittsburgh, Pa., Aug. 1–3, 1978.
27. H. B. Palmer and J. M. Beer, ed., *Combustion Technology—Some Modern Developments*, Academic Press, Inc., New York, 1974.
28. L. P. Harris, *Combustion* **50**(9), 21 (1979).
29. J. G. Singer, *Combustion Engineering—A Reference Book on Combustion, A Reference Book on Fuel Burning and Steam Generation*, Combustion Engineering, Inc., Windsor, Conn., 1991.

General References

H. E. Barner, H. Beisswenger, and K. E. Barner, "Chemical Equilibrium Relationships Applicable in Fluid Bed Combustion," *Proceedings of the Ninth International Conference on Fluidized-Bed Combustion*, Boston, Mass., May 4–7, 1987.
M. Bashar and T. S. Czarnecki, "Design and Operation of a Lignite-Fired CFB Boiler Plant," *Proceedings of the Tenth International Conference on Fluidized Bed Combustion*, San Francisco, May 1–4, 1989.
H. Beisswenger, R. Kittel, and L. Plass, *The 95.8 MWe CFB Utility Boiler of the Duisburg Municipal Power Company*, Lurgi Publication, Frankfurt, Germany, 1987.
R. J. Gendreau and D. L. Raymond, "Atmospheric Fluidized-Bed Combustion Update—Status and Applications," presented at the *1987 Joint ASME/IEEE Power Generation Conference*, Miami Beach, Oct. 4–8, 1987, ASME paper no. 87-JPGC-FACT-11.
E. J. Gottung and S. J. Sopko, "Design and Operation of a CFB Steam Generator Firing Anthracite Waste," presented at the *1988 Joint ASME/IEEE Power Generation Conference*, Philadelphia, Sept. 25–28, 1988.
R. L. Patel and co-workers, "Reactivity Characterization of Solid Fuels in an Atmospheric Bench-Scale Fluidized-Bed Combustor," presented at the *1988 Joint ASME/IEEE Power Generation Conference*, Philadelphia, Sept. 25–29, 1988; also as *Combustion Engineering* publication TIS-8391.
F. A. Sainz and co-workers, "Chatham Circulating Fluidized Bed Demonstration Project," *Proceedings of the First International Conference on Circulating Fluidized Beds*, Halifax, Nova Scotia, Canada, Nov. 18–20, 1985, Pergamon Press, Toronto.
W. Wein, *Flow Dynamics of Atmospheric Fluid Bed Combustion Systems and their Effect on SO_2 Capture and NO_x Suppression*, trans. Lurgi from *VGB Magazine*, Feb. 1985, pp. 119–123.
B. W. Wilhelm and co-workers, "100 MW Anthracite Culm CFB Small Power Producer," *Proceedings of the American Power Conference*, Vol. 50, Illinois Institute of Technology, Chicago, 1988.
R. Norris Shreve and J. A. Brink, "Chemical Process Industries," in *Perry's Chemical Engineers' Handbook*, 4th ed., McGraw-Hill Book Co., Inc., New York, 1977.

CARL R. BOZZUTO
ABB Kreisinger Development Laboratory

FUSION ENERGY

As far as is known, nuclear fusion, which drives the stars, including the Sun, is the primary source of energy in the universe. The process of nuclear fusion releases enormous amounts of energy. It occurs when the nuclei of lighter elements, such as hydrogen, are fused together at extremely high temperatures and pressure to form heavier elements, such as helium. Whereas practical methods for harnessing fusion reactions and realizing the potential of this energy source have been sought since the 1950s, achieving the benefits of power from fusion has proved to be a difficult, long-term challenge.

Fusion is widely held to be the ultimate resource for the world's long-term energy needs. The fuel reserves for fusion are virtually limitless and available to all countries. Fusion fuels can be extracted from water. Additionally, fusion promises to be an energy source which is potentially safe and environmentally benign. Radiological and proliferation hazards are much smaller than for fission power plants. The atmospheric impact is negligible compared to fossil fuels, and adverse impacts on the Earth's ecological and geophysical processes are smaller than for large-scale renewable energy sources (see also COAL; FUEL RESOURCES; FUELS, SYNTHETIC; GAS, NATURAL; PETROLEUM; RENEWABLE ENERGY RESOURCES). The economics and costs of fusion power plants are still being studied, but appear comparable to those for other medium- and long-term energy sources (see POWER GENERATION). The tantalizing promise of affordable essentially unlimited supplies of clean, safe energy, free of political boundaries, has motivated a worldwide research effort to develop this energy resource.

The nuclear burning mechanism of the Sun was elucidated in the 1930s (1). In a complex sequence of reactions starting with hydrogen, atomic nuclei are fused to form heavier species. Because of a mass deficit, Δm, exhibited by the reaction products, large amounts of energy, E, are released, as dictated by the well-known Einstein equivalence $E = \Delta mc^2$ where c is the speed of light. Large-scale fusion energy production was demonstrated dramatically on earth in the early 1950s with the explosion of thermonuclear fusion, ie, hydrogen, bombs. These weapons used the heat of nuclear fission (atomic bombs) to cause the fusion of deuterium [16873-17-9], D, and tritium [15086-10-9], ^{3}H or T. Subsequently, an international research effort was undertaken to harness this awesome power on a controllable scale for peaceful purposes. Several impressive advances in the 1980s and early 1990s have led to a well-founded feeling of optimism that fusion energy should become a practical energy source during the early twenty-first century.

In order to effect a fusion reaction between two atomic nuclei, it is necessary that these nuclei be brought together closely enough to experience an attractive nuclear force. All nuclei are positively charged and repel one another via Coulomb's law, the electrostatic law of the repulsion of like charges. This electrostatic barrier can be overcome by imparting sufficient kinetic energy to the reacting species so that the nuclei can approach closely enough together that quantum mechanical tunneling can occur. The repulsive forces increase rapidly with the magnitude of the nuclear charge; therefore, nuclear fusion research has concentrated on the lightest elements and the isotopes having the lowest atomic numbers.

The reactions of deuterium, tritium, and helium-3 [14762-55-1], ^{3}He, having nuclear charge of 1, 1, and 2, respectively, are the easiest to initiate. These have the highest fusion reaction probabilities and the lowest reactant energies.

Deuterium–Tritium Fusion

The D–T reaction involving the two heavy isotopes of hydrogen

$$D + T \longrightarrow {}^4He + n$$

is especially attractive to fusion scientists because of its relative ease of ignition. The products of this reaction are an alpha particle, ie, the helium nucleus, ^{4}He, and a free neutron, n, carrying kinetic energies of 3.5 and 14.1 MeV, respectively. In an electric power-generating facility the neutrons would be absorbed in a blanket surrounding the fusion region, and the kinetic energy converted into heat. Conventional power conversion systems could then be used to transform this heat into electrical energy. Fusion reactions are extremely energetic, and yields are measured in units of millions of electron volts, MeV (1 MeV $= 1.6 \times 10^{-13}$ J) (see DEUTERIUM AND TRITIUM; HELIUM GROUP, GASES).

Another set of reactions of practical interest involves only deuterons. The D–D reaction can proceed along either of two pathways with roughly equal probabilities:

$$D + D \to T + p + 4.0 \text{ MeV}$$

or

$$D + D \to {}^3He + n + 3.2 \text{ MeV}$$

Finally, the D–^{3}He reaction

$$D + {}^3He \to {}^4He + p$$

is noteworthy not only because of its high (18.3 MeV) energy release, but also because the reaction products are both charged particles, offering the possibility of high efficiency, direct energy conversion. Direct energy conversion would involve the extraction of the positively charged ions and negatively charged electrons from the reaction region directly onto collection electrodes having a potential difference of the same order of magnitude as the mean kinetic energy of the charged particles. A variation of the above reaction schemes is the catalyzed D–D reaction wherein the external feedstock is deuterium but the ^{3}He and T produced in the D–D reactions are recycled and burned in situ to enhance the net energy yield.

Because of its relatively high reactivity (2), the D–T fusion-fuel cycle is very likely to be employed in the first generation of fusion reactors. This implies the use of a neutron absorbing blanket and thermal (Carnot) conversion efficiencies. Deuterium, also known as heavy hydrogen, occurs naturally in the ratio of 1:6700 relative to ordinary hydrogen; 30,000 kg water contains one kilogram of deuterium. The separation of deuterium from water is a relatively simple and inexpensive process. Tritium, on the other hand, is a radioactive isotope of hydrogen found in nature only in trace amounts and has a half-life of only 12.36 yr. The initial inventory of tritium for a power-producing D–T fusion reactor is a few kilograms and could be supplied, for example, from heavy-water fission reactors where it is produced as a by-product. Further tritium needs can be met by breeding additional tritium in the fusion reactor itself, by absorbing the fusion-produced neutrons in a blanket of lithium and exploiting the reaction

$$n + {}^6\text{Li} \rightarrow \text{T} + {}^4\text{He} + 4.8 \text{ MeV}$$

${}^6\text{Li}$ accounts for about 7.5% of natural lithium and is abundantly available in the earth's crust and oceans. Detailed fusion-blanket designs incorporate additional isotopes, such as ${}^7\text{Li}$ and ${}^9\text{Be}$, which provide neutron-multiplying reactions, to compensate for the leakage and absorption losses of neutrons. A D–T fusion reactor, then, is in reality a consumer of deuterium and lithium. The estimated reserves of lithium should prove sufficient for at least several hundred years of D–T fusion reactor operation, even allowing for a significant increase in the world demand for energy (3). A common fusion evolution scenario relies on D–T fusion to fulfill the energy needs until the more difficult fuel cycle involving pure deuterium can be implemented. Then deuterium alone would be the fuel for the fusion energy economy. Because each liter of seawater contains enough deuterium to supply the energy equivalent of 300 L of gasoline, long-term energy needs would be assured.

Although the D–T reaction is the easiest route to fusion power production, it is no easy task to meet the conditions required to produce net fusion energy. Relative kinetic energies between the deuterons and tritons of 10 keV or more are necessary for practical energy generation, corresponding to relative particle velocities on the order of 10^6 m/s. Fusion-produced neutrons have, in fact, been created by impinging a beam of accelerated deuterons onto a solid target containing tritium. Unfortunately, a fusion reactor cannot be built around this concept because most of the incident beam energy is dissipated nonproductively through scattering and collision events in the target, and only a relatively small number of energy-producing fusion reactions occur. Other approaches, involving colliding beams of particles, have been proposed, but such schemes are inherently of very low power density and are not likely to yield practical energy sources.

Plasma Conditions Required for Net Energy Release

The most promising approach to attaining significant reaction rates is to heat the reacting species to a high temperature, thereby imparting large kinetic energies to the nuclei in the form of thermal motions. By doing so, the particles, eg, deuterons and tritons, may scatter among themselves many times before undergoing fusion reactions, without losing significant energy from the system. At any given temperature, a system of particles in thermal equilibrium is characterized by a

Maxwellian distribution of kinetic energies. The particles at the high energy end of this distribution account for most of the fusion reactions in fusion experiments.

The fusion fuel, when undergoing thermonuclear reactions, exists as an ionized gas called a plasma. In physics, the plasma state usually means a high temperature gas of net electrical neutrality consisting of free electrons and ions exhibiting collective behavior. The collection of charged particles exhibits characteristics of an electrically conducting fluid that can interact with electromagnetic fields. As such, its physical behavior is much more complex than that of an ordinary gas, and plasma confinement can be disrupted or reduced by many different kinds of plasma instabilities and other loss mechanisms. There exists a large literature and a number of outstanding books on plasma physics, such as Reference 4 (see also PLASMA TECHNOLOGY).

In a plasma undergoing fusion reactions, the reactivity, and thus the fusion-power output rate, increases with increasing temperature. However, over a wide range of temperatures, as the temperature of the plasma is raised, the radiation losses are also increased, primarily because of bremsstrahlung, or continuum, ie, braking radiation from the electrons. For any fusion-fuel system there exists a unique temperature at which the fusion power production is precisely balanced by the radiation losses. This temperature is called the ideal ignition temperature, and equals about 50 million K (5 keV) for a D–T plasma (1 keV $= 11.6 \times 10^6$ K). For a D–D plasma, this temperature is considerably higher, about 400×10^6 K (40 keV), a fact which considerably increases the difficulty of using pure deuterium fuel. Furthermore, a fusion system must be operated above the ideal ignition temperature for net power production, typically by a factor of 2–5.

Besides having to satisfy a minimum temperature requirement, the plasma must be sufficiently dense and contained for a long enough time to yield net power. If the plasma burns above the ideal ignition temperature for some time period, τ, the fusion energy released must at least equal the energy required to heat the plasma to that temperature plus the energy radiated during that period. It can be shown that this condition is met by requiring that the product of the plasma density, n, and confinement time, τ, exceed a characteristic value which depends only on the temperature. The minimum value of the product $n\tau$ represents the least stringent condition for the plasma to be a net producer of fusion energy. For D–T plasmas, this minimum occurs at a temperature of about 100×10^6 K, for which $n\tau \sim 10^{20}$ s/m^3. This minimum $n\tau$ product is called the Lawson criterion (5). For D–D, the minimum $n\tau$ product is about 10^{22} s/m^3 at a higher temperature, again indicating that a pure deuterium system requires a higher quality of confinement. A commonly used measure of the quality of plasma confinement is given by the triple product of the plasma density, n, ion temperature, T_i, and energy confinement time, τ, usually expressed in units of keV·s/m^3. A primary goal of fusion research is to achieve $n\tau T_i$ values of $\sim 10^{22}$ keV·s/m^3, as required for a D–T reactor. Experiments as of this writing (1993) have reached a value of 1.1×10^{21} keV·s/m^3 in the JT-60 tokamak in Japan (6).

Plasmas at fusion temperatures cannot be kept in ordinary containers because the energetic ions and electrons would rapidly collide with the walls and dissipate their energy. A significant loss mechanism results from enhanced radiation by the electrons in the presence of impurity ions sputtered off the container walls by the plasma. Therefore, some method must be found to contain the plasma at elevated temperature without using material containers.

Once a fusion reaction has begun in a confined plasma, it is planned to sustain it by using the hot, charged-particle reaction products, eg, alpha particles in the case of D–T fusion, to heat other, colder fuel particles to the reaction temperature. If no additional external heat input is required to sustain the reaction, the plasma is said to have reached the ignition condition. Achieving ignition is another primary goal of fusion research.

Paths to Fusion Power

Two diverse technical approaches to fusion power, magnetic confinement fusion, also known as magnetic fusion energy (MFE) and inertial confinement fusion, also known as inertial fusion energy (IFE) are being pursued worldwide. These form the basis of a large number of fusion research programs. Magnetic confinement techniques, studied since the 1950s, are based on the principle that charged particles such as electrons and ions, ie, deuterons and tritons, tend to be bound to magnetic lines of force. Thus the essence of the magnetic confinement approach is to trap a hot plasma in a suitably chosen magnetic field configuration for a long enough time to achieve a net energy release, which typically requires an energy confinement time of about one second. In the alternative IFE approach, fusion conditions are achieved by heating and compressing small amounts of fuel ions, contained in capsules, to the ignition condition by means of tightly focused energetic beams of charged particles or photons. In this case the confinement time can be much shorter, typically less than a millionth of a second.

Magnetic Confinement. In magnetic confinement, strong magnetic fields are used to confine the plasma. Electrons and ions in magnetic fields spiral in circles around the field lines but translate freely along the direction of the magnetic field. Thus the magnetic field of a long solenoid, for example, confines the plasma in two directions but does not prevent the particles from streaming from either end of the system. Furthermore, collisions between particles displace them from one field line to another, producing a net diffusion of plasma across the field toward the walls of the container. By employing more complex magnetic field configurations, fusion researchers have made significant progress toward solving the problem of magnetic confinement of plasmas by substantially reducing plasma losses.

One of the earliest configurations studied was the simple magnetic mirror. A simple mirror system is depicted in Figure 1. Particles gyrating about the field lines move freely along these lines until they enter regions of increased field strength at either end of the device. Conservation of angular momentum considerations dictate that, as the particles approach the end regions, they gyrate more energetically about the field lines and slow down in the direction of motion along the lines. Ultimately, their kinetic energy is completely converted into gyration energy, at which point the particles are reflected from these mirror points and return to the central, weaker field region. Particles having motion exactly along the axis of the device are not reflected and are lost through the ends. Although ingenious attempts have been made to reduce end losses from mirror machines

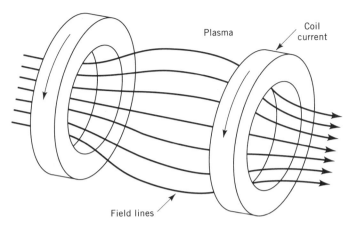

Fig. 1. Simple magnetic mirror open configuration.

and to make them stable against magnetohydrodynamic (MHD) and other instabilities, all single-cell mirror reactor designs have suffered from a high recirculating power fraction, ie, a lot of the output power has to be used to operate the
reactor itself. In single-cell mirror machines these losses are fundamentally too
high. The machines are referred to as too lossy, and the amount of injected power
required to maintain the plasma, usually in the form of high energy neutral
beams, has been too large to be practical.

A more advanced mirror approach involving multicells, called the tandem
mirror, has been studied as a means to overcome the leakage problem. One way
to view the tandem mirror is as a long uniform magnetic solenoid with two single-
cell mirrors installed at the ends to electrostatically plug the device. Plasma end
losses are impeded by electrostatic potentials developed by the plasma as the
electrons and ions attempt to leave the device at different rates.

Another mirror variation is the field-reversed mirror configuration, in which
the diamagnetic nature of the plasma is exploited to cause the interior magnetic
field lines within the central region of a singe-cell mirror to reverse and close on
themselves. The plasma current responsible for this field modification is at right
angles to the original field lines.

The problem with all the mirror approaches is that none has achieved the
degree of confinement quality that the closed systems have. Closed systems are
characterized by magnetic field lines that close on themselves so that charged
particles following the field lines remain confined within the system.

The simplest way of producing a closed configuration is to employ a torus or
doughnut-shaped container having current-carrying coils wrapped around the minor diameter as shown in Figure 2. In this geometry, the magnetic lines of force
are circles that traverse the torus and provide endless paths for the plasma ions
and electrons to spiral about. Unfortunately, such a simple toroidal configuration
is well known to have very poor confinement properties, because the magnetic
field strength is not constant across the plasma. Instead, it is stronger at the inner
wall and weaker at the outer wall of the toroidal chamber. As a result, the positive
ions and electrons drift in opposite directions across the field lines and establish

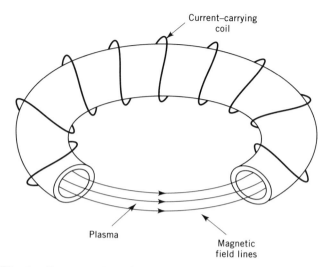

Fig. 2. Cutaway view of a simple toroidal field configuration.

an electric field within the plasma. This electric field, coupled with the applied magnetic field, then causes the plasma as a whole to move to the container wall and dissipate its energy (2).

This deleterious effect can be obviated by introducing additional components of magnetic field, causing the field lines to circumscribe the torus without ever closing on themselves. The net magnetic field is then composed of a major, or toroidal, field component produced by the current coils, plus a smaller poloidal component which gives the desired twist to the lines. Particle drifts weaken or nullify the harmful electrical field and the plasma no longer tends to move to the walls.

Several geometries for producing the required poloidal magnetic field component have been studied. The class of plasma devices called tokamaks generates the poloidal, ie, around the minor circumference, field component from a toroidal, ie, around the major circumference, current in the plasma itself, either induced by a pulse from an external transformer or driven by external current-drive systems. The basic components of a tokamak are shown in Figure 3. External current-drive systems, such as high energy neutral beam injection or radio-frequency (r-f) current drive, impart a net toroidal momentum to one of the charged species, ions or electrons. A toroidal system related to the tokamak, which has a higher plasma energy density, is called the reversed field pinch (RFP). The RFP is an inherently pulsed, or batch-burn device.

The poloidal field component can also be created externally by using current-carrying coils that wind helically around the outside surface of the torus. Such devices, called stellerators or torsatrons, have the advantage of not requiring a net toroidal current. The helical field windings, however, make these machines mechanically and magnetically more complex than the tokamak.

Tokamak. The design concept that has come the closest by far to achieving energy breakeven conditions is the tokamak. Invented in the 1950s by the Russian

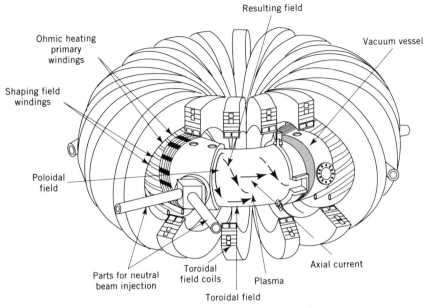

Fig. 3. The tokamak fusion approach.

physicists Andrei Sakharov and Igor E. Tamm (7), the tokamak derives its name from the Russian acronym for toroidal magnetic chamber. Technical progress in tokamaks was dramatic in the late 1980s and early 1990s (8,9). Central ion temperatures of 400×10^6 K have been reached, and energy confinement times have increased from 0.02 to about 1.4 seconds for strongly heated plasmas. The result has been $n\tau T_i$, triple products, of about 10^{21} keV·s/m³, compared to the value of $\sim 10^{22}$ keV·s/m³ required for a steady state D–T reactor.

Other important parameters have also shown dramatic gains. The normalized plasma pressure which is usually called beta, the plasma pressure divided by the confining magnetic field pressure, has been increased fourfold, to about 10%. This value is actually higher than that needed in a reactor. Bootstrap cur-

Table 1. Tokamak Plasma Parameters

Parameter	Achieved			Required for steady-state D–T reactor
	1971	1981	1991	
central ion temperature, T_i, keV	0.5	7	35	30
central electron temperature, T_e, keV	1.5	3.5	15	30
energy confinement time, τ, s	0.007	0.02	1.4	3
triple product, $n\tau T_i$, keV·s/m³	1.5×10^{17}	5.5×10^{18}	9×10^{20}	7×10^{21}
normalized plasma pressure, β, %	0.1	3	11	5
fusion reactivity, reactions per second				
D–D		3×10^{14}	1×10^{17}	
D–T			6×10^{17}	10^{21}

rents have been measured for the first time in several experiments. Bootstrap current is the name given to a toroidal current, theoretically predicted to arise spontaneously in tokamaks under near-reactor conditions. These in principle can eliminate much of the need for external current drive. Bootstrap currents open the possibility of a self-sustained steady-state tokamak reactor. Results from the large Japanese tokamak JT-60 are particularly interesting in this regard, where up to 80% of the 500,000 A of total plasma current is attributed to the bootstrap effect (8). Table 1 summarizes some of the progress made in the parameters of interest for magnetic fusion since 1971.

Some of the tokamaks in operation around the world, on which the data in Table 1 were obtained are

Designation	Tokamak	Location
ALC-A	Alcator-A	Plasma Fusion Center, Massachusetts Institute of Technology (MIT), Cambridge, Mass.
ALC-C	Alcator-C	Plasma Fusion Center, MIT
ASDEX	Axially Symmetric Divertor Experiment	Max Planck Institute for Plasma Physics, Garching, Germany
ATC	Adiabatic Toroidal Compressor	Princeton Plasma Physics Laboratory (PPPL), Princeton, N.J.
C-MOD	ALC-C Modified	MIT
DIII	Doublet III	General Atomics, San Diego, Calif.
DIII-D	Doublet III-D	General Atomics, San Diego, Calif.
ISX-B	Impurity Studies Experiment B	Oak Ridge National Laboratory (ORNL), Oak Ridge, Tenn.
JET	Joint European Torus	Abingdon, England
JFT-2M		Japan Atomic Energy Research Institute, Tokai, Japan
JT-60		Japan Atomic Energy Research Institute, Naka, Japan
ORMAK	Oak Ridge Tokamak	ORNL
PDX	Princeton Divertor Experiment	PPPL
PLT	Princeton Large Torus	PPPL
T-3	Tokamak-3	Kurchatov Institute, Moscow
T-10	Tokamak-10	Kurchatov Institute, Moscow
T-15	Tokamak-15	Kurchatov Institute, Moscow
TFR	Tokamak Fontenay-aux-Roses	Centre d'Etudes Nucleaire, Fontenay-aux-Roses, France
TFTR	Tokamak Fusion Test Reactor	PPPL

Additionally, two other reactors, the international thermonuclear experimental reactor (ITER) for which the location is under negotiation, and the Tokamak Physics Experiment at PPPL, Princeton, New Jersey, are proposed. The most impressive advances have been obtained on the three biggest tokamaks, TFTR, JET, and JT-60, which are located in the United States, Europe, and Japan, respectively. As of this writing fusion energy development in the United States is dependent on federal funding (10–12).

Until 1992, tokamak experiments were performed using deuterium or hydrogen only. The use of radioactive tritium greatly complicates the operation of experimental facilities, impeding the pace of research. Certain experiments, however, such as those directly involving D–T fusion, cannot be done without the use of tritium. A European research team in 1992 produced nearly 2 million watts of fusion power for about one second in the JET device, and opened the modern frontier of D–T fusion experiments (13). Only about half of the JET fusion energy release came from fusion in the thermal plasma, at temperatures of 15–20 keV. The other half came from fusion of the injected tritium beams striking the deuterium in the plasma. The ratio of tritium to deuterium was about 2% in JET. If a 50:50 mixture of tritium and deuterium had been used instead, an amount of fusion energy would have been released roughly equal to the energy required to heat and sustain the plasma, giving an energy gain, Q, of about unity. In December 1993, scientists at the Princeton Plasma Physics Laboratory initiated a series of experiments on the Tokamak Fusion Test Reactor (TFTR), introducing D–T fuel into the machine and producing over 6 MW of fusion power. For the first time in a tokamak experiment an approximately 50:50 mixture of deuterium and tritium was used as the fusion fuel. Preliminary analysis of the first 100 experimental runs indicated that the confinement in a D–T fuel mixture was better than in a pure deuterium plasma, the ion and electron temperatures were higher, and the plasma stored energy longer. No enhanced loss of alpha particles (the product of D–T fusion reactions) was observed as the fusion power was increased. These results are encouraging for tokamak-based power generation.

International Thermonuclear Experimental Reactor.   One of the largest obstacles to the development of fusion power has been that high powered, and correspondingly expensive, research facilities are needed at each step of the reactor development path. ITER (pronounced "eater") is a project supported by the United States, Japan, the European community, and Russia, wherein each party contributes equally to the effort and shares equally in the results (9). The main reason for making the ITER an international effort is cost sharing. The project is managed under the auspices of the International Atomic Energy Agency (IAEA), and the design is based on the tokamak concept. The central purpose of the ITER is to demonstrate the scientific and technological feasibility of fusion power by achieving, for the first time, controlled ignition and extended burn in a D–T plasma. ITER is expected to accomplish this by demonstrating technologies essential to a reactor in an integrated system, and by integrated steady-state testing of the high heat-flux and nuclear components (9).

A preliminary design of ITER, done in 1988–1990 by an international team (14), utilizes superconducting magnets. The heating and current drive are provided by a combination of 1.3 MeV negative-ion neutral beams, lower-hybrid frequency rf, and electron-cyclotron frequency rf. The negatively charged beams of

deuterons or tritons are to be accelerated to 1.3 MeV, neutralized, and then injected, unperturbed by the confining magnetic field, into the plasma. The design
is based on a conservative assessment of physics knowledge and allows for operational and experimental flexibility. The design calls for plasma major radius of
6 m, plasma minor radius of 2.1 m, plasma current of 2 MA, magnetic field of
4.85 T, average neutron wall loading of about 1 MW/m^2, and fusion power of about
1 GW thermal.

The second phase of ITER, the engineering design activity (EDA), was begun
in 1992 and is scheduled to be completed in 1998. The ITER engineering design
is being conducted at three cocenters: San Diego, California; Graching, Germany;
and Naka, Japan. At these cocenters, multinational teams focus on developing a
mature design in sufficient detail to allow the construction of the machine, with
industrial vendors able to bid on the fabrication and installation of ITER systems.
The first ITER plasma could be made as early as 2005. D–T operation could begin
a few years later.

Inertial Confinement.   Because the maximum plasma density that can be
confined is determined by the field strength of available magnets, MFE plasmas
at reactor conditions are very diffuse. Typical plasma densities are on the order
of one hundred-thousandth that of air at STP. The Lawson criterion is met by
confining the plasma energy for periods of about one second. A totally different
approach to controlled fusion attempts to create a much denser reacting plasma
which, therefore, needs to be confined for a correspondingly shorter time. This is
the basis of inertial fusion energy (IFE). In the IFE approach, small capsules or
pellets containing fusion fuel are compressed to extremely high densities by intense, focused beams of photons or energetic charged particles as shown in Figure
4. Because of the substantially higher densities involved, the confinement times

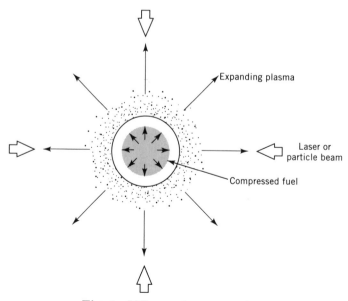

Fig. 4. ICF capsule compression.

for IFE can be much shorter. In fact, no external means are required to effect the confinement; the inertia of the fuel mass is sufficient for net energy release to occur before the fuel flies apart. Typical burn times and fuel densities are 10^{-10} s and $10^{31}-10^{32}$ ions/m^3, respectively. These densities correspond to a few hundred to a few thousand times that of ordinary condensed solids. IFE fusion produces the equivalent of small thermonuclear explosions in the target chamber. An IFE power plant design, therefore, must deal with very different physics and technology issues than an MFE power plant, although some requirements, such as tritium breeding, are common to both. Some of the challenges facing IFE power plants include the highly pulsed nature of the burn, the high rate at which the targets must be made and transported to the beam focus, and the interface between the driver beams and the reactor chamber (15).

Drivers. In inertial fusion the fuel is compressed and heated using driver beams. Achieving ignition requires a large amount of energy to be precisely controlled and delivered to the fuel target in a very short time, and the target must be capable of absorbing this energy efficiently. To produce net energy, the IFE system must have gain, ie, more energy output than was used to make, compress, and heat the fuel. Driver efficiency and capsule design and fabrication are therefore important issues for an IFE reactor (16).

The necessary energy can be delivered to the fuel by a variety of possible drivers. The four types of drivers receiving the most research attention are solid state lasers, KrF lasers, light-ion accelerators, and heavy ion accelerators. The leading driver for target physics experiments worldwide is the solid-state laser, and in particular the Nd:glass laser. The reason is that the irradiances required for IFE are in the $10^{18}-10^{19}$ W/m^2 range (17). The Nd:glass laser was the first driver which could produce these large power densities on target and it has remained in the forefront because of its high performance, reliable technology, and relative ease of maintenance. Low efficiencies and pulse rates have traditionally eliminated Nd:glass lasers from serious consideration in IFE reactor designs. However, new Nd:glass technology, replacing flash lamp pumping with higher efficiency diode pumping and utilizing crystalline disks and gas cooling, could change this view. Higher driver efficiencies are achievable in KrF lasers and particle beam accelerators. Particle beams have thus far had difficulty in achieving the low divergences and small focal spots required for IFE experiments, a technical area where lasers have a natural advantage. In IFE reactors, however, focal spots as large as 1 cm are permitted, and it appears that both light and heavy ion drivers could meet this requirement.

Targets. Two types of IFE targets have been investigated known as direct and indirect drive targets. Direct-drive targets absorb the energy of the driver directly into the fuel capsule, whereas indirect-drive targets use a cavity, called a hohlraum, to convert the driver energy to x-rays which are then absorbed by the fuel capsule. This latter method can tolerate greater inhomogeneities in driver illumination, albeit at the expense of the efficient delivery of energy to the capsule.

The extremely high peak power densities available in particle beams and lasers can heat the small amounts of matter in the fuel capsules to the temperatures required for fusion. In order to attain such temperatures, however, the mass of the fuel capsules must be kept quite low. As a result, the capsules are quite small. Typical dimensions are less than 1 mm. Fuel capsules in reactors could be larger (up to 1 cm) because of the increased driver energies available.

Laser Fusion. The largest and most powerful operating laser in the world is the NOVA 10-beam Nd:glass laser facility at the Lawrence Livermore National Laboratory in California. NOVA can deliver up to 40 kJ of 351-nm light in a 1-ns pulse onto the target. NOVA is primarily used for indirect-drive experiments. Other large Nd:glass laser facilities include the GEKKO XII laser at Osaka University in Japan, and the OMEGA laser at the Laboratory for Laser Energetics at the University of Rochester (Rochester, New York). The latter is used primarily for direct-drive experiments (see LASERS).

The krypton–fluoride (KrF) laser, which uses a gaseous lasing medium, can in principle operate at much higher pulse repetition rates and efficiencies than solid-state Nd:glass lasers. Moreover, the shorter (250 nm) wavelength and broad bandwidth, both of which improve coupling to the target, provide additional advantages. However, the use of KrF lasers is complicated by the long pulse length, which, for the 1 ns time scales of IFE, has to be shortened by a factor of about 100. At least two schemes have been proposed and demonstrated (15). In one method, angular multiplexing, many short, low power pulses are sent sequentially through the laser power amplifier stage for the entire duration of the pumping pulse, each at a different angle. After traversing paths of different optical length, these pulses are recombined at the target into a single high amplitude short pulse. In the second method, a long pulse is extracted and subsequently shortened in a Raman scattering cell filled with, for example, SF_6 gas (see INFRARED TECHNOLOGY AND RAMAN SPECTROSCOPY). Through Raman backscattering, the pulse can be shortened by the desired factor of 100. Both techniques have been demonstrated (15). KrF laser technology is not as well developed as the technology for Nd:glass lasers, however, and no KrF lasers have been constructed which are as powerful as NOVA. The efficiency of KrF lasers may also fall a little short of that needed for a power producing reactor.

Particle Beam Fusion. Advances in pulsed power technology have enabled large quantities of electrical energy to be generated in short pulses using relatively high efficiency and low cost. In a light-ion particle accelerator, an initial electrical pulse of the required energy is progressively shortened through a series of pulse forming steps to be delivered with an amplitude of several tens of megavolts to a diode which emits and accelerates the selected ions, eg, lithium, across a short gap to converge on the fuel capsule. The light-ion particle beam fusion accelerator II (PBFA II) at Sandia National Laboratory in New Mexico is the most energetic IFE driver, delivering up to 1 MJ on target. However, obtaining good beam divergence has been a challenge.

To survive the effects of the target explosion, the diode must be located at least several meters away from the target. The diode on PBFA II is only about 15 cm from the target. Long-lived, reliable diodes having 10 Hz repetition rates and beam-transport systems several orders of magnitude longer than those in use as of this writing are required to make a light-ion beam reactor feasible (15).

The Fusion Policy Advisory Committee of the Department of Energy has identified the heavy-ion accelerator as the leading candidate for an IFE reactor driver (16). The reasons include ruggedness, reliability, high pulse-rate capabilities, and potential for high efficiency. There are two different technologies being developed for heavy-ion accelerators: induction acceleration and radio frequency (rf) acceleration. The induction accelerator approach is pursued mainly in the United States, at the Lawrence Berkeley Laboratory. The rf accelerator approach is pursued primarily in Europe and Japan (15). The same types of heavy ions can be utilized in both approaches; typically cesium, bismuth, or xenon are chosen. To obtain the required 10^{18}–10^{19} W/m^2 on target in a reactor, using targets of 1 cm^2 size and accelerator energies limited to 5 GeV to provide the requisite stopping distance inside the target fuel, particle beam currents of around 100,000 A are required. These currents are quite large compared to traditional high energy physics accelerators, and in experiments where high currents have been achieved, the beam divergence has been unsatisfactorily large.

Environmental and Safety Aspects

Fusion reactors are expected to be relatively benign environmentally when compared to other sources of power. A 1989 National Research Council report cites the environmental issue as a persuasive reason for pursuing the fusion energy option (10). A general environmental advantage of nuclear power plants whether fission or fusion, compared to fossil fuel plants, is the minimization of mining requirements and no emission of noxious effluents. A further advantage of fusion, relative to fission, is the absence of meltdown dangers and avoidance of long-lived radioactive wastes (see NUCLEAR REACTORS). An accidental runaway reaction cannot occur in a fusion reactor for two reasons. First, the amount of deuterium and tritium in the reactor at any given time is small, and any uncontrolled burning quickly consumes all the available fuel and extinguishes itself. Second, a neutron chain reaction of the fission-reactor type is impossible in fusion, because fusion reaction rates are not sustained by neutrons.

The fusion of deuterium and tritium produces only energetic neutrons and alpha particles (helium nuclei), which are not themselves radioactive. The 14-MeV neutrons are absorbed in a blanket surrounding the reacting plasma, and the only unavoidable ash of the D–T reaction is ordinary helium gas. The main concern about radiation comes from a secondary process, namely activation of the reactor components by the fusion neutrons. The secondary nuclear reactions which result from the energetic neutrons depend on the materials selected for the reactor blanket and support structure (18). The materials aspects of fusion reactors have been reviewed (19), and the calculated decay of radioactivity following shutdown of D–T fusion reactors constructed of various materials is shown in Figure 5, together with that of a fission reactor (8,18). If advanced structural materials such as silicon carbide, SiC, can be used, fusion reactors are expected to reduce the amount of radioactive waste by six orders of magnitude or more.

A D–T fusion reactor is expected to have a tritium inventory of a few kilograms. Tritium is a relatively short-lived (12.36 year half-life) and benign (beta emitter) radioactive material, and represents a radiological hazard many orders of magnitude less than does the fuel inventory in a fission reactor. Clearly, however, fusion reactors must be designed to preclude the accidental release of tritium or any other volatile radioactive material. There is no need to have fissile materials present in a fusion reactor, and relatively simple inspection techniques

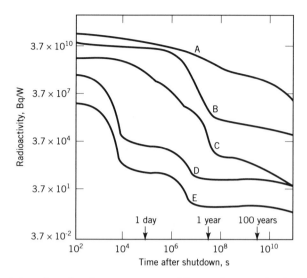

Fig. 5. Radioactivity after shutdown per watt of thermal power for A, a liquid-metal fast breeder reactor, and for a D–T fusion reactor made of various structural materials; B, HT-9 ferritic steel; C, V-15Cr-5Ti vanadium–chromium–titanium alloy; and D, silicon carbide, SiC, showing the million-fold advantage of SiC over steel a day after shutdown. The radioactivity level after shutdown is also given for E, a SiC fusion reactor using the neutron reduced D–^{3}He fuel cycle. To convert Bq to curie, multiply by 2.70×10^{-11}.

should suffice to prevent any clandestine breeding of fissile materials, eg, for potential weapons diversion.

Future Developments and Applications

The goal of fusion development is central station electrical power generation. Using the D–T fuel cycle, power would be extracted from the thermalization of the neutron kinetic energy deposited in the blanket. Pulsed systems such as inertial fusion require storage techniques to provide a continuous output of electrical power. In some cases, this storage medium may be simply the thermal blanket surrounding the reaction chamber. In MFE, significant technological challenges include the development of large superconducting magnets, efficient current drive systems, and adequate diverter plates and plasma facing components to handle the high particle and radiation heat loads. Provisions must also be made for the replacement and maintenance of components by remote handling techniques.

Potential fusion applications other than electricity production have received some study. For example, radiation and high temperature heat from a fusion reactor could be used to produce hydrogen by the electrolysis or radiolysis of water, which could be employed in the synthesis of portable chemical fuels for transportation or industrial use. The transmutation of radioactive actinide wastes from fission reactors may also be feasible. This idea would utilize the neutrons from a fusion reactor to convert hazardous isotopes into more benign and easier-to-handle species. The practicality of these concepts requires further analysis.

Fusion energy research is also the primary avenue for the development of plasma physics as a scientific discipline. The technologies and the science of plasmas developed en route to fusion power are already important in other applications and fields of science (see PLASMA TECHNOLOGY).

Cold Fusion

In the spring of 1989, it was announced that electrochemists at the University of Utah had produced a sustained nuclear fusion reaction at room temperature, using simple equipment available in any high school laboratory. The process, referred to as cold fusion, consists of loading deuterium into pieces of palladium metal by electrolysis of heavy water, D_2O, thereby developing a sufficiently large density of deuterium nuclei in the metal lattice to cause fusion between these nuclei to occur. These results have proven extremely difficult to confirm (20,21). Neutrons usually have not been detected in cold fusion experiments, so that the D–D fusion reaction familiar to nuclear physicists does not seem to be the explanation for the experimental results, which typically involve the release of heat and sometimes gamma rays.

Room temperature fusion reactions, albeit low probability ones, are not a new concept, having been postulated in 1948 and verified experimentally in 1956 (22), in a form of fusion known as muon-catalized fusion. Since the 1989 announcement, however, international scientific skepticism has grown to the point that cold fusion is not considered a serious subject by most scientists. Follow-on experiments, conducted in many prestigious laboratories, have failed to confirm the claims, and although some unexplained and intellectually interesting phenomena have been recorded, the results have remained irreproducable and, thus far, not accepted by the scientific community.

BIBLIOGRAPHY

"Fusion Energy" in *ECT* 3rd ed., Vol. 11, pp. 590–603, by G. M. Halpern, Exxon Research and Engineering Co.

1. H. A. Bethe, *Phys. Rev.* **55,** 103 (1939).
2. S. Glasstone and R. H. Lovberg, *Controlled Thermonuclear Reactions,* D. Van Nostrand, New York, 1960.
3. R. A. Gross, *Fusion Energy,* John Wiley & Sons, Inc., New York, 1984.
4. L. Spitzer, Jr., *Physics and Fully Ionized Gases,* 2nd rev. ed., John Wiley & Sons, Inc., New York, 1962.
5. J. D. Lawson, *Proc. Phys. Soc.* **B70,** 6(1957).
6. O. Naito and co-workers, *Plasma Phys. Control. Fusion* **35,** B215–B222 (1993); T. Kondo and co-workers, "High Performance and Current Drive Experiments in JA-ERI Tokamak-60 Upgrade," *Phys. Plasmas* (in print) (1994); H. Ninomiya and co-workers, "Recent Progress and Future Prospect of the JT-60 Program," in the *Proceedings of the 15th Symposium on Fusion Engineering,* Hyannis, Mass., 1993.
7. I. E. Tamm and A. D. Sakharov, in M. A. Leontovich, ed., *Plasma Physics and the Problem of Controlled Thermonuclear Reactions,* Vol. 1, Pergamon Press, New York, 1961.

8. H. P. Furth, *Science* **249,** 1522 (Sept. 1990); J. G. Gordey, R. J. Goldston, and R. R. Parker, *Phys. Today,* 22 (Jan. 1992).
9. R. W. Conn and co-workers, *Sci. Am.* **266,** 103 (Apr. 1992).
10. *Pacing the U.S. Magnetic Fusion Program,* National Academy Press, Washington, D.C., 1989.
11. *Fusion Policy Advisory Committee Final Report, DOE/S-0081,* Department of Energy, Washington, D.C., Sept. 1990.
12. *National Energy Strategy, First Edition 1991/1992,* Department of Energy, Washington, D.C., Feb. 1991.
13. The JET Team, *Nuc. Fusion* **32,** 187 (1992).
14. International Atomic Energy Agency, *ITER Conceptual Design Report,* IAEA, Vienna, 1991.
15. W. J. Hogan, R. Bangerter, and G. L. Kulcinski, *Phys. Today,* 42 (Sept. 1992).
16. *Review of the Department of Energy's Inertial Confinement Fusion Programs,* National Academy Press, Washington, D.C., Sept. 1990.
17. J. D. Lindl, R. L. McCrory, and E. M. Campbell, *Phys. Today,* 32 (Sept. 1992).
18. R. W. Conn and co-workers, *Nucl. Fusion* **30,** 1919 (1990).
19. J. P. Holdren and co-workers, *Report of the Senior Committee on Environmental, Safety, and Economic Aspects of Magnetic Fusion Energy,* report UCRL-53766, Lawrence Livermore National Laboratory, Livermore, Calif., Sept. 25, 1989.
20. *Cold Fusion Research, DOE Report S-0073,* U.S. Dept. of Energy, Washington, D.C., Nov. 1989.
21. F. Close, *Too Hot to Handle: The Role for Cold Fusion,* Princeton University Press, Princeton, N.J., 1991.
22. B. V. Lewenstein and co-workers, *Forum for Applied Research and Public Policy,* **7**(4), 67–107 (Winter 1992).

WILLIAM R. ELLIS
Raytheon Engineers & Constructors

G

G ACID. See Naphthalene derivatives.

GADOLINITE, GADOLINIUM. See Lanthanides.

GALENA, GALENITE. See Lead.

GALLIUM AND GALLIUM COMPOUNDS

Gallium [7440-55-3], atomic number 31, was discovered through a study of its spectral properties in 1875 by P. E. Lecoq de Boisbaudran and named from Gallia in honor of its discoverer's homeland. The first element to be discovered after the publication of Mendeleev's Periodic Table, its discovery constituted a confirmation of the Table which was reinforced shortly after by the discoveries of scandium and germanium.

Gallium is a scarce but not a rare element. It is found most commonly in association with its immediate neighbors in the Periodic Table, ie, zinc, germanium, and aluminum. It has also been found in a sphalerite from the Hautes-Pyrenees. The concentration of gallium in the earth's crust, 10–20 g/t (10–20 ppm), is comparable to that of lead and arsenic. Up to 1% of gallium has been found in a few rare ores, eg, South African germanite. Some geochemical data (1) are given in Table 1. The ratio of Ga:Al is quite constant. Additional data are available (2).

There is an abundance of aluminum ores and alumina plants (see Aluminum and aluminum alloys; Aluminum compounds, aluminum ox-

Table 1. Gallium in Rocks and Minerals[a]

Material	Number of studies	Ga, ppm	10^4 Ga:Al
igneous rocks	323	22	3.3
metamorphic rocks	90	19	2.2
sedimentary rocks	166	16	2.8
bauxites	76	55	2.2
sphalerites	31	35[b]	

[a]Data are mean values.
[b]Observed to be quite variable, eg, 3–100 or more.

IDE), and less importantly to gallium production zinc ores and plants (see ZINC AND ZINC ALLOYS); these are the main sources of gallium. Gallium is found also in the production of germanium from germanium minerals. In certain coals, the flue dust may contain 0.001–0.05 wt % Ga, sometimes as high as 1 wt % Ga (see COAL). Micas are also gallium-rich (up to 0.1%) and gallium extraction from a combined spodumene and lepidolite deposit (Bernic Lake, Ontario, Canada) was considered (see MICAS, NATURAL AND SYNTHETIC). Gallium has also been extracted from copper ores (Apex mine, St. George, Utah).

Properties

Physical Properties. Gallium, at. wt 69.717, has two stable isotopes, ^{69}Ga [*14391-02-7*], 60.4%, and ^{71}Ga [*14391-03-8*], 39.6%, and twelve unstable isotopes, from mass 63 through 76. The radius of the atom is 0.138 nm, and of the ions Ga^{3+} and Ga^+, is 0.062 nm and 0.133 nm, respectively. Solid gallium has a metallic, slightly bluish appearance.

The physical properties of gallium, especially its thermal properties, are exceptional (1). It has a low mp and vaporizes above 2200°C, ie, it has the longest liquid interval of all the elements. Also, it is easily supercooled. However, it expands during solidification by 3.2%, a property shared by only two other elements, germanium and bismuth. Its crystal structure is unusual for a metal. Gallium crystallizes in the orthorhombic system, and it is very anisotropic. This latter property is attributed to the existence of Ga–Ga covalent bonds along its [001] axis. Principal physical properties of the normal or alpha form are listed in Table 2. Only normal gallium (alpha or I) expands upon solidification. There are other isomorphic phases of gallium. Some, which are metastable at atmospheric pressure, are obtained from supercooled gallium. Some of the physical properties of the metastable forms are listed in Table 3.

Other forms of gallium are obtained at high pressure (3,4). The high pressure forms II, cubic, I$\bar{4}$3d; and III, tetragonal, I4/mm; are formed at 2.8 GPa (280 kilobar) and 2.6 GPa (260 kilobar), respectively. The mp of the latter is 48.7°C.

Gallium wets almost all surfaces, especially in the presence of oxygen which promotes formation of a gallium suboxide film. Gallium diffuses quickly into the crystal lattice of certain metals, particularly aluminum. If a line is traced with a

Table 2. Physical Properties of Normal Gallium

Property	Value
melting point, °C	29.7714
boiling point, °C	ca 2200
density at mp, g/cm^3	
solid	5.904
liquid	6.095
heat of fusion, J/g^a	79.8
heat capacity, J/(kg·K)^a	
30°C	381.5
−17°C	361.4
−36°C	352.8
thermal conductivity, W/(m·K)	
solid at 20°C	
a	88.4
b	16.0
c	40.8
liquid at 77°C	28.7
cubic coefficient of expansion, °C^{-1}	
solid at 0–20°C	5.95×10^{-5}
liquid	
at 103°C	1.20×10^{-4}
at 600°C	1.03×10^{-4}
vapor pressure, Pa^b	
1198 K	0.14
1249 K	0.44
1321 K	1.74
1478 K	16.5
crystallographic properties	orthorhombic Cmca space group
a, nm	0.45186
b, nm	0.76602
c, nm	0.45258
z	8
electrical resistivity, μΩ·cm	
solid at 20°C	
a	8.1
b	54.3
c	17.4
polycrystal	15.05
liquid at 30°C	25.795
debye temperature, °C	51.6
viscosity, dynamic, mPa·s(=cP)	
32°C	1.810
50°C	1.739
surface tension, mN/m(=dyn/cm)	
30°C	709
400°C	733

^{a}To convert J to cal, divide by 4.184.
^{b}To convert Pa to mm Hg, divide by 133.3.

Table 3. Physical Properties of Metastable Forms of Ga

Property	Metastable forms at normal pressure			
	β	λ	ε	γ
melting point, °C	− 16.3	− 19.4	− 28.6	− 35.6
heat of fusion, J/g[a]	38.0	37.0	36.2	34.9
density at melting point, g/cm^3	6.22	6.21		6.20
crystal structure	monoclinic	trigonal		orthorhombic
space group	C2/c	R$\bar{3}$m		Cmcm

[a]To convert J to cal, divide by 4.184.

piece of solid gallium on a sheet of aluminum, the aluminum quickly becomes brittle on that line as a result of the diffusion of gallium to the grain boundaries of the aluminum. Because of this property of embrittling aluminum, the DOT Office of Hazardous Materials has classified gallium in Hazard Class HM-181 and has placed restrictions on air shipment. IATA (international air transport regulation) classifies gallium as corrosive.

Chemical Properties. In accordance with its normal potential, gallium is chemically similar to zinc and is somewhat less reactive than aluminum. Just as aluminum, it is protected from air-oxidation at ambient temperature by a fine film of oxide. It is entirely oxidized by air and by pure oxygen at ca 1000°C. Normally, gallium is trivalent; however, it also may be monovalent. Gallium forms some compounds of mixed valence, eg, gallium(I) tetrachlorogallate(-1) [13498-12-9], Ga(GaCl$_4$). Gallium slowly reacts with water below 100°C, but it is completely oxidized by water in an autoclave at 200°C. At ambient temperature, concentrated inorganic acids and gallium react slowly. Oxidizing agents, such as aqua regia, H$_2$SO$_4$, H$_2$O$_2$, and perchloric acid, attack gallium effectively, especially when hot. The halogens readily react with gallium upon heating, as do sulfur, selenium, tellurium, phosphorus, arsenic, and antimony. Nitrogen reacts with gallium only under particular conditions, whereas boron seems not to do so.

Amphoteric character is one of gallium's principal properties. Gallium oxide is more acidic than alumina, and thus solutions of gallates(III) are not as unstable as those of aluminates. Solutions of alkali hydroxides readily attack gallium at 100°C. In alkaline solutions, eg, of sodium gallate [12063-93-3], NaGaO$_2$, as in acidic solutions, eg, of GaCl$_3$, the solubility in terms of Ga can be very high, on the order of 1000 g/L of solution. At a temperature of 500–1000°C or more, gallium is very corrosive to most metals. Only tungsten is unaltered; niobium, tantalum, and molybdenum are less resistant.

Extraction

A minor amount of gallium is extracted as a by-product from the zinc industry. The gallium content of sphalerites generally is concentrated in the residues of zinc distillation and in the iron mud resulting from purification of zinc sulfate solutions. Gallium is extracted from these streams by acidic solutions and gallium salts are recovered by liquid–liquid extraction using dialkylphosphates, trialkyl phosphates, hydroxyquinolines, or isopropyl ether, among others (see EXTRACTION, LIQUID–LIQUID) (5).

Recovery from Bayer Liquor. The significant amount of primary gallium is recovered from the alumina industry. The main source is the sodium aluminate liquor from Bayer-process plants that produce large quantities of alumina. Several methods have been developed to recover gallium from Bayer liquor.

Carbonation. Gallium can be extracted by fractional carbonation which consists of treating the aluminate solution with carbon dioxide in several controlled stages. This process is no longer under industrial operation (6).

Electrolysis. Gallium can be extracted by direct electrolysis of the aluminate solution at a strongly agitated mercury cathode. The recovery from a sodium gallate solution resulting from the carbonation process is another possibility. This process is probably no longer operative because of the environmental problems associated with the mercury.

Chemical Reduction. Reduction of gallium by aluminum has been developed in the former Soviet Union. This method is in operation (ca 1994). The Bayer liquor is contacted using a gallium–aluminum alloy named Gallam, and the gallium is deposited.

Liquid–Liquid Extraction. Among the various extractants available for the recovery of metals from aqueous streams, only 8-hydroxyquinoline derivatives are effective for the recovery of gallium from Bayer liquor. A process has been developed and patented (7–12). Production began in 1980 in Salindres, France, followed in 1989 in Pinjarra, western Australia.

Ion-Exchange Resins. Some attempts have been carried out to recover gallium by ion exchange (qv). Only the commercially available amidoxime resin has proved to be effective for the recovery of gallium from Bayer liquor. The process has been developed and patented (13) and is reported to be under operation in Japan.

Purification. Extraction from aluminum or zinc ores produces crude gallium metal or concentrates. These concentrates are transformed to sodium gallate, gallium chloride, or gallium sulfate solutions which are purified, then electrolyzed. Gallium is deposited as a liquid.

The purification of the gallium salt solutions is carried out by solvent extraction and/or by ion exchange. The most effective extractants are dialkylphosphates in sulfate medium and ethers, ketones (qv), alcohols, and trialkylphosphates in chloride medium. Electrorefining, ie, anodic dissolution and simultaneous cathodic deposition, is also used to purify metallic gallium.

Ultrahigh Purity Gallium. Many applications, particularly those in the electronics industry (see ELECTRONIC MATERIALS), require high (> 99.99999 % = 7.N) purity metallic gallium. This is achieved by a combination of several operations such as filtration, electrochemical refining, heating under vacuum, and/or fractional crystallization (see ULTRAPURE MATERIALS) (14).

Recycling. A large part of the wastes from the gallium arsenide [*12064-03-8*], GaAs, industry is recovered for both economical and environmental reasons. Several processes are effective and being used to recover both the gallium either as a metal, a salt, or a hydroxide for recycling, and the arsenic in some form for recycling or disposal. Thermal decomposition of gallium arsenide waste is one method which competes with the use of hydrometallurgical routes in caustic soda media (see RECYCLING, NONFERROUS METALS).

Production

About 60 metric tons of gallium were used throughout the world in 1992. Total worldwide gallium production capacity excluding the CIS, for which data are not available, is estimated to be at 250 t/yr.

The gallium either comes from mining sources or is recycled from scrap. Scrap-recycling capacity is taking a larger place each year and in 1992 represented about 30 t. In the United States, the main processors are Eagle-Picher Industries and Recapture Metals; in Japan, Rasa Industries, Mitsubishi Metals, and Sumitomo Metal Mining; in France, Rhône-Poulenc; and in Germany, Preussag (Metaleurop) and Ingal.

Recycled gallium changes the market. The scrap generated in the production of electronic devices is a very important source of gallium. Most of the scrap is in the GaAs form because of wafer production, but each stage of electronic device manufacturing generates its own waste. The gallium content may vary from less than 1% to over 98% in these scraps.

Analysis

Gallium is easily identified spectrographically, eg, using the lines 287–294 and 403–417 nm. It is somewhat difficult to separate from aluminum and zinc by chemical means. The best method is the extraction of gallium chloride in acidic medium using an organic solvent, eg, diethyl or diisopropyl ether. Gallium can also be identified by colorimetry (using rhodamine B), fluorometry (using 8-hydroxyquinoline), gravimetrically (using 5,7-dibromo-8-hydroxyquinoline), or by complexometry and titration (EDTA).

The conventional method for quantitative analysis of gallium in aqueous media is atomic absorption spectroscopy (qv). High purity metallic gallium is characterized by trace impurity analysis using spark source (15) or glow discharge mass spectrometry (qv) (16).

Alloys and Intermetallic Compounds

Alloys. Gallium has complete miscibility in the liquid state with aluminum, indium, tin, and zinc. No compounds are formed. However, these binary systems form simple eutectics having the following properties:

Metal	Ga, %	Melting point, °C
Al	96	26.4
In	76	15.7
Sn	91.5	20.6
Zn	96.3	25.0

The systems obtained when gallium is in the presence of bismuth, cadmium, germanium, mercury, lead, silicon, or thallium present miscibility gaps. No intermetallic compounds are formed.

Intermetallic Compounds. Numerous intermetallic gallium–transition element compounds have been reported (17). The principal compounds are listed in Table 4 (18–23). There are probably several Cs and Rb compounds; however, none is well known.

Lanthanides and Yttrium. The following compounds are known: MGa_2, hexagonal (M = Dy, La, Nd, Sm and Tb); MGa, orthorhombic; M_5Ga_3, tetragonal; MGa_2, hexagonal (M = Eu and Yb); MGa_3, cubic (M = Lu, Tm); MGa, orthorhombic; M_5Ga_3, hexagonal; $ErGa_3$ [12785-97-6], cubic; $ErGa_2$ [60874-25-1], hexagonal; Er_5Ga_3 [12324-45-7], hexagonal; $GdGa_2$ [12690-00-5], hexagonal; GdGa [12310-95-1], orthorhombic; Gd_3Ga_2 [12683-05-5], tetragonal; Gd_5Ga_3 [12324-83-3], tetragonal; $HoGa_3$ [67352-19-6], cubic; $HoGa_2$ [12599-74-5], hexagonal; HoGa [12310-96-2], orthorhombic; Ho_5Ga_3 [12324-84-4], hexagonal; $PrGa_2$ [12599-76-7], hexagonal; PrGa [12310-99-5], orthorhombic; Pr_5Ga_3 [12361-85-2], orthorhombic; YGa_2 [12435-21-1], hexagonal; YGa [12160-65-5], orthorhombic; and Y_5Ga_3 [12024-34-9], hexagonal.

Actinides. The following compounds are known: U_2Ga_3 [37381-84-3], orthorhombic; UGa_2 [12064-23-2], hexagonal; UGa_3 [12024-33-8], cubic (24); $PuGa_6$ [12160-80-4], tetragonal; $PuGa_4$ [12160-77-9], orthorhombic; $PuGa_3$ [12024-31-6], hexagonal; $PuGa_2$ [12160-71-3], hexagonal; PuGa [12160-58-6], tetragonal; Pu_5Ga_3 [12160-74-6], tetragonal; and Pu_3Ga [12160-59-7], tetragonal. Gallium is also used at low (ca 1%) concentrations to stabilize the delta phase of plutonium.

Compounds Other Than Intermetallic

Gallium compounds containing monovalent elements are described in Reference 25.

Hydrides. Gallium(I) hydride [13572-92-4], GaH, is detected only in the vapor phase. The hydride of Ga(III), gallane [13572-93-5], GaH_3, which decomposes above −15°C, can be obtained by reaction of trimethylaminegallane [19528-13-3], $(CH_3)_3N \cdot GaH_3$, and boron fluoride, BF_3, at −20°C. Halogen derivatives of gallane, especially dichlorogallane [13886-65-2], $GaHCl_2$, and dibromogallane [13886-66-3], $GaHBr_2$, are obtained by reaction of trimethylsilane and $GaCl_3$ or $GaBr_3$, respectively, below −20°C. The colorless crystals of dichlorogallane are stable up to 29°C. The dibromogallane decomposes at −5°C.

Mixed hydrides, $MGaH_4$, where M = Li, [17836-90-7]; Na, [32106-51-7]; K, [32106-52-8]; Rb, [32104-62-4]; and Cs [32104-63-4] also are known. The lithium gallium hydride [17836-90-7], is the most stable, and has been studied as a reagent for organic reactions. It is similar to the aluminum compound, $LiAlH_4$.

A number of organic derivatives of gallane, both addition and substitution compounds, are known. Some are much more stable than gallane, eg, $(CH_3)_3N \cdot GaH_3$, $(CH_3)_3P \cdot GaH_3$ [15279-03-5], and $(CH_3)_2N \cdot GaH_2$ [32342-80-6]. These are used for the preparation of other organic compounds of gallium.

Table 4. Intermetallic Compounds

Compound	CAS Registry Number	Crystal form	Compound	CAS Registry Number	Crystal form
Ag_3Ga	[57673-45-7]	hexagonal	$NbGa_3$	[12160-73-5]	tetragonal
$AgGa$	[12260-08-1]	hexagonal	Nb_5Ga_{13}	[12160-83-7]	orthorhombic
$AgGa_2$	[12044-77-8]	cubic	Nb_5Ga_4	[12024-35-0]	hexagonal
$AuGa$	[12006-53-0]	orthorhombic	Nb_3Ga_2	[12306-31-9]	trigonal
Au_2Ga	[54327-67-2]	orthorhombic	Nb_5Ga_3	[12024-30-5]	trigonal
Au_7Ga_2	[12536-48-6]	hexagonal	Nb_3Ga	[12024-05-4]	cubic
$BaGa_4$	[12230-66-9]	tetragonal	$NiGa_4$	[56627-21-5]	cubic
$BaGa_2$	[12258-56-9]	hexagonal	Ni_2Ga_3	[12629-62-8]	tetragonal
$BaGa$	[12258-55-8]	hexagonal	Ni_3Ga_4	[12435-24-4]	cubic
Ba_3Ga_2	[12258-59-2]	trigonal	$Ni_{13}Ga_9$	[12500-13-9]	monoclinic
Be_3Ga	[73156-95-3]		Ni_3Ga_2	[56541-04-8]	hexagonal
$CaGa_4$	[12177-60-5]	tetragonal	Ni_5Ga_3	[12401-05-7]	orthorhombic
$CaGa_2$	[12258-68-3]	hexagonal	Ni_3Ga	[12063-96-6]	cubic
$CaGa$	[12258-67-2]	orthorhombic	O_5Ga_3	[60862-21-7]	tetragonal
$CeGa_2$	[12157-58-3]	hexagonal	$PdGa_5$	[52935-27-0]	tetragonal
$CeGa$	[20328-37-4]	orthorhombic	Pd_3Ga_7	[73157-02-5]	cubic
$CeGa_3$	[12360-78-0]	tetragonal	$PdGa$	[59125-32-5]	cubic
Ce_5Ga_3	[12360-78-0]	tetragonal	Pd_5Ga_4	[73157-03-6]	cubic
Ce_3Ga	[73623-49-1]	cubic	Pd_5Ga_3	[52935-26-9]	orthorhombic
$CuGa_2$	[12443-57-1]	tetragonal	Pd_5Ga_2	[53095-62-8]	orthorhombic
$CuGa$	[12191-11-6]		Pd_2Ga	[12529-00-9]	orthorhombic
Cu_2Ga	[69847-96-7]	cubic	$PtGa_6$	[12411-32-4]	orthorhombic
Cu_9Ga_4	[12395-15-2]	cubic	Pt_3Ga_7	[12411-33-5]	cubic
Cu_3Ga	[68985-62-6]	cubic	$PtGa_2$	[12786-51-5]	cubic
$FeGa_2$	[12062-72-5]	tetragonal	Pt_2Ga_3	[12411-26-6]	tetragonal
Fe_3Ga_4	[53237-41-5]	monoclinic	$PtGa$	[12411-22-2]	cubic
$FeGa$	[71771-30-7]	cubic	Pt_5Ga_3	[12411-27-2]	orthorhombic
Fe_6Ga_5	[53262-40-1]	monoclinic	Pt_2Ga	[12401-01-3]	cubic
Fe_2Ga	[12160-14-4]	hexagonal	Pt_3Ga	[12411-23-3]	cubic
Fe_3Ga	[12063-30-8]	cubic	Rh_2Ga_9	[66703-44-4]	monoclinic
$HfGa_3$	[73156-96-4]	tetragonal	$RhGa_3$	[60862-22-8]	tetragonal
$HfGa_2$	[12186-72-0]	tetragonal	$Rh_{10}Ga_{17}$	[12064-38-9]	tetragonal
Hf_2Ga_3	[73156-97-5]	orthorhombic	$RhGa$	[73157-04-7]	cubic
Hf_5Ga_3	[73156-98-6]	hexagonal	$RuGa_3$	[60862-23-9]	tetragonal
Hf_2Ga	[12786-47-9]	tetragonal	$RuGa_2$	[12064-21-0]	orthorhombic
$IrGa_3$	[60682-20-6]	tetragonal	$RuGa$	[39388-93-7]	cubic
Ir_3Ga_5	[12064-32-3]	tetragonal	$SrGa_4$	[12160-78-0]	tetragonal
$IrGa$	[60921-95-1]	cubic	$SrGa_2$	[12259-26-6]	hexagonal
KGa_4	[12435-22-2]	orthorhombic	Sr_3Ga_2	[12259-27-7]	cubic
K_5Ga_8	[73156-99-7]		Ta_3Ga_2	[60862-19-3]	tetragonal
$LiGa$	[12519-03-8]	cubic	Ta_5Ga_3	[12160-75-7]	tetragonal
Li_3Ga_2	[63705-92-0]		$TiGa_3$	[12064-30-1]	tetragonal
Li_2Ga	[39343-49-2]		$TiGa_2$	[12186-73-1]	tetragonal
$LiGa_2$	[39343-50-5]		$TiGa$	[12398-41-3]	tetragonal
$LiGa_3$	[53570-25-5]		Ti_5Ga_4	[65453-84-1]	hexagonal
Mg_2Ga_5	[12411-31-3]	tetragonal	Ti_5Ga_3	[68565-51-5]	tetragonal
$MgGa_2$	[12411-25-5]	orthorhombic	Ti_2Ga	[12500-08-2]	hexagonal
$MgGa$	[12063-92-2]	tetragonal	Ti_3Ga	[12183-38-9]	hexagonal
Mg_2Ga	[12422-88-7]	hexagonal	V_8Ga_{41}	[55071-08-4]	tetragonal
Mg_5Ga_2	[12064-14-1]	orthorhombic	V_2Ga_5	[12024-48-5]	tetragonal
$MnGa_4$	[12160-76-8]	cubic	V_6Ga_7	[12024-50-9]	cubic
Mn_2Ga_5	[12763-96-1]	tetragonal	V_6Ga_5	[12024-49-6]	tetragonal
Mn_7Ga_6	[73157-00-3]	tetragonal	V_3Ga	[12024-15-6]	cubic

Table 4. (*Continued*)

Compound	CAS Registry Number	Crystal form	Compound	CAS Registry Number	Crystal form
Mn_3Ga_2	[12160-70-2]	tetragonal	$ZrGa_3$	[12064-31-2]	tetragonal
Mn_8Ga_5	[73157-01-4]	cubic	$ZrGa_2$	[12186-74-2]	
Mn_2Ga	[12160-55-3]	hexagonal	Zr_3Ga_5	[73156-91-9]	
Mn_3Ga	[12186-69-5]	hexagonal	Zr_2Ga_3	[73156-92-0]	
Mo_8Ga_{41}	[55071-07-3]	tetragonal	$ZrGa$	[73156-93-1]	tetragonal
Mo_6Ga_{31}	[52015-37-9]	monoclinic	Zr_5Ga_4	[73156-94-2]	hexagonal
Mo_3Ga	[37381-83-2]	cubic	Zr_3Ga_2	[66703-42-2]	tetragonal
$NaGa_4$	[12435-23-2]	tetragonal	Zr_5Ga_3	[66103-56-8]	hexagonal
Na_5Ga_8	[39297-66-0]		Zr_2Ga	[12786-50-4]	tetragonal

Halides   *Gallium(I).*   Halides of Ga(I) are known only in the vapor phase. Coordination complexes of GaCl and GaBr have been obtained with dioxane, morpholine, and acetylacetone.

Gallium(I)–Gallium(III).   Compounds having the formula Ga_2X_4 where X = Cl, Br, or I, are known. The structure of these mixed valance compounds, $Ga^+(GaX_4)^-$, has been better understood in recent years (26,27). Some properties are listed in Table 5. Other compounds such as $Ga^+(Ga_2X_7)^-$, where X = Cl or Br, have also been identified and the structures determined (28,29).

Gallium(III).   Anhydrous gallium(III) fluoride [7783-51-9], GaF_3, is obtained as colorless needles, which dissolve slowly in water, by thermal decomposition of ammonium hexafluorogallate(III) [14639-94-2] in a stream of argon. The hydrate, $GaF_3 \cdot 3H_2O$ [22886-66-4], obtained by dissolving metallic gallium or the hydroxide in hydrofluoric acid, is more soluble in water. Thermal decomposition of $GaF_3 \cdot 3H_2O$ yields the hydroxyfluoride, $16[Ga(OH,F)_3] \cdot 6H_2O$, which is fcc at 200°C (30). Between 350 and 500 MPa (3450–4940 atm) at 300°C (31), the reaction of HF and Ga_2O_3 yields another hydroxyfluoride, $Ga_8[(OH)_{0.45}F_{0.55}]_{24} \cdot 3H_2O$, which is of the pyrochloro cubic type.

Gallium trichloride [13450-90-3], $GaCl_3$, forms as colorless needles upon reaction of chlorine or HCl and gallium at ca 200°C in a stream of nitrogen. An airtight apparatus is used to exclude traces of moisture. $GaCl_3$ can be purified by distillation in a stream of chlorine or nitrogen. Its structure seems to be a bridged dimer molecular lattice. A hydrate, $GaCl_3 \cdot H_2O$ [23306-52-7], mp 44.4°C, is known. The gallium trichloride is extremly soluble in water, ie, >800 g/L.

Gallium hydroxychloride [73157-09-2], $GaCl_2OH$, can be obtained (32) by decomposition at 130°C of the complex, $GaCl_3 \cdot C_2H_5OH$ [19379-37-4]. The hydroxychloride melts at 120°C and is a cyclic tetramer.

Gallium tribromide [13450-88-9], $GaBr_3$, is obtained by reaction of a stream of bromine vapor and nitrogen on gallium. The tribromide has seven hydrates, $GaBr_3 \cdot xH_2O$, where x = 1, 2, 2.5, 3, 4, 6, or 15 (33). Gallium hydroxybromide [73157-08-1], $GaBr_2OH$, which melts at 175°C, can be obtained by the same method as used for the hydroxychloride.

Gallium triiodide [13450-91-4], GaI_3, is obtained by direct reaction of the elements or by reaction of iodine solution in carbon disulfide on gallium.

Table 5. Properties of Gallium Halides

Compound	CAS Registry Number	Mp,°C	Density, g/cm^3	Crystal structure	Space group
GaF$_3$	[7783-51-9]	subl. 800	4.47	trigonal	R$\overline{3}$c
Ga(GaCl$_4$)	[24597-12-4]	170.4		orthorhombic	Pnma
Ga(Ga$_2$Cl$_7$)	[24688-86-6]	dec 77.3	2.74	orthorhombic	Pna2$_1$
GaCl$_3$	[13450-90-3]	77.8	2.47	triclinic	P$\overline{1}$
Ga(GaBr$_4$)	[18897-61-5]	164.5	3.471		a
Ga(Ga$_2$Br$_7$)	[50647-37-5]	dec 81.2	3.631	orthorhombic	P2$_1$2$_1$2$_1$
GaBr$_3$	[13450-88-9]	124.1	3.74	orthorhombic	Pbca
Ga(GaI$_4$)	[17845-89-5]	211			
GaI$_3$	[13450-91-4]	212	4.15	monoclinic	P2$_1$/c

^{a}Two forms exist, having a transition at 152°C.

Gallium Oxyhalides. The compounds gallium oxychloride [15588-51-9], GaOCl, and galliumoxybromide [15605-97-7], GaOBr, are obtained by reaction of Ga$_2$O$_3$ on GaCl$_3$ or GaBr$_3$ in a sealed tube at ca 300°C. These materials crystallize in the orthorhombic system.

Gallium Halogenates. The anhydrous gallium(III) perchlorate [19854-31-0], Ga(ClO$_4$)$_3$, is obtained by reaction of chlorine peroxide, Cl$_2$O$_6$, on GaCl$_3$ at −180°C (34). The product is a white solid and is stable at room temperature.

Sulfohalides and Sulfohalogenates. Gallium(III) fluorosulfide [73157-10-5], GaSF, is obtained by heating Ga$_2$S$_3$ and GaF$_3$ under pressure at 310°C. The sulfohalides, Ga$_9$S$_8$Cl$_{11}$ [12268-03-0] and Ga$_9$S$_8$Br$_{11}$ [12049-26-2] are monoclinic solids, obtained by heating Ga$_2$S$_3$ and GaCl$_3$ in a sealed tube at 250°C, or GaBr$_3$ at 330°C, respectively (35). The gallium(III) chlorosulfite [30338-20-6], Ga(SO$_3$Cl)$_3$, and its addition compounds with LiCl and NaCl, were also obtained (36).

Compounds with Ammonia. Gallium halides give the following derivatives with ammonia: GaF$_3$·NH$_3$ [73157-06-9] and GaF$_3$·3NH$_3$ [36171-86-5]; GaCl$_3$·NH$_3$ [50599-24-1]; GaBr$_3$·NH$_3$ [54955-92-9]; GaI$_3$·NH$_3$ [58384-90-0]. Some poorly defined compounds of gallium chloride, bromide, and iodide that contain up to 14 NH$_3$ groups also exist.

Other Halides. Hexafluorogallates(III), M$_3$GaF$_6$, are known where M is a univalent element, M = Cs [16027-80-8], K [16061-04-4], Li [16061-07-7], Na [16061-62-4], Rb [16061-06-6], Tl [26846-24-2], and NH$_4$ [14639-24-2]. Tetrafluorogallates, MGaF$_4$, where M = Cs [15002-96-7], K [15002-94-5], Li [39210-67-8], Na [15002-93-4], Rb [15002-95-6], Tl [26667-61-8], and NH$_4$ [18532-60-0], also exist. There are also pentafluorogallates(III). Some typical examples are the compounds of MGaF$_5$, where M is Ca [27658-92-0], Cd [58984-51-3], Mn [35745-87-0], Pb [27658-94-2], or Sr [27658-93-1].

Two types of chlorogallates(III) are known (27,29): MGaCl$_4$, where M = Cs [21646-31-1], K [18154-89-7], or Rb [21646-29-7]; and MGa$_2$Cl$_7$, where M = Cs [12331-25-8], K [30617-31-1], or Rb [56531-09-0]. These are analogous to the mixed valence compounds, ie, M = Ga$^+$. Additionally, the following chlorogallates(III) are also known, but less precisely: MGaCl$_4$ where M = In [12432-57-4], Li [15955-98-3], Na [15007-28-0], or Tl [26490-70-0]; and several M(GaCl$_4$)$_2$ and MGaCl$_5$ compounds. The chlorogallate(III) GaSbCl$_6$ [73157-05-8] has been studied (19). Chlorogallates(III) are also obtained with $AsCl_3$, $AsCl_5$, PCl_3, PCl_5, $POCl_3$, $SeCl_4$, $TeCl_4$, $MoCl_5$, $NbCl_5$, and $TaCl_5$.

The following bromogallates(III) have been described (27,29): $MGaBr_4$ where M = Cs [52582-09-9], K [50328-30-8], or Rb [52582-08-8]; MGa_2Br_7 where M = Cs [52724-44-4], K [39681-16-8], or Rb [52582-11-3]; and MGa_3Br_{10} where M = Cs [62974-40-7], or Rb [62974-41-8]. The bromogallate(III) $SbGaBr_6$ [12514-56-6] also was studied (26). Three types of iodogallates (III) have been described (37): $MGaI_4$ where M = Cs [57146-51-7] or Rb [57146-50-6]; MGa_2I_7 where M = Cs [57143-24-5] or Rb [57143-25-6]; and $SbGaI_6$ [12519-02-7] (26).

Gallium trihalides, particularly $GaCl_3$ and $GaBr_3$, form a large number of complexes with organic Lewis bases. Complexes are also formed with other compounds, especially those containing, O, S, N, P, or As, eg, $LGaCl_3$, where L = $(C_2H_5)_2O$ or pyridine. Generally, because of its acceptor strength, the reactivity of trivalent Ga is ranked below aluminum and above indium and boron.

Gallium Oxides. The preparation conditions of gallium oxides and hydroxides are listed in Table 6.

Gallium(I). Gallium(I) oxide [12024-20-3], Ga_2O, is a dark brown powder, stable in cold, dry air, and has a density of 4.77 g/cm^3. Above 700°C, it decomposes into gallium and Ga_2O_3.

Gallium(III). Ga_2O_3 is the single gallium oxide that is stable under normal conditions. Like alumina, it exists in several crystalline forms. The most stable form is the beta oxide, mp ca 1725°C, density 5.88 g/cm^3. The density of α-Ga_2O_3 is 5.18 g/cm^3.

Gallium(III) oxide [12024-21-4] can be reduced at ca 600°C by hydrogen, carbon monoxide, or metallic gallium. It is more reactive than alumina and is strongly amphoteric, forming gallium salts with acids and gallates(III) with bases. There is only one well-defined hydroxide, gallium(III) oxide monohydrate [20665-52-5], GaOOH or $Ga_2O_3 \cdot H_2O$. The gallium(III) hydroxide [12023-99-3], $Ga(OH)_3$, probably has an amorphous structure.

Table 6. Gallium Oxides and Hydroxides

Compound	Means of preparation	Crystal structure
Ga_2O	reduction of Ga_2O_3 with Ga at 500–600°C, under vacuum	
α-Ga_2O_3	heating GaOOH at 300–500°C	trigonal, $R\bar{3}C$ (corundum α-AlO_3)
β-Ga_2O_3	calcination at $\geq$ 600°C of $Ga(OH)_3$ or other salts	monoclinic, C2/m (β-Al_2O_3)
γ-Ga_2O_3	rapid dehydration on gels at ca 400°C	cubic, Fd3m (spinel)
δ-Ga_2O_3	by decompositon of $Ga(NO_3)_3$ at ca 250°C	cubic, Ia3 (MnO_3)
ϵ-Ga_2O_3	brief heating of δ-Ga_2O_3 at ca 550°C	orthorhombic (κ-Al_2O_3)
GaOOH	oxidation of Ga with H_2O at ca 200°C, under pressure	orthorhombic Pbnm (diaspore)
$Ga(OH)_3$	neutralization of aqueous Ga(III) salts or alkaline gallates(III)	unstable gel

Gallates(III).   Use of the oxidation number (III) was recommended by the IUPAC to distinguish gallates(III) from the salts of gallic acid, 3,4,5-trihydroxybenzoic acid, which often are wrongly named gallates. The alkali metal gallates(III) are the only gallates(III) that are soluble in water; the former gives stable solutions at all concentrations, unlike aluminate solutions. They are obtained easily by direct reaction of metallic gallium with solutions of the hydroxides of Na, K, Rb, or Cs. The ratio M:Ga can be as low as unity; the solutions can be obtained in very high concentration by evaporation, eg, 800 g Ga/L solution for the sodium gallate(III). Lithium gallate(III) is less soluble than $NaGaO_2$.

The solid gallates(III) of numerous elements have been studied and include gallates(III) of alkaline and other metals. The former group includes Li_5GaO_4, $LiGaO_2$, and $LiGa_5O_8$; $MGaO_2$ where M = Cs, K, Na, or Rb; and $Na_2O_5 \cdot 4Ga_2O_3$ and $Na_2O \cdot 7Ga_2O_3$ (ionic conductors, of the type of β alumina). The latter group includes $MGaO_2$ (M = Cu or Ag); MGa_2O_4 (M = Ba, Sr, Co, Cu, Fe, or Zn); $CaO \cdot nGa_2O_3$ (n = 0.33, 1 or 2); $AlGaO_3$; and $LnGaO_3$ and $Ln_3Ga_5O_{12}$ with the lanthanides. Among these compounds, a number are studied for their useful magnetic or electric properties, especially spinels, perovskites, and above all, garnets, eg, $Ln_3Ga_5O_{12}$ (Ln = Dy, Sm, or Gd) (see FERROELECTRICS).

Gallium Chalcogenides.   Gallium forms two types of compounds by reaction with sulfur, selenium, and tellurium, and these are listed in Table 7. One type is represented by GaS, GaSe, and GaTe, in which the gallium is trivalent and the form is a lamellar crystal containing atomic groups such as S–Ga–Ga–S. In such a group, each gallium atom is surrounded by a tetrahedron containing 3 S and 1 Ga. This lamellar structure gives two-dimensional properties to these semiconductor compounds (see SEMICONDUCTORS). The other compound type includes Ga_2S_3 (38), Ga_2Se_3 (39), and Ga_2Te_3, which have vacancies in their structure that also give particular electronic properties.

Mixed Chalcogenides.   The mixed chalcogenides include $MGaS_2$ (or Se_2 or Te_2), where M = Ag, Cu, In, or Tl; among these compounds, some are semicon-

Table 7. Gallium Chalcogenides

Compound	CAS Registry Number	Mp, °C	Crystal structure	Density, g/cm³
GaS	[12024-10-1]	962	hexagonal (lamellar)	3.86
Ga_2S_3	[12024-22-5]	1090	monoclinic, Bb (superstructure of wurtzite type)	3.77
GaSe	[12024-11-2]	960–965	hexagonal (lamellar)	5.01
Ga_2Se_3	[12024-24-7]	1005–1010	monoclinic, B2/m (superstructure of deformed blend type) cubic by quenching of liquid compound (disordered sphalerite type)	4.95
GaTe	[12024-14-5]	825	monoclinic or hexagonal (lamellar structure)	5.44
Ga_2Te_3	[12024-27-0]	792	cubic (sphalerite type)	5.57
Ga_2Te_5	[73623-48-0]	stable	tetragonal only at 400–495°C	5.85

ductors, eg, AgGaS$_2$ [12002-67-4] or CuGaS$_2$ [12018-83-6]. Other mixed chalco-genides are MGa$_2$S$_4$ (or Se$_4$ or Te$_4$) where M = Cd, Hg, or Zn; MGaS$_3$, where M = In [12398-40-2] or Y [12018-83-6]; and M$_{18}$Ga$_{10}$S$_{42}$ (or Se$_{42}$), where M = lanthanides.

Sulfates and Selenates. The anhydrous gallium(III) sulfate [13494-91-2], Ga$_2$(SO$_4$)$_3$, is obtained by dissolving GaOOH in 50% H$_2$SO$_4$, evaporating, filtering, and drying to 360°C. It crystallizes from aqueous solutions as Ga$_2$(SO$_4$)$_3$·18H$_2$O. In the presence of alkali metal or ammonium sulfates, it gives alums, MGa(SO$_4$)$_2$·12H$_2$O, where M is Cs, K, Rb, or NH$_4$. Gallium selenate is obtained under the same conditions as the sulfate. Like the sulfate, it gives double sele-nates that are isomorphous with alums.

Gallium Nitrogen Compounds. Direct combination of gallium and nitrogen (40) involves a gas discharge at 750°C for the dissociation of N$_2$ into atomic N which combines with Ga vapor; a thin film of GaN is deposited. Gallium nitride [25617-97-4] can also be obtained by the reaction of ammonia on gallium at ca 1000°C, or on gallium compounds (oxide, chloride, sulfide, etc). GaN is also pre-pared by thermal decomposition of triammonium hexafluorogallate(III) in a stream of ammonia at ca 600°C. Gallium nitride, hexagonal, wurtzite type, den-sity, 6.10 g/cm^3, is very stable in air at ca 900°C; it can be sublimed in a stream of ammonia at 1150°C. It has a colorless crystal form when pure. It is a semicon-ductor and as such promises practical applications, especially as an electrolumi-nescent compound. Double nitrides with metals, particularly the alkali metals, are known, eg, Li$_3$GaN$_2$. The hydrated nitrate, Ga(NO$_3$)$_3$·9H$_2$O, is formed by dis-solving gallium in nitric acid.

Gallium(III) azide [73157-11-6], Ga(N$_3$)$_3$, is prepared by decomposition, at 250°C, of the addition compound GaF$_3$·NH$_3$ [73157-06-9].

Contrary to previous indications apparently gallium boride does not exist.

Gallium Compounds with Phosphorus, Arsenic, and Antimony. The 1:1 compounds, gallium phosphide [12063-98-8], GaP, gallium arsenide [1303-00-0], GaAs, and gallium antimonide [12064-03-8], GaSb, can be obtained by direct com-bination of the elements at high temperature. GaP and GaAs formation requires combination under pressure. The three compounds also can be prepared, mainly as thin films, by numerous exchange reactions in the vapor phase.

Gallium phosphide, density = 4.13 g/cm^3, melts at 1465°C under pressure; gallium arsenide, density = 5.360 g/cm^3, melts at 1238°C under pressure; and gallium antimonide, density = 6.096 g/cm^3 melts at 712°C. All three crystallize in the cubic system (sphalerite type) and possess semiconductor properties which make these materials, especially the arsenide, the most important in terms of practical applications of gallium. Many mixed compounds also are used, eg, GaAs$_{1-x}$P$_x$ or Ga$_{1-x}$Al$_x$As (see LIGHT GENERATION, LIGHT-EMITTING DIODES; OPTICAL DISPLAYS; SEMICONDUCTORS).

Gallium phosphates are obtained by the reaction of gallium hydroxide on the phosphorus oxoacids. Anhydrous gallium phosphate is prepared by heating the dihydrate or by the reaction of gallium on phosphoric acid in a sealed tube at 200°C. Anhydrous gallium phosphate [14014-97-2], GaPO$_4$, hexagonal (beta quartz type); hydrated phosphate, GaPO$_4$·2H$_2$O [23653-37-4], orthorhombic; and pyrophosphates, GaHP$_2$O$_7$ [34641-69-5] and Ga$_4$(P$_2$O$_7$)$_3$·xH$_2$O [19584-43-1] (x about = 23), are known.

Anhydrous gallium arsenate [13811-89-7], $GaAsO_4$, hexagonal (beta quartz type) is slightly soluble in water and can be obtained by the reaction of $GaCl_3$ on As_2O_3 at 800°C. It forms a hydrate, $GaAsO_4 \cdot 2H_2O$ [23653-38-5] which is orthorhombic, and it can form acidic salts, eg, $Ga(H_2AsO_4)_3 \cdot xH_2O$ with $x = 1$ [38296-94-5] or $x = 5$ [38296-97-8].

Gallium antimonate [28888-33-7], $GaSbO_4$, is tetragonal (rutile type) and has a density of 6.540 g/cm^3.

Carbon Compounds of Gallium. There is no binary gallium carbide; however, some ternary carbides have been studied, principally with regard to their structure and magnetic properties: Mn_3GaC [12069-65-7], Nd_3GaC [12127-26-3], and Mo_2GaC [12286-91-8] (see CARBIDES; MAGNETIC MATERIALS).

More than 1000 organic gallium compounds have been described in addition to the coordination compounds of the gallium hydride or halides already mentioned. These compounds include salts of organic acids, eg, acetate, oxalate, citrate, and numerous complexes, eg, oxalates, $Ga(C_2O_4)^+$, $Ga(C_2O_4)^{2-}$, $Ga(C_2O_4)_3^{3-}$; alkoxides or their derivatives by substitution, eg, $Ga(OCH_3)_3$ [2746-72-7], $Ga(OC_2H_5)_3$ [2572-25-0], $Ga(OC_6H_5)_3$ [2572-17-0], and $GaCl(OCH_3)_2$ [21907-51-7], and $GaCl_2(OCH_3)$ [22381-97-1]. Another group is the organogallium compounds, R_3Ga, which contain Ga–C bonds, where R is methyl [1445-79-0] (mp $-15.8°C$), ethyl [1115-99-7] (mp 82.3°C), n-propyl [29868-77-7], isopropyl [54514-59-9], propenyl (cis or trans), butyl [15677-44-8], pentyl [15677-45-9], hexyl, decyl, phenyl [55321-79-4], and α-naphthyl. In addition, a number of their substitution or coordination compounds have been described.

Toxicology

The toxicity of metallic gallium or gallium salts is very low. The corrosive, poisonous, or irritating nature of some gallium compounds is attributable to the anions or radicals with which it is associated. Gallium metal-organics, such as $Ga(CH_3)_3$, react vigorously with air, and can be explosive. The gallium halides, except the fluoride, hydrolyze in water to form corresponding halogen acids. Gallium phosphide, arsenide, selenide, and telluride react slowly with water, and more vigorously with acids and bases, to liberate toxic compounds. The LD_{50} of the solution of $[Ga(NO_3)_3 \cdot 9H_2O]$ for mice is ca 3–4 g/kg. The ^{72}Ga and ^{67}Ga isotopes were studied, eg, as citrate salts, for detection of tumors. ^{72}Ga concentrates in bone tissues and ^{67}Ga seem to have a tumor-specific affinity. Additional data are available (41,42).

Uses

Gallium Solar Neutrino Experiment   Among the few potential experiments for the detection of the overwhelming majority of solar neutrinos, which are low energy pp-neutrinos, is the radiochemical gallium solar neutrino experiment. It is the only experiment which has been demonstrated to be feasible. The feasibility was proved between 1979 and 1983. A Montecarlo simulation has shown that the minimum quantity of gallium must be 30–32 t to correspond to one neutrino captured per day. Experiments are conducted by the Max Plant Institute Gallex Project and a collaboration between Karlsruhe, Mossbader of Munich, France (Rhône-Poulenc, CEN Saclay, CNRS Nice), Italy (Milano, Roma) and Russia. Experiments should be finished in 1995 (43–45).

Gallium Using Epitaxial Technologies. *Diodes.* Most light-emitting diodes (LEDs) use vapor-phase epitaxy (VPE) or liquid-phase epitaxy (LPE) to grow GaAs, GaAsP, or GaP layers (46–49). The superbright red LED uses LPE to grow AlGaAs on GaAs or AlGaAs substrates. Epitaxial growth based on MOCVD technology is used in manufacturing some types of infrared LEDs used as optocouplers. MOCVD is also used to grow a GaAlAs or a GaAs layer (Buffer layer) on GaAs substrates for low cost optic fibers dedicated to local area networks (50) (see FIBER OPTICS; LIGHT GENERATION).

GaAs, GaAlAs, and GaP based laser diodes are manufactured using the LPE, MOCVD, and molecular beam epitaxy (MBE) technologies (51). The short wavelength devices are used for compact disc (CD) players, whereas the long wavelength devices, mostly processed by MBE, are used in the communication field and in quantum well structures.

Transitor Amplifiers. Most gallium-based field-effect transitor amplifiers (FETs) are manufactured using ion implantation (qv) (52), except for high microwave frequencies and low noise requirements where epitaxy is used. The majority of discrete high electron mobility transistor (HEMT) low noise amplifiers are currently produced on MBE substrates. Discrete high barrier transistor (HBT) power amplifiers use MOCVD and MBE technologies.

Integrated Circuits. For analogue integrated circuits (ICs) as frequencies increase, requirements for epitaxy grow at the same rate. For most microwave devices with frequencies over 20 GHz, an epitaxial GaAs layer is required. MBE is preferred for HEMT structures with better low noise, while MOCVD is used for HBT devices (see INTEGRATED CIRCUITS).

Photovoltaics. Photovoltaic applications require GaAs wafers and epitaxial layers. In most applications of photovoltaics in space, germanium substrates with GaAs active layers are used. Night vision system (NVS) devices use an epitaxial layer of GaAs applied to one end of a photomultiplier to enhance infrared images. The epitaxial layer consists of a 5 to 50 μm layer grown on a GaAs substrate using MOCVD techniques.

Other Applications. Lasers (qv) having wavelengths between 690 and 900 nm utilize GaAlAs. Main applications are fiber optic communication, digital audio disk, video disk, laser printer, optical recording, and test and measurement instruments. When using InGaAsP the wavelength changes from 0.7 μm up to 1.5 μm.

In photodetectors (qv) for photocouplers, optical switches, fiber optics, communication, and imaging, the materials used are InGaAs, InGaAsP, and GaSb.

Materials used in optical ICs are GaAs, InSb (for optoelectronic ICs), GaAlAs, InGaAsP (for optical array), and GaAlAs. Main applications are in supercomputers, design for very large-scale integration (VLSI) circuitry, nuclear fusion control, research of resources, high speed data system, control memories for large computers, memories for large-scale integration (LSI) testers, and direct broadcasting satellite receivers.

High voltage power transistors having high ($<$ ns) speed switching, very high (200 V) blocking voltage, working voltage of 1000 V, and a total current 10 times higher than Si MOSFET and twice as high as Si bipolar, use gallium. The materials used in magnetoelectric transducers are InSb, GaAs, and InAs. Applications are precision motors, direct drive motors, tension controllers (tape disk, disk deck, microcomputers), tape end sensors, and microcomputer-controlled machinery.

Magnet Applications.   For magnets of the type FeNdB, a small amount of gallium is effective in improving the intrinsic coercive force (53). It slows the pinning type recoil loop. The thermal stability is increased. The irreversible loss is less than 1.5% under 373 K (1000 h).

Medical Uses of Gallium.   Gallium can be used to detect such diseases as Hodgkin's disease, lymphomas, and interstitial lung disease (54). Gallium nitrate is also used as an anticancer drug for lymphomas and bones (55). It can reverse bone degeneration and/or hypercalcemia cancer, osteoporosis, and Paget's syndrome. In the case of therapeutic action gallium halts bone resorption, normalizes serum calcium levels, adds bone mass, and kills cancer cells. Some gallium drugs become acute renal toxins at 10 to 100 times the therapeutical dose. Gallium does not accumulate in tissues, nor does it cause mutations or show other signs of toxicity.

In dental applications gallium alloys are nonstaining and used in the fabrication and repair of dental protheses. These alloys exhibit a sufficiently high solder temperature enabling presoldering of the alloy without distortion of the restoration (56) (see DENTAL MATERIALS).

Catalysis Application.   Gallium is used in catalysts for aromatization in the petroleum (qv) industry (57). Benzene, toluene, and xylene can be produced by oligomerization and cyclization using zeolites containing 1 to 4% gallium as gallium nitrate (see BTX PROCESSING; MOLECULAR SIEVES). A gallium 99.999% is used and from 5 to 8 t/yr was forecasted for this application in 1993.

Economic Aspects

Despite very strong growth in the market for many gallium-containing devices used in electronics, achieving a balance in the gallium industry has been difficult economically. Because of a significant over-capacity in the gallium industry, prices have been weak and are decreasing. Light-emitting diodes (LED) must face increasing competition from liquid crystal displays (LCD) and other alternatives to cathode ray tubes (CRT) for display devices. However, developments in LED technology have created a new market in large area outdoor and semi-outdoor displays. In Japan, high brightness GaAlAs LEDs are used as rear braking lights in automobiles.

Gallium-based laser diodes have also lost some of the market in telecommunications to indium-based devices. Optical fiber systems moved to the use of longer wavelengths for long distance networks.

Nevertheless, the market in compact disc equipment is increasing, and computer data storage has expanded rapidly, as well as telecommunication by satellite and the use of high frequency devices.

In spite of the limited growth of gallium demand, the forecast for use in integrated circuits (qv) shows an increasing market. Vitesse Semiconductor, in the United States, announced the very large-scale integration (VLSI) products based on GaAs which should compete with emitter coupled logic (ECL) devices. Also in the United States, Gigabit Logic and Gazelle Microcircuits are marketing devices which can readily replace silicon-based products. The high speed of GaAs is a great advantage in the military, space and supercomputing, and general computer markets.

Another attractive application for gallium is the use in high performance photovoltaic cells (qv) in satellites, but owing to the weight of the substrates, these have been replaced by germanium. The solar cells are manufactured using multiple layers of gallium compounds deposited by metalloorganic chemical vapor deposition (MOCVD) on the Ge substrates.

Gallium prices, stable up to 1991, fluctuated in 1992 because of gallium from China and Eastern European Bloc countries. The market destablized owing to spot offers and dumping. The reliability of these products was a real problem, however, and the consequence of market variations was dramatic for small manufacturers.

BIBLIOGRAPHY

"Gallium" in *ECT* 1st ed., Vol. 7, pp. 53–58, by L. M. Foster, Aluminum Co. of America; "Gallium and Gallium Compounds" in *ECT* 2nd ed., Vol. 10, pp. 311–328, by P. de la Breteque, Sociétè Française pour l'Industrie de l'Aluminium, Division of Swiss Aluminium, Ltd; in *ECT* 3rd ed., Vol. 11, pp. 604–620, by P. de la Breteque.

1. P. de la Breteque, *Gallium, Bulletin d'Information et de Bibliographie*, N12, Alusuisse France SA, Marseille, 1974.
2. F. E. Katrak and J. L. Agarwal, *JOM* **33**, 33 (1981).
3. L. Bosio, A. Defrain, and I. Epelboin, *Colloq. Int. C.N.R.S.* **201**, 325 (1971).
4. L. Bosio, *J. Chem. Phys.* **68**(3), 1221 (1978).
5. R. Bautista, *JOM* **41**, 30 (1989).
6. L. K. Hudson, *JOM* **17**, 948 (1965).
7. U.S. Pat. 3,971,843 (July 27, 1976), J. Helgorsky and A. Leveque (to Rhône-Poulenc Industries).
8. U.S. Pat. 4,724,129 (Feb. 9, 1988), J. Helgorsky and A. Leveque (to Rhône-Poulenc Industries).
9. U.S. Pat. 4,169,130 (Sept. 25, 1979), J. Helgorsky and A. Leveque (to Rhône-Poulenc Industries).
10. U.S. Pat. 4,241,029 (Dec. 23, 1980), J. Helgorsky and A. Leveque (to Rhône-Poulenc Industries).
11. U.S. Pat. 4,485,076 (Nov. 27, 1984), D. Bauer, P. Fourre, and J. L. Sabot (to Rhône-Poulenc Specialítès Chimiques).
12. U.S. Pat. 4,559,203 (Dec. 17, 1985), D. Bauer, P. Fourre, and J. L. Sabot (to Rhône-Poulenc Specialítès Chimiques).
13. Eur. Pat. Appl. 076,404 (Apr. 13, 1983) (to Sumitomo Chemical Co.).
14. P. de la Breteque, *Mem. Sci. Rev. Met.* **67**(1), 57 (1970).
15. J. Allegre and B. Boudot, *J. Cryst. Growth* **106**, 139–142 (1990).
16. W. Vieth and J. C. Huncke, *Anal. Chem.* **64**, 2958–2964 (1992).
17. K. Yvon and P. Feschotte, *J. Less Common Met.* **63**, 1 (1979).

18. M. Puselj and K. Schubert, *J. Less Common Met.* **38**, 83 (1974).
19. N. E. Alekseevskii and V. M. Zakosarenko, *Dokl. Akad. Nauk SSSR* **208**(2), 303 (1973).
20. M. J. Philippe, B. Malaman, and B. Roques, *C. R. Acad. Sci. Ser. C* **278**(17), 1093 (1974).
21. S. P. Yatsenko and co-workers, *Izv. Akad. Nauk SSSR Metal.*, (1), 185 (1973).
22. K. Yvon, *Acta Crystallogr. Sect. B* **30**(4), 853 (1974).
23. P. Guex and P. Feschotte, *J. Less Common Met.* **46**, 101 (1976).
24. K. H. Buschow, *J. Less Common Met.* **31**, 165 (1973).
25. Ref. 1, N14, 1978, pp. 11–71.
26. E. Chemouni, *J. Inorg. Nucl. Chem.* **33**, 2317 (1971).
27. D. Mascherpa-Corral, dissertation, University of Montpellier, France, 1975.
28. Y. Dumas and A. Potier, *Bull. Soc. Chim. Fr.*, 2634 (1975).
29. D. Mascherpa-Corral and A. Potier, *Bull. Soc. Chim. Fr.*, 1912 (1973).
30. A. Baumer, R. Caruba, and G. Turco, *C.R. Acad. Sci. Ser.D* **271**(1), 1 (1970).
31. M. Rault, G. Demazeau, J. Portier, and J. Grannec, *Bull. Soc. Chim. Fr.*, 74 (1970).
32. L. Moegele, *Z. Naturforsch.B* **23**, 1013 (1968).
33. M. T. Roziere-Bories and co-workers, *Bull. Soc. Chim. Fr.*, 1285 (1974).
34. M. Chaabouni and co-workers, *J. Chem. Res.*, 80 (1977).
35. A. Hardy and D. Cottreau, *C.R. Acad. Sci. Ser.C* **262**, 739 (1966).
36. B. Vandorpe, M. Drache, and B. Dubois, *C.R. Acad. Sci. Ser.C* **276**, 73 (1973).
37. D. Mascherpa-Corral and A. Potier, *J. Inorg. Nucl. Chem.* **39**, 1519 (1977).
38. G. Collin and co-workers, *Mat. Res. Bull.* **11**, 285 (1976).
39. G. Ghemard, R. Ollitrault, and J. Flahaut, *C.R. Acad. Sci. Ser.C.* **282**, 831 (1976).
40. B. B. Kosicki and D. Kahng, *J. Vac. Sci. Technol.* **6**(4), 593 (1969).
41. N. I. Sax and R. J. Lewis, *Dangerous Properties of Industrial Materials*, 7th ed., Vol. III, Van Nostrand Reinhold, New York, 1989, pp. 1793–1795.
42. J. L. Domingo and J. Corbella, *Trace Elements in Medicine*, **8**, 56–64 (1991).
43. E. Henrich and K. H. Ebert, *Angew. Chem. Int. Ed. Engl.* **31**, 1283—1297 (1992).
44. GALLEX Collaboration, *Phys. Lett. B* **285,** 376–389 (1992).
45. *Ibid.*, pp. 390—397.
46. J. R. Knight, D. Effer, and P. R. Evans, *Sol. St. Electron.* **8,** 178 (1965).
47. D. Effer, *J. Electrochem. Soc.* **112,** 1020 (1965).
48. L. Hollan, J. P. Hallais, and J. C. Brice, *Current Topics Mater. Sci.* **5**, 1–217 (1980).
49. D. Elwell and H. J. Scheel, *Crystal Growth from High Temperature Solution*, Academic Press, London, 1975.
50. G. B. Stringfellow and H. T. Hall Jr., *J. Cryst. Growth* **43**, 47 (1978).
51. R. F. C. Farrow, *Current Topics Mater. Sci.* **2**, 237 (1977).
52. P. D. Townsend, J. C. Kelly, and N.E.W. Hartley, *Ion Implantation, Sputtering and their Applications*, Academic Press, New York, 1976.
53. R. Nakayama and T. Takeshita, *J. Alloys Comp.* **193**, 259–261 (1993).
54. C. L. Edwards and R. L. Hayes, *Nucl. Med.* **10,** 103–105 (1969).
55. R. H. Adamson, G. P. Canellos, and S. M. Sieber, *Cancer Chemother. Rep.* **59,** 599–610 (1975).
56. R. M. Waterstrat, *J. Am. Dental Assoc.* **78,** 536 (1969).
57. N. S. Gnep, J. Y. Doyemet, A. M. Seco, F. Ramoa Ribeiro, and M. Guisnet, *Appl. Catal.* **43,** 155–166 (1988).

General References

K. Wade and A. J. Banister, *The Chemistry of Al, Ga, In, and Tl*, Pergamon Press, New York, 1974.
A. J. Downs, ed., *Chemistry of Aluminium, Gallium, Indium, & Thalium*, Routledge, Chapman, and Hall, London, June 1993.

T. Ikegami, F. Hasegawa, and Y. Takeda, "Gallium Arsenide & Related Compounds, 1992," in the *Proceedings of the 19th International Symposium, Oct. 2, 1992, Kariuzawa, Japan*, Institute of Physics, Bristol, U.K., 1993.

G. B. Stringfellow, ed., "Gallium, Arsenide, and Related Compounds, 1991," in the *Proceedings of the Eighteenth International Symposium on Gallium, Arsenide, and Related Compounds, Sept. 9–12, 1991, Seattle, Wash.*, Institute of Physics, Bristol, U.K., 1992.

T. Ikoma and H. Watanabe, eds., "Gallium, Arsenide, and Related Compounds, 1989," in the *Proceedings of the 16th International Symposium on Gallium, Arsenide, and Related Compounds, Sept. 25–29, 1989, Kariuzawa, Japan*, No. 106, Institute of Physics, Bristol U.K., 1990.

J. S. Harris, ed., *Proceedings of the 15th International Symposium on Gallium, Arsenide, and Related Compounds, Sept. 11–14, 1988, Atlanta, Ga.*, No. 96, Institute of Physics, Bristol, U.K., 1989.

W. T. Lindley, ed., *Proceedings of the 13th International Symposium on Gallium, Arsenide, and Related Compounds, Sept. 28–Oct. 1, 1986, Las Vegas, Nev.*, No. 83, Institute of Physics, Bristol, U.K., 1987.

M. Fujimoto, ed., *Proceedings of the Third International Symposium, Sept. 23–26, 1985, Karuizawa, Japan*, No. 79, Institute of Physics, Bristol, U.K., 1986.

B. De Cremoux, ed., *Gallium, Arsenide, and Related Compounds*, Sept. 26–28, 1984, Biarritz, France, No. 74, Institute of Physics, Bristol, U.K., 1985.

G. E. Stillman, ed., papers contributed from the *Proceedings of the Tenth International Symposium of Gallium, Arsenide, and Related Compounds, Albuquerque, N.M., Sept. 19–22, 1982*, No. 65, Institute of Physics, Bristol, U.K., 1983.

T. Sugano, ed., *Gallium, Arsenide, and Related Compounds*, No. 63, Institute of Physics, Bristol, U.K., 1982.

D. F. Ferry, *Gallium, Arsenide Technology*, McMillan Co., New York, 1985.

M. J. Howes and D. V. Morgan, *Gallium Arsenide Materials, Devices, and Details*, John Wiley & Sons, Ltd., Chichester, U.K., 1985.

JEAN LOUIS SABOT
HUBERT LAUVRAY
Rhône-Poulenc

GAS CHROMATOGRAPHY. See CHROMATOGRAPHY.

GAS CLEANING. See AIR POLLUTION CONTROL METHODS; GAS, NATURAL.

GAS HYDRATES. See HYDROCARBONS; WATER, SUPPLY AND DESALINATION.

GASKETS. See PUMPS; VACUUM TECHNOLOGY.

GAS, MANUFACTURED. See FUELS, SYNTHETIC.

GAS, NATURAL

Natural gas is a mixture of naturally occurring hydrocarbon and nonhydrocarbon gases found in porous geologic formations beneath the earth's surface (see HYDROCARBONS). Methane is a principal constituent and the mixture may contain higher hydrocarbons such as ethane, propane, butane, and pentane. Gases such as carbon dioxide (qv), nitrogen (qv), hydrogen sulfide, various mercaptans, and water vapor along with trace amounts of other inorganic and organic compounds can also be present. Natural gas is found in a variety of geological formations including sandstones, shales, and coals (see COAL; OIL SHALE).

Discussions of natural gas can involve the following definitions (1):

Associated gas: free natural gas in immediate contact, but not in solution, with crude oil in the reservoir.

Dissolved gas: natural gas in solution in crude oil in the reservoir.

Dry gas: gas where the water content has been reduced by a dehydration process or gas containing little or no hydrocarbons commercially recoverable as liquid product.

Liquefied natural gas (LNG): natural gas that has been liquefied by reducing its temperature to 111 K at atmospheric pressure. It remains a liquid at 191 K and 4.64 MPa (673 psig).

Natural gas liquids (NGL): a liquid hydrocarbon mixture which is gaseous at reservoir temperatures and pressures, but recoverable by condensation or absorption (qv).

Nonassociated gas: free natural gas not in contact with, nor dissolved in, crude oil in the reservoir.

Sour gas: gas found in its natural state containing compounds of sulfur at concentrations exceeding levels for practical use because of corrosivity and toxicity.

Sweet gas: gas found in its natural state containing such small amounts of sulfur compounds that it can be used without purification with no deleterious effect on piping or equipment, and without the potential for health hazards.

Wet gas: unprocessed or partially processed natural gas produced from strata containing condensible hydrocarbons.

History

Natural gas and its combustion properties appear to have been known since early times (2). Some early temples of worship were located in areas where gas was seeping from the ground or from springs, and it is reported that Julius Caesar saw a phenomenon called the "burning spring" near Grenoble, France. Gas wells were drilled in Japan as early as 615 AD and in 900 AD the Chinese employed bamboo tubes to transport natural gas to their salt works, where the heat was used to evaporate water from salt brine. The existence of natural gas in the United

States was reported by early settlers who observed gas seeps and columns of fire in the Ohio Valley and the Appalachian area in 1775 (3).

The serious, practical use of natural gas occurred in the United States in 1821 when residents in the village of Fredonia, New York drilled the first well and piped the gas through hollowed-out logs to nearby houses for lighting. The depth of the initial well was 8.2 m and its drilling followed the accidental ignition of a gas seepage at nearby Canadaway Creek. The Fredonia Gas, Light, and Waterworks Company was formed in 1865 as the first natural gas company in the United States. The first large-scale, industrial use of natural gas in the United States was in the steel (qv) and glass (qv) works of Pittsburgh, Pennsylvania, in 1883.

The early use of natural gas relied on its availability from small, local, shallow fields. This frequently created a chaotic cycle of events consisting of the discovery of a field, followed by the development of a local distribution system which, in turn, attracted new industries and other customers to the locale and often resulted in a rapid depletion of the reserves. After a few years, curtailment, followed by a total cessation of supplies, occurred.

Later, larger reservoirs of natural gas were located and produced. The development of long-distance pipeline systems was initiated (see PIPELINES), natural gas storage systems were established, and gas processing technology evolved which could separate water vapor, noxious constituents, inert gases, and condensible hydrocarbons from the raw natural gas flowing from the wells. This provided an integrated infrastructure that allowed effective marketing of the natural gas. Natural gas is now used throughout the industrialized world as a source of energy in residential, commercial, industrial, and electric utility generation applications. It also represents an important feedstock to many segments of the chemical industry. Because of its clean burning characteristics, availability, and competitive cost, the use of natural gas continues to increase on a worldwide basis.

Gas Reserves and Production

Estimates of the amount of natural gas available are made within the context of definitions and are subject to revision as definitions change, as additional information becomes available, as resources are consumed, or as underlying assumptions are altered. These definitions include proved reserves where the resource is expected to be recoverable and marketable using known technology and prices; probable reserves where a resource has been identified but not completely characterized; and possible or potential gas where estimates are based on the available geological information, historical trends, and previous successes. There are variations in these definitions throughout the world.

Data compiled for 1992 placed the world's estimated proved natural gas reserves at approximately 1.24×10^{14} m^3 (4.38×10^{15} ft^3) (4). Data for the Confederation of Independent States (CIS) are denoted as explored reserves and include proved, probable, and some possible gas. The data for Canada also include some probable reserves. The worldwide natural gas reserves have continued to increase as the demand for gas has increased and exploration efforts have expanded. In 1976, the world natural gas reserves were estimated to be 6.58×10^{13} m^3.

Table 1. Natural Gas Reserves, Production, Imports, and Exports[a,b]

Country	Proved reserves,[c,d] $m^3 \times 10^{12}$		Production,[e,f] $m^3 \times 10^9$		Imports,[e,g] $m^3 \times 10^9$		Exports,[e,h] $m^3 \times 10^9$	
Abu Dhabi	5.18	(4.18)						
Algeria	3.30	(2.66)	55.79	(2.70)			34.52	(10.91)
Argentina			22.41	(1.09)				
Australia			22.15	(1.07)			3.10	(0.98)
Austria					4.87	(1.56)		
Belgium					10.80	(3.46)		
Bolivia							2.41	(0.76)
Brunei							8.90	(2.81)
Bulgaria					6.60	(2.11)		
Canada	2.74	(2.21)	112.78	(5.46)			43.40	(13.71)
China	1.00	(0.81)	16.78	(0.81)				
CIS (former USSR)	49.55	(39.97)	783.42	(37.95)			105.54	(33.35)
Czechoslo- vakia					13.62	(4.36)		
France					32.42	(10.38)		
Germany			16.31	(0.79)	55.99	(17.93)	1.46	(0.46)
Hungary					6.14	(1.97)		
India	0.73	(0.59)						
Indonesia	1.84	(1.48)	44.59	(2.16)			29.47	(9.31)
Iran	17.00	(13.71)	26.37	(1.28)			1.45	(0.46)
Iraq	2.69	(2.17)					2.26	(0.71)
Italy			16.92	(0.82)	26.69	(8.55)		
Japan					54.77	(17.53)		
Kuwait	1.36	(1.10)						
Libya	1.22	(0.98)						
Malaysia	1.67	(1.35)	14.06	(0.68)			10.31	(3.26)
Mexico	2.02	(1.63)	28.56	(1.38)				
the Nether- lands	1.97	(1.59)	71.77	(3.48)			33.95	(10.73)
Nigeria	2.97	(2.39)						
Norway	1.72	(1.39)	31.94	(1.55)			29.14	(9.21)
Poland					8.01	(2.56)		
Qatar	4.59	(3.70)						
Romania			27.10	(1.31)				
Saudi Arabia	5.21	(4.20)	31.11	(1.51)	6.89	(2.21)		
Spain					4.85	(1.55)		
United Arab Emirates			22.27	(1.08)			3.54	(1.12)
United Kingdom			53.77	(2.60)	8.12	(2.60)		
United States	4.79	(3.87)	495.54	(24.00)	43.11	(13.80)	2.42	(0.76)
Venezuela	3.11	(2.51)	22.20	(1.08)				
Yugoslavia					4.20	(1.34)		
Totals	[i]	(92.49)		(92.80)		(91.91)		(98.53)

[a]To convert m^3 to standard cubic feet (scf), multiply by 35.31. [b]Values in parentheses are percentages. [c]Data for 1992; Ref. 4. [d]Estimated values. [e]Data for 1990 from Ref. 5 assuming a specific energy content of 35.4 MJ/m^3. To convert from J to cal, divide by 4.184. [f]Total world production in 1990 was 2064.5×10^9 m^3. [g]Total world import volume in 1990 was 312.3×10^9 m^3. [h]Total world export volume in 1990 was 316.5×10^9 m^3. [i]Total estimated proved reserves for 1992 were 124×10^{12} m^3.

In 1987, the reserves were $1.06 \times 10^{14} \, m^3$, and by 1992 the reserve estimates had grown to $1.24 \times 10^{14} \, m^3$. The distribution of the reserves in 1992 by principal geographical areas is shown in Table 1. In 1992, the principal political/geographical entities of the United States, the Confederation of Independent States (CIS), and the Oil Producing and Exporting Countries (OPEC) held 3.87, 39.97, and 39.7% of the world's natural gas reserves, respectively.

Natural gas production on a worldwide basis has also continued to increase (5). Worldwide natural gas production increased at a rate of approximately 2.5%/yr for the period 1980–1985 and at a rate of approximately 3.7%/yr for the period 1986–1990. By 1990, the annual level had reached an energy equivalent of $73.1 \times 10^{18} \, J \, (1.75 \times 10^{16} \, kcal)$ or approximately $2.1 \times 10^{12} \, m^3 \, (7.42 \times 10^{13} \, ft^3)$ and provided 22.9% of the total energy used. The 1990 production levels and corresponding countries are also listed in Table 1 as are exports and imports of natural gas. Geographic consumption trends are shown in Figure 1.

Government-owned companies dominate the lists of international organizations holding and producing the available natural gas reserves. The top 20 holders of gas reserves in 1991 are listed in Table 2, and the top 20 natural gas producing organizations in 1991 are shown in Table 3 (6). Several of the international companies are based in the United States. The reserve holdings and the production data for these organizations reflect their activities throughout the world.

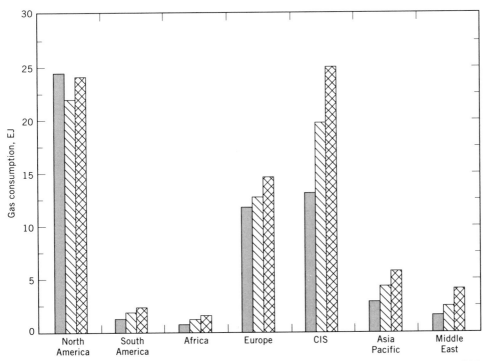

Fig. 1. Worldwide annual consumption of natural gas where ▢ represents 1980; ◺, 1985; and ⬚, 1990. 1 EJ = 10^{18} J. To convert J to cal, divide by 4.184.

Table 2. Natural Gas Reserve Holders[a]

Rank	Company	Country	1991 Reserves,[b] $m^3 \times 10^9$
1	National Iranian Oil Co.	Iran	17,000
2	Saudi Arabian Oil Co.	Saudi Arabia	5,212
3	Abu Dhabi National Oil Co.	Abu Dhabi	5,176
4	Qatar General Petroleum Corp.	Qatar	4,587
5	Petroleos de Venezuela SA	Venezuela	3,600
6	Sonatrach	Algeria	3,299
7	Nigerian National Petrol. Corp.	Nigeria	2,965
8	Iraq National Oil Co.	Iraq	2,690
9	Petroleos Mexicanos	Mexico	1,999
10	Pertamina	Indonesia	1,836
11	Petronas	Malaysia	1,672
12	Royal Dutch/Shell	the Netherlands	1,503
13	Kuwait Petroleum Corp.	Kuwait	1,359
14	National Oil Co.	Libya	1,218
15	China National Petroleum Co.	China	1,002
16	Oil & Natural Gas Commission	India	730
17	Exxon Corp.	United States	720
18	Yacimientos Petroliferos Fiscales SA	Argentina	566
19	Amoco Corp.	United States	530
20	Mobil Corp.	United States	526

[a]Ref. 6.
[b]To convert m^3 to ft^3, multiply by 35.31.

Properties

The composition of natural gas at the wellhead depends on the characteristics of the reservoir and is highly variable with respect to both the constituents present and the concentrations of these constituents. Compositions of various natural gases are given in Table 4.

Argon, oxygen, helium, hydrogen, organic sulfides, organic mercaptans, and organic disulfides may also be present and, using highly sensitive analytical instrumentation, trace concentrations of a variety of other constituents are being detected (see TRACE AND RESIDUE ANALYSIS). Mercury has been found in some wellhead gases at concentrations based on weight ranging from 1 to 230 ppm and arsenic compounds (qv) have been detected at the ppb level in gases obtained from a formation in New Mexico (U.S.). Arsines have been found in gases from West Virginia (U.S.) and gases produced from offshore Louisiana (U.S.) wells (8).

The physical properties of the principal constituents of natural gas are listed in Table 5. These gases are odorless, but for safety reasons, natural gas is odorized before distribution to provide a distinct odor to warn users of possible gas leaks in equipment. Sulfur-containing compounds such as organic mercaptans, aliphatic sulfides, and cyclic sulfur compounds are effective odorants at low concentrations and are added to natural gas at levels ranging from 4 to 24 mg/m^3.

Table 3. Natural Gas Producing Organizations[a]

Rank	Company	Country	1991 Production,[b] $m^3 \times 10^9$
1	Royal Dutch/Shell	the Netherlands	71
2	Pertamina	Indonesia	70
3	Mobil Corp.	United States	47
4	Sonatrach	Algeria	46
5	Saudi Arabian Oil Co.	Saudi Arabia	45
6	Exxon Corp.	United States	43
7	Petroleos de Venezula SA	Venezuela	42
8	Petroleos Mexicanos	Mexico	38
9	Amoco Corp.	United States	38
10	Chevron Corp.	United States	29
11	Texaco Inc.	United States	23
12	National Iranian Oil Co.	Iran	22
13	Yacimientos Petroliferos Fiscales SA	Argentina	21
14	Abu Dhabi National Oil Co.	Abu Dhabi	19
15	AGIP	Italy	19
16	ARCO	United States	17
17	Petronas	Malaysia	17
18	Unocal Corp.	United States	16
19	China National Petroleum Co.	China	15
20	Shell Oil Co.	United States	14

[a] Ref. 6.
[b] To convert m^3 to ft^3, multiply by 35.31.

The pressure–volume–temperature (PVT) behavior of many natural gas mixtures can be represented over wide ranges of temperatures and pressures by the relationship

$$PV = ZnRT$$

where P is the absolute pressure; V, the volume; Z, the compressibility factor for the mixture; n, the number of moles of gas; R, the universal gas constant; and T, the temperature in Kelvin. When natural gas contains high concentrations of H_2S and CO_2, along with water vapor, the PVT behavior deviates from the conventional compressibility relationship. The PVT properties of these mixtures can be described using empirically determined pseudocritical temperature and pressure adjustment factors which are a function of the total acid gas concentration and the H_2S to CO_2 ratio of the acid gases (9). More complex relationships such as the Benedict-Webb-Rubin relationship (BWR), an eight-constant parameter equation, provide a good representation of both liquid and vapor properties for mixtures containing nonpolar gases and light hydrocarbons. Modifications of the BWR equation extend the basic BWR approach to more of the complex mixtures encountered during the production and processing of natural gas (10) (see ENGINEERING, CHEMICAL DATA CORRELATION).

Table 4. Wellhead Compositions of North American Natural Gases[a]

Parameter	Rio Arriba County, N. Mex.	Terrell County, Tex.	Stanton County, Kans.	San Juan County, N. Mex.	Olds Field, Alberta, Canada	Cliffside Field, Amarillo, Tex.[b]
component, mol %						
methane	96.91	45.64	67.56	77.28	52.34	65.8
ethane	1.33	0.21	6.23	11.18	0.41	3.8
propane	0.19		3.18	5.83	0.14	1.7
butanes	0.05		1.42	2.34	0.16	0.8
pentanes and heavier	0.02		0.04	1.18	0.41	0.5
carbon dioxide	0.82	53.93	0.07	0.80	8.22	0.0
hydrogen sulfide		0.01			35.79	0.0
nitrogen	0.68	0.21	21.14	1.39	2.53	25.6
total sulfur, mg/m^3	0	0.27	0	0	984	
classification						
wet				x		
lean	x	x	x		x	x
sweet	x		x	x		x
sour		x			x	
gross heating value,[c] MJ/m^3	37.6	17.3	34.9	46.8	30.0	30.7
specific gravity	0.574	1.077	0.733	0.741	0.882	0.711

[a]Ref. 7.
[b]Also contains 1.8 mol % helium.
[c]To convert MJ/m^3 to Btu/ft^3, multiply by 26.86.

Processing

Because of the wide variation in the composition of natural gas as it is recovered at the wellhead and because natural gas can be used over a wide range of hydrocarbon contents, any specification for natural gas is usually broadly defined. However, the natural gas obtained at the wellhead usually undergoes some type of treatment or processing prior to its use for safety, economic, or system and material compatibility reasons.

In 1990, there were 1479 gas processing plants throughout the world having a combined capacity for treating 4.43×10^9 m^3 of natural gas per day. The operating level of these plants, in 1990, was 2.67×10^9 m^3/d of gas and 5.6×10^8 L/d of natural gas liquids. Eighty-seven percent of the gas processing plants, representing 63% of the worldwide processing capacity, were located in North America: 734 plants are in the United States; 540 plants are in Canada. The size of these plants, the elements making up the processing trains, and the operating conditions depend on the amount and the characteristics of the gas to be processed. Plant capacities range from small systems having feed rates less than 2.8×10^4 m^3/d to a 9.9×10^7 m^3/d plant located in the province of Alberta, Canada (11).

Dehydration.   Produced gas is usually saturated with water vapor at the wellhead temperature and pressure. Generally, these water-vapor levels are re-

Table 5. Physical Constants of Natural Gas Constituents

Compound	CAS Registry Number	Formula	Mol wt	Boiling point, K^a	Critical pressure, kPa^b	Critical temperature, K	Specific gravity, liquid^c	Gross heating value,^{d} MJ/m^3
methane	[74-82-8]	CH_4	16.043	111.64	4595	190.56	$(0.3)^e$	37.57
ethane	[74-84-0]	C_2H_6	30.070	184.55	4871	305.34	0.3562^f	65.83
propane	[74-98-6]	C_3H_8	44.097	231.08	4247	369.86	0.5070^f	93.60
2-butane	[75-28-5]	C_4H_{10}	58.123	261.37	3640	407.86	0.5629^f	120.98
n-butane	[106-97-8]	C_4H_{10}	58.123	272.65	3796	425.17	0.5840^f	121.37
2-pentane	[78-78-4]	C_5H_{12}	72.150	301.00	3381	460.44	0.6247	148.84
n-pentane	[109-66-0]	C_5H_{12}	72.150	309.23	3369	469.71	0.6311	149.14
n-hexane	[110-54-3]	C_6H_{14}	86.177	341.89	3012	507.38	0.6638	176.93
n-heptane	[142-82-5]	C_7H_{16}	100.204	371.58	2736	540.21	0.6682	204.71
n-octane	[111-65-9]	C_8H_{18}	114.231	398.83	2487	568.83	0.7070	232.47
n-decane	[124-18-5]	$C_{10}H_{22}$	142.285	447.32	2104	617.60	0.7342	288.05
nitrogen	[7727-37-9]	N_2	28.013	77.35	3400	126.21	0.8094^g	
oxygen	[7782-44-7]	O_2	31.999	90.20	5043	154.59	1.1421^g	
carbon dioxide	[124-38-9]	CO_2	44.010	194.68^h	7384	304.22	0.8180^f	23.70
hydrogen sulfide	[7783-06-4]	H_2S	34.076	212.88	8963	373.41	0.8014^f	
water	[7732-18-5]	H_2O	18.015	373.16	22055	647.14	1.0000	
air			28.963	78.83	3771	132.43	0.8748^g	

^{a}At atmospheric pressure, 101.3 kPa (1 atm).
^{b}To convert kPa to psi, multiply by 0.145.
^{c}At 288.72 K.
^{d}At 288.72 K, 101.325 kPa (1 atm).
^{e}Above critical point, estimated or extrapolated.
^{f}At saturation pressure, 288.72 K.
^{g}At normal boiling point.
^{h}Denotes sublimation temperature.

duced to concentrations no greater than 112 mg/m^3 (7 lbs/10^6 ft^3) gas to prevent condensation during transmission in high pressure pipelines and to reduce the possibility of corrosion. If the gas undergoes additional processing in a cryogenic plant to remove inert gases such as nitrogen or helium, the water-vapor levels must be reduced to a dew point corresponding to 172 K.

Usually the process selected for dehydration involves either liquid or solid desiccants (qv). Whereas solid desiccants, such as alumina, silica gel, or molecular sieves (qv), offer advantages of lower dew points and less susceptibility to corrosion, liquid systems based on glycols are frequently selected because of lower construction costs, lower operating costs, and greater economic effectiveness at larger scales. Glycol units can dehydrate natural gas to moisture contents of 8 mg/m^3 and, with the addition of other units, can achieve dehydration to a level of 4 mg/m^3. Triethylene glycol is frequently the liquid desiccant of choice. Tetraethylene glycol is also used.

Dehydration may also be accomplished by expansion refrigeration which utilizes the Joule-Thompson effect. This technique is normally used when the prime objective is hydrocarbon recovery.

Anhydrous calcium chloride absorbs water to a capacity of 3.5 kg/kg of calcium chloride and forms a nonreuseable brine. This technique is best suited for remote applications where modest dew point depressions are required and gas processing volumes are small.

The lower molecular weight hydrocarbon constituents of natural gas can react with liquid or free water to form crystalline solids called hydrates (8). These reactions can take place under many of the conditions associated with the recovery and movement of water-saturated natural gas. In solid hydrates, water and the hydrocarbon molecules interact through hydrogen bonding to form a stable cage-like structure known as a clathrate. These clathrates typically consist of 90 wt % water, the remaining 10 wt % being made up of methane, ethane, propane, isobutane, n-butane, nitrogen, carbon dioxide, and hydrogen sulfide. Hydrates involving only water and n-butane do not form; however, n-butane participates in hydrate formation in a mixture of other gases. Hydrate formation is prevented by reducing the water content of the gas below concentrations where hydrates form, by heating flow lines, or by the use of chemical additives such as ethylene glycol, diethylene glycol, or methanol (qv) to depress the temperature at which the hydrates or ice forms (see GLYCOLS; INCLUSION COMPOUNDS).

Natural Gas Liquids. Natural gases containing high concentrations of the higher hydrocarbons are processed both to reduce the potential for condensation of these higher molecular-weight compounds during transmission and subsequent use, and to recover the natural gas liquid (NGL) products which can be marketed in both the fuel and petrochemical feedstock market (see FEEDSTOCKS, PETROCHEMICAL). Natural gases are characterized with respect to the higher (ethane and above) hydrocarbon content as lean, moderately rich, or very rich. Lean gases have higher hydrocarbon contents that are less than 334 mL/m^3; very rich gases have hydrocarbon liquid contents greater than 668 mL/m^3. In 1990, 5.62 × 10^8 L of natural gas liquids were produced.

The market value of natural gas liquids is highly volatile and historically has been weakly related to the world price of crude oil. During the 1980s, the market value of natural gas liquids ranged from approximately 60% of the price of crude to 73% (12). In this 10-year interval, several fluctuations occurred in the natural gas liquid market. Because of the variability of the natural gas liquid market, the NGL recovery plants need to have flexibility. Natural gas liquid products compete in the following markets: ethane; propane; a liquefied petroleum gas (LPG); a C-3/C-4 mix; and n-butane all compete as petrochemical feedstocks. Propane and LPG are also used as industrial and domestic fuels, whereas 2-butane and natural gasoline, consisting of C-5 and heavier hydrocarbons, are used as refinery feedstocks.

Natural gas liquids are recovered from natural gas using condensation processes, absorption (qv) processes employing hydrocarbon liquids similar to gasoline or kerosene as the absorber oil, or solid-bed adsorption (qv) processes using adsorbents such as silica, molecular sieves, or activated charcoal. For condensation processes, cooling can be provided by refrigeration units which frequently use vapor-compression cycles with propane as the refrigerant or by using the Joule-Thompson expansion to lower the temperature of the feed gas, or through the use of expansion turbines which both reduce the temperature of the gas and derive work for use at other points in the recovery and separation process.

The condensation processes are generally favored for recovering natural gas liquids. If the feed gas is very rich in liquids, plants based on simple refrigeration cycles may be used. When the liquid content of the feed gas is relatively low, use of the expansion turbine may be favored. For conditions providing a very high inlet pressure, the Joule-Thompson expansion may be more economical. Low inlet pressures generally favor an expander plant or straight-refrigeration. Very low flow rates require a relatively simple process and may favor an automatically operated Joule-Thompson unit.

Absorber oil units offer the advantage that liquids can be removed at the expense of only a small (34–69 kPa (4.9–10.0 psi)) pressure loss in the absorption column. If the feed gas is available at pipeline pressure, then little if any recompression is required to introduce the processed natural gas into the transmission system. However, the absorption and subsequent absorber-oil regeneration process tends to be complex, favoring the simpler, more efficient expander plants. Separations using solid desiccants are energy-intensive because of the bed regeneration requirements. This process option is generally considered only in special situations such as hydrocarbon dew point control in remote locations.

Acid Gas Constituents. Hydrogen sulfide is both highly toxic and acidic enough to precipitate corrosion. Natural gases containing hydrogen sulfide are subjected to processing to reduce H_2S concentrations to less than 4 ppmv, or approximately 5.7 mg/m^3. Carbon dioxide, like hydrogen sulfide, forms acidic solutions in the presence of water and is also referred to as an acid gas. However, it is not toxic, and processing requirements to reduce CO_2 levels are less severe than those for H_2S. Carbon dioxide concentrations of 1 to 2% are usually tolerated. At these concentrations, the expenditure of energy to transport the inert constituent through pipelines is minimal, and the extent of the corrosion of materials used in transmission systems is acceptable (see CORROSION AND CORROSION CONTROL).

There are more than 30 processes available for removing the acid gas constituents such as hydrogen sulfide, carbon dioxide, and other organic sulfur compounds, ie, carbonyl sulfide, organic mercaptans, and disulfides (8,9). Because of the toxicity of hydrogen sulfide, requirements for removal are severe. In the

United States, natural gas is almost always processed to provide hydrogen sulfide levels not greater than 4 mg/m^3 and the specification can be as low as 1 mg/m^3 in some countries. Carbon dioxide levels of 2 to 3% are usually acceptable and in many instances it may be necessary to treat the gas for other sulfur-containing molecules. When the CO_2 content exceeds specifications, or when cryogenic processing follows, removal of carbon dioxide to levels of 100 ppmv can be achieved.

The process options reflect the broad range of compositions and gas volumes that must be processed. Both batch processes and continuous processes are used. Batch processes are used when the daily production of sulfur is small and of the order of 10 kg. When the daily sulfur production is higher, of the order of 45 kg, continuous processes are usually more economical. Using batch processes, regeneration of the absorbant or adsorbant is carried out in the primary reactor. Using continuous processes, absorption of the acid gases occurs in one vessel and acid gas recovery and solvent regeneration occur in a separate reactor.

Iron sponge is the oldest and most widely used batch process for removing sulfur compounds from natural gas. Iron sponge consists of wood chips or shavings impregnated with ferric oxide and sodium hydroxide. The chips are placed in a vertical contact tower where the hydrogen sulfide in the natural gas reacts with the iron oxide, forming ferric sulfide. Mercaptans also react with the iron oxide to form ferric mercaptides. When the chips are saturated, ie, no longer able to react with hydrogen sulfide, the addition of oxygen to the spent bed converts the iron sulfide back to iron oxide and elemental sulfur. The mercaptides are converted to iron oxide and a disulfide. These beds eventually deteriorate and must be replaced. The spent chips are considered toxic and disposal must be done in an environmentally acceptable manner. Other batch processes involve the use of an iron oxide slurry, zinc oxide particles suspended in a water solution of zinc acetate, or aqueous solutions of sodium nitrite.

There are numerous chemical and physical solvents available for use in continuous acid gas removal processes (8). The chemical absorbants include aqueous solutions of organic amines such as monoethanolamine, diethanolamine, triethanolamine, diglycolamine, or methyldiethanolamine. These solutions, which can be regenerated, react chemically with the acid gases and can be used to remove large quantities of hydrogen sulfide and carbon dioxide. Physical solvents are organic liquids that absorb carbon dioxide and hydrogen sulfide into solution at high pressures and ambient or low temperature. When physical solvents are used, the acid gases are recovered and the solvents regenerated by flashing at low pressures. The energy required to regenerate physical solvents is less than the regeneration requirements for chemical solvents. Physical solvents include methanol, mixtures of dimethylethers of polyethylene glycols, propylene carbonate, sulfolane, 1-acetylmorpholine, and other complex organic molecules. Solvent systems can include solutions of physical solvents and more chemically reactive materials. Solutions of potassium carbonate have been employed, and solutions involving metal–organic complexes are now being used in some small-scale applications. These metal–organic-complex-based solvents result in the direct oxidation of hydrogen sulfide to elemental sulfur as part of the regeneration process.

Adsorption systems employing molecular sieves are available for feed gases having low acid gas concentrations. Another option is based on the use of polymeric, semipermeable membranes which rely on the higher solubilities and diffusion rates of carbon dioxide and hydrogen sulfide in the polymeric material relative to methane for membrane selectivity and separation of the various constituents. Membrane units have been designed that are effective at small and medium flow rates for the bulk removal of carbon dioxide.

Whereas the list of process options for gas sweetening is extensive, the batch processes and the amine processes are used for over 90% of the wellhead applications. During the 1990–1991 period, there were about 700 amine units operating in the United States (8). The early amine processes were based on monoethanolamine [141-43-5] (MEA) as the absorbant. In many applications, the use of diethanolamine [111-42-2] (DEA) is frequently favored because of its lower heats of reaction, reduced corrosivity, and ability to be used at higher concentrations. These characteristics translate into process advantages which include higher acid gas loadings in the solvent, lower energy requirements for solvent regeneration, and lower solvent recirculation rates.

Subsequent to separating and recovering the acid gases from the raw natural gases, additional processing is undertaken to convert the hydrogen sulfide to elemental sulfur. A number of processes have been developed to accomplish this conversion, the process known as the Claus process being the most widely used in the natural gas industry. This process provides the basis for producing sulfur from acid gas streams having hydrogen sulfide concentrations ranging from approximately 20 to 100%. Economic considerations generally limit its application to plants that have a production capacity of sulfur greater than 30 t/d. Whereas there are several variations of the Claus process to accommodate specific feed stream conditions, the fundamental element is a reaction furnace operating at temperatures of approximately 1273 K where one-third of the hydrogen sulfide is burned with air (or oxygen-enriched air) to produce sulfur dioxide and water. A shift reaction then takes place between the remaining hydrogen sulfide and the sulfur dioxide to produce additional water and elemental sulfur which can be recovered. The reactor furnace is reported to achieve sulfur yields as high as 90% directly. Conversion of the remaining hydrogen sulfide to sulfur is accomplished in lower temperature catalytic reactors (see SULFUR REMOVAL AND RECOVERY).

In 1991, there were approximately 418 sulfur production plants associated with oil and gas production in operation throughout the world. Approximately 86% of these plants were based on the Claus process, and there were 118 Claus units operating in natural gas processing facilities (11).

Nitrogen. Natural gases containing nitrogen are processed to reduce the nitrogen levels to concentrations that do not detract significantly from the energy level of the gas and do not require the expenditure of excessive amounts of energy during transmission of the gas.

The separation of nitrogen from natural gas relies on the differences between the boiling points of nitrogen (77.4 K) and methane (91.7 K) and involves the cryogenic distillation of a feed stream that has been preconditioned to very low levels of carbon dioxide, water vapor, and other constituents that would form solids at the low processing temperatures.

Specifications

Whereas there is no universally accepted specification for marketed natural gas, standards addressed in the United States are listed in Table 6 (8). In addition to

Table 6. Natural Gas Pipeline Specifications[a]

Characteristic	Specification	Test method[b]
water content, mg/m^3	64–112	ASTM (1986) D1142
hydrogen sulfide, mg/m^3	5.7	GPA (1968) Std. 2265
gross heating value,[c] MJ/m^3	35.4	GPA (1986) Std. 2172
hydrocarbon dew point at 5.5 MPa,[d] K	264.9	ASTM (1986) D1142
mercaptan content, mg/m^3	4.6	GPA (1968) Std. 2265
total sulfur, mg/m^3	23–114	ASTM (1980) D1072
carbon dioxide, mol %	1–3	GPA (1990) Std. 2261
oxygen, mol %	0–0.4	GPA (1990) Std. 2261

[a]Gas must be commercially free of sand, dust, gums, and free liquid. Delivery temperature, 322.16 K; delivery pressure, 4.83 MPa. Ref. 8.
[b]ASTM = American Society for Testing Materials; GPA = Gas Processors' Association.
[c]To convert MJ/m^3 to Btu/ft^3, multiply by 26.86.
[d]To convert MPa to psi, multiply by 145.

these specifications, the combustion behavior of natural gases is frequently char-acterized by several parameters that aid in assessing the influence of composi-tional variations on the performance of a gas burner or burner configuration. The parameters of flash-back and blow-off limits help to define the operational limits of a burner with respect to flow rates. The yellow-tip index helps to define the conditions under which components of the natural gas do not undergo complete combustion, and the characteristic blue flame of natural gas burners begins to show yellow at the flame tip. These three parameters are determined experimen-tally using standardized test methods. An index referred to as the Wobbe Number is indicative of the combustion energy being delivered to a burner at a constant pressure drop across the burner. The Wobbe Number, defined as the square root of the ratio of the volumetric heat of combustion of the gas to the specific gravity of the gas, can be calculated directly from a knowledge of the gas composition.

Two gases characterized by the same Wobbe Number deliver approximately the same amount of chemical energy to the combustion zone of a burner without the need for adjustments in the pressure drop or orifice size of the burner even though the volumetric heating values and the specific gravities of the gases are different. For existing burners having fixed piping and dimensions, these per-formance indexes become important as variations in the composition of the gas being delivered are encountered. These parameters must also be considered dur-ing the development of new burner hardware.

Transmission and Storage

As exploration and production activities have expanded both the natural gas re-source base and the worldwide proven reserves, long-distance gas transmission pipelines have been constructed to link these resources to the industrialized areas and population centers of the world. The availability of high tensile-strength steel pipe and the development of techniques to construct, weld, and lay large diameter high pressure pipelines make it possible to economically transport natural gas to the marketplace. These transmission systems, coupled with localized, lower pressure distribution networks, bring gas to large segments of the world. Long-distance pipelines transport gas from the large fields in Siberia and the Ural Mountains of the Confederation of Independent States to both eastern and central Europe. Pipelines across the Mediterranean Sea connect the gas fields of North Africa with the European market. The European market is also served by long-distance pipelines transporting gas from the gas fields in the Middle East. Depending on the availability of the resource and the anticipated market volume, these long-distance transmission lines have used pipes having diameters ranging from approximately 36 to 142 cm (2,8). The gas transmission lines and distribution main systems existing as of 1989 are given.

Country	Pipeline, km
Belgium	35,200
Canada	219,000
Czechoslovakia	32,800
France	146,100
Germany	246,100
Italy	141,500
the Netherlands	96,100
Poland	55,200
CIS	432,800
United Kingdom	241,400
United States	1,700,000

As early as 1966, natural gas was available to all of the lower 48 states in the United States. During the period 1967–1990, the U.S. transmission system grew from 362,700 km to 450,800 km. Over this same time period, the distribution mains increased from 867,800 km to 1,347,000 km. As plastic pipe and reliable joining technology became available, the use of plastic pipe expanded to include the distribution of gas in low pressure systems. By 1990, approximately 24% of the U.S. distribution system was based on plastic pipe (1).

Natural gas production and transmission systems are complemented by underground storage systems. These systems provide the capability to respond to short-term gas demands which exceed the immediate production levels or transmission capabilities. They also provide an opportunity to sustain some production by refilling the storage areas when seasonal temperature variations lead to periods of reduced gas demand. In the United States in 1990, there were 397 storage pools having a combined capacity of 2.2×10^{11} m^3 (1).

Liquefied natural gas (LNG) also plays a large role in both the transportation and storage of natural gas. At a pressure of 101.3 kPa (1 atm), methane can be liquefied by reducing the temperature to about $-161°C$. When in the liquid form, methane occupies approximately 1/600 of the space occupied by gaseous methane at normal temperature and pressure. In spite of the very low temperature of the liquid, LNG offers advantages for both shipping and storing natural gas.

The U.S. Bureau of Mines, because of its interest in separating helium from natural gas, is credited with achieving the first liquefaction of natural gas on a practical scale in 1917 (2) (see HELIUM GROUP, GASES). The value of LNG as a means of providing natural gas to a system during periods of high demand was considered in the early 1940s and began to receive serious attention in the late 1950s and early 1960s, resulting in the commissioning of three LNG peak-shaving facilities in the United States in 1965. In 1978, there were 55 LNG peak-shaving plants (having liquefaction, storage, and regasification capabilities) in the United States. These were complemented by over 50 satellite plants (for storage and regasification) representing a combined supply–delivery capability of gas of 2.15 $\times$ 10^8 m^3/d (2). By 1987, the U.S. LNG storage capabilities had increased to a deliverability level of 2.66 $\times$ 10^8 m^3/d of gas (1). LNG peak-shaving capabilities, those systems that satisfy short-term demands not met by the primary supply system, have also been developed in many parts of the world, including Canada, Britain, Germany, France, Spain, the Netherlands, Belgium, and Australia.

In the early 1950s, projects were undertaken to construct barges and ships that could transport liquefied natural gas over long distances. In 1958, using a freighter that had been retrofitted with five storage tanks insulated with laminated balsa, 2000 metric tons of LNG were successfully transported from Lake Charles, Louisiana (U.S.), to a storage facility at Canvey Island, Essex (England). Subsequently, other international LNG projects were developed. These include LNG shipped from Algeria to Britain and France, from Alaska to Japan, from Libya to Spain and Italy, from Brunei (on the island of Borneo) to Japan, and from Australia to Japan. Construction of the liquefaction and shipping facilities for the Australia to Japan project was completed in mid-1989 and can provide an LNG capacity of 6 $\times$ 10^6 t/yr (13). LNG tankers having capacities ranging from 45,000 to 130,000 m^3 have been designed, constructed, and put into operation to serve the international LNG activities. The on-shore complexes that liquefy and store LNG include multitrain plants capable of liquefying in excess of 1.98 $\times$ 10^7 m^3/d of gas. The individual tanks for LNG storage can have capacities of 60,000 m^3 (2 $\times$ 10^6 ft^3) (2).

Uses

Fuel.   Natural gas is used as a primary fuel and source of heat energy throughout the industrialized countries for a broad range of residential, commercial, and industrial applications. The methane and other hydrocarbons react readily with oxygen to release heat by forming carbon dioxide and water through a series of kinetic steps that results in the overall reaction,

$$CH_4 + 2\ O_2 \rightarrow CO_2 + 2\ H_2O\ (g)$$

This exothermic reaction has an energy release of 50 kJ/g (12 kcal/g) of methane reacted. This energy can be released by raising the temperature of a methane–air mixture to its ignition temperature where the reaction becomes self-sustaining, producing high temperature reaction products. At atmospheric pressure, the combustion reactions can be sustained in methane–air mixtures for methane concentrations ranging from approximately 5.4 to 14 vol %. A methane–air mixture containing approximately 9.5% methane would be stoichiometrically balanced for CO_2 and H_2O (14). The adiabatic combustion temperatures for combustible methane–air mixtures are in the range of 1950 to 2325 K depending on the specific conditions. The overall reaction of methane and oxygen can also be promoted at lower temperature through a series of steps using catalysts and electrolytic cells that result in the direct conversion of the chemical energy to electrical energy without the use of an intermediate heat engine and generator. Heat energy is also derived from fuel-cell reactions and the overall energy utilization efficiencies of these units can exceed 80% (15) (see COMBUSTION SCIENCE AND TECHNOLOGY; FUEL CELLS).

Gas burner technology has been developed that permits natural gas to be used effectively as a primary fuel in both small and large applications. Small applications include furnaces, hot water heaters, clothes dryers, and cooking stoves for residential installations. A high performance, natural gas fueled furnace using a pulse-combustion process and operating at conditions resulting in the condensation of the water vapor in the combustion products has an overall energy efficiency exceeding 90%. This technology has been successfully incorporated into many of the residential or small-scale applications (16). Large-scale applications include the use of natural gas to supply process heat in the production of steel (qv), glass (qv), ceramics (qv), cement (qv), paper (qv), chemicals, aluminum, processed foods, fabricated metal products, etc. Natural gas fueled, indirect-fired metallic radiant tubes and ceramic radiant-tube burners have facilitated the expanded use of natural gas for heat-treating applications (17). Natural gas is also used as a primary fuel for the production of electrical energy.

In 1990, in the United States, the use of natural gas in the residential and commercial, industrial, and electricity-generating sectors corresponded to 37.5, 44.3, and 14.8%, respectively, of the 20.5 EJ (4.9×10^{18} cal) of gas energy consumed. In addition, 0.63 EJ (1.5×10^{17} cal) of natural gas energy was consumed in the transportation sector, mainly to transport the gas through the pipelines and distribution systems (1).

Because of its clean burning characteristics, natural gas is being strongly considered as a viable fuel for a larger segment of the transportation market throughout the world. Automobile manufacturers are developing engines to more effectively use compressed natural gas as a vehicular fuel (see GASOLINE AND OTHER MOTOR FUELS).

Natural gas is attractive as a fuel in many applications because of its relatively clean burning characteristics and low air pollution (qv) potential compared to other fossil fuels. Combustion of natural gas involves mixing with air or oxygen and igniting the mixture. The overall combustion process does not involve particulate combustion or the vaporization of liquid droplets. With proper burner design and operation, the combustion of natural gas is essentially complete. No unburned hydrocarbon or carbon monoxide is present in the products of combustion.

Natural gas combustion produces neither particulates nor significant quantities of SO_2. Natural gas, as delivered, usually contains only small amounts of sulfur (of the order of 0.3 mg/MJ) resulting from either residual H_2S or sulfur-containing odorants. In contrast, fuel oils contain between 50 and 500 mg/MJ and coal between 100 and 1500 mg/MJ (18). In addition, the individual constituents of natural gas do not contain nitrogen. Formation of nitrogen oxides, NO_x, through reactions involving the nitrogen in the air can be minimized by staging the mixing, ignition, and combustion processes. Burners specifically designed to be low NO_x burners have been developed.

Of all the fossil fuels, the use of natural gas results in the formation of the least amount of CO_2 per unit of heat energy produced. On a constant energy basis, natural gas combustion produces approximately 30% less CO_2 than liquid petroleum fuels and approximately 45% less CO_2 than coal and other solid fossil fuels.

Chemical Use.   Both natural gas and natural gas liquids are used as feedstocks in the chemical industry. The largest chemical use of methane is through its reactions with steam to produce mixtures of carbon monoxide and hydrogen (qv). This overall endothermic reaction is represented as

$$CH_4 + H_2O \rightarrow CO + 3\,H_2$$

In the presence of catalysts, the CO reacts with steam through the shift reaction to produce additional hydrogen and CO_2 as represented by

$$CO + H_2O \rightarrow CO_2 + H_2$$

Hydrogen is used mainly in ammonia synthesis, methanol synthesis, and petroleum refining.

In 1990, the global demand for methanol was 19.6×10^6 t. Increased demands for methanol are developing as a result of its use in the manufacture of methyl tertiary butyl ether (MTBE), an oxygenated additive to gasolines used to improve octane ratings while reducing the potential for hydrocarbon emissions from gasoline fueled vehicles (see OCTANE IMPROVERS) (19). Carbon monoxide, in addition to being used in the manufacture of methanol, is also used in the manufacture of many other products, including paints, plastics, pesticides, and adhesives.

Natural gas liquids represent a significant source of feedstocks for the production of important chemical building blocks that form the basis for many commercial and industrial products. Ethylene (qv) is produced by steam-cracking the ethane and propane fractions obtained from natural gas, and the butane fraction can be catalytically dehydrogenated to yield 1,3-butadiene, a compound used in the preparation of many polymers (see BUTADIENE). The n-butane fraction can also be used as a feedstock in the manufacture of MTBE.

Production

Natural gas is produced from reservoirs containing both oil and gas (associated gas) and from nonassociated reservoirs holding only gas. These reservoirs may be relatively shallow and require wells drilled to depths of a few hundred meters. However, production is also being realized from reservoirs located at substantial depths requiring wells drilled to depths in excess of 6100 m. Production takes place both at onshore installations and on offshore platforms which service wells drilled to provide access to reservoirs located below the floor of the ocean. Offshore facilities are operating in the continental shelf regions and shallow coastal waters of many parts of the world including the United States, Canada, Australia, Brazil, Norway, the North Sea, and the Persian Gulf. A gas field located in the Norwegian North Sea having recoverable reserves of 8000 m^3 had a water depth of 100 m. Production began in 1983 (20) and the daily production rate was 6×10^6 m^3/d.

The worldwide annual level of drilling, as represented by well completions, exceeded 47,000 to 53,000 wells per year from 1988 through 1990 (1). During this period, the annual number of well completions for gas was approximately 11,000 wells; approximately 22,000 wells were completed each year for oil. Of those wells completed for gas, 95% were in North America.

By 1990, there were 264,500 producing gas wells in the United States. During the period from 1970 through 1990, there was a continual increase in the production of natural gas from the offshore areas of Alaska, California, Louisiana, and Texas. United States offshore production increased from approximately 0.93 $\times 10^{11}$ m^3 annually in 1970, to 1.53×10^{11} m^3 in 1980, and 1.56×10^{11} m^3 in 1990 (1).

The cost of drilling natural gas wells depends on the drilling environment, the characteristics of both the overburden and the reservoir, and the depth of the well. In 1989, the average cost of drilling a well on the U.S. mainland ranged from $162/m for wells less than 380 m deep to $1910/m for wells drilled to depths of 6100 m and beyond. The average depth of the onshore gas wells drilled in 1989 was approximately 1585 m, drilled at an average cost of $285/m. In 1989, 233 wells were drilled offshore to an average depth of approximately 3100 m at an average drilling cost of $1010/m. These costs include all charges for drilling and equipping wells up to and including the Christmas trees, which are pipe and valve structures that offer access to the well, but exclude exploration costs, development and production costs, and expenditures for service wells (1).

Economic Aspects

Economically, natural gas represents an attractive energy option in those industrialized areas of the world where business infrastructures and delivery systems have been established to provided reliable service. Table 7 gives the 1988 natural

Table 7. 1988 Natural Gas Prices[a]

	Price, $/GJ[b]	
Country	Industrial	Residential
United States	2.72	5.05
Germany (West)	3.09	6.97
United Kingdom	3.55	6.67
Australia	2.55	5.21
Japan	10.48	21.42

[a]U.S. dollars; includes taxes. Ref. 6.
[b]To convert J to cal, divide by 4.184.

gas prices (including taxes) for selected industrialized countries (6). From 1979 to 1988, natural gas prices to users generally increased. However, in some geographic areas natural gas prices decreased during the latter part of the period. The geographic variations in the gas prices are a reflection of many factors, including differences in supply and demand, availability, differences in import requirements, and the extent of different governmental regulatory influences.

Figure 2 shows the acquisition price for U.S. natural gas at the wellhead, together with the acquisition prices of crude oil in both the international market and the U.S. domestic market (21). On a cost per unit energy basis, natural gas has been priced lower than crude oil by factors ranging from 1.4 to 3.5 for the period shown.

Figure 3 provides a comparison of the energy costs in the U.S. residential market for natural gas, electricity, or No. 2 fuel oil (1). The prices of all three forms of energy to residential users have increased for the period shown. Electrical energy has had the largest dollar increase.

Outlook. Since natural gas became a commodity of commerce and its potential as an energy form was recognized, its use on a worldwide basis has continued to increase. Adequate proven reserves have been developed and the necessary production and delivery infrastructures have been established to bring natural gas to the industrial markets on a basis that is economically competitive with other energy options. Technology has been developed to take advantage of the properties of natural gas, and sufficient confidence has been established in the long-range stability of the natural gas option to provide the necessary incentives for the continued investment of capital in order to maintain and expand existing capabilities. The outlook for natural gas depends on the ability of the international natural gas community to continue to demonstrate that the resource base and the economic competitiveness of natural gas are adequate to justify its use for short-term and long-term applications.

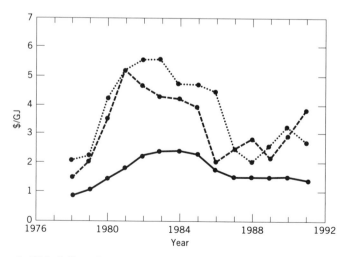

Fig. 2. Prices in U.S. dollars for (——), U.S. natural gas; (•––•), U.S. crude oil; and (•••••), Saudi 34 crude oil. To convert J to cal, divide by 4.184 (21).

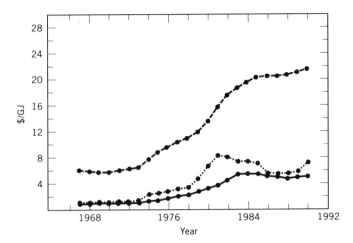

Fig. 3. Energy prices in U.S. dollars for the U.S. residential market for (——), natural gas; (•—–•), electrical energy; and (•••••), No. 2 fuel oil (21).

Reserves

During the period from 1950 to 1992, the worldwide proven reserves of natural gas increased at an average annual rate of approximately 6.7% (3–5). In this same period, worldwide production increased at an average annual rate of approximately 5.7% (5) so that the annual worldwide consumption of natural gas as of 1992 was equivalent to approximately 1.7% of the proven reserves. This suggests that the proven reserves, assuming the reserves could be delivered to the existing world markets, could satisfy the requirements of this market for more than 50 years. However, world population is expected to increase and the demand for natural gas is expected to expand both in areas that are heavily industrialized and in areas where industrialization activities are underway.

There are large concentrations of proven reserves in specific areas of the globe such as the Middle East, North America, and the Confederation of Independent States. Thus the future of natural gas, worldwide, is dependent on the flow of natural gas across international boundaries on a long-term basis. This future is also dependent on the maintenance of an economic balance between the price of oil and the price of natural gas.

In the United States as of this writing, the proven reserves inventory is being replenished at a rate approximately equal to the rate of gas consumption (22). During the 1970s, natural gas prices at the wellhead were federally regulated and low, and natural gas usage in the United States exceeded proven reserve replacements. However, reserve replacements in the 1980s were approximately equal to gas usage rates. In some years, replacement even exceeded the rate of use. This change is in part the result of technologies which make it possible to economically produce gas from resources previously considered too expensive to develop.

The gas reservoirs located in very deep waters, in coal beds, and in tight sands are now more accessible. Fifteen percent of the U.S. gas supply in 1992 was derived from tight sand formations and 1.4×10^{11} m^3 of coal-bed methane was added to the proven reserves (22). In 1992, U.S. proven reserves were placed at 4.67×10^{12} m^3 in the lower 48 states, and it was estimated that the identified gas resource in the United States and Canada exceeds 3.4×10^{13} m^3. Based on the 1992 rate of natural gas consumption, the United States has between 8 and 10 years of proven reserves and a domestic resource base containing between 50 and 70 years' supply.

In 1988, the U.S. Department of Energy completed a study indicating that more than 1.42×10^{13} m^3 of gas was recoverable at a wellhead price below \$2.84/GJ (in 1987 dollars) and 1.98×10^{13} m^3 could be available for \$4.75/GJ or less (23).

Numerous databases and models have been developed for the purpose of creating and examining future scenarios involving natural gas. These projections consider the primary elements of the gas use sector, the existing internal and external opportunities to provide natural gas to these sectors, the economics of competing energy alternatives, and the impact of technologies likely to effect the production, use, and economics of natural gas in the United States. The annual "Baseline Projection" for 1992 (16) showed the primary energy consumption in the United States growing from 88.8 EJ (21.2×10^{18} cal) in 1990 to 105.1 EJ (25.1×10^{18} cal) in the year 2010, to reflect an average annual growth rate of 0.84%. During this period, the use of natural gas was projected to increase from 20.5 EJ in 1990 to 25.6 EJ in 2010, an average annual growth rate of 1.13%. Only the residential sector was projected to have a gas demand requirement that was relatively constant during the 20-year period considered.

The results of research and development programs now in progress should impact the future availability and use of natural gas. In all industrialized countries, programs have been underway to improve the cost effectiveness of activities associated with natural gas. These programs are supported financially by funds from individual governments, funds derived from corporate activities, or monies collected from users of natural gas. The International Gas Union and the Gas Research Institute (17) both provide forums for the development of coordinated programs. These programs cover activities associated with exploration, production, processing, transmission, storage, and use of natural gas.

The development of diagnostic tools coupled with high speed computers is increasing the probability of locating specific gas-rich formations. Drilling technology such as directional drilling affords the opportunity to place a well-bore strategically with respect to either naturally occurring or induced fractures. More sophisticated representations of reservoirs and rock mechanics should improve well stimulations, and more cost-effective recovery of a specific resource. The continued development of coal seams as a source of natural gas is an example.

The more effective delivery of natural gas is being realized by the use of computerized operation centers that allow rapid responses to the variations in gas demand. Automated valves, more precise measuring systems, and high speed communication networks make it possible to closely monitor and manage the transmission and delivery of natural gas. This translates into improved service and cost effectiveness.

In the United States, the development of high performance, gas-fueled cogeneration systems that simultaneously generate electricity and produce steam has impacted gas use in both the commercial and industrial sectors (see POWER

GENERATION). The amount of gas used annually for fuel cogenerators increased from 0.2 EJ (5×10^{16} cal) in the early 1980s to 1.27 EJ (3.03×10^{17} cal) in 1991 (17). The phosphoric acid, natural gas-based fuel cell, a cogeneration technology which produces both electricity and heat electrochemically, is in the early stages of commercialization, whereas advanced fuel cells based on the use of molten carbonates and solid oxides are in the later stages of development. These technologies are expected to have broad applications in the commercial, industrial, and electricity-generating sectors of the natural gas market.

Combined-cycle power generation systems using natural gas and advanced gas turbine technology offer the potential for increasing power generation cycle efficiencies from 33% with conventional boilers to 48%. Opportunities are projected to retrofit existing power plants and to use the technology in new plant construction. Natural gas can also be used in co-firing and reburn modes to help coal-based electric power generation plants achieve lower emissions of sulfur dioxide and nitrogen oxides while protecting the investments in the original coal units. Natural gas can also be co-fired with solid wastes, providing for the more effective destruction of these materials, easing potential pollution problems, and assisting in converting the energy content of solid wastes to a more useable form (see FUELS FROM WASTE).

Technologies are being evaluated which could expand the opportunities for the use of natural gas in the general area of space conditioning. A heat pump being developed for residential applications uses a natural gas-fueled engine to drive the compressor in the refrigeration unit (24). On a larger scale, an advanced natural gas-fired engine has been developed and integrated with a compressor to form the basis for engine-driven chillers. Engine-driven chillers having capacities of 150 t are being marketed. The use of natural gas to regenerate desiccant-based dehumidifiers which could be used in individual family dwellings is also being explored. Similar technology has been successfully developed for larger-scale applications in the commercial market (23).

Natural gas, along with natural gas liquids, may also have an opportunity to provide energy as a transportation fuel. U.S. automakers are involved in limited production of natural gas-fueled vehicles, and approximately 500 refueling stations have been built as part of the infrastructure needed to support these vehicles (22).

BIBLIOGRAPHY

"Gas, Natural" in *ECT* 1st ed., Vol. 7, pp. 59–73, by F. S. Lott, Bureau of Mines, U.S. Department of the Interior; in *ECT* 2nd ed., Vol. 10, pp. 443–462, by D. E. Holcomb, United Gas Corp.; in *ECT* 3rd ed., Vol. 11, pp. 630–652, by J. H. Hillard, Mountain Fuel Supply Co.

1. *1991 Gas Facts*, The American Gas Association, Arlington, Va., 1991.
2. M. W. Peebles, *Evolution of The Gas Industry*, New York University Press, 1980.
3. *Gas Engineers Handbook*, 1st ed., Industrial Press, 1965.
4. *Oil and Gas Journal Data Book—1992 Edition*, PennWell Publishing Co., Tulsa, Okla., 1992.
5. *1990 Energy Statistics Yearbook*, Department of Economic and Social Development, United Nations, New York, 1992.

6. *Oil & Gas Journal Data Book—1993 Edition*, PennWell Publishing Co., Tulsa, Okla., 1993.

7. L. W. Brandt and L. Stroud, *Open File Information and Data Relating to the Extraction of Helium from Natural Gas by Low Temperature Processes*, U.S. Bureau of Mines, Amarillo, Tex., May 1, 1959.

8. F. S. Manning and R. E. Thompson, *Oil Field Processing of Petroleum*, Vol. 1, PennWell Publishing Co., Tulsa, Okla., 1991.

9. R. N. Maddox, *Gas Conditioning and Processing*, Vol. 4, Campbell Petroleum Series, Norman, Okla., 1985.

10. R. H. Perry and D. W. Green, *Perry's Chemical Engineers' Handbook*, 6th ed., McGraw-Hill Book Co., New York, 1984.

11. OJG Special, *Oil Gas J.*, 54–84 (July 22, 1991).

12. A. J. Tarbutton, Jr., *Oil Gas J.*, 50 (July 22, 1991).

13. J. Haggin, *Chem. Eng. News*, 23 (Aug. 1992).

14. B. Lewis and G. von Elbe, *Combustion, Flames, and Explosions of Gases*, Academic Press, Inc., New York, 1972.

15. H. Oman, *Energy Systems Engineering Handbook*, Prentice-Hall Inc., Englewood Cliffs, N.J., 1986.

16. T. J. Woods, *The Long-Term Trends in U.S. Gas Supply and Prices: 1992 Edition of the GRI Baseline Projection of U.S. Energy Supply and Demand to 2010*, Gas Research Insights, Gas Research Institute, Chicago, Ill., Dec. 1991.

17. *An Assessment of the Natural Gas Resource Base of the United States*, U.S. Department of Energy, DOE/W/31109-HL, Washington, D.C., 1988.

18. H. Richter, *Proceedings of the 18th World Gas Conference,* International Gas Union, Berlin, Germany, 1991, pp. 1–9.

19. *Hydrocarbon Process.*, 29 (July 1992).

20. Goodfellow Associates Ltd., *Applications of Subsea Systems*, PennWell Books, 1990.

21. *International Energy Statistics Sourcebook*, 2nd ed., PennWell Publishing Co., Tulsa, Okla., 1992.

22. *New Directions: Natural Gas Energy*, Natural Gas Council (American Gas Association, Independent Petroleum Association of America, Interstate Natural Gas Association of America, Natural Gas Supply Association), 1992.

23. W. M. Burnett, S. D. Ban, and D. A. Dreyfus, *Ann. Rev. Energy* **4**, 1–18 (1989).

24. *Gas Res. Inst. Digest,* 24 (Fall 1992).

KERMIT E. WOODCOCK
Consultant

MYRON GOTTLIEB
Gas Research Institute

GASOHOL.   See ALCOHOL FUELS; FUELS, SYNTHETIC–LIQUID FUELS; GASOLINE AND OTHER MOTOR FUELS.

GASOLINE AND OTHER MOTOR FUELS

Gasoline and other motor fuels comprise the largest single use of energy in the United States, and in 1988 accounted for 21% of all energy usage (1). The cost of this energy has been and is expected to continue to be a primary factor in the national economy. Moreover, the fraction of resources from which these fuels come that is provided by foreign sources is a matter of political concern (see FUEL RESOURCES). The fraction of total crude oil produced domestically shrunk from 73% in 1970 to 49% in 1991 and is predicted to continue to drop (2) (see PETROLEUM). In the 1970s, two Organization of Petroleum Exporting Countries (OPEC) embargoes resulted in rapid increases in the price of crude oil and therefore motor fuels. These increases triggered programs designed to develop alternative sources of fuels such as coal (qv), oil shale (qv), and natural gas (see FUELS, SYNTHETIC; GAS, NATURAL). In the 1990s, as a result of lower price volatility and improved energy efficiencies, the inflation adjusted cost of driving is about one-half of what it was in the 1960s (3,4). Alternative fuels are more important for the potential to reduce emissions and improve air quality than for securing energy self-sufficiency (see also AIR POLLUTION; EXHAUST CONTROL, AUTOMOTIVE).

General Aspects of Manufacture of Motor Fuels

All motor fuel in the United States is manufactured by private companies. Many of these are vertically integrated. That is, the same company finds the crude oil or buys it from a producing government, refines it into finished products, and then sells to independent retailers who specialize in that company's blended products or sells at company operated service stations. There are also a significant number of companies that participate in only some aspects of the business cycle such as refining or marketing.

Four groups are involved in the production or use of motor fuels in the United States: (1) manufacturers of the vehicles; (2) manufacturers and/or marketers of the fuels; (3) purchasers and users of fuels and vehicles; and (4) federal, state, and local regulatory agencies (qv). Each has a different role.

Vehicle manufacturers must build cars and trucks that operate well on available fuels. They also specify the fuel requirements of their vehicles. Fuel marketers must produce fuels that operate in both new and old vehicles. The consumer, the purchaser of the fuel and the vehicle, wants the fuel to be affordable, readily available, and able to provide a high level of performance. Regulators check that fuels are labeled properly and meaningfully and that no unwarranted claims are made. They also regulate emissions from vehicles as well as the fuels used.

All four groups, or stakeholders, must work together to guarantee that fuels and vehicles are well matched. The American Society of Testing and Materials (ASTM) was founded in 1902 to promote just such a need for all products. ASTM Committee D-2 provides a forum for regulators, vehicle manufacturers, fuel producers, and consumers to develop and recommend nonbinding standards for petroleum products. Among the types of ASTM standards are Specifications and

Test methods. Specifications define a precise set of requirements to be satisfied by a material. Test methods define procedures for measuring qualities and characteristics of a material. Specifications for gasoline are contained in ASTM D4814 (5) and for automotive diesel fuel in ASTM D975 (6).

ASTM committees must be balanced in that the number of voting producers must not be greater than the number of voting nonproducers. For petroleum products, nonproducers are regulators, consumers, and equipment manufacturers. Committee chairs must be nonproducers. Although standards must be approved by a majority, all negative votes must be carefully considered and a response made. In practice, because all issues are fully discussed and the discussions are based on hard data, very few negative votes are cast when standards are submitted for final approval.

Although ASTM specifies certain quality levels, there are a number of factors that contribute to other quality levels in the marketplace. At times, government regulations are more restrictive than ASTM specifications, especially with respect to environmental issues. Secondly, competitive forces may encourage companies to provide fuel quality that is better than that defined by ASTM. Thirdly, ASTM specifications do not have the force of law, and certain companies may decide to exceed or not meet their recommended values. In response to this last factor, some states have adopted ASTM fuel quality specifications as state regulations, thus forcing a minimum quality level in the field.

GASOLINE

Gasoline, the preferred fuel for the Otto engine, evolved from an unwanted by-product to an indispensable mainstay of modern life in only a few short years. As recently as 1900, gasoline was little more than a disposal problem for the infant petroleum industry in the production of stove and lamp oils, lubricating oils, and greases. The first prototype four-stroke, internal combustion, spark-ignited engine was built by Nicolaus Otto in 1876 in Germany. This engine, which ran on gasoline, provided enormous advantages over earlier internal combustion designs in terms of weight, power, and efficiency. By 1890, almost 50,000 of these engines had been sold. As the popularity of these engines grew at an exponential rate, the demand for gasoline grew accordingly. Gasoline demand grew by a factor of five between 1907 and 1915.

Figure 1 shows the growth in gasoline demand in the United States since 1948 (7). Demand is largely determined by the growth in the number of cars, the kilometers of paved roads available for driving, population, and economic growth. Figure 1 also shows the number of cars and the average yearly distance driven per car (8–12). Most Americans drive to work, to shop, and for recreation. In 1992, American consumers burned 422×10^9 L of gasoline. The average family car is driven 15,000 km/yr and Figure 1 shows that has not changed much since 1948. Despite stated goals of reducing the number of kilometers that people drive to work through increased use of mass transit and car-pooling, the public's preference for personal driving has remained constant.

A primary factor in moderating the demand for gasoline has been the dramatic improvement in automotive fuel economy since the mid-1970s. Figure 2

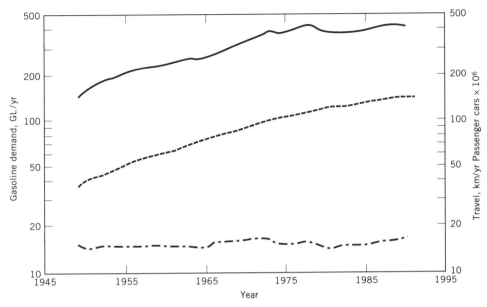

Fig. 1. Growth in annual (——) gasoline demand, in GL/yr; (————) millions of passenger cars; and (— · —) travel per vehicle, in km/yr.

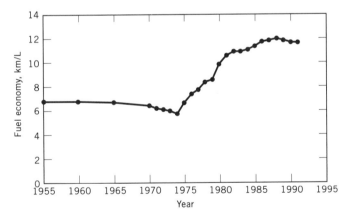

Fig. 2. Fleet average new car fuel economy.

shows the trend in new car fuel economy since 1955 (13). Improvement is the result of governmental regulations that were a direct outcome of the oil shortages of the 1970s.

Economic predictions for future gasoline demand are somewhat divided. Some economists argue that the number of cars have reached saturation levels, that the population is aging, and that future cars should have better fuel economy. These factors suggest that gasoline demand should decline over the next 15–20 years. Others believe that vehicle numbers and use should continue to increase and that fuel economy is not expected to improve. This scenario predicts gasoline demand growth. Projections of future gasoline demand range from an increase of 1.1% per year to a decrease of 0.4% per year over the time frame of 1990–2010 (14).

Gasoline demand in other industrialized regions follows a pattern similar to that in the United States. In western Europe, for example, gasoline consumption grew at an annual rate of 1.1% in the 1980s. Over the same period, consumption in nonindustrialized nations grew by 3.4% annually, reflecting economies that are less mature, less saturated with cars and drivers, and growing at a much faster rate (15).

Requirements of Good Gasoline

To satisfy high performance automotive engines, gasoline must meet exacting specifications, some of which are varied according to location and based on temperatures or altitudes. The fuel must evaporate easily and burn completely when the spark plug fires in each cylinder. Early detonation of the fuel in the cylinder can cause destructive engine knock. The fuel must be chemically stable. It should not form gums or other polymeric deposit precursors. There should be no particulate contaminants or entrained water. Contaminants must be prevented from the point of manufacture in the refinery all the way through the distribution system until the fuel is metered from the vehicle tank into the engine.

Octane. Octane is probably the single most recognized measure of gasoline quality. The octane value of gasolines are posted on service station dispensers, and most drivers recognize a fuel in which the octane is too low. Broadly speaking, octane is a measure of the combustion characteristics of gasoline. In an Otto cycle engine, gasoline vapor in the combustion chamber must start to burn only when the spark plug ignites after the charge has been compressed. Then it must burn with a well-defined flame front that travels all the way across the combustion chamber. Low octane gasoline has a tendency to preignite. As the flame front sweeps across the combustion chamber, the unburned portion of the fuel (end gas) heats up under rapidly rising temperature and pressure conditions. The fuel–air mixture undergoes chemical reactions which may cause it to autoignite and detonate the entire remaining mixture. Instead of being pushed down smoothly on the power stroke, the piston is given a hard instantaneous rap to which it cannot respond because of the large mechanical inertia present in the crankshaft and other pistons. This rapid energy release causes pressure fluctuations in the cylinder which result in a loud metallic noise commonly called knock. In addition to producing an objectionable sound, knock reduces the amount of useful work that can be extracted from the engine. Power is dissipated in the pressure waves and increased heat is radiated to the cylinder walls and into the cooling water. Under extreme conditions of prolonged knock, overheating and even engine damage can occur. The damage is typically caused by catastrophic melting of piston crowns or head gaskets.

Knock is caused by unwanted chemical reactions in the combustion chamber. These reactions are a function of the specific chemical species which make up the fuel and the environmental conditions to which the fuel is subjected during the compression and power stroke in the engine. Therefore, both the chemical makeup of the fuel and the engine design parameters must be considered when trying to understand knock.

Chemical Factors.   Because knock is caused by chemical reactions in the engine, it is reasonable to assume that chemical structure plays an important role in determining the resistance of a particular compound to knock. Reactions that produce knock are generally free-radical chain-type reactions which are different from those that occur in the body of the flame; the former occur at lower temperatures and are called cool flame reactions. Peroxy and hydroperoxy radicals play important roles in the knock process. A number of good reviews have discussed the details of the chemical mechanisms (16). Ignition delay (tau) has also been used for description of the chemical tendency to knock (17). The chemical factors affecting knock are

Change in structure	Knocking tendency
longer paraffin chains	+
isomerizing normal paraffins	−
aromatizing normal paraffins	−
alkylating aromatics	−
saturating aromatic rings	+

A lengthy and systematic investigation into the cause and prevention of knock, culminated in the use of tetraethyllead [*78-00-2*] (TEL) to decrease the knocking tendency of fuels dramatically. Other alkyllead compounds such as tetramethyllead [*75-74-1*] and dimethyldiethyllead [*1762-27-2*] were found to be effective as well. Lead alkyls function as antiknocks by decomposing in the combustion process to form lead oxides which terminate the free-radical chain mechanisms leading to knock. Lead additive packages contain, in addition to alkyllead compounds, scavengers such as ethylene dichloride [*107-06-2*] and ethylene dibromide [*106-93-4*]. These compounds react with lead oxide to form volatile lead halides which exit in the exhaust and do not form harmful lead deposits in the combustion chamber. The problem with lead additives is the presence of lead in the exhaust and the consequent potential health effects on the population, especially children, living near roadways. This exposure to lead and a second environmental issue involving catalytic convertors have led to a worldwide trend away from leaded gasoline.

Starting in 1975, in order to meet exhaust emission standards, most cars in the United States were built using noble metal-based catalytic converters. These converters reduce emissions of hydrocarbons (qv), carbon monoxide (qv), and oxides of nitrogen to levels that are less than 10% of those emitted in the 1960s. Even small amounts of lead in gasoline permanently poison a noble metal catalyst. Therefore, beginning in 1974, gasoline marketers were required to sell unleaded gasoline. As the number of cars which could use leaded fuel decreased, the market share of leaded gasoline dropped. In 1986, the Environmental Protection Agency (EPA) limited the lead content in leaded gasoline to 0.026 g/L. As of this writing, very few service stations sell leaded gasoline owing to its limited demand.

In 1957, Ethyl Corp. announced a new antiknock compound, methylcyclopentadienylmanganese tricarbonyl [*12108-13-3*] (MMT). MMT is almost as effective as lead on a per gram of metal basis, but because manganese was more expensive than lead, MMT was not widely used until limits were placed on the lead content of gasoline. MMT was used in unleaded fuel between 1975 and 1978. After a large fleet test suggested that MMT could increase exhaust emissions because it interfered with catalysts and oxygen sensors, EPA banned its use in unleaded fuel in 1978. MMT is used in Canada in unleaded fuel.

Vehicle Factors.   Because knock is a chemical reaction, it is sensitive to temperature and reaction time. Temperature can in turn be affected either by external factors such as the wall temperature or by the amount of heat released in the combustion process itself, which is directly related to the density of the fuel–air mixture. A vehicle factor which increases charge density, combustion chamber temperatures, or available reaction time promotes the tendency to knock. Engine operating and design factors which affect the tendency to produce knocking are

Increased knock	Decreased knock
higher compression ratio	increased turbulence
advanced spark schedule	exhaust gas recycle
higher coolant temperature	cooled air charge
turbocharging or supercharging	high altitudes
combustion chamber deposits	high humidity

Generally, anything which increases the efficiency or power output of the engine increases the tendency to knock. Higher compression ratio increases the initial temperature of the fuel–air charge and therefore the peak combustion temperatures. Combustion chamber deposits are highly insulating and reduce the amount of heat rejected to the coolant, thereby increasing combustion chamber temperatures. Advancing the time in the cycle at which the spark plug fires provides more time for the end gas to undergo preknock reactions. One of the easiest fixes for a knocking engine is to retard the spark timing. The engine loses a little power but the knocking can almost always be eliminated. In a similar manner, designing the combustion chamber to promote shorter burn times can effectively reduce knocking tendencies. Putting the spark plug near the center of the combustion chamber is one way. Another is to promote turbulence during burning thus increasing flame speed.

Ambient conditions can also affect knocking tendency. At high altitudes, air density is lower and the combustion of the fuel air charge generates less heat. High humidity dilutes the oxygen with water vapor and also interferes with some of the knock reactions leading to lower knocking tendencies.

Measuring Octane.   Two different values need to be considered when discussing octane measurements. One is the knocking tendency of the fuel, called the fuel octane number. The other is the knocking tendency of the vehicle, called octane number requirement.

It is impractical to measure a fuel's octane number in a full sized vehicle. Thus laboratory techniques have been developed through ASTM to measure these numbers in well-defined engines. The standard knock engine is a single cylinder cooperative fuels research (CFR) engine. This engine has a compression ratio which can be varied between 3 and 30. It also has four carburetor bowls so that reference and test fuels can be switched easily. It is equipped with a magneto-restrictive type detonation sensor that measures the rapid change in pressure characteristic of knock. Two primary reference fuels (PRFs) have been defined: 2,2,4-trimethylpentane [540-84-1], commonly referred to as isooctane, has an octane number of 100 and n-heptane [142-82-5] has an octane number of zero. Octane numbers of primary reference fuels between zero and 100 are formed by volumetric mixtures of isooctane and n-heptane. For example, a 50:50 mixture of the two has an octane number of 50. Primary reference fuels for measuring octane numbers above 100 contain specified amounts of tetraethyllead. For instance, isooctane containing 1 g/L of TEL has a defined octane number of 108.

The octane value of an unknown fuel sample is determined by comparing its knocking tendency to various primary reference fuels. Its measured octane is equal to the octane of the PRF which has the same knocking intensity. Knock intensity is controlled to an average value by varying the compression ratio of the CFR engine. In practice, the exact value of a fuel's octane number is determined to the nearest 0.1 octane number by interpolation from two PRFs that are no more than two octane numbers apart.

The CFR engine is operated at two conditions to simulate typical on-road driving conditions. The less severe condition measures research octane number (RON); the more severe one measures motor octane number (MON). Table 1 summarizes the operating conditions for each test.

The difference between RON and MON for a particular fuel is called the sensitivity. By definition, the RON and MON of the primary reference fuels are the same and the sensitivity is zero. For all other fuels, the sensitivity is almost always greater than zero. Generally, paraffins have low sensitivities whereas olefins and aromatics have sensitivities ranging up to 10 and higher.

The octane numbers of many pure compounds have been measured and reported in the literature. Probably the most comprehensive project was carried out under the auspices of the American Petroleum Institute (18). Table 2 lists RON and MON values for a number of representative compounds. Some aromatic compounds cannot be tested neat in the knock engine, so these are evaluated at levels of 20%, and the equivalent octane number is calculated. The values for oxygenates in Table 2 have been reported elsewhere (19).

Table 1. Octane Test Operating Conditions

Parameter	Research octane number (ASTM D2699)	Motor octane number (ASTM D2700)
engine speed, rpm	600	900
inlet air temperature, °C	51.7	38
mixture temperature, °C		149
spark advance,[a] °BTDC[b]	13	14–26

[a]Spark advance for motor method is a function of compression ratio.
[b]BTDC = before top dead center.

Table 2. Octane Numbers of Pure Compounds[a]

Compound	CAS Registry Number	RON	MON
Paraffins			
n-pentane	[109-66-0]	61.8	63.2
n-hexane	[110-54-3]	24.8	26.0
n-heptane	[142-82-5]	0	0
2-methylbutane	[78-78-4]	92.3	90.3
2-methylpentane	[107-83-5]	73.4	73.5
2-methylhexane	[591-76-4]	42.4	46.4
Olefins			
2-methyl-2-butene	[513-35-9]	97.3	85.5
1-pentene	[109-67-1]	90.9	77.1
1-hexene	[592-41-6]	76.4	63.4
1-heptene	[592-76-7]	54.5	50.7
Aromatics[b]			
toluene	[108-88-3]	124	112
o-xylene	[95-47-6]	120	102
ethylbenzene	[100-41-4]	124	107
propylbenzene	[103-65-1]	127	129
Oxygenates			
ethanol	[64-17-5]	130	96
methyl t-butyl ether (MTBE)	[1634-04-4]	118	100
ethyl t-butyl ether (ETBE)	[637-92-3]	118	102
methyl t-amyl ether (TAME)	[994-05-8]	111	98

[a]Refs. 18 and 19.
[b]Aromatics were measured in a 20% blend with a 60/40 mixture of 2,2,4-trimethylpentane and n-heptane.

These two test methods and the octane numbers of the fuels measured are ultimately used to evaluate the performance of vehicles. The two methods provide useful information about how fuels perform in cars. The RON corresponds to light load, low speed conditions, whereas the MON corresponds to heavier loads, and high speed severe driving conditions.

The octane number requirement (ONR) of a car is the octane number which causes barely audible, ie, trace knock when driven by a trained rater. The Coordinating Research Council (CRC), a research organization funded jointly by the American Petroleum Institute (API) and the American Automobile Manufacturers Association (AAMA), has defined test procedures for measuring ONR. Each car is driven under a set of light and heavy accelerations until the most sensitive driving mode is determined. Then a series of fuels is run in the car until trace knock is determined. Each year, CRC members measure ONR of more than 100 cars and publish the results.

It has long been known that the design of a car affects the way in which it responds to RON and MON. The response of a car to RON and MON can be characterized by a parameter called severity, which is a measure of its relative response to RON and MON. In the 1930s, most cars responded only to RON. That is, it didn't really matter what the MON of a fuel was. As long as the RON was above a minimum value, the car would not knock. Cars of the 1990s generally have a severity of about 1, meaning that a 1 number change in RON can be offset by a corresponding and opposite change in MON to provide equivalent knocking protection. Octane numbers which are posted on gasoline dispensers in all states show the antiknock index which is the average of the RON and MON values of the gasoline being sold.

In designing engines, automotive manufacturers must decide what kind of fuel that engine should use, from low to high octane. Compression ratio is the primary determining factor in a vehicle's octane requirements. High performance, high compressions ratio engines need high octane or premium fuel. It is also known that the octane requirement of a vehicle increases as a vehicle ages. This phenomena, known as octane requirement increase (ORI), is caused by the buildup of fuel and lubricant based deposits in the combustion chamber. These deposits are highly insulating and increase the combustion wall temperature, thereby increasing the tendency to knock. Generally, ORI reaches an equilibrium value of 6–9 octane numbers over the first 24–32,000 kilometers of operation. Increase in octane requirements as high as 12 are not uncommon. In some cases, as cars age and start to burn more oil, ORI can increase again when cars have 100,000+ kilometers on the odometer.

Over the years, engine compression ratios have undergone a number of changes. These increased rapidly through the 1950s and 1960s. The advent of unleaded gasoline and the demise of lead as a cheap source of octane, brought about a drop in fuel octane numbers and compression ratios. Starting in the early 1980s, both compression ratios and average octane numbers started to increase again as a result of newer refining capacity coming onstream and customer demand for more powerful and more fuel efficient cars.

Auto manufacturers have developed knock detecting and limiting devices called knock sensors which are installed in many high octane requirement engines. The sensor is a piezoelectric crystal which is tuned to the natural frequency of engine knock sound (see PIEZOELECTRICS; SENSORS). When knock is detected, a signal is sent to the on-board computer which retards the spark timing. In this way, knocking noise can be almost totally eliminated. The tradeoff, however, is reduced power. Full throttle acceleration times can be increased by up to 10% when these cars use low octane fuel. For these cars, customers value octane as a power enhancer instead of a noise reducer.

The particular fuel octane number which a gasoline marketer chooses to produce and sell is determined by a number of factors including the cost of producing the various levels of octane, the number of customers requiring the various octane levels, and competitive offerings. Determining customer requirements is the most difficult. The annual ONR survey by CRC only measures vehicle ONR using trained raters. The level of knock and octane requirement heard by an average customer is well below that considered to be trace knock by a CRC trained rater. This customer/rater delta has been the subject of a number of studies (20,21) and the average difference has been estimated to be between 3–5 octane numbers. Typically, a marketer decides that the premium grade should satisfy a certain fraction of customers, usually well over 90%. The octane values for the regular and/or intermediate grades are determined in a similar manner.

The optimum level of gasoline octane is a complex function of a number of factors involving the customer as well as the automotive and petroleum industries. Increasing octane values may require significant investment and higher operating costs on the part of the oil industry that could raise the price of gasoline to the consumer. On the other hand, vehicle manufacturers could take advantage of the availability of higher octane gasolines by raising the compression ratios of their engines which could in turn increase fuel economy. Higher compression ratio engines may cost more to build, but the consumer would save in fuel costs over the life of the car. Unless crude oil prices are considerably higher than $125/m^3 ($20/bbl), it is not economically attractive to increase compression ratio.

Volatility. The properties of a gasoline which control its ability to evaporate are critical to good operation of a vehicle. In an Otto cycle engine, the fuel must be in the vapor state for combustion to take place. The volatility or vaporization characteristics of a gasoline are defined by three ASTM tests: Reid vapor pressure (RVP), the distillation curve, and the vapor/liquid ratio (V/L) at a given temperature, ASTM D323, D86, and D2533, respectively. These three tests are sufficient to describe the volatility related characteristics of gasoline and predict its driveability performance. Moreover, these are the specifications that industry and government use to define volatility characteristics. Proper choice of these parameters ensures that cars operate well under all ambient conditions.

RVP is a vapor pressure measurement at a fixed air/liquid ratio of 4 and a temperature of 38°C. It is measured under conditions of water saturation. For samples which contain water-soluble components such as alcohols, ASTM D4953 is used.

The distillation (qv) test, D86, is a batch distillation using a specified heating rate. A 100-mL sample is distilled and the vapor temperatures at which various percentages of the sample have condensed are recorded. The data are reported as the temperature at which 10, 20, 30%, etc, of the sample has been evaporated. The amount of heavy material which doesn't evaporate is measured at the end of the test and the amount of light material which was not condensed is calculated by difference from the total condensate and the residue. The D86 test approximates a one-plate distillation with limited recycle. It has been compared to the kind of distillation the gasoline undergoes in a carburetor and intake manifold. For the more detailed description of the volatility characteristics needed to design and run a refinery there are other tools which can be used to generate true boiling point (TBP) curves. TBP are essentially distillation curves run at high efficiency and reflux. One method uses a glass distillation column with 15 theoretical plates and a reflux ratio of 5. Another uses a gas chromatography (gc) technique to identify most or all of the individual compounds in the sample and then uses a computer-based chemical library to mathematically construct a distillation curve using known or estimated vapor pressures, fugacities, etc.

Distillation data may be expressed in two ways: the percent evaporated at a given temperature (E_{xxx}); or the temperature for a given percent evaporation (T_{yy}). Because E_{xxx} values blend linearly, these are generally preferred by refiners and blenders. Gasoline performance specifications have been reported in both ways. ASTM specifications generally prefer the T_{yy} format.

The vapor/liquid ratio tests measure the amount of vapor formed from a given volume of liquid at a given temperature at atmospheric pressure. A common measure used in specifying gasoline is the temperature at which the vapor/liquid ratio is 20 ($T_{V/L = 20}$). Although V/L can be measured experimentally, it is a difficult and time consuming test to carry out, and techniques have been developed to calculate it from RVP and D86 values.

When designing fuel volatility targets, gasoline blenders must strike a balance between various driveability performance characteristics. Driveability refers to the ability of a car to start easily, accelerate and idle smoothly, and respond to changes in throttle position as expected. Too much volatility can cause as many problems as too little volatility. Targets must be matched to local ambient temperature conditions. As for other performance features, a good match between vehicle and fuel design is important to proper performance. Volatility requirements, like octane, are a strong function of vehicle design. A great deal of data on the volatility requirements of new cars is collected in cooperative programs run by CRC. Trained raters drive cars under strictly controlled conditions and evaluate the driveability performance on a range of fuels. Customers are generally less sensitive to problems than the raters. Each company decides which level of protection to provide. ASTM also uses these data to set target levels for the various fuel parameters.

Startability. In order to achieve combustion in an Otto cycle engine, the air/fuel ratio in the combustion chamber must be near the stoichiometric ratio. Unfortunately, when the engine is first started, the walls of the combustion chamber and the intake manifold are not hot enough to vaporize much fuel. Therefore, the vehicle is designed to meter extra fuel and less air to the engine upon start up so that there is adequate vapor in the engine to support combustion. From the fuel's perspective, a fuel which is easily vaporized and which contains a large percentage of light compounds such as butane is desirable to achieve good starting. Too much butane can be as much of a problem as too little, however. The ability of a fuel to achieve good starting can be correlated with RVP and a measure of the front end of the distillation curve, either E70 or T_{10}. Usually, minimum levels of RVP such as 60 kPa (0.6 bar) and E70 minimum of 10% (or a T_{10} maximum of 60°C) are satisfactory for good startability in most winter locations (22). Obviously, fuel specifications change with ambient temperatures. At higher temperatures, lower RVP and front-end volatilities are adequate to provide good starting characteristics.

Vapor Lock. At the other end of the spectrum from starting is vapor lock, a problem of too much volatility. Vapor lock occurs when too much of the fuel evaporates and either starves the engine for fuel or provides too much fuel to the engine. It occurs on days that are warmer than usual and when the car has reached full operating temperatures. A typical situation would be when a car is stuck in heavy traffic; the engine reaches high temperatures and the fuel reaches its boiling point. If the fuel pump is located near the engine, the suction side pulls only vapors and the engine becomes starved for fuel. Fuel-injected cars have fuel pumps located in the tank and the fuel lines are under positive pressure making them less prone but not totally immune to vapor lock. In carbureted vehicles, boiling of the fuel can occur in the carburetor bowl. This may dry out the bowl and stop the engine. It can also, interestingly enough, cause the same problem, ie, stopping the car, by causing too much fuel to flow to the engine. If there is foaming fuel in the carburetor bowl, the float sinks, causing the bowl to overfill and feed excess fuel into the carburetor main metering jet. The air/fuel ratio becomes overly rich and stalls the car.

Vehicle manufacturers minimize these problems by keeping the fuel system cool and under positive pressure. Fuel manufacturers minimize the problems by seasonal volatility blending. As might be expected, the front end of the gasoline controls vapor lock. The temperature to obtain a V/L ratio of 20 is referred to in ASTM D4814 as a way to control the vapor locking tendencies of gasoline. Another commonly used parameter is a combination of RVP and a measure of the front-end volatility such as E70 according to the formula:

$$\text{RVP} + n\text{E70}$$

This expression is known as the vapor lock index (VLI) or the front-end volatility index (FEVI). The value of n for U.S. cars is generally reported as 9 when RVP is in kPa (0.13 when pressure is in psi) (23). The maximum level of VLI is set by month and by region according to the ninetieth percentile daily maximum temperature.

Warm-Up. Warm-up refers to that period of operation beginning immediately after the car has started and continuing until the engine has reached normal operating temperatures, usually after 10 minutes or so of operation. During this period, the vehicle designer wants to get the vehicle equivalence ratio to stoichiometric as soon as possible to minimize emissions. On the other hand, if the mixture is leaned out too soon, the car experiences poor driveability during the warmup period. Fuel system design is critical during this period. With single point fuel metering systems such as carburetors and throttle body injectors, there is generally liquid fuel on the walls of the intake manifold during warm-up. If this liquid only reaches the cylinders in bursts the vehicle may experience unwanted surges. Manufacturers employ techniques to use the exhaust heat to rapidly heat up the walls of the intake manifold. Multipoint fuel injectors minimize many of these problems but make the scheduling of the fuel pulses critical.

From the fuel's perspective, the middle of the distillation curve plays the largest role in achieving good warm-up performance. Under the vehicle operating regime, the front end of the fuel totally evaporates. The back end of the fuel, or heaviest molecules, have trouble evaporating yet. The molecules boiling between about 100 and 150°C are the most important. Here again, the actual levels required are a strong function of temperature. The critical periods are usually the Spring and Fall, when temperatures are between 0 and 15°C. At very cold temperatures, the RVP and front-end volatility are so high that there is adequate fuel vaporization. Also, because cold weather emission control has not been a principal concern, vehicles can be calibrated to maintain rich air/fuel ratios longer than in warmer weather. Similarly, in hot weather, the engine heats up faster and maldistribution is not as much of a problem.

The most common expression for controlling driveability is known as the driveability index (DI), which has the form:

$$DI = 1.5T_{10} + 3T_{50} + T_{90}$$

It is generally felt that fuels which have values of DI below 570 when T is in °C (1200 when T is in °F) provide good warm-up driveability performance.

Icing. At temperatures within 5°C of freezing and under conditions of high humidity, ice can form in the intake system of vehicles with carburetors or throttle body fuel injectors. When the gasoline in these systems evaporates, it cools the incoming air because of the latent heat of vaporization. If the air is near its saturation point, ice crystals can form on the throttle plate, on the choke plate, or in the venturi throat. The ice can cause rough idle if it accumulates around the edges of the throttle plate when it is nearly closed. In the extreme, ice can clog the carburetor jets and stall the car completely. After the car is fully warmed-up there is generally enough heat in the intake system to prevent any ice buildup. The vehicle manufacturer minimizes icing problems by rapid heating of the intake system, which also helps lower emissions, and through the use of multipoint fuel injectors.

On the fuel side icing tendencies can be reduced by proper volatility blending and, if necessary, by the use of additives. The volatility parameters which control icing are the portions of the fuel evaporating below 100°C. If there is too much material in this boiling range, then the temperature drop caused by evaporation is large and icing might occur. Alternatively, fuel producers can add a few percent of a compound such as isopropyl alcohol, which depresses the freezing point of water, or a surfactant, which prevents ice crystals from sticking to and building up on metal surfaces. In the United States, very few cars on the road exhibit icing tendencies. Since 1972, manufacturers have been using heated intake manifolds in order to help reduce emissions of hydrocarbons and carbon monoxide. The intake manifold is heated by passing the exhaust manifold underneath. This causes more heat to be forced into the intake manifold and limits the ability of ice to form. More recently, the rapid proliferation of port fuel injection has almost eliminated this concern from newer cars.

Back-End Volatility. The portion of the gasoline that boils above 150°C is referred to as the back end. Molecules in this region have high energy density and provide a significant contribution to fuel economy. Too much material in this boiling range, however, can cause problems. This material is hard to volatilize and when the engine is cold, tends to accumulate on the walls of the cylinder. From there it can be washed into the oil sump and dilute the oil. Generally, as the engine heats up this material evaporates. However, if there are too many back ends in the gasoline, then not all may boil off and the performance of the lubricant may be degraded. Very heavy molecules, such as those having more than 12 carbon atoms, may contribute to combustion chamber deposits. Condensed ring aromatics are particularly effective contributors to these deposits (24). It has been reported that these same compounds reduce octane requirement increase associated with long-term gasoline use (25).

Vehicle volatility requirements are a strong function of ambient temperatures. ASTM has defined five volatility classes based on expected minimum and maximum daily temperatures. These classes and their ranges are shown in Table 3. Each month, each state is assigned a volatility class, depending on its temperature history. Table 3 also shows the ASTM volatility specifications for each class.

Cleanliness. Good gasoline must be both chemically and physically clean. Chemical cleanliness means that it does not contain nor react under conditions of storage and use to form unwanted by-products such as gums, sludge, and deposits. Chemical cleanliness is assured by controlling the hydrocarbon composition and by appropriate additives. Physical cleanliness means that there are no undissolved solids or large amounts of free water in the gasoline.

After production in the refinery, gasoline may undergo two types of oxidative degradation. The first occurs under ambient temperature conditions and long periods of time, as long as six months, and is reflected in storage stability. The second has an impact on high temperature stability and is a phenomenon of the high temperatures and short residence times encountered as the fuel makes its way through the vehicle intake system. Both mechanisms share some similar chemical features in that they involve oxidation of fuel compounds and free-radical polymerization to form high molecular weight gums. The gums and gum precursors formed under storage conditions can lead to damaging deposits in the carburetor, high pressure injectors, or intake valves.

Compounds associated with poor stability are olefins in general and conjugated diolefins in particular. A compound such as 1,3-cyclopentadiene is known to be particularly detrimental. Compounds having poor stability can be produced in cracking processes such as thermal cracking, including visbreaking and steam cracking, and catalytic cracking. The presence of compounds such as sulfur and nitrogen can also degrade stability and promote gum formation. Vehicle manufacturers know that copper promotes oxidation of the fuel and are careful to minimize the use of copper in the fuel system materials.

A number of laboratory tests are used to predict chemical stability. The amount of existent gum in a gasoline is determined by ASTM D381. This method involves evaporating a sample by a jet of heated air. The residue is weighed, solubles are extracted with n-heptane, and the sample is reweighed. The total is called unwashed gum and the insoluble portion is called existent gum. ASTM D4814 specifies that gasoline should contain less than 5 mg/100 mL of existent gum. Most gasolines contain less than 3 mg/100 mL.

Table 3. ASTM Volatility Specifications

Class	Ambient temperature, °C Min[a]	Max[b]	RVP, kPa	T_{10}, °C	T_{50}, °C	T_{90}, °C	End point, °C
A	>16	≥43	62	70	77–121	190	225
B	>10	<43	69	65	77–118	190	225
C	>4	<36	79	60	77–116	185	225
D	>−7	<29	93	55	77–113	185	225
E	≤−7	<21	103	50	77–110	185	225

[a]Tenth percentile 6-h minimum daily temperature.
[b]Ninetieth percentile maximum daily temperature.

Although ASTM D381 measures the gum existing in gasoline at a particular point in time, it does not indicate how much more might be formed during storage at the refinery, during various modes of transportation, in service station tanks, or in vehicle tanks. The oxidative stability test (ASTM D525) was developed to provide a rapid means of predicting future gum formation. This method consists of placing a sample of the gasoline in a bomb, pressurizing it to 690 kPa (100 psi) with oxygen, and maintaining it at a temperature of 100°C. The pressure inside the bomb is monitored continuously. The oxygen reacts with the gasoline slowly at first, but eventually the pressure drops sharply, 14 kPa (2 psi) or more within a 15 minute interval, indicating a breakpoint. The time from the start of the test to the breakpoint is the induction period and is a measure of the oxidative stability of the gasoline. ASTM D4814 specifies that gasoline induction period must exceed 240 minutes, although most good gasolines have induction periods in excess of 960 minutes, the duration of the test.

Another ASTM test method, Potential Gum (D873), combines the existent gum and the oxidation stability tests to measure potential gum. A sample of gasoline is subjected to the oxidation stability test for 960 min, filtered to remove particulates, and then subjected to an existent gum test. The potential gum is expressed as the total (unwashed) gum in this test.

Other tests to predict stability of gasoline have been developed and reported in the literature. One, developed by the U.S. military, stores gasoline at elevated (43°C) temperatures for up to 12 weeks and measures existent gum at the end of that period (26). Another measures existent gum in the presence of copper. The copper catalyzes oxidation and may be a better estimator of the stability of gasoline at high temperature/low residence time conditions.

Laboratory simulation tests have been developed which attempt to provide a better simulation of engine conditions than the analytical tests for predicting high temperature stability and intake system deposits, especially for gasolines containing additives. A test rig has been described (27,28) which could be used in a laboratory to predict the deposits formed in a carburetor and on intake valves. Such a test could be used for certification of fuels or for comparing the potency of competing additives. Nevertheless, these tests have not proven to be accurate measures of the deposit forming tendencies of fuels, and vehicle tests must always be carried out.

In summary, a number of tests are used to predict the chemical stability of gasolines. None, however, does an adequate job of predicting either storage stability or deposit forming tendencies. This may be a reflection of the various regimes that the fuel undergoes between production and combustion. Even in the intake system itself, the time/temperature history of a fuel in a carburetor is much different than it is in a fuel injector and on an intake valve. There is no *a priori* reason why one test should be expected to correlate with all of these cases. For this reason the best tests are not short laboratory tests but long expensive ones such as long-term storage, and engine/vehicle-based deposit tests for carburetor, fuel injector, and intake valve deposit formation. Much more fundamental work on the mechanisms of deposit formation must be carried out before short tests can be expected to be successful.

Other Requirements.   There are a number of other specific features that must be present (or absent) in gasoline. Some of these are achieved by blending or manufacturing; others are achieved through the use of additives. The sulfur content should be below 1000 ppm and the sulfur should not be present as foul smelling mercaptans. The gasoline should not promote rust in pipeline, service station tanks, or vehicle metals. It should be clear and bright in appearance and should not pick up water either as a haze or as an emulsion. Free water in gasoline contributes to rust and corrosion. Gasoline should not contain more than trace amounts of carbonyls which can dissolve elastomeric seals and diaphragms, nor should it contain phosphorus bearing additives which can damage sensitive catalytic converters.

Manufacture

Crude oil can be easily separated into its principal products, ie, gasoline, distillate fuels, and residual fuels, by simple distillation. However, neither the amounts nor quality of these natural products matches demand. The refining industry has devoted considerable research and engineering effort as well as financial resources to convert naturally occurring molecules into acceptable fuels. Industry's main challenge has been to devise new ways to meet the tremendous demand for gasoline without, at the same time, overproducing other petroleum products.

Distillation. Petroleum refining begins with the distillation of crude oil into a number of different fractions. In many cases, two distillations are carried out in units called pipestills. The first is at atmospheric pressure and temperatures up to 400°C. The second distillation is carried out at reduced pressure and fractionates the heaviest material from the atmospheric pipestill. Table 4 shows typical boiling ranges for the various crude oil fractions and typical yields from Arab Light, a common crude oil. The yield and properties of these straight-run or virgin fractions are a function of the type of crude oil being distilled. Virgin naphtha can be used as gasoline, except that its octane value is very low (78 RON/75 MON). When lead could be used, it was possible to add enough lead to bring the octane up to acceptable levels (>90 RON/MON). Whereas the potential yield of gasoline directly from crude oil is less than 20%, the demand is about 50%. The heavy material must be converted to lighter material and the octane of many of the existing streams improved by changing chemical composition.

Catalytic Cracking. Although it has long been known that heating crude oil fractions could break or crack the compounds into smaller molecules, the de-

Table 4. Properties of Crude Oil Fractions[a]

Fraction	Boiling range, °C	Yield, %
gas	<0	<1
virgin naphtha		
light	0–100	18
heavy	100–200	18
gas oil/kerosene	200–400	33
residue	>400	48

[a]From Arab Light crude.

velopment of suitable catalysts and processing designs has made catalytic cracking the premier refinery process for changing the molecular structure of the crude (see CATALYSIS). Catalytic cracking generates higher yields than thermal cracking as well as superior quality products. As of this writing, over 50% of the gasoline in the United States is obtained by catalytic cracking which uses a fluidized bed of powdered or small diameter catalysts that are continuously regenerated in an adjacent vessel called a regenerator (see CATALYSTS, REGENERATION; FLUIDIZATION). The fluidized-bed catalytic cracking (FCC) process was first commercialized in 1942 by Standard Oil Co. (New Jersey) at its Baton Rouge refinery (29) using powdered silica/alumina as a catalyst. Fluidized beds offered the ability to use powdered, high surface area catalysts, to operate continuously, to regenerate easily, and to use short contact times which increased yield and selectivity. Since 1942, many improvements have been made in process design, catalyst formulation, and the ability to handle heavier feeds (30–32). Catalysts of the 1990s are zeolites having highly controlled pore size and surface area (see MOLECULAR SIEVES).

The principal class of reactions in the FCC process converts high boiling, low octane normal paraffins to lower boiling, higher octane olefins, naphthenes (cycloparaffins), and aromatics. FCC naphtha is almost always fractionated into two or three streams. Typical properties are shown in Table 5. Properties of specific streams depend on the catalyst, design and operating conditions of the unit, and the crude properties.

Two undesirable aspects of FCC naphtha quality are that it may contain unacceptably high amounts of foul smelling mercaptans, and that its thermal stability may be too low. Mercaptans are usually found in the light FCC naphtha and may be removed or converted to sulfides and disulfides by a sweetening process such as Merox, developed by UOP. Thermal stability is improved in sweetening processes through removal of cresylic and naphthenic acids. It may be further improved by clay treating and by addition of oxidation inhibitors such as phenylene diamine.

Thermal Cracking. Certain cracking conversion processes are carried out without catalysts. Heavy residuum streams in the refinery can be cracked thermally to produce coke and a mixture of lighter products. If naphtha is produced, it may require extensive treating before it can be used directly in gasoline or used as feed to other processes. Depending on the specific process conditions and design these processes are known as delayed coking, visbreaking, Fluid Coking, and

Table 5. Properties of FCC Naphtha

Property	FCC naphtha		
	Light	Intermediate	Heavy
boiling range, °C	<105	105–160	160–220
RON	91	88	91
MON	79	77	81
aromatics, vol %	10	35	65
olefins, vol %	60	20	15

Flexicoking (the last two developed by Exxon). Heavy distillate may also be thermally cracked using high (~800°C) temperature steam. This process, steam cracking, is used to generate olefins for use in chemicals plants, but also generates material in the naphtha range which, if the quality is appropriate, may be used in gasoline.

Reforming. Catalytic reforming is a process to increase the octane of gasoline components. The feed to a reforming process is naphtha (usually virgin naphtha) boiling in the 80–210°C range. The catalysts are platinum on alumina, normally with small amounts of other metals such as rhenium. Because the catalysts contain noble metals, the feed must have very low levels of sulfur, nitrogen, and heavy metals such as lead and arsenic. Reactors may operate in continuous or semicontinuous mode, the difference being the method in which the catalyst is regenerated. Various designs have been developed by companies such as UOP (Platforming and CCR), Exxon (Powerforming), and Amoco (Ultraforming).

Depending on the catalysts and operating conditions, the following types of reactions occur to a greater or lesser extent: (*1*) heavy paraffins lose hydrogen and form aromatic rings; (*2*) cycloparaffins lose hydrogen to form corresponding aromatics; (*3*) straight-chain paraffins rearrange to form isomers; and (*4*) heavy paraffins are hydrocracked to form lighter paraffins.

Reformers generate highly aromatic, high octane product streams, and a great deal of hydrogen (qv). Reformate can have RON values over 100 and MON values of 90, but does not respond to the addition of lead as much as some other streams that are less aromatic. Because of the removal of lead from gasoline, reforming has become a much more important process. The hydrogen, which can be used to improve the quality of many other refinery streams, is an extremely valuable product of the reformer.

Some of the negative aspects of reformate are production of benzene, polynuclear or multiring aromatics (PNAs), and light gas (C_1–C_4) (see also BTX PROCESSING). Benzene is a recognized carcinogen and its concentration in gasoline is regulated in the United States, whereas PNAs can contribute to combustion chamber deposits. PNAs may be removed by distilling the whole reformate and discarding the heaviest fractions.

Alkylation. Alkylation (qv) is the chemical combination of two light hydrocarbon molecules to form a heavier one and involves the reaction of butenes in the presence of a strong acid catalyst such as sulfuric or hydrofluoric acid. The product is a heavier multibranched isoparaffin. Propene and the various pentenes may also be used, to produce C_7 or C_9 isoparaffins, respectively. Process designers need to be careful not to let three molecules combine to form C_{12} molecules because these have too high a molecular weight and probably would contribute to poor combustion and higher emissions. Octane numbers for alkylate are usually above 95 for both RON and MON, and the reaction has the effect of lowering the RVP of the available pool, a plus when refiners are trying to reduce it. The principal competitors to alkylation are the various etherification processes which also use C_4 and C_5 olefins as feeds (see also OLEFINS, LIGHT).

There are environmental concerns over the use of HF catalyst. The refining industry has taken steps to reduce the likelihood of an accidental release and to minimize the environmental impact in the event of a release. As a result of these environmental concerns, most new units use sulfuric acid catalysts.

Isomerization. Isomerization is a catalytic process which converts normal paraffins to isoparaffins. The feed is usually light virgin naphtha and the catalyst platinum on an alumina or zeolite base. Octanes may be increased by over 30 numbers when normal pentane and normal hexane are isomerized. Another beneficial reaction that occurs is that any benzene in the feed is converted to cyclohexane. Although isomerization produces high quality blendstocks, it is also used to produce feeds for alkylation and etherification processes. Normal butane, which is generally in excess in the refinery slate because of RVP concerns, can be isomerized and then converted to alkylate or to methyl *tert*-butyl ether (MTBE) with a small increase in octane and a large decrease in RVP.

Hydrogen Processing. Hydrogen is probably the most valuable refinery chemical in terms of its ability to improve the quality of refinery streams. It can be used to remove unwanted species such as sulfur and nitrogen. Hydrogen treating reduces the carbon-forming potential of diesel fuels and improves the ability to resist oxidative degradation. Under more severe conditions and with the proper catalysts, hydrogen can saturate olefinic and aromatic bonds. The largest producer of hydrogen in a typical refinery is the naphtha reformer, but hydrogen can also be produced by reforming methane into CO_2, CO, and H_2.

Blending Agents. Blending agents are components of gasoline that are used at levels up to 20% and which are not natural components of crude oil. As of this writing, all blending agents are oxygenated compounds such as ethers (qv) and alcohols and are used for one or more of a variety of reasons, including to increase the octane of the fuel, to reduce vehicle emissions, and/or to use renewable resources to reduce dependency on imported crude oil.

Ethers, such as MTBE and methyl *tert*-amyl ether (TAME) are made by a catalytic process from methanol (qv) and the corresponding isomeric olefin. These ethers have excellent octane values and compete on an economic basis with alkylation for inclusion in gasoline. Another ether, ethyl *tert*-butyl ether (ETBE) is made from ethanol (qv) and isobutylene (see BUTYLENES). The cost and economic driving forces to use ETBE vs MTBE or TAME are a function of the raw material costs and any tax incentives that may be provided because of the ethanol that is used to produce it.

Alcohols such as methanol, ethanol, and *tert*-butyl alcohol [75-64-9] (TBA) $C_4H_{10}O$, have also been used as gasoline blending agents. Methanol has limited miscibility in gasoline, and in the presence of even small amounts of water, the gasoline–methanol mixture separates into two phases. The two phases are unacceptable as fuel. In order to use methanol safely, it may be mixed with equal amounts of higher molecular weight alcohols having four or more carbon numbers. Common alcohols which have been proposed for this purpose are *sec*-butyl alcohol [78-92-2] and TBA. These C_4 alcohols have also been used by themselves. As of this writing, most interest in methanol is as a neat or near neat fuel. However, this use requires a modified or redesigned engine. Higher alcohols are generally not used because of cost. The only alcohol used in large volumes is biomass-derived ethanol (see CHEMURGY; FUELS FROM BIOMASS). When ethanol is used as a gasoline blending agent at a level of 10 vol %, the mixture is known as gasohol. Gasohol actually contains only 9.5 vol % pure ethanol because the ethanol must be denatured using 5% unleaded gasoline before it can legally be shipped anywhere. As of this writing, gasohol receives a 5.4 ¢/gal (1.4 ¢/L) federal tax subsidy in the United States. Some states, mostly farm states, also provide additional tax rebates for blenders of ethanol. These incentives significantly improve the economics of ethanol use.

Ethanol is more miscible with gasoline than methanol, but ethanol gasoline mixtures also have a tendency to separate into two phases in the presence of water. Additionally, ethanol is corrosive in its own right. Therefore, pipeline companies have refused to allow ethanol blends to be transported through the normal pipeline distribution system (see PIPELINES). Sales of gasohol are limited to areas where the ethanol can be transported in trucks or tank cars and can be blended at terminals. Despite this limitation, gasohol has achieved a significant market share. In 1991, gasohol had attained 7% of the total U.S. market. In large farm states, such as Illinois and Nebraska, gasohol sales were 27 and 46%, respectively, of the total gasoline market (33).

According to Section 211 of the Clean Air Act, EPA must approve use of any new blending agent in gasoline. "New" means that the agent is not "substantially similar" to unleaded gasoline used to certify 1974 vehicles. In 1981, EPA ruled that gasolines containing up to 2 wt % oxygen were substantially similar (34). All aliphatic ethers and alcohols could be used, except that methanol had to be blended with equal parts of higher alcohols. In 1991, EPA increased the approved level of oxygenates to 2.7 wt % oxygen for aliphatic ethers and alcohols, except methanol (35). Other waivers granted by EPA are listed in Table 6. For each of these waivers, a significant amount of data were generated to convince EPA and the auto manufacturers that the specified blending agent would not harm any of the vehicle components and would not increase emissions. Once a waiver is granted, all producers and marketers are free to use the material.

Another factor which must be considered when blending oxygenates is the effect of a large amount of a single component on the volatility and therefore the driveability characteristics of the fuel blend. In addition to desired cutpoints, the volatility curve of a good fuel should be smooth and not have any bumps or flat spots. Figure 3 shows the distillation curves for three fuels, one made from hydrocarbons, one containing 15% MTBE, and one containing 10% ethanol. The oxygenated fuels have distinct flat spots which may cause driveability problems in sensitive vehicles.

Table 6. Blending Agents Having EPA Section 211 Waivers

Compound	Max oxygen, wt %	Max oxygenate, vol %	Date waived
ethanol	3.5	10	1979, 1982
tert-butyl alcohol	3.5	15.7	1981
methanol + TBA (1:1)	3.5	9.4	1981
5% methanol (max) + 2.5% cosolvent (min)	3.7	[a]	1985[b,c]

[a]Varies with type of cosolvent.
[b]Ethanol, C_3, C_4 alcohols, and a corrosion inhibitor.
[c]Ethanol, C_3–C_8 alcohols, and a corrosion inhibitor. Date waived, 1988.

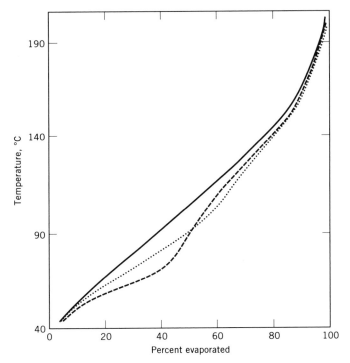

Fig. 3. Distillation curves for (——) nonoxygenated, and oxygenated, ie, (·····) 15% MTBE (2.7% oxygen), and (————) 10% ethanol (3.7% oxygen), fuels.

Additives. Gasoline additives are used to improve the performance of the fuel either because the hydrocarbon components themselves contain some deficiency or because it is more effective to add a small amount of additive than to change the composition of the gasoline. Lead antiknock additives were a good case in point. Octane can be increased either by processing steps such as reforming, which requires significant investment and increased operating costs, or by using additives which is considerably less expensive. Additives are added in parts per million levels to distinguish them from blending agents which are added in the percents.

Dyes. Dyes are added to gasoline to impart color for a number of reasons. Originally, these compounds were used to identify leaded gasoline so that it would not be used for other inappropriate purposes, such as solvents. Dyes are used to identify different gasoline grades so that pipeline companies can separate tenders, and so that service stations can easily check that the correct grade was placed in the underground tanks (aq). Dyes are usually based on azo chemistry and are added in concentrations below 10 ppm (see AZO DYES).

Antioxidants. Antioxidants (qv) are used in gasoline to improve storage stability. Poor storage stability may sometimes be traced to high concentrations of olefins, especially conjugated diolefins. In many cases, an antioxidant is added to the FCC naphtha in the sweetening process. The antioxidant may be added in large enough amounts so that the blended gasoline contains the proper concentration. Antioxidants are almost always one of two chemical types: phenylenedi-amines or hindered phenols (qv). These were first used in gasoline around 1930 (36) by Universal Oil Products.

Metal Deactivators. Small amounts of metals in gasoline can contribute to enhanced rates of oxidation. Copper is especially potent in this regard. In order to prevent metal catalyzed oxidation, chelating agents (qv) are often added to gasoline. The most commonly used additive is N,N'-disalicylidene-1,2-propane-diamine [94-91-7], $C_{17}H_{18}N_2O_2$ (37) at concentrations up to 10 ppm. Special care must be taken if any of the refinery streams are treated with a copper-based process. Automobile manufacturers sometimes use higher levels of metal deacti-vators in factory fill gasoline, because this gasoline may sit in vehicle tanks for a long period of time until the vehicle is sold.

Corrosion Inhibitors. Corrosion-causing free water is almost always present in at least small amounts in all parts of the gasoline distribution system. It can come about when hot, water saturated gasoline cools in storage tanks or pipelines. Salt water is found in ship holds, and rainwater is found in underground tanks. Acidic or caustic water may be carried over inadvertently from various processes in the refinery. Careful monitoring of service station storage tanks is an important part of good operating practice to prevent unwanted water from contaminating customers' vehicle tanks (see CORROSION AND CORROSION INHIBITORS).

All corrosion inhibitors in use as of this writing are oil-soluble surfactants (qv) which consist of a hydrophobic hydrocarbon backbone and a hydrophilic func-tional group. Oil-soluble surfactant-type additives were first used in 1946 by the Sinclair Oil Co. (38). Most corrosion inhibitors are carboxylic acids (qv), amines, or amine salts (39), depending on the types of water bottoms encountered in the whole distribution system. The wrong choice of inhibitors can lead to unwanted reactions. For instance, use of an acidic corrosion inhibitor when the water bot-toms are caustic can result in the formation of insoluble salts that can plug filters in the distribution system or in customers' vehicles. Because these additives form a strongly adsorbed impervious film at the metal liquid interface, low liquid con-centrations are usually adequate. Concentrations typically range up to 5 ppm. In many situations, pipeline companies add their own corrosion inhibitors on top of that added by refiners.

Rust protection is measured using any one of a number of tests: National Association of Corrosion Engineers (NACE) rust test (40), ASTM D665, and MIL-I-25017 B/C (41). These all involve immersing a steel spindle in a mixture of gasoline and water for a specified time and evaluating the resulting corrosion. ASTM 4914 lists the Copper Strip Corrosion Test which protects against the cor-rosive effects of fuel sulfur on copper.

Antiicing Additives. Additives to control icing are rarely used in the United States because vehicle design changes have made icing problems almost non-existent.

Detergent Additives. Deposits may build up in many parts of an engine: carburetor, fuel injector (throttle body and port), intake manifold hot spot, intake valves, and the combustion chamber itself. Both fuel and vehicle design contribute to deposit formation. Proper choice of additive or additives can prevent deposit buildup and in some cases can clean up existing deposits. The exact mechanisms of deposit formation are poorly understood. Moreover, the fuel undergoes a different time/temperature history in each engine location leading to different types of deposits. An additive that is effective in controlling one type of deposit may not work at all, or may even contribute to increased deposits, in another engine location. For instance, solvent oils can be particularly effective in preventing the buildup of intake valve deposits, but too much solvent oil causes excessive combustion chamber deposits and leads to high ORI.

Carburetor Detergents. Carburetors were the most common form of fuel metering device in automobiles until the mid-1980s when these were replaced to a large extent by throttle-body and port fuel injectors. Deposits can form on and around the throttle plate and adjacent throat area. Deposits in this region can cause poor fuel distribution and atomization at idle and low speed operation because the deposits interfere with air flow past the throttle. Deposits in the throat can clog air bleeds or idle jets and can upset the desired stoichiometry in the engine. The deposits are thought to be caused by the oxidation of unstable or partially oxidized fuel components reacting with materials found in the crankcase blowby and the exhaust gases which are recycled to control NO_x emissions. The deposit-forming tendency of fuels is determined using a bench engine which is run under strictly controlled conditions of temperature, humidity, crankcase blowby, and exhaust gas recycle (EGR) (42).

The first carburetor detergent used in gasoline was introduced by Standard Oil Co. of California in 1954 (43). It was an amide derived from a fatty acid and a polyamine (44) and effectively prevented deposits from forming on the throttle plate and in the fuel ports of the carburetors. It was discovered that carburetor detergents could also provide protection against icing and provide antirust protection. In 1972, Esso introduced a detergent package which, in addition to providing carburetor detergency, antiicing protection, and corrosion inhibition, also improved fuel distribution among the cylinders (45). A mixture of nonpolymeric amines derived from beef tallow was found to coat the intake manifold with a low surface energy film, creating small liquid droplets and promoting the rapid entrainment of the liquid film always present in the manifold walls. It was reported that driveability and fuel economy improved, whereas emissions were reduced. With the advent of catalysts starting in 1975, auto manufacturers stopped using such lean mixtures and maldistribution problems prevalent in the late 1960s and early 1970s disappeared.

Fuel Injector Detergents. Fuel injectors generally experience higher temperatures than carburetors, around 100°C. Deposits can accumulate in the small annulus in the tip of the injector through which fuel must flow. As the annulus becomes plugged, fuel flow becomes erratic, atomization degrades, and control of stoichiometry becomes more difficult. Although well-designed injectors can tolerate plugging levels of up to 30%, variability among injectors can cause problems even at much lower levels. This is because the vehicle computer controls stoichiometry for all cylinders on average. If one cylinder is running rich, then the other cylinders run lean so that average stoichiometry is correct. Plugged injectors lead to poor driveability and higher emissions.

Although the exact chemical mechanism which leads to injector deposits has not been identified, it is generally believed that heavy hydrocarbons oxidize and polymerize during vehicle hot soak. The resulting gums and resins trap other particles present in the crankcase blowby and in the EGR. Olefins and diolefins are known to be significant contributors, whereas the role of sulfur is still unresolved (46–48). Polar compounds have been shown to contribute to injector deposits. In one experiment, a dirty deposit forming fuel was clay filtered and the resulting filtrate had very low deposit-forming tendencies (49).

A number of additives have been developed which not only prevent injector deposits from forming, but which also remove deposits already in place. The best additives can unclog clogged injectors in less than one tankful of gasoline. Polybutene succinimides, polyether amines, and certain lower molecular weight amines have been shown to be effective injector detergents (50). Phenylenediamine, which is known to improve the oxidative storage stability of gasoline, does not prevent the reactions which lead to fuel injector deposits.

Intake Valve Detergents.   In the mid-1980s some vehicle manufacturers started to experience levels of intake valve deposits which caused driveability degradation, especially before the engine was fully warmed up. These problems were associated with deposit levels lower than those which had caused problems in the past. It has been hypothesized that the deposits act as hydrocarbon sponges, especially during acceleration when the mixture is usually enriched somewhat in order to provide adequate power. If the deposits, because of physical and chemical makeup, absorb some of the extra fuel that is passing by, the vehicle may stumble and hesitate because the mixture is too lean. Cars built to low emission standards generally have the stoichiometry set as lean as possible, and any upset which leans out the mixture even further is not well tolerated (51).

Intake valve deposits (IVD) have been studied extensively and a number of vehicle design and fuel factors have been determined which are principal contributors. As for other deposit-forming problems, the mechanism of deposit formation is not known in much detail, but the broad outlines are fairly well understood. The primary difference between injector deposits and intake valve deposits is the temperature regime. Intake valves generally operate at a temperature as high as 300°C (52), much higher than injectors. The hot soak portion of the driving cycle is not as important for development of IVD as it is for injector deposits. Also, there is a small but constant flow of lubricating oil down the stem of the intake valve which may influence deposit formation.

Vehicle factors that contribute to IVD are typically related to whether or not the valve is kept wet by liquid gasoline during operation. Design factors which encourage wetting of the valve tend to minimize deposits. Valve rotation, which rotates the valves slightly on each cycle, also tends to reduce deposits. Valves that do not rotate develop deposits on the back side which sees little fuel from the injectors. Oil flow down the valves can either help or hurt. If the oil flow is too low, the valve may stay dry, whereas oil flow that is too high encourages buildup of oil related deposits. Injector spray pattern is also important. An even, continuous spray is the best for minimizing deposits, whereas intermittent injectors tend to promote higher deposit levels.

Fuel factors which contribute to IVD have been shown to include olefins (53,54), alcohols, and cracked stocks (55).

Additive packages have been developed which do an excellent job of preventing IVD. The key to effective operation is to keep the valve wet so that the additive can prevent deposit buildup. Most packages include a combination of detergent/dispersant and a carrier oil or heavy solvent. If no carrier oil is present, then the fuel may evaporate off the valve too rapidly for the package to be effective. When the valves do not rotate, the portion of the valve which has the highest deposit level is the back side which is not constantly wet.

Additive effectiveness is measured a number of ways, but the most meaningful measures involve expensive vehicle testing. The first widespread test used a BMW 318i which was run on the road for 16,000 km, after which the engine head was removed and each valve weighed. More recently, the CRC has defined a test using a Ford engine (56). No laboratory simulation test has been shown to be a good predictor of IVD and additive effectiveness. Most gasoline sold in the United States contains additives which control IVD.

Demulsifiers. Because free water is present in many parts of the distribution system, special care must be taken to prevent emulsions (qv) from forming. Emulsions generally refer to thick oil in water emulsions that are very stable and form a separate phase at the fuel–water interface. Emulsions do not burn very well, and they always carry large quantities of dirt and rust particles with them. These particles plug vehicle fuel filters and cause a car to stop running. Haze is a dilute water-in-oil emulsion in which the water droplets are dispersed throughout the gasoline but which does not form a separate phase. Haze and emulsions may form during high shear conditions such as pumping, or during low shear conditions such as sloshing in the hold of a ship or barge. Gasoline without additives does not generally form emulsions, but the use of surfactants to control deposit formation and limit corrosivity can lead to gasoline having very severe emulsion forming behavior. Demulsifiers are highly surface active chemicals or mixtures having limited solubility in both fuel and water (57). The hydrophilic portion may be polyethylene or polypropylene and the hydrophobic portion may be a long-chain alkylphenol or alcohol. The specific demulsifiers chosen depend on the base gasoline, the rest of the additive package, and the type of water encountered in the distribution system. Demulsifiers may have to work for water bottoms that range from pure rainwater, to salty seawater, to caustic water bottoms (see DISPERSANTS).

Although there is no official ASTM test to measure the water handling properties of gasoline, there are a number of widely used industry tests. One uses a Waring Blender to simulate high shear (57), whereas another uses a wrist action shaker to simulate mild shear conditions (58).

Blending and Distribution

When blending gasoline from its components, refinery operators must balance a number of factors in the most economical way. First, the components must be used at the same rate at which these are produced or else the refinery either runs out of material or drowns in excess components. Secondly, each gasoline fuel grade must be produced to the specifications set by marketers and regulators. The specifications should not be exceeded in a way which increases manufacturing cost. Finally, blend targets must take into account the fact that the gasoline might not be sold immediately and may travel in pipelines, barges, or tankers and then sit in a distribution terminal. The time between production in the refinery and sale to the customer can be as long as one month.

Most refineries accomplish the difficult task of blending through the use of sophisticated linear programming algorithms. The linear programs are run in a few different time scales ranging from weekly to yearly to help refineries plan their seasonal operations. In the winter, gasoline demand is down and heating oil demand is up; in the summer, gasoline demand is at its peak.

Many gasoline properties, especially octane and RVP, do not blend linearly. Whereas the volatility properties of blends can be well approximated using non-ideal properties, it has not been possible to predict the octane values of blends from first principles. It is known that when various refinery streams are blended together, the octane of the blend is usually different from the volume weighted average octane of the individual components. This deviation can be either positive or negative. For instance, blending alkylate and reformate almost always results in an octane number that is lower than the weighted average of the individual components. This is known as an octane blending debit. Likewise, mixing FCC naphtha and virgin naphtha almost always results in a blended octane higher than expected, or an octane blending bonus (59). Experiments using blending agents show that the addition of most agents is nonlinear and depends strongly on the composition of the base gasoline.

Proper prediction of the octane of refinery blends is important because octane has traditionally been one of the most expensive gasoline properties and raising pool octane often entails significant investment and increased operating costs. Also, it is possible to meet targets for the different grades by properly choosing blend stocks to take advantage of octane bonuses available from the nonlinear blending characteristics.

Blending behavior of a binary mixture may be characterized by a linear blending value (LBV). Figure 4 shows the response curve of a hypothetical two-component mixture. The LBV for each of the components at any composition f_c is defined by the tangent at that point according to the formula

$$LBV_1 = ON_c + (1 - f_1)(dON/df_1)_c$$
$$LBV_2 = ON_c - f_1(dON/df_1)_c$$

where f_1 is the weight fraction of component 1, $(dON/df)_c$ is the slope of the blending curve at composition c, and ON_c is the octane number of the blend at composition c.

As can be seen from Figure 4, LBVs for these components are not constant across the ranges of composition. An interaction model has been proposed (60) which assumes that the lack of linearity results from the interaction of pairs of components. An approach which focuses on the difference between the weighted linear average of the components and the actual octane number of the blend (bonus or debit) has also been developed (61). The independent variables in this type of model are statistical functions (averages, variances, etc) of blend properties such as octane, olefins, aromatics, and sulfur. The general statistical problem has been analyzed (62) and the two approaches have been shown to be theoretically similar though computationally different.

Most of the octane blending values reported in the literature use a slight variation on this theoretically sound approach. The composition and octane of the

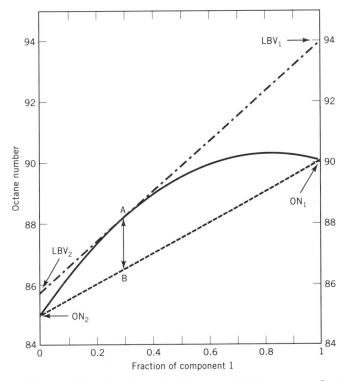

Fig. 4. Octane blending behavior where (——) represents the measured octane response curve, (—·—) the tangent to the curve, and (————) the linear connection between the octane number of components 1 and 2. The line AB corresponds to the octane blending bonus of this composition.

base fuel are assumed to be fixed and the second component is assumed to be added. Using the same nomenclature, the blending octane number (BON) of component 2 is defined as

$$BON_2 = (ON_c - ON_1 \cdot f_1)/f_2$$

The advantage of this definition is that it does not depend on measuring the tangent of the response curve, although the variation in the value of the blending octane number is greater. Typically, BONs are measured at an 80/20 mixture. This technique is also useful when trying to measure the octane of a compound such as butane or methanol that is difficult or impossible to measure in its pure state.

Most refineries develop individual octane blending equations which do a good job of predicting that refinery's blending behavior. In order to use these equations in refinery planning and operations, these may be linearized in a piecewise fashion.

Until the 1960s, most gasoline blending was done in batches. All the components were stored in large tanks, a recipe was defined by a master blender, and an empty tank was filled according to the recipe. After the blend was finished, a sample was sent to the laboratory for testing. If all properties were within specifications, the blend was released. If not, certain components would be added to fix the blend properties. In the extreme, a blend could be downgraded from premium to regular, it could be totally reblended, or it could be sold at a discount to a wholesale customer willing to use off-spec material. In the 1990s, this task is accomplished through in-line blending. Computers define preliminary recipes. As the blend is being made, special instruments sample the run-down pipe and measure critical properties such as octane and volatility every 20 minutes. If the integrated values of the blend are not within specifications, then the blend recipe is adjusted by the computer. Thus, it is very unusual to require reblending or post-blend adjustments. In a large refinery, an average blend might be 15 × 10^6 L, having an average flow rate though the blender of 10^6 L/h. As a result of the large number of repeat tests conducted on the blend as it is made, the properties of the blends may be known to a high degree of statistical confidence, and the blends may be pumped directly into a pipeline or a barge.

Gasoline blends are shipped from the refinery to a storage terminal. From the storage terminal, the gasoline is shipped by tank truck to individual service stations. The trucks, which have 40,000 L capacity, have 4–5 compartments so that they can deliver different grades at the same time. Service station tanks have a capacity of 12,000–15,000 L and are buried underground. Submerged turbine pumps transfer the gasoline from the tanks to dispensers at the dispensing islands. Tanks have been made of a number of materials, although the most popular is reinforced fiber glass. As a result of environmental concerns about gasoline leakage from underground tanks, the tanks have double-wall construction and have leak detectors between the two walls.

Fuel Economy

Fuel economy, typically expressed as distance driven per volume of fuel consumed, ie, in km/L (mi/gal), is measured over two driving cycles specified by the Federal Test Procedure (63). One cycle simulates city driving and consists of relatively low speed (~32 km/h) driving, and includes a portion where the car starts after having equilibrated at ambient conditions for 16 hours. The second simulates highway driving conditions and includes higher speeds and fewer starts, stops, and accelerations. The statutory fuel economy standards are based on a harmonic average of the city and highway tests assuming that 45% of distance is accumulated under highway conditions and 55% is accumulated under city driving conditions.

Fuel economy is measured using a carbon balance method calculation. The carbon content of the exhaust is calculated by adding up the carbon monoxide (qv), carbon dioxide (qv), and unburned hydrocarbons (qv) concentrations. Then using the percent carbon in the fuel, a volumetric fuel economy is calculated. If the heating value of the fuel is known, an energy specific fuel economy in units such as km/MJ can be calculated as well.

The most important fuel property in terms of volumetric fuel economy is the heat content of the fuel. The fuel economy in modern vehicles responds linearly and with a slope of almost unity to changes in energy content of the fuel. In turn, fuel energy content is most influenced by density and oxygen content. Density is positively correlated with heat content whereas oxygen is negatively correlated. Figure 5 shows the relationship between energy content and fuel economy for a group of twenty 1989 vehicles tested on a wide range of gasolines with and without oxygen (64).

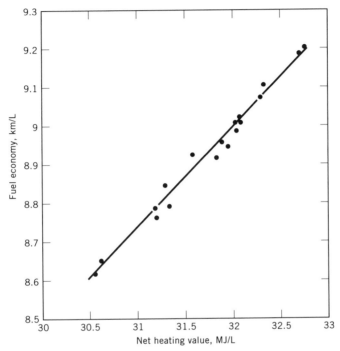

Fig. 5. Fuel economy vs energy content for various fuels. Each data point represents the average of twenty 1989 vehicles. To convert MJ/L to Btu/gal, multiply by 2.78×10^8.

Emissions

As a result of atmospheric pollution levels that exceed the National Ambient Air Quality Standards (NAAQS) in many parts of the United States, both the federal government and the State of California have implemented standards for exhaust and evaporative emissions from new vehicles (see EXHAUST CONTROL, AUTO-MOTIVE). The first of these standards went into effect in 1968 and mandated that the vapors from the vehicle crankcase be routed back through the engine and burned. Since then, the standards have continued to grow stricter. Table 7 shows the federal exhaust emission standards and Table 8 shows the more stringent California standards. California has mandated that starting in 1998 a certain percentage of new vehicles sales must be zero emissions vehicles (ZEV).

These ZEV vehicles are envisioned to be electric battery vehicles (see BATTERIES, SECONDARY CELLS–OTHER). The California fleet average standards

Table 7. Federal Light-Duty Exhaust Emission Standards

Year	Emissions, g/km[a]			Automotive hardware changes
	HC[b]	CO	NO$_x$	
precontrol	9.4	56	3.9	
1970	2.5	21		lean combustion
1972	1.8	17		
1973			1.9	exhaust gas recycle
1975	0.94	9.4		oxidation catalysts
1977			1.2	
1980	0.25	4.4		
1981		2.1	0.6	three-way catalysts, oxygen sensors
1994[c]	0.16[d]	2.1	0.25	
2003[e]	0.078[d]	1.1	0.12	

[a]To convert g/km to g/mi, multiply by 1.609.
[b]HC = hydrocarbons.
[c]Standards phase in over three years.
[d]Nonmethane hydrocarbons.
[e]If necessary and technically feasible.

Table 8. California Light-Duty Vehicle Exhaust Emission Standards

Year	Fleet average NMOG,[a] g/km	ZEV[b] sales restrictions
1994	0.155	
1995	0.144	0
1996	0.141	0
1997	0.126	0
1998	0.098	>2%
1999	0.071	>2%
2000	0.046	>2%
2001	0.044	>5%
2002	0.042	>5%
2003	0.039	>10%

[a]NMOG = nonmethane organic gases, ie, the total mass of exhaust hydrocarbon and oxygenated compounds, excluding methane.
[b]ZEV = zero emissions vehicle.

are to be met by selling combinations of various classes of vehicles. Manufacturers must certify their new cars in one of the categories shown. The sales weighted average for a given year must then be at or below the standard shown in Table 8.

The emissions standards have forced changes in vehicle hardware as shown in Table 7. Many of these changes in hardware have resulted in changes in fuel as well. For example, starting in 1975, vehicle manufacturers installed noble metal catalysts in order to meet federal standards and because the catalysts could not tolerate lead, all fuel manufacturers were mandated to sell at least one grade of unleaded gasoline. As demand for unleaded gasoline grew, new processing capacity was required and different blends developed.

In addition to setting standards for exhaust emissions, the government set standards for evaporative emissions. These refer to hydrocarbons that escape from the vehicle when fuel evaporates, either while the car is operating (running losses), while it is sitting and not being operated (diurnal emissions), or immediately after operation (hot soak). In order to control evaporative emissions, auto manufacturers have installed canisters of activated charcoal in their vehicles since 1972. The charcoal traps any hydrocarbon vapors from the fuel tank or from the carburetor bowl. During engine operation, the adsorbed hydrocarbons are stripped off the charcoal by a countercurrent air stream and fed into the engine and burned.

Although the charcoal canisters are about 95% effective, fuel volatility still impacts the mass of vapors that break through the canister. Therefore, EPA mandated that starting in the summer of 1992, RVP levels be reduced below the levels specified in ASTM D4814. Class C regions, generally the northern part of the country, are limited to a maximum RVP of 62 kPa (9.0 psi) vs an ASTM limit of 79 kPa (11 psi), and the southern Class B regions are limited to a maximum RVP of 54 kPa (7.8 psi) vs 69 kPa (9.0 psi) for ASTM.

The Clean Air Act Amendments of 1990 (65) introduced a new concept in emissions reduction: reducing exhaust emissions by controlling the composition of the fuel. Whereas previous regulations had lowered exhaust emissions by setting standards for new vehicles, this law mandated gasoline marketers to change gasoline composition so that emissions from existing vehicles would be reduced. Reduction targets of 15% and at least 20% were set for 1995 and 2000, respectively. Reductions are to be measured against 1990 vehicles and industry average gasoline. The reductions are for hydrocarbons (summer ozone season only), and for air toxics (year round). Air toxics are defined as the sum of the emissions of benzene [71-43-2], 1,3-butadiene [106-99-0], formaldehyde [50-00-0], acetaldehyde [75-07-0], and polycyclic organic material. Reformulated gasoline (RFG) may not result in any increase in NO_x emissions. Additionally, RFG must contain no more than 1% benzene and at least 2% oxygen.

Each refiner has the flexibility to choose the specific formulation to produce based on the economics of the individual refineries. RFG meeting the statutory requirements must be sold in the nine areas of the country which have the worst ozone (qv) problem. In addition, all other areas of the country which exceed the ozone NAAQS may elect the RFG regulations, and EPA has estimated that 40% of the nation's gasoline should be subject to RFG rules (66).

In the winter, the Clean Air Act mandates that gasoline in all areas which exceed the NAAQS for CO must contain at least 2.7% oxygen. This is based on the assumption that adding oxygen to the fuel reduces CO emissions.

Compliance with the RFG regulations is measured by using a formula to calculate the emissions from a given fuel. This formula, developed by EPA, predicts average exhaust and evaporative emissions as a function of gasoline chemical and physical parameters. In anticipation of the need to develop this equation, and because very little data were available, the three domestic auto companies and fourteen petroleum companies formed a joint research program in late 1989 called the Auto/Oil Air Quality Improvement Research Program (AQIRP). This four-year, $40 million program had the objective of "conducting a research and testing program to develop data on potential improvements in vehicle emissions and air quality, primarily ozone, from reformulated gasoline, various other alternative fuels, and developments in vehicle technology" (67).

The program developed a great deal of information on the effects of fuel properties on emissions. Table 9 lists some of the more important effects measured (68–70). RVP is the main fuel factor affecting evaporative emissions. The data collected by AQIRP represent the largest single body of information collected on this subject. Over two thousand individual emissions tests were carried out through the end of 1992. EPA used this and other data to generate an equation to predict emissions as a function of fuel properties. Final regulations were published after much consultation with the oil and automotive industries, and with consumer and environmental groups. Specific reduction targets were set for RFG, and gasoline manufacturers were given the option of complying with the regulations on every gallon or on average. In general, if averaging is used, the standard is more severe (1.5%) and the performance of any one blend may not be more than 2.5% below the regulatory level.

	Required RFG reductions, %		
Emission	1995	2000	Control season
volatile organic carbon	15.6	25.9	summer
NO_x	0	5.5	summer
toxics	15	25	year round

The state of California has taken a different conceptual approach to reducing emissions through control of gasoline composition. Instead of defining a performance target, ie, 25% reduction, the State has defined composition targets which are aimed at achieving emissions reductions. These targets, shown in Table 10

Table 9. Effect of Fuel Composition Change on Exhaust Emissions[a]

Fuel parameter	Change		Change[b] in exhaust emissions, %		
	From	To	HC	CO	NO_x
aromatics, %	45	25	-6.5 ± 1.9	-13.3 ± 3.2	2.0 ± 2.0*
MTBE, %	0	15	-5.5 ± 1.9	-11.1 ± 2.0	1.4 ± 2.1*
olefins, %	20	5	5.8 ± 2.0	1.5 ± 3.4*	-6.1 ± 1.9
T_{90}, °C	182	138	-21.7 ± 1.9	1.2 ± 3.9*	4.9 ± 2.3
sulfur, ppm	450	50	-16.1 ± 1.5	-12.9 ± 3.9	-9.0 ± 1.9
RVP, kPa[c]	60	53	-4.5 ± 3.2	-9.1 ± 6.2	-0.7 ± 4.9*

[a]Data from twenty 1989 light-duty vehicles.
[b]Effects are shown with their 95% confidence intervals. Effects that are smaller than their interval are not statistically significant and are shown with an asterisk.
[c]To convert kPa to psi, multiply by 0.145.

are to take effect in 1996. Gasoline producers must meet these targets on every liter of gasoline. If desired, a company may choose to meet yearly average targets, but these are slightly more severe than the per liter specifications, and the ranges which may be used for averagers are restricted.

Table 10. California Phase 2 Gasoline Composition

Property	Limit per liter	Values for averagers	
		Average	Cap
RVP, kPa[a]	48		48
sulfur, ppm	40	30	80
aromatics, vol %	25	22	30
olefins, vol %	6	4	10
T_{90}, °C	149	143	166
T_{50}, °C	99	93	104
oxygen, wt %	1.8–2.2		2.7
benzene, vol %	1.0	0.8	1.2

[a]To convert kPa to psi, multiply by 0.145.

Gasoline composition may also be regulated in Europe. A tripartite initiative is being carried out among the European Commission, the oil industry, and the automotive industry. Based on an analysis of the required improvements in air quality, new regulations are to be written that control vehicle emissions and fuel composition into the twenty-first century.

DIESEL FUEL

As a fuel for internal combustion engines, diesel fuel ranks second only to gasoline. In 1991, total U.S. demand for diesel fuel for transportation was 100 billion L, about 25% of the demand for gasoline (71). Use of diesel fuel for highway vehicles accounted for 67% of the total, off-highway vehicles for 12%, ships for 7%, locomotives for 10%, and military use for 4% (72). The volume of diesel fuel used relative to gasoline is expected to grow somewhat in the United States, although not as much as predicted in the 1980s. Diesel cars, thought at one time to be very promising, have encountered significant customer resistance in the United States. There are significant obstacles to overcome. Consumers do not like the diesel noise, the smoke, and the hard starting in winter. The fuel economy benefit that diesel engines can offer is not as attractive when gasoline prices are relatively low. Under the 1994 outlook, it is unlikely that there will be much growth in the use of diesel engines for passenger car use in the United States. In other countries, diesel engines have captured a much larger share of the passenger car market. In France, for example, diesel engines constituted 32% of new car passenger sales in 1990 (73), vs less than 0.1% in the United States (74).

Combustion in Diesel Engines

Unlike the spark-ignited gasoline engine, the diesel engine, first used by Rudolf Diesel in the 1890s to burn finely powdered coal dust, employs compression ignition. Liquid fuel was employed soon after. Through the action of the pistons, air alone is drawn into the cylinders and compressed. Near the end of the compression stroke, fuel is injected into the cylinder and after a short delay, is ignited by the high temperature generated during compression. The fuel must be finely atomized and well mixed with the air for complete combustion to take place. The ignition delay is useful because it allows the fuel to be injected and mixed before combustion starts. Its length depends largely on the composition of the fuel, and is described by cetane number.

Diesel engines offer a number of advantages over gasoline engines. Diesel engines are able to operate at high (up to 24 for light-duty diesels and 15–17 for heavy-duty diesels) compression ratios which improves energy efficiency. Because combustion starts at many different sites and there is no flame progressing across the combustion chamber, there are no knock concerns. The engine operates with unthrottled air flow, which also improves efficiency. Because the air flow is unthrottled, changing fuel flow results in a wide range of stoichiometry, always in the lean region. Lean operation also means that HC and CO emissions are generally low for diesels. If too much fuel is fed to the engine to boost power, visible black smoke and soot are emitted.

Diesels also suffer from a number of disadvantages relative to gasoline engines. Diesels are generally hard to start in cold weather. Many engines use glow plugs to heat up the combustion chamber so that the fuel can ignite. It used to be common for heavy-duty diesels to be left running in cold weather because they are so hard to restart. Particulate emissions from diesel engines contain materials such as benz-a-pyrene which are thought to be carcinogenic. Although diesel engine-out emissions of NO_x are not much different from those of equivalent gasoline engines, diesel engines do not have exhaust catalysts so the tailpipe emissions are much higher. Additionally, diesel engines tend to be heavier and more expensive than gasoline engines of the same power.

There are two categories of diesel engines: direct injection (DI) and indirect injection (IDI). In DI engines, the fuel is injected directly into the combustion chamber. In IDI engines, there is a small prechamber into which the fuel is injected. The fuel starts to ignite in the prechamber and the hot burning gases are forced out into the main combustion chamber through a small passage. IDI engines may operate at higher speeds and use lower pressure injector systems which tend to be less expensive. They are used mainly on passenger cars although progress has also been made in producing small, high speed DI engines for passenger car use.

Requirements for Good Diesel Fuel

Diesel fuel is used in a wide variety of vehicular engines ranging from small IDI powered passenger cars to large trucks and construction equipment. There are actually three grades of diesel fuel defined in ASTM D975, the specification for diesel fuels. The first is Grade 1-D, suitable for high speed engines which operate under widely varying conditions of speed and load. Grade 1-D also has excellent low temperature properties. Grade 2-D is a general-purpose diesel suitable for use either in automotive or nonautomotive applications. It can be used in high speed engines involving relatively high loads and uniform speeds. Grade 4-D is much more viscous and is used in low and medium speed engines having sustained loads at substantially constant speed. Most cars and trucks use 2-D, a general-purpose grade. 1-D, a more volatile, lower density, lower aromatic fuel, used in cold weather and in municipal buses.

Ignition Quality.   The ability of diesel fuel to burn with the proper characteristics is described by its cetane number, a measure of ignition delay. Excessively long ignition delays (low cetane number) cause rough engine operation, misfiring, incomplete combustion, and poor startability. Because the fuel starts to burn later in the cycle, pressure rise is more rapid without increasing the net work from the cycle. Power may be reduced and combustion may be incomplete leading to high emissions if the fuel does not have time to burn in the expansion stroke.

The procedure for measuring the cetane number of diesel fuel (ASTM D613) is similar to that used for measuring gasoline octane number. Cetane [544-76-3] (n-hexadecane), $C_{16}H_{34}$, is defined as having a cetane number of 100; α-methylnaphthalene [90-12-0], $C_{11}H_{10}$, is defined as having a cetane number of 0. 2,2,4,4,6,8,8-Heptamethylnonane [4390-04-9] (HMN), $C_{16}H_{34}$, which can be produced in high purity, is used as the low reference fuel and has a cetane number of 15. Blends of cetane and HMN represent intermediate ignition qualities according to the formula:

$$\text{cetane number} = \%\ \text{cetane} + 0.15\ (\%\ \text{HMN})$$

The cetane engine is a variable compression single cylinder engine very much like the octane engine. The engine is run at 900 rpm and injection is timed to start at 13° before top dead center (BTDC). The compression ratio is adjusted so that the test fuel starts to ignite at exactly top dead center (TDC), for an ignition delay of 13° or 2.4 ms. Reference fuels are chosen which bracket the sample and the cetane number of the sample is estimated by interpolation between the two reference fuels.

The cetane number of a fuel depends on its hydrocarbon composition. In general, normal paraffins have high cetane numbers, isoparaffins and aromatics have low cetane numbers, and olefins and cycloparaffins fall somewhere in between. Diesel fuels marketed in the United States have cetane numbers ranging between 35 and 65. Most manufacturers specify a minimum cetane number of 40–45.

Cetane number is difficult to measure experimentally. Therefore, various correlation equations have been developed to predict cetane number from fuel properties. One such equation may be found in ASTM D4737 to calculate a cetane index (CI). ASTM D975 allows use of CI as an approximation if cetane numbers are not available.

Cold Temperature Properties.   Diesel fuel must be able to be pumped and to flow through all filters and injectors at the lowest temperature that may be encountered in use. When the temperature is lowered, wax molecules in the fuel start to crystallize. This temperature is known as the wax appearance point or cloud point. These temperatures, which are generally the same, are measured by ASTM D2500 and D3117, respectively. If the temperature is lowered still further, the fuel gels and does not flow. This is the pour point and is measured by ASTM D97. These tests measure the ability of a fuel to operate in a diesel engine. Generally, the cloud point of a fuel is 4–6°C above the pour point, although fuels having differences of 11°C are not uncommon. The true operability temperature is somewhere in between the two; cloud point is too high and pour point is too low. Many engine manufacturers recommend fuels having pour points of 6°C below the lowest temperature at which the engine is expected to operate.

Some additives have the ability to lower the pour point without lowering the cloud point. A number of laboratory scale flow tests have been developed to provide a better prediction of cold temperature operability. They include the cold filter plugging point (CFPP), used primarily in Europe, and the low temperature flow test (LTFT), used primarily in the United States. Both tests measure flow through filter materials under controlled conditions of temperature, pressure, etc, and are better predictors of cold temperature performance than either cloud or pour point for additized fuels.

If the fuel temperature is below its pour point, the fuel has difficulty flowing out of the storage tank on the vehicle. Diesel powered vehicles generally do not have in-tank pumps. If, however, the fuel temperature is above its pour point but below its cloud point, the following situation may occur. The engine starts but as operation continues, wax crystals begin to collect on the fuel filter, plugging it after a few minutes and stopping the vehicle totally. Vehicle manufacturers can take a number of steps to minimize these problems. Some are relatively expensive, such as heated filters, tanks, and engine blocks. Others are relatively simple although not as effective. These include making sure that the fuel lines have no kinks or sharp bends in them and routing lines through sheltered locations.

Vehicle testing is the best way to determine low temperature requirements. These tests can be carried out in environmentally controlled facilities under strictly defined conditions or in actual use. Many companies and industry groups carry out customer tests in cold climates such as northern Canada or Finland.

Volatility. Volatile light fractions in diesel fuel help to provide easy engine starting but are generally low in cetane number and energy content. Heavy fractions, which have good cetane and energy content, can contribute to deposit formation and hard starting if present in too high concentrations. Desirable quality characteristics are obtained by careful blending of refinery streams. The temperatures at which 10, 50, and 90% of the fuel evaporate in the ASTM D86 distillation test are used as controls to provide good volatility. In general, most diesel fuels have an initial distillation temperature above 160°C and a 90% point of 290–360°C, depending on fuel grade.

Viscosity. For optimum performance of diesel engine injector pumps, the fuel should have the proper viscosity. Too low viscosity results in excessive injector wear and leakage. Viscosity that is too high may cause poor atomization of the fuel upon injection into the cylinders.

Diesel fuel kinematic viscosity is measured by ASTM D445, and is reported in units of mm^2/s at 40°C. Desired viscosity is a function of fuel grade and ranges from a minimum of 1.3 mm^2/s for 1-D to a maximum of 24 mm^2/s for 4-D.

Density. The greater the density of diesel fuel, the greater its heat content per unit volume and therefore the greater its power or fuel economy. Because diesel fuel is purchased on a volume basis, density is often stipulated in purchase specifications and measured on delivery. A common measurement of density is API gravity (ASTM D287), which is measured easily and accurately using calibrated hydrometers. The relationship between specific gravity (SG) and API gravity is

$$°API \text{ at } 15.6°C = (141.5/SG) - 131.5$$

Flash Point. The flash point of a fuel indicates the temperature below which the fuel can be handled without danger of fire. This is the temperature to which the fuel must be heated to create sufficient vapors above the surface of the liquid that they can be ignited in the presence of an ignition source. The flash point of diesel fuel is measured by ASTM D93, using a closed-cup Penske-Martens tester. The fuel is heated at a rate of 5.6°C/min and a test flame is introduced into the test chamber at 30 s intervals. The fuel temperature reached when the flame ignites the vapors is called the flash point. Specifications for flash point vary with grade; the lowest value is 38°C for grade 1-D. Controlling flash point is important in order to prevent the vapor space in storage and vehicle tanks from being in the explosive range. Setting the flash point at 38°C protects most storage vessels from exploding.

Carbon Residue. The tendency of a diesel fuel to form carbon deposits in an engine can be roughly predicted by one of two carbon residue tests: the Ramsbottom Coking Method (ASTM D524) or the Conradson Carbon Test (ASTM D189). Basically, both methods determine the amount of carbon residue left after evaporation and chemical decomposition of the fuel at elevated temperatures for a specified length of time. The Ramsbottom method involves heating the last 10% residue of the ASTM distillation at 549°C for 20 minutes. The result is reported as percent carbon on 10% residuum. For use in high speed diesel engines operating over a range of loads and speeds, ASTM specifications call for no more than 0.15% Ramsbottom carbon residue. Because ignition control additives may interfere with the tests without actually leading to deposit formation in the engine, these laboratory tests should be made on fuel before any addition of ignition control agent.

Sulfur. Sulfur in diesel fuel should be kept below set limits for both environmental and operational reasons. Operationally, high levels of sulfur can lead to high levels of corrosion and engine wear owing to emissions of SO_3 that can react with condensed water during start-up to form sulfuric acids. From an environmental perspective, sulfur burns to SO_2 and SO_3, the exact split being a function of temperature and time in the combustion chamber. SO_3 can react with water vapor in the exhaust which can further react to form sulfates, leading to high levels of particulates. Although the SO_3 fraction may be small, diesel fuel may contain high concentrations of sulfur (up to 5000 ppm) especially compared to gasoline (330 ppm average). The relationships between fuel sulfur content and total exhaust particulates has been well documented. As particulate emission standards for diesels have become more stringent, engine manufacturers have made the case that they could not meet the standards unless sulfur levels were reduced. Through negotiations between the Engine Manufacturers Association and API, it was agreed that sulfur content of highway diesel could be reduced, and EPA specified a maximum level of 0.05 wt % starting in October, 1993.

Ash Content.   The fuel injectors of diesel engines are designed to very close tolerances and are sensitive to any abrasive material in the fuel. Therefore, the maximum permissible ash content of the fuel is specified. This is measured by ASTM D482, which consists of burning a small sample of the fuel in a weighed container until all the carbonaceous matter has been consumed. The permissible amount of ash is between 0.01 and 0.1 wt % depending on the grade of diesel. Low speed engines operating at constant speed and load can tolerate higher levels of ash.

Aromatics Content.   Aromatic compounds have very poor ignition quality and, although they are not specifically limited in ASTM D975, there are practical limitations to using high aromatic levels in highway diesel fuel. In the United States, where gasoline demand represents about one-half of the crude barrel, and where heating oil demand is relatively low, there is great pressure to be able to blend aromatic FCC streams into diesel fuel. Average aromatic levels in the United States are about 30%. In addition to having poor fuel quality, aromatics also contribute to exhaust emissions. The federal government began effectively limiting aromatic content to below 40% starting in October, 1993 by specifying a minimum cetane index of 40. California limits aromatic levels below 10% beginning in the same time period, also because of emissions concerns.

Stability.   Diesel fuel can undergo unwanted oxidation reactions leading to insoluble gums and also to highly colored by-products. Discoloration is believed to be caused by oxidation of pyrroles, phenols, and thiophenols to form quinoid structures (75). Eventually, these colored bodies may increase in molecular weight to form insoluble sludge.

Gums can lead to deposits in critical injector orifices that can degrade atomization and combustion performance of the engine. The chemistry involved in the formation of these gums appears to be similar to that which occurs in gasolines. Stability is measured using ASTM D2274. In this test, a sample is heated at 95°C for 16 hours while oxygen is bubbled through the liquid. After cooling, the insoluble material is filtered and washed. The amount remaining is reported as gums.

ASTM specifications for different grades of diesel fuel are summarized in Table 11.

Diesel Fuel Manufacture

The biggest factors in determining how diesel fuel is blended in a given refinery are the availability of high cetane stocks. In order of decreasing ignition quality, the hydrocarbon types rank in the following order: normal paraffins, olefins, cycloparaffins, branched paraffins, and aromatics. Because straight-run distillates contain the greatest amount of normal paraffins and cycloparaffins and the least

Table 11. Specifications for Diesel Fuel[a]

Fuel property	Grade		
	1-D	2-D	4-D
flash point,[b] °C	38	52	55
T_{90}, °C	288, max	282–338	
kinematic viscosity, mm^2/s($=$cSt)	1.3–2.4	1.9–4.1	5.5–24.0
carbon residue,[c] %	0.15	0.35	
ash,[c] %	0.01	0.01	0.10
sulfur,[c] %	0.50	0.50	2.00
cetane number[b]	40	40	30

[a]From ASTM D975.
[b]Value given is minimum value.
[c]Value given is maximum permissible value.

amounts of branched paraffins and aromatics, these are the preferred stocks for diesel blending. Cracked stocks, which are relatively rich in aromatics, are less desirable from the standpoint of ignition quality. However, these have high energy density and good cold temperature properties. The amount of cracked material allowable in diesel fuel depends largely on the cetane number specification. In the United States, where a high level of cracking is necessary to meet gasoline demand, the large supply of cracked fractions and the relatively small supply of straight-run distillates make substantial use of cracked stocks economically necessary. This has been made possible through the use of cetane improvers to improve cetane and through the use of hydrogenation to improve stability.

Ignition Improvers. In order to meet the increasing demand for diesel fuel and to allow use of blendstocks having good low temperature properties but low cetane number, diesel manufacturers frequently use cetane improvers in the fuel. Cetane improvers work in just the opposite way that antiknock additives do. During the preignition period, the improvers generate free-radical species that promote faster onset of combustion and reduce the ignition delay properties of the base fuel. Various types of chemicals have been shown to be effective, including nitrates and nitrites, nitro and nitroso compounds, and peroxides (75). Alkyl nitrates are by far the most common and are used in concentrations up to 0.3%. Although the effectiveness varies with base fuel composition, 0.1% in a typical fuel would give an increase of five numbers (76).

Stability Improvers. Diesel fuels that contain high amounts of cracked stocks generally have poorer stability than virgin diesel. If the fuel is hydrotreated to remove sulfur, the stability is vastly improved. Traditional antioxidants such as hindered phenols are not particularly effective in preserving color although these still prevent other oxidation reactions. Stabilizers are amines or other nitrogen-containing compounds that prevent sediment formation by interfering with the oxidation reactions that occur between heteroatoms and available oxygen. These act as radical traps and/or peroxide decomposers.

Corrosion Inhibitors. The corrosion inhibitors used in diesel fuel are generally similar to those used in gasoline and, like the latter, produce an effect primarily by surface action. If amine additives are used for detergency, these may provide some corrosion protection as well.

Detergent Additives.   Diesel engine deposits are most troublesome in the fuel delivery system, ie, the fuel pump and both fuel side and combustion side of the injectors. Small clearances and high pressures mean that even small amounts of deposits have the potential to cause maldistribution and poor atomization in the combustion chamber. The same types of additives used in gasoline are used in diesel fuel. Low molecular weight amines can also provide some corrosion inhibition as well as some color stabilization. Whereas detergents have been shown to be effective in certain tests, the benefit in widespread use is not fully agreed upon (77).

Cold Flow Improvers.   The cold flow properties of diesel fuel may be improved by the use of additives or blending agents. As is the case for gasoline, additives are materials present in low concentrations (ppm), whereas blending agents are present in levels of a few percent or more. Blending agents work by diluting the concentration of wax crystal forming paraffins to a point below which these paraffins are no longer a problem. Kerosene and grade 1-D diesel fuel may be used as effective diluents. Gasoline, which also improves cold flow properties, is unsafe and dangerous to use a blending agent because it can drastically lower the flash point of the mixture.

Additives are often the most economical way to improve the cold flow properties of middle distillates. These additives have been called by a number of different names including wax crystal modifiers, mid-distillate flow improvers, and wax antisettling additives. All additives work by modifying the crystallization process in some way. Some change the shape of the wax crystals so that the crystals are less likely to plug fuel system filters. Others work by promoting nucleation of many small crystals instead of the growth of relatively few large crystals (78). Various types of chemistry have been shown to be effective in improving low temperature flow properties of diesel fuels. Examples are ethylene–vinyl ester copolymers, chlorinated hydrocarbons, and polyolefins. The exact nature of most commercially available additives is proprietary. Treatment rates and the specific additive formulation used are a strong function of the fuel composition and depend in large part on the distribution of molecular weights of the normal paraffins in the fuel and the concentration of high molecular weight paraffins (79). Fuels having a wide distribution of molecular weights are the most responsive to flow improver additives.

Three types of cold flow additives have been described (80). The first are pour point depressants, which are primarily low molecular weight ethylene–vinyl ester copolymers. These work by cocrystallizing with the paraffin wax and preventing further paraffin addition to existing crystals. The second type of additive is the cloud point depressant (CPD). CPDs are low molecular weight polymers where the paraffin segments have been designed to interact with the paraffins in the fuel to delay the onset of crystallization. The third and newest type of additive is called an operability additive (OA). OAs are multicomponent additives optimized for given fuel properties and show an improvement in the cold filter plugging point and low temperature flow tests as well as lowering the pour point.

Oxygenates.   Oxygenated materials have been considered for addition to diesel fuels for the same reasons these compounds are added to gasoline. Putting oxygen in the fuel should lead to better combustion and reduced emissions of CO and, more importantly, particulates. A number of different oxygenates have been tested, including esters made from vegetable oils (qv), eg, rapeseed methylesters (RSME); ethers (qv); and alcohols.

Diesel Environmental Regulations

Emission standards have been set for heavy-duty vehicles in much the same manner as they have been set for gasoline engines. Because heavy-duty vehicles are primarily diesels, the focus is on diesel engine emissions. Standards have been written in units of grams per brake-horsepower-hour (g/bhph) = g/kW·h × 1.34, which normalize the emissions according to the total energy output of an engine over the specified driving cycle. In contrast to light-duty vehicle testing where testing is carried out on the total vehicle, heavy-duty engines are certified in tests on an engine dynamometer. A series of accelerations is carried out and the emissions are measured. Table 12 shows U.S. emissions standards. For heavy-duty engines, the most difficult emissions standards to meet are total particulates and NO_x. When standards were relatively high, they were met by engine redesign without fuel changes. At 1994 and future levels, fuel changes have also been specified. Many studies have been published on the interactions between engine design, fuel composition, and emissions, and the conclusions are not always in agreement. It is generally agreed that sulfur contributes to particulate emissions and that lowering sulfur helps to meet the particulate standards. It is also generally agreed that in terms of engine design parameters, there is an inverse relationship between NO_x and particulates. Rapid, complete combustion reduces particulate emissions but also promotes the formation of NO_x. Results of a CRC program which measured emissions from nine fuels in a heavy-duty diesel engine showed that higher aromatics produced higher levels of both NO_x and particulates and that higher levels of cetane number produced lower levels of all four regulated

Table 12. Federal Heavy-Duty Truck Exhaust Emission Standards, g/(kW·h)[a]

Year	HC	CO	NO_x	HC + NO_x	Particulates
1970	4.89	47.4			
1974		30		11.9	
1979	1.1	18.7		7.5	
1984	0.97	11.6	8.0	3.7	
1985	0.97	11.6	8.0		
1988					0.45
1990			4.5		
1991			3.7		0.19
1994[b]					0.07
1998[b]			3.0		

[a]To convert g/(kW·h) to g/(bhph), multiply by 1.340.
[b]For 1994 standards and beyond, heavy-duty trucks must meet standards which reflect the greatest degree of emission reduction achievable through available technology.

pollutants HC, CO, NO_x, and particulates (81). Particulate emissions have also been shown to be affected by fuel density (82).

Diesel manufacturers have found it difficult to meet the stringent emissions targets. Development of exhaust treatment devices to reduce particulates and meet NO_x standards has been underway. These devices either trap or catalytically oxidize the particles or both.

California has taken a slightly different approach from that of the federal government. Starting in October, 1993, all diesel fuel in California must contain no more than 10% aromatics, 500 ppm sulfur, and meet all other ASTM specifications. Alternative formulations are possible if these are shown to have equivalent NO_x emissions to a base reference fuel. In addition to the specifications that apply to the commercial fuel, other aspects of the reference fuel composition are tightly controlled. The fuel must also have a minimum cetane number of 48 without cetane improvers. Fuel manufacturers test the emissions from proposed alternatives and compare them to the emissions from the reference fuel. For instance, manufacturers may be able to produce higher aromatic fuel if the fuel also has higher cetane number. In an analysis of the industry sponsored CRC program, NO_x emissions have been shown to be limiting in terms of aromatics and cetane levels (83). A curve relating aromatics and cetane number at constant NO_x levels shown in Figure 6 has been developed. This type of information is helpful in guiding refiners toward fuel specifications that can be tested and certified as equivalent. In Europe, the tripartite initiative may result in diesel fuel composition which would take effect in 2000.

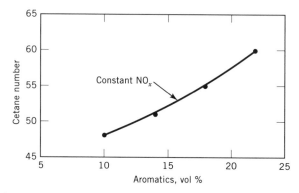

Fig. 6.   Trade-off between cetane number and aromatics for NO_x emissions (85).

ALTERNATIVE FUELS

Alternative fuels fall into two general categories. The first class consists of fuels that are made from sources other than crude oil but that have properties the same as or similar to conventional motor fuels. In this category are fuels made from coal and shale (see FUELS, SYNTHETIC). In the second category are fuels that are different from gasoline and diesel fuel and which require redesigned or modified engines. These include methanol (see ALCOHOL FUELS), compressed natural gas (CNG), and liquefied petroleum gas (LPG).

During World War II, Germany developed large-scale production of fuel from coal using Fischer-Tropsch synthesis. In the United States, in the 1970s, interest in producing alternative fuels was generated by upheavals in the price and availability of crude oil. In addition to coal, a great deal of research was carried out to develop processes to develop fuels from oil shale and natural gas.

Production of synthetic fuels from coal, oil shale, and methane involves changing the chemical structure of the raw material, especially the hydrogen–carbon ratio. In coal, the hydrogen–carbon ratio is too low and the molecular weight of the raw material is too high to burn cleanly. Coal can be converted to syngas (CO/H_2) and then converted to liquids via Fischer-Tropsch synthesis. Alternatively, it may be liquefied directly with the addition of hydrogen. Oil shale is retorted in large reactors or *in situ* to drive out the oil or kerogen. The oil is processed to resemble crude fractions. Methane may be converted to hydrocarbon liquids or to methanol via gasification to syngas (CO/H_2) followed by Fischer-Tropsch reactions. A principal factor in all of these processes is the amount of energy needed to run the process and the disposition of the waste products. In the case of shale oil, the spent rock took up more volume than the original shale and had to be disposed of in an environmentally safe manner.

These processes tend to produce liquids that are higher priced than those derived from crude oil. Timing for commercialization depends on a number of factors, including price of crude oil, need for energy self-sufficiency, and environmental considerations. The factors are different for each of the potential sources. Environmental considerations include polution from burning the fuel itself, pollution from the whole production cycle, and production of gases that are involved in the enhanced greenhouse effect (see AIR POLLUTION).

Compressed Natural Gas

Compressed natural gas (CNG), which contains mostly methane, is used in about 700,000 vehicles worldwide and offers a number of clear environmental advantages. It has a high octane number, has the potential to be clean burning, and is currently shipped through pipelines to many parts of the country. Optimized engines would have high compression ratios to take advantage of methane's high octane number. These vehicles store the CNG in high pressure cylinders made of lightweight spun carbon fiber (see CARBON AND GRAPHITE FIBERS) and use specially designed fuel injectors. Dedicated CNG vehicles are projected to have improved energy efficiency and lower emissions of hydrocarbons and carbon monoxide. Vehicle maintenance costs would probably be substantially lower than equivalent gasoline vehicles.

Unfortunately, CNG also has a number of disadvantages. The energy density of methane is very low and it cannot be compressed enough to provide equal volumetric energy density to gasoline. Producers of automobiles are faced with either eliminating trunk space to provide equal range or compromising on both range and storage space. New infrastructure would have to be put into place to refuel CNG vehicles. A service station refueling island would probably cost $300,000–500,000, and refueling a car would take about 10 minutes. A home compressor capable of overnight refueling would cost anywhere from $3,000–8,000.

CNG has some potential environmental debits as well. It is envisioned that dedicated CNG vehicles should operate on the lean side of stoichiometry, and this makes catalytic control of NO_x difficult. Methane is also a potent contributor to enhanced global warming, being anywhere from 20–50 times worse than CO_2. Not only do emissions from CNG vehicles contain large amounts of methane, but additional transmission of methane through pipeline systems increases the amount of leakage of methane into the atmosphere. The exact leakage rate is not well known but estimates range from less than 1% to almost 7%.

In the United States, CNG's initial use is in captive, centrally refueled fleets which require limited range, such as delivery vans, taxis, or school buses. Such vehicle fleets can afford the capital cost of refueling equipment and can tolerate the slightly longer refueling times. Many utility companies, such as Brooklyn Union Gas (New York), have purchased CNG vehicles as a way of encouraging the development of this fuel. All three domestic automobile manufacturers offer CNG vehicles for sale.

Liquefied Petroleum Gas

Liquefied petroleum gas (LPG) is a generally available fuel that is a mixture of mostly propane and butane. The exact proportion depends on marketplace conditions and alternative outlets for the two main components. LPG has a number of advantages over both gasoline and methane. It has a higher energy density than methane, because LPG is a liquid. It is easier starting than gasoline because of its higher vapor pressure. Generally, LPG vehicles vaporize the fuel before it is fed into the engine. In cold weather, if the butane content is too high, the fuel may have trouble vaporizing. The largest problem with LPG is that the supply is not large enough to make a significant dent in the total fuel demand picture.

Methanol

Methanol has been considered as a potentially attractive automotive fuel for many years. It has the advantage of a very high octane number, good combustion characteristics, and can be made from alternative energy sources such as coal and methane. Methanol can be thought of as a partially burned liquid form of methane. It has a higher energy density than methane and may be used in an engine with relatively minor modifications. On the negative side, methanol is quite corrosive, burns with a colorless flame, and is extremely toxic to humans. Amounts as small as a few milliliters can cause blindness and even death in humans. Methanol is not, however, as dangerous for other organisms and is believed to decompose fairly rapidly if spilled in water or on the ground.

Cars designed especially to run on methanol fuel generally use a mixture of 85% methanol and 15% gasoline, called M85. Small amounts of gasoline help these cars provide enough volatility for easy starting in cold weather and provide good flame luminosity in case of a spill and fire. Methanol vehicles require special metals and elastomers to stand up to methanol's corrosive and solvent properties. These vehicles also may exhibit significantly improved energy efficiency as a result of having high compression ratio engines and lean carburetion.

Emissions from methanol vehicles are expected to produce lower HC and CO emissions than equivalent gasoline engines. However, methanol combustion produces significant amounts of formaldehyde (qv), a partial oxidation product of methanol. Formaldehyde is classified as an air toxic and its emissions should be minimized. Formaldehyde is also very reactive in the atmosphere and contributes to the formation of ozone. Emissions of NO_x may also pose a problem, especially if the engine runs lean, a regime in which the standard three-way catalyst is not effective for NO_x reduction.

Initially, at least, methanol vehicles should be capable of operating on either M85, gasoline, or any mixture of the two. These vehicles are called flexible fueled (FFV) or variable fueled vehicles (VFV). It is expected that methanol could be sold in existing service stations out of tanks constructed of methanol-tolerant material such as carbon steel or certain fiber glass formulations. Eventually, if enough FFV/VFVs are sold and methanol becomes widely available, dedicated vehicles would likely be built and sold. Methanol has been used for years as a racing fuel.

Heavy-duty diesel engines can also be modified to operate on neat methanol. Emissions of NO_x and particulates are generally lower than the original engine. These types of engines have typically been used in urban buses to help reduce ambient pollution levels.

BIBLIOGRAPHY

"Gasoline" in *ECT* 1st ed., Vol. 7, pp. 113–117, by M. M. Hicks-Bruun, Consultant; "Gasoline and Other Motor Fuels" in *ECT* 2nd ed., Vol. 10, pp. 463–498, by J. C. Lane, Ethyl Corp.; in *ECT* 3rd ed., Vol. 11, pp. 652–695, by J. C. Lane, Ethyl Corp.; "Refinery Processes, Survey" under "Petroleum" in *ECT* 3rd ed., Vol. 17, pp. 183–256, by C. E. Jahnig, Consultant.

1. S. C. Davis and P. S. Hu, *Transportation Energy Data Book: Edition 11*, ORNL-6649, Oak Ridge National Laboratories for Office of Transportation Technologies, U.S. Dept. of Energy, Washington, D.C., 1991.
2. *Annual Energy Review 1991*, report no. DOE/EIA-0384 (91), Energy Information Administration, U.S. Dept. of Energy, Washington, D.C., 1992.
3. *Ibid.*, p. 161.
4. S. K. Skinner, *National Transportation Statistics, Annual Report, 1990*, U.S. Dept. of Transportation, Washington, D.C., 1990, p. 124.
5. *1992 Annual Book of ASTM Standards*, Vol. 5.3, American Society of Testing and Materials, Philadelphia, 1992, pp. 587–611.
6. *1992 Annual Book of ASTM Standards*, Vol. 5.01, American Society of Testing and Materials, Philadelphia, 1992, pp. 313–326.
7. *Annual Energy Review, 1992*, Report No. DOE/EIA-0384 (92), Energy Information Administration, U.S. Dept. of Energy, Washington, D.C., 1993, p. 139.
8. E. H. Dole and R. A. Barnhart, *Highway Statistics, Summary to 1985*, U.S. Dept. of Transportation, Federal Highway Administration, Washington, D.C., 1986, Table VM-201A, pp. 229–232.

9. R. E. Farris, *Highway Statistics 1987*, U.S. Dept. of Transportation, Federal Highway Administration, Washington, D.C., 1988, Table VM-1, p. 171.

10. T. D. Larson, *Highway Statistics 1989*, U.S. Dept. of Transportation, Federal Highway Administration, Washington, D.C., 1990, Table VM-1, p. 181.

11. T. D. Larson, *Highway Statistics 1991*, U.S. Dept. of Transportation, Federal Highway Administration, Washington, D.C., 1992, Table VM-1, p. 193.

12. T. D. Larsen, *Highway Statistics 1992*, U.S. Dept. of Transportation, Federal Highway Administration, Washington, D.C., 1993, Table VM-1, p. 193.

13. S. K. Skinner, in Ref. 4, p. 118.

14. *Annual Energy Outlook*, report no. DOE/EIA-0383 (92), Energy Information Administration, U.S. Dept. of Energy, Washington, D.C., 1992.

15. S. Meyer, ed., *Energy Statistics Sourcebook*, 6th ed., PennWell Publishing Co., Tulsa, Okla., 1991, p. 271.

16. J. B. Heywood, *Internal Combustion Engine Fundamentals*, McGraw-Hill Book Co., Inc., New York, 1988.

17. A. M. Douaud and P. Ezyat, *SAE Tans.* **87**, 294–308 (1978).

18. *Knocking Characteristics of Pure Hydrocarbons*, ASTM Special Publication No. 225, American Society of Testing Materials, Philadelphia, 1958.

19. G. H. Unzelman, *Oil Gas J.*, 33–37 (Apr. 10, 1989).

20. E. S. Corner, A. M. Hochhauser, and H. F. Shannon, *SAE Trans.* **87**, 1151–1527 (1978).

21. W. E. Bettoney and co-workers, *SAE Trans.* **89**(4), 4198–4212 (1980).

22. Y. Nakajima and T. Sato, *SAE Paper No. 700612*, 1970.

23. P. J. Clarke, *SAE Paper No. 830595*, SAE, Warrendale, Pa., 1983.

24. K. Owen and T. Coley, *Automotive Fuels Handbook*, Society of Automotive Engineers, Warrendale, Pa., p. 116.

25. U.S. Pat. 4,699,629 (Dec. 5, 1985), M. C. Croudace, T. Wusz, and S. G. Brass (to Union Oil Co. of California).

26. J. N. Bowden and D. W. Brinkman, *Stability Survey of Hydrocarbon Fuels*, Report BETC/17784, U.S. Dept. of Energy, Washington, D.C., 1979.

27. A. A. Johnston and E. Dimitroff, *SAE Paper No. 660783*, SAE, Warrendale, Pa., 1966.

28. H. W. Marbach and co-workers, *SAE Paper No. 790204*, 1979.

29. A. D. Reichle, *Chem. Eng. Progress*, 70–74 (Sept. 1990).

30. J. R. Murphy, *Oil Gas J.*, 49–58 (May 18, 1992).

31. A. A. Avidan, in Ref. 30, pp. 59–66.

32. A. A. Murcia, in Ref. 30, pp. 68–71.

33. *Nat. Pet. News* **84**(7), 94–101 (June, 1992).

34. U.S. Environmental Protection Agency, *Fed. Reg.* **46**(144) (July 28, 1981).

35. U.S. Environmental Protection Agency, *Fed. Reg.* **56**(28), 5352 (Feb. 11, 1991).

36. C. J. Pedersen, *Ind. Eng. Chem.* **48**(10), 1881 (Oct. 1956).

37. P. Polss, *Hydrocarbon Process.* **52**(2), 61 (Feb. 1963).

38. E. W. Unruh and F. M. Watkins, *Oil Gas J.* **47**63 (June 17, 1948).

39. L. M. Gibbs, *SAE Trans., J. Fuels Lubricants* **99**(4), 618–638 (1990).

40. National Association of Corrosion Engineers, *NACE Standard TM0172*, Houston, Tex., 1972.

41. U.S. Dept. of the Air Force, *Specification MIL-F-25017d for Fuel Soluble Corrosion Inhibitor*, Wright Patterson AFB, Dayton, Ohio, May 1981.

42. J. J. Malakar, J. B. Retzloff, and L. M. Gibbs, *SAE Paper No. 831708*, SAE, Warrendale, Pa., 1983.

43. H. W. Sigworth and J. Q. Payne, *SAE Paper No. 405*, SAE, Warrendale, Pa., 1954.

44. U.S. Pat. 2,839,373 (June 17, 1958), M. R. Barusch and E. G. Lindstrom (to Chevron Research and Technology Co.).

45. A. A. Zimmerman, L. E. Furlong, and H. F. Shannon, *SAE Trans.* **81**, 316–324 (1972).
46. J. D. Benson and P. A. Yaccarino, *SAE Trans.* **95**(6), 562–579 (1986).
47. R. C. Tupa, *SAE Paper No. 872113*, SAE, Warrendale, Pa., 1987.
48. D. L. Hilden, *SAE Trans.* **97**(3), 3.847–3.862 (1988).
49. C. Kim, S. I. Tseregounis, and B. E. Scruggs, *SAE Trans.* **96**, 7.617–7.629 (1987).
50. J. D. Benson and P. A. Yaccarino, in Ref. 46.
51. B. Bitting and co-workers, *SAE Trans.* **96**(7), 639–655 (1987).
52. G. T. Kalghatgi, *SAE Trans. J. Fuels Lubricants* **99**(4), 639–667 (1990).
53. T. Nishizaki and co-workers, *SAE Paper No. 790203*, SAE, Warrendale, Pa., 1979.
54. T. J. Bond, F. S. Gerry, and W. Wagner, *SAE Paper No. 892115*, SAE, Warrendale, Pa., 1989.
55. B. Bitting and co-workers, in Ref. 51.
56. S. A. Bannon, G. O. Sherer, and D. V. Swaynos, *SAE Trans. J. Fuels Lubricants* **101**(4), 1452–1484 (1992).
57. R. C. Tupa and C. J. Dorer, *SAE Trans* **95**(6), 340–374 (1986).
58. T. J. Sheahan, C. J. Dorer, and C. O. Miller, *SAE Paper No. 690516*, SAE, Warrendale, Pa., 1969.
59. W. E. Morris, *Oil Gas J.* **84**(3), 63 (Jan. 20, 1986).
60. W. E. Morris, *The Interaction Approach to Gasoline Blending*, paper presented at *NPRA Annual Meeting*, San Antonio, Tex., Mar. 23–25, 1975.
61. W. C. Healy, Jr., C. W. Maassen, and R. T. Peterson, *A New Approach to Blending Octanes*, paper presented at *24th Mid-year Meeting of the American Petroleum Institute's Division of Refining*, New York, May 27, 1959.
62. R. D. Snee, *Technometrics* **23**(2), 119–130 (May 1981).
63. *Fed. Reg. FR40CFR part 600* **51**(206), 37851 (Oct. 24, 1986).
64. A. M. Hochhauser and co-workers, *SAE Paper No. 930138*, SAE, Warrendale, Pa., 1993.
65. *Clean Air Act Amendments of 1990*, PL 1639, U.S. Environmental Protection Agency, Washington, D.C., Oct. 26, 1990.
66. *Fed. Reg. 57FR57980* **57**(236) (Dec. 8, 1992).
67. V. R. Burns and co-workers, *SAE Trans. J. Fuels Lubricants* **100**(4), 687–714 (1991).
68. A. M. Hochhauser and co-workers, in Ref. 67, pp. 748–788.
69. W. J. Koehl and co-workers, in Ref. 67, pp. 789–802.
70. R. M. Reuter and co-workers, *SAE Trans. J. Fuels Lubricants* **101**(4), 463–484 (1992).
71. Ref. 2, p. 143.
72. S. C. Davis and P. S. Hu, in Ref. 1, pp. 2–12.
73. *World Motor Vehicle Data, 1992 Edition*, Motor Vehicle Manufacturers Association of the U.S.A., Inc., Detroit, Mich., 1992.
74. J. W. Bush, ed., *Ward's Automotive Yearbook*, 54th ed., Ward's Communications, Detroit, Mich., 1992, p. 57.
75. R. C. Tupa and C. J. Dorer in Ref. 57.
76. T. R. Coley and co-workers, *SAE Paper No. 861524*, SAE, Warrendale, Pa., 1986.
77. J. E. Benethum and R. E. Windsor, in Ref. 67, pp. 803–809.
78. N. Feldman, *SAE Paper No. 730677*, SAE, Warrendale, Pa., 1973.
79. S. R. Reddy and M. L. McMillan, *SAE Trans.* **90**, 3598 (1981).
80. J. E. Chandler, F. G. Horneck, and G. I. Brown, *SAE Paper No. 922186*, SAE, Warrendale, Pa., 1992.
81. T. L. Ullman, R. L. Mason, and D. A. Montalvo, *SAE Trans. J. Fuels Lubricants* **99**(4), 1092–1103 (1990).
82. W. W. Lange, in Ref. 67, pp. 1233–1256.
83. B. J. Kraus, paper presented to *API/MVMA Detroit Advisory Panel Forum*, Apr. 29, 1992.

General References

J. B. Heywood, *Internal Combustion Engine Fundamentals*, McGraw-Hill Book Co., Inc.,
   New York, 1988.
K. Owen and T. Coley, *Automotive Fuels Handbook*, Society of Automotive Engineers,
   Warrendale, Pa., 1990.
L. M. Gibbs, *SAE Trans. J. Fuels Lubricants* **99**(4), 618–638 (1990).
G. T. Kalghatgi, *SAE Trans. J. Fuels Lubricants* **99**(4), 639–667 (1990).
R. Tupa and C. J. Dorer, *SAE Trans.* **95**(6), 340–374 (1986).

ALBERT M. HOCHHAUSER
Exxon Research and Engineering Co.

GASTROINTESTINAL AGENTS

Compounds used as gastrointestinal agents may be divided into several classes according to general activity, ie, agents for therapy of peptic ulcer and other gastric acid related diseases, laxatives, antidiarrheals, antiemetics, and gastric prokinetic drugs. Many of the compounds described in this article are discussed in detail elsewhere (1) and most drugs have been approved for marketing in the United States. Trade names are those used in the United States.

Antipeptic Ulcer Therapy

The primary aim in antiulcer therapy is either prevention of acid secretion from the gastric parietal cells or neutralization of the acid before it comes into contact with the ulcerated areas of the gastrointestinal tract. Long established therapy of duodenal ulcers, prior to the introduction of cimetidine [*51481-61-9*], $C_{10}H_{16}N_6S$ (**1**), included diet, antacids, and anticholinergics. Of these, only antacids have been shown conclusively to be effective. The antacids can produce healing, but only when given in large and frequent doses. Antacids, used mainly as adjunctive therapy and generally available without a prescription, are indicated for short-term therapy of esophageal reflux (heartburn) and gastric upset. Nonselective anticholinergics, such as atropine [*51-55-8*], $C_{17}H_{23}NO_3$, are only effective when given at doses high enough to produce side effects. Gastric acid selective anticholinergics have been developed, but as of early 1994 none had been licensed for use in the United States by the Food and Drug Administration (FDA).

The therapy for duodenal ulcers changed drastically upon the introduction of cimetidine (**1**) in the United States in 1977. Cimetidine was the first of a series of agents which work by antagonism of histamine responses at the site responsible for acid secretion (see HISTAMINE AND HISTAMINE ANTAGONISTS). This receptor is different from the H_1 receptor targeted for classical antihistaminic agents for treatment of allergies. Since the approval of cimetidine, three additional H_2 blockers have been approved for acute therapy of duodenal ulcers, ie, famotidine [*76824-35-6*], $C_8H_{15}N_7O_2S_3$ (**2**); nizatidine [*76963-41-2*], $C_{12}H_{21}N_5O_2S_2$ (**3**); and ranitidine [*66357-35-5*], $C_{13}H_{22}N_4O_3S$ (**4**).

A new class of agents, which specifically prevent the secretion of hydrochloric acid by the gastric parietal cells, acts by selective inhibition of the H^+ proton pump. This enzyme is found only in the gastric parietal cell and compounds inhibiting it must be selective antisecretory agents. The compound marketed in the United States (ca 1992), omeprazole [*73590-58-6*], $C_{17}H_{19}N_3O_3S$ (**5**), is a potent antisecretory agent and is used only for acute therapy of ulcerative disease and esophagitis. Several other agents are in development. However, it is not clear whether these have advantages over omeprazole.

Drugs acting by mechanisms not related to inhibition of gastric acid secretion have been introduced into therapy in the United States. The two compounds falling into this category are sucralfate (**6**), which works by an as yet to be determined mechanism, and misoprostil (**7**), a prostanoid derivative indicated only for prevention of gastrointestinal damage induced by nonsteroidal antiinflammatory agents such as aspirin. Therapy also has been changed by the acceptance that *Helicobacter pylori*, an organism able to withstand the acid media of the stomach, is associated with gastritis and ulcerative disease. It is not known as of this writing whether duodenal ulcers may be permanently cured by eradication of this organism. No specific agents have been approved for eradication of *Helicobacter*, but combinations of several antibiotics (qv) are effective and some bismuth-containing compounds may act by their effect against the organism. Trials have been published showing that other combinations such as omeprazole (**5**) and amoxylcillin [*26787-78-0*], $C_{16}H_{19}N_3O_5S$, and a bismuth complex with ranitidine, may be useful.

The market for antiulcer agents (Fig. 1) is large and is comprised of both prescription and over the counter (OTC) products. The estimated prescription market is over \$3 billion annually in the United States, whereas the more difficult to estimate OTC market is in the range of \$500 million annually. Several pharmaceutical companies are attempting to obtain approval for OTC use of prescription-only agents, and patent protection for several of the histamine antagonists runs out in the mid-1990s.

Antiulcer Agents. *Cimetidine.* This agent, also known as Tagamet [*51481-61-9*], $C_{10}H_{16}N_6S$ (**1**) is soluble in alcohol, slightly soluble in water, very slightly soluble in chloroform, and insoluble in ether. The hydrochloride is freely soluble in water, soluble in alcohol, very slightly soluble in chloroform, and practically insoluble in ether. The method of preparation of cimetidine is available (2).

Cimetidine was the first histamine-H_2 antagonist approved in the United States. Its mechanism of action is suppression of gastric acid secretion through the blockade of histamine H_2-receptors on the gastric parietal cells, thereby decreasing both basal and stimulated gastric acid secretion. It is indicated for short-term treatment of active duodenal ulcer, maintenance therapy for duodenal ulcer at reduced dosage, short-term treatment of active benign gastric ulcer and erosive gastroesophageal reflux disease, prevention of upper gastrointestinal bleeding in critically ill patients, and treatment of pathological hypersecretory condi-

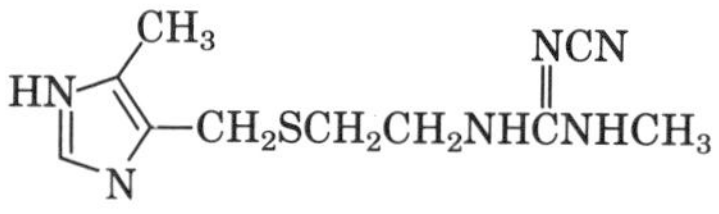

Fig. 1. Structures of prescription antiulcer agents.

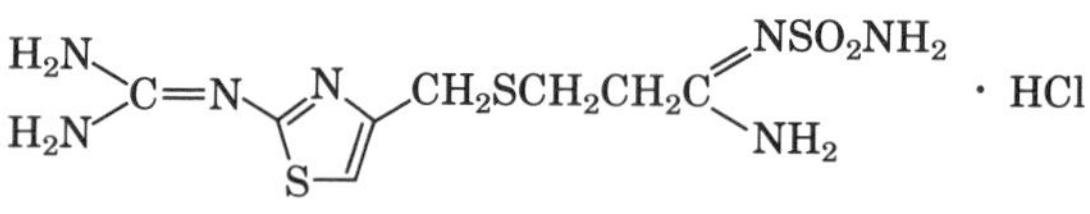

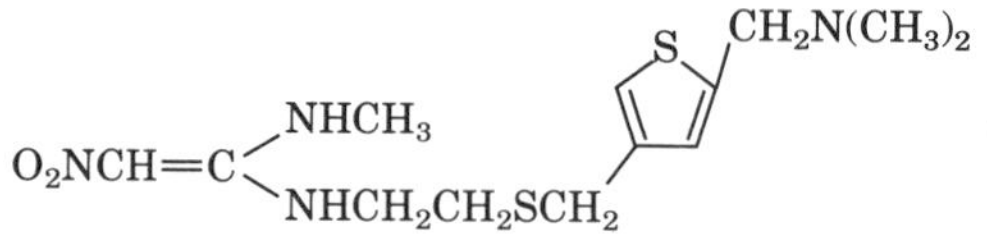

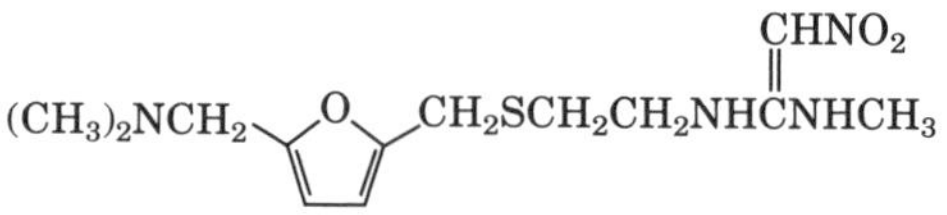

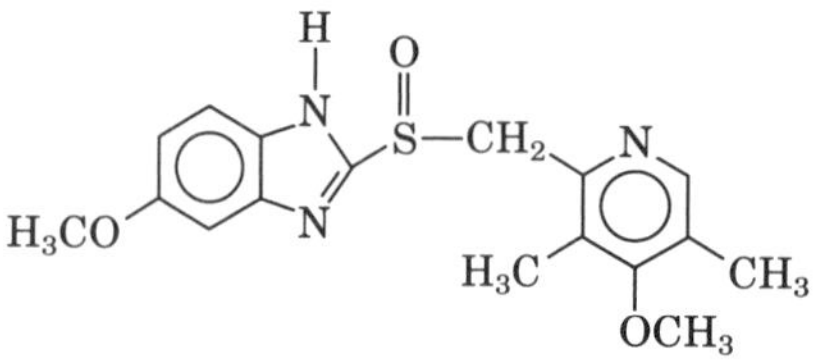

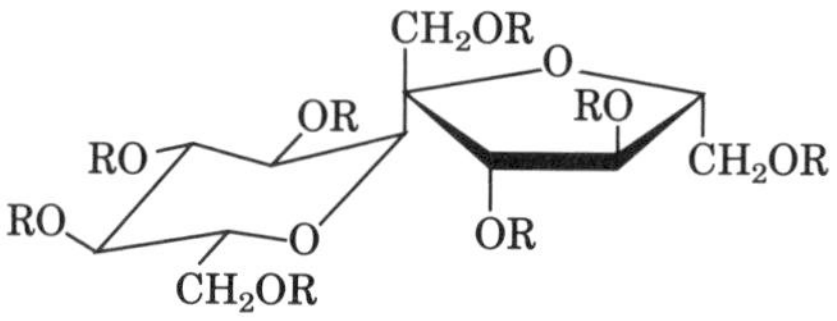

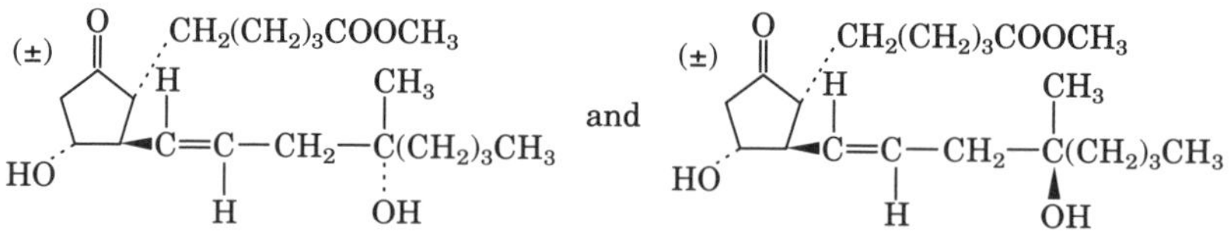

tions. Cimetidine has been reported to inhibit certain microsomal enzyme systems and reduce the liver metabolism of a large number of drugs. There should be careful monitoring of patients on concomitant therapy with other drugs, eg, antiasthmatic agents (qv).

Famotidine. Also known as Pepcid, famotidine [76824-35-6] (N'-(aminosulfonyl)-3-([[2-[(diaminomethylene) amino]-4-thiazolyl] methyl]thio)propanimidamide) (**2**) is a white to pale yellow crystalline compound, freely soluble in glacial acetic acid, slightly soluble in methanol, very slightly soluble in water, and practically insoluble in ethanol. It may be prepared by the method described in Reference 3.

Famotidine is a competitive histamine-H_2 antagonist that works primarily by inhibition of gastric acid secretion. It is three times more potent than ranitidine (**4**) and 20 times more potent than cimetidine (**1**). It is indicated for short-term treatment of active duodenal ulcers, maintenance therapy of duodenal ulcers using reduced doses, short-term treatment of benign gastric ulcer, treatment of gastroesophageal reflux disease and esophagitis, and treatment of pathological hypersecretory conditions such as Zollinger-Ellison syndrome and multiple endocrine adenomas.

Nizatidine. Nizatidine, or Axid [*76963-41-2*] (**3**), is an off-white to buff crystalline solid. It is soluble in water and its method of synthesis is available (4).

Nizatidine is a histamine-H_2 antagonist indicated for the treatment of active duodenal ulcer for up to eight weeks. It is equipotent with ranitidine (4). Nizatidine is also indicated at lower doses for maintenance therapy of duodenal ulcer disease and gastroesophageal reflux for up to 12 weeks. It may inhibit alcohol dehydrogenase in the gastric mucosa and produce higher alcohol levels in the blood. In patients given very high (3900 mg) doses of aspirin daily, increases in serum salicylate levels have been seen when nizatidine, at 150 mg bid, is administered concurrently.

Ranitidine. Ranitidine hydrochloride [*66357-59-3*] (Zantac) is a white to pale yellow granular substance. It is freely soluble in water and acetic acid, soluble in methanol, sparingly soluble in ethanol, and practically insoluble in chloroform. It has a slightly bitter taste and a sulfur-like odor. It may be made by the method described in Reference 5.

Ranitidine (**4**), the second H_2-receptor antagonist approved for use in the United States, is indicated by oral administration for short-term therapy of active duodenal ulcers, maintenance therapy for duodenal ulcers using reduced daily dosage, treatment of pathological hypersecretory conditions, short-term treatment of benign gastric ulcers, and treatment of gastroesophageal reflux. It has been useful by parenteral administration for some hospitalized patients with pathological hypersecretory conditions or intractable duodenal ulcers. Although it may interact weakly with cytochrome P-450, this has not been a significant effect seen in clinical use. Drug interactions are less frequent than for cimetidine; however, interactions with nifedipine [*21829-25-4*], $C_{17}H_{18}N_2O_6$; warfarin [*81-81-2*], $C_{19}H_{16}O_4$; theophylline [*58-55-9*], $C_7H_8N_4O_2$; and metoprolol [*37350-58-6*], $C_{15}H_{25}NO_3$, have been observed.

Omeprazole. Also known as Prilosec, omeprazole [*73590-58-6*] (5-methoxy-2-(((methoxy-3,5-dimethyl-2-pyridinyl)methyl)sulfinyl)-1*H*-benzimidazole)(**5**) is a white to off-white crystalline powder. It is a weak base freely soluble in ethanol and methanol, slightly soluble in acetone and isopropanol, and very slightly soluble in water. It is rapidly degraded in acid media but has acceptable stability under alkaline conditions. The method of preparation is available (6).

Omeprazole is supplied as a delayed-release formulation to prevent degradation in the acid environment of the stomach (see CONTROLLED RELEASE TECHNOLOGY, PHARMACEUTICAL). It is the first approved antisecretory agent that acts by suppression of gastric acid secretion by specific inhibition of the H^+/K^+ adenosine triphosphatase (ATPase) enzyme system in the gastric parietal cell. It blocks the final step of gastric acid secretion and works independently of the means of gastric acid stimulation. It is indicated for use for short-term treatment of duodenal ulcers but not maintenance. It also is used for the short-term therapy of gastroesophageal reflux disease and severe erosive esophagitis. Omeprazole has produced a dose-related increase in gastric carcinoid tumors in long-term tests in rats; the significance in humans is unknown. However, this agent should only be used for short-term therapy. Omeprazole inhibits microsomal P-450 monooxygenase and can be expected to interfere with the metabolism of some drugs.

Sucralfate. Sucralfate [54182-58-0] (Carafate) (**6**) is a white amorphous powder soluble in dilute hydrochloric acid and sodium hydroxide. It is practically insoluble in water, ethanol, and carbon tetrachloride. Dissolution of aluminum occurs at pH <3. It may be prepared by the method described in Reference 7.

The mechanism by which sucralfate accelerates healing of duodenal ulcers has not been determined. It does not have significant antisecretory, acid neutralizing activity or direct stimulation of ulcer healing. It is known that the mechanism is local rather than systemic. Binding of pepsin or bile salts may contribute to its effect. It is indicated for the short-term therapy of active duodenal ulcers and for maintenance at reduced dosage.

Misoprostil. Also known as Cytotec [59122-46-2], $C_{22}H_{38}O_5$, (a 1:1 mixture of (+/−)methyl-11-alpha,16-dihydroxy-16-methyl-9-oxoprost-13E-en-1-oate) (**7**), this agent is a light yellow liquid that is soluble in water. The methods for preparation are available (8).

Misoprostil is a stable prostaglandin analogue which has significant gastric antisecretory and cytoprotective activity (see PROSTAGLANDINS). As of this writing it does not have FDA acceptance for the treatment of gastric or duodenal ulcers, but is accepted only for the prevention of nonsteroidal antiinflammatory drug (NSAID) induced gastric ulcers in patients at high risk. It has not been shown to prevent duodenal lesions produced by NSAIDs, and has no effect on gastrointestinal pain or discomfort associated with NSAID usage. It is contraindicated in pregnant woman owing to abortifacient properties. There is also a significant incidence of diarrhea and abdominal pain associated with the drug.

Antacids. *Aluminum Hydroxide Gel.* Colloidal aluminum hydroxide [21645-51-2] (Amphogel), $AlOH_3$, is a suspension. Each 100 g contain the equivalent of 3.6–4.4 g aluminum oxide [1344-28-1]. Aluminum hydroxide gel may contain a variety of flavoring agents. It is a white viscous suspension from which small amounts of water may separate on standing and is translucent in thin layers.

Aluminum hydroxide gel may be prepared by a number of methods. The products vary widely in viscosity, particle size, and rate of solution. Such factors as degree of supersaturation, pH during precipitation, temperature, and nature and concentration of by-products present affect the physical properties of the gel.

In one manufacturing process, aluminum chloride is treated with a solution containing sodium carbonate and sodium bicarbonate. The product of this reaction is mixed with the precipitate obtained by reaction of a solution of aluminum chloride and ammonia. The mixed magma is dialyzed, the product mixed with glycerol (qv), sodium benzoate is added, and the mixture is then passed through a colloid mill.

Aluminum hydroxide gel is used as an antacid and for the treatment of phosphate nephrolithiasis. One gram of amphogel suspension raises the pH of 12.5–25 mL of simulated gastric juice, ie, 0.1 N HCl, to pH 3.5. The rate of neutralization is slow. Aluminum hydroxide gel is nontoxic, but has side effects, ie, constipation, nausea, or vomiting owing to astringent action or taste; hypophosphatemia and osteomalacia owing to interference with absorption; complexation with tetracycline; and interference with absorption of tetracycline and other classes of drugs (see ANTIBIOTICS, TETRACYCLINES).

Precipitated Calcium Carbonate. Calcium carbonate [471-34-1] (Tums), $CaCO_3$, is a fine white microcrystalline powder without odor or taste. It is stable in air. An aqueous suspension is close to neutrality. It is practically insoluble in water, insoluble in alcohol, and dissolves with effervescence in dilute acetic, hydrochloric, and nitric acids (see CALCIUM COMPOUNDS, CALCIUM CARBONATE).

Calcium carbonate can be prepared by the double decomposition of calcium chloride and sodium carbonate in aqueous solution. Its density and fineness are governed by the concentration of the solutions. Heavy and light forms are available. It is an effective gastric antacid for use in peptic ulcer disease in a dose of 2.5 g/h. Usual dose for OTC usage is 1–2 g as needed. One gram of $CaCO_3$ neutralizes 110 mL of 0.1 N HCl in 10 min, and 162 mL within 2 h. Long-term therapy, including large doses of $CaCO_3$ taken with milk or other sources of phosphate, may cause renal pathology, ie, milk–alkali syndrome, and systemic alkalosis, ie, 7–19% of calcium is absorbed; increased urinary calcium may favor calcific renal stones; and calcium carbonate may be constipating and rebound gastric secretion may occur with high doses that neutralize gastric contents, ie, greater than 2 g.

Magnesia and Alumina Suspension. A mixture of salts, available as Maalox, Mylanta, Gelusil, and Aludrox, contains magnesium hydroxide [1309-42-8], $Mg(OH)_2$, and variable amounts of aluminum oxide in the form of aluminum hydroxide and hydrated aluminum oxide, ie, 2.9–4.2% magnesium hydroxide and 2.0–2.4% aluminum oxide, Al_2O_3, for a mixture of 4.9–6.6% combined magnesium hydroxide and aluminum oxide. This mixture may contain a flavoring and antimicrobial agents in a total amount not to exceed 0.5% (see ALUMINUM COMPOUNDS, ALUMINUM OXIDE).

The magnesia and alumina suspension is prepared by treatment of an aqueous solution, containing aluminum and magnesium salt in the desired proportion, with sodium hydroxide. The coprecipitated aluminum and magnesium hydroxides are collected by filtration, washed free of soluble salts, and stabilized by the addition of a suitable hexatol.

Magnesia and aluminum suspension is useful for the therapy of duodenal ulcers when given at high doses at frequent intervals. It is available in both liquid and tablet formulations.

Magnesium Oxide. Magnesia [1309-48-4], MgO, is available in a very bulky white powder known as light magnesium oxide, or a relatively dense white powder known as heavy magnesium oxide. It absorbs moisture and carbon dioxide when exposed to air. It is practically insoluble in water, insoluble in alcohol, and soluble in dilute acids (see MAGNESIUM COMPOUNDS).

Light or heavy magnesium carbonate is exposed to a red heat, and carbon dioxide and water are expelled leaving light or heavy magnesium oxide. The density is also influenced by the calcining temperature; higher temperatures yield more compact forms.

Magnesium oxide is an effective nonsystemic antacid, ie, it is converted to the hydroxide. It does not neutralize gastric acid excessively nor does it liberate carbon dioxide. The light form is preferable to the heavy for administration in liquids because it is suspended more readily. One gram of magnesium oxide neutralizes 87 mL of 0.1 N HCl in 10 min, and 305 mL in 2 h.

Magnesium Trisilicate. Magnesium trisilicate [39365-87-2], $2MgO \cdot 3SiO_2 \cdot xH_2O$, is a compound of magnesium oxide and silicon dioxide with varying proportions of water. It contains not less than 20% magnesium oxide, nor less than 45% silicon dioxide, ie, magnesium silicate hydrate. It is a fine, white, odorless, tasteless powder free from grittiness, its suspension is neutral or only slightly alkaline, it is insoluble in water and alcohol, and it is decomposed readily by mineral acids with the liberation of silicic acid.

Magnesium trisilicate is prepared by precipitation of a solution of sodium silicate of the proper composition, ie, MgO to SiO_2 ratio equal to 1:1.5, using a solution of magnesium chloride or sulfate. The precipitate of the magnesium trisilicate is filtered, washed, and dried at a low temperature.

Magnesium trisilicate is a nonsystemic antacid and an adsorbent. It has a slow onset of activity and is a relatively weak antacid. It may cause diarrhea in large doses owing to soluble magnesium salts. One gram of magnesium trisilicate neutralizes 10 mL of 0.1 N HCl in 10 min, and 15 mL in 2 h.

Magaldrate. Aluminum magnesium hydroxide [*39366-43-3*], also known as magaldrate(tetrakis(hydroxymagnesium)decahydroxydialuminatedihydrate (Riopan)), $Al_2Mg_4(OH)_{14} \cdot 2H_2O$, contains the equivalent of 28–39% magnesium oxide and 17–25% aluminum oxide. It is a white, odorless, crystalline powder insoluble in water and alcohol, and soluble in dilute solutions of mineral acids.

Magaldrate is prepared by precipitation from aqueous solutions of sodium or potassium aluminate and a magnesium salt under controlled conditions of concentration and temperature. The precipitated product is collected by filtration, washed to remove soluble by-products, and dried.

Magaldrate is an antacid having somewhat more efficacy than aluminum hydroxide. It does not appear to disturb electrolyte balance or bowel function.

Sodium Bicarbonate. Sodium bicarbonate [*144-55-8*], $NaHCO_3$, is a white crystalline powder. It is odorless, has a saline and slightly alkaline taste, and is stable in dry air, but slowly decomposes in moist air. Its solubility is one gram in 10 mL water; in hot water it is converted into carbonate, and it is insoluble in alcohol.

Sodium bicarbonate may be prepared by the ammonia–salt (Solvay) process. Carbon dioxide is passed through a solution of sodium chloride in ammonia water. Sodium bicarbonate is precipitated and the ammonium chloride remains in solution. The ammonium chloride is heated with lime to regenerate ammonia (see ALKALI AND CHLORINE PRODUCTS).

Sodium bicarbonate is a gastric antacid that may cause systemic alkalosis on overdose and may contribute to edema owing to sodium retention. It is useful for systemic acidosis because both deficient ions are present in the same molecule, and it can be used topically as a moist paste or in solution as an antipruritic. Sodium bicarbonate also is an ingredient of many effervescent mixtures, alkaline solutions, etc. One gram of $NaHCO_3$ neutralizes 115 mL 0.1 N HCl.

Laxatives

Laxatives facilitate the passage and elimination of feces. These agents are used most commonly as self-treatment and are rarely prescribed by physicians. The use of laxatives should be limited to those few conditions in which they are warranted and should not be used chronically. Legitimate uses include the removal of hardened and dry stools causing partial or complete intestinal obstruction, the softening of stools in patients for whom straining could be harmful, and as preparation prior to diagnostic tests. Laxatives have traditionally been classified as bulking agents; contact, ie, stimulant; saline, ie, osmotic; and emollients, ie, lubricant. Emollient or lubricant laxatives such as mineral oil are not discussed herein. Bulking agents containing indigestible fiber appear to have the most physiological mechanism of action and should be the agent of choice unless purgation is needed (see DIETARY FIBER).

Bulk Laxatives. *Carboxymethylcellulose Sodium.* Carboxymethyl ether of cellulose sodium salt (Citrucel) (**8**) is a white granular substance soluble in water depending on the degree of substitution. It is equally soluble in cold and hot water and may be prepared by treating alkali cellulose with sodium chloroacetate.

When mixed with water, carboxymethylcellulose sodium [9004-32-4] makes a bulky hydrophilic colloid which is indigestible and nonabsorbable. It produces softening of formed stools within three days, but may cause fluid retention because of its sodium content.

$$R = H \text{ or } CH_2\overset{\displaystyle O}{\overset{\displaystyle \|}{C}}O^- Na^+$$

(**8**)

Polycarbophil. Polycarbophil [73038-24-1] (copolymer of acrylic acid and divinyl glycol (1,5-hexadiene-3,4-diol [1069-23-4])) consists of white-to-creamy white granules having a slight ester-like odor. It swells to contain a maximum of 1.5% water, but is insoluble in water and most organic solvents. It is prepared by copolymerization of acrylic acid and divinyl glycerol in a hot salt slurry using azobisisobutyronitrile as the initiator.

Polycarbophil binds free water and, therefore, increases the fluidity of stools. It is most active in the slightly acid or alkaline medium of the small bowel and colon. Polycarbophil's calcium salt [9003-97-8] is called Carbofil or Sorboquel.

Psyllium Hydrophilic Mucciloid. Metamucil, or psyllium hydrophilic mucciloid, is a white-to-cream colored, slightly granular powder having little or no odor and a slightly acidic taste. It is made from the mucilaginous portion, ie, outer epidermis, of blond psyllium seeds (*Plantago ovata*).

Metamucil is a popular bulking agent for use in the chronic therapy of several gastrointestinal disorders. It is also indicated for use when a high fiber intake is recommended.

Contact Laxatives. *Bisacodyl.* 4,4'-(2-Pyridylmethylene)bisphenol diacetate [603-50-9] (Dulcolax) (**9**) is a white to off-white crystalline powder in which particles of 50 μm dia predominate. It is very soluble in water, freely soluble in chloroform and alcohol, soluble in methanol and benzene, and slightly soluble in diethyl ether. Bisacodyl may be prepared from 2-pyridine-carboxaldehyde by condensation with phenol and the aid of a dehydrant such as sulfuric acid. The resulting 4,4'-(pyridylmethylene)diphenol is esterified by treatment with acetic anhydride and anhydrous sodium acetate. Crystalliza-tion is from ethanol.

Bisacodyl (**9**) is a contact laxative that may be given orally or rectally. It is often used for evacuation of the bowel prior to surgery or diagnostic examination. It may obviate the need for a cleansing enema.

Phenolphthalein. Alophen, Ex-Lax, Feen-a-Mint, Modane, and Phenolax are trade names for phenolphthalein [*77-09-8*] (3,3-bis(4-hydroxyphenyl)-1-(3*H*)-1 isobenzofuranone) (**10**). It is a white or faintly yellowish white crystalline powder, odorless and stable in air, and practically insoluble in water; one gram is soluble in 15 mL alcohol and 100 mL diethyl ether. Phenolphthalein may be prepared by mixing phenol, phthalic anhydride, and sulfuric acid, and heating at 120°C for 10–12 h. The product is extracted with boiling water, then the residue dissolved in dilute sodium hydroxide solution, filtered, and precipitated with acid.

Phenolphthalein is a cathartic drug and the basis of many OTC laxatives. Its action is mainly on the colon to produce a soft semifluid stool within six to eight hours. Its action may persist for several days owing to enterohepatic cir-culation and it may cause red urine if urine is alkaline.

(**9**)   (**10**)

Cascara Sagrada. Cascara sagrada, also known as sacred bark, chitten, dogwood, coffeeberry, bearberry, bitter bark, and bearwood, is the dried bark of *Rhamnus Purshiana* DeCandolle. It is in the form of brown, purplish brown, or brownish red flattened or transversely curved pieces, 1- to 5-mm thick, and has a characteristic odor and bitter taste. It should be collected at least one year prior to use. The active constituents are aloe-emodin [*481-72-1*], $C_{15}H_{10}O_5$; iso-emodin [*476-62-0*], $C_{15}H_{10}O_5$; purshianin [*1393-00-6*]; and several resins.

Cascara sagrada is used as a cathartic. It is most useful when prepared as a fluid extract, and tends to be a mild laxative causing little discomfort. However, on prolonged use it may result in characteristic melanotic pigmentation of the rectal mucosa. The bitter taste can be lessened, owing to neutralization of the acid constituents, if the ground substance is moistened and mixed with mag-nesium or calcium hydroxide. This treatment may lessen the potency of the preparation.

Castor Oil. Castor oil [*8001-79-4*] (qv) is the fixed oil from the seeds of *Ricinus communis* Linne. Pale yellowish or almost colorless, it is a transparent viscid liquid with a faint, mild odor and a bland taste followed by a slightly acrid and usually nauseating taste. Its specific gravity is between 0.945 and 0.965. Castor oil is soluble in alcohol, and miscible with anhydrous alcohol, glacial acetic acid, chloroform, and diethyl ether. It consists chiefly of the glycerides of ricinoleic acid [141-22-0], $C_{18}H_{34}O_3$, and isoricinoleic acid [73891-08-4], $C_{18}H_{34}O_3$, found in the small intestine. The seed contains a highly poisonous albumin (ricin) and base (ricinine).

Castor oil may be obtained by cold expression of the decorticated seed. The oil is steamed under vacuum to eliminate odors and coagulate the toxic albumin. Fuller's earth or activated charcoal may be used for further purification.

Castor oil is a cathartic only after lipolysis in the small intestine liberating ricinoleic acid. Ricinoleic acid inhibits the absorption of water and electrolytes. It is commonly used for preparation of the large bowel for diagnostic procedures.

Docusate Calcium. Dioctyl calcium sulfosuccinate [128-49-4] (calcium salt of 1,4-bis(2-ethylhexyl)ester butanedioic acid) (11) is a white amorphous solid having the characteristic odor of octyl alcohol. It is very slightly soluble in water, and very soluble in alcohol, polyethylene glycol 400, and corn oil. It may be prepared directly from dioctyl sodium sulfosuccinate dissolved in 2-propanol, by reaction with a methanolic solution of calcium chloride.

Docusate calcium is used both as a fecal softening agent and an emulsifier, ie, a wetting or dispersing agent for external preparations (see EMULSIONS).

(11)

Docusate Sodium. Aerosol OT, Colace, and Doxinate are trade names of docusate sodium [577-11-7] (dioctyl sodium sulfosuccinate, sodium salt of 1,4-bis(2-ethylhexyl)ester butanedioic). This white, wax-like, plastic solid, with a characteristic odor suggestive of octyl alcohol, is usually available in the form of pellets. One gram of the sodium salt slowly dissolves in about 70 mL water. Docusate sodium is freely soluble in alcohol and glycerol, very soluble in hexane, and decomposes in the presence of strong alkali. It may be prepared by treatment of maleic anhydride with 2-ethylhexanol to produce dioctyl maleate, which is allowed to react with sodium bisulfite under conditions conducive to saturation of the olefinic bond with simultaneous arrangement of the bisulfite to the sulfonate.

Docusate sodium is both a surface-active agent for use as a fecal softener, and a wetting agent in industrial, pharmaceutical, cosmetic, and food applications.

Saline or Osmotic Laxatives. *Lactulose.* 4-O-β-D-Galactopyranosyl-4-D-fructofuranose [4618-18-2] (Chronolac) (12) may be made from lactose using the method described in Reference 9. It is a synthetic disaccharide that is not hydro-lyzed by gastrointestinal enzymes in the small intestine, but is metabolized by colonic bacteria to short-chain organic acids. The increased osmotic pressure of these nonabsorbable organic acids results in an accumulation of fluid in the colon. Lactulose may not be tolerated by patients because of an extremely sweet taste. It frequently produces flatulence and intestinal cramps.

(**12**)

Magnesium Citrate Solution. The solution magnesium citrate (3:2) [*3344-18-1*] (citrate of magnesia), $C_{12}H_{10}Mg_3O_{14}$, contains in each 100 mL an amount of magnesium citrate corresponding to 1.55–1.9 g of MgO, as well as potassium bicarbonate and flavorings. It is a colorless to slightly yellow, clear, effervescent liquid with a sweet, acidulous taste and a lemon flavor.

In the preparation of a kilogram of magnesium citrate solution, anhydrous citric acid (49.4 g) is dissolved in 220 mL of hot purified water. Magnesium car-bonate is added slowly and stirred until dissolved into 220 mL of purified water. The syrup is added and the mixed liquids heated to the boiling point. The lemon oil is triturated with talc and added to the hot mixture, which is filtered into a strong sterile bottle of appropriate size. Boiled purified water is added to make 770 mL. The bottle is stoppered with cotton and allowed to cool, potassium bicar-bonate is added, and the bottle is closed securely. The mixture should be shaken occasionally and the bottle stored on its side in a cool place. Sodium bicarbonate (4.6 g) may be substituted for potassium bicarbonate, and the solution may also be further carbonated using carbon dioxide. The solution can be stabilized by sterilization or by the addition of 66 g citric acid and a quantity of magnesium carbonate equivalent to 13.2 g MgO. Precipitation on standing is increased by the presence of sucrose and CO_2, and decreased by sterilization. Magnesium citrate solution is a saline cathartic, usually packaged in a single-use container. It is used sometimes as a preoperative or prediagnostic cathartic.

Magnesium Sulfate. Magnesium sulfate heptahydrate [*10034-99-8*], $MgSO_4 \cdot 7H_2O$, also known as bitter salts and Epsom salts, are small, colorless crystals, usually needle-like, having a cooling, saline, and bitter taste, that efflo-resce in warm, dry air. The heptahydrate loses 5 molecules of water at 100°C. An aqueous solution of magnesium sulfate is neutral, and one gram is soluble in 1 mL water, 0.2 mL boiling water, or 1 mL glycerol; the heptahydrate is sparingly soluble in ethanol.

Magnesium sulfate heptahydrate may be prepared by neutralization of sul-furic acid with magnesium carbonate or oxide, or it can be obtained directly from natural sources. It occurs abundantly as a double salt and can also be obtained from the magnesium salts that occur in brines used for the extraction of bromine (qv). The brine is treated with calcium hydroxide to precipitate magnesium hydroxide. Sulfur dioxide and air are passed through the suspension to yield magnesium sulfate (see CHEMICALS FROM BRINE). Magnesium sulfate is a saline cathartic.

Sodium Phosphate. Disodium phosphate heptahydrate [7782-85-6] (Fleets Phospho-Soda), $Na_2HPO_4 \cdot 7H_2O$, is a colorless or white granular salt which effloresces in warm dry air. One gram of the heptahydrate dissolves in 4 mL water, and is only slightly soluble in alcohol. In the preparation, finely ground phosphatic material is mixed with a quantity of sulfuric acid a little in excess of the amount required to transform the phosphate into monobasic calcium phosphate. Bone phosphate or bone ash, obtained by heating bones to whiteness and which consists mainly of tribasic calcium phosphate and phosphorite (phosphate rock), may be used as a source of phosphatic materials. The mixture is leached with hot water, a concentrated solution of sodium carbonate sufficient to convert half of the phosphate into the dibasic sodium salt is added, and the mixture is boiled. After filtering, the solution is concentrated and the sodium phosphate allowed to crystallize. Disodium phosphate heptahydrate is used as a saline cathartic, mainly for preparation for diagnostic procedures (see PHOSPHORIC ACID AND THE PHOSPHATES).

Polyethylene Glycol Electrolyte Preparation. A mixture of osmotically balanced ingredients, made up of polyethylene glycol 3350 (a nonabsorbable osmotic agent), sodium chloride, sodium sulfate, potassium chloride, and sodium bicarbonate, is known as Colyte, GoLYTELY, and NuLytely. It is usually in a dry formulation to be made up to volume with cold water just prior to administration. The mixture provides complete gastrointestinal cleansing when given prior to endoscopic or x-ray diagnostic procedures. It is usually well tolerated with little change in blood electrolyte concentrations, but occasionally may cause nausea, vomiting, abdominal fullness, and cramps because of the large volume that must be ingested.

Antidiarrheal Therapy

Diarrhea is a common problem that is usually self-limiting and of short duration. Increased accumulations of small intestinal and colonic contents are known to be responsible for producing diarrhea. The former may be caused by increased intestinal secretion which may be enterotoxin-induced, eg, cholera and *E. coli*, or hormone and drug-induced, eg, caffeine, prostaglandins, and laxatives; decreased intestinal absorption because of decreased mucosal surface area, mucosal disease, eg, tropical sprue, or osmotic deficiency, eg, disaccharidase or lactase deficiency; and rapid transit of contents. An increased accumulation of colonic content may be linked to increased colonic secretion owing to hydroxy fatty acid or bile acids, and exudation, eg, inflammatory bowel disease or amebiasis; decreased colonic absorption caused by decreased surface area, mucosal disease, and osmotic factors; and rapid transit, eg, irritable bowel syndrome.

Diagnosis and alleviation of the cause, if possible, is of primary importance. Often, however, this is not possible and therapy is used to alleviate the inconvenience and pain of diarrhea. These compounds usually only mask the underlying factors producing the problem. Diarrhea may cause significant dehydration and loss of electrolytes and is a particularly serious problem in infants. Antidiarrheals do not usually prevent the loss of fluids and electrolytes into the large bowel and, although these may prevent frequent defecation, often the serious imbalance of body electrolytes and fluids is not significantly affected.

Commonly used antidiarrheals work by one of two mechanisms: effects on net intestinal secretion, or a decrease in intestinal propulsive motility. Narcotic analgesics are constipating and are antidiarrheal owing to the effect on intestinal propulsion. This is not a smooth muscle relaxing activity but results from the increase in nonpropulsive phasic smooth muscle contractions, and is apparently the mechanism of such commonly used compounds as codeine sulfate [76-57-3], diphenoxylate (**13**), difenoxin, and loperamide (**14**). Some opiates also have been found to have effects on intestinal secretion. Bismuth subsalicylate (Pepto-Bismol) is effective and works by several mechanisms. Although drugs such as atropine, which have antispasmodic activity, have been used as antidiarrheals in the past, they are no longer considered useful or effective at doses not causing significant side effects. Some drugs effective in the therapy of inflammatory bowel disease are also included even though they are not specifically antidiarrheal but are useful in chronic therapy of intestinal inflammation.

Bismuth Subsalicylate.   (2-Hydroxybenzoato-O^1)-oxobismuth [14882-18-9] (Pepto-Bismol) may be made by the process described in Reference 10. Bismuth subsalicylate has been shown to bind toxins produced by several bacterial strains. It may also act as a result of its salicylate component on prostaglandin formation and have specific intestinal antisecretory activity. It has been found to be effective in the prevention and therapy of traveler's diarrhea and diarrhea of nonspecific origin. It is available as a suspension or tablet, and because it is radiopaque it may interfere with radiological examination. It also may produce a grayish black discoloration of the stool which could be confused with melena.

Diphenoxylate Hydrochloride.   1-(3-Cyano-3,3-diphenylpropyl)-4-phenyl-4-piperidinecarboxylic acid monohydrochlorhydrate [3810-80-8] (Lomotil) (**13**) is a white, odorless, crystalline powder that melts at 220–226°C. It is soluble in methanol, sparingly soluble in ethanol and acetone, slightly soluble in water and isopropyl alcohol, freely soluble in chloroform, and practically insoluble in ether and hexane. The method of preparation for diphenoxylate hydrochloride is available (11). Diphenoxylate hydrochloride [3810-80-8] (**13**) is an antidiarrheal that acts through an opiate receptor. It has effects both on propulsive motility and intestinal secretion. Commercial forms are mixed with atropine to discourage abuse.

(**13**)

Loperamide.   4-(4-Chlorophenyl)-4-hydroxy-*N,N*-dimethyl-*a,a*-diphenyl-1-piperidinebutanamide monohydrochloride [43552-83-5] (Imodium) (**14**) is practically insoluble in water (0.002%) at physiological pH. Its crystals are not affected by light and it is not hygroscopic. It may be synthesized (12).

Loperamide is similar in action and use to diphenoxylate; however, it does not need to be formulated with atropine and is available by prescription and OTC. It is reported to have fewer central nervous system side effects than diphenoxylate.

(14)   (15)

Sulfasalazine. Salicylazosulfapyridine or Azulfadine [*599-79-1*] (2-hydroxy-5-[[4[(2-pyridylamino)sulfonyl]-phenyl]azo] benzoic acid) (**15**) is a light brownish yellow-to-bright yellow fine powder that is practically tasteless and odorless. It melts at ca 255°C with decomposition, is very slightly soluble in ethanol, is practically insoluble in water, diethyl ether, chloroform, and benzene, and is soluble in aqueous solutions of alkali hydroxides. Sulfasalazine may be made by the synthesis described in Reference 13. It is not used as an antidiarrheal as such, but is indicated for the treatment of inflammatory bowel diseases such as ulcerative colitis and Crohn's disease. Its action is purported to result from the breakdown in the colon to 5-aminosalicylic acid [*89-57-6*] (5-ASA) and sulfapyridine [*144-83-2*]. It may cause infertility in males, as well as producing idiosyncratic reactions in some patients; these reactions have been attributed to the sulfa component of the compound. The mechanism of 5-ASA is attributed to inhibition of the arachidonic acid cascade preventing leukotriene B_4 production and the ability to scavenge oxygen free radicals. The active component appears to be 5-aminosalicylic acid.

Mesalamine. Rowasa, Asacol, and Pentasa are trade names for mesalamine [*89-57-6*] (5-ASA, 5-amino-2-hydroxybenzoic acid). It is a white to pinkish crystalline substance that is slightly soluble in cold water and alcohol, more soluble in hot water, and soluble in hydrochloric acid. It may be prepared by the reduction of *m*-nitrobenzoic acid with zinc dust and HCl.

5-ASA appears to be the active component of sulfasalazine without the sulfa component, and is free of the serious side effects seen with sulfasalazine. It is used orally, in a delay-release formulation, as a retention enema, and as a suppository. It is well tolerated in most patients.

Antiemetics

Nausea and vomiting are frequent symptoms of disease and can be produced by a number of causes, eg, pregnancy, chemotherapy, motion sickness, radiation, and gastrointestinal infections. In most cases, it is self-limiting. However, in those cases in which fluid and electrolyte loss are significant, the control of emesis is needed. Drugs may act at a variety of physiological locations, eg, blockade of dopamine and effects at the chemoreceptor trigger zone, vestibular apparatus, or serotonin (5-HT$_3$) receptors (Fig. 2) (see HISTAMINE AND HISTAMINE ANTAGONISTS).

   Benzquinamide.   Emete-con [63-12-7] (*N,N*-diethyl-1,3,4,6,7,11*b*-hexahydro-2-hydroxy-9,10-dimethoxy-2*H*-benzo-(*a*)-quinolizine-3-carboxamide acetate hydrochloride) (**16**) may be made by the method described in Reference 14.

   Benzquinamide (**16**) is a nonamine-depleting benzoquinolizine derivative, chemically unrelated to phenothiazine and other antiemetics, with antiemetic, antihistaminic, anticholinergic, and sedative properties. It is given by parenteral administration for the prevention and treatment of nausea and vomiting associated with anesthesia and surgery. It acts directly at the chemoreceptor trigger zone and is not particularly effective against chemotherapy-induced vomiting.

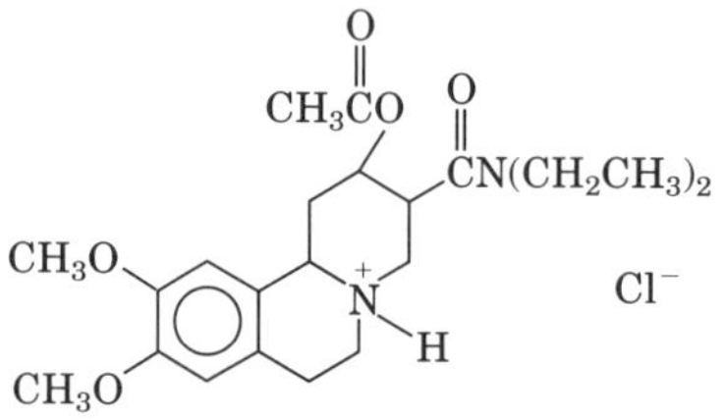

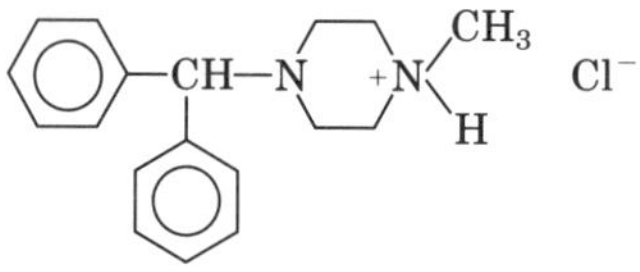

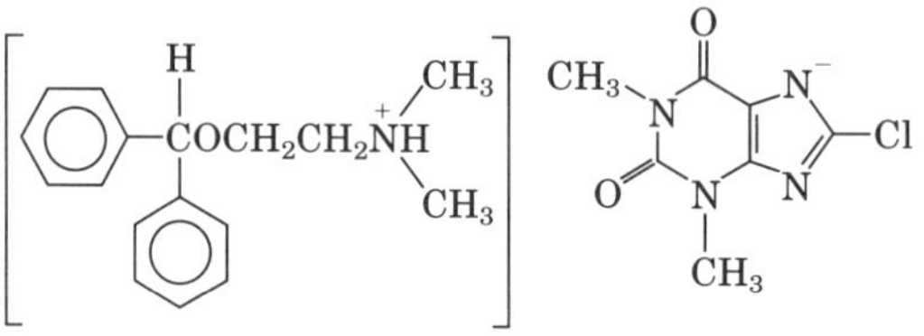

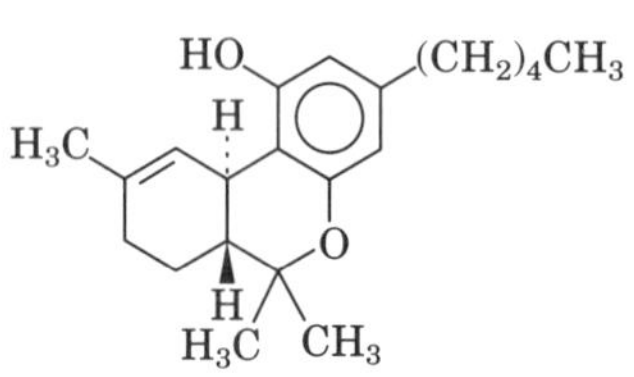

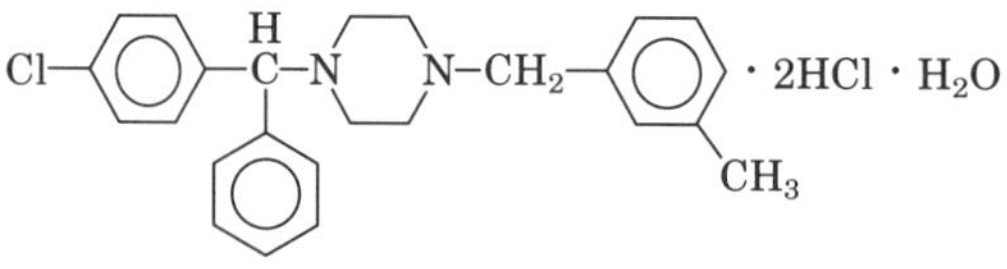

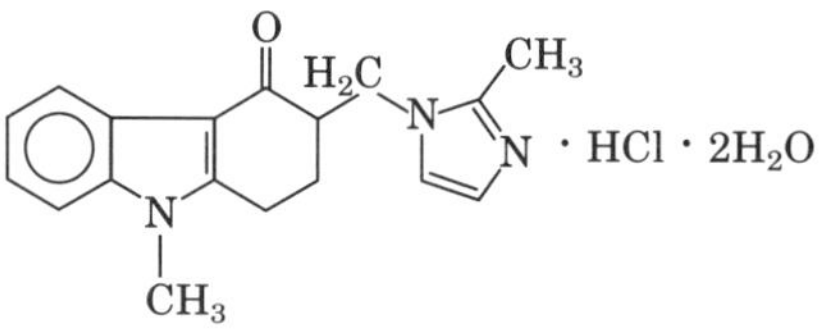

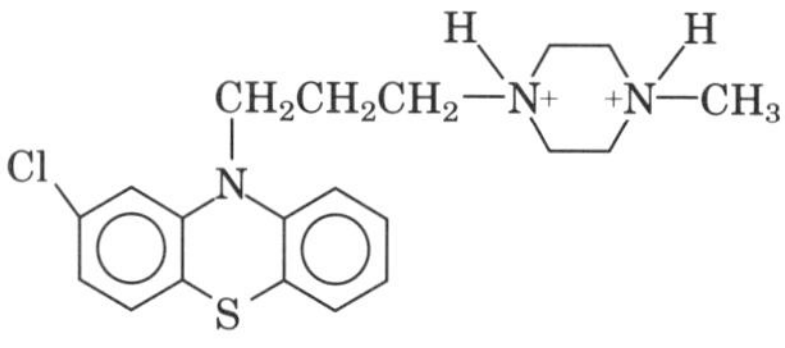

Fig. 2. Structures of antiemetic agents.

Cyclizine Hydrochloride.   1-(Diphenylmethyl)-4-methylpiperazine mono-hydrochloride [303-25-3] (Marezine) (**17**) is a white crystalline powder, or small colorless crystals, that is odorless or nearly so and has a bitter taste. It melts indistinctly and with decomposition at ca 285°C. One gram of cyclizine hydrochloride [303-25-3] is soluble in 115 mL water, 115 mL ethanol, and 5 mL chloroform; it is insoluble in diethyl ether. It may be made by the synthesis shown in Reference 15.

This compound has antihistaminic activity and is useful in the therapy of motion sickness. It may also be effective in the control of post-operative nausea and vomiting. It is classified as FDA Category B for Pregnancy, ie, no demonstrated risks shown in animal studies; however, no controlled trials in pregnant women. Large doses may cause drowsiness and dry mouth owing to decreased secretion of saliva.

Dimenhydrinate.   Dimenhydrinate [523-87-5] (Dramamine) (**18**) is a white crystalline, odorless powder that melts between 102 and 107°C. It is sparingly soluble in water, freely soluble in ethanol and chloroform, and sparingly soluble in diethyl ether. Dimenhydrinate is prepared by combining dimethylaminoethyl benzhydryl ester with 8-chlorotheophylline and refluxing in an isopropyl alcohol solution. The crystalline precipitate of dimenhydrinate that forms on cooling is collected by filtration, washed with cold ethyl acetate, and dried.

Dimenhydrinate is an antiemetic especially useful as an antinauseant in motion sickness, and for syndromes associated with vertigo such as Meniere's syndrome, radiation sickness, and vestibular dysfunction. It may produce mild drowsiness. It is FDA Category B for Pregnancy, and is available as an OTC preparation as well as by prescription.

Dronabinol.   Marinol (6aR-*trans*)-6a,7,8,10a-tetrahydro-6,6,9-trimethyl-3-pentyl-6H-dibenzo(B,D)pyran-1-ol is the principal psychoactive substance present in *Cannabis sativa* L., ie, marijuana. It is a controlled substance, formulated in sesame oil and encapsulated in soft gelatin capsules for oral administration.

Dronabinol is indicated for the treatment of the nausea and vomiting produced by cancer chemotherapy in patients who have failed to respond adequately to other conventional treatments. This agent may be habit forming and can be expected to produce disturbing psychomimetic reactions. It should only be used under close supervision.

Meclizine Hydrochloride.   Piperazine, Antivert, and Bonine are trade names for meclizine dihydrochloride monohydrate [31884-77-2] (**20**). It is a white or slightly yellowish crystalline powder with a slight odor, no taste, and a melting point of 217–224°C. The hydrochloride is practically insoluble in water and ether. It is freely soluble in chloroform, pyridine, methylacetamide, and mild acid alcohol–water mixtures, and is slightly soluble in dilute acids or alcohol. See Reference 16 for synthesis.

Meclizine is often used for the therapy of motion sickness and is available without prescription in a variety of formulations.

Ondansetron.   (+/−)1,2,3,9-Tetrahydro-9-methyl-3-((2-methyl-1H-imidazol -1-yl)methyl)-4H-carbazol-4-one, monohydrochloride, dihydrate [35727-72-1] (Zofran) (**21**) is a white to off-white powder that is soluble in water and normal saline. It may be prepared by the method in Reference 17. Ondansetron (**21**) is a new type of antiemetic, which has selective serotonin (5-HT_3) antagonistic activity. It is particularly useful in emesis induced by cytotoxic chemotherapeutic agents such as cisplatin. It lacks the high incidence of dystonic reaction seen with antiemetics that have dopamine-blocking activity. Side effects are infrequent other than a significant incidence of constipation. It is available for use by injection for the prevention of nausea and vomiting associated with initial and repeat courses of emetogenic cancer chemotherapy.

Prochlorperazine Maleate.   2-Chloro-10-[3-(4-methyl-1-piperazinyl)-propyl]-10H-phenothiazine maleate [*84-02-6*] (Compazine) (**22**) is a white or pale yellow crystalline powder. It is almost completely odorless, its saturated solution is acidic to litmus, it is practically insoluble in water and ethanol, and it is slightly soluble in warm chloroform. It may be made by the synthesis described in Reference 18. Prochlorperazine maleate [*84-02-6*] is an effective antiemetic and tranquilizing agent. It is not particularly effective for motion sickness. Adverse reactions that may occur include extrapyramidal reactions, motor restlessness, dystonias, tardive dyskinesia, and contact dermatitis. Prochlorperazine is also a significant phenothiazine antipsychotic.

Gastric Prokinetics

Gastric prokinetics are a new class of agent that can augment gastric emptying. In most cases these also increase the barrier pressure of the lower esophageal sphincter and increase acid clearance making them useful for therapy of esophageal reflux (heartburn) and esophagitis. They also may be effective in increasing gastric emptying in diabetic gastroparesis and other diseases in which there is a decrease in the ability for the stomach to empty. Cisapride [*81098-60-4*] (Propulsid) has been accepted by the FDA for the treatment of gastroesophageal reflux. It also has been shown to have significant effect on gastric emptying and may be useful in diabetic gastroparesis and other disease in which delayed gastric emptying is a problem. The prevalent hypothesis of its activity is that it has agonistic activity on the serotonin (5-HT$_4$) receptors on the nervous network in the gastrointestinal tract.

(**23**)

Metoclopramide.   Reglan (**23**) is a white crystalline powder soluble in water. It may be made by the method described in Reference 19.

Metoclopramide [*54143-57-6*] is an effective antiemetic which may work due to its central dopamine-blocking activity and/or serotonin (5-HT$_3$) blocking activity. It also stimulates propulsive motility of the upper gastrointestinal tract, probably due to stimulation of 5-HT$_4$ receptors. Interactions with concomitant medication may occur due to increased gastric emptying. It is indicated for the nausea and vomiting due to emetogenic cancer chemotherapy or prevention of postoperative nausea and vomiting. It is also indicated for symptomatic esophageal reflux, diabetic gastroparesis, small bowel intubation, and radiological examination. Use is often limited due to extrapyramidal symptoms presenting primarily as acute dystonic reactions. These side effects may be partially alleviated by ad-ministration of diphenhydramine or benztropine mesylate.

BIBLIOGRAPHY

"Antacids, Gastric" in *ECT* 1st ed., Vol. 1, pp. 930–934, by J. C. Krantz, Jr., University of Maryland; in *ECT* 2nd ed., Vol. 1, pp. 427–431, by J. C. Krantz, Jr., University of Maryland; "Gastrointestinal Agents" in *ECT* 3rd ed., Vol. 2, pp. 696–710, by H. I. Jacoby, McNeil Laboratories.

1. D. R. Bennett, ed. *Drug Evaluations Annual 1992*, American Medical Association, Chicago, Ill.
2. U.S. Pat. 3,950,333 (Apr. 13, 1976), G. J. Durant and co-workers (to SmithKline and French Ltd.).
3. U.S. Pat. 4,283,408 (Aug. 11, 1981), T. Hirata and co-workers (to Yamanuechi).
4. U.S. Pat. 4,375,547 (Mar. 1, 1983), R. Ploch (to Eli Lilly).
5. U.S. Pat. 4,128,658 (Dec. 5, 1978), B. Price and co-workers (to Allen and Hansberry).
6. U.S. Pat. 4,255,431 (Apr. 5, 1979), U. Jungren and co-workers (to Aktiebolget Hassle).
7. U.S. Pat. 3,432,489 (Mar. 11, 1969), Y. Nitta and co-workers (to Chugai Seiyako K.K.).
8. U.S. Pat. 3,965,143 (June 22, 1976), P. W. Collins and co-workers (to G. D. Searle).
9. J. H. Montgomery, *J. Am Chem Soc.* **52**, 2101 (1930).
10. G. Fischer, *Arch Pharm.* **231**, 680 (1893).
11. U.S. Pat. 2,898,340 (Aug. 18, 1959) P. A. J. Janssen, and co-workers (to Janssen Pharmaceutica).
12. U.S. Pat. 3,714,159 (Jan. 30, 1973), P. A. J. Janssen, and co-workers (to Janssen Pharmaceutica).
13. U.S. Pat. 2,396,145 (Mar. 12, 1946), Askelof and co-workers (to Aktiebolaset Pharmacia).
14. U.S. Pat. 3,053,485 (Sept. 11, 1962) Trettner (to Pfizer).
15. U.S. Pat. 2,630,435 (Mar. 30, 1953), R. Baltzly and co-workers (to Burroughs Welcome).
16. U.S. Pat. 2,709,169 (May 24, 1955), H. Murren (to Union Chemique Belge Societe Anonyme).
17. U.S. Pat. 4,695,578 (Sept. 22, 1987), R. Coates and co-workers (to Glaxo Group).
18. U.S. Pat. 2,902,484 (Sept. 1, 1959), R. J. Horclois and co-workers (to Societe des Unines Chemique Rhône-Poulenc).
19. U.S. Pat. 3,177,252 (Jan. 14, 1964), G. Thominet and co-workers (to Soc. d'Etudes Science and Industry).

General References

USP XXII, 22th rev., 1990, The U.S. Pharmacopoeial Convention, Rockville, Md.
Merck Index, 11th ed., Merck, Rahway, N.J., 1989.
C. A. Fleeger, ed., *USAN and USP Dictionary of Drug Names*, U.S. Pharmacopeial Convention Inc., Rockville, Md.

HENRY I. JACOBY
Discovery Research Consultants

GELATIN

Gelatin [9000-70-80] is a protein obtained by partial hydrolysis of collagen, the chief protein component in skin, bones, hides, and white connective tissues of the animal body (see PROTEINS). Type A gelatin is produced by acid processing of collagenous raw material; type B is produced by alkaline or lime processing. Because it is obtained from collagen by a controlled partial hydrolysis and does not exist in nature, gelatin is classified as a derived protein. Animal glue and gelatin hydrolysate, sometimes referred to as liquid protein, are products obtained by a more complete hydrolysis of collagen and thus can be considered as containing lower molecular-weight fractions of gelatin.

Use of animal glues was first recorded ca 4000 BC in ancient Egypt (1). Throughout subsequent centuries, glue and crude gelatin extracts with poor organoleptic properties were prepared by boiling bone and hide pieces and allowing the solution to cool and gel. Late in the seventeenth century, the first commercial gelatin manufacturing began. At the beginning of the nineteenth century, commercial production methods gradually were improved to achieve the manufacture of high molecular weight collagen extracts with good quality that form characteristic gelatin gels (1–3).

Uses of gelatin are based on its combination of properties; reversible gel-to-sol transition of aqueous solution; viscosity of warm aqueous solutions; ability to act as a protective colloid; water permeability; and insolubility in cold water, but complete solubility in hot water. It is also nutritious. These properties are utilized in the food, pharmaceutical, and photographic industries. In addition, gelatin forms strong, uniform, clear, moderately flexible coatings which readily swell and absorb water and are ideal for the manufacture of photographic films and pharmaceutical capsules.

Chemical Composition and Structure

Gelatin is not a single chemical substance. The main constituents of gelatin are large and complex polypeptide molecules of the same amino acid composition as the parent collagen, covering a broad molecular weight distribution range. In the parent collagen, the 18 different amino acids are arranged in ordered, long chains, each having ~95,000 mol wt. These chains are arranged in a rod-like, triple-helix structure consisting of two identical chains, called α_1, and one slightly different chain called α_2 (4–6). These chains are partially separated and broken, ie, hydrolyzed, in the gelatin manufacturing process. Different grades of gelatin have average molecular weight ranging from ~20,000 to 250,000 (7–14). Molecular weight distribution studies have been carried out by fractional precipitation with ethanol or 2-propanol and by complexing with anionic detergent molecules. The coacervates are isolated and recovered as gelatin fractions (15–17).

Analysis shows the presence of amino acids from 0.2% tyrosine to 30.5% glycine (see AMINO ACIDS). The five most common amino acids are glycine [56-40-6], 26.4–30.5%; proline [147-85-3], 14.8–18%; hydroxyproline [51-35-4], 13.3–14.5%, glutamic acid [56-86-0], 11.1–11.7%; and alanine [56-41-7],

8.6–11.3%. The remaining amino acids in decreasing order are arginine [74-79-3], aspartic acid [56-84-8], lysine [56-87-1], serine [56-45-1], leucine [61-90-5], valine [72-18-4], phenylalanine [63-91-2], threonine [72-19-5], isoleucine [73-32-5], hydroxylysine [13204-98-3], histidine [71-00-1], methionine [63-68-3], and tyrosine [60-18-4] (18–20).

Warm gelatin solutions are more levorotatory than expected on the basis of the amino acid composition, indicating additional order in the molecule, which probably results from Gly-Pro-Pro and Gly-Pro-Hypro sequences (21). The α-chain form of gelatin behaves in solution like a random-coil polymer, whereas the gel form may contain as much as 70% helical conformation (22). The remaining molecules in nonhelical conformation link helical regions together to form the gel matrix. Helical regions are thought to contain both inter- and intramolecular associations of chain segments.

Gelatin structures have been studied with the aid of an electron microscope (23). The structure of the gel is a combination of fine and coarse interchain networks; the ratio depends on the temperature during the polymer–polymer and polymer–solvent interaction leading to bond formation. The rigidity of the gel is approximately proportional to the square of the gelatin concentration. Crystallites, indicated by x-ray diffraction pattern, are believed to be at the junctions of the polypeptide chains (24).

Homogeneous α-chain gelatin has been prepared by pretreating collagen with pronase in the presence of 0.4 M CaCl$_2$ [10043-52-4], and extracting the gelatin with hot water at 80°C and pH 7.0 after inactivating the enzyme and removing the salts.

Stability. Dry gelatin stored in airtight containers at room temperature has a shelf life of many years. However, it decomposes above 100°C. For complete combustion, temperatures above 500°C are required. When dry gelatin is heated in air at relatively high humidity, >60% rh, and at moderate temperatures, ie, above 45°C, it gradually loses its ability to swell and dissolve (25,26). Aqueous solutions or gels of gelatin are highly susceptible to microbial growth and breakdown by proteolytic enzymes. Stability is a function of pH and electrolytes and decreases with increasing temperature because of hydrolysis.

Physical and Chemical Properties

Commercial gelatin is produced in mesh sizes ranging from coarse granules to fine powder. In Europe, gelatin is also produced in thin sheets for use in cooking. It is a vitreous, brittle solid, faintly yellow in color. Dry commercial gelatin contains about 9–13% moisture and is essentially tasteless and odorless with specific gravity between 1.3 and 1.4. Most physical and chemical properties of gelatin are measured on aqueous solutions and are functions of the source of collagen, method of manufacture, conditions during extraction and concentration, thermal history, pH, and chemical nature of impurities or additives.

Gelation. Perhaps the most useful property of gelatin solution is its capability to form heat reversible gel–sols. When an aqueous solution of gelatin with a concentration greater than about 0.5% is cooled to about 35 to 40°C, it first increases in viscosity, then forms a gel. The gelation process is thought to proceed through three stages: (1) rearrangement of individual molecular chains into ordered, helical arrangement, or collagen fold (27–30); (2) association of two or three ordered segments to create crystallites (29,31,32); and (3) stabilization of the structure by lateral interchain hydrogen bonding within the helical regions. The rigidity or jelly strength of the gel depends on the concentration, the intrinsic strength of the gelatin sample, pH, temperature, and additives.

Because the economic value of gelatin is commonly determined by jelly strength, the test procedure for its determination is of great importance. Commercially, gelatin jelly strength is determined by standard tests which measure the force required to depress the surface of a carefully prepared gel by a distance of 4 mm using a flat-bottomed plunger 12.7 mm in diameter. The force applied may be measured in the form of the quantity of fine lead shot required to depress the plunger and is recorded in grams. The measurement is termed the Bloom strength after the inventor of the lead shot device (33,34). In the early 1990s, sophisticated testing equipment utilizing sensitive load cells for the measurement are commonly used.

The conversion temperature for gelatin is determined as setting point, ie, sol to gel, or melting point, ie, gel to sol. Commercial gelatins melt between 23 and 30°C, with the setting point being lower by 2–5°C. Melting point determination, described in Reference 35, utilizes test tubes filled with gelatin solution that are gently chilled to form a gel. The tubes are tilted and colored carbon tetrachloride solution is placed on the gelatin surface. The tube is gradually warmed and the end point is determined when the descent of the colored solution is observed. Several methods have been used to determine the setting point of gelatin (36).

Solubility.   In most commercial applications, gelatin is used as a solution. Gelatin is soluble in water and in aqueous solutions of polyhydric alcohols such as glycerol [56-81-5] and propylene glycol [57-55-6]. Examples of highly polar, hydrogen-bonding organic solvents in which gelatin dissolves are acetic acid [64-19-7], trifluoroethanol [75-89-8], and formamide [75-12-7] (37). Gelatin is practically insoluble in less polar organic solvents such as acetone, carbon tetrachloride, ethanol, ether, benzene, dimethylformamide, and most other nonpolar organic solvents. Many water-soluble organic solvents are compatible with gelatin, but interfere with gelling properties (38). Dry gelatin absorbs water exothermally. The rate and degree of swelling is a characteristic of the particular gelatin. Swelled gelatin granules dissolve rapidly in water above 35°C. The cross-linking of gelatin matrix by chemical means is used extensively in photographic products, and this so-called hardening permanently reduces the solubility of gelatin (28,39–44).

Amphoteric Character.   The amphoteric character of gelatin is due to the functional groups of the amino acids and the terminal amino and carboxyl groups created during hydrolysis. In strongly acidic solution the gelatin is positively charged and migrates as a cation in an electric field. In strongly alkaline solution, it is negatively charged and migrates as an anion. The intermediate point, where net charge is zero and no migration occurs, is known as the isoelectric point (IEP) and is designated in pH units (45). A related property, the isoionic point, can be determined by utilizing a mixed-bed ion-exchange resin to remove all nongelatin cations and anions. The resulting pH of the gelatin solution is the isoionic point and is expressed in pH units. The isoionic point is reproducible, whereas the isoelectric point depends on the salts present. Type A gelatin has a broad isoionic region between pH 7 and pH 10; type B is in a lower, more reproducible region, reaching an isoionic point of 5.2 after 4 weeks of liming, which drops to 4.8 after prolonged or more vigorous liming processes (46–49). The isoelectric point can also be estimated by determining a pH value at which a gelatin solution exhibits maximum turbidity (50). Many isoionic point references are recorded as isoelectric points even though the latter is defined as a pH at which gelatin has net charge of zero and thus shows no movement in the electric field (51).

Viscosity. The viscosity of gelatin solutions is affected by gelatin concentration, temperature, molecular weight of the gelatin sample, pH, additives, and impurities. In aqueous solution above 40°C, gelatin exhibits Newtonian behavior. Standard testing methods employ use of a capillary viscometer at 60°C and gelatin solutions at 6.67 or 12.5% solids (34,36). The viscosity of gelatin solutions increases with increasing gelatin concentration and with decreasing temperature. For a given gelatin, viscosity is at a minimum at the isoionic point and reaches maxima at pH values near 3 and 10.5 (52). At temperatures between 30 and 40°C, non-Newtonian behavior is observed, probably due to linking together of gelatin molecules to form aggregates (53). Addition of salts decreases the viscosity of gelatin solutions. This effect is most evident for concentrated gelatin solutions (54,55).

Colloid and Emulsifying Properties. Gelatin is an effective protective colloid that can prevent crystal, or particle, aggregation, thereby stabilizing a heterogeneous suspension. It acts as an emulsifying agent in cosmetics and pharmaceuticals involving oil-in-water dispersions. The anionic or cationic behavior of gelatin is important when used in conjunction with other ionic materials. The protective colloid property is important in photographic applications where it stabilizes and protects silver halide crystals while still allowing for their normal growth and sensitization during physical and chemical ripening processes.

Coacervation. A phenomenon associated with colloids wherein dispersed particles separate from solution to form a second liquid phase is termed coacervation. Gelatin solutions form coacervates with the addition of salt such as sodium sulfate [7757-82-6], especially at pH below the isoionic point. In addition, gelatin solutions coacervate with solutions of oppositely charged polymers or macromolecules such as acacia. This property is useful for microencapsulation and photographic applications (56–61).

Swelling. The swelling property of gelatin is not only important in its solvation but also in photographic film processing and the dissolution of pharmaceutical capsules. That pH and electrolyte content affect swelling has been explained by the simple Donnan equilibrium theory, treating gelatin as a semipermeable membrane (62). This explains why gelatin exhibits the lowest swelling at its isoelectric pH. At pH below the isoelectric point, proper choice of anions can control swelling, whereas above the isoelectric point, cations primarily affect swelling. These effects probably involve breaking hydrogen bonds, resulting in increased swelling. The rate of swelling follows approximately a second-order equation (63). In photographic products, the swelling of the gelatin layer is controlled by coating conditions, drying conditions, chemical cross-linking, and the composition of the processing solutions (40–44). Conditioning at 90% rh and 20°C for 24 h greatly reduces swelling of hot dried film coatings (64). The ratio of lateral to vertical swelling is of great concern in the photographic industry since it can cause curling of photographic papers or films when changes in humidity or general moisture content take place.

Manufacture and Processing

Although new methods for processing gelatin, including ion exchange and cross-flow membrane filtration, have been introduced since 1960, the basic technology for modern gelatin manufacture was developed in the early 1920s. Acid and lime processes have separate facilities and are not interchangeable. In the past, bones and ossein, ie, decalcified bone, have been supplied by India and South America. In the 1990s, slaughterhouses and meat-packing houses are an important source of bones. The supply of bones has been greatly increased since the meat-packing industry introduced packaged and fabricated meats, assisted by the growth of fast-food restaurants. Dried and rendered bones yield about 14–18% gelatin, whereas pork skins yield about 18–22% (see also MEAT PRODUCTS).

Most type A gelatin is made from pork skins, yielding grease as a marketable by-product. The process includes macerating of skins, washing to remove extraneous matter, and swelling for 10–30 h in 1–5% hydrochloric [7647-01-0], phosphoric [7664-38-2], or sulfuric acid [7664-93-9]. Then four to five extractions are made at temperatures increasing from 55–65°C for the first extract to 95–100°C for the last extract. Each extraction lasts about 4–8 h. Grease is then removed, the gelatin solution filtered, and, for most applications, deionized. Concentration to 20–40% solids is carried out in several stages by continuous vacuum evaporation. The viscous solution is chilled, extruded into thin noodles, and dried at 30–60°C on a continuous wire-mesh belt. Drying is completed by passing the noodles through zones of successive temperature changes wherein conditioned air blows across the surface and through the noodle mass. The dry gelatin is then ground and blended to specification.

Type B gelatin is made mostly from bones, but also from bovine hides and pork skins. The bones for type B gelatin are crushed and degreased at the rendering facilities, which are usually located at a meat-packing plant. Rendered bone pieces, 0.5–4 cm, with less than 3% fat, are treated with cool, 4–7% hydrochloric acid from 4 to 14 d to remove the mineral content. An important by-product, dibasic calcium phosphate, is precipitated and recovered from the spent liquor. The demineralized bones, ie, ossein, are washed and transferred to large tanks where they are stored in a lime slurry with gentle daily agitation for 3–16 weeks. During the liming process, some deamination of the collagen occurs with evolution of ammonia. This is the primary process that results in low isoelectric ranges for type B gelatin. After washing for 15–30 h to remove the lime, the ossein is acidified to pH 5–7 with an appropriate acid. Then the extraction processing for type A gelatin is followed. Throughout the manufacturing process, cleanliness is important to avoid contamination by bacteria or proteolytic enzymes.

Bovine hides and skins are substantial sources of raw material for type B gelatin and are supplied in the form of splits, trimmings of dehaired hide, rawhide pieces, or salted hide pieces. Like pork skins, the hides are cut to smaller pieces before being processed. Sometimes the term calfskin gelatin is used to describe hide gelatin. The liming of hides usually takes a little longer than the liming of ossein from bone.

Most manufacturing equipment should be made of stainless steel. The liming tanks, however, can be either concrete or wood (qv). Properly lined iron tanks are often used for the washing and acidification, ie, souring, operations. Most gelatin plants achieve efficient processes by operating around the clock. The product is tested in batches and again as blends to confirm conformance to customer specifications.

Economic Aspects

World gelatin production in 1993 was believed to be about 200,000 t. The United States produced about 30,000 t, followed by France, Germany, Japan, Brazil, and Mexico. Of the gelatin produced in the United States, 55% is acid processed, ie, type A. The U.S. food industry consumes about 20,000 t/yr, with an annual growth rate of 0.5%; the pharmaceutical industry consumes about 10,000 t/yr; and the photographic industry about 7,000 t/yr. In the United States, the pharmaceutical gelatin market is expected to grow on the average of 2.5% per year. The photographic gelatin market has been stable or growing slightly. Color paper and x-ray products use over 55% of the photographic gelatin in the United States, with graphic arts and instant films using an additional 30%.

Analytical Test Methods and Quality Standards

Gelatin is identified by a positive test for hydroxyproline [51-35-4], turbidity with tannic acid [1401-55-4], or a yellow precipitate with acidic potassium dichromate [7778-50-9] or trinitrophenol [88-89-1]. A 5% aqueous solution exhibits reversible gel-to-sol formation between 10 and 60°C. Gelatin gives a positive color test for aldehydes and sugars that are considered undesirable impurities in photographic gelatin; nucleic acids are considered restrainers in photographic gelatins and their concentration is monitored closely for this application (65). Elemental analysis of commercial gelatin is reported as carbon, 50.5%; hydrogen, 6.8%; nitrogen, 17%; and oxygen 25.2% (66); a purer sample analyzed for 18.2–18.4% nitrogen (18,20). Regulations for quality standards vary from country to country, but generally include specifications for ash content, SO_2, heavy metals, chromium, lead, fluoride, arsenic, odor, and for the color or clarity of solutions (67). In addition, certain bacteriological standards, including E. coli and Salmonella, are specified. Restrictions on certain additives and preservatives are also listed. In the United States, the Food Chemicals Codex has been considering a new specification for food-grade gelatin; a final version should be issued soon (ca 1995). Standard testing procedures for viscosity, pH, ash, moisture, heavy metals, arsenic, bacteria, and jelly strength are described (67,68,33). Additional test procedures have been published by the photographic and gelatin industries including the Japanese PAGI Method (69). Specific tests for photographic gelatin have been devised by the International Working Group for Photographic Gelatin (IAG) (70) in Fribourg, Switzerland, and by individual photographic companies and gelatin companies.

Uses

Food Products. Gelatin formulations in the food industry use almost exclusively water or aqueous polyhydric alcohols as solvents for candy, marshmallow, or dessert preparations (see FOOD ADDITIVES). In dairy products and frozen foods, gelatin's protective colloid property prevents crystallization of ice and sugar. Gelatin products having a wide range of Bloom and viscosity values are utilized in the manufacture of food products, specific properties being selected depending on the needs of the application. For example, a 250-Bloom gelatin may be utilized at concentrations ranging from 0.25% in frozen pies to 0.5% in ice cream; the use of gelatin in ice cream has greatly diminished. In sour cream and cottage cheese, gelatin inhibits water separation, ie, syneresis. Marshmallows contain as much as 1.5% gelatin to restrain the crystallization of sugar, thereby keeping the marshmallows soft and plastic; gelatin also increases viscosity and stabilizes the foam in the manufacturing process. Many lozenges, wafers, and candy coatings contain up to 1% gelatin. In these instances, gelatin decreases the dissolution rate. In meat products, such as canned hams, various luncheon meats, corned beef, chicken rolls, jellied beef, and other similar products, gelatin in 1–5% concentration helps to retain the natural juices and enhance texture and flavor. Use of gelatin to form soft, chewy candies, so-called gummi candies, has increased worldwide gelatin demand significantly (ca 1992). Gelatin has also found new uses as an emulsifier and extender in the production of reduced-fat margarine products. The largest use of edible gelatin in the United States, however, is in the preparation of gelatin desserts in 1.5–2.5% concentrations. For this use, gelatin is sold either premixed with sugar and flavorings or as unflavored gelatin packets. Most edible gelatin is type A, but type B is also used.

Pharmaceutical Products. Gelatin is used in the pharmaceutical industry for the manufacture of soft and hard capsules. The formulations are made with water or aqueous polyhydric alcohols. Capsules are usually preferred over tablets in administering medicine (71). Elastic or soft capsules are made with a rotary die from two plasticized gelatin sheets which form a sealed capsule around the material being encapsulated. Methods have been developed to encapsulate dry powders and water-soluble materials which may first be mixed with oil. The gelatin for soft capsules is low bloom type A, 170–180 g; type B, 150–175 g; or a mixture of type A and B. Hard capsules consisting of two parts are first formed and then filled. The manufacturing process is highly mechanized and sophisticated in order to produce capsules of uniform capacity and thickness. Medium-to-high bloom type A, 250–280 g; type B, 225–250 g; or the combination of type A and B gelatin are used for hard capsules. Usage of gelatin as a coating for tablets has increased dramatically. In a process similar to formation of gelatin capsules, tablets are coated by dipping in colored gelatin solutions, thereby giving the appearance and appeal of a capsule, but with some protection from adulteration of the medication. The use of glycerinated gelatin (67) as a base for suppositories offers advantages over carbowax or cocoa butter base (72). Coated or cross-linked gelatin is used for enteric capsules. Gelatin is used as a carrier or binder in tablets, pastilles, and troches.

For arresting hemorrhage during surgery, a special sterile gelatin sponge known as absorbable gelatin sponge (23) or Gelfoam is used. The gelatin is partially insolubilized by a cross-linking process. When moistened with a thrombin or sterile physiological salt solution, the gelatin sponge, left in place after bleeding stops, is slowly dissolved by tissue enzymes. Special fractionated and prepared type B gelatin can be used as a plasma expander.

Gelatin can be a source of essential amino acids when used as a diet supplement and therapeutic agent. As such, it has been widely used in muscular disorders, peptic ulcers, and infant feeding, and to spur nail growth. Gelatin is not a complete protein for mammalian nutrition, however, since it is lacking in the essential amino acid tryptophan [73-22-3] and is deficient in sulfur-containing amino acids.

Photographic Products.   Gelatin has been used for over 100 years as a binder in light-sensitive products. The useful functions of gelatin in photographic film manufacture are a result of its protective colloidal properties during the precipitation and chemical ripening of silver halide crystals, setting and film-forming properties during coating, and swelling properties during processing of exposed film or paper. Quality requirements of photographic gelatin may be very elaborate and can include over 40 chemical and physical tests, in addition to photographic evaluation. Most chemical impurities are limited to less than 10 ppm. Aqueous solutions are employed for emulsions (see PHOTOGRAPHY). Photographic gelatins are manufactured to standard specifications since the testing is time-consuming and costly. A new gelatin product may require 6–12 months of testing, including extensive field testing prior to commercialization. Photographic products may have up to twenty gelatin layers grouped into three categories: (1) light-sensitive silver halide-bearing layers of 2–10 μm thickness, referred to as emulsion layers; (2) surface, spacer, filter, or protective layers of 1–2 μm thickness; and (3) backing, antihalo, or noncurl layers coated on the opposite side of the film substrate from the emulsion layer. The quality and uniformity standards are highest for emulsion gelatin because it controls silver halide nucleation, crystal growth, chemical sensitization, latent image stability, and numerous other factors affecting the total photographic response. Since the early 1970s, the photographic industry has switched from so-called active gelatins derived from hides to inert types derived from bones. The latter are very low or void of natural restrainers, reduction, and sulfur sensitizers. Other changes in techniques have been brought about by abandoning the lengthy noodle wash technique used to remove salts after silver halide precipitation in favor of precipitating, coagulating, or derivatizing gelatin and washing the precipitate by decanting or utilizing ultrafiltration techniques; by new coating techniques that allow simultaneous coating of several layers at one time at speeds 10 times as fast as before; and by short-time high temperature processing which may require new cross-linking agents unlike the aldehydes and metal salts previously used. Many new hardeners are extremely fast-acting and are metered into the solution during the coating operation. It is quite common to use a derivatized gelatin, such as phthalated gelatin, to precipitate silver halide (61). These materials with a low pH isoionic point form a coacervate at pH <4.0. Precipitation in this case is accomplished by lowering the pH, washing at low pH, and then increasing pH to above 6.5 to dissolve and redisperse the emulsion before reconstituting it with gelatin. Gelatin used in the auxiliary layers must be able to withstand high temperature processing and allow high speed coating.

Gelatin is also used in so-called subbing formulations to prepare film bases such as polyester, cellulose acetate [9004-35-7], cellulose butyrate, and polyethylene-coated paper base for coating by aqueous formulations. Solvents such as methanol [67-56-1], acetone [67-64-1], or chlorinated solvents are used with small amounts of water. Gelatin containing low ash, low grease, and having good solubility in mixed solvents is required for these applications (see COATINGS). In certain lithographic printing, light-sensitive dichromated gelatin is used. Light causes permanent cross-linking of gelatin in the presence of the dichromate; this phenomenon is used to make relief images for printing. Dichromated gelatin coatings are commonly used in production of high quality holographic images. In this application, the light sensitivity of the image-receiving medium is less important than the image-resolving power (73). Gelatin coatings in photographic products are further tested for brittleness, scratch resistance, friction, swelling rate, drying rate, curling tendency, dry adhesion, wet adhesion, and pressure sensitivity. These properties are becoming more critical with the development of more sophisticated cameras and printing and processing equipment. Photographic technology offers a rapidly changing, highly sophisticated, very competitive market for photographic gelatin manufacturers.

Derivatized Gelatin.   Chemically active groups in gelatin molecules are either the chain terminal groups or side-chain groups. In the process of modifying gelatin properties, some groups can be removed, eg, deamination of amino groups by nitrous acid [10024-97-2] (74), or removal of guanidine groups from arginine [74-79-3] by hypobromite oxidation (75); the latter destroys the protective colloid properties of gelatin. Commercially successful derivatized gelatins are made mostly for the photographic gelatin and microencapsulation markets. In both instances, the amino groups are acylated. Protein detergent is made by lauroylating gelatin. Phthalated gelatin is now widely used in the photographic industry (76). Arylsulfonylated gelatin has been patented for microencapsulation (qv) (77). Carbamoylated gelatin, made by treating gelatin with cyanate or nitrourea in neutral aqueous solution, is also used by the photographic industry (19,78–80). Active double bonds react with the amino groups in gelatin, and acrylic polymers have been grafted to gelatin (81). Gelatin has been derivatized by epoxides (82), cyclic sulfones (83), and cyanamide [420-04-2] (84). Cross-linking or hardening of gelatin attacks the same active groups, but an agent with two active sites is needed, eg, divinylsulfone [77-77-0], bis(isomaleimide) [13676-54-5], aziridines, bisepoxides, epichlorohydrin [106-89-8], polyisocyanates, and dichlorotriazine. Aldehydes such as formaldehyde [50-00-0] and glyoxal [107-22-2] are still used and to a small extent even potassium chromium alum [7788-99-0], $Cr_2K_2(SO_4)_4 \cdot 24H_2O$, and potassium aluminum alum [7784-24-9], $Al_2K_2(SO_4)_4 \cdot 24H_2O$.

BIBLIOGRAPHY

"Gelatin" in *ECT* 1st ed., Vol. 7, pp. 145–153, by C. E. Anding, Jr., Kind & Knox Gelatin Co.; in *ECT* 2nd ed., Vol. 10, pp. 499–509, by E. M. Marks, Kind & Knox Gelatin Co.; in *ECT* 3rd ed., Vol. 11, pp. 711–719, by F. Viro, Kind & Knox, Division of Knox Gelatine, Inc.

1. P. Koepff, in H. Ammann-Brass and J. Pouradier, eds., *Photographic Gelatin*, Proceedings of the Fourth IAG Conference, Internationale Arbeitsgemeinschaft für Photogelatine, Fribourg, Switzerland, 1983, 1985, pp. 3–35.

2. R. H. Bogue, *The Chemistry and Technology of Gelatine and Glue*, McGraw-Hill Book Co., Inc., New York, 1922.
3. P. I. Smith, *Glue and Gelatine*, Pitman Press, London, 1929.
4. A. H. Kang and co-workers, *Biochemistry* **5**, 509–515 (1966).
5. A. H. Kang and co-workers, *Biochem. Biophys. Res. Commun.* **36**, 345–349 (1969).
6. K. A. Piez and co-workers, *Brookhaven Symp. Biol.* **21**, 345–357 (1968).
7. A. Courts, *Biochem. J.* **58**, 70 (1954).
8. S. Aoyagi in Ref. 1, pp. 79–94.
9. D. Larry and M. Vedrines in Ref. 1, pp. 35–54.
10. C. Xiang-Fang and P. Bi-Xian in Ref. 1, pp. 55–64.
11. J. Butel in Ref. 1, pp. 65–78.
12. P. Koepff in H. Ammann-Brass and J. Pouradier, eds., *Photographic Gelatin Reports 1970-1982*, IAG, 1984, pp. 197–209.
13. I. Tomka in Ref. 12, p. 210.
14. J. Bohonek, A. Spühler, M. Ribeaud, and I. Tomka, in R. J. Cox, ed., *Photographic Gelatin II*, Academic Press, Inc., New York, 1976, pp. 37–55.
15. J. Pouradier and A. M. Venet, *J. Chim. Phys.* **47**, 381 (1950).
16. G. Stainsby, P. R. Saunders, and A. G. Ward, *J. Polym. Sci.* **12**, 325 (1954).
17. L.-J. Chen and A. Shohei, *J. Photogr. Sci.* **40**, 159 (1992).
18. J. E. Eastoe in G. N. Ramachandran, ed., *Treatise on Collagen*, Vol. 1, Academic Press, Inc., New York, 1967, pp. 1–72.
19. P. Johns, in A. G. Ward and A. Courts, eds., *The Science and Technology of Gelatin*, Academic Press, Inc., New York, 1977, pp. 475–506.
20. J. E. Eastoe, *Biochem. J.* **61**, 589 (1955).
21. J. Josse and W. F. Harrington, *J. Mol. Biol.* **9**, 269 (1964).
22. G. Stainsby in Ref. 19, p. 127.
23. I. Tomka, J. Bohonek, J. Spühler, and M. Ribeaud, *J. Photogr. Sci.* **23**, 97 (1975).
24. T. Fujii, *Bull. Soc. Sci. Photogr.* **16**, 274 (1966).
25. E. M. Marks, D. Tourtelotte, and A. Andux, *Food Technol.* **22**, 99 (1968).
26. R. T. Jones in K. Ridgway, ed., *Hard Capsules Development and Technology*, The Pharmaceutical Press, London, 1987, pp. 41–42.
27. *Standard Methods for Sampling and Testing Gelatins*, Gelatin Manufacturers' Institute of America, Inc., New York, 1986.
28. P. H. von Hippel and W. F. Harrington, *Biochim. Biophys. Acta* **36**, 427 (1959).
29. P. H. von Hippel and W. F. Harrington, *Brookhaven Symp. Biol.* **13**, 213 (1960).
30. P. J. Florey and E. S. Weaver, *J. Am. Chem. Soc.* **82**, 4518 (1960).
31. J. Engel, *J. Arch. Biochem.* **97**, 150 (1962).
32. H. Bredtker and P. Doty, *J. Phys. Chem., Ithica*, **58**, 968 (1954).
33. A. Veis in B. L. Horecker, N. D. Kaplan, and H. E. Sheraga, eds., *Molecular Biology*, Vol. V, Academic Press, Inc., New York, 1964, Chapt. 5.
34. E. Fuchs, *Adhesion* **5**, 225 (1961).
35. *Sampl. Test. Gelatins, Brit. Stand.* **6**, 757 (1975).
36. F. W. Wainwright, *GGRA Bull.* **17**(3), 10 (1966).
37. C. A. Finch and A. Jobling in Ref. 19, Chapt. 8; A. R. Krough in Ref. 19, Chapt. 14.
38. J. Q. Umberger, *Photogr. Sci. Eng.* **11**, 385 (1967).
39. K. M. Hornsby, *Brit. J. Photogr.* **103**, 17, 28 (1956).
40. J. Janus, A. W. Kechington, and A. G. Ward, *Research* **4**, 247 (1951).
41. N. Itoh, *J. Photogr. Sci.* **40**, 200 (1992).
42. B. E. Tabor, R. Owers and J. F. Janus, *J. Photogr. Sci.* **40**, 205 (1992).
43. T. Takahashi, *J. Photogr. Sci.* **40**, 212 (1992).
44. J. Rottman and H. Pietsch, *J. Photogr. Sci.* **40**, 217 (1992).
45. T. D. Weatherill, R. W. Henning, and K. A. Smith, *J. Photogr. Sci.* **40**, 220 (1992).
46. C. R. Maxey and M. R. Palmer in Ref. 14, pp. 27–36.

47. C. Li-juan in Ref. 1, pp. 136–144.
48. Y. Toda in Ref. 1, pp. 107–124.
49. Y. Toda, in S. J. Band, ed., "Photographic Gelatin," *Proceedings of the Fifth RPS Symposium*, Oxford, U.K., 1985, The Imaging Science and Technology Group of the Royal Photographic Society, 1987, pp. 28–37.
50. J. E. Eastoe and A. Courts, *Practical Analytical Methods for Connective Tissue Proteins*, Spon, London, 1963, Chapt. 6.
51. A. Veis "The Macromolecular Chemistry of Gelatin" in Ref. 33, p. 112.
52. G. Stainsby, *Nature (London)* **169**, 662–663 (1952).
53. G. Stainsby, in H. Sauverer, ed., *Scientific Photography*, Pergamon Press, London, 1962, p. 253.
54. C. W. N. Crumper and A. E. Alexander, *Aust. J. Sci. Res.* **A5**, 146 (1952).
55. G. Stainsby in Ref. 19, pp. 109–136.
56. V. Zitko and J. Rosik, *Chem. Zvesti* **17**, 109 (1963).
57. Brit. Pat. 930,421 (1963), (to Upjohn Co.).
58. P. D. Wood in Ref. 19, pp. 419–422.
59. B. Kuznicka, and J. Kuznicki, in Ref. 49, pp. 206–212.
60. R. J. Croome in Ref. 1, pp. 267–282.
61. D. L. Kramer, *J. Photogr. Sci.* **40**, 152 (1992).
62. H. R. Proctor and J. A. Wilson, *J. Chem. Soc.* **109**, 307 (1916).
63. A. Libicky and D. I. Bermane, in R. J. Cox, ed., *Photographic Gelatin*, Academic Press, Inc., New York, 1972, pp. 29–48.
64. D. W. Jopling, *J. Appl. Chem.* **6**, 79 (1956).
65. J. Russell and D. L. Oliff, *J. Photogr. Sci.* **14**, 9 (1966).
66. C. R. Smith, *J. Am. Chem. Soc.* **43**, 1350 (1921).
67. *The United States Pharmacopeia XXII, (USP XXII–NFXVII)*, The United States Pharmacopeial Convention, Inc., Rockville, Md., 1989.
68. K. Helrich, ed., *AOAC, Official Methods of Analysis, 15th Ed.*, Association of Official Analytical Chemists, Arlington, Va., 1990.
69. *PAGI Method, 7th ed.*, Photographic and Gelatin Industries, Japan, Tokyo, 1992.
70. *IAG Test*, Gelatin Manufacturers Institute of America, New York; see also Ref. 1.
71. *Consumer Survey*, A. C. Nielsen Co., Northbrook, Ill., 1976.
72. L. F. Tice and R. E. Abrams, *J. Am. Pharm. Assoc.* **14**, 24 (1953).
73. *Science J.* **4**(10), 27 (1968).
74. A. W. Kenchington, *Biochem. J.* **68**, 458 (1958).
75. P. Davis, in G. Stainsby, ed., *Recent Advances in Gelatin and Glue Research*, Pergamon Press, London, 1958, p. 225.
76. U.S. Pat. 3,184,312 (May 18, 1965), J. W. Gates and E. Miller (to Eastman Kodak Co.).
77. Brit. Pat. 1,075,952 (July 19, 1967), R. C. Clark and co-workers (to Gelatin and Glue Research Assoc.).
78. U.S. Pat. 2,525,753 (Oct. 10, 1950), H. C. Yutzky and G. F. Frame (to Eastman Kodak Co.).
79. U.S. Pat. 3,108,995 (Oct. 29, 1963), D. Tourtellotte and E. M. Marks (to Charles B. Knox Gelatin Co., Inc.).
80. U.S. Pat. 2,816,099 (Dec. 10, 1959), H. H. Young and E. F. Christophen (to Swift and Co.).
81. U.S. Pat. 3,291,611 (Dec. 13, 1966), J. J. Krajewski (to Swift and Co.).
82. Belg. Pat. 672,906 (Nov. 26, 1965), F. Dersch and S. L. Paniccia (to GAF Corp.).
83. Brit. Pat. 1,033,189 (June 15, 1956), J. W. Gates, Jr. and P. E. Miller (to Kodak Ltd.).
84. Brit. Pat. 1,100,842 (Jan. 24, 1968), R. J. Chamberlain (to American Cyanamid Co.).

THOMAS R. KEENAN
Kind & Knox Gelatine, Inc.

GEMSTONES

GEMSTONE MATERIALS

There are three types of gemstone materials as defined by the U.S. Federal Trade Commission (1): (*1*) natural gemstones are found in nature and at most are enhanced (see GEMSTONES, GEMSTONE TREATMENT); (*2*) imitation or simulated, fake, faux, etc, material resembles the natural material in appearance only and is frequently only colored glass or even plastic; and (*3*) synthetic material is the exact duplicate of the natural material, having the same chemical composition, optical properties, etc, as the natural, but made in the laboratory (2,3). Moreover, the word gem cannot be used for synthetic gemstone material. The synthetic equivalent of a natural material may, however, be used as an imitation of another, eg, synthetic cubic zirconia is widely used as a diamond imitation.

The first successful synthetic gemstone material was the 1885 Geneva ruby of unknown origin, misleadingly called reconstructed at the time and sold as natural (2). This was soon followed by Verneuil's flame-fusion synthetic ruby [*12174-49-1*] (4), which became an immediate commercial success. Early synthetics were usually the result of mineralogical studies or focused attempts to duplicate natural gemstones. More recently these materials have been by-products of technology-oriented studies, eg, the search for laser crystals (see LASERS). The historic availability of gemstone materials is summarized in Table 1. Most of these materials are made by crystal growing techniques (2,5,6) (see CRYSTALLIZATION). The U.S. patent literature has been summarized (7). There is also a large Japanese patent literature.

Synthetic gemstone materials often have multiple uses. Synthetic ruby and colorless sapphire are used for watch bearings, unscratchable watch crystals, and bar-code reader windows. Synthetic quartz oscillators are used for precision timekeeping, citizen's band radio (CB) crystals, and filters. Synthetic ruby, emerald, and garnets are used for masers and lasers (qv).

In the gemstone jewelry market, synthetics provide a less expensive alternative to natural gemstones, but of a better quality than that available in costume jewelry. In general, a synthetic should be available for no more than 10% of the cost of equivalent-quality natural gemstone to be commercially viable. Synthetics are frequently divided into three groups: (*1*) luxury synthetics, involving slow and difficult growth processes, produced in small quantities for a price-restricted market; (*2*) intermediates; and (*3*) low cost synthetics, produced on a large scale. In 1989 more than 10^9 carats (200 t) of both synthetic ruby and cubic zirconia [*1314-23-4*] were produced (3). Price ranges for these groups are summarized in Table 2.

Diamond simulants are usually included under synthetics, even though not all of these have been synthetic gemstone materials, such as the garnets. Many trade names have been used for each of these imitations (see CARBON–DIAMOND, SYNTHETIC).

Table 1. Availability of Synthetic Gemstone Materials

Year	Synthetic gemstone	CAS Registry Number	Manufacturing technique
1885	ruby	[12174-49-1]	Geneva
1905	ruby		Verneuil
1910	sapphire	[1317-82-4]	Verneuil
1910	spinel	[1302-67-6]	Verneuil
1947	star ruby and sapphire		Verneuil
1947	rutile	[1317-80-2]	Verneuil
1950	emerald	[1302-52-9]	flux
1950	quartz	[14808-60-7]	hydrothermal
1955	strontium titanate	[12060-59-2]	Verneuil
1965	ruby		flux
1965	emerald		hydrothermal
1970	diamond	[778-40-3][a]	high pressure
1973	alexandrite	[12252-02-7]	flux
1974	opal	[14639-88-4]	complex process
1974	citrine	[14832-92-9]	hydrothermal
1975	amethyst	[14832-91-8]	hydrothermal
1976	cubic zirconia	[1314-23-4]	skull melting
1978	sapphire		flux
1980	jadeite	[12003-54-2][a]	high pressure

[a]Grown for other purposes or experimental production.

Table 2. Price Ranges of Synthetic Gemstone Materials

Material	Cost, $/g[a]	
	Rough	Faceted[b]
Luxury		
emerald,[c] black opal, alexandrite, flux ruby,[c] sapphires[c]	50 to <500	500 to <5000
Intermediate		
star ruby[d] and sapphires,[d] white opal, amethyst	5 to <50	50 to <500
Low cost		
cubic zirconia, Verneuil ruby, sapphires, spinel, colorless and citrine quartz	0.50 to <2.50	5 to <50

[a]To convert $/g to $/carat, divide by 5.00.
[b]Usually at least five times the cost of the rough because cutting recovery is only 25% or even less. The cutting cost may become more important than the materials cost in small faceted stones.
[c]Flux crystal clusters sell for about the same price as faceting-quality rough.
[d]Star material has generally not been available as rough.

Distinguishing between natural and synthetic gemstone materials is one of the tasks of the trained gemologist. Whereas difficult at times, the importance derives from the huge differences in commercial value. For example, a natural flawless ruby could be valued at $20,000 a carat ($100,000/g); a synthetic pulled from the melt by the Czochralski technique goes for $2 a carat ($10/g). Gemology is taught by the Gemological Institute of America of Santa Monica, California, and New York, the Gemmological Association of Great Britain of London, and elsewhere. Textbooks are also available (8–10).

Properties

The important properties are those of importance in natural gemstones. First is hardness (qv), H. A value of 7 or greater on Mohs' scale is desirable to avoid scratches from the quartz (H = 7) sand present in dust. Next is color (qv) or a total lack of color, as in diamond and its simulants. A high refractive index (RI) permits the return by total internal reflection of most of the light falling onto a well-cut gemstone, giving brilliance, and a high dispersion (DISP) spreads the internally reflected light into spectral colors, resulting in fire. A large birefringence is undesirable, because that produces a doubling of facet junctions seen through the top of the stone.

Diamond is supreme among natural gemstones in H, RI, and DISP. Table 3 shows the steady improvement in the sequence of diamond imitations, the aim being to produce a colorless, adequately hard material having closely matching optical properties. The introduction of synthetic cubic zirconia in 1976 brought about a sufficiently close match.

Several gemstone species occur in various colors, depending on the presence of impurities or irradiation-induced color centers. Examples are the beryl, corundum, and quartz families. Quartz has poor optical properties (RI = 1.55, DISP = 0.013), but becomes of gemological interest when it exhibits attractive colors. Any material can have its color modified by the addition of various impurities: synthetic ruby, sapphires, and spinel are produced commercially in over 100 colors (2). Synthetic cubic zirconia has been made in essentially all colors of the spectrum (11), but only the colorless diamond imitation is produced commercially in any quantity.

Manufacture

The most frequently used techniques for the commercial manufacture of synthetic gemstone materials are summarized in Table 4. More details on these can be found in various texts (2,5–7). An overview including the various manufacturers is also available (3). Only rarely used for synthetics are such alternative growth techniques as the Bridgman technique of solidification in a crucible and the float zone technique, both involving growth from the melt (2,5–7).

The easiest and most rapid crystal growth techniques employ crystallization from the melt. Some materials are incongruently melting, ie, they decompose below their melting point, in which case solution techniques must be used, as for

Table 3. Properties of Diamond and Synthetic Gemstone Materials[a]

Material	CAS Registry Number	Composition	RI	DISP	H	Specific gravity	Year[b]	Disadvantages
spinel[c]	[1302-67-6]	$MgAl_2O_4$	1.73	0.02	8	3.64	1910	much less brilliant
sapphire, colorless	[1317-82-4]	Al_2O_3	1.77	0.018	9	4.00	1920	much less brilliant
rutile	[13463-67-7]	TiO_2	2.8	0.33	6	4.26	1947	yellowish, soft, birefringent, excessive fire
strontium titanate (tausonite)	[12060-59-2]	$SrTiO_3$	2.41	0.19	5½	5.13	1955	soft, excessive fire
YAG garnet	[12005-21-9]	$Y_3Al_5O_{12}$	1.83	0.028	8¼	4.55	1968	less brilliant
GGG garnet	[12024-36-1]	$Gd_3Ga_5O_{12}$	1.97	0.045	7	7.02	1974	slightly less brilliant, heavy
cubic zirconia[c]	[1314-23-4]	$9ZrO_2 \cdot Y_2O_3$	2.16	0.060	8¼	6.0	1976	somewhat heavy
diamond	[7782-40-3]	C	2.42	0.044	10	3.52		

[a]RI = refractive index; DISP = dispersion; H = hardness on Mohs' scale.
[b]Represents the start of widespread use.
[c]Composition property values can vary from batch to batch.

Table 4. Techniques for Commercial Gemstone Material Synthesis

Technique	Material
Crystal growth from the melt	
Verneuil (flame fusion)	ruby, sapphires, and stars; spinel; rutile; strontium titanate
Czochralski (pulling)	ruby and sapphire; alexandrite; garnets: YAG and GGG
float zone	ruby and sapphire; alexandrite
skull melting	cubic zirconia
Crystal growth from solution	
flux	alexandrite; emerald; ruby and sapphire; spinel
hydrothermal	colorless, amethyst, citrine, and smoky quartz; emerald; ruby and sapphire
high pressure	diamond;[a] jadeite[a]
Other techniques	
complex chemical[b]	opal

[a]Grown for other purposes or experimental production.
[b]See text for a description of this technique.

emerald. The same applies to those instances where there is a strong tendency for glass formation from the melt, as in the quartz family. Diamond is a special case in that it is only thermodynamically stable at high pressure. Whereas low pressure techniques have been successful for producing very thin single-crystal films or for bulk polycrystalline material, the growth of bulk single-crystal synthetic diamond has not been achieved (12).

Crystal Growth from the Melt. *The Verneuil Technique.* The Verneuil technique is also known as flame-fusion. A very pure feed powder is first made by chemical decomposition. For the growth of corundum, ammonium alum [7785-25-0], $NH_4Al/(SO_4)_2 \cdot 12H_2O$, is recrystallized from water containing added color-causing transition-metal impurities, such as Cr for ruby, again in the form of an alum. Decomposition at about 1200°C forms Al_2O_3 powder that is very fine and free-flowing. This feed powder is now sprinkled through a downward-pointing hydrogen–oxygen torch, as shown in Figure 1. The particles melt on impinging, forming a molten cap about 20 μm thick on the upper surface of the growing crystal boule, which is slowly lowered. Because of the small melt volume, even small irregularities in the feed rate and flame temperature produce irregular solidification, resulting in curved color striations and occasional incompletely melted grains or gas bubbles. The growth rate is one to several cm/h and boules over 4 cm in diameter and 40 cm in length have been grown.

For the Verneuil growth of rutile and strontium titanate it is necessary to maintain strongly oxidizing conditions to prevent excessive reduction of Ti^{4+} to Ti^{3+}. This is achieved by adding a third outer tube carrying extra oxygen to the Verneuil torch (Fig. 1) in the tricone modification. Annealing in O_2 at about 1100°C is subsequently used to achieve full oxidation.

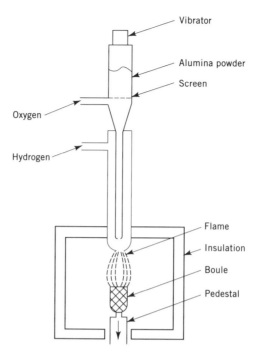

Fig. 1. The Verneuil technique, or flame-fusion growth, as used for synthetic ruby and sapphire.

The Czochralski Technique. Pulling from the melt is known as the Czochralski technique. Purified material is held just above the melting point in a crucible, usually of Pt or Ir, most often powered by radio-frequency induction heating coupled into the wall of the crucible. The temperature is controlled by a thermocouple or a radiation pyrometer. A rotating seed crystal is touched to the melt surface and is slowly withdrawn as the molten material solidifies onto the seed. Temperature control is used to widen the crystal to the desired diameter. A typical rotation rate is 30 rpm and a typical withdrawal rate, 1–3 cm/h. Very large, eg, kilogram-sized crystals can be grown.

Skull Melting. Skull melting, or the cold crucible technique, is used for high melting point materials. Zirconia [1314-23-4], ZrO_2, has such a high (about 2750°C) melting point, that no crucible material is available that does not react with the melt. In skull melting (2,5,13) a container of closely spaced water-cooled copper fingers, as shown in Figure 2, is filled with powder of the desired composition, usually about 90% ZrO_2 and 10% yttria [1314-36-9], Y_2O_3. A piece of Zr metal is added. Radio-frequency power from the coil readily passes through the narrow gaps between the fingers and melts the metal, which in turn melts the adjacent oxides, which then couple directly to the coil; the metal oxidizes in the air. All the contents of the container melt except for a relatively thin crust adjacent to the cold copper fingers, so that the melt is only in contact with a cold crucible of its own composition. On slowly lowering the container from out of the coil,

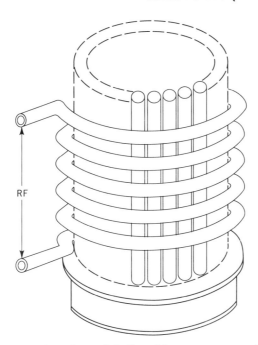

Fig. 2. Schematic drawing of one form of skull-melting apparatus; only some of the fingers are shown (2). RF = radio frequency.

random nucleation begins at the bottom and produces irregularly shaped crystals, which can be over 10 cm across.

Crystal Growth from Solution. *Flux Growth.* This procedure is called growth from the fluxed melt in the United Kingdom. A flux in this context is a high melting inorganic solvent containing substances such as PbO [*1317-36-8*], PbF$_2$ [*7783-46-2*], B$_2$O$_3$ [*1303-86-2*], etc. For example, for the growth of ruby, 3.6 kg Al$_2$O$_3$ and 30 g Cr$_2$O$_3$ [*1308-38-9*] are mixed with 22 kg of PbF$_2$ and melted in a 20-cm diameter and high, heavy-walled platinum crucible at about 1300°C and mixed by a period of rotation via the supporting pedestal as in Figure 3. A controlled, slow cooling to about 1000°C over a several-week period exceeds the solubility of ruby in the flux, and crystals self-nucleate and grow to several centimeters across on the slightly cooler crucible bottom. The unwanted flux may be drained off through the hollow pedestal. This is flux growth by slow cooling.

Flux reaction growth is one of several modifications used for the growth of emerald. Here one of several fluxes, possibly composed of Li$_2$O [*12057-24-8*] and MoO$_3$ [*1313-27-5*], is used with the constituents of emerald, ie, SiO$_2$, Al$_2$O$_3$, BeO [*12269-78-2*], and Cr$_2$O$_3$, diffusing toward thin seed plates in the central growth region, as shown in Figure 4.

Hydrothermal Growth. Hydrothermal growth is used in the form of solution transport for the growth of synthetic quartz. Crushed natural quartz is placed into the lower part of a high pressure steel vessel, called a bomb, and thin seed plates are located in the upper region, as seen in Figure 5. The vessel is filled, for example, to 80% capacity with a 4% NaOH [*1310-73-2*] solution; the NaOH acts

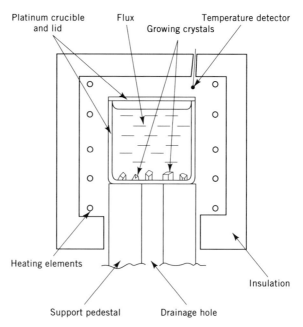

Platinum crucible and lid
Flux
Growing crystals
Temperature detector
Heating elements
Support pedestal
Drainage hole
Insulation

Fig. 3. Crucible inside a furnace as used for growth from the flux (2).

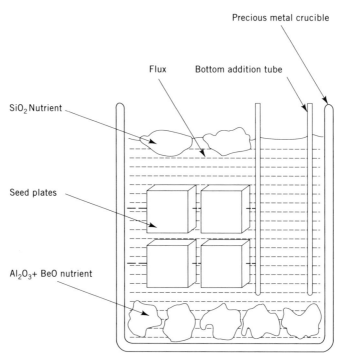

Precious metal crucible
Flux
Bottom addition tube
SiO_2 Nutrient
Seed plates
Al_2O_3+ BeO nutrient

Fig. 4. Schematic diagram of one of the arrangements for the flux-reaction growth of synthetic emerald (2).

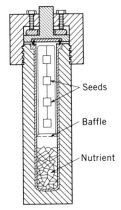

Fig. 5. A small silver-lined hydrothermal growth vessel, 35-cm long, as used for synthetic quartz. Courtesy of Bell Laboratories.

as a mineralizer to increase the solubility of SiO_2. When the closed vessel is heated to 390°C bottom and 330°C top, the resulting pressure is 170 MPa (25,000 psi) and a single fluid fills the vessel. It dissolves nutrient at the bottom and flows by convection, controlled by a baffle, to the upper region, where growth occurs at the lower temperature. A typical 4 cm × 15 cm × 1 mm thick seed plate grows to a 5-cm thickness in about 3 weeks in a 3-m long production vessel.

For emerald, a hydrothermal reaction arrangement is used, analagous to that of Figure 4. Here the mineralizer is strongly acid, containing $8N$ NH_4Cl [12125-02-9] and HCl [7647-01-1] to give pH between 0.2 and 4.5.

High Pressure Growth of Diamond. Ni dissolves graphite [7782-42-5] or diamond at about 7 GPa (1×10^6 psi) and 1800°C and crystallizes diamond in a temperature gradient onto a cooler seed plate (2,5–7,14). To reach such high pressures, a massive support apparatus using two pistons and a belt made of tungsten carbide [12070-13-2] bonded with cobalt [7440-48-4] and supported by shrunk-on steel rings is used in a uniaxial hydraulic press. In an alternative geometry, four pistons are arranged to form a tetrahedral space. Pyrophillite [12269-78-2] is used as an insulator and as a nonextruding gasket. Metals other than Ni and some alloys can also be used as solvents.

Materials

Alexandrite. Alexandrite [12252-02-7] is chrysoberyl [1304-50-3], $BeAl_2O_4$, when pure. The Cr-containing alexandrite form has a psychooptic color change, appearing blue-to-green when viewed in daylight or fluorescent-tube lighting, and red in the light from a candle or an incandescent light bulb. It is grown as a luxury synthetic from the flux and also more recently by the Czochralski technique.

Beryl. Beryl [1302-52-9], $Be_3Al_2Si_6O_{18}$, is called aquamarine [1327-51-1] when pale green or blue from inclusion of Fe, emerald [12415-33-7] when dark green from Cr or at times V, and morganite or red beryl when pink or red, respectively, from Mn. Only the synthetic emerald is in commercial production, although the other colors can also be grown. Both the flux and hydrothermal techniques are used to grow this luxury synthetic.

Corundum.   Crystalline Al_2O_3, corundum [1302-74-5], is called ruby [12174-49-1] when colored red by about 1% Cr, and sapphire [1317-82-4] for other colors, particularly when blue from charge transfer (2,11) between about 0.01% each of Fe^{2+} and Ti^{4+}. A wide range of other colors can be grown, including colorless (pure), yellow (containing Ni), orange (Ni + Cr), green (Co), alexandrite-like green/purple (V), and so on. Synthetic ruby is grown (1) for high optical-quality laser use by the Czochralski technique; (2) together with many colors of sapphire as a low cost synthetic by the Verneuil technique; and (3) together with blue sapphire as luxury synthetics by the flux technique. The hydrothermal technique has also been used. The colorless material has been used as a poor diamond imitation (Table 3), called Brillite, Diamondite, Thrilliant, etc. Synthetic ruby and variously colored sapphires are used in class rings as imitations of various gemstones, as well as for other jewelry uses.

The addition to Al_2O_3 of up to 1% TiO_2 [1317-80-2] results in the growth of clear crystals containing a solid solution. On annealing at about 1200°C, the exsolution of TiO_2 needles results in the formation of a star effect, as shown in Figure 6.

Cubic Zirconia.   As of this writing, cubic zirconia [1314-23-4], ZrO_2, is the best diamond imitation available (Table 3). It is marketed under such names as CZ, Cerene, Cubic zirconium, Diamonair II, Diamonique III, Fianite, etc, and grown by the skull melting technique (2,5,13). Pure ZrO_2 is monoclinic, changing through tetragonal to cubic at 2300°C. The addition of 10% yttria [1314-36-9] or other oxides such as calcia [1305-78-8] stabilizes the cubic form at room temperature. This material can also be made in almost any color. The pink color given by Co changes to a deep sapphire-like blue if the stabilizer concentration is raised to about 30%. The addition of any yellow-causing impurity can then produce an emerald-like green. Several techniques, including the thermal tester designed specifically for this purpose which recognizes the uniquely high thermal conduc-

Fig. 6.   Synthetic star ruby made by Linde (2).

tivity of diamond, provide ready distinctions between cubic zirconia and diamond (2,8–10).

Diamond.   The synthesis by a high pressure process of single-crystal diamond [7782-40-3] large enough for gemstone use was revealed by the General Electric Co. in 1971 (14). Commercial production was instituted (15) at DeBeers, Sumitomo, and others to produce gem-quality synthetic diamond crystals weighing over 11 carats (2.2 gm). The yellow color (containing N) is grown much more easily than colorless (pure) and blue (B). None of these is likely to be viable for use in jewelry in the near future. The low pressure deposition techniques used for growing diamond films (12) are not able to produce bulk single-crystal material.

Garnets.   Both YAG, yttrium aluminum garnet [12005-21-9], $Y_3Al_5O_{12}$, and GGG, gadolinium gallium garnet [12024-36-1], $Gd_3Ga_5O_{12}$, have the garnet structure and were used at one time as diamond imitations (Table 3) under trade names such as Diamonaire, Diamonique, Diamonite, Kimberly, Triamond, and YAIG for YAG and Diamonique II, Galliant, and Triple G for GGG. These have been supplanted by cubic zirconia.

Opal.   Opal [14639-88-4] is the only commercial synthetic gemstone material that is not a single crystal. As shown in Figure 7, it consists of a three-dimensional diffraction grating of geometrically aligned spheres of $SiO_2 \cdot xH_2O$, where x is usually $< 10\%$. The spheres have uniform size between 0.2 and 0.7 µm and are cemented together with a similar composition having a different x. A

Fig. 7.   Electron microscope view of Gilson synthetic opal; each sphere is ¼ µm in diameter. Courtesy of Ets. Ceramiques Pierre Gilson.

chemical precipitation process is used to make the spheres, followed by a lengthy settling process and some type of compaction step (2,7).

Quartz.   When colorless, quartz [*14808-60-7*] is also known as rock crystal. When irradiated, it becomes smoky from a color center associated with a ubiquitous Al impurity at about the 0.01% level. The name citrine [*14832-92-9*] is used when quartz is colored by Fe, and irradiation of this can produce the purple-colored amethyst [*14832-91-8*] under certain circumstances (2). Although not significantly lower priced than the natural materials, synthetic citrine and amethyst are used in jewelry because of the ability to provide matched sets of stones from large, up to 7-kg, hydrothermally grown crystals.

Rutile.   Rutile [*1317-80-2*], a form of TiO_2, was at one time used as a rather poor diamond imitation under trade names such as Kenya stone, Rutania, Titanic, and Ultimate. Small amounts of Al_2O_3 or Ga_2O_3 lighten the yellow color. Related is strontium titanate [*12060-59-2*], $SrTiO_3$, now more properly called synthetic tausonite. This latter was a great improvement over TiO_2 as a diamond imitation under trade names such as Brilliante, Diagem, Diamontina, Fabulite, and Wellington, and is still sometimes so used. The properties of both these materials, grown by the Verneuil technique, are listed in Table 3.

Spinel.   Although the composition of natural spinel [*1302-67-6*] is $MgAl_2O_4$, crystal growth is much eased by growing Al_2O_3-rich material in the solid solution region. Colorless (pure), blue (Co), and other colored synthetic spinels made by the Verneuil process are widely seen in class rings and in other jewelry uses, where the blue is often mislabeled as synthetic sapphire. Flux growth has also been used for stoichiometric spinel. Colorless synthetic spinel was once used as a poor diamond imitation (Table 3) under trade names such as alumag, radient, and strongite.

Other Synthetic Materials.   Many other natural gemstone materials have been duplicated in the laboratory on an experimental basis, often only in small sizes. Examples include tourmaline [*1317-93-7*], topaz [*1302-59-6*], and zircon [*1490-68-2*]. Of some potential is synthetic jadeite [*12003-54-2*], one of the two forms of jade. This crystallizes under medium pressure in polycrystalline form from an $NaAlSi_2O_6$ glass (qv) and can be colored green by Fe or lavender by Mn (16). Many gemstone-like materials have been grown for technological purposes and such material is sometimes faceted.

Discredited Synthetics.   There are several materials that have in the past been considered to be synthetics, but were found on closer examination not to deserve such a designation, being merely imitations. Examples include imitation coral, lapis lazuli, and turquoise, all made by ceramic processes. This same point has been raised (17) with respect to synthetic opal, which does contain some substances not present in natural opal and somewhat less water. However, the composition of natural opal is quite variable and is usually intermixed with significant amounts of rock-derived materials; hence the synthetic designation is usually retained.

BIBLIOGRAPHY

"Gems, Synthetic" in *ECT* 1st ed., Vol. 7, pp. 157–167, by A. K. Seemann, Linde Air Products Co., A Division of Union Carbide and Carbon Corp.; in *ECT* 2nd ed., Vol. 10, pp. 509–519, by C. Robert Castor, Electronics Division, Union Carbide Corp.; in *ECT* 3rd ed., Vol. 11, pp. 719–730, by L. R. Rothrock, Union Carbide Corp.

1. *Guides for the Jewelry Industry*, U.S. Federal Trade Commission, Washington, D.C., Feb. 27, 1979 (under revision in 1993).
2. K. Nassau, *Gems Made by Man*, Gemological Institute of America, Santa Monica, Calif., 1980.
3. K. Nassau, *Gems Gemol.* **26**, 50 (1990).
4. A. Verneuil, *Ann. Chim. Phys., Ser. 8*, **3**, 20 (1904).
5. K.-Th. Wilke and J. Bohm, *Kristall Züchtung*, Verlag H. Deutch, Thun, Frankfurt am Main, 1988.
6. J. C. Brice, *Crystal Growth Processes*, John Wiley & Sons, Inc., New York, 1986.
7. L. H. Yaverbaum, *Synthetic Gems*, Noyes Data Corp., Park Ridge, N.J., 1980.
8. R. T. Liddicoat, Jr., *Handbook of Gem Identification*, 12th ed., Gemological Institute of America, Santa Monica, Calif., 1989.
9. C. S. Hurlbut, Jr. and R. C. Kammerling, *Gemology*, 2nd ed., John Wiley & Sons, Inc., New York, 1991.
10. B. W. Anderson and E. A. Jobbins, *Gem Testing*, 10th ed., Butterworths, London, 1990.
11. K. Nassau, *The Physics and Chemistry of Color*, John Wiley & Sons, Inc., New York, 1983.
12. R. C. DeVries, *Ann. Rev. Mater. Sci.*, **17**, 161 (1987).
13. K. Nassau, *Gems Gemol.* **17**, 9 (1981).
14. K. Nassau, *J. Crystal Growth* **46**, 157 (1979).
15. J. E. Shigley and co-workers, *Gems Gemol.* **22**, 192 (1986); **23**, 187 (1987).
16. K. Nassau and J. E. Shigley, *Gems Gemol.* **23**, 27 (1987).
17. K. Schmetzer and U. Henn, *Gems Gemol.* **23**, 148 (1987).

Kurt Nassau
Nassau Consultants

GEMSTONE TREATMENT

Color and clarity are two of the attributes that give gemstones used in jewelry value. Gemstones deficient in either color or clarity can be enhanced (1). Almost worthless material can at times be converted into valuable-appearing gemstones. An estimated two-thirds of all colored gemstones used in jewelry have been treated. Accordingly, the identification of the use of treatments and the disclosure of enhancements to the purchaser are important. Table 1 lists the materials discussed herein.

Some treatments are practiced so widely that untreated material is essentially unknown in the jewelry trade. The heating of pale Fe-containing chalcedony to produce red-brown carnelian is one of these. Another example involves turquoise where the treated material is far superior in color stability. Such treatments have traditionally not been disclosed. Almost all blue sapphire on the market has been heat treated, but it is not possible to distinguish whether it was near-colorless corundum containing Fe and Ti before treatment, or whether it had already been blue and was only treated in an attempt at marginal improvement. The irradiation of colorless topaz to produce a blue color more intense than any occurring naturally is, however, self-evident, and treatments used on diamond are always disclosed.

Table 1. Gemstone Materials Discussed

Material	CAS Registry Number	Molecular formula
agate	[15723-40-7]	SiO_2
amber	[8002-67-3]	
amethyst	[14832-91-8]	$SiO_2 + Fe$
aquamarine	[1327-51-1]	$Be_3Al_2Si_6O_{18} + Fe$
azurite	[1319-45-5]	$Cu_3(CO_3)_2(OH)_2$
beryl	[1302-52-9]	$Be_3Al_2Si_6O_{18}$
carnelian		
chalcedony	[14639-89-5]	SiO_2
citrine	[14832-92-9]	$SiO_2 + Fe$
coral		
corundum	[1302-74-5]	Al_2O_3
diamond	[7782-40-3]	C
emerald	[12415-33-7]	$Be_3Al_2Si_6O_{18} + Cr^{3+}$
fluorite	[7789-75-5]	CaF_2
jade (jadeite)	[12003-54-2]	$Na(Al,Fe)Si_2O_6$
jade (nephrite)	[12172-67-7]	$Ca_2(Mg,Fe)_5Si_8O_{22}(OH)_2$
lapis lazuli	[1302-85-8]	complex and variable
malachite	[569-64-2]	$Cu_2(CO_3)(OH)_2$
marble	[471-34-1]	$CaCO_3$
opal	[14639-88-4]	$SiO_2 \cdot xH_2O$
pearl		
quartz	[14808-60-7]	SiO_2
ruby	[12174-49-1]	$Al_2O_3 + Cr^{3+}$
sapphire	[1317-82-4]	Al_2O_3
spodumene	[1302-37-0]	$LiAlSi_2O_6$
topaz	[1302-59-6]	$Al_2SiO_4(F,OH)_2$
tortoiseshell		
tourmaline	[1317-93-7]	complex and variable
turquoise	[1319-32-0]	$Al_6Cu(PO_4)_4(OH)_8 \cdot 4H_2O$
zircon	[1490-68-2]	$ZrSiO_4$
zoisite	[1319-42-4]	$Ca_2Al_3Si_3O_{12}(OH)$

The stability of a particular treatment is also important. The enhancement should survive during normal wear or display conditions. Whereas all the enhancements from heat treatments are stable, some produced by irradiation are not. There are also surface coatings which wear off, oilings which dry out, etc.

It is convenient to discuss enhancements in three groups: heat treatments, irradiations, and other treatments (1). Several types of treatments are at least 3000 years old; others, such as the filling of cracks with glass, arose only in the late 1980s.

Heat Treatments

The most commonly seen of the gemstones that have been enhanced by heat treatment are listed in Table 2. Parameters for specifying the conditions for heat treat-

Table 2. Gemstones Enhanced by Heating

Material	Change[a]	Product	Use[b]
amber	clarified, sun-spangled	amber	F
amber	reconstructed, aged	amber	R
beryl	green to blue	aquamarine	W
chalcedony	pale to red-brown or red	carnelian, agate, tiger's eye, etc	W
corundum	develop, intensify, or lighten blue	blue sapphire	W
corundum	develop or intensify yellow	yellow sapphire	W
corundum (ruby)	remove off-shades	ruby	F
corundum (ruby, sapphires)	remove silk, remove or develop asterism	starting material	W
corundum	diffuse in color or asterism	ruby, sapphires	R
diamond	change color after irradiation	starting material	R
quartz	amethyst to yellow citrine	starting material	W
quartz	crackled and dyed	various colors	R
zircon	brown to colorless or blue	starting material	W
zoisite	brown to deep purple-blue	starting material	W

[a]All product colors listed are stable.
[b]Prevalence of treatment occurring in product: R = rare to occasional; F = frequent; W = widespread or near-total.

ment of a gemstone material include the maximum temperature reached and the time for which the maximum temperature is sustained; the rate of heating to temperature, the rate of cooling down from temperature, and any holding stages while heating and cooling; the chemistry and pressure of the atmosphere; and any material in contact with the gemstone. Exact conditions for heat treatments vary widely according to the natural materials used.

The Effects of Heat. Heat can have many different effects, but to avoid cracking, gemstones are often preformed or even fully polished to eliminate any existing cracks and imperfections. Gemstones are then usually buried in an inert powder or placed into nested crucibles to avoid thermal shock. However, in some instances cracking may be desired as in amber or in the crackling of quartz by heating and dropping it into a dye solution. The stone resulting from the latter appears to be uniformly colored and may be used in imitation of another gemstone such as ruby or emerald. A cracking process has also been used on synthetic corundum to introduce natural-appearing fingerprint inclusions (2).

Heat is used to darken amber, ivory, and jade to simulate age. Pieces of amber and tortoiseshell can be reconstructed, ie, joined under heat and moderate pressure. By careful heating in oil, milky amber can be clarified when the gas and water within small bubbles diffuse out of the stone. If heating is rapid, the attractive sun-spangle cracking shown in Figure 1 results.

The destruction of color centers (1,3) by heating can result in bleaching or fading. Examples are brown or blue topaz, red tourmaline, smoky quartz, and some yellow sapphire. In other instances there may be a color change as when

Fig. 1. An amber bead containing sun spangles produced by a special heat treatment. Courtesy of Gemological Institute of America.

amethyst turns into yellow citrine, or when the heating of a brown topaz reveals the presence of a previously hidden Cr-derived color in a pinked topaz. These changes can usually be reversed by an irradiation treatment.

The yellow color center in some sapphire, designated Type 1 and usually from Sri Lanka, is quite unusual. This color fades gradually over the range 60–600°C. Irradiation restores and intensifies the color to a vivid orange. However, daylight restores the stable color, either from the heat-faded extreme or from the irradiated extreme (1,4). A color change derived from hydration in chalcedony involves the irreversible change from a yellow iron oxide or hydroxide or hydrate such as goethite [1310-14-1], FeO(OH), to brown-to-red hematite [1317-60-8], Fe_2O_3.

The most widely practiced enhancement is the heat treatment to produce a blue sapphire. This occurs from charge transfer involving about 0.01% each of Fe^{2+} and Ti^{4+} in Al_2O_3, corundum (1,3). Some pale corundum contains these impurities, but Fe is present mostly as Fe^{3+}. The desired redox balance can be reached by an appropriate heat treatment in a reducing atmosphere. An oxidizing atmosphere is used if the material is too dark from too much Fe^{2+} or reduced Ti. Temperatures in the 1800°C range for some hours are typically required. Depending on conditions, colors ranging from almost colorless through yellow, green, blue, to black can be obtained. If enough Ti is present, a star effect can be seen by reflection from microscopic needles of rutile [1317-80-2], TiO_2.

When synthetic Al_2O_3 containing about 0.2% TiO_2 is crystallized from the melt, the titanium, mostly as Ti_2O_3, remains in solid solution when cooled relatively rapidly, or it can crystallize as TiO_2 needles and form asteriated material on annealing at about 1200°C for some hours. Natural rubies and sapphires may contain the necessary titanium, but an anneal may be needed to develop the star. If the TiO_2 needles are too coarse or irregular, forming silk, a high temperature may be used to dissolve the needles. This is then followed by rapid cooling for clear material or a lower temperature anneal for asterism. If corundum contains Fe but no Ti, a high temperature heating produces a deep yellow-to-orange color.

If enough Ti to form a star or enough Fe and Ti for a blue color are not present in corundum, it is possible to diffuse these into the surface of a gemstone by heating for many hours close to the melting point in contact with an Al_2O_3 powder containing the missing impurities.

Irradiation Treatments

The process of irradiation involves the exposure of a specimen to one of a variety of radiations (1). A summary is given in Table 3.

There are three significant possible effects when radiation interacts with matter (5,6). First, the radiation can interact with the nucleus and induce radioactivity as in the case of neutrons. Second, displacement of atoms can occur. This

Table 3. Rays and Particles Commonly Used for the Irradiation of Gemstones

Irradiation type	Average energy, eV	Coloration uniformity	Induced radioactivity	Localized heating
x-rays	1×10^4	poor	none	none
γ-rays from Co-60 or Cs-137	1×10^6	good	none	none
neutron beam	1×10^6	good	strong	none
electron beam	1×10^6	poor	none	strong
electron beam	2×10^7	good	some	strong

has happened in a number of uranium- and thorium-containing minerals over geological periods. The outstanding example is zircon, which can contain over 10% Th and 2% U. The internal bombardment from these materials and their decay products over geological periods produces low or metamict zircon, where the disorder gives an amorphous state having a low density.

The third possible effect, a displacement of the outermost electrons in atoms, is of the most interest herein. This displacement can lead to the formation of color centers or to valence state changes. Many gemstones have been colored by irradiation over geological periods from radioactive elements in the earth's crust and from cosmic radiation. Blue and brown topaz, smoky, rose, and amethyst quartz, and purple fluorite are a few examples. All topaz is not blue or brown, because some of these materials have lost color either from relatively recent geological heating or by other means so that there has not been enough time for the color to reform. Additionally, in some localities the rock or soil is particularly free of radioactivity from geochemical depletion and acts as shielding.

With the exception of diamond coloring and the turning of topaz blue, the source of the irradiation is immaterial. Gamma rays are the preferred source because of uniformity of coloration and the absence of heating and induced radioactivity. The most commonly seen gemstones enhanced by irradiation are summarized in Table 4.

Color Centers.   Characteristics of a color center (1,3,7) include production by irradiation and destruction by heating. Exposure to light or even merely time in the dark may be sufficient to destroy these centers. Color arises from light absorption either from an electron missing from a normally occupied position, ie, a hole color center, or from an extra electron, ie, an electron color center. If the electron is a valence electron of a transition element, the term color center is not usually used.

Any material which can form a color center contains two types of precursors as shown in Figure 2a. The hole center precursor is an atom, ion, molecule, impurity, or other defect which contains two paired electrons, one of which can be ejected by irradiation, leaving behind a hole center (Fig. 2b). The electron center

Table 4. Gemstones Enhanced by Irradiation

Material	Change or product	Comments[a]	Use[b]
corundum	colorless to yellow	S,U,R	R
diamond	near colorless to black, blue, green, yellow, or red	S,R	F
pearl	darken to black	S	F
quartz	colorless to smoky	S,R	W
quartz	amethyst to amethyst–citrine	S,R	F
spodumene	pink kunzite to deep green	U,R	R
topaz	colorless or pale to blue	S,R	W
topaz	colorless or pale to brown	S,U,R	R
tourmaline	colorless or pale to red or multicolor	S,R	F

[a]S = stable; U = unstable, may fade; R = can be reversed by another treatment.
[b]Prevalence of treatment occurring in product: R = rare or occasional; F = frequent; W = widespread to near-total.

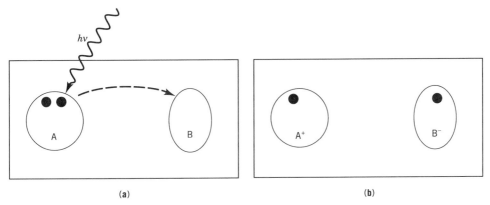

Fig. 2. (**a**) Irradiation of a material containing A, a hole precursor having an electron pair, and B, an electron precursor, to form (**b**) a color center having A^+, a hole center, and B^-, an electron center (1,3).

precursor is an atom, ion, etc, which can produce an electron center by trapping the electron ejected from the hole center precursor. A hole and an electron center are thus formed simultaneously. Either or both can be the color center. Almost all materials have hole center precursors. If there is no electron center precursor, however, the displaced electron returns to its original place and the material remains unchanged.

When light or heat liberates the trapped electron from the electron center, the hole center can be restored to the original state and color is lost, ie, fading or bleaching occurs. If the trap is shallow, room temperature alone can supply enough energy (about 0.1 eV) to release the trapped electron. For a deeper trap, the 1 to 3 eV of visible light is needed to produce fading as in blue Maxixe beryl, unstable yellow sapphire, and unstable brown topaz. A deep trap holds the electron so strongly that the coloration is stable. Smoky quartz, amethyst, blue topaz, stable yellow sapphire, and stable brown topaz are examples. Heating to as high as 600°C may then be required for bleaching.

In some materials the same color center can exist in both fading and nonfading forms, which are derived from the presence of different electron traps. Examples are yellow sapphire and brown topaz. When any colorless sapphire is irradiated to become yellow, the color usually fades on exposure to light. At the same time nonfading yellow sapphire that contains a color center, but no Fe, occurs naturally. Although both fading and nonfading color centers lose color on heating, irradiation restores both, and light exposure removes only the unstable color. The same pattern occurs in brown topaz. Some amethyst fades very slowly.

Color Changes in Quartz. All quartz contains small amounts of substitutional Al, typically 0.01%, as well as similar amounts of interstitial hydrogen or alkali metal ions. When an Al^{3+} replaces a Si^{4+}, one positive charge is missing and electroneutrality is maintained by H^+ or Na^+, etc. Irradiation ejects one of a pair of electrons on an O adjacent to an Al:

$$AlO_4^{5-} \rightarrow AlO_4^{4-} + e^-$$ (1)

The electron can be trapped, for example by an interstitial H^+, which is converted to an H atom. The AlO_4^{4-} is the hole color center which absorbs light and gives the color to smoky quartz. Bleaching is the result of thermal energy releasing the trapped electron, which then produces the reverse of reaction 1.

The amethyst color center in quartz is exactly like the smoky, except that Fe^{3+} replaces Al^{3+}.

$$Fe(III)O_4^{5-} \rightarrow Fe(III)O_4^{4-} + e^- \tag{2}$$

The $Fe(III)O_4^{4-}$ hole color center gives the purple color. On being heated, the trapped electron is released and the reverse of reaction 2 occurs producing $Fe(III)O_4^{5-}$, which provides the pale yellow color of citrine.

Color Changes in Topaz.   Essentially all colorless topaz can be turned either brown or blue by irradiation. There are two brown color centers formed at quite low irradiation doses. One is stable to light, the other fades. A stable blue color center (1,8,9) is formed also at higher irradiation doses and is revealed when heating at about 300°C removes the browns. Different shades of blue are produced by different irradiation procedures. Gamma rays give a medium blue color but lengthy exposure is needed. Electrons in the 10- to 20-MeV range produce a satisfactory blue within a reasonable time, as do neutrons, all at similar dose levels. The best color is obtained by a double treatment, using first neutrons, then electrons.

Color Center-Like Color Changes.   Green aquamarine contains Fe in two different sites. A blue color, which is not affected by heat, is produced by Fe^{2+} in interstitial sites. Most specimens also contain some Fe^{3+} substituting for Al^{3+}, which by itself produces the yellow color of golden beryl or, together with the blue Fe^{2+}, gives green aquamarine. On heating, the substitutional Fe^{3+} changes to Fe^{2+} from an electron released from a trap. This substitutional Fe^{2+} contributes no significant color. Heat thus bleaches yellow beryl to colorless and converts green aquamarine to blue aquamarine. Irradiation restores the yellow, frequently to a deeper intensity than that originally exhibited.

The irradiation darkening to produce black pearls is merely a charring of the organic conchiolin.

Radioactive Gemstones.   Zircon can contain radioactive elements, but the amount in jewelry-grade material is insignificant. Some of the treatments of Table 3 may leave irradiated material radioactive. Such gemstones have been released on rare occasions without the required cooling-off period (10).

Other Treatments

Other treatments fall into the three groups (1) summarized in Table 5. There are chemical treatments and impregnations which penetrate below the surface of the gemstone. Then there are surface coatings of various types. Finally, there are composite stones.

Chemical Treatments and Impregnations.   *Bleaching.*   Diluted hydrogen peroxide [7722-84-1], H_2O_2, and sunlight are employed to bleach essentially all pearls. The organic conchiolin is thus lightened. Similar processes can be applied

Table 5. Gemstones Enhanced by Other Processes

Process	Material[a]	Use[b]
Impregnations		
bleaching	chalcedony, coral, ivory, pearl, petrified wood, tiger's-eye, etc	W
impregnation, colorless oil/wax/ plastic	agate, chalcedony, fluorite, jade, lapis lazuli, malachite, marble, opal, turquoise, etc	W
impregnation, colored oil/wax/ plastic	same	F
crack filling, colorless plastic	emerald	R
crack filling, colorless oil	emerald	W
crack filling, colorless oil	ruby, sapphire	R
crack filling, colored oil	beryl, emerald, ruby, quartz	R
crack filling, colorless glass	diamond, ruby, sapphire	R
dyeing	agate, chalcedony, marble, onyx, etc	W
	amber, carnelian, coral, ivory, jade, malachite, opal, pearl, turquoise, etc	F
Surface modifications		
surface color coating	amber, carnelian, diamond, pearl, etc	R
foil back, mirror back, star back	used on any gemstone	R
lasering	diamond inclusions	F
Composite gemstones		
doublets, triplets	opal	F
doublets, triplets, artifact-included, gel-filled	amber, beryl, emerald, ruby, sapphire, etc	R
synthetic overgrowth	emerald on beryl	R

[a]Most of the enhanced products listed are unstable.
[b]Prevalence of treatment occurring in product: R = rare or occasional, F = frequent, W = widespread to near-total.

to other organic gem materials such as coral, producing gold from black, and to lighten ivory that has become dark with age. Brown tiger's-eye is frequently bleached to give the desired honey color.

Colorless Impregnations and Crack Filling. There may be several different aims in applying a colorless wax, oil, polymer, or glass to a gem material. First, cracks may be hidden, as in oiling an emerald. Ordinary lubricating oil has been used, but this may seep out of the cracks when the stones are warm, or the oil may dry up. Canada balsam [8007-47-4] and polymers have also been used; a vacuum and gentle heating improve penetration (11,12). A material such as turquoise may be stabilized using paraffin wax [8002-74-2] or polymers to prevent perfume and perspiration from entering the porous surface and producing a color change. At the same time, low grade turquoise which is white because of light scattering from its porosity may show an improvement from an impregnation which fills the pores. This same process may convert worthless chalky opal into a brilliantly colored form. A more recent development is the use of glass (qv) to fill cracks and even holes in ruby (13), using a high melting glass, and in diamond (14), using low melting glasses containing oxides and halides of Pb, Bi, etc.

Colored Impregnations. Colored oil to fill cracks is used on gemstones primarily to improve color, most frequently on emerald. Colored oil is also used to simulate other stones, most frequently quartz. Depending on the dye used, the colors may fade, in addition to the problems associated with colorless oiling.

Dyeing Porous Chalcedony. Many gemstone materials of the cryptocrystalline quartz family collectively known as chalcedony are porous and resistant to heat and acids. Thus many dyeing techniques are applied. The use of stable inorganic precipitates or decomposition products has been perfected (1). The oldest process, going back to Roman times, is the honey- or sugar-acid technique used to produce deep yellow, brown, or black. After cleaning, the material is soaked in honey or a sugar (qv) solution. Heat then gives yellow to brown colors. Alternatively, boiling in concentrated sulfuric acid [7664-93-9], H_2SO_4, carbonizes the sugars, giving a permanent black; essentially all black onyx is made this way. Many colors are produced by forming insoluble pigments within the pores, eg, $Fe_4(III)(Fe(II)(CN)_6)_3$, Prussian blue [14038-43-8] or Turnbull's blue [25869-98-1]. Metallic silver [7440-22-4] is precipitated for black. Heating to decompose water-soluble substances is used to produce chromic oxide [1308-38-9], Cr_2O_3, for green, iron oxides for yellow, brown, red, etc.

Dyeing Materials Other than Chalcedony. Most other gemstone substances cannot tolerate the acids or the temperatures used on chalcedony. Natural or synthetic organic dyes, even ordinary inks (qv), fabrics dyes, etc, are frequently used. Fading is, however, a problem (see DYES AND DYE INTERMEDIATES). Pearls are frequently dyed pink using an organic dye such as eosin [15086-94-9]. Exposure of pearls to silver salts followed by light exposure, or using a dyed pearl seed for culturing, give black. The honey- or sugar-acid process has also been used on porous opal to give a black background to the flashes of color. The older process, smoking, of opal packs the stones with oil or manure in paper, followed by heating.

Surface Modifications. A wide variety of surface modifications have been used on gemstones. These include treatment with wax, paints, varnishes, interference filters, foil backs, mirror backs, inscribings, selective decorations, and synthetic overgrowth. A clear varnish or wax coating is frequently applied to poorly finished tumbled stones, cabochons, or carved objects to improve the appearance or even to avoid completely the polishing step. Foil backs, shiny or colored metal sheets or metallic paints, behind a gemstone reflect light. These modifications were widely used in the days before faceting for total internal reflection. Most frequently such modifications are seen on glass or even plastic stones for costume and stage jewelry. Metal foil that has been embossed or inscribed with scratches is used behind a clear cabochon to simulate a star; the scratches also may be made directly on the back of the stone itself.

Drilling diamonds using a focused laser beam to burn out dark inclusions or make the inclusions accessible to a chemical treatment is a frequent enhancement. In a potential deception, a cubic zirconia diamond imitation was laser drilled to make it more convincing (15).

Composite Gemstones. Many types of composite or assembled gemstones have been made (1). Some are shown in Figure 3. In the United States a doublet has two pieces combined using a colorless cement. If three pieces are used, or if two pieces are assembled using a colored cement, the gemstone is a triplet. The use of composite stones has declined rapidly with the rise of inexpensive synthetics. Frequently seen are opal doublets, where precious opal is backed by a black material. In opal triplets a thin slice of precious opal is cemented between a black backing and a clear cover, usually of quartz. Additionally, insects and even fish have been inserted into amber.

A thin layer of dark green beryl had been grown by a hydrothermal technique over the surface of a pale beryl to imitate emerald. It has been suggested that such stones should be called synthetic emerald–beryl doublets (16). The ability to grow thin, but not thick, single-crystal diamond on the surface of natural diamond (17) leads to the possibility of growing such a thin film colored blue with boron; this has been done experimentally (18).

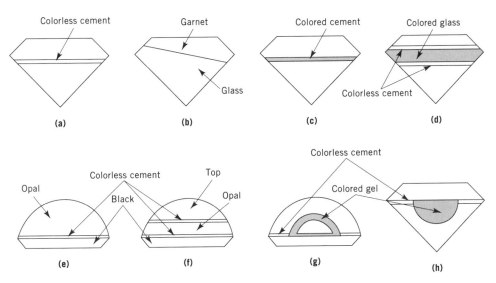

Fig. 3. Forms of composite gemstones: (**a**) doublet; (**b**) fused garnet-top doublet; (**c**) and (**d**) triplets; (**e**) opal doublet; (**f**) opal triplet; (**g**) and (**h**) gel-filled triplets (1).

Identification of Treated Gems

A trained gemologist, taught by the Gemological Institute of America of Santa Monica, Calif., and New York, the Gemmological Association of Great Britain of London, or elsewhere, is needed for identification of treated gems. This topic is also discussed in textbooks (19–21). In some materials the induced change is the exact equivalent of a process that also occurs naturally, so that such treatments cannot be identified.

BIBLIOGRAPHY

1. K. Nassau, *Gemstone Enhancement*, Butterworths, Boston, Mass., 1984; 2nd ed., 1994 (in press).
2. J. I. Koivula, *Gems Gemol.* **19**, 220 (1983).
3. K. Nassau, *The Physics and Chemistry of Color*, John Wiley & Sons, Inc., New York, 1983.
4. K. Nassau and G. K. Valente, *Gems Gemol.* **23**, 222 (1987).
5. J. Shapiro, *Radiation Protection: A Guide for Scientists and Physicians*, 2nd ed., Harvard University Press, Cambridge, Mass., 1981.
6. K. Z. Morgan and J. E. Turner, eds., *Principles of Radiation Protection*, John Wiley & Sons, Inc., New York, 1967.
7. E. Fritsch and G. R. Rossman, *Gems Gemol.* **24**, 3 (1988).
8. K. Nassau and B. E. Prescott, *Am. Mineral.* **60**, 705 (1975).
9. K. Nassau, *Gems Gemol.* **21**, 26 (1985).
10. R. Crowningshield, *Gems Gemol.* **17**, 215 (1981).
11. R. Ringsrud, *Gems Gemol.* **19**, 149 (1983).
12. R. C. Kammerling and co-workers, *Gems Gemol.* **27**, 70 (1991).
13. R. E. Kane, *Gems Gemol.* **20**, 187 (1984).
14. J. I. Koivula and co-workers, *Gems Gemol.* **25**, 68 (1989).
15. C. Fryer, *Gems Gemol.* **19**, 172 (1983).
16. H. Bank, *Z. Dt. Gemmol. Ges.* **29**, 197 (1980); **30**, 118 (1981).
17. R. C. DeVries, *Ann. Rev. Mater. Sci.* **17**, 161 (1987).
18. J. I. Koivula and R. C. Kammerling, *Gems Gemol.* **27**, 118 (1991).
19. R. T. Liddicoat, Jr., *Handbook of Gem Identification*, 12th ed., Gemological Institute of America, Santa Monica, Calif., 1989.
20. C. S. Hurlbut, Jr. and R. C. Kammerling, *Gemology*, 2nd ed., John Wiley & Sons, Inc., New York, 1991.
21. B. W. Anderson and E. A. Jobbins, *Gem Testing*, 10th ed., Butterworths, London, 1990.

KURT NASSAU
Nassau Consultants

GENETIC ENGINEERING

PROCEDURES

Genetics is the science dealing with the information content of an organism, especially the hereditary information passed on from one generation to another. The related discipline of molecular biology deals with the informational macromolecules that replicate and express genetic information in living systems. The term genetic engineering implies the deliberate manipulation in the laboratory of the hereditary information content (genotype) of a cell in order to alter the observable properties (phenotype) of an organism. In some sense, therefore, genetic engineering is as old as agriculture. Selective breeding in prehistoric times led to the introduction of maize and wheat as well as to the propagation of fermentative microbes. This Mendelian engineering is a principal industry in the 1990s as agricultural crops and animals are bred for yield, product composition, and disease resistance.

The more contemporary meaning of genetic engineering implies a use of the techniques of molecular biology, especially recombinant deoxyribonucleic acid (DNA) techniques, rather than breeding in the formation of new genotypes. Recombinant DNA molecules are composed of two parts: first, a vector where the function is to provide the biochemical functions necessary for replication of the recombinant DNA molecule, and secondly, the passenger DNA which is joined to the vector and is replicated passively under control of the vector. Recombinant DNA technology allows the construction *in vitro* of DNA molecules that are not found in nature and the subsequent introduction into organisms, resulting in new genotypes and phenotypes of the recipient. Particularly in plant and animal science, the two techniques are often complementary: a gene may be introduced into an organism by recombinant DNA technology and the line with the desired properties further manipulated by breeding (see GENETIC ENGINEERING, ANIMALS; GENETIC ENGINEERING, PLANTS).

Analysis of DNA Information

Molecular biology is an information-based science. In this context information can be defined as the negative logarithm of the probability of a system occupying a particular state, given the total number of states available to it. In a DNA sequence there are four possible states at each position, corresponding to the four nucleic acid bases, adenine [73-24-5] (A), $C_5H_5N_5$; guanine [73-40-5] (G), $C_5H_5N_5O$; thymine [65-71-4] (T), $C_5H_6N_2O_2$; and cytosine [71-30-7] (C), $C_4H_5N_3O$ (see NUCLEIC ACIDS). A DNA of chain length n therefore has 4^n possible arrangements. Although the cost in free energy terms of maintaining it is relatively small, this is an enormous amount of information. Even a small linear virus having a genome 5000 nucleotides long would have 4^{5000} (approximately 10^{3010}) potential sequences; by comparison, the number of elementary particles in the universe is estimated at 10^{80}. Because there are so many potential sequences of a DNA molecule, and because DNA molecules of the same base composition can have similar biochemical properties but very different sequences, standard biochemical techniques cannot address the most biologically important property of DNA, its information content. Genetic engineering techniques allow the analysis and manipulation of genetic information based on its nucleotide sequence.

Sequence-Dependent Cleavage of DNA by Restriction Enzymes. Bacteria in nature are constantly exposed to exogenous DNA, primarily from bacteriophage (viruses) in the environment. Probably as a defense system, many bacteria contain a two-part DNA restriction and modification system. Restriction enzymes are of several types; the most useful for cloning, the Type II restriction enzymes, recognize specific sequences, usually 4–8 base pairs (bp) in length, and cut DNA molecules within these sequences. In nature, restriction enzymes serve as a sort of immune mechanism. Invading viruses are inactivated by restriction enzyme digestion of their DNA.

The recognition sequences of restriction enzymes are short enough to be present in the host's genomic DNA many times; for example, a sequence of 6 bp would be present once every $4^6 = 4096$ bp by random chance. In the case of a standard bacterium, *Escherichia coli*, which has a genome of 4.7×10^6 bp, chromosomal DNA would be cleaved at over 1000 sites by a resident restriction enzyme, rendering it nonfunctional. In order to prevent this inactivation, bacteria encoding restriction enzymes also synthesize modification enzymes which modify DNA, usually by methylation, at the sequence recognized by the restriction enzyme, rendering the DNA refractory to digestion and thereby serving to distinguish host from invading DNA. A given restriction enzyme is useful in the analysis of DNAs prepared from hosts that do not contain an interfering modification enzyme. Because a variety of bacteria are used as sources for restriction enzymes, virtually all DNA from laboratory microorganisms or eukaryotic cells can be cut *in vitro* for recombinant DNA experiments. Table 1 lists the recognition sequences and corresponding sites of methylation for several restriction enzymes used in molecular cloning and DNA mapping experiments (1). All of these enzymes cleave DNA prepared from laboratory strains of *E. coli*, the most common bacterial host for genetic engineering experiments.

Restriction sites provide physical markers on a DNA molecule. The fragments resulting from restriction digestion are commonly separated by electrophoresis in gel supports of polyacrylamide or agarose and visualized by staining with the dye ethidium bromide, which fluoresces strongly when intercalated into DNA (see ELECTROSEPARATIONS, ELECTROPHORESIS). The length in bp of a DNA fragment is estimated by comparison of the mobility with that of a fragment of known length. In general, the mobility of a DNA fragment during electrophoresis is inversely proportional to its chain length; this relationship holds over a range of DNA size that depends on the concentration of the gel and the conditions of

Table 1. Sequence Specificities of Restriction Endonucleases

Bacterial source	Enzyme	Sequence specificity[a]
E. coli/RI	*Eco*RI	G—A—A—T—T—C
Bacillus amyloliquefaciens H	*Bam*HI	G—G—A—T—C—C
B. globigii	*Bgl*II	A—G—A—T—C—T
Xanthomonas malvacearum	*Xma*I	C—C—C—G—G—G
Providencia stuarti	*Pst*I	C—T—G—C—A—G

[a]The left end of each sequence is the 5′ end and the right end is 3′. Only one strand is shown for convenience, although the enzymes break duplex DNA. The arrow shows the position of the bonds broken.

electrophoresis. It is usually fairly simple to derive the restriction map of a small DNA, eg, a virus or plasmid.

When mapping longer DNA species such as those on the order of a whole chromosome, other techniques are used. A few enzymes are known which have recognition sequences of 8 bp in length. These longer sequences occur relatively more rarely in a chromosome than do 6-bp sequences. In some cases, naturally occurring DNAs contain fewer sites for these rare cutters than would be predicted by random chance, allowing the determination of relatively simple restriction maps of whole bacterial or eukaryotic chromosomes. Separation of the very large pieces of DNA resulting from restriction digestion in these experiments requires special electrophoresis conditions to resolve the fragments (2).

Location of Specific Sequences to DNA Restriction Fragments. A second technique that is universally applied to DNAs large and small is that of southern blotting (Fig. 1). In these experiments, DNA fragments separated by gel electrophoresis are denatured *in situ* and transferred by capillary action to a nitrocellulose or nylon membrane, thereby making a contact print of the DNA in the gel. The single-stranded DNA fragments are bound irreversibly to the filter which is then immersed in a solution containing a single-stranded nucleic acid probe. The probe forms double helical base-paired hybrid regions with filter-bound DNA of complementary sequence. Solution conditions of hybridization can be set up to

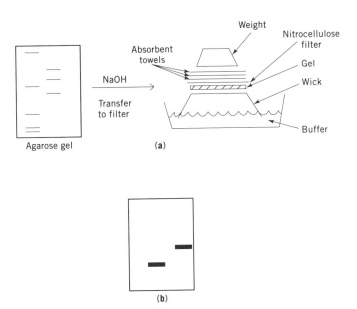

Fig. 1. Southern blot analysis of DNA showing (**a**) step 1, an agarose gel containing separated restriction fragments of DNA, denoted by (—), which is immersed in NaOH to denature the double-stranded structure of DNA, and then transferred by capillary flow to a nitrocellulose filter. In step 2, the bound DNA is allowed to hybridize to a labeled nucleic acid probe, and the unbound probe is washed off. In step 3, the filter is placed into contact with x-ray film resulting in (**b**) bands of exposure on the film which are detected after development and correspond to regions where the restriction fragment is complementary to the probe.

distinguish exact from inexact matches. The hybridized probe is detected by directly exposing the filter to x-ray film, if the probe is radioactive, or by enzymatic staining, if the probe is labeled by chemical modification. In either case the contact print of the gel shows the mobilities of the DNA fragments complementary to the probe. The specificity of Watson-Crick base pairing in DNA allows single fragments to be detected in the midst of a large excess of noncomplementary DNA. This specificity is the basis for, among other techniques, genetic fingerprinting of individual human DNAs and the use of species-specific gene probes for detection of bacterial species.

Restriction Sites as Genetic Markers. Classical genetic analysis uses observable phenotypes to infer genetic information. In order to derive a genetic map of a genotype, geneticists look for mutations which are linked, ie, carried on a chromosome. If two genes are carried on different chromosomes these cosegregate, ie, appear together in the progeny of a genetic mating statistically half the time. (Consider the probability (0.50) of getting two heads or two tails when two coins are flipped.) If two genes are located near each other on the same chromosome, the frequency of cosegregation is greater, somewhere between 50 and 100%.

In the early 1980s workers recognized that the presence of a restriction enzyme site is a chromosomal marker that can be assayed by southern blotting of genomic DNA. Thus, if the pattern of a southern blot is different for the DNA of different individuals in the population, and if one or more patterns cosegregate with a mutant gene causing a disease, the restriction pattern is a surrogate diagnostic marker for the disease. This phenomenon, termed restriction fragment length polymorphism (RFLP), can be used to predict an inheritance pattern in the absence of any other information about the disease other than its pattern of heredity. The probes used to detect RFLPs can be used as the starting point to isolate clones of the gene encoding the disease itself (3). Similar logic forms the basis of genetic fingerprinting of individuals for forensic analysis (see also FORENSIC CHEMISTRY) (4).

Gene Isolation by Recombinant DNA Techniques

Workers in the early 1970s recognized that restriction enzymes provided tools not only for DNA mapping but also for construction of new DNA species not found in nature. A collection of recombinant DNA species consisting of many passenger sequences joined to identical vector molecules is called a library. Individual recombinant DNAs are isolated from single clones of the library for detailed analysis and manipulation.

Plasmid DNAs. Plasmids are nucleic acid molecules capable of intracellular extrachromosomal replication. Usually plasmids are circular DNA species, but linear and RNA plasmids are known. In nature, plasmids can assume a variety of lifestyles. Plasmids can recombine into the host chromosome, be packaged into virus particles, and replicate at high or low copy number relative to the host chromosome. Additionally, their information can affect the host phenotype. Whereas no single plasmid is usually capable of all these behaviors, the properties of various plasmids have been used to construct vectors for a variety of purposes.

Ultimately a plasmid is defined by its mode of DNA replication. DNA replication is initiated at a single, characteristic sequence, termed the origin. The origin sequence determines the copy number of the plasmid relative to the host chromosome and the host enzymes that are involved in plasmid replication. Two different plasmids that contain the same origin sequence are termed incompatible. This term does not refer to the active exclusion of one plasmid by another from the cell but rather to a stochastic process by which the two plasmids are partitioned differentially into progeny cells. A cell which contains two plasmids of the same incompatibility group segregates two clonal populations, each of which has one of the two plasmids in it.

Plasmids can be introduced into cells by several methods. The most common method is transformation, where the recipient cells are made competent to receive DNA by washing with a solution of Ca^{2+} or other inorganic ions. Then the naked DNA is added directly; a fraction of the cells take up the DNA and replicate it. These cells are then selected by growth in media containing an antibiotic. In many cases, discharge of a high voltage capacitor across a solution of cells renders them permeable to DNA. This phenomenon, termed electroporation, can increase the efficiency of transformation substantially. Some but not all plasmids also transfer by conjugation, a sexual process where the DNA is donated from one cell to another after physical contact.

Most plasmids are topologically closed circles of DNA. They can be separated from the bulk of the chromosomal DNA by virtue of their resistance to alkaline solution. The double-stranded structure of DNA is denatured at high pH, but because the two strands of the plasmid are topologically joined they are more readily renatured. This property is exploited in rapid procedures for the isolation of plasmid DNA from recombinant microorganisms (5,6).

Plasmid Vectors for Facile Introduction of Passenger DNA and Selection of Recombinants.   The map of a commonly used plasmid vector, pUC19 (7), is shown in Figure 2. Three parts of the vector are key to its utility. The origin sequence, ori, allows the replication of plasmid DNA in high copy number relative to the chromosome. A gene, amp, encoding the enzyme beta-lactamase, which hydrolyzes penicillin compounds, allows growth of plasmid-containing cells in media containing ampicillin (see ANTIBIOTICS, β-LACTAMS–β-LACTAM INHIBITORS). The third region of the plasmid allows the introduction of passenger DNA. The polylinker sequence is a chemically synthesized sequence of DNA which contains recognition sequences for a variety of restriction enzymes. A large number of these sites are unique, occurring only once in the vector. Associated with the polylinker sequence is a portion of the gene encoding *E. coli* beta-galactosidase. In the appropriate genetic background, beta-galactosidase enzymatic activity can be detected by the use of a chromogenic substrate (x-gal) which is hydrolyzed to yield a blue compound. Bacterial colonies containing the intact polylinker sequence express beta-galactosidase activity and stain blue. When the polylinker is disrupted by insertion of a passenger DNA sequence, beta-galactosidase is no longer made and the colonies do not stain to a blue color. Clones containing recombinant plasmids therefore can be identified visually in a background of nonrecombinants.

Construction of a Recombinant Plasmid by Joining Vector and Passenger DNA.   The unique restriction sites in the plasmid vector DNA provide sites in the molecule for insertion of restriction-digested DNA fragments. Figure 3 shows an example: vector and passenger DNAs are digested separately using the enzyme *Eco*RII, and the digested DNAs are mixed in the presence of an energy donor, adenosine triphosphate [56-65-5] (ATP), and the enzyme DNA ligase. A fraction

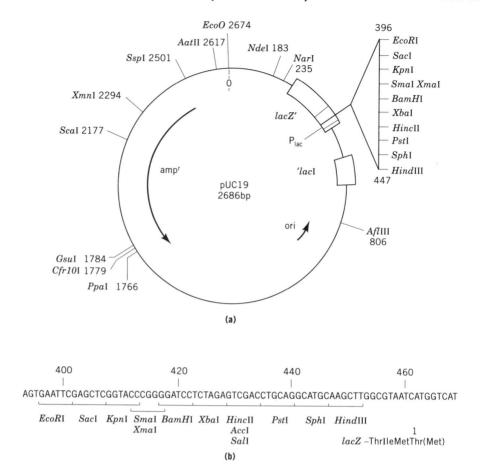

Fig. 2. (**a**) Map of pUC19, a commonly used plasmid vector where the numbers correspond to the positions of the various restriction enzyme cuts; and (**b**) nucleic acid composition of pUC19 from position 393 (5′-end) through position 469 (3′-end) (5,7).

of the complementary ends left by *Eco*RI in the vector and passenger DNAs form Watson-Crick base pairs and the DNA chains are then joined covalently by DNA ligase. The recombinant DNA formed in this example is then introduced into recipient *E. coli* cells. Recipients containing the recombinant DNA are recognized by their ability to grow in the presence of ampicillin, this is diagnostic of the vector, and their lack of beta-galactosidase enzymatic activity, which indicates the insertion of foreign DNA into the polylinker, ie, sequence B in Figure 3.

    Identification of the Desired Passenger Sequence in Plasmid Cloning Experiments. The objective of recombinant DNA construction is to obtain a clone of a single DNA sequence. If more than a single restriction fragment is ligated into different vector molecules, the result is a library of clones, all of which have the same vector sequence but with different passengers. Libraries are often described by the source of the passenger DNA. Genomic DNA libraries contain the total chromosomal DNA inserted into a vector. Copy DNA (cDNA) libraries con-

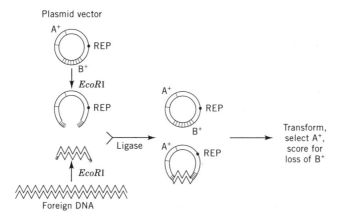

Fig. 3. Construction of a recombinant DNA by joining vector and passenger fragments where (⊔⊔⊔⊔) represent sticky ends. Both A and B genes represent selectable traits, so that introduction of foreign DNA in B gene leads to the loss of an identifiable function. REP represents replication and maintenance genes. Courtesy of CRC Press.

tain passenger DNA derived by copying messenger RNA (mRNA) into DNA using the enzyme RNA-dependent DNA polymerase. This process is known as reverse transcriptase.

The number of independent sequences in a DNA population is defined as its complexity. In a cloning experiment, the complexity of a library reflects the complexity of the passenger DNA population. In order to find a gene in a complex library, it is necessary to screen a large number of clones. The relationship between the complexity of the passenger DNA population, the size of the inserted passenger DNA fragments (I) and the number of clones (N) that must be screened to have a defined probability (P) of finding the correct sequence is given by Poisson statistics (8):

$$N = \ln(1 - P)/\ln (1 - I/G)$$

where G is the genome complexity in bp. For cloning a mammalian genomic sequence (complexity = 3×10^9 bp) using fragments of 1×10^4 bp of DNA, a library of 1.38×10^6 independent clones is required to find a sequence having a probability of 99%. The number of clones required to identify a gene could be reduced by the insertion of larger DNAs or by the use of a less complex initial population of passenger DNA. For example, a mammalian organism synthesizes fewer than 10^5 mRNAs, representing only a few percent of its genomic DNA information. Thus, fewer clones would need to be screened to identify a particular coding sequence in a cDNA library.

The complexity attainable in construction of a plasmid library is limited by the efficiency of introducing the DNA into recipient bacteria. Libraries of sufficient complexity to clone mammalian genes are not normally feasible in plasmid vectors. Usually, these complex DNAs are cloned in bacteriophage or cosmid vectors which can accept larger fragments of passenger DNA. Plasmid libraries have been used to clone microbial genomes (complexity approximately 4×10^6 bp) because correspondingly smaller library sizes are required.

Given a library of sufficient complexity it is then necessary to find the clone of interest against the background of recombinant clones containing other passenger sequences. Three strategies are generally employed. The simplest method is to use genetic complementation of a mutant in the host. Thus, for example, a plasmid genomic library from *Streptomyces coelicolor* was transformed into a mutant *E. coli* host deficient in the metabolism of galactose (gal). The transformants were selected for the ability to metabolize galactose. The cloned genes were shown to direct galactose metabolism in *S. coelicolor* (9). Alternatively, genes encoding antibiotic resistance have been identified by direct phenotypic selection. A wide variety of genes from bacteria and yeast have been detected in this fashion. In cloning a gene for which no *E. coli* phenotype can be selected, it is necessary to screen for the desired sequence. The most common screening method uses a radioactively labeled probe to hybridize to DNA from the recombinant bacteria. The agar plate containing colonies of recombinant bacteria is blotted with a sheet of nitrocellulose or nylon filter paper, thereby transferring some of the bacteria in the colony to the filter. These transferred colonies are lysed with alkali *in situ*, thereby also making the DNA single-stranded. This DNA contact print is hybridized to the probe in the same way as restriction fragments are hybridized in southern blotting (see Fig. 1). Colonies containing DNA complementary to the probe are identified by autoradiography. Because only a portion of the original colony is transferred to the filter paper, the pattern on the autoradiogram identifies the appropriate colonies on the master plate (see RADIOACTIVE TRACERS).

The limiting factor in identifying a clone of interest is the availability of a probe. Probes may be obtained by using a homologous sequence from a previously identified clone. Thus, for example, a sequence encoding mouse beta-globin can be used to identify beta-globin genes from the human. Similarly, sequences complementary to *E. coli* ribosomal RNA (rRNA) identify yeast rRNA sequences. In most screenings using heterologous probes it is necessary to hybridize the probe under conditions of lesser stringency to obviate the effects of DNA sequence variations across species.

A potentially general method of identifying a probe is, first, to purify a protein of interest by chromatography (qv) or electrophoresis. Then a partial amino acid sequence of the protein is determined chemically (see AMINO ACIDS). The amino acid sequence is used to predict likely short DNA sequences which direct the synthesis of the protein sequence. Because the genetic code uses redundant codons to direct the synthesis of some amino acids, the predicted probe is unlikely to be unique. The least redundant sequence of 25–30 nucleotides is synthesized chemically as a mixture. The mixed probe is used to screen the library and the identified clones further screened, either with another probe reverse-translated from the known amino acid sequence or by directly sequencing the clones. Whereas not all recombinant clones encode the protein of interest, reiterative screening allows identification of the correct DNA recombinant.

If an antibody to the protein of interest is available, it is sometimes possible to use vector sequences, eg, the beta-galactosidase promoter sequence, to direct the transcription of the passenger DNA into messenger RNA and the translation of that mRNA into protein which can be recognized by the antibody. Although this method is somewhat less reliable than the use of nucleic acid probes, specialized vectors are available for this purpose.

Vectors for Cloning Larger Fragments of DNA. Plasmid DNAs used in molecular cloning have a practical limit in the amount of DNA that can be inserted into them. Fragments longer than this limit often accumulate deletion variants where replication is favored over the original molecular species. This leads to loss of the original clone. In addition, the introduction of plasmids into recipient cells by transformation is relatively inefficient. When complex libraries are needed, for example to isolate a mammalian gene, other cloning strategies are needed. These strategies are based on the replication of the bacteriophage lambda.

Lambda infects *E. coli* in either of two modes. It can lyse the cell to release more virus particles in a short time or it can insert itself into the bacterial chromosome and be replicated passively with the host, a phenomenon termed lysogeny. The DNA encoding lysogenic functions, about 40% of the bacteriophage chromosome, can be replaced by foreign DNA without interfering with lytic growth of the phage. Lambda-derived vectors use this phenomenon for cloning purposes. An example is shown in Figure 4. The vector DNA is prepared from phage grown lytically and the central stuffer fragment is removed and discarded, leaving the two arms of the vector. The vector arms are ligated to fragments of the DNA. Conditions of ligation are such that long tandem molecules of DNA are formed, containing several vector-insert combinations linked together. The DNA is packaged *in vitro* into virus particles and the recombinant particles are mixed with host bacteria. Infection of the bacteria by recombinant packaged phage is essentially 100% efficient; this provides an advantage over the use of plasmid vectors.

Selection for phage clones containing recombinant DNA is provided by the packaging reaction. DNA packaging involves cutting of the catenated DNA at specific sequences termed cos for cohesive. If two cos sequences are separated by approximately 50 kilobases (kb) of DNA, the cut DNA fills the phage head and a viable phage results. If the sequences are closer together, as in the case of two arms being joined without an insert, no viable phage results. Similarly, if two passenger sequences are inserted between arms, the resulting DNA is too large to be packaged in the phage head.

Isolation of DNA for Phage Cloning. Because lambda-derived cloning vectors accept only a narrow size range of DNA inserts, a library constructed from completely restriction-digested DNA is unlikely to be representative of the total passenger DNA population. Because a restriction enzyme recognizing a 6-bp sequence cleaves DNA on the average of once in 4×10^3 bp and, to a first approximation, restriction sites are randomly located, many of the restriction sites are located significantly more or less than 20 kb apart. The fragments resulting from digestion at these sites would not be packaged and therefore would not be found in the library. In order to construct representative libraries the passenger DNA population is partially digested using a frequently cutting, ie, 4 bp recognition sequence, restriction enzyme under conditions where the average size of the products is close to 20 kb. The passenger DNA fragments are then separated by agarose gel electrophoresis or by sucrose density gradient centrifugation to eliminate those smaller and larger fragments in the digestion products. Only one passenger DNA molecule from this preparation yields a viable phage after the two steps of

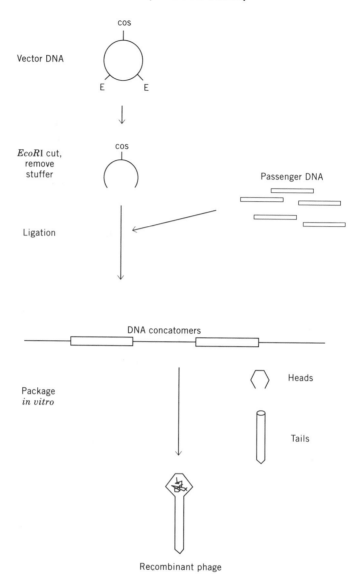

Fig. 4. Construction of recombinant phage in vectors derived from bacteriophage lambda where E represents the enzyme *EcoRI*. Other terms are defined in text.

ligation to vector arms and *in vitro* packaging. If size-fractionation of the passenger DNA population were not done before ligation, two or more small fragments might ligate to each other and be cloned together, a linkage which would not reflect the true relationship.

The vector arms in phage cloning experiments are prepared in such a way as to prevent self-annealing. First, the cos sequences of the phage DNA are ligated together. This step ensures that each recombinant phage DNA molecule encodes all of the phage functions essential for growth. Then the central stuffer fragment is removed by restriction digestion, separated from the arms by density gradient centrifugation, and discarded. Finally, the arms are treated with alkaline phosphatase to remove the 5′-phosphate from the sticky ends, preventing self-ligation of the vector arms.

Screening of Recombinant Phage by DNA Hybridization or Antibody Recognition. The result of a phage infection of bacteria growing in a nutrient agar plate is a plaque or hole in the lawn of host bacteria. The plaque contains a clonal population of phage that result from initial infection of a single cell by a single phage. Infection typically results in a burst of 100–200 progeny phage which then infect neighboring bacteria to continue the infection process. As the bacteria reach saturation and stop growing the process stops, leaving the plaque of lysed bacteria and infectious phage particles. In screening a recombinant phage library, the phage are plated at high density so that plaques are nearly contiguous. Then the plate is blotted with nitrocellulose or nylon filter paper. Dipping the filter into alkaline solution lyses the phage particles and denatures the DNA, making it ready for hybridization with a DNA probe. The phage from the hybridizing region of the plate are diluted to a lower density, plated, and rescreened. After two or three screenings, a clonal population of recombinant phage are present. The passenger DNA can be analyzed by southern blotting, restriction mapping, and sequencing.

If a DNA probe is not available, the recombinant phage can be screened by antibody methods. This has been used primarily in screening inserts derived by copying messenger RNA into cDNA using the enzyme reverse transcriptase. The cDNA population is then cloned into specialized phage that facilitate transcription and translation of the inserted cDNA. Antibody screening of the clones is done as described above. It should be noted that most eukaryotic genes contain noncoding introns inserted in the midst of the coding sequence. Therefore genomic DNA libraries usually cannot be screened by antibody technology.

Cosmid Vectors. Whereas the amount of information required for growth of a phage is large, that required for replication and selection of a plasmid is much less. Addition of cos sequences to a plasmid (hence the name cosmid) allows recombinant molecules of the appropriate size to be packaged into infectious particles. Because infection is much more efficient than transformation, use of recombinant particles overcomes a primary disadvantage of plasmid vectors. In addition, the cosmid can accept over 40 kb of passenger DNA without loss of infectivity. This means that smaller library sizes are required to represent a complex passenger DNA population. Once the recombinant cosmid DNA is inserted into a bacterial host cell it replicates like a plasmid, in some cases allowing the use of phenotypic selection to identify genes. Although artifacts can arise because of the instability of recombinant cosmids, their use has been valuable in the preparation of ordered libraries representing entire bacterial or eukaryotic genomes (9).

Yeast Artificial Chromosomes for Insertion of Passenger DNA Molecules Larger than 10^5 bp. The sizes of the passenger DNA molecules inserted into cosmid and phage vectors are limited by the requirement for packaging of the recombinant DNA into the phage head. In addition, recombinant plasmids in *E. coli* are often unstable if these are larger than 50–100 kb. A system for constructing libraries from complex DNAs has been developed (10) that overcomes many of these limitations. Reasoning that individual eukaryotic chromosomes are extremely large ($> 10^6$–10^7 bp in most organisms) DNA molecules, a vector that provides the sequences necessary for faithful replication and segregation of an individual yeast chromosome to the recombinant was constructed. Passenger DNA inserted into these vectors is replicated by yeast as an extra, linear chromosome in the nucleus. In addition to a cloning site and sequences allowing the preparation of the vector from *E. coli*, these yeast artificial chromosome (YAC) vectors (Fig. 5) contain sequences required for chromosomal propagation: CEN, the sequence that provides the information for segregation of the artificial chromosome through mitosis and meiosis; ARS, an autonomous replication sequence, ie, an origin of DNA replication; TEL, sequences directing the formation of telomeres, the ends of the chromosome. In YAC cloning experiments, large fragments of chromosomal DNA are prepared by partial digestion with a restriction enzyme and size fractionation. Then the passenger DNA is ligated into the vector and the ligation mixture is transformed into yeast. The YACs are replicated as linear molecules, just as are the resident chromosomes of yeast. Screening for desired sequences is done by standard protocols, as described above.

A YAC library can represent all the sequences of the human genome (3×10^9 bp) with a 99% probability of finding an individual sequence in as few as 50,000 clones. This library size is small enough to be carried as individual cultures in microtiter dishes and screened by automated means. Whereas for many genes the use of cosmid and phage libraries is sufficient to locate the appropriate information, mapping and sequencing of complex genomes, as in the Human Genome Project, is expected to use YAC libraries preferentially.

Clones Linked Together to Form Larger Maps. Many eukaryotic genes are larger than the carrying capacity of a single phage or cosmid vector. Genomic DNA from eukaryotes contains noncoding sequences termed introns within the

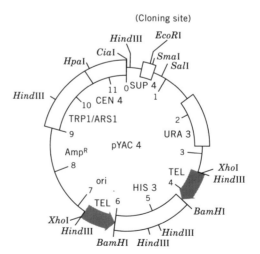

Fig. 5. Restriction map of the yeast artificial chromosome (YAC) vector used for cloning very large fragments of eukaryotic DNA. Terms defined in text (10).

coding sequence; therefore, the DNA required to encode a protein can be much longer than that predicted from the size of the corresponding mRNA. For example, the gene for the human blood clotting protein Factor VIII is over 2×10^6 bp long, but the corresponding mRNA is about one-tenth the size.

As a result of these considerations it is usually necessary to identify clones in the library that are linked to the first clone obtained by screening. The library is rescreened using a restriction fragment from one end of the passenger DNA segment in the first clone as a probe. The process, termed chromosome walking, may be reiterated several times until the contiguous inserts, termed contigs, cover the entire chromosomal region of interest. Complete contig maps have been determined for single chromosomes of several common laboratory organisms (eg, 10) as a guide to genomic sequencing.

Chromosome walking, in principle capable of determining large maps, is tedious, especially when assembling phage or cosmids into contigs. In addition, the presence of repeated sequences in the probe and/or genome of interest can confound the analysis. The use of YAC clones can lead to more efficiency. More specialized jumping techniques can also be applied, especially in cloning genes from nearby linked restriction fragment polymorphisms (11).

Experimental Protocols.   In many cases, optimized reaction buffers, nucleotides, and enzymes are packaged in kits along with detailed protocols by the manufacturers. Beyond the information provided by reagent manufacturers, laboratories need a collection of experimental protocols (5,6). One such collection is distributed on a subscription basis and updated quarterly (5).

Analysis of DNA Sequences

After a desired clone is obtained and mapped with restriction enzymes, further analysis usually depends on the determination of its nucleotide sequence. The nucleotide sequence of a new gene often provides clues to its function and the structure of the gene product. Additionally, the DNA sequence of a gene provides a guidepost for further manipulation of the sequence, for example, leading to the production of a recombinant protein in bacteria.

The sequence of a gene predicts the sequence of the protein it encodes. The relationship between nucleotide sequence of a DNA or its mRNA and the amino acid sequence of the protein it encodes is given by the genetic code. Several strategies are available for identifying protein-coding regions. The frequency at which synonomous codons are used varies in different organisms. Sequences are searched using a database of codon frequency in the organism of interest to identify the most likely coding regions. In addition, a DNA sequence can be translated in all three reading frames. A database of known protein coding sequences is then searched using the predicted amino acid sequences as a query. Statistically significant homologies can provide a clue to the structure and function of the protein(s) encoded by the cloned DNA.

Determination of DNA Sequence Information.   Almost all DNA sequence is determined by enzymatic methods (12) which exploit the properties of the enzyme DNA polymerase. Whereas a chemical method for DNA sequencing exists, its use has been supplanted for the most part in the initial determination of a sequence. Chemical or Maxam-Gilbert sequencing (13) is more often used for mapping functional sites on DNA fragments of known sequence.

DNA polymerase enzymes all synthesize DNA by adding deoxynucleotides to the free 3'-OH group of an RNA or DNA primer sequence. The identity of the inserted nucleotide is determined by its ability to base-pair with the template nucleic acid. The dependence of synthesis on a primer oligonucleotide means that synthesis of DNA proceeds only in a 5' to 3' direction; if only one primer is available, all newly synthesized DNA sequences begin at the same point.

DNA polymerases normally use 3'-deoxynucleotide triphosphates as substrates for polymerization. Given an adequate concentration of substrate, DNA polymerase synthesizes a long strand of new DNA complementary to the substrate. The use of this reaction for sequencing DNA depends on the inclusion of a single 2',3'-dideoxynucleoside triphosphate (ddNTP) in each of four polymerization reactions. The dideoxynucleotides are incorporated normally in the chain in response to a complementary residue in the template. Because no 3'-OH is available for further extension, polymerization is terminated. Thus, for a sequencing reaction initiated at a primer with a reaction mixture containing ddATP, at each T in the template, the enzyme incorporates either a deoxy-A or a dideoxy-A. In the first instance, polymerization continues; in the second, the chain terminates. The result at the completion of the reaction is a series of nested fragments each containing an identical 5' end and having different lengths. In a similar fashion, inclusion in other reaction mixtures of either ddGTP, ddCTP, or ddTTP produces nested fragments where the termini correspond to C, G, and A residues, respectively, in the template sequence (Fig. 6).

The nested oligonucleotides of a sequencing experiment are separated on the basis of chain length by electrophoresis in polyacrylamide gels. The gels separate the fragments on the basis of size, roughly proportional to the logarithm of the chain length. This means that shorter fragments are more widely separated than are long fragments. Practically, this means that sequences longer than 350–400 nucleotides are difficult to determine from a single experiment, although longer gels can resolve fragments up to 600 nucleotides long. The primer or newly synthesized chain is usually labeled isotopically, and the fragments are detected by autoradiography. In practice, reading the gels is the limiting step in DNA sequence analysis. An automated sequencing system is commercially available which uses fluorescent dyes to detect the fragments after electrophoresis; output from the instrument is stored as a computer file (14). Sequences are usually determined from both template strands to ensure against mistakes. It is also necessary to sequence across all the restriction sites used for subcloning portions of the fragment.

Computer Analysis of DNA Sequence Information

The amount of information from a single DNA sequencing project can be staggering. Therefore, it is almost always necessary to analyze these data by computer methods. A number of commercial systems are available for analysis of DNA se-

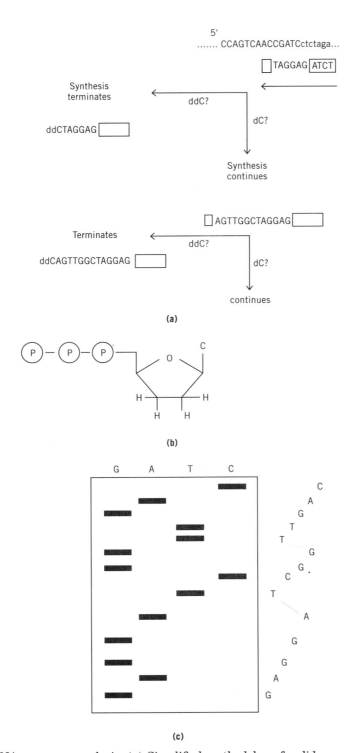

Fig. 6. DNA sequence analysis. (**a**) Simplified methodology for dideoxy sequencing. A primer, 5′-TCTA, hybridized to the template, is used to initiate synthesis by DNA polymerase. (**b**) Structure of 2′,3′-dideoxy CTP. When no 3′-OH functionality is available to support addition of another nucleotide to the growing chain, synthesis terminates once this residue is incorporated into the synthetic reaction. (**c**) Representation of a DNA sequencing gel and the sequence, read from bottom to the top of the gel, gives sequence information in the conventional 5′ to 3′ direction.

455

quence information, operating on a variety of platforms, ranging from personal computers through workstations and supercomputers, depending on the intensity of the task. The field is not fully developed and research is ongoing in algorithm development, database manipulation, and network applications, among others.

Assembly and Analysis of the Results from a Sequencing Project. The size of a gene almost always exceeds the data available from a single DNA sequencing experiment. It is necessary to identify contiguous regions of sequence information and assemble these into completed projects. This can be done by relatively simple string matching, followed by highlighting points where two sets of overlapping information disagree. Resolution of discrepancies is then a matter of the investigator's judgment.

The string of nucleotides in a DNA file is then translated into an amino acid sequence using the genetic code. Because amino acid residues are specified by a genetic code of three nucleotides, any nucleic acid sequence can be read in three separate reading frames. The three predicted sequences are then used as arguments to search the database of previously determined genes and gene products. The databases (qv), either Genbank, maintained by the National Library of Medicine (United States) or EMBL, maintained at the European Molecular Biology Laboratory (Heidelberg) share information, and a search of one is now equivalent to the search of the other. Databases are distributed either by electronic mail, on physical media (CD-ROM or tape) from the repositories, or by commercial software vendors as part of an analysis package. Virtually all journals in the field require deposition of the sequence file into a database before accepting a manuscript for publication.

Searching the database for a string of nucleotides or amino acids is extremely simple but normally unenlightening. Usually DNA and protein sequences vary among organisms so that, eg, an enzyme from the mouse and one from bacteria may have the same catalytic mechanism and function but different amino acid sequences. Search programs use statistically based scoring tables to evaluate the probability of two protein or DNA sequences being closely related. Sequences which share a large fraction of chemically related or identical amino acids are predicted to define biochemically related proteins (15).

Other database searches may be used to predict active sites or secondary structures in proteins, likely points of mRNA initiation (promoters), translation initiation, and restriction sites that could be useful in further manipulation. When more than two homologous sequences are aligned, a statistically most probable consensus can be generated as a guide to functional residues; however, finding the best multiple alignment of several sequences, for example, in defining the information necessary for promoter function, from first principles is computationally difficult and has not been implemented rigorously.

Uses of Sequence Information

DNA sequence information is the starting point for other applications, including the expression of a gene product, the search for related sequences in biological samples, *in vitro* mutagenesis of the sequence, and structure–function studies of gene expression.

Specific Amplification of Related Sequences by the Polymerase Chain Reaction. If the sequence of a gene is known, primers that are unique to the gene can be synthesized. An oligonucleotide longer than 15–18 residues is likely to be unique even in a complex genome. Such an oligonucleotide, if hybridized to a single-stranded DNA, can be used to prime DNA polymerase so that the DNA is replicated. If two primers are made, each complementary to one strand of a gene, then each strand of the DNA located between them can be specifically replicated by DNA polymerase (Fig. 7). The newly replicated DNA strands can be separated by heating and can then serve as template for another round of primed synthesis, leading to another doubling in the concentration of the original ampli-

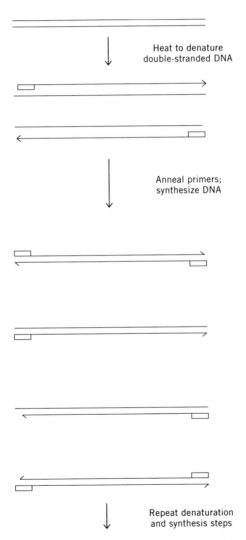

Fig. 7. The PCR reaction, showing amplification of a sequence located between two primers where (▢) represents a primer, (—) a single strand of DNA, and the arrow the 3′ end. A double-stranded DNA is replicated fourfold.

fied sequence. Because the concentration of the DNA of interest doubles with each cycle, at least a 50,000-fold increase in its concentration is achievable within a few hours. This polymerase chain reaction (PCR) (16) is used in a variety of experimental manipulations and diagnostic procedures. Sequences less than a few hundred base pairs long are most efficiently amplified by PCR but even this relatively limited information can be fruitful.

PCR amplification of a DNA sequence is facilitated by the use of a heat-stable DNA polymerase, Taq polymerase (TM), derived from the thermostable bacterium *Thermus aquaticus*. The thermostable polymerase allows the repeated steps of strand separation, primer annealing, and DNA synthesis to be carried out in a single reaction mixture where the temperature is cycled automatically. Each cycle consists of a high temperature step to denature the template strands, a lower temperature annealing of the primer and template, and a higher temperature synthesis step. All components of the reaction are present in the same tube.

It should be emphasized that the polymerase chain reaction requires specific primers for synthesis; therefore, the sequence flanking the sequence to be amplified must be known. PCR reactions are also very susceptible to contamination by other DNA. Precautions against contamination need to be rigorous, especially when PCR is used for diagnosis of disease, eg, in testing for the presence of viral nucleic acid in blood samples.

Applications of PCR.   The ability of PCR to specifically amplify DNA sequences leads to a wide variety of applications. It is possible to identify microorganisms in pathological samples without the need for culturing them, eg, in testing for the human immunodeficiency virus that causes AIDS; identify mutant genes in analyzing genetic disease; construct mutant genes for structure–function studies; find specific clones without the need for screening (sib selection); genetically fingerprint DNA from forensic samples; synthesize proteins by coupled transcription and translation *in vitro*; and clone rare messages from mammalian cells by amplification of specific sequences. In analysis of complex genomes, specific PCR primers form sequence tagged sites for physical mapping. Protocols for many of these uses of PCR have been collected (16) and new ones appear in the literature regularly.

Combinatorial or Affinity-Selected Libraries.   A recent application of PCR is the construction of new types of libraries consisting of randomized, synthetic sequences which are then selected for some biochemical property. The selected sequences are amplified by PCR and rescreened. The result is a set of one or more sequences where the members share the biochemical property. After cloning, the sequences of individual members of the set are determined. An example is shown in Figure 8. A hypothetical protein binds to a recognition sequence six nucleotides long. In an effort to determine the sequences required for this binding, a library is made consisting of synthetic DNA oligonucleotides randomized at six positions and having a common sequence in all other positions. The complexity of the initial library is therefore $4^6 = 4096$. The oligonucleotide library is then bound to purified protein *in vitro*. Bound oligonucleotides are separated from the unbound sequences by physical means, such as electrophoresis or entrapment on a nitrocellulose filter. Because DNA-binding proteins normally bind to a number of sequences with different affinities the initial population of selected DNAs is somewhat heterogeneous, consisting of 20 members in this example. These oligonu-

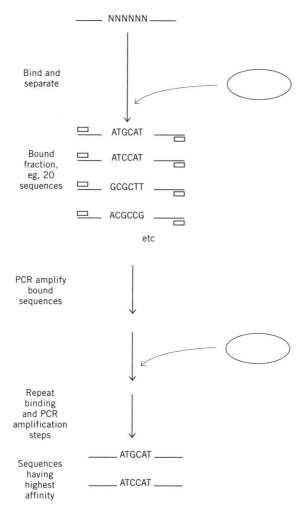

Fig. 8. A simplified combinatorial approach to identify the binding sequence for a (○) protein of interest, within 4096 sequences where (□) represents the PCR primers, N a nucleotide, ie, A, G, C, or T. An essential aspect of this experiment is the ability to separate protein-bound from free DNA prior to amplification. See text.

cleotides are then amplified by the PCR reaction to yield a population of DNAs representative of the initially selected sequence.

DNA sequencing of the selected population at this stage would show bands in more than one lane of the sequencing gel indicating that several positions in the sequence population consisted of more than one residue. Reiteration of the selection and amplification process would yield the sequences in the initially selected population that had highest affinity for the binding protein. In the example shown, this would consist of two members. If the finally selected sequence population were cloned and the DNA sequences of individual clones determined, two sequences would be found to be equally represented and to have identical affinities. Because selection is stochastic the representation of different sequences in the final, mixed population would reflect the affinity of the sequences for the protein with sequences having lowest bonding affinity K_d values being represented most often (17,18).

By similar logic, protein affinity libraries have been constructed to identify protein–protein combining sites, as in antibody–antigen interaction (19) and recombinant libraries have been made which produce a repertoire of antibodies in *E. coli* (20). In another case, a potential DNA-based therapeutic strategy has been studied (21). DNAs from a partially randomized library were selected to bind thrombin *in vitro*. Oligonucleotides, termed aptamers that bound thrombin shared a conserved sequence 14–17 nucleotides long.

Combinatorial libraries are limited by the number of sequences that can be synthesized. For example, a library consisting of one molecule each of a 60-nucleotide sequence randomized at each position, would have a mass of $>10^{14}$ g, well beyond the capacity for synthesis and manipulation. Thus, even if nucleotide addition is random at all the steps during synthesis of the oligonucleotide only a minority of the sequences can be present in the output from a laboratory-scale chemical DNA synthesis reaction. In analyzing these random but incomplete libraries, the protocol is efficient enough to allow selection of aptamers of lowest dissociation constants (K_d) from the mixture after a small number of repetitive selection and amplification cycles. Once a smaller population of oligonucleotides is amplified, the aptamer sequences can be used as the basis for constructing a less complex library for further selection.

Expression of Genes in a Heterologous Host

In many cases it is possible to synthesize the product of a gene in a different organism, eg, bacteria, yeast, or higher eukaryote. Recombinant DNAs directing the synthesis of the gene product must contain information specifying a number of biochemical processes.

Replication of the Recombinant DNA. In bacteria, replication of the recombinant DNA is provided by origin sequences, derived usually from plasmids indigenous to the host. Often, an ori sequence from one bacterium does not function in another. Therefore the vector can contain two origins of replication, each functioning in a different host. These vectors are termed shuttle vectors. In some cases, the origin is provided by integrating the foreign DNA into the chromosome of the host.

Selection of Recombinants. Selection of recombinants is provided either by a gene specifying antibiotic resistance or the ability to allow growth of recombinants in the absence of a particular nutrient.

Transcription of the Foreign Gene. Promoters are sequences preceding the start of transcription that direct RNA polymerase action. In general, these are specific to an organism and must be supplied, eg, when expressing a mammalian DNA sequence in bacteria. Often, expression of a heterologous gene is deleterious to the growth of the host. In these cases, strategies are available for the conditional transcription of the gene. For example, transcription of the gene might occur only at 42°C.

Translation of the Foreign Gene.   The translation of a mRNA into a protein is governed by the presence of appropriate initiation sequences that specify binding of the mRNA to the ribosome. In addition, not all the codons of the genetic code are used equally frequently by all organisms. Efficient translation depends on matching the preferred pattern of host codon usage in the heterologous gene.

Stability and Purification of the Recombinant Protein.   There are no hard and fast rules specifying, eg, whether a recombinant protein is available in a soluble state in the cell. In some cases, the expression system must be engineered by *in vitro* mutagenesis to optimize overall yield of the protein.

Mutagenesis of Cloned DNA

Genetics begins with mutants; indeed, the primary definition of a gene is a unit of mutation. Mutational analysis of a cloned gene is often essential for identifying structure–function relationships in its expression or in the protein encoded by the cloned gene. Alternatively, expression of a recombinant protein is often dependent on the codon usage optimal for the host. A number of techniques are available for mutagenesis. Randomized treatment of DNA using a chemical mutagen continues to be useful. In addition, a short synthetic DNA can be made having specific or random mutations introduced during synthesis. This altered information can then be incorporated into the cloned gene. A few of the more general approaches are described herein.

Mutagenesis by Synthetic DNA.   Synthetic DNA is the basis for many other techniques and was the first procedure realized (Fig. 9) (22). A synthetic oligonucleotide having one specific alteration is hybridized to a single-stranded DNA, usually from a subcloning of the gene to be mutagenized into a specialized bacteriophage. The oligonucleotide is then used to prime synthesis of a complementary strand. The double-stranded recombinant is then transformed or electroporated into a recipient cell. If the phage DNA is prepared from a mutant *E. coli* host that inserts deoxyuracil rather than thymidine [50-89-5] into its DNA during growth, then subsequent growth in a wild-type host discriminates against the template and mutant clones are obtained with high efficiency. An alternative procedure uses two complementary mutagenic oligonucleotides and amplifies the rest of the sequence by the polymerase chain reaction (23).

It is possible to make a number of mutations in a small sequence by synthesizing the mutagenic oligonucleotide with a small amount of incorrect nucleotide at each position, such that each oligonucleotide contains an average of one or two incorrect bases. This is the so-called dirty bottle synthesis. The oligonucleotide population is then hybridized to single-stranded DNA and primed synthesis is carried out as described above.

Linker-Scanning Mutagenesis.   Using linker-scanning mutagenesis (24) small sequences of DNA are removed and replaced with a synthetic restriction fragment or linker. This technique is commonly used in analysis of promoters and other control sequences in DNA, while preserving the spatial relationship between the sequences.

Mutagenic PCR.   More recently, methods have been developed to use the PCR reaction to randomly mutagenize a defined sequence (25). The Taq polymer-

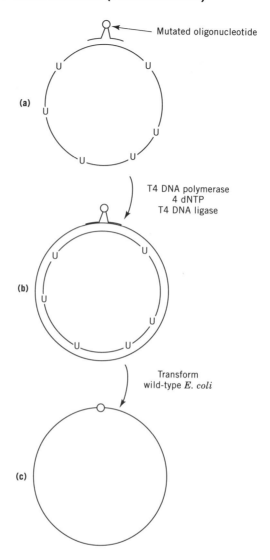

Fig. 9. Mutagenesis by a synthetic oligonucleotide of a cloned sequence available in single-stranded form: (**a**) single-stranded M13 template containing uracil (U) residues; (**b**) double-stranded product, uracil residues are not mutagenic; (**c**) strong selection for M13 phages containing mutation of interest (23).

ase used in PCR misincorporates nucleotides in a random fashion if manganese dichloride [*7773-01-5*], $MnCl_2$, is included in the reaction buffer during PCR. The library of mutagenized PCR products can be screened for the desired phenotype.

Protein Pharmaceutical Products

Development of recombinant proteins for pharmaceutical use has grown exponentially since 1982, when recombinant human insulin received approval from

Table 2. Approved Biotechnology Drugs, 1991[a]

Product	Company	Indication
human insulin	Lilly	diabetes (1982)
murine monoclonal antibodies	Ortho	reversal of acute kidney transplant rejection (1985)
interferons-alpha	Schering-Plough; Hoffmann-LaRoche; Interferon Sciences	hairy cell leukemia (1986); genital warts (1988, 1989); AIDS-related Kaposi's sarcoma (1988); non-A/non-B hepatitis (1991)
somatotropin	Genentech; Eli Lilly	human growth hormone deficiency in children (1985, 1987)
tissue plasminogen activator, alteplase	Genentech	acute myocardial infarction (1987); acute pulmonary embolism (1990)
erythropoietin	Amgen	anemia associated with chronic renal failure (1989)
hepatitis B vaccine	Merck	hepatitis B prevention (1989); interferon gamma (Genentech); management of chronic granulomatous disease (1990)
colony stimulating factors	Immunex; Hoechst-Roussel; Amgen	bone marrow transplantation (1991); treatment of chemotherapy-induced neutropenia (1991)

[a]Ref. 26.

the United States Food and Drug Administration (see INSULIN AND OTHER ANTI-DIABETIC DRUGS). Paralleling the development of small-molecule pharmaceuticals, previously approved proteins are being tested for new applications. For example, recombinant interferon alpha-2B (IntronA; Schering Plough) was initially approved in 1986 for treatment of hairy cell leukemia. Subsequent applications were approved for treatment of genital warts, AIDS-related Kaposi's sarcoma, and non-A/non-B hepatitis. Table 2 lists recombinant proteins approved by FDA as of August 1991. At that time, twice as many applications were pending and 10 times as many were in clinical trials (26).

The majority of the proteins in clinical trials in August 1991 were monoclonal antibodies for treatment of sepsis or neoplasia (see IMMUNOTHERAPEUTIC AGENTS). Monoclonal antibodies are derived in most cases from mouse cell lines. This leads to the problem that the antibodies are recognized as foreign by the human immune system. Research efforts are directed toward humanizing antibodies so that these molecules can escape immune surveillance, for example, by the systematic replacement of murine-specific peptide sequences with human homologues while still maintaining the binding specificity of the original monoclonal antibody (27).

Regulatory and Safety Issues

In the early days of recombinant DNA research, there were serious concerns about crossing assumed species barriers that prevented the exchange of DNA between different bacteria. Since that time, however, the exchange of DNA by conjugation has been demonstrated between numerous bacteria, including between gram-negative and gram-positive genera. Similar conjugal transfer has been shown to occur between bacteria and yeast. Thus, the DNA information in the biosphere can best be thought of as a continuum. In this background, safety regulations have been modified to recognize that no new hazards are created by recombinant DNA research, eg, DNA introduction does not make a pathogen out of a nonpathogen. Reference 28 contains a discussion of these views.

BIBLIOGRAPHY

"Genetic Engineering" in *ECT* 3rd ed., Vol. 11, pp. 730–745, by A. M. Chakrabarty, University of Illinois at the Medical Center, Chicago; in Suppl. Vol., pp. 495–513, by E. Jaworski and D. Tiemeier, Monsanto Co.

1. R. J. Roberts and D. Macelis, *Nucleic Acids Res.* **20**, 2167–2180 (1992).
2. G. Chu, D. Vollrath, and R. W. Davis, *Science* **234**, 1582–1585 (1986).
3. J. E. Richards and co-workers, *Proc. Natl. Acad. Sci.* **85**, 6437–6481 (1988).
4. A. J. Jeffreys and co-workers, *Nature* **354**, 204–209 (1991).
5. F. M. Ausubel and co-workers, eds., *Current Protocols in Molecular Biology*, Wiley-Interscience, New York.
6. J. Sambrook, E. F. Fritsch, and T. Maniatis, *Molecular Cloning: A Laboratory Manual*, Cold Spring Harbor Laboratory Press, Cold Spring Harbor, N.Y., 1989.
7. J. Norrander, T. Kempe, and J. Messing, *Gene* **26**, 101–106 (1983).
8. L. Clarke and J. Carbon, *Cell* **9**, 91–99 (1976).
9. J. Sulston and co-workers, *Nature* **356**, 37–41 (1992).
10. D. T. Burke and M. V. Olson, *Meth. Enzymol.* **194**, 251–270 (1991).
11. M. C. Ianuzzi and co-workers, *Am. J. Hum. Genet.* **44**, 695–703 (1989).
12. F. Sanger, S. Nicklen, and A. R. Coulson, *Proc. Natl. Acad. Sci.* **74**, 5463–5467 (1977).
13. A. M. Maxam and W. Gilbert, *Meth. Enzymol.* **65**, 499–559 (1980).
14. T. Hunkapiller and co-workers, *Science* **254**, 59–67 (1991).
15. W. R. Pearson, *Meth. Enzymol.* **183**, 63–98 (1990).
16. M. A. Innis and co-workers, *PCR Protocols: A Guide to Methods and Applications*, Academic Press, Inc., New York, 1990.
17. A. D. Ellington and J. W. Szostak, *Nature* **346**, 818–822 (1990).
18. C. Tuerk and L. Gold, *Science* **249**, 505–510 (1990).
19. J. K. Scott and G. P. Smith, *Science* **249**, 386–390 (1990).
20. A. Plueckthun, *Immunol. Rev.* **130**, 151–189 (1992).
21. L. C. Bock and co-workers, *Nature* **355**, 564–566 (1992).
22. T. A. Kunkel, *Proc. Natl. Acad. Sci.* **82**, 488–492 (1985).
23. D. H. Jones and B. H. Howard, *Biotechniques* **10**, 62–66 (1991).
24. S. L. McKnight and R. Kingsbury, *Science* **217**, 316–324 (1982).
25. R. C. Cadwell and G. F. Joyce, *PCR Methods and Applications* **2**, 28–33 (1992).
26. Pharmaceutical Manufacturers Association, *In Development: Biotechnology Medicines*, PMA, Washington, D.C., 1991.
27. J. R. Adair, *Immunol. Rev.* **130**, 5–40 (1992).
28. B. D. Davis, ed., *The Genetic Revolution*, Johns Hopkins University Press, Baltimore, Md., 1991.

FRANCIS J. SCHMIDT
University of Missouri, Columbia

ANIMALS

Most cells within an animal contain a complete copy of genetic information. This information is encoded by nucleic acids (qv), eg, deoxyribonucleic acid (DNA). Specific sequences of nucleotides are paired with their complementary base within the helically arranged double-stranded DNA molecule. A complete sequence of DNA is known as a genome, and the sequence of DNA dictates an animal's genotype. For humans and most mammals, a single cell contains about six billion base pairs (bp) of DNA. This six billion base pairs is composed of two copies of the genome, one from each parent, of approximately three billion base pairs each. Hundreds of thousands of genes that encode cellular ribonucleic acids (RNAs) and proteins (qv) are nested within these six billion base pairs of DNA. These genes are transcribed into messenger RNAs (mRNAs) that encode amino acid sequences for cellular, extracellular, and secreted proteins (see AMINO ACIDS). The amount and composition of these proteins within a cell determine its function within the organism. All proteins have some function within a cell and certain proteins are directly involved in growth and development of the animal.

Transgenic Animals

A transgenic animal is an animal that has a modified gene inserted into its DNA (1). This modified or foreign gene is called a transgene. Transgenic animals can be produced using techniques in embryo manipulation and molecular biology that include microinjection, retroviral infection, and embryonic stem cell production. Transgenic animals are produced that either over or under express specific proteins within certain cells. This leads to animals having unique characteristics.

Agricultural uses of this technology include insertion of genes into farm animals for improved milk production, growth rate, and disease resistance. Biomedical uses of this technology include the development of lines of transgenic laboratory animals as experimental models for human diseases and the production of farm animals that produce recombinant pharmaceutical proteins in their milk. The production and use of transgenic laboratory and farm animals represents an evolving technology of engineering animal species for specific roles in science and agriculture.

Methods of Gene Transfer in Animals. Transfer of foreign genes into animals is done at an early stage of embryonic development (one cell to blastocyst stage) prior to implantation or placentation. Embryos at this stage of development can be grown outside the uterus of the mother (*in vitro* culture) in specialized medium containing nutrients that support their growth. For best results, micromanipulation and gene transfer are performed on one-cell embryos because integration of the transgene into the DNA of a one-cell embryo theoretically assures that all the cells of the adult animal carry the foreign gene. In practice, gene transfer into one-cell embryos does not always result in an adult that has the transgene in every cell. This is because the transgene may not actually integrate into the embryonic DNA until several cleavages of the embryo have occurred. This results in an adult animal that is genetically mosaic, harboring mixtures of cells having different sites of transgene integration or populations of transgenic and nontransgenic cells. Gene transfer into later stage embryos, ie, two-cell, four-cell, or greater, is less desirable because the different cells would not be expected to integrate the transgene equivalently. Therefore, genetic mosaicism of the adult is more likely than if single-cell embryos are used. Three methods of gene transfer are used to insert foreign DNA into the early embryo. They are microinjection, retroviral infection, and embryonic stem cells.

Microinjection. The process of genetic engineering by microinjection of the one-cell embryo is illustrated in Figure 1. Foreign genes are assembled by splicing together several pieces of DNA from selected genes. The goal of this assembly is to construct a gene that becomes active within specific tissues or in response to specific signals. These gene assemblies or DNA constructs contain two important parts: a promotor and a gene coding for a protein. The promotor is a DNA sequence that dictates the activity as well as tissue specificity for the expression of the protein coding portion of the transgene. Examples of promotors include the mouse mammary tumor virus promotor (active in mammary tissue) and the mouse metallothionine promotor (activated by heavy metals). A DNA sequence that codes for a protein is spliced next to the promotor so that it is transcribed when the promotor region is activated within certain cells. Activation of the transgene results in the synthesis of mRNA and the production of a foreign protein within the transgenic animal. The protein from the transgene has the same biological activity and structure as naturally occurring animal proteins. However, it may be

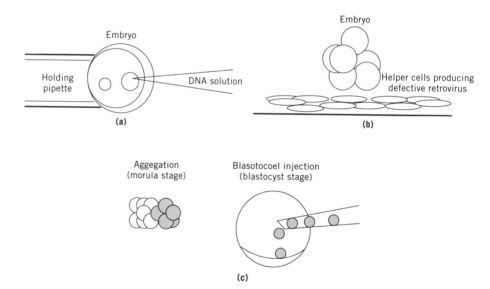

Fig. 1. Methods for the production of transgenic animals by introducing foreign genes into early embryos. (**a**) Microinjection. A DNA solution containing the foreign gene is injected into the pronucleus of a one-cell embryo. (**b**) Retroviral infection. An eight-cell embryo is cultured on top of a layer of helper cells that produce replication-defective retroviruses that carry the foreign gene into the embryonic cells and insert the foreign DNA into the embryonic DNA (see text). (**c**) Embryonic stem cells. The foreign gene is incorporated into stem cells (O) *in vitro* and then the stem cells are either fused to a morula stage embryo or injected into the blastocoele of a blastocyst stage embryo.

expressed in a greater amount or synthesized within a different tissue because it is controlled by the adjoining promotor region. An example of a DNA construct is the mouse growth hormone (GH) gene linked to the metallothionine promotor. This construct causes the expression of GH in animal tissues when the promotor is activated by heavy metals provided in the drinking water of the transgenic animal. Other commonly used constructs contain milk protein gene promotors. These cause the synthesis of foreign proteins in the milk of the transgenic animal.

Microinjection is performed on single-cell embryos that are either collected following natural insemination or are produced *in vitro* by maturing and fertilizing oocytes from slaughtered animals. Several thousand copies of the DNA construct are dissolved in a small quantity of buffered solution and loaded into a needle specifically designed for microinjection. The needle is then inserted into the pronuclei where the DNA solution is injected (2). The pronuclei of the mouse embryo is readily visible, but embryos of farm animals are briefly centrifuged to polarize the opaque cytoplasm and reveal the pronuclei prior to injection (3). The process through which the foreign DNA integrates into the embryonic genome is unknown and a greater understanding of this process may improve the efficiency of production of transgenic animals. Once the transgene is integrated into the DNA of the embryo, it is replicated and becomes a permanent part of the embryonic DNA. Microinjected embryos are generally cultured *in vitro* for a short period of time and then transferred to the oviduct or uterus of a surrogate mother.

Following pregnancy and parturition, samples of DNA are extracted from tissues of newborn animals and analyzed for the foreign DNA. The methodology for DNA analysis varies according to the original DNA construct. In general, the DNA is either analyzed by a southern blot, foreign DNA detected using a radiolabeled probe (4); or by a polymerase chain reaction (PCR), foreign DNA detected by specific amplification (5) (see GENETIC ENGINEERING, PROCEDURES). In practice, only a small percentage of microinjected embryos develop into transgenic animals. This is because many microinjected embryos do not develop within the uterus and are lost. Other (70 to 90%) embryos develop normally but fail to integrate the foreign DNA. These inefficiencies do not mean necessarily that the production of transgenic mice is prohibitively expensive. Mouse embryos can be collected, microinjected, and transferred in a short time. In addition, mice are conveniently housed within limited space, produce numerous embryos in response to superovulation, carry a large litter of potentially transgenic offspring, and have a short gestation period. In contrast, production of transgenic farm animals is a tedious process that can be attempted only at great cost. This is because fewer microinjected embryos of farm animals develop eventually into transgenic animals. Embryos from farm animals have a lower (4 to 8%) frequency of integration compared to mice (10 to 30%) and are less likely to survive after microinjection. These limitations combine to reduce the efficiency of production of transgenic farm animals to less than 1% of those embryos injected. Farm animals also need more space, consume more feed, have poorer superovulatory responses, carry smaller litters, and have a longer gestation. These factors increase the expense of production for transgenic farm animals.

Retroviral Infection of Embryos. The low incorporation rate of microinjected DNA can preclude the efficient production of some types of transgenic animals. Therefore, experimental methods that employ retroviruses have been developed

(6). Retroviruses are naturally equipped to perform gene transfer because they infect cells and integrate their genes into the DNA of the infected cell. Once inserted, the proviral DNA of the retrovirus may remain dormant in the host genome. However, when critical signals are received, the retroviral DNA is activated and new viral particles are produced that spread the infection. By slightly modifying a retrovirus, it is possible to infect embryonic cells with the retrovirus but prevent its subsequent replication. These replication-defective retroviruses are used to carry DNA constructs, ie, promotor and protein coding sequences, into developing embryos (7). Replication-defective retroviruses contain sequences for DNA integration called long terminal repeat regions (LTRs) that flank the transgene, ie, the promotor and protein coding sequences. Most of the other sequences of the retrovirus that are needed for replication and packaging of the viral particle are deleted. The defective retroviruses are propagated in a specially constructed helper cell line that expresses those proteins that are needed to make a mature retroviral particle. Therefore, the replication defective retrovirus can only multiply within the helper cell line. The embryo, which is usually at the four- to eight-cell stage, is incubated on top of the helper cells that are producing the replication-defective virus (Fig. 1**b**). The retrovirus infects the cells of the embryo and inserts the foreign DNA into the embryonic DNA. The foreign viral DNA is not harmful to the embryo because the virus cannot replicate.

Once the embryo has been infected, it is transferred to a surrogate mother for gestation. The provirus is carried throughout development as a foreign element in the animal's DNA and the presence of the provirus is confirmed by genetic testing of the newborn animal. Transgenic animals produced by retroviral infection are usually mosaic because the original retroviral infection is performed on four- to eight-cell embryos and all the cells are not necessarily infected with virus. Therefore, founder animals that test positive for the transgene are used to produce offspring that are selected based on the presence of the foreign genes.

Embryonic Stem Cell Chimeras.   Cells that have been isolated from developing embryos and allowed to multiply in tissue culture are called embryonic stem cells (8). Foreign genes can be inserted into the DNA of embryonic stem cells while the cells are being grown *in vitro*. These transgenic stem cells can then be used to produce transgenic animals. The procedure for producing transgenic animals from embryonic stem cells is presented in Figure 1**c**. Embryonic stem cells do not develop into an embryo if they are placed by themselves in the uterus. Therefore, cells from the embryonic stem cell line are either fused to developing morula stage embryos or injected into the blastocoele of blastocyst stage embryos. The stem cells and the cells of the embryo attach to each other and intermingle to form a chimeric embryo. Chimeric embryos are composed of a mixture of cells from different sources (original embryo and the stem cell) and can survive within the uterus. Chimeric offspring derived from these procedures must be mated to produce a second generation of animals, some of which will be direct decedents of the embryonic stem cell line and carry the desirable transgene.

The use of embryonic stem cell lines has many advantages when compared to other methods of transgenic animal production. First, methods for insertion of foreign DNA into cells in tissue culture are simpler and more efficient than methods for inserting DNA into embryos. Second, the cells that contain the foreign genes can be selected before they are fused to embryos and any chimeric offspring must carry the foreign gene. Finally, specialized cell lines that carry the transgene in proper location, ie, replacing endogenous gene sequences, can be selected. These advantages make the use of embryonic stem cells a possible alternative to either microinjection or retroviral infection for future production of transgenic farm animals. However, as of this writing, this technology has not been perfected and has not been successful for farm animal species. Improved methods for establishment, *in vitro* culture, and genetic engineering of embryonic stem cells are needed. In addition, genetic factors that control the development of the embryo need to be more clearly understood.

Expression of Foreign DNA in Transgenic Animals.   Animals that carry a transgene do not have equivalent levels of expression of the foreign protein (9). In some animals the transgene can be detected in the DNA but no foreign protein is detected in the body. This suggests that the gene is nonfunctional, ie, not expressing mRNA and not producing protein. Other animals carry the transgene and express the foreign protein at much too high levels. This can cause abnormal development and pathological conditions.

Regulation of expression of the protein coding portion of the transgene should be under the control of the promotor region of the DNA construct. However, other factors, including the flanking DNA sequences at the site of transgene insertion, can influence the level of expression of the foreign DNA. Poor reproducibility of expression of foreign genes in transgenic animals may be partially corrected by dominant control regions. These DNA sequences can be inserted into DNA constructs to provide more consistent expression of proteins in transgenic animals.

Transgenic Farm Animals.   Farm animals that carry foreign genes are being produced for two reasons. First, agricultural productivity may be increased if lines of transgenic farm animals can be developed that express specific genes for increased growth (10) or immunity (11). Therefore, several different growth-promoting genes have been inserted into farm animals and evaluated for potential benefit to animal productivity (see GROWTH REGULATORS, ANIMALS). Second, transgenic farm animals are being used as bioreactors for the production of pharmaceutical proteins in their milk (12,13). This is because pharmaceutical proteins extracted from the milk of transgenic farm animals can have greater biological activity when compared to similar proteins produced in bacteria.

Transgenic Farm Animals Engineered for Increased Growth or Muscling.   The first transgenic livestock carried genes designed to increase growth, milk production, or muscling (Table 1). These included growth hormone (GH), growth hormone releasing factor (GRF) and insulin-like growth factor-I (IGF-I). Growth hormone releasing factor causes the release of GH from the pituitary gland and GH causes the release of IGF-I from the liver (21). These genes were chosen because increased concentrations of GH and IGF-I in the blood have been associated with increased growth and milk production in farm animals. In addition, cattle or swine injected with GH have increased IGF-I in the blood and grow more efficiently. Furthermore, dairy cows injected with GH produce more milk (22). Therefore, a transgenic animal carrying an additional GRF, GH, or IGF-I gene should grow faster and more efficiently, and produce more milk. This concept was initially tested in mice.

Table 1. Transgenic Farm Animals having Genes Designed to Increase Growth or Productivity

Species	Promotor	Gene	Reference
bovine	metallothionine	growth hormone	14
ovine	metallothionine	growth hormone	15,16
ovine	metallothionine	GH releasing factor	15
porcine	metallothionine	growth hormone	17
porcine	mMLV	growth hormone	18
porcine	metallothionine	IGF-I	19
porcine	MSV-LTR	c-ski	20

Transgenic mice where produced that carried a GH gene hooked to a metallothionine promotor. These mice had increased concentrations of GH in the blood and grew considerably larger than the nontransgenic mice (23). The experiment suggested great promise for this technology in agricultural species; however, farm animals carrying the GH transgene have not shown similar responses. Transgenic swine with high levels of expression of the GH gene have increased feed efficiency and rate of gain but suffer from various illnesses including gastric ulcers, synovitis, and dermatitis. Apparently, the overproduction of GH cannot be tolerated by the animal.

Transgenic swine that carry the c-ski gene have also been produced (20). This gene is not related to GH, GRF, or IGF-I. The c-ski gene codes for a protein that is involved in muscle fiber development and transgenic mice harboring the c-ski gene have larger muscles. Possibly, farm animals with greater muscle mass could be produced if the c-ski gene were appropriately expressed in their muscles. As with the GH constructs, however, appropriate expression of the c-ski gene has not been achieved in farm animals. Although some transgenic swine having the c-ski gene have increased muscling, others have muscle weakness in fore and hind legs. Productive lines of swine that carry the c-ski gene have not been developed.

As of this writing, the illnesses and/or physical abnormalities reported for transgenic swine carrying the GH or c-ski genes severely limit the usefulness of these genes. Transgene expression must be tightly controlled in farm animals. Transgenes need to be active at specific periods of growth and inactive during other periods. This can be partially achieved through the selection of an appropriate promotor for the DNA construct. However, few promotors tightly control the expression of the foreign protein. Therefore, transgenes are not expressed correctly and the foreign protein is produced at all times, leading to many of the physical ailments of transgenic animals. Promotors that give better control over gene expression are important to the future success of these technologies. In addition, the actual site of insertion of the transgene within embryonic DNA may have to be controlled. This would prevent the complicating effects of flanking DNA sequences which can either increase or diminish the activity of promotor within the DNA construct.

Transgenic Immunity from Disease.   One method to increase productivity of farm animals is to increase their resistance to common diseases. Transgenic farm animals can be produced that carry genes for immunoglobulins directed against viruses or bacteria (11). Animals carrying the foreign immunoglobulin genes would not need vaccinations for certain diseases because the expression of the transgene would cause the production of immunoglobulins that convey immunity to disease. Production of transgenic rabbits, sheep, and pigs that carry mouse immunoglobulins has been accomplished (24,25). Furthermore, the foreign immunoglobulins are correctly assembled and secreted into the blood. Therefore, the use of genetic engineering to increase resistance to bacterial and viral diseases is a future possibility for farm animal species.

Production of Pharmaceuticals in Transgenic Farm Animals. Transgenic farm animals such as sheep, goats, and cattle can produce human pharmaceutical proteins in their milk (12,13). Farm animals are used because they produce large quantities of milk from which pharmaceuticals (qv) can be purified. A single cow can easily produce enough milk to supply the needs of most companies for certain specialized drugs. Several recombinant proteins have been produced in the milk of farm animals (Table 2). The gene coding for the protein is linked to a promotor that is active within the mammary gland, resulting in production and secretion of the desired protein into the milk. The milk is then collected and the drug extracted and purified (31). Examples of mammary-specific promotors are the whey acidic protein promotor and the β-lactoglobulin promotor. Both control the expression of proteins normally found in milk and can be adapted to direct the expression of pharmaceutical proteins.

Farm animals produce recombinant proteins less expensively than bacteria or cells in culture because the farm animals produce large volumes of milk containing up to 5 g/L of recombinant protein. In addition, modifications to the proteins that can be performed only by mammalian cells are made by the cells of the mammary gland. Therefore, numerous pharmaceuticals that previously could only be made by cells in culture or extracted from human tissue or blood are being produced by lactating farm animals.

Transgenic Mice in Biomedical Research. Transgenic mice are used widely in the field of biomedical research. Human diseases, including cancer, can be studied by producing transgenic mice that mimic closely the symptoms found in people. Examples are Alzheimer's disease (32), atherosclerosis (33), autoimmune insulitis (34), bone marrow graft rejection (35), breast cancer (36), chronic myeloid disease (37), Cushing's disease (38), Gaucher's disease (39), hypercholesterolemia (40), hypertension (41), hypoxanthine phosphoribosyltransferase deficiency (42), liver cancer (43), Menetrier's disease (44), polycystic kidney disease

Table 2. Pharmaceuticals Produced in Milk or Blood of Transgenic Farm Animals

Species	Pharmaceutical	Reference
bovine	human lactoferrin	26
caprine	human tissue plasminogen activator	27
ovine	human antihemophilic factor IX	28
ovine	human alpha-1-antitrypsin	29
porcine	human hemaglobin	30

(45), pulmonary adenocarcinoma (46), retinitis pigmentosa (47), retinoblastoma (48), rheumatoid arthritis (49), and Sjogren's syndrome (50).

Many human diseases are caused when certain proteins are either over- or underexpressed. For example, breast cancer can be induced by overexpressing certain cellular oncogenes within mammary tissue. To study the disease, researchers produce a line of transgenic mice that synthesize an abnormal amount of the same protein. This leads to symptoms of the disease in mice that are similar to what is found in humans. A protein can be overexpressed by inserting a DNA construct with a strong promotor. Conversely, underexpression of a protein can be achieved by inserting a DNA construct that makes antisense RNA. This latter blocks protein synthesis because the antisense RNA binds and inactivates the sense mRNA that codes for the protein. Once a line of mice is developed, treatments are studied in mice before these therapies are applied to humans.

Transgenic disease models are also developed using gene knock-out experiments (51). Instead of simply inserting the transgene randomly into the mouse genome, the goal of a knock-out experiment is to replace the animal's normal gene with a mutated gene sequence. Alternatively, a mutated gene may be replaced with a normal gene. This procedure, known as gene therapy, may someday be applied to humans to correct genetic defects. Knock-out experiments are preformed using embryonic stem cells. The transgene is inserted into the cell nucleus and replaces the normal gene by a process called homologous recombination. Embryonic stem cells that have undergone this gene replacement process can be selected and grown in culture. These cells are then used to produce chimeric embryos. The chimeric mice that result from these procedures are founder animals for new lines of mice derived from the original embryonic stem cells. These new lines of mice have their normal genes knocked out and replaced with a different gene. Therefore, the effects of the new gene can be studied independently from the old gene.

Production of Multiple Animals From Single Embryos

One goal of animal agriculture is to exploit the superior genetics of animals. A technology that accomplishes this goal is artificial insemination. Semen is collected from superior males, diluted, frozen in liquid nitrogen, and shipped to farms. This means that an average female can be mated to a superior male to produce above average progeny. Artificial insemination is so widely used in the dairy industry that very few bulls breed nearly all the dairy cows in the United States. It is much more difficult to exploit the genetics of superior female cattle because they produce few gametes (ova). Female cattle can be treated with stimulatory drugs that cause the release of numerous ova (superovulation), and these treatments can be used to maximize the production of embryos from a single female. However, it is impossible for female animals to match the gamete production of male animals. Therefore, it is necessary to develop methods to mass produce identical copies of embryos that are collected from a superior female animal. Two methods are available to do this: embryo splitting and embryo cloning. These are illustrated in Figure 2.

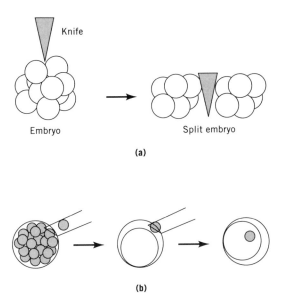

Fig. 2. Methods for production of identical cattle. (**a**) Embryo splitting. A microdissection knife is used to cut an early embryo into two identical halves. (**b**) Embryo cloning. A cell from a morula stage embryo is removed and then inserted into an oocyte that has had its nucleus removed. Following fusion of the two cells the newly produced one-cell embryo is identical to the original morula. The procedure can be repeated to give multiple copies of the original morula.

Embryo Splitting. The procedure for embryo splitting is not complicated but requires delicate instrumentation. Using microdissection tools (either a knife or needle), embryos are separated into two to four pieces (52) (Fig. 2**a**). The split embryos are then transferred into the uterus of foster mothers for development. Animals that are produced from the same original embryo are identical twins and should be genetically equivalent. The highest success rate for the production of identical offspring occurs when embryos are split into halves resulting in identical twins. Splitting embryos into four pieces can yield identical quadruplets, but the probability of pregnancy is much lower. This is because the quarter-sized embryos are less likely to develop within the uterus.

Embryo Cloning. A principal limitation to embryo splitting is that an embryo can be split only a few times before the pieces are too small to continue development. In contrast, embryo cloning can be used theoretically to make an unlimited number of copies of the same embryo (53). Cloned embryos are not limited by size because cells, housing a complete copy of the embryonic genome, are transplanted back into an oocyte that has had its nucleus removed. Therefore, the cell nucleus can become part of a new one-cell embryo. This process is illustrated in Figure 2**b**. A single cell from a 32-cell embryo is removed using a wide-bore micropipette. Single-cell oocytes (unfertilized) are isolated from ovaries collected from slaughtered cattle and are used as recipients for the cell from the 32-cell embryo. Prior to the nuclear transfer, the oocyte nucleus is removed using a micropipette. The cell from the 32-cell embryo is inserted next to the oocyte and the cytoplasmic membranes of the oocyte and cell are fused using a mild pulse of electricity. When the cell and the oocyte are joined, a single-cell embryo is formed that proceeds with normal development. If all the cells from the original 32-cell embryo are transplanted, then 32 genetically identical embryos can be produced. These embryos can be cultured *in vitro* and transferred to foster mothers or allowed to develop and used to make more embryo clones. Therefore, numbers of identical cloned animals that can be produced from a single embryo are theoretically limitless.

Unfortunately, many complications are associated with the cloning procedures (53). First, the procedure is very inefficient. As of this writing, only 1 to 4% of cloned embryos from farm animals complete development and yield live offspring. Most embryos are lost because of inherent inefficiencies of each step in the cloning procedure. There are additional problems associated with the offspring that are produced from cloned embryos (53). Genetic abnormalities have been found in calves from cloned embryos. In addition, calves of cloned embryos are often extremely large leading to complications at birth. Both inefficiency of production and genetic/developmental problems need to be solved.

Nucleic Acid Analysis of Individual Animals

The sequence of nucleotides within DNA varies across individuals within a species. This genotypic variation leads to phenotypic variation among animals. Some genes are not economically important. However, other genes, associated with milk production in dairy cattle and growth rate in meat-producing animals, ie, cattle, swine, and poultry, are of great economic importance. Most of these traits are believed to be quantitatively inherited. In other words, several (possibly 25 to 100) different genes are inherited and the additive effect of these genes determine the ultimate phenotype, ie, level of milk production or growth rate. The location of a gene that contributes to quantitative inheritance is known as a quantitative trait loci (QTL). If the genetic value of a gene is known, then animals at or before birth can be tested for their complement of genetic material. The genetic value of an animal, based on the additive inheritance of numerous genes, can then be calculated, and superior animals can be selected at an early age before time and money are invested. For this reason, finding the location of the genes that control production, as well as the alleles, ie, DNA sequences, for superior production, is a significant strategy for engineering improved animal species through biotechnology.

Deciphering the genetic information within species or individuals has important applications for human and animal sciences. The tremendous size (six billion base pairs) of the average genome makes the determination of the sequence of nucleotides within an individual nearly impossible. Indeed, hundreds of scientists are involved in determining the sequence of nucleotides from just one human genome. Therefore, nucleotide sequencing of the genome of multiple individuals from various species is not practical. Other methods have been developed to locate the genetic information associated with superior productivity in farm animals. Two possibilities exist. First, although relatively few genes controlling quantitative traits have been discovered or completely understood, multiple alleles are found for some of these genes. Therefore, alleles associated with superior production can be identified and animals having the superior allele selected. Unfortunately the vast majority of genes that control production have not yet been discovered. Therefore, a second method of genetic selection can be performed without knowledge of the specific genes that control quantitative traits. For this method, genetic markers, polymorphic gene sequences with unknown function, located close to QTL, are used to select superior farm animals. The use of these methods, ie, either using markers located within known QTL or using those markers located near undiscovered QTL to find superior animals is known as marker assisted selection (54,55).

Identification of Genetic Markers.   Most genes that control production are as yet unknown, and the discovery of these genes may take many years of scientific research. Therefore, selection of animals that carry desirable alleles for quantitative traits is carried out using genetic markers that are closely linked to the undiscovered QTL (54,55). This concept is illustrated in Figure 3. The process begins by mapping the genome of a particular species and finding DNA markers at evenly spaced intervals across the entire genome. A marker is a polymorphic region of DNA that is inherited in a mendelian fashion and can be used to monitor the segregation (during meiosis) of regions of the chromosome. By following the segregation of markers in offspring it is possible to follow the movement of portions of the chromosome that may contain QTLs that have desirable alleles.

There are three types of DNA markers used: restriction fragment length polymorphisms (RFLPs), minisatellites, and microsatellites (55). The RFLPs are DNA sequence differences (polymorphisms) among individuals that are detected by restriction enzyme digestion of DNA. These may be found in known genes or may be found outside or closely linked to known genes. The second type of marker is a variable number of tandem repeat marker (VNTR) or minisatellite. This is a small repeated region of DNA. The VNTRs are interspersed throughout the ge-

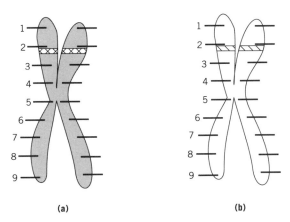

(a)   (b)

Fig. 3. Homologous chromosomes having evenly spaced genetic markers (1–9). The segregation of the ⬚⬚⬚ or ⬚⬚⬚ quantitative trait loci (QTL) can be followed using marker 2. Polymorphisms within marker 2 can be used to follow the segregation of the genes within the QTL, even though the exact location or nature of the QTL is unknown.

nome and there are several alleles for each VNTR. As is true for the RFLPs, the segregation of alleles for each VNTR can be used to mark chromosomal regions. The VNTRs are detected using specific DNA probes by southern blotting that hybridize to DNA that has been enzymatically digested and size-fractionated. The final type of marker is microsatellite DNA. Microsatellites are similar to VNTRs but are shorter in length. Microsatellites are interspersed throughout the genome and several alleles are found for each microsatellite. Unlike VNTRs, microsatellites can be best detected by PCR. Different alleles, ie, DNA of variable lengths, are amplified by selecting specific PCR primers for microsatellites.

Construction of the Genomic Map. Chromosomal locations are assigned to markers after they are identified. This is accomplished by a combination of approaches involving linkage analysis of pedigrees and somatic cell hybrid analysis. The goal of these procedures is to develop a genetic map that has markers interspersed at regular intervals throughout the genome (see Fig. 3, markers 1–9). Therefore, any QTL should segregate with one of the markers. Initial genetic maps contain 400–500 markers that are separated by about 10 centimorgans. A centimorgan is the physical distance between two genes or markers that would result in separation of the two genes in one percent of genetic recombinations. Therefore, in a 10 centimorgan map, markers and their associated QTLs would be separated in 10% of the offspring.

Selecting Superior Individuals Using Markers. A goal is to find markers that are linked to QTLs and to identify superior alleles for production. A parent, usually a sire with many offspring, is selected that is heterozygous (AB) for a given marker. His offspring are then compared based on whether they inherited the A allele or the B allele. If genetic superiority is associated with one allele or the other then the marker is probably located near a QTL. This process is repeated for all the markers to determine what markers are useful in identifying QTL for traits of significant economic importance. Progeny are then selected that carry the favorable marker. Other progeny that carry unfavorable markers, ie, genetically inferior animals or embryos, are not used in future animal breeding programs.

Use of Genetic Markers. Polymorphisms in DNA sequence exist within the prolactin gene locus in cattle (56). These are used as genetic markers for increased milk production (57). The genetic marker is classified as an RFLP and is not associated with any differences in the amino acid sequence of prolactin. The marker can be visualized by digestion of genomic DNA with AvaII enzyme. A 200 base-pair deletion near the prolactin gene can be detected using this approach. Offspring that inherit the favorable allele at the prolactin locus have milk production that averages 283 kg/yr more than offspring that inherit the unfavorable allele. This marker is used to select superior female cattle for milk production or to select superior sires for the artificial insemination industry.

A second marker is associated with porcine stress syndrome (PSS) and is used to breed swine for leaner carcasses (58). Porcine stress syndrome has been linked to a single base mutation in the ryanodine receptor gene that causes changes in calcium release channels in muscle. When swine that have PSS are exposed to stress prior to slaughter, a muscle reaction is elicited that damages portions of the carcass. An Hgi A1 polymorphism in the ryanodine receptor gene can be used to detect pigs that are homozygous for PSS. Hogs could be selected that do not carry the mutant allele; however, genetic advantage for leanness would be lost. Therefore, in order to take advantage of increased leanness, without risking potential PSS, homozygous boar lines carrying the defective receptor associated with PSS are being developed to breed normal sow lines having the normal receptor to produce heterozygous offspring for market. The market hogs are genetic carriers of the mutant receptor that should not suffer from PSS during slaughter because heterozygous pigs are only slightly susceptible to the disease. The heterozygous pigs should, however, have increased carcass leanness of 1 to 3%.

Genetic Analysis of Embryos

It is possible to select superior embryos during their initial development. As an example, the procedures for simultaneously testing embryos for gender, milk protein genotype, and somatotropin genotype are described below and illustrated in Figure 4. An embryo is allowed to develop *in vitro* to the blastocyst stage. Several cells are dissected from the embryo for genetic analysis. This procedure is not harmful to the embryo. The DNA within these cells is analyzed by PCR (5). In the polymerase chain reaction, DNA is denatured into single strands and specific primers are allowed to anneal to the denatured DNA. The sequence of DNA between the primers is then filled in by a DNA polymerase enzyme. This process is repeated over and over until the sequence between the two primers is amplified nearly one billion-fold. Each genetic test involves different PCR primers and a different PCR reaction. At the end of the genetic analysis, only embryos with the desired genotype are removed from *in vitro* culture and transferred to recipient females. Embryos of the undesirable genotypes can be frozen indefinitely in liquid nitrogen or discarded.

Embryo Sexing. The method for determining the sex of an embryo is presented in Figure 4. Female animals have identical sex chromosomes (XX), whereas male animals have an X chromosome as well as a Y (male specific) chromosome. The DNA sequences from the Y chromosome (male specific) are amplified using PCR to identify male embryos (59). Short pieces of DNA (primers) specific for a male gene are included in the PCR. Annealing of the DNA primers and DNA polymerization occurs only in male embryos. After several cycles of PCR, cells from male embryos generate a DNA product. In contrast, cells from female embryos do not give a DNA product because these lack the male specific genes. The male-specific DNA product is visualized by electrophoresis and DNA staining using ethidium bromide and ultraviolet illumination (see ELECTROSEPARATION, ELECTROPHORESIS). Depending on the needs of the producer, a male or female embryo is selected for transfer into surrogate mothers.

Milk Protein Genes. The milk proteins κ-casein and β-lactoglobulin are both found in A and B forms. An embryo can be selected based on the milk protein gene that is found in its DNA. This is advantageous because certain forms of these proteins are better for milk compared to cheese manufacturing (60). For example, there is a distinct advantage of the B form of κ-casein in the production of cheese but greater milk production is found in cows that carry the A form. Embryos are tested by PCR and restriction enzyme digestion of DNA (Fig. 4, PCR no. 2) A

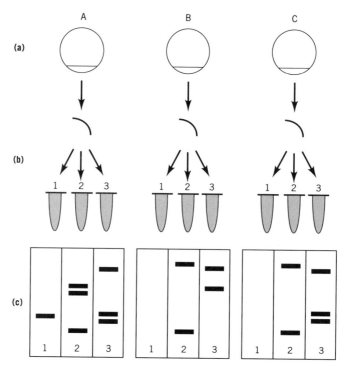

Fig. 4. Analysis of blastocyst stage bovine embryos for three genes by polymerase chain reaction. (**a**) Blastocyst stage embryos from which a piece of the wall of the blastocyst is surgically removed; (**b**) the DNA undergoes PCR and enzyme digestion using markers for (1) the Y chromosome, (2) κ-casein gene allele, and (3) somatotropin gene allele; and (**c**) the material is then placed on gels and undergoes electrophoresis and DNA staining of RFLPs. Embryo A is a male (DNA band in lane 1), and embryos B and C are females (no DNA band in lane 1). Embryo A is homozygous for the A allele of κ-casein (lane 2 has a three-band pattern), and embryos B and C are homozygous for the B allele (lane 2 has a two-band pattern). Embryos A and C are homozygous for the leucine form of somatotropin (lane 3 has a three-band pattern), and embryo B is homozygous for the valine form of somatotropin (lane 3 has a two-band pattern). When selecting an embryo for eventual use as a lactating cow, embryo C would be selected because it is a female, having superior genes for milk and cheese production (allele B for κ-casein and leucine allele for somatotropin).

portion of the milk protein gene that contains the polymorphic DNA sequences, giving rise to either A or B forms, is amplified using PCR. The amplified DNA from the milk protein gene is then digested with restriction enzymes that recognize and cut specific sequences of nucleotides within the DNA molecule. These different sequences of nucleotides give rise to different milk proteins. An enzyme is selected that cuts differently the amplified DNA from the two alleles. For example, in the case of κ-casein, a 350 bp section of the gene is amplified by PCR and then digested with the *Hinf*I restriction enzyme (61). The DNA of the A allele is cut twice by *Hinf*I, whereas the DNA of the B allele is cut once. After digestion and electrophoresis, embryos that have different genotypes show different patterns of DNA fragments, known as restriction fragment length polymorphisms (RFLP) as seen in lane 2 of Figure 4c. Embryos that are homozygous for the A allele have a three-band pattern and embryos that are homozygous for the B allele have a two-band pattern. Heterozygous embryos (AB) have a mixture of the two patterns. The embryo for transfer is selected based on the ultimate use of milk from the cow that develops from the embryo, eg, AA genotype for milk production and BB genotype for cheese production.

Somatotropin Genes.   Somatotropin is found as two different forms in cattle, and milk production may be greater for cows that synthesize a specific type of somatotropin. The two forms are different because a leucine or a valine amino acid is found at the 127th amino acid position of somatotropin. This difference is caused by a single nucleotide change in the somatotropin gene sequence. It is possible to amplify a 428 base pair piece of the somatotropin gene from cattle using similar techniques to those described above for the milk protein genes. After digestion with the *AluI* restriction enzyme, the DNA fragment for the leucine form of somatotropin is cut twice and the DNA fragment for the valine form of somatotropin is cut once (62). This differential digestion of *AluI* results from the single base mutation in the valine allele. As in the case of the κ-casein gene alleles, the RFLP can be used to determine what form of somatotropin gene the embryo carries (Fig. 4, PCR no. 3). A cow homozygous for the leucine allele is selected for milk production because the leucine gene is associated with superior milk production.

BIBLIOGRAPHY

"Genetic Engineering" in *ECT* 3rd ed., Vol. 11, pp. 730–745, by A. M. Chakrabarty, University of Illinois at the Medical Center, Chicago; in Suppl. Vol., pp. 495–513, by E. Jaworski and D. Tiemeier, Monsanto Co.

1. R. D. Palmiter and R. L. Brinster, *Cell* **41**, 343 (1985).
2. N. D. Allen and co-workers, in M. Monk, ed., *Mammalian Development: A Practical Approach*, IRL Press, Oxford, 1987, p. 217.
3. R. J. Wall and co-workers, *Biol. Reprod.* **32**, 645 (1985).
4. E. M. Southern, *J. Mol. Biol.* **98**, 503 (1975).
5. H. A. Erlich, D. Gelfand, and J. J. Sninsky, *Science* **252**, 1643 (1991).
6. D. Jähner and co-workers, *Proc. Natl. Acad. Sci. USA* **82**, 6927 (1985).
7. J. L. R. Rubenstein, J. F. Nicholas, and F. Jacob, *Proc. Natl. Acad. Sci. USA* **83**, 366 (1986).
8. E. Robertson, in E. Robertson, ed., *Teratocarcinomas and Embryonic Stem Cells: A Practical Approach*, IRL Press, Oxford, 1987, p. 71.
9. R. E. Hammer aand co-workers, *Nature* **315**, 680 (1985).
10. K. A. Ward and C. D. Nancarrow, *Experientia* **47**, 913 (1991).
11. M. Müller and G. Grem, *Experientia* **47**, 923 (1991).
12. I. Wilmut and co-workers, *Experientia* **47**, 905 (1991).
13. J. Jänne and co-workers, *Ann. Med.* **24**, 273 (1992).
14. T. G. McEvoy and J. M. Sreenan, *Theriogenology* **33**, 819 (1990).
15. C. E. Rexroad, Jr., and co-workers, *Mol. Reprod. Dev.* **1**, 164 (1989).
16. J. D. Murray and co-workers, *Reprod. Fertil. Dev.* **1**, 147 (1989).
17. P. D. Vize and co-workers, *J. Cell Sci.* **90**, 295 (1988).
18. K. M. Ebert and co-workers, *Anim. Biotech.* **1**, 145 (1990).
19. V. G. Pursel and co-workers, *Science* **244**, 1281 (1989).
20. V. G. Pursel and co-workers, *Theriogenology* **37**, 278 (1992).

21. P. D. Gluckman, B. H. Breier, and S. R. Davis, *J. Dairy Sci.* **70**, 442 (1987).
22. C. J. Peel and D. E. Bauman, *J. Dairy Sci.* **70**, 474 (1987).
23. R. D. Palmiter and co-workers, *Nature* **300**, 611 (1982).
24. D. Lo and co-workers, *Eur. J. Immunol.* **21**, 1001 (1991).
25. U. H. Weidle, H. Lenz, and G. Brem, *Gene* **98**, 185 (1991).
26. P. Krimpenfort and co-workers, *Biotechnology* **9**, 844 (1991).
27. J. Denman and co-workers, *Biotechnology* **9**, 839 (1991).
28. A. J. Clark and co-workers, *Biotechnology* **7**, 487 (1989).
29. G. Wright and co-workers, *Biotechnology* **9**, 830 (1991).
30. M. E. Swanson and co-workers, *Biotechnology* **10**, 557 (1992).
31. T. D. Wilkins and W. Velander, *J. Cell. Biochem.* **49**, 333 (1992).
32. A. Kammesheidt and co-workers, *Proc. Natl. Acad. Sci. USA* **89**, 10857 (1992).
33. R. A. Hegele, *Can J. Cardiol.* **8**, 116 (1992).
34. T. Miyazaki and co-workers, *Proc. Natl Acad. Sci. USA* **89**, 9519 (1992).
35. K. Kikly and G. Dennert, *J. Immunol.* **149**, 3489 (1992).
36. C. T. Guy and co-workers, *Proc. Natl. Acad. Sci. USA* **89**, 10578 (1992).
37. J. Groffen and co-workers, *Leukemia* **6** (Suppl. 1), 44 (1992).
38. A. Helseth and co-workers, *Am J. Pathol.* **140**, 1071 (1992).
39. E. Sidransky, D. M. Sherer, and E. I. Ginns, *Pediatr. Res.* **32**, 494 (1992).
40. A. S. Plump and co-workers, *Cell* **71**, 343 (1992).
41. S. Bachmann and co-workers, *Kidney Int.* **41**, 24 (1992).
42. D. J. Williamson, M. L. Hooper, and D. W. Melton, *J. Inherit. Metab. Dis.* **15**, 665 (1992).
43. H. Takagi and co-workers, *Caner Res.* **52**, 5171 (1992).
44. H. Takagi and co-workers, *J. Clin. Invest.* **90**, 1161 (1992).
45. M. Trudel and V. D'Agati, *Contrib. Nephrol.* **97**, 47 (1992).
46. K. A. Wikenheiser and co-workers, *Cancer Res.* **52**, 5342 (1992).
47. J. E. Olsson and co-workers, *Neuron* **9**, 815 (1992).
48. D. M. Albert and co-workers, *Invest. Ophthalmol. Vis. Sci.* **33**, 2354 (1992).
49. F. M. Brennan, R. N. Maini, and M. Feldmann, *Br. J. Rheumatol.* **31**, 293 (1992).
50. B. C. Johnson, J. I. Morton, and D. R. Trune, *Otolaryngol. Head Neck Surg.* **106**, 394 (1992).
51. M. A. Frohman and G. R. Martin, *Cell* **56**, 145 (1989).
52. J. M. Robl and N. L. First, *J. Reprod. Fertil. Suppl.* **33**, 101.
53. R. S. Prather, T. T. Stumpf, and L. F. Rickords, *Anim. Biotech.* **3**, 67 (1992).
54. M. R. Dentine, *Anim. Biotech.* **3**, 81 (1992).
55. J. M. Massey and M. Georges, *Anim. Biotech.* **3**, 95 (1992).
56. C. M. Cowen and co-workers, *Anim. Genet.* **20**, 157 (1989).
57. C. M. Cowen and co-workers, *Theor. Appl. Genet.* **79**, 577 (1990).
58. D. H. MacLennan and M. S. Phillips, *Science* **256**, 789 (1992).
59. T. Peura and co-workers, *Theriogenology* **35**, 547 (1991).
60. H. Bovenhuis, J. A. M. van Arendonk, and S. Korver, *J. Dairy Sci.* **75**, 2549 (1992).
61. J. F. Medrano and E. Aguilar-Cordova, *Biotechnology* **8**, 144 (1990).
62. M. C. Lucy and co-workers, *Dom. Anim. Endocrinol.* **10**, 325 (1993).

MATTHEW LUCY
University of Missouri, Columbia

ROBERT COLLIER
Monsanto Company

MICROBES

The field of genetic engineering of microbes is an active area of research experiencing constant improvements in technology. Successful commercialization of a variety of recombinant proteins (qv) having applications ranging from therapeutics to food processing (qv) has been accomplished using genetically engineered microbes. A comprehensive review of all the applications of engineered microbes is a formidable task, thus this article describes methodologies relevant to microbial genetic engineering and the potential impact of some genetically engineered microbes in commercial processes (see also GENETIC ENGINEERING, PROCEDURES).

Gene Technologies

The techniques involved in isolation and manipulation of genes from several microbes have become routine (1). Genes are cloned into plasmid or phage-based vectors and introduced into the microbe of choice. The number of vectors that have been engineered to be versatile is on the increase for several systems. Some vectors are commercially available for gene manipulations from several vendors. In addition, construction of gene libraries and screening are also performed by small biotechnology companies. Most of the reagents needed for gene manipulation such as oligonucleotides, peptides, and antibodies are commercially available. In addition, kits are also accessible for some of the routine tasks such as plasmid isolation, deoxyribonucleic acid (DNA) sequencing, etc.

Gene Transfer Methods.   An efficient method of DNA uptake by microbes is essential for gene manipulations. Certain microbes are naturally competent for DNA uptake during a particular phase of life cycle. Other bacteria are made competent upon treatment with calcium chloride or rubidium chloride. Alternatively, the cell wall of bacteria and fungi can be removed and the resulting spheroplasts are capable of DNA uptake in the presence of polyethylene glycol. Protoplasts are later grown on regeneration medium. However, none of the above methods has been very useful for a variety of industrial strains and a newer method called electroporation has been developed (2).

Electroporation.   When bacteria are exposed to an electric field a number of physical and biochemical changes occur. The bacterial membrane becomes polarized at low electric field. When the membrane potential reaches a critical value of 200–300 mV, areas of reversible local disorganization and transient breakdown occur resulting in a permeable membrane. This results in both molecular influx and efflux. The nature of the membrane disturbance is not clearly understood but bacteria, yeast, and fungi are capable of DNA uptake (see YEASTS). This method, called electroporation, has been used to transform a variety of bacterial and yeast strains that are recalcitrant to other methods (2). Apparatus for electroporation is commercially available, and constant improvements in the design are being made.

Gene Alteration and Amplification.   A divergent set of technologies converged in the mid-1980s. These included efficient methods of oligonucleotides synthesis and the use of these oligonucleotides as primer *in vitro* to amplify DNA. A combination of these technologies resulted in two powerful methodologies, ie, site-directed mutagenesis and polymerase chain reaction (PCR) amplification.

Site-Directed Mutagenesis.   This method enables the alteration of a specific amino acid in a protein to another desired amino acid by changing the genetic coding sequence (3). A mutagenic oligonucleotide is annealed to a single-stranded template and elongated in the presence of ligase so that a circular molecule is generated *in vitro*. This *in vitro* generated molecule is used to transform bacteria and the mutant molecule is amplified *in vivo*. The desired mutation is screened by a variety of techniques and verified by DNA sequencing. The original method has been improved and several site-directed mutagenesis commercial kits are available (Biorad, Amersham). Random alterations can also be introduced *in vitro* on DNA template by using chemical reagents such as bisulfite or hydroxylamine. This methodology is a key player on structure–function studies of proteins. One successful commercial application of this technology is the production of a thermostable subtilisin having a pH tolerance such that it can be used in laundry detergents (see DETERGENCY; SURFACTANTS).

Gene Amplification.   The technique to amplify DNA *in vitro*, known as polymerase chain reaction (PCR), has turned out to be extremely powerful and is widely used for gene manipulations (4,5). PCR is carried out in a series of cycles, and oligonucleotides having >20 bases that flank the target are used as primers. The primers are designed as follows: to anneal to both sides (5′ and 3′) of the target sequence; to anneal to a complementary strand outside of the target sequence; to be lacking in any self-complementary. Each cycle has three steps: denaturation, annealing, and elongation, and these are routinely performed at different temperatures. Denaturation of double-stranded DNA results in the target DNA being single stranded. This is followed by annealing of primers to complementary strands with the 3′ end of the primer facing the target sequence. The elongation step in which DNA polymerase elongates from the free 3′ end of the primer occurs usually at 72°C. The DNA polymerase from a thermophile allows the elongation to be carried out at a high temperature. Thermocyclers and the reagents needed for PCR are sold commercially by Perkin-Elmer.

PCR is an efficient method and low amounts of target DNA are sufficient for amplification. In addition, because the annealing temperature is variable, amplification of homologous target sequences is feasible under low stringency conditions, ie, lower temperature of annealing. The 5′ end of the primer need not be homologous to the target sequence, and can be designed to code a restriction site convenient for gene manipulation. PCR, used in a variety of experiments that hitherto involved laborious techniques, can also be used for site-directed mutagenesis, for domain swapping in proteins, and for isolating homologous genes. Additionally, PCR is used in diagnostics and forensic science (see FORENSIC CHEMISTRY) (4).

Gene Expression

The methods involved in the production of proteins in microbes are those of gene expression. Several plasmids for expression of proteins having affinity tails at the C- or N-terminus of the protein have been developed. These tails are useful in the isolation of recombinant proteins. Most of these vectors are commercially available along with the reagents that are necessary for protein purification. A majority of recombinant proteins that have been attempted have been produced in *E. coli* (1). In most cases these recombinant proteins formed aggregates resulting in the formation of inclusion bodies. These inclusion bodies must be denatured and refolded to obtain active protein, and the affinity tails are useful in the purification of the protein. Some of the methods described herein involve identification of functional domains in proteins (see also PROTEIN ENGINEERING).

Display of Peptides and Proteins on Phages.   The use of phages to display peptides and proteins is an extremely versatile method allowing rapid screening of large random libraries in a relatively short time for a specific function, eg, binding of a hormone (see HORMONES) to a receptor (6,7). For example, 1 in 10^{12} random hexapeptide sequences having a specific affinity to a ligand can be identified in a short (ca one week) time.

Bacteriophages are a class of viruses that are specific to bacteria. Bacteriophage consist of a proteinaceous capsid containing packaged genetic material, ie, DNA or RNA. The proteinaceous capsid consists of several proteins having a defined molecular shape and architecture. In the phage display method a random peptide is fused to a minor capsid protein and is exposed to the outside. Thus in theory this method can be used on any virus. However, bacteriophages are better suited because of the ability to replicate and package rapidly. Vectors are available in which random DNA sequences coding for a short peptide are fused to the coding region of a capsid protein. The ligated DNA is transformed and phage is propagated using standard methods. The random nature of the library can be verified by DNA sequencing. This is followed by selecting phages displaying peptide sequences that bind to a specific ligand by a protocol called biopanning. Certain variations in the basic methodology have been developed (7). The area of displaying peptides on the surface of phage is making rapid progress. For examples, peptide sequences that bind receptors or mimic other nonproteinaceous compounds have been identified (7).

The ability to select a random peptide having specific affinity to any ligand such as 1 in 10^{12} in a relatively short time makes this an extremely powerful technology. This method, still in its infancy, has attracted the attention of several biotechnology companies, because of the potential to provide much information for rational drug design and vaccine development (see PHARMACEUTICALS; VACCINE TECHNOLOGY). This method can also be used for the development of crop protection chemicals (see FERTILIZERS; FUNGICIDES, AGRICULTURAL; INSECT CONTROL TECHNOLOGY).

Expression of Functional Antibodies.   Monoclonal and polyclonal antibodies are used in both diagnostics and research (see IMMUNOASSAY; IMMUNOTHERAPEUTIC AGENTS). Isolation of a specific monoclonal antibody having high affinity to antigen is expensive and laborious. A combination of gene amplification technology along with phage display provides an excellent method to express a repertoire of antibodies in a short time (8–10).

Antibodies are large Y-shaped molecules having a molecular mass of $\sim$ 150,000 in which the domains forming the tips of the arm bind to the antigen. The functional domains carrying the antigen-binding activities and the constant region, called Fc fragments, can be swapped between antibodies. The business end of the molecule is the Fab fragment or Fv fragment which is rather small (see also BIOSENSORS). Both have been expressed in bacteria and are functional (8,9).

Immunization of an animal results in the production of antibodies in lymphocytes and a repertoire (10^{10}) of antibodies having different affinities. Hybridoma cell lines are constructed using these lymphocytes and identification of a high affinity monoclonal antibody is rather time consuming. Antibody genes from lymphocytes can be isolated by reverse-PCR. RNA from the spleen of an immunized mouse is used to create a complimentary DNA (cDNA) library by reverse-PCR. Bacterial clones containing antigen-binding activity can be screened. The bacteria have a doubling time of 30 min compared to 18 h for hybridoma cells and the reagents for bacterial growth are cheaper than mammalian cell culture. Microbial expression of antibodies is an active area of research in several biotechnology companies to generate novel antibodies for diagnostics and therapeutic purposes. In addition, catalytic antibodies, ie, antibodies that carry out catalysis, can also be screened easier in bacteria (11). Catalytic antibodies have potential in synthesis of novel organic compounds (11).

Gene Expression Systems. One of the potentials of genetic engineering of microbes is production of large amounts of recombinant proteins (12,13). This is not a trivial task. Each protein is unique and the stability of the protein varies depending on the host. Thus it is not feasible to have a single omnipotent microbial host for the production of all recombinant proteins. Rather, several microbial hosts have to be studied. Expression vectors have to be tailored to the microbe of choice.

Escherichia coli. A broad genetic and biochemical database combined with a critical mass of academic scientists made *E. coli* the preferred host for gene manipulations in the 1970s. Some of the therapeutic proteins such as insulin and human growth hormone are commercially manufactured from *E. coli* (see INSULIN AND OTHER ANTIDIABETIC DRUGS; HORMONES, HUMAN GROWTH HORMONE). The FDA approval process for recombinant proteins for therapeutic purposes involves elaborate tests which are both time consuming and expensive. Thus, once a recombinant protein from *E. coli* has been approved, it is then less expensive for companies to use the same expression system for new products. However, *E. coli* produces an endotoxin and thus cannot be used as a host for the production of food processing enzymes in certain European countries. Another limitation of *E. coli* is the lack of glycosylation of proteins.

Several expression vectors for *E. coli* are commercially available (1,12). Some of these vectors have been designed so that the recombinant protein can be expressed as a fusion protein having an affinity tail at the C-terminus that can be used for purification. High levels of intracellular expression of proteins in *E. coli* often result in the formation of inclusion bodies which can be easily separated from debris and soluble cellular proteins, solubilized, and refolded to biological activity (12). The denaturation and refolding process is expensive. Therefore, vectors to produce properly folded protein in *E. coli* have been developed. This is accomplished by targeting the recombinant protein to the periplasm. For example, active Fv antibody fragments (1 to 2 g/L) have been expressed in the *E. coli* periplasm at high levels (14). Media and growth optimization of recombinant *E. coli* fermentation has resulted in cell density yields up to values of optical density at 550 nm (OD_{550}) of 100 to 150.

Erwinia. *Erwinia* are gram-negative bacteria that are significant as plant pathogens causing soft-rot of vegetable mainly after harvest during storage. However, these bacteria have the potential to be used for production of specialty chemicals and enzymes. The principal attention on *Erwinia* has been toward understanding the mechanism of plant pathogenesis, which primarily results from the extracellular cell wall degrading enzymes such as pectinase, cellulases, and proteases. However, these enzymes can be used for biomass degradation (see ENZYME APPLICATIONS). Vectors to express genes in *Erwinia* have been developed, and *Erwinia* has been engineered for the conversion of glucose to 2-ketol-gulonic acid, a precursor in vitamin C production (see MICROBIAL TRANSFORMATIONS; VITAMINS, ASCORBIC ACID) (15).

Pseudomonas. These gram-negative bacteria are a diverse group of microbes that inhabit plants, water, and soil. *Pseudomonads* are metabolically versatile, capable of carrying out chemical transformations, mineralization of organic compounds, and colonization on plant roots (16). The use of *Pseudomonads* strains in the clean up of chemical wastes and oil spills has drawn considerable attention.

Plasmid vectors for regulatable gene expression are available for *Pseudomonas* strains. *Pseudomonas* is not commonly studied for the production of recombinant proteins by fermentation. However, because they can colonize on leaves and roots the attention on *Pseudomonas* species has been toward the release of genetically engineered microorganisms for crop protection. For example, *P. syringea* has been engineered to have a trait that prevents ice nucleation called Ice$^-$, and these bacteria when sprayed compete with Ice$^+$ bacteria that occur in nature. This has resulted in prevention of ice formation which results in a decrease of frost damage on certain vegetable plants. Insecticidal endotoxin from *B. thurengeisis* has also been expressed in *Pseudomonas fluorescens* and the recombinant strain expressing endotoxin was capable of normal root colonization. In addition, the recombinant *P. fluorescens* also showed increased toxicity toward corn root worms. A lot of the use of engineered *Pseudomonas* relies on the release of these genetically engineered microorganisms. Extensive studies on the potential benefits and risks have been addressed. For example, vectors with herbicide resistance genes have been constructed to easily trace the released microorganism (17).

Pseudomonads also have the ability for xenobiotic metabolism and are capable of carrying out diverse sets of chemical reactions. *Pseudomonas* species is used in the commercial production of acrylamide (qv) (18). Several operons involved in the metabolism of xenobiotic compounds have been studied. Use of *Pseudomonads* for the clean up of the environment and for the production of novel chemical intermediates is likely to be an area of active research in the 1990s.

Bacillus sp. These bacteria are gram-positive soil microbes. Members of the *Bacillus* species supply 58% of industrial enzymes sold (19). For example, proteases from *B. amyloliquefaciens* and amylases from *B. licheniformis*, glucose isomerase from *B. coagulans* are used in a variety of industrial processes (see ENZYME APPLICATIONS–INDUSTRIAL). The proteinaceous inclusions produced by *B. thuringiensis* are useful as insect toxins. Thus extensive fermentation technology has been developed for *Bacillus* species and low cost media are available (19).

B. subtilis remains the paradigm organism for *Bacillus* sp. for genetics and biochemistry (20). *B. subtilis* is naturally competent and thus gene manipulations are easy. Electroporation has made it feasible to transform some of the industrial strains such as *B. licheniformis, B. amyloliquefaciens,* and *B. thuringiensis.* A variety of vectors are available for gene expression *B. subtilis.* High level (3 to 6 g/L) secretion of *Bacillus* proteins has been accomplished in *B. subtilis,* but the export of certain non-*Bacillus* proteins has not reached commercially acceptable levels. A significant problem associated with protein secretion in *B. subtilis* is the presence of host proteases. *B. subtilis* strains lacking most of the principal extracellular proteases are available and these should help in stabilizing recombinant proteins. Most of the attention in *B. subtilis* has been focused on the extracellular protein production. However, it is feasible to obtain high levels of intracellular protein expression. Efforts are underway to produce certain toxin of pathogens as intracellular proteins in *B. subtilis* for vaccine production.

Genes for most of the *Bacillus* exoenzymes that are commercially used have been cloned. Thus recombinant proteases and amylases should replace the traditional enzymes. Extensive protein engineering of subtilisin has resulted in the creation of an enzyme that is tolerant of higher temperature and pH (19). *Bacillus* species continue to be attractive sources as host for the production low cost bulk enzymes. High levels of interleukin-3, secreted from *B. licheniformis,* is in clinical trails in Europe. *B. brevis* has been studied for the production of a variety of proteins in Japan (19). *B. thuringiensis* produces an intracellular proteinaceous crystal which is an insect toxin (20). The genes for several of these toxins have been cloned and the x-ray structure of the toxin has been solved. Several biotechnology companies (Ecogen, Mycogen) are actively exploring the development of the next generation of toxins based on the knowledge of toxin gene sequences.

Clostridium.   This genus is comprised of a heterogeneous assemblage of obligate, anaerobic, gram-positive, endospore-forming bacteria (21). Some of the members of this genus such as *C. perfringes, C. tetani, C. botulinum,* and *C. difficile* are pathogens. The toxin genes from the various species have been cloned and are being studied toward understanding pathogenesis. The availability of both the tetanus and botulism toxin genes should both improve vaccines and help in human medicine and food diagnostics. *Clostridium acetobutylicum* has the ability to ferment acetone, butanol, and ethanol. Some of the genes involved in the solventogenesis have been isolated. The physical and genetic maps of *C. perfringes* have been determined. Plasmid vectors for gene manipulations for *C. acetobutylicum* and *C. perfringes* are available and gene transfer by electroporation is feasible. The cellulases from *C. thermocellum* have been widely studied and the cellulase genes have been expressed in *E. coli* and *B. subtilis.* The increased interest in the use of biomass for the production of solvents such as ethanol (qv) should result in more use of *Clostridium* (see CHEMURGY) (21).

Streptomyces.   These gram-positive bacteria have been studied traditionally for the production of antibiotics (qv) and have advanced fermentation technology. *Streptomyces* species also secrete a variety of extracellular enzymes such as xylanases and ligninases (20). Plasmids and cloning vectors for heterologous gene expression are available for certain *Streptomyces* species. *Streptomyces lividans* has been studied as a host for heterologous protein expression. Expression of active soluble human T cell receptor CD4 into the growth medium of *S. lividans* has been demonstrated (22). A problem with heterologous protein expression in

Streptomyces results from the presence of extracellular proteases. However, low protease strains have been genetically constructed (20).

Rhodococcus.   *Rhodococcus* are gram-positive, aerobic, nonmotile actinomycetes having a growth cycle in which they range from cocci to rods to filaments with short projections. These bacteria are common throughout nature and some are pathogenic to humans, plants, and animals. These bacteria exhibit a broad metabolic diversity and are capable of degrading hydrocarbons, halogenated phenols, aromatic amino acids, and halogenated alkanes. The nitrile hydratase of *Rhodococcus* sps. is used in the commercial production of acrylamide in Japan (18). The genetics and molecular biology in these bacteria have not been well studied. Cloning vectors based on indigenous rhodoccocal plasmid have been constructed and *E. coli*–rhodococcal shuttle vectors are available (23). Protoplast transformation of several species of *Rhodococcus* sps. have been accomplished. The development of cloning vectors and gene transformation make it feasible to engineer *Rhodococcus* species for a variety of purposes. The nitrile hydratase gene has been cloned and expressed in *E. coli*. The metabolic potential of this class of microbes can be exploited for both bioremediation and synthesis of novel chemicals.

Lactic Acid Bacteria.   The lactic acid bacteria are ubiquitous in nature from plant surfaces to gastrointestinal tracts of many animals. These gram-positive facultative anaerobes convert carbohydrates (qv) to lactic acid and are used extensively in the food industry, for example, for the production of yogurt, cheese, sour dough bread, etc. The sour aromatic flavor imparted upon fermentation appears to be a desirable food trait. In addition, certain species produce a variety of antibiotics.

Lactococcus and *Lactobacillus* are members of this group and gene transfer methods are available (24). Plasmids vectors that function in several different species of lactic acid bacteria are available. In addition, some of the well-characterized staphylococcal plasmids are also functional in these bacteria. Heterologous genes such as amylases and proteases have been expressed in certain lactic acid bacteria. The endogenous proteases produced by these bacteria act on caesin resulting in small peptides. These peptides are thought to play a role in the flavor generation. The genes for some of the proteases has been cloned and expressed.

Yeasts.   Among the eukaryotic microbes yeast has drawn a lot of attention for the production of heterologous proteins (see YEASTS) (25). *Saccharomyces cerevisiae* is well characterized biochemically and genetically and was the organism of choice for most of the early experiments. However, heterologous expression seems to be better in some of the industrial strains of yeasts such as *Pichia pastoris, Hansenula polymorpha, Kluyveromyces lactis,* and *Yarrowia lipolytica* (25–28).

A set of regulatable gene expression vectors are available for *S. cerevisiae* and these vectors are based on autonomous replicating plasmids or by integration into the chromosome. Several therapeutic proteins such as insulin, growth hormone, and interlukins have been secreted from *S. cerevisiae*. A sizeable limitation of *S. cerevisiae* for the production of some of therapeutic proteins is the extensive glycosylation of the heterologous protein. This hyperglycosylation, which may be responsible for the highly antigenic nature of some of the secreted protein, may be advantageous for nontherapeutic purposes. For example, an increase in thermostability of *Aspergillus* glucose oxidase in *S. cerevisiae* was observed. High level (3 g/L) accumulation of a fungal glucose oxidase has also been accomplished in *S. cerevisiae* (28).

Methylotrophic yeasts *P. pastoris* and *H. polymorpha* have been used to secrete a variety of heterologous proteins. Stable multicopy integration into the chromosome occurs in both *P. pastoris* and *H. polymorpha*. *P. pastoris* appears to be a promising host, and several heterologous proteins, eg, human serum albumin and human interleukin, have been secreted at high levels (27). The extent of glycosylation in *P. pastoris* is different than that of *S. cerevisiae*. *K. lactis* has also been engineered to secrete high levels of human serum albumin (26). A limitation of the methylotrophic yeasts is the lack of basic research on the mechanism of secretion and gene regulation. In addition, extensive clonal variation in the level of gene expression have been observed, making multiple clone screening a necessity. The clonal variation seems to be a result of the original transformation, and lack of understanding of this mechanism makes it hard to target genes to specific sites in the chromosome.

Filamentous Fungi.   Fungi are highly versatile and generate a wide range of commercial products (29) including organic acids such as citric acid (qv), secondary metabolites such as antibiotics, and a variety of industrial enzymes. *Aspergillus nidulans* and *Neurospora crassa* have been the prototypes for genetics and biochemistry. However, some of the commercial strains that have been developed for heterologous protein expression are *Aspergillus niger* and *Trichoderma reesi* (30,31). *A. niger* is used in several commercial processes; *T. reesi* appears to have a very high capacity to secrete proteins to the level of 35 g/L. These filamentous fungi do not produce any pyrogens and thus do not pose any threat for the production of food processing enzymes and therapeutic proteins. However, the secreted proteins may be glycosylated and the glycosylation is different from that of mammalian cells which may present an increased antigenic problem for therapeutic proteins.

Vectors have been developed for regulated gene expression in both *Aspergillus niger* and *Trichoderma reesi*. Heterologous genes are maintained in these organisms by integration into the chromosome rather than on autonomous plasmids. Commercial levels of 1 g/L of chymosin (rennin) production have been accomplished in *A. niger* var. *awamori* (32). The recombinant chymosin has properties that are similar to the native calf rennin and can be used for cheese processing. The recombinant chymosin has been approved by FDA for use in food processing (cheese). Chymosin from *E. coli* has been available from Pfizer since the early 1990s. Attempts to secrete high levels of chymosin in other microorganisms such as *B. subtilis* and yeast were not successful. Efficient production of antibody fragments (150 mg/L) in *T. reesi* has been reported. Mechanism of protein secretion and genetic tools needed for improving yields are not available in *A. niger* and *T. reesi*.

Other Microbial Systems.   In addition to the systems described, gene cloning is routinely performed in several other bacterial strains including *Streptoccocus, Staphylococcus, Brevibacterium, Rhodobacter, Cornyebacterium, Glucanobacter, Acetobacter,* and *Zanthomonas* species.

Commercial Products from Genetically Engineered Microbes

Genetically engineered microbes are used to produce several commercial products. The number of products is likely to increase exponentially throughout the 1990s.

Therapeutics. Therapeutic materials represent a class of polypeptides that are a low volume, high value product. The production system need not be very efficient but the quality of the recombinant protein has to be extremely pure (33,34). Thus high cost mammalian production systems can be tolerated. However, some of the therapeutic proteins such as insulin, human growth hormone, interleukins, interferon, and streptokinase are produced microbially.

Bulk Enzymes. Enzymes such as proteases, amylases, glucose isomerases, and rennin are used in food processing. Similarly proteases and lipases are used in detergents. Cellulases and xylanases are used in the paper pulp industry. The genes for most of the enzymes used in the various commercial processes have been cloned and overexpressed. Rennin (chymosin) produced from *E. coli* and *A. niger* has been approved by FDA for use in the dairy industry.

Antibiotics. The genes involved in the synthesis of a variety of antibiotics have been isolated (34,35). These include antibiotics such as erythromycin, streptomycin, and also peptide antibiotics such as gramicidin and tyrocidin. Characterization of these gene products facilitates the design of novel antibiotics. In addition, overexpression of some of these gene products is also expected to improve the yield of the antibiotic (34,35).

Chemicals. Gluconic acid is produced by fermentation using *A. niger*. The genes for glucose oxidase and catalase from *A. niger* have been isolated. Gluconic acid production can be improved with the use of cloned genes, and alternative economical host systems can be developed. Acrylamide (6000/yr) is produced using the nitrile hydratase from *Rhodococcus* sp. N-774 or *Pseudomonas chloraphis* B23 (18). The genes for different nitrile hydratases from several *Rhodococcal* sps. and *P. chloraphis* have been isolated. The comparison of the various protein structures should aid in protein engineering of this enzyme and thus improve the economics of the acrylamide process.

Some chemicals such as indigo, tryptophan, and phenylalanine are overproduced in bacteria by pathway engineering (36–38). In this method, the enzymes involved in the entire pathway are overproduced. In addition, the host bacterium is also altered such that the carbon flow is directed toward the engineered pathway (38). *E. coli* has been modified to overproduce indigo and tryptophan and phenylalanine. *Corynebacterium glutamicum* has been engineered to overproduce tryptophan from 28 to 43 g/L. Similarly, attempts are underway to overproduce several vitamins by pathway engineering (34,38).

Poly-β-hydroxybutyrate (PHB) is a biodegradable thermoplastic that is produced by several microorganism. The PHB synthesis has been characterized in *Alcaligens eutrophus* and the operon involved in PHB production has been cloned. Recombinant *E. coli* strains that can produce high levels of PHB have been constructed (39).

Crop Protection. Microbes naturally produce a variety of fungicides and insecticides. In addition, certain bacteria such as *Pseudomonads* can help in frost prevention. *B. thuringienesis* toxins have been used for a long time as efficient insecticides. The genes for several of the insect toxins have been isolated, and it is feasible to construct hybrid toxins with higher potency. However, release of genetically engineered microorganisms remains a sensitive issue. The advantages of these engineered microbes in most cases far outweigh the potential risks.

Future Prospects.   One of the advances in the recombinant DNA technology is the rapidity with which most of the experiments can be performed. There has also been constant improvement in the quality of reagents. Automation in DNA sequencing should also have an impact on characterization of genes. Isolation of genes for a variety of enzymes having useful properties from thermophilic bacteria and methanogens should evolve (40,41). As rapid progress is made in this arena microbes are expected to be tailored to produce useful commercial products.

BIBLIOGRAPHY

"Genetic Engineering" in *ECT* 3rd ed., Vol. 11, pp. 730–745, by A. M. Chakrabarty, University of Illinois at the Medical Center, Chicago; in Suppl. Vol., pp. 495–513, by E. Jaworski and D. Tiemeier, Monsanto Co.

1. J. Sambrook, E. F. Fritsch, and T. Maniatis, *Molecular Cloning* Vols. 1, 2, 3, Cold Spring Harbor Laboratory Press, Cold Spring Harbor, New York, 1989.
2. B. M. Chassy, A. Merciener, and J. Flickinger, *Trends. Biotechnol.* **6**, 303–309 (1988).
3. M. Smith, *Ann. Rev. Genet.* **19**, 423–461 (1985).
4. N. Arnheim and H. Erlich, *Ann. Rev. Biochem.* **61**, 131–136 (1992).
5. M. A. Innis and co-workers, *PCR Protocols: Guide to Methods and Applications*, Academic Press, Inc., New York, 1990.
6. J. J. Devin, L. C. Panganiban, and P. E. Devlin, *Science* **249**, 404–406 (1990).
7. R. H. Hoess, *Curr. Opion. Struct. Biol.* **3**, 572–578 (1993).
8. J. McCafferty and co-workers, *Nature* **348**, 552–554 (1990).
9. G. Winter and C. Milstein *Nature* **349**, 293–299 (1991).
10. L. Sastry and co-workers, *Proc. Natl. Acad. Sci. USA* **86**, 5728–5732 (1989).
11. R. A. Lerner, S. A. Benkovicand, and P. G. Schultz, *Science* **252**, 659–667 (1991).
12. D. V. Goeddel, *Methods Enzymol.* **185** (1990).
13. J. Hodgson, *Biotechnology*, **11**, 887–893 (1993).
14. P. R. Carter and co-workers, *Biotechnology* **10**, 163–167 (1992).
15. J. Robert-Baudoy, *Trends. Biotechnol.* **9**, 325–326 (1991).
16. E. Galli, S. Silver, and B. Witholt, *Pseudomonas Molecular Biology and Biotechnology*, ASM, Washington, D.C., 1992.
17. M. Wilson and S. E. Linsow, *Ann. Rev. Microbiol.* **47**, 913–944 (1993).
18. M. Kobayashi, T. Nagasawas, and H. Yamada, *Trends. Biotechnol.* **10**, 402–408 (1992).
19. R. H. Doi and M. McGloughlin, *Biology of Bacilli Applications to Industry*, Butterworth-Heinemann, Stoneham, Mass., 1992.
20. A. L. Sonenshein, J. A. Hoch, and R. Losick, *Bacillus subtilis and other Gram-positive Microorganisms*, ASM, Washington, D.C., 1993.
21. M. Young and T. Cole, in Ref. 20, pp. 65–83.
22. J. A. Fornwald and co-workers, *Biotechnology* **11**, 1031–1036 (1993).
23. W. R. Finnerty, *Ann. Rev. Microbiol.* **46**, 193–218 (1993).
24. M. J. Gasson, *FEMS Microbiology Reviews* **12**, 3–20 (1993).
25. R. Fleer, *Curr. Opin. Biol.* **3**, 486–496 (1992).
26. R. Fleer and co-workers, *Biotechnology* **9**, 968–975 (1991).
27. J. F. Tschopp and co-workers, *Biotechnology* **5**, 1305–1308 (1987).
28. A. De Baetselier and co-workers, *Biotechnology* **9**, 559–561 (1991).

29. D. B. Finkelstein and C. Bell, *Biotechnology of Filamentous Fungi*, Butterworth-Heinemann, Stoneham, Mass., 1992.
30. R. W. Davies, in J. A. H. Murray, ed., *Transgenesis*, John Wiley & Sons, Inc., 1992, pp. 82–104.
31. E. Nyyssonen and co-workers, *Biotechnology* **11**, 591–595 (1993).
32. N. S. Dunn-Coleman and co-workers, *Biotechnology* **9**, 976–980 (1991).
33. T. J. R. Harris, *Protein Production by Biotechnology*, Elsevier Science Publishers Ltd., New York, 1990.
34. R. H. Baltz, G. D. Hegeman, and P. L. Skatrud, *Industrial Microorganisms: Basic and Applied Molecular Genetics*, ASM, Washington, D.C., 1993.
35. P. L. Skatrud, *Trends. Biotechnol.* **10**, 324–329 (1992).
36. R. Katsumata and M. Ikeda, *Biotechnology* **11**, 921–925 (1993).
37. D. Murdoick and co-workers, *Biotechnology* **11**, 381–385 (1993).
38. J. E. Bailey, *Science* **282**, 1668–1675 (1991).
39. B. S. Kim, S. Y. Lee, and H. N. Chang, *Biotechnol. Lett.* **14**, 811–816 (1992).
40. J. Koninsky, *Trends. Biotechnol.* **7**, 88–92 (1989).
41. W. W. M. Adams, *Ann. Rev. Microbiol.* **47**, 627–658 (1993).

VASANTHA NAGARAJAN
E. I. du Pont de Nemours & Co., Inc.

General Reference

S. Fahnestock, B. Moldover, and C. Perrotto, E. I. du Pont de Nemours, Inc., Wilmington, Del., 1993.

PLANTS

Several discoveries in the 1980s and 1990s permitted the transition of plant molecular biology from a fledgling science to commercial reality. These discoveries ranged from the identification of biologically important genes to the development of methods to introduce new genes into plants and regulate gene expression. The former process is commonly referred to as transformation. Nearly five dozen plant species have been transformed and the list of plant species subject to transformation include principal field crops such as corn, cotton (qv), rape, rice, soybean, and wheat (see SOYBEANS AND OTHER OILSEEDS; WHEAT AND OTHER CEREAL GRAINS). In addition, several horticultural species such as tomato, potato, petunia, chrysanthemum, apple, walnut, melons, etc, have been subject to transformation. More than 500 field tests have been conducted and transgenic plants such as transgenic tomato, soybean, corn, rape, potato, petunia, melons, and cucumbers are in the advanced stages of commercial development and regulatory process.

Four methods have been extensively investigated for the introduction of transferred deoxyribonucleic acid (T-DNA) into plants. These include agrobacterium mediated T-DNA transfer (1,2), direct uptake of DNA by protoplasts (3), particle acceleration techniques such as electrostatic discharge or biolistics gun technology (4–7), and DNA uptake into partially digested immature embryos (8). By far the most commonly used method for gene introduction into dicotyledonous plants is the agrobacterium technology. This bacterium delivers genes contained in the T-DNA region of the Ti plasmid to the nucleus of several dicotyledonous species. Within the nucleus, the T-DNA is randomly inserted into the chromosome of the recipient cell. The clonal progenies of the cell containing the inserted gene show a high degree of stability. The gene is transmitted in a Mendelian fashion during sexual stages of cell division and development (1,2,9).

Although agrobacterium mediated gene introduction into plants is highly efficient and routinely used, its primary limitation is that several plant species are recalcitrant to transformation via this bacterium. This is particularly so for monocotyledonous species such as corn, rice, and wheat. In these instances, particle gun technology is routinely used for the introduction of genes. Whereas most genes introduced into plants via the gun technology appear to be nuclear localized, this technology also has been reported to be useful in tranforming the chloroplast of plant cells (10). Several reviews documenting the progress in plant transformation during the early 1990s are available (6,11,12).

Expression of genes that have been introduced into plants is regulated by promoters, although the extent of regulation of gene activity by the promoter is influenced at least to some extent by the insertion site of the gene within the chromosome. As of this writing methods for DNA transfer cause random insertion of the DNA into the chromosome. Techniques for precise introduction of the transgene to specific sites with the plant genome are being developed. Numerous promoters have been used for gene expression in plants. The choice of promoters is dictated by the tissue and developmental specificity required for gene expression. By far the most commonly used promoter for constitutive gene expression in both mono- and dicotyledonous plants is the Cauliflower mosaic virus (CaMV) 35S promoter. This promoter appears to be expressed in several plant organs and cell types; however, it is not truly constitutive in that it is not uniformly expressed in all plant tissues. DNA elements within the 35S promoter, which cause tissue specific expression of genes, have been described (13,14). The activity of the 35S promoter may be enhanced by use of multiple copies of enhancer elements located within the 35S promoter (15).

For tissue regulated gene expression, promoters have been described which are expressed in a tissue specific manner (16). Examples of such promoters include the promoter for patatin which causes tuber specific expression of genes in potato (17–20), the 7S promoter of soybean (21), or the napin promoter of *Brassica* (22), which cause seed specific expression of genes, and the RB7 promoter which causes root specific expression of genes (23). These promoters may not only be spatially regulated in terms of cell and tissue specificity but may also be temporally regulated in that the promoters are active only at certain developmental stages of the cells and tissues in which the promoters are expressed.

In order to determine which plant cells have been transformed, selectable marker genes are introduced during tranformation. These marker genes permit selective growth of transgenic cells on the medium used for tissue propagation whereas the nontransgenic cells are killed. Examples of selectable marker genes include antibiotic resistance genes such as neomycin phosphotransferase (NPT-II), hygromycin phosphotransferase, and chloramphenicol acetyl transferase, as well as herbicide resistance genes such as phosphinothricin acetyl transferase, bromoxynil nitrilase, 2,4-D-oxygenase, etc. Plant cells expressing the NPTII gene are able to survive kanamycin [8063-07-8] and addition of kanamycin permits selection of those cells receiving and expressing the NPTII gene during transformation (see ANTIBIOTICS; HERBICIDES).

Herein two specific applications of plant biotechnology are discussed. The first is concerned with 5-enolpyruvylshikimate 3-phosphate synthase (EPSPS), the enzyme which is the target for the widely used herbicide glyphosate [1071-83-6], $C_3H_8NO_5P$. The second is directed toward a discussion of increasing starch biosynthesis in plants. The first application deals with a trait which directly impacts the farmer during the production phase of agriculture; the second application deals with a trait that impacts the consumer of agricultural products. These traits may be referred to as agronomic and quality traits, respectively.

A number of other agronomic and quality traits are being investigated. These include insect, virus, disease, and nematode resistance, fertilizer-use efficiency, ripening control, fruit firmness, etc. Of these traits the most advanced agronomic trait for bioengineering is insect resistance. Insect resistant cotton and corn have been obtained by introduction and expression of a *Bacillus thurigiensis* kurastaki gene (BtK gene) (24). The BtK protein encoded by this gene is selectively toxic to the lepidopteran pests, ie, cotton boll worm, pink boll worm, and European corn borer, which attack these crops. Insect-resistant potato has been obtained by expression of a BtT gene which encodes a protein, selectively toxic to the Colorado potato beetle, a principal pest of potato (25,26). This topic has been reviewed (26–28) (see also INSECT CONTROL TECHNOLOGY). Virus resistance, conferred by expression of the viral coat protein (CP) gene in transgenic plants, has also received considerable attention. Products such as potato, squash, melons, etc, based on this technology are in advanced stages of development and commercialization (29).

Both tomato fruit ripening and fruit firmness are among the advanced quality traits that are being investigated. A variety of approaches, based on inhibition of ethylene production (30,31) are being pursued for enhancement of shelf life of tomato. For enhancing fruit firmness, cell wall hydrolytic enzymes such as polygalacturonidase and pectin methylesterase are being investigated (32–34).

Bioengineering of Glyphosate Tolerance

The enzyme 5-enolpyruvylshikimate 3-phosphate (EPSP) synthase catalyzes the transfer of a carboxyvinyl moiety of phosphoenol pyruvate (PEP) to shikimate 3-phosphate (S3P), yielding inorganic phosphate and EPSP as reaction products. EPSPS has received considerable attention in recent years, in view of the demonstration that glyphosate (N-(phosphonomethyl)glycine [1071-83-6]), the active ingredient of the herbicide Roundup, kills plants by inhibition of this enzyme. EPSPS catalyzes the sixth reaction during aromatic amino acid biosynthesis via the shikimate pathway which exists only in plants and microorganisms. EPSP is the immediate precursor of chorismate, the first important branch point during aromatic amino acid and vitamin biosynthesis.

Perhaps the most important inhibitor of the EPSPS reaction is glyphosate which inhibits the EPSPS reaction via formation of a ternary complex with either S3P or EPSP and enzyme. Glyphosate is a competitive inhibitor with respect to PEP and an uncompetitive inhibitor with respect to S3P. Glyphosate, however, is not a structural analogue of PEP because the glyphosate does not inhibit any other PEP-dependent reaction. Whereas it has been suggested that glyphosate may be a transition-state analogue of the carbonium ion intermediate of PEP formed during catalysis, the bulk of the evidence suggests that this is unlikely. Nevertheless, glyphosate inhibits a wide range of EPSPS enzymes of bacterial, fungal, and plant origin. A number of structural analogues of glyphosate have also been tested for the ability of inhibit EPSPS. Only a few such analogues, eg, N-amino and N-hydroxy glyphosate, have been found to be inhibitors.

Roundup is a nonselective, post-emergent herbicide having activity against a wide range of annual and perennial grasses as well as broadleaf weeds. Because Roundup has no selectivity for weeds, use for weed control during active growth period of crops is fairly limited. Despite its nonselectivity, glyphosate, is extensively used in weed management because of broad-spectrum, systemic herbicidal activity; rapid inactivation in the soil (does not sterilize the soil); decomposition in the soil to the natural products, ie, carbon dioxide (qv), ammonia (qv), and phosphate; no toxicity to animal, aquatic, and avian species; it binds tightly to soil and does not contaminate ground water; and its cost effectiveness in weed control. In view of all the desirable features of glyphosate, the engineering of glyphosate tolerance in crop plants has the potential to open up new frontiers in weed management during cultivation. A substantial effort has been directed toward introducing Roundup tolerance to crop plants (35–38).

Engineering Roundup Tolerance. Knowing that the mode of herbicidal action of glyphosate is mediated via inhibition of EPSPS, at least two mechanisms can be considered for the introduction of Roundup tolerance to plants. The first option is to simply overproduce EPSPS so as to leave sufficient EPSPS enzymatic activity within the plant cells to satisfy the flux through the shikimate pathway. Alternatively, a gene encoding a glyphosate tolerant EPSPS enzyme can be used so that the EPSPS reaction is unaffected even in the presence of glyphosate. In addition, other approaches which are not related to the mode of action of glyphosate, such as glyphosate inactivation and inhibition of uptake, can be considered. These last are not discussed herein.

Overproduction of EPSPS. Overproduction of EPSPS has been demonstrated to confer glyphosate tolerance to both bacteria (39) and plant cells (40–42). Glyphosate tolerant plant cells have served as an excellent starting material for the isolation and purification of the EPSPS protein to homogeneity. N-Terminal amino acid sequence of the resulting protein provided the requisite information for synthesis of oligoprobes which were used for screening a complementary DNA (cDNA) library of petunia cells tolerant to glyphosate. From the library, the cDNA encoding petunia EPSPS was isolated and sequenced. The protein encoded by the cDNA had an N-terminal extension of 72 amino acids compared to the protein sequence obtained from the purified EPSPS enzyme. This N-terminal extension is necessary and sufficient to direct the EPSPS protein into the chloroplasts of plant cells (43). These studies also led to the conclusion that aromatic amino acid biosynthesis occurred primarily in the chloroplast of plant cells.

Whereas plant cells overproducing EPSPS could be generated by stepwise selection on glyphosate and the cells were glyphosate tolerant, these cells could not be regenerated into intact plants. Availability of the cDNA clone for EPSPS, however, provided a convenient tool for generating trangenic plants capable of overproducing EPSPS. Using an *Agrobacterium tumefaciens* transformation system, the EPSPS gene was introduced into both petunia and tobacco plants. Petunia plants overproducing EPSPS were thus produced and shown to be tolerant to Roundup (44). However, the extent of tolerance was not adequate for commercial use.

   Glyphosate-Tolerant EPSPS.   Several groups have tried to introduce Roundup tolerance into plants using genes encoding glyphosate-tolerant EPSPS enzymes. A mutant glyphosate-tolerant EPSPS enzyme, fivefold less sensitive to glyphosate, was isolated from *Salmonella typhimurium* (45). The introduction of the *S. typhimurium* mutant EPSPS gene into tobacco plants resulted in expression of the mutant gene such that plants were tolerant to glyphosate, but the extent of tolerance was not commercial (46).

   Other bacterial mutants, such as a mutant *Escherichia coli* enzyme tolerant to glyphosate, have been described (47). The *E. coli* mutant had a single amino acid change from the wild type, resulting in substitution of glycine 96 with alanine. An identical mutation was reported in glyphosate-tolerant *Klebsiella pneumoniae* (48). The nature of changes in the kinetic constants of the *K. pneumoniae* enzyme is similar to that of the *E. coli* enzyme.

   The *E. coli* mutant EPSPS was fused to the chloroplast transit peptide (CTP) sequence of petunia EPSPS in order to target the bacterial protein to the chloroplast (49). *In vitro* uptake experiments confirmed that the bacterial enzyme could indeed be imported and processed to a mature protein by chloroplast preparations. Introduction into petunia and tobacco plant cells led to regenerated plants expressing the bacterial gene either targeted to chloroplast or the cytosol. Tobacco plants containing the *E. coli* mutant EPSPS targeted to the chloroplast had higher levels of Roundup tolerance compared to either plants overproducing wild-type EPSPS or the control nontransgenic plants, but the level of tolerance was not sufficient for commerical use. The level of Roundup tolerance of plants having the *E. coli* enzyme targeted to the cytosol was only slightly higher than that of control plants, suggesting that the cytosolic EPSPS reaction was unable to complement the chloroplastic deficiency of EPSPS.

   The glycyl 96 (G96) and prolyl 101 (P101) residues occur in a conserved region of EPSPS which is present in bacterial, fungal, and plant EPSPS enzymes. Replacement of G96 with an amino acid other than alanine (A) results in an inactivation of the EPSPS activity of the protein. However, the G96 to A mutation can be transferred to other bacterial and plant EPSPS enzymes, and in every case the alanyl enzyme has a higher glyphosate tolerance compared to the glycyl enzyme (50). This suggests that there is a high degree of conservation of the active site of EPSPS between bacterial, fungal, and plant enzymes.

   Mutation of the conserved P101 to a serine residue also results in glyphosate tolerance of the EPSPS enzyme. This mutation was introduced into petunia EPSPS by site-directed mutagenesis and the seryl enzyme was demonstrated to be glyphosate tolerant. Analogous to the *S. typhimurium* enzyme, this mutation confers only marginal (sevenfold) glyphosate tolerance and no significant changes in the kinetic constants for the substrates. The petunia cDNA containing the prolyl to seryl mutation and the targeting sequence was introduced into tobacco plants. The Roundup tolerance of the tobacco plants expressing the seryl mutant was intermediate between plants expressing the wild-type and alanyl mutant enzymes.

   Numerous bacteria which utilize glyphosate as a growth substrate were screened for the presence of EPSPS enzymes having binding constants for PEP close to those of the wild-type enzyme, but at least a 100–10,000-fold increase in affinity for glyphosate (37). The EPSPS from agrobacterium CP4 is perhaps the best studied. These enzymes are referred to as class II in order to distinguish them from the class I enzyme already described. The class II EPSPS enzymes have natural resistance to glyphosate and a low binding constant for PEP whereas the class I EPSPS enzymes are highly sensitive to glyphosate. The agrobacterium CP4 enzyme has 28% identity to the *E. coli* enzyme and has the conserved glycyl to alanyl change. Antibodies which recognize the *E. coli* EPSPS recognize petunia EPSPS but do not recognize agrobacterium CP4 EPSPS. Similarly antibodies reacting with the agrobacterium CP4 EPSPS do not show immune reaction with either *E. coli* or petunia EPSPS.

Transgenic soybean plants expressing the CTP-CP4 EPSPS display commercial levels of Roundup tolerance. These results validate the importance of substrate kinetics of EPSPS in order to maintain adequate rates of aromatic biosynthesis. Furthermore, the fact that glyphosate tolerance can be obtained by expression of a glyphosate-tolerant EPSPS illustrates that the herbicidal mode of action of glyphosate is related solely to inhibition of the EPSPS reaction.

As described earlier, translation of the EPSPS mRNA of plants results in the formation of a protein which has an *N*-terminal extension. The *N*-terminal extension, referred to as the chloroplast transit peptide, is necessary and sufficient for the import of the preprotein by the chloroplast. Once imported by the chloroplast, the transit peptide is cleaved releasing the mature enzyme. As expected, introduction of the EPSPS transit peptide to other protein sequences results in the importation of the fusion protein by the chloroplast.

The three-dimensional structure of EPSPS from *E. coli* has been established by crystallographic techniques (51). A number of amino acid residues have been modified to establish the necessity of these residues for enzymatic activity. At its *N*-terminus, the lysyl residue at position 22 of the *E. coli* enzyme has been shown to be highly reactive and essential for enzymatic activity (52,53). It is likely that the lysyl residue is involved in substrate recognition. In addition to the lysyl residue at position 22, the arginyl residue at position 28 of EPSPS is conserved in all EPSPS enzymes studied to date. This arginyl residue is highly reactive, and its reaction with arginine reagents is inhibited by S3P and to a higher extent by a mixture of S3P and glyphosate. By site-directed mutagenesis, the arginyl residue has been replaced by lysyl, histidinyl, and glutaminyl residues. The latter two replacements appear to be detrimental for EPSPS activity, whereas the lysyl enzyme retains substantial activity (54). The roles of histidinyl, glutamyl, and cysteinyl residues of EPSPS have been probed by reaction with chemical modification reagents. These studies suggest that a glutamyl and histidinyl residue are critical for EPSPS activity (55). Similar studies with cystein modification suggest that cys-408 of *E. coli* EPSPS, although in a conserved region, is not essential for activity but is proximal to the active site (56).

Bioengineering of Increased Starch Content

The primary form of carbohydrate reserve in plants is starch (qv), entirely composed of the six-carbon sugar (qv) glucose. Starch typically is deposited in the form of water-insoluble granules, and is synthesized and stored in chloroplasts in photosynthetic tissues or in amyloplasts. Starch is a generic term used to describe a very heterogeneous class of molecules which differ in size and structure between different plants, different tissues within a plant, and at different stages of plant development. The heterogeneity of starch has proven useful in a number of different applications; for example, pea starch is widely used as a sizing agent in paper (qv) manufacture, and corn and potato starch are widely used to give viscosity, freeze–thaw tolerance, and body to a number of processed foods (see CARBOHYDRATES; FOOD PROCESSING).

The primary and likely sole pathway of starch biosynthesis is the adenosine diphosphate (ADP) glucose pathway (57). In this pathway the first enzyme, ADPglucose pyrophosphorylase (ADPGPP), catalyzes the conversion of glucose-1-phosphate to ADPglucose. In plants, it has been proposed that sucrose synthase is involved in the production of the ADPglucose used in starch biosynthesis (58). This model is not considered to be accurate given a number of mutants characterized affecting both starch and sucrose biosynthesis, and this topic has recently been reviewed (59). Another route for starch biosynthesis is through the action of starch phosphorylase. This enzyme is involved in the degradation of starch, forming glucose-1-phosphate from successive removal of glucose units from the polymer. The reaction is reversible *in vitro*; thus this enzyme potentially plays a role in the formation of starch. Through expression of antisense RNA, this enzyme has been eliminated in the amyloplast of potato tubers with no effect on starch content; thus any role in biosynthesis is proposed to be very minor (60).

Enzymes Involved in Starch Biosynthesis. Much of the early data dealing with starch biosynthesis in plants are derived from the study of various mutants. The shrunken-2 and brittle-2 mutants of maize have greatly reduced levels of ADPGPP activity owing to the absence of one of the two subunits of this enzyme, and result in a shrunken seed appearance. Mendel's early work on inheritance of traits was performed with a pea mutant deficient in branching enzyme activity (61). Mutations in plants affecting starch biosynthesis can have severe results to plant morphology and viability.

ADP Glucose Pyrophosphorylase. The rate-limiting reaction in both bacterial glycogen and plant starch biosynthesis is the first step, catalyzed by the enzyme ADPGPP. In bacteria the enzyme functions as a homotetramer subject to tight allosteric regulation by effector molecules that reflect the energy state of the cell, and is the only enzyme in the pathway of glycogen biosynthesis subject to such regulation. The enzyme is activated by glycolytic intermediates and inhibited by adenosine monophosphate (AMP), ADP, and/or inorganic phosphate (Pi). Fructose 1,6-bisphosphate is typically the primary activator and AMP the primary inhibitor (57,62,63). The role of the activator is to increase the affinity of the enzyme for its substrates, adenosine triphosphate (ATP) and glucose-1-phosphate, and increasing amounts of the activator relieves inhibition caused by AMP, ADP, or Pi. The allosteric regulation of this enzyme has been shown to regulate the flux of carbon through this pathway and control the level of glycogen that is produced. Much of this work has been performed with mutants of *E. coli* and *S. typhimurium* affected in their ability to accumulate glycogen.

The bacterial ADPGPP enzymes each have subunits that contain allosteric activator and inhibitor binding regions, substrate binding sites, and a site for binding Mg^{2+}. A series of chemical modification experiments lead to the elucidation of amino acid residues responsible for interacting with the various effector and substrate molecules (64,65). The ADPGPP enzymes in plants function as heterotetramers consisting of two distinct subunits encoded by two different genes (57). These subunits differ in molecular weight, amino acid composition and sequence, and antigenic properties. Antibodies made against the large subunit only weakly react with the small subunit from a given plant (and vice-versa); but antibodies against the large (or small) subunit recognize the corresponding subunit from different plant species (66), ie, certain sequences are conserved between widely devergent plant species. As in bacterial glycogen biosynthesis, ADPGPP catalyzes the rate-limiting step in starch biosynthesis. The levels of control are primarily via allosteric regulation, but regulation of gene expression also plays a role in controlling ADPGPP activity. The primary effector molecules differ from those in bacteria. For every plant system studied, the plant enzymes are activated by 3-phosphoglycerate (3-PGA) and inhibited by inorganic phosphate (Pi). One possible exception is the wheat endosperm enzyme which appears not to be activated by 3-PGA (67). However, this enzyme is inhibited by Pi, and the presence of 3-PGA overcomes the inhibition.

The importance of allosteric regulation to *in vivo* ADPGPP activity and starch content in plants has been demonstrated (68). The gene encoding the ADPGPP enzyme from *E. coli* strain 618, which is relatively insensitive to allosteric control, was isolated and inserted into transgenic potato plants via *Agrobacterium tumefaciens* transformation. The gene was designed to express the active protein only in the potato tuber, and such expression resulted in a 25–50% increase in starch content. In contrast, expression of the ADPGPP gene from a wild-type *E. coli* K12 strain, which encodes an enzyme subject to normal allosteric regulation, had little effect on starch content. These results showed the importance of allosteric control to ADPGPP activity, and circumvention of this control increases the flux of carbon through this pathway and results in an increase in starch biosynthesis and composition.

It is of interest to determine why the plant enzyme is composed of two distinct subunits and the bacterial enzymes only one. Because the enzyme must have binding sites for the allosteric activator and inhibitor, the substrates, and a catalytic site, it is possible that these sites are located on different subunits. The shrunken-2 and brittle-2 mutants of maize endosperm lack the large and small subunits, respectively, of the ADPGPP enzyme. These mutants have 12% and 17% of the wild-type ADPGPP activity and about 25% of wild-type levels of starch (69), demonstrating that both subunits are required for normal levels of enzyme activity and starch content, but that a single subunit by itself can form an active enzyme. This is supported by a starch-deficient mutant of *Arabidopsis* which lacks the large subunit, has about 5% wild-type levels of ADPGPP activity, and about 40% wild-type levels of starch (70). In addition, elimination of one of the ADPGPP subunits in transgenic potato through expression of antisense RNA results in a reduction in ADPGPP activity to 1.5–17% of wild type, and starch content to 4–35% of wild type (71). These results suggest that allosteric, substrate, and catalytic sites reside on each of the subunit types.

ADPGPP genes in plants are also controlled at the level of gene expression. In potato, the transcripts corresponding to the large and small subunits differ in their accumulation profiles in different organs (72). The steady-state levels of transcripts corresponding to the large subunit of ADPGPP are highest in tubers and stolons and are inducible by sucrose. In contrast, the steady-state levels of transcripts corresponding to the small subunit of ADPGPP are relatively equivalent in tubers, stolons, and aerial portions of the plant and are not strongly influenced by carbohydrates. Why the gene encoding the large subunit of ADPGPP is more tightly regulated than that encoding the small subunit is unknown.

Starch Synthase. In contrast to the bacterial systems where a single synthase is responsible for the elongation of the glucose chain, in plants several synthases are involved in building the starch granule. These synthases are either soluble or granule-bound. The soluble synthases are divided into two forms, designated as Type I and Type II, distinguished by size, kinetic properties, and immunological properties (63). These forms are encoded by separate genes which may show tissue and developmental regulation. Given these differences, the two types of enzymes likely play distinct roles in the formation of the starch granule, although this role is thought to be primarily involved in the synthesis of amylopectin, the branched form of starch. The granule-bound starch synthases are immunologically, physically, and kinetically distinct from the soluble synthases, and are encoded by one or more distinct genes. In maize endosperm, two forms of granule-bound synthase have been identified, bringing the total number of synthases identified in this tissue up to four. Unlike the situation for the ADPGPP gene in potato, the potato granule-bound starch synthase gene has been shown to be regulated solely at the level of gene expression (73).

The primary role of granule-bound starch synthase may be in the formation of amylose, the linear fraction of starch. Waxy-like mutations which are devoid of amylose and granule-bound starch synthase have been characterized in a number of plant systems, including maize, rice, barley sorghum, and potato (74). The waxy mutation was obtained in transgenic potato through expression of antisense RNA to granule-bound starch synthase (75) providing strong evidence that the waxy locus encodes the granule-bound starch synthase enzyme, and that this enzyme is responsible for the synthesis of amylose *in vivo*.

Branching Enzyme. Multiple forms of branching enzyme have been found in a number of plant species. These enzymes are all soluble and catalyze essentially the same reaction, but differ in physical, immunological, and kinetic properties, and like the synthases probably play different functional roles in the synthesis of the starch molecule (57). Branching enzymes are also encoded by multigene families which may show developmental and tissue-specific expression profiles (76,77). The most detailed studies involve the isoforms from maize endosperm (78), where three different forms of branching enzyme have been purified and designated BEI, BEIIa, and BEIIb. Polyclonal antibodies against BEI do not react against either form of BEII, and vice-versa, but forms BEIIa and IIb appear to be closely related. Monoclonal antibodies have been produced which react with all three isoforms, showing that the enzymes share a few common epitopes but are otherwise divergent (79). Each endosperm-branching enzyme has been highly purified and the branching characteristics studied (80). BEI was found to have high activity on amylose but little on amylopectin, and was found to preferentially transfer long chains. These chains would represent the B chains in the cluster model proposed for the structure of amylopectin (81). BEIIa and IIb were found to have low activity on amylose and high activity on amylopectin, and transferred preferentially short, or A chains. Differences between these two isoforms in the types of branches produced were not noted, and these enzymes appear to be very similar (80).

Branching enzymes have been characterized from a variety of other plant tissues. Only a single isoform has been detected in potato tubers. The gene for potato branching enzyme is regulated in a manner similar to the potato large subunit ADPGPP gene and is expressed most abundantly in the potato tuber (82). Antisense RNA expression in transgenic potatoes has resulted in a 90% decrease in branching enzyme activity, but with no discernable effect on starch content or structure (60). This implies that either branching enzyme activity is present in vast excess, or a second enzyme indeed exists. The former seems to be the case. In pea, the wrinkled seed phenotype has been linked to the *r* locus and results in a 66–75% reduction in total starch, and an increase in amylose from 33% in wild-type pea up to 60–70% in the mutant (83). Branching enzyme activity is reduced to 14% of wild-type levels because of the complete absence of one isoform of branching enzyme. The decrease in total starch levels is caused by a similar mechanism as in bacteria lacking branching enzyme activity, ie, as the glucose chain is elongated, it becomes a poorer substrate for the synthase enzyme.

One function of branching enzymes is to clip the elongating chain and provide additional substrate to the synthase enzymes. In this model, the synthase and branching enzyme work in concert, whereas the synthase elongates the chain, the branching enzyme cleaves, transfers a maltodextrin, and forms a new branch, which is then further elongated by the synthase. The dependence of starch synthase on branching enzyme has been shown in *in vitro* systems where the activity of starch synthase is observed to be greatly enhanced by the addition of branching enzyme (84). This model of concerted activity also provides the rationale for the existence of multiple isoforms of starch synthase and branching enzyme in plants. Amylopectin is an asymmetric molecule formed of both short (12–42 residues) and long (>49 residues) glucose chains (85). Synthesis of such an asymmetric structure requires starch synthases and branching enzymes having different specificities for elongation and for insertion of branch points at different distances along A- and B-chains. Further evidence for this comes from the study of a low starch mutant of *Chlamydomonas* that lacks soluble starch synthase II and shows a decrease in intermediate length chains in the amylopectin fraction (86). Thus the structure of the starch granule can be influenced by the properties of both starch synthases and branching enzymes, and further controlled by regulation of gene expression in different tissues or during plant organ development.

BIBLIOGRAPHY

"Genetic Engineering" in *ECT* 3rd ed., Vol. 11, pp. 730–745, by A.M. Chakrabarty, University of Illinois at the Medical Center, Chicago; in 3rd ed., Suppl. Vol., pp. 495–513, by E. Jaworski and D. Tiemeier, Monsanto Co.

1. K. Baron and co-workers, *Cell* **32**, 1033 (1983).
2. R. B. Horsch and co-workers, *Science* **223**, 496 (1984).
3. R. D. Shillito and co-workers, *Bio/Technology* **3**, 1099 (1985).
4. T. M. Klein and co-workers, *Nature* **327**, 70 (1987).
5. P. Christou *Plant J.* **2**, 257 (1992).

6. P. Christou *Curr Opin. Biotechnol.* **4**, 135 (1993).
7. T. M. Klein and co-workers, *Biotechnology* **10**, 286 (1992).
8. K. D'Halluin and co-workers, *Plant Cell* **4**, 1495 (1992).
9. M. Wallroth and co-workers, *Mol. Gen. Genet.* **202**, 6 (1986).
10. P. Maliga, *Trends Biotechnol.* **11**, 101, (1993).
11. R. M. Morrish and M. E. Fromm, *Curr. Opin. Biotechnol.* **3**, 141 (1992).
12. J. H. Oard, *Biotechnol. Adv.* **9**, 1 (1991).
13. P. Benfey, L. Ren, and N.-H. Chua, *EMBO J.* **8**, 2195 (1989).
14. P. Benfey, L. Ren, and N.-H. Chua, *EMBO J.* **9**, 1677 (1990).
15. R. Kay and co-workers, *Science* **236**, 1299 (1987).
16. J. W. Edwards and G. M. Coruzzi, *Annu. Rev. Genet.* **24**, 275 (1990).
17. M. Rocha-Sosa and co-workers, *EMBO J.* **8**, 23 (1989).
18. H. C. Wenzler and co-workers, *Plant Mol. Biol.* **12**, 41 (1989).
19. D. Twell and G. Ooms, *Plant Mol. Biol.* **9**, 365 (1987).
20. M. Koster-Topfer and co-workers, *Mol. Gen. Genet.* **219,** 390 (1989).
21. S. Naito, P. H. Dube, and R. N. Beachy, *Plant Mol. Biol.* **11**, 109 (1988).
22. S. E. Radke and co-workers, *Theor. Appl. Genet.* **75**, 685 (1988).
23. Y. T. Yamamoto and co-workers, *Plant Cell* **3**, 371 (1991).
24. F. J. Perlak and co-workers, *Biotechnology* **8**, 939 (1990).
25. R. T. Fraley, *Biotechnology* **10**, 40 (1992).
26. M. G. Koziel and co-workers, *Biotechnology* **11**, 194 (1993).
27. F. J. Perlak and D. A. Fischhoff, *Advances in Engineered Pesticides*, Marcel Dekker Inc., New York, 1993, pp. 199–211.
28. K. A. Barton and M. J. Miller, in Kung and Wu, eds., *Transgenic Plants*, Vol. 1, Academic Press, Inc., New York, 1993, pp. 297–315.
29. W. K. Kaniewski and co-workers, *Biotechnology* **8**, 750, (1990).
30. H. J. Klee, *Plant Physiol.* **102**, 911 (1993); A. J. Hamilton, G. W. Lycett, and D. Grierson, *Nature* **346**, 284 (1990).
31. P. W. Oeller and co-workers, *Science* **254**, 437 (1991); H. J. Klee and co-workers, *Plant Cell* **3**, 1187 (1991).
32. C. J. S. Smith, A. J. Hamilton, and D. Grierson, D. P. S. Verma, ed., in *Control of Plant Gene Expression*, 1993, pp. 535–546.
33. R. E. Sheehy, M. Kramer, and W. R. Hiatt, *Proc. Natn. Acad. Sci. U.S.A.* **85** 8805 (1988).
34. D. H. Tieman and co-workers, *Plant Cell* **4**, 667 (1992).
35. L. Comai and co-workers, *Science* **221**, 370 (1983).
36. G. M. Kishore and D. M. Shah, *Ann. Rev. Biochem.* **57**, 627 (1988).
37. G. F. Barry and co-workers, *Biosynthesis and Molecular Regulation of Amino Acid Biosynthesis in Plants*, Vol. 7, American Society of Plant Physiologists, 1992, pp. 139–145.
38. M. A. W. Hinchee and co-workers, *Transgenic Plants* **1**, 243 (1993).
39. S. R. Padgette and co-workers, *Arch Biochem. Biophys.* **258**, 564 (1987).
40. H. Sollandev-czytco, I. Sommer, and N. Amrhein, *Plant Mol. Biol.* **20**, 1029 (1992)
41. H. C. Steinrucken and co-workers, *Arch. Biochem. Biophys.* **244**, 169 (1986).
42. R. M. Hauptmann and co-workers, *Mol. Gen. Genet.* **211**, 357 (1988).
43. G. della-Cioppa and co-workers, *Proc. Natl. Acad. Science U.S.A.* **83**, 6873 (1986).
44. D. M. Shah and co-workers, *Science* **233**, 478 (1986).
45. D. M. Stalker, W. R. Hiatt, and L. Comai, *J. Biol. Chem.* **260**, 4724 (1985).
46. L. Comai and co-workers, *Nature* **317**, 741 (1985).
47. G. M. Kishore and co-workers, *Fed. Proc. Am. Soc. Exp. Biol.* **45**, 1506 (1986).
48. J. Jost and N. Amrhein, *Arch. Biochem. Biophys.* **282**, 433 (1990).
49. G. della-Cioppa and co-workers, *Bio/Technology* **5**, 579 (1987).

50. S. R. Padgette and co-workers, *J. Biol. Chem.* **266**, 22364 (1991).
51. W. C. Stallings and co-workers, *Proc. Natl. Acad. Sciences U.S.A* **88**, 5046 (1991).
52. Q. K. Huynh, G. M. Kishore, and G. S. Bird, *J. Biol. Chem.* **263**, 735 (1988).
53. Q. K. Huynh and co-workers, *J. Biol. Chem.* **263**, 11636 (1988).
54. S. R. Padgette and co-workers, *Arch. Biochem, Biophys.* **266**, 254 (1988).
55. Q. K. Huynh, *J. Biol. Chem.* **263**, 11631 (1988).
56. S. R. Padgette and co-workers, *J. Biol. Chem* **263**, 1798 (1988).
57. J. Preiss, *Oxford Survey of Plant Molecular and Cellular Biology*, Vol. 7, Oxford University Press, U.K., 1991, pp. 59–114.
58. J. Pozueta-Romero and co-workers, *Proc. Natl. Acad. Sci. U.S.A.* **88**, 5769 (1991).
59. T. W. Okita, *Plant Physiol.* **100**, 560 (1992).
60. J. Kobmann and co-workers, paper presented at the *Third International Symposium on the Molecular Biology of the Potato*, July 25–30, 1993, Santa Cruz, Calif.
61. M. K. Bhattacharyya and co-workers, *Cell* **60**, 115 (1990).
62. J. Preiss and T. Romeo, *Advances in Microbial Physiology*, Vol. 30, Academic Press, Inc., New York, 1989, pp. 183–238.
63. J. Preiss, *The Biochemistry of Plants*, Academic Press, Inc., Orlando, Fla., 1988, pp. 184–249.
64. T. F. Parsons and J. Preiss, *J. Biol. Chem.* **253**, 6197 (1978).
65. *Ibid.*, p. 7638.
66. J. Preiss and co-workers, *Plant Physiol.* **92**, 881 (1990).
67. M. R. Olive, R. J. Ellis, and W. W. Schuch, *Plant Mol. Biol.* **12**, 525 (1989).
68. D. M. Stark and co-workers, *Science* **258**, 287 (1992).
69. B. D. Dickinson and J. Preiss, *Plant Physiol.* **44**, 1058 (1969).
70. T. P. Lin and co-workers, *Plant Physiol.* **88**, 1175 (1988).
71. B. Muller-Rober, U. Sonnewald, and L. Willmitzer, *EMBO J.* **11**, 1229 (1992).
72. B. T. Muller and co-workers, *Mol. Gen. Genet.* **224**, 136 (1990).
73. R. G. F. Visser and co-workers, *Plant Mol. Biol.* **17**, 691 (1991).
74. O. E. Nelson and H. W. Rines, *Biochem. Biophys. Res. Commun.* **9**, 297 (1962).
75. R. G. F. Visser and co-workers, *Mol. Gen. Genet.* **225**, 289 (1991).
76. M. Bhattacharyya and co-workers, *Plant Mol. Biol.* **22**, 525 (1993).
77. T. Kawasaki and co-workers, *Mol. Gen. Genet.* **237**, 10 (1993).
78. B. K. Singh and J. Preiss, *Plant Physiol.* **79**, 34 (1985)
79. C. Takeda, Y. Takeda, and S. Hizukuri, *Cereal Chem.* **66**, 22 (1989).
80. H. P. Guan and J. Preiss, *Plant Physiol.* **102**, 1269 (1993).
81. S. Hizukuri, *Carbohydr. Res.* **147**, 342 (1986).
82. J. Kobmann and co-workers, *Mol. Gen. Genet.* **230**, 30 (1991).
83. J. Edwards, J. H. Green, and T. ap Rees, *Phytochemistry*, **27**, 1615 (1988).
84. C. D. Boyer and J. Preiss, *Plant Physiol.* **64**, 1039 (1979).
85. Z. Gunja-Smith and co-workers, *FEBS Lett.* **12**, 101 (1970).
86. T. Fontaine and co-workers, *J. Biol. Chem.* **268**, 16223 (1993).

JANICE W. EDWARDS
GANESH M. KISHORE
DAVID M. STARK
Monsanto Company

GEOTEXTILES

The American Society for Testing and Materials Committee D35 on Geosynthetics defines a geosynthetic as a planar product, manufactured from polymer material, used with soil, rock, earth, or other geotechnical engineering-related material, as an integral part of a "man-made" project, structure, or system. A geotextile, a subset of geosynthetics, is defined as any permeable textile material used with foundation soil, rock, earth, or any other geotechnical engineering-related material, as an integral part of a "man-made" project, structure, or system.

There are several different processes and polymers used in the manufacture of geotextiles which affect their appearance and physical properties. Geotextiles are produced in various weights and thicknesses, which also determine their physical properties and ultimately the performance of the material when installed on a project (see also HIGH PERFORMANCE FIBERS).

When first used in civil engineering projects, the materials now known as geotextiles were called filter fabrics, civil engineering fabrics, or construction fabrics. Their primary use in the early- to mid-1960s was as erosion control materials, and as an alternative for granular soil filters. Since the late 1970s, the use of these materials has expanded almost limitlessly. The varied uses of geotextiles include drainage, dissimilar materials separation, erosion control, environmental protection, highway pavement rehabilitation, and as component materials of geocomposites used in a variety of applications. Within each of these categories of use there exist numerous types of installations using the geotextiles.

Manufacturing Process

Polymers used in the manufacturing process in decreasing order of use include polypropylene, polyester, polyamide (nylon), polyethylene, and others to a much lesser extent (1). The fibers used in manufacturing the geotextiles are made by melt-spinning, ie, melting the polymer material and forcing it through a spinneret. Hardening of the fiber filaments is accomplished for the geotextile materials mostly through a process of simultaneous stretching and cooling of the fibers. The stretching of the fibers produces a more orderly arrangement of the molecules in the fibers, resulting in increased strength of the fiber (see FIBERS, SURVEY). Monofilament fibers are further processed to form the various types of yarns used in the manufacturing of geotextiles. Multifilament yarn is formed by several monofilaments being twisted together to form the yarn. Staple fibers are formed from a rope-like bundle of monofilaments being crimped and cut into 2.5- to 10-cm (1- to 4-in.) lengths. Staple yarns are formed from these staple fibers by twisting into longer fibers (yarns). Another type of material used in geotextiles is called slit film. A continuous flat sheet of the polymer is cut into fibers by slitting with knives, or through the use of air jets. The resulting ribbons are slit-film fibers. Figure 1 shows the types of fibrous components in geotextiles.

Geotextiles may be woven, nonwoven, or knitted. All types, woven, nonwoven, or knitted, are susceptible to degradation owing to the effects of ultraviolet light and water. Thus stabilizing agents are added to the base polymeric material to lessen the effects of exposure to ultraviolet light and water.

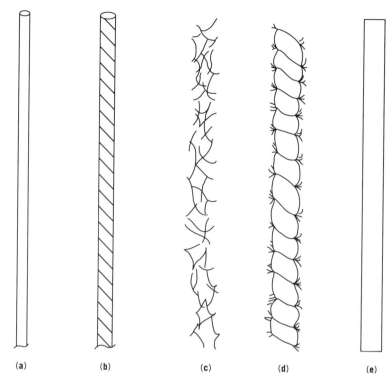

Fig. 1. Schematic drawings of five types of geotextile fibers: (**a**) monofilament, (**b**) multifilament, (**c**) staple fibers, (**d**) staple yarn, and (**e**) slit film.

Woven and nonwoven geotextiles are the most common types presently in use in geotechnical engineering. A woven geotextile is formed through the conventional textile weaving process resulting in a screen-like or mesh material with various sizes of mesh openings, depending on the tightness of the weave.

Nonwoven Geotextiles.   There are two basic manufacturing processes by which a nonwoven geotextile is produced: a one-stage or continuous process where the fibers are spun and bonded in one continuous process, and a two-stage process where fibers are laid down and then bonded (see NONWOVEN FABRICS). The most common types of bonding processes are the spun-bonded process, melt- or heat-bonded, resin-bonded, and needle-punched (Fig. 2). Each of the processes results in a geotextile that may have vastly different characteristics than one formed otherwise. Depending on the end use of the material, this may affect the choice of the geotextile to be used on a specific project.

Spun-Bonded Fabrics.   In this continuous process, one or more polymers are fed into an extruder, through the spinneret onto a continuous moving conveyor. As a result, a continuous web of material is formed. As the polymer(s) are being placed on the conveyor, fiber orientation is achieved by spinneret rotation, varying conveyor speed, or other actions. Bonding of the fabric is achieved through thermal, mechanical, or chemical treatment. Some of the notable characteristics of a geotextile produced by this means include high performance, low weight materials

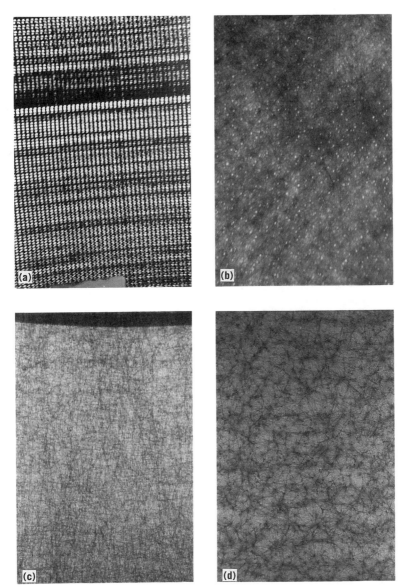

Fig. 2. Photomicrographs of geotextiles made by various methods: (**a**) woven geotextile, (**b**) needle-punched, (**c**) heat-bonded, and (**d**) resin-bonded.

as a result of the continuous character of the fibers. Product weights generally range between 10 and 200 g/m^2 (0.3 to 6 oz/yd^2). Product thicknesses range between 0.08 and 0.64 mm (3 to 25 mils).

Melt- or Heat-Bonded Fabrics.   In the two-stage melt or heat-bonding process, the fibers, which may be continuous or long staple fibers, are melted together at fiber crossover points. Bonding is achieved by calendering, ie, passing the geotextile between two rotating hot rollers, melting the fibers together at the crossover points. Resulting fabric characteristics depend on whether one or two different polymers are used for the fibers and on the intensity of the heat and roller pressure used in melting the fibers together. Fabric weights generally range between 70 and 400 g/m^2 (2 to 12 oz/yd^2).

Resin-Bonded Fabrics.   To achieve this type of bonding, an acrylic resin is either sprayed onto or impregnated into a fibrous web. The resulting structure is allowed to cure or is run through the calendering process, which develops strong bonds between the fibers. Often the fabric undergoes a forced-air drying process which redevelops the open pores of the structure. This type of bonding may not be used as frequently today as it was in the early years of geotextiles.

Needle-Punched Fabrics.   In the two-stage needle-punching process, a fibrous web is placed into a machine which is equipped with a series of specially designed reversed-barbed needles. The web of material is passed between two plates. As the fabric passes between the plates, the needles punch down through the top plate and fabric, reorienting the fibers and resulting in entanglement of the fibers. Fabric weights from this process generally range between 60 and 700 g/m^2 (1.7 to 20 oz/yd^2). Fabric thickness ranges between 0.4 and 5 mm (15 to 200 mils).

Physical Properties

The polymer used, the manufacturing process, and additives used in producing the geotextile determine the appearance and physical properties of the product. Properties may be classified as index properties or design and performance properties. Index properties provide a means of material differentiation, quality control in the manufacturing process, and quality assurance for the specifying agency. They generally are determined by relatively simple and quick physical tests that are not performed under the conditions in which the geotextile will be used. That is, they are tested in isolation, without the soil or material in which they are to be placed. Index properties include certain types of strength tests, mass per unit area, thickness, apparent opening size (pore openings), permittivity (volumetric flow rate), and temperature stability.

Design or performance properties define how the geotextile performs under specific installation conditions. Ideally testing is performed under the anticipated installation conditions, but it should not be expected that the manufacturer perform these tests. It is virtually impossible to duplicate every condition under which a geotextile will be placed. The design engineer should determine the specific conditions under which the geotextile will be installed and expected to perform, and select appropriate design and performance tests. There are few standard design and performance tests available. Generally an attempt to interpolate performance requirements based on index properties is made which is very difficult to do and is sometimes misleading. Some design or performance tests that do exist include wide-width strength, filtration efficiency, soil-clogging potential, and biological clogging.

Applications

Fabrics perform one or more function in each installation; generally there is one primary function. The five basic primary functions have been identified as separation, stabilization, reinforcement, filtration, and drainage. When the geotextile is impregnated with an impermeable material such as an asphaltic emulsion, it may function as a moisture barrier. Within these basic functions, there are over one hundred different application areas (1–4).

Separation. In this function, the geotextile serves to separate two dissimilar materials (Fig. 3), eg, two different soils, landfill material and the native soil, stone material and subgrade soil, old and new pavement, foundation soils and various types of walls, or one of many other similar situations. In some instances, it is difficult to distinguish between the separation and stabilization functions because in both situations the geotextile is serving as a separator. However, in stabilization some additional phenomena occur.

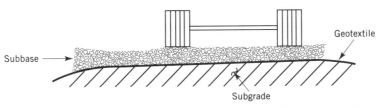

Fig. 3. Separation of subbase from subgrade soil by a geotextile.

In separation, the primary function of the geotextile is to prevent intrusion of one material into another in order to prevent contamination of either material by the other. In the case of an aggregate being placed over a firm foundation soil, the purpose is to maintain the drainage integrity of the aggregate; wall application is similar. In landfill application, the purpose is to prevent intrusion of waste material into the leachette collector system.

Even though separation may not be the primary function for which the geotextile is installed, in almost every case the geotextile does perform as a separator. For the separation function, the physical properties of concern are primarily strength-related, including wide width, puncture resistance, and tear resistance.

Stabilization. Figure 4 shows stabilization accompanying separation. In this application, the natural soil on which the geotextile is placed is usually a wet, soft, compressible material, exhibiting very little strength. By acting as a separator, the geotextile allows water from the soft natural soil to pass from this soil into a free-draining construction soil, which in turn allows consolidation of the natural soil to take place. As a result of the consolidation process, there is a strength gain in the natural soil, which then provides an adequate foundation for construction to take place.

The geotextile may also act to bridge over the soft soil in which case drainage of water from the soft, wet soil is not the foremost function served by the geotextile. The geotextile must have sufficient strength to support the construction soil. The area over which the geotextile is spread in this stabilization function is much

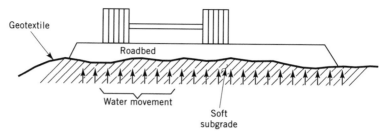

Fig. 4. Stabilization of a roadbed by a geotextile.

smaller than the area when it is used in the first-described stabilization application. For stabilization, strength is again of concern, but drainage characteristics must also be considered.

Reinforcement. The key difference between stabilization and reinforcement is that stabilization is accomplished by providing for drainage of water from the unstable soil, while in reinforcement the strength characteristics (stress–strain) of the geotextile provide added strength to the whole system. Another difference is that in stabilization the geotextile is placed on or around the area being stabilized and thereby also acts as a separator, whereas in the reinforcement application the geotextile is placed within the material being reinforced (Fig. 5). This is in line with reinforcement concepts in concrete and other materials.

In order for it to perform the reinforcement function, the geotextile must be allowed to deform to develop its strength. When stabilization of a site occurs, there is consolidation of the soil, and with this comes deformation of the geotextile. Due to the deformation of the geotextile, strength is required to ensure that a site failure does not occur, ie, there can be a reinforcement component in the stabilization process.

Areas where geotextiles are used as reinforcement include embankment construction, reinforced soil wall construction, and slope improvement. In reinforcement, the physical properties of importance are primarily related to strength, that is, a combination of the stress–strain characteristics of the material.

Filtration. Here the prime function is to retain soil or other fine materials, while allowing water to pass through. Again, it is seen that more than one function

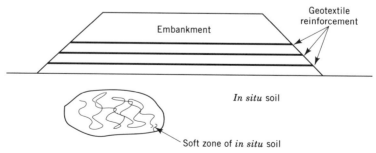

Fig. 5. Reinforced embankment.

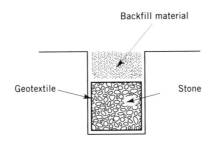

Fig. 6. Use of a geotextile in filtration.

is being performed (Fig. 6). If there were no drainage of water taking place, movement, and therefore retention of the soil, would not be of concern.

Part of the mechanism by which filtration occurs is through the development of a soil filter behind the geotextile. As the water passes through, soil is filtered out and collects behind the geotextile. As buildup takes place, a natural soil filter is developed. If the geotextile is improperly designed for the site, plugging or clogging of the geotextile will take place as a result of this buildup. If the geotextile is plugged or clogged to the point where no water is able to pass through it, excess pore pressure may develop and eventually lead to a failure of the site. It is therefore critical to design an installation so that there is a balance between the soil-retention characteristics and the drainage characteristics of the geotextile–soil system.

Filtration installations include wrapping the trench of a pavement-edge drain system to prevent contamination of the underdrain; placement behind retaining walls and bridge abutments to prevent contamination of the sand blanket placed against the structure to allow dissipation of pore pressures in order to avoid failure of the structure; as silt fences to allow surface runoff from a site while retaining the soil suspended in the runoff; and on earth slopes beneath larger stone or other overlay materials to prevent erosion of the slope as water escapes from the interior of the slope.

Drainage. In the previous sections, drainage was discussed as taking place in a direction perpendicular to the plane of the geotextile. Here, drainage parallel to the plane of the geotextile is described (Fig. 7). The property called transmissivity is defined as flow parallel to the plane of the geotextile. This type of flow can occur to some extent in all geotextiles, but is best achieved in needle-punched nonwoven materials. This class of geotextiles can be manufactured in a range of thicknesses such that this characteristic is optimized.

The drainage that is accomplished in this function takes place by one of two mechanisms: gravity or pressure flow. In order for gravity drainage to take place, the geotextile is installed at some incline from the horizontal, up to the point of being vertical. Installations using gravity drainage include chimney drains within earth dams, pore-pressure dissipators behind retaining walls, and for transport of water or air beneath a geomembrane. In pressure flow, fluid is flowing from a point of high pressure to a point of low pressure and the orientation of the geotextile is not critical. It can be placed horizontal, or at a specified incline from the

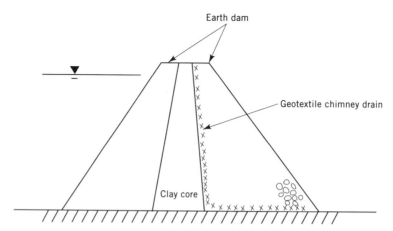

Fig. 7. Drainage parallel to the plane of a geotextile.

horizontal. This type of drainage occurs when the geotextile is placed as a vertical drain to increase the rate of consolidation of a soil, placed within reinforced earth walls, in earth embankments or dams, and beneath surcharge fills.

The physical property of primary concern in this application is the transmissivity and, depending on the type of installation, the clogging potential and strength properties.

Moisture Barrier. When impregnated with an asphaltic emulsion, geotextiles become impermeable and can then be used as moisture barriers. The primary application for this type of geotextile is in pavement rehabilitation (Fig. 8).

One of the primary factors in the deterioration of a pavement structure is the intrusion of surface water into the support structure of the pavement. When rehabilitating a pavement, the installation of a moisture barrier between the old, existing pavement surface and the new overlain surface acts to retard moisture intrusion, thus prolonging the life of the overlay.

In this application, the primary property concern is asphalt absorption and retention. If the geotextile has poor retention capabilities, then the necessary waterproofing cannot be achieved.

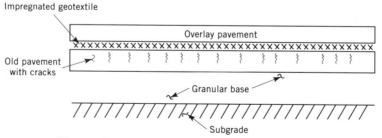

Fig. 8. Use of a geotextile as a moisture barrier.

Economic Aspects

Geotextiles are a relatively new concept for solving problems in geotechnical engineering. They have gained wide use as not only an economical solution to these problems, but in many instances as the only viable solution to a complex engineering problem. This is evidenced by the fact that over a seven-year period from 1976 to 1983 sales of geotextiles in North America alone rose from 5 to 115 m^2 (6 to 138 $\times$ $10^6/yd^2$).

The installed price of geotextiles ranges from less than $1.20/m^2$ to over $12/m^2$ ($1–10/yd^2$), depending on the geotextile and geographic location. The savings compared to what in the past were considered to be conventional solutions to geotechnical problems can range anywhere from a few thousand dollars per project to well over a million dollars, depending on the magnitude of the project.

Use of geotextiles should follow only after careful evaluation of the site conditions and the problems to be solved. In some sites almost any type of geotextile could be installed, survive the installation procedures, and perform as desired. However, there are sites that require a very detailed subsurface exploration program, appropriate soils testing to determine the soil conditions, followed by a detailed design procedure to ensure that the proper geotextile is properly installed. An improperly designed and installed geotextile can result in the desired improvement not being achieved, or in a failure of the site. In some cases a geotextile may not be the appropriate solution.

BIBLIOGRAPHY

1. R. M. Koerner, *Designing with Geosynthetics*, Prentice-Hall, Inc., Englewood Cliffs, N.J., 1986.
2. B. R. Christopher and R. D. Holtz, *Geotextile Design and Construction Guidelines*, Participant Notebook, rev. ed., National Highway Institute, Washington, D.C., 1992.
3. B. R. Christopher and R. D. Holtz, *Geotextile Engineering Manual*, Course Text, National Highway Institute, Washington, D.C., 1984.
4. *A Design Primer: Geotextiles and Related Materials*, Industrial Fabric Association International, St. Paul, Minn., 1990.

L. DAVID SUITS
New York State Department of Transportation

GEOTHERMAL ENERGY

Heat emanating from within the earth is one source of geothermal energy. This vast repository of energy is generated from the decay of natural radioisotopes (qv) and heat from the molten core of the earth. Energy from the core is transported to the earth's mantle by conduction, an extremely slow process because crystalline rock has very low thermal conductivity. Active volcanos provide evidence of the enormous amount of energy present deep within the earth. Geysers, fumaroles, and hot springs, more benign than volcanos, also demonstrate geothermal energy brought to the surface.

In certain circumstances, such as volcanic activity and mountain building, large masses of molten rock may intrude into the mantle and even erupt through the surface. Where these conditions prevail, high geothermal energy levels can be found at relatively shallow depths. The rate at which the temperature of the earth increases with depth, the geothermal gradient, is not uniform. The average value of the geothermal gradient worldwide is 25–30°C/km, but it may be as much as 10 times greater in geologically active regions (1). In addition, the temperature of the earth does not always increase uniformly with depth. Factors such as rock type, porosity, fluid content, and the presence of aquifers may affect the local geothermal gradient.

Natural sources of geothermal fluids for heating and bathing have been utilized since prehistoric times. In the 1800s and 1900s applications of hydrothermal resources expanded widely to include space and district heating, agriculture, aquaculture, industrial processing, and, most recently, electric power generation (qv). Historically this energy was utilized by diverting surface hot water or steam sources. As technology progressed, wells were drilled to tap geothermal fluids more efficiently, and improvements in drilling technology have enabled access to and recovery of deeper and hotter fluids. In addition, development of techniques to extract geothermal heat from rock in which no natural mobile fluids exist is under way to bring this energy to the surface.

The useful applications of hydrothermal resources depend on the temperature of the extracted fluid. Figure 1 shows the distribution of thermal energy use in the United States as a function of temperature (2). It is clear that relatively low temperature fluids can be effectively applied for purposes such as greenhouse heating, fish farming, and especially space heating. Waters at higher temperatures can be used for a variety of industrial processes. All direct uses of geothermal energy require that the point of application be essentially co-located with the source of the hot water. Transportation of hot fluid over more than a few kilometers is impractical. The energy can, however, be converted to electricity. It is then possible to apply the power generated by hydrothermal energy in a variety of ways and at distant locations. The efficiency of electrical generation is directly related to the thermal quality of the resource. Using the most advanced power generating equipment, it may be economical to generate electricity from geothermal waters at temperatures as low as 150°C (3).

To successfully compete with the multitude of energy sources available, geothermal energy must be available and retrievable in both a convenient and an economical manner. As of this writing, these conditions have only been met using

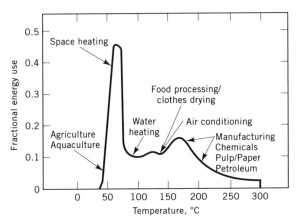

Fig. 1. Thermal energy use vs temperature (2). Electricity generation is practical from thermal energy sources hotter than 150°C.

high grade geothermal resources in the form of hot water and steam and these particular hydrothermal resources are limited. Most of the world's accessible geothermal energy is found in rock that is hot but dry. Although research and development have demonstrated success in extracting thermal energy on a limited basis, the vast hot dry rock resource has not yet been shown to be an economically feasible source of energy on a scale large enough for practical use.

Geothermal Energy Resources

Type. Figure 2 is a generalized view of a cross section of the earth indicating the various types of geothermal resources. Typically, the subsurface of the earth is rock that is dry, except for small amounts of immobile pore fluids. The earth becomes progressively hotter with depth constituting the ubiquitous hot dry rock (HDR) resource. In a few places, water has penetrated into faults, fissures, or porous regions of the hot rock. The water can be trapped there in the form of hot liquid under pressure or, much more infrequently, as steam (qv). In this manner, a reservoir of hot fluid is formed creating a hydrothermal resource. In some places these hydrothermal resources are visible in the form of hot springs, geysers, fumaroles, or similar geothermal features. Most often hydrothermal resources have no surface manifestations. A few of these hidden hydrothermal reservoirs have been discovered accidentally during drilling for oil, gas, or water wells. Oil and gas resources are widespread throughout the world (see FUEL RESOURCES). From the standpoint of geothermal energy technology, the only important combined geothermal and fossil energy resource type consists of hot water reservoirs which contain significant amounts of methane under pressure. This combination is known as the geopressured resource (Fig. 2).

Within the mantle of the earth, molten rock known as magma is found. Generally magma resources exist many kilometers below the earth's surface, far too deep to be accessed technologically. In a few places, however, magma bodies come

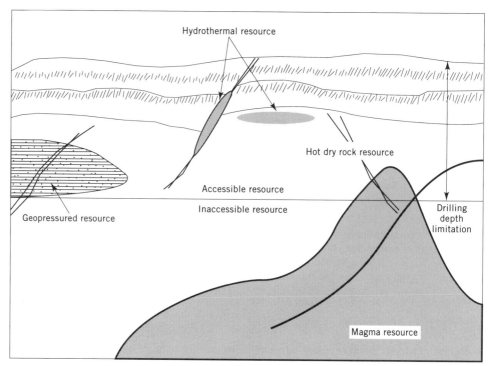

Fig. 2. Types of geothermal resources. Only the geopressured resource is typically found in sedimentary rock.

close to or actually penetrate the surface of the earth. Volcanic activity is an example. In such instances, it is possible to consider the thermal energy of the relatively shallow magma body as a potentially exploitable geothermal resource.

Magnitude. Whereas the total amount of energy stored within the earth is extremely large, only a very small fraction of that energy is accessible, in part because drilling and energy extraction costs escalate rapidly with depth. Commercial drilling can reach depths of about 9000 m, so all of the thermal energy in the earth to that depth can be considered part of the geothermal resource base. This resource base has been estimated to be on the order of 1×10^7 EJ (1×10^7 quads where 1 quad $= 10^{15}$ BTU) in the United States alone (4). This is equivalent to 3×10^7 m^3 (182×10^6 bbl) of oil. The amount of geothermal energy estimated to exist according to resource type is

Resource type	Accessible energy, EJ
hydrothermal	130,000
geopressured	540,000
hot dry rock	10,000,000
magma	500,000
Total	*11,170,000*

The total world consumption of energy in all forms is only about 300 EJ (300 quads); thus the earth's heat has the potential to supply all energy needs for the foreseeable future (5). Economic considerations, however, may preclude the utilization of all but a small part of this potential resource. Only a miniscule fraction of this energy supply has been tapped.

Hydrothermal Resources

Hydrothermal resources are characterized by the presence of heat relatively close to the earth's surface coincident with a trapped body of water to absorb that heat and provide a mechanism for its transfer to the surface. Hydrothermal resources occur, in general, throughout the world, in regions where continental plates meet, and the upwelling of magma and the earth's crust lead to rock temperatures that are higher than the worldwide average. Within these areas, the places where water is trapped in geologic formations are scattered and difficult to predict. Figure 3a is a map of geothermal power plant locations in the United States (6) where geothermal resources of commercially useful magnitudes have been developed. Most of these areas have been identified by surface exposure of the fluid resource.

Temperatures of hydrothermal reservoirs vary widely, from aquifers that are only slightly warmer than the ambient surface temperature to those that are 300°C and hotter. The lower temperature resources are much more common. The value of a resource for thermal applications increases directly with its temperature, and in regions having hotter water more extensive use of geothermal resources has been implemented. Resources in remote areas often go unused unless hot enough to be employed in generating electricity.

Drilling and Field Development. The techniques for drilling hydrothermal wells have been adapted from those in use in the oil and gas industry (7) (see GAS, NATURAL; PETROLEUM). Rotary drilling rigs are normally employed along with conventional drilling equipment such as steel casing, drilling lubricants, and casing cements. Drilling conditions encountered in geothermal areas are often more severe than those in oil fields, although, in some instances, soft sedimentary rock of the type common in oil and gas basins is encountered. Usually it is necessary to bore through extremely hard metamorphic or igneous rock, resulting in a slower drilling rate. Penetration rates of 5–13 cm/s (10–25 ft/h) are common, but frequently problems such as loss of circulation, caving, twist-off, and high pressure flow zones related to the rock formation cause interruptions. In addition, the temperatures encountered in drilling into hydrothermal reservoirs are usually considerably higher than those for oil and gas well drilling. Thus extra cooling procedures and special lubricant formulations must be employed. Moreover, geothermal drilling is subject to more stringent regulations than oil and gas drilling. The costs of drilling geothermal wells are from 2–4 times greater than those for oil and gas wells.

Hydrothermal drilling fluid, or mud, typically consists of a suspension of colloidal bentonite clay in water. It is circulated through the wellbore to lubricate and cool the drill bit as well as to carry the cuttings to the surface. Because drilling mud is relatively expensive, it is continually treated and reused to minimize the total volume of fluid consumed. Commonly, the mud is stable to about 150°C, but

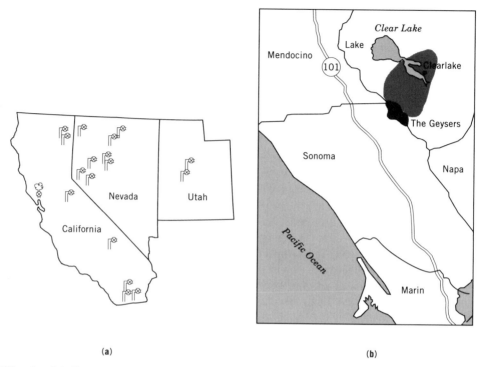

(a)   (b)

Fig. 3. (**a**) General locations of hydrothermal power plants in the continental United States (6). Power is produced directly from hydrothermal steam indicated by the steam plume at The Geysers in northern California. At all other locations, hot water resources are utilized for power production. In 1993, a hydrothermal power plant also came on line on the island of Hawaii. (**b**) Location of The Geysers steam-dominated hydrothermal field (▫) in Lake and Sonoma counties, within the boundaries of the Clearlake–Geysers thermal anomaly (■).

it tends to gel at higher temperatures as the clay particles flocculate. Boiling of the mud as it comes to the surface can also be a problem. The incorporation of air to increase the surface area of the mud (aeration), active cooling at the surface, pressure control so that any flashing to steam can be confined to a specially designed separator, or combinations of these techniques may be used to alleviate problems related to high wellbore temperatures. Sometimes special high temperature sepiolite clays or synthetic polymers may be substituted for bentonite, but these significantly increase the unit cost of drilling muds. In steam-dominated hydrothermal reservoirs, it is desirable to substitute air for the water-based drilling fluid in the region of the steam production in order to prevent the fluid from plugging fractures through which the steam issues. Not only is the air a poorer lubricant, but the cuttings can cause rapid abrasive degradation of drill string components as they are brought to the surface in the high velocity air stream. Therefore, air drilling is avoided except in special circumstances.

A phenomenon known as lost circulation is particularly troublesome in hydrothermal well drilling. Lost circulation occurs when the drill bit penetrates an open fracture in the rock. The drilling fluid then tends to flow into the fracture rather than back to the surface for recirculation. In some cases, lost circulation is self-limiting as the drilling mud solids clog and seal the fracture, but often it can only be stopped by removing the drill string and plugging the leaky zone with cement. Such cementing operations are time-consuming, expensive, and not always successful. In drilling into or through a fluid-producing fracture, the lost circulation problem becomes one of keeping the fracture open rather than sealing it. The primary goal in this situation is to assure that the productivity of the wellbore is not impaired by drilling mud solids that become lodged in the productive conduits to the wellbore. This is accomplished by diluting the circulating fluid to reduce its solids content or by reverting to air-cooling while the drill is passing through productive zones. In instances where test bores have been completed, it may sometimes be possible to predict the general locations of productive fractures. When these are encountered by surprise, however, the appropriate adjustments in the drilling strategy must be implemented rapidly, slowing the progress of the drilling operation and increasing the costs.

Geothermal cements are also employed to fix the steel wellbore casing in place and tie it to the surrounding rock (8). These are prepared as slurries of Portland cement (qv) in water and pumped into place. Additional components such as silica flour, perlite, and bentonite clay are often added to modify the flow properties and stability of the cement, and a retarder is usually added to the mixture to assure that the cement does not set up prematurely. Cements must bond well to both steel and rock, be noncorrosive, and water impermeable after setting. In hydrothermal applications, temperature stability is critical. Temperature cycling of wellbores as a result of an intermittent production schedule can cause rupture of the cement, leading to movement and, ultimately, failure of the wellbore casing.

Logging operations, in which drilling is temporarily suspended while instruments are lowered into the wellbore to make measurements, are very important in geothermal well drilling operations. The temperature, flow rate, and pressure of any fluid located can be determined and used as the basis for further drilling decisions. Hydrothermal drilling is often carried out in rough mountainous areas, and the terrain alone presents special problems in well and field development. Considerable costs can be incurred in preparing flat drilling pads; therefore, several wells may be directionally drilled from a single pad to reach different parts of the hydrothermal resource. Geothermal fluids have a low unit value relative to oil and gas; thus a geothermal well must be operated at a much higher flow rate to be profitable. This means that either the wells must be of greater diameter or flow rates must be considerably higher. Larger diameter wells become ever more expensive to drill, and high flow rates can lead to increased rates of abrasion and more rapid deterioration of piping. In addition, hydrothermal fluids that contain significant amounts of dissolved solids or corrosive gases can rapidly degrade piping. These effects can be mitigated by the use of pipe made from special steels or sometimes by adjusting operating conditions to minimize gas concentrations.

The geothermal drilling industry is much smaller than that of oil and gas drilling and the active geothermal rig count is generally less than 10. Thus, there is not a commercial basis for the development of specialized materials and equipment for geothermal drilling. For a number of years, the U.S. Department of Energy has sponsored the development of high temperature drilling fluids and cements especially designed for geothermal operations (9). Efforts have been concentrated on lightweight, carbon dioxide-resistant cements, thermally conductive and scale-resistant protective liners, improved materials to control lost circulation, and bonding agents.

Direct Uses of Hydrothermal Energy.   Use of low temperature hydrothermal energy for direct thermal applications is widespread (10). The largest volume use of hydrothermal fluid is also one of the simplest. In regions such as some parts of the state of Wyoming, where hydrothermal fluids are found in close proximity to partially depleted oil fields, the hot hydrothermal fluid is pumped down oil wells at the perimeter of the field to heat the remaining oil. The resultant decrease in the viscosity of the remaining oil makes it flow much more readily through the formation and enough added oil can thereby be pumped to the surface to make the process economically viable. In some areas, hydrothermal energy is used to provide central heating for all or part of a community as in Boise, Idaho (11). Hydrothermal energy is also employed to supply process heat for agriculture, primarily to heat greenhouses, and in aquaculture applications which involve warming the water in commercial ponds to enhance the rate of growth of fish.

Water sources for direct thermal uses range in temperature from less than 30°C to over 90°C. Resources in desirable locations can often be reached by simply drilling a few hundred feet into the earth. Hot water cannot be economically transported very far. All direct thermal uses of hydrothermal energy are tied to the quantity and quality of nearby hydrothermal resources.

Electric Power Generation.   Hydrothermal steam and hot water resources having temperatures in excess of about 150°C are generally suitable for the production of electricity (see Fig. 3a). Because electricity is easy to market and transport, it is the only product of hydrothermal energy which permits the resource to be utilized at some distance from its actual location.

Hydrothermal Steam: The Geysers.   In a few cases, the hydrothermal resource exists in the form of pressurized dry steam (qv) rather than hot water. Only a few significant hydrothermal steam fields are known to exist (12), ie, the Salton Sea area of California. The first commercial production of electricity from hydrothermal energy occurred in 1927 using steam from a large field at Larderello in northern Italy. Commercial steam reservoirs have also been developed in Japan and Indonesia. In 1960 electricity production from hydrothermal energy was begun in the United States at The Geysers steam field in northern California. In 1990, The Geysers was responsible for well over 50% of geothermal electricity generation in the United States.

The Geysers.   The Geysers steamfield, located about 120 km (75 mi) northwest of San Francisco, is about 50 km² (20 mi²) in extent (Fig. 3b). It is in an extremely rugged portion of the coastal range. The altitude varies from about 300 m to over 1200 m and the terrain has northwest-to-southeast trending ridges enclosing rather steep valleys. Precipitation and temperatures vary widely. This geographical setting presents special problems for geothermal development. Drilling pads must often be bulldozed flat and landslides during the rainy season can destroy wellheads and piping if proper precautions are not taken.

The Geysers is at the southern end of the Geysers-Clearlake thermal anomaly, a region of some 700 km² (270 mi²) which exhibits an extremely high heat flow (13). Numerous hot springs, mud pots, and fumaroles are found throughout the area. The source of the heat appears to be a large magma chamber centered about 10 km under Mt. Hannah, a few kilometers northeast of the northern edge of The Geysers. The basement rock, a granitic or granodiorite felsite, is overlain with a highly fractured graywacke. The graywacke is covered in most places by a surface layer of greenstone, serpentinite of the Franciscan formation, or metamorphic melanges.

The steam at The Geysers, believed to arise from meteoric or formation water at great depth, is found in steeply angled fractures in the felsite and, more often, in highly interconnected random fractures in the graywacke. Many of the latter are oriented at low angles to the horizontal. Numerous wells have been drilled into the Clearlake volcanics. In a number of cases, either no fluid was found or only hot water was produced. Producing wells vary in depth from a few hundred to about 2400 m. Static steam pressure is typically a few MPa (a few hundred psi). A Geysers well may produce steam at a rate of several thousand kg/h at temperatures in the range of 250°C.

Electricity Production. Depending on steam conditions and conversion efficiency, each megawatt of electricity production requires from 100–200 kg (15,000–30,000 psi) of steam (14). There are several hundred producing wells in the field and steam reserves are estimated to be on the order of a few million kg per well. Electricity production at The Geysers expanded rapidly during the 1970s and 1980s and capacity peaked at nearly 2000 MW in 1988. Since that time, there has been a significant decline in the steam pressure at The Geysers and a concomitant reduction in the production of electricity. By the summer of 1992, electricity production from The Geysers had declined to only about 1220 MW and the area was troubled by significant idle generating capacity (15).

A simplified schematic of an electricity generating plant using hydrothermal steam is shown in Figure 4. After the production of electricity, it is possible to condense and recover much of the fluid. Since the decline in the steam pressure at The Geysers, a more concerted effort has been made to reinject the condensate in order to recharge the hot rock from which the steam arises. This has met with some success.

Recharge has become a primary issue in the attempts to mitigate the steam decline problem and proposals have been put forward to obtain water from other sources to contribute to the recharge effort. Although reinjection may help stabilize the level of energy production from The Geysers, such artificial recharge techniques are not expected ever to restore The Geysers to the production levels reached in 1988.

Geochemistry and Environmental Considerations. The geochemistry of steam-dominated geothermal resources is concerned primarily with condensable and noncondensable gases in the steam. The amounts and composition of noncondensable gases in Geysers steam vary rather widely within the steamfield as shown in Table 1. The predominant gas is carbon dioxide (qv) in all cases. The most important noncondensable gas, however, is hydrogen sulfide, because H_2S can present both corrosion and environmental problems.

Another important gas in The Geysers steam is hydrogen chloride (qv), which forms hydrochloric acid in the presence of liquid water and therefore is

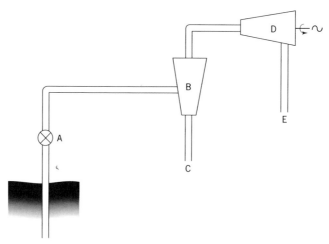

Fig. 4. Simplified schematic of a generating plant using hydrothermal steam. The steam simply issues from the well through a control valve, A, and passes through a preliminary separator, B, which removes particulates and any entrained liquid water, C. The steam drives a turbine, D. Then the spent fluid, E, may be treated to remove hydrogen sulfide, recondensed, or reinjected through another well depending on site-specific circumstances.

extremely corrosive. Significant amounts of hydrogen chloride are found in very high temperature ($ 300°C) steam arising either from a fluid source rich in chloride salts or from steam coming in contact with halite (17). Lower temperature steam may originally contain significant amounts of moisture, which tends to dissolve and wash out the hydrogen chloride before the fluid enters the energy-production cycle. Extended operations may in fact lead to drying out of produced steam and thus to increased concentrations of hydrogen chloride over time. It was demonstrated at Larderello that water reinjected into the reservoir could be used to scrub hydrogen chloride from the steam production flow (18). Studies suggest that there is a positive correlation between the level of hydrogen chloride and the total noncondensable gas concentrations.

Table 1. Noncondensable Gas Contents of The Geysers Steam[a]

Location within steamfield	Composition, ppmwt								Steam:gas, mol ratio
	Total gas	CO_2	H_2O	CH_4	NH_3	N_2	H_2	Ar	
northwest	65,223	55,560	1,710	2,580	576	560	347		31
central–southwest	13,524	11,500	662	851	223	153	133	<2	133
southeast	982	734	116	43	30	46	12	1.0	1,724

[a] Ref. 16.

Water-Dominated Hydrothermal Resources.   Hydrothermal resources in which the dominant or exclusive component is hot water are much more widespread than steam fields. Hot geothermal waters were first exploited to produce electricity in the Wairakei area of New Zealand in 1958. In the United States, electricity production from hot water did not begin until 1979. By 1992 water-dominated geothermal resources were being tapped to make electricity at a number of locations in California, Nevada, Utah, and Hawaii (Fig. **3a**). The Cascade Mountains and the basin and range area of Utah and Nevada contain identified but as yet undeveloped hot water resources which may be suitable for the generation of electricity. Hot water fields exist at depths of a few hundred to 2400 m and deeper. Often the hot water contains large amounts of dissolved solids. In the Salton Sea area of California, for example, fluids having dissolved solids levels in excess of 300,000 ppm (30%) are being commercially exploited.

Several types of plants are in use for converting the energy in hot water resources into electricity (19). The simplest design is a single flash unit similar to that shown in Figure 4. The hot fluid is simply separated (flashed) into liquid and vapor fractions in a cyclone separator in which the pressure is reduced. The vapor is then used to drive a turbine. The liquid fraction, which often contains high concentrations of dissolved solids, is typically disposed of by underground injection. This may or may not recharge the resource.

A second option is employed in double-flash generating plants, as shown in Figure 5. In a double-flash plant, liquid from the first flashing step is fed to a second separator at a lower pressure to produce more steam. The lower pressure steam from the second separation step is then admitted to the turbine at an appropriate point to act as a booster. Although double-flash plants may recover 15–20% more energy, these plants are obviously more complex and costly to build and operate. In addition, the final fluid by-product of a double-flash plant may be saturated or even supersaturated with solids, leading to extensive scaling, corrosion, and disposal concerns. The choice between a single- and double-flash plant

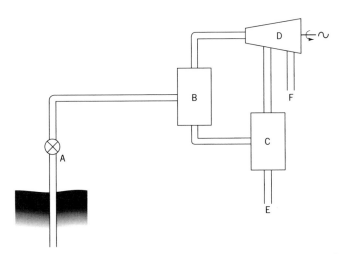

Fig. 5. In a double-flash plant for producing electricity from hydrothermal water, superheated water is delivered from the well, A, to an initial flashing unit, B, where the pressure is reduced to release steam which drives a turbine, D. The liquid fraction is then delivered to a second flashing unit, C, where further pressure reduction produces more steam which is introduced to the turbine at an intermediate stage. The waste fluid from the second flashing stage, E, may contain very high concentrations of dissolved or suspended solids, presenting significant disposal problems. The spent steam can be recondensed and reinjected, F.

depends on the characteristics of the resource and on capital investment considerations.

Another method, known as binary technology, utilizes two fluids to convert
hydrothermal energy to electricity. In such an operation, it is possible to extract
the energy from hot water without any vaporization of the hydrothermal water.
Figure 6 shows a schematic of a binary plant. Heat from the geothermal water is
transferred to a working fluid in a heat exchanger. The working fluid is thereby
vaporized and drives a generator. It is then recondensed and recirculated to be
heated again by additional hydrothermal water. Typical working fluids are very
volatile, low molecular weight hydrocarbons such as isobutane or fluorinated
hydrocarbons (see HEAT-EXCHANGE TECHNOLOGY, HEAT-TRANSFER MEDIA
OTHER THAN WATER). Binary power generators are often only 1–2 MW in capacity
and are usually employed as ganged units in multimegawatt geothermal power
stations.

Single-flash, double-flash, and binary generating plants are all being used
in the United States to produce energy from liquid-dominated hydrothermal resources. A fourth type of power plant, which is really a sophisticated version of a
binary system, is being evaluated. In this last design, often referred to as the
Kalina cycle, the working fluid is an adjustable proportion fluid mixture, most
often consisting of water and ammonia (20).

The production of electricity from hot-water resources is expanding in the
United States, but the individual production plant sizes are often small, ranging

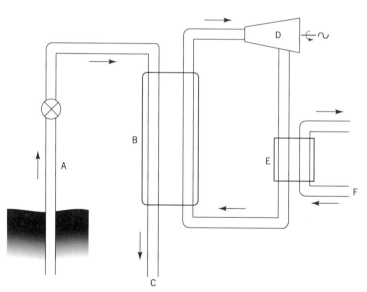

Fig. 6. In a binary electricity generation plant, the hydrothermal water from the well, A,
is passed through a heat exchanger, B, where its thermal energy is transferred to a second,
more volatile working fluid. The second fluid is vaporized and delivered to a turbine, D.
After exiting the turbine the spent working fluid is cooled and recondensed in another heat
exchanger, E, using water or air as the coolant, F. It is then fed back to the primary heat
exchanger to repeat the cycle. Waste hydrothermal fluid, C, can be reinjected into the
producing field.

from 1–10 MW. These systems can be highly automated so that little daily labor is needed. Plant availability is also generally very high, on the order of 95% or greater. The utilization of hot-water resources in the United States is expected to grow in response to an increasing need for small-to-moderate increments of clean power in the American West.

Most of the developed hot-water fields are located by significant surface indications, particularly in the form of hot springs. Once a resource has been identified, a variety of techniques can be used to map the system and determine whether it is of a size sufficient to justify commercial development. Hidden hot-water resources are much more difficult to locate, but geologic indicators such as volcanic activity and evidence of hydrothermal alteration can be used.

Worldwide Hydrothermal Development.   Electric generation capacity from hydrothermal energy outside the United States was more than 3000 MW in 1990 (21). Hydrothermal resources have been especially well developed in countries on the Pacific rim. Nearly 900 MW are on line in the Philippines, 700 MW in Mexico, 283 MW in New Zealand, and 215 MW in Japan. Lesser amounts of hydrothermal electric capacity have been installed or planned in Indonesia, El Salvador, Nicaragua, and a number of other developing nations. Italy, which has 545 MW at Larderello, is the chief producer of hydrothermal-based electric power in Europe. The small island nation of Iceland generates 44 MW of electric power from geothermal energy and has perhaps the most extensive direct-use applications of geothermal waters of any nation in the world. Electricity generation from hydrothermal resources is rapidly increasing in a number of developing countries. The relatively simple engineering, straightforward components, and ease of repair of hydrothermal plants make these plants ideal for application in nations with unsophisticated economies.

Economics.   In the early 1990s, the cost of electricity generated from geothermal energy varied from about 3.5–10¢/kWh (22). The cheapest electricity came from The Geysers, where steam can be delivered to the power plant for less than 2¢/kWh of electricity generating potential (23). Electricity costs from hot-water resources using flashed steam and binary plants, respectively, are progressively higher. In addition to the usual power plant costs, other significant up-front capital costs, including resource exploration, drilling, and field development, must be covered before a geothermal plant can begin producing revenue. Accordingly, capital financing carries an added risk in geothermal projects because continued supply of fuel is not assured until considerable capital has been expended. Environmental concerns may also add to the capital costs of a hydrothermal plant. Hydrogen sulfide abatement systems can run several million dollars. The exact costs vary with the composition of the gas to be treated. Liquid-dominated resources which are high in dissolved solids incur added capital and operating costs to pay for the collection and disposal of the spent brine and any precipitated solid residues.

The development of hydrothermal resources was given a strong push by a variety of governmental regulations instituted in response to the oil crisis of the mid-1970s. Perhaps the most significant law was the Public Utilities Regulatory Policies Act (PURPA), passed in 1978. This law required utility companies to purchase power from qualified independent power plants and cogeneration facilities at a cost equivalent to the avoided expense the utility saved by not constructing the facility itself. States were free to implement PURPA as they saw fit, and in California the Standard Offer-4 contract was passed in 1982 (24). These contracts allowed for payments based on capacity over a period of 30 years and energy delivery for the first 10 years of a project, thus providing a firm basis for financing construction of the power generating facility. These regulations proved to be a boon for all alternative energies, including cogeneration, wind, solar, and biomass, as well as geothermal. In 1990, 8440 MW of firm capacity was being generated by qualified suppliers in California. The share of this power coming from geothermal plants is approximately 9%, or about 760 MW. The growth of the geothermal energy industry during the 1980s can be attributed in large part to such governmental policies, especially the Standard Offer-4 regulations. Financial incentives played a key role in expanding hydrothermal technology from the utilization of steam at The Geysers to hot-water resource development. Commercial exploitation of lower temperature, but more abundant, hot-water resources has been made possible by continued technological improvements coupled with government environmental and fiscal policies designed to encourage the development of alternative energy sources (see RENEWABLE ENERGY RESOURCES).

Whereas there are significant known geothermal reserves and an estimated large amount of undiscovered geothermal energy, the future growth of the industry is tied closely to energy prices and environmental regulations. Hydrothermal energy utilized for the production of electricity is expected to remain exclusively a western U.S. resource.

Environmental Issues. Hydrothermal energy, recognized as one of the clean power sources for the twenty-first century, is not entirely free of environmental problems.

Atmospheric Emissions. The hydrogen sulfide found in many hydrothermal fluids is extremely toxic and has an unpleasant odor (25). In The Geysers area of California, special equipment has been installed at geothermal installations to remove hydrogen sulfide from the waste stream by the Stretford process. This technology utilizes a vanadium catalyst to reduce about 95% of the hydrogen sulfide to elemental sulfur. The product would be salable except that it is usually contaminated with vanadium and other traces of heavy metals which make it a hazardous material (see SULFUR REMOVAL AND RECOVERY). Hydrogen chloride, found in some geothermal steam, has proven to be more of a problem from a corrosion standpoint than as an atmospheric contaminant. Other gases are present in relatively small quantities. Geothermal plants release only about 5% as much carbon dioxide, and less than 1% of the nitrous and sulfur oxides emitted from fossil power plants in generating an equivalent amount of electricity (see AIR POLLUTION; AIR POLLUTION CONTROL METHODS).

Aquatic Pollution. Aquatic pollution is of some concern from hydrothermal resources. The primary problem is the disposal of highly saline fluids from water-dominated reservoirs. This is generally overcome by pumping into deep reservoirs situated well beneath potable water sources. The fluid is returned to the earth minus the thermal energy content. One other potential environmental effect is the use of water for cooling in arid locations where hydrothermal resources are often found. If the water resources of the region cannot support the volume of water needed, then air cooling, which is less efficient and more susceptible to the vagaries of the climate, must be used.

Terrestrial Problems. In some hot-water operations, solid wastes accumulate and present a disposal problem. These wastes are usually salt cakes and sulfides which must be disposed of in accordance with government regulations. The land use of a geothermal facility may be significant when a number of wells are used to feed a central power facility, but these can be compatible with agricultural use if piping is carefully routed. Subsidence of land has never been noted from steam reservoirs, but like oil fields, it has occurred above hot-water reservoirs (26). Reinjection of the cooled water can mitigate potential subsidence problems. Seismic hazards do not appear to be significant. Geothermal operations can sometimes affect the flow of nearby springs, an issue which raises serious concern when the affected springs are used for recreational or medicinal purposes. Other potential problems such as noise pollution (especially during drilling), cultural and archeological disturbances, and visual degradation of the landscape are common to geothermal as well as other large industrial development in a relatively untouched environment.

Geopressured Resources

The Resource. Geopressured resources consist of highly overpressured mixtures of hydrocarbons, predominantly methane, and water, in sedimentary formations (27). The potentially useful energy in geopressured resources exists as three components: fossil chemical from the methane, heat from the water, and mechanical from the high pressure of the fluid. Geopressured resources are generally found very deep in the earth at levels of 3600 to 6000 m or more. It is thought that these were formed when incompletely dewatered organic sediments were covered by layers of clay. Over time, the clay was converted from the smectite to the impervious illite form, effectively isolating the sediments and setting the stage for the formation of a geopressured compartment (see CLAYS). Increasing temperatures of the buried strata led to pressures above hydrostatic owing to aquathermal pressuring and the decomposition of the organic material into volatile low molecular weight compounds, particularly methane. The distinction between an oil or gas resource and a geopressured resource is somewhat arbitrary, as some water and pressure are often encountered in petroleum (qv) deposits. It was estimated that in 1983 about 2% of the more than 50,000 oil and gas wells along the Texas Gulf coast were producing from geopressured reservoirs (28). In these operations, however, only the oil and gas were recovered.

Figure 7 shows the locations of geopressured resources in the United States. The characteristics of the various basins vary significantly with regard to degree of overpressure, amount of entrained gases, and dissolved solids levels. The geopressured region along the Gulf Coast has been the most thoroughly studied. This area is estimated to contain 6000 EJ (6×10^{18} Btu) of energy in the form of methane gas and 11,000 EJ more in the form of the thermal energy in hot water at 121–260°C (29). The mechanical energy recoverable from the high (14–35 MPa (2000–5000 psi)) pressure at the surface, is as yet undetermined. The worldwide geopressured resource base, considering all three energy forms, is extremely large, amounting to more than 500,000 EJ. Tables 2 and 3 summarize the gas and salt analyses, respectively, of geopressured fluids from the Pleasant Bayou,

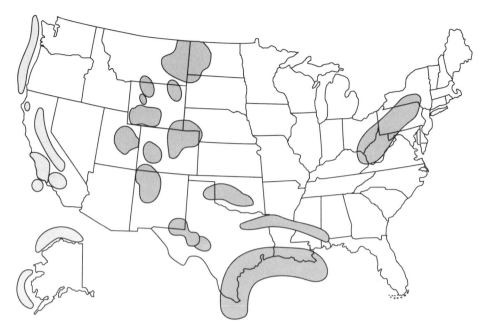

Fig. 7. Geopressured ( )–geothermal (▪) resources in the United States.

Table 2. Geopressured Gas Composition[a]

Component	Gas composition, mol %
methane	87.7
carbon dioxide	10.4
ethane	2.9
propane	1.0
nitrogen	0.5
isobutane	0.15
N-butane	0.14
pentanes	0.06
hexanes and higher hydrocarbons	0.06
hydrogen	0.02
helium	0.01

[a]Data from Pleasant Bayou well, Feb. 1990 (30).

Texas, well. The salts found in geopressured reservoirs are predominantly sodium and calcium chlorides but the composition and salinity vary widely. These salts appear to have multiple origins including seawater residues and dissolved sedimentary salt layers.

   The Technology.   Owing to the large overpressure, geopressured wells flow freely in high volumes. Production levels of 3000–5000 m³/d (20,000–30,000 bbl/d) have been achieved in some wells. At high flow rates, clogging of the pores near

Table 3. Geopressured Brine Analysis[a]

Component	Concentration, mg/L
chloride	72,400
sodium	37,100
calcium	7,950
strontium	843
barium	769
magnesium	603
potassium	561
silica as SiO_2	106
bromide	77
iron	46
lithium	31
boron	25
iodide	22
manganese	16
sulfate	5
fluoride	1.7

[a]Data from Pleasant Bayou well, May 8, 1988 (31).

the wellbore can occur when the structure of the formation sand is disturbed by the turbulent flow, greatly reducing the energy production capacity of the well. The high salinity of geopressured water leads to a spent fluid disposal problem. The most common solution is to pump the saline water down a nearby well into a formation at a shallower depth than the geopressured resource. The formation of calcium carbonate scale creates significant operational difficulties in utilizing highly saline geopressured fluids. Scale inhibitors and the requirement for frequent removal of accumulated scale from piping and equipment can both add substantially to the maintenance cost of geopressured facilities. In one proposed power plant design, the mechanical power is first utilized in a pressure-reduction turbine, then the hydrocarbon and aqueous fluids are separated, and the water is fed to the heat exchanger of a binary power plant. The gas is used to produce electricity through conventional technology or sold directly to off-site users.

Most of the work on developing techniques to exploit geopressured resources has been carried out under the auspices of the U.S. Department of Energy (DOE). Wells in Texas and Louisiana that were originally drilled for oil and gas production, but instead struck geopressured resources, were operated in a demonstration mode to evaluate the feasibility of utilizing geopressured resources in a commercially viable manner. A hybrid power plant was constructed at Pleasant Bayou, Texas, along the lines of that shown in Figure 8, except that the pressure reduction turbine was excluded (32). The plant was fed by 1600 m³/d (10,000 bbl/d) of geopressured brine at a temperature of 64°C. The contained gas consisted primarily of methane (87%). The principal noncombustible impurity was carbon dioxide. During operation, the system produced power at a rate of 1.225 MW. Of the electricity generated, about 56% came from the gas and 44% from the thermal energy of the water. Geothermal energy conversion, which required a far more complicated mechanical plant, produced a smaller percentage of the energy.

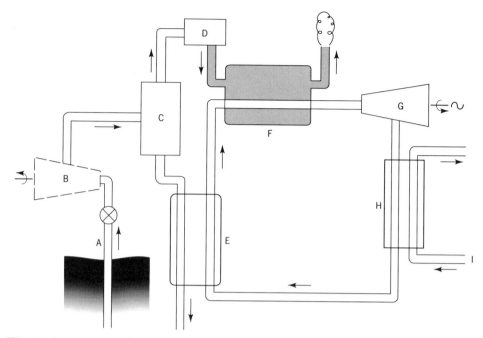

Fig. 8. A power plant for utilizing geopressured resources (32). The high pressure fluid is fed from the well, A, to a turbine, B, where its mechanical energy is utilized to generate electricity. The fluid then proceeds to a separator, C, where the methane is separated and used to generate electricity in a combustion generator, D. The liquid fraction is delivered to a heat exchanger, E, of binary plant. The binary plant loop contains a turbine, G, and a cooling heat exchanger, H, which uses water, I, to recondense the working fluid. It also contains an additional heat exchanger, F, to extract the thermal energy from the exhaust gas of the combustion generator. A pilot plant run for a year included all these elements except the initial turbine, B.

Economics.   The cost of energy from the Pleasant Bayou plant, at 12–18¢/kWh, was not competitive with the 4–6¢/kWh power produced from higher quality fossil fuels which are abundant in the Gulf Coast area (33). The high costs can be related to a number of factors. The multiple energy forms, each effectively requiring its own generating plant, result in higher capital costs. Additionally, the salinity of the fluid leads to problems in corrosion and scaling as well as in disposal. Finally, the depth at which the resource exists means that drilling costs are high. These factors are fundamental characteristics of the technology which, except for corrosion and scaling, are not readily amenable to solutions that are economical with the technology available today. Commercial utilization of geopressured resources for electricity production is not expected in the foreseeable future.

Direct Uses of Geopressured Fluids.   Many of the uses typical of hydrothermal energy, such as greenhouse, fishfarm, and space heating, have been proposed for geopressured resources, but none has been commercially developed (34). Hydrothermal fluids are widely used in enhanced oil recovery, however, to increase production from depleted oil fields.

Hot Dry Rock

The Resource. The largest quantity of accessible geothermal energy exists in the form of hot rock which contains insufficient natural fluids to allow the transport of its energy to the surface. Because hot dry rock (HDR) is widely distributed, it also has the greatest potential for widespread application and is the only technology capable of making geothermal energy available on a worldwide basis. Whereas the HDR resource is found almost everywhere, it is not equally easy or economic to reach at every location. The typical geothermal gradient worldwide is about 25–30°C/km, but in many places it is much higher. Figure 9 is a geothermal gradient map of the United States, showing that areas of high geothermal gradient are found in the western part of the country. Hydrothermal reservoirs are typically located in regions of high geothermal gradient.

The amount of thermal energy stored within hot dry rock at accessible depths is enormous. Estimates have placed the energy at more than 10^7 EJ (34). Even a minute fraction of this resource could supply all the world's energy needs for decades or even centuries. However, to utilize HDR resources, a practical means of accessing the hot rock and transporting its energy to the surface must be developed. In effect, an underground heat exchanger must be created to transfer the thermal energy of the rock to a mobile fluid. Because of the low thermal conductivity of hard rock, the surface area of the heat exchanger would have to be extremely large. It is not sufficient to simply circulate water through under-

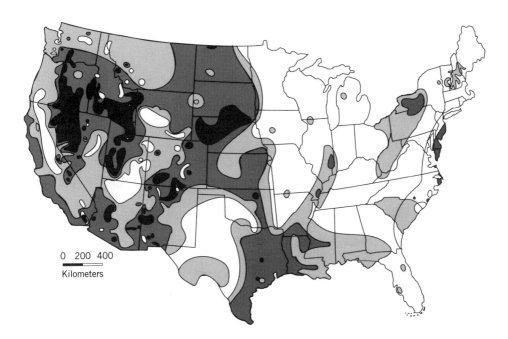

Fig. 9. A geothermal gradient map of the United States, where (□) represents a geothermal gradient of <20°C/km; (▫), 20–30°C/km; (▪), 30–50°C/km; (▪), 50–70°C/km; and (▪), >70°C/km.

ground pipes because, as the relatively small amount of rock in direct contact with the pipe cooled, the efficiency of heat transfer would rapidly decline.

The Technology.   The basic technique for extracting energy from HDR was conceived and patented in the early 1970s (35). It is based on drilling and hydraulic fracturing technologies developed in the petroleum and geothermal industries. Figure 10 shows an idealized HDR heat mine. The first step in constructing a heat mine is to drill a well into sufficiently hot and impervious rock, with the exact depth of the well to be determined by local heat-flow and thermal conditions. Wells drilled for HDR applications are similar in many aspects to

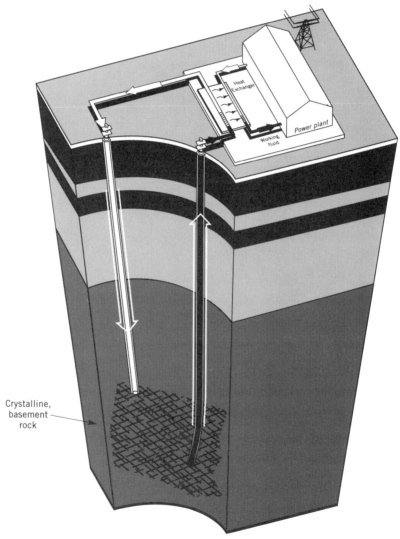

Fig. 10.   An idealized view of an HDR heat mine. Water circulating in a closed loop is heated as it passes through fractures in the hot rock. The energy is extracted at the surface where electricity is generated using binary technology.

hydrothermal wells except that these wells are deeper and sometimes penetrate into a much greater depth of hard, crystalline rock. For this reason, specialized drilling and logging equipment which can withstand extended exposure to high temperatures is required. After the well has been completed, a segment of the bottom portion of the well is blocked off using a packer which provides pressure isolation. Water under high pressure is pumped through the packer and forced into joints in the surrounding rock body to form a reservoir consisting of a relatively small amount of water in a very large volume of rock. The extent of the reservoir region may be controlled by the pressure applied via the injected fluid and the length of time the process is continued. The shape and orientation of the reservoir are functions of the natural jointing features of the host rock.

Thousands of tiny earth tremors are initiated by the hydraulic fracturing process as the rocks move apart. These microearthquakes can be detected by sensitive seismic instruments placed at nearby locations in relatively shallow wells and on the surface (36). By correlation of these data, the HDR reservoir can be mapped, ie, the number of microearthquakes detected in small (typically 1000-m^3) units of volume are compiled and mapped to give a view of the density of microearthquakes in the reservoir region and, by implication, the size and shape of the reservoir. To complete the system, a second well is drilled into the reservoir at some distance from the first, using the map as a guide. In operation, water pumped down one well heats as it flows through the joints in the reservoir rock and returns to the surface through the second well, where its thermal energy is extracted using binary technology (see Fig. 6). The water can then be recycled.

Issues related to operation of a HDR geothermal energy system are the efficiency of energy extraction from the rock in the reservoir region; the impedance to flow as the water traverses the reservoir body; and the water losses resulting from leakage from the reservoir. Because of the low thermal conductivity of rock, the efficiency with which the water extracts the energy from the rock is directly related to the number and geometry of the open joints. If the injection and production wells are directly connected by one or very few joints, the surface area of rock accessed by most of the circulating water is only a small fraction of the total volume of the reservoir. The rock in the region of these joints would then cool rapidly and the temperature of the produced water soon drop to the point where it is no longer hot enough to be useful. In order for sustained energy production to be achieved, a complex series of joints providing multiple pathways between the wells with access to a large volume of the reservoir rock is required.

As of this writing, it has not been possible to use the seismic data which defines the volume of the reservoir to also determine the joint structure. Extended flow testing is the most direct measure of the efficiency and sustainability of energy recovery from the reservoir. The use of chemical tracers in the circulating fluid can also provide valuable supporting data with regard to the multiplicity of flow paths and the transit time of fluid within the reservoir (37).

The impedance to fluid flow through the reservoir determines the pressure which must be applied to move the water from the injection to the production point. Impedance greatly influences both the pumping power required and the ultimate volume of water which can be pushed through the system per unit time. It is thus a primary factor in determining the absolute rate at which energy can be extracted as well as the cost per unit of energy produced. Impedance is at least in part a function of the injection pressure. At pressures below those required to open the joints at depth, the impedance is essentially infinite. After an HDR reservoir has been created, there is an injection pressure range high enough so that measurable flow through the reservoir takes place, but not so high that additional reservoir expansion occurs. Finally, at some pressure, active reservoir growth begins, as evidenced by microearthquake activity and a large increase in water consumption. Because the maximum possible flow rate with the least possible water consumption is desired for efficient performance of an HDR system, the ideal injection condition for continuous energy production from a stable reservoir is at a pressure just below the point at which active reservoir growth takes place.

Water loss in operating an HDR facility may result from either increased storage within the body of the reservoir or diffusion into the rock body beyond the periphery of the reservoir (38). When a reservoir is created, the joints which are opened immediately fill with water. Micropores or microcracks may fill much more slowly, however. Figure 11 shows water consumption during an extended pressurization experiment at the HDR facility operated by the Los Alamos National Laboratory at Fenton Hill, New Mexico. As the microcracks within the reservoir become saturated, the water consumption at a set pressure declines. It does not go to zero because diffusion at the reservoir boundary can never be completely eliminated. Of course, if a reservoir joint should intersect a natural open fault, water losses may be high under any conditions.

It is imperative that any HDR reservoir be created in rock which is free of natural faults. This can be accomplished by a thorough geologic study of a rock body prior to creation of an HDR reservoir within it, by close control of the hydrofracturing operation, and through real-time analysis of microearthquake data arising from joint opening to assure that the HDR reservoir stays within known bounds.

A four-month long flow test of the reservoir at Fenton Hill, New Mexico conducted in 1992 provides an illustration of the present stage of development of HDR energy production technology (39). As shown in Figure 12a, the test was

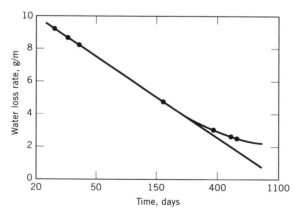

Fig. 11. Water consumption during extended pressurization of an HDR reservoir. The amount of water required to maintain a constant pressure declines with the logarithm of time as the microcracks in the reservoir rock are slowly filled with the pressurized fluid.

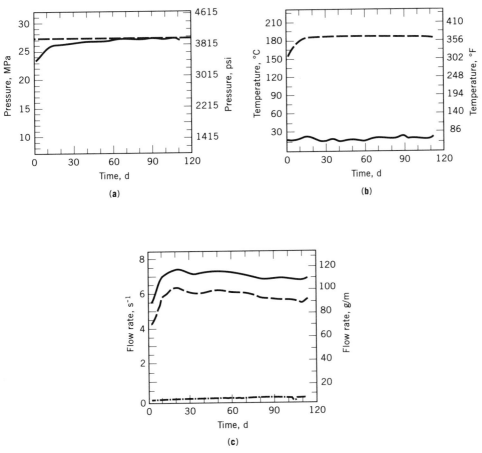

Fig. 12. HDR reservoir flow test at Fenton Hill, New Mexico (1992) (39). (**a**) Injection pressure profile (——) and a seismic limit (————); (**b**) injection (——) and production (————) fluid temperatures; and (**c**) injection (——) and production (————) flow rates. Tracer testing showed that the decline in flow rates was a result of the fluid following longer flow paths through the body of the reservoir as the test progressed. The bypass flow (—·—·—) arose from a leak in the injection wellbore.

carried out at the highest pressure that previous experience had shown could be tolerated without leading to reservoir growth, as evidenced by microseismic activity. The temperature of the rock in the Fenton Hill reservoir region is known to be in the range of 220°C. Considerable thermal loss to the surrounding rock occurred as the fluid passed up the 3.6-km wellbore resulting in surface production temperatures in the range of 180°C. As shown in Figure 12**b**, the temperature of the fluid produced at the surface declined very slowly over the course of this experiment, but temperature logging indicated no decline in temperature at depth. Figure 12**c** indicates that the flow rate declined by about 10% over the course of the test, indicating a significant increase in the impedance of the system. The lower flow rate and the resultant longer residence time moving up the wellbore led to the slight decline in the surface temperature of the production fluid over the span of the test. Tracer analyses in turn indicated that the lower flow rate could be attributed to increased distribution of the circulating water in long flow path channels through the hot rock. These results indicate that the access to hot reservoir rock improves as an HDR reservoir is operated. Because of the limited flow testing conducted, it is not possible to predict the extent of reservoir cooldown over the multiyear period of operation required to make the commercial development of HDR systems practical.

Economics.   The costs of developing HDR resources are closely tied to the depth at which sufficiently hot rock is found. This is most readily expressed in terms of the geothermal gradient shown in Figure 9. High gradient resources are located primarily west of the Mississippi River. In the eastern United States, it is generally necessary to drill much deeper to reach hot rock. Because drilling is the most expensive single factor in HDR development, the first HDR electric plants are expected to be built in the western United States.

A number of studies have been conducted to assess the economics of producing electricity from HDR. Estimates of busbar electricity costs range from 4–10¢/kWh. A 1990 report gave cost figures based on the quality of the HDR resource as reflected in the local geothermal gradient (40). Direct heating applications of HDR were also addressed (41). Results of that study are summarized in Table 4. Electric power from high grade HDR resources could be competitive in the 1990s, whereas that derived from medium grade resources is likely to be only marginally so. Because direct heating applications do not require temperatures as high as those needed for electricity production and do not suffer the losses inherent in conversion of thermal to electric energy, even lower grade HDR resources may prove to be competitive in site-specific space or industrial heating applications. No commercial HDR facility has been built as of this writing.

Geochemistry and Environmental Aspects.   The environmental characteristics of HDR make it potentially the cleanest geothermal energy source and thus place it among the most promising of all energy resources. When operated as a closed loop, no significant amounts of air, water, or terrestrial pollutants are produced. Because the active reservoir is located thousands of feet below the water table, there is no danger of ground or surface water contamination. Finally, when a plant is decommissioned at the end of its useful life, the underground system could be permanently shut in by techniques already well known and proven in the oil, gas, and geothermal industries.

Because the energy production and storage zone is far underground, HDR plants should occupy only a minimal space on the surface. In addition, siting

Table 4. Costs of Energy from HDR Resources[a]

Geothermal gradient of resource, °C/km	Breakeven electricity price, $/kWh	Cost of thermal energy, $/GJ[b]
80	0.05–0.06	2–4
50	0.08–0.09	4–7
30	0.16–0.18	10–17

[a]Ref. 40.
[b]To convert from GJ to Btu, multiply by 9.49×10^5.

opportunities are predicted to be extremely versatile, and the locations of facilities can be chosen for minimal visual impact or to eliminate the need for long runs of high tension lines.

Natural geothermal fluids contain widely varying levels of dissolved solids and gases resulting from the extended contact of these natural fluids with the underground environment. In contrast, the water circulating in a hot dry rock system is drawn from surface or groundwater sources. As the water is recycled, it rapidly approaches an equilibrium level of dissolved species which tends to be remarkably low. Table 5 is a geochemical analysis of water from the Fenton Hill HDR system after several months of circulation. The low levels of dissolved solids indicate that HDR fluids are not likely to pose significant environmental problems even if it should be necessary to dispose of them on the surface. These data seem to be a reasonable representation of a typical HDR reservoir developed in deep granitic rock.

HDR Outside the United States. The extraction of geothermal energy from HDR is being evaluated at a number of locations around the world (42). All work is based on the same general technical approach as that employed in the United States. In Japan, reservoirs have been created in rock at about 200°C at two locations. The most advanced work has been done in an area that had been previously explored for hydrothermal resources. Water losses to natural faults ini-

Table 5. Geochemical Analysis of HDR Plant Water[a]

Component	Concentration in production fluid, ppmwt
chloride	953
sodium	900
bicarbonate	588
silicate as SiO_2	424
sulfate	378
potassium	89
boron	35
calcium	18
fluoride	17
lithium	16
arsenic	7.2
bromide	5
ammonium	1.1
iron	0.8
strontium	0.8
bisulfate	0.2
magnesium	0.1
Total dissolved solids	*3434*
carbon dioxide	2370
nitrogen	71
oxygen	0.35
hydrogen sulfide	0.33

[a]From the Fenton Hill hot dry rock plant.

tially proved to be a significant problem, but in a flow test in 1990 more that 80% of the injected water was recovered. An extensive HDR effort was also begun in England in 1978. An HDR system was created 2 km below the surface, but because of the low thermal gradient, the rock temperature was less than 100°C. Extensive flow testing and rock mechanics studies were carried out for more than a decade. HDR wells have been drilled in both France and Germany, but no circulation systems have yet been established. Exploratory HDR work has also been carried out in Sweden, Switzerland, and Russia.

Magma

The Resource.   The core of the earth is generally believed to consist of molten rock known as magma. The energy content of the core is essentially boundless, but it is unreachable from a practical standpoint because it lies many kilometers below the surface. The technology to drill to those depths does not exist. In volcanically active regions, however, magma intrusions can be found relatively close to the surface in some localities. In these areas magma is a potentially useful geothermal resource, but relatively little is known about intrusive magma chambers.

The United States Geological Survey estimates that the total amount of magma energy existing within 10 km of the surface is on the order of 50,000–500,000 EJ in molten or partially molten magma (43). These calculations have been based primarily on indirect evidence obtained by drilling into granitic plutons, the frozen remnants of former liquid rock, and on studies of recent volcanism, cooling models, and plausible assumptions as to the size of magma chambers. The data imply that magma may exist at reachable depths in several regions of the United States, most notably in the large calderas at Long Valley in east-central California, the Valles Caldera of north-central New Mexico, and the Yellowstone region of northwestern Wyoming. The sizes of these magma bodies may be in excess of 1000 km^3 of fluid rock at temperatures in excess of 650°C. It has been estimated that only 2 km^3 of magma could provide enough energy to operate a 1000-MW electric power plant for 30 years.

Work on the extraction of useful energy from magma has been limited primarily to paper and laboratory studies aimed at understanding the formation, extent, cooling, and other facets of magmatic bodies. Field drilling has been limited. The earliest work was carried out at the Kilauea Iki Lava Lake on the island of Hawaii where lava exists very close to the surface. A significant amount of drill core was extracted from a partial melt zone having temperatures in excess of 1000°C. Models based on the Kilauea experience suggest that a single well into magma may be able to produce electricity at a rate of 30 MW.

The Long Valley Magma Experiment.   An exploratory effort to drill toward a magma chamber is being conducted at Long Valley, California. The intent of this project is not to penetrate the magma body, but simply to come close enough to confirm its existence. Downhole thermal and seismic techniques are then planned to demonstrate unequivocally that a liquid magma chamber exists there.

The project at Long Valley has been designed to proceed in four phases. Work on the first two stages of the project, which entailed drilling to 2313 m to reach rock at temperatures of 104°C, was completed in late 1991 (44). These depths and temperatures are typical of those seen in conventional hydrothermal drilling operations and the problems encountered by the project were not unlike those of other geothermal operations. The third phase of the work should extend the hole to a depth of 4267 m and temperatures of 300°C. Again, this drilling will not push beyond the frontiers of geothermal drilling technology. Phase four, however, is planned to complete the well at 6096 m into rock at 500°C, a temperature which is beyond the capacity of present technology. Drilling fluid additives are not stable at high temperatures, nor are casing cements. In addition, high temperatures cause more severe corrosion and available logging tools are expected to be limited. Insulated drill pipe, however, could allow water to be kept relatively cool as it is pumped to the bottom of a deep, hot magma well, and the development of insulated drill pipe has become a focus in the magma effort. The pipe envisioned would need to be double-walled. The mechanical strength of such a pipe should not be a problem, but fabrication at reasonable costs and the development of an efficient insulated tool joint to connect the pipe segments present formidable development problems (see HIGH TEMPERATURE ALLOYS).

Economics. A cost estimate of power from magma has been developed for the California Energy Commission in conjunction with the Long Valley Project, estimating that a 50-MW electricity-producing facility could produce power for 5.6¢/kWh (45). These costs hinge on completion of the Long Valley well, actual fabrication of insulated drill pipe from special corrosion-resistant alloys, design and testing of special drill bits to penetrate magma, the construction of a small-scale pilot facility, and successful drilling of a magma penetrating borehole. Utilization of magma energy will take a concerted effort in resource identification and verification, drilling technology, and materials development.

BIBLIOGRAPHY

"Geothermal Energy" in *ECT* 3rd ed., Vol. 11, pp. 746–790, by J. Tester and C. O. Grigsby, Los Alamos Scientific Laboratory.

1. C. Otte, *Geothermal Energy in Developing Countries, Energy and the Environment in the 21st Century*, MIT Press, Cambridge, Mass., 1991, p. 755.
2. G. M. Reistad, paper presented at the *Proceedings of the 2nd UN Symposium on the Development and Use of Geothermal Resources*, Vol. 1, Lawrence Berkeley Laboratory, San Francisco, 1975, pp. 2155–2164; "Analysis of the Economic Potential of Solar Thermal Energy to Provide Industrial Process Heat," *U.S. ERDA Contract*, EY-No-C-02-2829, Intertechnology Corp., 1977.
3. R. DiPippo, *Geothermal Energy as a Source of Electricity*, U.S. Department of Energy, Washington, D.C., 1980, p. 3.
4. *The Potential for Renewable Energy: An Interlaboratory White Paper*, Office of Policy Planning and Analysis, U.S. Dept. of Energy, SERI/TP-260-3674, Washington, D.C., 1990, p. C-2.
5. P. E. Gray, J. W. Tester, and D. O. Wood, *Energy Technology: Problems and Solutions, Energy and the Environment in the 21st Century*, MIT Press, Cambridge, Mass., 1991, p. 120.
6. L. McLarty, and R. J. Reed, *The U.S. Geothermal Industry: Three Decades of Growth, Energy Sources*, Vol. 14, No. 4, Taylor and Francis, Philadelphia, Pa., 1992, pp. 443–455.

7. L. E. Capuano, Jr., *Geotherm. Res. Counc. Bull.*, **21**(4), 113–116 (1992).
8. J. Evanhoff, and K. Harris, *Geotherm. Res. Counc. Bull.*, **21**(4), 108–112 (1992).
9. L. E. Kukacka, *Materials for Geothermal Production*, Proceedings of the Geothermal Program Review X, U.S. Dept. of Energy, Washington, D.C., 1992, p. 97.
10. D. H. Freeston, *Geothermal Resources Council Bulletin*, **19**(7), 192 (1990).
11. J. C. Austin, L. M. Fettkether, and B. J. Chase, *Geotherm. Res. Counc. Trans.*, **8**, 135–138 (1984).
12. L. M. Edwards, G. V. Chilingar, H. H. Rieke III, and W. H. Fertl, eds., *Handbook of Geothermal Energy*, Gulf Publishing Co., Houston, Tex., 1992, p. 59.
13. M. A. Walters and J. Combs, *Geotherm. Res. Counc. Monogr. The Geysers Geotherm. Field* **17**, 43–53 (1991).
14. J. Kestin, ed., *Sourcebook on the Production of Electricity from Geothermal Energy*, U.S. Dept. of Energy, Washington, D.C., 1980, p. 944.
15. K. F. Stelling, *The Geysers Production, The Geothermal Hot Line* **21**(1), 3 (1992).
16. A. H. Truesdell, J. R. Haizlip, W. T. Box, Jr., and F. D. Amore, in Ref. 13, p. 127.
17. Ref. 13, pp. 121–132.
18. J. R. Hazlip, Ref. 13, p. 143.
19. R. DiPippo, *Energy and the Environment in the 21st Century*, MIT Press, Cambridge, Mass., 1991, pp. 742–747.
20. C. H. Marston, *Mech. Eng.*, 81 (Sept. 1992).
21. G. W. Huttrer, *Geotherm. Res. Counc. Bull.* **19**(7), 175–187 (1990).
22. *Geothermal Energy—Heat from the Earth, Power for the Future*, U.S. Department of Energy Brochure, Meridian Corp., Alexandria, Va., 1992.
23. W. P. Short III, *Geotherm. Res. Counc. Bull.* **20**(9), 246 (1991).
24. D. J. Falcone and W. P. Short III, *Geotherm. Res. Counc. Bull.* **21**(10), 325–328 (1992).
25. M. Brower, *Cool Energy*, MIT Press, Cambridge, Mass., 1992, p. 151.
26. Ref. 19, p. 752.
27. *Geopressured–Geothermal Energy, the Untapped Resource*, U.S. Department of Energy, Washington, D.C., 1992, p. 2.
28. J. Negus-de Wys, *The Geopressured Habitat—A Selected Literature Review*, U.S. Department of Energy, Washington, D.C., 1992, pp. 117–118.
29. Ref. 27, p. 3.
30. Ref. 28, P. A-7.
31. Ref. 28, p. A-14.
32. R. G. Campbell and M. M. Hattar, *Proceedings of the Geothermal Program Review IX*, U.S. Department of Energy, Washington, D.C., 1991, pp. 163–166.
33. K. Taylor, in Ref. 32, p. 161.
34. Ref. 12, p. 83.
35. U.S. Pat. 3,786,858 (Jan. 22, 1974), R. M. Potter, E. S. Robinson, and M. C. Smith (to Los Alamos National Laboratory).
36. L. House, *Geophys. Res. Lett.* **14**(9), 919–921 (1987).
37. N. E. V. Rodrigues, B. A. Robinson, and E. Counce, "Tracer Experiment Results During the Long-Term Flow Test of the Fenton Hill Reservoir", *Proceedings of the Eighteenth Workshop on Geothermal Reservoir Engineering*, Stanford University, Stanford, Calif., 1994, in press.
38. D. W. Brown, in Ref. 32, pp. 153–157.
39. D. W. Brown and R. DuTeau, "Progress Report on the Long-Term Flow Testing of the HDR Reservoir at Fenton Hill, New Mexico", in Ref. 37.
40. J. W. Tester and H. J. Herzog, *Economic Predictions for Heat Mining: A Review and Analysis of Hot Dry Rock (HDR) Geothermal Energy Technology*, MIT Energy Lab. Report #MIT-EL 90-0001, Massachusetts Institute of Technology, Cambridge, Mass., 1990, p. 59.

41. Ref. 40, p. 137.
42. D. V. Duchane, *Geotherm. Res. Counc. Bull.* **20**(5), 135–142 (1991).
43. J. C. Eichelberger and J. C. Dunn, *Geotherm. Resourc. Counc. Bull.* **19**(2), 53–56 (1990).
44. J. T. Finger, *Proceedings of the Geothermal Program Review X,* U.S. Department of Energy, Washington, D.C., 1992, pp. 127–128.
45. R. A. Crewdson, W. F. Martin, Jr., D. L. Taylor, and K. Bakhtar, *An Evaluation of the Technology and Economics of Extracting Energy from Magma Resources for Electric Power Generation,* report by Mine Development and Engineering Co., Bakersfield, Calif., 1991, p. 3.

General References

R. DiPippo, *Geothermal Energy as a Source of Electricity,* U.S. Department of Energy, Washington, D.C., 1980.
J. Kestin, ed., *Sourcebook on the Production of Electricity from Geothermal Energy,* U.S. Department of Energy, Washington, D.C., 1980.
L. M. Edwards, G. V. Chilingar, H. H. Rieke III, and W. H. Fertl, eds., *Handbook of Geothermal Energy,* Gulf Publishing Co., Houston, Tex., 1982.
C. Stone, ed., *Monograph on The Geyser's Geothermal Field,* Geothermal Resources Council, Davis, Calif., 1992.
J. Negus-de Wys, *The Geopressured Habitat, A Selected Literature Review,* U.S. Department of Energy, Washington, D.C., 1992.
H. C. H. Armstead and J. W. Tester, *Heat Mining,* E. & F. N. Spon, London, 1987.
J. W. Tester, D. W. Brown, and R. M. Potter, *Hot Dry Rock Geothermal Energy—A New Energy Agenda for the 21st Century,* Los Alamos National Laboratory Report LA-11514-MS, Los Alamos, N.M., 1989.
R. A. Crewdson, W. F. Martin, Jr., D. L. Taylor, and K. Bakhtar, *An Evaluation of the Technology and Economics of Extracting Energy from Magma Resources for Electric Power Generation,* report by Mine Development and Engineering Co., Bakersfield, Calif., 1991.
M. Brower, *Cool Energy,* MIT Press, Cambridge, Mass., 1992.

DAVID DUCHANE
Los Alamos National Laboratory

GERANIOL. See TERPENOIDS.

GERANIUM OIL. See OILS, ESSENTIAL; PERFUMES.

GERMANIUM AND GERMANIUM COMPOUNDS

Germanium [7440-56-4], Ge, at. no. 32, having electronic configuration [Ar] $3d^{10}4s^24p^2$, is a semiconducting metalloid element found in Group 14 (IVA), Period 4 of the Periodic Table. Although it looks like a metal, it has a diamond cubic crystal structure and is fragile like glass (qv). Its electrical resistivity is about midway between that of metallic conductors and good electrical insulators. It was first discovered by Winkler in 1886 (1), but no commercial application was found until the early 1940s, when its interesting electrical properties were discovered. The first significant use was in solid-state electronics, and, using germanium, the transistor was invented. Indeed, the entire modern field of semiconductors (qv) owes its development to the early successful use of germanium. Whereas germanium is still used in the field of electronics, its use in infrared optics surpassed electronic applications in the 1970s (see INFRARED AND RAMAN SPECTROSCOPY, INFRARED TECHNOLOGY). Germanium has also found widespread use in the fields of γ-ray spectroscopy, catalysis, and fiber optics (qv).

Occurrence

The crust of the earth is estimated to contain Ge at a concentration of 1–7 g/t. Germanium, which usually occurs widely dispersed in minerals such as sphalerite [12169-28-7], rarely occurs in concentrated form. Almost all germanium production has been from zinc smelters (see ZINC AND ZINC ALLOYS). Copper (qv) smelters are the second largest source. There are only a few minerals of germanium, some having germanium concentrations up to about 8%. Most of these occur in Africa, where the highest concentration is near Tsumeb, Namibia. Nineteen germanium minerals found near Tsumeb have been reviewed (2). A discussion of the geochemistry and mineralogy of germanium is also available (3).

     Germanium also occurs in significant concentrations in many coals around the world. When coal (qv) is burned in power-generating or coking plants, the germanium tends to concentrate in the fly ash or flue dust produced. Significant germanium recovery from coal in Great Britain was reported in the 1950s (4), and amounts have been reported from other countries since then. Production from coal in the 1990s is reported in the former USSR and in China.

Properties

The physical, thermal, and electronic properties of germanium metal are shown in Table 1. Optical properties are given in Table 2.

Chemical Properties

     Germanium Metal.   Germanium is quite stable in air up to 400°C where slow oxidation begins. Oxidation becomes noticeably more rapid above 600°C. The metal resists concentrated hydrochloric acid, concentrated hydrofluoric acid, and

Table 1. Properties of Germanium

Parameter	Value
Physical properties	
atomic weight	72.59
density at 25°C, g/cm^3	5.323
atomic density at 25°C, at/cm^3	4.416×10^{22}
lattice constant at 25°C, a_o, nm	0.565754
surface tension, liquid at mp, mN/m($=$dyn/cm)	650
modulus of rupture, MPa[a]	110
Mohs' hardness	6.3
Poisson's ratio at 125–375 K	0.278
isotopes	
mass number	70 72 73 74 76
natural abundance, %	20.4 27.4 7.8 36.6 7.8
Thermal properties	
melting point, °C	937.4
boiling point, °C	2830
heat capacity at 25°C, J/(kg·K)[b]	322
latent heat of fusion, J/g[b]	466.5
latent heat of vaporization, J/g[b]	4602
heat of combustion, J/g[b]	7380
heat of formation, J/g[b]	4006
vapor pressure, kPa[c]	
at 2080°C	1.33
at 2440°C	13.3
at 2710°C	53.3
at 2830°C	101.3
coefficient of linear expansion, 10^{-6}/K	
at 100 K	2.3
at 200 K	5.0
at 300 K	6.0
thermal conductivity, W/(m·K)	
at 100 K	232
at 200 K	96.8
at 300 K	59.9
at 400 K	43.2
Electronic properties	
intrinsic resistivity at 25°C, Ω·cm	53
intrinsic conductivity type	N (negative)
intrinsic drift mobility at 25°C, cm^2/(V·s)	
electron	3800
hole	1850
band gap, direct, minimum eV	
at 25°C	0.67
at 0 K	0.744
number of intrinsic electrons at 25°C, cm^{-3}	2.12×10^{13}

[a]To convert MPa to psi, multiply by 145.
[b]To convert J to cal, divide by 4.184.
[c]To convert kPa to mm Hg, multiply by 7.5.

Table 2. Optical Properties of Germanium

Wavelength, μm	Refractive index at 25°C	Absorption coefficient, cm^{-1}
1.8	4.134	7.0
1.9	4.120	0.68
2.0	4.108	0.010
4.0	4.0255	0.0047
6.0	4.0122	0.0068
8.0	4.0074	0.0150
10.0	4.0052	0.0215
10.6	4.0048	0.0270
11.0	4.0045	0.0295
11.9	4.0040	0.200
12.0	4.0039	0.170

concentrated sodium hydroxide solutions, even at their boiling points. It is not attacked by cold sulfuric acid but does react slowly with hot sulfuric acid. Nitric acid attacks germanium at all temperatures more readily than does sulfuric acid. Germanium reacts readily with mixtures of nitric and hydrofluoric acids and with molten alkalies and more slowly with aqua regia. The principal reaction route for the mixed acids is the oxidation of the germanium by one constituent, then dissolution of the oxide by the other constituent. The reaction with fused alkalies is a direct oxidation with the release of hydrogen.

In compounds, germanium can have a valence of either 2 or 4. Although the divalent compounds tend to be less stable than the tetravalent ones, most can be stored at room temperature for years with no change in composition. At higher temperatures, most of the divalent compounds decompose. The syntheses and properties of many germanium compounds have been reviewed (5) and properties of germanium bonds discussed (5). There is also an excellent earlier review of inorganic germanium compounds (6).

Germanium Halides. Germanium tetrachloride [10038-98-8], $GeCl_4$, is made by the reaction of hydrochloric acid on germanium concentrates containing oxides or germanates. It can also be made by the reaction of chlorine on heated metallic germanium. The properties of $GeCl_4$ are shown in Table 3.

$GeCl_4$ is soluble in solvents such as acetone, absolute ethanol, benzene, carbon disulfide, carbon tetrachloride, chloroform, and diethyl ether. It is only slightly soluble in concentrated hydrochloric acid, and the solubility drops with acid normality, reaching a minimum at 5 N. At HCl concentrations below 5 N, the tetrachloride begins to hydrolyze to GeO_2. Germanium tetrachloride is insoluble in concentrated sulfuric acid and does not react with it. The solubility of free chlorine in $GeCl_4$ can reach as high as 4 wt %, especially at low temperatures.

Germanium tetrabromide [13450-02-5], $GeBr_4$, and germanium tetraiodide [13450-95-8], GeI_4, can be prepared easily by the reaction of the respective halogen and germanium metal or by the reaction of GeO_2 with HBr and HI solutions, respectively. The preparation of germanium tetrafluoride [7783-58-6], GeF_4, is not

Table 3. Properties of Germanium Tetrachloride

Property	Value
molecular weight	214.40
color	colorless
density at 25°C, g/cm^3	1.874
melting point, °C	-49.5
boiling point, °C	83.1
refractive index at 25°C, 0.5893 μm	1.464
heat capacity, C_p, of vapor at 25°C, J/(kg·K)^a	449
heat of vaporization at bp, J/g^a	137
heat of formation at 25°C, J/g^a	-3318
vapor pressure, Pa^b	
at 225 K	10^2
at 253 K	10^3
at 294 K	10^4
at 356 K	10^5
at 462 K	10^6
at $T_c = 550$ K	3.850×10^6

^{a}To convert J to cal, divide by 4.184.
^{b}To convert Pa to mm Hg, divide by 133.3.

so straightforward, and pure GeF$_4$ is usually made by decomposing barium hexa-fluorogermanate [60897-63-4], BaGeF$_6$, at ca 700°C (7) (see FLUORINE COM-POUNDS, INORGANIC–GERMANIUM).

Germanium Oxides.   Germanium dioxide [1310-53-8], GeO$_2$, is usually made by the hydrolysis of GeCl$_4$. GeO$_2$, also made by the ignition of germanium disulfide, exists in soluble, insoluble, and vitreous forms. The properties of these three forms are given in Table 4. The soluble form is the usual product of GeCl$_4$ hydrolysis. The insoluble form can be prepared by heating soluble oxide at 300–900°C, especially in the presence of about 0.5 wt % alkali halides. The glassy, or vitreous, form is prepared by melting either of the other forms and then cooling the melt.

Table 4. Properties of Germanium Dioxide

Property	Form		
	Soluble	Insoluble	Vitreous
structure	hexagonal	tetragonal	amorphous
density at 25°C, g/cm^3	4.228	6.239	3.637
melting point, °C	1116	1086	
solubility in water, g/L soln			
at 25°C	4.53	insoluble	5.18
at 100°C	13	insoluble	
solubility in HCl, HF, NaOH solutions	soluble	insoluble	soluble

Germanium monoxide [20619-16-3], GeO, can best be prepared in pure form by heating a mixture of Ge and GeO_2, in the absence of oxygen. At temperatures above 710°C, GeO sublimes from the mixture and condenses as a glassy deposit in the cooler part of the reaction vessel. Germanium monoxide is stable at room temperature.

Germanates. Germanates are usually prepared by the fusion of GeO_2 with alkali oxides or carbonates in platinum crucibles. Sodium heptagermanate [12195-31-2], $Na_3HGe_7O_{16} \cdot 4H_2O$, is precipitated by the neutralization of a sodium hydroxide solution of GeO_2 with hydrochloric acid to a pH above 7.

Germanides. Germanides can be formed by melting other metals with germanium in the proper stoichiometric concentrations and then freezing the melt. These compounds can also be prepared by vacuum-sintering the two metals together, usually followed by long annealing. Other procedures include the thermal dissociation of one germanide into another and the electrolysis of fused mixed salts. The preparation and properties of about 200 germanides have been tabulated and reviewed (8). One of the germanides that has been prepared most often is magnesium germanide [1310-52-7], Mg_2Ge.

Germanes. Germanium hydrides can be prepared by the reaction of a germanide, such as Mg_2Ge, with hydrochloric acid. Germane [7782-65-2], GeH_4, can also be produced by the reduction of $GeCl_4$ using lithium aluminum hydride or by the reduction of GeO_2 by sodium borohydride in water solution. The preparation and properties of the germanes have been reviewed (9,10).

Miscellaneous Inorganic Compounds. Germanium nitride [12065-36-0], Ge_3N_4, is about as inert as tetragonal GeO_2. It is prepared most easily from germanium powder and ammonia at 700–850°C. The nitride does not react with most mineral acids, aqua regia, or caustic solutions, even when hot. Germanium disulfide [12025-34-2], GeS_2, is an unusual and useful compound because it is insoluble in strong acids such as 6 N HCl and 12 N H_2SO_4. This insolubility permits the recovery of germanium from acid solutions by gassing with H_2S. The disulfide can also be made by the reaction of GeO_2 with sulfur.

Organogermanium Compounds. The field of organogermanium chemistry has drawn widespread interest for many years. Organogermanium compounds are generally characterized as having low chemical reactivity and relatively high thermal stability. Many syntheses begin with a Grignard reaction (qv). Many excellent reviews of the organogermanium literature have been published (11–23). Several organogermanium compounds have been produced in commercial quantities. These include spirogermanium [41992-22-7] (2-aza-8-germaspiro-[4,5]-decane-2-propanamine-8,8-diethyl-N,N-dimethyl   dihydrochloride), $C_{17}H_{36}GeN_2 \cdot 2HCl$, and carboxyethyl germanium sesquioxide [27031-31-8] (3,3'-germanoic anhydride dipropanoic acid), $C_6H_{10}Ge_2O_7$. These compounds have been studied extensively for possible anticancer and blood pressure effects (24) (see CARDIOVASCULAR AGENTS; CHEMOTHERAPEUTICS, ANTICANCER).

Alloys. Many Ge alloys have been prepared and studied. Most have been made by melting Ge with another metal, much as germanides are made. Collections of binary phase diagrams and comments about many Ge alloys are available (25–28).

Manufacturing and Processing

Ore Processing. No mineral is mined solely for its germanium content. Almost all of the Ge recovered worldwide is a by-product of other metals, primarily zinc, copper, and lead (qv). The enriched copper–lead concentrates from Tsumeb, Namibia, have been treated in a vertical retort from which germanium sulfide is sublimed and separated (4). The copper–zinc ores of Katanga, Zaire, have been treated by roasting with H_2SO_4, followed by leaching and selective precipitation of the germanium with MgO (4). In the United States, zinc concentrates have been roasted and then sintered for zinc recovery (29). The sinter fume is chemically leached, and the germanium is selectively precipitated from the leach solution by fractional neutralization and sent to the germanium refinery. Because of the low solubility of germanium sulfide and tannate in acid solutions, germanium has been recovered by precipitating it from acid solutions using H_2S or tannic acid. Sulfide precipitates are usually oxidized using sodium chlorate or permanganate, followed by, or concurrent with, dissolution in concentrated HCl and distillation of the resulting $GeCl_4$. Tannic acid precipitates are usually upgraded by igniting the precipitate to the oxide and dissolving the oxide in concentrated HCl, with subsequent distillation of the $GeCl_4$. Germanium can also be recovered from the still residue in the distillation of zinc metal (30).

From 1986 to 1990, germanium was recovered intermittently from the Apex Mine near St. George, Utah. This was an unusual operation in that germanium was the principal product from the mine and gallium was a by-product. Copper was also recovered because the Apex Mine was an old abandoned copper mine. The process involved dissolution of the screened ore in sulfuric acid, followed by cementation of the copper, solvent extraction of the gallium, and precipitation of the germanium with H_2S. The GeS_2 precipitate was oxidized with sodium chlorate and dissolved in HCl. The $GeCl_4$ formed was distilled and then hydrolyzed to a crude GeO_2. The operation was not a financial success and the producer was forced into bankruptcy. The assets were purchased in 1989 by Hecla Mining Co., which made significant changes in the processing circuit, including the addition of a solvent extraction system for germanium recovery. However, even these changes were not enough to provide profitability to the operation in a declining market, and production was again stopped in August 1990.

In electrolytic zinc plants, which have become increasingly important for environmental reasons, germanium is precipitated, usually along with iron, during the purification of the $ZnSO_4$ electrolyte prior to electrolysis. Germanium is one of several impurities that have an adverse effect on zinc electrolysis. If the germanium concentration is high enough in the separated solids, economic recovery of germanium is possible. Several solvent extraction and ion-exchange (qv) processes have been developed (31–37) that provide better germanium separations, primarily from $ZnSO_4$ electrolytes.

Purification. A simplified flow diagram for a germanium refinery is shown in Figure 1. Regardless of the source of Ge, all Ge concentrates are purified by similar techniques. The ease with which concentrated germanium oxides and germanates react with concentrated hydrochloric acid and the convenient boiling point (83.1°C) of the resulting $GeCl_4$ make chlorination a standard refining step.

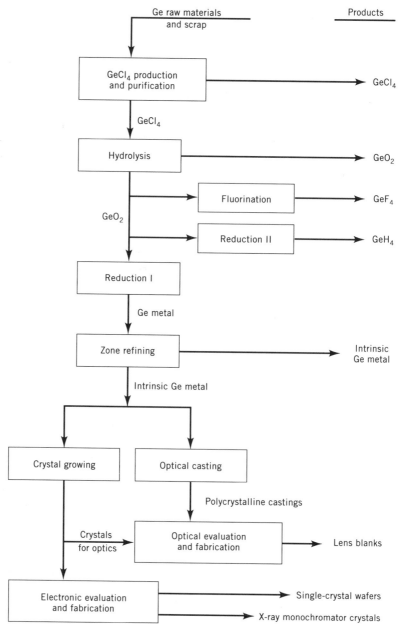

Fig. 1. Simplified germanium flow diagram.

An oxidizing agent is often added to the primary distillation or to the subsequent fractionation, or both, to suppress the volatility of arsenic (38). Other purification steps are used to separate whatever objectionable impurities are present. The fractionation is usually done in glass or quartz because most subsequent uses of GeCl$_4$ require metallic impurity levels of no more than about 1 mg/kg (1 ppm).

The purified $GeCl_4$ is hydrolyzed using deionized water to produce GeO_2, which is removed by filtration and dried. The dried GeO_2 is reduced with hydrogen at about 760°C to germanium metal powder, which is subsequently melted and cast into so-called first-reduction, or as-reduced, bars. These bars are then subjected to zone refining to produce intrinsic or electronic-grade germanium metal. Zone refining is ideally suited for the refining of germanium; in fact, the original development of this procedure was prompted by the need for ultrapure germanium (39).

Zone refining results in polycrystalline germanium, usually containing total impurities of <100 ng/g of Ge and electrically active impurities of <0.5 ng/g of Ge. For extremely high purity applications, such as for uncompensated γ-ray detector crystals, germanium has been refined to impurity concentrations of less than 0.0003 ng/g (0.0003 ppb). However, zone refining (qv) is not equally efficient in removing all impurities. Boron and silicon are not removed easily with this method and must be removed from germanium before the zone-refining step.

For use as a semiconductor, refined germanium is grown into single crystals from the melt. The electronic properties are controlled by the addition of selected impurities (dopants) to the melt before crystal growth begins. The grown crystal is sliced and fabricated into devices, and the germanium scrap generated is recycled through the refinery into intrinsic germanium.

For use in infrared optics, zone-refined germanium is recast or grown into forms suitable for lens and window manufacture. Polycrystalline castings of up to 115 kg and single crystals of up to 90 kg are routinely made. After the germanium is annealed, it is cut and ground into lens- or window-blanks, which are then polished, coated, and assembled into an ir system (see INFRARED AND RAMAN SPECTROSCOPY, INFRARED TECHNOLOGY).

Environmentally, the production of germanium is quite benign. Among the products produced, the chemicals consumed, and the by-products made, only the arsenic impurity would normally be considered a problem. The bulk of the arsenic is separated from the germanium at the smelter, and the small amounts entering the germanium refining plant are easily controlled. The acids, bases, and chlorine used in processing and in gas scrubbing can be neutralized and held in permanent containment ponds or disposed of in hazardous waste landfills. No hazardous substances are known to be discharged into surface waters or municipal treatment facilities from germanium refining plants.

Economic Aspects

World reserves of germanium have been estimated at 4000 metric tons, but it is impossible to discuss reserves without considering price. In many applications, the cost of germanium is but a small part of the overall cost of the product. Thus substantially higher germanium prices would have little impact on such uses but would expand germanium reserves significantly.

Because germanium is almost always recovered as a by-product, its price and availability over the long range are subject to supply and demand considerations for its host products, usually zinc and copper, as well as for itself. This is not the case over the short term (6–12 months) because producers often recover germanium from stockpiles of smelter residues that can last for many months. Therefore, short-term pricing is largely controlled only by demand. Figure 2 plots the price levels, adjusted for inflation, of intrinsic germanium in the United States since the mid-1970s. The November 1978 adjusted price of $467.50/kg was an all-time low. The price increases during 1979–1981 resulted from the increased demands for Ge ir optics and for $GeCl_4$ for fiber optics. The inability of Ge producers to fill all demands during this period from regular sources required the use of lower grade residues which incurred higher treatment costs and, hence, higher prices. Supply and demand were essentially in balance during 1982–1984, and the metal price remained at $1060/kg, unadjusted. During 1985–1986, demand for $GeCl_4$ and Ge optics softened significantly, and prices fell accordingly, reaching $750/kg ($684 adjusted) in August 1986. The U.S. price for intrinsic Ge remained at $750 through 1992 although inflation decreased the adjusted price to about $525/kg at the end of that year. There is also a free-market price for Ge metal. This price applies to Ge of variable quality usually exported from China or the Commonwealth of Independent States (CIS). The price of such metal is usually $100–150/kg less than U.S. or European producer metal.

World production of germanium exceeded 100 t/yr during the period of peak demand for Ge semiconductors in the 1950s and early 1960s. Production varied between 30 and 80 t/yr during the 1970s and between 80 and 110 t/yr in the 1980s.

An annual review of germanium is published by the U.S. Bureau of Mines (40) and a broader survey is published every five years (41). The *Engineering and Mining Journal* also published an annual germanium review through 1986 (42). The economics of germanium, with special emphasis on world trade, have been reviewed (43).

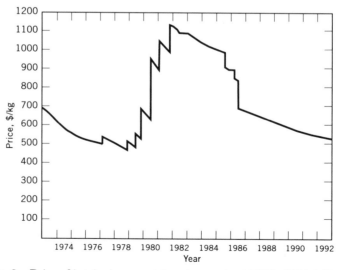

Fig. 2. Price of intrinsic germanium in constant 1982–1984 dollars.

Specifications

Germanium tetrachloride refined for use in making optical fibers is usually specified to contain less than 0.5 to 5 ppb of each of eight impurities: vanadium, chromium, manganese, iron, cobalt, nickel, copper, and zinc. Limits are sometimes specified for a few other elements. Also of concern are hydrogen-bearing impurities; therefore, maximum limits of 5 to 10 ppm are usually placed on HCl, OH, CH_2, and CH_3 contents.

Most electronic-grade GeO_2 is specified to have 99.9999% minimum purity (6 N) as measured spectrographically and to have 0.1–0.2% maximum volatile content, a bulk density of 1.5–2.0 g/cm^3, a metal billet resistivity of at least 5 $\Omega \cdot cm$ average on the last third of the billet to freeze, and an N-type (negative) billet conductivity type. The GeO_2 intended for chemical use, such as for use as a catalyst, is at least 5 N pure spectrographically, at least 98% finer than 325 mesh (44 μm), and at least 99.9% soluble in water or ethylene glycol.

All intrinsic germanium metal sold is specified to be N-type with a resistivity of at least 40 $\Omega \cdot cm$ at 25°C or 50 $\Omega \cdot cm$ at 20°C. Germanium metal prepared for use in infrared optics is usually specified to be N-type with a resistivity of 4–40 $\Omega \cdot cm$, to be stress-free and fine annealed, and to have certain minimum transmission (or maximum absorption) characteristics in the 3–5 or 8–12 μm wavelength ranges. Either polycrystalline or single-crystal material is specified.

Germanium single crystals intended for electronic applications are usually specified according to conductivity type, dopant, resistivity, orientation, and maximum dislocation density. They may be specified to be lineage-free unless the specified resistivity is below about 0.05 $\Omega \cdot cm$. Minority carrier lifetime and majority carrier mobility are occasionally specified.

Analytical and Test Methods

The analysis of ores of germanium is usually done with an emission spectrograph but can be done in the field using the phenylfluorone method (44). Analysis of germanium refinery samples is usually done after fusion of the sample with KOH or NaOH in nickel crucibles. Following distillation of the $GeCl_4$ from HCl solution of the fusion, the Ge can be determined gravimetrically, usually by precipitation of GeS_2 from acid solution and by ignition to GeO_2; titrimetrically, usually by reduction with sodium hypophosphite and titration with KIO_3 solution; or spectrally, using an atomic absorption spectrophotometer. The last procedure is not considered as accurate as either of the first two. Excellent reviews of the analytical chemistry of germanium have been published (45,46).

Analysis of refined germanium products is done in a wide variety of ways, including several methods that have become ASTM standards (47). Electronic-grade GeO_2 is analyzed using an emission spectrograph to determine its spectrographic purity. Its volatile content is measured in accord with ASTM F5 and its bulk density with F6. Other ASTM standards cover the preparation of a metal billet from a sample of the oxide (F27), and the determination of the conductivity type (F42) and resistivity (F43) of the billet.

The type and resistivity of all grades of germanium metal are also measured in accord with F42 and F43. The transmission characteristics of optical-grade germanium are determined with an infrared spectrophotometer, and the measurement of the interstitial oxygen content of the metal is covered in F120 and F122. Germanium single crystals can be further evaluated in accord with ASTM F26, F28, F76, F389, F398, and F673.

The overall spectrographic purity of $GeCl_4$ can be determined by using an emission spectrograph to examine a sample of GeO_2 produced from the $GeCl_4$ by hydrolysis with deionized water. The trace metal impurities of concern to fiber optic producers are determined by flameless atomic absorption, and the hydrogen-bearing impurity concentrations are measured by infrared absorption of the $GeCl_4$.

Toxicology

Germanium compounds generally have a low order of toxicity (24). Only germane [7782-65-2], GeH_4, is considered toxic, having a maximum time-weighted average 8-h safe exposure limit of only 0.2 ppm (48). The lethal dose median for GeO_2 is 750 mg/kg, and that of germanium is 586 mg/kg (49). The toxicity of specific germanium compounds usually must be considered more from the standpoint of the other part of the compound than from the Ge content. The biological activity of germanium has been reviewed (24).

Uses

The earliest commercial use of germanium was as a solid-state diode that performed as a detector in radar systems in 1941. This device was adapted to radio circuits as a radio frequency detector, and the germanium transistor was invented in 1947. This invention revolutionized the electronics industry and caused a sharp increase in the demand for germanium. The use of germanium in these conventional semiconductor roles reached a peak in the early 1960s, when world consumption averaged over 100 t/yr. Except for a brief upsurge in this demand during 1969, there has been a general decline in this application of germanium since the early 1960s.

The use of germanium as a semiconductor substrate deserves special mention. In this application, single-crystal wafers of germanium are used as substrates for the epitaxial deposition of gallium arsenide [1303-00-0], GaAs, or gallium arsenide phosphide [12044-20-1], GaAsP, for use as light-emitting diodes or solar cells. These substrates take the place of more expensive gallium arsenide wafers. Many metric tons of germanium were consumed in the mid-1970s for the production of GaAsP/Ge light-emitting diodes for calculators and watches (50). Large-scale production of GaAs/Ge solar cells did not begin until the late 1980s (51), but production is expected to increase in the 1990s. As a substrate for GaAs solar cells, Ge provides stronger, lighter, and cheaper cells while maintaining the high conversion efficiency of gallium arsenide (see LIGHT GENERATION, LIGHT-EMITTING DIODES; SOLAR ENERGY).

Many of the uses of germanium and Ge compounds have been reviewed (52). The largest use of germanium as of this writing is in the field of infrared optics. In this application, the transparency of germanium to infrared wavelengths longer than 2 μm and its high refractive index are utilized. Other advantageous properties of Ge for this use are low dispersion; easy machinability; reasonable strength; low price compared to other infrared materials; good resistance to atmospheric oxidation, to moisture, and to chemical attack; and availability in large sizes. It had been estimated in 1978 that the world demand for germanium for infrared devices would increase to 55–70 t/yr during the 1980s (53). From 55–65 t/yr was indeed consumed in this application from 1987 to 1989.

Infrared devices are principally used for military applications. Indeed, germanium should share in the credit, along with new ir detector materials, for revolutionizing modern warfare. Germanium infrared systems usually operate in the 8–12 μm range and usually contain several germanium lenses, a germanium window, and a color-correcting lens made from a Ge–Sb–Se glass (Texas Instruments, Inc., Dallas, Texas; and Amorphous Materials, Inc., Garland, Texas), a Ge–As–Se glass (Amorphous Materials, Inc.), or ZnSe. Some of the viewers also utilize a germanium infrared detector. The U.S. government accumulated 69 t of intrinsic Ge metal in its strategic stockpile during 1987–1991.

Nonmilitary infrared applications for germanium include CO_2 lasers (qv), intrusion alarms, and police and border patrol surveillance devices. Germanium is used as a thin-film coating for infrared materials to decrease reflection losses or to provide heavy filtering action below 2 μm.

Germanium metal is also used in specially prepared Ge single crystals for γ-ray detectors (54). Both the older lithium-drifted detectors and the newer intrinsic detectors, which do not have to be stored in liquid nitrogen, do an excellent job of spectral analysis of γ-radiation and are important analytical tools. Even more sensitive Ge detectors have been made using isotopically enriched Ge crystals. Most of these have been made from enriched ^{76}Ge and have been used in neutrino studies (55–57).

The primary application of germanium dioxide is in the preparation of germanium metal. However, there are several other uses that provide significant markets for the oxide. The largest of these is its use in place of antimony oxide as a catalyst in the reaction of ethylene glycol with terephthalic acid in the production of polyester fibers and poly(ethylene terephthalate) (PET) resins (see FIBERS, POLYESTER). Although considerably more expensive than Sb_2O_3, GeO_2 produces a polyester fiber that does not yellow with age, which is especially attractive to makers of white shirts and other white fabrics. The PET resins are used almost entirely in making beverage bottles such as 2- and 3-liter soft drink bottles (see CARBONATED BEVERAGES). Germanium dioxide produces a stronger, clearer bottle than those made with Sb_2O_3, but the change to GeO_2 is also being driven by the concern over the long-term effects of trace contamination of the bottle contents with antimony. The use of GeO_2 in this application is more widespread in the Pacific Rim countries than in other parts of the world.

Another significant use of GeO_2 is the production of bismuth germanium oxide [12233-56-6], $Bi_4Ge_3O_{12}$, crystals. These scintillation crystals are used primarily in positron emission tomography scanners, which are expected to find increased use as suitable radioisotopes become less expensive. Germanium dioxide is also used in significant quantities in the production of spirogermanium and carboxyethyl germanium sesquioxide (24), both of which have potential medical applications. In addition, GeO_2 is included in a few special glass formulations, primarily to increase the refractive index of the glass (qv).

The only significant application of $GeCl_4$, besides its use in the production of GeO_2, is in optical fibers (see FIBER OPTICS). In this application, $GeCl_4$ is converted to GeO_2 which is deposited, along with SiO_2 and sometimes B_2O_3 or P_2O_5, on the inside of a pure quartz tube which is subsequently collapsed to form a solid rod or preform (58,59). Alternatively, the mixed oxides can be deposited on the outside of a quartz rod and then covered with a second layer consisting only of SiO_2 (60) to form a preform. The preform is then drawn into fine fibers that can be used as optical waveguides, primarily in the 0.8–1.6 μm wavelength region. Such fibers have proved very useful, especially in long-distance telephone lines. Phone signals have been transmitted over 100 km (65 miles) through such fibers without amplification. Germanium dioxide provides the higher refractive index fiber core, which prevents signal loss; also, it produces an extremely clear glass that provides very low signal absorption at those wavelengths at which infrared emitters are available. $GeCl_4$ has also been used as an oil-cracking catalyst promoter.

Germane is used, along with silane, SiH_4, to make amorphous or crystalline silicon solar cells having an extended solar energy absorption range to increase conversion efficiency.

Several Ge alloys have found commercial applications. Examples include the following: Ge–Au alloys are used in precision castings, dental alloys, and electronic connection solders; Ag–Cu–Ge and Pd–Ge alloys are also used for dental castings; Ge–Cu alloys exhibit good resistance to acid corrosion; Ge is used in the precipitation-hardening of aluminum, and Ge–Al alloys are used in soldering of aluminum; Ge–Ag alloys have been found to decrease abrasion losses in commutators; Ge–Si alloys are useful in several thermoelectric applications; and Ge–Te alloy films are used in rewritable phase-change optical storage disks (61).

Other uses of germanium include the use of magnesium germanate as a phosphor; the addition of lead germanate to barium titanate capacitors; the use of niobium germanide [12025-22-8], Nb_3Ge, as a superconductor which has a T_c of 23 K; the use of germanium single crystals as x-ray monochromators for high energy physics applications (62); and the use of Ge metal for phase-change energy storage in satellites (63).

BIBLIOGRAPHY

"Germanium" in *ECT* 1st ed., Vol. 7, pp. 168–174, by A. W. Laubengayer, Cornell University, and A. E. Middleton, Battelle Memorial Institute; "Germanium and Germanium Compounds" in *ECT* 2nd ed., Vol. 10, pp. 519–527, by H. R. Harner, The Eagle-Picher Co., and A. W. Laubengayer, Cornell University; in *ECT* 3rd ed., Vol. 11, pp. 791–802, by J. H. Adams, Eagle-Picher Industries, Inc.

1. C. Winkler, *Ber. Deut. Chem. Ges.* **19**, 210–211 (1886).

2. W. E. Wilson, ed., *Tsumeb! The World's Greatest Mineral Locality*, The Mineralogical Record, Inc., Bowie, Md, 1977.
3. L. R. Bernstein, *Geochim. Cosmochim. Acta* **49**, 2409–2422 (1985).
4. *Eng. Min. J.* **157**, 75 (1956).
5. F. Glocking, *The Chemistry of Germanium*, Academic Press, Inc., London, 1969.
6. O. H. Johnson, *Chem. Rev.* **51**, 431 (1952).
7. C. J. Hoffman and H. S. Gutowsky, *Inorg. Synth.* **4**, 147 (1953).
8. G. V. Samsonov and V. N. Bondarev, *Germanides*, trans. A. Wald, Primary Sources, New York, 1970.
9. F. G. A. Stone, *Hydrogen Compounds of the Group IV Elements*, Prentice-Hall, Inc., Englewood Cliffs, N.J., 1962, pp. 63–76.
10. E. G. Rochow, in J. C. Bailar, Jr., and co-workers, eds., *Comprehensive Inorganic Chemistry*, Vol. 2, Pergamon Press Ltd., Oxford, U.K., 1973, pp. 1–41.
11. O. H. Johnson, *Chem. Rev.* **48**, 259 (1951).
12. E. G. Rochow, D. T. Hurd, and R. N. Lewis, *The Chemistry of Organometallic Compounds*, John Wiley & Sons, Inc., New York, 1957.
13. F. Rijkens, *Organogermanium Compounds*, Germanium Research Committee, Utrecht, the Netherlands, 1960.
14. D. Quane and R. S. Bottei, *Chem. Rev.* **63**, 403 (1963).
15. F. Rijkens and G. J. M. Van der Kerk, *Investigations in the Field of Organogermanium Chemistry*, Germanium Research Committee, Utrecht, the Netherlands, 1964.
16. F. Glocking, *Quart. Rev. Chem. Soc.* **20**, 45 (1966).
17. M. Dub, *Organometallic Compounds*, 2nd ed., Vol. 2, Springer-Verlag, Berlin, 1967.
18. K. A. Hooton, in W. L. Jolly, ed., *Preparative Inorganic Reactions*, Vol. 4, John Wiley & Sons, Inc., New York, 1968, pp. 85–176.
19. N. Hagihara, ed., *Handbook of Organometallic Compounds*, W. A. Benjamin, Inc., New York, 1968, pp. 449–467.
20. Ref. 5, pp. 58–199.
21. M. Lesbre, P. Mazerolles, and J. Satgé, *The Organic Compounds of Germanium*, John Wiley & Sons, Ltd., London, 1971.
22. Ref. 10, pp. 29–39.
23. R. C. Weast, ed., *CRC Handbook of Chemistry and Physics*, 59th ed., CRC Press, West Palm Beach, Fla., 1978, pp. C-692–697.
24. A. Furst, *Toxicol. Ind. Health* **3**, 167–204 (1987).
25. M. Hansen and K. Anderko, eds., *Constitution of Binary Alloys*, 2nd ed., McGraw-Hill Book Co., Inc., New York, 1958.
26. R. P. Elliot, ed., *Constitution of Binary Alloys*, First Supplement, McGraw-Hill Book Co., Inc., New York, 1965.
27. F. A. Shunk, ed., *Constitution of Binary Alloys*, Second Supplement, McGraw-Hill Book Co., Inc., New York, 1969.
28. W. G. Moffatt, *The Handbook of Binary Phase Diagrams*, Genium Publishing Corp., Schenectady, N. Y., 1989, updated.
29. J. A. O'Connor, *Chem. Eng.* **59**, 158 (1952).
30. U.S. Pat. 4,090,871 (May 23, 1978), A. Lebleu, P. Fossi, and J. Demarthe (to Societe Miniere et Metallurgique de Penarroya).
31. U.S. Pat. 3,883,634 (May 13, 1975), A. DeSchepper and A. Van Peteghem (to Metallurgie Hoboken-Overpelt).
32. U.S. Pat. 4,389,379 (June 21, 1983), D. Rouillard and co-workers (to Soc. Miniere et Metallurgique de Penarroya).
33. U.S. Pat. 4,432,951 (Feb. 21, 1984), A. DeSchepper and co-workers (to Metallurgie Hoboken-Overpelt).

34. U.S. Pat. 4,432,952 (Feb. 21, 1984), A DeSchepper and co-workers (to Metallurgie Ho-boken-Overpelt).

35. U.S. Pat. 4,525,332 (June 25, 1985), D. Boateng and co-workers (to Cominco, Ltd).

36. U.S. Pat. 4,568,526 (Feb. 4, 1986), G. Coat and co-workers (to Soc. Miniere et Metal-lurgique de Penarroya).

37. U.S. Pat. 4,666,686 (May 19, 1987), W. Krajewski and K. Hanusch (to Preussag AG Metall).

38. H. R. Harner, in C. A. Hampel, ed., *Rare Metals Handbook*, 2nd ed., Reinhold Publishing Co., 1961, pp. 188–197.

39. W. G. Pfann, *Zone Melting*, 2nd ed., John Wiley & Sons, Inc., New York, 1966, p. 134.

40. T. O. Llewellyn, *Germanium Minerals Yearbook*, U.S. Department of the Interior, Washington, D.C., 1989.

41. P. A. Plunkert, "Germanium" in *Mineral Facts and Problems*, U.S. Department of the Interior, Washington, D.C., 1985.

42. J. H. Adams, *Eng. Min. J.* **187**, 46 (1986).

43. *The Economics of Germanium*, 6th ed., Roskill Information Services, Ltd., London, 1990.

44. H. J. Cluley, *Analyst* **76**, 523 (1951).

45. J. R. Musgrave, in I. M. Kolthoff, P. J. Elving, and E. B. Sandell, eds., *Treatise on Analytical Chemistry*, Part II, Vol. 2, John Wiley & Sons, Inc., New York, 1962, pp. 208–245.

46. V. A. Nazarenko, *Analytical Chemistry of Germanium*, trans. N. Mandel, John Wiley & Sons, Inc., New York, 1974.

47. *1992 Annual Book of ASTM Standards, 10.05*, American Society for Testing and Materials, Philadelphia, Pa., 1992.

48. R. J. Lewis, Sr., and D. V. Sweets, eds., *Registry of Toxic Effects of Chemical Substances, 1983 Suppl. to 1981–1982 ed., Vol. 1*, NIOSH, Rockville, Md., 1984, p. 786.

49. H. E. Christensen, *Toxic Substances List*, NIOSH, Rockville, Md., 1972.

50. J. H. Adams, *Eng. Min. J.* **178**, 181 (Mar. 1977).

51. G. C. Datum and S. A. Billets, *Conference Record of the Twenty-Second IEEE Photovoltaic Specialist Conference, 1991*, **2**, 1422–1428 (1991).

52. J. H. Adams and D. W. Thomas in A. E. Torma and I. H. Gandiler, eds., *Precious and Rare Metal Technologies*, Elsevier Science Publishing Co., New York, 1989, pp. 577–587.

53. J. R. Piedmont and R. J. Riordan, *Proc. Soc. Photo-Opt. Instrumen. Eng.* **164**, 216–222 (1979).

54. E. E. Haller, H. W. Kraner, and W. A. Higinbotham, eds. *Nuclear Radiation Detector Materials*, North-Holland, New York, 1983.

55. H. V. Klapdor and co-workers, *Proceedings of the International Symposium on Weak and Electromagnetic Interactions in Nuclei (WEIN 89), Montreal*, Max Planck Institute fur Kernphysics, Heidelberg, Germany, 1989.

56. H. V. Klapdor, *Proceedings of the XXIII Yamada Conference on Nuclear Weak Process and Nuclear Structure, Osaka, Japan*, Max Planck Institute fur Kernphysics, Heidelberg, Germany, 1989.

57. F. T. Avignone, III and co-workers, *Physics Letters B*, **256**, 559 (1991).

58. U.S. Pat. 3,737,293 (June 5, 1973), R. D. Maurer (to Corning Glass Works).

59. U.S. Pat. 4,217,027 (Aug. 12, 1980), J. B. MacChesney and P. B. O'Conner (to Bell Telephone Labs).

60. U.S. Pat. 3,737,292 (June 5, 1973), D. B. Keck, P. C. Schultz, and F. Zimar (to Corning Glass Works).

61. B. Ryan, *Byte*, 289–296 (Nov. 1990).

62. D. M. Mills, *Physics Today*, 22–29 (Apr. 1984).
63. *Photonics Spectra*, 50–52 (Mar. 1992).

JACK H. ADAMS
DENNIS THOMAS
Eagle-Picher Industries, Inc.

GERMANIUM OIL. See OILS, ESSENTIAL; PERFUMES.

GETTERING. See VACUUM TECHNOLOGY.

GILSONITE. See COAL.

GIN. See BEVERAGE SPIRITS, DISTILLED; OILS, ESSENTIAL.

GLASS

Glass was formed naturally from common elements in the earth's crust long before anyone ever thought of experimenting with its composition, molding its shape, or putting it to the myriad of uses that it enjoys in the world today. Obsidian, for instance, is a naturally occurring combination of oxides fused by intense volcanic heat and vitrified (made into a glass) by rapid air cooling. Its opaque, black color comes from the relatively high amounts of iron oxide. Its chemical durability and hardness compare favorably with many commercial glasses (1). Pumice, a naturally occurring foam glass, is replete with tiny pockets of the gaseous products of the decomposition of many compounds. These gases were trapped by the viscous glass while it was cooling.

The origin of the first synthetic glasses is lost in antiquity and legend. Faience was made by the Egyptians who molded figurines from sand, SiO_2, the most popular glass-forming oxide. They coated them with natron, the residue left by the flooding Nile river, which was composed principally of calcium carbonate, $CaCO_3$; soda ash, Na_2CO_3; sodium chloride, NaCl; and copper oxide, CuO. Heating below 1000°C produced a glassy coating by the diffusion of the fluxes, CaO and Na_2O, into the sand and their subsequent solid-state reaction with the sand. The copper oxide gave the article an appealing blue color. Glass technology has evolved for 6000 years and some principles date back to early times. This includes what is known about the structure of glass, its composition, properties, method of manufacture, and uses (see also GLASS-CERAMICS; GLASSY METALS.)

Common usage of the term glass follows the definition of Morey (2): "Glass is an inorganic substance in a condition which is continuous with, and analogous to, the liquid state of that substance, but which, as the result of a reversible change in viscosity during cooling, has attained so high a degree of viscosity as to be, for all practical purposes, rigid." Similarly, ASTM (3) defines glass as "an inorganic product of fusion that has cooled to a rigid condition without crystallizing." Both organic and inorganic materials may form glasses if their structure is noncrystalline, ie, if they lack long-range order. This includes some plastics, metals (4–7), and organic liquids (8,9). In principle, rapid cooling could prevent crystallization of any substance if the final temperature is sufficiently low to prevent structural rearrangement. Thus glasses are formed primarily for kinetic reasons.

Glass is not merely a supercooled liquid. This distinction is illustrated by the volume–temperature diagram shown in Figure 1. When a liquid that normally does not form a glass is cooled, it crystallizes at or slightly below the melting point (path A). If there are insufficient crystal nuclei or if the viscosity is too high to allow sufficient crystallization rates, undercooling of the liquid can occur. However, the viscosity of the liquid rapidly increases with decreasing temperatures, and atomic rearrangement slows down more than would be typical for the supercooled liquid. This results in the deviation from the metastable equilibrium curve which is depicted by paths B and C in Figure 1. This change in slope with temperatures is characteristic of a glass. Structural rearrangement is too slow to be detected experimentally, and additional volume changes are virtually linear with continued cooling, the same as for any other single-phase solid. The cooling rate determines when the deviation begins to occur. Slower cooling (path B), for instance, results in less of a deviation from the extrapolated liquid curve. Figure 1 shows that the point of intersection of the two slopes defines a transformation point, the glass-transition temperature (T_g), for a given cooling rate. Practical limitations on cooling rate define the transformation range, $T_g \rightarrow T_g'$, as the temperature range in which the cooling rate can affect the structure-sensitive properties such as density, refractive index, and volume resistivity (10). The structure, which is frozen in during the glass transformation, persists at all lower temperatures. Thus a glass has a configurational or fictive temperature which may differ from its actual temperature. The fictive temperature is the temperature at which the glass structure would have been the equilibrium structure (11). It describes the structure of a glass as it relates to the cooling rate; a fast-quenched glass would have a higher fictive temperature than a slowly cooled glass.

Glasses can be prepared by methods other than cooling from a liquid state, including solution evaporation, sintering of gels, reactive sputtering, vapor deposition, neutron bombardment, and shock-wave vitrification (12,13). These tech-

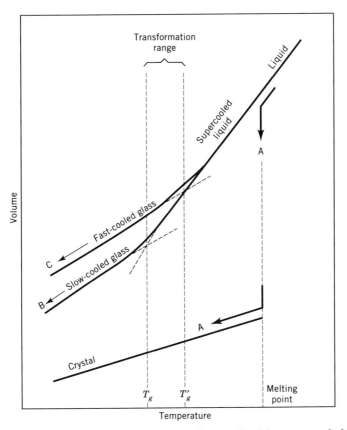

Fig. 1. Volume–temperature relationships for glasses, liquids, supercooled liquids, and crystals. Courtesy of Corning Inc.

niques suggest that the purely kinetic explanation of the glassy state is subject to question and that the previous definitions need modification. Extrapolation of the thermodynamic properties of the supercooled liquid gives the paradoxical results that entropies, heat contents, and volumes become less than those of the perfect crystal at the same temperature (14). Considering the glass transformation as a second-order transition (15) has yielded a satisfactory explanation for the properties of some organic systems, but this theory is still subject to confirmation for inorganic glasses. The dependence of transformation temperature on relaxation time has been considered and a definition based on structural factors proposed (16). Isotropic materials with long structural-relaxation times, eg, $>10^3$ s, would be defined as glasses. The determination of either the isotropy of the material or structural relaxation times distinguishes whether or not the material is a glass. This definition requires information regarding structure and does not consider previous thermal history as a distinguishing characteristic of a vitreous material.

Structure

The basic structural unit of silicate glasses is the silicon-oxygen tetrahedron in which a silicon atom is tetrahedrally coordinated to four surrounding oxygen atoms. Oxygens shared between two tetrahedra are called bridging oxygens. In pure vitreous silica, SiO_2, virtually all oxygens are bridging. Those that are not shared, for one reason or another, can be referred to as nonbridging oxygens. The geometric relationship between these tetrahedra is controversial and has yet to be completely resolved. The earlier crystallite theory has been modified by proponents of the random-network theory. Modern structural methods point to a compromise theory (see SILICA).

In 1921, a discontinuous index of refraction of vitreous SiO_2 near the $\alpha-\beta$ transition of quartz (crystalline SiO_2) was noted (17). These data and subsequent x-ray investigations of vitreous silica led to the suggestion (18) that crystallites on the order of 1.5 nm were present. It was demonstrated, however, that the crystal size would be less than 0.8 nm, and it was suggested that the term crystal loses meaning for these dimensions (19,20).

The random-network theory of glass was formulated in 1932 (21). It proposes that atoms present in glass form a three-dimensional connected structure without periodic order and with energy content comparable to that of the corresponding crystalline material. According to this theory, the coordination number of an atom determines its role in a glass structure, and the following four rules should be fulfilled for an oxide to form glass: (1) each oxygen atom must be linked to no more than two cations; (2) the number of oxygen atoms around any one cation must be small, ie, three or four; (3) the oxygen polyhedra must share corners, not edges or faces, to form a three-dimensional network; and (4) at least three corners must be shared. For one-component glasses, each polyhedron shares corners with at least three other polyhedra in such a way that the network is continuous in three dimensions. In multicomponent glasses, additional cations are distributed throughout holes in the network.

X-ray structural work strongly supported the random network theory (19). The x-ray scattering pattern of glass after Fourier analysis gives radial distribution curves that indicate the distribution of neighboring atoms about a central atom. No evidence of discrete particles or voids supporting an ordered structure was observed based on data for the first coordination shell (19). There were both theoretical and experimental limitations to this early work. Later, more complete data were obtained using fluorescence excitation to eliminate Compton scattering (22). Not only was a silicon–oxygen distance of 0.162 nm observed, but also peaks at 0.265 nm for the oxygen–oxygen distance, 0.312 nm for the silicon–silicon distance, 0.415 nm silicon–second silicon, and 0.64 nm for the silicon–third oxygen peak. X-ray scattering investigation of silica (23,24) suggested a structural ordering beyond the distances first reported (22). Data analysis resulted in a shorter silicon–oxygen bond distance (0.1595 nm) than reported at first (22). However, a similarity in bonding topology between tridymite and silica glass does not imply microcrystallinity of vitreous silica in a crystallographic sense. The similarity between crystalline and vitreous structure on the basis of silicon–oxygen–silicon bond angles and, hence, the relative orientation of the silicon tetrahedrons is pointed out in Figures 2 and 3.

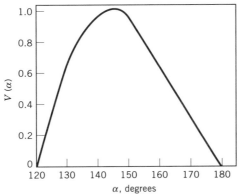

Fig. 2. The distribution of silicon–oxygen–silicon bond angles in vitreous silica (22,25). The function $V(\alpha)$ is the fraction of bonds with angles normalized to the most probable angle, 144°. This distribution gives quite a regular structure on the short range, with gradual distorting over a distance of 3 or 4 rings (2–3 nm). Crystalline silica such as quartz or cristobalite would have a narrower distribution around specific bond angles.

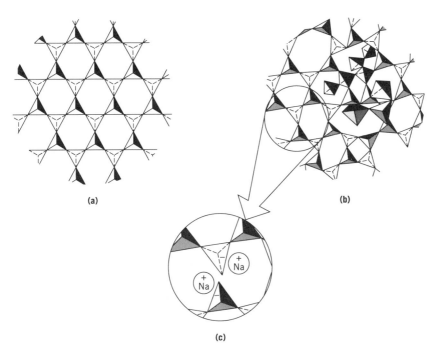

Fig. 3. The tetrahedra in the schematics represent four oxygens clustered around a silicon. Schematic representation of (**a**) an ideal crystalline structure (Si—O—Si bond angles = 180°); (**b**) a simple glass (Si—O—Si bond angles = 144° ± according to Fig. 2); and (**c**) the addition of a modifier, in this case one molecule of Na_2O, causes the breaking of one Si—O—Si bond to form two Si—ONa linkages. Courtesy of Corning Inc.

In addition to the question of long-range ordering (1–2 nm), there still are aspects of the random-network theory that are often criticized. It is possible, for example, to form glasses when no three-dimensional network is possible. Glassy orthosilicates, SiO_4^{4-}, of lead or sodium and calcium have been prepared (26). Furthermore, modifying cations have been shown to occur at regular interatomic distances ranging over several coordination shells (27). Dark-field transmission electron microscopy has been used to infer density fluctuations of silica which suggest ordered regions of approximately 1.0 nm in size. However, these results have been criticized (28).

Glass-forming systems other than silica have been examined. The fraction of three- and four-coordinated boron in borate glasses can be determined by nmr (29). Both nmr and x-ray diffraction (30) results led to the suggestion that the boroxyl ring is the structural unit of vitreous B_2O_3 (22,29). The intermediate-size boroxyl ring represents a compromise between the crystallite and the random-network theory (29) (see ANALYTICAL METHODS).

Composition

Conditions favorable for glass formation may be deduced from either geometric or bond strength considerations. On the basis of the rules (21) discussed above, the following oxides should be glass formers: B_2O_3, SiO_2, GeO_2, P_2O_5, As_2O_5, P_2O_3, As_2O_3, Sb_2O_3, V_2O_5, Sb_2O_5, Nb_2O_5, and Ta_2O_5. In fact, they are all so used. The only fluoride that fulfills the rules of glass formation is BeF_2, which readily forms a glass (31).

Glass formers generally have cation–oxygen bond strengths greater than 335 kJ/mol (80 kcal/mol). In multiple-component systems, oxides with lower bond strengths do not become part of the network and are called modifiers. Oxides with energies of ca 335 kJ/mol may or may not become part of the network and are referred to as intermediates. The dissociation energies used to predict glass formation are calculated, taking into account the coordination number of the cation (Table 1). In multiple-component glasses, the terms formers, modifiers, and intermediates are frequently used to define the role of the individual oxides. However, an element such as lead may be either a modifier or intermediate, depending on its coordination and the glass system considered.

Glass formation of individual oxides can be predicted from the melting point, and individual bond energies can be normalized by dividing by the melting point of the oxide (33). This ratio is relevant because the melting point is related to the amount of thermal energy available to rupture bonds. If the bond energy is large and the melting point low, glass formation is favored. This explains the ease of glass formation of B_2O_3 and from low melting eutectics in which neither oxide forms a glass separately, eg, $CaO–Al_2O_3$.

Other correlations of glass formation and properties have been offered. For example (34): (1) cation valence should be three or greater; (2) glass formation should increase with decreasing cation size; and (3) the Pauling electronegativity should be between 1.5 and 2.1. Using these criteria, four types of oxides are described: (1) strong glass formers such as Si, B, Ge, As, and P; (2) intermediate formers that require rapid cooling, such as Sb, V, W, Mo, and Te; (3) oxides that

Table 1. Coordination Number and Bond Strengths of Oxides[a]

Name	Formula	CAS Registry Number	Dissociation energy, kJ/mol[b]	Coordination number	Single-bond strength, kJ/mol[b]
			Formers		
boron oxide	B_2O_3	[1303-86-2]	1489	3	496
silicon oxide	SiO_2	[10097-28-6]	1774	4	443
germanium oxide	GeO_2	[1310-53-8]	1803	4	450
aluminum oxide	Al_2O_3	[1344-28-1]	1682–1326	4	420–332
boron oxide	B_2O_3	[1303-86-2]	1489	4	372
phosphorus oxide	P_2O_5	[1314-56-3]	1849	4	462–370
vanadium oxide	V_2O_5	[1314-62-1]	1879	4	469–376
arsenic oxide	As_2O_5	[1303-28-2]	1460	4	365–292
antimony oxide	Sb_2O_5	[1314-60-9]	1418	4	354–284
zirconium oxide	ZrO_2	[1314-23-4]	2029	6	338
			Intermediates		
titanium oxide	TiO_2	[13463-67-7]	1820	6	303
zinc oxide	ZnO	[1314-13-2]	602	2	301
lead oxide	PbO	[1317-36-8]	606	2	303
aluminum oxide	Al_2O_3	[1344-28-1]	1682–1326	6	280–221
thorium oxide	ThO_2	[1314-20-1]	2159	8	269
beryllium oxide	BeO	[1304-56-9]	1046	4	261
zirconium oxide	ZrO_2	[1314-23-4]	2029	8	253
cadmium oxide	CdO	[1306-19-0]	498	2	248
			Modifiers		
scandium oxide	Sc_2O_3	[12060-08-1]	1514	6	252
lanthanum oxide	La_2O_3	[1312-81-8]	1699	7	242
yttrium oxide	Y_2O_3	[1314-36-9]	1669	8	208
tin oxide	SnO_2	[18282-10-5]	1163	6	193
gallium oxide	Ga_2O_3	[12024-21-4]	1117	6	186
indium oxide	In_2O_3	[1312-43-2]	1083	6	180
thorium oxide	ThO_2	[1314-20-1]	2159	12	179
lead oxide	PbO_2	[1309-60-0]	970	6	161
magnesium oxide	MgO	[1309-48-4]	929	6	154
lithium oxide	Li_2O	[12057-24-8]	602	4	150
lead oxide	PbO	[1317-36-8]	606	4	151
zinc oxide	ZnO	[1314-13-2]	602	4	150
barium oxide	BaO	[1304-28-5]	1088	8	135
calcium oxide	CaO	[1305-78-8]	1075	8	134
strontium oxide	SrO	[1314-11-0]	1071	8	133
cadmium oxide	CdO	[1306-19-0]	498	4	124
sodium oxide	Na_2O	[1313-59-3]	502	6	83
cadmium oxide	CdO	[1306-19-0]	498	6	82
potassium oxide	K_2O	[12136-45-7]	481	9	53
rubidium oxide	Rb_2O	[18088-11-4]	481	10	48
mercury oxide	HgO	[21908-53-2]	284	6	47
cesium oxide	Cs_2O	[20281-00-9]	477	12	39

[a] Ref. 32.
[b] To convert J to cal, divide by 4.184.

form glasses in binary mixtures with nonglass formers, such as Al, Ga, Ti, Ta, Nb, and Bi; and (4) oxides that do not form glasses.

Glass composition work starts with the application of structural and bonding rules of glass formation. Numerous ternary systems and their glass-forming regions have been investigated (35). There are three types of ternaries: Type A, single former and two modifiers; Type B, two formers and one modifier; and Type C, three glass formers. Type A is shown in Figure 4. The structural rules suggested in Reference 21 can also define likely regions for glass formation. Additions of several percent of other oxides for property adjustments are usually made to each system to give commercially useful glasses.

A parallel but more historically comprehensive discussion of glass structure and composition has been given (36). Prediction of structural parameters and consequent properties from theoretical principles has increased with the advent of supercomputers. Of particular interest to glass scientists are those studies which have focused on crystalline and vitreous silica (37,38).

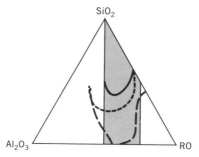

Fig. 4. Glass-forming region in Type A system (35). The shaded area represents the predicted glass-forming region based on Zachariasen's rules (21). RO = BaO (——); SrO (---); CaO (— — —).

Single-Phase Glasses

Vitreous Silica. Vitreous silica is the most important single-component glass. Highly cross-linked vitreous silica is viscous and has a thermal expansion coefficient within the 0–300°C range of about $5.5 \times 10^{-7}/°C$. It is an excellent dielectric and resists attack by most chemicals, except fluorides or strong alkali. Fused silica has a high spectral transmission, and in addition, is resistant to radiation damage which results in browning of other glasses. It is the ideal glass for space-vehicle windows, wind-tunnel windows, ultrasonic delay lines, crucibles for growing ultrapure silicon or germanium crystals, and for optical systems in spectrophotometric and photolithographic devices.

The same properties that make transparent fused silica useful also make it expensive and difficult to produce. Vitreous silica may be made by several processes. Fused quartz made by electrically fusing quartz crystal gives a product containing very little moisture and having good ir transmission. However, mineral impurities of natural quartz, including alumina, iron, and some chlorides, reduce uv transmission. Flame fusion of quartz or flame hydrolysis of $SiCl_4$, on the other hand, gives glasses containing larger amounts of water, which decreases the ir transmission. Long heat treatments of vitreous silica below 1723°C may cause crystallization. Stable crystalline forms of silica at atmospheric pressure are cristobalite [14464-46-1] (1723–1470°C), tridymite [15468-32-3] (1470–867°C), and quartz [14808-60-7] (below 867°C) (see SILICA, VITREOUS SILICA; SILICA, SYNTHETIC QUARTZ).

Multicomponent Silicate Systems. Most commercial glasses fall into the category of silicates containing modifiers and intermediates. Addition of a modifier such as sodium oxide, Na_2O, to the silica network alters the structure by cleaving the Si—O—Si bonds to form Si—O·Na linkages (see Fig. 3c).

Separating the silica tetrahedra from each other makes the glass more fluid and therefore more amenable to conventional melting and forming methods. Modifiers (or fluxes) also cause a decrease in resistivity, an increase in thermal expansion, and generally lower chemical durability. Glasses with a SiO_2:Na_2O molecular ratio less than one have so many nonbridging oxygens that they lack a continuous, three-dimensional structure (Zachariasen's rule 4). Such glasses, referred to as invert glasses, have been made containing Li_2O, Na_2O, and K_2O oxides. Alkali silicates that have a silica:alkali ratio of 0.5–3.4 are the basis of the soluble silicate glass industry. The composition and uses of sodium silicates for adhesives, cleansers, desiccants, abrasives, cement deflocculants, and surface coatings are discussed in Reference 39 (see SILICON COMPOUNDS).

The effectiveness of an alkali oxide, eg, Li_2O or Cs_2O, as a flux increases with the size of the cation and therefore with its polarizability. Large ions such as cesium are easily polarized and thus more likely to give up their oxygen to break the Si—O—Si bonds as discussed above. Lithium, on the other hand, is more likely to keep its oxygen, and therefore its fluxing power is less. This is consistent with the ease of glass formation as the size:charge ratio of the modifier is increased. Phase separation occurs often when less polarizable oxides are present. Lithium or magnesium silicates have a tendency to phase-separate during heat treatment (40). Ionic mobility is related both to charge and to size. Large alkalies are expected to be more mobile because of their greater polarizability. Increased size, however, tends to reduce mobility. Alkaline-earth silicates behave similarly to alkali silicates, but the fluxing power of alkaline earths is less than that of the alkalies. Mobility of divalent ions is less than that of monovalent ones; hence, resistivities of alkaline-earth glasses are usually higher. Divalent oxides increase the resistivity of alkali-containing glasses.

Alumina is frequently used in silicate glasses. It often adopts a four-coordinated structure with alkalies giving a $NaAlO_2$ tetrahedral unit which substitutes into the SiO_2 network. The extra negative charge associated with the four bridging oxygens surrounding Al^{3+} is offset by the Na^+ ion. A maximum in viscosity often occurs when the Al_2O_3:Na_2O ratio equals one.

Boron oxide often behaves as a flux. Boron softens glass for easier melting, but unlike alkalies, boron oxide increases expansion only slightly. This is the basis of the easily melted, low expansion commercial glasses known as borosilicates.

Soda–Lime Glasses. Mixtures of alkali and alkaline earths give glasses of higher durability than the alkali silicates. The actual compositions are usually more complex than the term soda–lime suggests. In addition to Na_2O, CaO, and

SiO_2, these glasses may contain MgO, Al_2O_3, BaO, or K_2O and various colorants. Alumina increases durability, whereas MgO prevents devitrification. Soda–lime glass accounts for nearly 90% of all the glass produced. The batch materials are inexpensive and relatively easy to melt. Soda–lime glass is used for containers, flat glass, pressed and blown ware, and lighting products where exceptional chemical durability and heat resistance are not required.

Borosilicate Glasses.   Replacement of alkali by boric oxide in a glass network gives a lower expansion glass. The fluxing action of the boron facilitates melting by weakening the network. This has been attributed to the presence of planar three-coordinate borons that weaken the silicate network at high temperature. Phase separation of borosilicate glasses often occurs during heat treatment, which may be useful for certain applications. However, most commercial borosilicate glasses have compositions that are miscible and homogeneous. Borosilicate glass is applied as ovenware, laboratory equipment, piping, sealed-beam headlights, and pharmaceutical ampuls.

Aluminosilicate Glasses.   Structural rules suggest that if the $R_2O:Al_2O_3$ or the $RO:Al_2O_3$ molar ratio is unity, an aluminosilicate glass has a silica-like structure in which nearly all oxygens are bridging oxygens. This is true of other silicate minerals and appears to be the case with glasses. Alumina is expected to be four-coordinated when the alkali to alumina molar ratio is greater than one, but if the ratio is less than one, sixfold coordination of alumina has been suggested.

Aluminosilicate glasses are used commercially because they can be chemically strengthened and withstand high temperatures. Thus applications include airplane windows, frangible containers, lamp envelopes, and flat panel display devices.

Lead Glasses.   Lead oxide is usually a modifier, although at times it may act as a network former. Lead glasses may be easily melted and have a long working range and a high refractive index which makes them useful for lead crystal, optical glass, and hand-formed art ware. Lead-containing glasses effectively shield high energy radiation and are therefore used commercially for radiation windows, fluorescent-lamp envelopes, and television bulbs. Low melting solder glasses and frit or decorative enamels are usually based on low melting lead compositions (see ENAMELS, PORCELAIN OR VITREOUS).

Borate Glasses.   Borates, including vitreous B_2O_3, have been studied more than any other glass-forming system with the exception of silicates (41). Vitreous boric oxide has a three-coordinate structure consisting of six-membered rings of alternating boron and oxygen atoms. Many physical properties of alkali borate glasses show a minimum or maximum at 15–30 mol % modifier (boron anomaly). Coordination changes of boron are detected by nmr, ir, Raman, and esr techniques. Broad quadripolar coupling typical of triangular boron coordination is readily distinguished from the sharp coupling of four-coordinate boron (29). The fraction of tetrahedral borons present appears directly proportional to the alkali-to-boron ratio as long as this ratio is less than 0.5 (42). The very low durability of borate glasses precludes their use in all except the most special applications. Low molecular weight Lindemann glasses, $Li_2O \cdot BeO \cdot B_2O_3$, were developed as x-ray transmitting glasses. Rare-earth borate glasses have optical uses because of their high refractive indexes and low dispersion. Additions of Gd_2O_3 to the latter increase the index but not the dispersion (42).

Phosphate Glasses. The structure of phosphate glass appears to be based on the phosphorus–oxygen tetrahedron (see PHOSPHORIC ACID AND THE PHOSPHATES). Like the borates, they tend to have low durability. Important commercial applications of phosphate glasses do exist, however. Because the absorption bands of iron oxide in phosphate glasses are sharper in the uv and ir than in silicate glasses, iron-containing phosphate glasses are nearly transparent to visible light. Almost clear heat-absorbing glasses with several percent iron oxide are possible. Phosphate-based glasses also are more resistant to fluoride than silicate glasses. Some of the optical glasses produced by Schott, Hoya, and Corning use phosphate as the primary glass former. Fluorophosphate glasses, designated FK-5 or FK-50 by Schott, have very low optical dispersion with Abbe-numbers of 70.4 and 81.5, respectively.

Other Oxide Glasses. Germanium, arsenic, and antimony oxides all form stable glasses and their structures have been predicted. The germania glass structure is quite similar to silica. Infrared transmission of germania glasses is higher than that of silica. Tellurium-containing lead glasses with a very high refractive index (>2.0) are also used commercially. Heavy-metal oxide glasses (>50 cation % bismuth and/or lead) possess the best infrared transmission and the highest nonlinear optical susceptibilities known to be available in oxide glasses (43).

Chalcogenide Glasses. Glasses based on sulfur, selenium, or tellurium rather than oxygen are well known. These glasses, although often opaque to visible light, transmit ir radiation of a much longer wavelength than oxide systems, and many are also semiconductors (qv). Conductivity usually increases with increasing atomic number. The most studied chalcogenide glasses contain the Group V elements arsenic and antimony.

Halide Glasses. Although examples of zinc chloride glasses are known, BeF_2-containing glasses are more common (31). Vitreous beryllium fluoride [7787-49-7] has a tetrahedral structure analogous to silica. Its unique spectral properties including transmission from <160 nm to 5500 nm, low refractive index, and very low dispersion have aroused considerable interest in beryllium fluoride glasses. Vitreous BeF_2 is very hygroscopic but addition of other fluorides, such as AlF_3, or alkali and alkaline-earth fluorides, increases durability. Selected compositions also resist devitrification and have optical properties that are not obtainable with oxide glasses. A refractive index of 1.3 with a dispersion of 100 is not uncommon for beryllium fluoride glasses. The toxicity of beryllium is, however, a concern associated with melting and forming operations. Fluoroberyllate, fluorophosphate, and fluorozirconate glasses are being evaluated for application as optical fiber materials, and doped with rare-earth cations as optical fiber amplifiers (44,45).

Metallic Glasses. Under highly specialized conditions, the crystalline structure of some metals and alloys can be suppressed and they form glasses. These amorphous metals can be made from transition-metal alloys, eg, nickel–zirconium, or transition or noble metals in combination with metalloid elements, eg, alloys of palladium and silicon or alloys of iron, phosphorus, and carbon.

Vacuum evaporation or sputtering techniques can produce thin films of amorphous metal elements as well as a wide variety of amorphous alloys (46). Liquid quenching at rates greater than 10^5 K/s is limited to alloys containing metalloids (47). Although glass-forming ability cannot be predicted with certainty, a low temperature eutectic in a system of high melting metals often forms a metallic glass (48,49) (see also GLASSY METALS).

Glass-Ceramics and Phase-Separated Glasses

Glasses are metastable and, under the appropriate conditions, revert to a thermodynamically more stable state. Glasses particularly susceptible to uncontrolled crystal growth or phase separation have traditionally created problems for their manufacturer. However, glass is also a good medium for controlled crystallization (50), and has become the basis for a number of unique crystalline materials known as glass-ceramics (qv). The separation of a single glass into multiple glassy phases can make the article cloudy and adversely affect its chemical durability. Controlled phase separation, however, can produce opaque, white opal glass or, after a leaching step, lead to new materials such as porous glass or 96% silica glass. The colloidal suspension of multiple phases in transparent glass produces precise colors for products such as optical filters (qv). Furthermore, photosensitive and photochromic glasses change their optical transmission and color, sometimes reversibly, upon stimulus by a combination of light and heat treatment (see CHROMOGENIC MATERIALS, PHOTOCHROMIC). All these transformations generally depend on the phenomena of diffusion, nucleation, and growth.

The thermodynamic driving force for phase separation depends on the amount of undercooling from some appropriate equilibrium temperature such as the liquidus temperature (path A, Fig. 1). That temperature is more correctly defined as the point where the free energy of formation of the new substance(s) changes from positive to negative as the glass cools. The free energy of formation depends on the standard free energies of all the products and reactants involved, as well as on each of their concentrations, or more correctly, upon their activities. Therefore, as the glass cools below the equilibrium temperature, eg, the liquidus, the free energy of formation becomes more negative and the driving force for phase separation becomes greater. However, as glasses cool they become more viscous, which further slows the diffusion process. Diffusion is necessary for sufficiently concentrating the elements to allow phase separation. Thus, upon cooling, the rate of reaction increases with the thermodynamic driving force until it reaches a maximum at some optimum temperature. Then it slows down as the diffusion rate becomes the dominant controlling factor. Both the nucleation and growth rates are affected by this balance between driving force and diffusion rate. However, because they are different phenomena, their optimum temperatures for maximum rate may or may not coincide. In general, the optimum temperature for maximum nucleation rate is lower than that for maximum growth rate. Because critical nucleus size is determined by a balance between volume free energy (pronucleation) and surface free energy (antinucleation), nucleation is more sensitive to driving force and less sensitive to diffusion rate than is growth. Consequently, just below the equilibrium temperature only a few nuclei form, but the growth rate is rapid and the glass has a few relatively large islands of a second phase.

At lower temperatures, the morphology is different. More nuclei form but do not grow as fast, and the other extreme occurs where the glass has many fine islands of the second phase. A single glass composition can have several phases competing for the elements that it needs to nucleate and grow, and one phase can dominate the others through the choice of a heat treatment which maximizes the nucleation and growth rates of the desired phase at the expense of the unwanted phases. Furthermore, the size and amount of a second phase can be controlled by balancing the heat treatment for nucleation with the heat treatment for growth. Finally, the nature of the second phase can be affected by nucleating agents which are either added to the glass or encouraged to grow in the glass. Nucleating agents affect either the thermodynamics of the components or the surface energy of the nuclei or both.

Glass Ceramics.   Glass-ceramics are melted and formed by conventional glass manufacturing techniques, and then given a subsequent heat treatment (ceram) to transform them into fine-grained crystalline materials. Table 2 lists some commercial glass-ceramics. By definition, glass-ceramics are more than 50% crystalline after heat treatment; frequently, the final product is more than 95% crystalline. The operation is generally accompanied by an increase in viscosity which increases the product's use temperature. For example, both Corning and NEG manufacture a $PbO-B_2O_3-ZnO$ glass-ceramic frit for joining color television bulb parts. The glassy frit flows, seals, and crystallizes, thereby becoming quite rigid above 440°C. The bulbs can be further processed at temperatures over 400°C without the parts shifting. In most commercial applications the thermal expansion of the glass-ceramic is lower than that of the uncerammed parent glass which increases its dimensional stability and thermal shock resistance. Telescope mirror blanks, missile radomes, and cooking utensils are examples. The intrinsic strength of glass-ceramics is less than that of glass, but if the crystal grains are of the optimum size and shape they inhibit crack propagation, yielding an actual strength greater than that of glass. In Corning's machinable glass-ceramic material, mica crystals occupy 55% of the volume in the glassy matrix. During machining, the interlocking intersections of the randomly oriented crystals stop the microscopic fractures that, in mass, make up the actual cut. Cracks do not propagate throughout the matrix because they are repeatedly deflected, branched, blunted, and finally arrested. This property also keeps the product strong during service regardless of whatever surface flaws might occur. Glass-ceramics can be further strengthened by application of a lower expansion glaze, by ion exchange, or by differential crystallization.

The formation of a glass-ceramic is extremely complex, but generally follows a four-step sequence (50–53). (1) A dispersed amorphous phase, structurally incompatible with the host glass, usually highly unstable, and enriched in one or two key oxides, eg, TiO_2 or ZrO_2, forms on either cooling or reheating the glass. (2) Primary crystalline nuclei, which are often titanates or zirconates, form either heterogeneously at the phase boundaries or homogeneously within the second phase. (3) A metastable crystalline phase heterogeneously nucleates on the primary crystallites and grows, generally at the expense of the glassy second phase. This produces a typically fine-grained, metastable, solid-solution material. (4) The metastable material breaks down to the final, stable, fine-grained crystalline structure by means of isochemical phase transformations, reaction between metastable phases, or exsolution.

Devitrification.   This process is the uncontrolled formation of crystals in glass during melting, forming, or secondary processing. Devitrification can ad-

Table 2. Commercial Glass-Ceramics[a]

Commercial designation	Principal crystalline phases	Properties	Application
Schott Zerodur	β-quartz solid solution, SiO_2	zero expansion	electric range tops, telescope mirrors
Corning 9632[b]	β-quartz	low expansion, high strength, thermal stability, chemical durability	electric range tops (transparent)
Corning 8603	lithium metasilicate, $Li_2O \cdot SiO_2$; lithium disilicate, $Li_2O \cdot 2SiO_2$	photochemically machinable	fluid amplifiers
Corning 9608	β-spodumene solid solution, $Li_2O \cdot Al_2O_3 \cdot (SiO_2)_{4-10}$	low expansion, high chemical durability	cooking utensils
Corning 9617[b]	β-spodumene solid solution; anatase, TiO_2	low expansion, high strength, thermal stability, chemical durability	electric range tops (opaque)
Corning 0330[c]	β-spodumene solid solution; rutile TiO_2	mechanical, chemical, thermal stability	exterior, interior cladding, laboratory bench tops
Corning 9455	β-spodumene solid solution; mullite, $3Al_2O_3 \cdot 2SiO_2$	low expansion, high thermal and mechanical stability	heat exchangers
Neoceram (Japan)	β-spodumene	low expansion	cooking ware
Corning 9606	cordierite, $2MgO \cdot 2Al_2O_3 \cdot 5SiO_2$; spinel, $MgO \cdot Al_2O_3$; MgO-stuffed β-quartz; quartz, SiO_2	low expansion, high transparency to radar	missile radomes
Corning 9625[d]	α-quartz solid solution, SiO_2; spinel, $MgO \cdot Al_2O_3$; enstatite, $MgO \cdot SiO_2$	very high strength	classified
Corning 9658	fluorphlogopite solid solution, $KMg_3AlSi_3O_{10}F_2$; mullite, $3Al_2O_3 \cdot 2SiO_2$	machinable, high dielectric strength, thermal stability	precision dielectric components, insulators, high vacuum components
High K (Corning)	$(Ba,Sr,Pb)Nb_2O_6$	high dielectric constant	capacitors

[a]Ref. 51. [b]Surface strengthening by differential crystallization.
[c]Surface strengthening by $Na^+ \Leftrightarrow Li^+$ ion exchange. [d]Surface strengthening by $2 Li^+ \Leftrightarrow Mg^{2+}$ ion exchange.

versely affect the optical properties, mechanical strength, and sometimes the chemical durability of the glass. These unwanted crystals nucleate and grow homogeneously within the glass or heterogeneously at the air–glass or refractory–glass interface. Devitrification is most likely to occur in glasses in which the optimum temperatures for maximum nucleation rate and for maximum growth rate nearly coincide. If these glasses are held too long in this critical temperature range or are cooled too slowly through it, the glass starts to crystallize. The critical temperature range is below the liquidus temperature, but above the temperature where the glass is sufficiently viscous to retard devitrification (10^5–10^6 Pa·s or 10^6–10^7 P). In soda–lime glasses, for example, the crystal phase known as devitrite, $Na_2O \cdot 3CaO \cdot 6SiO_2$, forms between 850–900°C (7.5–25 kPa·s or 7.5–25 $\times$ 10^4 P). However, since the glass is already quite viscous, the critical temperature range is short (ca 50°C), and devitrification is not much of a problem. Some optical glasses have liquidus temperatures as high as 1100°C (20–30 Pa·s or 200–300 P) and, therefore, must be cooled very quickly for several hundred degrees to avoid devitrification. Some forming operations require viscous glass (drawing) and, therefore, require glasses with low liquidus temperatures. Other operations (pressing of small articles) can tolerate glasses with much higher liquidus temperatures.

Phase-Separated Glass. Glasses that derive their color, optical transparency, or chemical durability from a small amount of a finely dispersed second phase are termed phase-separated glass. They are distinguished from glass-ceramics by virtue of their predominantly glassy character. Therefore, the control of the nucleation and morphology of the second phase is not so critical. Opal glasses are either translucent or opaque, depending on the particle size and quantity of the second phase, and on the refractive index difference between the second phase and the glass matrix. Incident light is refracted and scattered by the dispersed phase and the glass appears opaque. Fluorine-containing glasses, which separate into NaF- or CaF_2-containing phases, for instance, are commonly made into pressed items such as tableware. As the glass cools from the molten state, the fluoride phases become immiscible in the glass and separate as fine droplets (<0.5 µm). Generally, these droplets comprise only 5–7% of the volume of the opal glass. On further cooling, the fluoride may or may not crystallize. These glasses generally behave as spontaneous opals, opacifying or striking-in while they are being formed. Reheat or restrike opals are clear glass after forming, but like glass-ceramics, must be reheated for phase separation. These opals are not as economical to produce as the spontaneous opals, but their production is more controllable. Bubble opals can be formed with many tiny pockets of gas which make them opaque. At present, they have no commercial application.

The development of colloidal colors depends on controlled nucleation and growth of absorbing particles in glass. Colored filter glasses, eg, Corning Code 2403 sharp-cut red filter glass, can be made by precipitating colored crystals such as CdS or CdSe. Colloidal gold, copper, and silver are most often used. The particles have a selective absorption and a complementary selective reflection which depend on particle size. The metal to be precipitated is dissolved in the glass in conjunction with a reducing agent such as tin oxide or antimony oxide. After melting, the glass is rapidly cooled to a temperature region where the nucleation rate is high and then heated to a temperature where crystal growth dominates. By this technique, the size of the crystals, and therefore the color, can be controlled. Porous glass and reconstructed 96% silica glass are made with this processing technique. Porous glass is used in the biological sciences as a support for immobilized enzymes and antibodies (see ENZYME INHIBITORS).

Light-Sensitive Glasses. Photosensitive glasses (54) employ structural transformations, ie, precipitation and crystallization, which are sensitized by exposure to electromagnetic radiation, usually uv, but sometimes visible or x-ray radiation. The transformation is generally initiated by the photoreduction of a metal ion. Subsequent or simultaneous heat treatment morphologically transforms the glass in the regions exposed to the radiation leading, in general, to a change in optical or physical properties (see CHROMOGENIC MATERIALS, PHOTOCHROMIC).

The simplest photosensitive glasses are those colloidally colored with gold, silver, or copper. A sequence of exposure and heat treatments leads to the development of a controlled spatial pattern of color in the glass (55), eg, Corning Code 8600. In some light-sensitive glasses, this photosensitized precipitation of metal colloids can act as nucleation sites for subsequent phase changes in the glass. For example, an opalizing phase such as sodium fluoride may be so precipitated in a controlled manner to give an opal pattern within the glass. Corning Code 8601 is such a photosensitive opal which has found commercial and architectural uses, for example, in decorative windows in the United Nations Assembly Building. Chemical machining of glass is made possible by the heterogeneous precipitation of Li_2SiO_3 from lithium silicate-based glasses. In this system, the crystallized material is much more soluble in 10% hydrofluoric acid than the residual glass. If, after dissolution of Li_2SiO_3, the glass is again exposed to uv light or x-rays, subsequent heat treatment can produce a hard, nonporous, strong, high temperature glass-ceramic such as Corning Code 8603. Except for less than 1% shrinkage this can be done without distortion. The final material is almost completely crystalline, and is superior to the original glass in strength, hardness, thermal shock resistance, and electrical resistivity. It is used for printed circuit boards for high temperatures (400–500°C) and high humidities, for laminated dielectric structures, fluid amplifiers, etc.

Corning developed polychromatic glass, which is a special photosensitive material. A typical polychromatic glass before processing has properties similar to those of window glass. However, it has the potential, when activated by uv light and heat, to produce within the same glass all hues and a wide range of transparent colors or combination of colors, any desired pattern of white or colored opacity to employ either two- or three-dimensional geometry in the patterns, and to produce permanent full-color photographic images. The mechanism (56) of this variable, controlled coloration involves a photosensitized formation of metallic silver on a sodium halide microcrystalline phase, which itself has been precipitated by a photosensitive process from the glass.

Because absorption and scattering characteristics are due to silver particles and inorganic crystals in a stable glass matrix, they are as permanent as the glass. The only known way to destroy the colors without destroying the glass is to heat the material to about 400 to 450°C, where atomic diffusion can occur and the colors deteriorate to brown or yellow. The capability of reproducing color or opacity in three dimensions within a solid transparent medium may lead to new applications in photographic contour mapping, modeling, and sculpture, including holography (qv).

Photochromic glasses are those whose transmission of visible light decreases when the glass is exposed to uv or visible light and increases when the exposure is stopped. Photochromic glasses are commercially available as ophthalmic lens blanks, manufactured by Corning, Chance-Pilkington, Schott Optical Glass, and others. At one time, Corning also manufactured sagged photochromic lens blanks for nonprescription sunglasses. These blanks were cut from a continuous sheet made by the fusion process. Photochromic sheet has also found use in architectural applications and automobile sunroofs.

The commercially available photochromic glasses contain a fine dispersion of silver halide crystallites which are about 10 nm in diameter and about 100 nm apart. These crystallites are precipitated thermally without exposure to radiation. Hence, they are not photosensitive as defined earlier. Darkening can be likened to the latent image formation step in silver halide photography. The light causes the formation of very small amounts of silver metal within or at the surface of the silver halide crystallite. Because the glass traps the reaction products close to each other, the reaction reverses and the silver is reabsorbed into the silver halide when the irradiation ceases. Figure 5 shows the percent luminous transmittance of Corning Code 8111 ophthalmic lens glass under various conditions.

Experimental photochromic glasses have shown a wide variety of darkening and clearing rates. Some glasses clear in seconds, whereas others are essentially stable in their darkened state at room temperature. The fading process is temperature dependent; hence, glasses tend to darken less and fade faster at higher temperatures (darken more and fade slower at low temperatures). The temperature dependence of the darkened transmittance is generally greater for the faster-fading glasses. All commercial glasses are designed for particular applications and involve balances between maximum available darkening and fade rate and temperature dependence.

Deforming a photochromic glass at a temperature slightly below its softening point, by extrusion or redraw processes, elongates the silver halide particles within it. Such a glass polarizes light in the darkened state. If stretched photochromic glass is subjected to heat treatment in a reducing atmosphere, the resultant elongated silver particles produce permanent light polarization properties. Precise control of the process has yielded the family of Polarcor polarizing glass filters manufactured by Corning (57).

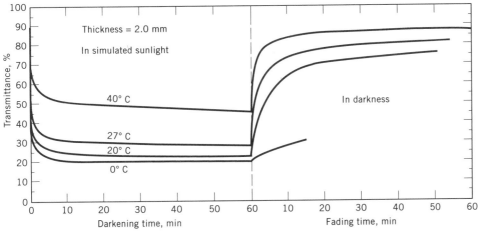

Fig. 5. Darkening and fading rate curves for Corning Code 8111. Courtesy of Corning Inc.

Properties

Rheological. The viscosity of a glass determines its melting, forming, and annealing procedures as well as the limitations of its use at high temperature. Viscosity is ordinarily measured between 10^{13} and 10 Pa·s (10^{14} and 100 P); at room temperature it is greater than 10^{19} Pa·s (10^{20} P). The rapid but smooth change of viscosity with temperature is shown for several glasses in Figure 6. Compositions and properties are shown in Tables 3, 4, and 5. The effect of modifiers on the viscosity of glass at high temperature depends on their polarizability or ionic field strength (58). Low field-strength modifiers decrease the viscosity of silica more than high field-strength ones. At low temperature, the effects of a modifier on viscosity are largely controlled by its coordination number. Modifiers with higher coordination numbers tend to increase low temperature viscosity as a result of packing restraints (see RHEOLOGICAL MEASUREMENTS).

Reference points on the viscosity–temperature curve in Figure 7 have been chosen to characterize properties of an individual glass and to facilitate comparisons of similar glasses. Those used most frequently are the working, softening, annealing, and strain points. Definitions of these selected points are given by ASTM. The working point of a glass is the temperature at which its viscosity is exactly 1 kPa·s (10^4 P). At this viscosity, glass is sufficiently fluid for forming by

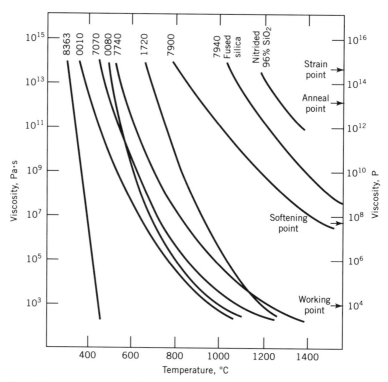

Fig. 6. Viscosity vs temperature for some commercial glasses designated by glass codes (see Table 3). Courtesy of Corning Inc.

most common methods. Viscosity between 10^6-1 Pa·s $(10^7-10$ P) is measured with a Margules viscometer, a calibrated instrument which measures the force exerted by molten glass upon a rotating spindle. The softening point of a glass cannot be defined by a precise viscosity. However, ASTM C338 defines the softening point as the temperature at which a uniform fiber of glass 0.55–0.75 mm diameter and 23.5 cm long elongates under its own weight at 1 mm/min when the upper 10 cm is heated at 5°C/min. For glass having a density of 2.50 g/cm^3, this corresponds to a viscosity of $10^{6.6}$ Pa·s $(10^{7.6}$ P). The exact viscosity is a function of density and surface tension. The annealing point is the temperature that corresponds to a viscosity of approximately 10^{12} Pa·s $(10^{13}$ P) (ASTM C336 and C598). At the annealing point, internal stress is reduced to an acceptable value in about 15 min. The strain-point temperature is the lower end of the annealing range and is approximately $10^{13.5}$ Pa·s $(10^{14.5}$ P) (ASTM C336 and C598). Internal stresses are reduced to an acceptable level in approximately four hours at the strain-point temperature.

Glass is usually melted and fined at viscosities between 5 and 50 Pa·s (50–500 P) but forming and final viscosity requirements vary greatly. The ranges of viscosity for various forming methods are compared in Figure 7.

Viscosities of glass are compared qualitatively. A hard glass has a high softening point and a soft glass has a lower softening point. Long and short refers to the temperature difference between the softening and strain point of glass. A long glass has a large temperature difference between its softening and strain point, ie, it solidifies (sets up) slower than a short glass as temperature decreases.

Thermal. *Expansion.* The thermal expansion of a glass determines the range of materials to which it can be safely sealed. It also affects the ability of the glass to survive thermal shock or cycling. The usefulness of glass as a heat exchanger or a thermal barrier, and its ease of melting and forming, depend on its heat-transfer properties and emissivity. The upper use temperature is a function of all of these properties.

Like most materials, glass expands when it is heated and contracts when it is cooled. If the thermal cycling is slow enough, there is virtually no hysteresis effect. Linear expansion, $\Delta L/L$, is the change in length per initial length. Generally, the expansion is proportional to temperature up to 300°C or more, depending on the glass. The slope of the linear expansion vs temperature curve, the linear thermal expansion coefficient, α, is therefore virtually constant between 0 and 300°C for most glasses, and thus α_{0-300} is a useful property for comparisons (see Table 4). As the temperature of the glass rises to near the set point (strain point + 5°C), the thermal expansion increases more rapidly (compare Figure 1, which is a volume–temperature curve for a general glass). The volume–temperature curve is analogous to linear expansion (length–temperature) curves. The density, ρ, which depends on the temperature, is easily calculated from the density at room temperature and the appropriate linear thermal expansion coefficient.

The sealing of glasses to either metals or other glasses depends on the thermal expansion mismatch of the two materials as well as their rheological and elastic properties (59). The materials must be heated to a point where the softer glass flows and intimately contacts the other material. In the case of glass-to-glass seals, the glasses flow together and bond extremely well. For glass-to-metal seals, however, a complicated and rigorously controlled oxidation-reduction re-

Table 3. Approximate Compositions of Commercial Silicate Glasses, wt %

Use	Glass code	SiO_2	Al_2O_3	B_2O_3	Li_2O	Na_2O	K_2O	MgO	CaO	BaO	PbO	ZnO	Others
lamp tubing	0010	63	1			8	6				22		
electronic and lamp tubing	0070	71	3			13	1	5	7				
lamp bulbs	0080	73	1			17		4	5				
electronic and lamp bulbs	0091	73.5	1			15	0.5	4	6				
lamp tubing	0120	56	2			4	9				29		
color TV neck	0137	52.5	1			0.5	13				28		5% SrO
color TV funnel	0138	54	2			6	8	2	3		23		
microsheet	0211	65	2	9		7	7	2				7	3% TiO_2
bench tops	0330	65	20	2	4			2				2	1% As_2O_3, 4% TiO_2
ignition tubes	1720	62	17	5		1		7	8				
electron tubes	1723	57	16	4				7	10	6			
iron sealing	1990	41			2	5	12				40		
optical filter	2405	70	1	12		5						11	CdS,Se
lamp bulbs	2473	67	1			16						12	1.5% CdO, SnO, Se, S^{2-}
tungsten sealing	3320	76	2	15		4	2						
table ware	6720	60	10	1		8	2					10	1% U_3O_8
lighting ware	6750	61	11			15			5	9			4% F^-
Kovar sealing	7040	67	3	23		4	3						3% F^-
series sealing	7050	68	2	24	1	6							
Kovar sealing	7052	64	8	19	1	2	3			3			
Kovar sealing	7056	68	3	18		1	9						F^-
substrate	7059	49	10	15						25			
low loss electrical	7070	72	1	25	0.5	0.5	1						1% As_2O_3
sealed beam lamps	7251	82	2	12		4							
laboratory ware	7280	71			1	12							16% ZrO_2
solder sealing	7570	3	11	12							74		
tungsten sealing	7720	74	1	15		4					6		

application	glass	SiO_2	Al_2O_3	B_2O_3	Li_2O	MgO	CaO	BaO	PbO	Na_2O	K_2O	ZnO	other
general	7740	81	2	13						4			1% As_2O_3
general	7760	78	2	15						3	1		
pharmaceutical	7800	72	6	11						7	1		
high temperature	7900	96.5	0.5	3				1				2	0.1% H_2O
optical	7940	99.9											7% TiO_2
optical	7971	93											0.7% F^-
electron tubes	8160	56	2		3			1	23	5	10		2% Rb_2O
electron tubes	8161	39							51	2	6		
Kovar sealing	8830	65	5	23						7	6		
CRT panel	9025	68	4	12	3			2	2	5	9		1% TiO_2, 1% CeO_2
color TV panel	9061	65	2				2	2	2	7	1		10% SrO, F^-, TiO_2, CeO_2, As_2O_3, Sb_2O_3
missile nose cones	9606	56	20			15	1						9% TiO_2
cooking ware	9608	70	18		3	3						1	5% TiO_2
smooth cooktop	9617	74	20		3	2						1	5% TiO_2, F^-
machinable glass-ceramic	9658	45	17	8		14					10		6% F^-
uv transmitting	9741	65	5	27				1		2			F^-
optical	N16B	5		34				2				6	6% ZrO_2, 41% La_2O_3, 6% CdO
laboratory ware	G20	76	5	7				4		6	1		
laboratory ware	N-51a	74.5	6	10				2		6	0.5		
laboratory ware	KG33	81	2	13						4			
reinforcement fiber	E-glass	54	14	10		4.5	17.5			0.5			
insulation fiber	T-glass	59	4.5	3.5		5.5	16			0.5			
chemical-resistant fiber	C-glass	65	4	5.5		3	14			8	0.5		
insulation fiber	SF-glass	59.5	5	7			14.5						4% ZrO_2
high strength fiber	S-glass	65	25			10							8% TiO_2, F^-

Table 4. Properties of Glasses[a]

Glass code[b]	Forms usually available[a,c]	Corrosion resistance[d] Weathering	Water	Acid	Thermal expansion ×10^{-7}/°C 0–300°C	25°C to setting point	Upper working temperatures[e] (Mechanical considerations only) Annealed Normal service, °C	Annealed Extreme service, °C	Tempered Normal service, °C	Tempered Extreme service, °C	Thermal shock resistance[f] plates 15 × 15 cm Annealed 3.2 mm thick, °C	6.4 mm thick, °C	12.7 mm thick, °C
0010	T	2		2	93.5	99.8	110	380			65	50	35
0070	BT	2		2	91.0	100							
0080	BMT	3	2	2	93.5	105	110	460	220	250	65	50	35
0091	B				91.0	100							
0120	TM	2		2	89.5	97	110	380			65	50	35
0137	T				97.0	102							
0138	P	1		3	98.5	109							
0211	S	1		2	74.0	84.0							
0330	RS[k]	1	1	3	9.7		538						
1720	BT	1	1	3	42	52	200	650	400	450	135	115	75
1723	BT	1	1	3	46	54	200	650	400	450	125	100	70
1990		3	3	4	124	136	100	310			45	35	25
2405	BPU[l]				43	53	200	480			135	115	75
2473	B[l]	2		2[n]	91		110	460			65	50	35
3320	m	1[n]	1[n]	2[n]	40	43	200	480			145	110	80
6720	P[o]	p	1	2	78.5		110	480	220	275	70	60	40
6750	BPR[o]	p	2	2	88	90	110	420	220	220	65	50	35
7040	BT	3[n]	3[n]	4[n]	47.5	54	200	430					
7050	T	3[n]	3[n]	4[n]	46	51	200	440	235	235	125	100	70
7052	BMPT	2[n]	2[n]	4[n]	46	53	200	420	210	210	125	100	70
7056	BTP	2	2	2	51.5	56	200	460					
7059	S	1		4	46.0	50.5							
7070	BMPT	2[n]	2[n]	2[n]	32	39	230	430	230	230	180	150	100
7251	P	1[n]	2[n]	2[n]	36.5	41.4	230	460	260	260	160	130	90
7280	BT	1		2	65.0	92							
7570		3	3	4	84	92	100	300					

7720	BPT	1	2[n]	2[n]	36	43	230	460	260	260	160	130	90
7740	BPSTU	1[n]	1[n]	1[n]	32.5	35	230	490	260	290	160	130	90
7760	BP	2	2	2	34	37	230	450	250	250	160	130	90
7800	T	1	1	1	50	53	200	460					
7913	BPRST	1	1	1	7.5	5.5[r]	900	1200					
7940	U	1	1	1	5.6	3.5[r]	900	1100					
7971	U	1	1	1	0.5	−2	800	1100					
8160	PT	2	2	3	91	100	100	380			65	50	35
8161	PT	2	1	4	90	99	100	390					
8830	PT	2		4	49.5								
9025	P	3		2	90.0	98.0							
9061	P	2		2	99.0	111							
9606	C[s]		1	4	57		700				200	170	130
9608	BP[s]		1	2	4–20[t]		700	800					
9617	R[s]	1		1	9.0								
9658	PRT[s]	1		4	89.0								
9741	BUT	3[n]	3[n]	4[n]	39.5	50	200	390			150	120	80
N16B					87								
G20	BPT												
N51a	T				50						115	95	65
KG33	BPSTU				32						180	150	100

Table 4. *(Continued)*

| Glass code[b] | Thermal stress resistance,[g] °C | Viscosity data[h] | | | | Knoop hardness, HK_{100}[i] | Density, g/cm³ | Young's modulus, GPa | Poisson's ratio | Log₁₀ of volume resistivity, Ω·cm | | | Dielectric properties at 1MHz, 20 | | | |
		Strain, °C	Annealing, °C	Softening, °C	Working, °C					25°C	250°C	350°C	Power factor, %	Dielectric constant	Loss factor, %	Refractive index[j]
0010	19	395	435	628	985	360	2.79	62	0.21	>17	8.9	7.0	0.16	6.7	1	1.540
0070		487	527	715			2.50									1.513
0080	16	473	514	696	1005	465	2.47	70	0.22	12.4	6.4	5.1	0.9	7.2	6.5	1.512
0091		485	523	705			2.48									
0120	20	395	435	630	985	382	3.05	59	0.22	>17	10.1	8.1	0.12	6.7	0.8	1.560
0137		436	478	661	977		3.18				10.1	8.4				1.570
0138		450	490	670			3.02					7.7				1.563
0211	20	508	550	720	1008		2.57	74	0.21							1.52
0330	178					522	2.54	86	0.26							
1720	28	667	712	915	1202	513	2.52	87	0.24	>17	11.4	9.5	0.38	7.2	2.7	1.530
1723	26	665	710	908	1168	514	2.64	86	0.24	>17	13.5	11.3	0.16	6.3	1.0	1.547
1990	14	340	370	500	756		3.50	58	0.25	>17	10.1	7.7	0.04	8.3	0.33	
2405	37	501	537	765	1083		2.48	68	0.21							1.507
2473	19	466	509	697			2.65	66	0.22							1.52
3320	43	493	540	780	1171		2.27	65	0.19		8.6	7.1	0.30	4.9	1.5	1.481
6720	20	505	540	780	1023		2.58	70	0.21							1.507
6750	18	447	485	676	1040		2.59									1.513
7040	37	449	490	702	1080		2.24	59	0.23		9.6	7.8	0.20	4.8	1.0	1.480
7050	39	461	501	703	1027		2.24	60	0.22	16	8.4	6.8	0.33	4.9	1.6	1.479
7052	41	436	480	712	1128	375	2.27	57	0.22	17	9.2	7.4	0.15	5.1	1.3	1.484
7056	33	472	512	718	1058		2.29	64	0.21		10.2	8.3	0.27	5.7	1.5	1.487
7059	32	593	639	844	1160		2.76	68	0.28		13.1	11.0		5.9		1.53
7070	66	456	496		1068		2.13	51	0.22	>17	11.2	9.1	0.06	4.1	0.25	1.469
7251	48	521	565	808	1192	451	2.26	64	0.19	18	8.0	6.5	0.45	4.9	2.18	1.476
7280	20	576	624	873	1234		2.62	83	0.22		6.0	4.8		6.12	0.07	1.54
7570	21	342	363	440	558		5.42	55	0.28	>17	10.6	8.7	0.22	15	3.3	1.86
7720	49	484	523	755	1146		2.35	63	0.20	16	8.9	7.3	0.23	4.6	1.3	1.487
7740	54	510	560	821	1252	418	2.23	63	0.20	15	8.1	6.6	0.39	4.6	2.6	1.474
7760	52	478	523	780	1198	442	2.24	62	0.20	17	9.4	7.7	0.16	4.6	0.79	1.473

Glass[a]	(1)	(2)	(3)	(4)	(5)	(6)	(7)	(8)	(9)	(10)	(11)	(12)	(13)	(14)	(15)	(16)
7800	33	576	533	795	1189	487	2.36	68	0.19	>17	7.0	5.7	0.04	3.8	0.15	1.491
7913	220	1020	890	1530			2.18	73	0.16	>17	9.7	8.1	0.001	3.8	0.0038	1.458
7940	286	1084	956	1580		489	2.20			>17	12.4	10.7	<0.002[q]		<0.008[q]	1.459
7971	3370		1000	1500			2.21	68	0.17	20.3	12.2	10.1		4.0[f]		1.484
8160	18	397	438	632	973		2.98			>17	10.6	8.4	0.09	7.0	0.63	1.553
8161	22	400	435	600	862		3.99	54	0.24	>17	12.0	9.9	0.06	8.3	0.50	1.659
8830	39	460	501	708	1042		2.24	56	0.22		7.8	6.3		6.5		
9025		421	460	647			2.60				8.5	6.8				1.529
9061		503	462	688	988		2.68				9.3	7.4				
9606	16					657	2.60	118	0.24	16.7	10.0	8.7	0.30	5.6	1.7	
9608						593	2.50	86	0.25	13.4	8.1	6.8	0.34	6.9	2.3	
9617							2.54				7.2	5.7		6.75	0.6	1.468
9658												9.8	8.4		5.9	0.2
9741	54	408	450	705	1161		2.16	49	0.23	>17	9.4	7.6	0.32	4.7	1.5	
N16B	19		540	720	1065		2.48				5.8	4.7		7.7	1.4	
G20		524	569	794	1190		2.39	72			7.4	6.0	0.51	4.9	2	1.49
N51a		538	580	798	1190		2.36				6.8	5.4	0.96	5.9	5.7	1.47
KG33		513	565	827	1240		2.23	74			8.1	6.6	0.46	4.6	2.1	1.548
							2.60							6.4		1.541
							2.54	70						7.3		1.549
							2.61							7.8		1.537
							2.57	87						8.3		1.523

[a]B = blow ware; M = multiform; U = panels; P = pressed ware; R = rolled sheet; C = castings; S = plate glass; T = tubing and rod; F = fiber. [b]Glasses 7905, 7910, 7911, 7912, 7913, and 7917 for special uv and ir applications. Glass 1720 is available with improved uv transmittance (designated glass 9730). Glass 7760 is also available with special transmission suitable for sun lamps. [c]Color is clear unless otherwise noted. [d]Weathering is defined as corrosion by atmospheric-borne gases and vapors. [e]Normal service: no breakage from excessive thermal shock is assumed. Extreme limits; glass is very vulnerable to thermal shock. Recommendations in this range are based on mechanical stability considerations only. Tests should be made before adopting final designs. These data are approximate only. [f]These data approximate only, based on plunging cold sample into cold water after oven heating. Resistance of 100°C means no breakage if heated to 110°C and plunged into water at 10°C. Tempered samples have over twice the resistance of annealed glass. [g]Resistance in °C is the temperature differential the two surfaces of a tube or a constrained plate that will cause a tensile stress of 6.9 MPa (1000 psi) on the cooler surface. [h]These data subject to normal manufacturing variations. [i]Determined by revised ASTM standard: number of standard not yet assigned. [j]Refractive index may be either the sodium yellow line (589.3 nm) or the helium line (587.6 nm). Values at these wavelengths do not vary in the first three places beyond the decimal point. [k]Gray. [l]Red. [m]Canary. [n]These borosilicate glasses may rate differently if subjected to excessive heat treatment. [o]Opal. [p]Since weathering is determined primarily by clouding which changes transmission, a rating for the opal glasses is omitted. [q]At 10 kHz. [r]Extrapolated values. [s]White. [t]Code 9608 may be produced in a range of expansion values depending on intended application.

Table 5. Properties of Glass Fibers

Property	Value for				
	E-glass	T-glass	C-glass	SF-glass	S-glass
thermal expansion $\times 10^7/°C^a$	60	80	72	75	34
strain point,[b] °C	507		435		760
annealing point,[b] °C	657		585		810
softening point,[b] °C	846	715	752	675	970
density, g/cm³	2.60	2.54	2.61	2.57	
Young's modulus, GPa	74		70		87
dielectric constant, 1 MHz	6.4	7.3	7.8	8.3	
refractive index,[c] 1 MHz	1.548	1.541	1.549	1.537	1.523

[a]At 0–300°C.
[b]These data subject to normal manufacturing variations.
[c]Refractive index may be either the sodium yellow line (589.3 nm) or the helium line (587.6 nm). Values at these wavelengths do not vary in the first three places beyond the decimal point.

action must take place at the interface to form a good bond. As the composite cools, each material shrinks along its linear thermal expansion curve. At high temperatures, the softer glass deforms by viscous flow to compensate for any thermal expansion difference. However, when the composite reaches the set point of the softer material, viscous deformation stops and both members are constrained to shrink equally. Therefore, an expansion mismatch is divided between the two materials, producing stresses that must satisfy force and moment balances consistent with the appropriate elastic constants and geometrical cross sections of the materials (59). Other factors also affecting the stresses set up in the glass sealing include the geometry of the seal, the cooling rate used in annealing the seal, and the difference in annealing ranges.

The upper use temperature for annealed ware is below the temperature at which the glass begins to soften and flow (about 10^{14} Pa·s or 10^{15} P). The maximum use temperature of tempered ware is even lower, because of the phenomenon of stress release through viscous flow. Glass used to its extreme limit is vulnerable to thermal shock, and tests should be made before adapting final designs to any use. Table 4 lists the normal and extreme temperature limits for annealed and tempered glass. These data are approximate and assume that the product is not subject to stresses from thermal shock.

Stresses.   Stresses caused by steady-state thermal gradients may or may not cause failure, depending on the degree of constraint imposed by some parts of the item upon others or by the external mounting. Thus, under minimum constraint and maximum uniformity of gradient through the thickness, very large temperature differences can be tolerated. If a plate is completely restrained, the tensile stress on the cool side is given by

$$\sigma = E\alpha \, \Delta T/2(1 - v)$$

where ΔT is the temperature difference across the glass, α is the expansion coefficient, E is Young's modulus, and v is Poisson's ratio. Normally, the maximum

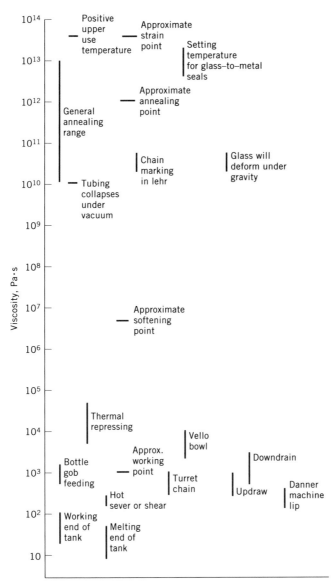

Fig. 7. Approximate viscosity values for forming and processing methods (Pa·s × 10 = P). Courtesy of Corning Inc.

design stress in glass is about 7 MPa (1000 psi). The face-to-face temperature differentials (thermal stress resistance) that cause this tensile stress on the cool face of cylinders or otherwise constrained shapes, are given in Table 4. The thermal stress resistance of glass-ceramic 9608 is calculated to be 480°C, considerably higher than that of most glasses and ceramics. This is due to the higher break resistance of the material and its low coefficient of thermal expansion.

When glass is suddenly cooled, as by removal from a hot oven, tensile stresses are introduced in the surfaces and compensating compressional stresses in the interior. Conversely, sudden heating leads to surface compression and internal tension. In either case, the stresses are temporary (transient) and disappear on attainment of temperature uniformity. Since glass fractures only in tension, usually at the surface, the temporary stresses from sudden cooling are much more damaging than those resulting from sudden heating, assuming, of course, that all surfaces are heated or cooled at the same time.

Because the strength of glass is greater under momentary stress than under prolonged load, thermal shock endurance cannot be directly calculated, but is generally determined by empirical testing. Resistance to breakage is determined by heating the ware to some appropriate temperature and then plunging it into cold water. Table 4 lists the thermal shock resistance of 15 × 15 cm annealed plates of three thicknesses. A resistance of 100°C means that no breakage occurs on heating to 100°C and plunging into water at 10°C. When other cooling media, eg, air or oil, are used, much higher values of thermal shock resistance are recorded. Tempered samples have more than twice the thermal shock resistance of annealed glass.

Heat Transmission. At room temperature, the thermal conductivity of glasses ranges from 0.67 to 1.21 W/(m·K), with the most common compositions near the upper end of the range. Thermal conductivity of glass-ceramics ranges from 1.7 to 3.8 W/(m·K). At a mean temperature of 200°C, the values are greater by approximately 25%. Above 500°C, conductivity increases rapidly because of radiation within the glass. For Corning Code 7740, used frequently in heat-transfer applications, the thermal constants are thermal conductivity (25°C): 0.96 W/(m·K); mean specific heat (25–175°C): 84 J/(kg·K) (20 cal/(kg·K)); thermal diffusivity: 0.56 mm²/s; and emissivity coefficient, radiant energy: 0.94.

Liquidus Temperature. The liquidus temperature determines the susceptibility of a glass to devitrification and therefore influences its forming limitations and often its heat-treating requirements.

Gas Permeation. Gas permeation through glass is of crucial importance to any high vacuum system. With the advent of high intensity tungsten–halogen lamps, the lamp envelopes must contain gas under pressure for a long time at high temperatures. The permeation rate varies with temperature, glass type and composition, and gas. The volume of gas (at STP) which diffuses through glass is proportional to the time, the area of the glass, and the pressure differential, and it is inversely proportional to the thickness of the glass (60). Figure 8 shows the constant of proportionality as a function of temperature for various gases permeating a 96% silica glass. Helium is the most mobile of the gases and silica is the most permeable glass. The addition of a modifier blocks the openings in the glass network and slows the permeation rate (60).

Mechanical. Because of its amorphous structure, glass is brittle, reasonably abrasion resistant, and nearly perfectly elastic as long as its temperature is low enough to prevent viscous flow. Among other things, glasses do not contain crystallographic planes which slip relative to each other and therefore permit the material to deform plastically when stress is applied (61). Slip tends to relieve stress and allow broken bonds to form new bonds before breakage occurs. Since no such phenomena take place in glass, a bond that is broken because of excessive stress quickly forms a crack. As the crack grows, the stress intensity at its tip increases; those bonds that were intact quickly break in its path, and the glass

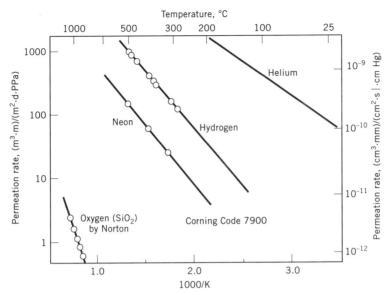

Fig. 8. Permeation of helium, hydrogen, neon, and oxygen for Corning Code 7900 (60). Courtesy of Corning Inc.

fractures. For this reason, a glass usually breaks because of an applied tensile stress and generally from a surface flaw. Several steps are commonly taken to make the glass more break resistant. The elimination of surface flaws, eg, micro-cracks or bruise checks, either by careful handling during and after forming, an-nealing, etc, or by acid etching the surface to remove these flaws, prevents a crack from starting. Secondary treatments, eg, ion exchange, thermal tempering, glaz-ing, or differential crystallization, provide a compressive stress on the surface of the glass that must be overcome before the applied tensile stress causes breakage. Figure 9 compares the strengths of various glass types and treatments. The depth of this compressive layer governs the resistance of glass to deep bruising or abrasion. Glass-ceramics resist fracture because the tiny microcrystals present in their structure arrest crack growth.

Strength. Theoretical calculation places the intrinsic strength of glass as high as 35 GPa (5×10^6 psi); the strength of flame-polished silica rods (25) is 14 MPa (2.0×10^6 psi). However, the practical strength of glass is a small fraction of these figures because of stress concentrations introduced by surface imperfec-tions. Strength of fibers may be measured in bending or pure tension; strength of more massive forms is measured only in bending.

A glass rod 6.4-mm dia with pristine surfaces may show a tensile strength of 1750 MPa (250,000 psi). However, normal handling of the severity experienced by many glass products in service introduces surface imperfections which reduce this strength to approximately 70 MPa (10,000 psi). Glass also exhibits a time-load effect, so that a glass article breaks at a lower stress under prolonged loading than under momentary loading.

For a safety factor, the allowable stress for annealed (stress-free) soda-lime glass is 7 MPa (1000 psi) under sustained loading for 1000 h or more. The chemical

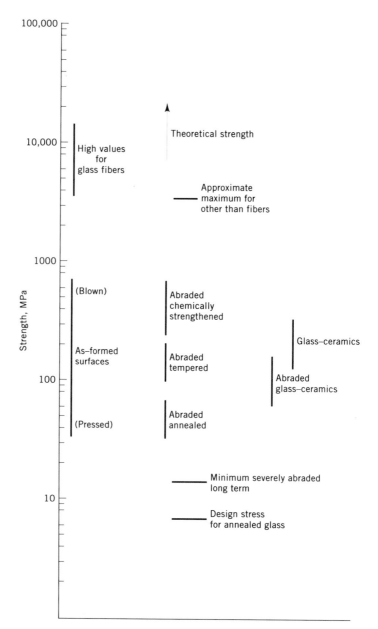

Fig. 9. Strength ranges for glass and glass-ceramic products. To convert MPa to psi, multiply by 145.

composition of glass has a significant effect on its design strength. The allowable stress for glasses with greater resistance to stress corrosion, eg, fused silica, can be twice as high as that for soda-lime glass (14 MPa or 2000 psi).

Elasticity. Glasses, like other brittle materials, deform elastically until they break in direct proportion to the applied stress. The Young's modulus E is the constant of proportionality between the applied stress and the resulting strain. It is about 70 GPa (10^7 psi) [(0.07 MPa stress per μm/m strain = (0.07 MPa·m)/μm)] for a typical glass.

A high modulus glass is one in which the atoms are closer together or the bonds are stronger and therefore more resistant to rupture. Articles made from brittle materials break when they exceed their strain tolerance which depends directly on size and configuration of the article as well as on surface condition. For bulk glass, it is on the order of 1000 μm/m, resulting in a strength of (0.07 MPa·m/μm) $\times$ (1000 μm/m) = 70 MPa or 10^4 psi. For fibers, the strain tolerance can be one or two orders of magnitude greater. Thus, for a given strain tolerance, the higher the Young's modulus, the greater the strength.

Young's modulus and Poisson's ratio v for many glasses are listed in Table 4. The modulus of rigidity G is $E/2(1 + v)$; the bulk modulus K is $E/3(1-2v)$, and the compressibility C is $1/K$. For a typical glass where E and v are 70 GPa (10^7 psi) and 0.22, respectively, G = 29 GPa (42 $\times$ 10^5 psi), K = 42 GPa (6.1 $\times$ 10^6 psi), and C = 24 $\times$ 10^{-9}/MPa (1.6 $\times$ 10^{-10}/psi).

Hardness. Glass hardness tests usually measure the resistance to abrasion by grinding or grit-blasting, resistance to scratching, or penetration by an indenter. The method to be used depends on expected service conditions. Knoop hardness (Table 4) is commonly used, because other methods usually fracture the glass.

On the Mohs' scale of scratch hardness, glass lies between apatite, 5, and quartz, 7. Some common materials hard enough to scratch glass include agate, sand, silicon carbide, hard steel, and emery. Glasses are harder than mica, mild steel, copper, aluminum, and marble. A typical glass-ceramic exhibits diamond penetration hardness ca 40% greater than borosilicate glass.

Electrical. Glasses are used in the electrical and electronic industries as insulators, lamp envelopes, cathode ray tubes, and encapsulators and protectors for microcircuit components, etc. Besides their ability to seal to metals and other glasses and to hold a vacuum and resist chemical attack, their electrical properties can be tailored to meet a wide range of needs. Generally, a glass has a high electrical resistivity, a high resistance to dielectric breakdown, and a low power factor and dielectric loss.

Resistivity. When a voltage is applied across a piece of glass, a minute current flows, partly over the surface and partly through the interior. Thus the insulation resistance is determined jointly by the surface resistivity measured in Ω/sq and the volume resistivity measured in Ω·cm. Surface electrical resistivity is of primary importance in all electrical insulation problems with glasses or ceramics, since it greatly affects performance under service conditions involving high humidity. Because of the adsorption of a moisture film, the surface electrical resistivity of glasses and ceramics, although high in comparision to that of many other materials, is markedly lower than volume resistivity. Thus chemical durability is an essential requirement in good electrical glass, because film thickness and film conductivity are both increased by the soluble products of unstable glass compositions (62) (see INSULATION, ELECTRIC).

Volume resistivity, when measured in terms of current flowing through unit path per unit of applied voltage, in accordance with Ohm's law, is markedly affected in glass by the amount of alkali present. Conduction in glass is generally by means of ionic transport and often by the movement of sodium ions, one of the ions most mobile in a silicate network (63). Glasses also conduct by means of potassium or lithium migration, but, when all three ions are present, they tend to interfere with one another and the electrical conductivity decreases. This is called the mixed-alkali effect. Volume resistivity of common glasses at ordinary room temperature varies widely with composition from as low as 10^8 to as high as 10^{19} $\Omega\cdot$cm. The resistivity of a given composition at a fixed temperature may vary by a factor of 3, depending on the degree of annealing or strain which the glass has received; strained glass has lower resistance than properly annealed glass which is practically strain free.

Dielectric Strength. Dielectric failure may be thermal or disruptive. In thermal breakdown, applied voltage heats the sample and thus lowers its electrical resistance. The lower resistance causes still greater heating and a vicious circle, leading to dielectric failure, occurs. However, if applied voltage is below a critical value, a stabilized condition may exist where heat input rate equals heat loss rate. In disruptive dielectric failure, the sample temperature does not increase. This type of failure is usually associated with voids and defects in the materials.

The intrinsic dielectric strength of glass is high. Thus in many applications this property is relatively unimportant compared with the problem of design to prevent flashover. Dielectric breakdown voltage decreases with increase in both frequency and temperature. At elevated temperatures, breakdown is governed mainly by the volume resistivity of the glass. Dielectric breakdown voltages for soda–lime (Corning Code 0080) glass plates are about 75% of those for Corning Code 7740 borosilicate glass.

Dielectric Loss. The loss tangent of glass is the ratio of the real to the imaginary current through a glass that is being subjected to an alternating voltage in a capacitor configuration. The loss tangent depends on the composition and structure of the glass. Material with a certain amount of alkali may exhibit certain loss properties, whereas another material with equal alkali content may be substantially different. On the other hand, a fixed composition with minor changes in alkali may show significant dielectric loss variation. Thus the loss properties can be controlled with variations in alkali and alterations in structure.

Several glasses have been developed with exceptionally low power factors of less than 0.015% at room temperature and 1 MHz. Figure 10 shows the variation of power factor with temperature.

Dielectric Constant. The dielectric constant determines electric energy storage in a polarized dielectric, thus serving to evaluate the performance of a capacitor. In d-c fields at ordinary room temperatures, dielectric absorption also affects dielectric constant measurements, giving values that vary with time. At $-75°C$ and below, the dielectric constant is independent of charge time. At ordinary room temperature and above, it increases with temperature at rates that vary with glass composition. This effect is less in high silica and borosilicate glasses and greater in those glasses containing substantial amounts of alkali.

In addition to the obvious effect of high dielectric constant glasses on the capacitance of the circuit elements into which they enter, their dielectric strengths may be more important. Since the amount of energy a capacitor can store varies as the first power of the dielectric constant and the second power of the voltage,

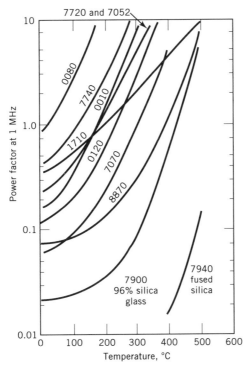

Fig. 10. Power factor at 1 MHz vs temperature for some commercial glasses. Courtesy of Corning Inc. The power factor is the ratio of the power in watts dissipated in a material to the product of the effective sinusoidal voltage and current in volt-amperes. When the dissipation factor is less than 0.1, the power factor differs from the dissipation factor by less than 0.5%.

a glass with twice the dielectric strength is as effective as one with four times the dielectric constant.

Microwave Compatibility. Microwave cooking has given rise to questions concerning the effects of microwave energy on various materials used in cooking vessels. Since the suitability of a cooking vessel for microwave oven use is affected by such factors as the configuration of the vessel, the distribution of food in it, and energy distribution in the oven, as well as the intrinsic properties of the material itself, there is no clear demarcation between suitable and nonsuitable materials. Nevertheless, the properties of some glasses and glass-ceramics make them more suitable for this application than others. The relative suitability of a glass or glass-ceramic material is expressed by a factor of merit which is proportional to the material's strength and inversely proportional to its Young's modulus, expansion coefficient, and rate of microwave energy absorption. Table 6 reflects the microwave energy absorption and the factor of merit of various types of glass and glass-ceramics (see MICROWAVE TECHNOLOGY).

Chemical Stability. *Corrosion Resistance.* The resistance of glass to chemical corrosion is frequently the reason for its use. However, the durability of a glass varies from highly soluble to highly durable, depending on its composition

Table 6. Microwave Energy Absorption and the Factor of Merit for Various Types of Glasses and Glass-Ceramics

Material	Energy absorbed, %	Factor of merit
fused silica	0.008	1500
glass-ceramic		
cordierite	0.03	41
β-spodumene	0.38	20
nepheline, celsian	6.1	0.7
tempered aluminosilicate glass	0.71	7
borosilicate glass	0.40	5
opal		
high strength	0.90	5
tempered	1.1	2
tempered lime	1.0	2

and the solvent considered. Comparisions are usually based on measurements of weight loss, changes in surface quality, or the analysis of solutions that were in contact with a glass.

The reaction of acids with glass may be either a leaching or a complete dissolution process (25). Hydrofluoric acid [7664-39-3] attacks silicate glasses by dissolving the silica network. Other acids such as hydrochloric or nitric acid [7697-37-2] may dissolve certain glasses, but most frequently they react by selective extraction of alkali and the substitution of protons in a diffusion-controlled process. As a layer of glass is leached, the rate of material extraction decreases as the square root of time. The temperature coefficient of leaching reactions of glass increases about 1.5 times for every 10 K increase in temperature. An etching or dissolution process constantly exposes a fresh glass surface, and thus the reaction rate is usually constant with time. Etching may give a smooth or a pitted glass surface, depending on the glass composition and quality and the solution used. The rate of dissolution increases about 2.5 times with each 10 K increase in temperature. Ratings of acid durability of various glasses are based on standard 5% hydrochloric acid solution test. Results for several glasses are shown in Table 4. A rating of 1 implies little or no visible damage, whereas a rating of 4 represents severe deterioration of the glass during the test (1 in. = 2.54 cm).

Classification	Thickness loss, in.
1	$<10^{-6}$
2	10^{-6} to 10^{-5}
3	10^{-5} to 10^{-4}
4	$>10^{-4}$

Reaction of bases with most commercial silicate glasses produces dissolution rates of 7.5–30 μm/d when tested in 5% NaOH solution at 95°C. The mechanism involves a complete dissolution process as described for acid corrosion. Weaker alkaline solutions, however, may both leach and dissolve and sometimes show a greater dependence on glass composition. For strong alkali solutions, the rate of attack ca doubles for each 10 K increase in temperature or each increase in pH unit. Higher alkali durability glasses such as Corning's Code 7280 zirconia–silicate glass are available for laboratory ware. As with acids, reaction products accumulating on the surface slow alkali attack. Glass-ceramics may exhibit preferential attack of various crystalline phases.

The attack of water is related to the leaching mechanism described for acids. Table 4 rates glasses based on their resistance to water attack. Low alkali, high alumina, or borosilicate glasses generally have high water durability.

Weathering of glass is the result of the action of water, carbon dioxide, and other atmospheric constituents. Water, initially absorbed by the glass, is exchanged for alkalies that form alkali salt solutions. If left in contact with the glass, these may cause additional surface damage. For this reason, weathering resistance may not correlate with acid durability. Test methods have been designed to accelerate the weathering process; for example, samples are weathered in a chamber at 50°C and 98% rh. Visual comparisons allow rating of glasses as shown in Table 4.

Element or metal halide attack of silica glass at elevated temperatures is of special interest to lighting ware manufacturers (64). At high temperatures, reaction occurs between vitreous silica and various metals or halides. Staining of glass by alkali metal vapors can occur at relatively low temperatures (<300°C), but at higher temperatures (ca 900°C) failure of the glass may occur as a result of severe devitrification that is induced. Rates of attack based on one-hour exposure indicate sodium causes failure of silica tubes because of devitrification below 1000°C. Other elements that induce devitrification are Mn, Sn, Fe, and Zn which cause failure at 1250 to 1350°C. Metal halides that react with silica glass below 1200°C are LiCl, NaCl, $SnCl_2$, NaBr, and KBr. At 1200–1400°C, RbCl, CsCl, NaI, KI, $MgCl_2$, $CaCl_2$, $BaCl_2$, and $AlCl_3$ cause failure.

Optical. *Refraction.* Optical glasses are usually described in terms of their refractive index at the sodium D line (589.3 nm), and their v value (or Abbe number) which is a measure of the dispersion or variation of index with wavelength, ie, $v = (n_D - 1)/(n_F - n_C)$, in which n_F is the refractive index at the hydrogen F line (486.1 nm) and n_C is the refractive index at the hydrogen C line (656.3 nm). These data have been incorporated into a six-digit numbering system used to identify optical glasses. For example, the glass commonly used to produce eyeglasses has the number 523–586. This means $n_D = 1.523$ and $v = 58.6$ for this glass. Glasses with index n_D less than 1.60 and a v value of 55 or above are defined as crown glasses; those with a v value below about 50 are defined as flint glasses. Glasses with n_D greater than 1.60 are defined as crown glasses if v is 50 or above. The Schott nomenclature often used for optical glasses is defined by Figure 11. The labeled areas represent commercially available glasses. Crowns are usually alkali silicate glasses and flints are lead alkali silicates. Phosphate and borate-based glasses are used to obtain high index or low dispersion glasses or both. Compositions of numerous optical glasses are given in Reference 2. Table 4 also lists values of n_D for several glasses.

Reflectance is dependent on the condition of the glass surface. It can be specular, as in polished or precision molded surfaces, or diffuse for ground or irregularly etched surfaces. For normal incidence, the fraction of incident light

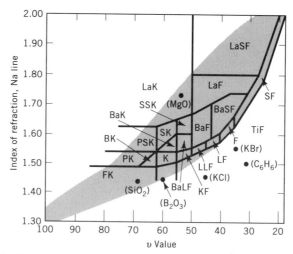

Fig. 11. Index of refraction vs dispersion and optical classification of glasses. The shaded area indicates the region of glass formation. BaF = barium flint; BaK = barium crown; BaLF = light barium flint; BaSF = heavy barium flint; BK = borosilicate crown; F = flint; FK = fluorcrown; K = crown; KF = crown flint; LaF = lanthanum flint; LaSF = heavy lanthanum flint; LaK = lanthanum crown; LF = light flint; LLF = very light flint; PK = phosphate crown; PSK = heavy phosphate crown; SF = heavy flint; SK = heavy crown; SSK = very heavy crown; TiF = titanium flint.

reflected by a single surface is given by $R = [(n - 1)/(n + 1)]^2$, in which n is the index of refraction at the wavelength of interest. For silicate glasses having an index of refraction of 1.5 for green light, the reflection loss per surface is about 4%. A perfectly clear glass therefore transmits ca 92% of the incident light. As the index of refraction increases above 1.5, transmission decreases. Indexes of refraction of 1.7 and 1.9 predict the transmission of clear glasses to be 87 and 81%, respectively. Low index coatings or special surface treatments of certain glasses may significantly reduce reflection losses (65).

Transmission. The spectral transmission of glass is determined by reflection at the glass surfaces and the optical absorption within the glass. Overall transmission of a flat sample at a particular wavelength is equal to $(1 - R)^2 e^{-\beta t}$, where β is the absorption coefficient, t the thickness of glass, and R the air–glass reflection coefficient. Transmission is a function of wavelength. In silicate glasses, it is limited by the absorption of silica at approximately 150 nm in the uv and at 6000 nm in the ir. Iron impurities further reduce transmission in the uv and near ir. Dissolved water absorbs at 2700 nm and is a serious problem in making ir transmitting glasses.

Transmission may be controlled by the type of glass used, eg, silicate, phosphate, borate, etc, or by the control or addition of coloring additives, melting atmosphere, melting temperatures, and cooling schedules (66,67). In glasses containing suspended particles, transmission is diffuse. Opal glasses, used extensively for lighting products, exemplify diffuse transmission. Oxides that are frequently used to control color in glass are listed in Table 7.

Table 7. Coloration of Glasses[a]

Effect	Additive[b]
colorless, uv transmitting	SiO_2 glass
	96% SiO_2 glass
	P_2O_5 glass
colorless, uv absorbing	CeO_2
	TiO_2
	Fe_2O_3 (P_2O_5 glasses)
blue	Co_3O_4
	Cu_2O + CuO
	$S(B_2O_3$ glasses)
purple	Mn_2O_3
	$NiO(K_2O$ glasses)
green	Cr_2O_3
	Fe_2O_3 + Cr_2O_3 + CuO
	V_2O_3
	CuO
	MoO_3 (P_2O_5 glasses)
brown	MnO (reduced)
	MnO + Fe_2O_3
	TiO_2 + Fe_2O_3
	NiO (Na_2O glasses)
	MnO + CeO_2
amber	Na_2S (reduced)
yellow	CdS
	CeO_2 + TiO_2
	Ag staining
	UO_3
orange	CdS + Se
red	CdS + Se
	Au
	Cu or Cu staining
	UO_3 (PbO glasses)
	Sb_2S_3
black	Co_3O_4 (+ Mn, Ni, Fe, Cu, Cr oxides)
colorless, heat absorbing	FeO (P_2O_5 glasses)
colorless, ir transmitting	PbO glasses
	$CaO \cdot Al_2O_3$ glasses
	Te, Ge, or Sb glasses
	As_2S_3 glass
	Ge–Sb–Se or Ge–As–Te glasses

[a]Refs. 68 and 69.
[b]Glass former is silica unless otherwise noted. See Table 1.

Stress-Optical Coefficients.   When strained, glass becomes doubly refracting, or birefringent. Stress optical effects are presumed to be elastic; therefore, strain is proportional to stress. The stress optical coefficient, B, or Brewster's constant, is a measure of this proportionality, and varies greatly with glass composition (70). Values of B range from about -1 for extra dense flints containing 80% PbO to ca 3.6 for 96% silica glass. For soda–lime glass, B is about 2.5.

Radiation Effects.   Interaction of glass with low energy visible and uv radiation may result in alteration of the electronic states. These changes may cause coloration or luminescence effects, eg, photochromic or photosensitive glass. When changes in color are produced by sunlight, the effect is frequently referred to as solarization (71,72). Effects produced by ions, gamma rays, and x-rays are relevant to glasses used as electronic tube envelopes, radiation-shielding windows, and dosimeters. The stability of glass in neutron fields depends on the relative absence of boron in the glass composition. Boron-free glasses can, in general, withstand fluxes of 10^{19} neutrons/cm^2. Ion bombardment disrupts the glass structure, producing discoloration and an evolution of oxygen. This is troublesome in electronic tube envelopes where the oxygen can react with the cathode, reducing the tube life. Electron browning is also troublesome in cathode ray tubes containing oxides of lead or bismuth, which are easily reduced by the impinging electrons, producing a brown tint in the glass. Relatively low doses (2.6 C/kg $= 10^4$ roentgens) of gamma rays tend to discolor dense lead-containing glasses. The glass most resistant to this effect is fused silica; it shows virtually no discoloration at doses above 2.6×10^5/kg (10^9 roentgen). This glass, however, is of low density (2.20 g/cm^3) and it is expensive. Glasses of intermediate density and relatively low cost can be protected against darkening in γ-ray fields by the addition of multivalent oxides such as cerium dioxide to give color stability up to about 2.6×10^4 C/kg (10^8 roentgen). These glasses have a density of about 3.30 g/cm^3, and can be combined with lead glasses of higher density to make composite windows. These windows are remarkably resistant to radiation damage, and can be used extensively with little or no maintenance in radiation flux fields of more than 2.6×10^2 C/(kg·h) (10^6 roentgen/h). They match the absorption of the heavy concretes used for shielding walls. Conversely, glasses containing cobalt have been developed that exhibit color changes on absorption of radiation. More sensitive silver phosphate glasses are used as dosimeters to measure earlier x-ray or γ ray exposure.

Ce^{3+} can either increase or decrease the visible absorption induced in glasses by irradiation. It has a weak absorption band at about 310 nm. Irradiation into this band, or electron beam irradiation, can cause the Ce^{3+} to trap a hole to become a Ce^{4+} which absorbs very strongly in the wavelength region between 300–320 nm. The hole trapping efficiency is very high compared to many other impurities or defects which trap holes. Therefore, it protects the glass against the induced absorption due to holes trapped by other defects. On the other hand, the Ce^{3+} is a good source of electrons. It, therefore, promotes induced darkening if the glass contains defects which trap electrons. The violet color produced in glasses containing cerium and vanadium is an example of such an induced visible color (73,74).

Manufacture and Processing

Most glass articles are manufactured by a process in which raw materials are converted at high temperatures to a homogeneous melt that is then formed into the articles. The flow diagram in Figure 12 summarizes the details of glass container manufacturing. The vapor deposition of SiO_2 from a flame fed with $SiCl_4$ and oxygen is the basis for manufacturing high purity glass used for blanks which are redrawn into optical fibers (see FIBER OPTICS). Fused silica items that cannot be formed from viscous melts of SiO_2 or quartz are prepared by vapor deposition.

Raw materials are selected according to purity, supply, pollution potential, ease of melting, and cost. Sand is the most common ingredient. In the United States, approximately 90% of the quality sand produced is consumed by the glass industry. Both purity and grain size are important. Iron oxide, titania, and zirconia are the primary contaminants, but high concentration of feldspars from sands in the western United States account for large amounts of alumina and potash in glasses from that area. Acid-washed varieties offer the lowest iron concentration. Container-glass manufacturers generally use sand between 590 and 840 μm (20–30 mesh) for the best compromise between the high cost of producing fine sand and melting efficiency. Fiber glass manufacturers, however, use a fine grain sand, <70 μm (200 mesh). Both agglomerated fine sand grains and undissolved coarse sand grains may cause melting problems. Shipping costs are often 3–4 times the original cost of the sand, and therefore the plant should be near the sources of raw materials.

Limestone is the source of calcium and magnesium. It is available as a high calcium limestone and consists primarily of calcite (95% $CaCO_3$) or as a dolomitic limestone, a mixture of dolomite, $CaMg(CO_3)_2$, and calcite. High quality limestones contain less than 0.1% Fe_2O_3 and approximately 1% silica and alumina. The mineral aragonite (98% $CaCO_3$) is another source of CaO. Large deposits of high purity aragonite exist near the coast of the Bahamas (see LIME AND LIMESTONE).

The amount of soda ash, Na_2CO_3, produced by the Solvay process has decreased, and most soda ash now comes from the Trona, Wyoming deposits (trona, $Na_2CO_3 \cdot NaHCO_3 \cdot 2H_2O$). Caustic soda, NaOH, solutions may be used in wet batching processes as a source of soda (see ALKALI AND CHLORINE PRODUCTS).

Other raw materials include boron, generally from deposits located in California or Turkey. Either anhydrous borax or boric acid is used, but their consumption is decreasing because of the energy required to produce them. Mineral ores such as colemanite, rasorite, or ulexite are now used in addition to boric acid or pentaaquo borax, $Na_2B_4O_7 \cdot 5H_2O$, when possible. Feldspar and nepheline syenite are common mineral sources of alumina. Litharge is used as a source of lead oxide for the manufacture of lead glasses. Sulfates or nitrates oxidize other oxides in the glass and control fining reactions.

Powdered anthracite coal is a common reducing agent in glass manufacture. Fining agents remove the bubbles in the molten glass and include sulfates, halides, peroxides, chlorates, perchlorates, CeO_2, MnO_2, As_2O_3, and Sb_2O_3. They react by release of oxygen or sulfur trioxide, or by vaporization as in the case of halides. Controlled decomposition of sodium sulfate with powdered coal is used

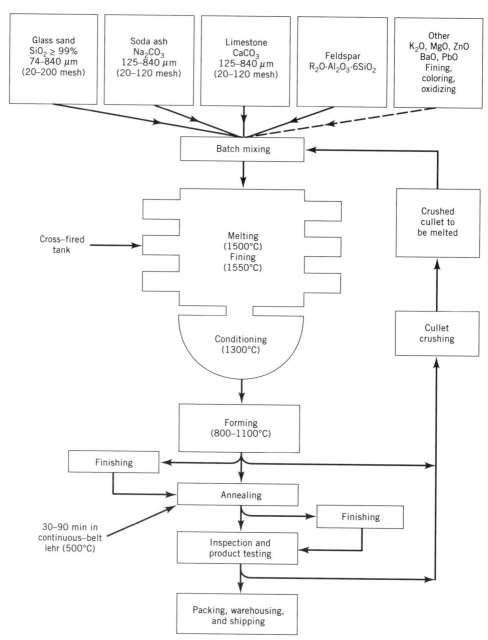

Fig. 12. Glass manufacture. Temperatures are for common soda–lime glass. Other glasses may require appreciably different temperatures.

to fine many soda–lime compositions. Arsenic acid, used as an aqueous solution in place of arsenic trioxide, reduces airborne contamination of As_2O_3 in the batching and melting areas; solutions are dispensed on a volume or weight basis and cover the batch with a more even distribution.

Common colorants for glass include iron, chromium, cerium, cobalt, and nickel (75). Small amounts of iron are sometimes desirable for color and controlled radiant heat transfer during melting. Ferrous sulfide or iron pyrites produce amber-colored glass used in the container industry. Sodium dichromate is a source of chromium for green container glasses, whereas small amounts of cobalt and nickel oxides added to flint glass decolorize the yellow-green color that results from iron contamination. Selenium with iron and cobalt yields a bronze color. Ceria is used to increase uv absorption in optical or colorless soda–lime glass and to protect glasses from x-ray browning (see COLORANTS FOR CERAMICS).

Cullet, or broken glass, is used as a batch material to enhance glass melting and to reduce the amount of dust and other particulate matter that often accompanies a batch made exclusively from raw materials. Some forming operations, eg, the ribbon machine, generate as much as 70% waste glass which must be recycled as cullet. More efficient operations, as used in the container industry, may require the purchase of cullet from flat-glass companies or recycled-glass distributors. Typically, 10–50% of a glass batch is comprised of cullet (see RECYCLING).

Modern analytical techniques have not only aided glass producers, but also enabled suppliers of raw materials to produce a more consistent product. Closer monitoring of purity and on-site blending of raw materials are now used by many producers. A stable and uniform raw material is the key to the manufacture of high quality glass.

Preparation of Raw Materials.   Melting and fining depend on the batch materials interacting with each other at the proper time and in the proper order. Thus, extreme care must be taken to obtain materials of optimum grain size, to weigh them carefully, and to mix them together intimately. The efficiency of the melting operation and the uniformity and quality of the glass product are often determined in the mix house. Batch handling systems vary widely throughout the industry, from manual to fully automatic. Large tanks that melt glass of the same composition for years can easily be fully automated. Each batch material is stored in its own bin and the correct amount, determined and controlled by a computer, is weighed directly from the bin onto a conveyor belt which is fed directly into a mixer. Alternatively, sometimes only the gross weighing is made automatically and the final dribble feed is controlled manually. For small tanks and pot furnaces, where the composition changes frequently, the batches are sometimes weighed into a bin or hopper. Often the weighing step is the most crucial in small tank operations, which are the most sensitive to fluctuations, and often melt the types of glasses that must meet the most rigid specifications.

The method by which the batch is mixed depends more on the type of glass than on the size of the tank. High SiO_2 glasses (soda–lime, borosilicate, aluminosilicate) tend to be batch-mixed in pan-type mixers. The batch is first dry blended and then small amounts of liquid are sometimes added for wet blending. The whole operation takes 3 to 8 min; longer treatment may unmix or segregate the batch. Many specialty glasses, especially those with high PbO, are mixed in

Muller-type mixers, where a smearing action coats the batch particles with each other.

Wet mixing and batch agglomeration (pelletizing or briquetting) are coming into vogue for several reasons, partially because of the increased use of arsenic acid (see SIZE ENLARGEMENT). A wet batch (3–4% water) prevents dusting, controls air pollution, ensures homogeneity, and therefore increases melting efficiency and glass quality (76). Especially in batches with raw materials of varying grain size, a wet batch prevents particle segregation because of fine particles settling to the bottom of the bin. The homogeneous batch melts more efficiently because all of the batch materials are more intimately in contact with each other and also in the correct proportion. Homogeneity can be assured by agglomeration into pellets or application of pressure to form briquettes.

Table 8 shows by example the problem of batch homogenization, where 1.4 kg pyrites and 4 kg salt must be uniformly dispersed throughout a 1000 kg batch, nearly one-third of which is 300 μm (50 mesh) sand and nearly one-half of which is cullet (1–2 cm) glass.

Melting. Ideally, when the intimately mixed batch is charged into the hot furnace, a series of melting, dissolution, volatilization, and redox reactions takes place between the materials in a particular order and at the appropriate temperature (78). Preheating raises the batch as rapidly as possible to the melting temperature where significant reactions occur which generally result in a distinct change in the flow characteristics of the batch. Pellets or briquettes heat up to melting temperature more efficiently than a granular batch.

Table 8. Typical Batch Mixture for Amber Soda–Lime Container Glass[a]

Batch materials	Weight, kg	Oxides supplied, kg					
		SiO_2	Al_2O_3	CaO	Na_2O	FeO	LOI,[b]kg
sand, SiO_2	300	299.3	0.2			0.03	0.5
soda ash, Na_2CO_3	100				58.3		41.7
aragonite, $CaCO_3$[c]	90			49.0		0.02	40.7
feldspar, $SiO_2 \cdot Al_2O_3$ mineral[d]	40	26.4	7.6	0.4	1.3	0.03	0.1
salt cake, NaCl	4				2.1		1.9
powdered coal[e]	9						9
iron pyrites, FeS_2	1.4					0.84	0.6
cullet	460	333.7	9.2	48.8	67.2	1.03	0.1
Total	*1004.4*	*659.4*	*17.0*	*98.2*	*128.9*	*1.95*	*94.6*
yield of glass, kg, and wt % oxides present	909.8	72.48	1.87	10.79	14.17	0.21	

[a]Ref. 77.
[b]Loss on ignition. Generally, the oxides of carbon and sulfur (plus some chlorine, depending on the fining agent) volatilize during melting.
[c]Also 0.2 kg MgO or 0.02 wt %.
[d]Also 4.1 kg K_2O or 0.45 wt %.
[e]Used primarily to reduce the Fe_2O_3 to FeO to give the characteristic amber color, although the redox state of the glass melt also influences the fining reactions.

Dissolution of the more refractory (higher melting) grains, such as sand, is accelerated by fluxes (lower melting materials), eg, Na_2CO_3. The decomposition of alkali carbonates (Li_2CO_3, Na_2CO_3, K_2CO_3) or alkaline-earth carbonates ($MgCO_3$, $CaCO_3$, $SrCO_3$, $BaCO_3$) results in a similar fluxing action on sand and other minerals, notably alumina-containing ones such as nepheline syenite and feldspars. Undissolved grains (stones) can be introduced either when the refractory grains fail to react completely or because the reactants are not intimately mixed. Increasing the melting temperature aids in dissolving stones and compensates for minor deviations from an ideally prepared and delivered batch.

Fining is the physical and chemical process of removing gas bubbles (seeds, blisters) from the molten glass melt. Gas is evolved during the first stages of melting because of (*1*) the decomposition of the carbonates or sulfates, or both; (*2*) air trapped between the grains of the fine-grained batch materials; (*3*) water evolved from the hydrated batch materials; and (*4*) the change in oxidation state of some of the batch materials, eg, red lead:

$$2Pb_3O_4 \rightarrow 6\ PbO + O_2$$

Table 8 shows how much gas may be evolved from a typical amber soda–lime container glass batch. Fining agents are employed that generally react at higher temperatures than are needed for melting; thus the fining reactions continue after dissolution and volatilization have taken place. Only materials that can release gases without delay through the formation of boiling bubbles can act as fining agents. As the gases are released, the bubbles rise to the surface roughly according to Stokes' law. Each bubble, during its trip through the glass melt, attracts new quantities of gas by diffusion from neighboring layers and by coalescence with other bubbles. High temperatures make the glass more fluid and increase the diffusivity which speeds up the fining process greatly. The most common fining agents are the sulfates, followed by sodium or potassium nitrates in combination with arsenic or antimony trioxides. Arsenic trioxide is used for higher melting glasses, 1450–1500°C, and antimony for lower melting glasses, 1300–1400°C. As the glass cools, dissolution fining may occur, ie, oxygen bubbles are removed by reaction with the arsenic or antimony trioxide to form the pentoxide.

Melt homogenization followed by cooling to working temperatures completes the melting process. Batch segregation, melt segregation, volatilization, and temperature fluctuations, as well as refractory corrosion from tank-lining material, cause compositional differences (cord, stria) within the melt. These inhomogeneities must be removed by diffusion and flow before the glass is cooled in the forehearth prior to delivery and forming. Vigorous fining action, as well as convection current mixing help break up cord. Mechanical and static mixers continuously shear the glass to help active homogeneity.

Melting Units. Melting units range from small pot furnaces for manual production to large, continuous tanks for rapid machine forming (79). After mixing, the raw materials are charged into a furnace for melting. Pot furnaces are used for melting smaller quantities of glasses below 1400°C. Pots, whether single or multiple port, are inefficient fuel consumers and have poor temperature control. However, because they can be heated from the sides as well as the top, they are useful for melting heat-absorbing specialty glasses. Hand producers may still use pots for experimental melts where the time between filling the pot and production can be as short as 24 h. However, for experimental melts day tanks are usually preferred because they have better refractories and higher attainable temperatures. Day tanks burn gas or oil and use a single opening for both charging and gathering of glass. The glass quality is better than that of pot-melted glass, but not as good as that of glass melted in a commercial tank.

Small, continuous-melting tanks, filled automatically or by hand, produce high quality glass, such as optical and ophthalmic glasses, at low volume. They are used especially when many glass changes are necessary. Capacity is in the 7 t/d range and two melting zones are used. The premelt section may be heated by electricity, gas, or a combination of the two and operates at a lower temperature (eg, 1400°C). The finer is an electrically heated platinum tube approximately 20 cm dia. Some of these tanks operate as cold crown units with the unmelted batch covering the surface of the molten glass in the premelt region. These units are heated electrically from the bottom or side, and their small size makes possible quick changes of glass. Draining and flushing are usually preferred. Pollution from glass melting is substantially reduced by using cold crown units.

The largest furnaces are continuous regenerative furnaces that recover waste heat from burned gases (80). They produce large quantities of quality glass and are either cross- or end-fired. The latter type is usually smaller. The hot combustion gases are passed through a chamber filled with refractory checkers. Combustion air and waste gas flow are reversed at 15–30 min intervals to alternately heat each chamber. Incoming heated air mixes with fuel and burns over the surface of the glass. Figure 13 shows the design of a cross-fired regenerative furnace typically employed for glass container manufacturing. The two ports of an end-port furnace produce a U-shaped flame over the glass that enters and exhausts from the back wall, whereas cross-firing from side to side allows more even heating across a larger surface area. Each type of furnace has a melting portion and a conditioning portion which is separated by a refractory bridge wall. The opening between the two areas, called the throat, is beneath the surface and allows glass to flow, but reduces surface contamination. Flat-glass furnaces may not have a bridge wall, but are longer than container furnaces to give greater output of high quality glass. Scum, foam, or surface imperfections are removed by a skimming or floater device near the surface of the glass. The depth of glass ranges from 1 to 1.3 m and must be closely controlled by subsequent forming operations. Melting efficiency is determined by the tank area necessary to melt a given quantity of glass. Soda–lime furnaces for flat and container glass melt approximately one metric ton of glass per 0.41 m^2 of melting area per day $(500 \text{ lb/ft}^2 \cdot \text{d})$. Other glasses, such as borosilicates, may require as much as $1.0–1.5 \text{ m}^2/\text{t}$ $(9.8–14.6 \text{ ft}^2/2000 \text{ lb})$ of glass melted.

Continuous recuperative furnaces employing metallic recuperators (heat exchangers) have been in use since the 1940s. Operation of these furnaces is simplified and the combustion process is more precisely controlled; no reversal of air flow causes temperature variations. The recuperator metal must be carefully selected because of chemical attack at high temperature. Recuperative furnaces are often used in the production of textile fiber glass because they maintain a constant temperature.

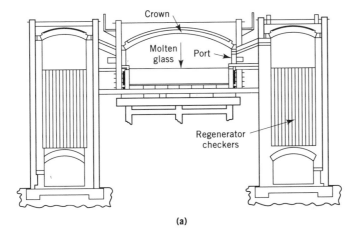

(a)

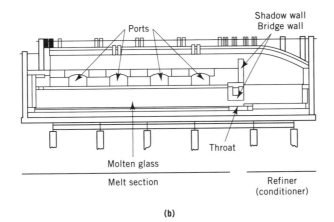

(b)

Fig. 13. Cross-fired regenerative furnace: (**a**) end view; (**b**) side view.

Development of molybdenum electrodes in the 1950s permitted the use of electrically assisted melting in regenerative furnaces (81). In the 1990s, approximately one-half of all regenerative tanks are electrically boosted. Operating practice has shown that effective use of electricity near the back end of the furnace, where the batch is added, can reduce fossil fuel needs. This lowers surface temperature and reduces batch volatilization.

All-electric melting of glass has been patented since the beginning of the twentieth century. Electric furnaces produce large quantities of commercial glass including fiber glass, lead, soda–lime, borosilicate, and fluoride opal types. Furnace designs range from shallow rectangular types to round and vertical (82) (see FURNACES, ELECTRIC). Submersible electrodes of rod or plate design are placed on the bottom or side of tanks. High energy efficiency by conversion of electrical energy into useful heat and low volatilization are primary advantages of all-electric melting. For example, in a fluoride-containing batch as much as 40% volatilization might occur with gas firing compared to 2% with electric. High quality glass, cleanliness of operation, and small space requirements are additional advantages. Disadvantages, especially to the container industry, include the difficulty in maintaining a constant glass level in the tank when melting cold crown, the lack of ability to use alternative fossil fuels, and a loss of pull rate flexibility. Most glasses, in principle, can be melted electrically. However, glass compositions with steep resistivity–temperature curves, such as an alkali-free glass, present problems, since a cold batch surface is quite difficult to maintain because of the rapid decrease in resistivity with increasing temperature.

The choice of refractories (qv) depends on the type of glass and furnaces used and the position of the refractory in the furnace.

Batch Feeding Systems. The two common feeding designs used are the screw feeder and the reciprocating pusher. The former delivers the batch from a hopper to the furnace by tube and helical gear drive, whereas the latter forces a layer of the batch from the feed chutes onto the molten glass. The enfolding feeder uses a vibrating chute to deliver the batch as a rotating device dips into the surface of the molten glass, which causes the batch to be enfolded into the surface. Blanket feeding is usually done by conveyor. Regardless of the type of charger used, all empty into the backwall area of the furnace or into an extension known as the doghouse. A batch charged into a doghouse usually is glazed over before entering the hotter melting area. All-electric melting relies on the complete coverage of the molten glass surface by several centimeters of the batch.

Control Devices. Control devices have advanced from manual control to sophisticated computer-assisted operation. Radiation pyrometers in conjunction with thermocouples monitor furnace temperatures at several locations (see TEMPERATURE MEASUREMENT). Batch filling is usually automatically controlled. Combustion air and fuel are metered and controlled for optimum efficiency. For regeneration-type units, furnace reversal also operates on a timed program. Data acquisition and digital display of operating parameters are part of a supervisory control system. The grouping of display information at the control center is typical of modern furnaces.

Benefits resulting from better control of glass melting are lower fuel consumption, better glass quality, more efficient production time, and better pollution control. Fewer operators are needed and working conditions are better.

Fuels and Efficiency. Natural gas, oil, and electricity are the primary sources of energy; propane is used as backup reserve in emergencies. Natural gas is the least expensive and most frequently used fuel, with heat content ranging from 34–45 MJ/m^3 (900–1200 Btu/ft^3) for raw gas and approximately 3 MJ/m^3 (80 Btu/ft^3) for air–gas mixtures. Fuel oil has heat content between 39–43 MJ/L (139,600–153,000 Btu/U.S. gal). Fuel oil is viscous at low temperature and must be heated before being fed to atomizing burners where it is mixed with air for combustion.

The efficiency of gas-fired regenerative furnaces is about 30%. Oxygen, when substituted for air, reduces the fuel required to melt a unit of glass. For a well engineered soda–lime glass furnace, fuel reduction is typically 30%. Fuel savings can be greater with higher melting temperature silicate glasses (83). Direct application of electrical energy to molten glass by electrodes melts glass more efficiently. The efficiency of electric melting is 2 to 3.5 times that of fossil fuels, but production of electricity from fossil fuel at the power plant is only about 30%

efficient. Electric furnaces absorb less heat and there are no costly regenerators to repair or replace. However, the most flexible furnaces are still the electrically boosted fossil fuel tanks where a combination of energy sources is used rather than a single fuel (see FURNACES, ELECTRIC; FURNACES, FUEL-FIRED).

Pollution (84) from soda–lime glass production is minimal, mostly caused by sodium sulfate particulates. Other soda–lime contaminants include SO_2, NO_x, CO_2, and hydrocarbons which are reduced by scrubbers, baghouses, and precipitators, or sometimes controlled by optimizing operating conditions. Volatilization of lead, fluorine, and arsenic for specialty glass manufacture has to be carefully controlled. All-electric melting and new batching procedures have been helpful in reducing volatilization. Electrostatic precipitator dust (EP dust) collected from a stack can be recycled and may, in some cases, offset the operating cost of the pollution control device. Powdered coal as a reducing agent controls the decomposition of sodium sulfate more efficiently than increased temperature. Therefore, less sulfate can be used which results in lower emissions with equivalent glass fining. Higher cullet ratios are also known to reduce emissions. New operating procedures or furnace modifications, such as smaller ports near the rear (batching end) of a furnace rather than near the bridge wall, reduce volatilization.

Forming. Molten glass is either molded, drawn, rolled, or quenched, depending on desired shape and use. Bottles, dishes, optical lenses, television picture tubes, etc, are formed by blowing, pressing, casting, and/or spinning the glass against a mold to cool it and to set its final shape. Window glass, tubing, rods, and fiber are formed by drawing the glass in air (or on a bath of molten tin as in the float process) until it sets up and can be cut to length. Art glass is usually hand-formed by blowing and shaping it while soft. Glass that is intended to be crushed into powder, called frit, is quenched between water-cooled rollers or ladled or poured directly into water and then dried (dry-gauged).

Molding. Many glass articles are made by shaping a glass gob by forcing it under pressure against a mold. The source of pressure and type of mold vary with the application.

Blowing. Deep items such as bottles, jars, or light bulb envelopes are formed by the use of air pressure either from a pair of lungs or an air compressor. Hard blowing is still practiced by highly skilled craftsmen for specialty items for scientific, industrial, lighting, and electrical applications. Items up to a kilogram in weight can be made with a single ball of glass gathered on the end of a punty iron, preshaped, and then blown into a mold.

Articles of circular cross section may be made in iron paste molds. To keep the inner surface of the paste mold moist, it is coated with shellac or varnish and a mixture of charcoal and linseed oil is baked on. Hot iron molds are used for ware of any shape, particularly for screw threading, multiple decoration, or raised lettering.

The Hartford-Empire individual section machine is an in-line machine of up to 12 sections (eg, H.E. IS-12) used to blow small-neck containers (Fig. 14). Each section is an individually functioning hollow glass machine which can handle one to four gobs at a time, depending on the size of the ware. The rate of production is about 10 pieces per section per minute. In this blow-and-blow process there is no movement (other than opening and closing) of the molds under the glass delivery system. Although this machine usually makes narrow-neck bottles, it can

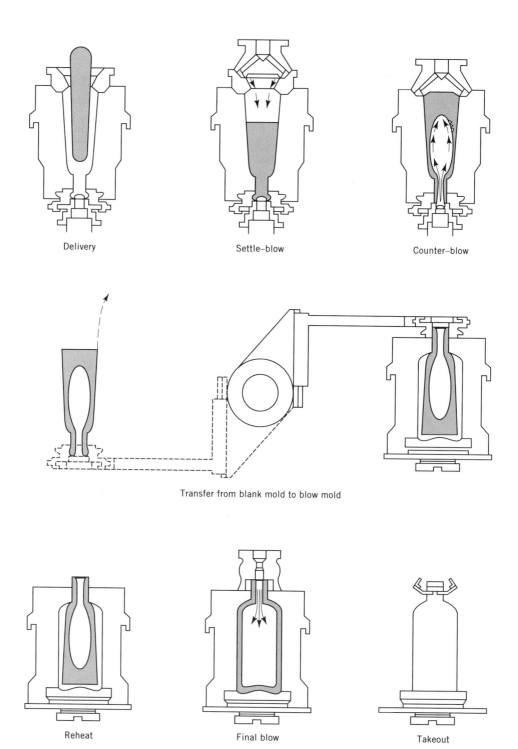

Delivery   Settle–blow   Counter–blow

Transfer from blank mold to blow mold

Reheat   Final blow   Takeout

Fig. 14. The H.E. IS blow-and-blow machine (85). The gob is delivered into a blank mold, settled with compressed air, and then preformed with a counter-blow. The parison or preform is then inverted and transferred into the blow mold where it is finished by blowing.

be converted readily to a press-and-blow machine for making wide-mouth jars with screw threads. In general, iron molds are used.

The ribbon machine employs a puff-and-blow method to make incandescent light bulb envelopes (1200/min), flashbulb envelopes (2000/min), Christmas tree ornaments, and the like (Fig. 15). For larger articles, the turret chain machine also uses the puff-and-blow method, but produces less cullet and uses individual glass gobs rather than ribbon.

The Hartford-Empire 28 is a press-and-blow machine used to make articles such as drinking glasses (tumblers). It uses paste molds and the ware is rotated to avoid the mold seams; jars with screw threads cannot be produced. The product leaves the machine as an almost closed, hollow object and is finished by severing and fire-polishing with a burn-off machine.

Pressing.   Flat items such as dinnerware, optical and sealed-beam lenses, filter glasses, and television tube panels are pressed between a plunger and a mold, most commonly with an automatic rotary press. Typically, there are 8–16 mold positions on the periphery of a horizontally rotating table. Cast iron, bronze, steel, and some specialty alloys are common mold materials. A gob of glass at about 200–400 Pa·s (2000–4000 P) is fed into the mold at the first station. The table rotates and a retaining ring and plunger press the article at the next station. The glass cools for several more stations until it sets up at 10^5–10^8 Pa·s (10^6–10^9 P) and is taken from the mold. The mold is cooled and ready to take its turn and receive another gob. Pressed ware ranges from 5 g to 15 kg, pressing pressures range from 0.5 to 0.8 MPa (75–100 psi). Solid shapes are pressed in font molds which are small cavity-split molds into which molten glass is extruded by the plunger (Fig. 16). Automatically produced pressed ware has about 0.5% variation in bottom thickness when made from the same mold, and with accurate gobbing, the manufacturing tolerance for diameter is about ±0.05%.

Casting.   This is the process of shaping glass by pouring it into a mold. The largest piece ever cast was a borosilicate telescope mirror 600 cm in diameter weighing 36 t. Dimensional tolerances as low as 0.1 mm can be achieved. Funnel-shaped ware, such as missile radomes and television bulbs, are formed by centrifugal casting (88). Gobs from a feeder are dropped into a spinning mold that creates centrifugal forces making the glass flow upward, forming a wall of relatively uniform thickness.

Drawing.   Drawing molten glass from a specially designed orifice in a continuous manner and letting it set up before it touches a solid mold gives the glass a uniform cross section and a fire-polished surface. The surface rarely has to have a secondary finishing operation. There are few, if any, moving parts in the forming equipment, and the ware can be cut to any practical length at the end of the draw.

By the 1950s the drawn-sheet processes such as the Fourcault process, the LOF-Coburn process, and the PPG Pennvernon process manufactured cheap, fire-polished window glass eminently suitable for domestic glazing and greenhouses (85). High quality glass was made by polished-plate processes, which incorporated a costly grinding and polishing step. However, about 20% of the glass was waste which was ground off in order to make the surfaces flat and parallel (89).

The advent of the Pilkington float process at the end of the 1950s revolutionized the flat-glass industry, and by 1976 the United States was producing about 85% of its flat glass by the process (90). It combines the low cost and fire

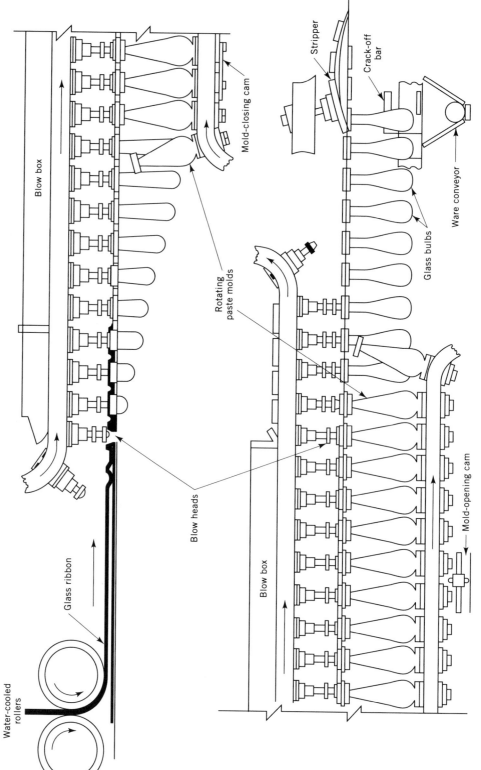

Fig. 15 The Corning ribbon machine (86). Courtesy of Corning Inc.

604

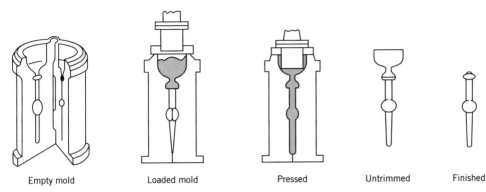

| Empty mold | Loaded mold | Pressed | Untrimmed | Finished |

Fig. 16. Font-mold pressing of glass (87). Courtesy of McGraw-Hill Book Co., Inc.

polish of drawn-sheet glass with the distortion-free quality of polished glass. A continuous ribbon of glass moves out of the melting furnace and floats along the surface of an enclosed bath of molten tin (Fig. 17). The ribbon is held in a chemically controlled atmosphere at a high enough temperature for a long enough time for the irregularities to even out and for the surfaces to become flat and parallel. Because the surface of the molten tin is flat, the glass also becomes flat. The ribbon is then cooled while still advancing across the molten tin until the surfaces are hard enough for it to be removed from the bath without the rollers marking the bottom surface (91). In 1975, PPG Industries developed a novel delivery system in which a stream of molten glass of the desired production width, normally about 4 m, formed at the exit end of the melting furnace and remained that width as it flowed onto the bath. By contrast, glass melted in the Pilkington process enters the bath through a restricted opening and is deposited, unshaped, on the molten tin. The improved PPG delivery system minimizes flow effects which introduce optical distortion in the finished glass (90). At present, the float process is limited to soda–lime/silica glass and is capable of producing ca 50,000 m^2 (500–800 t) of glass per day that is typically 3–3.5 m wide. The equilibrium thickness of glass on the molten tin is 67 mm, depending on the various glass–tin–atmosphere

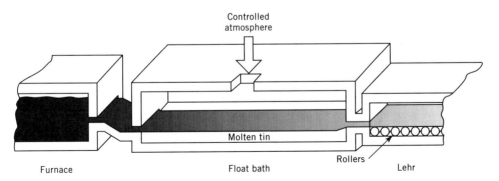

Fig. 17. Diagram of the float process.

interfacial tensions and on the densities of the liquid glass and tin. These parameters, of course, change as the glass moves along the tin from 1050°C (ca 1 kPa·s or 10^4 P) to 600°C (ca 10^{10} Pa·s or 10^{11} P). In practice, thinner glass can be made (down to 2 mm) by stretching the glass ribbon on the molten tin. However, imperfections, which become elongated and thus more serious, must now be more closely controlled. Float glass is highly vulnerable to chemically produced effects (89), such as molten metal or a reducing atmosphere of nitrogen and hydrogen.

The fusion process, developed by Corning (92) in the late 1960s is a down-draw process for specialty applications (Fig. 18). It can handle a wide variety of glasses (working points) ranging from soft (950°C) photochromic glasses, glasses for liquid crystal displays (1160°C), to hard (1265°C) aluminosilicate chemically strengthenable glasses for automobiles and aircraft windows. Molten glass flows into a trough, then overflows and runs down either side of a pipe, fuses together at the root, and is drawn downward until cool. Both sides have a fire-polished surface comparable to polished-plate or float glass. This is a higher viscosity process (20 kPa·s or 200,000 P) than float (1 kPa·s or 10^4 P). The thickness of the sheet is dependent on the flow rate and therefore is controlled directly by the velocity of pulling. Very thin glass (0.4–0.8 mm) can be formed routinely, and thick glass (up to 1 cm) has been produced. Although scale-up is possible, present production widths are not quite half the width of float-process glass. Currently, production rates are less than those of the float process.

Glass Rod and Tubing.   These are also made by hand drawing (85). About 15 m of 140-mm OD, 2-mm wall thickness tubing can be made on one draw by a gaffer and two helpers. Gathers of glass as heavy as 30–35 kg are made with a bubble in their center at the end of the blowpipe. This is rotated in place by the gaffer and rotated and stretched out by one of the helpers walking away from it. The size and diameter of the tubing is a function of the helper's walking speed, and is further controlled by a second helper who cools the tubing by fanning at the appropriate time. As much as 150 m of the much smaller diameter thermometer tubing can also be made in this manner from one gather.

In the Danner process (Fig. 19), a continuous stream of glass flows onto a rotating mandrel. Tubing is drawn off the end while air blown through the man-

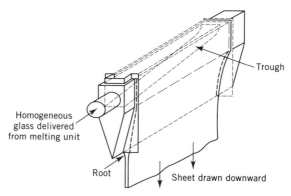

Fig. 18. Schematic diagram of the fusion process (92). Courtesy of Corning Inc.

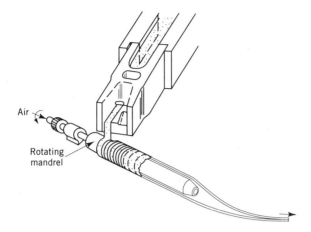

Fig. 19. Working scheme of Danner rotating mandrel.

drel helps to maintain tubing dimensions. The glass can be heat-treated on the mandrel. Because of the circular symmetry of the glass it is more thermally uniform. However, hairpin cords of inhomogeneity and volatilization from the surface may occur. In the Vello process the molten glass passes down through the annulus between a horizontal ring and a vertical bell. The moving stream is then brought around to a nearly horizontal runway and, as with the Danner process, is drawn off at the appropriate rate for dimensional specifications and stability. The Vello process can form tubing faster because the glass can be cooled more quickly to the appropriate forming viscosity in a stationary forehearth of any desired dimensions. Cooling in the Danner process is carried out on the rotating mandrel itself. Although the Vello process is more difficult to operate, it is better suited to longer production runs with few size changes.

Fiber Glass.   There are two basic forms of fiber glass being manufactured: continuous filaments (415 μm dia) and fibers (6–9 μm dia and 20–40 cm long) (85,93). Single filaments can be combined into a strand that is easily unwound from a spool. The product is twisted textile yarn. Staple fibers that meet diameter and minimum length requirements are twisted together into a tow that can also be easily unwound from a spool. The end product is staple yarn. Glass fibers that do not meet these requirements are not suited for textile applications, but may be used for thermal and sound insulation in the form of glass wool (85). Nearly all continuous filament yarn is made from E-glass which is well-suited for electrical insulation (see INSULATION, ELECTRIC), heat-resistant applications (see INSULATION, THERMAL) (93), and plastic reinforcement and mats (see LAMINATED AND REINFORCED PLASTICS). Most staple fibers are spun from C-glass which is more resistant to chemical corrosion and is used for chemical filtration fabrics, etc. A low cost soda–lime glass, T-glass, is used when high durability is not required, eg, coarse fiber mats for air filters and thermal and acoustical insulation (see INSULATION, ACOUSTIC). For weathering resistance, SF-glass is used in fine and ultrafine wool products such as very low density thermal and acoustic insulators, paper additives, and high efficiency all-glass filter papers.

Formerly, molten glass was formed into marbles which were remelted in an electric furnace and formed into filaments. However, many manufacturers now use the direct-melt method, where the molten glass is formed directly into filaments. Continuous filament yarn is drawn from the molten glass as it comes through small holes in a high temperature alloy bushing. The high speed winders, operating as fast as 3000 m/min, pull the streams into filaments. Staple fibers are formed by jets of air that pull the fibers from the furnace and deposit them on a revolving vacuum drum.

The Owens steam-blowing process produces most of the fibers for the glass wool application. Glass issues from slot-type openings in a platinum–rhodium bushing and is fiberized by a steam blower mounted just below the bushing. The fibers fall through a spout into a hood and deposit uniformly onto a conveyor chain belt which transports the fibers in the form of a felt or mat from the hood (85). In the rotary wool-forming process, molten glass is delivered to a rotating cylinder the face of which is perforated with a large number of holes. Glass streams, projected laterally from the holes by centrifugal force, are then attenuated into discontinuous fibers by a high velocity gas stream (air, steam, or combustion gases) (94). The strength of glass fibers (strength-to-weight ratio) is over twice that of the strongest textile fibers, and much stronger than ordinary glass because of the relatively pristine surface.

Fritting.   Cooling a glass quickly, directly from the melting unit, creates a high thermal stress that shocks and often breaks it into small pieces suitable for charging into a ball mill. The molten glass is either quenched in water (dry gauged) or between water-cooled metal rollers to make a thin, friable ribbon, depending on its viscosity and sensitivity to water. The thin-rolled ribbon often provides more uniform mill feed. Dry gauging sometimes forms clinker-like, tempered pieces that are difficult to mill.

Protective and decorative enamels, solder glasses for glass-to-glass or glass-to-metal seals, and a variety of sintered glass products (multiform) are made from frit. Enamels (qv) and solder glasses are designed to be fluid (viscosity ca 300 Pa·s or 3000 P) at their processing temperatures, whereas the substrates to which they are joined remain relatively rigid (viscosity = 10^8–10^{13} Pa·s or 10^9–10^{14} P). The resultant product is a glaze/substrate composite. Protective coatings such as porcelain enamels and glazes for both metal and glass are ball-milled with a variety of clays, electrolytes, and fluids to achieve the correct particle size (< 74 to < 44 μm or -200 to -325 mesh), shape and distribution prior to dipping, spraying, or dusting on the ware for final firing (glazing) (95–97) (see CERAMICS).

Small, complex-shaped glass articles such as thread guides for the textile industry and television gun mounts for the electronics industry are made by the multiform process. The dry-milled powder is mixed with an inorganic binder and a fluid vehicle, and then atomized by a spray dryer into small, dried agglomerates of glass powder and binder with good flow characteristics. They are subsequently pressed to the desired shape and fired.

Foam Glass.   This is made from powdered glass that has been mixed with an oxidizing agent and lampblack. When heated close to melting temperatures, the CO_2 evolved produces a cellular structure in the glass which can be controlled to nearly any density below that of the base glass. Foam glass, made by Pittsburgh-Corning, is used as insulation and has a density of 0.14 g/cm^3.

Slip Casting.   Complex shapes can be made cheaply and easily by slip or drain casting from compositions that cannot be formed by the usual glass-forming techniques. Glass is finely ground and suspended in water. This slip, which also contains a binding agent, is poured into plaster of Paris molds which absorb the water by capillary action. The resulting partially dried articles are removed from the molds and fired to produce consolidated products. Honeycomb structures are made by a modification of this process. Thin sheets of glass or ceramic powder on a paper substrate can be corrugated, stacked, and fired to produce bodies with continuous channels. Ceramic burners, catalyst supports, and heat exchangers are formed by this technique.

Lamination.   Simultaneous forming of two glasses to produce a single article is possible if the two have similar viscosities in the appropriate temperature range. Thermometer tubing is a composite of a white opal-glass stripe embedded in a clear glass tube parallel to the hollow bore. The opal glass is melted separately from the clear glass, but, while still fluid, it is fed into the Vello tubing machine and applied during the forming operation to the clear glass. The glasses are then drawn out the orifice and down the runway together.

In the late 1960s, Corning (98) developed a vertical rotating turret machine which, when used with a special orifice, forms a laminated glass article consisting of a core glass and a skin glass (Fig. 20). This process is used by Corning for the manufacture of moderately priced plates, bowls, cups, and saucers. The piece of ware is formed and glazed in one operation. The skin is of a lower thermal expansion than the body; hence, when the entire article cools, the skin has a residual compressive stress and the body a residual tensile stress. Surface compression adds both thermal shock resistance and mechanical strength to the ware (see LAMINATED MATERIALS, GLASS).

Extrusion.   This forming process is used extensively in metals and plastics industries, but infrequently in glass manufacture. The main reason for the lack of acceptance is the inherent low rate of production when employed to shape glass. The maximum extrusion rate is tens of centimeters per minute (99). Depending on the type of glass being processed, temperatures of up to 950°C and pressures up to 1 MPa (1000 bars) are utilized. Through careful control of the processing parameters and the selection of nonwetting die materials, sharp edged complicated rods and tubes can be made to close tolerances.

Advantages of the extrusion process include the relatively low forming temperatures, the manufacture of products from glasses with steep viscosity temperature curves and/or a high tendency to crystallize, the ability to form shapes not available through other glass-forming processes, and the ability to form laminated and reinforced glass products.

Lehring.   Glass articles formed at high temperature must be cooled in order to reduce the strain and associated stress caused by temperature gradients in the glass to a low level to prevent damage during finishing and subsequent use. The continuous solidification of glass on cooling and the methods of reducing strain are unique to glasses and do not apply to crystalline materials that undergo phase transitions. Lehring processes include annealing, tempering, densifying or compacting, and post-heat treatments (100,101). The term annealing generally refers to removal of stress, and terms such as fine annealing or trimming of optical glass infer structural changes associates with lehring.

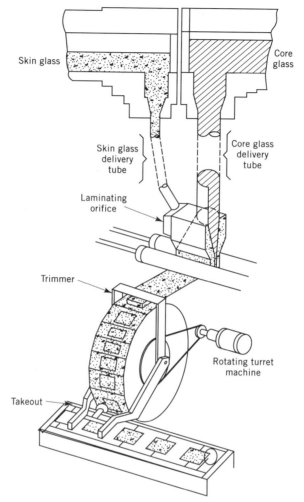

Fig. 20. Schematic of the rotating turret machine (98). Courtesy of Corning Inc.

The cooling of a flat piece of glass creates a parabolic temperature distribution (87) within the glass. Depending on the original temperature of the glass, there are two possibilities: (*1*) if the temperature is high enough to allow structural relaxation, then after the temperature gradient is removed, a strain profile results. Tensile stress produced by this strain occurs in the interior with compressive stress on the surfaces. In the ideal case of infinitely slow cooling, ie, $\Delta T = 0$, stress does not develop within the glass; or (*2*) if the original temperature of the glass was so low that the structure could not relax, then the temporary tension at the surface and compression in the interior, which were caused by the temperature gradient, dissipate completely when the temperature gradient is removed.

Stress annealing is the process of removing or preventing the objectionable stresses in glassware. The glass is brought to a temperature within the annealing range (usually defined as the temperature range between the anneal and strain points) and held there until the stresses are relieved. Cooling rates are determined by stress levels permitted in the ware being made. Generally, cooling rates increase as the temperature nears the strain point, and strain is directly proportional to the cooling rate (102). When the glass has cooled to about 50°C below the strain point, faster cooling is possible since the strain is only temporary. Schedules are worked out based on the item being manufactured, and depend on (1) shape, eg, thick or thin, flat or curved, etc; (2) glass composition, ie, the thermal expansion curve, Young's modulus, and thermal diffusivity; and (3) the product and requirements such as cutability, finishing, and refractive index. The stress in glass can be calculated from the measured birefringence and the glass's stress optical properties.

Glass manufacturers using these principles for annealing flat glass have found that the maximum cooling rate generally cannot be achieved because of technical difficulties in cooling 3-m wide sheets of glass uniformly without surface damage. For pressed and blow items (hollow ware), the variations in thickness of a single article makes the problem considerably more complex. After forming, hollow ware receives a soak at constant temperature to assure that strain is reduced to a suitable level. Soaking is followed by a cooling through the anneal range and finally more rapid cooling to room temperature. A significant difference between flat and hollow ware is that during cooling the inside surface of the hollow ware is hotter than the outside. This results in compression on the outer surfaces, as with flat glass, but tension at the inner surface. Poorly annealed items may be subject to breakage if the tension is high or the inner surface is bruised.

Annealing of optical glass (fine annealing or refractive-index trimming) is important because it changes the refractive index and increases the density (103,104). Residual strain can also cause a nonuniform index or distortion during finishing operations. The fine annealing of large or thick optics may require days or even weeks to achieve low stress and ensure the maximum homogeneous, stable refractive index.

Post-heat treatment of glass induces phase separation and crystallization. A very precise secondary heat treatment is necessary to develop and control photochromic properties. Glass-ceramics receive a thermal ceramming process that induces crystallization.

Tempering refers to the rapid air chilling of the surface of a glass that has been heated to reach its softening point. This produces a temperature gradient, which, in turn, produces a high surface-compressive layer balanced by an internal tensile stress. Tempering increases the strength of the glass because applied forces must overcome the compression on the surface before fracture occurs. After tempering, items usually cannot be cut or ground without breaking, but the thickness of the compressive layer allows the article to retain much of its strength even if the surface is slightly damaged. Top-of-stove ware, drinkware, auto side windows, and safety glass for doors are representative items that are thermally tempered. Air cushions have been used to support flat glass during tempering to avoid surface damage. The PPG Gas Hearth (105) or the Pennglass Air Float systems (106) support flat glass over a series of heated air jets in a conveyor system. Thinner, lightweight door glass or automobile windows can be made by air flotation. Bending of glass over a template in a glass-bending lehr is also possible by a modification of the air-float process (107). The greater precision of this method allows use of thinner glass and the production of tempered laminated ware.

Secondary Operations. In secondary forming, a piece of preformed glass is reheated and reworked into the finished product. Repressing of optical blanks in precision molds below the primary pressing temperature is sometimes done by lens manufacturers who do not melt their own glass. Redrawing of tubing and cane (rods) is called for when manufacturing bundles of thin fibers, as in channel amplifiers, or long, thin single fibers, as in optical waveguides. Sagging flat glass, eg, automotive windshields, is accomplished at temperatures near the softening point so that the glass bends to fit a form made from a refractory or metal. Lamp-working and sealing are often either hand or automated operations in which the glass (rod or tubing) is heated near its working point (10^3–10^4 Pa·s or 10^4–10^5 P) and formed to the desired shape. Novelties, labware, and glass-to-metal wire seals for electrical and electronic applications are common lampworked articles. Precision tubing with inside-diameter variations as small as 30 μm/m length are made by shrinking. This is done by vacuum-collapsing the hot tube over a mandrel which is withdrawn when the glass has cooled (87).

Mechanical Finishing. Glass can be cut by five methods. (*1*) Flame cut-off is accomplished with pinpointed flames which heat the glass until it is soft enough to separate. (*2*) Sawing is done with band, wire, or circular blades used in conjunction with a loose or bonded abrasive. (*3*) To score-break glass, the piece is scored with a tool such as a file, a diamond, or a steel or carbide wheel, and then bent to apply tension to the scored area. (*4*) Score-thermal crack-off is done by applying heat opposite a mechanically produced score on the glass surface. (*5*) Localized heating with pinpointed burners followed by rapid chilling, usually with water, accomplishes a thermal crack-off. Production rates, tolerances, sizes, etc, vary, depending on the type of glass and the specific application.

Drilling. Glass is drilled with carbide or bonded-diamond drills under a suitable coolant such as water or kerosene. Other drilling processes include a metal tube rotating about its axis (core drilling), an ultrasonic tool in combination with an abrasive slurry, or an electron beam. Tolerances less than ±0.1 mm are readily obtained with diamond-core drilling and, if required, holes smaller than 25 μm-dia can be made with the electron-beam method.

Grinding. Glass is ground with sand, garnet, corundum, silicon carbide, boron carbide, or diamond. These materials are used loose, as in the grinding process employed for plate glass, or they are bonded in grinding wheels or coated on fabric belts. Grinding rates vary, depending on the glass and the abrasive used, but removal rates near 0.75 mm/min are common. In general, any shape that is ground in metal can be produced in glass on the same machine.

Mechanical Polishing. This process is similar to grinding, but the polishing compound, usually cerium oxide, zirconium oxide, or ferric oxide, is finer. The polishing tool may be plastic, cellulose, felt, or pitch. A mixture of hydrofluoric and sulfuric acids is used for acid polishing. By directing flames on a glass surface, fire polishing can produce surface finishes as good, except for flatness, as those obtained with mechanical polishing.

Copper-wheel engraving is used for decoration and in artware production. Abrasive compounds are applied in water suspension to the spinning copper wheel held in a chuck. The glass is brought into contact with the wheel to produce the design (see ABRASIVES).

Chemical Finishing. Treatment of glass surface with a chemical may alter its strength, appearance, or durability. For example, ion exchange alters the strength of a glass article. A glass, eg, a sodium aluminosilicate, is immersed in a bath of potassium nitrate at a temperature about 50°C below its strain point (ca 500°C) for 6–10 h. The small sodium ion diffuses from the glass and is replaced by the larger potassium ion. This stuffing of the first 100 μm of the surface produces a surface compression of 450 MPa (65,000 psi) and a corresponding central tension of 10–30 MPa (1450–4350 psi), depending on the article's total thickness. Stress release is minimized because of the low process temperature. Chemically strengthened glass is used in auto and aircraft windows and eyeglasses. Chemical strengthening produces higher strength in thinner glass than does thermal tempering, but it is more susceptible to weakening by abrasion or scratching because of the thinner compression layer. Copper- and silver-containing mixtures can be applied in the same way to glass to produce colors by ion-exchange staining.

Acid etching or frosting with dilute hydrofluoric acid produces articles with good light-diffusing properties, eg, a lamp envelope. In the process to make Vycor glassware, leaching produces 96% silica glasses (108). A suitable heat treatment is first given to the glass to separate it into a soluble and insoluble phase and soluble sodium borate is precipitated. Leaching removes this phase and leaves a skeleton of silica-rich glass. On subsequent firing, a 14% linear shrinkage is obtained in a consolidated, transparent, nonporous glass. Chemically machinable photosensitive glass is acid-machined after patterns are photochemically induced into the glass.

Quality Control. Quality control is governed by the uses for which a glass is designated. For example, color-television panel glass must pass a wide variety of chemical and property specifications. Its chemical composition is checked spectrographically by comparison with a standard. The x-rays generated inside the picture tube during operation must be absorbed by the glass; the absorption coefficient at 0.06 nm is calculated from the composition, which in turn is usually maintained to 0.1–0.2% for each oxide. The physical properties, such as low temperature viscosity (softening, annealing, and strain points) and thermal expansion, must be controlled to within 1% or less to ensure a good glass-to-metal seal for electrical leads and a good glass-to-glass seal when joined to the funnel. Electrical resistivity must be maintained above a minimum to prevent electrical fields from being generated around the tube during operation. The color of the panel must give optimum enhancement of the color-generating phosphors, maximum contrast with surrounding light, and uniform transmission. Other products have other requirements. Refractive index ($\pm$ 0.0001) and dispersion ($\pm$ 0.5) are closely controlled in optical and ophthalmic glass. Tinted optical glass, eg, sunglasses and filters, must maintain a color as well as a visible, ir, and uv transmission standard. Glazes, decorations, and food service items in general must maintain rigid standards of chemical durability and heavy-metal release. FDA regulations require less than 0.5–3 ppm Pb and less than 0.25–0.5 ppm Cd released depending on the use of the article. The test conditions are 4% acetic acid at room temperature for 24 h. Generally, each manufacturer has strength and abrasion standards; a notable exception is ophthalmic lenses which the FDA specifies must withstand a 15.88 mm-dia steel ball dropped from 1.27 m.

Melting defects must be held to a minimum. Solid inclusions (stones) in the form of refractory particles, unmelted batch, or devitrification, affect the strength as well as optical integrity. Gaseous inclusions (seeds and blisters) caused by improper fining or electrochemical reboil, have the same effect as striae (cords) from improper homogenization.

Forming defects are numerous and have been alluded to in the discussion of the forming machines and techniques. Tolerances in dimensions and capacities of glass bottles are independently negotiated between manufacturer and customer. Containers for expensive products like perfume and liquor carry the most precise tolerances. Champagne and soft-drink bottles must meet hydraulic pressure specifications which are affected by design and glass quality as well as dimensional tolerances. Flat glass, depending on the use, but especially in the case of automobile windshields, has precise melting defect, thickness, color, and optical distortion requirements. Dimensional tolerances, especially thickness, are monitored by an x-ray scanning device that measures relative thickness on-line. Lasers (qv) can also be used.

Absolute thickness is measured by a micrometer. The center thickness of a ribbon of float glass can be held to within 250 μm for window glass and to within 100 μm for high quality windshields. The allowable thickness variation or thickness profile of a 2-mm thick piece of sheet glass, for example, is much less than 250 μm. Distortion is measured by the angle between the piece of glass and a black-and-white striped board which first produces optical distortion (zebra angle). An angle of 50–60° for windshields corresponds to about 150 μm/m thickness variation.

Cleaning. Methods of cleaning glass are closely associated with the type of glass and its particular use (109,110). Cleaning requires consideration of (*1*) the soil to be removed; (*2*) the methods available; (*3*) the interaction between soil and glass; and (*4*) the effects of cleaning on the glass surface (110). Aqueous solvents frequently used range from chromic–sulfuric acid mixtures for cleaning borosilicate laboratory ware to common detergent solutions. Care must be taken to prevent leaching or etching of silicate glasses with very acidic or basic solutions. Organic solvents usually have no harmful effect on most glasses. Some organic solvents such as 2-propanol may be used alone or as aqueous mixtures in commercial cleansers. Hydrocarbon or halocarbon solvents or simply a hot flame are often used to remove nonpolar organic compounds. Mechanical cleaning may be required such as scrubbing, wiping, brushing, or repolishing and buffing the surface. Ultrasonic and electron-, ion-, or atomic-discharge cleaning methods are available for special cleaning problems usually associated with removal of organic materials (see ULTRASONICS).

Other Methods. *Vapor Deposition. Optical Fibers.* Special applications require nonstandard methods of manufacture (111). Vapor deposition is principally used in the manufacture of high quality glass such as optical fibers and optical mirrors (112,113). More than 90% of the optical fiber market is comprised of single-mode fiber. Corning, the world's largest producer, sells its most popular single-mode fiber as Corning SMF-28 optical fiber. It is a thin glass fiber that typically transmits information via 1310 nm laser light with 100–1000 times the capacity of conventional copper cables. Single-mode fiber typically has an attenuation at 1310 nm of 0.35 dB/km (92% incident light transmitted per km). By contrast, optical glass and window glass have attenuations of 1000 dB/km (0.35 dB/35 cm) and 100,000 dB/km (0.35 dB/3.5 mm), respectively. Hence, an optical waveguide can transmit as much light over 1 km as optical glass can over about 35 cm and ordinary window glass can over 4 mm. Sufficient freedom from light-absorbing chemical impurities, such as iron and water, and light scattering physical defects, such as seeds and compositional inhomogeneities, is very difficult to achieve for glass made by conventional glassmaking methods. Therefore, optical waveguide communication systems use glass fibers made by a vapor-phase process (112).

Compositionally, the typical single-mode fiber contains a core with a diameter of about 8 μm (less than six times the wavelength of the light that it carries) and a cladding of 125 μm ± 1 μm diameter. The core, which has a step refractive index profile, is an 8 wt % GeO_2 + 92 wt % SiO_2 glass. The germania raises the refractive index to about 1.4585. The refractive index of the pure silica cladding is about 1.4534. That difference in refractive index is sufficient to guide the laser light with minimum distortion. Besides facilitating strength and handleability, the thickness of the cladding is required for minimum attenuation. The outside of the cladding is coated with two types of polymer to protect the pristine glass surface, theoretically maintaining its strength at upward of 4.8 GPa (700,000 psi). Another popular single-mode product is dispersion-shifted fiber, with attenuations of 0.20 dB/km at 1550 nm combined with minimum dispersion. This is the highest capacity fiber type and is used for carrying longer wavelength laser light over longer distances without amplifications. Some multimode fibers have larger cores (eg, 62.5 μm), parabolic index profiles, and higher refractive indexes.

Commercial optical fibers (qv) use high vapor pressure metal chlorides such as $SiCl_4$ and $GeCl_4$, which are processed to provide bulk high purity oxide glasses (called blanks or preforms) from which optical fibers can be drawn. Because the fiber must exhibit negligible optical attenuation, the reaction mixture must be free from optically absorbing transition metals, eg, $FeCl_3$, VCl_4, $CuCl_2$. The pure chlorides are entrained in an oxygen carrier-gas system, accurately metered, transported, and then react at temperatures above 1500°C. The chloride reaction with oxygen, to form the desired oxides plus chlorine gas, is virtually homogeneous and produces a finely divided particulate glass material commonly called soot. This high purity, inclusion-free glass soot has a high surface area which provides a powerful driving force for rapid, thermally activated, viscous sintering. The glass soot is formed into solid inclusion-free glass bodies, which are then heated to temperatures where the glass is fluid enough to be drawn into optical fibers.

Numerous fabrication processes utilize this basic scheme. For example, Outside Vapor Deposition (OVD) is a Corning developed and patented process that uses a flame heat source, a removable rod target, and a perpendicular orientation for soot deposition (114). It is illustrated in Figure 21. A hot soot stream of the desired glass composition is generated by passing the vapor stream through a methane–oxygen flame directed at a rotating and traversing ceramic target rod. The glass soot sticks to the rod in a partially sintered state and, layer by layer, a cylindrical porous glass preform is built up; first the core, and then continuing

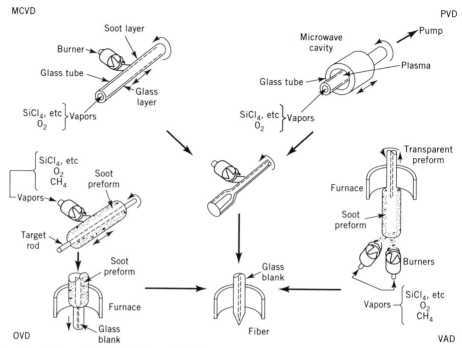

Fig. 21. Schematic illustration of the four primary vapor-phase deposition processes used in optical-fiber fabrication: outside vapor deposition (OVD), modified chemical vapor deposition (MCVD), plasma vapor deposition (PVD), and vapor axial deposition (VAD) (115).

into the cladding. By properly controlling the metal halide vapor stream composition during the soot deposition process, the desired glass composition profile is built into this porous preform. The preform is consolidated (sintered) between 1400 and 1550°C to a solid, bubble-free, glass blank. The consolidation process uses chlorine to rid the blank of water and trace impurities. The drying/chlorination process drives water down to the 10 ppb range. Sometimes a two-step process can be employed for efficiency. A preform is made which is roughly half core and half cladding. The sintered preform is then drawn into rod and then overclad with pure silica soot to obtain the appropriate core/clad ratio.

The Blanks can also be made using Vapor Axial Deposition (VAD) (Fig. 21). The process involves simultaneous flame deposition of both core- and cladding-glass soots onto the end (ie, axially) of a rotating fused-silica target rod. The finished perform is then consolidated in a process similar to the OVD process.

The Modified Chemical Vapor Deposition (MCVD) process (Fig. 21) involves reactions within a glass tube and glass deposition on the inside walls of this tube. A rotating, conventionally fabricated, silica glass tube is heated with an oxy–hydrogen torch to cause vapor-phase oxidation of the metal halide gases which are flowing inside. Ultimately, the tube becomes the outer part of the fiber cladding, and the deposited soot becomes the fiber core. Using a conventional glass tube as part of the fiber can produce fiber of lower strength and higher attenuation. As the hot soot flows downstream, it is attracted to the cold walls of the tube where it is deposited as a thin porous layer, not unlike the porous OVD preform structure described earlier. The torch is steadily moved toward this downstream portion of the tube and, as it passes over the soot deposit, it zone-sinters it to a clear, bubble-free glass layer. When the torch reaches the exhaust end of the tube, it is quickly returned to the inlet end, and the process is repeated. After the required number of passes, enough glass is deposited on the inside wall of the tube. Then the torch traverses the tube, which softens and collapses to a solid rod blank. This is then drawn into fiber.

Vapor Deposition. *Bulk-Fused Silica.* Bulk-fused silica is commercially produced by a variety of techniques, including vapor deposition. The physical properties of the product depend strongly on the method used. If the target is kept above 1800°C during the SiO_2 soot deposition, simultaneous sintering of the soot occurs, yielding in a single step a solid, bubble-free glass. This is achieved if the heat from the soot-generating burners also sinters the soot as it hits a hot-fused sand target. Layer by layer, a boule of solid fused silica is deposited which, because hydrogen-containing fuels are used, contains ca 1200 ppm OH. Very large boules, weighing over 500 kg, can be obtained by using many soot deposition burners and/or running furnaces for many days at a time. This approach is used to manufacture numerous high silica glasses. Optical blanks, windows, crucibles, tubing, and mirrors for large telescopes are produced by further processing these boules with conventional cutting, grinding, polishing, and flameworking techniques. This material is also used extensively for windows in spacecraft, because of its refractory nature, thermal shock resistance, and optical homogeneity.

Glass Films. Glass films are formed by both reactive and nonreactive deposition methods. Examples of nonreactive deposition methods are evaporating, sputtering, and ion-implantation or ion-plating. Reactive deposition methods are described below. Glass films are used in the semiconductor industry because of their dielectric properties, and they are used for encapsulating integrated circuits and other electronic devices because they provide a reasonably hermetic seal. Deposition techniques for dielectric films are given in Table 9.

Reactive sputtering is a process in which highly active metal atoms react with a gas before their deposition on cold substrate. In the presence of oxygen or nitrogen, for instance, combined with a high purity silicon metal cathode in a sputtering chamber, high purity amorphous silicon dioxide or silicon nitride is deposited. The high quench rate and the absence of impurities have made possible the production of numerous glasses of $PbTeO_3$, GeO_2, SiO_2, Al_2O_3–SiO_2, and many other oxides.

Thermal oxidation is one of the oldest and most commonly used methods of forming a primary passivating film of SiO_2 on silicon. The metal is heated in dry oxygen, in H_2O-containing oxygen, or in steam. A silica layer grows inwardly from the surface by a thermal oxidation mechanism. Although this process occurs slowly even at room temperature, yielding a layer ca 1-nm thick after long exposures, the silicon wafer is usually heated to 600–1200°C to achieve 1-μm thick films in ca 1 h.

A glass layer can also be made by anodic oxidation on a metal or semiconductor surface, such as silicon, by making it the anode in an electrolytic cell, immersing this anode in a suitable electrolyte (often aqueous), and passing a current through it. By applying a sufficient overvoltage, the metal surface oxi-

Table 9. Deposition Techniques for Dielectric Films[a]

Deposition process	Source material	Typical deposition rate	Typical substrate temperature, °C	Sources of impurities	Typical films[b] formed	Typical electronics applications
Low pressure depositions						
evaporation	high vapor-pressure solid (eg, SiO)	1–2 μm/min	25–200	filament/crucible walls	SiO	optical coatings
reactive evaporation	solids, eg, Al in gas (eg, O_2)	1–2 μm/min	25–200	filament/crucible	Al_2O_3	passivation layers
sputtering; reactive sputtering	low vapor-pressure solids, eg, SiO_2 with gas (eg, Ar)	1–10 nm/min	25	walls, gases used, sources	SiO_2,[c] Si_3N_4, TaN	passivation layers
plasma deposition	gases (eg, SiH_4–NH_3 mixtures)	1–10 nm/min	25–400	walls, gases, electrodes	Si_3N_4, Si	passivation layers, solar cells
low pressure CVD	gases (eg, SiH_4–NH_3 mixtures)	0.1–1 μm/min	600–1000	walls, gases, susceptors	SiO_2, Si_3N_4, ORPS[d]	passivation layers
101-kPa (1-atm) depositions						
thermal oxidation	substrate plus gas	nonlinear ca 1 μm in 1 h	800–1200	walls, substrate surface	SiO_2, Al_2O_3	channel oxides (MOS[e]), diffusion marks, etch marks, passivation layers
chemical vapor deposition[f]	gases (eg, SiH_4–NH_3 mixtures, SiH_4–O_2 mixtures)	0.1–1 m/min	800–1200 250–500	source gases, walls	SiO_2,[e] Si_3N_4, PSG,[e] BSG[e]	passivation layers, diffusion sources, dielectrics for multilevel metal
anodization	substrate plus electrolyte	nonlinear 10–100 nm in 1 min	25	electrolyte	Al_2O_3, Ta_2O_5	passivation layers
ion implantation	high purity ion source (mass-spectrometer)		25	substrate surfaces	nitride layers	etch-resists

[a] Ref. 116. [b] Crystalline nature of the deposit is amorphous unless otherwise noted. [c] Stoichiometric composition usually not obtained. [d] ORPS = oxygen-rich polycrystalline silicon. [e] BSG = borosilicate glass; PSG = phosphosilicate glass; MOS = metal oxide semiconductor. [f] Deposit is polycrystal amorphous.

dizes, forming, in some cases, a glass film up to several hundred nanometers thick. In one form of chemical vapor deposition (CVD), heterogeneous reaction and deposition of organometallic or metal halide vapors occur at a heated solid substrate surface. This is the most widely used and versatile method for forming dielectric films in the fabrication of semiconductor devices and for depositing layers of glass or crystalline oxides or nitrides over silicon, etc. For example, by passing gas mixtures such as SiH_4, PH_3, oxygen (or $SiCl_4$, $POCl_3$, and oxygen) over a heated silicon wafer, the vapors react heterogeneously at the wafer surface, depositing a thin layer of P_2O_5–SiO_2 glass over all heated exposed surfaces.

Miscellaneous. Unconventional melting includes splat cooling, which has been used extensively to produce amorphous metals; laser-spin melting, which can produce 100–800 μm spheres of refractory oxide glasses; and melting under high pressure so that enormous quantities of volatile components may be incorporated into the glass structure. Solid-state transformations produce amorphous materials from bulk crystalline phases by destroying the lattice structure at low temperatures. This can be achieved by neutron bombardment in a reactor, and is found in nature in certain minerals containing radioactive elements. Solid-state transformation also occurs upon exposure to shock waves, as in a meteorite impact. Solution methods have been used to form silica gels which can be sintered to clear vitreous silica at low temperatures (117). Such gels can be doped with a variety of cations to impart useful modifications. The technique is also being employed to produce glasses having compositions which cannot be obtained via conventional glass forming methods.

Economic Aspects

Glass manufacture is classified according to the product into flat, container, fiber, or specialty glass. U.S. manufacturers of flat and container glass produced over 17×10^6 t soda–lime glass products in 1992 (118). Specialty glass manufacturers of pressed and blown ware, television bulbs, lighting, and optical glasses melt hundreds of glass compositions to fulfill the need for a large variety of products.

Table 10 summarizes sales statistics for the principal segments of the industry. The data were gathered by a survey conducted annually by the editors of *Ceramic Industry* magazine. In 1991, the survey scope was changed from U.S. to a worldwide basis (119,124). Tables 11 and 12 present data gathered by U.S. Department of Commerce agencies.

Flat Glass. In the United States the main producers of flat glass are PPG Industries, Libbey-Owens-Ford (LOF), Guardian Industries, Ford Motor Co., and AFG Industries. Growth of this industry segment depends on the construction and automotive markets, 57 and 25% of the market, respectively (119). The float process produces more than 85% of all flat glass. The remainder is thin sheet for picture glass or rolled and patterned glass. A small but growing market is that for thin sheet glass for liquid crystal displays. In 1992, approximately 4.27×10^8 m^2 (4.6 billion ft^2) of flat glass was shipped from U.S. factories. Production of flat glass is cyclical but increases in volume by approximately 2–3% per year. Since 1990, the value of exports has exceeded that of imports; the balance is strongly influenced by the variations in current exchange rates.

Table 10. Distribution of Glass Sales, %[a]

Glass type	U.S., 1987	U.S., 1989	World, 1990	World, 1992
flat glass	23.7	21.3	30.0	32.0
fiber glass	23.4	25.0	17.0	17.0
container glass	22.8	22.5	12.0	17.0
specialty				
lighting	18.7	18.0	18.0	18.0
consumer	4.9	6.4	9.0	5.0
TV and cathode ray	3.2	3.4	4.0	9.0
other	3.3	3.4	10.0	2.0
total sales $\times\ 10^6$ \$	17,574	18,558	41,240	48,260

[a]Refs. 120–123.

Table 11. Glass and Related Mineral Industry Group Statistics[a]

	1982		1987		1990
Industry group	Number of establishments	Value of shipments, 10^6 \$	Number of establishments	Value of shipments, 10^6 \$	Value of shipments, 10^6 \$
stone, clay, and glass products	16,545	45,181	16,166	61,477	63,468
flat glass	69	1,666	81	2,549	2,279
glass and glassware, pressed or blown	459	7,941	522	8,339	8,918
products of purchased glass	1,337	2,977	1,432	5,429	6,141
cement, hydraulic	237	3,542	215	4,335	4,251
structural clay products	628	1,868	598	2,915	3,087
pottery and related products	910	1,762	1,006	2,416	2,613
concrete, gypsum, and plaster products	9,933	14,947	9,814	24,427	24,595
cut stone and stone products	711	555	745	841	989
miscellaneous nonmetallic mineral products	na	na	1,753	10,226	10,595

[a]Ref. 124.

Table 12. Trends and Forecasts for Flat Glass, 10^6 \$[a]

Item	1987	1988	1989	1990[b]	1991[c]
	Industry data				
value of shipments[d]	2549	2442	2477	2307	2265
value of shipments, 1987 \$	2549	2410	2495	2355	2300
capital expenditures	151	151	144		
	Product data				
value of shipments[e]	3509	3413	3405	3170	3043
value of shipments, 1987 \$	3509	3368	3426	3269	3140
	Trade data				
value of imports	502	513	510	497	460
value of exports	356	416	506	676	685

[a]Ref. 119.
[b]Estimated, except exports and imports.
[c]Estimate.
[d]Value of all products and services sold by establishments in the flat glass industry.
[e]Value of products classified in the flat glass industry produced by all industries.

The conversion of the industry to the float process during the 1970s caused concern over the possibility of overcapacity, but the continued demand by both the automotive and construction markets allowed a relatively easy transition. Float processing is amenable to surface treatments that produce reflective coatings. Because the cost of the glass represents a relatively small part (1–2%) of the total construction costs, the use of solar-efficient glazing appears economical compared to more expensive mechanical heating or cooling equipment. Solar energy (qv) conversion also represents a potentially large market for flat-glass producers.

Container Glass.   Statistics for food, beverage, drug and cosmetic, and household and industrial containers are compiled and published by the Glass Packaging Institute (GPI) (125). About 75% of the containers are narrow neck. Beverage containers, including soft drink, beer, wine, and liquor, constitute the largest segment, followed by food, drug and cosmetic, and household and industrial containers. Approximately 85% of container glass is clear, the remainder is mostly amber. Other tints make up a small percentage of the total. In the United States, principal producers are Owens-Illinois, Anchor, and Kerr glass companies. Over 41×10^9 containers weighing more than 13×10^6 t were shipped in 1992 with an estimated value of over $\$4 \times 10^9$ (126). Imports are of very minor importance because of high transportation costs. Export of container glass is a very small percent of the total amount manufactured for the same reason.

Fiber Glass.   Fiber glass is classified as either textile or wool. Many companies produce textile fibers for draperies, upholstery, tires, reinforced plastics, paper, and tape, including Manville/Schuller, Owens-Corning Fiberglas, PPG Industries, and Nicofibers. Glass wool is primarily used for building insulation, industrial equipment and pipe insulation. Growth of the insulation market has been faster than the textile area because of the demand for additional insulation in both new and existing buildings.

Specialty Glass. The pressed-and-blown or hollow-ware industry is comprised of over one hundred companies in the United States, including Corning Glass Works, Owens-Illinois, General Electric, and Anchor Hocking. The wide variety of products is divided into categories of pressed-and-blown glass for table, kitchen, art, and novelty applications and products of purchased glass. The latter consists of items for scientific, technical, and industrial uses such as electrical and electronic products, laboratory glassware, optical and ophthalmic glass, etc.

Information Sources

There are several professional societies worldwide that publish technical journals relating to glass science and technology. Prominent among them are The American Ceramic Society, The Society of Glass Technology (U.K.), Deutshe Glastechnische Gesellschaft, Nordiska Glastekniska Foreningen, Institute Du Verre (France), Ceramic Society of Japan, and the Indian Ceramic Society. The societies also sponsor conferences for which proceedings are published. The foremost international conferences on glass are the International Glass Congresses presented every three years by the International Commission on Glass, which is a union of scientific and technical organizations. The most recent Glass Congress was the 16th, held in Madrid, Spain in 1992. The proceedings of these conferences are full of information about the latest developments in glass science, technology, and art.

Additional periodicals of special interest to workers in the field of glass are the *Journal of Non-Crystalline Solids*, North-Holland/Elsevier Science Publishers BV; *Glass Production Technology International*, Sterling Publications Ltd. annual; and two Russian publications translated by Consultants Bureau, New York, *Glass and Ceramics* and *Glass Physics and Chemistry*. There are many other journals published by technical societies and organizations in fields such as chemistry, physics/optics, and materials science and engineering that frequently contain important papers on glass. These can be found most easily through a search of the *Chemical Abstracts* and/or the *Ceramic Abstracts* on-line databases. Another database of importance is available on CD-ROM from the New Glass Forum of Japan (127). These databases contain references to patents as well as to publications.

Industry periodicals of significance are *Ceramic Industry*, Business News Publishing Co., Troy, Mich.; *Glass, Monthly Journal of the European Glass Industry* and *Glass International*, both by FMJ International Publications, Surrey, U.K.; *Glass Digest* and *Glass Industry*, both by Ashlee Publishing Co. Inc., New York. An excellent, and comprehensive, directory is published annually by *Glass Industry*. The 1993 directory is divided into nine parts. They are (*1*) primary glass manufacturers listed alphabetically by company name; (*2*) secondary glass manufacturers listed alphabetically by company name; (*3*) glass manufacturers listed alphabetically by country, except the United States; (*4*) glass manufacturers listed alphabetically by product; (*5*) suppliers listed alphabetically by company name; (*6*) suppliers listed alphabetically by product or service; (*7*) associations and unions involved with the glass industry; (*8*) laboratories or organizations offering research programs or services to the glass industry; and (9) educational institutions offering training in glass technology and other glass-oriented subjects (128–130).

BIBLIOGRAPHY

"Glass" in *ECT* 1st ed., Vol. 7, pp. 175–206, by H. G. Vogt, Corning Glass Works; in *ECT* 1st ed., Suppl. 2, pp. 435–454, by S. D. Stookey, Corning Glass Works; in *ECT* 2nd ed., Vol. 10, pp. 533–604, by J. R. Hutchins III and R. V. Harrington, Corning Glass Works; in *ECT* 3rd ed., Vol. 11, pp. 807–880, by D. C. Boyd and D. A. Thompson, Corning Glass Works.

1. J. E. Ericson and co-workers, *J. Noncryst. Solids.* **17**, 129 (1975).
2. G. W. Morey, *The Properties of Glass*, 2nd ed., Reinhold Publishing Corp., New York, 1954, p. 28.
3. *Standard Terminology of Glass and Glass Products*, ASTM Standards 1993, ASTM C162-92, Vol. 15.02, pp. 29–40.
4. H. A. Davies, *Phys. Chem. Glasses* **17**(5), 159 (1976).
5. L. B. Davies and P. J. Grundy, *Phys. Stat. Sol.* **8a**, 189 (1971); *J. Noncryst. Solids* **11**, 179 (1972).
6. D. E. Polk, *J. Noncryst. Solids* **5**, 365 (1971).
7. G. S. Cargill, *J. Appl. Phys.* **41**, 2248 (1970).
8. F. G. A. Stone and W. A. G. Graham, *Inorganic Polymers*, Academic Press, Inc., New York, 1962.
9. G. Tammann, *Der Glaszustand*, Leopold Voss, Leipzig, 1933.
10. H. N. Ritland, *J. Am. Ceram. Soc.* **37**, 370 (1954).
11. A. Q. Tool, *J. Res. Natl. Bur. Stand.* **37**, 73 (1946).
12. N. J. Kreidl, *Glass Ind.* **58**, 26 (1977).
13. D. R. Secrist and J. D. Mackenzie, *Glass Ind.* **45**, 408, 451, (1964).
14. W. Zauzmann, *Chem. Rev.* **43**, 219 (1948).
15. J. N. Gibbs, in J. D. Mackenzie, ed., *Modern Aspects of the Vitreous State*, Part 1, Butterworth, Inc., Washington, D.C., 1960, pp. 152–157.
16. A. R. Cooper and P. K. Gupta, *J. Am Ceram. Soc.* **58**, 350 (1975).
17. A. A. Lebedev, *Arb. Staatl. Opt. Inst. Leningrad* **2**(10), (1921).
18. J. T. Randall, H. P. Rooksby, and B. S. Cooper, *J. Soc. Glass Technol.* **14**, 219 (1930); N. Valenkov and E. Pori-Koshitz, *Z. Krist.* **45**, 195 (1936).
19. B. E. Warren and I. Biscoe, *J. Am. Chem. Soc.* **21**, 49 (1938).
20. N. Valenkov and E. Porai-Koshits, *Z. Krist.* **95**, 195 (1936).
21. W. H. Zachariasen, *J. Am. Chem. Soc.* **54**, 3841 (1932).
22. R. L. Mozzi and B. E. Warren, *J. Appl. Crystallogr.* **2**, 164 (1969).
23. J. H. Konnert, J. Karle, and G. A. Gerguson, *Science* **179**, 177 (1973).
24. A. C. Wright and A. J. Leadbetter, *Phys. Chem. Glasses* **17**(5), 122 (1976).
25. R. H. Doremus, *Glass Science*, John Wiley & Sons, Inc., New York, 1973.
26. K. H. Sun and A. Silverman, *J. Am. Ceram. Soc.* **25**, 101 (1942).
27. M. E. Milberg and C. R. Peters, *Phys. Chem. Glasses* **10**, 46 (1969).
28. P. Chaudhari and co-workers, *Phys. Rev. Lett.* **29**, 425 (1972).
29. P. J. Bray and J. G. O'Keefe, *Phys. Chem. Glasses* **4**, 37 (1963).
30. R. L. Mozzi and B. E. Warren, *J. Appl. Crystallogr.* **3**, 251 (1970).
31. V. M. Goldschmidt, *J. Soc. Glass Technol.* **11**, 337 (1927).
32. K. H. Sun, *J. Am. Ceram. Soc.* **30**, 277 (1947).
33. H. Rawson, *Inorganic Glass-Forming Systems*, Academic Press, Inc., London, 1967.

34. J. E. Stanworth, *J. Am. Ceram. Soc.* **54**, 61 (1971).
35. M. Imaoka and T. Yamosaki, *J. Ceram. Assoc. (Japan)* **71**, 215 (1963).
36. H. W. S. De Jong, *Ullmann's Encyclopedia of Industrial Chemistry*, VCH Verlagsgesellschaft mbH, Weinheim, Germany, 1989, pp. 365–432.
37. D. C. Allen and M. P. Teter, *J. Am. Ceram. Soc.* **73** (11), 3247–3250 (1990).
38. X. Gonze, D. C. Allen, and M. P. Teter, *Phys. Rev. Lett.* **68**(24), 3603–3606 (1992).
39. J. G. Vail, *Soluble Silicates, Their Properties and Uses*, Reinhold Publishing Corp., New York, 1952.
40. V. I. Aver'yanov and E. A. Porai-Koshits, *The Structure of Glass*, Vol. 6, Part 1, Consultants Bureau, New York, 1966, p. 98.
41. L. D. Pye and co-workers, *Borate Glasses: Structure, Properties, Applications*, Vol. 12, Materials Science Research, Plenum Press, New York, 1978.
42. Ger. Pats. 63,126 (Aug. 5, 1978), and 62,408 (June 20, 1969), W. Heindorf.
43. W. H. Dumbaugh and J. C. Lapp, *J. Am. Ceram. Soc.* **75**, 2315–2326 (1992).
44. P. A. Tick and P. L. Bocko, "Optical Fiber Materials," *The Handbook of Optical Materials*, Vol. 3, Marcel Dekker, New York, 1988.
45. S. Zemon and co-workers, *Proc. S.P.I.E.-Int. Soc. Opt. Eng.* **1789** 58–65 (1993).
46. G. S. Cowgill, III, *NY Acad. Sci.* **279**, 208 (1976).
47. H. S. Chen, *Acta Metall.* **22**, 1505 (1974).
48. M. H. Cohen and D. Turnbull, *Nature (London)* **189**, 131 (1961).
49. M. Marcus and D. Turnbull, *Mater. Sci. Eng.* **23**, 211 (1976).
50. S. D. Stookey and R. D. Maurer, in J. E. Burke, ed., *Progress in Ceramic Science*, Vol. 2, Pergamon Press, New York, 1962, pp. 77–101.
51. G. H. Beall, in L. L. Hench and S. W. Friedman, eds., *Nucleation and Crystallization in Glasses*, American Ceramic Society, 1972, pp. 251–261.
52. J. F. MacDowell and G. H. Beall, *J. Am. Ceram. Soc.* **52**(1), 17 (1969).
53. D. R. Stewart, in L. D. Pye and co-eds, *Introduction to Glass Science*, Plenum Press, New York, 1972, pp. 237–271.
54. T. P. Seward, III, private communication, Corning Inc., 1980.
55. S. D. Stookey, *Ind. Eng. Chem.* **41**, 856 (1949).
56. S. D. Stookey, G. H. Beall, and J. E. Pierson, *J. Appl. Phys.* **49**, 5114 (1978).
57. N. F. Borrelli and T. P. Seward, in S. J. Schneider, ed., *Engineered Materials Handbook*, Vol. 4, *Ceramics and Glasses*, ASM International, Materials Park, Ohio, 1991, pp. 439–444.
58. A. G. F. Dingwall and H. Moore, *J. Soc. Glass Technol.* **37**, 316 (1953).
59. H. E. Hagy, *Electron. Packag. Prod.* **18**(7), 182 (July 1978).
60. V. O. Altemose, "Gas Permeation Through Glass," *Seventh Symposium on the Art of Glassblowing*, The American Scientific Glassblowers Society, Wilmington, Del., 1962.
61. W. C. LaCourse in Ref. 53, pp. 451–512.
62. W. H. Barney, private communication, Corning Inc., 1980; various Corning public information bulletins.
63. L. L. Hench and H. F. Schaake, in L. D. Pye and co-eds., *Introduction to Glass Science*, Plenum Press, New York, 1972, pp. 583–659.
64. *Ceramic Data Book, Lamp Components Sales Operation*, General Electric Co., Cahners Publishing Co., Boston, Mass., 1979, p. 304.
65. M. J. Minot, *J. Opt. Soc. Am.* **66**, 515 (1976).
66. W. A. Weyl, *Colored Glasses*, Society of Glass Technology, Sheffield (Reprinted by Dawson's of Pall Mall, London) 1951.
67. J. Wong and C. A. Angell, *Glass: Structure by Spectroscopy*, Marcel Dekker, Inc., New York, 1976.
68. N. J. Kreidl in F. V. Tooley, ed., *Handbook of Glass Manufacture*, Books for Industry, Inc., New York, 1984, pp. 957–998.

69. W. Vogel, *Chemistry of Glass*, American Ceramic Society Inc., Columbus, Ohio, 1985, p. 325.

70. E. J. Coker and L. N. G. Filon, *Treatise on Photoelasticity*, Cambridge University Press, London, 1931.

71. G. E. Rindone, *Transactions of the International Conference on Glass*, Paris, 1956, p. 373.

72. N. J. Kreidl and J. Rood, in R. Kinglake, ed., *Optical Materials, Applied Optics and Optical Engineering*, Academic Press, Inc., New York, 1965.

73. E. J. Friebele and D. L. Griscome, in M. Tomozawa and R. Doremus, eds., *Treatise on Materials Science and Technology*, Vol. 17, Academic Press, Inc., 1979, pp. 257–351.

74. E. J. Friebele, in D. R. Uhlmann and N. J. Kreidl, eds., *Optical Properties of Glass*, American Ceramics Society, 1991, pp. 205–262.

75. C. R. Bamford, *Colour Generation and Control in Glass*, Elsevier Scientific Publishing Co., New York, 1977, pp. 141–164.

76. W. C. Bauer and J. E. Bailey, in *Engineered Materials Handbook*, Vol. 4, ASM International, 1991, pp. 378–385.

77. E. D. Spinosa, P. M. Stephan, and J. R. Schorr, *Review of Literature on Control Technology which Abates Air Pollution and Conserves Energy in Glass Melting Furnaces*, to Corning Incorporated, EPA-600/2-77-005,/2-76-269,/2-76-032b, Battelle, Columbus, Ohio, Nov. 11, 1977.

78. F. E. Woolley in Ref. 76, pp. 386–393.

79. W. Trier, *Glass Furnaces (Design, Construction and Operation)*, K. L. Loewenstein, trans., Society of Glass Technology, U.K., 1987.

80. R. S. Arrandale and B. L. Schmidt, in F. V. Tooley, ed., *The Handbook of Glass Manufacture*, Books for Industry, Inc., New York, 1984, pp. 248–386.

81. A. G. Pincus and G. M. Diken, *Electric Melting in the Glass Industry*, Books for Industry, New York, 1976.

82. U. S. Pat. 3,524,206 (Aug. 18, 1970), G. B. Boettner (to Corning Inc.).

83. J. T. Brown, "100% Oxygen - Fuel Combustion for Glass Furnaces," *51st Conference on Glass Problems*, Amer. Ceram. Soc., Columbus, Ohio, Nov. 1, 1990.

84. J. Levins, in *Glass Production Technology International*, Sterling Publications Limited, London, 1992, pp. 47–50.

85. W. Giegerich and W. Trier, *Glass Machine Construction and Operation of Machines for the Forming of Hot Glass*, trans. N. J. Kreidl, Springer-Verlag, Berlin, 1969.

86. U. S. Pat. 1,790,397 (Jan. 27, 1931), W. J. Woods and D. E. Gray (to Corning, Inc.).

87. E. B. Shand, *Glass Engineering Handbook*, 2nd ed., McGraw-Hill, Book Co., Inc., New York, 1958.

88. U. S. Pat. 2,662,346 (Dec. 15, 1953), J. W. Giffen (to Corning, Inc.).

89. A. Pilkington, *Glass Technol.* **17**, 182 (Oct. 1976).

90. PPG Industries, Inc., press release, Feb. 17, 1975, Sept. 17, 1976, Feb. 17, 1977, May 19, 1977, Nov. 17, 1977, Public Relations Dept., Pittsburgh, Pa.

91. L. A. B. Pilkington, *Proc. Roy. Soc. London A* **314**, 1 (1969).

92. U. S. Pat. 3,338,696 (Aug. 29, 1967), S. M. Dockerty (to Corning Inc.).

93. R. F. Caroselli in H. F. Mark, S. M. Atlas, and E. Cernia, eds., *Man-Made Fibers*, Wiley-Interscience Publishers, Inc., New York, 1968, pp. 425–454.

94. G. R. Machlan in F. V. Tooley, ed., *Handbook of Glass Manufacture*, Books for Industry, Inc., New York, 1984, pp. 715–734.

95. A. I. Andrews, *Procelain Enamels*, The Garrard Press, Champaign, Ill., 1961.

96. J. A. Pask, in J. D. MacKenzie, ed., *Modern Aspects of the Vitreous State*, Vol. 3, Butterworth & Co., Washington, D. C., 1964, pp. 1–28.

97. C. W. Parmelee, *Ceramic Glazes*, 3rd ed., Cahners Publishing Co., Inc., Boston, 1973.

98. U. S. Pat. 3,231,356 (Jan. 25, 1966), J. W. Giffen (to Corning, Inc.).
99. Erwin Roeder, in Ref. 84, pp. 205–209.
100. H. A. McMaster, in Ref. 94, pp. 799–832.
101. A. G. Pincus and T. R. Holmes, *Annealing and Strengthening in the Glass Industry*, Magazines for Industry, Inc. New York, 1977.
102. L. H. Adams and E. D. Williamson, *J. Franklin Inst.* **190**, 597,835 (1920).
103. H. R. Lillie, *Glass Ind.* **21**, 355,382 (1950).
104. H. N. Ritland, *J. Am. Ceram. Soc.* **37**, 370 (1954).
105. U. S. Pat. 3,223,501 (Dec. 14, 1965), J. C. Fredley and G. E. Sleighter (to Pittsburgh Plate Glass Co.).
106. U. S. Pat. 3,332,762 (July 25, 1967), H. A. McMaster and N. C. Nitschke (to Perma-glass, Inc.).
107. U. S. Pat. 3,468,645 (Sept. 23, 1969), H. A. McMaster and N. C. Nitschke, and J. J. Kawecka (to Permaglass, Inc.).
108. U. S. Pat. 2,106,744 (Feb. 1, 1938), H. P. Hood and M. E. Nordberg (to Corning Inc.).
109. *ASTM Standard Practices for Designing a Process for Cleaning Technical Glasses*, ASTM C 912, pt. 17, 1980.
110. L. Holland, *The Properties of Glass Surfaces*, Chapman and Hall, London, 1966.
111. G. W. Scherer and P. C. Schultz, *Glass: Science and Technology, Vol. 1, Glass Forming Systems*. Academic Press, Inc., 1983, Chapt. 2, pp. 49–103.
112. *Fiber Opt. Commun. Newsl.* **1**, 8 (Aug. 1978).
113. U.S. Pat. 3,737,292 (June 5, 1973), D. B. Keck and co-workers (to Corning Inc.).
114. R. V. VanDewoestine and A. J. Morrow, *J. of Lightwave Tech.* **LT-4**(8), pp. 1020–1025, (Aug. 1986).
115. D. B. Keck and A. J. Morrow, *Phil. Trans. R. Soc. Lond. A 329*, 71–81 (1989).
116. J. A. Amick, G. L. Schnable, and J. L. Vossen, *J. Vac. Sci. Technol.* **14**, 1053 (1977).
117. I. Strawbridge and P. F. James, in *High Performance Glasses*, Blackie and Sons Ltd., Glasgow, 1992, Chapt. 2, pp. 20–49.
118. C. P. Ross, *Ceram. Ind.* **141**(5), 28–35 (1993).
119. *U.S. Industrial Outlook '92*, U.S. Dept. of Commerce, International Trade Administration, Washington, D.C., pp. 7–11.
120. "Giants in Glass," *Ceram. Ind.* **131**(2), 29 (1988).
121. "Giants in Glass," *Ceram. Ind.* **135**(2), 45 (1990).
122. "Giants in Glass," *Ceram. Ind.* **137**(3), 41 (1991).
123. "Giants in Glass," *Ceram. Ind.* **141**(3), 53 (1993).
124. *Statistical Abstract of the United States 1992*, U.S. Dept. of Commerce, Bureau of the Census, Washington, D.C., 1991, Table 1244, p. 736.
125. *Annual Reports*, Glass Packaging Institute, Washington, D. C.
126. "Business Outlook," *Ceramic Ind.* **141**(6), 14 (1993).
127. Y. Suzuki, *Ceram. Bul.* **70**(2), (1991).
128. *Dictionary of Glass-Making* (English, French, German), International Commission on Glass, Elsevier Scientific Publishing Co., New York, 1983.
129. L. S. O'Bannon, *Dictionary of Ceramic Science and Engineering*, Plenum Press, New York, 1984.
130. W. W. Perkins, ed., *Ceramic Glossary 1984*, The American Ceramic Society, Inc., Columbus, Ohio.

General References

F. V. Tooley, ed., *The Handbook of Glass Manufacture*, Ashlee Publishing Co., New York, 1984.
O. V. Mazurin and co-workers, *Handbook of Glass Data*, Elsevier Science Publishing Co. Inc., New York, 1991.

N. P. Bansal and R. H. Doremus, *Handbook of Glass Properties*, Academic Press, Inc., New York, 1986.
Ibid., Engineered Materials Handbook, Ceramics and Glass, Vol. 4, ASM International, New York, 1991.

DAVID C. BOYD
PAUL S. DANIELSON
DAVID A. THOMPSON
Corning Incorporated

GLASS-CERAMICS

Glass-ceramics are polycrystalline materials formed by the controlled crystallization of glass. Most commercial glass-ceramic products are formed by highly automated glass-forming processes and converted to a crystalline product by the proper heat treatment. Glass-ceramics can also be prepared via powder processing methods in which glass frits are sintered and crystallized. The range of potential glass-ceramic compositions is therefore extremely broad, requiring only the ability to form a glass and control its crystallization.

Glass-ceramics can provide significant advantages over conventional glass or ceramic materials, by combining the ease and flexibility of forming and inspection of glass with improved and often unique physical properties in the glass-ceramic. They possess highly uniform microstructures, with crystal sizes on the order of 10 micrometers or less; this homogeneity ensures that their physical properties are highly reproducible.

Unlike conventional ceramic materials, glass-ceramics are fully densified with zero porosity. They generally are at least 50% crystalline by volume and often are greater than 90% crystalline. Other types of glass-based materials that possess low amounts of crystallinity, such as opals and ruby glasses, are classified as glasses and are discussed elsewhere (see GLASS).

Glassmakers have traditionally aimed to formulate highly stable glass compositions that are resistant to devitrification in order to minimize the crystal growth rate in the working range. However, it was reasoned that carefully controlled crystallization might provide glass with greatly improved mechanical properties. In an early attempt (1) to crystallize soda-lime glasses only surface crystallization similar to that observed in popsicles was achieved, wherein the crystals grow in a columnar fashion from the surfaces, resulting in deformation and poor strength.

The development of the principles of nucleation and growth early in the twentieth century (2) ultimately led to the discovery that certain nucleating agents can induce a glass to crystallize with a fine-grained, highly uniform microstructure that offers unique physical properties (3). The first commercial glass-ceramic products were missile nose cones and cookware.

More than $500 million in glass-ceramic products are sold yearly worldwide. These range from transparent, zero-expansion materials with excellent optical properties and thermal shock resistance to jade-like highly crystalline materials with excellent strength and toughness. The highest volume is in cookware and tableware, architectural cladding, and stovetops and stove windows. Glass-ceramics are also referred to as Pyrocerams, vitrocerams, devitrocerams, sitalls, slagceramics, melt-formed ceramics, and devitrifying frits.

Processing

Glass-ceramic articles can be fabricated by means of either bulk or powder processing methods. Both methods begin with melting a glass of the desired composition.

Bulk Glass-Ceramic Processing. In this most common method of glass-ceramic manufacture, articles are melted and fabricated to shape in the glass state. Most forming methods may be employed, including rolling, pressing, spinning, casting, and blowing. The article is then crystallized using a heat treatment designed for that material. This process, known as ceramming, typically consists of a low temperature hold to induce nucleation, followed by one or more higher temperature holds to promote crystallization and growth of the primary phase. Because crystallization occurs at high viscosity, article shapes are typically preserved with little or no shrinkage (1–3%) or deformation during the ceramming.

Nucleation and Crystallization. Nucleation commonly begins with phase separation, whereby an amorphous, homogeneous glass unmixes into two immiscible phases of different compositions. Although some glass compositions are self-nucleating, more commonly certain nucleating agents are added to the batch to promote phase separation and internal nucleation. These melt homogeneously into the glass, but promote very fine-scale phase separation on reheating. The dispersed phase, which can be a metal, titanate, zirconate, phosphate, sulfide, or halide, is structurally incompatible with the host glass and is normally highly unstable as a glass. It therefore precipitates tiny crystalline nuclei on heating at temperatures about 30–100°C above the annealing point of the host glass. These crystals serve as the sites for subsequent nucleation of the primary crystalline phases.

Nucleation is followed by one or more higher temperature treatments to promote crystallization and development of the desired microstructure. During this stage of the heat treatment, typically at 750 to 1150°C, the primary crystalline phases nucleate and grow on these nuclei until they impinge on neighboring crystals. Given a high nucleation density, the resulting microstructure is highly uniform, consisting of fine-grained, randomly oriented crystals in a matrix of minor residual glass. Depending on composition and heat treatment, crystal size ranges from less than 0.1 μm to upward of 10 μm.

A typical heat treatment cycle, as illustrated in Figure 1, comprises both nucleation and crystallization temperature holds, but some glass-ceramics are

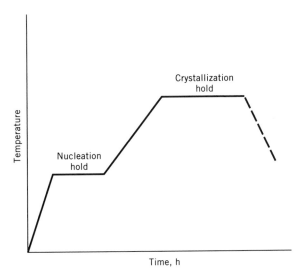

Fig. 1.   Heat treatment cycle for a glass-ceramic material.

designed to nucleate and/or crystallize during the ramp itself, eliminating the
need for multiple holds.

Powder Glass-Ceramic Processing.   The manufacture of glass-ceramics
from powdered glass, using conventional ceramic processes such as spraying, slip-
casting, or extrusion, extends the range of possible glass-ceramic compositions by
taking advantage of surface crystallization. In these materials, the relict surfaces
of the glass grains serve as the nucleating sites for the crystal phases. The glass
composition and processing conditions are chosen such that the glass softens prior
to crystallization and undergoes viscous sintering to full density just before the
crystallization process is completed. Given these conditions, the final crystalline
microstructure is essentially the same as that produced from the bulk process.

The precursor glass powders may be produced by various methods, the simplest being the milling of quenched glass to an average particle size of 3–15 μm.
Sol gel processes, in which highly uniform, ultrafine amorphous particles are
grown in a chemical solution, may be preferable for certain applications.

Such so-called devitrifying frits are employed extensively as sealing frits for
bonding glasses, ceramics, and metals. Other applications include cofired multi-
layer substrates for electronic packaging (4), matrices for fiber-reinforced composite materials (5), refractory cements and corrosion-resistant coatings, and honeycomb structures in heat exchangers. These products take advantage of the
complete densification, thermal stability, and wide range of physical properties
possible in glass-ceramic systems, and provide a number of advantages over conventional glass frits. Products incorporating glass-ceramic seals, for example, can
be reheated to their soldering temperature without deformation.

Secondary Processing (Strengthening).   Because their polycrystalline
microstructures provide resistance to crack propagation, glass-ceramics possess
mechanical strength inherently superior to that of glass. This strength may be
augmented by a number of techniques which impart a thin surface compressive

stress to the body. These techniques induce a differential surface volume or expansion mismatch by means of ion-exchange, differential densification during crystallization, or by employing a lower expansion surface glaze.

Design

There are three key variables in the design of a glass-ceramic: the glass composition, the glass-ceramic phase assemblage, and the nature of the crystalline microstructure.

The glass composition controls much of the workability of the material, including glass viscosity as well as the effectiveness of nucleation and rapidity of crystallization. A glass-ceramic base glass must meet the seemingly contradictory requirements of resistance to uncontrolled devitrification during melting and forming, while simultaneously being amenable to high nucleation density and homogeneous crystallization during subsequent heat treatment.

The glass-ceramic phase assemblage, ie, the types of crystals and the proportion of crystals to glass, is responsible for many of the physical and chemical properties, such as thermal and electrical characteristics, chemical durability, elastic modulus, and hardness. In many cases these properties are additive; for example, a phase assemblage comprising high and low expansion crystals has a bulk thermal expansion proportional to the amounts of each of these crystals.

Finally, the nature of the crystalline microstructure, ie, crystal size and morphology and the textural relationship among the crystals and glass, is the key to many mechanical and optical properties, including transparency/opacity, strength and fracture toughness, and machinability. These microstructures can be quite complex and often are distinct from conventional ceramic microstructures (6).

Properties

Given the three key structural variables discussed under design, glass-ceramics can be engineered to provide a broad range of physical properties.

Thermal Properties. Many commercial glass-ceramics have capitalized on their superior thermal properties, particularly low or zero thermal expansion coupled with high thermal stability and thermal shock resistance: properties that are not readily achievable in glasses or ceramics. Linear thermal expansion coefficients ranging from -60 to $200 \times 10^{-7}/°C$ can be obtained. Near-zero expansion materials are used in applications such as telescope mirror blanks, cookware, and stove cooktops, while high expansion frits are used for sealing metals.

Glass-ceramics have high temperature resistance intermediate between that of glass and of ceramics; this property depends most on the composition and amount of the residual glass in the material. Generally, glass-ceramics can operate for extended periods at temperatures of 700 to over 1200°C. Thermal conductivities of glass-ceramics are similar to those of glass and much lower than those of conventional oxide ceramics, ranging from 0.5 to 5.5 W/(m·K).

Mechanical Properties. Like glass and ceramics, glass-ceramics are brittle materials which exhibit elastic behavior up to the strain that yields breakage. Because of the nature of the crystalline microstructure, however, strength, elasticity, toughness (resistance to fracture propagation), and abrasion resistance are higher in glass-ceramics than in glass. The modulus of elasticity ranges from 80–140 GPa (12–20×10^6 psi) in glass-ceramics, compared with about 70 GPa (10^7 psi) in glass. Abraded modulus of rupture values in glass-ceramics range from 50 to 300 MPa (7250–43,500 psi), compared with 40 to 70 MPa (5800–10,000 psi) in glass. Strengths can be increased further by employing a compressive layer. Fracture toughness values range from 1.5 to 5.0 MPa$\sqrt{m}$ in glass-ceramics, compared with less than 1.5 MPa$\sqrt{m}$ in glass. Knoop hardness (qv) values of up to 900 can be obtained in glass-ceramics containing particularly hard crystals such as sapphirine.

Optical Properties. Glass-ceramics may be either opaque or transparent. The degree of transparency is a function of crystal size and birefringence, and of the difference in refractive index between the crystals and the residual glass (7,8). When the crystals are much smaller than the wavelength of light, as in some mullite and spinel glass-ceramics, or when the crystals have low birefringence and the index of refraction is closely matched, as in some Mg-stuffed β-quartz glass-ceramics, excellent transparency can be achieved.

Certain glass-ceramic materials also exhibit potentially useful electro-optic effects. These include glasses with microcrystallites of Cd-sulfoselenides, which show a strong nonlinear response to an electric field (9), as well as glass-ceramics based on ferroelectric perovskite crystals such as niobates, titanates, or zirconates (10–12). Such crystals permit electric control of scattering and other optical properties.

Chemical Properties. The chemical durability is a function of the durability of the crystals and the residual glass. Generally, highly siliceous glass-ceramics with low alkali residual glasses, such as glass-ceramics based on β-quartz and β-spodumene, have excellent chemical durability and corrosion resistance similar to that obtained in borosilicate glasses.

Electrical Properties. The dielectric properties of glass-ceramics strongly depend on the nature of the crystal phase and on the amount and composition of the residual glass. In general, glass-ceramics have such high resistivities that they are used as insulators. Even in relatively high alkali glass-ceramics, alkali migration is limited, particularly at low temperatures, because the ions are either incorporated in the crystal phase or they reside in isolated pockets of residual glass. Loss factors are low, typically 0.01–0.02 at 1 MHz and 20°C. The fine-grained, homogeneous, nonporous nature of glass-ceramics also gives them high dielectric breakthrough strengths, especially compared with ceramics, allowing them to be used as high voltage insulators or condensers.

Most glass-ceramics have low dielectric constants, typically 6–7 at 1 MHz and 20°C. Glass-ceramics comprised primarily of network formers can have dielectric constants as low as 4, with even lower values (K<3) possible in microporous glass-ceramics (13). On the other hand, very high dielectric constants (over 1000) can be obtained from relatively depolymerized glasses with crystals of high dielectric constant, such as lead or alkaline earth titanate (11,14).

Glass-Ceramic Families

All commercial as well as most experimental glass-ceramics are based on silicate bulk glass compositions. Glass-ceramics can be further classified by the composition of their primary crystalline phases, which may consist of silicates, oxides, phosphates, or borates.

Glass-Ceramics Based on Silicate Crystals.   The principal commercial glass-ceramics fall into this category. These can be grouped by composition, simple silicates, fluorosilicates, and aluminosilicates, and by the crystal structures of these phases.

Simple silicates and fluorosilicates are composed primarily of alkali and alkaline-earth silicate or fluorosilicate crystals, whose crystal structures are based on single or multiple chains of silica tetrahedra or on two-dimensional hexagonal arrays (layers) of silica and alumina tetrahedra. Their crystal morphologies tend to reflect the anisotropy of their structures, with chain silicates typically occurring as blades or rods, and layer silicates occurring as plates. With the exception of the lithium silicate glass-ceramics, these materials are valued most for their mechanical properties. Their microstructures of randomly oriented, highly anisotropic crystals typically give these glass-ceramics superior strength and toughness, for in order for a fracture to propagate through the material, it generally will be deflected and blunted as it follows a tortuous path around or through cleavage planes of each crystal. Indeed, glass-ceramics based on chain silicate crystals have the highest toughness and body strength of any glass-ceramics.

Simple Silicates.   The most important simple silicate glass-ceramics are based on lithium metasilicate [*10102-24-6*], Li_2SiO_3; lithium disilicate [*13568-46-2*], $Li_2Si_2O_5$; enstatite [*14681-78-8*], $MgSiO_3$; diopside [*14483-19-3*], $CaMgSi_2O_6$; and wollastonite [*14567-57-2*], $CaSiO_3$. A number of silicate glass-ceramic compositions are given in Table 1.

There are two groups of commercially important lithium silicate glass-ceramics. One group, which is nucleated with P_2O_5, yields high expansion lithium disilicate glass-ceramics which match the thermal expansion of several nickel-based superalloys and are used in a variety of high strength hermetic seals, connectors, and feedthroughs (15).

The second group is nucleated with colloidal silver, which in turn is photosensitively nucleated. By suitably masking the glass and then irradiating with uv light, it is possible to nucleate and crystallize only selected areas. Although this process can be used to make decorative products, an even more interesting group of products takes advantage of the large difference in chemical durability between the crystals and the residual glass. The crystallized portion consists of dendritic lithium metasilicate crystals, which are much more soluble in hydrofluoric acid than is the glass. These crystals can thus be etched away, leaving the uncrystallized portion intact. The resulting photoetched glass can then be flood-exposed to uv rays and heat-treated at higher temperature, producing the stable lithium disilicate phase. The resulting glass-ceramic is strong ($\sim$140 MPa = 20,000 psi), tough, and faithfully replicates the original photoetched pattern. These chemically machined materials have been used as fluidic devices, cellular display screens, lens arrays, magnetic recording head pads, and charged plates for ink jet printing.

Table 1. Silicate Glass-Ceramic Compositions, Wt %

Oxide	Fotoform/Fotoceram[a,b] Corning 8603	Enstatite Corning[c]	Slagsitalls[d,e] Slagsitall White	Minelbite Gray
SiO_2	79.6	58.0	55.5	60.9
Li_2O	9.3	0.9		
MgO		25.0	2.2	5.7
CaO			24.8	9.0
Al_2O_3	4.0	5.4	8.3	14.2
Na_2O	1.6		5.4	3.2
K_2O	4.1		0.6	1.9
ZnO			1.4	
MnO			0.9	2.0
Fe_2O_3			0.3	2.5
crystal phases	lithium metasilicate lithium disilicate	enstatite β-spodumene tetragonal zirconia	wollastonite	diopside
abraded MOR,[f] MPa		193 ± 15	65–100	80–120

[a]Also Ag, 0.11; Au, 0.001; CeO_2, 0.014; SnO_2, 0.003; and Sb_2O_3, 0.4.
[b]Commercial applications include substrates, fluidic devices, fine mesh screens, and magnetic head pads.
[c]Also ZrO_2, 10.7.
[d]White is from Russia, gray from Hungary. White has 0.4 and gray 0.6 wt % S.
[e]Used as cladding and in industrial products.
[f]MOR = modulus of rupture; to convert MPa to psi, multiply by 145.

No commercial applications have yet been found for glass-ceramics based on the chain silicate enstatite, $MgSiO_3$, but these materials are interesting because the phase undergoes a martensitic transformation on cooling. This in turn provides an additional toughening mechanism as fracture energy is absorbed by fine lamellar twinning (Fig. 2). Highly crystalline, fine-grained enstatite glass-ceramics can be produced in the SiO_2–MgO–ZrO_2 and SiO_2–MgO–Al_2O_3–Li_2O–ZrO_2 systems (16). Fracture toughness values as high as 5 $MPa\sqrt{m}$ and use temperatures approaching 1525°C can be obtained in these materials.

Blast furnace slags, with added sand and clay, have been used in Eastern Europe for over 25 years to manufacture inexpensive nonalkaline glass-ceramics called slagsitall (17). The primary crystalline phases are wollastonite ($CaSiO_3$) and diopside ($CaMgSi_2O_6$) in a matrix of aluminosilicate glass. Metal sulfide particles serve as nucleating agents. The chief attributes of these materials are high hardness, good to excellent wear and corrosion resistance, and low cost. The relatively high residual glass levels of these materials (typically >30%), coupled with comparatively equiaxial (less rod-shaped) crystals, confers only moderately high mechanical strengths of ~100 MPa (14,500 psi). Slagsitall materials have found wide use in the construction, chemical, and petrochemical industries. Applications include abrasion- and chemical-resistant floor and wall tiles, industrial machinery parts, chimneys, plungers, parts for chemical pumps and reactors, grinding me-

Fig. 2. Replica electron micrograph of the fracture surface of enstatite–β-spodumene–zirconia glass-ceramic, showing twinning in the enstatite grains (white bar = 1 μm).

dia, and coatings for electrolysis baths. These materials presently constitute the largest volume applications for crystallized glass.

More recently, attractive translucent architectural panels of wollastonite glass-ceramics have been manufactured by Nippon Electric Glass and sold under the trade name Neopariés. A sintered glass-ceramic with about 40% crystallinity, this material can be manufactured in flat or bent shapes by molding during heat treatment. It has a texture similar to that of marble, but with greater strength and durability than granite or marble. Neopariés is used as a construction material for flooring and exterior and interior cladding.

Fluorosilicates.   Compared to the simple silicates, these crystals have more complex chain and sheet structures. Examples from nature include hydrous micas and amphiboles, including hornblende and nephrite jade. In glass-ceramics, fluorine replaces the hydroxyl ion; fluorine is much easier to incorporate in glass and also makes the crystals more refractory. Four commercial fluorosilicate glass-ceramic compositions and their properties are listed in Table 2.

Sheet Fluorosilicates.   Machinable glass-ceramics based on sheet silicates of the fluorine–mica family have unique microstructures composed of interlocked platy mica crystals. Because micas can be easily delaminated along their cleavage planes, fractures propagate readily along these planes but not along other crystallographic planes. The random intersections of the crystals in the glass-ceramic, therefore, cause crack branching, deflection, and blunting, thereby arresting crack growth. This provides the material with high intrinsic mechanical strength, but,

Table 2. Fluorosilicate Glass-Ceramic Compositions and Properties

Components, wt %	Macor (Corning 9658)	Dicor (Dentsply)	Pyroceram (Corning)	Code 9634 (Corning)
SiO_2	47.2	56–64	67.3	54–62
Al_2O_3	16.7	0–2	1.8	1–4
B_2O_3	8.5			
MgO	14.5	15–20	14.3	0–2
CaO			4.7	17–25
Na_2O			3.0	6–10
K_2O	9.5	12–18	4.8	6–12
Li_2O			0.8	
BaO			0.3	
F	6.3	4–9	3.5	4–8
P_2O_5			1.0	
ZrO_2		0–5		
CeO_2		0.05		
crystal phases	F-phlogopite	tetrasilicic fluormica	F–K–rich-terite cristobalite	canasite
abraded MOR,[a] MPa^b	100	~150	150–200	250–300
fracture toughness, $MPa\sqrt{m}$			3.2	4–5
Young's modulus, GPa^b	65		95	80
hardness, $KNH_{100}{}^c$	250	~360		500
coefficient of thermal expansion	12.9×10^{-6} (25–600°C)	7.2×10^{-6} (25–600°C)	11.5×10^{-6} (0–300°C)	12.5×10^{-6} (0–300°C)
commercial applications	machinable components	dental restorations	tableware	rigid disk substrates

[a]MOR = modulus of rupture.
[b]To convert MPa to psi, multiply by 145.
[c]Knoop Hardness (qv).

in addition, the combination of ease of fracture initiation with almost immediate fracture arrest enables these glass-ceramics to be readily machined.

The commercial glass-ceramic Macor (Corning Code 9658), based on a fluorophlogopite [12003-38-2], $KMg_3AlSi_3O_{10}F_2$, mica, is capable of being machined to high tolerance (± 0.01 mm) by conventional high speed metal-working tools. By suitably tailoring its composition and nucleation temperature, relatively large mica crystals with high two-dimensional aspect ratios are produced, enhancing the inherent machinability of the material. This "house-of-cards" microstructure is illustrated in Figure 3. In addition to precision machinability, Macor glass-ceramic has high dielectric strength, very low helium permeation rates, and is

Fig. 3. Replica electron micrograph showing "house-of-cards" microstructure in a machinable fluormica glass-ceramic (white bar = 10 μm).

unaffected by radiation or oxygen–acetylene flames. The glass-ceramic has been employed in a wide variety of applications including high vacuum components and hermetic joints, precision dielectric insulators and components, seismograph bobbins, sample holders for field ion microscopes, boundary retainers for the space shuttle, and gamma-ray telescope frames.

Another machinable glass-ceramic, Dicor, has been developed for use as dental restorations (18). Based on a tetrasilicic fluormica, $KMg_{2.5}Si_4O_{10}F_2$, this material has higher strength 150 MPa (~22,000 psi), and improved chemical durability over that of Macor. The glass-ceramic has a hardness, radiographic density, and translucency closely matching natural tooth enamel. The desired translucency is achieved by maintaining a fine-grained (~1 μm) crystal size and by roughly matching the refractive indices of the crystals and glass. Ceria is added to simulate the fluorescent character of natural teeth. The glass-ceramic may be accurately cast using a lost wax technique and conventional dental laboratory molds. The material's high strength, low thermal conductivity, and transparency to x-rays provide working advantages over conventional metal–ceramic systems.

Chain Fluorosilicates.   Interlocking blade- or rod-like crystals can serve as important strengthening or toughening agents, much as fiber glass is used to reinforce polymer matrices. Naturally occurring, massive aggregates of chain silicate amphibole crystals, such as nephrite jade, are well known for their durability and high resistance to impact and abrasion.

Two families of chain fluorosilicate glass-ceramics have found commercial application (16). Glass-ceramics based on the amphibole potassium fluorrichterite ($KNaCaMg_5Si_8O_{22}F_2$) have a microstructure consisting of tightly interlocked, fine-grained, rod-shaped amphibole crystals in a matrix of minor cristobalite, mica, and residual glass. The flexural strength of these materials can be further enhanced (150 to 200 MPa = 22,000 to 29,000 psi) by employing a compressive glaze. Richterite glass-ceramics have good chemical durability, are usable in microwave ovens, and, when glazed, resemble bone china in their gloss and translucency. These glass-ceramics are currently utilized for high performance institutional tableware and as mugs and cups for the Corelle line.

An even stronger and tougher microstructure of interpenetrating acicular crystals, as shown in Figure 4, is obtained in glass-ceramics based on fluorcanasite ($K_{2-3}Na_{4-3}Ca_5Si_{12}O_{30}F_4$) crystals. Cleavage splintering and high thermal expansion anisotropy augment the intrinsic high fracture toughness of the chain silicate microstructure. Highly crystalline canasite glass-ceramics have flexural strengths up to 300 MPa (43,500 psi) and fracture toughness of 5 MPa$\sqrt{m}$. Fluorcanasite glass-ceramics have been developed for use as thin, rugged magnetic memory disk substrates. Other potential applications include architectural materials and consumer ware.

Aluminosilicates. These silicates consist of frameworks of silica and alumina tetrahedra linked at all corners to form three-dimensional networks; familiar examples are the common rock-forming minerals quartz and feldspar. Framework

Fig. 4. Fracture surface (replica micrograph) of fluorcanasite glass-ceramic showing interlocking blade-shaped crystals and effects of cleavage splintering (white bar = 1 μm).

silicates generally form blocky crystals, more isotropic in nature than the crystals discussed previously. The nature of their tetrahedral linkages often results in low thermal expansion coefficients.

Commercial glass-ceramics based on framework structures comprise compositions from the $Li_2O-Al_2O_3-SiO_2$ (LAS) and the $MgO-Al_2O_3-SiO_2$ (MAS) systems. The most important, and among the most widespread commercially, are glass-ceramics based on β-spodumene, cordierite, and the various structural derivatives of high (β-) quartz. These silicates are important because they possess very low bulk thermal expansion characteristics, with the consequent benefits of exceptional thermal stability and thermal shock resistance. Thus materials based on these crystals can suffer large thermal upshock or downshock without experiencing strain that can lead to rupture, a critical property in products like missile nose cones and cookware. Representative aluminosilicate glass-ceramic compositions are given in Table 3.

Glass-ceramics in the LAS system have great commercial value for their very low thermal expansion and excellent chemical durability. These glass-ceramics are based on essentially monophase assemblages of either β-quartz or β-spodumene (keatite) solid solution, with only minor residual glass or accessory phases. Partial substitution of MgO and ZnO for Li_2O improves the working characteristics of the glass while lowering the materials cost. Figure 5 illustrates the range of thermal expansion coefficients of some β-quartz glass-ceramics, compared with those of fused silica and borosilicate glassware (19).

Table 3. Aluminosilicate Glass-Ceramic Compositions,[a] Wt %

Components	Visions[b]	Zerodur[c]	Ceran[c]	Neoceram[d]	Corningware 9608[e]	Code 9606[b,f]
SiO_2	68.8	55.5	63.4	65.1	69.7	56.1
Al_2O_3	19.2	25.3	22.7	22.6	17.8	19.8
Li_2O	2.7	3.7	3.3	4.2	2.8	
MgO	1.8	1.0		0.5	2.6	14.7
ZnO	1.0	1.4	1.3		1.0	
CaO						0.1
BaO	0.8		2.2			
P_2O_5		7.9		1.2		
Na_2O	0.2	0.5	0.7	0.6	0.4	
K_2O	0.1			0.3	0.2	
F				0.1		
Fe_2O_3	0.1	0.03		0.03	0.1	0.1
TiO_2	2.7	2.3	2.7	2.0	4.7	8.9
ZrO_2	1.8	1.9	1.5	2.3	0.1	
As_2O_3	0.8	0.5		1.1	0.6	0.3

[a]Crystal phase is β-quartz $ZrTiO_4$ unless otherwise noted.
[b]Corning.
[c]Schott.
[d]NEG.
[e]Crystal phase is β-spodumene rutile.
[f]Crystal phase is cordierite, rutile, and $MgTi_2O_5$.

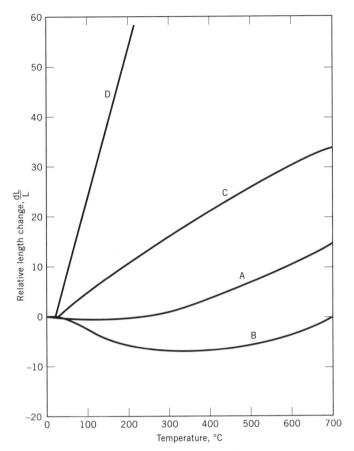

Fig. 5. Relative change in length with temperature of β-quartz glass-ceramics designed for A, precision optics and B, cooktop panels. Curves for C, fused silica glass and D, borosilicate laboratory glassware, are given for comparison (19).

Glass-ceramics containing β-quartz or β-spodumene can be made from glass of the same composition by modifying its heat treatment: β-quartz is formed by ceramming at or below 900°C, and β-spodumene by ceramming above 1000°C.

β-*Quartz Solid Solution.*   A mixture of ZrO_2 and TiO_2 produces highly efficient nucleation of β-quartz [*14808-60-7*], resulting in very small (<100 nm) crystals. This fine crystal size, coupled with low birefringence in the β-quartz phase and closely matched refractive indices in the crystals and residual glass, results in a transparent yet highly crystalline body, as illustrated in Figure 6. The combination of near-zero thermal expansion behavior with transparency, optical polishability, excellent chemical durability, and strength greater than that of glass, has made these glass-ceramics highly suitable for use as telescope mirror blanks (eg, Zerodur), thermally stable platforms, ring laser gyroscopes, infrared-transmitting rangetops (eg, Ceran) woodstove windows (eg, Neoceram) and transparent cookware (eg, Visions).

β-*Spodumene Solid Solution.*   Opaque, low expansion glass-ceramics are obtained by ceramming these LAS materials at temperatures above 1000°C. The

Fig. 6. Microstructure of transparent β-quartz solid solution glass-ceramic as revealed by transmission electron microscopy (white bar = 0.1 μm).

β-quartz to β-spodumene [*1302-37-0*] transformation takes place between 900 and 1000°C and is accompanied by a five- to ten-fold increase in grain size (to 1–2 μm). When TiO_2 is used as the nucleating agent, rutile development accompanies the silicate phase transformation. The combination of larger grain size with the high refractive index and birefringence of rutile gives the glass-ceramic a high degree of opacity. Secondary grain growth is sluggish, giving these materials excellent high temperature dimensional stability. Because of their larger grain size, these glass-ceramics are stronger than those based on β-quartz, and they are also amenable to additional surface strengthening techniques. β-spodumene glass-ceramics, such as Corningware 9608, have found wide utilization as cookware, architectural sheet, benchtops, hot plate tops, valve parts, ball bearings, sealing rings, heat exchangers, and matrices for fiber-reinforced composite materials.

  Cordierite Glass-Ceramics. These glass-ceramics in the MAS system combine high strength and good thermal stability and shock resistance with excellent dielectric properties at microwave frequencies. Corning Inc. produces a silica-rich, cordierite-based glass-ceramic with a complex phase assemblage that includes the high expansion cristobalite phase. The surface of the glass-ceramic can be selectively leached to provide an abrasion resistant layer (about 0.38 mm thick) that inhibits flaw initiation and increases the strength from 120 to 240 MPa (17,400 to 34,800 psi). This material, with its high transparency to radar, is the standard glass-ceramic used for missile radomes (Code 9606).

More recently, sintered glass-ceramics of cordierite [12182-53-5] (with minor clinoenstatite) have been developed by IBM for use as high performance multi-layer ceramic packaging (20). These higher magnesia cordierite compositions provide a number of advantages over conventional alumina packaging: lower dielectric constant and thermal expansion, superior dimensional control, and low coprocessing temperatures compatible with copper. Small additions of B_2O_3 and P_2O_5 enhance sintering and densification prior to crystallization; the higher MgO levels aid sintering and can be adjusted to provide a thermal expansion match for either silicon or GaAs.

Other Aluminosilicates. Transparent mullite glass-ceramics can be produced from modified binary $Al_2O_3–SiO_2$ glasses (21). In these materials, the bulk glass phase separates into tiny alumina-rich droplets in a siliceous matrix. Further heat treatment causes these droplets to crystallize to mullite spherulites less than 0.1 μm in size. When doped with ions such as Cr^{3+}, transparent mullite glass-ceramics can be made to absorb broadly in the visible while fluorescing in the near-ir (22,23), thereby making them potentially useful for luminescent solar collectors.

Glass-Ceramics Based on Nonsilicate Crystals. *Oxides.* Although not widespread commercially, glass-ceramics consisting of various oxide crystals in a matrix of siliceous residual glass offer properties not available with more common silicate crystals. In particular, glass-ceramics based on spinels and perovskites can be quite refractory and can yield useful optical and electrical properties.

Spinels have cubic crystal structures, with the general chemical formula AB_2O_4, where A is a tetrahedrally coordinated, typically divalent metal such as Mg, Zn, Fe, or Mn, and B is an octahedrally coordinated, usually trivalent metal such as Al, Fe, or Cr. Glass-ceramics based on spinel compositions ranging from gahnite ($ZnAl_2O_4$) toward spinel ($MgAl_2O_4$) can be crystallized from glasses in the system $ZnO–MgO–Al_2O_3–SiO_2$, with ZrO_2 or TiO_2 as a nucleating agent (7,24). These glass-ceramics can be made highly transparent, with spinel crystals on the order of 50 nm, and can be Cr-doped for luminescent properties. They possess high thermal stability, maintaining transparency up to 1200°C. Possible applications include high temperature lamp envelopes, solar collector panels, and liquid crystal display screens.

Perovskites have the chemical formula ABO_3, where A is an 8- to 12-coordinated cation such as an alkali or alkaline earth, and B is a small, octahedrally coordinated high valence metal such as Ti, Zr, Nb, or Ta. Glass-ceramics based on perovskite crystals are characterized by their unusual dielectric and electro-optic properties. Examples include highly crystalline niobate glass-ceramics which exhibit nonlinear optical properties (12), as well as titanate and niobate glass-ceramics with very high dielectric constants (11,14).

Phosphates. Many phosphates claim unique material advantages over silicates that make them worth the higher material costs for certain applications. Glass-ceramics containing the calcium orthophosphate apatite, for example, have demonstrated good biocompatibility and, in some cases, even bioactivity (the ability to bond with bone) (25). Recent combinations of fluorapatite with phlogopite mica provide bioactivity as well as machinability and show promise as surgical implants (26).

A number of other phosphate-based glass-ceramic systems have been investigated. Fine-grained glass-ceramics based on crystals isostructural with NZP $(NaZr_2(PO_4)_3)$ can be prepared from certain transition-metal orthophosphate and silicophosphate glasses (27). The extensive solid solution possible in the NZP phase provides for a wide range in thermal expansion, with coefficients ranging from -20 to $60 \times 10^{-7}/°C$.

Glass-ceramics containing BPO_4 can be obtained by heat treating B_2O_3–P_2O_5–SiO_2 glasses having molar ratios between 1:1:1 and 1:1:3 (28). This phase is isostructural with β-cristobalite, a high temperature polymorph of silica. The material can be opaque or transparent depending on the crystallite size. Crystallization of the BPO_4 phase produces a three-fold increase in chemical durability over that of the bulk glass, as the residual glass becomes more and more siliceous and depleted in B_2O_3 and P_2O_5. These glass-ceramics also have exceptional dielectric properties, with d-c resistivity of 10^{16} at 250°C, a loss tangent less than 10^{-3} above 1 KHz and below 200°C, and dielectric constants of 3.8 to 4.5.

Heat treatment of related glasses melted under reducing conditions can yield a unique microfoamed material, or "gas-ceramic" (29). These materials consist of a matrix of BPO_4 glass-ceramic filled with uniformly dispersed 1–10 μm hydrogen-filled bubbles. The hydrogen evolves on ceramming, most likely due to a redox reaction involving phosphite and hydroxyl ions. These materials can have densities as low as 0.5 g/cm³ and dielectric constants as low as 2.

Borates. Little work has been carried out in this area, but some frit-derived glass-ceramics based on alkaline-earth aluminoborate $(RAl_2B_2O_7)$ crystals have been investigated (30). These nonsilicate glass-ceramics can be processed at temperatures below 1000°C and can yield thermal expansion coefficients of less than $10 \times 10^{-7}/°C$.

New and Potential Applications

Although glass-ceramics have been employed for many years as cookware and dinnerware, as cladding and other architectural materials, and in a wide range of industrial components, their future roles are likely to involve entirely new technologies. Glass-ceramics will find numerous applications in the rapidly growing field of optoelectronics. Zero-expansion β-quartz glass-ceramics are used for ring laser gyroscopes, which require a transparent medium of critical dimensional stability. The investigation of unusual optical properties in transparent crystalline materials is still at the early stages of research, but these materials show great promise for applications ranging from optical switches to tunable lasers. The potential coupling of nonlinear optical properties, lasing capability, or luminescence with thermal and dimensional stability promises to yield unique combinations of properties not available in glass or ceramics.

Glass-ceramics will also play a role in the burgeoning field of information storage (qv) and display. Canasite glass-ceramics have been developed for use as disk substrates in hard disk drives; these are thinner, flatter, and more rugged than traditional aluminum alloy substrates, and their precise, uniform surfaces permit higher data storage capacity and improved reliability. Glass-ceramics have also been used for the read/write heads used in disk drives. Other glass-ceramics may find potential use in liquid crystal and electroluminescent displays.

Finally, glass-ceramics will play a key role in the growing arsenal of advanced materials, both alone and in combinations with other materials. For example, glass-ceramic composites incorporating ceramic fibers yield high temperature strengths superior to metal alloys, and also exhibit gradual failure behavior similar to that of metals and plastics, as opposed to the normal catastrophic behavior common to brittle ceramics (5) (see COMPOSITE MATERIALS, CERAMIC-MATRIX). Low dielectric-constant glass-ceramics are projected as the best candidates for high performance multilayer packaging materials. Other potential products include superconducting glass-ceramics, bioceramics for bone implants and prostheses, durable glass-ceramics for waste disposal, and refractory, corrosion-resistant glass-ceramic coatings for superalloys.

BIBLIOGRAPHY

"Slagceram" in *ECT* 2nd ed., Suppl. Vol., pp. 876–889, by S. Klemantaski, British Steel Corp. (Corporate Laboratories); "Glass-Ceramics" in *ECT* 3rd ed., Vol. 11, pp. 881–893, by D. R. Stewart, Owens-Illinois.

1. A. Reaumur, *Mem. Acad. Sci.*, 370 (1739).
2. G. Tammann, *The States of Aggregation*, D. Van Nostrand Co., New York, 1925.
3. S. D. Stookey, *Ind. Eng. Chem.* **51**, 805 (1959).
4. G. Partridge, C. A. Elyard, and M. I. Budd, in M. H. Lewis, ed., *Glasses and Glass-Ceramics*, Chapman and Hall, London, 1989, pp. 226–271.
5. K. M. Prewo, J. J. Brennan, and G. K. Layden, *Am. Ceram. Soc. Bull.* **65**, 305–313 (1986).
6. G. H. Beall, *Ann. Rev. Mater. Sci.* **22**, 91–119 (1992).
7. G. H. Beall and D. A. Duke, *J. Mater. Sci.* **4**, 340–352 (1969).
8. G. Partridge and S. V. Phillips, *Glass Technology* **32**, 82–90 (1991).
9. N. F. Borrelli and co-workers, *J. Appl. Phys.* **61**, 5399–5409 (1987).
10. N. F. Borrelli, *J. Appl. Phys.* **38**, 4243–4247 (1967).
11. A. Herczog, *J. Am. Ceram. Soc.* **47**, 107–115 (1964).
12. N. F. Borrelli and M. M. Layton, *J. Non-Cryst. Sol.* **6**, 197–212 (1971).
13. J. F. MacDowell and G. H. Beall, *Ceram. Trans.* **15**, 259–277 (1990).
14. B. G. Aitken, *16th Int. Cong. Glass* **5**, 33–38 (1992).
15. T. J. Headley and R. E. Loehman, *J. Am. Ceram. Soc.* **67**, 620–625 (1984).
16. G. H. Beall, *J. Non-Cryst. Sol.* **129**, 163–173 (1991).
17. A. I. Berezhnoi, *Glass-Ceramics and Photo-Sitalls*, Plenum Press, New York, 1970.
18. K. A. Malament and D. G. Grossman, *J. Prosthetic Dentistry* **57**, 62 (1987).
19. H. Scheidler and E. Rodek, *Am. Ceram. Soc. Bull.* **68**, 1926–1930 (1989).
20. R. R. Tummala, *J. Am. Ceram. Soc.* **74**, 895–908 (1991).
21. J. F. MacDowell and G. H. Beall, *J. Am. Ceram. Soc.* **52**, 17–25 (1969).
22. U.S. Pat. 4,396,720 (Aug. 2, 1983) G. H. Beall and co-workers (to Corning Glass Works).
23. A. J. Wojtowicz and co-workers, *IEEE J. Quantum Electronics* **24**, 1109 (1988).
24. U.S. Pat. 4,687,750 (Aug. 18, 1987), L. R. Pinckney (to Corning Glass Works).
25. L. L. Hench, *J. Am. Ceram. Soc.* **74**, 1487–1510 (1991).
26. W. Vogel and co-workers, in T. Yamamuro, L. L. Hench, and J. Wilson, eds., *Handbook of Bioactive Ceramics, Vol. 1, Bioactive Glasses and Glass-Ceramics*, CRC Press, Boca Raton, Fla., 1990.

27. B. G. Aitken, *15th Int. Cong. Glass* **3a**, 96–101 (1989).
28. J. F. MacDowell, in Ref. 27, pp. 90–95.
29. U.S. Pat. 4,666,867 (May 19, 1987), G. H. Beall and J. F. MacDowell (to Corning Glass Works).
30. J. F. MacDowell, *J. Am. Ceram. Soc.* **73**, 2287–2292 (1990).

General References

P. W. McMillan, *Glass-Ceramics*, 2nd ed., Academic Press, Inc., New York, 1979.
R. Morrell, *Handbook of Properties of Technical and Engineering Ceramics, Part 1: An Introduction for the Engineer and Designer*, HMSO, London, 1985.
Z. Strnad, *Glass-Ceramic Materials, Glass Science and Technology*, Vol. 8, Elsevier, Amsterdam, 1986.
W. Vogel, *Chemistry of Glass*, The American Ceramic Society, Columbus, Ohio, 1985.
M. H. Lewis, ed., *Glasses and Glass-Ceramics*, Chapman and Hall, London, 1989.

LINDA R. PINCKNEY
Corning Incorporated

GLASSES, ORGANIC–INORGANIC HYBRIDS

Organic glasses are amorphous structures with relatively low glass-transition temperatures, T_g, ie, <400°C. Their applications are limited to additives such as dyes, stains, and processing aids for foods, polymers, metals, etc. High molecular weight organic glasses, ie, polymers, are generally associated with plastics. Typically, organic polymer glasses possess low modulae owing to the low bond energy associated with carbon–carbon bonds. However, a wide variety of properties can be engineered into these materials by variations in processing, additives, and second-phase reinforcement. Inorganic glasses, on the other hand, which are high energy oxide structures that are useful at elevated temperatures (>400°C) have high modulae. Traditionally, a glass can be defined as an inorganic product of fusion which has cooled to a rigid condition without crystallizing. However, this definition has become somewhat obsolete. Low temperature synthetic routes, such as the sol–gel process, have been developed which allow the production of high purity, multicomponent inorganic glasses at temperatures significantly less than those required for traditional fusion, ie, much less than 1000°C.

The search for new high performance materials has spurred the development of composites combining high modulus/high thermal stability inorganic glasses and low modulus/low thermal stability polymeric glasses. Research has resulted in a novel class of amorphous polymer–glass composites referred to as organic–

inorganic hybrids or inorganic–organic hybrids, depending on the component with the highest volume fraction. These materials are synthesized in a variety of ways but ultimately exhibit near-molecular-level mixing of the matrix and the filler. Hence the term hybrid. Typically, this high degree of mixing results in transparent materials which exhibit significant increases in thermomechanical properties owing to extensive interaction between the polymeric and inorganic phases. However, the relatively high volume fraction of polymer included in these materials normally limits their service temperatures to well below 400°C. One route to promoting mixed, interactive phases is the sol–gel processing of metal alkoxides, which allows the development of an inorganic, oxygen-bridging network during composite consolidation and promotes the formation of polymeric inorganic glasses with a morphology, and consequently a resultant polymer–glass interface, that is a function of the traditional sol–gel processing variables and other controllable factors. In essence, this approach results in materials that can be broadly defined as amorphous, interpenetrating network structures.

Numerous combinations of polymers–oligomers and oxide glasses have been mixed in the attempt to create hybrid composites. Similarly, variations such as end capping the polymer with functional groups prior to sol–gel processing and also the *in situ* precipitation of the inorganic glass into an existing polymer film have been tried. It has been demonstrated that variables other than the traditional sol–gel processing variables of the water-to-alkoxy ratio, acid-to-alkoxy ratio, reaction medium, and catalyst exist. For example, the molecular weight of the polymer, its ability to stabilize the developing inorganic network, and the miscibility of the solvent used to dissolve the polymer and the alkoxysilane monomer used are of paramount importance. These new factors control the solubility between the organic and inorganic materials developing networks and hence the degree of mixing between the components, which ultimately governs composite properties. This article addresses material systems, processing considerations, and the properties of these transparent (unless otherwise specified), hybrid organic–inorganic glasses.

Organic–Inorganic Glass Systems

Many polymers and oligomers have been utilized in the synthesis of these hybrids, including poly(alkenes, acrylates, ethers, esters, amides, imides, and dienes). Some novel inorganic precursors have been used, such as clay. The use of metal alkoxides including methoxides, ethoxides, isopropoxides, as well as phenyl and other organically substituted derivatives of silicon, titanium, zirconium, and aluminum, have been reported. The research concerning the sol–gel processing of silica and its gel-to-glass transition far outweighs that done on the other alkoxides. The systems presented more or less follow the chronological development of hybrid amorphous composites.

Polydimethyl Siloxane. The development of organic–inorganic networks began in the early to mid-1980s using polydimethylsiloxane (PDMS) oligomers. This polymer was chosen because of the close structural similarity between the siloxane chains and the expected sol–gel-derived silica. Hence, it is also an ideal organic–inorganic glass. Similar synthetic approaches were employed independently by two groups. One approach focused on improving the mechanical properties of PDMS by using the sol–gel processing of metal alkoxides to generate inorganic "fillers" (silica, titania, alumina, and zirconia) within the PDMS network during or after the development of a cross-linked PDMS network (1–9). In all cases, the hybrids generated by either the simultaneous curing and filling or the precipitation of the inorganic component within an existing swollen network (*in situ* precipitation) showed significantly better reinforcement than the network containing fumed silica, an alternate reinforcing agent. Electron microscopy of the hybrids synthesized using the *in situ* precipitated silica shows that the inorganic phase does not agglomerate into large particles within the original PDMS network (2,4). Rather, most particles fall in the range of 20 to 30 nm and are very finely dispersed. Small-angle x-ray scattering (saxs) experiments of these reinforced networks demonstrated that under strongly basic conditions and large excesses of water, uniformly dense nonfractal particles resulted (6). Conversely, under acidic or neutral conditions, more polymeric structures resulted. The thermal stability of *in situ* filled PDMS networks was also studied using thermogravimetric analysis (tga), and it was demonstrated that this type of reinforcement enhances the thermoxidative stability of ordinary PDMS and fumed silica-reinforced PDMS (9). This reference also provides an excellent review of the different types of sol–gel-derived reinforcement studied (9).

In another approach, PDMS oligomers were used as the polymeric component and sol–gel-derived silica was used as the reinforcing phase. The materials that were developed were termed ceramers to denote the contributions of both the ceramic, inorganic glass and the polymeric, organic glass (10–12). As the name implies, the sol–gel glasses contained more inorganic than organic material. The intent was to engineer glasses with controllable flexibility and optical properties, rather than reinforcement of rubbery systems. This research examined the effect of silanol-terminated PDMS (as a function of wt %) upon the mechanical properties of these sol–gel-derived glasses. The principle means of characterization were mechanical tensile testing, dynamic mechanical spectrometry (dms), and small-angle x-ray scattering. Dms provided new insights, such as the degree of silica-siloxane interaction, into the structure–property behavior of these mixed-phase glasses. The effects of varying the acid (catalyst) concentration employed, the amount of water and tetraethoxysilane [78-10-4] (TEOS) added, and the molecular weight of the PDMS used were reported (10–12). These results indicate that the extent of hydrolysis and condensation the network undergoes is a function of the amount of water added. Therefore, in all likelihood there are residual ethoxy groups that may still be present even after the inorganic–organic network has gelled. Similarly, the simultaneous hydrolysis and condensation of all alkoxysilanes may result in preferential reactions between inorganic TEOS monomer and itself (and the silanol-terminated PDMS and itself), thereby affecting network morphology. Lastly, the degree of solubility between the oligomers, the metal alkoxides, and any solvents added greatly affects morphology and the degree of phase separation present upon final consolidation.

"Rubbery ORMOSILs" have been prepared using PDMS and TEOS (13,14). An ORMOSIL is an organically modified silicate. Traditionally, such materials are comprised of organosilanes which, as the name implies, possess one or more organic pendent groups on the tetrafunctional silicon atom. When the materials react via the sol–gel process, these pendent groups remain in the structure and impart flexibility to the resulting hybrid. These materials have been thoroughly studied and a good review of this work is available (15).

Other PDMS–silica-based hybrids have been reported (16,17) and related to the ceramer hybrids (10–12,17). Using differential scanning calorimetry, dynamic mechanical analysis, and saxs, the microstructure of these PDMS hybrids was determined to be microphase-separated, in that the polysilicate domains (of ca 3 nm in diameter) behave as network cross-link junctions dispersed within the PDMS oligomer-rich phase. The distance between these domains increases from 5.1 to 15.4 nm as the molecular weight of the PDMS chains increases from 850 to 1950, respectively. Phase separation was due to thermodynamic incompatibility between the siloxane backbone, the siloxane end-caps, and the hard segments (urethane–urea linkages) used to cap the PDMS chains with multifunctional alkoxysilanes. Hence, a three-phase system exists. Additionally, hydrogenated polybutadiene (H-PBD) based hybrids were also synthesized and studied. However, these H-PBD based hybrids could be considered two-phase systems because of good miscibility between the polybutadiene backbone and urethane–urea linkages (17).

Poly(tetramethylene oxide). It was hypothesized that the tendency of the PDMS and inorganic glass to phase-separate may be due to the immiscibility of the PDMS with the water added for hydrolysis and generated during condensation (18). Therefore, new polymer systems were sought which have better miscibility with water. Poly(tetramethylene oxide) (PTMO) was chosen for precisely this reason. However, the lack of the siloxane backbone to compatibilize the organic polymer and inorganic glass was a concern. Previously, silanol-terminated PDMS was all that was necessary to react with the *in situ* generated silica. To improve the compatibility of the PTMO with silica, the oligomers were end-capped with isocyanatopropyltriethoxysilane, thereby boosting the polymer functionality to 6. Similarly, it was assumed that the reactivities of the ethoxy groups on the TEOS and the PTMO end caps would be the same, thereby eliminating any preferential condensations, as discussed above in the case of PDMS systems. With these concerns addressed, work was begun on the new hybrids focusing on the variables of oligomer molecular weight and functionality, TEOS content, other inorganic constituents, gel age, acid concentration and type (organic vs inorganic), and processing temperatures and methods (18–33). By using characterization techniques such as saxs, dms, mechanical tensile testing, infrared (ir) spectroscopy and nuclear magnetic resonance (nmr), the structure–property relations of this class of hybrids were determined. In particular, saxs still revealed the presence of microphase-separation within the transparent hybrids. However, the mechanical and dynamic mechanical properties were greatly enhanced, confirming a high degree of mixing, and overall the PTMO-based systems were much better than those of the PDMS hybrids.

It was expected that the mixed region, the diffuse glass–polymer interface, would be responsible for many of the composites properties. Therefore, more recently, methods of modifying the microstructure of these hybrids *in situ* have been examined (34,35). It was found that by using a strongly basic solution of ethylamine and water, the solubility of silica at high pH could be used to selectively modify the interface between the organic and inorganic phases. Essentially, syneresis and ripening of the polysilicate domains just like that experienced for traditional sol–gel-derived inorganic glasses aged at high pH could be induced, resulting in phase separation. This modification resulted in systematic changes in the strength of the hybrid. Additionally, this process has been used in the synthesis of organic–inorganic hybrid interpenetrating networks (IPN). By controlling the degree of interaction (and hence PTMO restriction), the ethylamine procedure was used to control the equilibrium mass uptake of various vinylic monomers absorbed by the original hybrid network. Subsequent γ-radiation-induced polymerization of this absorbed monomer resulted in the formation of an IPN. Indeed, the properties of the new IPN were a function of the preirradiation, ethylamine exposure processing. These new materials were synthesized using methacrylic acid (MAA), N-vinylpyrrolidinone (NVP) and cyclohexyl methacrylate (CHMA) monomers. In particular, the PMAA and poly(N-vinylpyrrolidinone) (PVP) IPN exhibited hydrogel-like behavior. This approach opens new potential applications for these hybrid glasses.

Nitrile Rubber.   Vulcanized rubber sheets of NBR and montmorillonite clay intercalated with Hycar ATBN, a butadiene acrylonitrile copolymer have been synthesized (36). These rubber hybrids show enhanced reinforcement (up to four times as large) relative to both carbon black-reinforced and pure NBR. Additionally, these hybrids are more easily processed than carbon black-filled rubbers.

Nylon-6.   Nylon-6–clay nanometer composites using montmorillonite clay intercalated with 12-aminolauric acid have been produced (37,38). When mixed with ε-caprolactam and polymerized at 100°C for 30 min, a nylon clay–hybrid (NCH) was produced. Transmission electron microscopy (tem) and x-ray diffraction of the NCH confirm both the intercalation and molecular level of mixing between the two phases. The benefits of such materials over ordinary nylon-6 or nonmolecularly mixed, clay-reinforced nylon-6 include increased heat distortion temperature, elastic modulus, tensile strength, and dynamic elastic modulus throughout the − 150 to 250°C temperature range.

Poly(ethylene oxide).   The synthesis and subsequent hydrolysis and condensation of alkoxysilane-terminated macromonomers have been studied (39,40). Using ^{29}Si-nmr and size-exclusion chromatography (sec) the evolution of the silicate structures on the alkoxysilane-terminated poly(ethylene oxide) (PEO) macromonomers of controlled functionality was observed. Also, the effect of vitrification upon the network cross-link density of the developing inorganic–organic hybrid using percolation and mean-field theory was considered.

The successful synthesis of a transparent solid polymer electrolyte (SPE) based on PEO and alkoxysilanes has been reported (41). The material possessed good mechanical properties and high electrical conductivity (around 1.8×10^{-5} S/cm at 25°C) dependent on the organic–inorganic ratio and PEO chain length.

Poly(ethyloxazoline).   A transparent, inorganic–organic glassy material composed of a 50:50 volume ratio of SiO_2 to poly(ethyloxazoline) has been synthesized (42). The two components are not covalently bonded, as is usually the intent of other researchers. The report cites no evidence of any second-phase behavior using saxs, differential scanning calorimetry (dsc), tem, and swelling in water and THF. This morphology is expected to give rise to composites with different properties than those hybrids utilizing covalent bonding.

Poly(2-methyl-2-oxazoline).   Polymers containing strong electron-donor groups can be incorporated and well-dispersed into an evolving inorganic glass (43,44). These hybrids can function as precursors for porous silica glass of controllable pore size generated by hybrid pyrolysis.

Poly(vinyl acetate).   The dielectric and mechanical spectra of hybrids produced by mixing a poly(vinyl acetate)–THF solution with TEOS, followed by the addition of HCl have been investigated (45). Mixtures were made which were believed to be 0, 5, 10, 15, and 20 wt % SiO_2, respectively. These composites were transparent and Fourier transform infrared spectroscopy (ftir) revealed hydrogen bonding between the silicate network and carbonyl units of the poly(vinyl acetate) (PVAc). No shift in the T_g of the composites from that of the pure PVAc was observed. Similarly, the activation energies were calculated and shown to be independent of SiO_2 loading. However, the breadth of the tan δ-associated T_g relaxations did increase with increased filler. Dynamic mechanical and dielectric studies suggested a "microheterogeneous" environment within the hybrid.

Transparent, homogeneous hybrids using a 50:50 PVAc-to-TEOS mixture and an acid-catalyzed reaction have been produced and characterized by dsc and dms (46). Dsc indicated only a slight increase in the T_g of the hybrid with incorporation of silica. Dynamic mechanical tan δ responses indicate a strong interaction between the organic and inorganic phases and, hence, well-dispersed phases that lead to high modulus rubbery plateaus.

Poly(p-phenylene vinylene).   An organic–inorganic hybrid of poly(p-phenylene vinylene) (PPV) has been made utilizing a very ingenious approach (47). By mixing a sulfonium polyelectrolyte precursor for PPV and tetramethoxysilane (TMOS) together and subsequently polymerizing the organic precursor via thermolysis, a hybrid network was formed. The catalyst required for the polymerization of the TMOS was supplied as a by-product of the thermolysis. The polymer-doped glass exhibited improved optical quality over that of pure PPV. Similarly, initial studies showed that the hybrid has promise as a waveguide material, although the sol–gel processing of the TMOS may result in a shortening of the conjugation length in the PPV. This is important for future nonlinear optical applications.

Poly(N-vinylpyrrolidinone).   Transparent, well-dispersed organic–inorganic hybrids can be produced by the sol–gel processing of a metal alkoxide in the presence of an amide carbonyl group by utilizing hydrogen bonding between the silanol group of silica and the carbonyl (43,44,48). By using atomic-force microscopy and BET analysis, it was demonstrated that the hybrid possesses a very dense microstructure exhibiting little porosity, ie, the silica domains in this material have much less pore volume than pure sol–gel-derived silica. Hydrogen bonding was confirmed by ftir and cross-polarized, magic angle spinning ^{13}C-nmr. A good review of this work as well as of sol–gel-derived silica hybrids and their uses is available (44).

It has also been found that polymers possessing functional groups such as amines and pyridines are soluble in pregelled sol solutions, especially, poly(2-vinylpyridine) and poly(N-vinylpyrrolidinone) (PVP) (49). There, materials were made as part of a study of the synthesis of nonshrinking sol–gel-derived networks (49).

Transparent, homogeneous hybrids using a 50:50 PVP-to-TEOS mixture and an acid-catalyzed reaction have been reported, but only tga data were presented in the way of characterization (46).

Polymethacrylates. The synthesis of poly(allyl methacrylate) (PAMA) homopolymer- and methyl methacrylate copolymer-containing hybrids using group-transfer polymerization has been reported (50). The allyl methacrylate homopolymers were then functionalized by hydrosilylation of the allylic segments using Speier's catalyst. When mixed with tetrafunctional metal alkoxides in the presence of HCl or methanesulfonic acid and water, transparent, hybrid sol–gel glasses were produced over a wide range of TEOS:triethoxysilyl groups. However, when copolymers of the functionalized PAMA and poly(methyl methacrylate) (PMMA) were used, only the methanesulfonic acid produced transparent hybrid glasses. Similarly, lower fractions of the functionalized PAMA, ie, $\approx 10\%$, tended to produce grainy and opaque materials. The lack of a well-defined T_g for the polymethacrylate using dsc suggested good dispersion of the organic and inorganic phases. Titanium alkoxides were also employed, but met with less success.

Poly(methyl methacrylate). Extensive work has been done at the Eastman Kodak Corporate Research Labs toward determining the chemical nature of the organic–inorganic interaction and the thermomechanical properties that result from the noncovalently bonded phases interacting within PMMA hybrid glasses (46,51,52). The ability to produce well-dispersed networks was attributed to hydrogen bonding between the silanol groups and the carbonyl units in the PMMA (46). Hence, the highly hydrated, "open" polysiloxane chains resulting from the acid-catalyzed sol–gel process were shown to form more homogeneous, transparent hybrids which exhibited enhanced mechanical properties beyond the T_g. Similarly, the mechanical properties were also affected by curing time and temperature (52). Transparency was also a function of the temperature of the substrate the films were cast on, eg, $\geq 30^\circ$C resulted in films exhibiting no macrophase separation. Base-catalyzed systems, films cast on cooler substrates and polymers incapable of forming hydrogen bonds produced cloudy, poorly dispersed glasses.

The properties of porous, sol–gel-derived glass impregnated with benzoyl peroxide (BPO)-initiated PMMA have been studied. The effect of silane coupling agents was also evaluated (53). It was found that the density, elastic modulus, modulus of rupture, and the compressive strength of the material decreased as the volume fraction of PMMA increased. Conversely, the refractive index increased as the volume fraction PMMA increased. Methacryloxypropyltrimethoxysilane was used as a coupling agent to improve the interaction between the silica glass and the PMMA. The result was an increase in the modulus of rupture as the amount of coupling agent employed increased.

PMMA-impregnated sol–gel-derived silica gels have also been examined (54). Long-wave uv illumination was employed in addition to benzoyl peroxide for PMMA polymerization. This method prohibited the degradation of the silica xerogel from moisture adsorption and desorption. Overall the material behaved more like bulk PMMA than bulk silica, with the exception of hardness.

Poly(n-butyl methacrylate). Homogeneous, translucent organic–inorganic "alloys" combining poly(n-butyl methacrylate) ($M_n = 75{,}000$) with tetraethyl titanate (TET) and *tert*-butyl titanate (TBT), respectively, have been synthesized

(55). The thermal stability of these networks was studied using tga and it was determined that the thermal onset of significant network degradation could be increased by the addition of the titania phase. It was revealed by dsc that crystallization of the amorphous TiO_2 to anatase, in the TET case, occurred at temperatures above the degradation temperature of the polymer. Mechanical reinforcement was also observed by the addition of the titania.

Poly(arylene ether) Ketone. Transparent hybrids of varying organic contents using triethoxysilane end-capped poly(arylene ether) ketone (PEK) and TEOS have been produced. (56). The T_g dependence on the curing temperature behaved as predicted by the time–temperature transformation behavior for network development (57). Of particular importance was the demonstrated ability of the oligomeric, high T_g PEK to covalently bond to and disperse within the inorganic network, thus producing an optically clear gel. However, as is the case with other hybrids, saxs experiments confirmed the presence of microphase separation. The correlation distance increased as the ratio of PEK:TEOS decreased in a manner similar to that reported for TEOS–PTMO systems.

Polyimide. Hybrid glassy materials have been made using polyamic acid and TEOS (58). The resulting films consist of noncovalently bonded polyimide–silica species. Transparent, organic–inorganic networks result when the wt % silica is less than or equal to 8 (assuming 100% conversion to an oxygen-bridging network). Beyond the 8% limit, only opaque films result. However, all of the 0 to 70 wt % SiO_2-containing films exhibit considerable flexibility.

The process known as transimidization has been employed to functionalize polyimide oligomers, which were subsequently used to produce polyimide–titania hybrids (59). This technique resulted in the successful synthesis of transparent hybrids composed of 18, 37, and 54% titania. The effect of metal alkoxide quantity, as well as the oligomer molecular weight and cure temperature, were evaluated using differential scanning calorimetry (dsc), thermogravimetric analysis (tga) and saxs.

Molecularly mixed composites of montmorillonite clay and polyimide which have a higher resistance to gas permeation and a lower coefficient of thermal expansion than ordinary polyimides have been produced (60). These polyimide hybrids were synthesized using montmorillonite intercalated with the ammonium salt of dodecylamine. When polymerized in the presence of dimethylacetamide and polyamic acid, the resulting dispersion was cast onto glass plates and cured. The cured films were as transparent as polyimide.

Hydrogenated Polybutadiene. The mechanical properties of alkoxysilane-terminated hydrogenated polybutadiene (H-PBD) macromonomers (and PDMS oligomers) of varying molecular weight cross-linked via the inorganic silicate phase induced by the sol–gel process have been studied (17). Using a point-source x-ray apparatus, saxs investigations were made to determine the morphology of these materials. The data, in conjunction with dynamic mechanical analysis and dsc measurements, allowed development of a qualitative morphological model for this new hybrid system, which the researchers compare to the hybrids synthesized using PDMS oligomers. In particular, the two-phase nature of this hybrid system was demonstrated. Related work is discussed in the section on poly(ethylene oxide) hybrids (40).

Sodium Poly(4-styrene sulfonate).   The sol–gel processing of TMOS in the presence of sodium poly-4-styrene sulfonate (NaPSS) has been used to synthesize inorganic–organic amorphous complexes (61). These sodium silicate materials were then isothermally crystallized. The processing pH, with respect to the isoelectric point of amorphous silica, was shown to influence the morphology of the initial gel structures. Using x-ray diffraction, the crystallization temperatures were monitored and were found to depend on these initial microstructures. This was explained in terms of the electrostatic interaction between the evolving silicate structures and the NaPSS prior to heat treatment at elevated temperatures.

Applications

There are several areas of application for organic–inorganic hybrids.

Microelectronic Usage.   Organically modified ceramics (ORMOCERS) have been developed based on the sol–gel processing of phenyl and vinyl-substituted organosilanes and silica (62). The systems can be cured either thermally or using photopolymerization of the vinyl groups. The properties of these coatings include excellent electrical properties, such as low ϵ and high surface and bulk resistance, even after weathering. Additionally, the materials are stable up to 260°C.

Several coating compositions based on organically substituted silicon and aluminum alkoxides, eg, vinyltrimethoxysilane, aluminum tri-*sec*-butylate, etc, have been developed. After mixing these solutions with photoinitiators, the resulting solutions are spin-coated onto various substrates and can undergo photocuring to produce patterned microelectronic devices (63). Polymerization can be induced using either a high wattage uv light or a frequency-doubled argon laser.

Abrasion-Resistant Coatings.   Melamine, tris(*m*-aminophenyl)phosphineoxide (TAPO), diethylenetriamine (DETA), polyethyleneimine (PEI), 4,4-diaminodiphenylsulfone (DDS) and bis(3-aminophenoxy-4-phenyl) phosphine oxide (BAPPO), epoxy resins and other oligomers have all been functionalized using a triethoxysilane coupling agent to produce molecules which can react via the sol–gel process (64–66). These functionalized species may or may not be mixed with additional metal alkoxides. Once the sol–gel process is initiated, but before significant gelation occurs, these sols can be spin-coated onto substrates where they are allowed to finish gelling. Additionally, curing at elevated temperatures may also be employed. The resulting coatings are glasses composed of organic moieties or oligomers covalently bonded via siloxane linkages. These coatings have been shown to significantly enhance the abrasion resistance of a Lexan substrate, and do so with increasing effectiveness as the curing temperature increases (64,65). Similarly, when additional metal alkoxides such as titanium or zirconium are employed in conjunction with the functionalized moieties, significant increases in the refractive index of the coatings result (64).

The ability of organically modified ceramics based on alumina, zirconia, titania, or silica (and mixtures of each) to function as abrasion-resistant coatings has also been studied (62). For example, polycarbonate, when coated with an epoxy–aluminosilicate system, experiences a significant reduction in the degree of hazing induced by an abrader, as compared to uncoated polycarbonate.

Thermal–Oxidative-Resistant Coatings. The thermal stability of coatings produced by either covalently or noncovalently incorporating 2,4-dinitroaniline into an inorganic silicate network and coating it onto a sapphire substrate has been examined (67). Although some increase in the thermal stability of the chromophore was observed using uv-vis spectroscopy, the authors conclude that this sol–gel method of chromophore encapsulation does not provide any real thermal or oxidative protection in either the covalently or noncovalently bonded state.

Nonlinear Optical Devices. A transparent, optically active, sol–gel-derived organic–inorganic glass has been synthesized (68). This hybrid consists of a 2,4-dinitroaminophenylpropyl-triethoxysilane covalently bound to a silicon alkoxide-derived silica network. This hybrid exhibits a strong electric field-induced second harmonic signal and showed no signs of crystallization.

Hydrogels. A new approach has been taken to produce organic–inorganic hybrid interpenetrating networks using the sol–gel process, as described previously for PTMO, and γ-radiation (69–70). By swelling existing organic–inorganic networks with either methacrylic acid or N-vinylpyrrolidinone and subsequent γ-polymerization of the monomers in situ, IPNs which exhibit some degree of hydrogel behavior, ie, the ability to absorb significant amounts of moisture or preferential swelling dependent upon pH, are produced, and they possess the optical transparency of the original gels. Additionally, very significant increases in the strength (up to a factor of 50) and elongation at break, as well as the elastic and dynamic storage modulus were induced.

Biosensors. Novel biosensors have been created by incorporating several types of biochemically active proteins into the pore structures of sol–gel silica to produce transparent, colored, organic–inorganic glasses (71). The porous structure of the silica matrix enables the transport of small molecules into and out of the sensor at reasonable rates. Additionally, these materials have proved remarkably stable in that the proteins remain within the pores. Using optical spectroscopy (and changes in the color of the glass) to monitor the changes in the protein structures, the ability to remove the copper ion from bovine copper–zinc superoxide dismutase in situ by exposing the hybrid to a solution of the chelating reagent EDTA has been shown. Similarly, once removed, the ion can be "reinstalled" and the original color and optical spectrum returned by exposing the material to a solution containing the original metal ion. Also, it was possible to reduce horse heart myoglobin (Mb) to deoxy Mb and subsequently react this with dioxygen or carbon monoxide to make either oxy Mb or carbonyl Mb, respectively. This work has shown that these encapsulated proteins are capable of reversible reactions within the glassy matrix. Similarly, such materials are ideally suited for biosensors.

Cross-Linking Agents. A review article addresses the use of polyfunctional organosilanes as cross-linking agents for a variety of polyolefins (72). Most techniques employ a grafting initiator, usually an organic peroxide, which is mixed with the polyolefin prior to extrusion. Upon heating, the initiator thermally decomposes to free radicals which abstract hydrogen from the polyolefin backbone, thereby encouraging grafting of the organosilane onto the backbone. Subsequent hydrolysis and condensation of the alkoxysilanes, with or without a catalyst, after polymer processing and shaping induces cross-linking in the preformed product. These methods have the advantage of not requiring treatment at elevated temperatures (eg, above the crystalline melting point) to induce the cross-linking in the final part, as is the case for peroxide-induced cross-linking. Hence, dimensional stability is maintained throughout the cross-linking operation. Traditionally, primarily peroxide and radiation are used to induce cross-linking. However, numerous patents exist regarding this novel organosilane cross-linking method.

BIBLIOGRAPHY

1. J. E. Mark, C. Y. Jiang, and M. Y. Tang, *Macromolecules* **17**, 2613–2616 (1984).
2. Y. P. Ning, M. Y. Tang, C. Y. Jiang, W. C. Roth, and J. E. Mark, *J. Appl. Polym. Sci.* **29**, 3209–3212 (1984).
3. G. S. Sur and J. E. Mark, *Eur. Polym. J.,* **21**, 1051–1052 (1985).
4. J. E. Mark, Y. P. Ning, C. J. Jiang, and M. Y. Tang, *Polymer,* **26**, 2069–2072 (1985).
5. J. E. Mark and S. B. Wang, *Polym. Bull.* **20**, 443–448 (1988).
6. D. W. Schaefer and co-workers, *Polym. Prep., Am. Chem. Soc. Div. Polym. Chem.* **30**, 102–103 (1989).
7. C. C. Sun and J. E. Mark, *Polymer* **30**, 104–106 (1989).
8. S. B. Wang and J. E. Mark, *Polym. Prep., Am. Chem. Soc. Div. Polym. Chem.* **32**, 523–524 (1991).
9. G. B. Sohoni and J. E. Mark *J. Appl. Polym. Sci.* **45**, 1763–1775 (1992).
10. G. L. Wilkes, B. Orler, and H. H. Huang, *Polym. Prep., Am. Chem. Soc. Div. Polym. Chem.* **26**, 300–302 (1985).
11. H. H. Huang, B. Orler, and G. L. Wilkes, *Polym. Bull.* **14**, 557–564 (1985).
12. H. H. Huang, B. Orler, and G. L. Wilkes, *Macromolecules* **20**, 1322–1330 (1987).
13. Y. J. Chung, S. J. Ting, and J. D. Mackenzie, *Mater. Res. Soc. Symp. Proc.* **180**, 981–986 (1990).
14. Y. Hu and J. D. Mackenzie, *J. Mater. Sci.* **27**, 4415–4420 (1992).
15. H. Schmidt, *Mater. Res. Soc. Symp. Proc.* **32**, 327–334 (1984).
16. S. Kohjiya, O. Kenichiro, and S. J. Yamashita, *Non-Cryst. Solids* **119**, 132–35 (1990).
17. F. Surivet, T. M. Lam, J. P. Pascault and C. Mai, *Macromolecules* **25**, 5742–5751 (1992).
18. H. H. Huang and G. L. Wilkes, *Polym. Prep., Am. Chem. Soc. Div. Polym. Chem.* **28**, 244–245 (1987).
19. H. H. Huang and G. L. Wilkes, *Poly. Bull.* **18**, 455–462 (1987).
20. H. H. Huang, R. H. Glaser, and G. L. Wilkes, in M. Zeldin, K. J. Wynne, and H. R. Allcock, eds., *Inorganic and Organometallic Polymers,* ACS Symposia Series Vol. 360, American Chemical Society, Washington, D.C., 1987, pp. 355–376.
21. A. B. Brennan, H. H. Huang, and G. L. Wilkes, *Polym. Prep., Am. Chem. Soc. Div. Polym. Chem.* **30**, 105–106 (1989).
22. B. Wang, H. H. Huang, A. B. Brennan, and G. L. Wilkes, *Polym. Prep., Am. Chem. Soc. Div. Polym. Chem.* **30**, 146–147 (1989).
23. D. E. Rodrigues and G. L. Wilkes, *Polym. Prep., Am. Chem. Soc. Div. Polym. Chem.* **30**, 227–228 (1989).
24. H. H. Huang, G. L. Wilkes, and J. G. Carlson, *Polymer* **30**, 2001–2012 (1989).
25. G. L. Wilkes, A. B. Brennan, H. H. Huang, D. E. Rodrigues, and B. Wang, *Mater. Res. Soc. Symp. Proc.* **171**, 15–29 (1990).
26. B. Wang, A. B. Brennan, H. H. Huang, G. L. Wilkes, *J. Macromol. Sci. Chem.* **A27**, 1447–1468 (1990).
27. A. B. Brennan, B. Wang, D. E. Rodrigues, and G. L. Wilkes, *J. Inorgan. Organomet. Polym.* **1**, 167–187 (1991).

28. B. Wang and G. L. Wilkes, *J. Polym. Sci. Polym. Chem. Ed.* **29**, 905–909 (1991).
29. A. B. Brennan and G. L. Wilkes, *Polymer* **32**, 733–739 (1991).
30. D. E. Rodrigues and co-workers, *Polym. Prep., Am. Chem. Soc. Div. Polym. Chem.* **32**, 525–527 (1991).
31. D. E. Rodrigues and co-workers, *Chem. Mater.* **4**, 1437–1446 (1992).
32. A. B. Brennan and F. Rabbani, *Polym. Prep., Am. Chem. Soc. Div. Polym. Chem.* **32**, 496 (1991).
33. A. B. Brennan and F. Rabbani, *North American Thermal Analysis Society Conference Proceedings*, Vol. 20, 1991, pp. 164–170.
34. T. M. Miller and A. B. Brennan, *Polym. Prep., Am. Chem. Soc. Div. Polym. Chem.* **34**, 641–642 (1993).
35. A. B. Brennan and T. M. Miller, *Chem. Mater.*, **6**, 262–267 (1994).
36. A. Okada and co-workers, *Polym. Prep., Am. Chem. Soc. Div. Polym. Chem.* **32**, 540–541 (1991).
37. A. Okada and co-workers, *Mat. Res. Soc. Symp. Proc.* **171**, 45–50 (1990).
38. A. Okada, *Polym. Prep., Am. Chem. Soc. Div. Polym. Chem.* **28**, 447–448 (1987).
39. F. Surivet, T. M. Lam, and J. P. Pascault, *J. Polym. Sci. Polym. Chem. Ed.* **29**, 1977–1986 (1991).
40. F. Surivet, T. M. Lam, J. P. Pascault, and Q. T. Pham, *Macromolecules* **25**, 4309–4320 (1992).
41. M. Fujita and K. Honda, *Polym. Comm.* **30**, 200–201 (1989).
42. I. A. David and G. W. Scherer, *Polym. Prep., Am. Chem. Soc. Div. Polym. Chem.* **32**, 530–531 (1991).
43. T. Saegusa, *J. Macromol. Sci. Chem.* **A28**, 817–829 (1991).
44. Y. Chujo and T. Saegusa, *Advances in Polymer Science* **100**, 11–29 (1992).
45. J. J. Fitzgerald, C. J. T. Landry, R. V. Schillace, and J. M. Pochan, *Polym. Prep., Am. Chem. Soc. Div. Polym. Chem.* **32**, 532–533 (1991).
46. C. J. T. Landry and co-workers, *Polymer* **33**, 1496–1506 (1992).
47. C. J. Wung, Y. Pang, P. N. Prasad, and F. E. Karasz, *Polymer* **32**, 605–608 (1991).
48. M. Toki and co-workers, *Polym. Bull.* **29**, 653–660 (1992).
49. B. M. Novak and M. W. Ellsworth, *Mater. Sci. Eng.* **A162**, 257–264 (1993).
50. Y. Wei, R. Bakthavatchalam, D. Yang, and C. K. Whitecar, *Polym. Prep., Am. Chem. Soc. Div. Polym. Chem.* **32**, 503–505 (1991).
51. C. J. T. Landry, and B. K. Coltrain, *Polym. Prep., Am. Chem. Soc. Div. Polym. Chem.* **32**, 514–515 (1991).
52. C. J. T. Landry, and B. K. Coltrain, *Polym. Prep., Am. Chem. Soc. Div. Polym. Chem.* **32**, 514–515 (1991).
53. E. J. A. Pope, M. Asami, and J. D. Mackenzie, *J. Mater. Res.* **4**, 1018–1026 (1989).
54. L. C. Klein and B. Abramoff, *Polym. Prep., Am. Chem. Soc. Div. Polym. Chem.* **32**, 519–520 (1991).
55. K. A. Mauritz and C. K. Jones, *J. Appl. Polym. Sci.* **40**, 1401–1420 (1990).
56. J. L. W. Noell, G. L. Wilkes, D. K. Mohanty, and J. E. McGrath, *J. Appl. Polym. Sci.* **40**, 1177–1194 (1990).
57. J. K. Gillham, *Polym. Eng. and Sci.* **16**, 353–356 (1976).
58. A. Morikawa, Y. Iyoku, M. Kakimoto, and Y. Imai, *Polym. J.* **24**, 107–113 (1992).
59. A. B. Brennan, Ph.D. thesis, Virginia Polytechnic Institute and State University, Blacksburg, Va., 1990.
60. K. Yano, A. Usuki, A. Okada, T. Kurauchi, and O. Kamigaito, *Polym. Prep., Am. Chem. Soc. Div. Polym. Chem.* **32**, 65–66 (1991).
61. K. Nakanishi and N. Soga, *J. Non-Cryst. Solids* **108**, 157–162 (1989).
62. H. Schmidt, and H. Wolter, *J. Non-Cryst. Solids* **121**, 428–435 (1990).

63. M. Popall, H. Meyer, H. Schmidt, and J. Schulz, *Mater. Res. Soc. Symp. Proc.* **180**, 995–1001 (1990).
64. B. Wang and co-workers, *Polym. Prep., Am. Chem. Soc. Div. Polym. Chem.* **32**, 521–522 (1991).
65. C. Betrabet and G. L. Wilkes, *Polym. Prep., Am. Chem. Soc. Div. Polym. Chem.* **32**, 286–287 (1992).
66. B. Tamami, C. Betrabet, and G. L. Wilkes, *Polym. Bull.* **30**, 39–45 (1993).
67. C. L. Schutte, K. W. Williams, and G. M. Whitesides *Polymer* **34**, 2609–2614 (1993).
68. G. Puccetti, E. Toussaere, I. Ledoux, and J. Zyss, *Polym. Prep., Am. Chem. Soc. Div. Polym. Chem.* **32**, 61–62 (1991).
69. A. B. Brennan and R. Vinocur, *North American Thermal Analysis Society Conference Proceedings* Vol. 21, 1992, pp. 105–108.
70. T. M. Miller, R. Vinocur, and A. B. Brennan, *PMSE Prep., Am. Chem. Soc. Div. Polym. Mater.* **69**, 374–375 (1993).
71. L. M. Ellerby, C. R. Nishida, and F. Nishida, *Science* **255**, 1113–1115 (1992).
72. D. Munteanu, in J. E. Sheats, C. E. Carrahar, C. U. Pittman, eds., *Metal Containing Polymeric Systems*, Plenum Press, New York, 1985, pp. 479–509.

A. B. BRENNAN
T. M. MILLER
University of Florida, Gainesville

GLASS, POLYMER-MODIFIED. See GLASSES, ORGANIC–INORGANIC HYBRIDS.

GLASSY METALS

Materials can be classified in several ways, one of which is by atomic structure, ie, whether the material is periodic (crystalline) or nonperiodic (amorphous) in space. Most metals, minerals, and ceramics are crystalline having the atoms arranged in periodic form and having translational symmetry. The bulk of materials that are nonperiodic is composed of various oxide glasses but also includes amorphous or glassy metals. These metallic glasses do not occur naturally but are produced by various techniques, the oldest of which is by cooling molten metals so rapidly that the atoms do not get to form regular crystalline structures. The atoms are frozen in random or nonrepeating atomic patterns similar to those found in organic glasses (see GLASS). By metallic it is meant that the amorphous material is composed primarily, but not necessarily exclusively, of metallic elements that exhibit the properties of metals such as electrical or magnetic behavior.

The first synthesis of a metallic glass drawing wide attention among material scientists occurred in 1960 (1). A liquid gold–silicon alloy, when rapidly quenched to the temperature of liquid nitrogen, was reported to form a glass. Numerous metallic alloys have been amorphized using this melt-quenching technique. These have a wide variety of elements and compositions (2,3). A drawback of the early techniques is that one or more of the dimensions had to be thin, often as thin as 15 μm, making it unlikely that metallic glasses would be used for structural applications. Primarily through modification of the chemistries, glasses have been produced that can truly be considered bulk, even exceeding 1 cm in all dimensions (4). Warm extrusion and consolidation of metallic glass powders (5) produce blocks of glassy metals that can later be machined into components. Besides rapid solidification into thin ribbons or flakes, there exists a wide range of techniques to produce metallic glass. Methods available include chemical means (6), mechanical alloying (7,8), vaporization (9), and solid-state reactions (10).

Interest is maintained in these materials because of the combination of mechanical, corrosion, electric, and magnetic properties. However, it is their ferromagnetic properties that lead to the principal application of glassy metals. The soft magnetic properties and remarkably low coercivity offer tremendous opportunities for this application (see MAGNETIC MATERIALS, BULK; MAGNETIC MATERIALS, THIN FILM).

A limitation of metallic glass is that it exists in metastable form, which means that it tends to crystallize if heated with sufficient thermal energy to allow the kinetics of crystallization, ie, both nucleation and growth, to occur. If glassy metal alloys were all intrinsically unstable, however, they would be much less promising as an engineering material. Understanding the solid-state structure, the unusual mechanical properties, the liquid-like electrical properties, and the ferromagnetic properties of metallic glasses has been the focus of research and development efforts since the late 1950s.

Structure of Metallic Glass

An understanding of glassy metal alloys begins with an understanding of conventionally processed metal alloys. Macroscopically crystalline metals and metallic glasses look and feel the same. Atomistically, under normal conditions, metals and metal alloys are crystalline, ie, the atoms are arranged in a periodically repeated three-dimensional pattern (Fig. 1a). In amorphous materials the atoms are arranged more or less randomly (Fig. 1b), although there may be regions of local order owing to constraints imposed by the close packing of atoms as the melt solidifies. This order, however, is not repeated over large distances in the solid, and in general, these regions are very localized and widely spaced. Various experimental diffraction techniques are customarily used to determine the structure of metals. These include x-ray, neutron, and electron diffraction techniques (see Fig. 1). X-ray diffraction is the most popular and historically most important method in the structural analysis of metallic glasses (see X-RAY TECHNOLOGY).

Diffraction occurs when the Bragg equation is satisfied, resulting in well-known diffraction patterns having sharp intensity peaks. In amorphous materials, the unordered assemblages of atoms interact with radiation in such a way as

Structure   Electron   X-ray

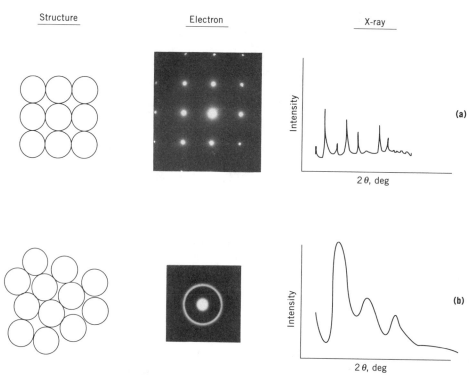

Fig. 1. Structures of (○) atoms and corresponding electron and x-ray diffraction patterns for (**a**) a periodic arrangement exhibiting translational symmetry where the bright dots and sharp peaks prove the periodic symmetry of the atoms by satisfying the Bragg condition, and (**b**) in a metallic glass where the atoms are nonperiodic and have no translational symmetry. The result of this structure is that the diffraction is diffuse.

to produce a dispersed, nondistinctive pattern. Because the range of the core electrons distributed around the nucleus is nearly the same as the x-ray wavelength, the phase of the scattered x-rays depends on the position of the electron corresponding to the origin of scattering in the atoms. Thus the atomic scattering factor of x-rays drastically decreases with an increase in the scattering angle. In contrast, the wavelength of the thermal neutrons used in general diffraction experiments is considerably larger than the size of the nucleus ($\simeq 10^{-12}$ cm); ie, the scattering amplitude is independent of the scattering angle. Whereas electron diffraction as recorded on photographic film has relatively poor quantitative intensity accuracy, this can be improved by employing digitally recorded images in modern electron microscopes (see MICROSCOPY). Advantages of electron diffraction are that the difference in angular dependence is smaller than that for x-ray diffraction. This is important in systems that contain both light and heavy elements, in that the effect of the light elements plays a significant role in scattering intensity in the low scattering angle regions.

   The description of the atomic distribution in noncrystalline materials employs a distribution function, $g(r)$, which corresponds to the probability of finding another atom at a distance r from the origin atom taken as the point $r = 0$. In a system having an average number density $\rho = N/V$, the probability of finding another atom at a distance r from an origin atom corresponds to $\rho_o g(r)$. Whereas the information given by $g(r)$, which is called the pair distribution function, is only one-dimensional, it is quantitative information on the noncrystalline systems and as such is one of the most important pieces of information in the study of non-crystalline materials. The interatomic distances cannot be smaller than the atomic core diameters, so $g(r)_{r\to 0} = 0$. Because the correlation of atomic positions decreases as $r\to\infty$, $g(r)_{r\to\infty} = 1$. The function $4\pi^2 \rho g_o(r)$, the radial distribution function (RDF), may also be used in the discussion of noncrystalline systems. This function corresponds to the number of atoms in the spherical shell between r and $r + dr$.

Because glasses do not possess a crystal structure, atomic positions can only be described on a statistical basis. Assessment of the amorphous structure involves the determination of the radial (or pair) distribution structure and the use of modeling (2,11). Complementary experiments, eg, Mössbauer spectroscopy or nuclear magnetic resonance (12), are useful because the radial distribution function does not provide a total picture of the amorphous structure. The models that are compared with experimental results include the dense random packing (DRP) model (13); an extension of the DRP model obtained by introducing chemical ordering (2,11); and a model based on prism packing of small identifiable units of a stable crystalline structure (14). The models become increasingly sophisticated depending on the experimental detail considered.

The DRP model can be appreciated by considering the packing and kneading of hard balls in a rubber bag. Subsequent analysis of the resulting structure reveals basic structural polyhedra units. Tetrahedral and octahedral units make up the majority of the structure, the ratio of frequency of various polyhedra in the structure, however, has not been agreed on. The model does agree with computed RDF for some metallic glasses but for metal–metalloid systems the smaller atom is sometimes not included. An improvement of this model results when the condition of hard spheres is removed. Use of soft atoms has led to better agreement. Many approximations to the model remain, however, leading to other difficulties.

It is generally accepted that pure metals cannot be quenched from the liquid state into a metallic glass using the cooling rates available as of this writing. Many alloys cannot be turned into glasses either. Alloy phase theory is not developed sufficiently to predict the composition of a metallic alloy that can be quenched into a glass, but there are experimental guidelines (2,15). Systems having a deep eutectic lying in the glass-forming composition region and systems composed of elements undergoing strong interactions, ie, negative free energy of mixing, often form glasses. Many glasses contain both a strongly metallic element including Fe, Co, Ni, Pd, Pt, Cu, and a nonmetallic element or metalloid, eg, B^{3+} forms Fe–B; C, Si, or Ge of valence 4, forms Au–Si for example; or P of valence 5, forms Cu–P. Binary metal alloys include Cu–Zr, Nb–Ni, Ta–Ni, Y–Cu, and Ti–Ni (2). Ternary alloys that can produce glass may include two metals and one metalloid such as Pd–Ni–P, or one metal and two metalloids, such as Fe–P–B. For ternaries, the metallic elements make up to 70–85 atomic % of the material. The remaining atomic % is metalloid (15). A simple list of glass-forming systems includes (1) intertransition metals, (2) transition-metal–metalloid (semimetal), (3) systems based on alkaline-earths, and (4) the actinide–transition-metal system (8). Examples of the various categories of metallic glasses are given in Table 1.

A family of glasses discovered in 1988 (16–19) is based on aluminum, a rare-earth (RE) element, and a transition metal (TM) (see ALUMINUM AND ALUMINUM ALLOYS). Aluminum contents may be as high as 90 atomic %. The empirical rules of predicting metallic glass compositions became even more difficult after this discovery because of the unusual glass formability in these Al-based metallic glasses. For conventional amorphous alloys, the glass-formation range usually coincides with a deep eutectic region where the liquid is stable to a lower temperature than in other regions of the phase diagram. In principle, the degree of supercooling required to form a glassy state is reduced in the eutectic region, and crystallization can be suppressed more easily, thus allowing the formation of metallic glass. A significant difference from the usual glass-forming systems is that liquidus temperatures of the Al–RE binary alloys increase rapidly as minority RE is added to Al; yet the glass formation in $Al_{100-x}RE_x$ has been demonstrated for $x = 8$–16 for Gd, $x = 1$–10 for La or Ce, $x = 9$–13 for Y, and $x = 8$–16 for Sm (see LANTHANIDES). This was totally unexpected (18,19).

Table 1. Metallic Glasses

Alloy	Useful properties/applications
$Fe_{78}B_{13}Si_9$	Metglas 2605S2,[a] good magnetic properties
$Fe_{80}B_{20}$	Metglas 2605[a]
$Pd_{80}Si_{20}$	easy to form glass, thick samples produced
$Pt_{60}Ni_{15}P_{25}$	
$Cu_{84.3}P_{15.7}$ eutectic	brazing foil
$Al_{85}Ni_5Fe_2Gd_8$	high strength low density thick ribbons
$Mg_{80}Cu_{15}Sn_5$	low density cast 4-mm dia rods
$Co_{83}Gd_{17}$	sputtered sample can support magnetic bubbles
$Mo_{64}Re_{16}P_{10}B_{10}$	superconducting below 8.7 K
$W_{60}Ir_{20}B_{20}$	crystallization temperature above 1200 K
$Zr_{41.2}Ti_{13.8}Cu_{12.5}Ni_{10}Be_{22.5}$	14-mm rod produced

[a]Produced by AlliedSignal.

Formability

A molten metal alloy would normally be expected to crystallize into one or several phases. To form an amorphous, ie, glassy metal alloy from the liquid state means that the crystallization step must be avoided during solidification. This can be understood by considering a time–temperature–transformation (TTT) diagram (Fig. 2). Nucleating phases require an incubation time to assemble atoms through a statistical process into the correct crystal structure which is capable of surmounting an activation barrier ΔG^* (Fig. 3). Incubation times can vary from fractions of a second to many seconds. The shape of the TTT curve is in the form of a C because of competing phenomena. As temperature is lowered, the free energy available to nucleate and grow a crystalline phase increases but the kinetic ability

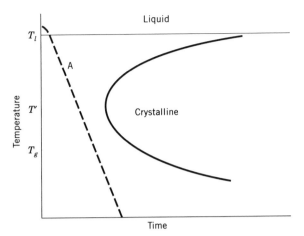

Fig. 2. Time–temperature–transformation(TTT) diagram where A represents the cooling curve necessary to bypass crystallization. The C-shaped curve separates the amorphous solid region from the crystalline solid region. Terms are defined in text.

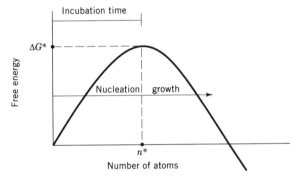

Fig. 3. Curve illustrating the activation energy (barrier) to nucleate a crystalline phase. The critical number of atoms needed to surmount the activation barrier of energy ΔG^* is n^* and takes time equal to the incubation time. One atom beyond n^* and the crystallite is in the growth regime.

to do so through atomic diffusion decreases, resulting in a nose at T' (see CRYSTALLIZATION; ZONE REFINING). The glass-forming ability of a material is determined by the kinetics of this process followed by the initial stages of crystal growth (20). If the liquid alloy is cooled from above the melting temperature to a temperature below the nose of the TTT curve at T' within a time less than the time where crystallization begins (eg, Fig. 2, part A), the alloy's liquid-like structure becomes frozen in when the temperature drops below the glass-forming temperature, T_g, whereby an amorphous solid is obtained. Once the alloy is below T_g, diffusion processes are such that growth of a crystalline embryo (Fig. 3) is essentially halted, and the liquid structure is preserved.

A reduced glass temperature can be defined as $T_{rg} = T_g/T_l$ which represents a measure of glass-forming ability. The higher the T_g and the lower the liquidus temperature, T_l, the easier it is to supercool the metal melt to a glassy state. Conventional theory predicts $T_{rg} = 0.65$–0.7 for good glass formers, eg, $T_{rg} \simeq 0.63$ for $Pd_{77}Cu_6Si_{17}$, and $T_{rg} \simeq 0.67$ for $Pd_{40}Ni_{40}P_{20}$, where a quenching rate as low as 100–1000 K/s is sufficient to produce the glassy state (21–23). However, substantially smaller T_{rg} values were observed for Al–Ni–Fe–Gd metallic glasses. For example T_{rg} for $Al_{85}Ni_5Fe_2Gd_8$ is about 0.44, which is close to the lowest limit of T_{rg} in glass-forming metallic systems previously observed. In chalcogenide alloys such as Ge_xSe_{100-x}, easy glass formation by air or water quenching has been observed in the region where the system exhibits only low (0.4–0.5) T_{rg} (24). However, high (10^9–10^{10} K/s) cooling rates are generally required to melt-quench metallic alloys having such low T_{rg} into amorphous states. Unexpectedly, despite the low T_{rg} for Al–Ni–Fe–Gd alloys, only low cooling rates are necessary for the glass formation. The observed good glass formability combined with a low reduced glass-transition temperature strongly suggests that the formation of metallic glass in Al–TM1–TM2–RE alloys, where TM1 and TM2 represent two different transition metals, is unique, and to some degree is similar to that in chalcogenide systems. This unique glass formability makes it possible to produce thick amorphous ribbons based on aluminum.

The good glass formability of Al-based glasses must ultimately be related to the atomic interactions and the structure of the amorphous state. Neutron and x-ray scattering have been applied to study the atomic structure of amorphous $Al_{90}Fe_xCe_{1-x}$ (25,26). In these amorphous alloys, the distance of an Fe–Al pair is 0.02 nm shorter than the sum of the atomic radii of Al and Fe atoms. Further structural analysis of these data reveals that Fe atoms are surrounded by approximately 10 Al atoms (0.3 nm), and of these, six are in close contact with Fe. Also, the rare-earth atoms in Al–Fe–Ce metallic glasses form a dilute dense random packing substructure, and repel each other, whereas the substructure of Fe atoms and the surrounding Al atoms are substantially different from the random packing. The strong Al–Fe interaction suggested from structural studies is consistent with the large negative volume additivity of the Al–Fe alloys in the liquid or solid state (27). Furthermore, an Al melt actually contracts with a small addition of Fe, and more remarkably the shear viscosity of molten Al is sharply increased by dissolution of small amounts of Fe (28). Based on these analyses, the unusual glass formability in Al–TM–RE may be explained by an increase in the shear viscosity of the molten alloy through Al–TM interactions (29) resulting in a high resistance to nucleation and crystallization. The amorphous arrangement of Al–TM clusters is further stabilized by the randomly distributed rare-earth atoms. This behavior occurs through the entire supercooling process to temperatures lower than glass temperature T_g assuring the glass formation.

Among metallic glasses, only a few alloys have high resistance to nucleation and crystallization during the quenching process; examples are Pd–Cu–Si and Pd–Ni–P alloys which can be water quenched into amorphous rods of 2-mm diameter (30). In this case, the critical cooling rate for amorphous phase formation is reduced to a few hundred degrees per second. It has been demonstrated that upon the elimination of heterogeneous nucleation sites, critical cooling rates as low as a few degrees per second are sufficient to allow the formation of an amorphous phase in $Pd_{40}Ni_{20}P_{20}$ alloy (31). Adding a second transition metal to Al-based glasses greatly enhanced the glass-forming ability and submillimeter thick

Al–Ni–Co–Y (32) and Al–Ni–Fe–Gd (29) amorphous ribbons can be produced at rather low quenching rates. The search for metallic glasses having superb glass-forming ability and high resistance to nucleation and growth was quickly extended to Mg-based ternary alloys containing rare earths where amorphous rods of $Mg_{65}Cu_{25}Y_{10}$ having a diameter of 4 mm could be produced (33). Replacing part of the transition metal in La–Al–TM (34) and Nd–Al–TM (35) alloys by one or more other transition-metal elements drastically enhances the glass formability (36). Using a high pressure die-casting method, the maximum diameter of an amorphous alloy rod in $La_{55}Al_{25}Ni_{10}Cu_5Co_5$ could be increased to 9 mm, and the resulting critical cooling rate (Fig. 2) for glass formation is estimated to be less than 100 K/s. More recently, the formation of amorphous Zr–Ti–Cu–Ni–Be alloy rod having a diameter of up to 14 mm was produced using a water quenching method at a critical cooling rate less than 10 K/s (4). The diversity of atomic species and the large variation in atomic size plays an important role in the amorphous phase formation in these alloys. These results demonstrate that a wide variety of glasses can be produced in bulk form that can then be cast and later machined into useful components, such as small gears (37). The ability to produce bulk metallic glasses gives added confidence that these materials may find structural uses.

Processing

Traditionally, production of metallic glasses requires rapid heat removal from the material (Fig. 2) which normally involves a combination of a cooling process that has a high heat-transfer coefficient at the interface of the liquid and quenching medium, and a thin cross section in at least one-dimension. Besides rapid cooling, a variety of techniques are available to produce metallic glasses. Processes not dependent on rapid solidification include plastic deformation (38), mechanical alloying (7,8), and diffusional transformations (10).

Splat quenching or gun techniques involve rapid solidification through atomizing molten metal by blowing it out a tube. The resulting liquid vapor is quenched by impingement upon a metal substrate having a high thermal coefficient. Development of this simple technique opened the window to a whole new world of novel alloys, ie, metallic glasses and other metastable crystalline and quasicrystalline alloys. Because of the undesireable characteristics of the small size of the resulting material and its irregular thickness, other techniques followed. The piston and anvil (39) and twin pistons (Fig. 4a) are two of these. Splat quenching results in an irregular foil of variable thickness. The approach is suitable for research but not for commercial production.

Melt-spinning can produce large quantities of very uniform ribbons, filaments, or tapes. The product is made in continuous form. Most simply, the metal is melted by induction in a chamber (Fig. 4b) and then expelled onto a rotating wheel of high thermal conductivity, eg, copper. The rapidly formed ribbon is then collected in a catch chamber. The quench rate (10^5–10^8 K/s) determines the thickness and width of the final product which usually can be from 20–300 μm thick and from a few to hundreds of mm wide. For example, increasing wheel speed from 27 to 47 m/s decreases the ribbon thickness from 37 to 22 μm for an

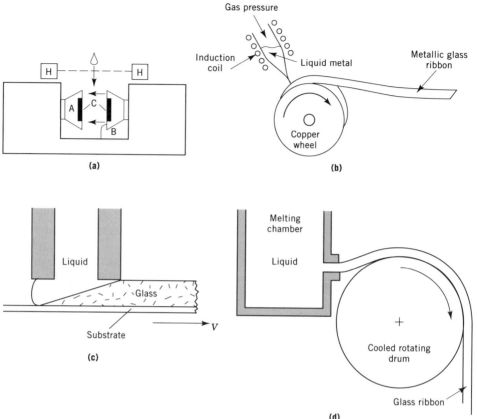

Fig. 4. Schematic diagrams of (**a**) piston-anvil quenching (39), where A is the fixed anvil; B, the fast-moving piston; C, copper disks; and H, photo cells; (**b**) melt-spinning apparatus; (**c**) planar flow casting; and (**d**) melt drag process.

$Fe_{40}Ni_{40}B_{20}$ alloy (40). This technique has the drawback of short contact times between the molten alloy jet and the wheel. This in turn limits the available cooling rate. A more recent development is centrifuge melt spinning (41). Here the alloy is induction melted and centrifugal force impinges the melt onto the inner surface of a copper rim. This technique can produce cooling rates of 10^8 K/s (42) and the ribbon has a more uniform thickness than using conventional melt spinning.

The production of wide (>5 mm) ribbons requires a technique whereby a rectangular melt is forced through a slotted nozzle very close (0.5 mm) to the cooling substrate (Fig. 4**c**) rotating at high speed. This technique is called planar flow casting (43) and is responsible for large-scale production of continuous metallic glass tapes. Ribbons as wide as 300 mm have been produced (44) using this technology.

The melt drag process drags molten metal from an orifice onto a cooled drum (Fig. 4**d**) (45). Ribbons in excess of 20 cm can be produced having thicknesses from 25 to 1000 μm (46). Gravity is used to force the molten liquid from the orifice so that it touches the rotating drum. The partially solidified alloy is then dragged onto the drum forming wire or ribbons.

Lasers (qv) can be used to obtain very fast quench rates up to 10^8 K/s. A thin layer is melted on the surface of a material. The high energy density generated can melt small areas rapidly where the resulting amorphous layer is up to 400 μm in $Pd_{91.7}Cu_{4.2}Si_{5.1}$ (47,48). Techniques of altering the surface chemistry by feeding in elemental or alloy powders on the surface prior to melting also exist using this approach (49). Electron-beam surface melting is a similar technology (see SURFACE AND INTERFACE ANALYSIS).

Metallic glass powders can be made in various sizes through atomization and comminution processes. Atomization can be accomplished by a variety of techniques (50–54) including gas (Fig. 5), gas–liquid (55,56), and ultrasonic gas atomization (57,58). Cooling speeds vary from 10^5–10^9 K/s depending on the approach used and the diameter of the powder produced (59,60). After the powder is sized, it must then be consolidated by other techniques such as warm pressing or extrusion. Using these techniques, crystallization and embrittlement can be difficult to avoid. Dynamic or quick processing has been applied to consolidate the amorphous powders (61,62). Atomization of $Al_{85}Ni_{10}Mm_5$, where Mm is misch metal (see CERIUM AND CERIUM COMPOUNDS), into powders of 13 μm average diameter has been successfully consolidated, and warm (453 K) material has been extruded with a cross-sectional reduction area of 40% (63). Samples $2 \times 2 \times 4$ mm^3 were compressively tested from the compact having strengths of 1060 MPa (147,700 psi). This was the first demonstration that bulk amorphous materials could be produced by powder metallurgy techniques.

Ion implantation (qv) has a large (10^{14} K/s) effective quench rate (64). This surface treatment technique allows a wide variety of atomic species to be introduced into the surface. Sputtering and evaporation methods are other very slow approaches to making amorphous films, atom by atom. The processes involve deposition of a vapor onto a cold substrate. The buildup rate (20 μm/h) is also sensitive to deposition conditions, including the presence of impurity atoms which can

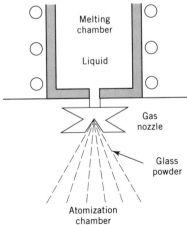

Fig. 5. Schematic diagram of the gas atomization process.

facilitate the formation of an amorphous structure. An approach used for metal–metalloid amorphous alloys is chemical deposition and electrodeposition.

The first solid-state amorphization reaction observed (65) was the loading of a binary metallic Zr–Rh metastable crystalline phase with hydrogen. The result of raising the free energy of this metastable phase through hydrogen loading is decomposition of the phase to form an amorphous hydride. The metastable hydride is more stable than the hydrogen-loaded crystalline phase under the conditions of the experiment. This amorphous phase is only stable in a limited high temperature range. More work on solid-state amorphization (10) led to the conclusion that transformation to an amorphous phase is possible in any binary metallic system demonstrating anomalous diffusion where one type of atom has a much higher mobility than the other within the working temperature range. By producing specially made binary structures having very thin elemental layers, which can then be annealed to provide activation energy for diffusion, such systems can be used to form bulk amorphous samples of almost any size as long as the elemental layers are thin enough for thorough interdiffusion (66).

Not long after the solid-state work, it was noticed that very thin-layered microstructures were obtainable by mechanical milling elemental powders, a process which could be directly applicable to solid-state amorphization (7,8). More recently, investigations have been carried out on the formation of amorphous phase through solid-state reactions, typically by the use of mechanical alloying and interdiffusion between thin films (qv). Because these solid-state transformations do not rely on high heat-transfer rates, there is the possibility of forming bulk amorphous material having much larger dimensions through powder metallurgy than through rapid solidification techniques. Mechanical alloying can be performed by any of several methods, but the most common is ball milling. Using this method, alloys are formed by placing the constituent elemental powders into a hardened metal jar with a number of hard metal balls, then sealing the jar, usually under some inert atmosphere to prevent powder oxidation, and placing the jar into a machine that shakes it for a long period of time, until the elemental powders are so intimately cold-welded and interdeformed that they transform into powdered alloy. In the case of easily amorphizable alloy systems, the powder product is often seen to be the amorphous phase. Never during the process does any material melt, therefore eliminating all consideration of rapid solidification phenomena. At short milling times, when the elemental powders have not yet interdeformed to the extent that they have lost their individual identities, a fine structure of highly deformed cold-welded elemental layers is observed, analogous to kneading balls of different colored modeling clay together. It has been shown that amorphous powders can be formed simply by mechanically milling, and thereby mechanically alloying, the elemental powders of an appropriate binary metallic system such as Ni–Zr (67). Particularly useful is the fact that the milling process itself produces enough localized thermal energy to drive the interdiffusion and lead to amorphization. The structure of the amorphous phase formed in this way is essentially the same as that produced through rapid-quenching (67). Application of this technique to $Al_{80}Ni_8Fe_4Gd_8$ (68) where the elements are mixed and ball milled up to 80 hours is shown in Figure 6 where x-ray diffraction traces as a function of milling time illustrates the amorphizing of this alloy.

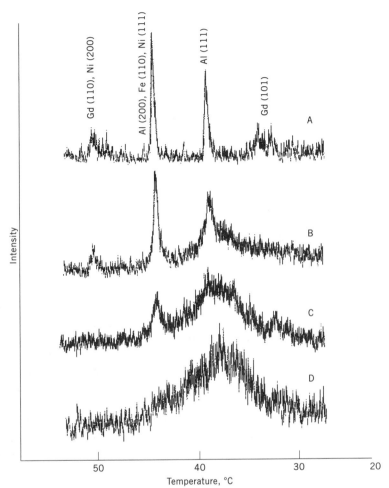

Fig. 6. X-ray diffraction traces vs ball milling (mechanical alloying) for $Al_{85}Ni_5Fe_2Gd_8$ at A, 2 h; B, 5 h; C, 20 h; and D, 80 h. After 80 hours of milling the original elemental powders are alloyed into a metallic glass (29).

Crystallization

Metallic glasses are metastable. A combination of thermal energy and time leads to crystallization. At room temperature, this may require a very long time but at moderate temperatures of 373 K or more (depending on the alloy) devitrification can occur in minutes. Usually the crystallization temperature is given as the temperature at which crystallization begins as the alloy is heated at a constant rate. Figure 7 is a differential scanning calorimeter (dsc) trace for an $Al_{85}Ni_5Fe_2Gd_8$ glass. T_x, the crystallization temperature, is defined as the onset temperature or the peak temperature of the first exotherm in the dsc scan. In this case crystallization begins at 543 K. Crystallization leads to an increase in density of about 1%. In general, T_x is somewhere between 0.4 and 0.65 T_m where T_m is the melting temperature, which for specific alloys is 400 K for Mg-based glasses

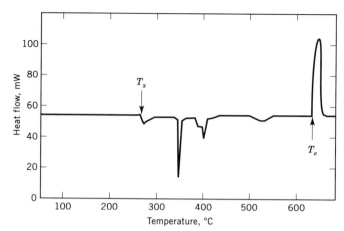

Fig. 7. Dsc scan for glassy $Al_{85}Ni_5Fe_2Gd_8$ alloy where T_x is the crystallization temperature and T_e the eutectic temperature.

and 1200 K for refractory amorphous material. The commercially important ferromagnetic iron-based glasses have T_x values of about 700 K (69).

Crystallization, by definition, implies that the initial structure be a glass, followed by the nucleation and growth of a crystalline phase, be it the equilibrium one or a metastable phase. The process is a first-order transformation and involves atomic diffusion. Types of crystallization reactions that occur include polymorphous crystallization, which is a composition invariant transformation such as that in Fe–B (70), and eutectic crystallization, T_e, in FeNiPB glass, where fine lamellae of iron–nickel austenite and metastable $(FeNi)_3PB$ phases grow cooperatively. In the primary crystallization reaction, the initial crystals formed have an overall composition different from the bulk glass (Fig. 8). The kinetics are then dependent on the rate of diffusion in the remaining glassy matrix. A more complicated mechanism is for the glassy metal to separate first into two distinct glassy phases followed by crystallization separately in each of these phases, eg, the $Zr_{36}Ti_{24}B_{40}$ system (71).

An area of some concern is whether metallic glasses are indeed amorphous (72). Amorphous materials are generally characterized by a broad diffuse x-ray or electron diffraction pattern. However, it has been a subject of intensive debate whether a material exhibiting a broad diffuse x-ray diffraction pattern has a truly amorphous structure or whether it consists of randomly oriented microcrystallites or microquasicrystallites. Thus techniques other than diffraction experiments are necessary to distinguish between the two. High resolution electron microscopy (hrem) and differential scanning calorimetry can provide direct and indirect observations of the structural nature of metallic glasses (see MICROSCOPY; THERMAL, GRAVIMETRIC, AND VOLUMETRIC ANALYSES). Modern medium voltage high resolution electron microscopy having a resolution < 0.2 nm enables identification of microcrystallinity in a sample. Using calorimetric techniques to study the isothermal crystallization kinetics allows a differentiation to be made between amorphous and microcrystalline structures (73). Figure 9 shows amorphous aluminum glass and microcrystalline glass (74). The $Al_{90}Fe_5Gd_5$ samples observed in hrem

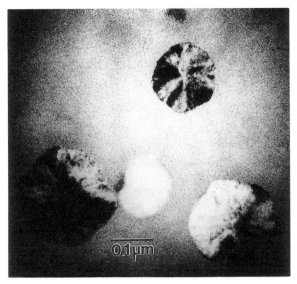

Fig. 8. Electron micrograph showing crystallization of icosahedral phase from glassy Pd–U–Si alloy.

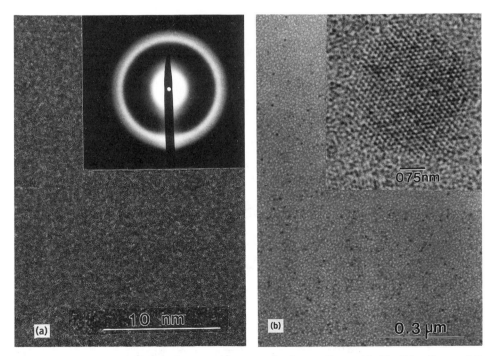

Fig. 9. (**a**) A high resolution electron micrograph of amorphous $Al_{90}Fe_5Gd_5$ alloy and its Fourier transform (inset). (**b**) Conventional electron micrograph of same alloy after annealing at 447 K for 150 minutes. The small particles are crystallites and can be imaged with high resolution electron microscopy (inset) that clearly shows the periodic nature of the crystallites which are aluminum imaged in a 110 direction.

do not reveal any crystalline structure in the sample (Fig. 9**a**), and the Fourier transform of the hrem image shows structureless features, providing direct evidence that the glass is truly amorphous. Annealed samples illustrate crystallization in the glass (Fig. 9**b**). The periodic structure is a structural image of an aluminum crystal embedded in a glassy matrix. X-rays would not differentiate crystallites this small.

To confirm that the matrix is amorphous following primary solidification, isothermal dsc experiments can be performed. The character of the isothermal transformation kinetics makes it possible to distinguish a microcrystalline structure from an amorphous structure assuming that the rate of heat released, dH/dt, in an exothermic transformation is proportional to the transformation rate, dx/dt, where H is the enthalpy and $x(t)$ is the transformed volume fraction at time t. If microcrystals do exist in a grain growth process, the isothermal calorimetric signal dH/dt is proportional to $1/r^{m+2}$, where r is the average grain size and m is a positive exponent (73). The heat-release rate decreases monotonically with time, which is the case in a sputtered Al–Mn film (73).

In an amorphous material, the alloy, when heated to a constant isothermal temperature and maintained there, shows a dsc trace as in Figure 10 (74). This trace is not a characteristic of microcrystalline growth, but rather can be well described by an isothermal nucleation and growth process based on the Johnson-

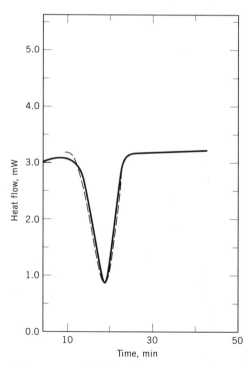

Fig. 10.   An isothermal dsc trace of $Al_{90}Fe_5Gd_5$ sample at 605 K. The solid line ——— is experimentally determined and the dashed line (– – –) is a computer simulation according to the Johnson-Mehl-Avrami equation. The degree of overlap demonstrates that the sample was truly amorphous.

Mehl-Avrami (JMA) transformation theory (75). The transformed volume fraction at time t can be written as

$$x = 1 - \exp\{-[K(t - \tau)]^n\}, \, t > \tau$$

where $K = K_o \exp(-E_x/RT)$ is the reaction rate constant and E_x is the activation energy in the crystallization process, which incorporates the activation energy for both nucleation and growth of the new phases. E_x generally has a value between 40 and 400 kJ/mol (10 and 100 kcal/mol) (69). T is the absolute temperature, R is the gas constant, and n, the Avrami exponent, is > 0 and varies between 1.5 to 4.0 (75).

Crystallization need not always be deleterious to properties such as strength. In Al-based glasses, partially crystallized material actually increased the fracture strength by 30%. A wholly amorphous $Al_{90}Fe_5Gd_5$ material obtained by melt-spinning and then annealed at 445 K for 20 minutes resulted in the nucleation and growth of nanocrystals (Fig. 9b) (76). These 5-nm precipitates are face-centered cubic and aluminum rich. Because of the very high ($10^{23}/m^3$) density and uniform distribution, the efficacy of the nanocrystalline amorphous matrix material in preventing the onset of fracture is high in this system.

Mechanical Properties

Of the various physical properties, it is the mechanical properties that make metallic glasses so unique when compared to their crystalline counterparts. A metallic glass obtains its mechanical strength in quite a different way from crystalline alloys. The disordered atomic structure increases the resistance to flow in metallic glasses so that these materials approach their theoretical strength. Strengths of $E/50$ where E is Young's modulus, are common (Table 2). An attractive feature is that metallic glasses are equally strong in all directions because of the random order of their atomic structure.

The ductility of glassy metals varies according to the kind of stress applied. For example, glasses are ductile when they are bent (flexible) or rolled in compression, but have little overall ductility in tension. Owing to the large intrinsic ductility, metallic glasses can be plastically deformed into useful shapes at no loss of mechanical strength. Wire drawing from $Fe_{29}Ni_{49}P_{14}B_6Si_2$ ribbons is an excel-

Table 2. Mechanical Properties of Metallic Glasses[a]

Alloy	Fracture strength, σ, GPa[b]	Density, ρ, mg/m³	Specific strength, σ/ρ	Young's modulus, E, GPa [b]
$Al_{85}Ni_6Fe_3Gd_6$	1.3	3.51	0.37	72.7
$Fe_{80}B_{20}$	3.6	7.4	0.49	170
$Ti_{50}Be_{40}Zr_{10}$	2.3	4.1	0.56	105
$Cu_{50}Zr_{50}$	1.8	7.3	0.25	85

[a] Refs. 29 and 77.
[b] To convert GPa to psi, multiply by 145,000.

lent example of this capability (78). The moduli of metallic glasses is generally less than crystalline forms of the same material. This is in part because of the atomic bonding in the metallic glass which does not have the overall symmetry of the crystalline metal. Upon supplying thermal energy, the moduli increase in association with structural relaxation. Typical ranges of elastic moduli ratios are 1.2–1.4 for crystalline to amorphous (79). Poisson's ratios are higher for metallic glasses: ~0.4 compared to crystalline metal values of about 0.33.

Plastic deformation in crystalline metals is accomplished by dislocation motion on specific atomic planes of atoms. Because of the random packing of atoms in amorphous metals, this mechanism is not available, and plastic flow occurs by homogeneous and inhomogeneous flow (80). The former, in which the metal glass deforms uniformly, is prevalent when stresses are less than 1/50 of the shear modulus. This occurs in creep situations. The microscopic mechanisms for homogeneous flow are based on models of free volume, microscopic shear transformation, structural relaxation, configurational entropy, and isoconfigurational flow (80). Each of these mechanisms has a temperature range where it is applicable. At temperatures between T_g and $T_g - 20$ the free volume and configurational entropy models are employed. The free volume model accounts for the temperature dependence of flow by incorporating the changes in free volume. The alteration of atomic configuration and size of flow areas are examined by the configurational entropy. At temperatures lower than $T_g - 20$, models based on structural relaxation take on increased importance. Almost by definition, the structure of the metal glass is not the equilibrium one. At temperatures below T_g, the atomic flow resulting from stress can result in increased flow resistance which may result from an increase in structural order.

Inhomogeneous flow, which occurs at higher stresses, is > 1/50 of the shear modulus (80), and its resulting strain is in the shear band. The formation of these shear bands, which are regions of highly localized deformation, are randomly spaced. Sometimes only a single band forms prior to fracture. Strains within the shear bands are as high as 10 and the bands are very narrow (<5 nm). In tension, only one band may form, but in more complex processes such as bending, rolling, or extrusion, multiple shear bands result with many orientations (Fig. 11). The internal structure of a single shear band, examined using high resolution electron microscopy (81), revealed nanocrystals in an $Al_{90}Fe_5Gd_5$ amorphous ribbon. The ribbon was bent through 180 degrees generating a high density of slip bands that were measured to be 5 nm in width. The deformation process resulted in aluminum-rich nanocrystals forming, all within slip band and surrounded by the glassy matrix. Whereas the fracture strength of a metallic glass may nearly equal its yield strength in tension, thus labeling the material brittle, compound deformation modes may enable the ductility of a metallic glass to be as high as 100% (bending).

The fracture of metallic glasses occurs by the formation of shear bands having a 45° orientation relative to the tensile axis. Typical veining on the fracture surface is a characteristic of almost all metallic glasses (82). The smooth curvature of the veins results from necking of the material (Fig. 12) and involves considerable viscous flow which has been described as the result of shear displacements along the shear band implying that the glassy metal is tough because the crack tip is dull (82). However, it is along the shear band that microcracks nucleate and

Fig. 11. Scanning electron micrograph showing the intersection of primary shear bands with the glassy ribbon surface produced by simple bending.

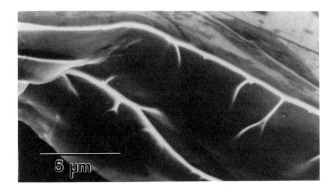

Fig. 12. Typical fracture surface of metallic glass.

propagate until impinging upon other cracks. The result of this intersection is that necking begins leading to the veining pattern.

The deleterious embrittlement of a glassy metal during annealing is also accompanied by a change in fracture mode. Almost all metallic glasses containing Fe, Co, and Al show this behavior. Figure 13 illustrates the transition to brittle fracture. When an Al–Fe–Gd glass is tested in tension, the ribbons fail by fracture along shear bands oriented at 45° to the tensile axis. The fracture surface has the typical veining pattern. After annealing at 543 K for 20 minutes, the fracture mode begins changing to a brittle fracture at 90° to the tensile axis, and a much different fracture surface is seen.

Fracture toughness experiments have been accomplished using tear tests to determine the critical stress intensity factor, K_{IIIC} (83). Iron-based glasses have a lower fracture toughness than crystalline materials such as AISI 4340 and maraging steels (84) (see STEEL). This is a result of the substantially increased yield

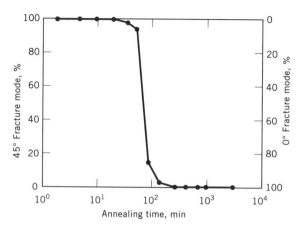

Fig. 13. Transition between ductile fracture and brittle fracture when $Al_{90}Fe_5Gd_5$ metallic glass is annealed at 170°C.

strengths of the metallic glasses which in turn is a result of the lack of plasticity in glassy materials.

At low applied stress levels crystalline alloys display fatigue properties superior to metallic glasses. At large loads approaching the fracture strength, metallic glasses can survive many more cycles than comparable strength crystalline alloys. This is because the metallic glass deforms elastically up to the high stress levels, whereas crystalline alloys generate localized regions of plastic deformation which lead to nucleation of fatigue cracks. In a $Pd_{80}Si_{20}$ alloy the fatigue crack propagation rate follows the same dependence as crystalline alloys (85):

$$da/dN = A\Delta K^m$$

where a is the crack length, N the number of cycles, ΔK the cyclic stress intensity, and A and m the material constants.

Properties

Chemical. Along with magnetic and mechanical behavior, high corrosion resistance is one of the most desirable properties of metallic glasses. The lack of crystalline defects, including grain boundaries and dislocations, helps to ensure that metallic glasses have attractive electrochemical properties. Coupled with this is the induced passivity owing to elements such as Cr and Al, and rapid repassivation following damage to the surface film. Corrosion rates can be increased by several orders of magnitude by introducing second phases by further processing steps including annealing or mechanical working, eg, warm extrusion, which could have a chemical potential difference. When elements such as Cr are added to the alloy, these enrich the surface film through the formation of complex hydroxides, eg, $CrO_x(OH)_{3-2x} \cdot nH_2$ (86). The amount and type of metalloid also influences the corrosion resistance of the passivating film. Adding P and C together in an $Fe_{70}Cr_{10}P_{13}C_7$ dramatically increases the concentration of Cr that appears on the surface film (87). In most instances the general overall corrosion resistance is determined by the presence of a passivation element rather than the amorphous nature of the alloy (88).

Magnetic. More experimental research and developmental work has been done on magnetic effects in metallic glasses than any other property (89). This is, of course, because of the focus of the technological and commercial importance (90,91). The first important glassy ferromagnet was $Fe_{83}P_{10}C_7$ (92). Alloy compositions that produce useful magnetic amorphous structures stable at room temperatures contain large fractions of transition-metal or rare-earth elements. Combinations of transition metals (primarily Fe, Ni, and Co), between 75 and 84 atom % and metalloid or glass-forming elements including C, P, B, Si, and Ge, around 20%, produce glasses strongly magnetic at room temperature. Adding a second transition metal allows ribbons of greater thicknesses to be produced. This benefit leads to multiple elements being employed. Table 3 (93) compares some properties of amorphous alloys with a crystalline-oriented FeSi transformer steel.

The Curie temperature for several magnetic glasses has been experimentally determined (89,94). This temperature represents the point where thermal agitation of the mutual interactive forces of atomic dipoles, which tend to align neighboring dipoles to one another, are perturbed. The net average moment is zero at the Curie temperature. Metallic glasses have a low Curie temperature relative to crystalline alloys. The results show that substituting Ni for Fe raises the Curie temperature in glasses such as $(Fe_xNi_{1-x})_{80}B_{20}$·FeCo alloys which have a broad maximum in Curie temperature near a Fe/Co ratio of 1. As Fe content is further increased, the Curie temperature is decreased which may reflect the atomic packing. Alloying the metallic glasses with metalloid and additional transition-metal additions has predictable effects. For example, when transition metals Cr, Mo, or V are substituted for Fe, the Curie temperature is lowered precipitously. The addition of P has a tendency to effect lower values of Curie temperature than the addition of B. Alloys having two or more transition metals have a less sharply defined Curie temperature than those having only one. The reason for this may be chemical inhomogeneity and phase separation.

For binary iron-based amorphous alloys, the saturation magnetization, ie, maximum magnetic moment per unit volume, is linear with metalloid content at

Table 3. Magnetic Properties[a]

Alloy	Coercive field, H_c, A/m[b]	Curie temp, θ_c,°C
$Fe_{80}B_{20}$[c]	3.1	374
$Fe_{80}P_{16}B_1C_3$[d]	4.0	292
$Fe_3Co_{72}P_{16}B_6Al_3$	1.2	260
$Fe_{96.8}Si_{3.2}$[e]	20–40	730

[a] Ref. 93.
[b] To convert A/m to oersted, multiply by 0.0126.
[c] Metglas 2605.
[d] Metglas 2615.
[e] Material is crystalline.

low temperatures (95). As temperature is raised, the saturation magnetization increases, reaches a peak value, then drops suddenly owing to the Curie temperature relationship with increasing metalloid content. Values for saturation magnetization have been collected for a large number of ferromagnetic amorphous alloys. Fe_xB_{100-x} alloys have shown the largest magnetization values (96). Saturation magnetization values for metallic glasses are generally lower than for crystalline alloys because of the presence of metalloid atoms which contribute to a higher volume of atoms without magnetic moments. One of the primary parameters for soft magnetic materials is the static coercivity, H_c. For metallic glasses this value is very low (Table 3).

Amorpheous alloys are not perfectly homogeneous because these materials do not show isotropic magnetic properties. Real glasses have varying amounts of magnetic anisotropy. The reasons are plentiful and include not only nonhomogeneous segregation of alloying elements but also internal stresses, both chemical and mechanical, resulting from cooling differences, and final processing of the material.

Electrical.   Unlike their crystalline counterparts, amorphous metals generally have high electrical resistivity not only at room temperature, but also, because of a very small temperature coefficient, near absolute zero. This can be understood by the atomic nature of the glass; that is, the randomness of the structure which efficiently scatters electrons. Figure 14 illustrates the difference between amorphous and crystalline FePC (97). The resistivity of the glass has little temperature dependence until it crystallizes at 675 K. Certain metallic glasses, eg, $La_{80}Au_{20}$ (98), do show superconductivity. The critical temperature is as high as 8.7 K. Metallic glass superconductors are also relatively insensitive to composition (99) and have very short electron mean-free paths.

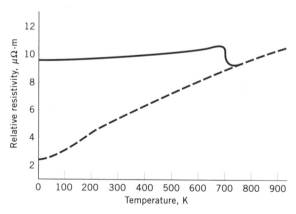

Fig. 14.   Relative resistivity of FePC as (———) a glass, and (— — —) in crystalline form.

Uses

The magnetic properties of glassy metals provide the only commercial use in bulk quantities, although brazing foils provide another niche for metallic glasses. Metallic glasses have yet to find their way into commercial products in structural applications in spite of the great strength of some glasses. This is related to their shearing instability and size limitations of the glassy product which requires a fabrication step to obtain bulk material.

Magnetic Applications. Metallic glasses are utilized in electric transformers and this provides a significant commercial success for these materials. Alloys such as MetGlas 2605 SC, $Fe_{81}B_{13.5}Si_{3.5}C_2$, are prepared in sheets having superior soft magnetic properties. These alloys can be magnetized and demagnetized with low remnant magnetizations. This is manifested in a much narrower hysteresis loop than found in an Fe–Si alloy which is the prime competing crystalline alloy to metallic glasses. The large magnetization response coupled with low magnetic losses and low excitation power make amorphous magnetic alloys attractive in applications such as transformers and inductors (100). In the United States and Japan amorphous metal-cored transformers are replacing distribution transformers where the cores are based on Fe–Si. The amorphous metals provide much lower energy losses under alternating current excitation. Amorphous metals can be also woven into sheets for magnetic shielding applications.

The combination of properties found in metallic glasses make them good candidates for recording heads where magnetic properties combined with wear resistance and mechanical hardness are attractive. Metallic glasses have often been considered for mechanical applications because of their combination of high strength and toughness. Upon discovery of aluminum glasses, these materials can also be lightweight. Suggestions for applications include rivets, razor blades, and tire cords. As of this writing, none have come to realization (101). The progress made in casting large ingots of metallic glass (4) opens up the possibility of casting and machining small components such as gears.

BIBLIOGRAPHY

"Glassy Metals" in *ECT* 3rd ed., Vol. 11, pp. 893–910, by T. Egami, Max-Planck Institut für Metallforschung.

1. P. Duwez, R. H. Willens, and W. Klement, Jr., *J. Appl. Phys.* **31**, 1136 (1960).
2. R. W. Cahn, in R. W. Cahn and P. Haasen, eds., *Physical Metallurgy*, 3rd ed., Elsevier Science Publishers, Amsterdam, 1983, p. 1779.
3. P. Duwez, in R. G. Lerner and G. L. Trig, eds, *Concise Encyclopedia of Solid State Physics*, Addison-Wesley, Reading, Mass., 1983, p. 112.
4. A. Peker and W. L. Johnson, *Appl. Phys. Lett.* **63**, 2342 (1993).
5. Y. Kawamura, A. Inoue, and T. Masumoto, *Scripta Metall. Mater.* **29**, 25 (1993).
6. H. Jones, *J. Mater. Sci.* **19**, 1043 (1984).
7. E. Hellstern and L. Schultz, *Appl. Phys. Lett.* **48**, 124 (1986).
8. E. Hellstern and L. Schultz, *J. Appl. Phys.* **63**, 1408 (1988).
9. G. K. Wehner and G. S. Anderson, in L. I. Maissel and R. Glang, eds., *Handbook of Thin Film Technology*, McGraw-Hill, New York, 1970, pp. 1–3.
10. R. B. Schwartz and W. L. Johnson, *Phys. Rev. Lett.* **51**, 415 (1983).
11. J. L. Finney, in F. E. Luborsky, ed., *Amorphous Metallic Alloys*, Butterworths, London, 1983, p. 42.
12. F. Spaepen, in M. B. Bever, ed., *Encyclopedia of Material Science and Engineering*, Pergamon Press, Oxford, U.K., 1986, p. 2976.

13. J. D. Bernal, *Proc. R. Soc. London A* **284**, 299 (1964).
14. P. H. Gaskell, *J. Non-Cryst. Solids* **32**, 207 (1979).
15. S. R. Elliott, *Physics of Amorphous Materials*, Longman, London, 1983, p. 311.
16. Y. He, S. J. Poon, and G. J. Shiflet, *Science* **241**, 1640 (1988).
17. G. J. Shiflet, Y. He, and S. J. Poon, *J. Appl. Phys.* **64**, 6863 (1988).
18. A. Inoue, K. Ohtera, A. P. Tsai, and T. Masumoto, *Jpn. J. Appl. Phys.* **27**, L479 (1988).
19. A. Inoue, K. Ohtera, T. Zhang, and T. Masumoto, *Jpn. J. App. Phys.* **27**, L1583 (1988).
20. M. G. Scott, in Ref. 11, p. 144.
21. H. A. Davis, *Phys. Chem. Glasses* **17**, 159 (1976).
22. H. A. Davis, in Ref. 11, p. 8.
23. F. Spaepen and D. Turnbull, *A. Rev. Phys. Chem.* **35**, 241 (1984).
24. R. Azoulay, H. Thibierge, and A. Brenac, *J. Non-Cryst. Solids* **18**, 33 (1975).
25. H. Y. Hsieh and co-workers, *J. Mater. Res.* **5**, 2807 (1990).
26. H. Y. Hsieh, T. Egami, Y. He, S. J. Poon, and G. J. Shiflet, *J. Non-Cryst. Solids* **135**, 248 (1991).
27. D. Turnbull, *Acta Metall. Mater.* **38**, 243 (1990).
28. E. Gebhardt, M. Becker, and S. Dorner, *Z. Metallk.* **44**, 510 (1953); *ibid.*, **44**, 573 (1953).
29. Y. He, G. M. Dougherty, G. J. Shiflet, and S. J. Poon, *Acta Metall. Mater.* **41**, 337 (1993).
30. H. Chen, *Acta Metall.* **22**, 1505 (1974).
31. A. J. Drehman, A. L. Greer, and D. Turnbull, *Appl. Phys. Lett.* **41**, 716 (1982).
32. A. Inoue, N. Matsumoto, and T. Masumoto, *Mater. Trans. JIM* **31**, 493 (1990).
33. A. Inoue, A. Kato, T. Zhang, S. Kim, and T. Masumoto, *Mater. Trans. JIM* **32**, 609 (1991).
34. A. Inoue, T. Zhang, and T. Masumoto, *Mater. Trans. JIM* **31**, 425 (1990).
35. Y. He, C. E. Price, S. J. Poon, and G. J. Shiflet, *Phil. Mag. Lett.* in press (1994).
36. A. Inoue, T. Nakamura, T. Sugita, T. Zhang, and T. Matsumoto, *Mater. Trans. JIM* **34**, 351 (1993).
37. L. Greer, *Nature* **366**, 303 (1993).
38. D. Kulmann-Wilsdorf and M. S. Bednar, *Scripta Metall. Mater.* **28**, 371 (1993).
39. P. Duwez, *ASM Trans. Q.* **60**, 608 (1967).
40. H. H. Liebermann, *Mater. Sci. Eng.* **43**, 203 (1980).
41. G. Rosen, J. Avissar, J. Baram, and Y. Gefen *Internat. J. Rapid Solidif.* **2**, 67 (1986).
42. J. Baram, *J. Mater. Sci.* **23**, 3656 (1988).
43. U.S. Pat. 4,142,571 (1979), M. C. Narasimhan (to Allied Chemicals).
44. W. A. Heineman, in, S. Steeb and H. Warlimont, eds., *Rapidly Quenched Metals V*, Elsevier Science Publishers, Amsterdam, 1985, p. 27.
45. D. King and W. La, *Metals* **2**, 32 (1967).
46. J. C. Hubert, F. Mollard, and B. Lux *Z. Metakkde.* **64**, 835 (1973).
47. E. M. Breinan, B. H. Kear, and C. M. Banas, *Phys. Today* **29**, 44 (1976).
48. S. M. Complex, in B. Cantor, ed., *Proceedings of the 3rd International Conference on Rapidly Quenched Metals*, Vol. 1, The Metals Society, London, 1978, p. 147.
49. E. M. Breinan, D. Snow, C. O. Brown, and B. H. Kear, in R. Mehrabian, B. H. Kear, and M. Cohen, eds., *Rapid Solidification Processing: Principles and Technology*, Claitor's Publishing Development, Baton Rouge, La., 1978, p. 440.
50. R. J. Grandzol and J. A. Tallmadge, *Int. J. Powder Metall. Powder Technol.* **11**, 103 (1975).
51. M. R. Glicksman, R. J. Patterson, and N. E. Schockey, in Ref. 50, p. 46.
52. P. R. Holiday, A. R. Cox, and R. J. Patterson, in Ref. 50, p. 98.
53. A. R. E. Singer, A. D. Roche, and L. Day, *Powder Metall.* **23**, 81 (1980).

54. A. R. E. Singer and A. D. Roche, in E. N. Aqua and C. I. Whitman, eds., *Modern Developments in Powder Metallurgy*, Metal Powder Industries Federation, Princeton, N. J., 1977, p. 127.

55. S. A. Miller and R. J. Murphy, *Scripta Metall.* **13**, 673 (1979).

56. S. A. Miller and R. J. Murphy, in R. Mehrambian, B. H. Kear, and M. Cohen, eds., *Rapid Solidification Processing: Principles and Technologies II,* Claitor's Publishing Division, Baton Rouge, La., 1980, p. 385.

57. N. Grant, in Ref. 50, p. 230.

58. V. Anand, A. J. Kaufman, and N. J. Grant, in Ref. 57, p. 273.

59. T. Yamaguchi and K. Narita, *IEEE Trans. Magn.* **MAG-13**, 1621 (1977).

60. A. E. Berkowitz and J. L. Walter, in Ref. 57 p. 294.

61. C. F. Cline and R. W. Hopper, *Scripta Metall.* **11**, 1137 (1977).

62. C. F. Cline, J. Mahler, F. Milton, W. Kuhl, and R. Hopper, in Ref. 50, p. 380.

63. Y. Kawamura, A. Inoue, and T. Masumoto, *Scripta Metall. Matr.* **29**, 25 (1993).

64. L. Mendoza-Felis, *Phys. Rev. B* **26**, 1306 (1982).

65. X. L. Yeh, K. Samwer, and W. L. Johnson, *Appl. Phys. Lett.* **49**, 146 (1983).

66. L. Schultz, in M. von Allmen, ed., *Proc. Materl. Res. Soc. Europe Meeting on Amorph. Metals and Non-Equil. Processing*, Les Ulis, Les Editions de Physique, Strasbourg, 1984, p. 135.

67. L. Schultz, *Phil. Mag. B* **61**, 453 (1990).

68. G. M. Dougherty, G. J. Shiflet, and S. J. Poon, *Acta Metall. Mater.* **41** (1993).

69. M. G. Scott, in Ref. 12, p. 2968.

70. A. L. Greer, *Acta Metall.* **30**, 171 (1982).

71. L. Tanner and R. Ray, *Scripta Metall.* **14**, 1657 (1980).

72. G. S. Cargill, *Solid St. Phys.* **30**, 227 (1975).

73. L. C. Chen and F. Spaepen, *Nature* **336**, 366 (1988).

74. Y. He, H. Chen, G. J. Shiflet, and S. J. Poon, *Phil Mag. Lett.* **61**, 297 (1990).

75. J. W. Christian, *The Theory of Transformations in Metals and Alloys*, 2nd ed., Pergamon, New York, 1975.

76. H. S. Chen, Y. He, G. J. Shiflet, and S. J. Poon, *Scripta Metall. Mater.* **25**, 1421 (1991).

77. R. W. Cahn, *Contemp. Phys.* **21**, 43 (1980).

78. S. Takayama, *J. Mater. Sci.* **16**, 269 (1981).

79. D. Weaire, M. F. Ashby, J. Logan, and M. J. Weins, *Acta Metall.* **19**, 779 (1971).

80. F. Spaepen and A. J. Taub, in F. E. Luborsky, ed., *Amorphlus Metallic Alloys*, Butterworths, London, 1983, p. 231.

81. H. Chen, Y. He, G. J. Shiflet, and S. J. Poon, *Nature* **367**, 541 (1994).

82. H. J. Leamy, H. S. Chen, and T. J. Wang, *Metall. Trans.* **3**, 699 (1972).

83. H. Kimura and T. Masumoto, *Scripta Metall.* **9**, 211 (1975).

84. L. A. Davis, in *Metallic Glasses*, American Society for Metals, Metals Park, Ohio, 1978, p. 190.

85. T. Ogura, K. Fukushima, and T. Masumoto, *Scripta Metall.* **9**, 979 (1975).

86. K. Asami, K. Hahimoto, T. Masumoto, and S. Shimodaira, *Corrosion Sci.* **16**, 909 (1976).

87. T. Masumoto and K. Hadhimoto, *Ann. Rev. Mater. Sci.* **8**, 215 (1978).

88. K. Hashimoto, in Ref. 81, p. 471.

89. C. D. Graham, Jr. and T. Egami, *Ann. Rev. Mater. Sci.* **8**, 423 (1978).

90. F. E. Luborsky and L. A. Johnson, *J. de Physq. (Paris)* **41**, 820 (1980).

91. D. Raskin and C. H. Smith, in Ref. 81,

92. P. Duwez and S. C. H. Lin, *J. Appl. Phys.* **38**, 4096 (1967).

93. F. E. Luborsky, in R. A. Levy and T. Hasegawa, eds., *Amorphous Magnetism II*, Plenum Publishing Corp., New York, 1977.

94. F. E. Luborsky, J. D. Livingston, and G. Y. Chin, *Physical Metallurgy*, North Holland Publishers, Amsterdam, 1983, p. 1674.
95. F. E. Luborsky, *J. Magn. Mater.* **7**, 143 (1978).
96. N. S. Kazama, M. Mitera, and T. Masumoto, *Proceedings of the Third International Conference on Rapidly Quenched Metals*, The Metals Society, London, 1978, p. 164.
97. P. Duwez, *Ann. Rev. Mat. Sci.* **61**, 83 (1976).
98. T. Masumoto, K. Hashimoto, and M. Naka, in B. Cantor, ed., *Rapidly Quenched Metals III*, Vol. 2, The Metals Society, London, 1978, p. 435.
99. M. M. Collver and R. H. Hammond, *Phys. Rev. Lett.* **30**, 92 (1972).
100. F. E. Luborsky, *Amorphous Metallic Alloys*, Butterworths, London 1983, p. 360.
101. K. Ohtera, A. Inoue, and T. Masumoto, *Mater. Sci. Eng.* **A134**, 1212 (1991).

GARY J. SHIFLET
University of Virginia

GLOBULINS.   See FRACTIONATION, BLOOD.

GLUCOSE.   See CARBOHYDRATES; SUGAR; SYRUPS.

GLUE.   See ADHESIVES.

GLUTAMIC ACID.   See AMINO ACIDS.

GLUTARIC ACID.   See DICARBOXYLIC ACIDS.

GLUTEN.   See BAKERY PROCESSES AND LEAVENING AGENTS.

GLYCERIDES.   See DRYING OILS; FATS AND FATTY OILS.

GLYCEROL

Glycerol [56-81-5], propane-1,2,3-triol, glycerin (USP), a trihydric alcohol, is a clear, water-white, viscous, sweet-tasting hygroscopic liquid at ordinary room temperatures above its melting point. Glycerol was first discovered in 1779 by Scheele, who heated a mixture of litharge and olive oil and extracted it with water. Glycerol occurs naturally in combined form as glycerides in all animal and vegetable fats and oils, and is recovered as a by-product when these oils are saponified in the process of manufacturing soap, when the fats are split in the production of fatty acids, or when fats are esterified with methanol in the production of methyl esters. Since 1949 it has also been produced commercially by synthesis from propylene [115-07-1]. The latter currently accounts for ca 30% of United States production.

The uses of glycerol number in the thousands, with large amounts going into the manufacture of drugs, cosmetics, toothpastes, urethane foam, synthetic resins, and ester gums. Tobacco processing and foods also consume large amounts either as glycerol or glycerides.

Occurrence

Glycerol occurs in combined form in all animal and vegetable fats and oils (see FATS AND FATTY OILS). It is rarely found in the free state in these fats but is usually present as a triglyceride combined with such fatty acids as stearic, oleic, palmitic, and lauric acids, and these are generally mixtures or combinations of glycerides of several fatty acids. Coconut and palm kernel oils containing a high percentage (70–80%) of C-6–C-14 fatty acids yield larger amounts of glycerol than do fats and oils containing mostly C-16 and C-18 fatty acids, such as animal fats, cottonseed, soybean, olive, and palm oil. Glycerol also occurs naturally in all animal and vegetable cells in the form of lipids such as lecithin (qv) and cephalins. These complex fats differ from simple fats in that they invariably contain a phosphoric acid residue in place of one fatty acid residue.

Nomenclature

The term glycerol applies only to the pure chemical compound 1,2,3-propanetriol, $CH_2OHCHOHCH_2OH$. The term glycerin applies to the purified commercial products normally containing ≥95% of glycerol. Several grades of glycerin are available commercially. They differ somewhat in their glycerol content and in other characteristics such as color, odor, and trace impurities. The ending -ol in glycerol connotes the presence of hydroxyl groups. The three hydroxyl positions in glycerol are designated 1,2,3- (formerly the designations α, β, and γ were used).

One method of naming the esters of glycerol with organic acids that have simple names is to replace the -ic acid ending by -in, eg, 1-monobutyrin (glycerol 1-butyrate); tributyrin (glycerol tributyrate). The degree of esterification is indicated by the prefixes mono-, di-, and tri-. A mixed triglyceride can be named in three ways, as illustrated for the compound of the formula

$$CH_3(CH_2)_{16}COOCH_2$$
$$CH_3COOCH$$
$$CH_3(CH_2)_{14}COOCH_2$$

by the names 2-aceto-3-palmito-1-stearin, 2-aceto-3-stearo-1-palmitin, and 3-palmito-1-stearo-2-acetin. Any one of the three acids can be given the -in termination; if possible, this acid receives the locant 1. The other acids are cited in alphabetical order (regardless of carbon content) and numbered, if there is a choice, to give the lowest possible numbering (see also CARBOXYLIC ACIDS; ESTERS, ORGANIC).

Properties

Physical properties of glycerol are shown in Table 1. Glycerol is completely soluble in water and alcohol, slightly soluble in diethyl ether, ethyl acetate, and dioxane, and insoluble in hydrocarbons (1). Glycerol is seldom seen in the crystallized state because of its tendency to supercool and its pronounced freezing point depression when mixed with water. A mixture of 66.7% glycerol, 33.3% water forms a eutectic mixture with a freezing point of −46.5°C.

Glycerol, the simplest trihydric alcohol, forms esters, ethers, halides, amines, aldehydes, and such unsaturated compounds as acrolein (qv). As an alcohol, glycerol also has the ability to form salts such as sodium glyceroxide (see also ALCOHOLS, POLYHYDRIC).

Synthesis

A variety of processes for synthesizing glycerol from propylene are shown in Figure 1. The first glycerol process, put on stream in 1948, followed the discovery that propylene could be chlorinated in high yields to allyl chloride [107-05-1] (see CHLOROCARBONS AND CHLOROHYDROCARBONS, ALLYL CHLORIDE). Since allyl chloride could be converted to glycerol by several routes, the synthesis of glycerol from propylene [115-07-1] became possible. Propylene can also be oxidized in high yields to acrolein [107-02-8]. Several routes for conversion of acrolein to glycerol are shown in Figure 1.

In the traditional allyl chloride route, the allyl chloride may be converted into glycerol by two processes. The allyl chloride may be treated with aqueous chlorine, and the resulting mixture of glycerol dichlorohydrins dehydrochlorinated to epichlorohydrin [106-89-8], which is then hydrolyzed to glycerol. In the second process, allyl chloride is hydrolyzed to allyl alcohol [107-18-6] (see ALLYL ALCOHOL AND MONOALLYL DERIVATIVES). The allyl alcohol is chlorohydrinated with aqueous chlorine solution to yield a mixture of monochlorohydrins which are hydrolyzed to glycerol in 90% yield based on allyl alcohol. The product from either of the above procedures is a dilute aqueous solution containing 5% or less of glyc-

Table 1. Physical Properties of Glycerol

Property	Value
mp, °C	18.17
bp, °C	
at 0.53 kPa[a]	14.9
at 1.33 kPa[a]	166.1
at 13.33 kPa[a]	222.4
at 101.3 kPa[a]	290
sp gr, 25/25°C	
in vacuum	1.2617
100% glycerol in air	1.2620
95% glycerol in air	1.2491
n_D^{20}	1.47399
vapor pressure, Pa[b]	
at 50°C	0.33
at 100°C	26
at 150°C	573
at 200°C	6100
surface tension at 20°C, mN/m($=$ dyn/cm)	63.4
viscosity at 20°C, mPa·s($=$ cP)	1499
heat of vaporization, J/mol[c]	
at 55°C	88.12
at 195°C	76.02
heat of solution to infinite dilution, kJ/mol[c]	5.778
heat of formation, kJ/mol[c]	667.8
thermal conductivity, W/(m·K)	0.28
flash point, °C	
Cleveland open cup	177
Pensky-Martens closed cup	199
fire point, °C	204

[a]To convert kPa to mm Hg, multiply by 7.5.
[b]To convert Pa to mm Hg, multiply by 0.0075.
[c]To convert J to cal, divide by 4.184.

erol. High purity glycerol is obtained in several steps: the crude glycerol is concentrated to ca 80% in multiple-effect evaporators and salt is removed by centrifuging; additional concentration of the product, followed by desalting, yields 98% glycerol; colored substances are removed by solvent extraction; and the product is refined by steam-vacuum distillation (see CHLOROHYDRINS).

Acrolein-based glycerol manufacture via no-chlorine processing proceeds by epoxidation (2) and reduction (Fig. 1), in either order, followed by hydration. The epoxidation to glycidaldehyde [765-34-4] can proceed through treatment of acrolein with aqueous sodium hypochlorite solution (3) or with hydrogen peroxide. If desired, glycidol [556-52-5] can be separated as an intermediate. Alternatively, allyl alcohol from the reduction of acrolein can be hydroxylated with aqueous hydrogen peroxide to yield directly a glycerol solution in 80–90% yield.

Propylene oxide-based glycerol can be produced by rearrangement of propylene oxide [75-56-9] (qv) to allyl alcohol over trilithium phosphate catalyst at

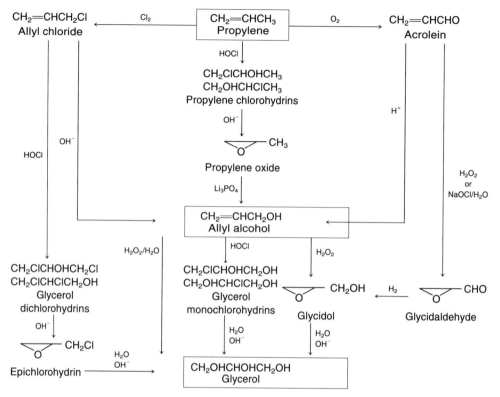

Fig. 1. Routes for the manufacture of glycerol.

200–250°C (yield 80–85%) (4), followed by any of the appropriate steps shown in Figure 1. The specific route commercially employed is peracetic acid epoxidation of allyl alcohol to glycidol followed by hydrolysis to glycerol (5). The newest international synthesis plants employ this basic scheme.

Manufacture

Until 1949 all glycerol was obtained from the glycerides in fats and oils. Currently ca 70% of U.S. production is from natural glycerides. The production of synthetic glycerol peaked in the 1960s and 1970s, when it accounted for 50–60% of the market.

Glycerol from glycerides (natural glycerol) is obtained from three sources: soap manufacture, fatty acid production, and fatty ester production. In soap manufacture, fat is boiled with caustic soda solution and salt. Fats react with the caustic to form soap (qv) and glycerol. The presence of salt causes a separation into two layers: the upper layer is soap and the lower layer, referred to as spent lye, contains glycerol, water, salt, and excess caustic. Continuous saponification (consap) processes for producing soap are now common and produce a spent lye similar to batch or kettle processes. In producing fatty acids, the most common process is continuous, high pressure hydrolysis where a continuous, upward flow of fat in a column flows countercurrent to water at 250–260°C and 5 MPa (720 psi). The fat is split by the water into fatty acids and glycerol. The fatty acids are withdrawn from the top of the column, and the glycerol-containing aqueous phase (called sweet water) falls and is withdrawn from the bottom. Concentration of the sweet water by evaporation results in a product called hydrolysis crude. The fatty acids from splitting are used to make soap, reduced to the corresponding fatty alcohol, or marketed as fatty acids. A third source of natural glycerol is the esterification of fats with alcohol to produce fatty esters. A fat usually reacts with methanol in the presence of an alkali catalyst such as sodium methoxide to produce methyl ester and glycerol, which is separated from the methyl ester by water washing. Acidulation with hydrochloric acid and removal of residual methanol produces a crude glycerol with a few percent salt content. The methyl esters are reduced to the corresponding fatty alcohols, marketed as fatty esters, or used as an emission reducing component of diesel fuels.

Recovery. The spent lyes resulting from current soapmaking processes generally contain 8–15% glycerol; sweet waters from hydrolysis of fats contain as much as 20% glycerol; crude glycerol from esterification contains 80% or more glycerol. The grade of fat used directly affects the treatment required to produce glycerol of an acceptable commercial quality. The chemicals most commonly used to remove impurities from spent lye and sweet water are hydrochloric acid and caustic soda.

The treatment of spent lye consists of a series of operations designed to remove nearly all of the organic impurities (6,7). The spent lye commonly is treated with mineral or fatty acids to reduce the content of free caustic and soda ash and to reduce the pH to 4.6–4.8 (8). Sulfates are to be avoided since they are associated with foaming and heat exchanger fouling during subsequent refining. After cooling, the solid soap is skimmed, and an acid and a coagulant are added, followed by filtration. Addition of caustic soda removes the balance of coagulant in solution and adjusts the pH to a point at which the liquor is least corrosive to subsequent process treatment. Spent lyes from modern liquid–liquid countercurrent extraction used with continuous saponification systems require little treatment other than reduction of free alkali by neutralization with hydrochloric acid. The dilute glycerol is now ready for concentration to 80% soap lye crude glycerol.

The sweet water from continuous and batch autoclave processes for splitting fats contains little or no mineral acids and salts and requires very little in the way of purification, as compared to spent lye from kettle soapmaking (9). The sweet water should be processed promptly after splitting to avoid degradation and loss of glycerol by fermentation. Any fatty acids that rise to the top of the sweet water are skimmed. A small amount of alkali is added to precipitate the dissolved fatty acids and neutralize the liquor. The alkaline liquor is then filtered and evaporated to an 88% crude glycerol. Sweet water from modern noncatalytic, continuous hydrolysis may be evaporated to ca 88% without chemical treatment.

Ester crude glycerol is usually of high quality; however, salt residue from the esterification catalyst is typically present at a concentration of one percent or higher. Crude glycerol originating from esterification or splitting of 100% vegetable oils is segregated from other glycerols throughout processing to produce kosher glycerin.

Concentration.   The quality of crude glycerols directly affects the refining operation and glycerin yield. Specifications for crude glycerols usually limit ash content, ie, a measure of salt and mineral residue; matter organic nonglycerol (MONG), which includes fatty acids and esters; trimethylene glycol (TMG), ie, propane-1,3-diol; water; arsenic; and sugars (8).

Dilute glycerol liquors, after purification, are concentrated to crude glycerol by evaporation. This process is carried out in conventional evaporation (qv) under vacuum heated by low pressure steam. In the case of soap–lye glycerol, means are supplied for recovery of the salt that forms as the spent lye is concentrated. Multiple effort evaporators are typically used to conserve energy while concentrating to a glycerol content of 85–90%.

Refining.   The refining of natural glycerol is generally accomplished by distillation, followed by treatment with active carbon. In some cases, refining is accomplished by ion exchange (qv).

Distillation.   In the case of spent-lye crude, the composition is ca 80% glycerol, 7% water, 2% organic residue, and less than 10% ash. Hydrolysis crudes are generally of a better quality than soap–lye crudes with a composition of ca 88% glycerol, <1% ash (little or no salt), and <1.5% organic residue.

Distillation equipment for soap–lye and esterification crude requires salt-resistant metallurgy. The solid salt which results when glycerol is vaporized is removed by filtration or as bottoms from a wiped film evaporator. The Luwa scraped wall evaporator is capable of vaporizing glycerol very rapidly and almost completely, such that a dry, powdery residue is discharged from the base of the unit (8). Distillation of glycerol under atmospheric pressure is not practicable since it polymerizes and decomposes glycerol to some extent at the normal boiling point of 204°C. A combination of vacuum and steam distillation is used in which the vapors are passed from the still through a series of condensers or a packed fractionation section in the upper section of the still. Relatively pure glycerol is condensed. High vacuum conditions in modern stills minimize glycerol losses due to polymerization and decomposition (see DISTILLATION).

Bleaching and Deodorizing.   The extensive use of glycerol and glycerol derivatives in the food industry (see FOOD PROCESSING) stresses the importance of the removal of both color and odor (also necessary requirements of USP and extra-quality grades). Activated carbon (1–2%) and diatomite (qv) filter aid are added to the glycerol in the bleach tank at 74–79°C, stirred for 1–2 h, and then filtered at the same temperature, which is high enough to ensure easy filtering and yet not so high as to lead to darkening of the glycerol.

Ion Exchange.   Most natural glycerol in the United States is refined by the methods described above. However, several refiners employ or have employed ion-exchange systems. When ionized solids are high, as in soap–lye crude, ion-exclusion treatment can be used to separate the ionized material from the nonionized (mainly glycerol). A granular resin such as Dowex 50WX8 may be used for ion exclusion. For ion exchange, crude or distilled glycerol may be treated with a resin appropriate for the glycerol content and impurities present. Macroreticular resins such as Amberlite 200, 200C, IRA-93, and IRA-90 may be used with undiluted glycerol. However, steam deodorization is often necessary to remove odors imparted by the resin. Ion exchange and ion exclusion are not widely used alternatives to distillation (8).

Grades. Two grades of crude glycerol are marketed: (*1*) soap–lye crude glycerol obtained by concentration of lyes from kettle or continuous soapmaking processes contains ca 80% glycerol; and (*2*) hydrolysis crude glycerol resulting from hydrolysis of fats contains ca 88–91% glycerol and a small amount of organic salts. Since glycerol from methyl ester production contains salt, it is usually marketed as soap–lye crude.

Several grades of refined glycerol, such as high gravity, dynamite, and USP, are marketed; specifications vary depending on the consumer and the intended use. USP-grade glycerol is water-white, and meets the requirements of the USP (see FINE CHEMICALS). It is classified as GRAS by the FDA, and is suitable for use in foods, pharmaceuticals, and cosmetics, or when the highest quality is demanded or the product is designed for human consumption. It has a minimum specific gravity (25°C/25°C) of 1.249, corresponding to no less than 95% glycerol. Kosher glycerin meets all USP requirements and is produced synthetically or from 100% vegetable glycerides. The *European Pharmacopoeia* (PH.EUR.) grade is similar to the USP, but the common PH.EUR. grade has a minimum glycerol content of 99.5%. The chemically pure (CP) grade designates a grade of glycerol that is about the same as the USP but with the specifications varying slightly as agreed by buyer and seller. The high gravity grade is a pale-yellow glycerol for industrial use with a minimum specific gravity (25°C/25°C) of 1.2595. The dynamite grade has the same specific gravity but is more yellow (see EXPLOSIVES AND PROPELLANTS, EXPLOSIVES). All these grades satisfy the federal specifications for glycerol (0-G491B-2).

Economic Aspects

Commercial production and consumption of glycerol has generally been considered a fair barometer of industrial activity, as it enters into such a large number of industrial processes. It generally tends to rise in periods of prosperity and fall in recession times.

Glycerol production in the United States (Table 2) rose from 19,800 metric tons in 1920 to a peak of 166,100 t in 1967 (10). World production of glycerol is ca 600,000 t/yr. Synthetic glycerol accounts for ca 30% of the United States annual production. Widespread consolidation of glycerin refining sites occurred between

Table 2. Glycerol Production[a] in the United States, t

Year	Production of crude[b]	Year	Production of crude[b]
1920	19,800	1985	145,500
1940	71,600	1987	139,300
1950	102,300	1988	134,600
1960	136,900	1989	133,200
1970	153,900	1990	133,200
1980	136,577	1991	133,800

[a]100% Glycerol basis.
[b]Synthetic included on a crude basis since June 1949.

1975 and 1992. In North America there were 20 natural and four synthetic sites in 1975. By 1992 there were only 14 natural sites and one synthetic site. During this time natural or glyceride-derived capacity increased by 55,000 t/yr, while synthetic capacity fell by 85,000 t/yr (11). In 1929, the largest single industrial use was as an automotive antifreeze (qv). This use has completely disappeared and been replaced by outlets in food, drug, cosmetics, tobacco processing, and urethane foams (see URETHANE POLYMERS). The uses in alkyd resins (qv) (owing to the increase in use of water-based paints) and cellophane (owing to the popularity of other transparent wraps) continue to decrease. The use distribution for the years 1986 and 1990 is compared to that in 1977 in Table 3 (10).

Since 1920, the price of refined glycerol in the United States has varied from a low of $0.22/kg in the early 1930s to a high of nearly $1.95/kg in 1987. In 1992 glycerin prices for USP grade ranged from $1.28 to $1.65/kg. Since glycerol is a by-product of fatty acid, ester, and alcohol production, prices are quite sensitive to changes in oleochemical demand. Glycerol prices also respond to prices for tallow, coconut oil, and petroleum, the feedstocks from which it is prepared.

Table 3. Glycerol Use,[a] t

Use	1977	1986	1990
alkyd resins	21,300	6,480	2,870
cellophane and meat casings	9,880	3,650	3,340
tobacco	15,900	22,670	21,870
explosives and military use	2,690	930	2,030
drugs, including toothpaste	19,820	24,730	28,330
cosmetics	4,670	6,410	9,170
monoglycerides and foods	13,740	13,480	18,340
urethane foams	13,620	13,730	15,290
miscellaneous	11,710	7,140	6,770
distributor sales	14,520	27,840	35,880
Total	*127,850*	*127,060*	*143,880*

[a]Ref. 10.

Identification and Analysis

The methods of analysis of the American Oil Chemists' Society (AOCS) are the principal procedures followed in the United States and Canada and are official in commercial transactions. When the material is for human consumption or drug use, it must meet the specifications of the USP (12). Commercial distilled grades of glycerol do not require purification before analysis by the usual methods. The determination of glycerol content by the periodate method (13), which replaced the acetin and dichromate methods previously used, is more accurate and more specific as well as simpler and more rapid.

Glycerol is most easily identified by heating a drop of the sample with ca 1 g powdered potassium bisulfate and noting the very penetrating and irritating odor of the acrolein that is formed. Owing to the toxicity of acrolein, a preferred method is the Cosmetics, Toiletry, and Fragrance Association (CTFA) method GI-1, an ir spectrophotometric method. Glycerol may be identified by the preparation of crystalline derivatives such as glyceryl tribenzoate, mp 71–72°C; glycerol tris(3,5-dinitrobenzoate), mp 190–192°C; or glycerol tris(p-nitrobenzoate), mp 188–189°C (14).

The concentration of distilled glycerol is easily determined from its specific gravity (15) by the pycnometer method (16) with a precision of ±0.02%. Determination of the refractive index also is employed (but not as widely) to measure glycerol concentration to ±0.1% (17).

The preferred method of determining water in glycerol is by the Karl Fischer volumetric method (18). Water can also be determined by a special quantitative distillation in which the distilled water is absorbed by anhydrous magnesium perchlorate (19). Other tests such as ash, alkalinity or acidity, sodium chloride, and total organic residue are included in AOCS methods (13,16,18).

Handling and Storage

Most crude glycerol is shipped to refiners in standard tank cars or tank wagons. Imported crude arrives in bulk, in vessels equipped with tanks for such shipment, or in drums.

Refined glycerol of a CP or USP grade is shipped mainly in bulk in tank cars or tank wagons. These are usually stainless steel-, aluminum-, or lacquer-lined. However, pure glycerol has little corrosive tendency, and may be shipped in standard, unlined steel tank cars, provided they are kept clean and in a rust-free condition. Some producers offer refined glycerol in 4.5-kg (3.8-L or 1-gal) tinned cans and more commonly in 250–259-kg (208-L or 55-gal) drums of a nonreturnable type (ICC-17E). These generally have a phenolic resin lining.

Storage. For receiving glycerol from standard 30.3-m^3 (8000-gal) tank cars (36.3-t), a storage tank of 38–45-m^3 ((10–12) × 10^3-gal) capacity should be employed. Preferably it should be of stainless steel (304 or 316), of stainless- or nickel-clad steel, or of aluminum. Certain resin linings such as Lithcote have also been used. Glycerol does not seriously corrode steel tanks at room temperature but gradually absorbed moisture may have an effect. Therefore, tanks should be sealed with an air-breather trap.

Handling Temperatures. Optimum temperature for pumping is in the 37–48°C range. Piping should be stainless steel, aluminum, or galvanized iron. Valves and pumps should be bronze, cast-iron with bronze trim, or stainless steel. A pump of 3.15-L/s (50-gal/min) capacity unloads a tank car of warm glycerol in ca 4 h.

Health and Safety Factors

Glycerol, since 1959, is generally recognized as safe (GRAS) as a miscellaneous or general-purpose food additive (qv) under the CFR (20), and it is permitted in certain food packaging (qv) materials.

Oral LD$_{50}$ levels have been determined in the mouse at 470 mg/kg (21) and the guinea pig at 7750 mg/kg (22). Several other studies (23–25) have shown that large quantities of both synthetic and natural glycerol can be administered orally to experimental animals and humans without the appearance of adverse effects. Intravenous administration of solutions containing 5% glycerol to animals and humans has been found to cause no toxic or otherwise undesirable effects (26).

The aquatic toxicity (TLm96) for glycerol is >1000 mg/L (27), which is defined by NIOSH as an insignificant hazard.

Uses

Glycerol is used in nearly every industry. The largest single use is in drugs and oral care products including toothpaste, mouthwash, and oral rinses (Table 3). Its use in tobacco processing and urethane foams remains at a fairly even consumption level. Use in foods and cosmetics is growing.

Foods.   Glycerol as a food is easily digested and nontoxic, and its metabolism places it with the carbohydrates, although it is present in combined form in all vegetable and animal fats. In flavoring and coloring products, glycerol acts as a solvent and its viscosity lends body to the product. Raisins saturated with glycerol remain soft when mixed with cereals. It is used as a solvent, a moistening agent, and an ingredient of syrups (qv) as a vehicle. In candies and icings, glycerol retards crystallization of sugar. Glycerol is used as a heat-transfer medium in direct contact with foods in quick freezing, and as a lubricant in machinery used for food processing (qv) and packaging. The polyglycerols and polyglycerol esters have increasing use in foods, particularly in shortenings and margarines.

Drugs and Cosmetics.   In drugs and medicines, glycerol is an ingredient of many tinctures and elixirs, and glycerol of starch is used in jellies and ointments. It is employed in cough medicines and anesthetics (qv), such as glycerol–phenol solutions, for ear treatments, and in bacteriological culture media. Its derivatives are used in tranquilizers (eg, glyceryl guaiacolate [93-14-1]), and nitroglycerin [55-65-0] is a vasodilator in coronary spasm. In cosmetics (qv), glycerol is used in many creams and lotions to keep the skin soft and replace skin moisture. It is widely used in toothpaste to maintain the desired smoothness, viscosity, and lending a shine to the paste (see DENTIFRICES).

Tobacco.   In processing tobacco, glycerol is an important part of the casing solution sprayed on tobacco before the leaves are shredded and packed. Along with other flavoring agents, it is applied at a rate of ca 2.0 wt % of the tobacco to prevent the leaves from becoming friable and thus crumbling during processing; by remaining in the tobacco, glycerol helps to retain moisture and thus prevents drying out of the tobacco, and influences the burning rate of the tobacco. It is used also in the processing of chewing tobacco to add sweetness and prevent dehydration, and as a plasticizer in cigarette papers.

Wrapping and Packaging Materials.   Meat casings and special types of papers, such as a glassine and greaseproof paper, need plasticizers (qv) to give them pliability and toughness; as such, glycerol is completely compatible with the base materials used, is absorbed by them, and does not crystallize or volatilize appreciably.

Lubricants.   Glycerol can be used as a lubricant in places where an oil would fail. It is recommended for oxygen compressors because it is more resistant to oxidation than mineral oils. It is also used to lubricate pumps and bearings exposed to fluids such as gasoline and benzene, which would dissolve oil-type lubricants. In food, pharmaceutical, and cosmetic manufacture, where there is contact with a lubricant, glycerol may be used to replace oils (see LUBRICATION AND LUBRICANTS).

Glycerol is often used as a lubricant because its high viscosity and ability to remain fluid at low temperatures make it valuable without modification. To increase its lubricating power, finely divided graphite may be dispersed in it. Its viscosity may be decreased by addition of water, alcohol, or glycols, and increased by polymerization or mixing with starch; pastes of such compositions may be used in packing pipe joints, in gas lines, or in similar applications (see PACKAGING MATERIALS, INDUSTRIAL PRODUCTS). For use in high pressure gauges and valves, soaps are added to glycerol to increase its viscosity and improve its lubricating ability. A mixture of glycerin and glucose is employed as a nondrying lubricant in the die-pressing of metals. In the textile industry, glycerol is frequently used in connection with so-called textile oils, in spinning, knitting, and weaving operations.

Urethane Polymers. An important use for glycerol is as the fundamental building block in polyethers for urethane polymers (qv). In this use it is the initiator to which propylene oxide, alone or with ethylene oxide, is added to produce trifunctional polymers which, on reaction with diisocyanates, produce flexible urethane foams. Glycerol-based polyethers (qv) have found some use, too, in rigid urethane foams.

Gaskets and Cork Products. Sheets and gaskets made with ground cork and glue require a plasticizer that has some humectant action in order that they may be pliable and tough. Glycerol is used because it has low vapor pressure, is not readily extractable by oils and greases, is readily absorbed by the cork, and is compatible with glue. With crown sealers and cork stoppers that come into contact with foods, it fulfills the additional requirement of nontoxicity.

Other Uses. Glycerol is used in cement compounds, caulking compounds, lubricants, and pressure media. It is also used in embalming fluids, masking and shielding compounds, soldering compounds, and compasses; cleaning materials such as soaps, detergents, and wetting agents; emulsifiers and skin protectives used in industry; asphalt (qv); ceramics (qv); photographic products; leather (qv) and wood (qv) treatments; and adhesives (qv).

Derivatives

Glycerol derivatives include acetals, amines, esters, and ethers. Of these the esters are the most widely employed. Alkyd resins (qv) are esters of glycerol and phthalic anhydride. Glyceryl trinitrate [55-63-0] (nitroglycerin) is used in explosives (qv) and as a heart stimulant (see CARDIOVASCULAR AGENTS). Included among the esters also are the ester gums (rosin acid ester of glycerol), mono- and diglycerides (glycerol esterified with fatty acids or glycerol transesterified with oils), used as emulsifiers and in shortenings. The salts of glycerophosphoric acid are used medicinally.

Mixtures of glycerol with other substances are often named as if they were derivatives of glycerol; eg, boroglycerides (also called glyceryl borates) are mixtures of boric acid and glycerol. Derivatives, such as acetals, ketals, chlorohydrins, and ethers, can be prepared but are not made commercially, with the exception of polyglycerols.

The polyglycerols, ethers prepared with glycerol itself, have many of the properties of glycerol. Diglycerol, $HOCH_2CHOHCH_2OCH_2CHOHCH_2OH$ [627-82-7], is a viscous liquid (287 $mm^2/s(=cSt)$ at 65.6°C), about 25 times as viscous as glycerol. The polyglycerols offer greater flexibility and functionality than glycerol. Polyglycerols up to and including triacontaglycerol (30 condensed glycerol molecules) have been prepared commercially; the higher forms are solid. They are soluble in water, alcohol, and other polar solvents. They act as humectants, much like glycerol, but have progressively higher molecular weights and boiling points. Products based on polyglycerols are useful in surface-active agents (see SURFACTANTS), emulsifiers (see EMULSIONS), plasticizers, adhesives, lubricants, antimicrobial agents, medical specialties and dietetic foods (see FOOD ADDITIVES; SWEETENERS).

Esters.   The mono- and diesters of glycerol and fatty acids occur naturally in fats that have become partially hydrolyzed. The triglycerides are primary components of naturally occurring fats and fatty oils.

Mono- and diglycerides are made by the reaction of fatty acids or raw or hydrogenated oils, such as cottonseed and coconut, with an excess of glycerol or polyglycerols. Commercial glycerides are mixtures of mono- and diesters, with a small percentage of the triester. They also contain small amounts of free glycerol and free fatty acids. High purity monoglycerides are prepared by molecular or short-path distillation of glyceride mixtures.

The higher fatty acid mono- and diesters are oil-soluble and water-insoluble. They are all edible, except the ricinoleate and the erucinate, and find their greatest use as emulsifiers in foods and in the preparation of baked goods (28) (see BAKERY PROCESSES AND LEAVENING AGENTS). A mixture of mono-, di-, and triglycerides is manufactured in large quantities for use in superglycerinated shortenings. Mono- and diglycerides are important modifying agents in the manufacture of alkyd resins, detergents, and other surface-active agents. The monoglycerides are also used in preparation of cosmetics (qv), pigments (qv), floor waxes (see POLISHES), synthetic rubbers (see RUBBER CHEMICALS), coatings (qv), textiles (qv) (29), etc.

Tailored triglycerides with unique nutritional properties have grown in importance in recent years. These compounds are produced from glycerol esterification with specific high purity fatty acids. A triglyceride consisting primarily of C-8, C-10, and C-22 fatty acid chains designated "caprenin" has been marketed as a low calorie substitute for cocoa butter (30) (see FAT REPLACERS). By starting with behenic monoglyceride made from glycerol and behenic acid, the shorter caprylic and capric acids can be attached to the behenic monoglyceride to deliver a triglyceride having only one long fatty acid chain (31,32).

Acetins.   The acetins are the mono-, di-, and triacetates of glycerol that form when glycerol is heated with acetic acid. Physical properties are shown in Table 4; they are all colorless.

Monoacetin (glycerol monoacetate [26446-35-5]), is a thick hygroscopic liquid, and is sold for use in the manufacture of explosives, in tanning, and as a solvent for dyes. Diacetin (glycerol diacetate [25395-31-7]) is a hygroscopic liquid, and is sold in a technical grade for use as a plasticizer and softening agent and as a solvent. Its n_D^{20} is 1.44.

Table 4. Physical Properties of Acetins

Property	Monoacetin	Diacetin	Triacetin
CAS Registry Number	[26446-35-5]	[25395-31-7]	[102-76-1]
bp, °C			
at 22 kPa^a	158^b		
at 101.3 kPa^a		259	258–259
d_4^{20}, g/cm^3	1.206		1.160
d_4^{16}, g/cm^3		1.184	
solubility			
soluble in	water, ethanol	water, ethanol	ethanol, diethyl ether and other organic solvents
sl sol in	diethyl ether	diethyl ether, benzene	water
insoluble in	benzene	carbon disulfide	

^{a}To convert kPa to mm Hg, multiply by 7.5.
^{b}At 0.4 kPa,^{a} bp = 130°C.

Triacetin, mp = −78°C, has a very slight odor and a bitter taste. Glycerol triacetate [102-76-1] occurs naturally in small quantities in the seed of *Euonymus europaeus*. Most commercial triacetin is USP grade. Its primary use is as a cellulose plasticizer in the manufacture of cigarette filters, and its second largest use is as a component in binders for solid rocket fuels. Smaller amounts are used as a fixative in perfumes, as a plasticizer for cellulose nitrate, in the manufacture of cosmetics, and as a carrier in fungicidal compositions (see FUNGICIDES, AGRICULTURAL).

BIBLIOGRAPHY

"Glycerol" in *ECT* 1st ed., Vol. 7, pp. 216–229, by N. N. Dalton and J. C. Kern, Association of American Soap & Glycerin Producers, Inc., and C. S. Miner, Jr., The Miner Laboratories; in *ECT* 2nd ed., Vol. 10, pp. 619–631, by J. C. Kern, Glycerin Producers' Association; in *ECT* 3rd ed., Vol. 11, pp. 921–932 by J. C. Kern, Glycerin Producers' Association.

1. *Physical Properties of Glycerin and Its Solutions*, Glycerin Producers' Association, New York, 1975.
2. G. B. Payne, *J. Am. Chem. Soc.* **80,** 6461 (1958); **81,** 4901 (1959).
3. C. Schaer, *Helv. Chim. Acta* **41,** 560, 614 (1958); U.S. Pat. 2,887,498 (May 9, 1959), G. Hearne, D. S. la France, and H. D. Finch (to Shell Development Co.).
4. U.S. Pat. 2,426,264 (Aug. 26, 1947), G. W. Fowler and J. T. Fitzpatrick (to Carbide and Carbon Chemicals Corp.).

5. K. Yamagishi and O. Kageyama, *Hydrocarbon Process.* **55**(12), 139 (1976).
6. T. M. Patrick, Jr., E. T. McBee, and H. B. Haas, *J. Am. Chem. Soc.* **68,** 1009 (1946).
7. W. E. Sanger, *Chem. Met. Eng.* **26,** 1211 (1922).
8. E. Woollatt, *The Manufacture of Soaps, Other Detergents and Glycerin*, John Wiley & Sons, Inc., New York, 1985, pp. 296–357.
9. J. L. Trauth, *Oil Soap* **23,** 137 (1946).
10. *SDA Glycerin and Oleochemicals Statistics Report*, The Soap and Detergent Association, New York, 1992.
11. E. T. Sauer, *World Glycerin Conference*, Paris, Oct. 1992, H. B. International and C. A. Houston and Assoc.
12. *The United States Pharmacopeia XX (USP XX-NF XV)*, The United States Pharmacopeial Convention, Inc., Rockville, Md., 1980.
13. *Official and Tentative Methods*, 3rd ed., American Oil Chemists' Society, Chicago, Ill., 1978, Ea6-51.
14. C. S. Miner and N. N. Dalton, *Glycerol, ACS Monograph 117*, Reinhold Publishing Corp., New York, 1953, pp. 171–175.
15. L. W. Bosart and A. O. Snoddy, *Ind. Eng. Chem.* **19,** 506 (1927).
16. Ref. 13, Ea7-50.
17. L. T. Hoyt, *Ind. Eng. Chem.* **26,** 329 (1934).
18. Ref. 13, Ea8-58.
19. C. P. Spaeth and G. F. Hutchinson, *Ind. Eng. Chem. Anal. Ed.* **8,** 28 (1936).
20. *Code of Federal Regulations,* Title 21, Sect. 182.1320, Washington, D.C., 1993.
21. H. F. Smyth, J. Seaton, and L. Fischer, *J. Ind. Hyg. Toxicol.* **23,** 259 (1941).
22. R. C. Anderson, P. N. Harris, and K. K. Chen, *J. Am. Pharm. Assoc. Sci. Ed.* **39,** 583 (1950).
23. V. Johnson, A. J. Carlson, and A. Johnson, *Am. J. Physiol.* **103,** 517 (1933).
24. C. H. Hine, H. H. Anderson, H. D. Moon, M. K. Dunlap, and M. S. Morse, *Arch. Ind. Hyg. Occup. Med.* **7,** 282 (1953).
25. W. Deichman, *Ind. Med. Ind. Hyg. Sec.* **9**(4), 60 (1940).
26. H. A. Sloviter, *J. Clin. Inv.* **37,** 619 (1958).
27. W. Hann and P. A. Jensen, *Water Quality Characteristics of Hazardous Materials*, Texas A&M University, College Station, 1974, p. 4.
28. N. H. Nash and V. K. Babayan, *Food Process. (Chicago)* **24**(11), 2 (1963); *Baker's Dig.* **38**(9), 46 (1963).
29. A. E. Parolla and C. Z. Draves, *Am. Dyestuff Rep.* **46,** 761 (Oct. 21, 1957); **47,** 643 (Sept. 22, 1958).
30. *Caprenin*, U.S. FDA GRAS Petition 1G0373, U.S. Food and Drug Administration, Washington, D.C., 1990.
31. U.S. Pat. 5,142,071 (Aug. 25, 1992), B. W. Kluesener, G. K. Stipp, and D. K. Yang (to Procter & Gamble).
32. U.S. Pat. 5,142,072 (Aug. 25, 1992), G. K. Stipp and B. W. Kluesener (to Procter & Gamble).

LOWEN R. MORRISON
Procter & Gamble

GLYCOLIC ACID.   See HYDROXYCARBOXYLIC ACIDS.

GLYCOLS

ETHYLENE GLYCOL AND OLIGOMERS

Glycols are diols, compounds containing two hydroxyl groups attached to separate carbon atoms in an aliphatic chain. Although glycols may contain heteroatoms, those discussed here are composed solely of carbon, hydrogen, and oxygen. These are adducts of ethylene oxide and can be represented by the general formula, $C_{2n}H_{4n}O_{n-1}(OH)_2$.

Ethylene glycol, the adduct of water and ethylene oxide, is the simplest glycol and is the principal topic of this article. Diethylene, triethylene, and tetraethylene glycols are oligomers of ethylene glycol. Polyglycols are higher molecular weight adducts of ethylene oxide and are distinguished by intervening ether linkages in the hydrocarbon chain. These polyglycols are commercially important; their properties are significantly affected by molecular weight. They are water soluble, hygroscopic, and undergo reactions common to the lower weight glycols (see also POLYETHERS, ETHYLENE OXIDE POLYMERS).

Ethylene glycol, EG, is a colorless, practically odorless, low viscosity, hygroscopic liquid of low volatility. It is completely miscible with water and many organic liquids. EG was first prepared by Wurtz in 1859 by hydrolysis of ethylene glycol diacetate. It did not achieve commercial interest until World War I, when it was used in Germany as a substitute for glycerol (qv) in explosives manufacture (1). The uses for ethylene glycol are numerous. Some of the applications are polyester resins for fiber, PET containers, and film applications; all-weather automotive antifreeze and coolants, defrosting and deicing aircraft; heat-transfer solutions for coolants for gas compressors, heating, ventilating, and air-conditioning systems; water-based formulations such as adhesives, latex paints, and asphalt emulsions; manufacture of capacitors; and unsaturated polyester resins. The oligomers also have excellent water solubility but are less hygroscopic and have somewhat different solvent properties. The number of repeating ether linkages controls the influence of the hydroxyl groups on the physical properties of a particular glycol.

Glycols undergo reactions common to monohydric alcohols forming esters, acetals, ethers, and similar products. For example, both simple and polyesters are produced by reaction with mono- or dibasic acids (eqs. 1 and 2):

$$HOCH_2CH_2OCH_2CH_2OCH_2CH_2OH + 2\ CH_3COOH \longrightarrow \tag{1}$$
$$\text{triethylene glycol}$$

$$CH_3COOCH_2CH_2OCH_2CH_2OCH_2CH_2OOCCH_3 + 2\ H_2O$$
$$\text{triethylene glycol diacetate}$$

$$HOCH_2CH_2OH + HOOC\!-\!R\!-\!COOH \longrightarrow H\!\!-\!\!(OOC\!-\!R\!-\!COOCH_2CH_2)_{\overline{x}}OH + H_2O \tag{2}$$
$$\text{polyester}$$

Physical Properties

Ethylene glycol and its lower polyglycols are colorless, odorless, high boiling, hygroscopic liquids completely miscible with water and many organic liquids. Physical properties of ethylene glycols are listed in Table 1. Vapor-pressure curves of the ethylene glycols at various temperatures are illustrated in Figure 1. Ethylene glycols markedly reduce the freezing point of water (Fig. 2). Some important physical constants of ethylene glycol are given in Table 2.

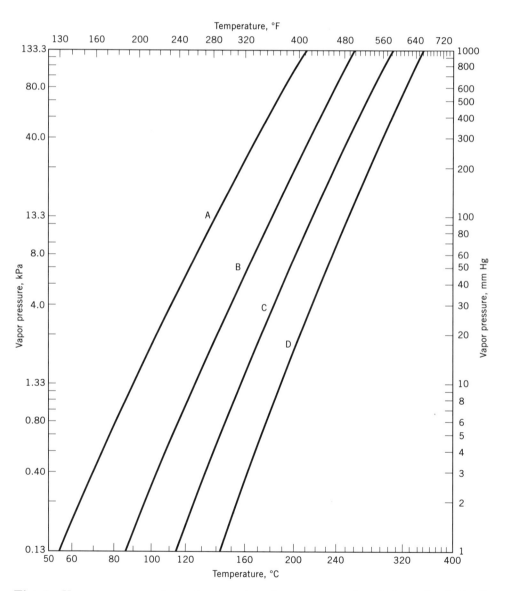

Fig. 1. Vapor pressures of glycols at various temperatures. A, ethylene glycol; B, diethylene glycol; C, triethylene glycol; and D, tetraethylene glycol.

Table 1. Properties of Glycols[a]

Property	Ethylene glycol	Diethylene glycol	Triethylene glycol	Tetraethylene glycol
CAS Registry Number	[107-21-1]	[111-46-6]	[112-27-6]	[112-60-7]
formula	$HOCH_2CH_2OH$	$HO(CH_2CH_2O)_2H$	$HO(CH_2CH_2O)_3H$	$HO(CH_2CH_2O)_4H$
mol wt	62.07	106.12	150.17	194.23
sp gr, 20/20°C	1.1155	1.1185	1.1255	1.1247
bp at 101.3 kPa,[b] °C	197.6	245.8	288	dec
mp, °C	− 13.0	− 6.5	− 4.3	− 4.1
viscosity at 20°C, mPa·s(=cP)	20.9	36	49	61.9
refractive index, n_D^{20}	1.4318	1.4475	1.4561	1.4598
heat of vaporization at 101.3 kPa,[b] kJ/mol[c]	52.24	52.26	61.04	62.63
flash point of commercial material, °C	116[d]	138[e]	172[e]	191[e]

[a]Ref. 2–5.
[b]To convert kPa to mm Hg, multiply by 7.5.
[c]To convert kJ to kcal, divide by 4.184.
[d]Determined by ASTM D56, using the Tag closed cup.
[e]Determined by ASTM D92, using the Pensky-Martens closed cup.

Chemical Properties

The hydroxyl groups on glycols undergo the usual alcohol chemistry giving a wide variety of possible derivatives. Hydroxyls can be converted to aldehydes, alkyl halides, amides, amines, azides, carboxylic acids, ethers, mercaptans, nitrate esters, nitriles, nitrite esters, organic esters, peroxides, phosphate esters, and sulfate esters (6,7).

The largest commercial use of ethylene glycol is its reaction with dicarboxylic acids to form linear polyesters. Poly(ethylene terephthalate) [25038-59-9] (PET) is produced by esterification of terephthalic acid [100-21-0] (1) to form bishydroxyethyl terephthalate [959-26-2] (BHET) (2). BHET polymerizes in a transesterification reaction catalyzed by antimony oxide to form PET (3).

$$HOOC-\bigcirc-COOH \; + \; 2\,HOCH_2CH_2OH \xrightarrow{-2\,H_2O} HOCH_2CH_2OOC-\bigcirc-COOCH_2CH_2OH$$

$$\text{(1)} \qquad\qquad\qquad\qquad\qquad\qquad\qquad\qquad \text{(2)}$$

$$\text{(3)}$$

$$\xrightarrow{Sb_2O_3} H{-}\!\Big[OOC-\bigcirc-COOCH_2CH_2\Big]_n\!{-}OH \; + \; HOCH_2CH_2OH$$

$$\text{(3)}$$

Ethylene glycol esterification of BHET is driven to completion by heating and removal of the water formed. PET is also formed using the same chemistry start-

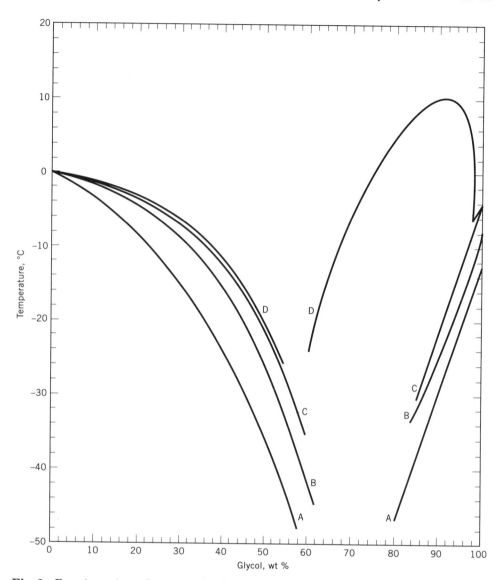

Fig. 2. Freezing points of aqueous glycol solutions. A, ethylene glycol; B, diethylene glycol; C, triethylene glycol; and D, tetraethylene glycol. Ethylene glycols form a slush with water in the apparent discontinuous ranges.

ing with dimethyl terephthalate [120-61-6] and ethylene glycol to form BHET also using an antimony oxide catalyst.

Glycols may undergo intramolecular cyclization or cyclically condense with other molecules to form a number of ring structures. Transesterification of carbonates with ethylene glycol produces ethylene carbonate [96-49-1] (eq. 4). Numerous materials catalyze carbonate transesterifications.

Table 2. Physical Properties of Ethylene Glycol[a]

Property	Value
flash point	
Pensky-Martens closed cup, °C	127
Cleveland open cup, °C	127
autoignition temperature, °C	400
density at 20°C, g/mL	1.1135
surface tension at 20°C, mN/m(= dyn/cm)	48.4
specific heat, J/(g·K)[b]	
as liquid, 19.8°C	2.406
as ideal gas, 25°C	1.565
electrical conductivity at 20°C, S/m	1.07×10^{-4}
solubility in water at 20°C, % by wt	100.0
solubility of water in EG at 20°C, % by wt	100.0
heat of combustion at 25°C, kJ/mol[b]	-1189.595
heat of formation at 25°C, kJ/mol[b]	-392.878
heat of fusion, kJ/mol[b]	11.63
onset of initial decomposition, °C	165
critical constants	
temperature, °C	446.55
pressure, kPa[c]	6515.73
volume, L/mol	0.186
compression factor, Z_c	0.2671
viscosity, mPa·s(= cP)	
at 0°C	51.37
at 40°C	9.20

[a]Ref. 2; see also Table 1.
[b]To convert J to cal, divide by 4.184.
[c]To convert kPa to mm Hg, multiply by 7.5.

$$CH_3O-\overset{\overset{\displaystyle O}{\|}}{C}-OCH_3 \ + \ HOCH_2CH_2OH \ \xrightarrow{Na_2CO_3} \ \left[\text{cyclic carbonate}\right] \ + \ 2\ CH_3OH \qquad (4)$$

Diethylene glycol readily dehydrates using an acid catalyst to make 1,4-dioxane [123-91-1] (eq. 5).

$$HOCH_2CH_2OCH_2CH_2OH \ \xrightarrow{H^+} \ \left[\text{1,4-dioxane}\right] \ + \ H_2O \qquad (5)$$

Ethylene glycol also produces 1,4-dioxane by acid-catalyzed dehydration to diethylene glycol followed by cyclization. Cleavage of triethylene and higher glycols with strong acids also produces 1,4-dioxane by catalyzed ether hydrolysis with subsequent cyclization of the diethylene glycol fragment. Diethylene glycol condenses with primary amines to form cyclic structures (eq. 6), eg, methylamine [74-89-5] reacts with diethylene glycol to produce N-methylmorpholine [109-02-4].

$$HOCH_2CH_2OCH_2CH_2OH \ + \ CH_3NH_2 \ \longrightarrow \ O \underset{}{\overset{}{\diagup}} N{-}CH_3 \ + \ 2\ H_2O \tag{6}$$

Ketones and aldehydes react with ethylene glycol under acidic conditions to form 1,3-dioxolanes (cyclic ketals and acetals) (eq. 7).

$$HOCH_2CH_2OH \ + \ RCOR' \ \xrightarrow{H^+} \ \underset{R}{\overset{R'}{\diagup}} \overset{O}{\underset{O}{\diagdown}} \ + \ H_2O \tag{7}$$

Manufacture

In 1937 the first commercial application of the Lefort direct ethylene oxidation to ethylene oxide [75-21-8] followed by hydrolysis of ethylene oxide became, and remains in the 1990s, the main commercial source of ethylene glycol production (1) (see ETHYLENE OXIDE). Ethylene oxide hydrolysis proceeds with either acid or base catalysis or uncatalyzed in neutral medium. Acid-catalyzed hydrolysis activates the ethylene oxide by protonation for the reaction with water. Base-catalyzed hydrolysis results in considerably lower selectivity to ethylene glycol. The yield of higher glycol products is substantially increased since anions of the first reaction products effectively compete with hydroxide ion for ethylene oxide. Neutral hydrolysis (pH 6–10), conducted in the presence of a large excess of water at high temperatures and pressures, increases the selectivity of ethylene glycol to 89–91%. In all these ethylene oxide hydrolysis processes the principal by-product is diethylene glycol. The higher glycols, ie, triethylene and tetraethylene glycols, account for the remainder.

The large excess of water from the hydrolysis is removed in a series of multiple-effect evaporators (8), and the ethylene glycol is refined by vacuum distillation. Figure 3 depicts a typical process flow diagram.

Ethylene glycol was originally commercially produced in the United States from ethylene chlorohydrin [107-07-3], which was manufactured from ethylene and hypochlorous acid (eq. 8) (see CHLOROHYDRINS). Chlorohydrin can be converted directly to ethylene glycol by hydrolysis with a base, generally caustic or caustic/bicarbonate mix (eq. 9). An alternative production method is converting chlorohydrin to ethylene oxide (eq. 10) with subsequent hydrolysis (eq. 11).

$$CH_2{=}CH_2 + HOCl \longrightarrow HOCH_2CH_2Cl \tag{8}$$

$$HOCH_2CH_2Cl + NaOH \longrightarrow HOCH_2CH_2OH + NaCl \tag{9}$$

$$HOCH_2CH_2Cl + Ca(OH)_2 \longrightarrow \overset{O}{\overset{\diagup\diagdown}{CH_2 - CH_2}} + CaCl_2 + H_2O \tag{10}$$

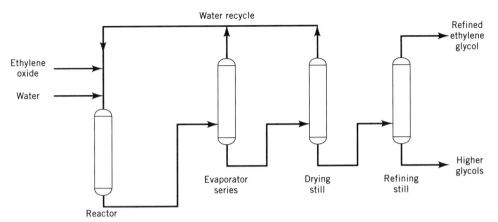

Fig. 3. Simplified glycol process flow diagram.

$$CH_2 \overset{O}{-} CH_2 + H_2O \longrightarrow HOCH_2CH_2OH \tag{11}$$

This process allows the purification of glycols without the difficulties of salt separation because the manufacturing procedure is done in two discrete steps with ethylene oxide distillation prior to hydrolysis. The hydration step is either uncatalyzed at high temperatures and pressures or utilizes an acid catalyst. A U.S. Industrial Chemicals, Inc. process uses a sulfuric acid catalyst at moderate temperatures producing an aqueous solution of glycol-containing acid (9). This process requires an additional step in the purification to remove the catalyst.

Du Pont commercially produced ethylene glycol from carbon monoxide, methanol, hydrogen, and formaldehyde (qv) until 1968 at Belle, West Virginia. The process consisted of the reaction of formaldehyde, water, and carbon monoxide with an acid catalyst to form glycolic acid. The acid is esterified with methanol to produce methyl glycolate. Subsequent reduction with hydrogen over a chromate catalyst yields ethylene glycol and methanol (eq. 12–14). Methanol and formaldehyde were manufactured on site from syngas.

$$CO + CH_2O + H_2O \xrightarrow{H^+} HOOCCH_2OH \tag{12}$$

$$HOOCCH_2OH + CH_3OH \longrightarrow CH_3OOCCH_2OH + H_2O \tag{13}$$

$$CH_3OOCCH_2OH + H_2 \xrightarrow{Cr_2O_3} HOCH_2CH_2OH + CH_3OH \tag{14}$$

Coal was the original feedstock for syngas at Belle; thus ethylene glycol was commercially manufactured from coal at one time. Ethylene glycol manufacture from syngas continues to be pursued by a number of researchers (10).

Ethylene glycol can be produced from acetoxylation of ethylene (eq. 15). Acetic acid, oxygen, and ethylene react with a catalyst to form the glycol mono- and diacetate. Catalysts can be based on palladium, selenium, tellurium, or thallium. The esters are hydrolyzed to ethylene glycol and acetic acid (eq. 16). The net reaction is ethylene plus water plus oxygen to give ethylene glycol. This technology has several issues which have limited its commercial use.

$$CH_3COOH + CH_2{=}CH_2 + O_2 \xrightarrow{\text{TeO}_2,\text{Br}_2} CH_3COOCH_2CH_2OH + CH_3COOCH_2CH_2OOCCH_3 \quad (15)$$

$$CH_3COOCH_2CH_2OH + CH_3COOCH_2CH_2OOCCH_3 + \xrightarrow{\text{3 H}_2\text{O}} 2\ HOCH_2CH_2OH + 3\ CH_3COOH \quad (16)$$

The catalysts and acetic acid are highly corrosive, requiring expensive construction materials. Trace amounts of ethylene glycol mono- and diacetates are difficult to separate from ethylene glycol limiting the glycol's value for polyester manufacturing. This technology (Halcon license) was practiced by Oxirane in 1978 and 1979 but was discontinued due to corrosion problems.

Ethylene glycol can be manufactured by the reaction of ethylene oxide with carbon dioxide to form ethylene carbonate (eq. 17) which can be hydrolyzed to ethylene glycol (eq. 18).

$$CH_2 - CH_2 + CO_2 \xrightarrow{\text{KI}} \quad (17)$$

$$+ H_2O \xrightarrow{\text{Na}_2\text{CO}_3} HOCH_2CH_2OH + CO_2 \quad (18)$$

Catalysts for the reaction of ethylene oxide and carbon dioxide to produce ethylene carbonate are alkali halides, quaternary ammonium halides, and quaternary phosphonium halides; conversion of the ethylene carbonate to ethylene glycol is catalyzed by basic materials (11,12). A significant advantage of the carbonate process is almost complete conversion of ethylene oxide to ethylene glycol with only around 1% diethylene glycol and higher glycols formed. The hydrolysis of ethylene carbonate requires a ratio of <2:1 water to carbonate by weight thus allowing significant reductions in distillation evaporators and concentrators.

Ethylene glycol can be manufactured by the transesterification of ethylene carbonate. A process based on the reaction of ethylene carbonate with methanol to give dimethyl carbonate and ethylene glycol is described in a Texaco patent (13); a general description of the chemistry has also been published (14) (eq. 19). Selectivities to ethylene glycol are excellent with little diethylene glycol or higher glycols produced. A wide range of catalysts may be employed including ion-exchange resins, zirconium and titanium compounds, tin compounds, phosphines, acids, and bases. The process produces a large quantity of dimethyl carbonate which would require a commercial outlet.

$$+ 2\ CH_3OH \xrightarrow{\text{ZrCl}_4} HOCH_2CH_2OH + CO(CH_3O)_2 \quad (19)$$

Oxalic acid produced from syngas can be esterified (eq. 20) and reduced with hydrogen to form ethylene glycol with recovery of the esterification alcohol (eq. 21). Hydrogenation requires a copper catalyst giving 100% conversion with selectivities to ethylene glycol of 95% (15).

$$\text{HOOCCOOH} + 2\ \text{ROH} \longrightarrow \text{ROOCCOOR} + 2\ \text{H}_2\text{O} \tag{20}$$

$$\text{ROOCCOOR} + 4\ \text{H}_2 \xrightarrow{\text{Cu}} \text{HOCH}_2\text{CH}_2\text{OH} + 2\ \text{ROH} \tag{21}$$

Ethylene glycol can be produced by an electrohydrodimerization of formaldehyde (16). The process has a number of variables necessary for optimum current efficiency including pH, electrolyte, temperature, methanol concentration, electrode materials, and cell design. Other methods include production of valuable oxidized materials at the electrochemical cell's anode simultaneous with formation of glycol at the cathode (17). The compound formed at the anode may be used for commercial value directly, or coupled as an oxidant in a separate process.

An early source of glycols was from hydrogenation of sugars obtained from formaldehyde condensation (18,19). Selectivities to ethylene glycol were low with a number of other glycols and polyols produced. Biomass continues to be evaluated as a feedstock for glycol production (20).

Conventional uncatalyzed hydrolysis of ethylene oxide to ethylene glycol also forms diethylene glycol, triethylene glycol, and higher weight glycols. Although the market demands for ethylene glycol and diethylene glycol are mostly independent of each other, the current manufacturing of ethylene glycol and diethylene glycol gives dependent production. Selectivity to the different glycols is currently controlled by varying the ratio of water to ethylene oxide with a large excess of water promoting ethylene glycol selectivity. Removing the excess water is energy intensive and requires capital investment in evaporators. These factors limit the amount of excess water which can be used for control of the uncatalyzed selectivity to ethylene glycol.

Although catalytic hydration of ethylene oxide to maximize ethylene glycol production has been studied by a number of companies with numerous materials patented as catalysts, there has been no reported industrial manufacture of ethylene glycol via catalytic ethylene oxide hydrolysis. Studied catalysts include sulfonic acids, carboxylic acids and salts, cation-exchange resins, acidic zeolites, halides, anion-exchange resins, metals, metal oxides, and metal salts (21–26). Carbon dioxide as a cocatalyst with many of the same materials has also received extensive study.

Economic Aspects

The domestic price remained stable from the early 1950s through the mid-1960s (27). A sharp decline occurred in the late 1960s and early 1970s because of decreased glycol consumption associated with the demise of double-knit polyester apparel (28). The oil embargo and escalating feedstock costs caused the ethylene glycol price to soar in the mid-1970s (29). During the 1980s through the early 1990s glycol prices have varied widely due to capacity variations and supply and

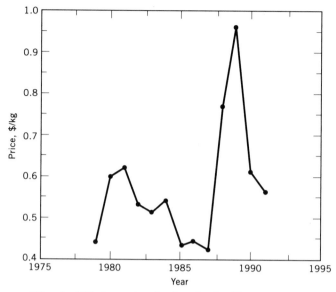

Fig. 4.   Ethylene glycol prices in the United States.

demand imbalances. These fluctuations are shown in Figure 4. The January 1993 price for industrial grade ethylene glycol was reported to be 44¢/kg (30).

During 1950–1978 the United States glycol demand roughly doubled with each succeeding decade. The current North American ethylene glycol capacity estimate is 3.5×10^6 t (35). A listing of producers and their capacity is shown in Table 3. Although accurate figures are not available, worldwide ethylene glycol nameplate capacity is estimated at 8.8×10^6 metric tons.

Table 3. 1993 North American Ethylene Glycol Capacity[a]

Producer	Location	Capacity, 10^6 kg/yr
BASF	Geismer, La.	164
Dow	Fort Saskatchewan, Can.	205
	Plaquemine, La.	211
Eastman	Longview, Tex.	105
Hoechst-Celanese	Clear Lake, Tex.	250
Oxy Petrochemicals	Bayport, Tex	264
PD Glycol	Beaumont, Tex.	359
Quantum	Morris, Ill.[b]	100
Shell	Geismer, La.	239
Texaco	Port Neches, Tex.	341
Union Carbide	Montreal, Can.[b]	70
	Prentiss, Can.	273
	Seadrift, Tex.	250
	Taft, La.	636
Total		*3467*

[a]Ref. 30.
[b]These units have been shut down.

Health, Safety, and Environmental Factors

Biodegradability of a product may be evaluated by extended-term biochemical oxygen demand (BOD) tests. This procedure permits comparison of the amount of oxygen consumed by microorganisms in the oxidation of the test material to the theoretical oxygen required to completely oxidize the chemical to carbon dioxide and water. Laboratory BOD tests using unacclimated biomass show that ethylene glycol (2,31,32) is readily biodegraded in a system which attempts to simulate the dilute biological conditions of a river or lake. The mean of several BOD determinations on ethylene glycol for 5, 10, and 20 days are 51, 78, and 97% biooxidation, respectively (2). Ethylene glycol can be treated effectively in conventional wastewater treatment plants and does not persist in the environment under expected conditions. For the higher molecular weight glycols (32) the laboratory BOD tests, using acclimated biomass which should occur in a wastewater treatment plant, show the 20-day value for diethylene glycol (3), triethylene glycol (4), and tetraethylene glycol (5) to be 67, 86, and 88% biooxidation, respectively. Consequently, in an acclimated environment, the higher molecular weight glycols (tetraethylene glycol is the highest molecular weight studied) readily biodegrade and do not persist in the environment.

Acute aquatic toxicity tests on ethylene glycol (33,34), diethylene glycol (3), and triethylene glycol (4,35) indicate no toxicity to *Daphnia magna* and fathead minnows at concentrations through 10,000 mg/L during the 48- and 96-h test duration. Also, bacterial inhibition tests indicate no bacterial growth inhibition at concentrations through 10,000 mg/L. Tetraethylene glycol (5) is relatively nontoxic to *Daphnia magna* and fathead minnows. The measured 48-h LC_{50} value with *Daphnia magna* was 7750 mg/L, whereas the 96-h LC_{50} value with fathead minnows was greater than 10,000 mg/L. Bacterial inhibition tests for tetraethylene glycol indicate no bacterial growth inhibition at concentrations through 5000 mg/L.

None of these glycols are highly irritating to the mucous membranes or skin (36). A splash of these neat materials in the eye may produce marked irritation, but permanent damage should not be expected. In general, each glycol is believed to have low acute dermal toxicity and vapor and inhalation toxicity when exposure is in an industrial situation at normal room temperature, since the amount absorbed through the skin is believed to be minimal and the vapor pressures are relatively low (37). Table 4 presents ranges of values for acute oral LD_{50} tests in rats (calculated dose that kills 50% of the treated animals) performed on the different glycols.

Ethylene glycol is not recommended for use as an ingredient in food or beverages, or where there is a significant contact with food or potable water. For adult humans, death has occurred after as little as 30 mL ethylene glycol has been ingested (37); yet survival has been reported following one liter and more (41). The clinical sequence of toxic response has been described as follows: Phase I: central nervous system (CNS) and metabolic abnormalities; Phase II: cardio and pulmonary abnormalities; Phase III: renal insufficiency. Urinary oxalate crystals are a common, but not invariable, feature of ethylene glycol poisoning. For adults, the reported probable oral acute lethal dose for diethylene glycol is between 1–2 g/kg which is in the same range as ethylene glycol (42). Some of the

Table 4. Rat Acute Oral LD$_{50}$ Values of Ethylene Glycols

Glycol	LD$_{50}$ rats, g/kg[a]		
	Ref. 38	Ref. 39	Ref. 40
ethylene	6.1	8.5	13.4
diethylene	16.6	20.8	
triethylene	16.8	22.1	
tetraethylene		32.8	

[a]Single oral dose.

same clinical signs have been noted for diethylene glycol poisoning as in ethylene glycol poisoning. However, hepatotoxicity may be more common in diethylene glycol poisoning. There are no recorded human fatalities by any route of exposure to triethylene or tetraethylene glycols.

Human volunteers tolerated continuous (20–22 h/d) inhalation exposure to ethylene glycol aerosols up to 75 mg/m^3, but considered 140 mg/m^3 irritating and 200 mg/m^3 intolerable for more than 15 minutes (43). Repeated skin contact with undiluted ethylene glycol has a very low potential to cause the development of allergic contact dermatitis. There is no evidence for a skin sensitizing potential in a repeated insult test in guinea pigs or humans for diethylene or triethylene glycols.

There is no evidence of genetic toxicity for ethylene, diethylene, and triethylene glycols from a battery of *in vitro* tests (44–46). Tetraethylene glycol is not believed to be mutagenic. Questionable results were obtained for *in vitro* tests; however, tetraethylene glycol did not induce mutagenic effects in tests with higher levels including an *in vivo* dominant lethal test in rats. Ethylene glycol was tested under the National Toxicology Program to determine its chronic toxicity and neoplastic (tumor causing) potential. After two years of feeding mice diets with ethylene glycol as high as 50,000 ppm, there was no evidence of carcinogenic activity (44). Likewise in a chronic feeding study in rats, no treatment-related neoplastic (tumorigenic) effects were observed (47).

Effects of repeated ethylene glycol peroral overexposure in treated rats and mice can result in kidney, liver, and nervous system damage. The most sensitive indicators of ethylene glycol toxicity are disturbances in acid–base balance and nephrotoxic (kidney) effects. Effects of repeated chronic peroral overexposure of diethylene glycol in treated rats result in kidney and liver damage (48).

There were no treatment-related effects on the reproductive system or performance in a three-generation reproduction study in which rats were fed diets containing ethylene glycol as high as 1000 mg/kg. When given to pregnant laboratory animals in the diet or by cutaneous application, ethylene glycol does not induce fetal malformations. Studies conducted through the National Toxicology Program have shown teratogenic and fetotoxic effects resulting from administration to mice of high doses by gavage (49) or in drinking water (50); however, no teratogenic or fetotoxic effects were observed when rabbits were given doses high enough by gavage to cause maternal death (51). Nose-only exposure with aerosols is minimally effective in producing developmental toxicity and only at levels that would be intolerable for humans, even for one or two breaths. Occluded cutaneous application of ethylene glycol does not demonstrate a potential to cause developmental toxicity in animal tests.

In animal studies, it has been concluded that high exposure to diethylene glycol and triethylene glycol do not result in a teratogenic response. Tetraethylene glycol has not been tested for teratogenic effects, but it and ethylene glycol have been shown to have no effect on reproduction when the males alone are exposed to high doses (52).

Following administration by various routes, ethylene glycol is rapidly absorbed, distributed, and cleared. Metabolic pathways and rates of metabolism and excretion vary with species, dose, and route of administration. The principal excretory end product at low doses is carbon dioxide in expired air and at higher doses is glycolate and unchanged ethylene glycol by urinary elimination (52). The metabolic pathway for the higher glycols has not been defined.

Uses

About 42% of the ethylene glycol produced domestically is used as a nonvolatile antifreeze for liquid-cooled motor vehicles (35) (see ANTIFREEZES AND DEICING FLUIDS). With smaller cars and reduced change frequency, antifreeze demand is stable.

In contrast to the antifreeze market, polyester markets are growing. The largest market, polyester fibers, consumes 27% of the ethylene glycol; PET bottles and other packaging materials, 12.5%; and PET film, 6% (35) (see FIBERS, POLYESTER; POLYESTERS, THERMOPLASTIC). Poly(ethylene terephthalate) provides unique properties for each end use. Wrinkle resistance, strength, durability, and stain resistance are enhanced when polyester is combined with natural fibers to produce apparel, home furnishing fabrics, carpeting, and fiberfill. Polyester's strength is an asset in industrial products such as tire cord, seat-belts, flexible belts, rope, and industrial fabric. Polyester containers are preferred for packaging soft drinks, food, personal care, and other consumer products where clarity, shatter resistance, and recyclability are important. Special grades of container resin are used to produce dual-oven containers suitable for microwave or conventional ovens and thermal-formed items such as cups. The strength of thin films makes polyester ideal for manufacturing video, audio, and computer tapes as well as various packing materials.

Miscellaneous industrial and PET resin uses consume the remaining 12.5% of the ethylene glycol. These markets typically utilize the freeze point depressing, polarity, and reactivity of ethylene glycol. Specially formulated fluids are used for defrosting and deicing aircraft; anti-icing or deicing airport runways and taxiways; and heat-transfer solutions for a wide temperature range (-51 to $135°C$). Ethylene glycol is used in adhesive, latex paint, and asphalt-emulsion water-based formulations to provide freeze protection. High purity ethylene glycol is a solvent and suspending medium for ammonium perborate, the conductor used in most electrolytic capacitors. Polyester resins based on maleic and phthalic anhydrides, ethylene glycol and higher glycols, and vinyl-type monomers are important in the low pressure laminating of glass fibers for furniture, suitcases, boat hulls, aircraft parts, and automobile bodies.

Derivatives

In addition to oligomers ethylene glycol derivative classes include monoethers, diethers, esters, acetals, and ketals as well as numerous other organic and organometallic molecules. These derivatives can be of ethylene glycol, diethylene glycol, or higher glycols and are commonly made with either the parent glycol or with sequential addition of ethylene oxide to a glycol alcohol, or carboxylic acid forming the required number of ethylene glycol subunits.

Ethylene glycol monoethers are commercially manufactured by reaction of an alcohol with ethylene oxide. The resulting glycol ether contains an ether and an alcohol functional group in the same molecule, thus providing unique solvency characteristics for diverse applications. Glycol monoethers having longer hydrocarbon-like alkoxide groups display solubility more characteristic of hydrocarbons. Thus glycol ethers produced from higher molecular weight alcohols have limited water solubility. The ether groups introduce additional sites for hydrogen bonding with improved hydrophilic solubility performance. These glycol ethers are miscible with a wide range of polar and nonpolar organic solvents and miscible with water in most cases (53). These properties allow applications as mild-ordered solvents for many resins, oils, waxes, fats, and dyestuffs, and as coupling agents for many water/organic systems.

Applications for glycol monoethers are found in a broad array of end uses (Table 5). Glycol monoethers are widely used in cleaning formulations to facilitate the removal of grease and greasy soils, to aid in solubilizing other components in the cleaner formulation, and to improve the storage stability of the formulations (54). Glycol monoethers are used as jet fuel additives to inhibit icing in fuel systems (55). They are used as solvents and cosolvents for conventional solvent-based lacquer, enamel, and wood stain industrial coating systems as well as cosolvents for waterborne industrial coating systems (56). Other applications include dye solvents in the textile, leather, and printing industries; solvents for insecticides and herbicides for agricultural applications; couplers and mutual solvents for soluble oils, hard-surface cleaners, and other soap–hydrocarbon systems; semiconductor manufacture; printed circuit board laminating formulations; freeze–thaw agents in latex emulsions; diluents for hydraulic brake fluids; chemical reaction solvents; and chemical intermediates (53).

Ethylene glycol diethers are made by derivatizing both glycol hydroxyl groups. Glycol diethers contain only the ether functional group; thus they are aprotic polar molecules which are relatively inert chemically with excellent solvent properties (57). Glycol dimethyl ethers are miscible in water, ethanol, acetone, benzene, diethyl ether, and octane. Glycol diethers with larger alkoxide end groups show more hydrophobic character and less water solubility. Strong solvating and stability properties allow numerous applications for glycol diethers including adhesives and coatings, ink formulations, cleaning compounds, batteries, electronics, polymer solvents, polymer plasticizers, gold refining, and gas purification.

Table 5. Applications for Glycol Monoalkyl Ethers, $RO(CH_2CH_2O)_nH^a$

Glycol monoether[b]	CAS Registry Number	n	R	Agriculture	Chemical	Food	Cleaning[d]	Textile	Transportation
methyl Cellosolve[e]	[109-86-4]	1	CH_3	X					X
methyl Carbitol	[111-77-3]	2	CH_3	X		X		X	X
methoxytriglycol	[112-35-6]	3	CH_3		X			X	X
Cellosolve[f,g]	[110-80-5]	1	C_2H_5						
Carbitol	[111-90-0]	2	C_2H_5	X		X	X	X	X
ethoxytriglycol	[112-50-5]	3	C_2H_5		X			X	X
propyl Cellosolve	[2807-30-9]	1	C_3H_7				X		X
propyl Carbitol	[6881-94-3]	2	C_3H_7				X	X	X
butyl Cellosolve	[111-76-2]	1	C_4H_9	X	X	X	X		X
butyl Carbitol[e,f]	112-34-5	2	C_4H_9	X	X	X	X	X	X
butoxytriglycol	[143-22-6]	3	C_4H_9						
hexyl Cellosolve	[112-25-4]	1	C_6H_{13}				X		
hexyl Carbitol[g]	[112-59-4]	2	C_6H_{13}				X		

[a] Ref. 53.
[b] Monoethylene glycol derivatives are termed "cellosolves"; diethylene glycol derivatives, "carbitols"; and triethylene glycol derivatives, triglycols. CELLOSOLVE and CARBITOL are registered trademarks of Union Carbide Corp.
[c] All but methyl Cellosolve are used for coatings.
[d] Household and institutional.
[e] Also used in electronics.
[f] Used in pharmaceuticals.
[g] Also used in printing.

Cyclic polyethers, or crown ethers, are cyclic structures containing ethylene glycol units as $+CH_2CH_2O+_n$ with $n > 2$ and generally between 4 and 8. The oxygen atoms are arranged in a cavity which is efficient in complexing cations. The number of glycol units determines the size of the cavity and effects the ability to complex specific cations. The closer the fit in size between the cavity and cation, the more stable the complex. Crown ethers greatly improve the solubility of salts in organic solvents and have a wide range of applications in organic reactions, especially as phase-transfer reagents (58) (see CATALYSIS, PHASE-TRANSFER).

Glycols can be used in the manufacture of poly(ethylene glycol) (PEG). Ethylene glycol, ethylene oxide, and a base catalyst react to form a family of linear polymers containing repeating $+CH_2CH_2O+$ ether groups with hydroxyl functions on the ends. Length of the chains is controlled by addition of ethylene oxide. PEGs find applications in agriculture products, ceramics, chemical intermediates, coatings and adhesives, cosmetics and toiletries, electronics, foods and feeds, household products, lubricants, metal processing, mining, paper, petroleum, pharmaceuticals, photography, plastics and resins, printing, rubber chemicals, textiles and leather, and wood products (59). Ethylene glycols can also react with other alkylene oxides as well as other alkylene glycols reacted with ethylene oxide to form mixed oxide polyglycols which can be used in many applications similar to those already mentioned (see POLYETHERS).

Ethylene glycol (as well as the higher glycols) can be esterified with traditional reagents such as acids, acid chlorides, acid anhydrides, and via transesterification with other esters. Preference between mono- and diesters can be controlled by the molar ratio of reactants. Low molecular weight glycol esters are good solvents for cellulose esters and printing inks, and are employed in industrial extraction processes and the protective coating industry. Fatty acid esters, together with other surfactants, are good-emulsifying, stabilizing, dispersing, wetting, foaming, and suspending agents.

Ethylene glycol monoethers contain a hydroxyl function which is readily derivatized to glycol ether–esters utilizing the same chemistry as glycol esters. Ether–ester applications are in cellulose acetate, acrylic, urethane, and polyester coatings, generally in conjunction with other oxygenated solvents (57). Other uses include retarder solvents for organic coatings and coalescing aids for latex paints.

Ethylene glycol in the presence of an acid catalyst readily reacts with aldehydes and ketones to form cyclic acetals and ketals (60). 1,3-Dioxolane [646-06-0] is the product of condensing formaldehyde and ethylene glycol. Applications for 1,3-dioxolane are as a solvent replacement for methylene chloride, 1,2-dichloroethane, 1,1,1-trichloroethane, and methyl ethyl ketone; as a solvent for polymers; as an inhibitor in 1,1,1-trichloroethane; as a polymer or matrix interaction product for metal working and electroplating; in lithium batteries; and in the electronics industry (61). 1,3-Dioxolane can also be used in the formation of polyacetals, both for homopolymerization and as a comonomer with formaldehyde. Cyclic acetals and ketals are used as protecting groups for reaction-sensitive aldehydes and ketones in natural product synthesis and pharmaceuticals (62).

Glyoxal [107-22-2] is a highly reactive dialdehyde produced commercially by the vapor-phase oxidation of ethylene glycol. Glyoxal has utility in paper manufacture as a wet-strength agent, to confer water resistance to starch-clay paper coatings, and with other pigment coating binders. Other applications include improving resistance to shrinkage and creasing of cotton, rayon, and other cellulosics; leather tanning; intermediate for biologically active molecules; reaction with hydrocolloids for modification of adhesives containing starch, carbohydrates, and other reactive materials; and in hardening gelatin (qv), either alone or condensed with diols or amino alcohols in photography.

Diethylene Glycol. Physical properties of diethylene glycol are listed in Table 1. Diethylene glycol is similar in many respects to ethylene glycol, but contains an ether group. It was originally synthesized at about the same time by both Lourenco and Wurtz (63) in 1859, and was first marketed by Union Carbide in 1928. It is a coproduct (9–10%) of ethylene glycol produced by ethylene oxide hydrolysis. It can be made directly by the reaction of ethylene glycol with ethylene oxide, but this route is rarely used because more than an adequate supply is available from the hydrolysis reaction. The 1993 capacity was estimated to be 359,000 t/yr with a price of 46¢/kg, fob Gulf (64).

Manufacture of unsaturated polyester resins and polyols for polyurethanes consumes 45% of the diethylene glycol. Approximately 14% is blended into antifreeze. Triethylene glycol from the ethylene oxide hydrolysis does not meet market requirements, which leads to 12% of the diethylene glycol being converted with ethylene oxide to meet this market need. About 10% of diethylene glycol is converted to morpholine. Another significant use is natural gas dehydration, which uses 6%. The remaining 13% is used in such applications as plasticizers for paper, fiber finishes, compatiblizers for dye and printing ink components, latex paint antifreeze, and lubricants in a number of applications.

Triethylene Glycol. Physical properties of triethylene glycol are listed in Table 1. Triethylene glycol is a colorless, water-soluble liquid with chemical properties essentially identical to those of diethylene glycol. It is a coproduct of ethylene glycol produced via ethylene oxide hydrolysis. Significant commercial quantities are also produced directly by the reaction of ethylene oxide with the lower glycols. The price for tank car quantities of technical-grade triethylene glycol was $1.35/kg in February 1990 (65).

Triethylene glycol (65) is an efficient hygroscopicity agent with low volatility, and about 45% is used as a liquid drying agent for natural gas (see DRYING AGENTS). Its use in small packaged plants located at the gas wellhead eliminates the need for line heaters in field gathering systems (see GAS, NATURAL). As a solvent (11%) triethylene glycol is used in resin impregnants and other additives, steam-set printing inks, aromatic and paraffinic hydrocarbon separations, cleaning compounds, and cleaning poly(ethylene terephthalate) production equipment. The freezing point depression property of triethylene glycol is the basis for its use in heat-transfer fluids.

Approximately 13% triethylene glycol is used in some form as a vinyl plasticizer. Triethylene glycol esters are important plasticizers (qv) for poly(vinyl butyral) resins, nitrocellulose lacquers, vinyl and poly(vinyl chloride) resins, poly(vinyl acetate) (see VINYL POLYMERS), and synthetic rubber compounds and cellulose esters. The fatty acid derivatives of triethylene glycol are used as emulsifiers, demulsifiers, and lubricants. Polyesters derived from triethylene glycol are useful as low pressure laminates for glass fibers, asbestos, cloth, or paper. This application and polyols account for 12% of the triethylene glycol. Triethylene glycol is used in the manufacture of alkyd resins (qv) used as laminating agents and in adhesives. It is reported (65) that 8% of the U.S. produced triethylene glycol is exported and 5% goes to other applications, some of which are noted above.

Tetraethylene Glycol. Physical properties of tetraethylene glycol are listed in Table 1. Tetraethylene glycol has properties similar to diethylene and triethylene glycols and may be used preferentially in applications requiring a higher boiling point, higher molecular weight, or lower hygroscopicity (5).

Tetraethylene glycol is miscible with water and many organic solvents. It is a humectant that, although less hygroscopic than the lower members of the glycol series, may find limited application in the dehydration of natural gases. Other possibilities are in moisturizing and plasticizing cork, adhesives, and other substances.

Tetraethylene glycol may be used directly as a plasticizer or modified by esterification with fatty acids to produce plasticizers (qv). Tetraethylene glycol is used directly to plasticize separation membranes, such as silicone rubber, poly(vinyl acetate), and cellulose triacetate. Ceramic materials utilize tetraethylene glycol as plasticizing agents in resistant refractory plastics and molded ceramics. It is also employed to improve the physical properties of cyanoacrylate and polyacrylonitrile adhesives, and is chemically modified to form polyisocyanate, polymethacrylate, and to contain silicone compounds used for adhesives.

Tetraethylene glycol has found application in the separation of aromatic hydrocarbons from nonaromatic hydrocarbons (BTX extraction) (66) (see BTX PROCESSING). In general, the critical solution temperature of a binary system, consisting of a given alkyl-substituted aromatic hydrocarbon and tetraethylene glycol, is lower than the critical solution temperature of the same hydrocarbon with triethylene glycol and is considerably lower than the critical solution temperature of the same hydrocarbon with diethylene glycol. Hence, at a given temperature, tetraethylene glycol tends to extract the higher alkylbenzenes at a greater capacity than a lower polyglycol.

BIBLIOGRAPHY

"Glycols" in *ECT* 1st ed., Vol. 7, pp. 238–263, by C. P. McClelland and P. R. Rector, Union Carbide and Carbon Corp.; "Ethylene Glycol, Propylene Glycol, and Their Derivatives" under "Glycols" in *ECT* 2nd ed., Vol. 10, pp. 638–660, by P. H. Miller, Union Carbide Corp.; "Ethylene Glycol and Propylene Glycol" under "Glycols" in *ECT* 3rd ed., Vol. 11, pp. 933–956, by E. S. Brown, C. F. Hauser, B. C. Ream, and R. V. Berthold, Union Carbide Corp.

1. G. O. Curme and F. Johnston, *Glycols*, ACE Monograph No. 114, Reinhold Publishing Corp., New York, 1952, Chapt. 2.
2. *Ethylene Glycol Brochure F-49193B-ICD*, Union Carbide Chemicals and Plastics Co. Inc. Danbury, Conn., 1991.
3. *Diethylene Glycol Brochure F-49181A-ICD*, Union Carbide Chemicals and Plastics Co. Inc., Danbury, Conn., 1989.
4. *Triethylene Glycol Brochure F-49191A-ICD*, Union Carbide Chemicals and Plastics Co. Inc., Danbury, Conn., 1989.
5. *Tetraethylene Glycol Brochure F-80066-ICD*, Union Carbide Chemicals and Plastics Co. Inc., Danbury, Conn., 1990.
6. R. C. Larock, *Comprehensive Organic Transformations*, VHC Publishers, Inc., New York, 1989.

7. J. March, *Advanced Organic Chemistry*, 4th ed., John Wiley & Sons, Inc., New York, 1992.
8. U.S. Pat. 3,970,711 (July 20, 1976), C. R. Reichle and J. A. Heckman (to PPG Industries, Inc.).
9. U.S. Pat. 2,409,441 (Oct. 15, 1946), F. J. Metzger (to U. S. Industrial Chemicals, Inc.).
10. H. Papp and M. Baerns, *Surf. Sci. Catal.* **64**, 430 (1991).
11. U.S. Pat. 4,314,945 (Feb. 9, 1982), C. H. McMullen and co-workers, (to Union Carbide Corp.).
12. U.S. Pat. 4,400,559 (Aug. 23, 1981), V. S. Bhise (to Halcon SD Group, Scientific Design).
13. U.S. Pat. 4,661,609 (Apr. 28, 1987), J. F. Knifton (to Texaco Inc.).
14. J. F. Knifton and R. G. Duranleau, *J. Mol. Catal.* **67**, 389 (1991).
15. U.S. Pat. 4,677,234 (June 30, 1987), W. J. Bartley (to Union Carbide Corp.).
16. U.S. Pat. 4,478,694 (Oct. 23, 1984), N. L. Weinberg (to SKA Associates).
17. U.S. Pat. 4,950,368 (Aug. 21, 1990), N. L. Weinberg, J. D. Genders, and D. J. Mazur (to The Electrosynthesis Co. and SKA Associates).
18. Ger. Pat. 373,975 (Mar. 19, 1923), A. Wohl and K. Braunig.
19. U.S. Pat. 2,060,880 (Nov. 17, 1936), W. A. Lazier (to E. I. du Pont de Nemours & Co., Inc.).
20. E. Tronconi and co-workers, *Chem. Eng. Sci.* **47**(9–11), 2451 (1992).
21. U.S. Pat. 4,822,925 (Apr. 18, 1989), J. R. Briggs and J. H. Robson (to Union Carbide Corp.).
22. U.S. Pat. 4,277,632 (July 7, 1981), T. Kumazawa, T. Yamamoto, and H. Odanaka (to Nippon Shokubai Kaguku Kogyo Co., Ltd.).
23. U.S. Pat. 4,551,566 (Nov. 5, 1985), J. H. Robson and G. E. Keller (to Union Carbide Corp.).
24. J. R. Briggs, A. M. Harrison, and J. H. Robson, *Polyhedron* **5**, 281 (1986).
25. U.S. Pat. 4,982,021 (Mar. 28, 1985), R. D. Best and co-workers, (to Union Carbide Corp.).
26. U.S. Pat. 4,967,018 (Oct. 30, 1990), H. Soo, B. C. Ream, and J. H. Robson (to Union Carbide Chemicals and Plastics Co. Inc.).
27. "Chemical Pricing Patterns, 1952–1967," *Oil Paint Drug Rep.* (1968).
28. *Ibid.* (July issues 1968–1971).
29. *Chem. Mark. Rep.* (July issues 1972–1978).
30. "Chemical Profiles," *Chem. Mark. Rep.* (Jan. 25, 1993).
31. H. F. Lund, *Industrial Pollution Control Handbook*, McGraw-Hill Book Co., Inc., New York, 1971, pp. 14–21.
32. W. H. Evans and E. J. David, *Water Res.* **8**(2), 97–100 (1974).
33. M. A. Mayes, H. C. Alexander, and D. C. Dill, *Bull. Environ. Contam. Toxicol.* **31**, 139–147 (1981).
34. U. M. Cowgill, I. T. Takahaski, and S. L. Applegath, *Environ. Toxicol. Chem.* **4**(3), 415–422 (1985).
35. G. A. LeBlanc, and D. C. Surpenant, *Arch. Environ. Contam. Toxicol.* **12**(3), 305–310 (1983).
36. E. Browning, *Toxicology and Metabolism of Industrial Solvents*, 2nd ed., Elsevier, Amsterdam, 1965, p. 625.
37. *Patty's Industrial Hygiene and Toxicology*, Vol. 2, 3rd rev. ed., Wiley-Interscience, New York, 1981.
38. Laug and co-workers, *J. Ind. Hygiene Toxicol.* **21**, 173 (1939).
39. Smyth and co-workers, *J. Ind. Hygiene Toxicol.* **23**, 259 (1941).
40. K. Bove, *Am. J. Clin. Path.* **45**, 46 (1966).
41. J. B. Stokes, III and F. J. Aueron, *J. Am. Med. Assoc.*, **243**, 2065, 198 (1980).

42. R. E. Gosselin and co-workers, *Clinical Toxicology of Commercial Products*, 4th ed., The Williams & Wilkins Co., Baltimore, 1976, p. II-119.
43. J. H. Wills and co-workers, *Clin. Toxicol.* **7**, 463 (1974).
44. *NTP Technical Report on the Toxicology and Carcinogenesis Studies of Ethylene Glycol (CAS No. 107-21-1) in B6C3F1 Mice (Feed Studies)*, NIH Publication No. 91-3144, U.S. Department of Health and Human Services, Washington, D.C., Public Health Service, 1991.
45. R. S. Slesinski and co-workers, *The Toxicologist* **6**, 228 (1986).
46. C. R. Clark and co-workers, *Toxic. Appl. Pharmac.* **51**, 529 (1979).
47. L. R. DePass and co-workers, *Fund. Appl. Toxic.* **7**, 547 (1986).
48. O. G. Fitzhugh and A. A. Nelson, *J. Ind. Hygiene Toxic.* **28** (1946).
49. C. J. Price, R. W. Kimmel, and M. C. Marr, *Toxicol. Appl. Pharmacol.* **81**, 113–127 (1985).
50. J. C. Lamb and co-workers, *Toxicol. Appl. Pharmacol.* **81**, 100–112 (1985).
51. R. W. Tyl and co-workers, *Fund. Appl. Toxicol.* **20**, 402–412 (1991).
52. Union Carbide internal reports, Conn., 1993.
53. *Glycol Ethers, Brochure F-60617A*, Union Carbide Corp., Conn., 1989.
54. *Glycol Ethers for Cleaners, Brochure F-60669*, Union Carbide Corp., Conn., 1990.
55. *Petrochemicals: Methyl Cellosolve Fuel Additive Grade*, brochure, Union Carbide Canada Ltd., 1991.
56. *UCAR Solvents Selection Guide for Coatings, Brochure F-7465z*, Union Carbide Corp.
57. *Glymes: The Grant Family of Glycol Diethers*, brochure, Grant Chemical Div., Ferro Corp., La., 1992.
58. *Crown Ethers, Technical Bulletin*, PCR Research Chemicals, Inc., Fla.
59. *Carbowax Polyethylene Glycols, Brochure F-4772M-ICD*, Union Carbide Corp., Conn., 1986.
60. *UCAR Esters for Coating Applications, Brochure F-48589A*, Union Carbide Corp., Conn., 1988.
61. *1,3-Dioxolane*, Grant Chemical Div., Ferro Corp., La., 1993.
62. T. W. Greene, *Protective Groups in Organic Synthesis*, John Wiley & Sons, Inc., New York, 1988, p. 114.
63. Ref. 1, Chapt. 7.
64. "Chemical Profiles," *Chem. Mark. Rep.* (Feb. 1, 1993).
65. "Chemical Profiles," *Chem. Mark. Rep.* (Feb. 5, 1990).
66. J. A. Vidueira, in T. C. Lo, M. H. I. Baird, and C. Hanson eds., *Handbook of Solvent Extraction*, John Wiley & Sons, Inc., New York, 1983, p. 531.

M. W. FORKNER
J. H. ROBSON
W. M. SNELLINGS
Union Carbide Corporation

PROPYLENE GLYCOLS

The propylene glycol family of chemical compounds consists of monopropylene glycol (PG), dipropylene glycol (DPG), and tripropylene glycol (TPG). The proper IUPAC chemical name for PG is 1,2-propanediol and it is listed on the U.S. EPA Toxic Substances Control Act (TSCA) inventory of chemical substances by its CAS Registry Number, [57-55-6]. The IUPAC name and TSCA listing for DPG is oxy-bispropanol [25265-71-8] and for TPG [(1-methyl-1,2-ethanediyl)bis(oxy)]bispro-panol [24800-44-0]. These chemicals are manufactured as coproducts and are used commercially in a large variety of applications. They are available as highly purified products which meet well-defined manufacturing and sales specifications. All commercial production is via the hydrolysis of propylene oxide. A fourth propylene glycol product, trimethylene glycol or 1,3-dihydroxypropane [504-63-2], has been available in commercial quantities in the past, but is not an important product in the 1990s (1). It was obtained as a by-product in the production of glycerol by either saponification or fermentation of animal fats (see GLYCEROL).

The propylene glycols are clear, viscous, colorless liquids that have very little odor, a slightly bittersweet taste, and low vapor pressures. The most important member of the family is monopropylene glycol, also known as 1,2-propylene glycol, 1,2-dihydroxypropane, 1,2-propanediol, methylene glycol, and methyl glycol. The more common commercial names are Propylene Glycol Industrial (PGI) and Propylene Glycol USP (PG USP), which designates the grade for general industrial as opposed to the food and drug grade. All of the glycols are totally miscible with water.

Propylene glycol, when produced according to the U.S. Food and Drug Administration good manufacturing practice guidelines at a registered facility, meets the requirements of the U.S. Food, Drug, and Cosmetic Act as amended under Food Additive Regulation CFR Title 21, Parts 170–199. It is listed in the regulation as a direct additive for specified foods and is classified as generally recognized as safe (GRAS). In addition, it meets the requirements of the *Food Chemicals Codex* and the specifications of the *U.S. Pharmacopeia XXII*. Because of its low human toxicity and desirable formulation properties it has been an important ingredient for years in food, cosmetic, and pharmaceutical products.

Manufacturing

Wurtz (2) first prepared propylene glycol in 1859 by hydrolysis of propylene glycol diacetate, and it was commercialized in 1931 by Carbide and Carbon Chemicals Corp. (3). This first commercial production used the chlorohydrin process to make propylene oxide, which was subsequently hydrolyzed to the glycol. In the mid-1930s Du Pont Co. operated a high pressure coconut oil hydrogenation plant which yielded propylene glycol as a by-product. Propylene glycol was gaining acceptance as a substitute for glycerol in pharmaceuticals, and shortages during World War II led to new production facilities by Dow Chemical Co. in 1942 and Wyandotte Chemical Corp. in 1948.

All commercial production of propylene glycol is by high pressure, high temperature, noncatalytic hydrolysis of propylene oxide (qv). A large excess of water is used in the conversion of propylene oxide to a mixture of mono-, di-, and tripropylene glycols. Typical product distribution is 90% PG and 10% coproducts. Hydration reactor conditions are 120–190°C at pressures up to 2170 kPa. After the hydration reaction is completed, excess water is removed in multieffect evaporators and drying towers, and the glycols are purified by high vacuum distillation.

Propylene oxide [75-56-9] is manufactured by either the chlorohydrin process or the peroxidation (coproduct) process. In the chlorohydrin process, chlorine, propylene, and water are combined to make propylene chlorohydrin, which then reacts with inorganic base to yield the oxide. The peroxidation process converts either isobutane or ethylbenzene directly to an alkyl hydroperoxide which then reacts with propylene to make propylene oxide, and t-butyl alcohol or methylbenzyl alcohol, respectively. Table 1 lists producers of propylene glycols in the United States.

Table 2 shows the production, sales, and value for the glycols in the United States in 1990. Production of monopropylene glycol peaked in 1988 at 404,000 t. Imports have been minimal except for the mid-1980s when 25,600 t were imported in 1984. In the period 1986–1989 U.S. exports averaged about 75,000 t per year.

Table 1. U.S. Producers of Propylene Glycols

Producer	Location	Annual capacity, 10^3 t
Arco[a]	Bayport, Tex.	163
Dow[a]	Freeport, Tex.	113
Dow[a]	Plaquemine, La.	68
Eastman[a]	S. Charleston, W.Va.	36
Olin[a]	Brandenburg, Ky.	32
Texaco[b,c]	Beaumont, Tex.	68
Total		*480*

[a] Ref. 4.
[b] Ref. 5.
[c] Projected 1994 startup.

Table 2. Production and Sales of Propylene Glycols, 1990[a]

Glycol	Production, 10^3 t	Sales, 10^3 t	Average sales, $/kg
propylene	342.2	258.3	0.94
dipropylene	35.6	29.4	0.90
tripropylene	10.4	[b]	[b]

[a] Ref. 6.
[b] Not reported.

Chemistry

Monopropylene glycol (1,2-propanediol) is a difunctional alcohol with both a primary and a secondary hydroxyl. Chemically, the presence of the secondary hydroxyl group differentiates propylene glycol from ethylene glycol, which has two primary hydroxyl groups. Coproducts dipropylene glycol and tripropylene glycol have several possible structural and stereochemical isomers. Examination of the mechanisms for addition of an alcohol to an oxirane ring under various reaction conditions explains the distribution of the various isomers in the product mix (7). In the high pressure, high temperature process for hydrolysis of propylene oxide to propylene glycol and the subsequent formation of dipropylene and tripropylene glycol, the neutral to slightly acidic conditions dictate a nonspecific opening of the oxirane ring. The nonspecific nature of the acid-catalyzed reaction is seen in the approximately 50:50 product distribution for the primary to secondary alcohol isomers in the dipropylene glycol produced in the process. In the base-catalyzed propoxylation of an alcohol, for example in poly(propylene glycol) manufacture, attack is at the less substituted position of the oxirane ring, leading to the preferential formation of the secondary alcohol.

The primary and secondary alcohol functionalities have different reactivities, as exemplified by the slower reaction rate for secondary hydroxyls in the formation of esters from acids and alcohols (8). 1,2-Propylene glycol undergoes most of the typical alcohol reactions, such as reaction with a free acid, acyl halide, or acid anhydride to form an ester; reaction with alkali metal hydroxide to form metal salts; and reaction with aldehydes or ketones to form acetals and ketals (9,10). The most important commercial application of propylene glycol is in the manufacture of polyesters by reaction with a dibasic or polybasic acid.

$$CH_3CHOHCH_2OH + HOOCRCOOH \rightarrow HO \overset{\overset{\displaystyle CH_3}{|}}{(CHCH_2OOCRCOO)_n} H + n\ H_2O$$

In the manufacture of unsaturated polyester resins the polyester is synthesized and then diluted with a vinyl reactive monomer such as styrene (see POLYESTERS, UNSATURATED). A portion of the dibasic acid of the polyester is maleic or some other vinyl reactive diacid that can be polymerized with the styrene to yield a highly cross-linked, high performance polymer system. Other esters made with propylene glycol, dipropylene glycol, and tripropylene glycol are used as emulsifiers in foods, as plasticizers in polymer systems, and as part of acrylate resin systems.

Polyethers are also products of commercial importance. Ethers can be formed by thermal dehydration, as shown for the formation of dipropylene glycol from propylene glycol. Cyclic ethers can form by elimination of water from di- or tripropylene glycol.

$$2\ CH_3CHOHCH_2OH \rightarrow CH_3CHOHCH_2OCH_2CHOHCH_3 + H_2O$$

The principal product obtained when heating propylene glycol in the presence of aluminum silicate at 200 to 400°C is the cyclic ether 2,5-dimethyl-1,4-dioxane (11).

The synthesis practiced in industry to make the important class of polyethers called polyols or polyglycols is the acid- or base-catalyzed addition of an epoxide such as propylene oxide to an active hydrogen compound such as a glycol. Any of the propylene glycols may be used for this purpose. Polyglycols can be tailored to meet rigorous application specifications for use in polyesters, polyurethanes, or other systems where active hydrogen compounds are needed.

$$CH_3CHOHCH_2OH + NaOH \rightarrow CH_3CHOHCH_2O^- \ Na^+$$

$$CH_3CHOHCH_2O^- \ Na^+ \ + \ \underset{\underset{O}{\diagdown\diagup}}{CH_2CHCH_3} \rightarrow CH_3CHOHCH_2OCH_2\overset{\overset{\displaystyle CH_3}{|}}{CHO^-} \ Na^+$$

Oxidation of a glycol can lead to a variety of products. Periodic acid quantitatively cleaves 1,2-glycols to aldehydes and is used as an analysis method for glycols (12,13). The oxidation of propylene glycol over Pd/C modified with Pb, Bi, or Te forms a mixture of lactic acid, hydroxyacetone, and pyruvic acid (14). Air oxidation of propylene glycol using an electrolytic crystalline silver catalyst yields pyruvic aldehyde.

Certain bacterial strains convert propylene glycol to pyruvic acid in the presence of thiamine (15); other strains do the conversion without thiamine (16). Propylene oxide is the principal product of the reaction of propylene glycol over a cesium impregnated silica gel at 360°C in the presence of methyl ethyl ketone and xylene (17).

Aldehydes and ketones react with glycols to form acetals and ketals which are easily hydrolyzed, making this a convenient method for protecting aldehyde or ketone functionality in organic synthesis. Propylene glycol in the presence of an acid catalyst reacts with aldehydes and ketones with concurrent removal of water to give 4-methyl-1,3-dioxolanes (18). The reaction of chloroacetaldehyde with propylene glycol in the presence of an acid catalyst gives 2-(chloromethyl)-4-methyl-1,3-dioxolane (R = CH_2Cl; R′ = H) (19).

$$CH_3CHOHCH_2OH + R\overset{\overset{\displaystyle O}{||}}{C}R' \longrightarrow CH_3\text{—}\underset{}{\bigg\langle}\ \text{structure}$$

Lactones are prepared from formaldehyde and carbon monoxide by cyclocondensation with propylene glycol in the presence of a strong acid and a Cu(I) or Ag carbonyl catalyst (20).

$$CH_3CHOHCH_2OH + H\overset{\overset{\displaystyle O}{||}}{C}H + CO \longrightarrow \text{structure}$$

Cyclic carbonates are made by treating 1,2-diols with dialkyl carbonates using an alkyl ammonium and tertiary amine catalyst. The combination of propylene glycol and dimethyl carbonate has been reported to result in a 98% yield of propylene carbonate (21).

$$CH_3CHOHCH_2OH + CH_3O\overset{\overset{\displaystyle O}{\|}}{C}OCH_3 \longrightarrow \underset{\underset{\displaystyle CH_3}{}}{\overset{\overset{\displaystyle O}{}}{\big\langle}} +2\ CH_3OH$$

Stereochemical and Structural Isomers

Propylene glycol, dipropylene glycol, and tripropylene glycol all have several isomeric forms. Propylene glycol has one asymmetric carbon and thus there are two enantiomers: (R)-1,2-propanediol and (S)-1,2-propanediol. 1,3-Propanediol is a structural isomer. Dipropylene glycol exists in three structural forms and since each structural isomer has two asymmetric carbons there are four possible stereochemical isomers per structure or a total of twelve isomers. These twelve consist of four enantiomer pairs and two meso- compounds. Tripropylene glycol has four structural isomers and each structural isomer has three asymmetric carbons so each structural isomer has eight possible stereochemical isomers or a total of 32 isomers of tripropylene glycol. Table 3 gives a listing of the IUPAC names, unique structures, and CAS Registry Numbers for the various isomers of propylene glycol, dipropylene glycol, and tripropylene glycol that have been reported in the literature.

Physical and Chemical Properties

Table 4 lists various physical and chemical properties and constants for the propylene glycols. A comprehensive source for additional physical and chemical properties is Reference 25.

Consumption and Use

Consumption of propylene glycol follows an erratic pattern in the United States which reflects domestic economic conditions. Table 5 gives a breakdown by percentage for the principal uses of propylene glycol in the United States.

Propylene Glycol. Propylene glycol is unique among the glycols in that it is safe for humans to take internally. Propylene glycol intended for human use is designated as USP grade and is commonly found in foods, pharmaceuticals, cosmetics, and other applications involving possible ingestion or absorption through the skin. The U.S. Food and Drug Administration has approved the use of propylene glycol in various food categories, as shown in Table 6.

An industrial grade of propylene glycol is usually specified for other uses. In common with most other glycols, propylene glycol is odorless and colorless, and has a wide range of solvency for organic materials, besides being completely water

Table 3. Propylene Glycol Isomers

IUPAC name	Formula	CAS Registry Number	Boiling point,[a] °C
Propylene glycol			
1,2-propanediol	$CH_3CH(OH)CH_2OH$	[57-55-6]	189, 115.9[b,c]
(±)-1,2-propanediol		[4254-16-4]	
(R)-1,2-propanediol		[4254-14-2]	
(S)-1,2-propanediol		[4254-15-3]	
Dipropylene glycol			
oxybispropanol		[25265-71-8]	152.5[b,d]
1,1'-oxybis-2-propanol	$[CH_3CH(OH)CH_2]_2O$	[110-98-5]	222.2[c], 229–32[d]
(R*,R*)-(±)-1,1'-oxybis-2-propanol		[55716-55-7]	
(R*,S*)-1,1'-oxybis-2-propanol		[55716-54-6]	
[S-(R*,R*)]-1,1'-oxybis-2-propanol		[61217-63-8]	
2,2'-oxybis-1-propanol	$[HOCH_2CH(CH_3)]_2O$	[108-61-2]	225.7[c]
(R*,R*)-(±)-2,2'-oxybis-1-propanol		[20753-87-1]	
(R*,S*)-2,2'-oxybis-1-propanol		[20753-88-2]	
[S-(R*,R*)]-2,2'-oxybis-1-propanol		[125948-51-8]	
2-(2-hydroxypropoxy)-1-propanol	$HOCH_2CH(CH_3)OCH_2CH(CH_3)OH$	[106-62-7]	224.0[c]
(R*,R*)-(±)-2-(2-hydroxypropoxy)-1-propanol		[62376-49-2]	
(R*,S*)-2-(2-hydroxypropoxy)-1-propanol		[62376-50-5]	
[S-(R*,R*)]-2-(2-hydroxypropoxy)-1-propanol		[110813-96-2]	
Tripropylene glycol			
[(1-methyl-1,2-ethanediyl)bis(oxy)]bispropanol	$HOCH(CH_3)CH_2OCH(CH_3)CH_2OCH_2CH(CH_3)OH$	[24800-44-0]	181.6[b,d]
1,1'-[(1-methyl-1,2-ethanediyl)bis(oxy)]bis-2-propanol	$HOCH_2CH(CH_3)OCH(CH_3)CH_2OCH_2CH(CH_3)OH$	[1638-16-0]	
2,2'-[(1-methyl-1,2-ethanediyl)bis(oxy)]bis-1-propanol	$HOCH_2CH(CH_3)OCH_2CH(CH_3)OCH_2CH(CH_3)OH$		
2-[1-(2-hydroxypropoxy)-2-propoxy]-1-propanol	$HOCH_2CH(CH_3)OCH(CH_3)CH_2OCH(CH_3)CH_2OH$		
2-[2-(2-hydroxypropoxy)propoxy]-1-propanol		[45096-22-8]	268[e]

[a]At 101.3 kPa = 1 atm, unless otherwise noted. [b]At 6.7 kPa (50 mm Hg). [c]Ref. 22. [d]Ref. 23. [e]Ref. 24.

Table 4. Properties of Glycols

Physical properties	Propylene glycol	Dipropylene glycol	Tripropylene glycol
formula	$C_3H_8O_2$	$C_6H_{14}O_3$	$C_9H_{20}O_4$
molecular weight	76.1	134.2	192.3
boiling point at 101.3 kPa,[a] °C	187.4	232.2[b]	265.1[b]
vapor pressure, kPa,[a] 25°C	0.017	0.0021	0.0003
density, g/mL			
25°C	1.032	1.022	1.019
60°C	1.006	0.998	0.991
freezing point, °C	supercools	supercools	supercools
pour point, °C	< -57	-39	-41
viscosity, mPa·s($=$cP)			
25°C	48.6	75.0	57.2
60°C	8.42	10.9	9.7
surface tension, mN/m($=$dyn/cm), 25°C	36	35	34
refractive index at 25°C	1.431	1.441	1.442
specific heat at 25°C, J/(g·K)[c]	2.51	2.18	1.97
flash point, °C, PMCC[d]	104	124	143
coefficient of expansion $\times 10^4$, 0–60°C	7.3	7.0	8.1
thermal conductivity at 25°C, W/(m·K)	0.2061	0.1672	0.1582
heat of formation, kJ/mol[c]	-422	-628	-833
heat of vaporization at 25°C, kJ/mol[c]	67.0	45.4	35.4

[a]To convert kPa to mm Hg, multiply by 7.5.
[b]Varies with isomer distribution.
[c]To convert J to cal, divide by 4.184.
[d]PMCC = Penskey-Martens closed cup.

Table 5. Uses of Propylene Glycols in the United States, 1992

Application	Consumed, %
unsaturated polyester resins	37
cosmetics, pharmaceuticals, foods	17
pet food	3
tobacco humectant	4
functional fluids	16
paints and coatings	5
liquid detergents	11
others	7

soluble. Propylene glycol is also a known antimicrobial and is an effective food preservative (26).

Propylene glycol is an important solvent for aromatics in the flavor concentrate industry, enabling manufacturers to produce low cost flavor concentrates of high quality. It is also an excellent wetting agent for natural gums, greatly sim-

Table 6. U.S. FDA-Approved Use of Propylene Glycol in Foods

Food category	Maximum content, wt %
alcoholic beverages	5.0
confections and frosting	24.0
frozen dairy products	2.5
seasonings and flavorings	97.0
nuts and nut products	5.0
all other foods	2.0

plifying the compounding of citrus and other emulsified flavors. PG also finds use as a solvent in elixirs and pharmaceutical preparations containing some water-soluble ingredients, and as a solvent and coupling agent in the formulation of sunscreen lotion, shampoos, shaving creams, and other similar products. Certain esters of propylene glycol such as propylene glycol monostearate [1323-39-3] are also popular as an emulsifier in cosmetic and pharmaceutical creams.

Aqueous solutions of propylene glycol display excellent antifreeze properties and are therefore valuable as low temperature heat-transfer fluids. For applications involving indirect food contact, heat-transfer fluids formulated with the USP grade product are preferred, since there could be inadvertent contact with a food product. These fluids are commonly used in the brewing and dairy industries as well as in refrigerated display cases in retail grocery stores.

Propylene glycol is also an effective humectant, preservative, and stabilizer and is found in such diverse applications as semimoist pet food, bakery goods, food flavorings, salad dressings, and shave creams. Humectancy, or the capability of retaining moisture in a product, is a result of the vapor–liquid equilibria of the glycol–water system and can be estimated from tables provided by suppliers (27).

The industrial grade of propylene glycol is an important intermediate in the production of alkyd resins for paints and varnishes. It is the preferred glycol for manufacturing high performance, unsaturated polyester resins for many uses, eg, reinforced plastic laminates for marine construction, gel coats, sheet molding compounds (SMC), and synthetic marble castings. It is also used as a solvent and plasticizer in printing inks, as a preservative in floral arrangements, and as a stabilizer in hydraulic fluids. Heat-transfer fluids used in commercial and industrial building heating and cooling systems, chemical plants, stationary engines, and solar heat recovery can be formulated with the industrial grade of propylene glycol. More recently, propylene glycol-based coolants for automobiles and heavy-duty diesel engine trucks have been introduced which compete with traditional ethylene glycol-based products (28,29). The newly published ASTM standard D5216 specifies aqueous propylene glycol-based engine coolants for automobile and light-duty truck service. All heat-transfer applications require corrosion inhibitor additives and are designed for specific operating temperature ranges and types of materials of construction. Operation at low temperature without freezing and at high temperature without excessive pressure are the principal features of these systems.

Due in large part to its lower toxicity and the concomitant lesser concern about its environmental impact, propylene glycol use in the air transportation industry as an airplane and runway deicing agent has grown substantially in recent years (see ANTIFREEZES AND DEICING FLUIDS). Other glycols, such as ethylene glycol, have historically been used in this industry.

Dipropylene Glycol. Dipropylene glycol is similar to the other glycols in general properties, and its fields of use are comparable. However, its greater solvency for certain materials and higher viscosity make it of interest in certain applications for which the other glycols are not as well suited. The greater solvency of dipropylene glycol for castor oil indicates its usefulness as a component of hydraulic brake fluid formulations; its affinity for certain other oils has likewise led to its use in cutting oils, textile lubricants, and industrial soaps. It is also used as a reactive intermediate in manufacturing polyester resins, plasticizers, and urethanes. Fragrance or low odor grades of dipropylene glycol are established standard base formulating solvents in the fragrance industry and for some personal care products such as deodorants.

Tripropylene Glycol. Tripropylene glycol is an excellent solvent in many applications where other glycols fail to give satisfactory results. Its ability to solubilize printing ink resins is especially marked, so much so that it finds its way into creams designed to remove ink stains from the hands. A combination of water solubility and good solvent power for many organic compounds plus low volatility and a high boiling point also have led to its use by formulators of textile soaps and lubricants, cutting oil concentrates, and many similar products. Tripropylene glycol is also used as a reactant to produce acrylate resins which are useful in radiation-cured coatings, adhesives, and inks. Polyethers used in the manufacture of urethane rigid foam insulation are made by alkoxylation of tripropylene glycol.

Toxicology

All of the propylene glycols display a low acute oral toxicity in laboratory rats as shown in Table 7 (30). Information for sucrose is shown for comparison.

Studies in which rats were fed drinking water containing as much as 10% propylene glycol over a period of 140 days (31,32) showed no apparent ill effects. Other investigations have revealed that rats can tolerate up to 4.9% propylene glycol in the diet for two years without significant effects on growth rate (33). However, minor liver damage was observed. In a more recent study (34), dogs

Table 7. Acute Oral Toxicity of Propylene Glycols[a]

Compound	Oral rat LD_{50}, g/kg
propylene glycol	20.0
dipropylene glycol	14.9
tripropylene glycol	3.0
sucrose	29.7

[a]Ref. 30.

were fed a diet containing 8% propylene glycol for two years and were unaffected, as judged by mortality, body weight changes, diet utilization, histopathology, organ weights, and blood, urine, and biochemical parameters. Because of its low chronic oral toxicity, propylene glycol is considered safe for use in foods and pharmaceuticals.

Rats showed no adverse effects from 5.0% dipropylene glycol in their drinking water for 77 days, but at a dose of 10.0% in the drinking water, kidney and liver injury and some deaths occurred (35). A sufficient number of studies have not been carried out on tripropylene glycol to permit conclusions to be drawn regarding its chronic oral toxicity.

Propylene glycols produce a negligible degree of irritation upon eye or skin contact. From tests on New Zealand white rabbits (36) in 1982 it was concluded that propylene glycol is a slight eye irritant. Other tests conducted both *in vitro* and *in vivo* have shown propylene glycol to be a nonirritant to the eye. Both dipropylene glycol and tripropylene glycol have been tested for skin and eye irritation in rabbits and have been found to be nonirritating. The expert panel of the *Cosmetic Ingredient Review*, after conducting a comprehensive safety assessment, has concluded that propylene glycol may be used in cosmetic products in concentrations up to 50% (37).

Inhalation of the vapors of any of the propylene glycols appears to present no significant hazard in ordinary applications and this is reflected in the fact that OSHA has not found it necessary to establish a permissible exposure level in the workplace. However, in 1985 the American Industrial Hygiene Association reviewed human experience and animal data and established a Workplace Environmental Exposure Level (WEEL) guideline for propylene glycol at 50 ppm total vapor and aerosol averaged over an eight-hour period. Although not legally binding, the WEEL guidelines have been adopted by industry as good industrial hygiene practice. Limited data indicate that breathing mists of di- and tripropylene glycol may be harmful. Prolonged inhalations of saturated vapors of propylene glycol have produced no ill effects in animals, but such concentrations would likely be irritating to the upper respiratory tract and possibly the eyes of humans. Only limited work has been done on the vapor toxicity of the other glycols; however, because of their very low vapor pressures and low systemic toxicities, it is unlikely that injury would occur as a result of limited vapor inhalation.

Environmental Considerations

The propylene glycols vary in biodegradability, as shown in Table 8. The tests involved were conducted with standard municipal inoculum, and other studies (31) have shown that biodegradability can be greatly enhanced when using an acclimated bacteria. For example, tripropylene glycol has shown 66% of theoretical oxygen demand at 20 days with an industrial seed. Thus it is expected that all of the propylene glycols will exhibit moderate to high biodegradability in a natural environment.

All of the propylene glycols are considered to be practically nontoxic to fish on an acute basis ($LC_{50} > 100$ mg/L) and practically nontoxic to aquatic invertebrates, also on an acute basis. Acute marine toxicology testing (38) on propylene

Table 8. Biodegradation of Propylene Glycols With Standard Municipal Inoculum

Glycol	Theoretical O_2 demand	5-Day O_2 demand	10-Day O_2 demand	20-Day O_2 demand
PG	1.68	1.16	1.18	1.45
DPG	1.91	a	0.14	0.71
TPG	1.38	a	a	a

[a] Data not available.

glycol showed that the 96-h LC_{50} for fathead minnows was 54,900 mg/L and the 48-h LC_{50} for *Daphnia magna* was 34,400 mg/L. A 24-h NOEL of 50,000 mg/L was also observed for fingerling trout. Similar results were observed for guppies and rainbow trout (39).

BIBLIOGRAPHY

"Glycols" in *ECT* 1st ed., Vol. 7, pp. 238–263, by C. P. McClelland and P. R. Rector, Carbide and Carbon Chemicals Co., a Division of Union Carbide and Carbon Corp.; "Ethylene Glycol, Propylene Glycol, and Their Derivatives" under "Glycols" in *ECT* 2nd ed., Vol. 10, pp. 638–660, by P. H. Miller, Union Carbide Corp.; "Ethylene Glycol and Propylene Glycol" under "Glycols" in *ECT* 3rd ed., Vol. 11, pp. 933–956, by E. S. Brown, C. F. Hauser, B. C. Ream, and R. V. Berthold, Union Carbide Corp.

1. J. W. Lawrie, *Glycerol and the Glycols*, Monograph Series, American Chemical Society, 1928, New York, p. 388.
2. A. Wurtz, *Ann. Chim. Phys.* **55**(3), 438 (1859).
3. I. Mellan, *Polyhydric Alcohols*, Spartan Books, Washington, D.C., 1962, p. 46.
4. *Chem. Mark. Rep.*, 45 (Jan. 4, 1993).
5. D. J. Caney, *Chem. Mark. Rep.*, 13 (July 20, 1992).
6. *Synthetic Organic Chemicals*, United States Production and Sales, United States International Trade Commission, Washington, D.C., 1990.
7. R. E. Parker and N. S. Isaacs, *Chem. Rev.* **59**, 737 (1959).
8. G. Reginato, A. Ricci, S. Roelens, and S. Scapecchi, *J. Org. Chem.* **55**, 5132 (1990).
9. J. A. Monick, *Alcohols—Their Chemistry, Properties, and Manufacture*, Reinhold Publishing Corp., New York, 1968.
10. G. O. Curme and F. Johnston, *Glycols*, Reinhold Publishing Corp., New York, 1953.
11. E. Swistak, *Compt. Rend.* **240**, 1544 (1955).
12. S. Siggia, *Quantitative Organic Analysis via Functional Groups*, John Wiley & Sons, Inc., New York, 1949, p. 8.
13. Ref. 10, p. 329.
14. T. Tsujino, S. Ohigashi, S. Sugiyama, K. Kawashiro, and H. Hayashi, *J. Mol. Catal.* **71**(1), 25–35, 1992.
15. Y. Izumi, Y. Matsumura, Y. Tani, and H. Yamada, *Agric. Biol. Chem.* **46**(11), 2673–2679 (1982).
16. T. Shigeno and T. Nakahara, *Biotechnol. Lett.* **13**(11), 821–826 (1991).
17. Eur. Pat. Appl. EP 145447 (June 19, 1985), J. Y. Ryn (to Exxon Research and Engineering Co.).
18. T. H. Chan, M. A. Brook, and T. Chaly, *Synthesis*, 203 (1983).
19. Eur. Pat. 456157 A1 (Nov. 13, 1991), M. Ishizuka and T. Wakasugi (to Kureha Chemical Industry Co., Ltd., Japan).

20. U.S. Pat. 4,990,629 (Feb. 5, 1991), Y. Souma (to Agency of Industrial Science and Technology Japan).

21. U.S. Pat. 5,091,543 (Feb. 25, 1992), R. A. Grey (to Arco Chemical Technology, Inc.).

22. A. R. Sexton and E. C. Britton, *J. Am. Chem. Soc.* **75**, 4357 (1953).

23. *A Guide To Glycols, Brochure 117-00991-92HYC,* The Dow Chemical Co., Chemicals and Metals Department, Midland, Mich., 1992.

24. R. C. West and J. G. Grasselli, *Handbook of Data on Organic Compounds*, 2nd ed., Vol. VI, CRC Press Inc., Boca Raton, Fla., 1989.

25. *Physical and Thermodynamic Properties of Pure Chemicals,* American Institute of Chemical Engineers, New York, 1991.

26. M. Burr and L. Tice, *J. Am. Pharm. Assoc.* **XLVI**(4), 217 (1957).

27. *A Guide to Glycols,* The Dow Chemical Co., Midland, Mich., 1992.

28. G. E. Coughenour and L. K. Hwang, Paper 930584, Society of Automotive Engineers, Inc., Warrendale, Pa., 1993.

29. W. J. Kilmartin and D. C. Dehm, Paper 930588, Society of Automotive Engineers, Inc., Warrendale, Pa., 1993.

30. R. J. Lewis, Sr., *Dangerous Properties of Industrial Materials*, 8th ed., Van Nostrand Reinhold, New York, 1992.

31. M. A. Seidenfeld and P. J. Hanzlik, *J. Pharmacol. Exp. Therap.* **44**, 109 (1932).

32. J. H. Weatherby and H. G. Haag, *J. Am. Pharm. Ass.* **27**, 446 (1938).

33. H. J. Morris and co-workers, *J. Pharmacol. Exp. Therap.* **74**, 266 (1942).

34. C. S. Weil and co-workers, *Food Cosmet. Toxicol.* **9**, 479 (1971).

35. H. D. Keston, N. G. Mulinos, and L. Pomerantz, *Arch. Pathol.* **27**, 447 (1939).

36. J. P. Guillot, and co-workers, *Food Chem. Toxicol.* **20**(5), 573–582 (1982).

37. *Tentative Report of the Safety Assessment of Propylene Glycol, Cosmetic Ingredient Review,* The Cosmetic, Toiletry, and Fragrance Association, Washington, D.C., 1992.

38. K. Verschueren, *Handbook of Environmental Data,* 2nd ed., Van Nostrand Reinhold Co., New York, 1983.

39. F. L. Mayer, and M. R. Ellersieck, *Manual of Acute Toxicity: Interpretation & DB*, FWS, Resource Publ. 160, U.S. Department of the Interior, Washington, D.C., 1986.

ALTON E. MARTIN
FRANK H. MURPHY
The Dow Chemical Company

OTHER GLYCOLS

Glycols such as neopentyl glycol, 2,2,4-trimethyl-1,3-pentanediol, 1,4-cyclohex-anedimethanol, and hydroxypivalyl hydroxypivalate are used in the synthesis of polyesters (qv) and urethane foams (see FOAMED PLASTICS). Their physical properties are shown in Table 1 (1–6).

Neopentyl Glycol

Neopentyl glycol, or 2,2-dimethyl-1,3-propanediol [126-30-7] (**1**) is a white crystalline solid at room temperature, soluble in water, alcohols, ethers, ketones, and toluene but relatively insoluble in alkanes (1). Two primary hydroxyl groups are provided by the 1,3-diol structure, making this glycol highly reactive as a chemical

Table 1. Physical Properties of Several Glycols

Properties	Neopentyl glycol (1)	2,2,4-Trimethyl-1,3-pentanediol (7)	1,4-Cyclohexane-dimethanol[a] (8)	Hydroxypivalyl hydroxypivalate (9)
CAS Registry Number	[126-30-7]	[144-19-4]	[105-08-5]	[1115-20-4]
molecular formula	$C_5H_{12}O_2$	$C_8H_{18}O_2$	$C_8H_{16}O_2$	$C_{10}H_{20}O_4$
mol wt	104.2	146.2	144.2	204.3
melting range, °C	124–130	46–55	45–50[b]	46–50
sublimation temp, °C	128			
boiling point, °C, at kPa[c]				
at 0.13			118	
at 0.45	93–94			
at 1.33			160	
at 101.3	212	236	286	290
boiling range, °C		215–235		
assay (commercial grade), wt % min	97	96	99	98
density, g/cm³				
at 20°C	1.06		1.02	1.02
at 15°C		0.937		
pour point, supercooled, °C			10	
crystallization point, °C			35	
viscosity at 50°C, mPa·s(= cP)			675	70
heat of combustion, kJ/mol[d]	−3100	−5050	−4849[e]	
flammability				
fire point, COC,[f] °C	135	118	174	
flash point, COC,[f] °C	129	113	167	161
autoignition temp,[g] °C	388	346	316	404
heat of fusion,[h] kJ/mol[d]	21.77	8.63		
heat of vaporization, kJ/mol[d]				
at 32 kPa,[c] 170°C	67.1			
at 101.3 kPa[c]			95.6	
at 101.3 kPa,[c] 204°C	56.5			
hygroscopity,[i] wt % H_2O				
at 50% rh		0.1–0.2		
at 51% rh	0.3			
at 78% rh	11.3			

[a]Mixture of isomers, cis/trans ratio (wt %) = ~ 32/68.
[b]Mp of cis isomer [3236-47-3] = 41°C; mp of trans isomer [3236-48-4] = 70°C.
[c]To convert kPa to mm Hg, multiply by 7.5.
[d]To convert kJ to kcal, divide by 4.184.
[e]Paar bomb.
[f]Cleveland open cup.
[g]ASTM D286.
[h]Estimated.
[i]At equilibrium; neopentyl glycol at 25–38°C, and 2,2,4-trimethyl-1,3-pentanediol at 25°C.

intermediate. The *gem*-dimethyl configuration is responsible for the exceptional hydrolytic, thermal, and uv stability of neopentyl glycol derivatives.

Chemical Properties. Neopentyl glycol can undergo typical glycol reactions such as esterification (qv), etherification, condensation, and oxidation. When basic kinetic studies of the esterification rate were carried out for neopentyl glycol, the absolute esterification rate of neopentyl glycol with *n*-butyric acid was approximately 20 times that of ethylene glycol with *n*-butyric acid (7).

Manufacture.   Commercial preparation of neopentyl glycol can be via an alkali-catalyzed condensation of isobutyraldehyde with 2 moles of formaldehyde (crossed Cannizzaro reaction) (2,8). Yields are ~70%.

$$(CH_3)_2CHCHO + 2\ CH_2O + KOH \rightarrow HOCH_2 \overset{\displaystyle CH_3}{\underset{\displaystyle CH_3}{-\overset{|}{\underset{|}{C}}-}} CH_2OH + HCOOK$$

(1)

Neopentyl glycol is manufactured by Eastman Chemical Co., BASF, Perstorp, Hoechst, Mitsubishi Gas, Polioli, and Hüls. In 1993, the bulk U.S. price was $1.61/kg.

Toxicity.   Acute toxicity data for neopentyl glycol (1) are reported in Table 2.

Table 2. Toxicity Data for Various Glycols

Parameter	Structure number[a]			
	(1)	(7)	(8)	(9)
oral LD_{50} (rat), mg/kg	6,400–12,800	3,730	3,200–6,400	>3,200
oral LD_{50} (mouse), mg/kg	3,200–6,400	1,600–3,200	1,600–3,200	1,600–3,200
inhalation LC_{50} (rat),[b] mg/L/6 h[b]	168	73.3[c]		>1.18[d]
dermal LD_{50} (guinea pig), g/kg	14[e]	slight	>1	>1
eye irritation (rabbit)	slight[f]	moderate–strong	slight	moderate
Reference	9	10	11	12

[a]See text and Table 1.
[b]Unless otherwise noted.
[c]Also mouse.
[d]1.18 mg/L/6 h.
[e]No skin sensitization (guinea pig).
[f]Skin irritation (rabbit) is moderate.

Uses.   Neopentyl glycol is used extensively as a chemical intermediate in the manufacture of polyester resins (see ALKYD RESINS), polyurethane polyols (see URETHANE POLYMERS), synthetic lubricants, polymeric plasticizers (qv), and other polymers. It imparts a combination of desirable properties to properly formulated esterification products, including low color, good weathering and chemical resistance, and improved thermal and hydrolytic stability.

The weatherability and hydrolytic stability of unsaturated polyesters based on neopentyl glycol have made it a popular intermediate for use in formulations exposed to severe conditions, eg, in gel coats for cultured marble and marine applications (see COATINGS, MARINE) (13).

Reactive saturated polyester resins (oil-free alkyds) based on neopentyl glycol are produced for use in formulating premium-quality surface coatings (14–16). These coatings exhibit excellent water, detergent, and stain resistance, and excellent weatherability, acid rain resistance, and gloss retention. (17). They may be formulated as conventional solvent-borne coatings, as high solids coatings, or as dry powders for electrostatic coatings applications.

A comparison of coatings formulations based on various glycols to determine the effects of the various glycol structures on the performance properties of the coatings has been made. Properties compared included degree of cure, flexibility, hardness, hydrolytic stability, processibility, chemical and stain resistance, and viscosity (18,19).

The polyurethane industry provides other uses for neopentyl glycol as an intermediate in the manufacture of hydroxy-terminated polyester polyols. Beginning with basically the same ingredients, products with a wide range of properties, varying from soft to rigid foams to elastomers (qv) and adhesives (qv) may be produced from polyols based on neopentyl glycol. This glycol also is employed to improve thermal, hydrolytic, and uv stability (20–23).

Synthetic lubricants are made with neopentyl glycol in the base-stock polyester (24). Excellent thermal stability and viscosity control are imparted to special high performance aviation lubricants by the inclusion of polyester thickening agents made from neopentyl glycol (25,26) (see LUBRICATION AND LUBRICANTS). Neopentyl glycol is also used to manufacture polymeric plasticizers that exhibit the improved thermal, hydrolytic, and uv stability necessary for use in some exterior applications (27).

Neopentyl glycol can be used for thermal energy storage by virtue of its solid-phase transition, which occurs at 39–41°C, a temperate range useful for solar heating and cooling (28–31).

Derivatives. A number of derivatives of neopentyl glycol have been prepared; some show promise for commercial applications.

$$H_2N-CH_2-\underset{\underset{CH_3}{|}}{\overset{\overset{CH_3}{|}}{C}}-CH_2-NH_2$$

(2)   (3)   (4)   (5)   (6)

Organophosphorus Derivatives. Neopentyl glycol treated with pyridine and phosphorus trichloride in anhydrous dioxane yields the cyclic hydrogen phosphite, 5,5-dimethyl-1,3-dioxaphosphorinane 2-oxide (2) (32,33). Compounds of this type may be useful as flameproofing plasticizers, stabilizers, synthetic lubricants, oil additives, pesticides, or intermediates for the preparation of other organophosphorus compounds (see FLAME RETARDANTS; PHOSPHORUS COMPOUNDS).

Acetals and Ketals. Acetals of 1,3-diols are prepared by refluxing the diol with the aldehyde in the presence of an acid catalyst, even in an aqueous medium. The corresponding ketals are more difficult to prepare in aqueous solution, but cyclic ketals of neopentyl glycol, eg, 2-butyl-2-ethyl-5,5-dimethyl-1,3-dioxane (3), can be prepared if the water of reaction is removed azeotropically (34).

Cyclopropane Derivatives. 2,2-Dimethylcyclopropanenitrile [5722-11-2] (4) has been made by preparing the di-*p*-toluenesulfonate of neopentyl glycol and treating the diester with potassium cyanide (35).

Diamine. 2,2-Dimethyl-1,3-propanediamine [7328-91-8] (5) has been prepared by amination of neopentyl glycol by treating the glycol with ammonia and hydrogen at 150–250°C at 10–31 MPa (1500–4500 psig) over a Ni catalyst. The diamine is useful for preparation of crystalline polyureas by reaction with diisocyanates (36).

Esters. Neopentyl glycol diesters are usually liquids or low melting solids. Polyesters of neopentyl glycol, and in particular unsaturated polyesters, are prepared by reaction with polybasic acids at atmospheric pressure. High molecular weight linear polyesters (qv) are prepared by the reaction of neopentyl glycol and the ester (usually the methyl ester) of a dibasic acid through transesterification (37–38). The reaction is usually performed at elevated temperatures, *in vacuo*, in the presence of a metallic catalyst.

Cyclic carbonates are prepared in satisfactory quality for anionic polymerization by catalyzed transesterification of neopentyl glycol with diaryl carbonates, followed by tempering and depolymerization. Neopentyl carbonate (5,5-dimethyl-1,3-dioxan-2-one) (**6**) prepared in this manner has high purity (99.5%) and can be anionically polymerized to polycarbonates with mol wt of 35,000 (39).

2,2,4-Trimethyl-1,3-Pentanediol

2,2,4-Trimethyl-1,3-pentanediol (**7**) is a white, crystalline solid. It is used in surface coating and unsaturated polyester resins. It also appears promising as an intermediate for synthetic lubricants and polyurethane elastomers and foams.

Trimethylpentanediol is soluble in most alcohols, other glycols, aromatic hydrocarbons, and ketones, but it has only negligible solubility in water and aliphatic hydrocarbons (4).

Chemical Properties. Trimethylpentanediol, with a primary and a secondary hydroxyl group, enters into reactions characteristic of other glycols. It reacts readily with various carboxylic acids and diacids to form esters, diesters, and polyesters (40). Some organometallic catalysts have proven satisfactory for these reactions, the most versatile being dibutyltin oxide. Several weak bases such as triethanolamine, potassium acetate, lithium acetate, and borax are effective as stabilizers for the glycol during synthesis (41).

Manufacture and Processing. 2,2,4-Trimethyl-1,3-pentanediol can be produced by hydrogenation of the aldehyde trimer resulting from the aldol condensation of isobutyraldehyde [*78-84-2*].

(**7**)

Eastman Chemical Co. is the only manufacturer of this glycol. Prices in 1993 were $1.20/kg for bulk quantities.

Toxicity. Acute toxicity data for this glycol (**7**) are reported in Table 2.

Uses. The versatility of trimethylpentanediol as an intermediate is reflected by the diversity of its commercial applications.

Unsaturated polyesters derived from trimethylpentanediol are characterized by a low exotherm, low shrink curing, and a product that has good electrical properties and excellent hydrolytic stability and chemical resistance (10,19,40,41). These unsaturated resins exhibit low viscosity, low density, and good glass–fiber or filler wetting (10,41). They are resistant to hydrolytic action, probably because of the protection of the ester linkage by the pendent methyl groups, and by virtue of the lesser number of ester groups in a given weight of material than would be afforded with lower molecular weight glycols (42). Chemically resistant applications where trimethylpentanediol-based unsaturated resins are used include reinforced polyester storage tanks and pipelines for hot, concentrated acids, some solvents, mild bases, hypochlorite solution, and ammonia (see POLYESTERS, UNSATURATED).

Saturated polyester resins based on trimethylpentanediol are used in various coating applications, most notably in water-borne (43) and high solids coatings (44–45). Resins manufactured with this diol are characterized by low viscosities, which permits formulation of enamels with 85% nonvolatiles when sprayed. Such formulations are cross-linked with isocyanates or melamines to give premium coatings useful for industrial applications, appliance coil coatings, and the like (44–46). Other saturated polyester resins based on trimethylpentanediol are useful in high gloss hypoallergenic nail polishes (47) (see POLYESTERS, THERMOPLASTIC).

The monoisobutyrate ester of trimethylpentanediol is especially useful as a coalescing aid in flat and semigloss (48) latex paint formulations (see PAINT). This product is commercially available from Eastman as Texanol ester alcohol.

The diisobutyrate ester of trimethylpentanediol is an economical, low color primary plasticizer for use in surface coatings, vinyl flooring, moldings, and other vinyl products. This diester is commercially available from Eastman as Kodaflex TXIB plasticizer (49).

Various other diesters, mixed esters, and polyesters of trimethylpentanediol are useful as monomeric or polymeric plasticizers for coatings and plastic film and sheeting (49). They are compatible with, and useful in, cellulosics, vinyls, polystyrenes, and some other plastics.

Trimethylpentanediol is used in hard-surface cleaners as a coupling agent (50) and in temporary or semipermanent hair dyes (51). Other applications involving trimethylpentanediol, or a derivative, are in urethane elastomers (52), in foams (53), as a reactive diluent in urethane coatings (see URETHANE POLYMERS) (54), as a sound-insulating, glass–laminate adhesive (see ADHESIVES) (55), as a bactericide–fungicide (56), and as a cross-linking agent in poly(vinyl chloride) adhesive (57).

Derivatives. *Esters.* The monoisobutyrate ester of 2,2,4-trimethyl-1,3-pentanediol is prepared from isobutyraldehyde in a Tishchenko reaction (58,59). Diesters, such as trimethylpentane dipelargonate (2,2,4-trimethylpentane 1,3-dinonanoate), are prepared by the reaction of 2 mol of the monocarboxylic acid with 1 mol of the glycol at 150–200°C (60,61). The lower aliphatic carboxylic acid diesters of trimethylpentanediol undergo pyrolysis to the corresponding ester of 2,2,4-trimethyl-3-penten-1-ol (62). These unsaturated esters reportedly can be epoxidized by peroxyacetic acid (63).

Ketals. Trimethylpentanediol reportedly forms a cyclic ketal by heating it with benzophenone in the presence of sulfonic acid catalysts at reflux temperatures in toluene (64). These are said to be useful as aprotic solvents for ink-jet printing and as inflammation inhibitors for cosmetic preparations (65).

1,4-Cyclohexanedimethanol

1,4-Cyclohexanedimethanol, 1,4-dimethylolcyclohexane, or 1,4-bis(hydroxymethyl) cyclohexane (**8**), is a white, waxy solid. The commercial product consists of a mixture of cis and trans isomers (6). This diol is used in the manufacture of polyester fibers (qv) (64), high performance coatings, and unsaturated polyester molding and laminating resins (5).

$$HOCH_2-\langle\ \rangle-CH_2OH$$

(**8**)

1,4-Cyclohexanedimethanol is miscible with water and low molecular weight alcohols and appreciably soluble in acetone. It has only negligible solubility in hydrocarbons and diethyl ether (6).

Chemical Properties. The chemistry of 1,4-cyclohexanedimethanol is characteristic of general glycol reactions; however, its two primary hydroxyl groups give very rapid reaction rates, especially in polyester synthesis.

Manufacture. The manufacture of 1,4-cyclohexanedimethanol can be accomplished by the catalytic reduction under pressure of dimethyl terephthalate in a methanol solution (47,65). This glycol also may be prepared by the depolymerization and catalytic reduction of linear polyesters that have alkylene terephthalates as primary constituents. Poly(ethylene terephthalate) may be hydrogenated in the presence of methanol under pressure and heat to give good yields of the glycol (see POLYESTERS) (66,67).

1,4-Cyclohexanedimethanol is produced commercially by Eastman Chemical Co. The price within the United States in bulk truckloads as of June 1993 was 99% purity, $2.49/kg.

Toxicity. Acute toxicity data are reported in Table 2 (11).

Uses. The most important application for 1,4-cyclohexanedimethanol is in the manufacture of linear polyesters for use as polyester fibers such as the Kodel polyester fibers (68) (see FIBERS, POLYESTER). Compared with fibers made from poly(ethylene terephthalate), fibers made from poly(1,4-cyclohexanedimethanol terephthalate) have lower densities and higher melting points. Linear polyesters produced from dicarboxylic acids, eg, terephthalic acid, generally have greater hydrolytic stability and better electrical properties than similar polyesters made from other glycols. The high dielectric strength, good dielectric constant, and low dielectric loss make them useful for many electrical applications, eg, in capacitors, wire coatings, and magnetic tape coatings (69,70).

Unsaturated resins based on 1,4-cyclohexanedimethanol are useful in gel coats and in laminating and molding resins where advantage is taken of the properties of very low water absorption and resistance to boiling water (6). Thermal stability is imparted to molding resins, both thermoplastic (71,72) and thermoset (73–76), enabling retention of physical and electrical properties at elevated temperatures (77). Additionally, resistance to chemical and environmental exposure is characteristic of products made from these resins (78).

High performance polyester enamels are manufactured from saturated resins containing 1,4-cyclohexanedimethanol. Such enamels may be formulated as electrostatically applied powder coatings (79–82), water-borne (83,84), cationic electrodeposition coatings (85), or solvent-based coatings (see COATING PROCESSES). These coatings are characterized by an extremely hard, durable, stain- and detergent-resistant finish (88–88). Other polyester resins based on 1,4-cyclohexanedimethanol are also used as plasticizers (89), in hot-melt adhesives (90,91), in elastomers (92,93), and urethane elastomeric coatings (94) (see COATINGS).

Another area in which 1,4-cyclohexanedimethanol is commercially important is in the manufacture of polyurethane foams (see FOAMED PLASTICS). The two primary hydroxyl groups provide fast reaction rates with diisocyanates, which makes this diol attractive for use as a curative in foams. It provides latitude in improving physical properties of the foam, in particular the load-bearing properties. Generally, the ability to carry a load increases with the amount of 1,4-cyclohexanedimethanol used in producing the high resilience foam (95). Other polyurethane derivatives of 1,4-cyclohexanedimethanol include elastomers useful for synthetic rubber products with a wide range of hardness and elasticity (96).

Derivatives. *Mixed Phosphonate Esters.* Unsaturated, mixed phosphonate esters have been prepared from monoesters of 1,4-cyclohexanedimethanol and unsaturated dicarboxylic acids. For example, maleic anhydride reacts with this diol to form the maleate, which is treated with benzenephosphonic acid to yield an unsaturated product. These esters have been used as flame-retardant additives for thermoplastic and thermosetting resins (97).

Diesters. Diesters prepared from the diol and monocarboxylic acids are useful as antioxidants (qv) for polypropylene (98), and as plasticizers (qv).

Polyesters. Polyesters containing carbonate groups have been prepared from this diol (see POLYCARBONATES) (99). Films of this polymer, formed from an acetone or ethyl acetate solution, exhibit excellent adhesive properties.

Hydroxypivalyl Hydroxypivalate

Hydroxypivalyl hydroxypivalate or 3-hydroxy-2,2-dimethylpropyl 3-hydroxy-2,2-dimethylpropionate (**9**) is a white crystalline solid at room temperature. It is used to manufacture polyester resins for use in surface coatings where good resistance to weathering and acid rain are of particular importance (6).

$$HO-CH_2-\underset{\underset{CH_3}{|}}{\overset{\overset{CH_3}{|}}{C}}-\overset{\overset{O}{\|}}{C}-O-CH_2-\underset{\underset{CH_3}{|}}{\overset{\overset{CH_3}{|}}{C}}-CH_2-OH$$

(9)

Hydroxypivalyl hydroxypivalate is soluble in most alcohols, ester solvents, ketones, and aromatic hydrocarbons. It is partially soluble in water (6).

Chemical Properties.   Both hydroxy groups on hydroxypivalyl hydroxypivalate are primary, which results in rapid reactions with acids during esterification. The absence of hydrogens on the carbon atom beta to the hydroxyls is a feature this glycol shares with neopentyl glycol, resulting in excellent weatherability. The relatively high molecular weight of this glycol requires lower levels of aromatic acid to produce polyester resins, thus contributing to the improved weatherability of polyesters made from this glycol.

Manufacture.   Hydroxypivalyl hydroxypivalate may be produced by the esterification of hydroxypivalic acid with neopentyl glycol or by the intermolecular oxidation–reduction (Tishchenko reaction) of hydroxypivaldehyde using an aluminum alkoxide catalyst (100,101).

Eastman Chemical Co., BASF, Mitsubishi Gas, and Union Carbide are manufacturers of this glycol. The U.S. price in June 1993 was $2.97/kg.

Toxicity.   Acute toxicity data for (9) appear in Table 2.

Uses.   Saturated polyesters made from hydroxypivalyl hydroxypivalate are most often used for formulating coatings which have very low initial color and which retain the low color exposure to weathering. The most typical example is in clear topcoat useful in automotive finishes (102). This glycol is often used as a partial replacement for neopentyl glycol in polyester resins to provide better resin solubility, reduced crystallinity, lower glass-transition temperatures, and lower melt viscosity. These characteristics make this glycol particularly useful for resins used in coil coatings, powder coatings, waterborne coatings, and unsaturated polyester gel coats (6,103).

BIBLIOGRAPHY

"Other Glycols" under "Glycols" in *ECT* 2nd ed., Vol. 10, pp. 676–680, by H. C. Twiggs and J. C. Hutchins, Eastman Chemical Products, Inc.; in *ECT* 3rd ed., Vol. 11, pp. 963–971, by P. Von Bramer and J. H. Davis, Eastman Chemical Products, Inc.

1. *Publication No. N-154*, Eastman Chemical Products, Inc., Kingsport, Tenn., 1985.
2. Fischer and Winter, *Monatsch. Chem.* **21**, 301 (1900).
3. *Publication No. N-307*, Eastman Chemical Products, Inc., Kingsport, Tenn.
4. *Publication No. N-153*, Eastman Chemical Products, Inc., Kingsport, Tenn.
5. *Publication No. N-199*, Eastman Chemical Products, Inc., Kingsport, Tenn.
6. *Publication No. N-332*, Eastman Chemical Products, Inc., Kingsport, Tenn., 1991.
7. *Publication No. N-115*, Eastman Chemical Products, Inc., Kingsport, Tenn., 1963.
8. U.S. Pat. 3,920,760 (Nov. 18, 1975), J. B. Heinz (to Eastman Kodak Co.).
9. *Publication MSDS No. 100000043/F/USA*, Eastman Chemical Co., Kingsport, Tenn., 1992.
10. *Publication MSDS No. 100000455/F/USA*, Eastman Chemical Co., Kingsport, Tenn. 1992.

11. *Publication MSDS No. 100000410/F/USA*, Eastman Chemical Co., Kingsport, Tenn., 1992.
12. *Publication MSDS No. 100004745/F/USA*, Eastman Chemical Co., Kingsport, Tenn., 1993.
13. *Mod. Plast.* **50**(1), 70 (1973).
14. J. R. Eiszner, R. S. Taylor, and B. A. Bolton, *Paint Varn. Prod.* **49**(3), 54 (1959).
15. L. Beth, *Am. Paint Coat. J.* **60**(8), 54 (1975).
16. U.S. Pat. 5,120,415 (Jan. 15, 1992), San C. Yuan (to E. I. du Pont de Nemours & Co., Inc.).
17. Jpn. Pat. 02,206,669 (Aug. 16, 1990), H. Koneko, T. Yoshida, and T. Hirayama (to Hitachi Chemical Co., Ltd.).
18. *Publication No. N-330*, Eastman Chemical Co., Kingsport, Tenn., 1990.
19. D. J. Golob, T. A. Odom, R. L. Whitson, *Polym. Mater. Sci. Eng.* **63**, 826–832 (1990).
20. *Res. Discl.* **143**, 35 (1976).
21. PRC Pat. 1,057,849 (Jan. 15, 1992), F. Zhan, W. Huang, and J. Yu (to Ministry of Chemical Industry, Ocean Paint Institute).
22. Jpn. Pat. 03,064,310 (Mar. 19, 1991), T. Takemoto, M. Saito, and H. Akiyama (to Sanyo Chemical Industries, Ltd.).
23. Jpn. Pat. 02,274,789 (Nov. 8, 1990), H. Iwasaki, T. Tajiri, M. Ito, and K. Kido (to Mitsubishi Rayon Co., Ltd.).
24. L. A. Sadovnikova and co-workers, *Neftekhimiya* **16**(2), 316 (1976).
25. E. L. Niedzielski, *Ind. Eng. Chem. Prod. Res. Dev.* **15**(1), 54 (1976).
26. Jpn. Pat. 04,164,993 (June 10, 1992), K. Tsuruoka and H. Kobashi (to Nippon Oil and Fats Co., Ltd.).
27. Jpn. Pat. 04,132,756 (May 7, 1992), T. Tanaka, K. Yamamoto, K. Shiotani, and T. Hirose (to Dainippon Ink and Chemicals, Inc.).
28. C. H. Son and J. H. Morchouse, *J. Thermophys. Heat Trans.* **5**(1), 122–124 (1991).
29. F. Walnut, M. Ribet, P. Bermir, and L. Elegant, *Solid State Commun.* **83**(12), 961–964 (1992).
30. C. H. Son and J. H. Morchouse, *J. Sol. Energy Eng.* **113**(4), 244–249 (1991).
31. F. Walnut, M. Ribet P. Bermir, and P. Girault, *Solid State Commun.* **76**(5) 621–626 (1990).
32. U.S. Pat. 2,952,701 (Sept. 13, 1960), R. L. McConnell and H. W. Coover (to Eastman Kodak Co.).
33. U.S. Pat. 4,956,406 (Sept. 11, 1990), G. L. Myers and R. H. S. Wang (to Eastman Kodak Co.).
34. W. E. Conrad and co-workers, *J. Org. Chem.* **26**, 3571 (1961).
35. E. R. Nelson, M. Maienthal, L. A. Lane, and A. A. Benderly, *J. Am. Chem. Soc.* **79**, 3467 (1957).
36. U.S. Pat. 5,099,070 (Mar. 24, 1992), G. Luce and A. McCollum (to Eastman Kodak Co.).
37. E. R. Alexander, *Principles of Ionic Organic Reactions*, John Wiley & Sons, Inc., New York, 1950, p. 231.
38. V. V. Korshak and S. V. Vinograndova, *Polyesters*, Pergamon Press, Inc., New York, 1965.
39. Ger. Pat. 4,109,236 (Sept. 24, 1992), N. Schoen, H. J. Buysch, E. Leitz, and K. H. Ott (to Bayer AG).
40. W. W. Blount, *Soc. Plast. Eng. Tech. Pap.* **21**, 26 (1975).
41. *Publication No. N-206*, Eastman Chemical Products, Inc., Kingsport, Tenn., 1975.
42. P. J. Trent, D. L. Edwards, and P. Von Bramer, *Mod. Plast.* **43**, 172 (Apr. 1966).
43. U.S. Pat. 3,979,352 (Sept. 7, 1976), J. W. Brady, F. D. Strickland, and C. C. Longwith (to Shanco Plastics & Chemicals).

44. L. Gott, *J. Coat. Technol.* **48**, 52 (July 1976).
45. J. D. Bailey, *Chem. Ind. NZ* **9**(4), 15 (1975).
46. F. Sheme, S. Belote, and L. Gott, *Mod. Paint Coat.* **65**, 31 (Apr. 1975).
47. U.S. Pat. 4,301,046 (Nov. 17, 1981), M. L. Schlossman (to Tevco, Inc.).
48. *Publication No. M-205*, Eastman Chemical Co., Kingsport, Tenn., 1991.
49. *Publication No. L-151D*, Eastman Chemical Products, Inc., Kingsport, Tenn., 1978.
50. U.S. Pat. 5,108,660 (Apr. 28, 1992), D. W. Michael (to The Procter and Gamble Co.).
51. Jpn. Pat. 01,050,812 (Feb. 29, 1989), (to Bristol Myers Co.).
52. *Res. Discl.* **148**, 55 (1976).
53. *Res. Discl.* **143**, 35 (1976).
54. *Res. Discl.* **138**, 13 (1975).
55. Brit. Pat. 1,367,977 (Sept. 25, 1974), (to Saint-Gobain).
56. U.S. Pat. 3,671,654 (June 20, 1972), H. G. Nosler and H. Schnegelberger (to Henkel and Cie, GmbH).
57. L. Foster, D. Beeler, and D. L. Valentine, *Def. Publ. U.S. Pat. Off. T*, 912,016 (1973).
58. H. J. Hagemeyer and G. C. DeCroes, *The Chemistry of Isobutyraldehyde and Its Derivatives*, Tennessee Eastman Co., Kingsport, Tenn., 1953.
59. F. J. Villani and F. F. Nord, *J. Am. Chem. Soc.* **68**, 1674 (1946).
60. Jpn. Kokai, 7 494,621 (Sept. 9, 1974), T. Kojima, T. Hirai, T. Tachimoto, and M. Nakamura (to Chisso Corp.).
61. Jpn. Kokai, 7 494,620 (Sept. 9, 1974), T. Kojima, T. Hirai, T. Tachimoto, and M. Nakamura (to Chisso Corp.).
62. U.S. Pat. 2,941,011 (June 14, 1960), H. J. Hagemeyer, D. C. Hull, and M. A. Perry (to Eastman Kodak Co.).
63. W. V. McConnell and H. W. Moore, *J. Org. Chem.* **28**, 822 (1963).
64. Jpn. Pat. 62,004,280 (Jan. 10, 1987), K. Mascoka and co-workers (to Neos Co., Ltd).
65. Ger. Pat. 2,526,312 (Dec. 30, 1976) and Ger. Pat. 2,526,675 (Dec. 30, 1976) H. Moella and co-workers (to Henkel and Cie, GmbH).
66. Jpn. Kokai 75142,537 (Nov. 17, 1975), T. Mizumoto and H. Kamatani (to Toyobo Co., Ltd.).
67. Jpn. Kokai 75130,738 (Oct. 16, 1975), T. Mizumoto and H. Kamatani (to Toyobo Co., Ltd.).
68. *Publication No. K-192*, Eastman Chemical Products, Inc., Kingsport, Tenn., 1976.
69. U.S. Pat. 4,374,958 (Feb. 22, 1983), A. Barmabeo (to Union Carbide Corp.).
70. Ger. Pat. 3,929,650 (Aug. 19, 1990), K. Tamazaki and co-workers (to TDK Corp.).
71. Ger. Pat. 2,544,069 (Apr. 8, 1976), H. Inata (to Teijin, Ltd.).
72. U.S. Pat. 3,668,157 (June 6, 1972), R. L. Combs and R. T. Bogan (to Eastman Kodak Co.).
73. S. Oswitch, *Reinf. Plast.* **17**, 308 (1973).
74. Ger. Pat. 2,151,877 (May 10, 1972), F. Fekete and J. S. McNally (to Koppers Co., Inc.).
75. U.S. Pat. 3,674,727 (July 4, 1972), F. Fekete and J. S. McNally (to Koppers Co., Inc.).
76. Jpn. Kokai 6805,911 (Mar. 4, 1968), M. Izumi, S. Matsumura, and N. Asano (to Sumitomo Electric Industries, Ltd.).
77. J. Litwin, H. H. Beacham, and C. W. Johnson, paper 1B-18, *Rp/C Conference*, Society of the Plastics Industry, Inc., Feb., 1973.
78. E. H. G. Sargent and K. A. Evans, *Plastics* **34**, 721 (June 1969).
79. Ger. Pat. 2,542,191 (Apr. 1, 1976), G. Slinckx (to UCB SA).
80. Ger. Pat. 2,351,176 (Apr. 25, 1974), R. C. Harrington, J. D. Hood, and P. M. Grant (to Eastman Kodak Co.).
81. P. M. Grant and H. R. Lyon, *Def. Publ. U.S. Pat. Off. T*, 914,001 (1973).
82. Ger. Pat. 2,454,880 (May 13, 1976), J. Rueter and H. Scholten (to Chemische Werke Huels AG).

83. Fr. Pat. 1,567,254 (May 16, 1969), (to N. F. Chemische Industrie Synres).
84. Jpn. Pat. 03,124,779 (May 28, 1992), O. Iwase and co-workers (to Kansai Paint Co., Ltd.).
85. Jpn. Pat. 04,219,177 (Aug. 10, 1992), O. Iwase and co-workers (to Kansai Paint Co., Ltd.).
86. Jpn. Kokai 7413,852 (Apr. 8, 1974), T. Kimura, T. Ohzeki, S. Kobayashi, H. Naka-moto, and Y. Maeda (to Mitsubishi Rayon Co., Ltd.).
87. *Publication No. N-217*, Eastman Chemical Products, Inc., Kingsport, Tenn., 1975.
88. Jpn. Pat. 02,053,881 (Aug. 19, 1988), N. O. Komoto and co-workers (to Towa Kasei Koggo Co., Ltd.).
89. J. R. Caldwell and J. M. McIntire, *Def. Publ. U.S. Pat. Off. T*, 939,013 (1975).
90. Jpn. Kokai 75100,123 (Aug. 8, 1975), Y. Niinami and K. Mizuguchi (to Toyobo Co., Ltd.).
91. U.S. Pat. 3,931,073 (Jan. 6, 1976), W. J. Jackson and W. R. Darnell (to Eastman Kodak Co.).
92. Fr. Pat. 1,596,552 (July 31, 1970), J. R. Caldwell and R. Gilkey (to Eastman Kodak Co.).
93. Jpn. Kokai 7696,890 (Aug. 25, 1976), H. Sakai, Y. Takeuchi, and S. Kuris (to Teijin, Ltd.).
94. Ger. Pat. 2,241,413 (Mar. 21, 1974), D. Stoye, W. Andrejewski, and A. Draexler (to Chemische Werke Hüls AG).
95. U.S. Pat. 4,338,407 (July 6, 1982), K. Chandalia and co-workers (to Union Carbide Corp.).
96. U.S. Pat. 4,522,762 (June 11, 1985), R. W. Ortel and co-workers (to Upjohn Co.).
97. U.S. Pat. 3,810,960 (May 14, 1974), W. T. Gormley and M. C. Russ (to Koppers Co., Inc.).
98. U.S. Pat. 3,962,313 (June 8, 1976), M. Dexter and D. H. Steinberg.
99. Ger. Pat. 1,568,342 (July 18, 1974), R. Nehring, K. H. Hornung, and W. Seeliger (to Chimische Werke Huels AG).
100. U.S. Pat. 5,041,621 (1991), D. Mores and G. Luce (to Eastman Chemical Co.).
101. U.S. Pat. 5,024,772 (1991) L. Thurman, J. Dowd, and K. Fischer (to BASF Corp.).
102. U.S. Pat. 230,774 (Aug. 5, 1987), I. Kordomenso, A. Dervan and T. Semanision (to E. I. du Pont de Nemours & Co., Inc.).
103. L. Johnson and W. Sade, *Proceedings of the 18th Waterborne, Higher Solids, Powder Coating Symposium*, pp. 65–77.

T. E. PARSONS
Eastman Chemical Company

GLYCOLS, 1,4-BUTYLENE GLYCOL AND BUTYROLACTONE. See ACETYLENE-DERIVED CHEMICALS.

GLYOXAL. See GLYCOLS, ETHYLENE GLYCOL AND OLIGOMERS.

GOLD AND GOLD COMPOUNDS

Gold [7440-57-5], Au, is presumably the first metal known and used by humans. It occurs in nature as a highly pure metal and is treasured because of its color, its extraordinary ductility, and its resistance to corrosion. Early uses in medicine and dentistry date to the ancient Chinese and Egyptians. In the Middle Ages the demand for gold led to the intense, unsuccessful efforts of alchemists to convert base metals into gold. These pursuits became the basis for chemical science. The search for gold has been an important factor in world exploration and the development of world trade.

Properties

Gold, atomic number 79, is a third row transition metal in Group 11 (IB) of the Periodic Table. It occurs naturally as a single stable isotope of mass 197, ^{197}Au, which is also formed via the decay of ^{197}Pt (half-life 20 min) formed in the irradiation of platinum with slow neutrons (1). The electronic configuration of gold is [Xe]$4f^{14}5d^{10}6s^1$. Common oxidation states are 0, 1, and 3. Selected properties are shown in Tables 1 and 2 (2). Gold is characterized by high density, high electrical and thermal conductivities, and high ductility. One gram of gold can be drawn to 165 m of wire having a 0.02-mm diameter. Gold leaf 0.14 μm thick still exhibits the shiny color of gold. At least 26 unstable gold isotopes have been made; the most frequently used is ^{198}Au which has a half-life of 2.7 d.

Gold is the most noble of the noble metals. Other than in the atomic state (3), the metal does not react with oxygen, sulfur, or selenium at any temperature. It does, however, react with tellurium at elevated (ca 475°C) temperatures (4) to produce gold ditelluride [12006-61-0], $AuTe_2$, which is also found in the naturally occurring mineral, calaverite [37043-71-3], AuTe. Gold reacts with the halogens, particularly in the presence of moisture. At low (≤ 200°C) temperatures, chlorine is adsorbed on the gold surface with formation of surface chlorides. The rate of further chlorination is limited by the rate of diffusion of Cl or, more likely, the diffusion of gold through this surface chloride layer. At higher temperatures, the reaction is kinetically controlled as gold chlorides sublime and fresh surface is continually exposed. At 700–1000°C, adsorbed Cl_2 dissociates into Cl atoms which then react with the surface (5). The chlorides are, however, unstable in this region.

Gold reacts with various oxidizing agents at ambient temperatures provided a good ligand is present to lower the redox potential below that of water. Thus, gold is not attacked by most acids under ordinary conditions and is stable in basic media. Gold does, however, dissolve readily in 3:1 hydrochloric–nitric acid (aqua regia) to form $HAuCl_4$ [16903-35-8] and in alkaline cyanide solutions in the presence of air or hydrogen peroxide to form $(Au(CN)_2)^-$. These reactions are important to the extraction and refining of the metal. Similarly, gold anodes are solubilized in chloride or cyanide baths (see METAL ANODES). Gold dissolves slowly in concentrated selenic acid at 68°C.

At high temperatures, attack by concentrated sulfuric and nitric acids is slow and is negligible for phosphoric acid. Gold is very resistant to fused alkalies and

Table 1. Gold Properties

Property	Value
atomic weight	196.9665
melting point, K	1337.59
boiling point, K	3081
atomic radius, Au lattice, nm	0.1422
crystal structure	fcc
atoms/unit cell	4
lattice constant at ambient temperature, nm	0.407
interatomic distance at ambient temperature, nm	0.2878
density[a] at 273 K, g/cm^3	19.32
Brinell hardness (10/500/90), annealed at 1013 K, kgf/mm^2	25
modulus of elasticity at 293 K, annealed at 1173 K, MPa[b]	7.747×10^4
Poisson's ratio, as drawn	0.42
tensile strength, annealed at 573 K, MPa[b]	123.6–137.3
elongation, annealed at 573 K, %	39–45
compressibility at 300 K, Pa^{-1} [c]	6.01×10^{-12}
heat of fusion, J/mol[d]	1.268×10^4
heat of evaporation at 298 K, J/mol[d]	3.653×10^5
vapor pressure, Pa[c]	
at 1000 K	5.5×10^{-8}
at 1500 K	8.5×10^{-2}
at 2000 K	82
at 2500 K	4.9×10^3
at 3000 K	7.1×10^5
specific heat at 298 K, J/(g·K)[d]	1.288×10^{-1}
thermal conductivity at 273 K, W/(m·K)	311.4
thermal expansion at 273–373 K, K^{-1}	1.416×10^{-7}
electrical resistivity at 273 K, Ω·cm	2.05×10^{-6}
temperature coefficient of resistivity at 273–373 K, K^{-1}	4.06×10^{-3}
work function, J[d]	
thermionic	$7.69–7.85 \times 10^{-19}$
photoelectric	$8.17–8.76 \times 10^{-19}$
thermal emf,[e] mV, at K	
at 373 K	0.92
at 773 K	6.40
at 1073 K	12.35
total emissivity at 493–893 K	0.018–0.035
susceptibility (magnetic) at 291 K, cm^3/g(= emu/g)	1.43×10^{-7}
Hall coefficient at 295 K, (Ω·cm)/T[f]	-6.97×10^{-17}
entropy at 298 K, J/K[d]	47.33
standard reduction potential, V	1.69
Au$^+$ + e^- → Au	

[a]The commercially accepted value has been given. Measured values and density calculations from x-ray data show some variations.
[b]To convert MPa to psi, multiply by 145.
[c]To convert Pa to mm Hg, divide by 133.3.
[d]To convert J to cal, divide by 4.184.
[e]Emf values are given vs the reference-grade platinum of NIST (NBS Pt27) with the cold junction at the ice point.
[f]To convert T to gauss, divide by 1.0×10^{-4}.

Table 2. Optical Properties of Gold

Property	Thickness, μm			
	0.40	0.55	0.70	1.00
reflectance, %	38.7	81.6	96.7	98.1
refractive index		0.331	0.131	0.179
extinction coefficient		2.324	3.842	6.044

to most fused salts except peroxides. Gold readily amalgamates with mercury. Gold is very corrosion and tarnish resistant and imparts corrosion resistance to most of the commonly used gold alloys, especially to alloys containing 50 or more atom % of gold. Although gold is resistant to organic acids, some of the base metal containing alloys used in jewelry may become tarnished by perspiration and, in rare instances, cause allergic reactions.

Gold alloys also are subject to stress–corrosion cracking, especially alloys below 14 carat (58% gold). Jewelry items may have areas of high local stress that can induce cracking and corrosion in normally harmless environments such as solutions of hydrochloric or nitric acid. Stress–corrosion cracking occurs essentially as intergranular corrosion at unstable structural sites by an electrochemical mechanism (6) whereby the nongold components of the alloy become preferentially oxidized. Stress–corrosion cracking occurs primarily in single-phase alloys. A remedy against it is stress relief by annealing; however, in the case of multiphase, low carat alloys, care must be exercised so that annealing does not lead to the formation of homogeneous solutions (see CORROSION AND CORROSION INHIBITORS; FRACTURE MECHANICS).

Extraction and Refining

Placer mining is the oldest form of gold mining. Auriferous sand or gravel is swirled with water in a pan and gold, which settles out owing to its high density, is collected. On a large scale, a mixture of sand, gravel, and water is passed through a sluice box fitted with transverse riffles behind which the gold accumulates. The technique is still in use in places where appropriate alluvial or marine deposits exist, such as Alaska, but requires large quantities of water (7). Air classification also is possible but less efficient (8) (see SEPARATION, SIZE). At present, most gold is obtained either by deep mining, most notably in South Africa, or by open pit mining such as in the United States.

Mined gold ore is milled sufficiently to allow separation of the gold; recoveries usually are 92–96% of the ore's gold content, ca 5–6 ppm in South African ore and ca 1 ppm in U.S. ores (7,9). After size reduction (qv), various pretreatment and recovery processes may be used such as gravity concentration, flotation (qv), roasting, chlorination, and cyanidation, the choice depending on the metallurgical characteristics of the ore (see METALLURGY, EXTRACTIVE; MINERAL RECOVERY AND PROCESSING).

Amalgamation, which was once widely used, has been largely discontinued because of inefficiency and environmental concerns. Gold in association with metals such as copper, nickel, and lead, generally follows these in the concentration process and eventually can be separated and recovered.

In cyanidation, the ground ore is leached with a solution of sodium cyanide (0.02–0.05%) or an equivalent of calcium cyanide together with some lime. The leaching solution is aerated to provide oxygen and gold is dissolved with formation of sodium dicyanoaurate(−1) [15280-09-8] $Na(Au(CN)_2)$ (see CYANIDES).

The rate of dissolution is limited by oxygen availability rather than by cyanide concentration. When oxygen solubility is reduced by water salinity or by consumption by ore constituents such as sulfide minerals, enrichment of the air with oxygen or addition of hydrogen or calcium peroxide improves leaching kinetics and decreases cyanide consumption (10).

The cyanide solution also dissolves silver and some of the base metals which may lead to an intolerably high cyanide consumption. An economic cyanide usage would be 1–2.5 kg NaCN, or its equivalent in impure calcium cyanide, per metric ton of ore. Heap leaching, in which the cyanide solution is distributed over the top of an open mound or leveled heap of coarsely crushed or pelleted ore, is gaining in popularity, particularly when low grade ores or mine tailings are involved. Models which predict the performance of heap leaching as a function of diffusion in porous ore particles and dissolution rates have been developed (11) but may not be generally applicable.

Conventionally, gold and most other metals are recovered from solution by treating with zinc or aluminum dust. A newer technique is the removal of gold from solution using activated carbon which permits collection of ca 9–12 kg gold per metric ton of carbon, along with silver. In this, the solution is slurried with granular activated carbon after addition of suitable thickeners to maintain the carbon in suspension (carbon-in-pulp). Alternatively, carbon and ore can be slurried directly in a cyanide solution (carbon-in-leach). This is particularly advantageous when the ore contains carbonaceous material or clays (qv) which can reabsorb gold (12). The loaded carbon is separated by screening and gold and silver are stripped using concentrated alcoholic alkaline cyanide and recovered from the resulting solution by electrolysis. The carbon is reactivated by controlled roasting. Development to improve the extraction and stripping processes is under way. For example, resin-in-pulp ion-exchange (qv) techniques have been introduced to recover gold from cyanide pulp. These techniques eliminate the need for thermal regeneration and result in improved gold recoveries (13). Ion-exchange and complexing resins also have been shown to be more effective than other methods in recovering gold from very low grade ores or tailings and in reducing the amount of cyanide discharged in wastewater (14).

Environmental concerns have led to the installation of various special precautions, monitoring, and leak detection techniques at mines employing cyanidation. Processes to detoxify wastewater and spent ore prior to discharge or disposal have been developed (7). The same concerns also have provided an added impetus to efforts to develop extraction processes in which cyanide is replaced by thiourea, iodine, malononitrile, or by bioextraction. Whereas none of these has as yet been established as a cost-effective alternative (7), the use of bacteria to extract gold (15,16) or to improve extraction by oxidizing interfering ore constituents such as sulfides, arsenic, iron, copper, and carbonaceous materials (17–22) is receiving increasing attention. The largest plant employing biological preoxidation of gold-bearing sulfide ores is in Ghana (22).

A proposed method which avoids cyanide consists of treating gold ore with gaseous chlorine at elevated (≤250°C) temperatures to volatilize gold as chloride, Au_2Cl_6 [12446-79-6] or $AuMCl_6$, (M = Fe [12523-43-2], Al [73334-09-5], or Ga [73334-08-4]) and recovering it by condensation (23).

Refractory or difficult to treat ores, ie, those containing excessive amounts of sulfur as pyrites or pyrrhotites, arsenic, tellurides, or carbonaceous material, are pretreated prior to extraction. Pretreatment may include fine grinding, roasting, or pressure oxidation (autoclaving) (18–27), as well as biological preoxidation. Roasting achieves oxidation of sulfides, tellurides, and carbon, and creates porosity (17). The roasting temperature must be carefully controlled. The economics of processing refractory ores have been reviewed (28).

The impure gold concentrates from any of the primary recovery processes are melted under oxidizing conditions to remove most of the copper (qv) and other base metals, leaving gold plus silver. If the silver content is very low, the gold can be recovered by the Wohlwill electrolysis process in a chloride solution (29). Gold in the anodes dissolves and is deposited in pure form at the cathode. The silver is converted to chloride, which tends to coat the anode; however, coating is sharply reduced by superimposing alternating current on the system. Typically, the electrolyte would contain 90 g/L of gold as $HAuCl_4$ plus a small amount of free HCl. The bath is operated at 70°C using a cathode current density of 10–15 A/dm^2. The resulting cathode deposit should contain 99.95% gold after melting.

Treatment of impure gold is largely via the Miller process (30) in which chlorine is bubbled through the molten metal and converts the base metals to chlorides, which volatilize. Silver is converted to the chloride, which is molten and can be poured. The remaining gold is less pure (99.6%) than that produced by the Wohlwill process and may require additional treatment such as electrolysis. If platinum-group metals (qv) are present, the chlorine process is unsuitable.

In the refining of gold, associated silver and the platinum-group metals are difficult to remove. In the refining of scrap gold, silver is encountered frequently because gold–silver alloys, named doré metal, are often used in the fabricated products which generate the scrap. Doré metal that contains moderate amounts of gold can be treated by electrolysis in a nitrate solution at room temperature and at an anode current density of ca 4.5 A/dm^2. The gold does not dissolve but is retained in canvas anode bags. The silver deposits at the cathode as very pure crystals; 99.9% is typical for melted cathode silver. The salt content of the electrolyte typically is comprised of ca 60% silver nitrate and 40% copper nitrate.

Silver also can be removed from doré metal by treatment using hot sulfuric acid. The gold remains undissolved but is lower in purity than that resulting from most other processes. A purity of 94% is reasonable for a single treatment with sulfuric acid (31) but this can be raised to 99.5% by repeated treatment and washing. This process works best with doré metal containing 20–25% gold.

For doré metal containing up to 33% gold, silver can be removed by treatment with warm nitric acid. The gold remaining after washing and melting may have a purity of 99% which is considerably higher than that using the hot sulfuric acid. However, the process is more expensive.

In both the sulfuric and nitric acid processes, the doré metal must be in shot form prior to treatment to secure a reasonably rapid reaction. A number of steps also may be required in processing the doré metal to remove miscellaneous impurities, particularly in treating material from copper-anode slime (31).

In refining precious metal scrap and some concentrates, the gold is converted to $HAuCl_4$ by treatment with aqua regia. After heating to remove nitrogen oxides, gold is precipitated from solution by reduction with sulfur dioxide or ferrous sulfate. Alternatively, $HAuCl_4$ can be extracted with a water immiscible solvent such as dibutyl carbitol from which gold is recovered by reduction with oxalic acid (32). Any platinum-group metals can be recovered after the complete removal of the gold. If any gold remains in solution, explosive compounds may form in the palladium recovery process, because reaction of ammonia or ammonium salts and gold chlorides produces so-called fulminating gold, the precise composition of which is not known.

A number of special problems arise in the treatment of precious metal wastes of various types, eg, fabrication scrap and dusts (sweeps), and in treating scrap comprising gold-clad base metals. The scraps may contain copper, nickel, zinc, iron, tin, and lead, as well as gold, silver or platinum-group metals (see also RECYCLING, NONFERROUS METALS). The assay of incoming materials also is difficult. When the scrap can be melted to a homogeneous liquid, a dip sample converted to shot probably is the most reliable assay technique but care must be taken to avoid two-phase melts which can result if considerable silver and iron-group metals are present. Dilution with copper or high copper scrap may be required to secure a single-phase melt. In refining, much of the zinc can be fumed away as zinc oxide, and iron, lead, and some of the tin may be removed as slag. The precious metals remain with the copper and most of the nickel. This product can be used as the anode in a sulfate solution and most of the copper and nickel removed, the precious metals remaining as an anode slime or mud. The latter is treated as described earlier. The presence of nickel in the copper solution introduces complications, and arrangements must be made to remove considerable amounts of electrolyte periodically for purification.

Sweeps and related materials containing nonmetallic particles can be treated by adding appropriate fluxes to produce a low melting slag. Litharge [1317-36-8], PbO, should be present in the mixture and some of this is reduced to produce metallic lead, which dissolves the fine, precious metal particles. The resulting noble metal–lead alloy, which should contain a reasonable amount of silver, is oxidized in a later step, producing litharge which is decanted, and doré metal which is treated as described.

Production

Gold is widely distributed and the average content in the earth's crust is estimated to be 3.5 ppb (8). The gold content of ocean water varies considerably with location. Reported values range from 40 ppb to 3 parts per trillion (ppt) (33). However, later investigations indicate that the average value is of the order of 10 ppt (34) which is well below the concentration (3 ppm) required for economic recovery. Larger local concentrations of gold (ca 75–90% pure) in the metallic state occur in sedimentary and igneous rocks in combination with gangue minerals, usually quartz, and with silver and base metal sulfides, selenides, and tellurides. Gold is further obtained as a by-product in the electrolytic refining of copper (qv) and nickel (see NICKEL AND NICKEL ALLOYS). The anode slimes may contain silver and platinum-group metals as well as gold.

From primary lodes in rocks, gold has been released by weathering in conjunction with erosion by flowing streams in the form of fine grains or nuggets or as residual or stream placers. Original placers have often become buried by substantial layers of rock; eg, at the Vaal Reefs Mine in South Africa, gold of residual origin is recovered from a depth of 2200 m.

As of 1991, land resources of gold were estimated to amount to 7.5×10^4 metric tons and total world production through 1991 was 1.09×10^5 t (35). Gold is produced in many countries, the leading one being the Republic of South Africa, followed by the United States, the former Soviet Union, Australia, and Canada. A survey of new mine production worldwide from 1982–1991 is given in Table 3 (9). Whereas production by South Africa and the Soviet Union has declined somewhat during this period, that by other countries has risen significantly and total production in 1991 was nearly double that in 1978 (36). Future growth is expected to be at a lower rate because of decreased exploration resulting from depressed pricing and the inability of existing mines to respond rapidly if prices increase.

In the United States, about 90% of gold production originates from ores and placer deposits. The remainder is recovered primarily as a by-product of the refining of base metals, chiefly copper. The principal gold producing states are Nevada (60%) and California (10%) followed by Montana, Utah, S. Dakota, Washington, Colorado, Alaska, Idaho, Arizona, and New Mexico (7).

Economic Aspects

The price of gold has declined substantially from its historic high of U.S. $27,328/kg ($850/troy oz) reached in January 1980. The average annual price was as follows:

Year	$/kg	$/troy oz
1987	14,402	448
1988	14,092	438
1989	12,300	383
1990	12,376	385
1991	11,680	363
1992	11,091	345

During this same period, the demand for fabricated gold and the cost of production, although fluctuating, have increased steadily. The price of gold reflects gold's status as both a commodity and an investment vehicle. In real terms, ie, in constant 1991 currency, the average price of gold in the early 1990s differs little from average levels since the beginning of the century (9).

Table 3. Mine Production of Gold, Metric Tons

Location	1982	1983	1984	1985	1986	1987	1988	1989	1990	1991
Canada	66.5	73.0	86.0	90.0	105.7	116.5	134.8	159.5	167.0	176.7
United States	45.3	62.6	66.0	79.5	118.3	154.9	201.0	265.5	294.2	300.0
Republic of South Africa	664.3	679.7	683.3	671.7	640.0	607.0	621.0	607.5	605.1	601.1
other Africa										
Zimbabwe	13.4	14.1	14.5	14.7	14.9	14.7	14.8	16.0	16.9	17.8
Ghana	13.0	11.8	11.6	12.0	11.5	11.7	12.1	15.3	17.3	25.8
Zaire	4.2	6.0	10.0	8.0	8.0	12.0	12.5	10.6	9.3	8.8
other	15.0	15.0	15.0	17.0	18.2	25.0	27.5	25.2	25.0	31.5
Total other Africa	*45.6*	*46.9*	*51.1*	*51.7*	*52.6*	*63.4*	*66.9*	*67.1*	*68.5*	*83.9*
Latin America										
Brazil	34.8	58.7	61.5	72.3	67.4	84.8	102.2	101.2	84.1	80.0
Chile	23.0	23.5	22.2	22.8	24.0	23.3	26.7	29.0	33.3	32.5
Colombia	15.5	17.7	21.2	26.4	27.1	32.5	33.4	31.7	32.5	32.2
Peru	6.9	9.9	10.5	10.9	10.9	10.8	10.0	12.6	14.6	15.1
Venezuela	2.0	6.0	9.5	12.0	15.0	16.0	20.0	17.1	14.2	13.2
Bolivia	2.5	4.0	4.0	6.0	6.0	6.0	9.0	11.5	10.4	10.0
other	25.1	24.2	25.2	27.9	30.4	32.0	33.2	33.8	30.4	27.5
Total Latin America	*109.8*	*144.5*	*154.1*	*178.3*	*180.8*	*205.4*	*234.5*	*236.9*	*219.5*	*210.5*

Table 3. (Continued)

Location	1982	1983	1984	1985	1986	1987	1988	1989	1990	1991
Asia										
Philippines	31.0	33.3	34.3	36.9	38.7	39.5	39.2	38.0	37.2	30.5
Indonesia	2.0	2.1	3.6	5.6	8.4	12.2	12.3	10.8	13.3	18.4
Japan	3.2	3.1	3.2	5.3	10.3	8.6	7.3	6.1	7.3	8.3
other	5.4	5.4	7.6	5.7	8.6	8.1	11.1	13.5	13.0	13.5
Total Asia	*41.6*	*43.9*	*48.7*	*53.5*	*66.0*	*68.4*	*69.9*	*68.4*	*70.8*	*70.7*
Europe	*17.4*	*19.1*	*20.1*	*21.5*	*20.3*	*21.9*	*23.2*	*29.5*	*32.2*	*32.4*
Oceania										
Australia	27.0	30.6	39.1	58.5	75.1	110.7	157.0	203.6	243.1	234.2
Papua/New Guinea	17.8	18.4	18.7	31.3	36.1	33.9	36.6	33.8	33.6	60.6
other	1.7	1.5	2.4	2.8	4.1	4.0	6.6	9.4	10.1	11.5
Total Oceania	*46.5*	*50.5*	*60.2*	*92.6*	*115.3*	*148.6*	*200.2*	*246.8*	*286.8*	*306.3*
other countries[a]										
former Soviet Union	266.0	267.0	269.0	271.0	275.0	277.0	280.0	285.0	270.0	242.0
China	56.0	58.0	59.0	59.0	65.0	72.0	78.0	86.0	95.0	110.0
North Korea								9.5	13.0	13.0
Mongolia	7.0	7.0	9.0	10.0	10.0	8.5	9.5	10.0	10.0	10.0
Total other countries	*329.0*	*332.0*	*337.0*	*340.0*	*350.0*	*357.5*	*367.5*	*390.5*	*388.0*	*357.0*
Total world	*1366.0*	*1452.2*	*1506.5*	*1578.8*	*1649.0*	*1743.6*	*1919.0*	*2071.7*	*2132.1*	*2156.6*

[a]Estimated.

Specifications

Dentistry. Most casting alloys meet the composition and properties criteria of specification no. 5 of the American Dental Association (37) which prescribes four types of alloy systems constituted of gold–silver–copper with addition of platinum, palladium, and zinc. Composition ranges are specified, as are mechanical properties and minimum fusion temperatures. Wrought alloys for plates also may include the same constituents. Similarly, specification no. 7 prescribes nickel and two types of alloys for dental wires with the same alloy constituents (see DENTAL MATERIALS).

Analytical and Test Methods

Analysis. The method used to determine gold depends on the state of the sample to be analyzed and on its expected gold content and interfering impurity levels. Many of the available methods require conversion of gold in the sample to a soluble form, eg, $[AuCl_4]^-$. This is accomplished by digestion of the sample with aqua regia, evaporating to remove oxides of nitrogen, and dissolving the residue in concentrated HCl. The resulting solution is reevaporated and redissolved until removal of nitrogen oxides is complete (generally twice). In place of aqua regia, Cl_2—HCl or H_2O_2—HCl can be employed. Prior to analysis, the gold in solution can be concentrated or separated from impurities by solvent extraction, eg, using diethyl ether, ethyl acetate, methyl isobutyl ketone or dichloromethane–tetrahydrofuran, or by chromatography on ion-exchange or complexing resins. Extraction with a chloroform solution of nonylpyridine oxide permits separation of gold from platinum-group metals (38).

Gravimetric Methods. In the fire assay (39), used traditionally for the determination of the gold content in ores, the finely ground ore sample is mixed with a flux containing carbon and litharge, and the mixture is fused for at least one hour. After cooling, a button of lead containing gold and other precious metals is separated from the slag. The button is cupelled on bone ash or magnesia which absorb the lead oxides. The remaining bead is treated with nitric acid to remove silver, leaving gold and any platinum metals present in the sample.

Gravimetric methods more suitable for general use involve the precipitation of metallic gold from tetrachloraurate solutions by reduction with oxalic acid, SO_2, or hydroquinone. Formaldehyde, hydrazine, ferrous sulfate, and hypophosphorous acid also have been used but are considered less efficient (40).

Gold can be precipitated in the form of complexes of uncertain composition which then are ignited. Reagents suitable for this purpose are dimethylglyoxime (41), tetramisole (citarin) (42), thiophenol (43), and 8-quinolinol (44) (see CHELATING AGENTS). A number of insoluble tetrachloro- or tetrabromoaurates can be precipitated quantitatively and gold then can be determined by weighing. Suitable cations are *N*-(*N*-bromotetradecylbetainyl)-*C*-tetradecylbetaine (45), bis(trimethyl)hexamethylenediammonium ion and 2,4,6-triphenylpyrilium ion (46) as well as complexes containing sulfur ligands formed by reactions with thioglycolic acid (47) or di-2-thienylketoxime. A method suitable for determining gold in cyanide plating solutions involves adding silver nitrate, which reacts with free cyanide, followed by fuming with sulfuric acid (48).

Titrimetric Methods.   Frequently, the reduction of gold solutions leads to finely divided precipitates which are difficult to recover quantitatively. In such cases, the reduction of Au(III) to Au(0) by, eg, hydroquinone, can be followed potentiometrically (49). The end point in such titrations also can be determined with indicators such as benzidine (50) or o-anisidine (51). Alternatively, the reduction can be effected with excess hydroquinone which is then back-titrated with Ce(IV) (52). Iodometric determination of Au(III) also is useful (53).

Spectrophotometric Methods.   The gold content of aqueous solutions of $[AuCl_4]^-$ or $[AuBr_4]^-$ can be determined directly from the absorbance of these species at 380 nm. Interferences by other metal species, particularly tetrahalo-anions of the Pt group, can be minimized by extraction of the Au(III) into organic solvents (54). The visible spectra of colored complexes formed by reactions with such reagents as stannous chloride (55), rhodanine (56), rhodamine B (57), and malachite green (58) also can be used to determine gold. The methods generally involve extraction of the colored complexes into organic solvents which overcomes problems of aqueous solubility and facilitates separation from interfering metal ions and concentration. The electronic and vibrational spectra of gold complexes have been determined and have been used to elucidate structure and bonding.

Emission Spectrography, Atomic Absorption, and Neutron Activation.   Instrumental techniques are particularly useful for the determination of trace quantities of gold, especially when used in conjunction with concentration via solvent extraction or chromatography. Emission spectrography is used to determine both small quantities of gold in metals and gold purity. Emission lines at 242.78 or 267.60 nm are useful for determining small (0.5–20 ppm) concentrations and a line at 312.29 nm for large (10–160 ppm) amounts in platinum (59), whereas the lines at 314.8 and 310.5 nm serve as internal standards in the determination of trace impurities in gold (60). Improved sensitivity (to ca 6 ppb) is achievable via d-c argon plasma emission spectroscopy (see PLASMA TECHNOLOGY) (61).

Determination of gold concentrations to ca 1 ppm in solution via atomic absorption spectrophotometry (62) has become an increasingly popular technique because it is available in most modern analytical laboratories and because it obviates extensive sample preparation. A more sensitive method for gold analysis is neutron activation, which permits accurate determination to levels ≤ 1 ppb (63). The sensitivity arises from the high neutron-capture cross section (9.9×10^{-27} m^2 = 99 barns) of the only natural isotope, ^{197}Au. The resulting isotope, ^{198}Au, decays by β and γ emission with a half-life of 2.7 d. Gold is a useful calibration standard for this method (see RADIOACTIVE TRACERS). Whereas similar sensitivities can be achieved by inductively coupled plasma mass spectrometry (qv), the latter requires more extensive sample preparation to overcome interference by other metals such as copper (64).

Health and Safety Factors

The discovery in 1890 that potassium dicyanoaurate(−1) [14265-59-3], KAu(CN)$_2$, was toxic to the tuberculosis-causing bacterium, mycobacterium tuberculosis, *in vitro* and was somewhat effective in the treatment of the disease led to a search for other effective gold compounds with less serious side effects. These efforts resulted in the preparation of less toxic sulfur-containing compounds, such as disodium gold(I) thiomalate [42722-04-3] (Myocrysin) and gold(I) thioglucose [12192-57-3] (Solganal), which were used extensively in the treatment of tuberculosis from 1925–1935. During this period, the utility of these compounds in the treatment of rheumatoid arthritis was discovered, and chrysotherapy, ie, gold therapy, along with some immunosuppressive and antimalarial drugs, remains as one of the few treatments capable of slowing or halting the damage caused by this crippling disease (65,66).

Excretion of gold during chrysotherapy is relatively slow (ca 20% within two weeks and less thereafter) and toxic side effects (such as kidney and liver damage, dermatitis, stomatitis, thrombocytopenia, bone marrow suppression, and other haematopoietic disorders) can be long-lasting and sufficiently severe to require cessations of treatment in about 35% of patients (67). The excretion of gold can be accelerated by administration of chelating thiols such as DL-penicillamine (of which the D enantiomer has antiarthritic activity of its own) and British antilewisite (BAL) (68) (see CHELATING AGENTS).

Toxicity and the requirement for parenteral administration provided an impetus to synthesize less toxic gold drugs that can be administered orally. The effectiveness of tertiary phosphine–gold complexes, in particular, triethylphosphinegold(I) chloride [15529-90-5], $(C_2H_5)_3PAuCl$, in suppressing adjuvant arthritis in rats without reaching harmful concentrations in the kidney when administered orally (69) led to further development culminating in the introduction of the thioglucose acetate derivative, (2,3,4,6-tetra-*O*-acetyl-1-thio-beta-D-glucopyranosato-*S*)-(triethylphosphine)gold [34031-32-8] having the generic name, Auranofin, into clinical practice in a number of West European countries in 1982 and in the United States in 1985. Clinical results indicate that Auranofin therapy results in a lower total body burden of gold than does injectable gold, that patient tolerance is considerably greater, and that effectiveness in responding patients is comparable (66,70). However, although Auranofin is less toxic than other second-line treatments, the overall response rate achieved is lower (71). Although the mechanism of action of gold compounds is not fully understood, there is some speculation that inhibition of thiol-containing enzymes involved in the inflammatory process may be responsible (72).

Radioactive ^{198}Au, prepared by irradiating natural gold in a nuclear reactor, has been used in radiotherapy either as grains which can be implanted in cancerous tissue such as of the prostate (73) or nasopharynx (74) or infused in colloidal form as in the treatment of bladder cancer (75) (see RADIOPHARMACEUTICALS). Colloidal ^{198}Au also has been employed in radiosynovectomy, which is the destruction of all or part of the synovial membrane in knees badly damaged by rheumatoid arthritis (76), and in diagnostic applications such as bone marrow scanning and visualization of organs such as the lungs and liver (see MEDICAL IMAGERY TECHNOLOGY). The ultrashort (30.6 s) half-life of ^{195m}Au has been used to advantage in multiple sequential evaluation of ventricular function (77) (see RADIOACTIVE DRUGS).

The labeling of various biological macromolecules with gold using such soluble species as $[Au(CN)_2]^-$, $[AuCl_4]^-$, and $[AuI_4]^-$, eliminates phasing problems in the determination of three-dimensional structures by x-ray diffraction because the high x-ray scattering power of Au permits its ready location in electron-density maps (78) (see X-RAY TECHNOLOGY). Conjugates of colloidal gold with various proteins have been used as an aid to visualization in immunoelectron microscopy (79).

Uses

Gold Metal.   Besides its use for monetary reserves, gold is used in the private sector principally for investment and fabrication. A breakdown by use of fabricated gold from 1982–1991 is given in Table 4 (9). United States consumption of fabricated gold during the last five years of this period ranged from ca 113 t in 1987 to ca 114 t in 1991 (80).

By far, the largest commercial use is jewelry. In jewelry, the weight fraction of gold is either expressed as the carat where 24 carat represents 100% or in fineness where 1000 fine represents 100%. Typical carat ranges are from 22–8, the most popular grades in the United States, Europe, and Japan being 18 and 14 carat.

In the electronics industry, gold is used as fine wires or thin film coatings and frequently in the form of alloys to economize on gold consumption and to impart properties such as hardness. Gold has properties that satisfy specific requirements not achievable with less expensive metals (see ELECTRICAL CONNECTORS; ELECTRONICS COATINGS; THIN FILMS).

In dentistry, gold is used for a variety of restorations (see DENTAL MATERIALS). Other industrial-decorative uses include inexpensive jewelry and watches either as gold-filled or as rolled-gold materials, in which gold is bonded to a base-metal backing, or as electroplated articles having gold coatings of up to 10 μm thickness; decorative coatings on glass or chinaware; reflective or absorptive optical coatings; electrical contacts and contactors; brazing alloys for high temperature applications (see SOLDERS AND BRAZING ALLOYS); instrumentation for measurement of temperatures near absolute zero, or for jet-engine temperature control; spinnerets for rayon manufacture, corrosion-resistant materials in the chemical industry; and as catalysts and for medicinal purposes.

Alloys.   *Jewelry.*   Except for white golds, the carat golds used in jewelry are alloys of gold, silver, and copper. Frequently these are modified by other metals, mainly zinc, as well as cadmium, nickel, platinum, or palladium. Varying compositions of the basic alloy produce shades and colors of gold ranging from yellow to green. The dependence of color on composition is shown in Figure 1. A newer jewelry alloy, 99 Au-1 Ti, has the mechanical properties of 18 carat gold but lacks the allergic response occasionally caused by base metal alloy constituents (82).

Liquidus isotherms of the gold–silver–copper system and the binary phase diagrams of the constituents are given in Figure 2. For binary gold and silver and for gold and copper alloys, the liquidus and solidus temperatures are close. For copper alloys (qv), a congruent minimum appears at 910°C at 20% copper. These alloys form continuous solid solutions. Copper and silver form a simple eutectic at 779°C having a silver content of 72%. In the copper–gold system, ordered

Table 4. World Fabrication of Gold,[a] Metric Tons

Fabrication	1982	1983	1984	1985	1986	1987	1988	1989	1990	1991
jewelry	935.6	842.4	1092.9	1187.1	1167.8	1214.5	1531.8	1907.0	2036.7	2111.1
electronics	88.8	106.4	130.4	114.4	123.0	124.0	132.7	136.7	147.8	146.9
dentistry	60.6	51.3	52.4	53.3	51.1	47.7	50.5	50.9	52.7	55.3
other industrial and decorative uses	57.9	52.7	55.6	54.5	56.5	56.5	59.2	63.6	65.4	65.2
medals, medallions, and unofficial coins	21.8	31.7	44.1	14.4	11.8	16.0	19.0	18.8	19.3	23.0
official coins	130.6	165.1	130.5	119.2	348.3	202.6	129.5	135.4	117.7	141.7
Total	*1295.3*	*1249.6*	*1505.9*	*1542.9*	*1758.5*	*1661.2*	*1922.8*	*2312.5*	*2439.6*	*2543.2*

[a]Former Soviet Union and China and bullion purchases not included.

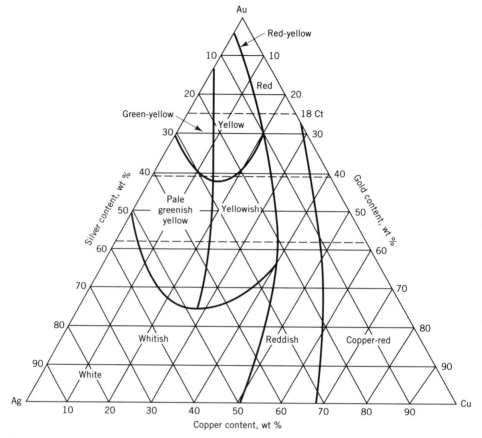

Fig. 1. Gradation of color of gold–copper–silver alloys (81). Courtesy of Academic Press, Inc.

phases appear below 400°C: Au_3Cu [12044-96-1], AuCu [12006-51-8], and $AuCu_3$ [60748-60-9], whereas the gold and silver system forms only one ordered AgAu [63717-64-6] phase. The liquidus surface of the ternary system accordingly exhibits a shallow minimum at the copper–gold binary edge and a eutectic gutter at the copper–silver binary edge. The ternary system also forms duplex systems at lower temperatures so that it is possible to age-harden rapidly quenched alloys. This is important in that a workable alloy may be substantially hardened by a final heat treatment after the forming operation.

Zinc and gold form a eutectic at 642°C having a zinc content of 15%. The alpha phase contains 6–13% Zn, depending on temperature. A number of ordered phases, including the compositions Au_3Zn [12256-74-5] and $AuZn_3$ [11089-97-7], exist at lower temperatures. Despite these complexities, zinc is used widely in jewelry alloys. At low concentrations (≤5%), zinc acts essentially as a deoxidizer without appreciably affecting the properties of the alloys. In concentrations of up to 10% and more, especially in alloys of 14 carat or less, zinc lowers the liquidus and solidus temperatures but also suppresses phase segregation by restricting

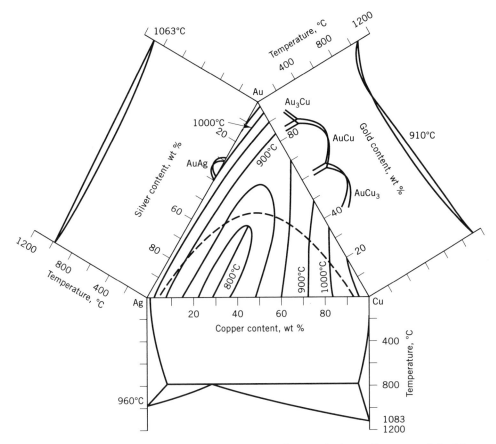

Fig. 2. Liquidus isotherms of gold–copper–silver alloys and phase diagrams of the binary constituents (83). Courtesy of Academic Press, Inc.

the range at which duplex phases coexist in gold–silver–copper alloys or suppresses formation of ordered phases. Preserving the single-phase state improves the tarnish-and-corrosion resistance of the alloy. Zinc or cadmium (84) also may be used to maintain the gold color of a lower carat alloy or to improve mechanical and other properties.

White Golds.   As shown in Figure 1, the silver-rich phases in the gold–silver–copper system are white but lack tarnish resistance. Therefore, two other alloy systems are those used primarily as white golds: gold–nickel–copper (often together with zinc) and gold–palladium alloys which usually also contain other metals such as silver, nickel, platinum, and zinc. Figure 3 shows liquidus isotherms of the gold–nickel–copper system and the binary phase diagrams of its constituents. Gold and nickel form continuous solid solutions at elevated temperatures with a congruent minimum at 950°C and a nickel content of 17.5%. At lower temperatures, the solid solutions decompose into two phases: nearly pure gold and a nearly pure nickel. This phase separation together with the development of ordered solid phases of the copper–gold system results in age-hardening

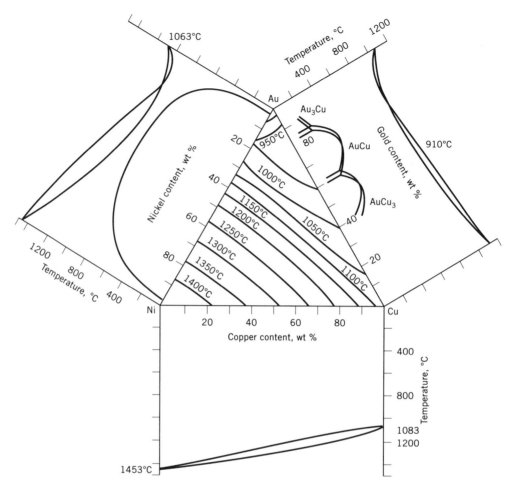

Fig. 3. Liquidus isotherms of gold–copper–nickel alloys and phase diagrams of the binary substituents (85). Courtesy of Academic Press, Inc.

accompanied by lower corrosion resistance (85). This can be reversed by homogenizing the alloy by annealing and quenching. Palladium-based white golds are expensive but also are malleable, ductile, and corrosion resistant, and can be cast (86). A number of other metals have been reported to bleach gold; however, these have had little application.

Solders and brazes for jewelry gold alloys are usually gold–silver–copper alloys that can be modified to yield a desired melting range by adjusting the silver-to-copper ratio. Additions of zinc, cadmium, tin, or nickel compensate for the changes in color introduced by compositional modification (87).

Dentistry. High purity gold in the form of foil or powder is used to a minor extent. Generally, gold alloys are employed for either castings, or as castings bonded to porcelain enamel, or for plates and wires. The presence of palladium and especially of platinum increases the hardness of the alloys considerably. Furthermore, age-hardening can be carried out. Zinc in small (0.5–1.0%) concentrations acts not only as a deoxidizer but contributes to age-hardening. Small amounts of metals, such as iridium, ruthenium, rhodium, and rhenium, act as grain refiners (88).

Alloys suitable for castings that are to be bonded to porcelain must have expansion coefficients matching those of porcelain as well as solidus temperatures above that at which the ceramic is fired. These are composed of gold and palladium and small quantities of other constituents: silver, calcium, iron, indium, tin, iridium, rhenium, and rhodium. The readily oxidizable components increase the bond strength with the porcelain by chemical interaction of the oxidized species with the oxide system of the enamel (see DENTAL MATERIALS).

Industry and Technology.   Gold and its alloys have many technological applications, in most cases in the form of thin films produced by electro- or chemical deposition. In instrumental and similar applications, gold alloys are employed in bulk as resistors, conductors, potentiometers, resistance thermometers, and contacts. The alloys are chosen because of specific mechanical, electrical, and thermal properties and because of high corrosion resistance (see TEMPERATURE MEASUREMENT). The gold–palladium system, which forms continuous solid solutions, offers a wide range of resistivities useful in resistors and potentiometers. The alloys 45 Au-55 Pd and 70 Au-20 Pd-10 V, have resistivities of 28 $\mu\Omega$·cm (88) and 217 $\mu\Omega$·cm (89), respectively, and minimal temperature coefficients of resistivity. A 50 Au-45 Pd-5 Mo alloy also is suitable for potentiometer use (90). For resistance thermometry at cryogenic temperatures, 99.6 Au-0.4 Mn has been recommended (91). For the measurement of elevated temperatures in the range of 1000°C, notably for temperature control of aircraft jet engines, alloy thermocouples (Platinel) of the type 55 Pd-31 Pt-14 Au (positive leg)-65 Au-35 Pd (negative leg) (92) have been developed. For cryogenic temperatures approaching 0 K, gold–iron thermocouples with an iron content of 0.07% in the United States and 0.03% in the United Kingdom are used (93). The other leg of the couple normally is chromel. Gold–platinum thermocouples show stabilities comparable to that of platinum resistance thermometers over a wide temperature range (94). For contacts, wipers, or sliprings, a number of hard alloys have been developed and are used frequently in clad or inlay form which allows a broader range of composition. The alloys involved are gold–copper, gold–silver, gold–nickel, gold–silver–platinum, gold–silver–nickel, gold–silver–copper, or gold–rhenium. An alternative method of gaining improved hardness for this application is by dispersion strengthening gold with small amounts of an oxide such as ceria (95). An overlay of 95 Au-5 Ni on telephone relay contacts containing palladium eliminates the disturbing catalytic action of palladium, causing the formation of polymeric material from organic vapors on the contact surface (96). An alloy 60 Au-40 Pd is used in spark plugs for special internal combustion engines, eg, in snowmobiles (97).

Gold and gold-based alloys are used for corrosion-resistant equipment. Gold–platinum alloys, 75 Au-25 Pt or 84 Au-15 Pt-1 Rh, are used as crucible material for many molten salts (98). Spinnerets for rayon manufacture are based on the Au–Pt system which exhibits a broad miscibility gap in the solid state so that the alloys can be age-hardened. Spinneret alloys contain 30–40% or more platinum modified by small additions of usually rhodium (99). Either gold or gold–platinum alloys are used in rupture disks for service with corrosive gases (100).

Fine (dia 0.0127–0.0381 mm) wires of gold, purity 99.975%, are used in the circuitry of semiconductors (qv) and transistors (101) (see INTEGRATED CIRCUITS). Many metallurgical problems arise in the bonding of these wires, eg, formation of embrittling eutectics if the wires are in contact with silicon or germanium (102). Using silicon, the eutectic point is at 370°C at 6% Si, and using germanium, at 350°C at 12% Ge. During bonding, gold diffuses into silicon or germanium causing p-type conductivity and requires countermeasures. On the other hand, gold that has diffused into silicon greatly accelerates the rate of minority carrier recombination which is utilized in high speed switching devices (103). Intermetallic phases form above 250°C if gold is in contact with aluminum surfaces: Au_4Al [12003-05-3], Au_5Al_2 [12522-6-6], Au_2Al [12250-39-4], AuAl [12250-38-3], $AuAl_2$ [12004-03-4]. Resulting embrittlement, however, is not ascribed to the formation of these phases but to development of Kirkendall voids (104).

Brazes and Solders. A multitude of alloys has been developed for applications in the electronics, aerospace, and nuclear power industries. These are used at given temperature conditions and impart specific wetting characteristics, eg, for graphite and diamond, and corrosion resistance adequate for the application. A principal requirement is a small differential between solidus and liquidus temperatures. Alloy systems for brazes and solders are: gold–copper; gold–silver; gold–silver–copper (for lower temperatures: 12–30 Au in Ag–Cu–Cd–Zn) (105); gold–nickel–copper; gold–nickel–silver; gold–nickel–chromium; gold–nickel–molybdenum, gold–nickel–tantalum; gold–palladium (most oxidation resistant, 92 Au-8 Pd) (106); gold–palladium–nickel (brazing for tungsten, molybdenum, nickel, and stainless steel) (107); gold–tin–germanium (solder for electronic circuits, eg, 80–85 Au, 5–19 Sn, 1–10 Ge) (108); gold–tin (solder for electrical connections to gallium arsenide, eg, 80 Au-20 Sn and 77 Au-19 Sn-3.4 Ga); gold–silicon (solder for connections to silicon eg, Au-2 Si) (109); gold–antimony; gold–arsenic (solder for semiconductors, eutectic at 665°C, 32% As) (110); gold–indium; and gold–indium–bismuth.

Electrodeposition. Electrodeposition is the most widely practiced technique for producing gold or gold alloy coatings for jewelry and for decorative and industrial purposes (111–113) (see ELECTROPLATING). In the electronics industry, for instance, essentially the entire gold consumption (Table 4) is in the form of coatings, and the majority of these are prepared by electrodeposition. A vast number of plating baths has been developed. These baths are based principally on two complexes of monovalent gold, predominantly the cyanide complex $(Au(CN)_2)^-$ which usually is employed as the potassium salt [1397-50-5] and, to a lesser extent, sulfite complexes (presumably $(Au(SO_3)_2)^{3-}$) usually employed as the sodium [39394-92-8] or potassium [19153-99-2] salt and stabilized with organic molecules such as amines (114). The sulfite complex is more suitable for high speed plating, eg, 5 μm in 2–4 min instead of 10–20 min (115). Plating speeds as high as 3.5 μm/sec have been achieved by jetting the electrolyte onto the area to be plated and passing a current through the workpiece. This technique is highly selective and capable of producing line widths as low as 150 μm without masking (116). In order to obtain bright smooth deposits of desired physical properties (hardness, ductility) and with minimum porosities, the bath composition, temperature, current density, and agitation must be controlled. Techniques such as pulse plating, reversed plating, and superposed alternating current may be used. In addition to the gold complex employed for electrodeposition, the baths contain other salts which enhance conductivity, buffers which adjust pH, or chelating agents which define the concentrations of other metal ions to be codeposited or which tie up undesired impurities, surface-active agents, and agents that act as brighteners, hardeners, and grain refiners.

Pure gold coatings can be deposited from alkaline, neutral, or acid cyanide baths. Alkaline, neutral, and acid pH ranges are achieved using excess cyanide (117), additions of citrates (118), and citric acid (119), respectively. Various organic additions and small quantities of agents, eg, arsenic and antimony, are added for grain refining, leveling, and brightening (120). For plating of electrical contacts, bright hard gold deposits are obtained by codepositing 0.1–0.2% nickel or cobalt (121). A great number of baths for gold alloy deposition have been formulated, mainly for jewelry and decorative purposes, and yield a variety of color shades. Rose-pink and red shades are obtained from gold–copper codeposits (122), yellow gold from gold–nickel or gold–cobalt alloys which lighten the color of gold (123), green colors from gold–silver or gold–cadmium codeposits (124), and white gold from gold–nickel–cadmium–copper codeposits or from nickel–zinc codeposits (125). In the industrial area, an 18-carat electrodeposited alloy of gold–copper–cadmium has been recommended for contact use (126), and a gold alloy containing 1.5–2% antimony has been recommended for coating electrical contacts (127). Gold films containing antimony also have been developed for coating germanium and silver transistors, and gold films containing 0.1–0.5% indium are used in the production of p-type germanium resistors (128). Recently, cadmium-free low carat gold plating baths have been developed (129).

Electroless Plating.   Surface coatings that are obtained from baths comparable to those used for electrodeposition but that are applied without current are produced by electroless plating (qv). The process operates either by electrochemical displacement (atomic exchange) or by chemical reduction. The former is restricted to metals less noble than gold; after a thin (1–2 μm) film of gold has been formed, no further gold deposition is possible. The latter permits deposition of heavier (1–15 μm) films on metallic or nonmetallic substrates, eg, glass, ceramics, and plastics.

Gold in a medium of ammonium citrate and chelating agents is suitable for electrochemical displacement (130). For electroless plating by reduction, a variety of complexing agents for gold have been formulated such as imides (131), amines, thiourea, and citrates (132), and reducing agents such as hypophosphite, hydrazine, borohydrides, amine boranes, and formaldehyde. The process is autocatalytic. The initial formation of metallic gold on the surface accelerates the rate of subsequent reduction. Therefore, preplating with gold, eg, by electrochemical displacement, is beneficial. In the case of nonconducting substrates, the surfaces are first rendered catalytic by applying a small amount of palladium solution which reacts with a layer of predeposited stannous chloride. In some cases other catalytic agents are employed, eg, cobaltous chloride (133). Spray applications, in which a gold solution is sprayed onto the substrate simultaneously with a reducing solution, also have been developed.

Electroforming, which is used in the production of art objects or jewelry is a combination of electroless plating and electrodeposition. A wax mold of the object to be produced is made conductive by electroless gold plating, a thick layer of gold or gold alloy is then electrodeposited and, finally, the wax is removed by melting (134).

Deposition from Liquid Gold.   Liquid gold refers either to organic suspensions or emulsions containing finely divided gold powder or to solutions of organogold compounds in organic solvents. Liquid golds are formulated as paints or inks which are applied by painting, brushing, screen printing, etc, and that yield well-bonded gold films after subsequent drying and firing. The thickness of films derived from dispersed metal is ca 25 μm. The thickness of films derived from organogold compounds ranges from $(5-20) \times 10^{-2}$ μm.

Initially, the gold compounds employed consisted of poorly defined gold sulforesinates prepared from gold chloride and sulfurized natural terpenes, eg, sulfurized Venetian turpentine. Subsequently, synthetic compounds have been developed for this use. Secondary-alkyl or aralkyl mercaptogold (135) and t-alkyl mercaptogold compounds (136) in suitable vehicles have been formulated to yield bright, semibright, and matte gold films or, when combined with other organometallic compounds, gold alloy films (137). A preferred compound is t-dodecyl gold mercaptide [26403-08-7] which has a decomposition temperature of 150°C and, therefore, allows coating of heat-sensitive substrates. The addition of iodide or bromide to gold mercaptide solutions further decreases the decomposition temperature to 80–90°C and yields highly reflective and conductive films (138). For use at temperatures $\geq$ 800°C, an intermediate thin film diffusion barrier, eg, ceria, is applied prior to gold coating to prevent interdiffusion of gold and the supporting metal (139). Gold and gold alloy films prepared from liquid gold are used widely for decorative and industrial purposes, particularly in electronics, and for optical and heat-concentrating and heat-shielding applications.

Catalysis.   Although the literature contains numerous references to catalysis (qv) by gold and gold compounds, particularly gold alloys or bimetallic clusters (140–143), practical applications in this area remain extremely meager. Thus, although gold-containing catalysts frequently display improved selectivities, their activities are generally far lower than those of the Group VIII metals or of the other transition metals in Group I. Although lack of activity has been ascribed to the lack of d-band vacancies in gold, it is by no means clear that this restriction necessarily applies to surface atoms, particularly in highly dispersed systems. In fact, extremely high activity for CO oxidation over gold catalysts highly dispersed on various transition metal oxides has been reported when these are prepared by coprecipitation from aqueous solutions of tetrachloroauric acid, $HAuCl_4$, and corresponding metal nitrates followed by calcination (144). Whereas these materials have generated substantial interest for their potential as low temperature CO oxidation catalysts, their stability under reaction conditions appears open to question (145,146).

More conventionally, preparation of supported gold catalysts usually involves impregnation of a carrier with a solution of $HAuCl_4$, followed by chemical reduction or thermal decomposition (140,147). Organogold complexes also have been used for this purpose (148). Olefin hydrogenation (149) and selective hydrogenation of dienes and acetylenes to monoolefins (150) over such catalysts has been reported. However, severe conditions are required in these cases and the catalysts rapidly lose activity, presumably owing to coking. Another use of supported gold catalysts is in oxidative dehydrogenation where the weak adsorption of oxygen on gold presumably prevents complete oxidation, making such catalysts highly selective in this application (151). For example, in the oxidative dehydrogenation of ketones, α,β-unsaturated ketones are produced with selectivities of 70–94% at conversions of 22–48% at 400–600°C (152). Similar results, 94% selectivity at 53% conversion at 700°C, have been reported in the oxidative dehydrogenation of ethylbenzene to styrene (153). Although superior selectivities have been described for catalysts containing gold and other catalytically active metals (primarily the Group VIII noble metals) in such diverse applications as hydrogenation (154), isomerization (155), hydrocracking (156), reforming (157), and partial oxidation of olefins (158), the only commercial use of such catalysts is in the manufacture of vinyl acetate. The catalysts typically contain palladium and gold on an inert support (eg, SiO_2 or TiO_2) and have a total metal content ca 2–3% with a Pd–Au ratio of 2:1–3:1 (158) (see VINYL COMPOUNDS).

The metals are impregnated together or separately from soluble species, eg, Na_2PdCl_4 and $HAuCl_4$ or acetates (159), and are fixed by drying or precipitation prior to reduction. In some instances sodium or potassium acetate is added as a promoter (160). The reaction of acetic acid, ethylene, and oxygen over these catalysts at ca 180°C and 618–791 kPa (75–100 psig) results in the formation of vinyl acetate with 92–94% selectivity; the only other product is CO_2. The role of gold in improving selectivity and life in these catalysts is unclear. However, as in other cases involving bimetallic gold catalysis (142), gold may modify and/or stabilize the active Pd centers in the surface layer (161).

Derivatives

Gold Compounds. The chemistry of nonmetallic gold is predominantly that of Au(I) and Au(III) compounds and complexes. In the former, coordination number two and linear stereochemistry are most common. The majority of known Au(III) compounds are four coordinate and have square planar configurations. In both of these common oxidation states, gold preferably bonds to large polarizable ligands and, therefore, is termed a class b metal or soft acid.

Although it has been proposed that gold(II) species are intermediate in the reduction of Au(III) to Au(I) chlorides (162), the only stable mononuclear gold(II) complexes which have been isolated are tetrabutylammonium maleonitriledithiolategold(II) [33637-83-1] (163) and bis(1,4,7-trithiacyclononane)gold(II) bis(tetraflouroborate) [125438-78-0] (164). Compounds of Au(V) are equally rare. However, both the fluoride, AuF_5 [57542-85-5], and salts of the complex fluoride AuF_6^-, have been prepared (165). Numerous binary and complex gold compounds have been prepared and characterized (166); however, only a few have assumed practical importance. Foremost among these are: the Au(III) halides, which are involved in the purification and analysis of gold as well as being important starting materials for the preparation of other gold compounds, and the alkaline cyanides, which are of importance in the extractive metallurgy and electroplating of the metal.

For the most part, the chemistry of gold is more closely related to that of its horizontal neighbors in the Periodic Table, platinum and mercury, than to the other members of its subgroup, copper and silver. Comprehensive treatments of gold chemistry can be found in the literature (see *General References*).

Halides.   Gold(III) chloride [13453-07-1] can be prepared directly from the elements at 200°C (167). It exists as the chlorine-bridged dimer, Au_2Cl_6 in both the solid and gas phases under an atmospheric pressure of chlorine at temperatures below 254°C. Above this temperature in a chlorine atmosphere or at lower temperatures in an inert atmosphere, it decomposes first to AuCl [10294-29-8] and then to gold. The monochloride is only metastable at room temperature and slowly disproportionates to gold(0) and gold(III) chloride. The disproportionation is much more rapid in water both for AuCl and the complex chloride, $[AuCl_2]^-$, formed by interaction with metal chlorides in solution.

The tetrachloroaurate ion, $[AuCl_4]^-$, is most conveniently prepared by dissolving gold in aqua regia. The acid, $HAuCl_4$, can be crystallized as the trihydrate by evaporating HCl solutions, whereas the addition of metal halides leads to various salts of the composition $M(AuX_4)_n$. The chlorides in $[AuCl_4]^-$ can be replaced by a variety of charged and neutral ligands. Thus, the tetrabromo- and tetra-cyanoaurates (168) are formed when a solution of the tetrachloroaurate is treated with bromide or cyanide and evaporated. Reaction with fluorine leads to the tetra-fluoroaurates. Similarly, complexes of the type $[AuX_3L]$ can be prepared where L represents nitrogen donors, such as pyridine (169) or nitriles (170). The addition of ammonia leads to the formation of $[Au(NH_3)_4]^{3+}$ (171). Reaction of gold(III) chlorides with easily oxidized ligands such as organophosphines (172), dialkyl-sulfides (173), or mercaptans (174), leads to the corresponding Au(I) complexes, LAuX, which can be oxidized to the corresponding gold(III) complexes with chlorine (175) or converted to other gold(I) complexes, particularly when L is a stabilizing ligand such as a tertiary phosphine.

The chemistry of the bromides is completely analogous to that of the chlorides. Gold(III) iodide [31032-13-0], on the other hand, is unstable, loses iodine, and converts to AuI [10294-31-2] (176).

Cyanides.   Salts of the complex ion, $[Au(CN)_2]^-$, can be formed directly from gold, ie, gold dissolves in dilute solutions of potassium cyanide in the presence of air. Additionally, a gold anode dissolves in a solution of potassium cyanide. The potassium salt can be isolated by evaporation of the solution and purified by recrystallization from water (177). Boiling of the complex cyanide in hydrochloric acid results in formation of AuCN [506-65-01]. Halogens add oxidatively to $[Au(CN)_2]^-$ to yield salts of $[Au(CN)_2X_2]^-$ which are converted to the tetracy-anoaurates using excess cyanide (178). These last can also be prepared directly from the tetrahaloaurates.

Oxides and Hydroxides.   The existence of Au_2O [1303-57-7] and its hydrated oxides is doubtful. Gold(III) hydroxide [1303-52-2], $Au(OH)_3$, is precipitated from solutions of $(AuCl_4)^-$ by addition of alkali hydroxides. It is dehydrated to the hydrated oxide, Au(O)OH [30779-22-7], and on heating to 140°C, to the oxide Au_2O_3 [1303-58-8] which decomposes to gold and oxygen above 160°C. The hydroxide and hydrated oxide are soluble in both strongly acidic and basic media. In the latter case, salts, eg, $KAuO_2\cdot3H_2O$ [12256-44-9], are formed which decompose on heating.

Sulfides.   Gold(I) sulfide [30695-60-4], Au_2S, is formed when acidified solutions of $KAu(CN)_2$ are treated with H_2S. It is highly insoluble in water and dilute acids but reacts with strong oxidizing agents, such as aqua regia or chlorine, or with strongly complexing anions such as cyanide. The less stable Au_2S_3 [1303-61-3] can be prepared from $AuCl_3$ or complex chloraurates by reaction with hydrogen sulfide in anhydrous ether. The reaction in aqueous solutions leads to formation of metallic gold as does heating of the compound in air.

Organogold Compounds. Both alkyl and aryl complexes of Au(I) and Au(III) as well as olefin and acetylene complexes have been prepared and studied. In general, the preparation of alkyls and aryls involves the reaction of a gold halide with a Grignard or organolithium reagent; halides can be displaced directly by unsaturated hydrocarbons. Stabilization of the σ-bonded compounds of Au(I) is facilitated by the presence of donor ligands such as tertiary phosphines or arsines. For example, complexes of the type, R_3PAuCH_3, are prepared by the reaction of R_3PAuCl and CH_3Li (179). Addition of a second mole of CH_3Li leads to formation of $Li(Au(CH_3)_2)$ [53863-37-9] (180) which decomposes when recovery from diethyl ether solution is attempted, but can be isolated as a stable adduct of $(CH_3)_2NCH_2CH_2N(CH_3)CH_2CH_2N(CH_3)_2$ (181). The alkyl tertiary phosphine gold(I) complexes decompose thermally to gold(0) and a dimer of the alkyl. Thus, heating of R_3PAuCH_3 to 100°C leads to formation of ethane (182). Reaction with acids cleaves the gold–carbon bond with formation of hydrocarbon, eg, CH_4 in the case of R_3PAuCH_3 (179), whereas halogens react to yield alkyl halides and the corresponding Au(I) or Au(III) halides (183). Reaction with Hg(II) salts leads to exchange of the alkyl group for halide (184). Addition of methyl iodide leads initially to formation of the dimethylgold(III) complex which reacts rapidly with more R_3PAuCH_3 to yield $R_3PAu(CH_3)_3$. This, in turn, reacts more slowly to yield ethane (185).

Dialkylgold(III) halides are prepared by the reaction of a Grignard reagent with a gold(III) halide (186). They are readily converted to other dimethylgold(III) complexes via reaction with the appropriate silver salt, eg, AgCN (186), or by replacement of the halide with a donor ligand (187). Dialkylgold(III) halides are inert to acids and metal halides and react slowly with halogens (188). Trialkylgold(III) complexes can be prepared from Au(III) halides by reaction with alkyllithium reagents, but these compounds are unstable unless a donor ligand also is added. The stable compounds are prepared more conveniently by treating the appropriate Au(I) halide with alkyllithium and methyl iodide (189). The tetramethylgold(III) anion can be prepared and stabilized in the same way as described for the dimethylgold(I) anion (181). The tri- and tetraalkylgold(III) complexes react with acids, halogens, and metal halides to yield the dialkyl species.

Olefin and acetylene complexes of Au(I) can be prepared by direct interaction of the unsaturated compounds with a Au(I) halide (190,191). The resulting products, however, are not very stable and decompose at low temperatures. Reaction with Au(III) halides leads to halogenation of the unsaturated compound and formation of Au(I) complexes or polynuclear complexes with gold in mixed oxidation states (see ORGANOMETALLICS).

Cluster Compounds. More recently, an increasing amount of interest has developed in gold-containing bimetallic cluster compounds (143) which permit investigation at the molecular level of the metal–metal interactions thought to occur in bimetallic catalysts or alloys. Most often, these compounds are prepared from organophosphine stabilized Au(I) compounds such as halides or alkyls and generally contain one to three gold phosphine fragments bonded to one or more transition metal atoms, most often as carbonyl species.

BIBLIOGRAPHY

"Gold and Gold Alloys" in *ECT* 1st ed., Vol. 7, pp. 274–284, by F. E. Carter, Baker & Co., Inc.; "Gold Compounds" in *ECT* 1st ed., Vol. 7, pp. 284–288, by E. F. Rosenblatt, Baker & Co., Inc.; "Gold and Gold Compounds" in *ECT* 2nd ed., Vol. 10, pp. 681–694, by E. M. Wise, Consultant; in *ECT* 3rd ed., Vol. 11, pp. 972–995, by J. G. Cohn and E. W. Stern, Engelhard Minerals and Chemicals Corp.

1. K. E. Zimen, *Strahlende Materie*, Ullstein, Frankfurt, 1990, p. 110.
2. J. G. Cohn, *Gold Bull.* **12**, 21 (1979).
3. D. McIntosh and G. A. Ozin, *Inorg Chem.* **15**, 2869 (1976).
4. R. J. L. Audon, J. F. Martin, and K. C. Mills, *J. Chem. Soc. A*, 1788 (1971).
5. M. N. Zyranox, G. A. Khlebinka, and V. A. Krenev, *Zh. Neorg. Khim.* **18**, 918 (1973).
6. L. Graf in W. D. Robertson, ed., *Stress Corrosion Cracking and Embrittlement*, John Wiley & Sons, Inc., New York, 1956, pp. 48–60.
7. J. M. Lucas, *Gold, Annual Report 1990*, U. S. Bureau of Mines, Washington, D.C., Apr. 1992.
8. J. M. West, *Bur. Mines. Prepr. Bull.* **667** (1975); W. C. Butterman, *Mineral Commodity Profiles MCP-25*, U. S. Bureau of Mines, Washington, D.C., 1978, p. 14.
9. S. Murray and co-workers, *Gold 1992*, Gold Fields Minerals Services, Ltd., London, 1992.
10. S. R. La Brouy, D. Muir, and T. Komosa, *World Gold '91, Austral. Min. Metall.*, Parkville, Australia, 1991, p. 165; A. Nugent, K. Brackenbury, and J. Skinner, *Ibid.*, p. 173.
11. D. G. Dixon and J. L. Hendrix, in Ref. 10, p. 201.
12. K. Chandrasekaran and R. E. Ott, *Mining Eng.* **43**, 1124 (1991).
13. C. Glynn, C. Michalopoulos, and T. Green, *Gold*, Consolidated Gold Fields, Ltd., London, 1979.
14. *IPMI Precious Metals News and Review*, **16**(6), 4 (1992).
15. Can. Pat. Appl. 2,030,900 (May 28, 1991), D. G. Kleid, W. J. Kohr, and F. R. Thibodeau (to Geobiotics, Inc.).
16. D. M. Noel, M. C. Fuerstenau, and L. Hendrik, in Ref. 10, p. 65.
17. J. Avramides and co-workers, in Ref. 10, p. 45.
18. B. Peinemann, in Ref. 10, p. 3.
19. P. A. Spencer, J. R. Budden, and M. K. Rhodes, in Ref. 10, p. 59.
20. C. Kenna and P. Montz, in Ref. 10, p. 133.
21. P. A. Spencer, J. R. Budden, and M. K. Rhodes, *Miner Eng.* **4**, 1143 (1991).
22. *U.S. Bureau of Mines Mineral Survey*, Washington, D.C., Oct. 1992.
23. U.S. Pats. 3,825,651 (July 23, 1974), H. J. Heinen, J. A. Eisele, and D. Fisher (to United States of America); 3,834,896 (Sept. 7, 1974), H. J. Heinen and J. A. Eisele (to United States of America).
24. K. S. Fraser, R. H. Walter, and J. A. Wells, *Miner Eng.* **4**, 1029 (1991).
25. P. M. Afenya, in Ref. 24, p. 1043.
26. C. T. O'Connor and R. C. Dunne, in Ref. 24, p. 1057.
27. B. J. Scheiner, *High Temp. Mater. Processes* **9**, 249 (1990).
28. K. G. Thomas, *Can. Inst. Mining Bull.* **84**, 33 (1991).
29. A. E. Richards, *Symposium on Refining Non-Ferrous Metals*, Institute of Mining and Metallurgy, London, 1950, pp. 73–118.
30. D. Thompson, *Min. J.* **232**, 80, 90 (1949).

31. J. W. Laist, in M. C. Sneed, J. L. Maynard, and F. C. Beasted, eds., *Comprehensive Inorganic Chemistry*, Vol. 2, D. Van Nostrand Co., Inc., Princeton, N.J., 1954.

32. R. I. Edwards, W. A. M. te Riele, and G. J. Bernfeld, in *Gmelin Handbook of Inorganic Chemistry*, 8th ed., Supp. A1, Springer-Verlag, Berlin, 1986, p. 16.

33. G. L. Putnam, *J. Chem. Ed.* **30**, 576 (1953).

34. J. B. Rosenbaum, J. T. May, and J. M. Riley, *AIME Preprint 69-As-82*, Society of Mining and Engineering, New York, 1969.

35. J. M. Lucas, *Mineral Commodity Summaries 1992*, Bureau of Mines, Washington, D. C., 1992, p. 75.

36. W. C. Butterman, *Mineral Commodity Summaries 1978*, Bureau of Mines, Washington, D.C., 1979, p. 13.

37. *Guide to Dental Materials and Devices*, 7th ed., American Dental Association, Chicago, 1974–1975.

38. M. G. B. Drew and M. J. Hudson, *J. Chem. Soc., Dalton*, 771 (1985).

39. F. E. Beamish, *The Analytical Chemistry of the Noble Metals*, Pergamon Press, New York, 1966, p. 162.

40. F. E. Beamish, J. G. Sen Gupta, and A. Chow, in E. Wise, ed., *Gold Recovery Properties and Applications*, D. Van Nostrand Co., Inc., Princeton, N.J., 1964, p. 325.

41. S. O. Thomson, F. E. Beamish, and M. Scott, *Ind. Eng. Chem. Anal. Ed.* **9**, 420 (1937).

42. L. Vanino and O. Guyot, *Arch. Pharm.* **264**, 98 (1926).

43. J. E. Currah and co-workers, *Ind. Eng. Chem. Anal. Ed.* **18**, 120 (1946).

44. R. Berg, *The Analytical Uses of o-Oxyquinoline and its Derivatives*, 2nd ed., Ferd. Enke, Stuttgart, Germany, 1938, p. 34.

45. A. E. Harvey, Jr. and J. H. Yoe, *Anal. Chim. Acta.* **8**, 246 (1953).

46. T. C. Chadwick, *Anal. Chem.* **46**, 1326 (1947).

47. A. K. Mukherjee, *Anal. Chim. Acta* **23**, 325 (1960).

48. J. B. Kushner, *Ind. Eng. Chem. Anal. Ed.* **10**, 641 (1938).

49. A. Czaplinski and J. Trokowicz, *Chem. Anal. (Warsaw)* **4**, 463 (1959).

50. R. Belcher and A. J. Nutten, *J. Chem. Soc.*, 550 (1951).

51. W. B. Pollard, *Analyst* **62**, 597 (1937).

52. S. C. S. Rajan and N. A. Raju, *Talanta* **22**, 185 (1975).

53. F. A. Gooch and F. H. Morley, *Am. J. Sci.* **8**, 261 (1899); V. E. Herschlag, *Ind. Eng. Chem. Anal. Ed.* **13**, 561 (1941).

54. W. A. E. McBride and J. H. Yoe, *Anal. Chem.* **20**, 1094 (1948).

55. C. G. Fink and G. L. Putnam, *Ind. Eng. Chem. Anal. Ed.* **14**, 468 (1942).

56. N. R. Day and S. N. Bhattarcharya, *Talanta* **23**, 535 (1976).

57. B. J. MacNulty and B. D. Woollard, *Anal. Chim. Acta.* **22**, 192 (1960).

58. L. M. Kul'berg, *Zavosd. Lab.* **5**, 5170 (1936).

59. A. J. Lincoln and J. C. Kohler, *Anal. Chem.* **34**, 1247 (1962).

60. A. J. Lincoln and J. C. Kohler, in Ref. 40, p. 345.

61. K. Smolander and M. Kauppinen, *Anal. Chim. Acta* **248**, 569 (1991).

62. F. J. M. Maessen, F. D. Posma, and J. Balke, *Anal. Chem.* **46**, 1445 (1974).

63. V. P. Guinn, in A. Weissberger and B. W. Rossiter, eds., *Physical Methods of Chemistry*, Wiley-Interscience, New York, 1972, p. 484.

64. G. E. M. Hall, J. C. Pelchat, and C. E. Dunn, *J. Geochem. Explor.* **37**, 1 (1990).

65. T. J. Constable and co-workers, *Lancet I*, 1176 (May 24, 1975).

66. B. M. Sutton, in E. D. Zysk and J. A. Bonnuci, eds., *Precious Metals 1985, IPMI Conf. Proc.* Int. Precious Metals Inst., Allentown, Pa., 1986, p. 387.

67. N. O. Rothermich and co-workers, *Arthritis Rheum.* **19**, 1321 (1976).

68. M. Rubin and co-workers, *Proc. Soc. Exp. Biol. Med.* **124**, 290 (1967).

69. D. T. Walz and co-workers, *J. Pharm. Exp. Ther.* **181**, 292 (1972).

70. R. C. Blodgett and R. G. Pietrusko, *Scand. J. Rheumatol. Suppl.* **63**, 67 (1986).

71. D. T. Felton, J. J. Anderson, and R. F. Meenan, *Arthritis Rheum.* **33**, 1449 (1990).
72. J. S. Lawrence, *Ann. Rheum. Dis.* **20**, 341 (1961).
73. E. D. Kwon, S. A. Loening, and C. E. Hawtrey, *J. Urol.* **145**, 524 (1991).
74. J. M. Sham and co-workers, *Br. J. Radiol.* **62**, 355 (1989).
75. E. E. Rogoff, R. Romano, and E. W. Hahn, *Radiology* **114**, 225 (1975).
76. R. M. Van Soesbergen and co-workers, *Clin. Rheumatol.* **7**, 224 (1988).
77. M. Radice and co-workers, *Acta Cardiol.* **42**, 49 (1987).
78. R. M. Burnett and co-workers, *J. Biol. Chem.* **249**, 4383 (1973); F. R. Salenime and co-workers, *J. Biol. Chem.* **248**, 3910 (1973); B. C. Wang, C. S. Yoo, and M. Sax, *J. Mol. Biol.* **87**, 505 (1974).
79. M. I. Herrera and co-workers, *Ultrastruct. Pathol.* **12**, 439 (1988); D. R. Sparkman and C. L. White, *J. Electron Microsc. Tech.* **13**, 152, (1989); S. Narayaswami, K. Lundgren, and B. A. Hamkalo, *Scanning Microsc. Suppl.* **3**, 65 (1989).
80. J. M. Lucas, *Gold, Ann. Rept. 1991*, U.S. Bureau of Mines, Washington, D.C., 1992.
81. W. S. Rapson and T. Groenewald, *Gold Usage*, Academic Press, Inc., New York, 1978, p. 37.
82. Ref. 14, p. 6.
83. Ref. 81, p. 32.
84. *Ibid.*, p. 35.
85. *Ibid.*, p. 45.
86. *Ibid.*, p. 48.
87. *Ibid.*, p. 83.
88. A. S. Darling, *Gold Bull.* **5**, 74 (1972).
89. F. Sperner, *Gold Bull.* **6**, 72 (1973).
90. A. S. Darling, *Platinum Met. Rev.* **12**, 54 (1968).
91. *Gold Bull.* **6**, 39 (1973).
92. U.S. Pat. 3,066,177 (Nov. 27, 1962), J. F. Schneider and D. Accinno (to Engelhard Industries, Inc.).
93. R. Berman, *Gold Bull.* **6**, 34 (1973).
94. M. Gotoh, K. D. Hill, and E. G. Murdock, *Rev. Sci. Inst.* **62**, 2778 (1991).
95. U.S. Pat. 4,018,599 (Apr. 19, 1977), J. S. Hill and co-workers, (to Engelhard Minerals and Chemicals Corp.).
96. Ref. 81, p. 121.
97. Ref. 91, p. 69.
98. Ref. 81, pp. 150, 167.
99. *Ibid.*, p. 148.
100. *Ibid.*, p. 167.
101. *Ibid.*, p. 292.
102. *Ibid.*, p. 144.
103. *Ibid.*, p. 125.
104. E. Philofsky, *Solid State Electron.* **13**, 1391 (1970).
105. *Gold Bull.* **5**, 13 (1972).
106. Ref. 72, p. 141.
107. *Aerospace Materials Specs. AMS 4*, SAE, 1968, pp. 784–786.
108. U.S. Pat. 3,579,312 (May 18, 1971) O. A. Shoit (to E. I. Dupont de Nemours & Co., Inc.).
109. D. M. Jacobson and G. Humpson, *Interdisciplanary Sci. Revs.* **17**, 244 (1992).
110. Ger. Offen. 1,508,311 (Jan. 8, 1970), L. Piffle (to Brown Boverie et cie).
111. S. Helmsley, L. Mayer, and T. Wildman, *Proc. AESF 77th Ann. Tech. Conf.* **1**, 1 (1990); J. F. Di Nunzio, *Ibid.*, 11; J. A. Lochet, *Ibid.* **2**, 983.
112. M. Antler, *Plat. Surf. Finish.* **78**, 58 (1991).
113. M. Dettke, *Galvanotechnik* **82**, 1238 (1991).

114. Ref. 81, p. 249.
115. *Gold Bull.* **4**, 72 (1971).
116. C. C. Bocking, *Trans. Inst. Met. Finish* **69**, 119 (1990).
117. Ref. 81, p. 224.
118. D. G. Foulke, in F. H. Reid and W. Goldie, eds., *Gold Plating Technology*, Electrochemical Publ. Ltd., Ayreshire, Scotland, 1974, p. 42.
119. *Ibid.*, p. 46.
120. Ref. 81, pp. 218–219.
121. M. Antler, in Ref. 118, p. 269.
122. Ref. 81, p. 231.
123. *Ibid.*, p. 234.
124. *Ibid.*, p. 233.
125. *Ibid.*, p. 235.
126. D. R. Mason, *Gold Bull.* **7**, 104 (1974).
127. N. P. Fedot'ev, P. M. Vyacheslavov, and G. A. Volyanyuk, *Zh. Prikl. Khim.* **40**, 1759 (1967).
128. E. A. Parker, *Plating* **45**, 631 (1958).
129. R. Green and S. Peary, *Galvano-Organ.* **60**, 379 (1991).
130. U.S. Pat. 3,230,098 (Jan. 18, 1966), H. W. Robinson (to Engelhard Industries, Inc.).
131. U.S. Pat. 3,917,885 (Nov. 4, 1975), K. D. Baker (to Engelhard Minerals & Chemicals Corp.).
132. Y. Okinaka, in Ref. 109, p. 82.
133. U.S. Pat. 3,506,462 (Apr. 14, 1970), T. Oda and J. Hayashi (to Japan Electric Co.).
134. G. Desthomas and K. Miscioscio, *Proc. AESF 77th Ann. Tech. Conf.* **1**, 27 (1990).
135. U.S. Pat. 3,163,665 (Dec. 29, 1964), H. M. Fitch (to Engelhard Industries, Inc.).
136. U.S. Pat. 2,984,575 (May 16, 1961), H. M. Fitch (to Engelhard Industries, Inc.).
137. U.S. Pat. 3,313,632 (Apr. 11, 1967), R. C. Langley and H. M. Fitch (to Engelhard Industries, Inc.).
138. U.S. Pat. 3,391,010 (July 2, 1968), A. P. Hauel (to Engelhard Industries, Inc.).
139. U.S. Pats. 3,176,678 (Apr. 6, 1965) and 3,445,662 (May 20, 1969), R. C. Langley (to Engelhard Industries, Inc.).
140. F. H. Lancaster and W. S. Rapson, *Gold in Catalysis: A Bibliography*, International Gold Corp. Ltd., Marshalltown, S. Africa, 1978.
141. J. Schwank, *Gold Bull.* **16**, 103 (1983).
142. J. Schwank, *Gold Bull.* **18**, 2 (1985).
143. P. Braunstein and J. Rose, *Ibid.*, 17.
144. M. Haruta and co-workers, *J. Catal.* **115**, 301 (1989).
145. S. D. Gardner and co-workers, *J. Catal.* **129**, 114 (1991).
146. A. Knell and co-workers, *J. Catal.* **137**, 306 (1992).
147. G. Parravano, *J. Catal.* **18**, 320 (1970).
148. G. C. Bond and co-workers, *J. Chem. Soc. Chem. Commun.*, 444 (1973).
149. Y. L. Lam and M. Boudart, *J. Catal.* **50**, 530 (1977).
150. D. A. Buchanan and G. Webb, *J. Chem. Soc. Farad. Trans. 1* **71**, 134 (1975).
151. G. C. Bond, *Gold Bull.* **5**, 11 (1972).
152. Brit. Pat. 1,152,817 (May 21, 1969), (to Mobil Oil Corp.).
153. U.S. Pat. 3,742,079 (June 26, 1973), R. W. Etherington (to Mobil Oil Corp.).
154. S. Inami and H. Wise, *J. Catal.* **26**, 92 (1972); H. G. Rushford and D. A. Wharl, *Trans. Farad. Soc.* **67**, 3577 (1971); B. J. Joice and co-workers, *Discuss. Farad. Soc.* **41**, 223 (1966).
155. U.S. Pat. 3,974,102 (Aug. 10, 1976), G. L. Kaiser (to SCM Corp.).
156. U.S. Pats. 3,576,736 (Apr. 27, 1971), J. R. Kittrell (to Chevron Research Co.); 3,617,489 (Nov. 2, 1971), S. M. Csicsery (to Chevron Research Co.).

157. U.S. Pats. 3,953,368 (Apr. 27, 1976), J. H. Sinfelt (to Exxon Research and Engineering Co.); 3,785,960 (Jan. 15, 1974), H. E. Merrill and R. S. Lunt (to Esso Research and Engineering Co.); 3,899,413 (Aug. 12, 1975), J. W. Myers (to Phillips Petroleum Co.).
158. U.S. Pat. 3,989,674 (Nov. 2, 1976), J. H. Sinfelt and A. E. Barnett (to Exxon Research and Engineering Co.).
159. Ger. Offen. 2,601,154 (July 21, 1977), G. Scharfe (to Bayer, AG); U.S. Pat. 4,048,096 (Sept. 13, 1977), T. C. Bissot (to E. I. du Pont de Nemours & Co., Inc.); Brit. Pat. 1,177,515 (Jan. 14, 1970), C. W. Capp (to British Petroleum, Ltd.).
160. Ger. Offen. 2,057,087 (June 22, 1972), H. Fernholz, F. Wunder, and H. J. Schmidt (to Farbwerke Hoechst, AG).
161. E. G. Allison and G. C. Bond, *Catal. Rev.* **7**, 233 (1972).
162. R. L. Rich and H. Taube, *J. Phys. Chem.* **58**, 6 (1954).
163. J. H. Waters and H. B. Gray, *J. Am. Chem. Soc.* **87**, 3534 (1965).
164. A. J. Blake and co-workers, *Angew. Chem. Internat. Ed.* **29**, 197 (1990).
165. K. Leary and N. Bartlett, *J. Chem. Soc. Chem. Commun.*, 903 (1972); J. H. Holloway and G. J. Schrobligen, *J. Chem. Soc. Chem. Commun.*, 623 (1975).
166. C. E. Housecroft, *Coord. Chem. Revs.* **115**, 117 (1992).
167. L. Capella and C. Schwab, *Compt. Rend.* **260**, 4337 (1965).
168. H. T. S. Britton and E. N. Dodd, *J. Chem. Soc.*, 100 (1935).
169. A. Buraway and C. S. Gibson, *J. Chem. Soc.*, 860 (1934).
170. F. Calderazzo and D. B. Dell'Amico, *J. Organomet. Chem.* **76**, C59 (1974).
171. L. H. Skibsted and J. Bjerrum, *Acta Chem. Scand. Ser. A* **28**, 740 (1974).
172. F. G. Mann, A. F. Wells, and D. Purdie, *J. Chem. Soc.*, 1828 (1937).
173. D. deFillippo, F. Devillanova, and C. Preti, *Inorg. Chem. Acta* **5**, 103 (1971).
174. S. Akerstrom, *Arkiv. Kemi* **14**, 387 (1959).
175. F. G. Mann and D. Purdie, *J. Chem. Soc.*, 1235 (1940).
176. J. L. Ryan, *Inorg. Chem.* **8**, 2058 (1969).
177. A. R. Raper, in Ref. 10, p. 51.
178. L. H. Jones, *Inorg. Chem.* **3**, 1581 (1964).
179. G. Calvin, G. E. Coates, and P. S. Dixon, *Chem. Ind.*, 1628 (1959).
180. A. Tamaki and J. K. Kochi, *J. Organomet. Chem.* **57**, C39 (1973).
181. G. W. Rice and R. S. Tobias, *Inorg. Chem.* **15**, 489 (1976).
182. A. Tamaki and J. K. Kochi, *J. Organomet. Chem.* **61**, 441 (1973).
183. E. G. Perevalova and co-workers, *Izv. Akad Nauk SSSR Ser. Khim.*, 2148 (1970).
184. B. J. Gregory and C. K. Ingold, *J. Chem. Soc. B*, 276 (1969).
185. A. Tamaki and J. K. Kochi, *J. Organomet. Chem.* **64**, 411 (1974).
186. M. S. Kharasch and H. S. Isbell, *J. Am. Chem. Soc.* **53**, 2701 (1931).
187. C. F. Shaw, J. W. Lundeen, and R. S. Tobias, *J. Organomet. Chem.* **51**, 365 (1973).
188. F. H. Brain and C. S. Gibson, *J. Chem. Soc.*, 762 (1939).
189. A. Tamaki and J. K. Kochi, *J. Chem. Soc. Dalton Trans.*, 2620 (1973).
190. R. Huttel, H. Reinheimer, and H. Dietl, *Chem. Ber.* **99**, 462 (1966).
191. R. Huttel and H. Forkl, *Chem. Ber.* **105**, 1644 (1972).

General References

R. J. Puddephatt, *The Chemistry of Gold*, Elsevier Scientific Pub. Co., Amsterdam, 1978.
B. F. G. Johnson and R. Davis, in J. C. Bailar, Jr. and co-workers, eds., *Comprehensive Inorganic Chemistry*, Vol. 3, Pergamon Press, Oxford, 1973, p. 129.
G. Chaudron and O. Dimitrov, eds., *Monographies sur les Métaux de Haute Pureté*, Vol. 3, Masson, Paris, 1977, p. 491.
H. Schmidbauer, "Organogold Compounds," *Gmelin Handbook of Inorganic Chemistry*, Springer-Verlag, Berlin, 1980.
W. S. Rapson and T. Groenewald, *Gold Usage*, Academic Press, Inc., New York, 1978.

E. M. Wise, ed., *Gold: Recovery, Properties and Applications*, D. Van Nostrand Co., Inc., Princeton, N.J., 1964.

Gold Bulletin & Gold Patent Digest, quarterly journal published by the World Gold Council, Geneva.

Annual Reports on Gold, Gold Fields Mineral Services, Ltd., London.

Gold, Annual Reports; Mineral Commodity Summaries; Minerals Yearbooks; Mineral Industry Surveys; Mineral Trade Notes, U.S. Dept. of Interior, Bureau of Mines, Washington, D.C.

L. M. Gedansky and L. G. Helper, *Thermochemistry of Gold and Its Compounds, Engelhard Industries Technical Bulletin X(1)*, Engelhard Minerals & Chemicals Corp., Edison, N.J., 1969.

Metals and Alloys in the Unified Numbering System, 6th ed., SAE/ASTM, Warrendale, Pa., 1993.

J. G. COHN
ERIC W. STERN
Engelhard Corporation

GONADOTROPIC HORMONES. See HORMONES, ANTERIOR PITUITARY HORMONES.

GOSSYPOL. See VEGETABLE OILS.

GRAMICIDIN. See ANTIBIOTICS, PEPTIDES.

GRAPHITE. See CARBON, CARBON AND ARTIFICIAL GRAPHITE; CARBON, NATURAL GRAPHITE.

GRAVITY CONCENTRATION. See MINERAL RECOVERY AND PROCESSING; METALLURGY, EXTRACTIVE.

GREAT SALT LAKE CHEMICALS. See CHEMICALS FROM BRINE.

GREENHOUSE EFFECT. See AIR POLLUTION; ATMOSPHERIC MODELS; FLUORINE COMPOUNDS, ORGANIC−FLUORINATED ALIPHATIC COMPOUNDS; OZONE.

GRIGNARD REACTIONS

The term Grignard reaction refers to both the preparation of a class of organo-magnesium halide compounds and their subsequent reaction with a wide variety of organic and inorganic substrates. As such it has had a wide and profound influence on synthetic chemistry since its first elucidation by Victor Grignard at the beginning of the twentieth century.

Barbier reported (1) in 1899 that a mixture of methyl iodide, a methyl ketone, and magnesium metal in diethyl ether produced a tertiary alcohol. Detailed studies by his student Victor Grignard are documented in his now classical doctoral thesis, presented in 1901. Grignard established (2) that the reaction observed by Barbier could be separated into three distinct steps: Grignard reagent formation, Grignard reaction, and hydrolysis.

$$CH_3I \ + \ Mg \ \longrightarrow \ CH_3{-}Mg{-}I$$

As a consequence, the general nature of this sequence can be used to prepare an extraordinary variety of new compounds. For this Grignard was awarded the Nobel Prize in Chemistry in 1912 at the age of 39.

The general sequence of the reactions is now embodied in the following generic forms, where RX = an organic halide (most typically a chloride or bromide, although fluorides can be induced to react); S = a coordinating solvent (such as an ether or an amine); and AZ = a substrate with an electronegative group, Z:

$$RX \ + \ Mg \ + \ n\,S \to RMgX{\cdot}S_n$$
$$RMgX{\cdot}S_n \ + \ AZ \to RAZMgX{\cdot}S_n$$
$$RAZMgX{\cdot}S_n \to RA \ + \ ZMgX{\cdot}S_n$$

The heterolysis of AZ is dependent on the substrate and does not always occur. The final isolation of the product usually involves a hydrolysis step.

For many years the Grignard reaction was viewed as a generally useful synthetic method only for research (gram or kilogram scale). The drawbacks were the perceived cost limitations of the organic halides and magnesium metal, the well-known difficulties encountered with diethyl ether as a solvent, and the widespread belief that the reaction was inherently hazardous. The development of improved industrial procedures, including the substitution of tetrahydrofuran (THF) (3) for diethyl ether and the demonstration that the less reactive, but significantly less expensive, vinyl and aryl chlorides (4–7) could be successfully used, has greatly expanded the commercial possibilities of this reaction. In the flavor, fragrance, pharmaceutical, and fine chemical industries, its use can generally be regarded as routine. Tens of thousands of metric tons of Grignard reagents are produced annually for captive use or merchant sale.

For convenience and economy of text, the molecular formula of a Grignard reagent is typically written as RMgX. Grignard himself proposed this general formula. The actual structures of the reagents in solution are considerably more complex. The colligative properties of Grignard reagents led Schlenk and Schlenk (8) to propose the equilibria

$$2\ RMgX \rightleftarrows R_2Mg + MgX_2$$

$$R_2Mg + MgX_2 \rightleftarrows R_2MgMgX_2$$

In fact, the use of dioxane to precipitate MgX_2 from solutions of Grignard reagents is a standard method of preparation for solutions of diorganomagnesium compounds. Numerous methods have been used to establish the nature of Grignard reagents in solution and this work has been reviewed (9).

The great value of the Grignard reaction to the synthetic chemist is its general applicability as a building block for an impressive range of structures and functional groups. The Grignard reagent can act both as a prototypical carbon nucleophile that can undergo addition and substitution reactions and as a strong base that can deprotonate acidic substrates, resulting in the conjugate base or in some cases elimination reactions. Grignard reagents react with most functional groups containing polar multiple bonds (eg, ketones, nitriles, sulfones, and imines), highly strained rings (epoxides), acidic hydrogens (eg, alkynes), and certain highly polar single bonds (eg, carbon–halogen and metal–halogen). A brief but by no means all-inclusive list of typical reactions is given in Table 1. The richness of Grignard chemistry is illustrated by the fact that the reaction conditions permit the choice of converting an ester to either a tertiary alcohol (Table 1, entry 9) or a ketone (Table 1, entry 10). Likewise, the reaction of a Grignard reagent with cyclohexenone can result in either a 1,2-addition yielding a substituted cyclohexanol (Table 1, entry 18) or a 1,4-addition to give a 3-substituted cyclohexanone (Table 1, entry 19). The now dated but still impressive volume by Kharasch and Reinmuth (11) gives a systematic compilation of the synthetic possibilities.

Preparation of Grignard Reagents

A Grignard reagent is prepared by first adding magnesium and a partial charge of solvent to the reactor (11,12), followed by the addition of RX, in the remaining solvent, to the reaction flask. Initiation should occur within the first 10 wt % addition of the RX mixture. Evidence for the initiation is a large exotherm. Many chemists look for turbidity or change of color, but this can be misleading if the agitation is strong enough to shear the magnesium. Until initiation has been confirmed, no more than 20 wt % of the RX charge should be added to the reactor. Otherwise, there is a risk of initiation after all of the RX has been added. This is

Table 1. Reaction of RMgX with Various Reactants[a]

Entry[b]	Reactant	Product
1	H_2O	RH
2	O_2/H_2O	ROH
3	CO_2/H_2O	RCOOH
4	I_2	RI
5	R'X	R–R'
6[c]	MX_n	MR_n
7[d]	$COCl_2$	R_3COH
8	R'–C(=O)–R' (ketone)	R'–C(OH)(R)–R'
9[e]	R'–C(=O)–O–R" (ester)	R'–C(R)(OH)–R
10[f]	R'–C(=O)–O–R" (ester)	R'–C(=O)–R
11[e]	R'–C(=O)–OH (carboxylic acid)	R'–C(R)(OH)–R
12[e]	R'–C(=O)–Cl (acid chloride)	R'–C(R)(OH)–R
13	R'–C(=O)–H (aldehyde)	R–C(OH)(H)–R'
14	R'–C(=O)–N(R")(R") (amide)	R'–C(=O)–R
15	R'–C(=O)–O–C(=O)–R" (anhydride)	R'–C(=O)–R
16	R'–N=C=O (isocyanate)	R–C(=O)–N(H)–R'
17	R'–C(=O)–O–OC(CH$_3$)$_3$	$(CH_3)_3C$–O–R
18	2-cyclohexenone	1-R-2-cyclohexen-1-ol (HO, R)
19[g]	2-cyclohexenone	3-R-cyclohexanone

770

Table 1. (*continued*)

Entry[b]	Reactant	Product
20[e]	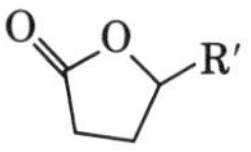 γ-butyrolactone with R' (O=C−O−CH(R')−CH2−CH2 ring)	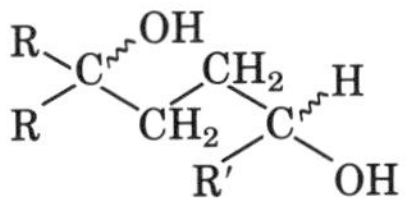 $\underset{R}{\overset{R}{C}}$(OH)−CH2−C(R')(OH)−CH2−CH2... (R2C(OH)CH2CH(R')CH2... diol product)
21	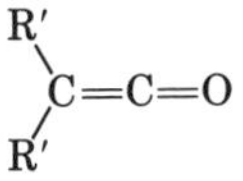 $R'_2C{=}C{=}O$	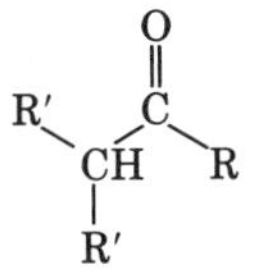 $R'CH(R')$−C(=O)−R (R−C(=O)−CH(R')R')
22	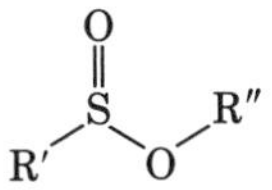 $R'{-}S(=O){-}O{-}R''$	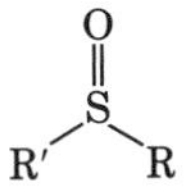 $R{-}S(=O){-}R$
23[e]	R'SSR'	RSR'
24	S	RSH
25[e]	S	RSR
26	CS_2	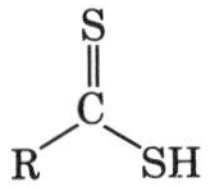 $R{-}C(=S){-}SH$
27	SO_2	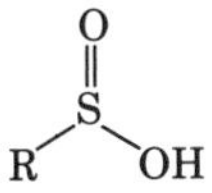 $R{-}S(=O){-}OH$
28	$R'{-}N{=}C{=}S$	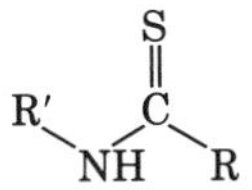 $R'{-}NH{-}C(=S){-}R$
29	$R'{-}N{=}C{=}O$	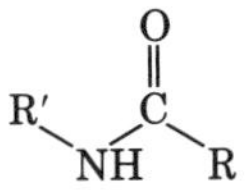 $R'{-}NH{-}C(=O){-}R$
30	$R'{-}C{\equiv}N$	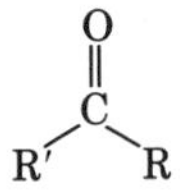 $R'{-}C(=O){-}R$
31	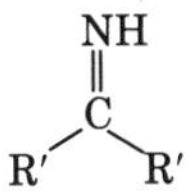 $R'{-}C(=NH){-}R'$	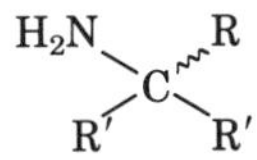 $H_2N{-}C(R)(R')R'$
32	 R'-substituted epoxide (R', R' on oxirane)	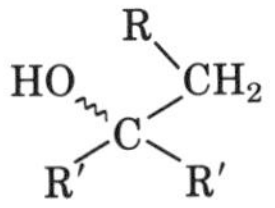 $R{-}C(R')(R')$... HO−C(R)(CH2)... (HO−C(R')(R')−CH2−R type)
33	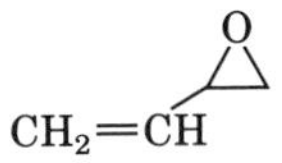 $CH_2{=}CH$−(epoxide)	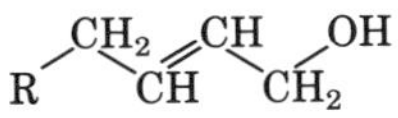 $R{-}CH_2{-}CH{=}CH{-}CH_2{-}OH$ type ($R{-}CH_2{-}CH{=}CH{-}CH_2OH$)
34	$R'{-}C{\equiv}C{-}H$	$R'{-}C{\equiv}C{-}MgX$

[a]Refs. 10 and 11.
[b]The reactant specified reacts with 1 mole of RMgX without catalyst unless otherwise specified.
[c]*n* RMgX.
[d]3 RMgX.
[e]2 RMgX.
[f]Catalyst.
[g]CuI < −40°C.

771

especially dangerous, because the resulting exotherm could cause a release of the reactor contents into the atmosphere, and possibly result in a fire.

The solvent, magnesium, and RX can have a deleterious effect on the preparation of the Grignard reagent. Some of the problems are a homocoupled product, formation of $RMgO_2X$, and noninitiated reaction of RX with Mg. Therefore, proper preparation and handling of each component must be carried out.

Solvent Preparation.   The most critical aspect of the solvent is that it must be dry (less than 0.02 wt % of H_2O) and free of O_2. If the H_2O content is above 0.02 wt %, then the reaction of Mg and RX does not initiate, except for an extremely reactive RX species, such as benzyl bromide. Although adventitious O_2 does not retard the initiation process, the O_2 reacts with the Grignard reagent to form a $RMgO_2X$ species. Furthermore, upon hydrolysis, the oxidized Grignard reagent forms a ROH species that may cause purification problems.

The peroxo species can oxidize other reactants, liquids, catalyst, or final product in the subsequent coupling reaction. One example of such oxidation is observed in the preparation of triphenylphosphine (13–15). If this reaction is hydrolyzed in air instead of an inert N_2 atmosphere, then the amount of triphenylphosphine oxide increases from less than 1 wt % to greater than 15 wt %.

$$Cl_3P + 3\ C_6H_5MgCl \xrightarrow{\text{air}} 0.15\ (C_6H_5)_3PO + 0.85\ (C_6H_5)_3P$$

Other considerations for the solvent are the solubility of the Grignard reagent and the temperatures required for initiation and adventitious reactions of the Grignard with the solvent (16–19). Based on these three considerations, the best general solvent for the preparation of a Grignard reagent is THF. However, other solvents that are commonly used are diethyl ether, methyl t-butyl ether, di-n-butyl ether, glycol diethers, toluene, dioxane (R_2Mg), and hexane.

Magnesium Preparation.   A surface coating resulting from the oxidation or hydration of the metal surface is the principal problem encountered for the magnesium reaction component. This coating prevents the Grignard reaction from initiating. Fortunately, there are dozens of methods to remove the inert coating, thus activating the magnesium (11,20–24). These methods include addition of I_2, $BrCH_2CH_2Br$, activated Mg (20,24), and ultrasound. However, for industrial use, the best method is using freshly chipped Mg turnings with a small quantity of the desired Grignard added to the reactor before addition of RX (11,25).

The Organohalogen Component.   Just as for Mg and the solvent, the organic halide must be dry (less than 0.02 wt % of H_2O) and free of O_2 for the reasons previously discussed. The relative reactivity of the halogens is reflected in the rate of disappearance of Mg which follows the general order I > Br > Cl >> F. Unfortunately, the rate of disappearance of Mg does not always correlate with the formation of active Grignard. Typically, the more reactive the RX is, the higher the probability of forming a homocoupled product. Therefore, when choosing X, the rate of reactivity, product selectivity, and cost must be taken into account.

The general effect of R on the rate of formation of RMgX is allyl, benzyl > 1°alkyl > 2°alkyl, cycloalkyl ≥ 3°alkyl, aromatic > vinyl. Again, the more reactive R is, the more probable the formation of homocoupled product. Another important consideration for the R group is the compatibility of functional groups in the organic halide. Because of the reactivity of the Grignard reagent, noncompatible functional groups need to be protected. However, some chemists design a reactive functionality into the organic halide to carry out an intramolecular cyclization.

Other Methods.   There are several common alternative methods for making Grignard reagents. Metal-exchange reactions are straightforward and MgR_2 can easily be prepared by this route (26–30). This method is typically used when Grignard reagent formation is difficult to initiate or a diorganomagnesium reagent is desired:

$$R_nM + n\ MgCl_2 \rightarrow n\ RMgCl + MCl_n$$

where M = Na, Li, or Hg. Acid–base preparation (21) works only if RH is more acidic than R′H. This method is typically used to make Grignard reagents of terminal alkynes:

$$RH + R'MgX \rightarrow RMgX + R'H$$

Hydromagnesation reactions allow for the preparation of a Grignard from an olefin (31–35).

$$CH_2{=}CHCH_2R + CH_3CH_2CH_2MgCl \xrightarrow{\text{catalyst}} ClMgCH_2CH_2CH_2R + CH_2{=}CHCH_3$$

This method provides an economic advantage if the olefin is significantly less expensive than the corresponding organic halide. Interestingly, when MgH_2 reacts with an alkyne in the presence of a Cu catalyst, only the cis-alkene Grignard reagent is formed.

Industrial Manufacturing Process

In spite of its industrial use for many years, the commercial-scale production of Grignard reagents has not been extensively described. The general features of a small production unit have been reported (36) and a pilot-size operation suitable for pharmaceutical applications has been described (37). A continuous system, practical for small- or medium-scale production, was patented by Hoffman-La Roche (38). However, the only practically important method is the batch method described by Grignard in 1900, namely formation of the Grignard reagent, reaction with a substrate, followed by hydrolysis of the reaction mixture. A simplified schematic diagram is shown in Figure 1.

The equipment can usually be constructed of carbon steel except for the hydrolysis vessel, which is usually glass-lined to avoid corrosion by aqueous acids. It is desirable to use stainless steel or, preferably, glass-lined vessels throughout. All vessels must be supplied with an inert gas (nitrogen or argon) for purging and blanketing and are vented to release off-gases. It is imperative that the reaction vessel be protected with a rupture disk.

Formation of the Grignard Reagent.   In the classical methodology for the start-up of a batch process (Fig. 1), magnesium turnings or chips are charged to

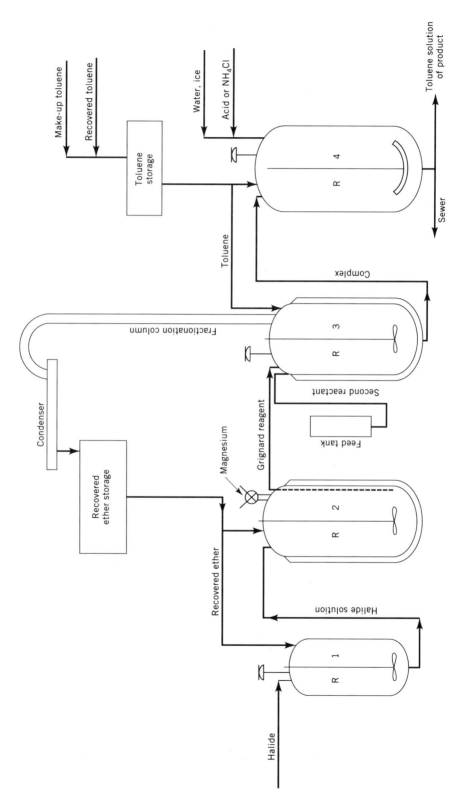

Fig. 1. Flow diagram of a Grignard plant.

774

vessel R-2. After thorough purging with an inert gas, a portion of the solvent is then added, followed by a fraction of the halide charge from vessel R-1. The chief hazard associated with the manufacture of Grignard reagents occurs at this stage in connection with the difficulty in initiating the reaction. If the addition of the halide is continued before initiation, then the presence of excess halide suddenly initiating generates such a large exotherm that the cooling system may be overwhelmed. This may result in a dangerous pressure buildup and a potential explosion.

To minimize the effects of this difficulty, an initiator is frequently employed. Among the numerous suggestions in the literature, the most satisfactory industrial procedure is to retain a portion of the Grignard from the preceding batch and to add this portion to the initial ether charge. The purpose of this procedure is to eliminate residual water and to clean the magnesium surface. Once this initiator has been added, the halide is added at a rate determined by the temperature and the pressure in the reaction vessel.

Cooling is routinely applied, either with ambient process water if THF is the solvent or with chilled brine if diethyl ether is used. Since Grignard reagents are particularly reactive with water, liquid hydrocarbon coolants may be preferred, to eliminate the risk that could arise from a cooling-system leak.

Reaction of the Grignard Reagent With a Substrate. For the process represented in Figure 1, the Grignard reagent is usually transferred from the reaction vessel R-2 to the coupling vessel R-3 for the reaction with the substrate. In most cases it is advisable to remove particles of residual magnesium by filtration. The substrate, dissolved in an ether or hydrocarbon solvent, is then added slowly from the feed tank (with cooling of the reaction vessel). The addition can also be reversed, ie, via addition of the Grignard reagent slowly from vessel R-2 to a solution of the substrate in vessel R-3. The choice of direct or reverse addition is dictated by several factors and is usually established by suitable process and pilot-plant studies before reaching the plant stage. After the reaction is complete, most of the ether is removed by distillation, with or without the addition of enough hydrocarbon to R-3 to maintain adequate agitation. At this point the contents of R-3 are pumped to R-4 for further processing.

Hydrolysis of the Reaction Mixture. Apart from the reaction of Grignard reagents with organic or inorganic halides, the reaction gives a complex that requires a quench step to release the desired product. Water has sufficient acidity to hydrolyze the complex, but the basic magnesium salts formed in this reaction generate a gel. This can be overcome through the addition of a stronger acid to neutralize the basic salts and produce water-soluble magnesium salts. Sulfuric and hydrochloric acids are commonly employed, but may lead, particularly in the case of some tertiary alcohols, to dehydration. In these cases, the use of weaker acids such as acetic acid or ammonium chloride permits the recovery of the desired alcohol. After the hydrolysis step is complete, the aqueous phase is separated from the organic phase and the product is then isolated.

In those cases where hydrolysis is not required, the metathetical reaction leads directly to the product (plus magnesium halide) in the solvent. Volatile products may be removed by direct distillation or the solid magnesium halide may be removed by filtration, before solvent evaporation. If the final product is not water sensitive, water-washing of the final reaction product mixture usually removes the salts conveniently.

Special handling procedures have been described (39) for those cases where the Grignard reagent is purchased commercially (either in drums or bulk) for use in a separate facility.

A number of continuous processes for the manufacture and use of Grignard reagents have been proposed (40), but are of no current commercial significance.

Analysis of Grignard Reagents

There are three potential problems that may occur during Grignard reagent preparation: oxidation by O_2, hydrolysis by H_2O, or homocoupling during the addition of alkyl or aryl halide. All three of these reactions decrease the active Grignard reagent while maintaining the same equivalents of base. Consequently, the concentration of a Grignard reagent should not be assumed, based on the reactants. The disadvantages of not analyzing the Grignard reagent are improper stoichiometry, potentially deleterious side reactions, highly exothermic quenching processes, phase splits, waste disposal, and cost problems. The analytical technique must be able to differentiate between active Grignard and total basicity.

Many methods are available to measure the active Grignard (41–50), ranging from titration to electrophilic quenching followed by gc analysis. Potentiometric titration (41,42) using 2-butanol (42) as the titrant is recommended as the general method for determining the activity of Grignard reagents. The advantages of this method are that it does not titrate Mg–OH or Mg–OR, results are reproducible (3 standard deviations = 0.6% of value), analysis of the solution is direct, and analysis takes less than 10 minutes. Standard titration techniques (43,44), using 2-butanol as the titrant, have greater error (3 standard deviations = 1–5% of value), but in most cases this is still acceptable.

Electrophilic quench followed by gc analysis can give the accuracy and precision of a potentiometric titration (46–50). However, each Grignard reagent and its product must be calibrated vs a gc standard, which is time-consuming. Also, these methods typically take a minimum of one hour from the start of the quench procedure to obtaining the final chromatogram. The advantage of the gc method is that a direct measurement of the homocoupled product, oxidized Grignard, hydrolyzed Grignard, and unreacted alkyl or aryl halide can all be made.

Total basicity is measured by standard acid–base titration techniques. The activity divided by the total basicity should be greater than 90%. If it is not, then the Grignard reagent should be checked for unreacted alkyl or aryl halide, homocoupled product, hydrolysis products, and oxidation products.

Spectroscopic techniques such as nmr and crystallography have been employed in structure elucidation (51–63). The nmr data are extremely dependent on the solvent, steric bulk of the organic group, temperature, and concentration. Typically, to observe the Schlenk equilibrium, low temperatures are required ($-70°C$), but it has been observed at ambient temperature with the bulky *tert*-butylmagnesium chloride (58,59) in THF. The high field resonance arises from R_2Mg and the low field resonance from RMgX. Similar findings have also been observed by ^{25}Mg-nmr (60).

Crystal structures of Grignard reagents do not necessarily correspond to their structure in solution. In general, the crystal structures (61–64) indicate the reagents are ligated with THF or diethyl ether and are frequently observed to be dimers. The Mg atoms in the dimers do not have a Mg–Mg bond; instead the dimers are typically held together by a halide bridge.

Economic Aspects

The Grignard reaction has been commercially important for a number of years, and for certain industrial processes it remains the favored (or only) practical route to construct various element-to-carbon bonds. Compared with other organometallic reagents, organomagnesium compounds offer considerably greater flexibility, efficiency, and degree of reactivity resulting from the choice of solvent and organic halide. The range of commercially available organoaluminum or organolithium compounds is limited to simple alkyl and aryl groups (methyl, butyl, and phenyl for lithium; butyl and octyl for aluminum). Grignard reagents, by contrast, can tolerate a wider range of substituents on the carbon framework (Table 2).

Table 2. Commercial Organometallic Reagents[a]

Parameter	R_3Al	RMgX	RLi
solvents	neat, hydrocarbons	ethers, hydrocarbons, amines	hydrocarbons, amines
R groups	alkyl	alkyl or aryl with substituents	alkyl
relative cost per mole of R^b transferred	1	2	15

[a]Available in bulk or in metric ton quantities.
[b]Calculations are based on R = C_4H_9.

For reasons of cost, the organic chloride is preferred as a starting material over the corresponding bromide or iodide, in spite of its generally lower reactivity. With the development of THF as a substitute for diethyl ether as a solvent, aryl (7) and vinyl (65) Grignard reagents can be made from the corresponding chlorides in high yield.

There are five components to the cost of using a Grignard reagent: *(1)* magnesium metal, *(2)* the halide, *(3)* the solvent, *(4)* the substrate, and *(5)* disposal of the by-products. The price of magnesium in mid-1992 was $3.20/kg, having risen from $1.20/kg in 1966 to $1.36/kg in 1970 and $2.90/kg in 1979. Prices for tetrahydrofuran and diethyl ether, the two most commonly used solvents, have also increased (Table 3) in the same period. The cost of the halide depends on its structure, but as a general rule the order of cost is chloride < bromide < iodide.

There is often a choice of two or more different substrates which can give rise to the same product with different Grignard reagents. Thus 1,1-diphenylethanol [599-67-7] can be prepared by three different routes, as shown in

Table 3. Costs of Components of Grignard Reagents, U.S. $/kg.

Component	1970	1979	1992
magnesium	1.36	2.90	3.20
tetrahydrofuran	0.81	1.65	2.30
diethyl ether	0.38	0.70	1.15
toluene	0.07[a]	0.29[a]	2.42

[a]$/L.

Figure 2. The choice depends on the yield and ease of purification as well as the cost of the substrate and the Grignard reagent.

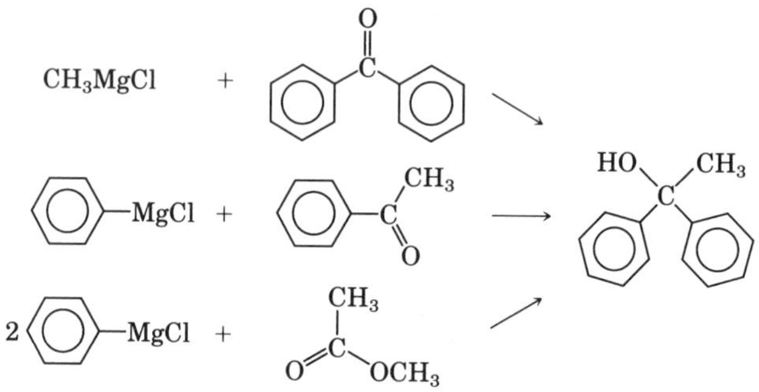

Fig. 2. Preparation of 1,1-diphenylethanol from methylmagnesium chloride and benzophenone, phenylmagnesium chloride and acetophenone, or phenylmagnesium chloride and methyl acetate.

Aqueous work-up of the typical Grignard reaction gives a mixed magnesium hydroxide–halide solution or suspension which must be disposed of. The cost of disposal of the acidic aqueous waste in accordance with local wastewater treatment regulations must also be considered.

For many fine chemical producers, including pharmaceutical manufacturers, using a Grignard reagent is often a make-or-buy decision, the choice depending on *(1)* the possibility of more cost-effective uses for the equipment being used to generate the Grignard reagent, *(2)* the facility for solvent recovery, and *(3)* waste disposal. For these reasons, custom synthesis by a company specializing in Grignard chemistry is often an option. Mid-1992 prices for commercially available Grignard reagents are given in Table 4.

Health and Safety Factors

Fire Hazards. The hazards associated with the manufacture, transport, and use of Grignard reagents are related to the flammability of the solvents employed and the exothermic reactions involved in their preparation and use. Historically, diethyl ether was the common solvent used in the preparation of Grignard reagents. Commercially, however, its use is limited by its extreme flammability and relatively low boiling point, which preclude its use in reactions with higher work-

Table 4. Prices for Commercially Available Grignards[a]

Reagent	Solvent	Concentration, mol/kg	Selling price, $/kg
CH_3MgCl	THF	3.0	7.80
CH_3MgBr	THF/toluene	2.0	8.60
n-C_4H_9MgCl	THF	2.5	9.10
C_6H_5MgCl	THF	2.0	10.00
$2\ ClC_6H_4MgCl$	THF	2.0	12.50
t-C_4H_9MgCl	THF	2.0	12.50
$C_6H_5CH_2MgCl$	THF	1.7	13.40

[a]Available in 1992.

ing temperatures. Since the 1950s it has been largely replaced by tetrahydrofuran (THF) in commercial-scale production. THF is a cyclic ether with a boiling point 30°C higher than that of diethyl ether and a flash point of -17°C, compared with -45°C for diethyl ether.

Because of the exotherm, Grignard reactions are carried out in jacketed, glass-lined steel reactors equipped with cooling, explosion-proof electrical wiring, explosion disks, and inert atmosphere for purging and blanketing. To control the heat of reaction when THF solvent is employed, ambient cold water may suffice. For reactions using diethyl ether, chilled water or brine is required to keep the reaction mass below the boiling point. Since Grignard reagents are particularly reactive with water, liquid hydrocarbon coolants, eg, kerosene, may be substituted to eliminate the risk of a water leak. Similar precautions apply for the use of commercial Grignard reagents.

Spills of Grignard reagents should be contained with inert absorbent material and transferred with nonsparking bronze, brass, or plastic tools. Because Grignard reagents are highly reactive; fire extinguishers containing carbon dioxide, chlorinated hydrocarbons, or water must not be used. Dry chemical fire extinguishers and full personal protective clothing, including self-contained breathing apparatus, are recommended.

Toxicology. Because of their high reactivity, there is little meaningful information on the health hazards of Grignard reagents per se. Rather, consideration needs to be given to the reagents employed, including the solvents and the products (or by-products) of the reaction. Some starting materials, such as organic halides (notably methyl bromide and vinyl chloride), are particularly toxic. Hydrolysis of a phenylmagnesium halide produces benzene.

Regulatory Considerations. Commercial use of a Grignard reagent in the United States requires that it appear on the Environmental Protection Agency (EPA) list of Chemical Substances in Commerce. A corresponding registration exists for the European Community and for Japan. Table 5 gives the CAS Registry Numbers of those currently listed.

Because they are classified as flammable liquids, Grignard reagents in the United States must be packaged in drums or other suitable containers bearing a red U.S. Department of Transportation label. The appropriate international des-

Table 5. Grignard Reagents Listed on the U.S. EPA Toxic Substances Control Act (TSCA) List

Name	RMgX formula	CAS Registry Number
methylmagnesium chloride	CH_3MgCl	*[646-58-4]*
methylmagnesium bromide	CH_3MgBr	*[75-16-1]*
methylmagnesium iodide	CH_3MgI	*[917-64-6]*
ethylmagnesium chloride	CH_3CH_2MgCl	*[2386-64-3]*
ethylmagnesium bromide	CH_3CH_2MgBr	*[925-90-6]*
n-propylmagnesium chloride	$CH_3CH_2CH_2MgCl$	*[2234-82-4]*
isopropylmagnesium chloride	$(CH_3)_2CHMgCl$	*[1068-55-9]*
n-butyl magnesium chloride	$CH_3(CH_2)_2CH_2MgCl$	*[693-04-9]*
sec-butylmagnesium chloride	$CH_3CH_2CH(CH_3)MgCl$	*[15366-08-02]*
n-pentylmagnesium bromide	$CH_3(CH_2)_4MgBr$	*[693-25-4]*
n-hexylmagnesium bromide	$CH_3(CH_2)_5MgBr$	*[3761-92-0]*
vinylmagnesium bromide	$CH_2{=}CHMgBr$	*[1826-67-1]*
phenylmagnesium bromide	C_6H_5MgBr	*[100-58-3]*
phenylmagnesium chloride	C_6H_5MgCl	*[100-59-4]*
benzylmagnesium chloride	$C_6H_5CH_2MgCl$	*[6921-34-2]*
p-tolylmagnesium bromide	$4\text{-}CH_3C_6H_4MgBr$	*[4294-57-9]*
o-tolylmagnesium chloride	$2\text{-}CH_3C_6H_4MgCl$	*[33872-80-9]*

ignation of Grignard reagents is UN-1993. Container size restrictions apply for ocean shipment of Grignard reagents which generate gaseous by-products (C-1 to C-4 hydrocarbons) upon hydrolysis.

Reactions and Applications of Grignard Reagents

There are several reviews and books that discuss the general chemistry of Grignard reagents (11,66,67). The focus here is the specific commercial growth areas in Grignard chemistry.

Asymmetric Syntheses Using Grignard Reagents. Grignard chemistry used in the pharmaceutical industry is typically a nucleophilic addition reaction. Asymmetric synthesis is possible using Grignard reagents. There are several successful synthetic strategies that have been exploited: preparation of a chiral Grignard reagent (68,69), internal chelation of a reactant with the Grignard reagent (70,71), and the use of a chiral metal catalyst (72,73) to generate chiral derivatives that have high % ee. For example, in the following the S/R ratio for the 3-ethylcyclopentene product is 96.8/3.2.

Grignard Reactions With Inorganic Chlorides. The principal advantage of using a Grignard reagent with an inorganic halide, usually a main group halide, is that a sterically hindered product can be made at low cost. Typically, the less expensive organoaluminum reagent does not deliver all three of its sterically hindered organic groups, or the actual organoaluminum reagent must be prepared from the corresponding Grignard reagent. In either case, the effect is to make the organoaluminum chemistry more expensive. The alternative sodium and lithium chemistries also have economic drawbacks, as compared to the Grignard chemistry. These comments are especially true for aromatic compounds. Commercial examples of sterically hindered products are tricyclohexylphosphine, cyclohexyldiphenylphosphine, dicyclohexylphenylphosphine, trineophyltin hydroxide, tricyclohexyltin hydroxide, bistrineophyltin oxide, and tetraneophylzirconium (74,75).

One of the largest commercially used Grignard reagents is phenylmagnesium chloride. Millions of kg per year of this Grignard react captively with inorganic halides. Some examples of these products are triphenylphosphine, triphenyltin hydroxide, sodium tetraphenylborate, and triphenylantimony.

Grignard Reagents as Bases. Typically, Grignard reagents are better nucleophiles and poorer bases, compared to organolithium reagents. However, there are many synthetic examples where Grignard reagents are used as bases (76,77). A commercial example is the preparation of the nuclear extractant n-octylphenyl-N,N-diisobutylcarbamoylphosphine oxide (78,79) in which two equivalents of octylmagnesium chloride [38841-98-4] are used, one to displace the ethoxide and the second to abstract the proton:

$$C_6H_5-\overset{\overset{\displaystyle O}{\|}}{\underset{\underset{\displaystyle OCH_2CH_3}{|}}{P}}-H \quad + \quad 2\ CH_3(CH_2)_7MgCl \quad \longrightarrow$$

$$\left[C_6H_5P\overset{\diagup O}{\diagdown(CH_2)_7CH_3} \right]^- \quad \xrightarrow{\ \underset{\overset{\|}{O}}{\overset{CH_2\diagdown N(i\text{-}C_4H_9)_2}{Cl-C}}\ } \quad C_6H_5\diagdown\underset{CH_3(CH_2)_7}{\overset{O\ \ O}{\overset{\|\ \ \|}{P-C}}}\diagup\underset{CH_2}{}\diagdown N\overset{i\text{-}C_4H_9}{\diagdown_{i\text{-}C_4H_9}}$$

Experiments (^{31}P nmr) using 0.8 and 2 equivalents of octylmagnesium chloride with ethyl benzenephosphinate indicate that the nucleophilic displacement occurs first, followed by proton abstraction (80). Interestingly, the order of the two steps is reversed when methylmagnesium chloride is used (81). This reaction demonstrates the difference in reactivity between the octyl and the methyl Grignard reagents.

Metal-Assisted/Modified Grignard Reactions. Over the last 15 years there has been a significant amount of work using catalysts to modify Grignard reactions. The focus has been on modifying the Grignard reagent's reactivity or using the Grignard reagent as the nucleophile in a catalytic system. Some of the specific reactions developed are asymmetric synthesis, a soft Grignard reagent, Ni cross-coupling reactions, and *gem*-dimetallic Grignard chemistry.

A soft (nucleophilic) Grignard reagent has been developed (82–84). The value of this reaction is demonstrated in acylation reactions (82).

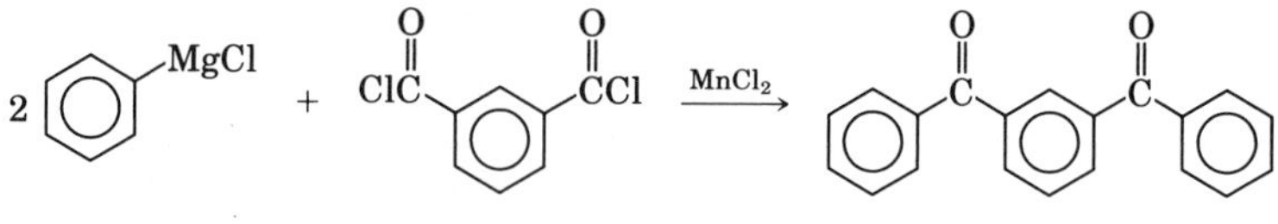

Using only the phenylmagnesium chloride without the $MnCl_2$ catalyst results in a mixture of products. This mixture includes the alcohol(s) resulting from the diaddition of the Grignard reagent to the carbonyl groups. Other catalysts, such as Fe(III) and Ni(II), have also been used to achieve similar results (85).

Another area of interest to the industrial sector is the development of a more efficient synthesis of biaryl compounds. This has been accomplished using a Ni(II)-catalyzed Grignard coupling reaction with an aryl halide (86–89).

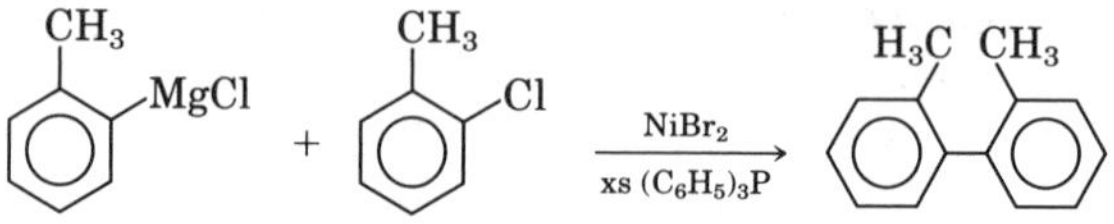

These aryl–aryl couplings are applications of the Ni(0) and Ni(II)-catalyzed cross-couplings of unreactive organic halides with Grignard reagents. This work has been extensively reviewed (90).

An interesting academic development that should have future applications is the advent of mixed *gem*-dimetallic Grignard reagents (91), such as the Mg/Zn reagent shown in Figure 3. The advantage of the mixed dianion is that two different electrophiles can be added simultaneously and react regiospecifically with the dianion. When the CH_3SSCH_3 is added alone, only the aldehyde is formed. This emphasizes the difference in reactivity between the two anions of this system.

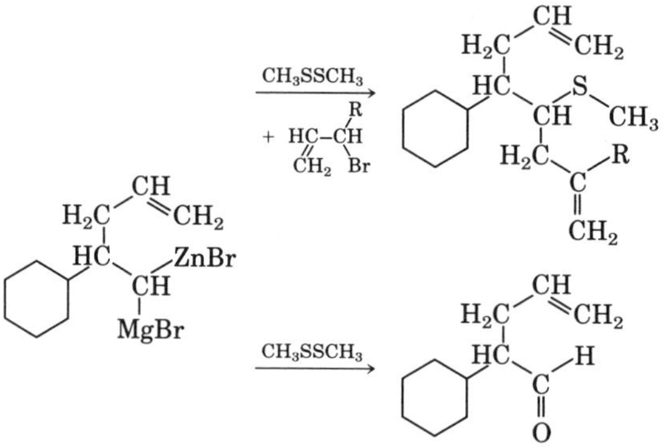

Fig. 3. Reactivity of a mixed *gem*-dimetallic Grignard reagent.

Intramolecular Grignard Reactions.   Some chemists use the reactivity of a substituent in the Grignard reagent to generate the desired product. For example, a one-step spiroannelation reaction has been developed by taking advantage of a dihalide or a cyano-halide reactant (92–94).

Other types of cyclization reactions have been demonstrated (95).

Grignards as Methacrylate Polymerization Catalysts. In this application the Grignard reagent is used to initiate the polymerization of methyl methacrylate (MMA) (96–98). The tacticity is affected by the amount of THF in the Grignard catalyst. The catalyst is isotactic-directing when the THF-to-Grignard mole ratio is in the range of 2:1 to 10:1. High levels of THF yield polymers that have a Bernoullian distribution of triads. A 4-centered transition state has been proposed involving RMgX·THF and MMA. The implication is that the Schlenk equilibrium has an effect on this reaction. However, further work needs to be carried out to confirm this hypothesis.

Grignard Reagents as Supports for the Ziegler-Natta Process. The Ziegler-Natta process is used to make polymers (polypropylene and polyethylene) from olefins via an olefin insertion mechanism on a transition metal, such as Ti. Since it would be expensive to have a solid Ti catalyst bed, a similar metal must be used as a filler. Mg^{2+} has an ionic radius similar to that of Ti^{4+} (0.068 nm). Magnesium chloride also has a similar crystal structure relative to the α and γ forms of $TiCl_3$. Therefore, $MgCl_2$ was initially chosen as a support for the Ti catalyst. In fact the initial work was done by grinding $MgCl_2$ with the other catalyst components. However, many of the current processes adsorb a Grignard reagent (butyl MgCl) in THF onto a support bed and quench the Grignard with an alcohol, followed by $TiCl_4$ treatment. The alcohol quench is not always used. In such a case, the active catalyst would be the reaction product of the Grignard reagent and $TiCl_4$. These current processes yield a catalyst with a higher active surface area (99–101).

BIBLIOGRAPHY

"Grignard Reaction" in *ECT* 1st ed., Vol. 7, pp. 314–324, by T. D. Waugh, Arapahoe Chemicals, Inc.; in *ECT* 2nd ed., Vol. 10, pp. 721–734, by T. D. Waugh, Arapahoe Chemicals, Division of Syntex Corp.; in *ECT* 3rd ed., Vol. 12, pp. 30–44, by T. E. McEntee, Arapahoe Chemicals, Inc.

1. P. Barbier, *Compt. Rend.* **128,** 110 (1899).
2. V. Grignard, *Compt. Rend.* **130,** 1322 (1900).
3. *Tetrahydrofuran as a Reaction Solvent*, E. I. du Pont de Nemours & Co., Inc., Wilmington, Del., 1960.
4. U. S. Pat. 2,795,628 (June 11, 1957), H. E. Ramsden (to Metal and Thermit Corp.).
5. U. S. Pat. 2,873,275 (Feb. 10, 1959), H. E. Ramsden (to Metal and Thermit Corp.).
6. U. S. Pat. 2,881,225 (Apr. 7, 1959), E. Kaiser and L. Sporar (to Armour and Co.).
7. U. S. Pat. 2,959,596 (Nov. 8, 1960), H. E. Ramsden and A. Balint (to Metal and Thermit Corp.).
8. W. Schlenk and W. Schlenk, Jr., *Chem. Ber.* **62B,** 920 (1929).
9. E. Ashby, J. Laemmele, and H. Neumann, *Acc. Chem. Res.* **7,** 272 (1974).
10. *The Handbook of Grignard Reactions*, 2nd ed., Elf Atochem NA, Philadelphia, Pa., 1992.

11. M. Kharasch and O. Reinmuth, *Grignard Reactions of Nonmetallic Substances*, Prentice-Hall, Inc., New York, 1954.

12. M. Fieser and L. Fieser, *Reagents for Organic Synthesis*, John Wiley & Sons, Inc., New York, 1967, pp. 415.

13. G. Silverman and M. DiFilipantonio, Elf Atochem NA internal publication, Philadelphia, Pa., 1987.

14. R. Han and G. Parkin, *J. Am. Chem. Soc.* **114,** 748 (1992).

15. S. Czernecki, C. Georgoulis, and E. Michel, *J. Organomet. Chem.* **140,** 127 (1977).

16. L. Jones, S. Kirby, D. Kean, and G. Campbell, *J. Organomet. Chem.* **284,** 159 (1985).

17. E. Bartmann, *J. Organomet. Chem.* **284,** 149 (1985).

18. A. Maercker, *Angew. Chem., Int. Ed. Engl.* **26,** 972 (1987).

19. F. Freijee, G. Schat, R. Mierop, C. Blomberg, and F. Bickelhaupt, *Heterocycles* **7,** 237 (1977).

20. B. Bogdanovic, *Acc. Chem. Res.* **21,** 261 (1988).

21. R. Rieke, *Acc. Chem. Res.* **10,** 301 (1977).

22. K. Klabunde, *Acc. Chem. Res.* **8,** 393 (1975).

23. B. Han and P. Boudjouk, *Tetrahedron Lett.* **22,** 2757 (1981).

24. H. Brown and U. Racherla, *J. Org. Chem.* **51,** 427 (1986).

25. U. S. Pat. 2,552,676 (May 15, 1951), J. Hill (to Cincinnati Milling Machine Co.).

26. L. Rosch, *Angew. Chem., Int. Ed. Engl.* **22,** 247 (1977).

27. H. Gilman and J. Swiss, *J. Am. Chem. Soc.* **62,** 1847 (1940).

28. E. Colomer and R. Corriu, *J. Organomet. Chem.* **133,** 159 (1977).

29. H. O. House and M. Lusch, *J. Org. Chem.* **42,** 183 (1977).

30. L. Costa and G. Whitesides, *J. Am. Chem. Soc.* **99,** 2390 (1977).

31. F. Sato, *J. Organomet. Chem.* **285,** 53 (1985).

32. G. Cooper and H. Finkbeiner, *J. Org. Chem.* **27,** 1493 (1962).

33. E. Ashby and T. Smith, *J. Chem. Soc., Chem. Commun.* **1,** 30 (1978).

34. J. Eisch and J. Galle, *J. Organomet. Chem.* **160,** C8 (1978).

35. E. Ashby, J. Lin, and A. Goel, *J. Org. Chem.* **43,** 757 (1978).

36. T. Waugh and R. Waugh, *Adv. Chem. Ser.* **23,** 73 (1959).

37. P. Colin and J. Gillin, *Chem. Engr. Prog.* **56,** 71 (1960).

38. J. Kollonitsch, *Annal N.Y. Acad. Sci.* **125**(1), 161 (1965).

39. P. Rakita, J. Aultman, and L. Stapelton, *Chem. Eng.* **97**(3), 110 (1990).

40. U. S. Pat. 3,911,037 (Oct. 7, 1975), G. Blackmar, C. Wright, and R. Smith (to Nalco Chemical Co.).

41. K. Kham, C. Chevrot, J. Folest, M. Troupel, and J. Perichon, *Bull. Soc. Chim. Fr.* **4,** 243 (1977).

42. K. Botlagudur and P. Branigan, Elf Atochem NA, internal standard test method, Rahway, N.J., 1971.

43. D. Bergbreiter and E. Pendergrass, *J. Org. Chem.* **46,** 219 (1981).

44. Y. Aso, H. Yamashita, T. Otsubo, and F. Ogura, *J. Org. Chem.* **54,** 5627 (1989).

45. M. Winkle, J. Lansinger, and R. Ronald, *J. Chem. Soc., Chem. Commun.* **1,** 87 (1980).

46. S. Watson and J. Eastham, *J. Organomet. Chem.* **9,** 165 (1967).

47. H. House and W. Respess, *J. Organomet. Chem.* **4,** 95 (1965).

48. R. Hollander and M. Anteunis, *Bull. Soc. Chim. Belg.* **72,** 77 (1963).

49. A. Wowk and S. Giovanni, *Anal. Chem.* **38,** 742 (1966).

50. M. Molinari, J. Lombardo, O. Lires, and G. Videla, *An. Assoc. Quim. Argent.* **48,** 140 (1960).

51. D. Hutchinson, K. Beck, R. Benkeser, and J. Gruntzner, *J. Am. Chem. Soc.* **95,** 7075 (1973).

52. P. Allen, S. Haglas, S. Lincoln, C. Mair, and E. Williams, *Ber. Bunsenges. Phys. Chem.* **86,** 515 (1982).

53. G. Westera, G. Schat, C. Blomberg, and F. Bickelhaupt, *J. Organomet. Chem.* **144,** 273 (1978).
54. F. Walker and E. Ashby, *J. Am. Chem. Soc.* **91,** 3845 (1968).
55. G. M. Whitesides, M. Witanowski, and J. D. Roberts, *J. Am. Chem. Soc.* **87,** 2854 (1965).
56. M. Witanowski and J. D. Roberts, *J. Am. Chem. Soc.* **88,** 737 (1965).
57. G. Frankel, C. Cottrell, and D. Dix, *J. Am. Chem. Soc.* **93,** 1704 (1971).
58. G. Paris and E. Ashby, *J. Am. Chem. Soc.* **93,** 1206 (1971).
59. E. Ashby, *Bull. Soc. Chim. Fr.* **59,** 2133 (1972).
60. R. Benn and A. Rufinska, *Angew. Chem., Int. Ed. Engl.* **25,** 861 (1986).
61. R. Cramer, P. Richmann and J. Gilje, *J. Organomet. Chem.* **408,** 131 (1991).
62. M. Vallino, *J. Organomet. Chem.* **20,** 1 (1969).
63. G. Stucky and R. Rundle, *J. Am. Chem. Soc.* **86,** 4825 (1964).
64. A. Spek, G. Voorbergen, G. Schat, C. Blomberg, and F. Bickelhaupt, *J. Organomet. Chem.* **77,** 147 (1974).
65. Fr. Pat. 1,137,707 (Mar. 2, 1959), H. Normant (to Société des Usines Chimiques Rhône-Poulenc).
66. C. Raston and G. Salem, *Chem. Met.–Carbon Bond* **4,** 159 (1987).
67. W. E. Lindsell in G. Wilkinson, G. Stone, and E. Abel, eds., *Comprehensive Organometallic Chemistry*, Vol. 4, Pergamon Press Ltd., Oxford, U.K., 1982, p. 476.
68. H. Schumann, B. Wassermann and F. Ekkehardt, *Organometallics* **11,** 2803 (1992).
69. H. M. Walborsky, *Acc. Chem. Res.* **23,** 286 (1990).
70. P. Wade, D. Price, J. McCauley, and P. Carroll *J. Org. Chem.* **50,** 2805 (1985).
71. A. I. Meyers, *Acc. Chem. Res.* **11,** 375 (1978).
72. G. Consiglio and A. Indolese, *Organometallics* **10,** 3425 (1991).
73. A. Hoveyda and Z. Xu, *J. Am. Chem. Soc.* **113,** 5079 (1991).
74. U.S. Pat. 2,912,465 (Nov. 10, 1959), H. Ramsden (to M&T Chemicals).
75. U.S. Pat. 4,011,383 (Mar. 8, 1975), R. A. Setterquist (to E. I. du Pont de Nemours & Co., Inc.).
76. P. Eaton, C. Lee, and Y. Xiong, *J. Am. Chem. Soc.* **111,** 8016 (1989).
77. Y. Sato, Y. Yagi, and M. Koto, *J. Org. Chem.* **45,** 613 (1980).
78. R. Gatrone and P. Horwitz, *Solv. Extr. Ion Exch.* **6,** 937 (1988).
79. G. Silverman and M. DiFillipantonio, internal publication, Elf Atochem NA, Rahway, N.J., 1986.
80. G. Silverman, internal publication, Elf Atochem NA, Rahway, N.J., 1986.
81. R. Gatrone, Argonne National Laboratories, private communication, 1989.
82. U.S. Pat. 4,827,044 (Mar. 8, 1989), P. Tozzolino and G. Chaiez (to Société Nationale Elf Aquitaine).
83. U.S. Pat. 4,983,774 (Jan. 8, 1991), G. Chaiez, B. Labone, and P. Tozzolino (to Société Nationale Elf Aquitaine).
84. G. Chaiez and B. Laboue, *Tetrahedron Lett.* **30,** 3345 (1989).
85. C. Cardellicchio, V. Fiandanese, G. Marchese, and L. Ronzini, *Tetrahedron Lett.* **28,** 2053 (1987).
86. U.S. Pat. 4,912,276 (Mar. 21, 1990), T. Puckette (to Eastman Kodak Co.).
87. S. Sirmans and G. Silverman, unpublished results, Elf Atochem NA, Rahaway, N.J., 1988.
88. U.S. Pat. 4,263,466 (Apr. 21, 1981), I. Colon, L. Maresca, and G. Kwiatkowski (to Union Carbide Corp.).
89. I. Colon and D. Kelsey, *J. Org. Chem.* **51,** 2627 (1986).
90. P. W. Jolly in Ref. 67, Vol. 8, p. 738.
91. P. Knochel and J. Normant, *Tetrahedron Lett.* **27,** 1043 (1986).
92. H. Xiong and R. Rieke, *Tetrahedron Lett.* **32,** 5269 (1991).

93. H. Xiong and R. Rieke, *J. Org. Chem.* **57,** 6560 (1992).
94. H. Xiong and R. Rieke, *J. Org. Chem.* **57,** 7007 (1992).
95. F. Bickelhaupt, *Angew. Chem., Int. Ed. Engl.* **26,** 990 (1987).
96. P. Allen and D. Williams, *Ind. Eng. Chem. Prod. Res. Dev.* **24,** 334 (1985).
97. K. Hatada, K. Ute, T. Kitayma, and M. Kamachi, *Polym. J.* **15,** 771 (1983).
98. S. Kanoh, N. Kawaguchi, and H. Suda, *Makromol. Chem.* **188,** 463 (1987).
99. P. Gacens, M. Bottrill, J. Kelland, and J. McMeeking in Ref. 67, Vol. 3, p. 476.
100. A. Munoz-Escalona, A. Fuentes, J. Liscano, A. Albornoz, *Stud. Surf. Sci. Catal.* **56,** 377 (1990).
101. C. Wilen, M. Auer, and J. Nasman, *J. Polym. Sci. Part A: Polym. Chem. Ed.* **30,** 1163 (1992).

GARY S. SILVERMAN
Elf Atochem North America

PHILIP E. RAKITA
Elf Atochem Japan

GRINDING.   See ABRASIVES; SIZE REDUCTION.

GROUNDWATER MONITORING

Groundwater monitoring is used to analyze the impact of a variety of surface and subsurface activities, including seawater intrusion, application of agricultural products such as herbicides (qv), pesticides, and fertilizers (qv), residential septic systems, and industrial waste ponds. Another focus of groundwater monitoring has been contamination associated with waste landfills and ruptured underground petroleum (qv) storage tanks (see TANKS AND PRESSURE VESSELS).

Groundwater monitoring is a necessary component in any investigation of subsurface contamination. A wide variety of information can be gleaned from the data including groundwater velocity and direction, and contaminant identification and concentration. These data can be combined with other observations to infer various characteristics of the contamination. Examples are source and timing of the release, and future location of the contaminant plume.

The design of a groundwater monitoring strategy requires a basic understanding of groundwater flow systems. The majority of groundwater flow occurs in formations known as aquifers. At least two types of data can be retrieved using groundwater wells, ie, groundwater pressure and groundwater quality. A monitoring well allows measurement of these properties at a specific point in an aquifer. Monitoring wells come in a variety of sizes and materials, but each is simply a pipe extending from the ground surface to a point in the aquifer at which the pressure or contaminant is to be assessed. Monitoring wells are functional only in the saturated zone of the subsurface. Within the unsaturated soil zone, tensiometers, soil moisture blocks, and psychrometers have been used to assess fluid pressures. Fluid samples are retrieved using suction cup lysimeters for subsequent quality analysis.

Aquifers

The term aquifer is used to denote an extensive region of saturated material. There are many types of aquifers. The primary distinction between types involves the boundaries that define the aquifer. An unconfined aquifer, also known as a phraetic or water table aquifer, is assumed to have an upper boundary of saturated soil at a pressure of zero gauge, or atmospheric pressure. A confined aquifer has a low permeability upper boundary that maintains the interstitial water within the aquifer at pressures greater than atmospheric. For both types of aquifers, the lower boundary is frequently a low permeability soil or rock formation. Further distinctions exist. An artesian aquifer is a confined aquifer for which the interstitial water pressure is sufficient to allow the aquifer water entering the monitoring well to rise above the local ground surface. Figure 1 identifies the primary types of aquifers.

Calculation of the flow in the saturated portion of the subsurface is generally much easier than that in the unsaturated zone. However, calculation of flow

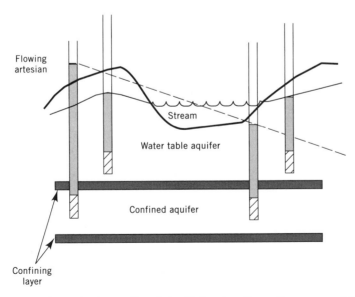

Fig. 1. Aquifers and monitoring wells where ▨ denotes the well screen and ▪ the water-filled space in the monitoring well. (—) denotes the water table level, (— —), the potentiometric surface, and (—) the ground surface. Terms are discussed in text.

in either requires a fundamental understanding of groundwater pressure and energy.

Groundwater Pressure and Energy

The energy state of soil water can be defined with respect to the Bernoulli equation, neglecting thermal and osmotic energy as

$$E = z + P/\gamma + v^2/2g \tag{1}$$

where E is the energy per unit weight, P the pressure, γ the specific weight, z the elevation, and v the average velocity. The three energy terms represented by the right-hand side of the equation are pressure energy, potential energy, and kinetic energy, respectively. In most groundwater applications, the kinetic energy term is much less significant than the other two and is neglected. Thermal gradients cause moisture to migrate toward colder regions. However, thermal energy has been neglected in the present formulation and the equation cannot be used to simulate problems where there is a significant temperature gradient present.

When the energy terms are expressed as energy per unit weight, the term head is often used. Therefore, the total head, h, is equal to the elevation head, z, plus the pressure head, P/γ:

$$h = z + \frac{P}{\gamma} \tag{2}$$

The total head is also often denoted as ψ. For a saturated soil having no vertical component of acceleration, the pore water pressure is calculated using basic hydrostatic principles, and the depth below the free surface defines the fluid pressure. This depth is evaluated using a monitoring well. Water passes through the screened portion of the well and rises in the casing until it reaches an elevation associated with the energy status of the fluid at the screened elevation. A variety of means can be used to determine the elevation of the fluid in the monitoring well, including electrical depth devices, sonar techniques, or steel tape and chalk (1).

In the unsaturated zone, measurement of the fluid head is a bit more complex, because the fluid pressures are less than atmospheric, and therefore, fluid does not rise above the point of measurement in a monitoring well, ie, ψ is negative. Instead, a tensiometer such as that shown in Figure 2 may be used to determine the soil water suction. A tensiometer consists of an airtight, water-filled tube having a porous cup at the base. After insertion into the soil, moisture exits the tensiometer while hydraulic equilibration is achieved with the surrounding unsaturated soil. As moisture exits, a vacuum is created in the evacuated space at the upper portion of the tensiometer. When the suction created in the tensiometer is equivalent to the negative pressure head in the surrounding soil, equilibration has been achieved and the corresponding pressure can be read on the tensiometer gauge. Other set-ups that may be used to evaluate the pore fluid

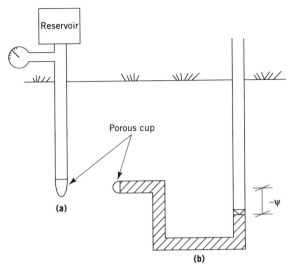

Fig. 2. (a) Schematic of a tensiometer and (b) a hydraulic analogy showing the negative value of the total head, ψ.

pressure in the unsaturated zone include soil moisture blocks, thermocouple psychrometers, γ-ray attenuation, and nuclear moisture logging (1–3).

Calculation of Groundwater Flow

The framework for the solution of porous media flow problems was established by the experiments of Henri Darcy in the 1800s. The relationship between fluid volumetric flow rate, Q, hydraulic gradient, and cross-sectional area, A, of flow is given by the Darcy formula:

$$Q = KA\frac{h_1 - h_2}{\Delta l} \tag{3}$$

The constant K, which maintains the equality, has been termed the hydraulic conductivity, permeability, or simply conductivity. The permeability is generally accepted to be a constant for a saturated soil, except for very small gradients (2–4). Here h_z represents the hydraulic head at location z, whereas Δl is the hydraulic length between points 1 and 2. A is an area perpendicular to the discharge vector. In differential form

$$Q = -KA\frac{\partial h}{\partial l} \tag{4}$$

The gradient, $\frac{\partial h}{\partial l}$, is often denoted i for simplicity. It is often convenient to analyze the discharge for a unit area using the specific discharge, q. The specific discharge represents the volumetric discharge divided by the total cross-sectional area, ie,

$$q = -Ki \tag{5}$$

In terms of fluid velocity,

$$v = \frac{Q}{An}$$

or

$$v = -K\frac{i}{n}$$

The porosity appears in the denominator of the right-hand side of the equality owing to the dependence of velocity on the available flow area, which is reduced from the total cross-sectional area by the factor n.

This form of Darcy's law is applicable only to saturated flow. As discussed earlier, there are distinctions between the state of soil water in the saturated and unsaturated regions. These distinctions lead to an alternative form of Darcy's law for the case of unsaturated flow (2,5).

Application of equation 5 requires caution. In this simplistic form, the equation can be used to find only one component of fluid velocity, namely that defined by the direction over which the gradient is measured, ie, the line between two monitoring wells. In general, however, the direction of groundwater flow at a point is fully characterized by assignment of values in three mutually orthogonal directions. Figure 3 provides an example of such a situation.

The vertical component of flow can be determined if a well is screened at two different elevations as shown in Figure 1. Frequently, nested wells are used instead of a single well and multiple screenings to determine the vertical component

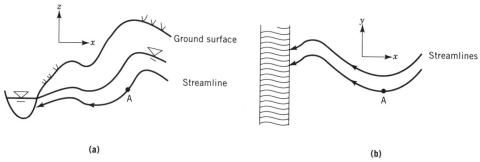

(a)   (b)

Fig. 3. Three-dimensional flow for stream recharge via a water table aquifer where (**a**) is the elevation view and (**b**) is the plan view.

of flow (2). Nested wells must be situated close enough to one another so horizontal gradients do not become a factor.

Nested wells can also be used to analyze multilayer aquifer flow. There are many situations involving interaquifer transport owing to leaky boundaries between the aquifers. The primary case of interest involves the vertical transport of fluid across a horizontal semipermeable boundary between two or more aquifers. Figure 4 sets out the details of this type of problem. Unit 1 is a phraetic aquifer, bound from below by two confined aquifers, having semipermeable formations at each interface.

Judging from the hydraulic heads, the vertical flow across the semipermeable interface 1 is in a downward direction, whereas across the semipermeable interface 2 it is in an upward direction. Therefore, unit 2 is being fed by fluid from the phraetic aquifer above it and the confined aquifer below.

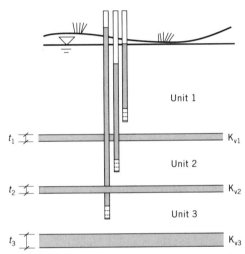

Fig. 4. Multilayered aquifer flow where t represents thickness of confining layers between units 1, 2, and 3.

Monitoring Well Design for Contaminant Transport Studies

There are a variety of contaminant problems that may prompt the development of a groundwater monitoring program. The specific details of the program depend on the situation prompting such monitoring. For example, groundwater monitoring may be required in the vicinity of a new or existing landfill, and would serve the purposes of clarifying groundwater flow conditions, identifying background water quality, and leak detection. Groundwater monitoring in the vicinity of known contamination is used to delineate the spatial extent of the contamination as well as to verify the chemicals present. Groundwater monitoring may be required in association with real estate transactions to verify the existence of a pristine water source for well development.

Monitoring wells are installed by first completing a soil boring to the approximate depth of groundwater measurements. Drilling methods for the borehole

include auger, mud rotary, cable tool, jetted wells, and driven wells (1,6). During
the drilling, a boring log is prepared that records details of the subsurface materials encountered as the depth progresses. A well casing is installed in the borehole with a well screen at or near the bottom of the borehole. The annular space
between the borehole and the casing must be filled properly to allow free passage
of groundwater from the monitored zone to the well screen and to preclude passage of moisture from the surface vertically along the sides of the casing. In the
vicinity of the well screen, a filter pack of natural, ie, typically sand or pea gravel,
or synthetic materials is used to preclude clogging of the well screen. The specific
design of the filter pack must take into consideration details of the aquifer soil.
Often, a secondary filter pack consisting of finer materials is placed above the
primary filter. Above this is the virtually impermeable bentonite seal. A neat
cement grout above this layer extends to the ground surface. Figure 5 illustrates
the primary components of a monitoring well.

A variety of techniques can be used to retrieve the groundwater sample once
the well is in place. Pumps, bailers, and syringes are among the devices used to
draw the sample to the surface. Typically the well is purged of three to ten casing
volumes of fluid prior to retrieval of the sample, to ensure standing water is not
being analyzed (1,6,7). Care must be taken during sampling and delivery to the
lab. The characteristics of the sample may be altered if protocols are not followed.
Volatile gas stripping, oxidation, and pH shifts are examples of modifications that
may occur owing to the introduction of oxygen or other gases to the samples (7).

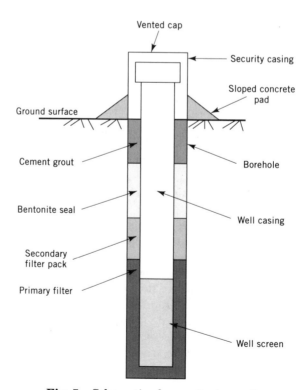

Fig. 5. Schematic of a monitoring well.

It is often important to quantify the contamination of pore fluid in the unsaturated soil zone, where monitoring wells are ineffective. In this region, suction cup lysimeters are useful (7). These samplers consist of a porous cup, typically ceramic, having two access tubes which are usually Teflon. One access tube provides a pressure-vacuum, the other discharges the sampled fluid to the surface. The porous cup, typically between 2 and 5 cm in diameter, is attached to a PVC sample accumulation chamber.

The installation of the probes should ensure good contact between the suction cup portion of the sampler and the surrounding soil, and minimize side leakage of liquid along the hole that has been cored for the sampler and access tube lines. Typically a clay plug of bentonite is used to prevent leakage down the core hole. A silica–sand filter provides good contact with the suction cup and prevents clogging of the cup. To retrieve a sample, the sample tube is clamped and suction is applied to the lysimeter through the air tube, which is then clamped. Moisture enters the accumulation chamber through the porous cup. The suction is released and pressure applied, forcing the sample to the surface through the sample collection tube. Figure 6 shows a sample installation.

Design of a groundwater monitoring program minimally includes consideration of materials, location, indicator parameters, and timing. Material selection is important for both the well casing and well screen. Materials of construction must be inert to the fluid being tested and to the ambient soil. The material must not release any type of chemical that could be interpreted as present in the groundwater. Typical inert materials include Teflon, polypropylene, PVC, and stainless steel (3,8,9). Material durability is also an issue, especially because many monitoring systems must be utilized for 50 years or more. The screens should also be evaluated regarding the potential for clogging, either via the porous media or biological activity.

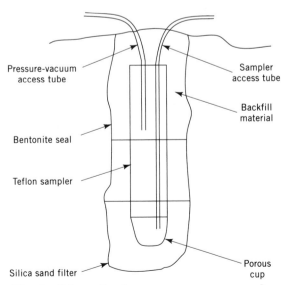

Pressure-vacuum
access tube

Sampler
access tube

Backfill
material

Bentonite seal

Teflon sampler

Porous
cup

Silica sand filter

Fig. 6. Schematic of a pressure-vacuum sampler.

Locational considerations include both surficial location and screened interval, ie, the sampling depth. The surficial location is selected based on whether the sample is to represent background quality or quality at the location of contamination, or potential leak location. In selecting the surficial location, the groundwater flow parameters, velocity and direction, are assumed to be known from other monitoring wells or borings already completed. The sampling depth is selected based on the type of contaminant monitored, ie, light or heavy, aqueous or nonaqueous, and/or the groundwater depth of interest (10,11). For example, if unit 2 of Figure 4 is used for groundwater well development as a drinking source, it is likely that a monitoring well is screened within that depth interval to assess water quality in that zone. Because it is possible for contaminants to migrate vertically, other zones may also be monitored.

Indicator parameters are those chemicals for which the water sample is analyzed. Often it is a simple matter to select the indicator parameter, if one suspects a discharge of a particular chemical. However, the situation is often much more complex. If monitoring wells are used to assess the occurrence of leachate leaks below a landfill, selection of the indicator parameters should be based on the expected chemical composition of the landfill leachate. In addition, the indicator parameters should be distinct from chemicals known to exist in the background groundwater (6). If the monitoring program is used for leak detection, the indicator parameter should be one that is expected to have an early arrival at the monitoring well, eg, the material having negligible adsorption.

Groundwater monitoring programs typically employ a routine schedule of sampling. Depending on the application, samples may be retrieved for analysis at weekly, monthly, quarterly, or other appropriate intervals. When the monitoring program serves the purpose of leak detection, as around the periphery of a landfill, wells are sampled quarterly. If contamination of an aquifer is known to exist, and monitoring wells are used to track movement of contaminants or the effectiveness of remediation efforts, sampling may occur more frequently. If monitoring wells are used in combination with a tracer test (4,6) to analyze flow characteristics, continuous sampling may be required.

Data analysis is aided by a variety of statistical techniques to assess significance, highlight trends, and form mathematical models of any correlations developed (12). It is never possible to design a groundwater monitoring program that samples an aquifer completely. Many pockets of unknown quality remain. A geostatistical technique, such as kriging (12), can, however, be used to determine an optimized estimate of groundwater quality at such an unsampled location using observed data from surrounding sampled locations.

BIBLIOGRAPHY

1. D. M. Nielsen, ed., *Practical Handbook of Ground-Water Monitoring*, Lewis Publishers, Inc., Chelsea, Mich., 1991.
2. R. A. Freeze and J. A. Cherry, *Groundwater*, Prentice Hall, Inc., Englewood Cliffs, N.J., 1979.
3. C. W. Fetter, *Applied Hydrogeology*, 2nd ed., Macmillan Publishers, New York, 1988.

4. P. Domenico and F. Schwarz, *Physical and Chemical Hydrogeology*, John Wiley & Sons, Inc., New York, 1990.

5. J. Bear, *Hydraulics of Groundwater*, McGraw-Hill Book Co., Inc., New York, 1979.

6. M. Barcelona, A. Wehrmann, J. Kelly, and W. Pettyjohn, *Contamination of Ground Water: Prevention, Assessment, Restoration*, Pollution Technology Review No. 184, Noyes Data Corporation, Park Ridge, N.J., 1990.

7. J. Devinny, L. Everett, J. Lu, and R. Stollan, *Subsurface Migration of Hazardous Wastes*, Van Nostrand Reinhold, New York, 1990.

8. E. A. Boettner, G. L. Ball, Z. Hollingsworth, and R. Aquino, *Organic and Organotin Compounds Leached from PVC and CPVC Pipes*, U.S. Environmental Protection Agency Report, EPA-600/1-81-062, Washington, D.C., 1981.

9. J. E. Voytek, *Ground Water Monitor. Rev.* **3**(1), 70–71 (1983).

10. J. F. Villaume, *Ground Water Monitor. Rev.* **5**(2), 60–74 (1985).

11. G. G. Hunkin, T. A. Reed, and G. N. Brand, *Ground Water Monitor. Rev.* **4**(1), 43–45 (1984).

12. J. Davis, *Statistics and Data Analysis in Geology*, 2nd ed., John Wiley & Sons, Inc., New York, 1986.

CAROL J. MILLER
Wayne State University

GROWTH REGULATORS

ANIMAL

The growth of animals can be defined as an increase in mass of whole body, tissue(s), organ(s), or cell(s) with time. This type of growth can be characterized by morphometric measurements; eg, skeletal muscle or adipose tissue growth can be described by observing temporal changes in cell number, ie, hyperplasia, and cell size, ie, hypertrophy. Growth also includes developmental aspects of function and metabolism of cells and tissues from conception to maturity.

Both types of growth are influenced by genotype, nutritional status, and gender of the animal. Studies conducted in the early 1900s described how these factors influenced allometric growth of tissues, through dissection or proximate composition measurements. Development of histological and histochemical methods allowed animal scientists to characterize cellular aspects of tissue growth, but other methods were needed to determine the mechanisms by which cell number or size were controlled. Within groups of animals that share similar genotype, nutritional regimen, or gender, differences in metabolic hormone concentrations and/or action bring about differential regulation of proteins (qv), lipid, carbohydrates (qv), and mineral metabolism (see MINERAL NUTRIENTS). These differences influence how consumed nutrients are used by growing animals. Change in partitioning of nutrients occurs coincident with normal allometric growth from birth to mature size. When animals are offered diets ad libitum (to appetite) the proportion of nutrients used for lipid accretion increases from birth to sexual maturity, or from birth to normal market weight or mature size, unless energy intake is restricted. This change is influenced by metabolic hormone action; rarely do any of the hormones or other influencing factors act independent of each other to regulate nutrient partitioning. Complex interactions allow for integration of influences to accommodate a coordinated chronic regulation of nutrient use for maintenance or growth so that an animal may adapt to its environment (see FEEDS AND FEED ADDITIVES).

Improved understanding of the control of metabolic aspects of growth has provided the opportunity to regulate animal growth. Improvement of rate and efficiency of growth benefits the producer. Improvement in composition of meat animals benefits the producer through more efficient gain and greater value, and benefits the processor through less labor requirement for trimming and removal of fat. The consumer benefits by receiving a quality, desirable food at a cost reflective of efficient production.

Four general classes (ca 1993) of growth regulators are approved by the Food and Drug Administration (FDA) for use in food-producing animals in the United States. These include naturally occurring and synthetic estrogens and androgens, ie, anabolic steroids (qv); ionophores; antibiotics (qv); and bovine somatotropin. Compounds in the first class, anabolic steroids, act as metabolism modifiers to alter nutrient partitioning toward greater rates of protein synthesis and deposition, thereby increasing the weight at which 25 to 30% lipid content in the body or carcass is achieved. Ionophores have highly selective antibiotic activity and appear to enhance feed conversion efficiency through effects on ruminal microbes. Antibiotics, administered at subtherapeutic doses, enhance growth through improving feed conversion efficiency and/or growth rate, with no consistent effect on body or carcass composition.

Two other classes of growth regulators, ie, somatotropin or somatotropin secretogogues, and select synthetic phenethanolamines, have been investigated for the ability to alter growth; no compound in either class has yet been approved by the FDA for use in animals raised for meat. In 1993, the FDA approved administration of recombinant bovine somatotropin for increasing milk production in dairy cows (see GENETIC ENGINEERING, ANIMALS). Administration of native or recombinant somatotropin (ST) to growing pigs, cattle, and lambs dramatically enhances rate, efficiency, and composition of gain. Likewise, experimental dietary administration of select synthetic phenethanolamines, most of which are β-adrenergic agonists, also has produced striking changes in rates of skeletal muscle and adipose tissue growth and accretion in growing cattle, lambs, pigs, and poultry.

Somatotropin, the β-adrenergic agonists, and the anabolic steroids are considered metabolism modifiers because these compounds alter protein, lipid, carbohydrate, mineral metabolism, or combinations of these; and they partition nutrient use toward greater rates of protein deposition, ie, muscle growth, and lesser rates of lipid accretion. Historical data leading to understanding of the mechanism(s) of action are found in reviews on anabolic steroids (1), somatotropin (2–4), and the phenethanolamines (5–7).

Anabolic Steroids

Naturally occurring and synthetic estrogens and androgens have been extensively and safely used to improve efficiency and carcass composition in growing beef cattle in the United States since the early 1950s. Several anabolic steroid implants have been approved for use in beef cattle in the United States, but only one, zeranol [55331-29-8], is approved for use in lambs. Anabolic steroids are not used for growth regulation in swine or poultry (see STEROIDS).

Commercial products approved by the Food and Drug Administration include the naturally occurring hormone estradiol [50-28-2] (Compudose) (1); the natural hormone progesterone [57-83-0] (2), used in combination with estradiol or estradiol benzoate, ie, Steer-oid, Synovex-S, and Synovex-C for calves; the fungal metabolite zeranol [55331-29-8] (Ralgro) (3) which has estrogenic properties; the synthetic progestin melengestrol acetate [2919-66-6] (MGA) (4); testosterone (5) propionate [57-85-2] in combination with estradiol benzoate, ie, Synovex-H or Heifer-oid; and a synthetic testosterone analogue, trenbolone acetate [10161-34-9] (TBA) (6) which is used alone, ie, Finaplix, or in combination with estradiol, ie, Revalor. Structures of these anabolic steroids are shown in Figure 1.

Classification of the anabolic steroids is based on chemical structures and associated actions. A review of the biosynthesis and metabolism of the naturally occurring estrogens and androgens is available (1). Names, descriptions, approval dates, and recommended doses of the commercial products are found in References

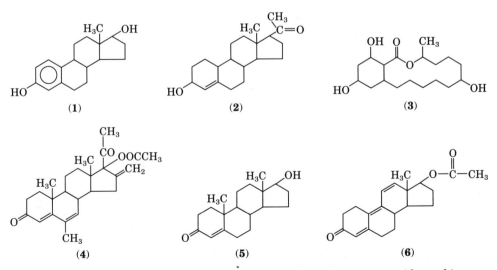

Fig. 1. Chemical structure of naturally occurring and synthetic sex steroids used in commercial anabolic steroid implant preparations.

1, 8, and 9. Although steroids may be orally active, the FDA approved mode of administration is the subcutaneous implant. Effective dose is lower with implant rather than oral administration.

Efficacy of these anabolic steroid implants has been summarized (1,8–12). Growth responses to anabolic steroids vary greatly, ranging from no response in feedlot bulls (13) to a 69.9% increase in average daily gain in heifers treated with trenbolone acetate (14). The choice of anabolic steroid depends on gender. The estrogenic compounds are generally more effective in steers. The response of females to both estrogens and androgens is more variable and less consistent, but superior responses are seen using the androgenic steroids. Use of a combination of anabolics generally produces an additive response compared to use of either estrogenic or androgenic implant alone. Response in bulls is generally less than that of steers, and implanted steers often achieve the growth performance observed in nonimplanted bulls (15).

Growth Performance Response. The consistent net effect of anabolic steroid implant use in growing ruminants appears to be increased rate of protein and live weight gain, and increased live weight at which carcass or empty body fat concentration equals that in nonimplanted cattle; thus increasing their potential mature size. Increased feed intake is frequently observed.

Rate of live weight gain is increased 10 to 20% on average with the use of anabolic steroids. Responses approaching 50% have been observed in lambs implanted with a combination of 35 mg trenbolone acetate (TBA) plus 5 mg estradiol-17β (16), and in beef steers housed in metabolism chambers and implanted with a combination of 140 mg trenbolone acetate and 20 mg estradiol-17β (17). Young animals may respond better to steroid implants than older animals (18–21). Greater responses have been observed during the initial period following implantation, which may be caused by the declining circulating concentration of the anabolics after the first few weeks (8,17,22,23). Trenbolone–estradiol combinations appear to be superior to the use of either implant alone. Dose-response efficacy trials conducted for FDA approval of TBA–estradiol combinations indicate that the average daily gain (ADG) plateaued at a dose of 118 mg TBA plus 24 mg estradiol, but the feed efficiency plateaued at 139 mg TBA plus 28 mg estradiol (24). ADG increased 18% and the feed:gain ratio reduced 9.5%; both exceeded the response to 30 mg of estradiol alone. Implants of 140 mg TBA alone did not improve growth performance in this and other studies (23,25). A TBA–estradiol ratio of 5:1 appears to be optimum for feedlot steers fed a high grain diet.

Efficiency of feed use for growth is usually improved with anabolic steroids, but the magnitude of the response is somewhat variable. Improvements of 5 to 14% have been reported (26–30). Trenbolone–estradiol combinations decrease feed:gain ratios 10–13% (24,31–33). The degree of response is clearly influenced by changes in dry matter intake, which are also variable, and by the degree of change in composition of gain. The majority of studies in which large increases in gain are observed also result in 10% increases in intake (30) and proportional increases in lean mass in cattle (18) and lambs (16,34). However, no significant changes in intake were observed in several studies (13,35–37).

Composition of Gain Response. Few studies investigating the effects of anabolic steroids on growth in ruminants in the United States include the direct measurement of carcass or empty body composition necessary to understand the mode of action and to define nutrient requirements. Total carcass lean, ie, muscle, increased 9.5 and 10.4% in steers implanted twice with 300 mg trenbolone acetate and 36 mg resorcylic acid lactone [26538-44-3] (Ralgro, Pitman-Moore) over the live weight range 250 to 400 kg (35). Separable fat was reduced two percentage points. Efficiency of gain was greater in implanted cattle fed at the higher of two levels of energy intake. Dressing percentage was higher in implanted cattle, which implies that neither organ weights nor gut fill were increased with treatment.

Long-term administration of trenbolone acetate and resorcylic acid lactone to heifers and steers fed to 491, 612, or 731 days of age exhibited greater absolute and relative amounts of carcass lean and lesser absolute and relative amounts of carcass fat than nonimplanted cattle (18). Sex-by-implant interactions were not significant. Cattle implanted the longest exhibited a greater increase in carcass lean than cattle implanted for shorter periods. The increase in carcass weight of implanted cattle was accounted for entirely by increase in carcass lean; decrease in fat accretion was offset by increased bone growth. Few studies on the effects of anabolic steroids in fed cattle have been conducted with degree of marbling as the end point. Growth performance and composition of gain responses to TBA–estradiol implants, compared in three breeds of steers representing different frame sizes, indicated little effect on carcass composition, carcass quality grade, or retail cut distribution (33). However, live weight required to reach the small degree of marbling end point was increased 25 to 45 kg using TBA–estradiol implant; fat gain was increased by 19% on average in these cattle. These results suggest that anabolic steroids stimulate growth without dramatic effects on composition of gain and increase the weight at which a common carcass or intra-muscular fat concentration is achieved.

The consistent improvement in rate of protein deposition observed in grow-ing ruminants indicates that anabolic steroid implants exert their primary in-fluence through altering protein metabolism. There are lesser effects on lipid metabolism.

Mechanism of Action. Few data are available that describe the effects of anabolic steroids on protein metabolism; even fewer data exist for assessment of direct effects of anabolic steroids on lipid metabolism in growing ruminants. The lack of any consistent change in somatotropin, prolactin, insulin, or other meta-bolic hormones (qv) in a total of 15 studies has been noted (1,38).

Protein metabolism studies suggest that rates of fractional protein synthesis and protein degradation may be reduced by trenbolone acetate; degradation rates may be reduced to a greater extent in rats (39,40) and lambs (41,42) to increase protein accretion rate. Combined TBA–estradiol implant treatment increased daily live weight gain 50–60% at similar feed intakes, and increased daily nitro-gen retention 100 and 146% in steers during the first seven weeks of treatment in two separate studies (17). Estimates of whole-body protein synthesis rate, based on metabolic body size, were similar throughout the 10-week experiment; amino acid oxidation was lower in treated steers at weeks two and five, compared to control animals. Urinary 3-methyl histidine excretion was slightly less and total energy retention was unaffected in treated steers, indicating that reduction of protein degradation rate may account for the bulk of the improvement in daily gain and nitrogen retention. Heat production was not increased in steers treated with the TBA–estradiol combination.

One possible mechanism responsible for the ability of trenbolone acetate to stimulate skeletal muscle hypertrophy may be through enhanced proliferation and differentiation of satellite cells as the result of increased sensitivity to insulin-like growth factor-I (IGF-1) and fibroblast growth factor (43).

Very little data are available regarding effects of anabolic steroid implants on the lipid metabolism in growing ruminants. Lipogenic enzyme activity and fatty acid synthesis *in vitro* were elevated in subcutaneous adipose tissue from bulls implanted with estradiol (44), which may account for the increase in fat content of carcasses reported in some studies. TBA implants have no effect on lipogenesis in intact heifers, and only tend to reduce lipogenic enzyme activities in ovariectomized heifers (45).

Economics.   Estimates of anabolic steroids in growing cattle indicate that savings associated with reduced feed costs are approximately $50.00 per animal. Increased value of the carcass resulting from the increased amount of saleable lean meat produced is estimated to range from $15.00 to $30.00 per animal.

Withdrawal from anabolic steroid treatment is not required before slaughter because residue levels in edible tissues are negligible, and are significantly lower than other sources of estradiol such as the normal endogenous production in humans and the phytoestrogens consumed in plant food sources (1).

Ionophores

An ionophore may be defined as an organic substance that binds a polar compound and acts as an ion-transfer agent to facilitate movement of monovalent, eg, sodium and potassium, and divalent, eg, calcium, ions through cell membranes (46). The change in electrical charge in membranes influences the transport of nutrients and metabolites across the cell membrane, but the exact mechanism by which ionophores improve growth performance in growing ruminants is not known. Several reviews of the proposed mode of action and efficacy of ionophores are available (46–52).

The FDA first approved use of a polyether ionophore as a feed additive for cattle in 1975. Ionophores were first isolated from bacteria generally of the *Streptomyces* genus, but are produced commercially by bacterial fermentation (qv). Monensin [17090-79-8] and other ionophores are being fed to over 90% of feedlot cattle grown for beef (53) to enhance efficiency of gain; improvements of 5–10% are common. Ionophores also are used as anticoccidial drugs in poultry production and have similar, but lesser, effects in ruminants (54).

Doses range from 6 to 33 ppm in the diet, but very little if any ionophore can be measured in the circulation after feeding. Monensin is absorbed from the gut, metabolized by the liver, and excreted into the bile and back into the gut. Thus tissue and blood concentrations are very low. Over 20 metabolites of monensin, which have little or no biological activity, have been identified (47,55).

Growth Performance Response.   Ionophores consistently improve feed conversion efficiency in growing cattle. In many cases feed intake is reduced without changing the rate of weight gain. When feed efficiency is improved, but intake is not changed, an increase in rate of weight gain is observed. Trials in which monensin was fed indicated that the gain-to-feed ratio increased 4–12%, rate of gain did not significantly increase (1.6%), and feed intake was reduced 6.4% (48). An examination of growth performance responses against monensin dose, for published data up to 1990, generally showed a similar response magnitude for rate and efficiency of gain (48,52). Effects on carcass yield as a percentage of live weight and on carcass composition were very small and of little economic importance in most cases.

Dietary administration of ionophores is coupled with the use of anabolic steroid implants to maximize rate and efficiency of gain in growing cattle. Effects of ionophores and anabolic steroid implants are generally additive.

Mechanism of Action. The positive effects of ionophores on growth performance in growing cattle have long been thought to result from changes in the digestive system, particularly those that occur in the rumen. Hydrogen production is reduced, which leads to reduced methane production and less energy lost in this form. A shift in fermentation products toward greater propionate and less acetate production, decreased ammonia production which may increase protein availability, and reduction of lactate-producing bacteria in the rumen to prevent rumen acidosis may all contribute to the increased efficiency of gain which occurs (52). Increased or improved amino acid composition of absorbed nitrogen could remove constraints on amino acid availability for protein synthesis, or could result in indirect effects on metabolic hormone secretion rates (56). Plasma concentrations of minerals are also altered upon feeding ionophores (57,58). The consequences of these changes are unknown. Other benefits of ionophores include maintenance of good general animal health through reducing the incidence or severity of legume and feedlot bloat (59), and pulmonary emphysema (60).

Antibiotics

Antibiotics used in livestock and poultry production improve growth rates and efficiency of gain. Subtherapeutic doses are used for these purposes, and effects are similar in magnitude to those achieved with ionophores in growing ruminants. However, antibiotics are efficacious in all livestock species and in poultry. Intermediate doses are used to prevent disease in exposed animals, and therapeutic doses are used to treat animals that are ill. Antibiotics, produced by microorganisms, and chemobiotics or chemotherapeutics, chemically synthesized, are drugs and therefore regulated by the Food and Drug Administration. Monitoring of proper use and avoidance of residues entering the human food chain is accomplished through joint monitoring and surveillance programs conducted by the FDA and the Food Safety and Inspection Services of the USDA. Certification programs among producer groups assure that appropriate withdrawal times and use guidelines are followed.

Antibiotics approved for use as growth enhancers in livestock and poultry include bacitracins, bambermycins, lincomycin [154-21-2], penicillin [69-53-4], streptomycin [57-92-1], tetracyclines, tiamulin [55297-95-5], tylosin [1401-69-0], and virginiamycin [11006-76-1] (61).

Chemically synthesized antimicrobials used in animal and poultry feeds include arsenicals, eg, arsanilic acid [98-50-0], sodium arsanilate [127-85-5], and roxarsone [121-19-7]; sulfa drugs, eg, sulfadimethoxine [122-11-2], sulfamethazine [57-68-1], and sulfathiazole [72-14-0]; carbadox [6804-07-5]; and nitrofurans, eg, furazolidone [67-45-8] and nitrofurazone [59-87-0] (see ANTIBACTERIAL AGENTS, SYNTHETIC; ANTIPARASITIC AGENTS).

Effects on Growth Performance. Effects of subtherapeutic use of antibiotics were documented as early as 1950 (62–65), and the efficacy in food-producing animals has been summarized (Table 1) (61,66–69). Effects in very young animals are greater than in older animals, presumably because significant benefits are achieved through inhibiting growth of bacteria that have adverse effects on growth. Conversely, effects are smaller when animals are exposed to environmental conditions that minimize exposure to pathogenic bacteria or that minimize stress and nutritional inadequacies.

Mechanism of Action. The mechanisms by which antibiotic administration at subtherapeutic levels enhance growth rate and efficiency of gain in growing animals have not been clarified. Possible modes of action include disease control, nutrient sparing, and metabolic effects. There is extensive evidence that the principal benefit from subtherapeutic use of antibiotics results from the control of harmful microorganisms.

Transport, intermingling of animals, and environmental stress can result in exposure to nonresident microorganisms or a greater predisposition to subclinical disease. The use of subtherapeutic levels of antibiotics can reduce this stress and result in improved, more cost-efficient production. The bacteriostatic or bacteriocidal effects are apparent in contaminated or previously used environments, where 5–10% improvements in growth rate or feed efficiency commonly are observed. Young animals in which the immune system is not yet fully developed also respond to a greater extent than older animals. Controlled experiments demonstrate that feeding antibiotics at subtherapeutic levels allows animals in these environments to perform closer to their genetic potential.

Table 1. Benefits of Subtherapeutic Level of Antibiotic in Food-Producing Animals,[a] % Improvement

Species	Number of experiments	Rate of gain	Feed/unit gain
pigs			
starter	378	16.1	6.9
grower	276	10.7	4.5
finisher	279	4.0	2.1
cattle			
calves	85	14.3	
feedlot	65	4.9	5.3
chicken			
broiler	286	2.9	2.5
layer hens	244	4.0[c]	4.7[b]
turkeys	126	7.0	3.8

[a]Refs. 61, 68, and 69.

[b]Egg production improvement; feed required per dozen eggs.

The nutrient sparing effect of antibiotics may result from reduction or elimination of bacteria competing for consumed and available nutrients. It is also recognized that certain bacteria synthesize vitamins (qv), amino acids (qv), or proteins that may be utilized by the host animal. Support of this mode of action is found in the observed nutritional interactions with subtherapeutic use of antibiotics in animal feeds. Protein concentration and digestibility, and amino acid composition of consumed proteins may all influence the magnitude of response to feeding antibiotics. Positive effects appear to be largest when protein intake is insufficient or optimum amino acid composition of absorbed nitrogen is not present in order to achieve optimal rates of weight gain.

Evidence for consistent, positive metabolic effects of feeding antibiotics is fragmented and inconclusive. Direct measurement of increased uptake of nutrients, ie, *in vivo* amino acids, glucose, or volatile fatty acids in ruminants, have not been reported.

Somatotropin

Growth and metabolism of tissues in domestic animal species are influenced or regulated by several metabolic hormones. Insulin [9004-10-8], the thyroid hormones, and the catecholamines are all important in maintaining homeostasis through acute regulation of protein, lipid, carbohydrate, and mineral metabolism. However, somatotropin [9002-72-6] (growth hormone) exerts its influence in a chronic, coordinated way to regulate metabolism and somatotropic development and growth of principal tissues and organs in the body during postnatal growth (70). Normal somatotropin (ST) concentrations in the circulation are essential for normal growth. Exogenous administration of ST accelerates growth of several tissues through stimulation of cell proliferation and accumulation of deoxyribonucleic acid (DNA). The increases in circulating levels that result also repartition nutrient use toward greater rates of protein synthesis and deposition and toward much reduced rates of lipogenesis and lipid accumulation in growing swine, sheep, and cattle. Body composition is markedly altered in growing animals administered ST for periods of several weeks or months. The mechanisms of action of ST on tissue growth and metabolism is discussed in detail elsewhere (2–4,71,72).

Effects on Growth and Composition of Gain.   Because ST is a protein, exogenous administration to influence growth must be by subcutaneous or intramuscular injection, or by long-term implant. Ingestion would destroy biological activity, as has been demonstrated in safety trials. Maximal increases of overall mean plasma or serum concentrations to approximately 10- to 13-fold control concentrations are achieved in a dose-dependent manner with daily doses up to 200 µg/kg body weight (BW) in pigs, cattle, or sheep. Elevated concentrations are maintained for approximately 8 to 12 hours after administration, depending on dose.

The dramatic effects of exogenous porcine ST (pST) administration are demonstrated by the results of dose-response studies, using growing pigs treated for 6 to 12 weeks, shown in Table 2. The maximum response is not achieved at the same dose for all response variables (2,73–77). Average daily gain is increased with increasing dose of pST, ie, up to 20% with 150 µg/kg body weight per day;

Table 2. Effects of Porcine Somatotropin (pST) Dose on Growth Performance[a]

Item	0	50	100	150	200	SEM[d]
	pST dose, μg/kg·d[b,c]					
number of pigs	10	10	10	10	10	
initial weight, kg	30.8	31.2	31.3	30.7	30.8	0.94
average daily gain, g	890	990	1030	1000	1040	30
daily feed intake, kg	2.86	2.51	2.35	2.20	2.15	0.90
feed:gain ratio	3.23	2.53	2.30	2.20	2.09	0.08
carcass tissue accretion rate, g/d						
protein	93	138	150	152	158	4.0
lipid	264	144	104	59	30	11.0
ash	24	31	34	34	34	1.0
chilled side weight, kg	33.7	34.0	33.4	32.5	32.3	0.41
muscle, kg	15.7	19.6	19.9	20.5	21.4	0.45
adipose, kg	11.6	7.2	5.0	3.8	3.0	0.47
bone, kg	3.7	4.0	4.1	4.2	4.3	0.10
skin, kg	1.8	2.1	2.5	2.5	2.5	0.08

[a]Pigs received daily injections of excipient or the specified dose of recombinant pST. Data are least square means summarized from Refs. 73 and 74.
[b]Kilograms of body weight.
[c]Confidence level > 95%.
[d]SEM = standard error of the means.

feed conversion efficiency is improved throughout an even greater dose range. The latter is explained in part by the continued reduction in feed intake with further dose increments. These relationships have been documented by numerous studies in market pigs fed ad libitum. Carcass protein accretion rates are increased up to 74%, coincident with an 82% decrease in lipid accretion rate when pST was administered from 30 to 90 kg body weight (BW). Water accretion rates paralleled protein accretion rates, and ash accretion rates were increased 26–40%. The observed stimulation of bone growth by ST is also dose-dependent. Near maximal response is achieved at pST dose of 100 μg/kg BW. Weight of bone in the carcass increased 10–17%, and skin mass increased 15–38% with increasing pST dose.

Two important aspects of the relationships between growth performance and ST administration are (1) the maximum increase in rate of body weight gain may be constrained not only by reduced feed intake, if nutrient density is inadequate, but also by reduction in adipose weight which more than offsets the increase in muscle mass; and (2) response in protein accretion to pST doses above 100 μg/kg BW is not parallel in lipid accretion rate. The reduction in lipid accretion rate is linear from 50 to 200 μg/kg BW of pST, suggesting that the physiological effects of ST on composition of gain reflect independent effects on skeletal muscle and adipose tissue. In general, exogenous ST administration does not significantly alter growth or composition in avian species.

Growing ruminants, eg, lambs and cattle, also respond to exogenous ST administration in a dose-dependent manner, but responses are generally of lesser magnitude than those observed in pigs (78,79). It has been unclear if this was the result of biological differences between species, or whether nutritional constraints of the more complex ruminant digestive system were responsible. Significant effects have, however, been demonstrated in lambs (80–83) and cattle (79,84). Typical responses of growing lambs to daily exogenous ST administration are shown in Table 3 (85). Average daily gain increased 12–19%, and feed conversion efficiency increased 20–22% in lambs. In contrast to the reduction observed in pigs, feed intake has generally not changed with ST treatment in growing lambs. Carcass protein and moisture accretion rates increased 36 and 33%, respectively, and lipid accretion rates were reduced 30%. These relative responses are approximately one-half those observed in growing pigs administered similar doses of ST for similar treatment periods. However, the 18% increase in individual hind leg muscle weights observed in Table 2, and the 24% increase in total dissected muscle observed in ewe lambs treated with ST (86) were not markedly different.

The more variable responses with growing cattle appear to result from lower doses, nutritional constraints, or lesser responsiveness of younger animals, ie, veal calves. A dose-dependent reduction in feed intake in finishing cattle, which also reduced average daily gain, has been observed (84). However, carcass composition was improved in a dose-dependent manner.

Table 3. Effects of Ovine Somatotropin (oST) and Human Growth Hormone-Releasing Factor (hGRF) on Growth and Composition of Gain in Lambs[a]

Response	Treatment				
	Control	oST[b]	5 µg hGRF[c]	10 µg hGRF[c]	SEM[d]
number of animals	18	19	20	20	
plasma variables					
oST, ng/mL	2.15	22.3	4.74	5.14	0.92
IGF-I, ng/mL[e]	278.4	469.0[c]	453.2	444.1	27
growth performance, % difference vs control					
average daily gain, g	304	14[c]	13	1.6	12
feed:gain ratio	4.99	−22.4[c]	−18	−19	0.24
composition of carcass gain					
number of animals	9	9	10	10	
protein accretion, g/d	17.2	36	30.8	34.9	1.0
water accretion, g/d	55.6	33.5	19.6	28.8	2.7
lipid accretion, g/d	79.9	−30.4	−21.2	−28.4	3.4
ash accretion, g/d	5.0	18	32[b]	42[b]	0.6
semitendinosus weight, g	91.6	20	10.5	15	2.1
semimembranosus weight, g	261.5	15[c]	10.7	7.6	5.8

[a]Lambs received saline, oST at 40 µg/kg BW, or the indicated dose of hGRF per kg BW four times per day for 42 or 56 days. Half of the lambs were withdrawn from treatment after 42 days. Carcass data shown are for lambs treated 56 days. Carcass composition data were analyzed by analysis of variance using carcass weight as the covariate. Data are summarized in Ref. 85.
[b]Confidence level >99% vs control, unless otherwise noted.
[c]Confidence level >95% vs control, unless otherwise noted.
[d]SEM = standard error of the means.
[e]IGF-I = insulin-like growth factor I.

Age, Gender, and Genotype Interactions. Young pigs, ie, birth to 15-kg live weight; bob veal calves, ie, newborn calves; and young lambs do not consistently exhibit improvement in growth performance or composition of gain in response to exogenous ST administration. This is explained in part by the apparent lack of the full complement of ST receptors in responsive tissues in very young animals. Alternatively, fractional rates of protein synthesis are highest in animals shortly after birth and decline with increasing weight gain. It may be that rates are near maximum in early development of the animal, and further increases may not be possible. Reduction of lipid accretion rate appears to be greatest when ST is administered during the later phases of growth, ie, when animals are approaching normal market weights and beyond (87). This is the stage of growth when lipid accretion rates are still increasing or are maximal in animals fed a high energy diet at ad libitum levels of intake.

Intact males exhibit faster rates of weight gain, more efficient conversion of feed-to-live weight gain, and leanest carcasses among genders of meat animals. However, intact males are not routinely used for pork or beef production in the United States. Exogenous administration of pST can reduce gender differences at moderately high (100 μg/kg) doses (88), although very high (200 μg/kg) doses were required to completely eliminate these differences in one study (73). Conversely, genotype differences in growth performance and composition of gain are not removed when these same dose ranges are used (73,89–92). The relative changes appear to be greatest in inferior genotypes, ie, those having lower protein accretion rates. Direct comparisons of the effects of ST among gender or genotypes of sheep and cattle are few. However, ewe lambs, which exhibit greater rates of lipid accretion than castrated males at the same live weight, exhibit greater reductions in fat accretion and greater responses in growth performance than wether lambs when either ST or growth hormone-releasing factor (GRF) was administered over an eight-week period prior to slaughter (85).

Nutritional Interactions. The large increases in protein deposition in growing animals administered ST may suggest that dietary protein and/or energy intake requirements may be increased. Protein accretion and growth of skeletal muscle may be constrained by inadequate intake of protein or energy. Nutrient requirements vary among growing animals of the same species and age, and protein and energy intake requirements are best defined by titration experiments in which whole-body protein accretion rates are used as the measured response variable (2,93–95). This approach was used to study the effects of ST administration (95–97). Results suggest that amino acid requirements are not changed in young pigs, ie, 20–55-kg live weight, when basal diets are adequate for the untreated pigs. However, amino acid requirements may be increased by a small amount in heavier pigs, ie, 55–110-kg live weight, when porcine ST is administered. The increase in protein accretion rate is accomplished in part by an increase in the percentage of absorbed protein (amino acids) which is deposited or retained. Increased efficiency of protein utilization is observed in both swine and growing ruminants administered ST (72), but the mechanisms by which this is achieved have not been clarified. The gain in lean tissue growth and efficiency of feed conversion achieved with ST or other growth promotants depend on adherence to the fundamental concepts of protein and energy nutrition.

Growth Hormone-Releasing Factor

Exogenous administration of the naturally occurring growth hormone-releasing factor ($GRF(1-44NH_2)$) stimulates ST secretion and increases circulating concentrations of ST in growing pigs, cattle, and sheep (98–100). Maximum elevation of ST concentration is achieved within approximately 5–15 minutes after GRF administration, depending on mode of administration (101). Duration of elevated ST concentration is short, approximately 30–45 minutes, and return to near basal ST concentrations occurs within 60–90 minutes. This is a much shorter duration than the 8–10 hours achieved with direct administration of ST. Therefore, to obtain chronic elevation of ST concentration in the blood, intermittent administration or continuous release, as from an implant of GRF, would be necesssary (see CONTROLLED RELEASE TECHNOLOGY).

Twice-daily sc injection of 10 or 20 µg human GRF (hGRF)$(1-44)NH_2$/kg BW for 36 days in barrows weighing 78 kg improved feed conversion efficiency and lean content of the ham (102). However, treatment with hGRF was less effective than pST injection of 20 or 40 µg/kg BW at the same frequency (103).

For growing wether and ewe lambs (85), four daily sc injections of synthetic hGRF at 5 or 10 µg/kg BW for eight weeks is nearly equivalent to injection of oST for improving growth performance and composition of gain (Table 3). Overall mean plasma ST concentration increases 2.5-fold when compared with controls, and lambs do not become refractory to the hGRF after 3, 6, or 8 weeks of administration. Although feed:gain ratios are reduced 18% with both doses of GRF, the higher dose reduces feed intake 6% and impairs an increase in daily gain. Carcass protein accretion rate increases 30–35% coincident with a 21–28% reduction in lipid accretion rate and 32–42% increase in ash accretion rate; the weights of two hind leg muscles show an increase of 10–15%. The overall mean plasma concentration increases to only half that achieved with oST administration, but IGF-I concentrations increase to an equivalent extent. Continuous sc administration of GRF for five weeks is as effective as GRF injection four times per day in significantly altering growth performance and carcass composition in wether lambs.

A shorter synthetic analogue of the native hGRF molecule, ie, hGRF(1-29)NH_2, has been shown to be as potent as native hGRF(1-44) in stimulating ST secretion in several species (104). Because the first 29 amino acids contain the active domain of the molecule for stimulating ST secretion, other even more potent (1-29) analogues have been synthesized and administered to growing pigs. Administration of a superactive analogue, ie, (desamino-Tyr[1], Ala[15]) hGRF(1-29)NH_2, by sc injection three times daily in pigs from approximately 50 to 105 kg BW increased serum pST in a dose-dependent manner (105). At a dose of 6.66 µg/kg BW, serum ST concentrations were elevated for a significantly longer period of time, over four hours total, than in other studies, which resulted in an approximate threefold elevation in mean ST concentration. Average daily gain was not significantly increased, but feed intake was reduced 15% and feed:gain ratios were reduced 20% using the GRF analogue. Treatment increased skeletal muscle mass 16%, reduced adipose tissue mass 25%, increased bone mass 19%, and increased skin mass approximately 30% (106). These changes were equivalent in magnitude to those observed using moderate doses of exogenous pST.

Because administration of GRF is presumed to act through the same mechanisms involved in ST mediation of metabolism and tissue growth, similar interactions with gender, genotype, and nutritional status are expected.

β-Adrenergic Agonists

Synthetic compounds called β-adrenergic agonists exhibit profound effects on growth and metabolism of skeletal muscle and adipose tissue in growing animals. Phenethanolamines have been categorized as β-adrenergic agonists because of the similar structural and pharmacological properties to the endogenous catecholamines, norepinephrine [51-41-2] (7) and epinephrine [51-43-4] (8). Among the most extensively studied compounds are clenbuterol [37148-27-9] (9), cimaterol [54239-37-1] (10), L-644-969 (11), ractopamine (12), and salbutamol [18559-94-9] (Fig. 2).

The β-adrenergic agonists are all orally active, and most have been shown to repartition nutrient use toward enhanced skeletal muscle growth, or protein deposition, and reduced lipid accretion. However, broad generalizations regarding efficacy and mode of action cannot be uniformly applied because differences exist in responsiveness among mammalian and avian species, and among dose-response relationships (5,6,107,108). For example, clenbuterol, cimaterol, and

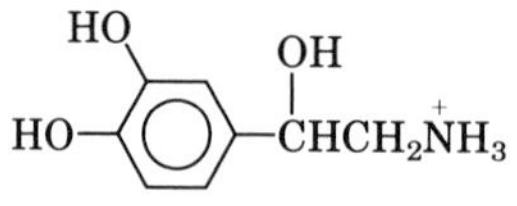

HO—, OH, HO—⟨O⟩—CHCH₂NH₃⁺ (7)

HO—, OH, HO—⟨O⟩—CHCH₂NH₂CH₃⁺ (8)

Cl, OH, H₂N—⟨O⟩—CHCH₂NH₂C(CH₃)₃⁺, Cl (9)

N≡C, OH, H₂N—⟨O⟩—CHCH₂NH₂CH(CH₃)₂⁺ (10)

OH, H₂N—⟨O,N⟩—CHCH₂NHCHCH₂CH₂—⟨O⟩, CH₃ (11)

OH, CH₃, HO—⟨O⟩—CHCH₂NHCCH₂CH₂—⟨O⟩—OH, H (12)

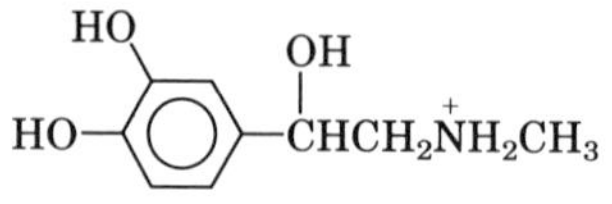

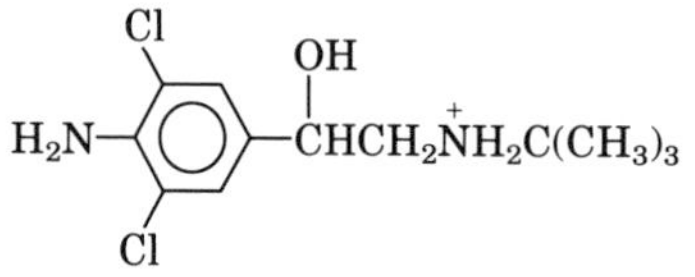

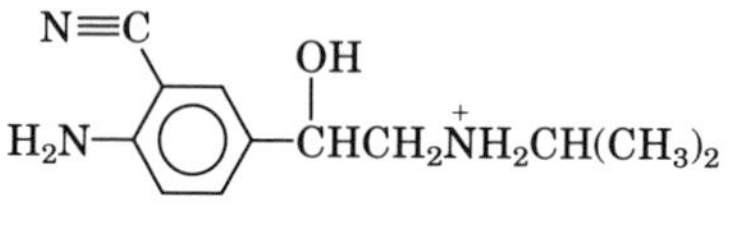

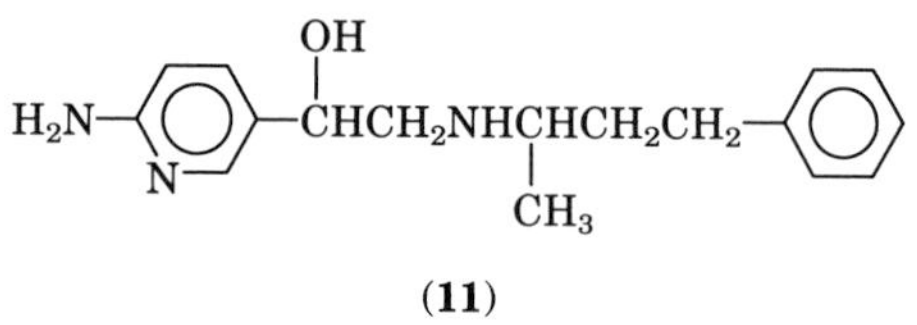

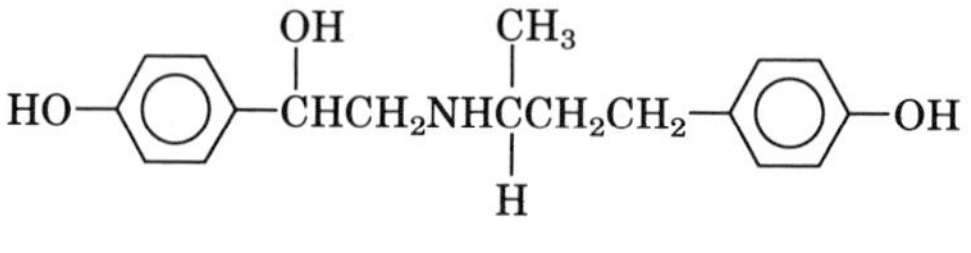

Fig. 2. Chemical structure of the endogenous catecholamines, epinephrine (8), and norepinephrine (7), and several synthetic phenethanolamines that alter animal growth.

L-644,969 are particularly effective in growing ruminants, ie, lambs and cattle, at doses of 1–10 ppm in the diet, whereas ractopamine is less effective, requiring administration at doses of 20–80 ppm for maximal effect on growth or body composition (7). The basis for these differences is not entirely clear, but may be related to receptor specificity, pharmacokinetics, or development of refractoriness with chronic administration.

Effects on Growth and Composition of Gain. The β-adrenergic agonists that alter skeletal muscle and adipose tissue growth in animals are orally active, unlike somatotropin, other peptide hormones, or growth factors. These compounds increase skeletal muscle mass and reduce lipid content of most adipose tissue deposits in a dose-dependent manner, with little or no effect on bone. These effects were first observed in rats (109), but have subsequently been described in all domestic farm animal species, ie, lambs, cattle, and pigs, and in poultry, ie, broiler chickens, turkeys, and ducks. Increased rates and efficiency of live weight gain are not consistently observed, and depend on the dose, treatment interval, and overall effect on composition of gain. Efficacy is reduced at extremely high doses (110–112). Largest, but typical, responses include 20–30% increases in average daily gain and 15–20% reductions in feed:gain ratios of lambs fed 1–10 ppm cimaterol, L-644,969, or L-655,871 in conventional mixed concentrate diets offered ad libitum (Table 4) (5). Skeletal muscle mass of individual muscles of the hind leg or total dissectable muscle mass in the carcass is increased 10–30%, and dissected adipose tissue may be decreased 15–30%. Similar responses have been observed in growing cattle, but responses in growing swine are generally smaller. However, when adequate nutriment is provided, similar changes in skeletal muscle and adipose tissue mass have been observed in pigs fed ractopamine (116,117). Responses in poultry are generally similar to or smaller than those observed in swine (118–120).

One striking feature common to all animal responses to these compounds is the lack of anabolic effects on visceral organ or bone growth. Another similarity among responses is that young animals that are nursing, are being reared on milk replacer diets, or have recently been weaned, exhibit little improvement in growth performance or body composition when fed these compounds. Evidence suggests that responses in young animals may be constrained by the lack of complete β-receptor differentiation in responsive tissues. This has not been unequivocally supported. Reductions in lipid accretion rates appear to be highest in animals that exhibit relatively high rates of lipid accretion, ie, those which are more physiologically mature, but are still approaching normal market weights.

The magnitude of the growth performance response is greatest during the early stages of administration, ie, the first few weeks, and in lambs the full effect on relative increases in skeletal muscle mass is achieved within three weeks when relatively high doses are fed (114). Direct infusion of very low doses of cimaterol into the external iliac artery in the hind leg of growing steers results in maximal increases, up to 260%, in amino acid uptake from the circulation at 14 days of administration, but the response is transient and amino acid uptake is returned to normal after 21 days of treatment (121). However, the relative differences in body composition observed in growing ruminants fed β-agonists for three to six weeks are not significantly diminished with continued administration for 10 to 12 weeks. Generalizations across species and the several compounds studied are

Table 4. Effects of β-Agonists on Growth and Carcass Composition of Growing Lambs

Treatment and dose, ppm	Treatment period, d	Control values and proportional responses, %		Carcass composition, %		Reference
		ADG,[a] g/d	Feed:gain[b]	Protein	Lipid	
cimaterol[c]	45					113
0		352^d	4.94	66.9	16.6	
0.57		3.7	0	6.4	-16.7^d	
2.29		17.9	-7.3	5.2	-16.3^d	
11.42		19.3	-14.7^d	9.0	-33.1^d	
cimaterol						114[e]
0	21	170	6.5	15.04	26.7	-14.7^d
10	21	25	-10.0	10.6^d	-25.0^d	
0	42	165	6.0	14.3	29.2	
10	42	20	-15.0^d	19.6^d	-20.0^d	
L-655,871	42					115
0		211	7.26	15.11	32.6	
0.25		23.7^f	-12.2^f	7.3	2.2	
1		26.1^f	-15.9^f	9^d	0	
4		29.4^f	-19.9^f	12.6^d	-6	

[a]ADG = average daily live weight gain.
[b]Kilograms feed per kg live weight gain.
[c]Data for carcass composition corresponding to protein and lipid are percent-dissected skeletal muscle and adipose, respectively.
[d]Confidence level >95%.
[e]Lambs were housed in metabolism crates.
[f]Confidence level >99%.

inappropriate because differential dose-response relationships are apparent. Very few detailed reports that characterize the pharmacokinetics of these compounds in domestic animals have been published (122,123).

Genotype, Gender, and Nutritional Interactions.   There have been relatively few specific gender or genotype interaction studies conducted in growing ruminants fed β-adrenergic agonists. Results available indicate little or no differential effect. Cimaterol and ractopamine increase skeletal muscle growth in both lean and obese swine (124–126), but anabolic responses to ractopamine were larger in genotypes that exhibited superior growth performance and carcass muscle and protein accretion rates (117,127,128). Genotype differences are not eliminated with β-agonist treatment in swine.

Adequate protein and energy intake are prerequisites for achieving maximal response to β-agonist administration. Inadequate protein intake constrains the nitrogen retention response in growing pigs fed 20 ppm ractopamine (129,130), but ractopamine does not increase the efficiency with which growing pigs utilize consumed protein (131). This is in contrast to the observed effect of ST administration. Studies have not been reported for evaluation of effects of β-agonists on the efficiency of protein utilization in growing ruminants. However, additive effects of rumen bypass protein and cimaterol on muscle growth have been demonstrated in lambs (132).

Mechanism of Action. β-Agonists stimulate skeletal muscle growth by accelerating rates of fiber hypertrophy and protein synthesis, but generally do not alter muscle DNA content in parallel with the increases in protein accretion (133–135). This is in contrast to the effects of anabolic steroids and ST on skeletal muscle growth. Both of the latter stimulate fiber hypertrophy and muscle protein synthesis, but also increase muscle DNA content coincident with increased protein accretion. Whether the β-agonists decrease muscle protein degradation is equivocal.

The short-term or acute effects of the β-agonists may be different from chronic effects. Acute lipolysis and glycogenolysis are not observed beyond the first day or two of treatment. Exact mechanisms of action on lipid metabolism may differ among species. Chronic effects of the β-agonists reduce circulating insulin concentrations; ST treatment causes an opposite change. Whereas residue levels may be of concern with administration of several of the β-agonists, such is not the case for ST or GRF.

Health and Safety Information

The U.S. Food and Drug Administration's Center for Veterinary Medicine thoroughly evaluates the proposed use of any compound, natural or synthetic, used in food-producing animals for human food safety, safety to the animal of intended use, and safety to the environment. A comprehensive review of the FDA approval process for compounds administered to food-producing animals is available (136). When a compound receives approval by the FDA, the efficacy and safety have been extensively investigated, and necessary labeling, handling, use, and withdrawal time requirements, if any, are determined. This information is provided by manufacturers of the compound to the food animal producers, giving appropriate handling, dose, mode of administration, and other use restrictions, guidelines, and procedures. Technical bulletins and reference manuals are available from the manufacturer of each approved product. The Food Safety and Inspection Service (FSIS) of the USDA is responsible for ensuring that USDA-inspected meat and poultry products are safe, wholesome, and free of adulterating residues. The FSIS conducts the National Residue Program (NRP) (137) to help prevent the marketing of animals containing unacceptable (violative) residues from animal drugs, pesticides, or potentially hazardous chemicals. The monitoring and surveillance activities of the NRP provide assurance that meat and poultry products produced from animals slaughtered under federal inspection are in compliance (see MEAT PRODUCTS). Not all animal growth regulators produce residue levels that may require withdrawal of the compound before the animal is marketed, eg, the anabolic steroid implants used in growing cattle. Only MGA carries a withdrawal requirement, ie, 48 hours.

BIBLIOGRAPHY

1. D. L. Hancock, J. F. Wagner, and D. B. Anderson, *Growth Regulation in Farm Animals*, *Advances in Meat Research*, Vol. 7, Elsevier Science Publishers Ltd., Essex, U.K., 1991, pp. 255–297.
2. R. D. Boyd and D. E. Bauman, *Animal Growth Regulation*, Plenum Publishing Corp., New York, 1989, pp. 257–293.
3. D. H. Beermann and D. L. DeVol, in Ref. 1, pp. 373–426.
4. T. D. Etherton and S. B. Smith, *J. Anim. Sci.* **69**(Suppl. 1), 2–26 (1991).
5. D. H. Beermann, *The Endocrinology of Growth, Development, and Metabolism in Vertebrates*, Academic Press, Inc., San Diego, Calif., 1993, pp. 345–366.
6. A. Moloney and co-workers, in Ref. 1, pp. 455–513.
7. D. B. Anderson and co-workers, *Advances of Applied Biotechnology Series, Fat and Cholesterol Reduced Foods: Technologies and Strategies*, Vol. 12, The Portfolio Publishing Co., The Woodlands, Tex., 1991, pp. 43–73.
8. B. D. Schanbacher, *J. Anim. Sci.* **59**, 1621 (1984).
9. L. A. Muir, *J. Anim. Sci.* **61**(Suppl. 2), 154 (1985).
10. H. Galbraith and J. H. Topps, *Nutr. Abstr. Rev. Ser.* **B52**, 521 (1981).
11. J. F. Roche and J. F. Quirke, *Control and Manipulation of Animal Growth*, Butterworths, London, 1986, pp. 39–51.
12. D. H. Beermann, *Animal Growth Regulation*, Plenum Press, New York, 1989, pp. 377–400.
13. C. R. Calkins, D. C. Clanton, T. J. Berg, and J. E. Kinder, *J. Anim. Sci.* **62**, 625 (1986).
14. J. C. Bouffault and J. P. Willemart, *Anabolics in Animal Production*, Office International des Epizooties, Paris, 1983.
15. A. V. Fisher, J. D. Wood, and O. P. Whelehan, *Anim. Prod.* **42**, 203 (1986).
16. A. H. Sulieman, H. Galbraith, and J. H. Topps, *Anim. Prod.* **47**, 65 (1988).
17. G. E. Lobley and co-workers, *Br. J. Nutr.* **54**, 681 (1985).
18. M. G. Keane and M. J. Drennan, *Anim. Prod.* **45**, 359 (1987).
19. T. L. Mader and co-workers, *J. Anim. Sci.* **61**, 546 (1985).
20. D. D. Simms and co-workers, *J. Anim. Sci.* **66**, 2736 (1988).
21. D. L. Whittington, *South Dakota Beef Report*, Animal and Range Sciences Department, South Dakota State University, Brookings, 1986, p. 92.
22. L. J. MacVinish and H. Galbraith, *Anim. Prod.* **47**, 75 (1988).
23. J. M. Hayden, W. G. Bergen, and R. A. Merkel, *J. Anim. Sci.* **70**, 2109–2119 (1992).
24. S. J. Bartle, R. L. Preston, R. E. Brown, and R. J. Grant, *J. Anim. Sci.* **70**, 1326–1332 (1992).
25. J. K. Apple, M. E. Dikeman, D. D. Simms, and G. Kuhl, *J. Anim. Sci.* **69**, 4437–4448 (1991).
26. T. S. Rumsey, *J. Anim. Sci.* **46**, 463 (1978).
27. J. R. Greathouse and co-workers, *J. Anim. Sci.* **57**, 355 (1983).
28. G. W. Mathison and L. A. Stobbs, *Can. J. Anim. Sci.* **63**, 75 (1983).
29. R. W. J. Steen, *Anim. Prod.* **41**, 301 (1985).
30. M. L. Thonney, *J. Anim. Sci.* **65**, 1 (1987).
31. A. Trenkle, *Feedstuffs* **59**, 43 (1987).
32. D. E. Eversole, J. P. Fontenot, and D. J. Kirk, *Nutr. Rep. Int.* **39**, 995 (1989).
33. T. C. Perry, D. G. Fox, and D. H. Beermann, *J. Anim. Sci.* **59**, 4696–4702 (1991).
34. A. H. Sulieman, H. Galbraith, and J. H. Topps, *Anim. Prod.* **43**, 109 (1986).
35. T. W. Griffiths, *Anim. Prod.* **34**, 309 (1982).
36. A. V. Fisher and J. D. Wood, *Anim. Prod.* **42**, 195 (1986).
37. W. Vanderwert and co-workers, *J. Anim. Sci.* **61**, 537 (1985).

38. P. J. Buttery and P. A. Sinnett-Smith, *Manipulation of Growth in Farm Animals*, Martinus Nijhoff Publisher, Boston, Mass., 1984, pp. 211–232.
39. B. G. Vernon and P. J. Buttery, *Br. J. Nutr.* **36**, 575 (1976).
40. B. G. Vernon and P. J. Buttery, *Anim. Prod.* **26**, 1 (1978).
41. B. G. Vernon and P. J. Buttery, *Br. J. Nutr.* **40**, 563 (1978).
42. P. A. Sinnett-Smith, N. W. Dumelow, and P. J. Buttery, *Br. J. Nutr.* **50**, 225 (1983).
43. S. H. Thompson, L. K. Boxhorn, W. Kong, and R. E. Allen, *Endocrinology* **124**, 2110 (1989).
44. R. L. Prior, S. B. Smith, B. D. Schnabacher, and H. J. Mersmann *Anim. Prod.* **37**, 81 (1983).
45. L. C. St. John and co-workers, *J. Anim. Sci.* **64**, 1428 (1987).
46. W. G. Bergen and D. B. Bates, *J. Anim. Sci.* **58**, 1465 (1984).
47. A. L. Donoho, *J. Anim. Sci.* **58**, 1528 (1984).
48. R. D. Goodrich and co-workers, *J. Anim. Sci.* **58**, 1484 (1984).
49. E. L. Potter, R. L. VanDuyn, and C. O. Cooley, *J. Anim. Sci.* **58**, 499 (1984).
50. G. T. Schelling, *J. Anim. Sci.* **58**, 1518 (1984).
51. G. C. Todd, M. N. Novilla, and L. C. Howard, *J. Anim. Sci.* **58**, 1512 (1984).
52. F. N. Owens, J. Zorrilla-Rios, and P. Dubeski, in Ref. 1, pp. 321–342.
53. M. L. Galyean and F. N. Owens, *ISI Atlas Sci.: Anim. and Plant Sci.* **1**, 71 (1988).
54. R. H. G. Stockdale, *Vet. Med. Small Anim. Clin.* **76**, 1575 (1981).
55. K. L. Davison, *J. Agri. Food Chem.* **31**, 1273 (1984).
56. G. M. Davenport, J. A. Boling, K. K. Schillo, and D. K. Aaron, *J. Anim. Sci.* **68**, 222 (1990).
57. J. Reffett-Stabel, J. W. Spears, R. W. Harvey, and D. M. Lucas, *J. Anim. Sci.* **67**, 2745 (1989).
58. J. W. Spears, B. R. Schricker, and J. C. Burns, *J. Anim. Sci.* **67**, 2140 (1989).
59. E. E. Bartley and co-workers, *J. Anim. Sci.* **5**, 1400 (1983).
60. D. C. Honeyfield, J. R. Carlson, M. R. Nocerini, and R. G. Breeze, *J. Anim. Sci.* **60**, 226 (1985).
61. *Antibiotics in Animal Feeds*, Rept. no. 88, Council for Agricultural Science and Technology (CAST), Ames, Iowa, 1981.
62. E. F. Bartley, F. C. Fountaine, and F. W. Atkeson, *J. Anim. Sci.* **9**, 646 (1950).
63. T. J. Cunha and co-workers, *J. Anim. Sci.* **9**, 653 (1950).
64. T. H. Jukes and co-workers, *Arch. Biochem.* **26**, 324 (1950).
65. J. McGinnis and co-workers, *Poult. Sci.* **29**, 771 (1950).
66. V. W. Hays, *The Use of Drugs in Animal Feeds*, National Academy of Science, National Research Council Publication no. 1679, Washington, D.C., 1969, p. 11.
67. R. G. Warner, unpublished transcript of presentation to the U.S. Food and Drug Administration Task Force on the use of antibiotics in animal feeds; Ref. 55, p. 25.
68. D. R. Zimmerman, *J. Anim. Sci.* **62**(Suppl. 3), 6 (1986).
69. V. W. Hays, in Ref. 1, pp. 299–320.
70. D. E. Bauman, J. H. Eisemann, and W. B. Currie, *Fed. Proc.* **41**, 2538–2544 (1982).
71. I. C. Hart and I. D. Johnson, *Control and Manipulation of Animal Growth*, Butterworths, London, 1986, pp. 135–159.
72. D. H. Beermann and R. D. Boyd, *Control of Fat and Lean Deposition*, Butterworth Heinemann, Oxford, U.K., 1992, pp. 249–275.
73. B. J. Krick and co-workers, *J. Anim. Sci.* **70**, 3024–3034 (1992).
74. L. F. Thiel, D. H. Beermann, B. J. Krick, and R. D. Boyd, *J. Anim. Sci.* **71**, 827–835 (1992).
75. T. D. Etherton and co-workers, *J. Anim. Sci.* **64**, 433–443 (1987).
76. D. G. McLaren and co-workers, *J. Anim. Sci.* **68**, 640–651 (1990).
77. C. D. Knight and co-workers, *J. Anim. Sci.* **69**, 4678–4689 (1991).

78. W. J. Enright, *Use of Somatotropin in Livestock Production*, Elsevier, London, 1989, pp. 132–156.
79. B. A. Crooker and co-workers, *J. Nutr.* **120**, 1256–1253 (1990).
80. I. D. Johnsson and co-workers, *Anim. Prod.* **44**, 405–414 (1987).
81. A. S. Zainur and co-workers, *Austral. J. Agric. Res.* **40**, 195–206 (1989).
82. J. M. Pell and co-workers, *Brit. J. Nutr.* **63**, 431–445 (1990).
83. C. L. McLaughlin and co-workers, *J. Anim. Sci.* **71** (in press) (1993).
84. W. M. Moseley and co-workers, *J. Anim. Sci.* **70**, 412–425 (1992).
85. D. H. Beermann and co-workers, *J. Anim. Sci.* **68**, 4122–4133 (1990).
86. I. D. Johnsson, I. C. Hart, and B. W. Butler-Hogg, *Anim. Prod.* **41**, 207–217 (1985).
87. J. P. McNamara and co-workers, *J. Anim. Sci.* **69**, 2273–2281 (1991).
88. R. G. Campbell and co-workers, *J. Anim. Sci.* **67**, 177–186 (1989).
89. R. G. Campbell and co-workers, *J. Anim. Sci.* **68**, 2674–2681 (1990).
90. C. L. McLaughlin and co-workers, *J. Anim. Sci.* **67**, 116–127 (1989).
91. E. Kanis and co-workers, *J. Anim. Sci.* **68**, 1193–1200 (1990).
92. J. P. Bidanel and co-workers, *J. Anim. Sci.* **69**, 3511–3522 (1991).
93. R. G. Campbell and co-workers, *J. Anim. Sci.* **66**, 1643–1655 (1988).
94. R. G. Campbell and co-workers, *J. Anim. Sci.* **68**, 3217–3225 (1990).
95. T. J. Caperna and co-workers, *J. Anim. Sci.* **68**, 4243–4252 (1990).
96. R. G. Campbell, *Nutr. Res. Rev.* **1**, 233–253 (1988).
97. R. D. Boyd and co-workers, *J. Anim. Sci.* **69**(Suppl. 2), 56–75 (1991).
98. L. A. Kraft and co-workers, *Domest. Anim. Endocrinol.* **2**, 133–139 (1985).
99. M. A. Della-Fera, F. C. Buonomi, and C. A. Baile, *Domest. Anim. Endocrinol.* **3**, 165–176 (1986).
100. D. Petitclerc and co-workers, *J. Anim. Sci.* **65**, 996–1005 (1987).
101. R. S. Kensinger and co-workers, *J. Anim. Sci.* **64**, 1002–1009 (1987).
102. J. L. Johnson and co-workers, *J. Anim. Sci.* **68**, 3204–3211 (1990).
103. T. D. Etherton and co-workers, *J. Anim. Sci.* **63**, 1389–1399 (1986).
104. A. M. Felix and co-workers, *Proceedings of the 19th European Peptide Symposium*, Chalkidiki, Greece, 1986, p. 481.
105. P. Dubreuil and co-workers, *J. Anim. Sci.* **68**, 1254–1268 (1990).
106. S. A. Pommier and co-workers, *J. Anim. Sci.* **68**, 1291–1298 (1990).
107. P. E. V. Williams, *Nutr. Abstr. Rev. (Series B)* **57**, 453–464 (1987).
108. D. H. Beermann, *Animal Growth Regulation*, Plenum Publishing, New York, 1989, pp. 377–396.
109. P. W. Emery and co-workers, *Biosci. Rep.* **4**, 83–91 (1984).
110. C. A. Ricks and co-workers, *J. Anim. Sci.* **59**, 1247–1255 (1984).
111. J. P. Hanrahan, *Recent Advances in Animal Nutrition*, Butterworths, London, 1986, pp. 125–138.
112. P. J. Reeds and co-workers, *Brit. J. Nutr.* **56**, 249–258 (1986).
113. J. P. Hanrahan and co-workers, *Beta-Agonists and Their Effects on Animal Growth and Carcass Quality*, Elsevier Applied Science, London, 1987, pp. 106–118.
114. R. M. O'Connor and co-workers, *Domest. Anim. Endocrinol.* **84**, 549–445 (1991).
115. E. L. Rickes and co-workers, *J. Anim. Sci.* **67**(Suppl. 1), 221 (1989).
116. L. E. Watkins and co-workers, *J. Anim. Sci.* **68**, 3588–3595 (1990).
117. L. J. Bark and co-workers, *J. Anim. Sci.* **70**, 3391–3400 (1992).
118. J. B. Morgan, S. J. Jones, and C. R. Calkins, *J. Anim. Sci.* **67**, 2646–2654 (1989).
119. R. H. Wellenreiter and L. V. Tonkinson, *Poult. Sci.* **69**(Suppl. 1), 143 (1990).
120. *Ibid.*, p. 142.
121. T. M. Byrem and co-workers, *FASEB J.* Abst. #3735, (1993).
122. H. H. D. Meyer and L. Rinke, *J. Anim. Sci.* **69**, 4538–4544 (1991).
123. T. M. Byrem and co-workers, *J. Anim. Sci.* **70**, 3812–3819 (1992).

124. J. T. Yen and co-workers, *J. Anim. Sci.* **68**, 3705–3712 (1990).
125. J. T. Yen and co-workers, *J. Anim. Sci.* **68**, 2698–2706 (1990).
126. J. T. Yen and co-workers, *J. Anim. Sci.* **69**, 4810–4822 (1991).
127. Y. Gu and co-workers, *J. Anim. Sci.* **69**, 2685–2693 (1991).
128. Y. Gu and co-workers, *J. Anim. Sci.* **69**, 2694–2702 (1991).
129. D. B. Anderson and co-workers, *Fed. Proc.* **46**, 1021 (1987).
130. A. Bracher-Jakob and J. W. Blum, *Anim. Prod.* **51**, 601–611 (1990).
131. F. R. Dunshea and co-workers, *J. Anim. Sci.* **69**(Suppl. 1), 302 (1991).
132. D. H. Beermann and co-workers, *J. Anim. Sci.* **62**, 370–380 (1986).
133. C. A. Maltin, M. I. Delday, and P. J. Reeds, *Biosci. Rep.* **6**, 293–299 (1986).
134. D. H. Beermann and co-workers, *J. Anim. Sci.* **63**, 1314–1524 (1987).
135. Y. S. Kim, Y. B. Lee, and R. H. Dalrymple, *J. Anim. Sci.* **63**, 1392–1399 (1987).
136. S. S. Collins and co-workers, *Nutr. Rev.* **47**, 238 (1989).
137. USDA, *Domestic Residue Data Book*, Food Safety and Inspection Service, National Residue Program, Washington, D.C., 1992; published annually.

General References

A. P. Moloney and co-workers, *J. Anim. Sci.* **68**, 1269–1277 (1990).
R. W. Jones and co-workers, *J. Anim. Sci.* **61**, 905–913 (1985).
A. Bracher-Jakob, P. Stoll, and J. W. Blum, *Livestock Prod. Sci.* **25**, 231–246 (1990).
E. L. Rickes and co-workers, *Poult. Sci.* **66**(Suppl 1), 166 (1987).

DONALD H. BEERMANN
Cornell University

PLANT

The arrival of new plant growth regulators on the market in the early 1990s, especially synthetic ones, is in a static state. Many growth regulators continue to be used experimentally, but the transition to approved usage is being delayed for several reasons. On the financial side, a number of mergers, buy-outs, and other dispositions of chemical companies has led to a decrease in the number of commercial compounds available. Moreover, the cost of registration has prompted some producers to withdraw chemicals from the market. Some older plant growth regulators have undergone several trade name changes, adding confusion to the field.

The ideal plant growth regulator should leave no harmful persistent residue in a finished product or crop and the paradigm compounds are ones that have high specific activity, are target specific, and are environmentally biodegradable.

Plant growth regulator compounds are listed herein by origin and in alphabetical order for ease of identification, rather than being classified by chemical groups. Numerous books and technical data sheets are available for more detailed information on time and rates of application. However, in many cases these compounds have dual uses, and because plant growth regulation is minor compared to various other applications, their mode of action from the perspective of growth regulation has not been determined. Furthermore, the economics may not allow for the elucidation of these mechanisms.

Natural Products

The most acceptable growth regulators appear to be those compounds that already occur in nature (Table 1) and elicit certain desirable responses in economic crops. Relative to the number of purely synthetic materials available, the natural products are a very small group that has not grown appreciably since the early 1950s.

Brassinosteroids. The brassinosteroids, especially brassinolide [72962-43-7] (**1**) and 24-epibrassinolide [78821-43-9] (**2**), are unique chemical structures representing a group of biologically highly active natural products (Fig. 1) that induce plant growth regulatory responses at exceedingly low concentrations. The fact that these compounds are applied at low rates and are effective at 10^{-11} M implies that they are environmentally safe. Toxicology studies show that 24-epibrassinolide has very low toxicity.

Brassinolide was originally isolated from the evergreen tree, *Distylium racemosum* Sieb. et Aucc, in Japan, where it bears the common name Isunoki. Initially, 430 kg of fresh leaves were gathered and extracted to give 751 mg of *Di-*

Table 1. Natural Plant Growth Regulators[a]

Product	Structure	Trade name	LD$_{50}$, g/kg
	Available natural products[b]		
brassinolide	(**1**)		rat[c]
24-epibrassinolide	(**2**)		rat, 1[c]
cytokinins, mixed[d]	(**3**), (**4**)	Cytogen	rabbit, 10
		Trigger	rat, 5
n-decanol	(**5**)	Off-Shoot-T, Royaltac,	rat, 12.8
		Sucker Plucker, Antak	
dikegulac	(**6**)	Atrimmec, Atrinal	mouse, 19.5; rat, 31
ethylene	(**7**)	Cerone, Prep, Ethrel,	rat, 4.22
ethephon	(**8**)	Chipcor, Florel	
gibberellins			
GA$_3$	(**9**)	Berelex, Gib-Tabs, Gib-Sol,	mouse, 15
		Pro-Gibb	
GA$_4$ and GA$_7$	(**10**), (**11**)	Pro-Gib 47, Regulex	no toxicity
lactic acid	(**12**)	Propel	4.94
	Available natural product derivatives[e]		
benzylamine purine	(**13**)	Promalin (as a mixture with	mouse, 1.69
		GA$_4$ and GA$_7$)	
indole-3-*n*-butyric acid	(**14**)	IBA, Hormodin, Rhizopon (AA),	mouse, 100
		Jiffy Grow	
N-(phosphonomethyl)-	(**15**)	Roundup, Glyphosate, Polado	rat, 3.9
glycine			

[a]Ref. 1.
[b]See Fig. 1.
[c]Oral.
[d]Kinetin (**3**); 6-(4-hydroxy-1,3-dimethylbut-*trans*-2-enylamino)-9-β-D-ribofuranosylpurine (**4**).
[e]See Fig. 2.

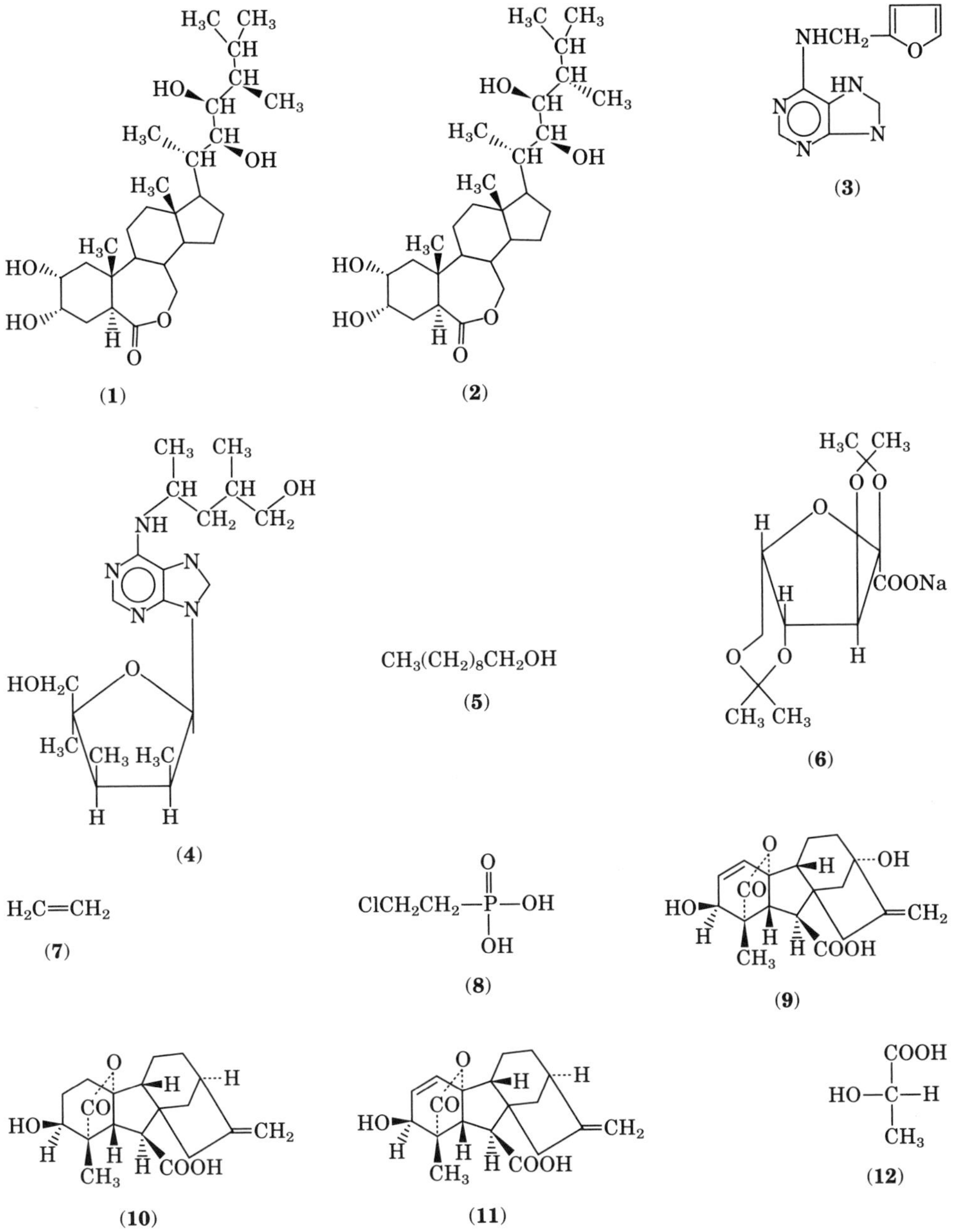

Fig. 1. Natural plant growth regulators. See Table 1.

stylium factor A, 50 µg of factor A$_2$, and 236 µg of factor B, all of which demonstrate biological activity. Later, factor B was identified as a mixture of brassinolide and 28-norbrassinolide and factor A contained brassinone and castasterone [80736-41-0] (2). However, none of these structures were known in 1968 and results were published only on the biological properties of the *Distylium* fac-

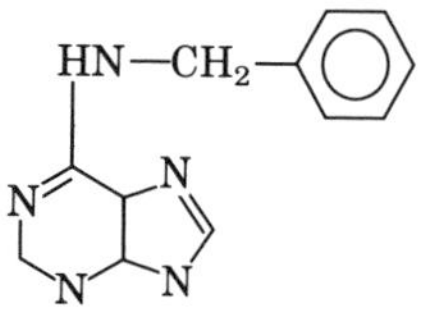

HN—CH$_2$—⟨○⟩ (purine structure) (indole) —CH$_2$CH$_2$CH$_2$COOH

$$HO-\overset{\overset{\displaystyle O}{\|}}{\underset{\underset{\displaystyle OH}{|}}{P}}-CH_2NHCH_2COOH$$

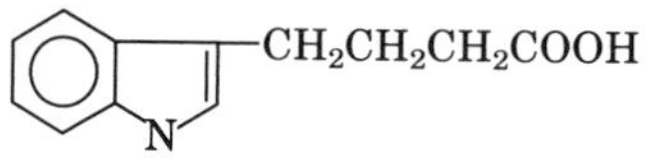

(14)

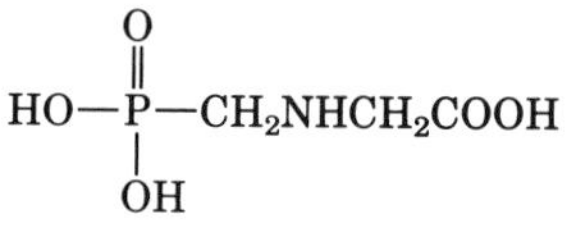

(15)

(13)

Fig. 2. Natural product derivative plant growth regulators. See Table 1.

tors (3). From 1963 to 1979, the Agricultural Research Service, USDA, concentrated its efforts on isolating a biologically active substance from rape pollen (*Brassica napus* L.). In order to obtain sufficient quantities of pollen, bees were used to gather material from flowering canola plants. Extraction and purification of a subsample consisting of 38 kg finally yielded ~4 mg of pure brassinolide (**1**) (4) which is difficult to obtain in large quantities from natural sources. Yields from chemicals synthesis are also low.

24-Epibrassinolide (**2**) can be obtained in relatively large quantity starting with brassicasterol, which is present as 10–20% of the sterol fraction of canola oil; it may then be synthesized in high yield. 24-Epibrassinolide has been used successfully to treat barley (*Hordeum vulgare*), lucerne (*Medicago sativa*), and some horticultural crops in Russia. Barley yields increased up to 25% using applications of 50 or 100 mg/hm^2 in 500 L of water. Concomitantly, lucerne seed yields increased 26% (5). In China, during a six-year period, 24-epibrassinolide was evaluated on 3333 hm^2 of wheat using concentrations that ranged from 0.1 to 0.001 ppm (6). There was a consistent increase in wheat yields of up to 15%. During 1990–1991 real usage was increased to a total of 4000 hm^2 with yield increases of 8–15%. The total land mass under treatment was expected to be increased to over 23,000 hm^2 between 1992–1994 and the total amount of 24-epibrassinolide to be used was a mere 100 g. Corn, cucumber, tobacco, and watermelon also have been treated in China with substantially increased yields although data are not available. Sesame yields have been increased 17% (7). Increased fruit set has been obtained by spraying at time of flowering in eggplant, grape, orange, canola, and strawberry, and this has led to increased yields. There was increased sugar content, fresh weight, and earlier maturity in grapes. Mushroom growth has also been increased using 24-epibrassinolide applications. This natural product, commercially available from Beak Consultants (Brampton, Ontario, Canada), is expected to have broad agronomic application.

The toxicity data for 24-epibrassinolide and the low dose at which it elicits growth regulatory responses in plants makes it appealing for agronomic and horticultural use.

Mixed Cytokinins. The first cytokinin, kinetin [*525-79-1*] (**3**), was isolated from stale herring sperm (8) but, like so many biologically active natural products, it was later found in the vascular system of tobacco stems and leaves (9). Yeast also proved to have a very high titre of kinetin (see YEASTS) (8). The compound is very active in increasing cell division in tobacco wound callus tissue that has been cultured on White's agar medium supplemented with 2 mg/L of indole-3-

acetic acid (IAA) [*87-51-4*]. The presence of IAA is mandatory to induce cell division in the presence of kinetin.

Other cytokinins discovered in microorganisms include *trans*-zeatin, *trans*-zeatin riboside, and 6-(4-hydroxy-1,3-dimethylbut-*trans*-2-enylamino)-9-β-D-ribofuranosylpurine (**4**), a new cytokinin from the phytopathogenic bacterium *Pseudomonas syringae* pr. *savastanoi*, that causes olive knot and galls on stems of ash, jasmine, oleander, and privet.

Despite the fact that a number of purines are available for possible plant growth regulator use, none have specifically found their way into the market. The exceptions are the so-called natural cytokinins which are mostly zeatin-like (10). One of these has the common name cytokinin and it is an aqueous extract from *Ascophyllum nodosum* which is marketed under a series of trade names, eg, Cytogen, and Arise seed dressing. Another entry, Soil Triggrr, contains zeatin, dihydrozeatin [*23599-75-9*], and the purine isopentyladenine [*97856-37-6*], all as free bases or as the riboside derivative. Cytogen has been used to boost yields in cereal grains, cotton (qv), soybeans, leafy and head vegetables, onions, peppers, and tomatoes (10). Soil Triggrr is used as a soil treatment to obtain yield increases in alfalfa, corn, cotton, jojoba, lupine, peanuts, rice, sorghum, soybeans, sugar beet, triticale, wheat, and certain fruits and vegetables. It is also used on ornamentals, trees, and turf. In using either product the amount of crude solution used falls between 3.65–43.9 mL/hm^2 (5–60 oz/A), depending on the crop or ornamental treated, and they may be added to most pesticides in a tank mix. The products are safe but care should be taken when handling them (see SOYBEANS AND OTHER OILSEEDS; WHEAT AND OTHER CEREAL GRAINS).

The difficulties in dealing with relatively crude plant or microbial extracts are twofold. First, whereas the putative compounds may be present in the extract, the solution may contain bacteria or fungi. This may lead to the biotransformation or breakdown of the supposed active metabolite, yielding yet another generation of secondary plant growth regulators. Second, a fermentation product utilizing fungi or bacteria may give rise to a number of unidentified and, therefore, undisclosed biologically active natural products. Proof of a cytokinin can only be established by rigorous chemical identification. In the case of Cytogen, the material is derived from the tissue of marine algae, specifically the *Laminarinaceae* and *Fucaceae* (10–12). Soil Triggrr is composed of microbial fermentation products, plant extracts, nutrients, surfactants, and preservatives (10). The efficacy of seaweed extracts relative to plant growth regulatory activity has been discussed (10).

n-Decanol. *n*-Decanol [*112-30-1*] (**5**) may be prepared by a number of routes; it also occurs naturally in certain vegetable oils (qv). It is classified as a plant growth regulator and is used to control axillary shoot growth in mature tobacco as a contact spray. As such, it is not a plant hormone and its mode of action is by plasmolysis of the tender axillary shoots. The substance is applied to tobacco plants when the flowers are in the button stage, ie, compact, immature racemes, before the corollas have burst through the calyx or just after the inflorescences have been mechanically removed. The length of the axillary shoot is important if the *n*-decanol is to be effective. If shoots are greater than 2.5 cm long they must be removed by hand. The substance is applied at rates of 2.45–12.78 kg/m^2, a rather heavy application, but it is nontoxic. It has an unpleasant, acrid fatty acid odor (10), hence it is carefully handled by applicators.

Dikegulac.   2,3:4,6-Bis-*O*-(methylethylidene)-α-L-xylo-2-hexulofuranosonic acid [*18467-77-1*] is a monosaccharide derivative that was isolated as an intermediate in the commercial synthesis of L-ascorbic acid. It possesses potent plant growth regulatory activity, occurs naturally as the free acid, but is sold as the sodium salt, dikegulac (**6**), which is the most active form. In initial experiments, plants treated with 6 g/L solutions included barley, perennial ryegrass (*Lolium perenne*), crabgrass, apple, grape, privet, bean, tomato, gerbera (*Gerbera jamesonii*), and rhododendron. The compound induced growth inhibition in the first five species but there were variable responses with the others. Bean plant petioles abscissed, tomato fruits were parthenocarpic, axillary shoots were increased in gerbera giving rise to more flowers, and rhododendrons were pinched so that axillary shoots broke dormancy and plants became bushy (13). The compound is used to control growth in shrubs and ornamentals (10).

Ethylene.   Ethylene [*74-85-1*], C_2H_4 (**7**) is considered to be one of the principal plant growth regulators in the natural products family. However, ethylene remains unique because it is a simple compound and in its natural state exists as a gas. It cannot be used in the field except in some bound form, but it may be used in gastight enclosures, provided it is not exposed to sparks or flames because of its extremely explosive nature. This characteristic has always posed a problem on banana boats where ethylene produced by the ripening fruit has collected below-deck in the hold. Smoking has been expressly forbidden because of the violent combustion that can result. The gas has been shown to be generated by a large number of fungi (14) and it is produced by many fruits during maturation and ripening, bananas being one of many such fruits.

The effects of ethylene on plant growth were first reported during the last quarter of the nineteenth century. As early as 1935, ethylene gas was used in Germany (15) to ripen, or color to a pleasing yellow, tobacco during the curing process, specifically at the coloring stage. It was used later in Italy, ie, during World War II (16). Ethylene also has been used to induce senescence and defoliation in cotton and other select plants (17). However, a practical form of ethylene did not arrive on the market until around 1970 when the experimental compound Ethrel [*16672-87-0*] (ethephon, 2-chloroethylphosphonic acid) (**8**) was distributed for experimental use. 2-Chloroethanol had been used as a sprout promoter in Irish potatoes (*Solanum tuberosum*). It had been extensively employed in the preparation of seed potatoes in the 1940s. The process, where seed potatoes were wrapped in cheesecloth and dunked in solutions of 2-chloroethanol, was clumsy, the chemical tended to volatilize in the sun, and the fumes were toxic. In 1946, the synthesis of 2-chloroethylphosphonic acid was described (18). In 1965, the compound, which is the phosphate ester of 2-chloroethanol, was discovered. 2-Chloroethylphosphonic acid has proven to be a highly utilitarian agrochemical for use as a ripening agent in many crops. Under physiological conditions it is readily translocated and breaks down to release ethylene, chloride ions, and phosphate. The release of ethylene is pH mediated with 3.5 being the threshold. Furthermore, the compound is highly water soluble. It is used to promote fruit maturity and loosening of fruit in apples, blackberries, black currants, cherries, coffee, pepper, and tomatoes. Other unique uses include increasing flower bud development in apples, uniform flower bud initiation in pineapples, stimulating latex flow in rubber trees, inhibiting arrowing (flowering) in sugarcane, increasing sucrose content in the barrels, and yellowing mature flue-cured tobacco to reduce curing time. The cost savings in tobacco curing alone are staggering and it has been estimated in the United States that there is an annual savings of 68.9 million L of LPG, 257 million L of fuel oil, and 110 million kW·h of electricity, thus the compound is an energy saver. Worldwide figures are not available. The compound is also used to loosen walnuts, and in the flower industry to shorten stems in daffodil bulbs that have been forced and to stimulate and increase branching on geraniums; greenhouse roses (*Rosa* sp.) also may be treated to increase basal budding. Application rates vary, depending on the crop being treated. Most importantly, the compound is nonflammable yet possesses many of the useful properties of ethylene.

The Gibberellins. The gibberellins are natural products that occur in fungi and higher plants and are responsible for causing cell elongation in plants. Diseased plants have been noted to be hyperelongated (19). The casual organism for the Bakanae disease of rice was *Gibberella fujikuroi*; culture filtrates of the organism induced elongation in rice and grass seedlings. In 1938, the isolation of gibberellin A and B, in pure crystalline form from the Bakanae fungus, was reported. This information was ignored until after 1945, when the gibberellins became an important series of chemical relatives because of their apparent potential to increase crop yields. By 1957 the USDA Agricultural Research Service had compiled the abstracts of 632 scientific articles dealing with gibberellins (19); since 1957 both the numbers of structurally different types of gibberellins and the literature surrounding them has burgeoned. However, the principal gibberellins of commerce are gibberellic acid [77-06-5], GA_3 (**9**) and a mixture of gibberellins, GA_4 and GA_7. Both GA_4 and GA_7 are used practically in the field in combination with 6-benzylaminopurine; GA_4 is 1α,2β,4aα,4bβ,10β-2,4A-dihydroxy-1-methyl-8-methylene gibbane-1,10-dicarboxylic acid, 1-4A-lactone [561-56-8] (**10**) and GA_7 is 1α,2β,4aα,4bβ,10β-2,4A-dihydroxy-1-methyl-8-methylene gibb-3-ene-1,10-dicarboxylic acid, 1,4A-lactone [510-75-8] (**11**).

All gibberellins used commercially are obtained from fermentation. Both GA_4 and GA_7 are sold as a mixture and are generally applied in 467.7 L of water per hm^2 containing ~50 ppm. The only agronomic use seems to be on gynoecious cucumber plants to induce the development of male flowers. This is done every five days for a total of three applications. Whereas the use for GA_4 and GA_7 is very selective, the chemical congener GA_3 exhibits a vast range of growth regulating responses in many crops. Its application amounts range from 10 mg/L for hops (*Humulus lupulus*) to 2 gm in 65.5 L of water/hm^2 for sugar cane. The material is used in blueberries (*Vaccinium* sp.) to increase fruit set when pollination is not efficiently carried out by honey bees. In cherries, depending on the variety, GA_3 is used to counteract the effect of cherry yellows virus; in sweet cherries, application of the chemical improves fruit size, texture, color, and prolongs harvest time. The compound may also be employed to reduce flowering and, consequently, fruiting in both tart and sweet cherries.

Before the advent of GA_3 it was common practice to girdle grape vines to increase yields; using GA_3, in the early 1950s, the girdlers job became redundant. Most applications to grapes are made either at berry shatter or normal girdling time. The results are an increase in berry size and, in the case of raisin grapes, maturity is hastened with a concomitant rise in quality.

The citrus industry also uses GA_3 to improve crop quality. In lemons (*Citrus* sp.), application delays yellowing and fruit size is increased. Treated lemons also have increased shelf life. Both Navel and Valencia oranges have delayed ripening when treated with GA_3 and the skins of both varieties are considerably improved. Whereas the same holds true for tangerines, one of the main uses in that crop is to improve fruit set, especially in those varieties in which pollen production is insufficient, eg, Minneola, Orlando, and Robinson varieties. If grapefruit are treated with GA_3, the fruit is retained on the tree longer and there are fewer losses due to early fruit drop; the skin remains firmer and takes on a less mature look. GA_3 has been used to control the attack of the Caribbean fruitfly (*Anastrepha suspensa*) by keeping the skin of the grapefruit firmer and in a less mature state; the reduction of chemical attractants in the peel or the color of the skin, or both, makes the fruit less attractive to the insect (20,21). In the Pome family, the only member that responds favorably to GA_3 treatment is the pear (*Pyrus communis*); the chemical is generally used to increase fruit set especially in those varieties with inefficient pollination. Apart from cherries, which have been discussed, the other stonefruit routinely treated with GA_3 is the Italian prune. Spraying with GA_3 increases yields and inhibits internal browning. Other horticultural crops also improved with standard GA_3 treatments include rhubarb (*Rheum* sp.), artichokes (*Cynara scolymus*), asparagus (*Asparagus officionalis*), beans (*Phaseolus vulgaris*), celery (*Apium graveolens*), cucumbers (*Cucumis sativus*), lettuce seed production (*Lactuca sativa*), parsley (*Petroselinum crispum*), peas (*Pisum sativum*), seed-potato (to break dormancy), barley (*Hordeum vulgare*) in the malting process to control malt production and thereby increase the relative concentration of neutral grain spirit in the manufacture of gin and vodka, oats (*Avena sativa*), cotton, rye (*Secale cereale*), soybeans, strawberries, and wheat (*Triticum aestivum*). The material also has application on golf courses where Bermuda grass (*Cynodon dactylon*) may be induced to start growth, or growth may be maintained throughout the season. Aesthetically, GA_3 may be used to keep the grass green at times of stress or when light frosts occur.

The effects of GA_3 are also responsible for its use in sugarcane where ground or aerial application just before harvest increases sucrose yields. The application rates are 2 g in 65.5 L of water/hm^2, which is relatively high compared to treatments in other crops. This effect on sucrose levels is only seen in Hawaii where sugar is a two-year crop when harvested and the rootstock is mechanically destroyed. In the West Indies, where sugarcane is harvested annually and each rootstock is viable for about seven years, ie, eight crops are cut from the same rootstock, GA_3 has no effect on sucrose content. There seems to be a direct correlation between altitude, temperature, and response. Sugarcane grown in the lowlands of Hawaii experiences some stress as the result of reduced seasonal temperatures with a concomitant slowing of growth; it is these varieties that respond well to GA_3 treatments (22).

Gibberellic acid is also used successfully in rice culture to promote seed germination and the growth of semidwarf varieties. The treated seed can be planted deeper than normal. In addition, the sprouting seedlings are much taller than the untreated ones; they compete well against weeds. The material is sold under the trade name Release.

Ornamentals are treated with GA_3 for a variety of purposes including more profuse flowering, increasing flower number, and in some circumstances flower size. It is a common practice in the southeastern United States to gib camellias prior to flowering, ie, a small amount of the potassium salt of GA_3 at 100 ppm is dropped into the floral bud at a very early stage of development to produce large showy flowers.

Until other gibberellins (~80) can be either fermented or synthesized in bulk, their future is uncertain for use as industrial plant growth regulators except in very special circumstances.

Lactic Acid. L-Lactic acid [79-33-4] (**12**) is a common natural product. It is a result of the metabolism of glucose in the Embden-Meyerhof pathway where glucose is split to yield phosphoglyceraldehyde by a series of steps. This, in turn, is acted upon by the enzyme phosphoglyceraldehyde dehydrogenase to produce phosphoglyceric acid. The latter comes under the influence of lactic dehydrogenase and L-lactic acid is formed. The practical use for lactic acid as a plant growth regulator is limited. It has been applied at the rate of 4.39 mL/hm^2, using a stock 80% solution, to increase fruit set, ripening, and promote shoot and root formation in members of the *Cruciferae*, stonefruit, *Solanaceae*, citrus, corn grapes, pineapples, strawberries, sugarcane, and walnuts (23). The mode of action is not known but it appears to be most efficacious when plants undergo stress. This indicates, possibly, that the material feeds into the Embden-Meyerhof pathway and from there into the Krebs cycle. Because lactic acid is a terminal metabolic product there are few alternative suggestions as to its ultimate fate in plants. Polymers of lactic acid appear to be the active form of this compound.

Natural Product Derivatives

Very few natural product derivatives (Fig. 2) have been marketed successfully. Of those that have, the profits relative to expenditure have been large. Certain natural products have been used as experimental templates upon which to construct marketable products; eg, 2,4-dichlorophenoxyacetic acid which was conceived from the indole ring.

Benzylamine Purine. The purine 6-benzylaminopurine [1214-39-7] (**13**) is an analogue of the natural product adenine, a component of both deoxyribonucleic and ribonucleic acid. It is not employed alone, but rather in combination with the natural products GA_4 and GA_7 to improve the size, weight, and thereby, yield per hm^2 of Red Delicious apples (10,24,25). Compounds with cytokinin activity were reported in 1913 (26) and asymmetric growth in apples was published in 1968 (27).

Indole-3-*n*-Butyric Acid. Among 500 compounds tested from the late 1920s to 1950 for plant growth regulatory activity, indole-3-butyric acid [133-32-4] (**14**) was one of the most interesting (28). It is a homologue of the naturally occurring plant growth regulator, indole-3-acetic acid, but differs in the side chain by the addition of two more carbon atoms. It is a relatively stable compound in solution, especially in comparison to indole-3-acetic acid, and has more utility as a practical chemical. It exhibits auxin-like activity and promotes the rooting of cuttings.

Indole-3-*n*-butyric acid can be prepared as a solution and used at a concentration of 0.1–1.0%. A convenient method for handling the chemical involves dissolving an appropriate quantity in acetone or similar solvent, and then adding the solution to a carrier such as talc. The material can be spun on a rotary distillation apparatus and the acetone evaporated away from the powder. The preparation, which has an evenly coated residue of indole-3-butyric acid, may be stored for many years in a cool place without the loss of activity. Cuttings of herbaceous or woody plants can be easily rooted by treating the basal ends with either a solution or the activated talc and placing them in moist sand in a misting chamber. The material is nontoxic at the concentrations used. The talc formulation is available commercially. Not all woody cuttings respond to treatment, eg, larch and cedar.

N-(Phosphonomethyl)Glycine. Glyphosate [*1071-83-6*] (Roundup, Polado) is a good example of a lucrative agrochemical that is a natural product derivative. Structurally, this allelochemical, ie, compound produced by one organism which detrimentally affects another, is composed of the simple amino acid glycine that has been phosphorylated. Because the compound is biologically active, the temptation exists to predict that homologous amino acid series, properly phosphorylated, would have herbicidal activity; they do not.

Glyphosate was first prepared in 1950, in Switzerland (29), but its potent herbicidal activity was not disclosed until 1970 by Monsanto Chemical Co.

Roundup is a broad-spectrum post-emergence herbicide that is nonselective, though certain crop plants are being genetically engineered to be resistant to the chemical (see GENETIC ENGINEERING, PLANTS). Whereas it is active against most annual and perennial plants, it is environmentally safe because it is nontoxic to mammals, fish, birds, insects, and many bacteria and therefore does not enter the food chain in any harmful way. It can be safely used around woody plants, which are not affected by the material, provided the actively growing vegetative parts are not directly treated. Roundup is bound to the soil where it rapidly undergoes microbial degradation into its component parts, glycine and phosphate. It is extremely useful in orchards to control broadleaved weeds and grasses, and it has been used successfully by homeowners to control weeds in vegetable gardens, provided the crops are screened from the chemical, and as an edging agent to control the borders of lawns. Many homeowners use the material to effectively control honeysuckle (*Lonicera japonica*), poison ivy (*Rhus toxicodendron*), and kudzu vine (*Pueraria thunbergiana*).

The mode of action is by inhibiting 5-enolpyruvyl-shikimate-3-phosphate synthase. Roundup shuts down the production of the aromatic amino acids phenylalanine, tyrosine, and tryptophane (30). Whereas all these amino acids are essential to the survival of the plant, tryptophane is especially important because it is the progenitor for indole-3-acetic acid, or auxin, which plays an important role in growth and development, and controls cell extension and organogenesis.

Polado is the sodium sesqui salt of *N*-phosphonomethyl glycine (**15**). The primary use of the compound is to accelerate ripening of sugarcane, bringing about an increase in the sucrose content in the barrel (internodes). It is especially used on green cane that is slow to mature because of climatic conditions. Nevertheless, timing is critical in applying the material and it must be applied at 4–10 weeks prior to harvest (23). An added advantage of treatment is that many of the green leaves are desiccated. This helps when the fields are fired prior to harvest to remove debris, trash, and snakes. As expected, the chemical also has herbicidal properties and may be used to inhibit seed head development in certain turf species (23). As with the parent material, glyphosate, it is environmentally benign and does not affect wildlife.

Synthetics

There are many synthetic plant growth regulators. Historically, it was common practice to generate a number of synthetics and test them in various biological systems. If a particular structure exhibited activity, it was used as a lead compound and synthetically elaborated in several ways. Much of the information concerning the development of synthetic plant growth regulators remains proprietary except in very successful cases. Many compounds were worked on secretly in the 1950s, but the information was never published. These compounds are presented as new discoveries by scientists working at universities and other institutions who do not have access to this earlier information. Table 2 offers added information on commercially available synthetic plant growth regulators.

Alar B. Butanedioic acid mono(2,2-dimethylhydrazide) [1596-84-5] (Alar B, Daminozide) (16) is water soluble, readily translocated throughout plant tissues, and nonpersistent in soil. In experimental trials, 50% of the compound could not be found in a variety of treated soils after one week and in greenhouse tests up to 90% of the material could not be found in two weeks. Furthermore, it is readily degraded microbially and has low toxicity to wildlife and fish. For example, the LD_{50} for mallard duck is 4.64 g/kg in 8-d trials, whereas the acute LD_{50} for rats is 8.4 g/kg. Even two-year feeding studies with rats and dogs demonstrated that 3.0 g/kg/d had no adverse effect. In the late 1980s residues of the chemical began to appear in apples where it had been used to develop the lobes on apples, especially the Delicious varieties. Because infants and children are given apple juice as part of their diet, it was insinuated that the plant growth regulator was hazardous to health and, subsequently, was voluntarily withdrawn from the U.S. market for use on agricultural crops by the manufacturer. There followed a reduction in use in Europe. However safe a growth regulator may appear from toxicity data, the compound may become a liability if the public perceives the material to be hazardous.

Daminozide still has multiple use outside the United States (Fig. 4) in ornamental crops where it is used to control vegetative growth, reduce plant height, and add to the general vigor of plants by stimulating resistance to stress conditions (23). There are certain metals with which it should not be mixed, including copper, presumably because there exists the possibility of forming a chelate with the carbonyl groups, thereby inactivating the substance. Its mode of action is by inhibiting gibberellin transport, as opposed to gibberellin biosynthesis, and this accounts for the inhibition of the internodes in ornamentals (31,32). In this respect it is unique among plant growth regulators.

Amidochlor. N-[(Acetylamino)methyl]-2-chloro-N-(2,6-diethylphenyl)acetamide [40164-67-8] (Amidochlor) (17) has been registered by the Environmental Protection Agency (Fig. 3) for nonresidential turf use only. It is not phytotoxic if

Table 2. Authorized Synthetic Plant Growth Regulators[a]

Product	Structure[b]	Trade name	LD$_{50}$, g/kg
alar B[c]	(16)	Alar, B-Nine, Daminozide, Dazide	rat, 8.4
amidochlor	(17)	Amidochlor, Limit	rat, 3.1
ancymidol	(18)	Ancymidol, A-Rest, Reducymol, Sleetone	mouse, 5; rat, 4.5
butralin[c]	(19)	Butralin, Tamex	rat, 1.26
chlormequat chloride	(20)	CCC, Cycogan, Hormocel, Chlormequat Chloride, Cycocel, Arotex 5C	rat (male), 31; rat (female), 18
chlorpropham	(21)	Chlorpropham, CIPC, Sprout NIP, Spud-NIC, Tater PIX	rat, 1.2
3-CPA	(22)	Fruitone CPA, 3-CP, Cloprop	rat, 10
4-CPA	(23)	PCPA, Tomato Fix, Sure-Set	rat, 0.85
2,4-dichlorophenoxyacetic acid	(24)	2,4-D, Citrus Fix, Hivol-44	mouse, 0.368; rat, 0.375
dimethipin	(25)	Harvade, Dimethipin "UBI-N252", Dimethipin "N252"	rat, 1.18
dormex	(26)	Dormex, Alzodep	rat, 1.25
etacelasil[c]	(27)	Alsol, CGA 13586	rat, 2.06
ethoxyquin	(28)	Stop Scald, Nix-Scald, Santoquin	rat, 1.92; mouse, 1.73
flumetralin	(29)	Prime+, CGA 41065	rat, 3.1
flurprimidol	(30)	Cutless, Cutless-TP	rat (male), 0.914; rat (female), 0.709
folcysteine[c]	(31)	Ergostim	rat (albino), 4.5
inabenfide[c]	(32)	Seritard, CGR 811	rat, 15
maleic hydrazide	(33)	Maleic hydrazide, Sucker-Stuff, MH-30, Retard, Fair-Plus, Royal MH-30	rat, 4.0
mefluidide	(34)	Embark 2-S	mouse, 1.92; rat, 4.0
mepiquat chloride[c,d]	(35)	PIX, BAS 08300, Ponnax, Terpal (including Ethephon)	rat, 1.42; rat (Terpal), 1.5
merphos	(36)	Folex	rat, 0.348
morphactin[c]	(37)	Morphactin, Maintain-A, Multiprop, Curbiset	rat, 4.0
naphthalene acetic acid	(38)	NAA-800, NAA	rat, 1.0
naphthalene acetamide	(39)	Furitone-N	rat, 6.4
Off-Shoot-O	(40)	Off-Shoot-O	rat, 20.5
paclobutrazol	(41)	Bonzi, Clipper, PP333, Proturf, Bounty Parlay	rat, 1.356
N-(phenylmethyl)-1-H-purine-6-amine	(42)	SD 8339, Proshear, ACCEL	mouse, 0.926; rat, 1.64
N-phenylthalamic acid[c]	(43)	Nevirol	rat, 9.0
sevin	(44)	Sevin, Carbaryl	rat, 0.560

[a]Ref. 1.
[b]See Fig. 3, unless otherwise noted.
[c]Not available in the United States; see Fig. 4.
[d]Terpal product available only outside the United States; other products available in the United States.

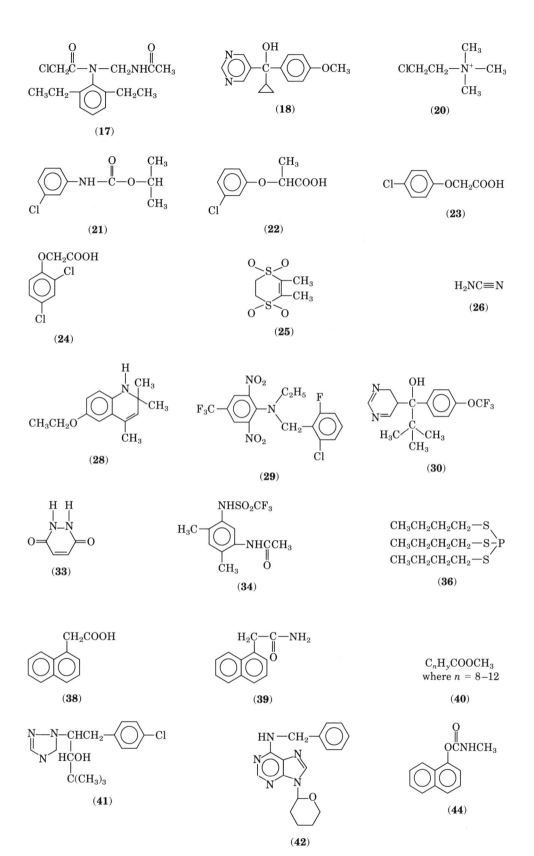

Fig. 3. Synthetic plant growth regulators authorized for use in the United States and elsewhere. See Table 2.

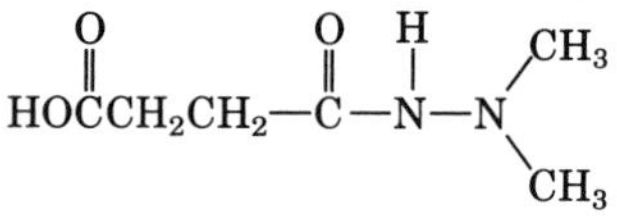

$$\underset{\text{(16)}}{\text{HOCCH}_2\text{CH}_2-\overset{\text{O}}{\overset{\|}{\text{C}}}-\overset{\text{H}}{\overset{|}{\text{N}}}-\text{N}\underset{\text{CH}_3}{\overset{\text{CH}_3}{<}}}$$

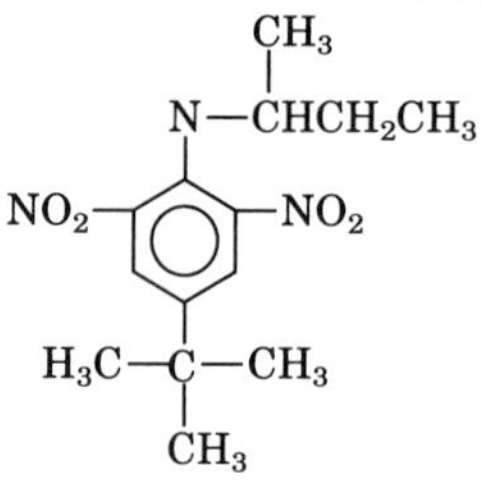

(19)

(CH$_3$OCH$_2$CH$_2$O)$_3$SiCH$_2$CH$_2$Cl

(27)

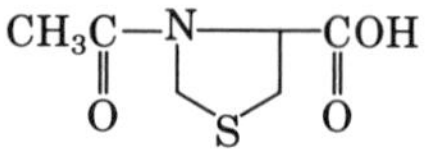

(31)

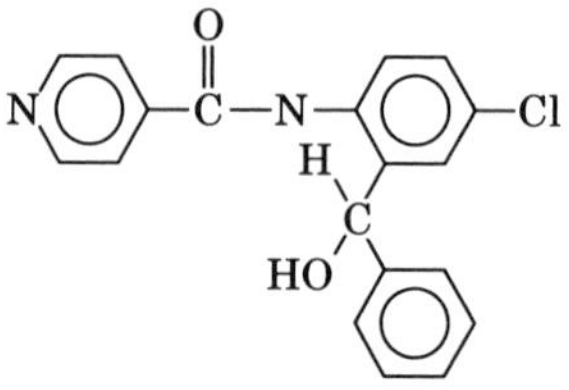

(32)

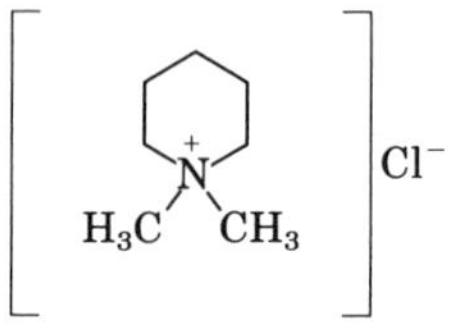

(35)

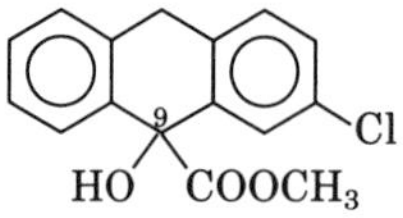

(37)

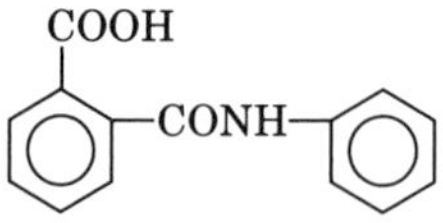

(43)

Fig. 4. Synthetic plant growth regulators available outside the United States. See Table 2.

used according to label instructions and only inhibits the height of grasses, thereby reducing vegetative growth, for approximately six weeks following treatment. The material is absorbed through the roots which means that it first has to reach the soil to be effective; from there it is translocated to the shoot apex. The mode of action appears to be the suppression of mitosis and may include inhibition of cellular differentiation in the meristematic zone (33); it does not inhibit either gibberellin biosynthesis or transport. It does not inhibit the growth of all grasses and it is active only on cool-season turf grasses, including Kentucky bluegrass, tall and fine fescues, and perennial rye grass; in these species vegetative growth may be cut by 50%. Amidochlor does not inhibit the growth of crabgrass, quack grass, zoysia, bermuda, or St. Augustine grass, neither does it control broadleaf or grassy weeds. Thus the antimitotic and antidifferentiation properties are quite selective. The anomalies in which some grasses are controlled whereas others are not is a fertile area for research.

The compound is relatively nontoxic. Because many golf courses and recreational grassy areas abut lakes and ponds that are used for fishing, the environmental toxicity data are important. The LC_{50} for trout (96 h) is 80 mg/L; for bluegill (96 h), 36 mg/L; and for Daphnia, 64 mg/L.

Ancymidol.   α-Cyclopropyl-α-(4-methoxyphenyl)-5-pyrimidinemethanol [*12771-68-5*] (Ancymidol) (**18**) is a heterocyclic nitrogenous compound used to control the height of container-grown ornamental plants such as azaleas (*Azalea* sp.), chrysanthemums (*Chrysanthemum* spp.), dahlias (*Dahlia* sp.), delphiniums (*Delphinium* sp.), lilies (*Lilium* sp.), poinsettias (*Euphorbia pulcherrima*), tulips (*Tulipa* sp.), and bedding plants (10,22,23). The growth arresting properties may be reversed by using gibberellic acid GA_3. The compound acts by specifically inhibiting the oxidative steps from *ent*-kaurene to *ent*-kaurenoic acid (31) where *ent*-kaurene is metabolized to *ent*-kaurenol then to *ent*-kaurenal and, in this sequence, to *ent*-kaurenoic acid. The synthesis of the compound is not difficult.

(**18**)

Butralin.   N-*sec*-Butyl-4-*tert*-butyl-2,6-dinitroanaline [*33629-47-9*] (Butralin, Tamex) (**19**) is an aniline that has a twofold purpose as a plant growth regulator. It is used as a contact agent to control axillary shoot growth in all tobacco types and, in that regard, to inhibit the growth of young shoots, or suckers, as effectively as maleic hydrazide or the long-chain C_8–C_{10} fatty acids. In contrast to the fatty acids, its mode of action is mainly by apical cell inhibition, though it seems not to be translocated throughout the plant. It has also been incorporated as a preplant herbicide in soil (see HERBICIDES). The principal breakdown product is 4-*tert*-butyl-2,6-dinitroaniline, which is the result of soil degradation by soil microflora, especially *Paecilomyces* spp. (34).

Chlormequat Chloride.   2-Chloroethyltrimethylammonium chloride [*999-81-5*] (chlormequat chloride), known as CCC (**20**), is an onium-type plant growth regulator that has variable use depending on where it is used. CCC by itself is registered for use in the United States by the Environmental Protection Agency, but the mixture of CCC with choline chloride has not been approved.

The early development of the compound took place in the 1960s, when the chemical structure and details about the activity of CCC as a plant growth inhibitor were revealed; the effects induced in plants were opposite to those induced by gibberellic acid (35). It was concluded that CCC and gibberellins were not in competition at a common enzymatic site (36). By 1981, CCC had been used experimentally as a growth inhibitor on a number of economic crops and ornamentals. Outside the United States the chemical has been used for a variety of reasons (22). In India, the material has been registered for use to promote fruit bud formation and increase yields in apples and pears; it has been used in grapes to inhibit floral drop, and increase fruit set and yields (22). Another effect in grapes has been to increase leaf disease resistance (22). India also uses the compound to ripen sugarcane; for a short period it was used for the same purpose in Hawaii. At one time India used CCC to reduce sugarcane lodging which, in turn, gave greater yields to improved tillering in ratoons.

On ornamental plants CCC is applied to azaleas, geraniums, and hibiscus (*Hibiscus* sp.) to make compact plants, and to poinsettias to reduce stem height and increase the red color of the bracts. A considerable amount of work has been carried out on cereals with CCC to reduce stem length and inhibit lodging. In Europe, the effect of CCC on shortening the culms of cereals is dependent upon the genotype. It has been demonstrated that the effect is as follows: wheat>triticale>durum wheat>rye>oats>barley>corn = millet = rice (37). In barley, culms are initially inhibited but later the plant overcomes the inhibition (37). This has been attributed to poor assimilation, translocation, and rapid breakdown in wheat (38).

The mode of action of CCC is attributed to the inhibition of *ent*-kaurene synthetase A, the enzyme that drives the biosynthesis of geranylgeranylpyrophosphate by copalyl pyrophosphate to *ent*-kaurene. The compound is registered in Europe to control lodging and is registered with the EPA.

Chlorpropham. (3-Chlorophenyl)carbamic acid 1-methyl ester [*101-21-3*] (Chlorpropham, CJPC) (**21**) was patented in the early 1950s and is a carbamate. Its only use in the United States is on stored Irish potatoes to inhibit bud development. The potatoes, which are generally stored at temperatures >10°C for maximum flavor, are treated by passing a stream of air laced with chloropropham over the potatoes for 48 hours after which the potatoes are purged with pure air.

The compound is not effective when applied as a preharvest dormancy agent on potatoes and it should not be used until injuries to the tuber coat have healed. Neither should it be applied until some suberization has occurred, generally two weeks after storage at >10°C (23). CIPC is widely used outside the United States on a number of crops as a herbicide.

3-CPA. 2-(3-Chlorophenoxy) propanoic acid [*5825-87-6*] (3-CPA, Fruitone CPA) (**22**) has been on the market for a number of years and its use in the United States primarily has been for thinning stonefruit. The material also is used in pineapples to inhibit vegetative crown growth and to increase the size of the fruit, presumably by partitioning minerals and growth factors. 3-CPA is not easily translocated from the foliage and during application it is important to contact the fruit. Application must be made ~15 weeks prior to harvest, or at dry petal fall (10).

4-CPA. A very close relative of 3-CPA is *p*-chlorophenoxyacetic acid [*122-88-3*] (4-CPA) (**23**). It has a limited but essential use as a specific plant growth regulator to improve fruit set in tomatoes; it also can replace pollination (23). The compound was reported as having unique activity in 1949 (39) when it was observed that the material induced the apex of kalanchoe to develop a pseudospathe that could then be excised and rooted. It has also been used to induce slip production in pineapples that have been forced with naphthalene acetic acid (40). 4-CPA is sold as an aerosol spray in either a 2 or 0.005% solution, as the diethanolamine or sodium salt, or it may be purchased as tablets. It is sprayed only on the corollas when they are fully open. The spray should be light, and because tomato plants flower and fruit asynchronously, care should be taken not to spray the fruit or leaves. Because it may replace pollination, 4-CPA should not be used on plants designated for seed production. The seeds do not have the same genetic complement as the normally pollinated germplasm. In addition to treating tomatoes, the compound is used to treat mung bean seed, bean sprouts, prior to germination. The seed is soaked for 5–8 hours in a solution so that 3 g of compound treats 450 kg (1000 lb) of seed. Following treatment, the seed is washed and sown, resulting in bean sprouts having inhibited root growth (23).

The 2-isopropylhydrazide derivative of 4-CPA is iproclozide [3544-35-2], a pharmaceutical that inhibits monoamine oxidase.

2,4-Dichlorophenoxyacetic Acid. The phenoxyacetic acids, specifically 2,4-dichlorophenoxyacetic acid [94-75-7] (2,4-D) (**24**) and 2,4,5-trichlorophenoxyacetic acid [35915-18-5], are the first principal synthetic plant growth regulators extensively used as herbicides. 2,4-D is lethal to all broadleaved weeds but not to grasses, although it has been used to control grasses when used at very high concentration in sugarcane in the West Indies. The first report of the synthesis of 2,4-D was made in 1941 (41) from the model, indole-3-acetic acid (42,43); 2,4-D was a substituted phenyl and IAA was a phenyl-pyrrole. For years, 2,4-D was the chemical paradigm for manufacturers of agricultural chemicals; p-chlorophenoxyacetic acid and 2-(3-chlorophenoxy) propanoic acid, among others, attest to this.

The commercial use of 2,4-D has decreased substantially and (ca 1993) it has general use for home lawns to control broadleaved weeds; it also is used on a limited basis to control broadleaved weeds in commercial moncotyledonous crops, eg, sugarcane. 2,4-D is used on citrus when the fruit is 1/3 to 1 inch in diameter to increase fruit size and to limit fruit drop on trees more than six years old. It should not be applied to trees that are in full flush. A further use includes treatment of harvested lemons at 500 mg/L to improve storage properties and to delay yellowing (23). It is used in certain parts of the world to increase latex flow in old rubber tree plantations.

Dimethipin. 2,3-Dihydro-5,6-dimethyl-1,4-dithiin-1,1,4,4-tetraoxide [55290-64-7] (dimethipin, oxidimethiin, UBI-N252, Harvard) (**25**) is used as a cotton defoliant and has been used as an experimental desiccant in potato vines. In addition, it defoliates nursery stock, grapes, dry beans, and natural rubber and is used as a desiccant for seed of canola, flax (Linum usitatissimum), rice, and sunflower (Helianthus annuus) (10). The product has been available since the mid-1970s and the experimental work was first reported in 1974 (44).

Oxidimethiin acts as a defoliant by eliciting the formation of an abscission layer at the middle lamella, and examination with cotton explants indicates that cellulase activity is increased at the abscission zone.

Dormex. Dormex [156-62-7] (Alzodep, hydrogen cyanamide) (**26**) should not be confused with the same term used for calcium cyanamide. Its other names include carbamide, cyanogenamide, and amidocyanogen. The compound is easily prepared by the continuous carbonation of calcium cyanamide in water (see CYANAMIDES).

The compound is used as a plant growth regulator and has shown remarkable responses in grapevines, sugarcane, and kiwifruit (Actinidia chinensis). On grapevines, the material is sprayed onto the buds prior to their breaking dormancy and causes swift bud break (37). In sugarcane, the setts, or short three-eyed pieces of sugarcane that are planted, may be treated with hydrogen cyanamide; this results in shoot formation approximately two weeks earlier than in controls (37). It has been stated that hydrogen cyanamide may be metabolized to arginine (45).

Dorex is very toxic (see Table 2) and must be handled with extreme care. Because it may produce severe dermatitis on moist skin, it is difficult to use in hot, humid climates; inhalation of the dust or spray may irritate the mucous membranes. Whereas symptoms may include a flushed face, tachycardia, headache, vertigo, and hypotension, it does not produce the typical cyanide effect.

Etacelasil. 2-Chloroethyl-tris (2'-methoxy-ethoxy)-silane [37894-46-5] (Etacelasil, Alsol) (**27**) is made by the reaction of three moles of 2-methoxyethanol with one mole of $Cl_3SiCH_2CH_2Cl$. The name Alsol should not be confused with aluminum acetotartrate, also known as Alsol. Etacelasil is sold in Europe for use as an abscissing agent in olives and it is applied 6–10 days before harvest. It is not sold for use in the United States. It acts by releasing ethylene which affects the abscission layer and allows for easier harvesting when the trees are either mechanically or hand shaken. The parent compound is rapidly degraded in a few hours. It is not readily absorbed by leaves and is not translocated (10).

Ethoxyquin. 6-Ethoxy-1,2-dihydro-2,2,4-trimethylquinoline [91-53-2] (Ethoxyquin) (**28**) has been available since the 1940s. The original structure was described in 1933 (46). The primary use, ie, as an antioxidant in food and feed (see FEEDS AND FEED ADDITIVES) and as an antidegradation compound for rubber, gives no clue as to its plant growth regulatory use. Ethoxyquin's only known use as a plant growth regulator is to control common scald in apples and Anjou pears (23). It is applied as a post-harvest dip or wax emulsion. The material should not be used on late harvested Winesap apples, it should never be used in hydrocooler water, temperatures must be below 27°C, and produce should not remain in the dip over 2–3 minutes. Treated fruit must be allowed to drip before being placed in cold storage, and treated fruit should be kept out of sunlight. Dip water must be completely chlorine free (23).

Flumetralin. 2-Chloro-N-[2,6-dinitro-4-(trifluoromethyl)-phenyl]-N-ethyl-6-fluorbenzenenemethanamine [62924-70-3] (Flumetralin, CGA 41065, Prime +) (**29**) is a dinitroaniline that has singular use in tobacco where it is employed to control axillary shoot formation. It was first prepared in 1977 (47); further reports followed (48,49). Flumetralin is applied at the early elongated button stage of flowering and controls axillary shoot growth if the shoots are shorter than one-inch long. Unlike other contact chemicals it has residual properties. In cases of highly variable tobacco plant growth it is advisable to apply contact axillary shoot inhibitors until there is uniform growth, and then apply Prime +. In order to be effective, the material must contact each sucker. Buds are not burned as they are with contact agents but the shoots may take on a chlorotic appearance.

The mode of action has not yet been elucidated but the manufacturer states that it probably behaves like the herbicide triflurolin and its congeners. These materials inhibit cell division by binding to tubulin thereby interrupting microtubule development. This, in turn, stops spindle fiber formation essential to mitosis and cell division. Experiments with ^{14}C-labeled Prime + show that it is acutely toxic to fish with estimated LC_{50} (96 h) of less than 100 ppb for rainbow trout and bluegill sunfish. However, channel catfish did not exhibit any toxic response at the maximum attainable water concentration (10).

Flurprimidol. α-(1-Methylethyl)-α-4-(trifluoromethoxy)phenyl-5-pyrimidinemethanol [56425-91-3] (Flurprimidol, Cutless) (**30**) is a pyrimidine compound which belongs to the class of nitrogen-containing heterocyclic compounds. It exhibits plant growth inhibiting properties and its main use is to control the growth of turf and ornamentals. The mode of action is by inhibiting the oxidative steps from *ent*-kaurene to *ent*-kaurenoic acid (31). The material is generally sprayed on turf after the second mowing which may take place in late spring or early summer, though application can be made in late summer.

Folcysteine. 3-Acetyl-4-thiazolidinecarboxylic acid [*8064-47-9*] (Folcysteine, Ergostim) (**31**) is used as a biostimulant to increase germination, promote plant growth, enhance fruit set, and increase yields in corn, rice, wheat, potatoes, sugar beets, grapes, strawberries, apples, and several other crops (10). It acts by stimulating the general chemistry of the plant, and making metabolic reserves available during development and organogenesis. Thiol groups are released in the plant which bind to specific substrates. These groups also play an important role in DNA, RNA, and, therefore, protein synthesis, resulting in the regulation of growth and development. The compound is readily synthesized by treating 4-thioazolidine with acetic anhydride.

Inabenfide. [4-Chloro-2-(α-hydroxybenzyl)]-isonicotinanilide) [*82211-24-3*] (Inabenfide) (**32**) is not for use in the United States, but is used in other countries to inhibit the growth of rice plants. The compound is applied to the soil 40–60 days prior to the heading up of plants, where it is absorbed through the roots and translocated throughout the stem. It inhibits the elongation of the lower internodes and this stops lodging. It is extremely toxic to fish.

Maleic Hydrazide. 1,2-Dihydro-3,6-pyridazinedione [*123-33-1*] (maleic hydrazide) (**33**) is one of the earliest organic plant growth regulators to be marketed; it is of approximately the same vintage as 2,4-dichlorophenoxyacetic acid. Maleic hydrazide is easily prepared by the treatment of maleic anhydride with hydrazide hydrate in alcohol. It inhibits the growth and development of plants, though not all plants respond to application. Most effective responses are seen in tobacco where it is used to control axillary shoot growth after the floral apices have been removed, an essential step in tobacco production. Applications to stored potatoes and onions stop shoot growth, and treatment of grass, trees, shrubs, and ivy inhibits growth; not all ivy species respond.

The mode of action has been a subject for research for a number of years. While it was originally thought that maleic hydrazide replaced uracil in the RNA sequence, it has been determined that the molecule may be a pyrimidine or purine analogue and therefore base-pair formation is possible with uracil and thymine and there exists the probability of base-pair formation with adenine; however, if maleic hydrazide occurs in an *in vivo* system as the diketo species, then there remains the possibility of base-pairing with guanine (50). Whatever the mechanism, it is apparent that the inhibitory effects are the result of a shutdown of the *de novo* synthesis of protein.

Mefluidide. *N*-[2,4-Dimethyl-5-[[(trifluoromethyl)sulfonyl]amino]phenyl]-acetamide [*53780-34-0*] (Mefluidide) (**34**) was the outcome of a synthetic program strongly oriented to the manufacture of *N*-aryl-1,1,1-trifluoromethanesulfonamides. The compound has been evaluated under several experimental situations to control seed head formation in grass and to improve protein, sugar, and digestibility. It has also been used to increase sucrose levels in sugarcane and is applied at the time that the cane matures (37). However, this is generally done in specific geographical areas, ie, Brazil, and it should be understood that a large percentage of the Brazilian sugar crop is converted to ethanol for motor fuel (see ALCOHOL FUELS) (51) and pesticide residues are not a problem. For the most part, mefluidide is used to regulate the growth of turf in public and commercial lands. It is a growth inhibitor and should be applied to grass at the height that is aesthetically appealing to the eye. The compound may be applied one day after mowing or the grass may be mowed three to seven days after application (10). Mefluidide inhibits the growth of the meristems and therefore probably inhibits mitosis or cell elongation. It is degraded into numerous metabolites which are biologically insignificant (10). Its life in the soil is short; the half-life is two days. Photo decomposition proceeds rapidly on wet soil surfaces but the element of moisture is critical for this reaction. In cow and sheep experiments, using radio labeled mefluidide, it has been shown that neither milk nor meat are affected (52).

Mepiquat Chloride. 1,1-Dimethylpiperidinium chloride [24307-26-4] (mepiquat chloride) (35) has been in existence since 1973. This quaternary ammonium is remarkably active as a growth regulator in cotton where it inhibits growth, boll set, and yields (10), and leads to early maturation (37). As the genesis of the compound progresses, it is applied as multisplit treatments in opposition to a single application. The material is ideally suited to cotton cultivation and produces many desired effects. These include reduction in vegetative growth, giving rise to compact plants that have healthy, dark green leaves. Rows become easily accessible and air currents circulate unhindered so that invasion by phytopathogens is reduced. Furthermore, there is less abscission of essential plant organs. Several morphological and anatomical studies have shown that the stems and branches are shorter. There is also an increase in the spongy mesophyll with a concomitant elongation of the palisade cells. The final product, cotton fiber, is not affected by treatment (53).

In basic soils mepiquat chloride is nonpersistent and is rapidly degraded. There is a half-life of two years in other soil types, but 86–93% is metabolized within 30 days (10). Breakdown does not appear to be a function of microbial activity.

Mepiquat chloride is mixed with 2-chloroethylphosphonic acid to control lodging of winter barley in Europe under the name Terpal. The plants are treated after the first nodes have become visible and treatments may be applied until the flag leaf develops. Terpal is not used in the United States.

Merphos. 5,5,5,-Tributylphosphorotrithioate [150-50-5] (Merphos, Folex) (36), a trithiophosphite, has limited use as a cotton defoliant. It is one of the older plant growth regulators on the market, making its first appearance in 1958. It may be used as a bottom defoliant, as a preconditioning defoliant, or as a complete defoliant. As a bottom defoliant, it is used as a directed bottom spray to selectively defoliate bottom leaves when the first set bottom bolls are sufficiently mature. This allows the bottom leaves to drop and increases air circulation among the rows so that boll rot is prevented. As a preconditioning agent, the compound is sprayed onto green, rank cotton leaves. This gives some leaf drop and slows the growth of the plant. It should not be used on varieties that have already been chopped-out (23). Application to induce total defoliation should be made when the upper bolls cannot easily be squeezed between the thumb and forefinger, while the boll content is still quite firm. The chemical acts by affecting the abscission zone; this is implied because the leaves are still green at drop.

Morphactin.   Methyl-2-chloro-9-hydroxyfluorene-(9)-carboxylate [2536-31-4] (**37**) was developed in Germany from 1960–1963 with many other homologues to control plant growth. All were fluorene derivatives; the parent fluorene consisted of the A, B, C rings only and substituents were synthesized on to the B ring at the C-8 position. One of the functional groups was a carboxylic acid making a convenient site for ester production (54). The morphactins were regarded as a novel group of structures that interfered with morphogenesis (55).

Morphactin is used (ca 1993) to control the growth of weeds along roadsides; to inhibit the growth of grass when mixed with maleic hydrazide; to control tree height, shrubs, and vines (23); and to make bean plants bushier. Past use has included pineapple where it improved the propagation of slips. It also has improved the size and yield of cucumbers (23). In early testing, morphactin induced odd effects when applied to lettuce and timothy grass (*Phleum pratense*). In both species the roots, hypocotyls, and coleoptiles, respectively, grew in all directions and exhibited negative geotropism (56). The mode of action appears to be the blocking of auxin transport (37). This compound is not used in the United States.

Naphthalene Acetic Acid and Naphthalene Acetamide.   Naphthalene acetic acid [26445-01-2] (**38**) is historically one of the first plant growth regulators. Reports concerning its activity in crops and plants have been a subject in much of the early literature (57). Consequently, it has been used as a starting material for other compounds, eg, *vide infra*, Sevin. Naphthaleneacetamide [31093-43-3] (**39**) has been used as a standard material to evaluate abscission prior to 1953 and its effect on apple drop was reported in 1953 (58). The substance is used as an internal standard in the abscission bioassay (59).

Naphthalene acetic acid is used to thin apple and pear blossoms and to control apple and pear preharvest drop (10). It also is used to induce flowering in pineapple, but conversely inhibits sprouting in potatoes, sweet potatoes, and turnips (*Brassica rapa*) (23). It also is used to promote rooting in cuttings and may substitute for indole-3-butyric acid in this regard. It is considered to be an auxin, though a synthetic one. Naphthalene acetamide has virtually the same properties as naphthalene acetic acid and is used for precisely the same crops.

Off-Shoot-O.   The methyl esters of the C_8–C_{12} fatty acids (**40**) are collectively sold under the name Off-Shoot-O and are closely related to 1-decanol, the fatty alcohol sold to control axillary shoots in tobacco. The material is a contact-type chemical used to pinch ornamental plants such as azaleas, cotoneaster, juniper (*Juniperus* sp.), privet, rhamnus, and taxus (*Taxus* sp.). As a result of treatment the shrubs become bushier. The mode of action is by plasmolysis of the young, sensitive tissues. Therefore, application timing may be critical.

Paclobutrazol.   1-(4-Chlorophenol)-4,4-dimethyl-2-(1*H*-1,2,4-triazol-1-yl)-pentan-3-ol [76738-62-0] (Paclobutrazol) (**41**) is a triazole plant growth regulator that has received a good deal of attention as a systemic growth inhibitor. Whereas it may be applied at any time to trees, injection through the trunk is recommended when they are transpiring. Trees that have been treated successfully include Norway maple, red maple, silver maple, sugar maple, Australian pine, sweet gum, laurel oak, water oak, willow oak, live oak, Chinese elm, and Siberian elm. Sugar maples should not be treated within one year of tapping for syrup; nut and fruit trees should not be harvested for fruit consumption until a year after application (10). The compound may be used for ornamentals including chrysanthemums, freesias (*Freesia* sp.), geraniums, and poinsettias. Outside the United States, it has been used to reduce lodging and to increase tillering and yields in rice. The main use in fruit trees is to halt vegetative growth and to promote spur growth (23). Its mode of action is inhibition of gibberellin biosynthesis (31). Translocation occurs in the xylem, but not in the phloem. Paclobutrazol is formulated in methanol; the latter cannot be made nonpoisonous, hence, it should be kept away from water and mammals.

N-(Phenylmethyl)-1,H-Purine-6-Amine. *N*-(Phenylmethyl)-1,*H*-purine-6-amine [1214-39-7] once had the trademark ACCEL, which should not be confused with the material registered as Accel. The former is the purine plant growth regulator (**42**); the latter is a lactic acid starter consisting of *Pediococcus cerevisia* for the fermentation of summer sausage. Except for older literature, the name ACCEL has been replaced by Proshear. The only use (ca 1994) for Proshear is as a plant growth regulator in white pine where it is applied two weeks before or after shearing to shape the trees (23). Formerly, it was used to enhance the number of lateral buds in carnations, chrysanthemums, and roses (10). It is considered to be a synthetic cytokinin that stimulates dormant lateral buds.

N-Phenylthalamic Acid. A product of Hungary, *N*-phenylthalamic acid [4727-29-1] (Nevirol) (**43**), is a benzoic acid derivative not sold in the United States. It is used to increase pollination and results in setting more fruit when weather conditions are unfavorable for normal fertilization. It is employed in both greenhouses and fields on apples, beans, cherries, lupine, peas, peppers, soybeans, and sunflower (23).

Sevin. 1-Naphthalenol methylcarbanate [63-25-2] (Sevin) (**44**) was developed as an insecticide. However, the conception of the molecule, in the mid-1950s, was as a possible herbicide. The compound ultimately was useless as a herbicide, but in routine testing it was discovered to be an excellent insecticide. Sevin was active in the oat mesocotyl assay and demonstrated weak auxin-like activity. During the development of Sevin, it caused massive apple drop in the western United States in an orchard being treated for insects. It is used (ca 1993) as an abscising agent to thin apples.

Experimental Synthetic Plant Growth Regulators. *AC 94377.* The compound 1-(3-chlorophthalimide) cyclohexanecarboxamide [51971-67-6] (Table 3) (**45**), also known as AC 94377, is a phthalamide that has plant growth regulating properties. It is relatively nontoxic so that its use in floricultural crops appears to be safe. It is mainly used to control stem length and stem numbers in hybrid tea roses (23). However, it has very limited use and is still experimental.

BAS 111. 1-Phenoxy-3-(1*H*-1,2,4-triazole-1-yl)-4-hydroxy-5,5-dimethylhexane [9003-11-6] (BAS 111) (**46**) is a triazole that has plant growth inhibiting properties. It exerts its influence by inhibiting the production of gibberellic acid in plants; this has been demonstrated in canola (31,37).

Other experiments with *Gibberella fujikuroi*, the fungus that produces gibberellin, indicate that GA_3 production is blocked by BAS 111. Very detailed and careful experiments conducted with enzymes in cell-free systems strongly support this mode of action, ie, using *ent*-kaurene oxidase and cinnamate 4-monooxygenase isolated from pea apices and soybean suspension cells, and flavanone-2-hydroxylase and dihydroxypterocarpane 6-hydroxylase from soybean suspension cells (31).

Table 3. Experimental Synthetic Plant Growth Regulators[a]

Product	Structure[b]	Trade name	LD_{50}, g/kg
AC 94377	(45)	AC 94377	mouse, 5
BAS 111	(46)	BAS 111	rat, 5
N-2-chloro-4-pyridinyl-N-phenylurea	(47)	CPPU, CN-11-3183	rat, 4.918
chlorthal-methyl	(48)	Razor, DCPA, Dacthal Rid	rat, 3
cimectacarb	(49)	Cimectacarb, CGA 163935	rat, 4.46
ethylchlozate[c]	(50)	Figaron	rat, 4.8
HOE 074 784	(51)	HOE 074 784	
lactidichlor-ethyl	(52)	PPG 1721, AC 310449	rat, 0.8

[a]Ref. 1.
[b]See Fig. 5.
[c]Experimental in the United States; approved for use in areas outside the United States.

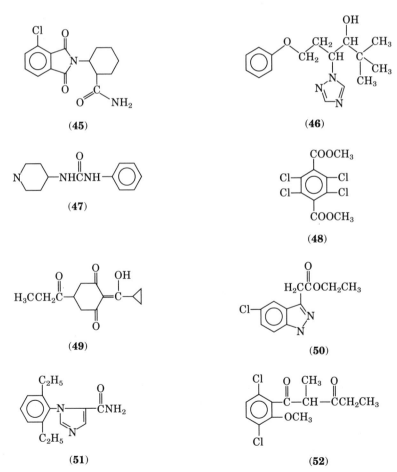

Fig. 5. Experimental synthetic plant growth regulators. See Table 3.

Based on these fundamental pieces of information, the material is used on ornamentals to reduce height, producing compact plants. Additional side effects are improved bud set and color. Experimentally, the compound has been used to dwarf fruit trees which has led to increasing the aveal planting number.

N-2-Chloro-4-pyridinyl-N-phenylurea.   CPPU [*68157-60-8*] (**47**) is a phenylurea plant growth regulator that has been used experimentally to increase fruit size in apples, grapes, kiwifruit, macadamia nuts (*Macadamia ternifolia*), cotton, small grains, and ornamentals (23).

Chlorthal-methyl.   2,3,5,6-Tetrachloro-1,4-benzenedicarboxylic acid dimethyl ester [*1861-32-1*] (chlorthal-methyl) (**48**) belongs to the class of chlorinated benzene compounds. It has very limited use on an experimental basis as an axillary shoot growth inhibitor in tobacco. Like many sucker control agents the compound is probably a contact one with its mode of action being most probably by plasmolysis. Each axillary shoot must be contacted and wetted by the spray solution in order to be killed (23). The compound, an old one, was originally called Dacthal, Rid, and DCPA. It was first developed and patented as a pre-emergence herbicide (60).

Cimectacarb.   4-(Cyclopropyl-α-hydroxymethylene)-3,5-dioxocyclohexanecarboxylic acid ethylester [*95266-40-3*] (Cimectacarb, CGA 163935) (**49**) belongs to the family of cychohexanetrione plant growth regulators that have growth inhibiting properties. In addition, the compound is used as the ethyl ester congener. Its primary use is as a foliar inhibitor of both cool and warm season turf with application during periods of active growth (10). It also is used experimentally as an antilodging compound in cereals and canola (23). The mode of action of cimectacarb has been examined in detail and its inhibitory action is due to its influence on GA biosynthesis beyond the GA_{12}-aldehyde step (31). One explanation is that the cyclohexatriones bear a similar structural relation to 2-oxoglutarate, the essential cosubstrate in GA biosynthesis beyond the GA_{12}-aldehyde step which, in addition, requires soluble dioxygenases as the catalyst (21). Further experiments using the model compound calcium 3,5-dioxo-4-propionylcyclohexane carboxylate, also known as prohexadione calcium, have shown that the effects of the cyclohexane-triones may be reversed by treatment with various GAs (61,62).

Ethylchlozate.   Ethyl-5-chloro-1*H*-3-indazolyl-3-ylacetate [*27512-72-7*] (Ethylchlozale, Figaron) (**50**) is an analogue of indole-3-acetic acid used in Japan to enhance the color of citrus. It is also being used experimentally to promote sugars in grapes, pineapples, and sugarcane. Additionally, it is used as a thinning agent in apples, peaches, and pears and may be used to increase maturation in apples and pears by one to two weeks. It also increases yields and promotes protein content in soybeans and wheat (23). Its mode of action is attributed to the release of ethylene. The substance is extremely toxic to fish and it is not available in the United States.

HOE 074 784.   [1-(2,6-Diethylphenyl)-imidazole-5-carboxamide] (HOE 074 784) (**51**) is an experimental imidazole analogue that has plant growth inhibiting activity; it is used to prevent lodging in cereal grains including barley, oats, rice, and wheat, and it is also effective in reducing height in canola and ornamentals. The compound is readily translocated throughout the plant and acts by inhibiting gibberellin biosynthesis (23).

Lactidichlor-ethyl.   Benzoic acid-3,6-dichloro-2-methoxy,2-ethoxy-1-methyl-2-oxoethyl ester [*87214-73-1*] (PPG-1721) (**52**) is a bichlorinated benzoic acid developed for use in horticultural crops. Benzoic acid and its derivatives have been used for decades as food preservatives and its antimicrobial properties are well documented. Its occurrence in nature is in both the free and bound form with berries containing ~0.05% and gum benzoin, ~20%. In combination with salicylic acid it has been used as a topical antifungal agent (60). It has also been combined with 2,4-D to make Sesin, ie, benzoic acid 2,4-dichlorophenoxy ethyl ester [*94-83-7*]. The chemical structure and growth activity of substituted benzoic acids has been a topic for research since the early 1940s (63), eg, 2,3,6-trichlorobenzoic acid has been found to be highly active in promoting cell elongation (63). It has been suggested that the nucleophilic nature of the substrate with which the substituted benzoic acids react has been demonstrated by the theoretical chemistry of molecular orbitals (63–65). Consequently, substituted benzoic acids have been assiduously worked as biologically active materials that possess auxin-like properties.

   Lactidichlor-ethyl has been used to increase yields, reduce fruit drop in apples, and increase maturity in grapes. It causes fruit drop in peaches which decreases actual number of peaches per tree while increasing yields due to larger fruit size. The mode of action is attributed to inhibition of ethylene production; this arrests development of the abscission layer. The compound is still undergoing experimental evaluation.

BIBLIOGRAPHY

"Plant Growth Substances" in *ECT* 1st ed., Vol. 10, pp. 723–736, by P. W. Zimmerman, Boyce Thompson Institute for Plant Research, Inc.; in *ECT* 2nd ed., Vol. 15, pp. 675–688, by H. M. Cathey, U.S. Dept. of Agriculture; "Plant-Growth Substances" in *ECT* 3rd ed., Vol. 18, pp. 1–23, by L. G. Nickell, Velsicol Chemical Corp.

1. *Merck Index*, 11th ed., Merck & Co., Rahway, N.J., 1989.
2. S. Marumo, *"The Advance of Brassinosteroid Researches in Japan,"* in the *Proceedings of the 14th Annual Meeting of the Plant Growth Society of America*, 1987.
3. S. Marumo and co-workers, *Agric. Biol. Chem* **32**, 528 (1968).
4. G. L. Steffens, *U.S. Department of Agriculture Brassins Project: 1970–1980*, ACS Symposium Series 474, Washington, D.C., 1991.
5. V. A. Khripach, V. N. Zhabinskii, and R. P. Litvinovskaya, *Synthesis and Some Practical Aspects of Brassinosteroids*, in Ref. 3.
6. N. Ikekawa and Y.-J. Zhao, *Application of 24-Epibrassinolide in Agriculture*, in Ref. 3.
7. N. Ikekawa, unpublished results, 1991.
8. C. O. Miller, F. Skoog, M. H. Von Saltza, and F. M. Strong, *J. Am. Chem. Soc.* **77**, 1392, (1957).
9. J. R. Jablonski and F. Skoog, *Physiol. Plantarum* **7**, 16 (1954).
10. H. G. Cutler and B. A. Schneider, *Plant Growth Regulator Handbook of The Plant Growth Regulator Society of America*, 3rd ed., Boyce Thompson Institute, New York, 1990.
11. K. R. Brain and co-workers, *Plant Sci. News.* **1**, 241 (1973).
12. E. F. Button and C. F. Noyes, *Agron. J.* **56**, 44 (1964).
13. P. F. Bocion and W. H. deSilva, *Nature* **258**, 142 (1975).

14. V.-L. Ilag, *Ethylene Production by Fungi*, Ph.D. dissertation, Purdue University, Lafayette, Ind., 1970, p. 93.

15. G. Pfutzer and H. Losch. *Umschau* **11**, 202 (1935).

16. La Rotunda, C. V. Rossi, and G. Petrosini, *Zeitschr. f. Untersuchung der Lebensmittel* **85**, 64 (1943).

17. F. T. Addicott, *Plant Physiol.* **43**, 1471 (1968).

18. M. I. Kabachnik and P. A. Rossiskaya, *Izvest. Akad. Nauk. S.S.S.R., O. ph.n*, 405 (1946).

19. E. Kurosawa, *J. Nat. Hist. Soc. Formosa* **16**, 213 (1926); Engl. transl. in F. H. Stodola, ed., *Source Book of Gibberellin 1828–1957*, ARS (ARS-71-11), USDA, 1958.

20. R. E. McDonald and co-workers, *Proceedings of the 6th International Citrus Congress*, Tel-Aviv, Israel, 1988.

21. P. D. Greany, R. E. McDonald, W. J. Schroeder, and P. E. Shaw, *Florida Entomol.* **74**, 570 (1991).

22. G. W. Elson, *The Gibberellins*, ICI, U.K.

23. W. T. Thompson, *Agricultural Chemicals, Book III—Miscellaneous Agricultural Chemicals, 1991–92 Revision*, Thompson Publications, Fresno, Calif., 1991.

24. C. R. Unrath, *J. Am. Soc. Hort. Sci.* **99**, 381 (1974).

25. M. W. Williams and E. A. Stahly, *J. Am. Hort. Sci.* **94**, 17 (1969).

26. G. Haberlandt, *Sitz Ber. K. Preuss. Akad. Wiss*, 318 (1913).

27. M. J. Bukovac and S. Nakagawa, *J. Am. Soc. Hort. Sci.*, **3**, 172 (1968).

28. W. J. Robbins, "The Expanding Concepts of Plant Growth Regulation," in *Plant Growth Regulation*, Iowa State University Press, Iowa, 1961.

29. A. Bader, *Aldrichemica Acta*, **21**, 15 (1988).

30. S. O. Duke, *Environ. Health Perspec.* **87**, 263 (1990).

31. W. Rademacher, "Biochemical Effects of Plant Growth Retardants," in *Plant Biochemical Regulators*, Marcel Dekker Inc, New York, 1991.

32. J. A. Riddell, H. A. Hageman, C. M. J. Anthony, and W. L. Hubbard, *Science* **136**, 391 (1962).

33. *Questions and Answers about Limit*,® Monsanto Agricultural Co., 1986.

34. P. C. Kearney and co-workers, *J. Agr. Food Chem.* **22**, 856 (1974).

35. N. E. Tolbert, *J. Biol. Chem.* **235**, 475 (1960).

36. B. Balden and A. Lang, *Am. J. Bot.* **52**, 408 (1965).

37. P. E. Schott and H. Walter, "Bioregulators: Present and Future Fields of Application," in Ref. 30.

38. J. Jung, H. Sturm, and W. Zwick, *Landwirtsch. Forsch. SD* **24**, 46 (1973).

39. P. W. Zimmerman and A. E. Hitchcock, *Contr. Boyce Thompson Inst.* **15**, 421 (1949).

40. D. P. Gowing, "Some Comments on Growth Regulators with a Potential in Agriculture," in Ref. 27.

41. R. Pokony, *J. Am. Chem. Soc.* **63**, 1768 (1941).

42. H. G. Cutler, *Natural Products and Their Potential in Agriculture*, ACS Symposium Series 380, Washington, D.C., 1988.

43. P. W. Zimmerman and A. E. Hitchcock, *Contr. Boyce Thompson Inst.* **12**, 321 (1942).

44. R. B. Ames, A. D. Brewer, and W. S. McIntire, "N-252-A New Harvest-Aid Chemical," *Proceedings of the Beltwide Cotton Production Research Conference*, 1974, p. 61.

45. *Plant Growth Regulator Dormex*, Technical Data Sheet, SKW, Trostberg, Germany, 1986.

46. Cliffe, *J. Chem Soc.*, 1327 (1933)

47. M. Wilcox and co-workers, *Proc. Plant Growth Reg. Work. Gr.* **4**, 194 (1977).

48. P. C. Kennedy and co-workers, *Proc. Plant Growth Reg. Work. Gr.* **5**, 172 (1978).

49. P. C. Kennedy, *Proc. Plant Growth Reg. Work. Gr.* **6**, 102 (1979).

50. P. D. Cradwick, *Nature* **258**, 774 (1975).
51. J. Goldemberg, *O Alcool Brasileira e a Gasoline Importada, uma Batalha Decisiva*, Vol. 1, ICARO, Brazil, 1983.
52. G. W. Ivie, *J. Agric. Food Chem.* **28**, 1286 (1980).
53. P. W. Schott and co-workers, *Melliand Textilber. Heidelberg SD* **64**, 380 (1983).
54. *Technical Informations for Crop Protection*, E. Merck, Darmstadt, Germany, 1964.
55. G. Schneider and co-workers, *Nature* **20**, 1013 (1965).
56. A. A. Khan, *Farm* **32**, 2 (1966).
57. A. C. Leopold, *Auxins and Plant Growth*, University of California Press, Berkeley, 1955.
58. L. J. Edgerton and M. B. Hoffman, *Am. Soc. Hort. Sci. Proc.* **62**, 159 (1953).
59. J. W. Mitchell and G. A. Livingston, *Methods of Studying Plant Hormones and Growth-Regulating Substances*, Agriculture Handbook No. 336, USDA, Washington, D.C., 1968.
60. *The Merck Index*, 10th ed., indent 2821, Merck & Co, Inc., Rahway, N.J., 1983.
61. I. Nakayama and co-workers, *Plant Cell Physiol.* **31**, 195 (1990).
62. W. Rademacher, K. E. Temple-Smith, D. L. Griggs, and P. Hedden, *Plant Physist.* **93**, abstr. 16 (1990).
63. R. M. Muir and C. Hansch, "Chemical Structure and Growth-Activity of Substituted Benzoic Acids," in Ref. 27.
64. C. Hansch, R. M. Muir, and R. L. Metzenberg, Jr., *Plant Physiol.* **26**, 812 (1951).
65. R. M. Muir and C. Hansch, *Plant Physiol* **26**, 369 (1951).

General References

L. G. Nickell, *Plant Growth Regulators*, Springer-Verlag, New York 1982, p. 173.
R. M. Sacher, "Strategies to Discover Plant Growth Regulators for Agronomic Crops," in *Chemical Manipulation of Crop Growth and Development*, 1982, p. 167.

HORACE G. CUTLER
USDA, ARS

GUAIAC. See RESINS, NATURAL.

GUAIACOL. See ETHERS.

GUMS

Over the years, the term gums has been used to denote a wide range of compounds including polysaccharides, terpenes, proteins, and synthetic polymers. In the 1990s, the term more specifically denotes a group of industrially useful polysaccharides or their derivatives that hydrate in hot or cold water to form viscous solutions, dispersions, or gels (1).

Gums are classified as natural or modified. Natural gums include seaweed extracts, plant exudates, gums from seed or root, and gums obtained by microbial fermentation (Table 1). Modified (semisynthetic) gums include cellulose and starch derivatives and certain synthetic gums such as low methoxyl pectin, propylene glycol alginate, and carboxymethyl and hydroxypropyl guar gum (see CELLULOSE; STARCH).

Gums are used in industry because their aqueous solutions or dispersions possess suspending and stabilizing properties. In addition, gums may produce gels or act as emulsifiers, adhesives, flocculants, binders, film formers, lubricants, or friction reducers, depending on the shape and chemical nature of the particular gum (2). Considerable research has been carried out to relate the structure and shape (conformation) of some gums to their solution properties (3,4).

Table 1. Classification of Gums

Algal	Botanical	Microbial
agar	*seed gums*	dextran
algin	guar gum	xanthan gum
carrageenan	locust bean gum	gellan gum
	plant exudates/extracts	welan gum
	gum arabic	rhamsan gum
	gum ghatti	
	gum tragacanth	
	karaya gum	
	pectin	

Economic Aspects

Gums fall into a category of specialty chemicals called thickeners and stabilizers. This market is dominated by starch, starch derivatives, and cellulosics (Table 2). Although the gums only represent approximately 5% of the sales by weight, they represent approximately 25% in dollars.

Production, application, and value of many industrial gums are not publicized because many of the uses are maintained as trade secrets. However, estimates of gum production, consumption, and cost may be obtained by reference to the U.S. Department of Commerce figures, market research reports, and trade journals. Estimates for the sales and prices of the principal gums are shown in Table 3.

Table 2. U.S. Market for Thickeners and Stabilizers[a]

Specialty chemicals	10^3 t	10^6 \$
starches	805	410
cellulosics	61	220
guar derivatives	16	53
gums	52	255
Total	*934*	*938*

[a]Ref. 5.

Table 3. Estimates of Markets for Gums[a]

Gum	Sales, 10^3 t		Price, \$/kg
	U.S.	World	
Thickeners			
gum arabic	4.5	18	5.5–7.7
gum tragacanth	0.23		33–88
guar	36	73	0.88–1.32
locust bean gum	2.3		07.7–10.5
xanthan	9.1	23	12.4–15.4[b]
			9.9[c]
Gelling agents			
agar	0.45		28.6–33
alginate	3.2	18	13.2–15.4
carrageenan	3.2–5		8.8–17.6[b]
pectin	2.3–3.2	16–20.5	11–15.4

[a]Ref. 6.
[b]Food.
[c]Industrial.

Algal (Seaweed) Gums

Agar. This gum is extracted from certain marine algae belonging to the class *Rhodophyceae*, red seaweed, which abound off the coasts of Japan, Mexico, Portugal, and Denmark. Important species include *Gelidium cartilagineum* and *Gracilaria confervoides*.

Structure. When the structure of agar [9002-18-0] was determined (7), the fraction with the greatest gelling ability was separated and named agarose (8). The other fractions were named agaropectin. Agarose is an alternating copolymer of 3-linked β-D-galactopyranose and 4-linked 3,6-anhydro-α-L-galactopyranose units (9). Agaropectin has essentially the same structure except that varying amounts of the units in the copolymer are replaced by 4,6-O-(1-carboxyethylidene)-D-galactopyranose or by sulfated or methylated sugar residues. The replacement occurs in such a manner that the alternating sequence of 3-linked β-D-units and 4-linked α-L-units is maintained. Originally the terms agarose and agaropectin have been used to describe the less and more ionic fractions of agar, respectively. However, recent research has indicated that agar contains a spectrum of molecules that have a similar but continuously varying chemical structure (9). The structure of agarose, as defined above, is one example, and it is apparent that the terms agarose and agaropectin represent an oversimplification of the structure of agar.

Agarose, the gelling portion of agar, has a double helical structure. Aggregation of double helices to form a three-dimensional framework is the proposed method of gel formation (3,10,11). The result is a thermally reversible gel that exhibits a large degree of hysteresis in the melting and gelling temperatures.

Properties.   Limited data are available for the types or grades of commercial agar which is usually in the form of chopped shreds, sheets, flakes, granules, or powder. The official specifications for agar are provided in the USP (12) and the *Food Chemicals Codex* (13).

Agar is insoluble in cold water, but is soluble in boiling water. On cooling to about 35°C, a firm gel forms that does not melt or liquefy below about 85°C. These setting and melting temperatures are characteristic for agar. Gels formed at agar concentrations greater than 0.5% are rigid, but gelation can take place at concentrations as low as 0.04%.

Sources and Processing.   Harvesting or collection of red seaweed is carried out by hand and is labor intensive. In some areas of the world this is accomplished by divers. In other places the seaweed can be collected at low tide by wading or from small boats. After collection, the seaweed is dried and bleached in the sun prior to baling.

The details of the commercial processes are proprietary. The agar can be extracted from the seaweed with hot water, followed by freezing and thawing for purification. Commercial extraction procedures involve washing, chemical extraction, filtration, gelation, freezing, bleaching, washing, drying, and milling.

Uses.   Agar is used primarily for its gelling properties based on its unique gel setting and melting temperatures and the heat resistance of the gels. It is also used for its emulsifying and stabilizing properties. Agar is nondigestible, but because of its colloidal and gelling properties, it has been used as a stabilizer in numerous food products, such as pie fillings, icings, toppings, and meringues (14). However, because of its high cost, agar has been replaced by other gums in most applications except in icings and bacteriological agar. At a use level of 0.2–0.5%, it can reduce the drying time and prevent the sugar from adhering to the wrapper. It increases the adherence of doughnut glazes, reducing setting time and producing a more flexible and stable glaze. Also a synergism with locust bean gum allows for a considerably lower agar use level.

Agar, which is low in metabolizable or inhibitory substances, debris, and thermoduric spores, is ideal for the propagation and pure culture of yeasts, molds, and bacteria. Agar also meets the other requirements of ready solubility, good gel firmness and clarity, and a gelation temperature of 35–40°C and a gel melting temperature of 75–85°C. A clarified and purified form of the bacterial polysaccharide, gellan gum, is the only known satisfactory substitute.

Algin.   Algin [*19005-40-7*], the generic description for salts of alginic acid [*9005-32-7*], was discovered about 1880 in the United Kingdom (15). Commercial production started in 1929 in California. The development of propylene glycol alginate [*9005-37-2*] in 1944 increased the utility of algin in both food and general industrial applications.

Algin occurs in all members of the class *Phaeophyceae*, brown seaweed, as a structural component of the cell walls in the form of the insoluble mixed calcium, magnesium, sodium, and potassium salt of alginic acid.

Structure. Alginate is a linear copolymer composed of two monomeric units, D-mannuronic acid [*1986-14-7*] and L-guluronic acid [*1986-15-8*]. These monomers occur in the alginate molecule as regions made up exclusively of one unit or the other, referred to as M-blocks or G-blocks or as regions in which the monomers approximate an alternating sequence. The calcium reactivity and gelling properties of alginates are a consequence of the particular molecular geometries of each of these regions (11).

Because of the particular shapes of the monomers and their modes of linkage in the polymer, the geometries of the G-block regions, M-block regions, and alternating regions are substantially different. Specifically, the G-blocks are buckled whereas the M-blocks have a shape referred to as an extended ribbon. If two G-block regions are aligned side by side, a diamond-shaped hole results. This hole has dimensions that are ideal for the cooperative binding of calcium ions. When calcium ions are added to a sodium alginate solution, such an alignment of the G-blocks occurs and the calcium ions are bound between the two chains like eggs in an egg box. Thus the reactivity of algins is the result of calcium-induced dimeric association of the G-block regions (16). The detailed chemical structure has been determined by nmr spectroscopy (17,18).

The block structure of alginates extracted from seaweeds species of commercial importance is shown in Table 4. These data can be used to predict the observed gelling characteristics of alginates from different sources. For example, the alginate from *Laminaria hyperborea*, with a large percentage of polyguluronate segments, forms rigid, brittle gels which tend to undergo syneresis, or loss of bound water. In contrast, alginate from *Macrocystis pyrifera* or *Ascophyllum*

Table 4. Composition[a] of Alginates from Different Seaweeds[b]

Source	M:G	M, %	G, %	MM, %	GG, %	MG, %	Sodium alginate content, % dry weed
Laminaria hyperborea							
stem	0.45	30	70	18	58	24	25–27
leaf	1.20	55	45	36	26	38	15–25
Laminaria digitata	1.20	55	45	39	29	32	20–26
Ecklonia maxima	1.20	55	45	38	28	34	40
Macrocystis pyrifera	1.50	60	40	40	20	40	26
Lessonia nigrescens	1.50	60	40	43	23	34	35
Ascophyllum nodosum	1.85	65	35	56	26	18	26–28
Durvillea potatorum	3.35	77	23	69	15	16	53

[a]M = D-mannuronic acid; G = L-guluronic acid; MM and GG indicate M- and G-block regions; MG indicates alternating regions.
[b]Ref. 19.

nodosum forms elastic gels that can be deformed and can have a markedly reduced tendency toward syneresis.

Properties. Alginates available for industrial use include the sodium alginate [*9005-38-3*], potassium alginate [*9005-36-1*], ammonium alginate [*9005-34-9*], mixed ammonium–calcium alginate [*9005-31-6*], and mixed sodium–calcium alginate [*12698-40-7*] salts of alginic acid and propylene glycol alginate. These water-soluble alginates are produced in many forms varying in molecular weight, calcium content, particle form (granular or fibrous), particle size distribution, and mannuronic-to-guluronic acid ratio. The propylene glycol ester can also vary in the degree of esterification.

The ability of alginates to form edible gels by reaction with calcium salts is an important property. Calcium sources are usually calcium carbonate, sulfate, chloride, phosphate, or tartrate (20). The rate of gel formation as well as the quality and texture of the resultant gel can be controlled by the solubility and availability of the calcium source.

In practice, alginate gels are obtained using three principal methods, namely, diffusion setting, internal setting, or setting by cooling (21). Diffusion setting is the simplest technique and, as the term implies, the gel is set by allowing calcium ions to diffuse into an alginate solution. Since the diffusion process is slow, this approach can only be effectively utilized to set thin strips of material (eg, pimiento strips), or to provide a thin gelled coating on the surface of a food product such as an onion ring.

In internal or bulk setting, which is normally carried out at room temperature, the calcium is released under controlled conditions from within the system. This method led to the development of structured fruits, structured pet foods, and a host of cold prepared desserts. Calcium sulfate (usually as the dihydrate) and dicalcium phosphate (calcium hydrogen orthophosphate) are the sources of calcium most commonly used.

The third method of preparing alginate gels involves dissolving the gelling ingredients, alginate, calcium salt, acid, and a sequestrant, in hot water and allowing the solution to set by cooling. The calcium salts used in this system are the same as those already mentioned for internal setting. Although the calcium ions required for the setting reaction are already in solution with the alginate, setting does not occur at elevated temperatures because the alginate chains have too much thermal energy to permit alignment. It is only when the solution is cooled that the calcium-induced interchain associations can occur.

Commercially available alginates dissolve in hot or cold water to produce solutions with viscosities ranging from a few to several hundred mPa·s($=$cP). The actual viscosity depends on the molecular weight and calcium content of the alginate.

The viscosity of alginate solutions decreases with increasing temperature, but provided the temperature is not maintained at high levels for extended periods, the viscosity decrease is reversible. Partial depolymerization of the alginate occurs if solutions are exposed to excessive temperatures or to sufficiently elevated temperatures for extended periods.

The viscosity of alginate solutions is independent of pH in the range 5–10, but below pH 4.5, the viscosity increases until the pH reaches 3 when insoluble alginic acid precipitates. Propylene glycol alginate is soluble and stable at pH 2–3 but not above 6.5.

The viscosity of sodium alginate solutions is slightly depressed by the addition of monovalent salts. As is frequently the case with polyelectrolytes, the polymer in solution contracts as the ionic strength of the solution is increased. The maximum viscosity effect is obtained at about 0.1 N salt concentration.

The flow properties of sodium alginate solutions depend on concentration. A 2.5% medium viscosity sodium alginate solution is pseudoplastic, especially at the higher shear rates in the range of 10–10,000/s.

Sources and Processing. All species of brown algae contain algin; however, most of the algin produced commercially is isolated only from a few species. In the United States, algin is extracted from the giant kelp, *Macrocystis pyrifera*, which grows along the west coast of the North American continent, Mexico to California. In Canada, algin is extracted from the rockweed, *Ascophyllum nodosum*, which grows along the shore of the southern portion of Nova Scotia. Harvesting is carried out using small mechanical harvesters.

In Europe, principally in the United Kingdom, Norway, and France, algin is extracted from *Ascophyllum nodosum*, *Laminaria hyperborea*, and *Laminaria digitata*. Harvesting in these countries is generally carried out manually. Japan and the Republic of Korea are the only other countries with a significant algin industry. Other important brown algae used for algin extraction include *Ecklonia maxima* and *Lessonia nigrescans*.

The alginic acid content of some of the commercially important brown algae is shown in Table 4. The commercial processes for the production of algin are proprietary (22,23).

Propylene glycol alginate is the only organic derivative of algin that is widely used in industrial applications, principally the food industry. This product is prepared by the reaction of partially neutralized alginic acid with propylene oxide (24). A typical manufacturing process for sodium alginate is shown in Figure 1.

Uses. Alginates are used in a wide range of applications, particularly in the food, industrial, and pharmaceutical fields (25–27). As shown in Table 5, these applications arise from the properties of gelation, thickening/water holding, emulsification, stabilization/binding, and film forming.

Health and Safety Factors and Environmental Considerations. The toxicological properties of alginates have been extensively investigated, and it has been established that alginates are safe to use in foods (28).

Ammonium alginate, calcium alginate [9005-35-0], potassium alginate, and sodium alginate are included in a list of stabilizers that are generally recognized as safe (GRAS) under 21 CFR 184. Propylene glycol alginate is approved as a food additive under 21 CFR 172.858 as an emulsifier, stabilizer, or thickener (see FOOD ADDITIVES).

Carrageenan. The term carrageenan [19000-07-1] is the generic description for a complex mixture of sulfated polysaccharides that are extracted from certain genera and species of the class *Rhodophyceae*, red seaweed.

Residents of County Carragheen on the south coast of Ireland are reported to have used a seaweed in foods and medicines about 600 years ago. This seaweed became known as Irish moss. The extraction and purification of the polysaccharide

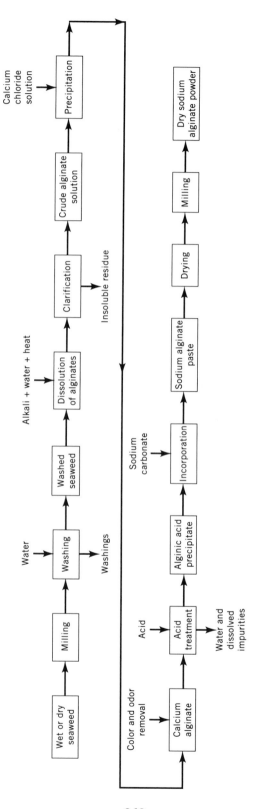

Fig. 1 Sodium alginate manufacturing process.

Table 5. Alginate Properties Utilized in Food and Industrial Applications

Property	Application
Food	
gel forming	pet food, restructured fruit and vegetables, restructured fish and meat, puddings and desserts, bakery creams and jellies
thickening/water holding	frozen foods, pastry fillings, syrups, bakery icings, dry mixes, meringues, and frozen desserts
emulsifying (propylene glycol alginate)	salad dressings, beverages
stabilization	beer (propylene glycol alginate), fruit juice, syrups and toppings, sauces and gravies
film forming	glazes for meat and fish, coatings for cakes and cookies
Industrial	
gel forming	air freshener gels, explosives, and hydromulching
thickening/water holding	paper coating and sizing, adhesives, textile printing and dyeing
binding	ceramics and welding rods
filming	warp sizing and paper sizing

from Irish moss was patented in 1871 (29). This polysaccharide eventually became known as carrageenan; it was not produced and marketed until 1937.

Commercial carrageenan has a molecular weight in the range of 100,000 to 1,000,000. It is sold as powder ranging from white to beige, depending on the grade. Carrageenan produces high viscosity solutions and gels in water which react with proteins, especially casein. This allows the preparation of high strength milk gels. Most of the uses for carrageenan are in the food industry. A small amount of carrageenan is used in cosmetics and pharmaceuticals.

Structure. Carrageenan is a mixture of galactans that carry varying proportions of half-ester sulfate groups linked to one or more of the hydroxyl groups of the galactose units (30,31). The repeating galactose units are joined by alternating α-1,3, β-1,4 glycosidic linkages. In addition, the β-1,4-linked galactose units often occur as 3,6-anhydro-D-galactose.

It is convenient to define three main types: kappa, iota, and lambda carrageenan; however, very few seaweeds yield these ideal structures. Most seaweeds yield a mixture of two or more or an irregular structure. Only the kappa, iota, and lambda carrageenans are available commercially. Composition depends on the type of seaweed from which the carrageenan is extracted and its geographical location. The idealized structures of these three carrageenan types are as follows: λ: D-galactose-2-sulfate–D-galactose-2,6-disulfate; κ: D-galactose-4-sulfate–3,6-anhydro-D-galactose; ι: D-galactose-4-sulfate–3,6-anhydro-D-galactose-2-sulfate (see CARBOHYDRATES).

The sulfate groups are believed to be important in gelation with potassium and in reaction with protein. The 3,6-anhydrogalactose units increase hydrophobicity and reduce solubility whereas the sulfate groups increase hydrophilicity and solubility.

The shape or conformation of carrageenan has been determined (11,32) and related to the gelation mechanism, including the formation of kappa and iota carrageenan gels with potassium and calcium ions, respectively; lambda carrageenan does not form gels. The structure of iota and kappa carrageenan allows the formation of double helices in the sol–gel transformation, whereas the structure of lambda carrageenan inhibits it. Furthermore, in the structure of iota and kappa carrageenan, incomplete conversion of galactose to anhydrogalactose causes the helical carrageenan chain to kink. This molecular kinking interrupts the helix formation between two chains and allows the chain to form helices with other chains and a three-dimensional network rather than a collection of chain pairs.

Sources and Processing.   Several genera and species of red algae contain carrageenan, including *Chondrus crispus*, *Eucheuma*, and *Gigartina stellata* (33). Red algae are abundant on the northeast coast of the North American continent. *Eucheuma* species grow in tropical waters extending from the Philippines to the east coast of Africa.

The seaweed is harvested by raking and hand-gathering. Mechanized harvesting or use of divers has met with limited success. The collected seaweed is dried mechanically in many areas and shipped to the processing plants.

Processing conditions are closely guarded trade secrets, but manufacturer's literature provides some information. The seaweed is extracted with hot water at slightly alkaline pH. The aqueous extract is filtered and recovered by alcohol precipitation, dried, and milled. Drum-drying provides a less pure product.

Uses.   The use of carrageenan in foods is dependent upon the properties of solubility, viscosity, gelation, reactivity with proteins, and synergism with the nongelling polysaccharide, locust-bean gum. Iota and kappa carrageenan act as gelling agents; lambda carrageenan is nongelling, but acts as a thickener (34). Because of its reactivity with certain proteins, carrageenan has found use at low concentrations (0.01–0.03%) in a number of milk-based products, such as chocolate milk, ice cream, puddings, and cheese analogues. It is also used in bakery gels and icings, low sugar jams and jellies. Because of its reactivity with proteins, it is used in ham pumping, an operation in which commercial hams are injected with a carrageenan and phosphate solution to retain moisture on cooking. Kappa carrageenan/locust bean gum blends have been used in room deodorant gels, and in Europe and Australia in canned, gelled pet foods.

Health and Safety Factors.   The Food and Drug Administration lists carrageenan as GRAS, an approved food additive in 21 CFR 172.620. Numerous tests show that carrageenan is not ulcerogenic, teratogenetic, or carcinogenic (35).

Botanical Gums

The botanical gums represent a family of polysaccharides obtained from a wide variety of plant sources. They are subdivided into exudate gums, seed gums, and gums obtained by extraction of plant tissue. For a gum to be used in commercial quantities, it must be present in the tissues or be readily extractable in relatively pure form which limits the number of commercial botanical gums.

The properties of a botanical gum are determined by its source, the climate, season of harvest, and extraction and purification procedures. Table 6 illustrates one of the important basic properties of all gums, ie, the relationship between concentration and solution viscosity. The considerable viscosity variation observed among gums from different sources determines, in part, their uses.

Table 6. Comparative Viscosities[a] of Botanical Gums, mPa·s(=cP)

Conc., %	Natural plant exudates				Seed gums	
	Gum tragacanth	Gum karaya	Gum ghatti	Gum arabic	Guar	Locust bean
1	50	3,000	2		4,500	100[b]
1.5					15,000	
2	900	8,500	35			1,600[c]
3	10,500	20,000				
5	111,000	45,000	290	5		
10			1,000	15		
20			2,250	40		
30				200		
40				950		
50				4,200		

[a]Measured in cold water.
[b]Hot water viscosity = 300 mPa·s(=cP).
[c]Hot water viscosity = 3400 mPa·s(=cP).

PLANT EXUDATES

Most plant families include species that exude gums, and those that produce copious quantities represent a ready supply of gums. These exudates were the first gums to be used commercially and still represent a significant, but diminishing segment of the natural gum market. The plants are usually shrubs or low growing trees. Collection is by hand and labor costs represent a large proportion of the cost of these gums. Raw gum prices have remained low and steady because of the low labor costs in the producing countries of the Middle East and North Africa.

Natural gums are exuded in a variety of shapes characteristic of the species of origin. These shapes include the globular shape of gum arabic and the flakes or thread-like ribbons of gum tragacanth (36).

The quality of individual gums is mainly determined by color and taste or odor. Many gums are colorless when secreted, but darken on aging. Most gums are usually tasteless unless contaminated by the bitter flavors of tannins which precludes their use in foods.

Although many plant gum exudates are known (37,38), only gum arabic, ghatti, karaya, and tragacanth have wide industrial use.

Gum Arabic.   Gum arabic [*9000-01-5*] is a dried exudate from a species of the acacia tree found in various tropical and semitropical areas of the world. Most of the commercial gum comes from a single species, *Acacia senegal*. The largest producers are the Republic of Sudan and several other West African countries, with over 75% of the world's production coming from the Sudan. The best grade comes from *Acacia senegal* and about 90% of the Sudan's production is from this source; the remainder comes from *Acacia seyal*.

The acacia trees produce gum arabic only under adverse conditions, lack of moisture, poor nutrition, and hot temperatures. Gum arabic is produced at wounded surfaces of the acacia trees. The wounds are generally produced deliberately in cultivated trees by stripping bark during the dry season. The gum is collected by hand over a period of several weeks with average yields of 250 grams per tree per year. Crude exudates are hand sorted and exported before processing and milling to various specifications.

Gum arabic is a complex mixture of calcium, magnesium, and potassium salts of arabic acid, a complex branched polysaccharide that contains galactose, rhamnose, glucuronic acid, and arabinose residues (39). Although the constituent sugars are the same, composition and molecular weight of the gum vary from species to species (39,40). Reported molecular weights are in the range of 260,000 to 1,160,000, suggesting a broad molecular weight distribution. The gum is highly soluble in water, and solutions of up to 50% gum concentration can be prepared. No appreciable viscosity is noted at concentrations below 30%. Solutions are slightly acidic with maximum viscosity obtained at neutral pH. Viscosity is stable over a wide range of pH, but is lowered by the addition of salts and increased temperatures.

Gum arabic has wide and varied industrial uses (1), but is mainly a stabilizer and thickener in foods. The largest use is as a crystallization inhibitor in sugar syrups and confections (41). Other uses are in the preparation of spray-dried flavors where it forms a thin protective layer around the flavor material. Gum arabic is widely used to stabilize beverage emulsions and for emulsifying the flavor oils used in these beverages (42). Standard specifications have been established for gum arabic, karaya, and tragacanth, and are published in the *Food Chemicals Codex* (13). Gum arabic has limited application in the cosmetic and pharmaceutical industries as a binder and emulsion stabilizer, but the main nonfood uses are in the formulations of inks and adhesives, and in lithography, where gum arabic is used widely in plating and etching solutions.

Gum Karaya.   Gum karaya [*19000-36-61*] or sterculia gum is the dried exudate of the *Sterculia urens* tree, which is now cultivated in India, the primary producing area. The best quality gum is collected by tapping the trees during the period April to June with a second collection of lower quality product later in the year. The gum is allowed to dry on the trees, and the crude gum is collected and sorted according to color and purity. It is further sorted and processed to powdered gum karaya in the country of use. The quality of supplies varies greatly (43).

Gum karaya is a partially acetylated high molecular weight polysaccharide which contains L-rhamnose, D-galactose, D-galacturonic acid, and D-glucuronic acid residues (44). Gum karaya is the least soluble gum exudate and does not dissolve completely in water to give true solutions, but swells in cold water to give viscous solutions. Gum concentrations up to 4% can be prepared in cold water and up to approximately 25% by heating. Viscosity decreases with increasing ionic strength and temperature or changes in pH. Gum karaya is compatible with most other plant hydrocolloids and proteins.

In foods, gum karaya has been used as a thickening and suspending agent and as a stabilizer in sauces and dressings, and in frozen desserts where it prevents the formation of large ice crystals. It has been used as an emulsifier and processing aid for cheese spreads, as a binder and emulsifier in sausages and meat products, and in baked goods where it retards staling and improves the tolerance of the dough to overmixing.

The majority of the imported gum karaya is used by the pharmaceutical industry as a bulk laxative, in dental adhesives, and in sealing gaskets for colostomy bags.

Gum Tragacanth. Gum tragacanth [9000-65-11] is an exudate from several species of tree, of the genus *Astralagus*, found in the dry, mountainous regions of Iran, Syria, and Turkey. Iran is the best commercial source and produces the highest quality gum (45). The plants are incised and the gum exudes; after drying, it is collected as ribbons, which are the highest quality gum, or as flakes. These two forms are obtained from two different shrubs. Gum tragacanth consists of a complex mixture of acidic polysaccharides containing galacturonic acid, galactose, fucose, xylose, and arabinose (46).

Solutions of gum tragacanth have extremely high viscosity, and the viscosity is stable over a wide pH range to about pH 2 (46). For this reason and because of its stabilizing and emulsifying properties, gum tragacanth was once widely used in food products (47). However, in many applications it has been increasingly replaced by propylene glycol alginate and more recently, xanthan gum. Gum tragacanth is compatible with other plant hydrocolloids as well as proteins and carbohydrates. Primary food applications include confectionery and icings, dressings and sauces, oil and flavor emulsions, frozen desserts and bakery fillings. The gum is used widely for the preparation of pharmaceutical emulsions and jellies and also in the cosmetic industry in face and hair lotions.

Gum Ghatti. Gum ghatti [9000-28-61] is an exudate from *Anogeissus latifolia*, a tree that is found in India and Sri Lanka. The exudations are natural, but yields can be increased by making artificial incisions. The sun-dried gum is classified according to color and impurities and processed by grinding to a fine powder.

Gum ghatti is the calcium and magnesium salt of a complex polysaccharide which contains L-arabinose, D-galactose, D-mannose, and D-xylose and D-glucuronic acid (48) and has a molecular weight of approximately 12,000. On dispersion in water, gum ghatti forms viscous solutions of viscosity intermediate between those of gum arabic and gum karaya. These dispersions have emulsification and adhesive properties equivalent to or superior to those described for gum arabic.

In foods and pharmaceuticals, gum ghatti has been used in many applications described for gum arabic, particularly as an emulsifier for oil and water emulsions (49). It has also been used as a waterproofing agent in liquid explosives, and to stabilize paraffin wax emulsions. However, in the 1990s, gum ghatti consumption has declined in most applications.

Pectin. Pectin [9000-69-5] is a generic term for a group of polysaccharides, mainly partially methoxylated polygalacturonic acids, which are located in the cell walls of all plant tissues. The main commercial sources of pectin are citrus peel and apple pomace, where it represents 20–40% and 10–20% of the dry weight respectively. The pectin is extracted, the extract purified, and the pectin precipitated (50); increased extraction times lead to the production of low methoxyl pectins.

Pectins are readily soluble in water to give viscous stable solutions. However, the importance of pectin to industry, in particular the food industry, is the ability of its solutions to form gels with sugar (ca 65% solids) and acid or calcium ion under suitable conditions (51).

Pectins are generally classed according to their ester content as high methoxyl pectins (>50% of the carboxyl groups esterified) or low methoxyl pectins (<50% of carboxyl groups esterified) (pectic acid, methyl ester [9049-34-1]). Low methoxyl pectins, like algins, require calcium for gelation.

Pectin preparations, particularly the high methoxyl ones, are used mainly for jams and jellies. The low methoxyl pectins are of increasing utility in the preparation of low calorie, low sugar jams and jellies as low methoxyl pectin–calcium gels (52). However, unlike high methoxyl pectins, low methoxyl pectins face competition from other gums such as carrageenan and gellan gum. Pectins are also used as stabilizers and thickeners in frozen desserts, fruit juice drinks, milk beverages, and in the preparation of restructured fruits and vegetables. Industrial applications, in particular for paper and textiles, have been limited because of high cost.

SEED GUMS

Although most seeds contain starch as the principal food reserve, many contain other polysaccharides and some have industrial utility. The first seed gums used commercially were quince, psyllium, flax, and locust bean gum. However, only locust bean gum is still used, particularly in food applications; quince and psyllium gums are only used in specialized applications.

Harvesting of these gums is expensive; however, harvesting from annual plants costs less than from perennial plants or trees. This is clearly demonstrated by the tremendous increase in the use of guar, a gum that is extracted from an annual leguminous plant. Since the beginning of commercial production in 1953, the use of guar has risen rapidly. More guar gum is consumed than all other gums combined.

Guar Gum.   Guar gum [9000-30-0] is derived from the seed of the guar plant, *Cyamopsis tetragonolobus*, a pod-bearing nitrogen-fixing legume grown extensively in Pakistan, India, and on a commercial scale since 1946, in the southwestern United States. During processing, the seed coat is removed by heating and milling. The endosperm, comprising approximately 40% of the seed, is then separated from the germ by various milling processes. The final milled endosperm, which is commercial guar gum, has a typical analysis of crude fiber, 2.5%; moisture, 10–15%; protein, 5–6%; and ash, 0.5–0.8%.

Structurally, guar gum comprises a straight chain of D-mannose with a D-galactose side chain on approximately every other mannose unit; the ratio of mannose to galactose is 2:1 (53). Guar gum has a molecular weight on the order of 220,000 (54).

Guar gum hydrates in either cold or hot water to give high viscosity solutions. Although the viscosity development depends on particle size, pH, and temperature, guar gum at 1% concentration fully hydrates typically within 24 h at room temperature and in 10 min at 80°C. Guar gum solutions are stable over the pH range of 4.0–10.5 with fastest hydration occurring at pH 8.0. Solutions are somewhat cloudy because of the small amount of insoluble fiber and cellulosic material present.

Guar gum is compatible with other common plant gums, starch, and water-soluble proteins such as gelatin, except for the synergistic viscosity increases given by blends with xanthan gum (55). The presence of salts, water-miscible solvents or low molecular weight sugars can affect the hydration rate. Small amounts of borate, however, cross-link the guar to form rubbery gels (56). These gels can be reversed by adjusting the pH to the acid ranges.

As a result of its properties and low cost, guar gum is the most extensively used gum, both in food and industrial applications. In the mining industry, guar gum is used as a flocculant or flotation agent, foam stabilizer, filtration aid, and water-treating agent. In the textile industry, it is used as a sizing agent and as a thickener for dyestuffs. The biggest consumer of guar gum is the paper industry where it facilitates wet-end processing and improves the properties of the product. Guar gum is used as a thickening and gelling agent for slurry explosives as it readily hydrates in concentrated ammonium nitrate solutions, is easily cross-linked, and acts as a waterproofing agent for the final gelled product (57).

In the food industry, guar gum is applied as a stabilizer in ice cream and frozen desserts (58), and it is also used as a stabilizer for salad dressings, sauces, frozen foods, and pet foods.

Many derivatives of guar, including cationic, carboxymethyl, carboxymethylhydroxypropyl and oxidized guar, have been prepared. One in particular, hydroxypropyl guar gum [39421-75-5], is of industrial importance. Derivatization leads to subtle changes in properties, eg, decreased H-bonding, increased solubility in alcohol–water mixtures, and improved electrolyte compatibility. These changes result in increased use in textile applications and liquid slurry explosives. In addition, the decrease in insolubles in the derivatized product has resulted in application of hydroxypropyl guar gum in oil-well fracturing applications (59).

Locust Bean Gum. Locust bean gum [9000-40-2] is produced by milling the seeds from the leguminous evergreen plant, *Ceratonia siliqua*, or carob tree, which is widely grown in the Mediterranean area. Pods produced by the carob tree consist of a husk, embryo, and endosperm. The latter, the source of the gum, is separated from the tough outer husk and the yellow embryo tissue by a variety of rolling and milling operations, and subsequently is milled into a fine powder (60).

Locust bean gum is a galactomannan, consisting of a main chain of D-mannose units with single-unit galactose side chains on approximately every fourth unit (54) and has a reported molecular weight of 310,000 (61). The side chains of guar are alternately disposed along the D-mannan backbone, whereas those of locust bean gum are disposed in uniform blocks along the backbone (62).

Locust bean gum is not completely soluble in cold water; it must be heated to 80°C and cooled to attain a stable solution that has high viscosity at low concentrations. The gum is compatible with other plant gums and the viscosity of solutions is not appreciably affected by pH or salts.

Concentrations above 0.3% form a gel with borate which is reversible upon the subsequent addition of mannitol (a sequestrant for borate) or of acid. Useful combinations are formed with carrageenan (63) and xanthan gum (64) and agar. In many applications, it is used in combination with these gums at considerable cost savings.

In the food industry, locust bean gum is used as a stabilizer in ice cream and in the preparation of processed cheese and extruded meat products. It is also used as an emulsifier and stabilizer of dressings and sauces and overall has similar properties to those outlined for guar gum.

Locust bean gum and its derivatives are excellent film formers and can be used either alone or in combination with starch as textile sizing agents and dye thickeners in textile printing, and as fiber bonding and beater additives in the papermaking industry. However, in most of these applications it has been replaced by guar.

Tamarind Gum. Tamarind gum [*39386-78-9*] is another seed gum with potential industrial application. It is obtained from the seed kernels of the tamarind tree, *Tamarindus indica*, which is cultivated in India and Bangladesh. The seeds are a by-product from the production of tamarind pulp which is used as a food flavor. Seed production is 150,000 t/yr.

Tamarind kernel powder is insoluble in cold water, but upon heating forms thick viscous colloidal dispersions at relatively low concentrations (2–3%). Upon drying, elastic films are formed. Tamarind seed polysaccharide is used as a low cost textile sizing agent in India.

Tamarind seed polysaccharide, the gum fraction obtained from tamarind kernel polysaccharide, forms gels over a wide pH range in the presence of high sugar concentrations (> 65 wt %), and it can therefore substitute for fruit pectins (65).

Psyllium Seed Gum. Psyllium seed gum [*8036-16-9*] is derived from plants of the genus *Plantago,* several species of which are used as commercial sources. However, most current production is from *Plantago ovata*, grown in India. The gum is located in the coat which is removed by cracking. The gum is then extracted with boiling water and separated from the insoluble residue by filtration. It consists of mixtures of both neutral and acidic polysaccharides, the composition of which is species dependent (66).

The purified gum hydrates slowly in water to give viscous solutions at concentrations up to 1% and clear gelatinous masses at higher concentrations. Psyllium seed gum is a laxative additive and is used in cosmetics and in hair-setting lotions.

Quince Seed Gum. Quince seed gum [*9011-85-2*] is obtained from the seeds of the quince fruit. Quince trees, *Cydonia vulgaris* or *Cydonia oblongs*, are cultivated in temperate regions throughout the world, with 75% of the total production of quince seeds coming from Iran. The gum is extracted from the seed with cold or hot water and separated from insoluble cellulosic residues by filtration through muslin. The gum is soluble in cold water but hydrates more rapidly in hot water to give highly viscous solutions at concentrations up to 1.5%.

As predictable from the similarity of the properties of the two gums, quince seed gum is used in the applications described above for psyllium seed gum. Specific applications are in cosmetics and hair-setting lotions. It has also been used as an emulsifier and stabilizer in pharmaceutical preparations.

Larch Gum. Larch gum [37320-79-9] (larch arabinogalactan) is obtained by water extraction of the western larch tree, *Larix occidentalis*, the heartwood of which contains 5–35% on a dry wood basis. In the early 1960s, a countercurrent hot water extraction system was developed, and the gum was produced commercially by the St. Regis Paper Co. under the trade name Stractan. The potential production capacity of this gum is 10,000 t/yr based on the wood residues from the lumber industry. However, the product could not compete with gum arabic, and commercial production is now limited to small batches for a specific medical application.

Structurally, arabinogalactan is a complex, highly branched polymer of arabinose and galactose in a 1:6 ratio (67). It is composed of one fraction with an average molecular weight of 16,000, and one of 100,000 (68).

Larch gum is readily soluble in water. The viscosity of these solutions is lower than that of most other natural gums and solutions of over 40% solids are easily prepared. These highly concentrated solutions are also unusual because of their Newtonian flow properties. Larch gum reduces the surface tension of water solutions and the interfacial tension existing in water and oil mixtures, and thus is an effective emulsifying agent. As a result of these properties, larch gum has been used in foods and can serve as a gum arabic substitute.

Larch arabinogalactan is approved in 21 CFR 172.610 as a food additive for use as an emulsifier, stabilizer, binder or bodying agent for essential oils and nonnutritive sweeteners, flavor bases, nonstandardized dressings, and pudding mixes. It has also been used in the preparation of cosmetic and pharmaceutical dispersions and as an emulsifier in oil–water emulsions (69). Industrially, the main use has been in lithography as a gum arabic substitute.

Microbial Gums

Dextran. This polysaccharide is produced from sucrose by certain species of *Leuconostoc* (70). Dextran [9004-54-0] was the first commercial microbial polysaccharide. It was used as a blood plasma extender in the U.S. Army during the late 1940s and early 1950s. This program was discontinued in 1955.

In the late 1950s and early 1960s, the possibility of producing fermentation gums, under controlled conditions, was investigated. Many polysaccharides produced by microorganisms have been studied, including xanthan gum (71), curdlan [54724-00-4] (72), bioalgin (73), pullulan [9057-02-7] (74), scleroglucan [39464-87-4] (75), and more recently, gellan, welan, and rhamsan gums (76). To date, only xanthan gum [11138-66-2] has achieved commercial sigificance.

Development of bioalgin has ceased. Curdlan, pullulan, and scleroglucan have found only limited application and gellan, welan, and rhamsan have only been recently introduced to the marketplace.

Xanthan Gum. As a result of a project to transform agriculturally derived products into industrially useful products by microbial action, the Northern Regional Research Laboratories of the USDA showed that the bacterium *Xanthomonas campestris* produces a polysaccharide with industrially useful properties

(77). Extensive research was carried out on this interesting polysaccharide in several industrial laboratories during the early 1960s, culminating in commercial production in 1964.

Structure and Conformation. As subsequently found for all microbial polysaccharides, the primary structure consists of regular repeating units. Each unit contains five sugars: two glucose units, two mannose units, and one glucuronic acid unit. The main chain of xanthan gum is built up of β-D-glucose units linked through the 1- and 4-positions, ie, the chemical structure of the main chain of xanthan gum is identical to the chemical structure of cellulose. A three-sugar side chain is linked to the 3-position of every other glucose residue in the main chain (78). About half of the terminal D-mannose residues contain a pyruvic acid residue linked to the 4- and 6-positions. The distribution of these pyruvate groups is unknown. The nonterminal D-mannose unit in the side chain contains an acetyl group at position 6.

Properties. Xanthan gum is a cream-colored powder that dissolves in either hot or cold water to produce solutions with high viscosity at low concentration. These solutions exhibit pseudoplasticity, ie, the viscosity decreases as the shear rate increases. This decrease is instantaneous and reversible. Solutions, particularly in the presence of small amounts of electrolyte, have excellent thermal stability, and their viscosity is essentially constant over the range 0 to 80°C. They are not affected by changes in pH ranging from 2 to 10.

Xanthan gum dissolves in acids and bases, and under certain conditions, the viscosity remains stable for several months. Xanthan gum has excellent stability and compatibility with high concentrations of many salts, eg, 15% solutions of sodium chloride and 25% solutions of calcium chloride (79).

The most unusual property of xanthan gum is the reactivity with galactomannans, such as guar gum and locust bean gum (80,81). The xanthan gum–locust bean gum combination at low gum concentration (less than 0.1%) has a significantly higher viscosity in solution than would be expected on the basis of the viscosity of the individual components. At higher gum concentrations (greater than 0.2%), a cohesive, thermoreversible gel is formed. Xanthan gum–guar gum combinations provide higher than expected viscosities, but do not form a gel.

Production. Xanthan gum is produced by the microorganism *X. campestris*, originally isolated from the rutabaga plant. The gum is produced commercially by culturing *X. campestris* purely under aerobic conditions in a medium containing commercial glucose, a suitable nitrogen source, dipotassium phosphate, and appropriate essential elements. When the fermentation is complete, the gum is recovered from the fermentation broth by precipitation with isopropyl alcohol, and dried, milled, tested, and packed.

Health and Safety Factors. The toxicological and safety properties of xanthan gum have been extensively investigated (82). On the basis of these studies, the FDA issued a food additive order in 1969 that allowed the use of xanthan gum in food products without specific quantity limitations.

Uses. The unique properties of xanthan gum make it suitable for many applications for the food, pharmaceutical, and agricultural industries (79).

Principal food applications include dressings, relishes, sauces, syrups and toppings, puddings, dry mix products, cake mixes, beverages, dairy products, and confectionery. It is also used in pharmaceutical suspensions and toothpaste. Industrial applications include textile printing, pigment and dye suspensions, cleaners, and polishes. Oilfield applications include drilling muds, workover and completion fluids, fracturing, and in enhanced oil recovery.

Gellan Gum. Gellan gum is the generic name for the extracellular polysaccharide produced by the bacterium, *Pseudomonas elodea* (ATCC 31461). Proprietary to Kelco Division of Merck & Co., Inc., gellan gum is manufactured in an aerobic, submerged fermentation (76).

Gellan gum is a linear anionic heteropolysaccharide with a molecular weight of 0.5×10^6 daltons. It is composed of tetrasaccharide repeat units comprising 1,3-β-D-glucose, 1,4-β-D-glucuronic acid, 1,4-β-D-glucose, and 1,4-α-L-rhamnose (83). In the solid state, the molecule forms a parallel, half-staggered intertwined double helix (84). The polymer, as secreted by the organism, contains *O*-acyl groups that are readily removed by alkali treatment. The native or acylated product produces elastic gels, whereas the low acyl product, resulting from treatment by heating at pH 10 or higher, forms firm brittle gels on heating and cooling. The native gum also contains cellular material that can be removed by centrifugation or filtration to give gels with excellent clarity (76).

Gellan dissolves in deionized water or low ionic strength solutions. Heating facilitates dissolution and sequestrants remove the inhibiting effect of divalent cations. Upon cooling, firm gels are produced, the strength and texture of which are determined by the gum concentration, cation concentration and makeup, degree of acylation, and the presence of sucrose (85). Higher concentrations of monovalent cations are required to produce a given gel strength. Although no synergism is observed, the gel texture can be modified by the addition of other thickening and gelling agents such as guar, xanthan, and locust bean gums (86). Gellan gum can also be used to modify the texture of starch and gelatin gels.

Gellan gum has been permitted for use in foods in Japan since 1988, and in 1992 received general food approval in the United States. Gellan gum is recognized as a food additive under the provisions of the U.S. Food and Drug Administration regulations (21 CRF 172.665) for use as a stabilizer and thickener in foods (87). Potential food applications include fabricated foods, water-based gels, pet foods, confectionery, jams and jellies, pie fillings and puddings, icings and frostings, and dairy products. Nonfood applications include deodorant gels and industrial films and gels.

Welan Gum. This gum is produced by a carefully controlled aerobic fermentation using an *Alcaligenes* strain (ATCC 31555) (88). The backbone of welan gum is identical to that of gellan gum, but it carries a single sugar sidechain which is either L-mannose or L-rhamnose (89).

Welan has similar properties to xanthan gum except that it has increased viscosity at low shear rates and improved thermal stability and compatibility with calcium at alkaline pH (90). The increased thermal stability has led to its use as a drilling mud viscosifier especially for high temperature wells. The excellent compatibility with calcium at high pH has resulted in its use in a variety of specialized cement and concrete applications.

Rhamsan Gum. Rhamsan gum, produced by *Alcaligenes* strain (ATCC 31961) (91), has the same backbone as gellan and welan gums, but it carries a disaccharide sidechain (92).

Solutions of rhamsan have high viscosity at low shear rates and low gum concentrations (90). The rheological properties and suspension capability combined with excellent salt compatibility, make it useful for several industrial applications including agricultural fertilizer suspensions, pigment suspensions, cleaners, and paints and coatings.

BIBLIOGRAPHY

"Gums and Mucilages" in *ECT* 1st ed., Vol. 7, pp. 329–339, by D. C. Beach, S. B. Penick & Co.; "Gums, Natural" in *ECT* 2nd ed., Vol. 10, pp. 749–754, by M. Glicksman, General Foods Corp.; "Gums" in *ECT* 3rd ed., Vol. 12, pp. 45–66, by I. W. Cottrell and J. K. Baird, Kelco Division of Merck & Co., Inc.

1. R. L. Whistler, *Industrial Gums*, 2nd ed., Academic Press, Inc., New York, 1973, pp. 6–7.
2. P. A. Sandford and J. K. Baird, in G. O. Aspinall, ed., *The Polysaccharides*, Vol. 2, Academic Press, Inc., New York, 1983, pp. 411–490.
3. D. A. Rees, *Biochem. J.* **126,** 257 (1972).
4. D. A. Rees, *Polysaccharide Shapes*, Halstead Press, a division of John Wiley & Sons, Inc., New York, 1977.
5. B. J. Spalding, *Chem. Week* **136**(15), 31 (1985).
6. *Chem. Mark. Rep.* (1990–1992).
7. C. Araki and K. Arai, *Bull. Chem. Soc. Jpn.* **40,** 1452 (1967).
8. C. Araki, *Bull. Chem. Soc. Jpn.* **29,** 543 (1956).
9. M. Duckworth, K. C. Hong, and W. Yaphe, *Carbohydr. Res.* **18,** 1 (1971).
10. I. C. M. Dea, A. A. McKinnon, and D. A. Rees, *J. Mol. Biol.* **68,** 153 (1972).
11. E. R. Morris, *Br. Polym. J.* **18**(1), 14 (1986).
12. *The United States Pharmacopeia XX (USP XX-NFXV)*, The United States Pharmacopeial Convention, Inc., Rockville, Md., 1980.
13. *Food Chemicals Codex*, 3rd ed., National Academy of Sciences, National Research Council, Washington, D.C., 1980.
14. M. Glicksman, *Gum Technology in the Food Industry*, Academic Press, Inc., New York, 1969, pp. 210–213.
15. Brit. Pat. 142 (July 9, 1881), E. C. C. Stanford.
16. G. T. Grant and co-workers, *Febs Lett.* **32,** 195 (1973).
17. H. Grasdalen, B. Larsen, and O. Smidsrød, *Carbohydr. Res.* **68,** 23 (1979).
18. H. Grasdalen, B. Larsen, and O. Smidsrød, *Carbohydr. Res.* **89,** 179 (1981).
19. E. Onsoyen, in A. Imeson, ed., *Thickening and Gelling Agents for Food*, Blackie Academic and Professional, London, 1992, Chapt. 1, p. 6.
20. U.S. Pat. 3,455,701 (July 15, 1969), A. Miller and J. Rocks (to Kelco Co.).
21. *Structured Foods with the Algin/Calcium Reaction,* Technical Bulletin F-83, Kelco Division of Merck & Co., San Diego, Calif., 1984.
22. U.S. Pat. 2,036,934 (Apr. 7, 1936), H. C. Green (to Kelco Co.).
23. U.S. Pat. 2,128,551 (Aug. 30, 1938), V. C. E. LeGloahec and J. R. Herter (to Algin Corp. of America).
24. U.S. Pat. 2,463,824 (Mar. 8, 1949), A. B. Steiner and W. H. McNeely (to Kelco Co.).
25. W. H. McNeely and D. J. Pettitt, in Ref. 1, Chapt. 4.
26. M. Glicksman, in Ref. 14, Chapt. 8.
27. E. Onsoyen, in Ref. 19, Chapt. 1, pp. 1–24.
28. W. H. McNeely and P. Kovacs, in A. Jeanes and J. Hodge, eds., *ACS Symposium Series 15*, American Chemical Society, Washington, D.C., 1975, pp. 269–281.

29. U.S. Pat. 112,535 (Mar. 14, 1871), G. Bourgade.
30. A. N. O'Neill, *J. Am. Chem. Soc.* **77,** 6324 (1955).
31. D. A. Rees, *J. Chem. Soc.*, 1821 (1963).
32. D. A. Rees, *British Food Manufacturing Industries Research Association Symposium Proceedings No. 13*, London, 1972, pp. 7–12.
33. E. Booth, in J. P. Riley and G. Skirrow, eds., *Chemical Oceanography*, Vol. 4, 2nd ed., Academic Press, Inc., New York, 1975, pp. 244–245.
34. W. R. Thomas, in Ref. 19, Chapt. 12, pp. 25–39.
35. D. J. Stancioff and D. W. Renn, in Ref. 28, pp. 282–295.
36. J. D. Dziezak, *Food Technol.*, 116 (Mar. 1991).
37. F. Smith and R. Montgomery, *The Chemistry of Plant Gum and Mucilages*, Van Nostrand Reinhold, New York, 1959.
38. F. N. Howes, *Vegetable Gums and Resins*, Chronica Botanica, Waltham, Mass., 1969.
39. D. M. W. Anderson and co-workers, *Food Add. Contam.* **8,** 405 (1990).
40. D. M. W. Anderson, *Proc. Biochem.* **12,** 24 (1977).
41. H. Reidel, *Confect. Prod.* **49,** 612 (1983).
42. F. Thevenet, in S. J. Risch and G. A. Reineccius, eds., *Flavour Encapsulation*, American Chemical Society, Washington, D.C., 1988, Chapt. 5, pp. 37–44.
43. W. Meer, in R. L. Davidson, ed., *Handbook of Water Soluble Gums and Resins*, McGraw-Hill Book Co., Inc., New York, 1990, Chapt. 10, pp. 1–14.
44. D. LeCerf, F. Irinei, and G. Muller, *Carbohydr. Polymers* **13,** 375 (1990).
45. S. R. J. Robbins, *A Review of Recent Trends in Selected Markets for Water Soluble Gums*, Overseas Development Natural Resources Institute, bulletin no. 2, Kent, U.K., 1987.
46. D. M. W. Anderson, *Food Add. Contam.* **6,** 1 (1989).
47. K. R. Stauffer, in Ref. 43, Chapt. 11, pp. 1–31.
48. G. O. Aspinall, E. L. Hirst, and A. Wick-Strom, *J. Chem. Soc.*, 1160 (1955).
49. M. Glicksman, in M. Glicksman, ed., *Food Hydrocolloids,* Vol. 2, CRC Press, Boca Raton, Fla., 1983, p. 31.
50. W. G. Hull, C. N. Lindsay, and W. E. Baier, *Ind. Eng. Chem.* **45,** 876 (1953).
51. R. M. Ehrlich, *Food Prod. Dev.* **2,** 36 (1968).
52. P. C. Fass, "Pectins in Confections," *Proceedings of the 21st Pennsylvania Manufacturing Confectioners Association*, Drexel Hills, Pa., 1977.
53. E. E. Heyne and R. L. Whistler, *J. Am. Chem. Soc.* **70,** 2249 (1948).
54. J. W. Hoyt, *J. Polym. Sci. Part B* **4,** 713 (1966).
55. J. K. Rocks, *Food Technol.* **25,** 22 (1971).
56. Ref. 1, pp. 317–318.
57. U.S. Pat. 3,355,336 (Nov. 28, 1967), W. M. Lyerly (to E. I. du Pont de Nemours & Co., Inc.).
58. W. S. Arbuckle, in *Ice Cream*, Avi Publishing Co., 1986, pp. 84–94.
59. *Oil Field Chemicals, North America*, C. H. Kline & Co., Fairfield, N.J., 1980.
60. M. Glicksman, *Adv. Food Res.* **11,** 162 (1962).
61. R. L. Whistler and C. L. Smart, *Polysaccharide Chemistry*, Academic Press, Inc., New York, 1953.
62. J. Hoffman, B. Lindberg, and T. Painter, *Acta Chem Scand. Ser. B* **30,** 365 (1976).
63. U.S. Pat. 2,669,519 (Feb. 16, 1954), G. L. Baker (to Seaplant Chemical Co.).
64. U.S. Pat. 3,557,016 (Jan. 19, 1971), H. R. Schuppner (to Kelco Co.).
65. G. R. Savur and A. Sreenivasen, *J. Soc. Chem. Ind. (London)* **67,** 190 (1948).
66. R. A. Laidlaw and E. G. V. Percival, *J. Chem. Soc.*, 1600 (1949).
67. H. A. Swenson and co-workers, *Macromolecules* **2,** 142 (1969).
68. G. Lystad-Borgin, *J. Am. Chem. Soc.* **71,** 2247 (1949).
69. M. R. Nazareth, C. E. Kennedy, and V. N. Bhatia, *J. Pharm. Sci.* **50,** 560 (1961).

70. A. Jeanes, in P. A. Sandford and A. Laskin, eds., *ACS Symposium Series No. 45*, American Chemical Society, Washington, D.C., 1977, p. 284.
71. J. K. Baird and D. J. Pettitt, in I. Goldberg and R. Williams, eds., *Biotechnology and Food Ingredients*, Von Nostrand Reinhold, New York, 1992, Chapt. 9, pp. 223–263.
72. T. Harada, in Ref. 70, pp. 265–283.
73. P. Gacesa, *Carbohydrate Polymers* **8,** 161 (1988).
74. K. Hannigan, *Food Engineering* **56,** 98 (1984).
75. N. E. Rodgers, in Ref. 1, Chapt. 22.
76. J. K. Baird, P. A. Sandford, and I. W. Cottrell, *Biotechnology* **1,** 778 (1983).
77. A. Jeanes, J. E. Pittsley, and F. R. Senti, *J. Appl. Polym. Sci.* **5,** 519 (1961).
78. P. E. Jansson, L. Kenne, and B. Lindberg, *Carbohydr. Res.* **45,** 275 (1975).
79. J. K. Baird, in J. I. Kroschwitz, ed., *Encyclopedia of Polymer Science and Engineering*, Vol. 17, 2nd ed., John Wiley & Sons, Inc., New York, 1989, pp. 901–918.
80. J. K. Rocks, *Food Technol.* **27**(5), 22 (1971).
81. P. Kovacs, *Food Technol.* **27**(3), 26 (1973).
82. *Food Chemicals Codex*, 3rd ed., National Academy Press, Washington, D.C., 1981.
83. P. E. Jannson, B. Lindberg, and P. A. Sandford, *Carbohydr. Res.* **124,** 135 (1983).
84. R. Chandrasekaran, R. P. Millane, and S. Arnott, *Carbohydr. Res.* **175,** 1 (1988).
85. G. R. Sanderson and R. C. Clark, *Food Technol.* **37**(4), 63 (1984).
86. G. R. Sanderson and co-workers, in G. O. Phillips, D. J. Wedlock, and P. A. Williams, eds., *Gums and Stabilizers for the Food Industry*, Vol. 4, IRL Press, Washington, D.C., 1988, pp. 301–308.
87. *Fed. Reg.* (Nov. 25, 1992).
88. U.S. Pat. 4,342,866 (1982), K. S. Kang and G. T. Veeder (to Kelco Division of Merck & Co., Inc.).
89. P. E. Jansson, B. Lindberg, and Widmalm, *Carbohydr. Res.* **139,** 217 (1985).
90. K. Clare, *Speciality Chem.*, 238 (Aug. 1989).
91. U.S. Pat. 4,401,760 (1983), J. A. Peik, S. M. Steenbergen, and H. R. Hayden (to Kelco Division of Merck & Co., Inc.).
92. P. E. Jannson and co-workers, *Carbohydr. Res.* **156,** 157 (1986).

JOHN K. BAIRD
Kelco Division of Merck & Co., Inc.

GYPSUM.   See CALCIUM COMPOUNDS, CALCIUM SULFATE.

HAFNIUM AND HAFNIUM COMPOUNDS

Hafnium [7440-58-6], Hf, is in Group 4 (IVB) of the Periodic Table as are the lighter elements zirconium and titanium. Hafnium is a heavy gray-white metallic element never found free in nature. It is always found associated with the more plentiful zirconium. The two elements are almost identical in chemical behavior. This close similarity in chemical properties is related to the configuration of the valence electrons, $4d^2 5s^2$ and $5d^2 6s^2$ for zirconium and hafnium, respectively; and to the close similarity in ionic radii of the M^{4+} ions, Zr^{4+}, 0.084 nm and Hf^{4+}, 0.083 nm. The latter is a consequence of the lanthanide contraction. Hafnium and zirconium have more similar chemical properties than any other pair in the Periodic Table apart from the inert gases.

Whereas zirconium was discovered in 1789 and titanium in 1790, it was not until 1923 that hafnium was positively identified. The Bohr atomic theory was the basis for postulating that element 72 should be tetravalent rather than a trivalent member of the rare-earth series. Moseley's technique of identification was used by means of the x-ray spectra of several zircon concentrates and lines at the positions and with the relative intensities postulated by Bohr were found (1). Hafnium was named after *Hafnia*, the Latin name for Copenhagen where the discovery was made.

Hafnium is obtained as a by-product of the production of hafnium-free nuclear-grade zirconium (see NUCLEAR REACTORS; ZIRCONIUM AND ZIRCONIUM COMPOUNDS). Hafnium's primary use is as a minor strengthening agent in high temperature nickel-base superalloys. Additionally, hafnium is used as a neutron-absorber material, primarily in the form of control rods in nuclear reactors.

Properties

Physical Properties.   Hafnium is a hard, heavy, somewhat ductile metal having an appearance slightly darker than that of stainless steel. The color of hafnium sponge metal is a dull powder gray. Physical properties of hafnium are summarized in Table 1. These data are for commercially pure hafnium which may contain from 0.2 to 3% zirconium. Although a number of radioactive isotopes have been artificially produced, naturally occurring hafnium consists of six stable isotopes (Table 2). Hafnium crystallizes in a body-centered cubic system which transforms to a hexagonal close-packed system below 2033 K.

Chemical Properties.   Hafnium's normal stable valence is also its maximum valence of four. Hafnium exhibits coordination numbers of six, seven, and eight in its compounds. The aqueous chemistry is characterized by a high degree of hydrolysis, the formation of polymeric species, a very slow approach to true equilibrium, and the multitude of complex ions that can be formed. Partially reduced di- and trihalides have been produced by reducing anhydrous hafnium tetrahalides with hafnium metal.

Hafnium is a highly reactive metal. The reaction with air at room temperature is self-limited by the adherent, highly impervious oxide film which is formed. This film provides oxidation stability at room temperature and resistance to corrosion by aqueous solutions of mineral acids, salts, or caustics. Thicker oxide films are formed at higher temperature, but slowly enough that forging or hot rolling of hafnium ingots is conducted in air at a temperature between 900 and

Table 1. Physical Properties of Hafnium

Property	Value	Reference
atomic number	72	
atomic weight	178.49	
density, at 298 K, kg/m^3	13.31×10^3	2
melting point, K	2504	3
boiling point, K	4903	3
specific heat, at 298 K, J/(kg·K)[a]	144	3,4
latent heat of fusion, J/kg[a]	1.53×10^5	3
electrical resistivity, at 298 K, Ω·m.	3.37×10^{-7}	5
Hall coefficient, at 298 K, V·m/(A·T)	-1.62×10^{-12}	6
work function, J[a]	6.25×10^{-19}	6
thermal conductivity, W/(m·K)		
at 273 K	23.3	
at 1273 K	20.9	
Young's modulus, at 293 K, GPa[b]	141	8
shear modulus, at 293 K, GPa[b]	56	8
Poisson's ratio, at 293 K	0.26	8
thermal expansion coefficient, linear,[c] from 293 to 1273 K, 10^{-6}/K	6.1	9

[a]To convert J to cal, divide by 4.184.
[b]To convert GPa to psi, multiply by 145,000.
[c]For random polycrystalline orientation.

Table 2. Hafnium Isotopes

Isotope mass number	Abundance, %	Thermal neutron cross section, $m^2 \times 10^{-28}$	Contribution to the total cross section
174	0.16	620	1.0
176	5.21	23	1.2
177	18.61	375	69.8
178	27.30	85	23.2
179	13.63	46	6.3
190	35.10	13	4.6
Total	*100.01*		*106.1*

1000°C, with subsequent removal of surface scale by sandblasting and then a nitric–hydrofluoric acid pickling. High surface area hafnium powder or porous sponge metal ignites quite easily in air. Clean hafnium metal ignites spontaneously in oxygen of about 2 MPa (300 psi).

Hafnium begins to react with nitrogen at about 900°C to form a surface nitride film, and reacts rapidly with hydrogen at about 700°C to form hydrides (qv). The hydrogen diffuses rapidly and converts the bulk metal into the brittle hydride.

Hafnium is readily soluble in hydrofluoric acid and is slowly attacked by concentrated sulfuric acid. Hafnium is unaffected by nitric acid in all concentrations. It is resistant to dilute solutions of hydrochloric acid and sulfuric acid. Hafnium is attacked by all mineral acids if traces of fluorides are present. Hafnium is very resistant to attack by alkalies.

Occurrence and Mining

Hafnium and zirconium are always present together in naturally occurring minerals. The primary commercial source is zircon [14940-68-2] (zirconium orthosilicate). Zircon sand is found in heavy mineral sand layers of ancient ocean beaches. Principal zircon sand producing countries are Australia, South Africa, the United States, and Russia. Other countries producing zircon include India, Sri Lanka, Malaysia, China, Thailand, Sierra Leone, and Brazil, and production is planned in Indonesia and Madagascar. Zircon is always a coproduct from the mining of rutile and ilmenite mineral sands to supply the titanium oxide pigment industry (see PIGMENTS, INORGANIC; TITANIUM COMPOUNDS, INORGANIC). Baddeleyite [1490-68-2], a naturally occurring zirconium oxide, is available from South Africa and Russia.

Most of the heavy mineral sands operations in the world are similar. Typically the quartz sand overburden is bulldozed away to reach the heavy mineral sand layer, which usually has 2 to 8% heavy minerals. The excavation is flooded and the heavy mineral sands layer is mined by a floating dredge with a cutter-head-suction. The sand slurry is pumped to a wet-mill concentrator mounted on a barge behind the dredge. Wet concentration using screens, cones, spirals, and sluices removes roots, coarse sand, slimes, quartz, and other light minerals. The tailings are returned to the back end of the excavation. Rehabilitation of worked-out areas is about a 10-year project which includes replacing the overburden and topsoil to pre-existing levels and contours, and reestablishing the natural vegetation, usually from company-owned nurseries (see MINERAL PROCESSING AND RECOVERY).

The heavy mineral sand concentrates are scrubbed to remove any surface coatings, dried, and separated into magnetic and nonmagnetic fractions (see SEPARATION, MAGNETIC). Each of these fractions is further split into conducting and nonconducting fractions in an electrostatic separator to yield individual concentrates of ilmenite, leucoxene, monazite, rutile, xenotime, and zircon. Commercially pure zircon sand typically contains 64% zirconium oxide, 34% silicon oxide, 1.2% hafnium oxide, and 0.8% other oxides including aluminum, iron, titanium, yttrium, lanthanides, uranium, thorium, phosphorus, scandium, and calcium.

Zircon sands containing 3% hafnium oxide (6% Hf/(Hf + Zr)) have been found in Brazil and Nigeria, and higher concentrations of hafnium oxide have been found in altered zircons such as cyrtolite, malacon, alvite, and naëgite. In general, zircon sands containing higher amounts of hafnium also contain higher amounts of radioactivity.

Manufacture

Decomposition of Zircon.
Zircon sand is inert and refractory. Therefore the first extractive step is to convert the zirconium and hafnium portions into active forms amenable to the subsequent processing scheme. For the production of hafnium, this is done in the United States by carbochlorination as shown in Figure 1. In the Ukraine, fluorosilicate fusion is used. Caustic fusion is the usual starting procedure for the production of aqueous zirconium chemicals, which usually does not involve hafnium separation. Other methods of decomposing zircon such as plasma dissociation or lime fusions are used for production of some grades of zirconium oxide.

Carbochlorination.   Milled zircon and coke are reacted with hot chlorine gas in a fluidized bed using chlorine as the fluidizing medium:

$$(ZrHf)SiO_4 + 4\ C + 4\ Cl_2 \xrightarrow{1100^\circ C} (ZrHf)Cl_4 + SiCl_4 + 4\ CO$$

This reaction is endothermic and additional energy must be provided to sustain it, usually by induction heating, or by adding silicon carbide grain which chlorinates exothermically. The product gases are cooled below 200°C to condense and collect the zirconium–hafnium tetrachloride as a powder. The offgas stream then is refrigerated to obtain by-product silicon tetrachloride liquid.

Fluorosilicate Fusion.   Milled zircon and potassium hexafluorosilicate are heated together to yield potassium hexafluorozirconate and silica (10):

$$(ZrHf)SiO_4 + K_2SiF_6 \xrightarrow{700^\circ C} K_2(ZrHf)F_6 + 2\ SiO_2$$

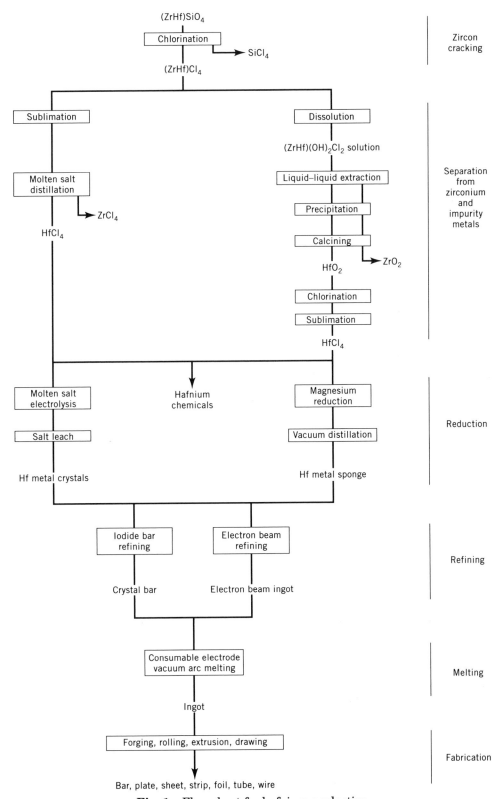

Fig. 1. Flow sheet for hafnium production.

The addition of potassium carbonate or chloride to the fusion mix reduces the loss of volatile silicon tetrafluoride, improving the conversion efficiency. The fused product is cooled, crushed, and leached with acidified hot water. The resulting hot solution of potassium hexafluorozirconate–hafnate is filtered to remove silica, then cooled to allow crystallization of the potassium hexafluorozirconate–hafnate. Many of the impurity metals remain in solution.

Separation of Hafnium. Many methods have been proposed for the separation of hafnium and zirconium; three different industrial methods are in use.

Liquid–Liquid Extraction. In the United States, both Teledyne Wah Chang Albany Corp. and Western Zirconium Division of Westinghouse use a liquid–liquid extraction process first proposed in Germany as an analytical procedure (11) (see EXTRACTION, LIQUID–LIQUID). This was modified into an industrial method at the Oak Ridge National Laboratory, and subsequently put into production at the U.S. Bureau of Mines, Albany, Oregon (12). In this process ammonium thiocyanate is added to an acidic solution of zirconium–hafnium hydroxide chloride [*22196-48-1*], $(ZrHf)(OH)_2Cl_2$. This solution is contacted with an immiscible methyl isobutyl ketone solution containing thiocyanic acid in a series of countercurrent stages. Both hafnium and zirconium thiocyanate complexes are soluble in both phases, but hafnium is preferentially distributed in the ketone phase. Hafnium is recovered from the ketone by scrubbing with dilute sulfuric acid which is then neutralized to precipitate hydrous hafnium oxide which is filtered and calcined to hafnium oxide. The reaction of milled hafnium oxide with carbon and chlorine in a fluidized bed at 900–1000°C produces hafnium tetrachloride. Before being reduced the hafnium tetrachloride is purified by sublimation in a nitrogen–hydrogen atmosphere.

Molten Salt Distillation. Hafnium tetrachloride is slightly more volatile than zirconium tetrachloride, but a separation process based on this volatility difference is impractical at atmospheric pressures because only solid and vapor phases exist. The triple point for these systems is at about 2.7 MPa (400 psia) and 400°C so that separation of the liquids by distillation would necessarily require a massive pressurized system (13).

In France, Compagnie Européene du Zirconium (CEZUS) now owned jointly by Pechiney, Framatome, and Cogema, uses a separation (14) based on the extractive distillation of zirconium–hafnium tetrachlorides in a molten potassium chloride–aluminum trichloride solvent at atmospheric pressure at 350°C. For feed, the impure zirconium–hafnium tetrachlorides from the zircon chlorination are first purified by sublimation. The purified tetrachlorides are again sublimed to vapor feed the distillation column containing the solvent salt. Hafnium tetrachloride is recovered in an enriched overhead fraction which is accumulated and reprocessed to pure hafnium tetrachloride.

Fluorozirconate Crystallization. Repeated dissolution and fractional crystallization of potassium hexafluorozirconate was the method first used to separate hafnium and zirconium (15), potassium fluorohafnate solubility being higher. This process is used in the Prinieprovsky Chemical Plant in Dnieprodzerzhinsk, Ukraine, to produce hafnium-free zirconium. Hafnium-enriched (about 6%) zirconium hydrous oxide is precipitated from the first-stage mother liquors, and redissolved in acid to feed ion-exchange columns to obtain pure hafnium (10).

Reduction. Hafnium oxide can be reduced using calcium metal to yield a fine, pyrophoric metal powder (see CALCIUM AND CALCIUM ALLOYS). This powder contains considerable oxygen contamination because of oxygen's high solubility in hot hafnium, and cannot be consolidated into ductile metal. To obtain low oxygen ductile hafnium, the feed must be an oxygen-free halide compound such as hafnium tetrachloride or potassium hexafluorohafnate [16871-86-6].

Kroll Process. Hafnium tetrachloride vapor is reduced by liquid magnesium in an inert atmosphere in a batch operation. Hafnium tetrachloride powder is charged into the annular upper chamber of a vertical cylindrical steel retort. Cast ingots of pure magnesium are placed inside a stainless-steel liner within the lower chamber. After the retort has been closed, evacuated, and backfilled with argon, heat is applied to the lower retort zone to melt the magnesium. Then the reaction is initiated by heating the upper zone to gradually sublime the hafnium tetrachloride. Hafnium tetrachloride vapors pass into the lower chamber and react with the liquid magnesium to form three micrometer-sized spheres of hafnium metal which settle to the bottom of the chamber as a hafnium–magnesium mud. The rate of this exothermic reaction is controlled by adjusting the upper zone temperature, thereby controlling the rate of tetrachloride sublimation. After the reaction, the reduction retort is cooled and unloaded. The reduction mass is placed in the upper portion of a vacuum distillation furnace which is then evacuated to a pressure of about 0.1 Pa (75 μm Hg) before heating. As the temperature passes above 700°C the magnesium chloride melts and drains into the cooler lower zone, and the magnesium begins to distill from the mud, allowing the hafnium spheres to slowly sinter into a porous sponge. The distillation is complete after several hours at 980°C. After the furnace is cooled, the sponge is broken by a hydraulic chisel operating in an argon atmosphere box. The sponge chunks are visually graded, then chopped to less than 19 mm and sampled (16).

Electrolysis. Electrowinning of hafnium, zirconium, and titanium has been proposed as an alternative to the Kroll process. Electrolysis of an all chloride hafnium salt system is inefficient because of the stability of lower chlorides in these melts. The presence of fluoride salts in the melt increases the stability of Hf^{4+} in solution and results in much better current efficiencies. Hafnium is produced by this procedure in France (17).

Refining. Kroll-process hafnium sponge and electrowon hafnium do not meet the performance requirements for the two principal uses of hafnium metal. Further purification is accomplished by the van Arkel-de Boer, ie, iodide bar, process (18) and by electron beam melting.

For van Arkel-de Boer refining, hafnium sponge or crystals are loaded around the interior periphery of a vertical cylindrical Inconel vessel. The vessel lid supports insulated electrodes from which hafnium wire filaments in hairpin shape are suspended. Iodine is added to the evacuated vessel which then is heated to 300°C. The iodine reacts with the sponge forming hafnium tetraiodide vapor which diffuses to the hafnium filament which is resistance-heated to 1300–1600°C. The tetraiodide thermally dissociates at the hot filament, depositing a film of hafnium and releasing iodine to react again with additional sponge.

$$Hf + 2\,I_2 \xrightarrow{\ 300°C\ } HfI_4 \xrightarrow{\ 1400°C\ } Hf + 2\,I_2$$

Oxygen and nitrogen stay gettered by the residual sponge, and other metals transfer poorly under these conditions so that the filament product, hafnium crystal bar, is much purer and therefore softer than the starting sponge. Crystal bar growth is slow and this refining procedure is costly.

Electron beam melting of hafnium in a high vacuum removes those impurities having partial pressure at the surface of the melt greater than the vapor pressure of hafnium, about 0.1 Pa (0.075 μm Hg) at 2500 K. Some oxygen is removable because hafnium monoxide is more volatile than hafnium. Only nitrogen and higher melting point metals, such as tantalum and tungsten, are not removed. To obtain the very purest hafnium, both refining procedures are used in sequence. Typical analyses of as-produced and refined hafnium are shown in Table 3.

Many of the impurities are much lower than the values shown in Table 3, but these analytical lower limits are typical and more than sufficient for all but special applications. Zirconium content can be from 0.01 to 4.5%, and is typically 0.5–2%, but this is a function of how far the separation process was carried, not a function of the reduction or refining processes.

Table 3. Analysis of Kroll Process, Electrowon, and Refined Hafnium, ppm

Impurities	Kroll process sponge	Electrowon crystals	Refined from Kroll sponge	
			Electron beam ingot	Iodide bar
Al	65	15	<10	<10
B	<0.2	<0.2	<0.2	<0.2
C	35	35	35	<20
Cd	<1	<1	<1	<1
Co	<5	<5	<5	<5
Cr	40	30	<10	<10
Cu	40	<40	<40	<40
Fe	250	100	<50	50
Mg	370	<10	<10	<10
Mn	<10	10	<10	<10
Mo	15	10	<10	<10
N	25	15	15	<10
Ni	15	35	<10	<10
Nb	<50	<50	<50	<50
Pb	<5	<5	<5	<5
O	900	600	300	90
Si	40	<40	<40	<40
Ta	<100	<50	<50	<50
Ti	95	30	<30	40
U	1	<1	<1	<1
V	<5	<5	<5	<5
W	<20	<20	<20	<20

Economic Aspects

Total hafnium available worldwide from nuclear zirconium production is estimated to be 130 metric tons annually. The annual usage, in all forms, is about 85 t. The balance is held in inventory in stable intermediate form such as oxide by the producers Teledyne Wah Chang (Albany, Oregon) and Western Zirconium in the United States; Cezus in France; Prinieprovsky Chemical Plant in Ukraine; and Chepetsky Mechanical Plant in Russia (crystal bar).

Demand for hafnium has not shown significant growth since the late 1980s, nor has pricing changed. Hafnium oxide is priced at $80–120/kg depending on quality; hafnium crystal bar about $180–220/kg; and wrought plate, sheet, tube, and wire, from $250–500/kg.

Specifications and Standards

Several ASTM Standard Specifications have been promulgated for hafnium materials: B737-90 for Hot-Rolled or Cold-Finished Hafnium Rod and Wire; B776-91 for Hafnium and Hafnium Alloy Strip, Sheet, and Plate; G2-88 and G2M-88 (metric) for Corrosion Testing of Products of Zirconium, Hafnium, and Their Alloys in Water at 680°F (633 K) or in steam at 750°F (673 K); C1076-92 for Nuclear-Grade Hafnium Oxide Pellets; and C1098-88 for Nuclear-Grade Hafnium Oxide Powder.

In addition, the following ASTM Standard Specifications are for the Nb–10Hf–1Ti–1Zr alloy known commercially as C103: B652-92 Niobium–Hafnium Alloy Ingots; B654-92 Niobium–Hafnium Alloy Foil, Sheet, Strip, and Plate; B655-92 Niobium–Hafnium Alloy Bar, Rod, and Wire.

Analytical Procedures

Analyses of alloys or ores for hafnium by plasma emission atomic absorption spectroscopy, optical emission spectroscopy (qv), mass spectrometry (qv), x-ray spectroscopy (see X-RAY TECHNOLOGY), and neutron activation are possible without prior separation of hafnium (19). Alternatively, the combined hafnium and zirconium content can be separated from the sample by fusing the sample with sodium hydroxide, separating silica if present, and precipitating with mandelic acid from a dilute hydrochloric acid solution (20). The precipitate is ignited to oxide which is analyzed by x-ray or emission spectroscopy to determine the relative proportion of each oxide.

Hafnium metal is analyzed for impurities using analytical techniques used for zirconium (19,21,22). Carbon and sulfur in hafnium are measured by combustion, followed by chromatographic or ir measurement of the carbon and sulfur oxides (19). Chromatographic measurement of liberated hydrogen follows the hot vacuum extraction or fusion of hafnium with a transition metal in an inert atmosphere (23,24).

Oxygen and nitrogen also are determined by conductivity or chromatographic techniques following a hot vacuum extraction or inert-gas fusion of hafnium with a noble metal (25,26). Nitrogen also may be determined by the Kjeldahl technique (19). Phosphorus is determined by phosphine evolution and flame-emission detection. Chloride is determined indirectly by atomic absorption or x-ray spectroscopy, or at higher levels by a selective-ion electrode. Fluoride can be determined similarly (27,28). Uranium and U-235 have been determined by inductively coupled plasma mass spectroscopy (29).

Inductively coupled plasma (icp) emission, direct current plasma (dcp), and inductively coupled plasma mass spectrometry (icp/ms) have taken over as the methods of choice for the simultaneous detection of metallic impurities in hafnium and hafnium compounds (29,30).

Health and Safety

High surface area forms of hafnium metal such as foil, fine powder, and sponge are very easily ignited, and fine machining chips can be pyrophoric. If ignited, hafnium can be extinguished with a blanket of argon or a layer of dry salt or dry sand. The use of water to extinguish burning hafnium is extremely hazardous. The resulting steam and hydrogen may disperse the burning fragments and spread the fire. Damp hafnium sponge or powder burns more quickly and completely than dry hafnium. Methods for safe handling of both hafnium and zirconium are discussed in the literature (32).

Most hafnium compounds require no special safety precautions because hafnium is nontoxic under normal exposure. Acidic compounds such as hafnium tetrachloride hydrolyze easily to form strongly acidic solutions and to release hydrogen chloride fumes, and these compounds must be handled properly. Whereas laboratory tests in which soluble hafnium compounds were injected into animals did show toxicity, feeding test results indicated essentially no toxicity when hafnium compounds were taken orally (33,34).

Uses

Nuclear. The first and primary market for many years has been for control rods in the pressurized light water reactors which power many naval vessels. Hafnium excels in this application because of its excellent hot water (450°C) corrosion resistance, good ductility and machinability, and large neutron absorption cross section. It absorbs both thermal and epithermal neutrons and its absorption cross section decreases slowly after long periods of neutron irradiation. The existence of several successive hafnium isotopes having large cross sections permits the neutron absorption by one hafnium isotope to result in the formation of a new hafnium isotope which also has a large absorption cross section (see Table 2). The ability of hafnium to absorb neutrons above thermal energies (resource energy absorption) makes it more effective as an absorbing material. Hafnium has not been used significantly in commercial nuclear power reactors.

Hafnium neutron absorption capabilities have caused its alloys to be proposed as separator sheets to allow closer spacing of spent nuclear fuel rods in interim holding ponds. Hafnium is the preferred material of construction for certain critical mass situations in spent fuel reprocessing plants where hafnium's excellent corrosion resistance to nitric acid is also important.

Alloying. Hafnium is used as an alloying element in high temperature application alloys such as superalloys and refractory metal alloys where hafnium additions improve the high temperature tensile and creep strength (see HIGH TEMPERATURE ALLOYS). Hafnium reacts with carbon, nitrogen, or oxygen in these alloys to form a fine dispersion of hafnium carbide, nitride, and oxide which provide second-phase particle dispersion strengthening. This effect is maintained at high temperature because of the high melting points and thermodynamic stability of these components. The largest use for hafnium as of 1994 is as a 1–2% addition in the superalloys used for cast vanes and turbine blades placed in the hottest stages following the combustion zone of jet aircraft engines. Small additions of hafnium and carbon react to form second-phase dispersions in tantalum, molybdenum, and tungsten alloys (35–37), including the tantalum alloys T-111, T-222, and Astar 811C, and the molybdenum alloy MHC.

In larger proportions, hafnium provides both dispersed particle strengthening and solid solution strengthening in niobium-base alloys (38), including such as C103 (Nb–10 Hf–1 Ti–1 Zr), C-129Y (Nb–10 W–10 Hf–0.7 Y), and WC 3015 (Nb–30 Hf–15 W–1.5 Zr). Hafnium-based alloys containing 20–27% tantalum and up to 2% molybdenum are unique refractory alloys which exhibit high oxidation resistance.

Hafnium is an effective solid solution strengthener at higher temperatures for other alloys such as nickel aluminides (39,40).

Other Uses as Metal. Very pure hafnium forms effective diffusion barriers for multilevel integrated circuits (qv). Hafnium metal stubs are being used in water-cooled copper holders as cathodes for plasma arc cutting, and hafnium is replacing tungsten in copper coated welding (qv) tips. Hafnium has been proposed as an ingredient in permanent magnets (see MAGNETIC MATERIALS).

Uses as Compounds. Hafnium oxide is a very refractory compound which has found some use in specialized refractories (qv) such as insulating sheaths for tungsten tungsten–rhenium thermocouples above 1500°C. Hafnium oxide [12055-23-1] sputtering targets are used for depositing coatings (qv) on optical components for lasers (qv) (41). Hafnium oxide is one of several hafnium compounds proposed as luminescents and phosphors during uv and x-ray excitation (42–46). Hafnium oxide is reactively sputtered on the surface of a magnetic recording disk as a wear-resistant coating (47).

Hafnium oxide 30–40 mol % titanium oxide ceramics (qv) exhibit a very low coefficient of thermal expansion over the temperature range of 20–1000°C. A 45–50 mol % titanium oxide ceramic can be heated to over 2800°C with no crystallographic change (48).

Hafnium nitride [25817-87-2], which has been used as a chemically stable surface coating on steel-grade cemented-carbide cutting tools (49,50), does not weld to the steel being cut, thereby reducing friction and tool wear (see TOOL MATERIALS). Hafnium nitride's gold coloring has been used as a decorative coating, and its optical properties (51) make it a candidate for wavelength-selective optical films that could be used on solar collectors (see SOLAR ENERGY). Hafnium nitride's high emissivity in the visible spectrum at high temperature has led to a

Table 4. Physical Properties of Some Hafnium Compounds

Property	HfB_2	HfC	HfO_2	HfN	HfP
CAS Registry Number	[12007-23-7]	[12069-85-1]	[12055-23-1]	[25817-87-2]	[12325-59-6]
melting point, °C	3370	3830	2810	3330	
sublimation point, °C					
specific gravity, g/cm³					
theoretical	11.2	12.7		13.84	9.78
measured	10.5	12.2	9.68		
resistivity at RT, $\mu\Omega\cdot cm$	8.8	37	$>10^8$	33	
color	gray	gray	white	gold	
coefficient of thermal expansion, 10^{-6}	5.7	6.59	6.1	6.9	
hardness,[b] kgf/mm²	2900[c]	2300[d]	1050[e]	1640[f]	
structure	hexagonal	face-centered cubic	monoclinic[f]	cubic	hexagonal
lattice parameters, nm					
a	0.3141	0.4640	0.51156	0.452	0.365
b			0.51722		1.237
c	0.3470		0.52948		

[a]At 3.34 MPa (33 atm).
[b]See HARDNESS; 1 kgf/mm² = 9.8 MPa.
[c]Vickers' hardness.
[d]Knoop hardness.
[e]Diamond pyramid hardness (DPH), 2 kg.
[f]Microhardness, 50 gf/mm² = 490 kPa.
[g]Tetragonal above 1600°C.

proposed use (52) as a coating on incandescent light filaments to improve visible light output.

Hafnium carbide [12069-85-1] can be used as surface coating on cemented-carbide cutting tools, shows promise as a stable field emission cathode (53), and has been alloyed with niobium carbide to produce a substitute for tantalum carbide as a constituent of cemented carbide tool bits (see CARBIDES) (54).

Hafnium tetrafluoride [13709-52-9] is one component in the cladding layer of a proposed zirconium fluoride glass optical waveguide fiber composition which is expected to have a lower intrinsic light absorption than fused quartz optical fiber (see GLASS; FIBER OPTICS; FLUORINE COMPOUNDS, INORGANIC–ZIRCONIUM).

After the discovery of isotactic polymerization of propylene using zirconocene catalysts, structurally analogous hafnium catalysts produced from hafnium tetrachloride [13499-05-3] were found to produce high yields of high molecular weight polypropylene (55), but not enough to lead to commercial development.

Hafnium Compounds

Most hafnium compounds have been of slight commercial interest aside from intermediates in the production of hafnium metal. However, hafnium oxide, hafnium carbide, and hafnium nitride are quite refractory and have received considerable study as the most refractory compounds of the Group 4 (IVB) elements. Physical properties of some of the hafnium compounds are shown in Table 4.

Table 4. (*continued*)

HfS$_2$	HfSe$_2$	HfSi$_2$	HfF$_4$	HfCl$_4$	HfBr$_4$	HfI$_4$
[18855-94-2]	[12162-21-9]	[12401-56-8]	[13709-52-9]	[13499-05-3]	[13777-22-5]	[13777-23-6]
		1750	>968	432^a	424^a	449^a
			968	317	322	393
		8.03			5.09	
6.03	7.46	7.2			4.90	
	20	60				
purple-brown	dark brown		white	white	white	yellow-orange
		930^f				
hexagonal	hexagonal	rhombic^g	monoclinic	monoclinic	cubic	cubic
0.364	0.375	0.3677	0.957	0.631	1.095	1.176
0.584	0.616	1.455	0.993	0.7407		
		0.3649	0.773	0.6256		

Hafnium Boride. Hafnium diboride [12007-23-7], HfB$_2$, is a gray crystalline solid. It is usually prepared by the reaction of hafnium oxide with carbon and either boron oxide or boron carbide, but it can also be prepared from mixtures of hafnium tetrachloride, boron trichloride, and hydrogen above 2000°C, or by direct synthesis from the elements. Hafnium diboride is attacked by hydrofluoric acid but is resistant to nearly all other reagents at room temperature. Hafnium dodecaboride [32342-52-2], HfB$_{12}$, has been prepared by direct synthesis from the elements (56).

Hafnium Tetrahydridoborate. Hafnium tetrahydridoborate [25869-93-6], Hf(BH$_4$)$_4$, is the most volatile compound of hafnium: mp, 29°C; bp, 118°C; and vapor pressure, 2 kPa (14.9 mm Hg) at 25°C. It is prepared by the reaction of hafnium tetrachloride with lithium tetrahydridoborate, followed by distillation of the hafnium tetrahydridoborate and separation from lithium chloride.

Hafnium Carbide. Hafnium carbide, HfC, is a dark gray brittle solid. This carbide can be prepared by heating an intimate mixture of the elements or by the reaction of hafnium tetrachloride with methane at 2100°C, but is commonly produced by the reaction of hafnium oxide with lampblack in graphite crucibles in hydrogen at 1900–2300°C or under vacuum at 1600–2100°C. The carbide is not a true stoichiometric compound, but a solution of carbon at preferred interstitial sites in a face-centered cubic hafnium lattice. The hafnium–carbon phase diagram shows that hafnium carbide is homogeneous over the composition range of 34–48 atomic % carbon at 2200°C (57).

Hafnium carbide is inert to most reagents at room temperature, but is dissolved by hydrofluoric acid solutions which also contain an oxidizing agent. Above 250°C, hafnium carbide reacts exothermically with halogens to form hafnium tetrahalide, and above 500°C, with oxygen to form hafnium dioxide. At higher temperatures in a flow of hydrogen, hafnium carbide slowly loses some of its carbon.

Hafnium Halides.   Hafnium tetrafluoride, HfF_4, can be prepared by careful thermal decomposition of ammonium fluorohafnate in an oxygen-free atmosphere, or by passing anhydrous hydrogen fluoride over hafnium tetrachloride at 300°C. The direct synthesis from the elements is incomplete because the product fluoride forms a film on the metal.

Ammonium fluorohafnate [16925-24-9], $(NH_4)_2HfF_6$, or potassium fluorohafnate [16871-86-6], K_2HfF_6, can be prepared by crystallization from an aqueous hydrofluoric acid solution by addition of ammonium fluoride or potassium fluoride, respectively.

Hafnium tetrachloride, $HfCl_4$, can be made by reaction of chlorine with hafnium above 317°C or with an intimate mixture of hafnium oxide and carbon above 700°C, or by reaction of carbon tetrachloride and hafnium oxide above 450°C. Hafnium tetrachloride reacts with water, forming hafnium oxide chloride and hydrochloric acid. It reacts with almost all hydroxide-containing organic compounds to form addition compounds with the subsequent evolution of hydrogen chloride. Hafnium tetrachloride vapor reacts with steam to form finely divided hafnium dioxide. It forms addition compounds with molten alkaline halides, eg, $2KCl \cdot HfCl_4$ [19381-63-6], iron chloride, and aluminum chloride. These addition compounds decompose at higher temperatures. Hafnium tetrachloride can be reduced to lower chlorides by reaction with hafnium metal or by reaction with aluminum metal in liquid aluminum chloride (58). Hafnium tetrachloride is the starting material used in preparation of hafnium-containing organometallic compounds (59).

Hafnium hydroxide chloride heptahydrate [93245-94-4], $Hf(OH)_2Cl_2 \cdot 7H_2O$ occurs as white tetragonal crystals. On heating, it first dissolves in its hydration water but rapidly loses both water and hydrogen chloride. On continued heating, further decomposition yields hafnium dioxide. The hydoxide chloride is produced by adding hafnium tetrachloride to water or by dissolving hydrous hafnium oxide in hydrochloric acid. Hafnium hydroxide chloride is soluble in water. The solubility in hydrochloric acid decreases with increasing acid strength to a minimum solubility at 8.5 M hydrochloric acid. The solubility in hydrochloric acid is increased greatly at elevated temperatures. Repeated crystallization of hafnium hydroxide chloride from hot concentrated hydrochloric acid is the classical method to obtain hafnium salts free of all metallic impurities except zirconium. The hydroxide chloride is the common starting material for the preparation of other aqueous hafnium compounds.

Hafnium tetrabromide [13777-22-5], $HfBr_4$, is very similar to the tetrachloride in both its physical and chemical properties. Hafnium tetraiodide [13777-23-6], HfI_4, is produced by reaction of iodine with hafnium metal at 300°C or higher. At temperatures above 1200°C, the iodide dissociates to hafnium metal and iodine. These two reactions are the basis for the iodide-bar refining process. Hafnium iodide is reported to have three stable crystalline forms at 263–405°C (60).

Hafnium Hydride.   Hafnium reacts reversibly with hydrogen to form hafnium hydride [12656-74-5], $HfH_{1.6-2.0}$. Below 250°C, the reaction rate becomes quite slow. The proportion of hydrogen depends on the temperature and pressure of hydrogen at which the hafnium is exposed. As hydrogen is absorbed, the hafnium transforms from hexagonal metal to face-centered cubic hydride and then to face-centered tetragonal hydride. At room temperature, the face-centered cubic form exists as a single phase at an H/Hf ratio of 1.7–1.8 (61). The face-centered tetragonal form exists as a single phase above H/Hf of 1.87 at room temperature. The face-centered tetragonal phase has been reported to be stable only below 407°C.

Hafnium hydride is brittle and easily crushed to very fine particle sizes. It is usually produced as an intermediate in the process of making hafnium powder from massive hafnium metal. The hydrogen can be removed by high vacuum pumping above 600°C.

Hafnium Nitride. Hafnium nitride, HfN, a gold-colored brittle solid, is prepared by heating hafnium metal to 1000–1500°C in an atmosphere of nitrogen or ammonia. The higher temperatures yield a product with a nitrogen-to-metal ratio approaching one, but the reaction is slow because of the slow diffusion of nitrogen through the protective nitride layer. Hafnium nitride also can be prepared by the reaction of hafnium tetrachloride vapor with nitrogen in a hydrogen atmosphere at >1000°C. This is the basis for a chemical vapor deposition process used to deposit a thin layer of hafnium nitride on steel-grade cemented-carbide cutting tools. The hafnium nitride layer reduces the frictional forces and wear, thereby increasing the life of the tool by a factor of 6–8. A straw-colored higher nitride, Hf_3N_4 [104382-33-4], is produced by dual ion beam sputter deposition (62) and by low temperature chemical vapor deposition at 250–400°C (63).

Hafnium Oxide. Two oxides of hafnium, hafnium monoxide [12029-22-0], HfO, and HfO_2, are known to exist but only the dioxide is stable under ordinary conditions. Gaseous hafnium monoxide can be present at >2000°C, especially when the partial pressure of oxygen is low. Hafnium monoxide is probably the compound form in which oxygen is evolved when hafnium metal is melted in an electron-beam melting furnace. HfO(g) is the species observed mass spectrometrically when hafnium dioxide vaporizes.

Hafnium dioxide is formed by ignition of hafnium metal, carbide, tetrachloride, sulfide, boride, nitride, or hydrous oxide. Commercial hafnium oxide, the product of the separation process for zirconium and hafnium, contains 97–99% hafnium oxide. Purer forms, up to 99.99%, are available.

Pure hafnium dioxide transforms into the tetragonal structure at about 1700°C. The difference between the heating transformation temperature and the cooling transformation temperature is 40–80°C, considerably less than for zirconia. The hafnium dioxide undergoes a shrinkage of about 3% upon transforming into the tetragonal phase. The tetragonal form converts to a cubic polymorph having the fluorite structure above 2600°C. The fluorite structure can be rendered stable at lower temperatures by addition of erbium oxide, yttrium oxide, calcium oxide, or magnesium oxide. Compared to zirconium oxide, the higher transformation temperature of hafnium oxide, pure or stabilized, has aroused considerable interest (64) and should lead to several specialized applications. Reference 64 is a thorough review of hafnium oxide and hafnium oxide-toughened ceramics.

At room temperature, hafnium dioxide is slowly dissolved by hydrofluoric acid. At elevated temperatures, hafnium dioxide reacts with concentrated sulfuric acid or alkali bisulfates to form various sulfates, with carbon tetrachloride or with chlorine in the presence of carbon to form hafnium tetrachloride, with alkaline fluorosilicates to form alkali fluorohafnates, with alkalies to form alkaline hafnates, and with carbon above 1500°C to form hafnium carbide.

The hydrous oxide, $HfO_2 \cdot xH_2O$, is precipitated from acidic solutions by addition of ammonium hydroxide or dilute alkaline solutions. However, the hydrous oxide exhibits a limited solubility in strongly alkaline solutions (65). The existence and relative stability of soluble alkaline peroxy compounds has been demonstrated (66).

Hafnium Sulfides.   Several sulfides of hafnium have been prepared, including Hf_2S, HfS, and HfS_2, by the reaction of the mixed elements at 500°C or by passing hydrogen sulfide over heated hafnium powder. Of these, hafnium disulfide [18855-94-2] is fairly well characterized. The disulfide is a layered compound reported to have good lubricating properties similar to other layered chalcogenide compounds such as NbS_2, MoS_3, and WS_3. The electrical resistance and sulfur vapor pressure are given as a function of temperature and composition in Reference 67.

Hafnium Carbonate.   Basic hafnium carbonate [124563-80-0], $Hf_2(OH)_4CO_3 \cdot XH_2O$, is prepared as a wet paste by reaction of a slurry of basic hafnium sulfate [139290-14-5], $Hf_5O_7(SO_4)_3 \cdot XH_2O$, and sodium carbonate, then filtering. The basic carbonate has a short shelf life and is preferably prepared as needed. It is a starting material for the preparation of various hafnium carboxylates.

Hafnium Acetate.   Hafnium acetate [15978-87-7], $Hf(OH)_2(CH_3COO)_2$, solutions are prepared by reacting the basic carbonate or freshly precipitated hydroxide with acetic acid. The acetate solution has been of interest in preparing oxide films free of chloride or sulfate anions.

BIBLIOGRAPHY

"Hafnium" in *ECT* 1st ed., Vol. 7, pp. 340–344, by E. G. Enck, Foote Mineral Co., and E. M. Larsen, University of Wisconsin; "Hafnium and Hafnium Compounds" in *ECT* 2nd ed., Vol. 10, pp. 754–768, by R. H. Nielsen, Wah Chang Corp.; in *ECT* 3rd ed., Vol. 12, pp. 67–80, by R. H. Nielsen, Teledyne Wah Chang Albany.

1. D. Coster and G. von Hevesy, *Nature* **3**, 252 (1923).
2. *CRC Handbook of Chemistry and Physics*, 73rd ed., CRC Press, Boca Raton, Fla., 1992, pp. 4–61.
3. K. L. Komarek, ed., *Hafnium: Physico-Chemical Properties of Its Compounds and Alloys*, International Atomic Energy Agency, Vienna, 1981, pp. 11,13,14,16. Covers thermochemical properties, phase diagrams, crystal structure, and density data on hafnium, hafnium compounds, and alloys.
4. Ref. 2, pp. 5–79.
5. Ref. 2, pp. 12–34.
6. *Hafnium*, brochure, Teledyne Wah Chang Corp., Albany, Oreg., 1987, p. 11.
7. C. Y. Ho, R. W. Powell, and P. E. Liley, *J. Phys. Chem. Ref. Data.* **3**(Suppl. 1), 747–754 (1974).
8. E. A. Brandes and G. B. Brooks, eds., *Smithells Metals Reference Book*, 7th ed., Butterworth-Heineman Ltd., Oxford, U.K., 1992, p. 15-2.
9. Ref. 8, p. 14-3.
10. N. P. Sajin and E. A. Pepelyaeva, in *Proceedings of the International Conference on Peaceful Uses of Atomic Energy*, Vol. 8, United Nations, New York, 1958, p. 559.
11. W. Fischer and W. Chalybaeus, *Z. Anorg. Allg. Chem.* **225**, 79 (1947).

12. J. H. McClain and S. M. Shelton, in C. R. Tipton, Jr., ed., *Reactor Handbook*, 2nd ed., Vol. 1, Interscience Publishers, Inc., New York, 1960, pp. 64–73.
13. U.S. Pat. 2,852,446 (Sept. 16, 1958) M. L. Bromberg (to E. I. du Pont de Nemours & Co., Inc.).
14. L. Moulin, P. Thouvenin, and P. Brun, in D. G. Franklin and R. B. Adamson, eds., *Zirconium in the Nuclear Industry: Sixth International Symposium*, ASTM STP 824, American Society for Testing and Materials, Philadelphia, 1984, pp. 37–44.
15. G. von Hevesy, *Chem. Rev.* **2**, 1 (1925).
16. H. P. Holmes, M. M. Barr, and H. L. Gilbert, *USBM Rep. Invest. 5169*, U.S. Bureau of Mines, U.S. Government Printing Office, Washington, D.C., 1955.
17. A. P. Lamaze and D. Charquet, in K. C. Liddell and co-workers, eds., *Refractory Metals: Extraction, Processing and Applications*, The Minerals, Metals & Materials Society, Warrendale, Pa., 1990, pp. 231–253.
18. E. M. Sherwood and I. E. Campbell, in D. E. Thomas and E. T. Hayes, eds., *The Metallurgy of Hafnium*, U.S. Government Printing Office, Washington, D.C., 1960, pp. 107–118. A complete treatise on all aspects of hafnium except chemical properties.
19. R. T. Van Santen and co-workers, in F. D. Snell and L. S. Ettre, eds., *Encyclopedia of Industrial Chemical Analyses*, Vol. 14, Wiley-Interscience, New York, 1971, pp. 103–148.
20. R. B. Hahn and E. S. Baginski, *Anal. Chim. Acta* **14**, 45 (1956).
21. R. B. Hahn, in I. M. Kolthoff, P. J. Elving, and E. B. Sandell, eds., *Treatise on Analytical Chemistry*, Part 2, Vol. 5, Interscience Publishers, New York, 1961, pp. 61–138.
22. *Suggested Methods for Analysis of Metals, Ores, and Related Material*, 9th ed., E-2 Sm 8-25, American Society for Testing and Materials, Philadelphia, 1992.
23. R. K. McGeary, in *A Symposium on Zirconium and Zirconium Alloys*, American Society for Testing and Materials, Philadelphia, 1953, p. 168.
24. W. G. Guldner, *Talanta* **8**, 191 (1961).
25. N. G. Smiley, *Anal. Chem.* **27**, 1098 (1955).
26. P. Elbing and E. G. Goward, *Anal. Chem.* **32**, 1610 (1960).
27. J. Surak and co-workers, *Anal. Chem.* **32**, 17 (1960).
28. J. Lingane, *Anal. Chem.* **39**, 881 (1967).
29. G. L. Beck and O. T. Farmer, *J. of Analytical Atomic Spectrometry* **3**, 771–773 (1988).
30. G. F. White and C. J. Pickford, AERE-M3235, Atomic Energy Research Establishment, Environmental and Medical Sciences Division, AERE, Harwell, U.K., 1982.
31. J. Schlewitz and M. Shields, *At. Absorpt. Newsl.* **10** 39, 43 (1971).
32. J. Schemel, *ASTM Manual on Zirconium and Hafnium*, ASTM STP 639, American for Testing and Materials, Philadelphia, 1977. Covers safe handling of hafnium metal.
33. *Documentation of the Threshold Limit Values and Biological Exposure Indices*, Vol. 2, 6th ed., American Conference of Governmental Industrial Hygienists, Inc., Cincinnati, Ohio, 1991, p. 719.
34. B. Venugopal and T. D. Luckey, *Metal Toxicity in Mammals: Chemical Toxicity of Metals and Metalloids*, Vol. 2, Plenum Press, New York, 1978.
35. A. Luo, K. S. Shin, and D. L. Jacobson, *Acta Metall. Mater.* **40**(9), 2225–2232 (1992).
36. B. H. Tsao and co-workers, *Proceedings of the 27th Intersociety Energy Conversion Engineering Conference*, Vol. 2, San Diego, Aug. 3–7, 1992, pp. 2.5–2.10.
37. J. A. Shields, *Adv. Mater. Proc.*, 28–36 (Oct. 1992).
38. U.S. Pat. 4,931,254 (June 5, 1990), M. R. Jackson (to General Electric Co.).
39. K. Vedula and co-workers, *High Temperature Ordered Intermetallic Alloys I*, Vol. 39, Materials Research Society, Pittsburgh, 1985, p. 411.
40. M. Takeyama and C. T. Liu, *J. Mater. Res.* **5**, 1189–1196 (1990).

41. M. Fritz and co-workers, *Proceedings of the 32nd Annual Technical Conference—Society of Vacuum Coaters*, Albuquerque, N.M., 1989, pp. 264–269.
42. U.S. Pat. 4,295,989 (Oct. 20, 1981) P. H. Klein, A. Addamiano, and R. Allen (to the United States of America as represented by the Secretary of the Navy).
43. P. S. Bryan and S. A. Ferranti, *J. Luminescence*, **31 & 32**, 117–119 (1984).
44. Can. Pat. 1,082,877 (Aug. 5, 1980), C. F. Chenot, J. E. Mathers, and F. N. Shaffer (to GTE Sylvania, Inc.).
45. L. H. Brixner and G. Blasse, *J. Solid State Chemistry* **91**, 390–393 (1991).
46. U.S. Pat. 5,112,700 (May 12, 1992), P. M. Lambert and co-workers (to Eastman Kodak Co.).
47. U.S. Pat. 5,078,846 (Jan. 7, 1992) M. S. Miller and R. L. Peterson (to Seagate Technology, Inc.).
48. S. R. Skaggs, *Rev. Int. Hautes Tempér.Réfractor* **16**, 157–167 (1979).
49. J. J. Oakes, *Thin Solid Films* **107**, 159–165 (1983).
50. A. J. Perry, M. Grösl, and B. Hammer, *Thin Solid Films* **129**, 263–279 (1985).
51. B. Karlsson and C. G. Ribbing, *Proceedings of the SPIE International Society of Optical Engineering* **324**, 52–57 (1982).
52. U.S. Pat. 5,148,080 (Sept. 15, 1992), R. J. Von Thyne (to Hilux Development).
53. W. A. Mackie and co-workers, *J. Vac. Sci. Technol. A* **10**(4), 2852–2856 (1992).
54. P. H. Booker and R. E. Curtis, *Cutting Tool Eng.* **30**(9/10), 18 (1976).
55. J. A. Ewen and co-workers, *J. Am. Chem. Soc.* **109**, 6544–6545 (1987).
56. J. F. Cannon and P. B. Farnsworth, *J. Less-Common Metals* **92**, 359–368 (1983).
57. Ref. 3, p. 67.
58. E. M. Larsen, in H. J. Emeléus and A. F. Sharpe, eds., *Advances in Inorganic Chemistry and Radiochemistry*, Academic Press, Inc., New York, 1988, pp. 92–97. Covers some aspects of chemical behavior of hafnium.
59. D. J. Cardin, M. F. Lappert, and C. L. Raston, *Chemistry of Organo-Zirconium and -Hafnium Compounds*, Hasted Press, Division of John Wiley & Sons, Inc., New York, 1986. Excellent for organometallic chemistry of zirconium and hafnium.
60. F. D. Stevenson and C. E. Wicks, *J. Chem. Eng. Data* **10**, 33 (1965).
61. W. M. Mueller, J. P. Blackledge, and G. G. Libowitz, *Metal Hydrides*, Academic Press, Inc., New York, 1968, p. 322.
62. B. O. Johansson and co-workers, *J. Mater. Res.* **1**, 442–451 (1986).
63. R. Fix, R. G. Gordon, and D. M. Hoffman, *Chem. Mater.* **3**, 1138–1148 (1991).
64. J. Wang, H. P. Li, and R. Stevens, *J. Mater. Sci.* **27**, 5397–5430 (1992). An excellent review on hafnium oxide.
65. F. K. McTaggart, *Rev. Pure Applied Chem. (Australia)* **1**, 152 (1951).
66. F. R. Duke and R. F. Bremer, *Iowa State Coll. J. Science* **25**, 493 (1951).
67. J. Rasneur, C. Cauchemont, and F. Marion, *C. R. Acad. Sci. (Paris) Ser. C* **283**(10), 409 (1976).

General References

References 3, 18, 32, 58, 59, and 64 are good general references.

K. L. Komarek, ed., *Hafnium: Physico-Chemical Properties of Its Compounds and Alloys*, International Atomic Energy Agency, Vienna, 1981.

D. E. Thomas and E. T. Hayes, eds., *The Metallurgy of Hafnium*, U.S. Government Printing Office, Washington, D.C., 1960.

J. Wang, H. P. Ki, and R. Stevens, *J. Mater. Sci.* **27**, 5397–5430 (1992).

D. J. Cardin, M. F. Lappert, and C. L. Ralston, *Chemistry of Organo-Zirconium and -Hafnium Compounds*, Hasted Press, Division of John Wiley & Sons, Inc., New York, 1986.

P. C. Wailes, R. S. P. Coutts, and H. Weigold, *Organometallic Chemistry of Titanium, Zirconium, and Hafnium*, Academic Press, Inc., New York, 1974. Excellent for organometallic chemistry of zirconium and hafnium.

E. Negishi and T. Takahashi, *Aldrichimica ACTA* **18**(2), 31–48 (1985). Excellent for organometallic chemistry of zirconium and hafnium.

RALPH H. NIELSEN
Teledyne Wah Chang Corporation

HAHNIUM.   See ACTINIDES AND TRANSACTINIDES.

HAIR PREPARATIONS

The phenomenal growth of the personal care market since the 1940s is a result of socioeconomic changes combined with an increasing focus on personal aesthetics, assisted by affordability of products. The attempt to satisfy the freshly awakened needs of the consumer and the drive for competitive advantages among markets has led to a variety of grooming aids and products; ie, shampoos to cleanse the hair, hair rinses and conditioners to make it soft and combable, hair colorants and permanent waves to impart to hair properties it did not have, and setting preparations and hair sprays to keep hair in the desired style.

Hair products are normally cosmetics and are thus subject to all laws and regulations that control the labeling and claims of all cosmetic products. There are, however, several significant variations to this premise, ie, hair colorants, professional use only products, and products that make drug claims.

Structure and Composition of Hair.   The essential growth structures of hair are follicles which are deeply invaginated in the scalp tissue in the tens of thousands. At the base of each follicle, the cells proliferate. As they stream upward, the complex processes of protein synthesis, structural differentiation, alignment, and keratinization transform their cytoplasm into the tough and resilient material which is known as hair. Hair grows at a rate of about 1 cm per month for a period of 3–5 years, followed by a resting period of 4–6 months, during which the old hair is shed and a new growth begins.

Scalp hair is typically 50–80 μm in diameter and its exterior consists of a layer of flat, imbricated cuticle scales pointing outward from root to tip. This arrangement of cuticle cells permits better mechanical retention of the fiber in the follicle and also serves as a self-cleaning feature. Although the individual scales are thin, ie, 0.5 μm, they are long and overlap each other to form a continuous multilayered shield (3–4 μm) around the fiber. Enveloped by the protective sheath of the cuticle is hair cortex which constitutes the bulk of the fiber. The cortical cells are fibrillar in nature, highly elongated, and oriented along the length of hair. Dispersed throughout the structure of cortex are pigment particles called melanin. Their number, chemical character, and distribution pattern determine the color of hair. In some hairs, centrally located vacuolated medulla cells are also present.

Chemically, hair is a biopolymer composed largely of cystine-cross-linked proteins termed keratins. Two principal protein fractions have been isolated from hair, ie, low and high sulfur proteins. The low sulfur fraction consists of protein of high molecular weight and high degree of molecular organization, ie, α-helical; the protein of the high sulfur group are of low molecular weight and of unknown structural pattern. Electron microscope studies reveal that both proteins participate in a biphase composite, filament-matrix texture which is the dominant structural element of hair cortex. The filaments are composed of low sulfur proteins and the surrounding matrix is made up of high sulfur proteins. The structure and chemical composition of the cuticle differs from that of the cortex, and cuticle cells do not seem to contain any organized low sulfur proteins. The distal zone of each cuticle is heavily cross-linked by cystine; this fact, in conjunction with the multilayered structure, makes the cuticle a formidable barrier to penetration of materials into the interior of the hair.

Although hair of different racial origin differs in shape, degree of waviness (curl), and color, there is very little difference in the underlying chemical composition and physical structure. The rate of reaction with a variety of chemical reagents and most physical properties are similar (1,2). Differences between hair from different ethnic groups are much smaller than the variation in the properties of hair taken from different individuals within one ethnic group. Compared to Caucasian hair, Negro hair is more oval in the shape of its cross-section, and is much curlier. The tight curls are occasionally associated with unevenness in fiber diameter, resulting in weak spots along the fiber length. These could cause problems during chemical treatments as well as during hot combing. Asian hair tends to be more perfectly round than Caucasian hair and somewhat thicker in diameter, on the average (1). The greater fiber diameter results in a slow uptake of dyes because the ratio of surface area to volume is smaller.

Shampoos

One of the largest segments of the hair care market is the shampoo category. With development of mass marketing, the number of shampoos available has reached immense proportions; this has been aided by the availability of synthetic detergents. Synthetic detergents, which have replaced (ca 1993) most soap-based products used in shampoos, allow for greater formulating flexibility and control, and meet new product standards. In the early 1990s a shampoo must not only cleanse and have tolerance to hard water, but it should also be able to provide different performance effects. Through the use of synthetic detergents, new technology, and novel additives, various options in formulating offer consumers a number of alternative shampoo types (see also DETERGENCY; SURFACTANTS).

Properties.   The primary purpose of a shampoo is to clean the hair and scalp. In its cleansing process the typical shampoo must be able to remove the various soils found on the hair and scalp, ie, natural oily exudates and scales; conditioners and setting products that may be applied; and airborne soils that accumulate on the hair and scalp. The shampoo should leave hair soft, lustrous, and in a manageable condition without leaving a harsh, dry, raspy feel. With the use of synthetic detergents there is no problem in providing good cleansing; in fact the development chemist is concerned with product over-cleaning. Thus, a good performing shampoo should not completely remove all the oils while cleaning the hair.

There are a number of factors in formulating an acceptable shampoo product. In addition to cleansing, the shampoo should have good lathering properties. Although it has nothing to do with cleansing, the lather connotes a pleasant consumer experience, allowing easy spreadability of the shampoo during the cleansing process. Lathering of the shampoo should take place in either hard or soft water and the lather should be easily and completely removed from the hair without leaving a residue. Further, an acceptable shampoo should be safe for repeated use, nontoxic and nonirritating, adequately preserved, chemically and physically stable, and not damaging to the eyes; it also should have a pleasant fragrance.

Product Forms. Shampoos generally consist of an aqueous solution or dispersion of one or more cleansing additives, together with other ingredients to enhance performance and consumer acceptability, ie, foaming enhancers, preservatives, colors, fragrances, and pearling agents in the case of opaque shampoos. Other additives found in shampoo compositions include thickeners, conditioners, antidandruff agents, lime soap dispersants, sequestrants, and buffering agents.

Shampoos have been prepared in various forms, and have included systems that are thick and thin, clear and opaque, pourable liquids, solids, gels, pastes, powders, flakes, and aerosol types. In many cases, shampoos have been prepared and directed for various hair types, eg, normal, dry, damaged, and color treated. Most marketed shampoos (ca 1993) are primarily clear liquid and opaque lotion types; gel and paste forms also are available. Aerosol shampoos are available to the consumer; however, their impact on the market has been limited to the dry shampoo aerosol.

Clear Shampoos. Aqueous solutions of one or more detergents, together with other water-soluble modifying additives, are the usual makeup of clear shampoos. The clarity of the shampoo offers the impression of superior cleansing and better rinsing qualities. In formulating, it is essential that components used in these products have low cloud points to maintain clarity at low temperature storage conditions. This is achieved through the careful selection of cleansing agents, additives, and solubilizers. The viscosity of these products is controlled through adjustment with salt and/or the use of alkanolamides or cellulosic thickeners.

Gel Shampoos. These shampoos are generally versions of the clear product but consist of higher concentrations of the cleansing and thickening agents. They are usually packaged in tube form for dispensing purposes.

Opaque Shampoos. Lotion shampoos are opaque in appearance, offering the consumer a range of rich consistencies. These products are usually formed through suspensions of opacifying agents of the glycol stearate type, which often lead to a pleasing pearlescent appearance. Since clarity is not of concern with these products, less soluble cleansing agents can be used. These shampoos also are prepared with dispersions of oils, silicones, and antidandruff additives for specific hair and scalp treatment effects. Thickening agents are used to increase viscosities of these shampoos in order to maintain the various opacifiers and dispersed agents in suspension.

Paste Shampoos.   These shampoos represent thickened versions of opaque shampoos. They have a somewhat firm, cream-like consistency and are packaged in jars and/or tubes. Thickening of these systems is usually accomplished through additions of stearate soaps and electrolytes.

Aerosol Shampoos.   These shampoos constitute a very small percentage of the market. They have been available in two versions, ie, liquid foam types and dry spray forms. The liquid foam type, despite its convenience and appealing appearance, did not attain high general use. Factors involved in its low acceptability include not only higher product cost but also serious stability issues with can corrosion.

Aerosol dry shampoos fill an important market for those unable to tolerate wet hair, such as the sick and infirm. These products are based on oil absorbing powders which include talc, starch, and/or clay. They can be sprayed onto hair and then brushed off after absorbing soils from the hair.

Synthetic Detergents.   Examples of shampoo formulations are given in Table 1. The names of the ingredients are those designated by the Cosmetics, Toiletry and Fragrance Association (CTFA).

Table 1. Formulations of Typical Shampoos

Ingredients	CAS Registry Number	Wt %	Function
Clear shampoo			
sodium lauryl sulfate	[151-21-3]	10.0–20.0	primary surfactant
lauramide DEA	[120-40-1]	4.5	foam booster
methylparaben	[99-76-3]	0.1	preservative
propylparaben	[94-13-3]	0.05	preservative
tetrasodium EDTA	[64-02-8]	0.05	sequesterant
fragrance		0.5	fragrance
D&C Yellow #10	[8004-92-0]	0.004	colorant
FD&C Blue #1	[3844-45-0]	0.0005	colorant
water	[7732-18-5]	to 100.00	diluent
Opaque, pearlescent shampoo			
sodium lauryl sulfate	[151-21-3]	10.0–20.0	primary surfactant
magnesium aluminum silicate	[1327-43-1]	0.5	suspending agent
glycol stearate	[111-60-4]	3.0	opacifier, pearl agent
cocamide DEA	[61791-31-9]	3.0	foaming viscosity aid
sodium chloride	[7647-14-5]	0.5	viscosity adjuster
methylparaben	[99-76-3]	0.1	preservative
propylparaben	[94-13-3]	0.05	preservative
fragrance		0.5	fragrance
deionized water	[7732-18-5]	to 100.00	diluent

At one time, soaps were the prime cleansing agents in shampoo products. Synthetic detergents are now (ca 1993) the backbone of most shampoo products, but soap systems are used in a few specialty brands. Soap shampoos provide good lathering and rinsing characteristics except in situations where hard water is available. In these cases, poor lathering results and complexing of the soap with the heavy alkali metals of hard water occurs, causing deposition of a dulling film on hair with rinsing. Soap shampoos continue to have followers who find that in their use the hair is left in a conditioned state (3). To capitalize on this effect of soap, some manufacturers of shampoos combine soap with synthetics to obtain the advantages of both cleansing and conditioning.

The use of synthetic detergent has become widespread in a number of product applications. These surfactants are classified according to the electrical properties of their hydrophyllic groups in aqueous solutions; ie, designated as anionic for those negatively charged, nonionic for those with no charge, cationic for those with a positively charged hydrophyll, and amphoteric for those having both positive and negative ionic features.

The anionics are used primarily in shampoo preparations because of their superior foaming and cleansing properties. The nonionics, although good cleansers, have low foaming properties and are not widely used in shampoo formulations. The amphoterics are low foaming detergents and are generally regarded as low cleansers. But they are very mild and often are found in baby-type shampoos. Cationic detergents are poor foamers; however, they are substantive to hair and, as a consequence, are used primarily in systems for hair conditioning purposes.

Anionic Surfactants. In terms of general usage in cosmetic products, the anionics are by far the most widely used and are chiefly found in shampoo systems. They provide the formulator with the basic conditions for preparing these products, ie, foaming, cleansing, and solubility.

Primary Alkyl Sulfates. These detergents were first developed in Germany to be less dependent on the use of fats and oils for soap preparation and to have more effective detergents to solve the precipitation problems of soaps. These detergents are prepared from the corresponding fatty alcohols which have been sulfated with either chlorosulfonic acid or sulfuric acid. The fatty alcohols are usually prepared by hydrogenation of fatty acids. The use of the cholorosulfonic acid process continues to be the prime method of preparing alkyl sulfates, although a continuous sulfation process using sulfur trioxide is gaining importance in this detergent's manufacture (4).

The most widely used alkyl sulfate in shampoo preparation is lauryl sulfate. The alkyl component of this sulfate ranges from C-10 to C-18 with a predominance of the C-12 (lauryl) component. By distillation of the fatty alcohol, certain cuts can be obtained which offer the best effects in foaming, cleansing, and rinsing properties for the alkyl sulfate preparation. The range which appears to be most desirable is between C-12 and C-16. Lauryl sulfate detergents are available in various salt forms with the sodium, ammonium, and triethanolamine types being used most frequently in shampoos.

Sodium lauryl sulfate is available in solution, paste, and solid forms. As a solution its activity ranges between 28–30%, and as a paste it is 55% active. With this detergent in a shampoo, inorganic salts can affect viscosity. In addition, the limited solubility of sodium lauryl sulfate requires its judicious use in low cloud point clear shampoo systems.

Ammonium lauryl sulfate [2235-54-3] is available as an approximately 28% active, clear, liquid form. It has greater solubility than the sodium salt and is more likely to be used in formulating clear shampoos. Systems using this detergent show their best stability at pH between 6 and 7. Lower pH would tend to hydrolyze the detergent, eventually releasing ammonia.

Of the lauryl sulfates, the triethanolamine form has the best water solubility. Because of this, it is available from suppliers as a clear solution at an active concentration of 40%. Its main disadvantage is discoloration during storage, ie, yellow to amber, which limits its use in clear shampoo systems.

Shampoos based on lauryl sulfates can range from 6–17% of the active surfactant. However, though they are effective cleansers, the alkyl sulfates tend to be defatting. In an effort to make these shampoos more mild, many shampoos are now based on blends of amphoterics and alkyl sulfates or the less irritating alkyl ether sulfates.

Alkyl Ether Sulfates.   These surfactants are also found in shampoo applications. They are prepared similarly to alkyl sulfates except that the fatty alcohol is first subjected to ethoxylation; ethoxylation may range from 2 to 3 moles per mole of fatty alcohol. Lauryl alcohol is the typical fatty alcohol reacted. Because of high water solubility, alkyl ether sulfates have low cloud points, making them suited for clear shampoo formulations (5). Viscosities based on the ether sulfates can be easily controlled through addition of inorganic salts such as sodium chloride.

Alkyl Sulfosuccinate Half Esters.   These detergents are prepared by reaction of maleic anhydride and a primary fatty alcohol, followed by sulfonation with sodium bisulfite. A typical member of this group is disodium lauryl sulfosuccinate [26838-05-1]. Although not known as effective foamers, these surfactants can boost foams and act as stabilizers when used in combination with other anionic surfactants. In combination with alkyl sulfates, they are said to reduce the irritation effects of the latter (6).

Fatty Acid–Sarcosine Condensates.   These surfactants are prepared by the reaction of fatty acid chlorides with methyl glycine; sodium lauroyl sarcosinate [137-16-6] is an example of this group. They are most effective at pH 5.5–6.0 for foaming activity in soft to moderately hard water. The action of these detergents is greatly reduced under severe hard water conditions. The sarcosinates exhibit compatibility with cationic surfactants and have been suggested for use in formulation of conditioning shampoos (7).

Fatty Acid–Peptide Condensates.   These protein detergents are reaction products of fatty acid chlorides and hydrolyzed proteins. They are used in shampoos because of their mildness on skin, hair, and to eyes when used alone or in combination with alkyl surfactants (8).

Alkyl Monoglyceride Monosulfates.   These detergents are among the earliest synthetic surfactants. For example, the sulfated coconut fatty acid monoglyceride in its ammonium salt form offers a shampoo with good foaming and detergency. Shampoos formulated with these monoglycerides are regarded as exceptionally mild.

Acyl Isethionates. These are among the oldest of the synthetic detergents and were developed in Germany to overcome problems of hard water. They are prepared by reaction of fatty acid chlorides with a salt of isethionic acid, ie, 2-hydroxyethanesulfonic acid [107-36-8]. These detergents have moderate foaming properties and have seen only limited use in shampoos.

Alpha-Olefin Sulfonates. Sulfonation of alpha-olefins yields a mixture of alkene sulfonates, hydroxyalkane sulfonates, and some amount of various disulfonates. These detergents are excellent foamers with good detergency properties. They are unaffected in hard water and their effects are considered superior to the alkyl ether sulfates (9).

Alkyl Sulfoacetates. These surfactants are prepared by esterification of sulfoacetic acid or by sulfonation of the alkyl chloroester. They are considered to produce good foaming and are less irritating to the eyes than the alkyl and alkyl ether sulfates (10).

Nonionic Detergents. Nonionic surfactants rarely are used as the primary cleansing additives in shampoos. They are generally poor foaming, but have value as additives to modify shampoo properties, eg, as viscosity builders, solubilizers, emulsifiers, and conditioning aids.

Alkanolamides. The fatty acid alkanolamides are used widely in shampoo formulations as viscosity and lather builders. They are formed by the condensation of a fatty acid with a primary or secondary alkanolamine. The early amides were compositions of 2:1 alkanolamine to fatty acid. Available technology allows the formation of amides with a 1:1 ratio of these additives. These amides are classified as superamide types. The typical amide used in shampoo preparations usually contains the mono- or diethanolamine adduct, eg, lauric diethanolamide [120-40-1] (see AMIDES, FATTY ACID).

Amine Oxides. These surfactants are formed by oxidation of tertiary fatty amines, eg, lauryldimethylamine oxide [1643-20-5]. They are used to modify foaming and also may find application as hair conditioning agents in shampoos, ie, acting as antistatic agents to provide manageability (see AMINE OXIDES).

Ethoxylated Nonionics. These are the largest group of nonionics. They consist of ethoxylated forms of alkylphenols, fatty alcohols, fatty esters, mono- and diglycerides, etc. Although these surfactants exhibit excellent cleansing properties they have poor foaming, and their application in shampoos has been limited. They can be found as solubilizers and emulsifiers. In some cases, nonionics are combined with certain shampoo surfactants to minimize eye irritation (11).

Amphoteric Detergents. These surfactants, also known as ampholytics, have both cationic and anionic charged groups in their composition. The cationic groups are usually amino or quaternary forms while the anionic sites consist of carboxylates, sulfates, or sulfonates. Amphoterics have compatibility with anionics, nonionics, and cationics. The pH of the surfactant solution determines the charge exhibited by the amphoteric; under alkaline conditions it behaves anionically while in an acidic condition it has a cationic behavior. Most amphoterics are derivatives of imidazoline or betaine. Sodium lauroamphoacetate [68647-44-9] has been recommended for use in non-eye stinging shampoos (12). Combinations of amphoterics with cationics have provided the basis for conditioning shampoos (13).

Cationic Detergents. These find little application in shampoo preparation, primarily because most compositions use anionics as the surfactant of choice. Generally, cationics are not combined with anionics in shampoo formulation because the opposing charge differences of these detergents result in a precipitated complex. Despite their substantivity to hair to provide conditioning, cationics have the deficiency of being severe eye irritants. They generally also have low foaming properties which would limit consumer acceptance of shampoos prepared with them. Cationic surfactants have been recommended for special applications, eg, with amphoterics for a hair conditioning shampoo (11).

Shampoo Additives. Although the primary function of a shampoo is for cleansing, a number of additives are included in their formulation to enhance and improve properties of the product.

Thickeners. These are used to increase viscosity of shampoos to achieve certain consistency characteristics in the product, from a thickened liquid to gels and pastes. Among the most important materials used for this purpose are the alkanolamides. The chain length of the amide alkyl group and its solubility in the shampoo system are important aspects to be considered in their use for effects on viscosity. In general, as the chain length increases the viscosity response improves. The viscosity increase also is related to the water solubility of the amide; the more water-soluble forms provide a lower viscosity response than the less soluble amides.

Inorganic salts are also used to promote shampoo thickening. These should be used sparingly since an excess may have a deleterious effect on a product's physical stability. Sodium chloride commonly is used in these cases. The additions of sodium stearate and stearic amides can be found in paste shampoos for thickening.

Other thickeners used include derivatives of cellulose such as methylcellulose, hydroxypropyl methylcellulose, and cellulose gum; natural gums such as tragacanth and xanthan (see CELLULOSE ETHERS; GUMS); the carboxyvinyl polymers; and the poly(vinyl alcohol)s. The magnesium aluminum silicates, glycol stearates, and fatty alcohols in shampoos also can affect viscosity.

Opacifiers. Opaque shampoos are produced by incorporating high melting, wax-like, dispersible materials into their preparation. Some of these materials crystallize in such a fashion that they effect a pearlescence in the product. Opacifying agents found in shampoos include the glycol mono- and diesters, higher fatty alcohols such as cetyl and stearyl forms, stearate soaps, and latex copolymer emulsions.

Conditioners. Conditioning agents are added to shampoos to provide manageability properties to hair, eg, ease of combing, detangling, and reduced static. Although the basic detergent is an important factor in the contribution to hair conditioning, additives can be included in the shampoo composition to minimize any ill-effects the detergents might impart. Materials found in many products for conditioning include lanolin and its derivatives, fatty amine oxides, cationic polymers, cationic guar gums, fatty amines and alcohols, alkanolamides, and quaternary ammonium compounds. Humectants, eg, glycerine, sugar, sorbitol; and oils, eg, fatty glycerides, esters, and silicones, also are used. Other additive products include beer, egg, honey, milk, and herb extracts for conditioning.

Preservatives. Shampoos present an ideal environment for microbial growth which can have a harmful effect on the physical/chemical properties of the shampoo and may pose a health hazard to the consumer. To prevent microbial growth, preservatives are added to shampoos. Among those used are methyl and propyl parabens, DMDM hydantoin, quaternium-15, phenoxyethanol, imidazolidinyl urea, and a mixture of methylchloroisothiazolinone and methylisothiazolinone. The selection of preservative is determined through challenge testing which subjects the product to the worst conditions encountered in manufacture, shelf storage, and use (14).

Other Additives. To provide and maintain the clarity of clear shampoos, the use of either ethyl or isopropyl alcohol may be employed. Perfumes are added to make shampoos more pleasing in terms of odor, while dyes are incorporated to give visual aesthetics to the products. Salts of ethylenediaminetetraacetic acid are found to sequester and prevent formation of insoluble alkaline-earth metal salts.

Baby Shampoos. These shampoos, specifically marketed for small children, feature a non-eye stinging quality. The majority of the products in this category are based on an amphoteric detergent system; a system combining the use of an imidazoline amphoteric with an ethoxylated nonionic surfactant has been successfully marketed (15,16). The sulfosuccinates also have been suggested for baby shampoo preparation because of their mildness (17). However, their widespread use for this purpose has been limited.

Medicated Dandruff Shampoos. Dandruff is a scalp condition characterized by the production of excessive cellular material (18). A number of shampoos have been marketed which are designed to control and alleviate this condition, and many additives have been included in shampoo compositions to classify them as treatment products for dandruff. These additives include antimicrobial additives, eg, quaternary ammonium salts; keratolytic agents, eg, salicylic acid and sulfur; heavy metals, eg, cadmium sulfide; coal tar; resorcinol; and many others. More recent (ca 1993) systems use selenium sulfide [7488-56-4] or zinc pyrithione [13463-41-7] as active antidandruff shampoo additives. Both of these additives are classified as drugs, but can be found in over-the-counter products. A stronger version, incorporating the use of higher levels of selenium sulfide in a shampoo, is available but requires a prescription for purchase.

Two-in-One Shampoos. These shampoos are combination cleansing and conditioning imparting products. The conditioning aspect is primarily the ease in wet combing, and a conditioning rinse product is not needed after shampooing. These shampoos were first introduced in the early 1970s. Because they did not perform to the satisfaction of the consumer, these early systems quickly faded from the market. In the late 1980s, new and more effective versions of these products appeared and have a strong presence in the shampoo market (ca 1993). These new products are based on conventional anionic detergents to provide desired physical shampoo properties combined with conditioning additives. An important factor in the re-emergence of two-in-one shampoos was efficient delivery of silicones to hair to make it comb easier and to impart a smooth and soft feel (19). In addition to silicones, quaternaries, cationic guar gums, and polymers are found in more recently introduced two-in-one products.

Manufacture, Evaluation, and Safety. The manufacture of shampoos is a relatively simple operation requiring a suitable stainless steel kettle with provisions for heating and cooling and equipped with appropriately sized mixers. Although shampoos are easily handled during preparation, precautions should be taken to not aerate the product. Cream shampoos are particularly sensitive to aeration and require more special care in their manufacture.

Evaluation. The performance evaluation of a shampoo is an important aspect in determining its acceptability for consumer use. A number of laboratory tests have been developed for this purpose. Although these give an indication of a product's acceptability, the final evaluation is still found through actual use testing. Laboratory methods are valuable to help assess such factors as foaming, cleansing, rinsing, and wet and dry combing effects in the development of a shampoo. These can be determined under standardized, controlled conditions. More critical evaluations can be made through half-head salon comparisons to competitive products and by panel tests under actual use conditions.

Safety. Shampoos generally do not represent a hazard with regard to skin and eye safety; once used, shampoos are almost immediately rinsed and have little contact time on sensitive areas. To assure this safety, provisions to test the finished product for skin and eye irritation should be made.

Hair Conditioners

Hair conditioning can be associated with almost any hair product sold in the marketplace because they all claim some benefit to the hair when used. Thus the term hair conditioner can be applied to such products as rinses, hair dressings, setting lotions, and hair sprays. Conditioners are used to provide different effects to the hair; primarily ease of combing, sheen, and soft feel.

Hair Rinses. These products generally are designed to be used in conjunction with shampoos to provide special benefits to hair, eg, wet and dry combing ease, antistatic effects, shine, manageability, and detangling. In years past, an after-shampoo acid rinse, such as a vinegar and lemon rinse, was considered essential to remove the soap film resulting from available products of that time. With the advent and use of synthetic detergents for shampoo preparation, the effect of soap is no longer a factor. However, shampoos made with synthetic detergents are more efficient cleansers, leaving hair less manageable and more difficult to comb, and hence requiring an after-treatment to alleviate these conditions. Creme rinses containing hair substantive cationic additives have been found useful for this purpose. Their after-shampoo use leaves hair smooth to the touch, easy to comb, and unsnarled. The active ingredients in most creme rinses are quaternary ammonium compounds such as steartrimonium chloride and cetrimonium chloride. Other additives useful in after-rinse hair conditioners include certain fatty amines, amine oxides, and cationic polymers. After-shampoo conditioners also are used to improve the finish of hair with respect to manageability, body, texture, etc. Additives used to obtain these effects include protein additives, silicones, and lanolin and its derivatives. Most rinses are opaque products, although clear versions can be found, and they range in consistency from pourable liquids to thick creams.

Hairdressings. Products associated with final grooming effects to hair are termed hairdressings. They are used to impart not only a holding effect to hair, but also provide an added benefit of giving hair a natural, healthy, lustrous appearance. Hairdressings can be found as liquid or cream emulsions, gels, or as hydroalcoholic preparations. They are usually applied by spreading the product through the hair with the fingers and then combing for an even distribution.

Brilliantines. The primary purpose of brilliantines is to add a level of grooming and to impart sheen attributes to hair. Historically, the main constituent of brilliantines is an oil, usually a mineral oil type, in a rather high concentration; in liquid brilliantines, the concentration of oil can run from 80 to 100% in the formula. Brilliantine action on hair is due to formation of a thin-film coating on the strands of hair fibers; the oil is not absorbed but provides a grooming effect. Higher viscosity mineral oils give better grooming than the lower viscosity oils, but have poorer spreadability properties. To offset this aspect, high viscosity oils are sometimes diluted with other hydrocarbons to enhance their spreading effects.

In certain brilliantine compositions, vegetable and animal oils are used as substitutes for mineral oil. In these systems, because of their potential for rancidity, antioxidants must be included. Other alternatives to mineral oils that have found utility in brilliantines are the polyethylene glycols which come in a variety of solubilities and spreading properties. Use of these materials offers the advantage of chemical stability to rancidity. Other additives found in brilliantines to improve their aesthetics include colorants, fragrance, medicated additives, lanolin, and fatty acid esters.

Solid brilliantines and pomades may be considered heavy-duty type hairdressings. These product types range in consistency from soft textured to waxlike and generally have poor spreading properties when compared to liquid versions. Their main component is petrolatum [8027-32-5] with additions of various waxes to obtain different consistency ranges. Early formulations were prepared with naturally occurring waxes such as spermaceti, beeswax, ceresin wax, and paraffin wax. Synthetic waxes are employed as substitutes for the natural waxes since they can be reproduced with greater uniformity to give more consistent end products.

Alcoholic Hair Tonics. Hairdressing products have been prepared by dilution of various oils with alcohol. This allows for good wetting action of the alcohol and, upon its evaporation, results in a deposition of a uniform thin layer of oil on the hair. Synthetics, which have all but replaced natural oils in formulating these products, provide uniformity of components, chemical stability, compatibility, and a range of emolliencies. Additives found in these products include ethoxylated and propoxylated glycols, ethoxylated ethers, various lanolin derivatives, and ethoxylated and propoxylated diols and triols. Because of the range in solubilities which these additives offer, products can be formulated that are less greasy and oily when compared to those made with mineral or vegetable oils. In addition, to offer more consumer-desired properties, alcoholic hair tonics may contain quaternary conditioners, keratolytic agents, hair setting resins, colorants, and fragrance.

Hairdressing Emulsions. Emulsified hairdressings have been formulated both in liquid and cream forms, and prove to be popular among consumers of hairdressings. They are either water-in-oil or oil-in-water emulsions. Mineral oil is commonly used in water-in-oil hairdressing emulsions. Emulsifiers include magnesium, zinc, or aluminum stearate, beeswax, borax, sorbitan sesquioleate, ethoxylated fatty alcohols, ethoxylated lanolin alcohols, polyglyceryl esters, and acetylated ether esters. Other emulsifiers include absorption bases which allow use of as much as 80% water in preparing these product forms.

Oil-in-water based hairdressings offer good spreading characteristics, have a less greasy feel, and are more easily rinsed than water-in-oil forms. Mineral oil is found in these products, although natural and synthetic oils have been used. The typical oil-in-water emulsifiers are used in preparing these emulsions.

Clear Gel Hairdressings. Clear gel hairdressings come in two forms, ie, microemulsions or setting gels. Microemulsions are systems containing mineral oil which can be blended with emollients, conditioning additives, lanolins, and protein compounds. The formation of these gels is achieved through use of high concentrations, ie, 10–40%, of ethoxylated emulsifiers based on fatty alcohols, lanolin alcohols, sucrose fatty acid esters, and oleyl ether phosphates. Additives used in these systems should not interfere with the clarity of the product.

Setting gels are formed through use of high molecular weight polymers such as the methylcellulose ethers and carboxyvinyl polymers. With these polymers, gels of soft to firm consistency can be prepared which provide hair holding properties. The gels are usually water or water/alcohol based to which soluble grooming agents such as poly(*N*-vinyl-2-pyrolidinone) [*9003-39-8*] (PVP) can be added. To help modify the film properties of these products, other additives such as lower molecular weight alcohols, lower alkoxypolyoxyalkylene glycols and diols, glycerin, and ethylene and propylene glycols may be used.

Fixatives

Fixatives are liquid products used to achieve a desired hairstyle and to temporarily hold that style in place. These products can be grouped into two classes, ie, styling products and finishing sprays. Styling products are used primarily on wet hair to make combing easier and to give the hair some tack so that the style remains in place as the hair is dried. Once dried, the products leave a nontacky coating of film-forming polymers which hold the hair in the desired style. Examples of this type of fixative include styling mousses, gels, lotions, and spray gels. Finishing sprays, eg, pump and aerosol hair sprays, are applied to the hair after the style is dried and set. They generally are not used to style the hair but rather to hold the hairstyle more firmly than styling products. Spritzes are sometimes marketed as styling products, but their technology most closely resembles finishing sprays.

The method used to apply and dry a fixative affects the degree of stiffness and hold it imparts to the hairstyle. If applied and then manipulated with a comb, brush, or fingers as the polymer forms its film, the adhesive bonds between the hair are broken, the film coating the hair is broken, and the end result is a soft feel with little set retention. If the fixative is allowed to dry undisturbed, then the result is a firmer feel and better hold.

When formulating a hair fixative, the balance between the two principal benefits, hold and styling ease, must be selected. Hold is characterized by the stiffness of the polymer film and its ability to remain stiff when exposed to high humidity. Styling ease is characterized by the product's ability to decrease surface friction of the hair during the combing, drying, and styling process. These benefits tend to be inversely proportional. If a product delivers a strong hold benefit, it will tend to be less suitable for improving styling ease.

Once the proper balance of hold and styling ease is attained, the products are tested for delivery problems, proper degree of tack when wet and no tack when dry, visible flaking of the polymer after it dries, dullness to the hair, excessively coated, heavy, or stiff-feeling hair, ease of wash out, stability problems, microbiological problems, and, for finishing sprays only, quickness of drying. Laboratory evaluations of hair fixatives are usually performed on human hair tresses. The hair is prepared to a standard length, weight, and density. For styling products, the hair is shampooed and rinsed with the excess water squeezed out. The product is applied at a standard amount and the hair combed to evaluate comb drag. The hair is rolled into a standard curl formation and allowed to dry and set. Comparisons are made to a control for stiffness, combing ease, curl resiliency after combing, flaking, and stickiness upon contact with moisture. Additionally, treated curls are evaluated for their ability to maintain a tight curl formation under various humidities.

Finishing sprays and spritzes are sprayed on clean, dry hair and tested for drying time, stiffness, combing ease, flaking after combing and stickiness upon contact with moisture. Additionally, hair holding properties are evaluated by measurements of the curl retention at various humidities. Standardized water-set curls are sprayed with standard amounts of product under controlled circumstances. They are then dried, placed into various humidity levels, and the curl fall monitored at various time periods.

Test salons are often used to evaluate hair fixatives. Half-head studies are performed, with the test product applied to one side of the head and a control product to the other in realistic use amounts. Similar properties as described in laboratory tests are measured. Finished products are often sent to testers' homes where they have an opportunity to evaluate the products in real use situations for extended periods.

Styling Products. Table 2 lists typical styling product formulations. Styling products use similar types of ingredients to provide their primary benefits, but differ in their physical form, ie, styling gels use thickening ingredients to increase viscosity, styling mousses use propellants and surfactants to create foams, and styling sprays often use alcohol to improve spraying and solubilize water-resistant holding polymers.

Ingredients. Holding Polymers. The setting or holding ingredient in styling products is the film-forming polymer. Its main functions are to hold the hair in a styled configuration and to stiffen the hair, which increases body and fullness. The dried polymer film should be clear, easy to remove by shampooing, tack-free, and not create visible flakes with combing. The polymer should be completely soluble in the product and show good stability under varying temperature conditions. Amounts of total polymer vary from product to product depending on the performance desired; levels can be found from 1 to 7% with the majority formulated from 3–4%.

Table 2. Fixative Styling Product Formulas

Component	CAS Registry Number	Wt %	Function
Styling gel, firm hold			
deionized water	[7732-18-5]	92.40	solvent
carbomer 940	[9007-17-4]	0.50	gelling agent
PVP/VA (65/35) copolymer	[25086-89-9][a]	2.00	holding polymer
PVP (K-90)	[9003-39-8]	1.50	holding polymer
triethanolamine	[102-71-6]	0.50	neutralizing agent
polyquaternium-11 (20%)		1.00	styling ease
laureth-23	[9002-92-0]	1.00	solubilizer
propylene glycol	[57-55-6]	0.50	solubilizer
disodium EDTA	[139-33-3]	0.05	chelating agent
fragrance		0.20	fragrance
methylparaben	[99-76-3]	0.15	preservative
diazolidinyl urea	[78491-02-8]	0.15	preservative
propylparaben	[94-13-3]	0.05	preservative
Styling mousse, firm hold			
deionized water	[7732-18-5]	84.35	solvent
polyquaternium-11 (20%)		6.00	styling ease
polyquaternium-4		0.50	holding polymer
isosteareth-10	[52292-17-7]	0.20	foam stabilizer
octoxynol-9	[42173-90-0]	0.30	foam stabilizer
dimethicone copolyol	[64365-23-7]	0.20	styling ease
fragrance		0.15	fragrance
DMDM hydantoin	[6440-58-0]	0.20	preservative
methylparaben	[99-76-3]	0.10	preservative
isobutane/propane blend	[75-28-5], [74-98-6]	8.00	propellants
Styling spray			
deionized water	[7732-18-5]	57.10	solvent
SD alcohol 40[b]	[64-17-5]	40.00	solvent
PVP/VA (35/65) copolymer	[25086-89-9][a]	2.00	holding polymer
polyquaternium-16		0.50	styling ease
dimethicone copolyol	[64365-23-7]	0.20	styling ease
fragrance		0.20	fragrance

[a]PVP/VA copolymer.   [b]190° proof = 95%.

Polyvinylpyrrolidinone (PVP), introduced in the early 1950s, was the first synthetic polymer to be used in styling products. It is soluble in both alcohol and water and exhibits excellent film forming properties. It is available in various molecular weights from 10,000–360,000, with the higher weights forming stiffer, more humidity-resistant films. Its principal negative is its hygroscopic nature. It becomes sticky and loses hold under humid conditions. Despite this, it is used widely in a variety of products.

Polyvinylpyrrolidinone/vinyl acetate copolymer (PVP/VA) was developed as an improved, less hygroscopic version of PVP. The monomer ratios control the stiffness and the resistance to humidity; however, too high a vinyl acetate monomer content requires another solvent in addition to water to completely solubilize it.

Polyquaternium-11 is the copolymer of N-vinylpyrrolidinone and dimethylaminoethyl methacylate quaternized with dimethylsulfate (20). It is used widely but provides little hold to the hair. Its primary advantage is its ability to improve the ease of wet combing and control static while giving a light set hold.

Polyquaternium-4 is the copolymer of hydroxyethylcellulose and diallydimethyl ammonium chloride. It provides a firmer hold than polyquaternium-11 and has better humidity resistance.

Natural Gums. These were used in early styling products, formerly called wave sets. Natural gums generally make hazy solutions and the dried film tends to flake. For these reasons they are rarely used. Some of these gums, eg, tragacanth and alginates, can create humidity-resistant films and have limited use for specific product concepts.

Conditioning Agents. These ingredients are added to make combing easier during styling by reducing friction between wet hair and the comb. Wet hair is much weaker than dry hair and can be damaged by the mechanical action of combing or brushing during styling. Conditioning ingredients in widespread use include the aforementioned polyquaternium materials, particularly numbers 6, 7, 10, 11, and 16; quaternary ammonium salts such as cetrimonium chloride; silicone compounds such as dimethicone copolyol and amodimethicone; and natural oils such as jojoba, corn oil, and safflower oil.

Solvents. The most widely used solvent is deionized water primarily because it is cheap and readily available. Other solvents include ethanol, propylene or butylene glycol, sorbitol, and ethoxylated nonionic surfactants. There is a trend in styling products toward alcohol-free formulas. This may have consumer appeal, but limits the formulator to using water-soluble polymers, and requires additional solvents to solubilize the fragrance and higher levels of preservatives.

Gelling Agents. These are used to build viscosity in styling gels. Desired properties include clarity, processing ease, and a plastic, shear-thinning rheology. Carbomer 940 is the most common thickening agent for styling gels and best provides these properties. Carbomers are high molecular weight polyacrylic acids. When dissolved in water at their natural pH of about 3, the polymer only slightly increases viscosity. When neutralized with a base, eg, triethanolamine, triisopropanolamine, etc, negative charges on the carboxylic groups mutually repel, causing a spontaneous stretching of the molecule, thus creating a viscous solution (21). Other thickening agents include hydroxyethylcellulose and hydroxypropyl methylcellulose.

Foaming Agents / Propellants. These ingredients are necessary in a styling mousse, the French word for foam. Styling mousses were introduced in Europe in the early 1980s and in the United States in 1983, and they are extremely popular.

The foam structure is the result of dispersed, liquefied gas bubbles (surrounded by the mousse concentrate containing the foam agents) suddenly being released from a pressurized system to atmospheric pressure. The propellant rapidly expands to a gas, thus blowing a foam. The stability of this foam depends primarily on the foaming agents selected and the percent of alcohol used in the liquid phase. Ideally, the foam should be stable in the hand for several minutes after dispensing, but break down when sheared to spread easily through the hair.

Many different types of foaming agents are used, but nonionic surfactants are the most common, eg, ethoxylated fatty alcohols, fatty acid alkanolamides, fatty amine oxides, nonylphenol ethoxylates, and octylphenol ethoxylates, to name a few (see ALKYLPHENOLS). Anionic surfactants can be used, but with caution, due to potential complexing with cationic polymers commonly used in mousses.

Hydrocarbon propellants, eg, propane, isobutane, butane, are the most commonly used in mousses. These are insoluble in the mousse concentrate; therefore, vigorous shaking of the can before use is required to properly disperse the propellants. Most products use a blend of two or more hydrocarbons. The more volatile the propellant blend, the faster a foam structure is formed and the less dense it is. An exception to this is the use of hydrofluorocarbon 152A [75-37-6] (HFC) which has partial solubility in mousse concentrates and a high volatility. The high volatility creates an immediate foam structure but the partial solubility leads to a creamy, dense foam. HFC is not an ozone depleting propellant as are its cousins the chlorofluorocarbons (CFCs); additionally, it is not considered a volatile organic compound by the states of California and New York due to its low activity in reactions that create ground level ozone.

Preservatives.   Most products must contain preservatives to ensure that yeasts, molds, and bacteria do not thrive in them. These preservatives include alcohol, methylparaben, propylparaben, DMDM hydantoin, diazolidinyl urea, and imidazolidinyl urea. The parabens have limited solubility in water, eg, 0.25% for methylparaben and 0.05% for propylparaben (22). If these levels are exceeded in the formula, then the addition of solvents is needed to ensure clear, stable products.

Miscellaneous.   Many other ingredients are used in styling products. Some have function; some are there to support marketing claims. Fragrances play a significant factor in the aesthetics of the product, but can be difficult to incorporate into the products due to being insoluble in water, thus requiring the use of additional solvents. Ultraviolet absorbers such as octyl methoxycinnamate, benzophenone-4, or DEA-methoxycinnamate are often used in styling products. In addition to their uv protection capability, they also can have a stabilizing effect on the viscosity of gels. Chelators, such as disodium EDTA, are used to improve stability of gels and to increase the efficiency of preservative systems. Proteins are film forming polymers and can act to improve the body and fullness of hair if used in levels of 1–4%. Vitamin derivatives, botanical extracts, marine extracts, and other exotic materials may create real benefits, but they rarely are used in high enough levels due to high cost.

Manufacture.   The manufacturing of styling products is relatively simple. Generally, a tank with simple agitation is sufficient for low viscosity products, ie, mousse, spray gels, and sculpting lotions. The use of powdered holding polymers and exotic materials can sometimes create manufacturing difficulties. The effect of the manufacturing procedure on the stability and performance of each formula must always be evaluated.

Styling gels are not as easy to manufacture. The gelling materials used are hygroscopic and tend to clump when added to water. This is particularly true of carbomers and can be avoided by using cool water and slowly sifting the carbomer into rapidly agitated water. Additionally, the use of an eductor or in-line powder disperser can decrease this clumping problem.

Mousses pose little manufacturing problem, but because they are aerosolized they must be filled with special equipment. The pressure fill technique requires the container to be filled with mousse concentrate, then a valve is crimped on and a vacuum of approximately 2.4 kPa (18 mm Hg) is pulled. The propellants are added through the valve. Another technique, the under-the-cup method, fills the container under pressure with propellant and crimps the valve, all in one step.

Packaging. There are many shapes, sizes, and forms of packaging used for styling products. Gels are usually sold in a low density polyethylene tube or in a bottle with a pump. Spray gels, heat activated sprays, scrunching sprays, etc, are generally in high density polyethylene bottles with pump spray devices. Mousses are generally packaged in aluminum cans because they are pressurized and the high water content tends to corrode tin-plated steel aerosol cans.

Finishing Sprays. Table 3 gives typical aerosol and pump finishing spray formulations.

Ingredients. Holding Polymer. The primary setting agent in finishing sprays is the film-forming polymer. Its principal function is to hold hair in a styled configuration for a period of time in all types of weather. It is also important that the polymer is easily removed with shampooing, does not dull the hair, is not sticky in humid weather, and does not flake appreciably with combing. The polymer should be completely soluble in the total system and in aerosols; the polymer's solubility should not be affected by the propellants. The polymer also must be compatible with the other additives in the product because any precipitation over time can lead to dispensing problems by clogging the pump or aerosol valve. The polymer's kinematic viscosity in the selected solvent system has a large impact on the type of spray that results. Polymer concentrations range from 2–7% in aerosols and up to 7% in pump versions. Generally, the more polymer in the product, the firmer the hold and the coarser the spray due to an increase in solution viscosity.

Vinyl acetate (VA)/crotonates copolymer became available in the late 1950s. It was the first polymer used in fixatives to contain carboxylic acid groups which, depending on neutralization percent, could produce variations in film properties; eg, stiffness, humidity resistance, resiliency, tack, and removability by shampoo. It has largely been replaced in hair sprays by newer polymers.

VA/crotonates/vinyl neodecanoate copolymer is the most used polymer in aerosol hair sprays (ca 1993). Like its precursor above, it has free carboxylic acid groups which can be neutralized to give various film properties. Recommended neutralizing agents include aminomethyl propanol, ammonium hydroxide, and dimethyl stearamine. Recommended percent neutralization is 90%, but products can be found in the 80–110% range.

Ethyl and butyl esters of poly(vinyl methyl ether)/maleic anhydride (PVM/ MA) copolymer were introduced in the early 1960s for use in hair sprays. These polymers also have free carboxy acid groups that can be neutralized. Recommended neutralization is 10%, but products can be found in the range of 5–30%,

Table 3. Fixative Finishing Spray Formulations

Component	CAS Registry Number	Wt %	Function
	Nonaerosol		
SD alcohol 40[a]	[64-17-5]	89.42	solvent
ethyl ester of PVM/MA copolymer	[50935-57-4]	10.00	holding polymer
aminomethyl propanol	[124-68-5]	0.23	neutralizer
cetearyl octanoate		0.15	plasticizer
fragrance		0.20	fragrance
	Nonaerosol[b]		
SD alcohol 40[c]	[64-17-5]	79.95	solvent
octylacrylamide/acrylates/ butylaminoethyl methacrylate copolymer		4.00	holding polymer
aminomethyl propanol	[124-68-5]	0.70	neutralizer
triethyl citrate	[77-93-0]	0.15	plasticizer
fragrance		0.20	fragrance
deionized water	[7732-18-5]	15.00	solvent
	Aerosol		
SD alcohol 40[c]	[64-17-5]	74.65	solvent
VA/crotonates/vinyl neodecanoate copolymer	[55353-21-4]	4.50	holding polymer
aminomethyl propanol	[124-68-5]	0.45	neutralizer
dimethicone copolyol	[64365-23-7]	0.25	plasticizer
fragrance		0.15	fragrance
isobutane/propane	[75-28-5], [74-98-6]	20.00	propellant
	Aerosol[b]		
SD alcohol 40[c]	[64-17-5]	55.00	solvent
VA/crotonates/vinyl neodecanoate copolymer	[55353-21-4]	4.50	holding polymer
aminomethyl propanol	[124-68-5]	0.45	neutralizer
dimethicone copolyol	[64365-23-7]	0.20	plasticizer
fragrance		0.15	fragrance
deionized water	[7732-18-5]	14.70	solvent
dimethyl ether	[115-10-6]	15.00	propellant
isobutane/propane blend	[75-28-5], [74-98-6]	10.00	propellant

[a]190° proof = 95%. [b]80% volatile organic chemicals (VOC). [c]200° proof = 100%.

and recommended neutralizers include ammonium hydroxide, aminomethyl propanol, and triisopropanolamine. These were the most widely used polymers in hair sprays before their use decreased dramatically in the early 1990s.

Octylacrylamide/acrylates/butylaminoethyl methacrylate copolymer became available in the late 1960s. It is a very stiff polymer which gives excellent holding properties in hair sprays. This copolymer is carboxylated at regular intervals and also has cationic groups. When the carboxyl groups are neutralized they take on an anionic character. Recommended neutralizing agents for this copolymer are aminomethyl propanol, triisopropanolamine, and triethanolamine (see ALKANOL-AMINES). The recommended neutralization level is 95%, but products are known in the range of 85–130%. The use of this copolymer in hair sprays increased dramatically in the 1980s and into the early 1990s.

Plasticizers and Other Film-Modifying Additives. The selection of a polymer neutralizer and the level used can greatly affect the properties of the film upon drying. Additionally, glycols, phthalates, and fatty alcohols tend to soften the polymer film, whereas silicone derivatives, lanolin derivatives, various oils, and various fatty esters plasticize the polymers and lessen moisture absorption; polymers can become over-plasticized if too much of these are added, causing the film to soften and feel slick. Fragrance also may have an affect on the film and should be considered in the total selection of plasticizers when formulating a finishing spray.

Solvent Systems. The principal solvent for finishing sprays is ethanol. Methylene chloride was used as a cosolvent in the past in aerosol hair sprays containing hydrocarbon propellants, but has been banned. Compatibility in the methylene chloride-free products is now ensured through careful selection of polymer and neutralizers. Traditionally, aerosol hair sprays use anhydrous ethanol as the principal solvent while almost all nonaerosol hair sprays contain some water, generally 6–10%; but beginning in the early 1990s the use of water became more prevalent. Water originally was added to lower the cost of the formulas, but with the states of California and New York regulating the total volatile organic compound (VOC) content, its use has become a necessity. The addition of water to hair spray formulas poses some problems. The more water added, the longer the product takes to dry and the stickier the polymer film during drying. The most serious problem is that the kinematic viscosity of a polymer solution increases dramatically. Table 4 shows the change in the viscosity of three hair spray polymers at various water levels. In aerosol hair sprays, the incompatibility of water with hydrocarbon propellants is an additional problem. Dimethyl ether can be used in combination with the hydrocarbons because of its good solubility in water.

Table 4. Effect of Water on Kinematic Viscosity,[a] cS

Polymer	Water content, %			
	0	6.8	15.0	22.7
ethyl ester of PVM/MA copolymer[b]	5.1	7.3	9.9	10.7
VA/crotonates/vinyl neodecanoate copolymer[c]	3.8	5.1	6.9	8.2
octylacrylamide/acrylates/ butylaminoethyl methacrylate copolymer[c]	7.5	10.9	14.6	18.0

[a]5.0% active polymer solutions at 25°C.
[b]Neutralized 10% with aminomethyl propanol (AMP).
[c]Neutralized 100% with AMP.

Propellants.   Aerosol hair sprays began using chlorofluorocarbons as propellants in 1948. In 1978, these propellants were banned in the United States for use in hair sprays due to the effect they had in depleting stratospheric ozone (23). To replace these, many gases were explored including carbon dioxide and nitrogen, but the hydrocarbons (propane, isobutane, and butane) were found to create the most consistent sprays throughout the can. Additionally, they did not interfere with performance of the hair spray polymers and were compatible with ethanol and properly neutralized polymers. The principal problem with using hydrocarbons was their flammability. With the advent of VOC content limits, dimethyl ether is used because of its compatibility with water. Another propellant, hydrofluorocarbon 152A, is used in a few products. It is not considered a VOC nor does it deplete stratospheric ozone; therefore, it can be used to create anhydrous hair sprays that comply with initial VOC limits. However, it is quite expensive and in short supply which limits its use (see AEROSOLS).

Manufacture.   Finishing sprays are easily prepared as simple solutions of the polymers, neutralizers, plasticizeres, fragrance, etc, in ethanol. If water is in the formulation, it must be added last. The aerosol products are filled by the methods described for styling mousses.

Packaging.   Evaluations of finishing sprays must include stability and performance testing in the finished state. With aerosol packages, pitting and corrosion of the can and flaking of the lining are real concerns. Standards of flammability and combustibility must be followed. Flame extension, flashbacks, and flashpoints must be determined for aerosol sprays and the shipping cases must be properly labeled according to U.S. Department of Transportation (DOT) standards. In both pump and aerosol products, spray rates, spray patterns, and particle size distribution have to be optimized. These can be controlled by variations in pump design in nonaerosols and by variations in valves, actuators, and concentrate/propellant ratio in aerosols.

Aerosol finishing sprays generally are packaged in tin-plated steel containers consisting of a dome, body, and base which may or may not be lined for protection against corrosion. For products containing higher water contents, aluminum is generally used for the package. Nonaerosol sprays are typically packaged in high density polyethylene.

Health and Safety Factors.   Finishing spray products generally have high alcohol contents which create a flammability hazard. This hazard is magnified in aerosolized sprays because hydrocarbon propellants used are very flammable. Deliberate inhalation of aerosols poses a potential health hazard to the consumer that could be fatal. Additionally, spraying a high alcohol content finishing spray into the eyes can cause severe irritation. Appropriate warnings must be displayed on the package. Detailed information covering labeling requirements of fixatives in general are published by the FDA (24).

Environmental Regulation.   In 1978, federal regulation banned the use of chlorofluorocarbons in hair sprays. This forced a dramatic change in the technology of aerosol hair sprays, requiring new formulations and new dispensing parts to accommodate hydrocarbon propellants. In the early 1990s, California and New York enacted strict limits on allowable VOC content; VOCs are defined as any compound containing at least one atom of carbon, but no more than 12 atoms, with a vapor pressure of 13.3 Pa (0.1 mm Hg) or more at 20°C. The regulations were enacted due to the role that VOCs play in creating ground level ozone. In 1992, California allowed a maximum of 80% VOC content in finishing sprays, with reduction to 55% maximum VOC content by 1998 (25). In New York the regulations are almost the same, and other states are expected to follow.

The primary VOCs in hair sprays are lower order alcohols and propellants. In nonaerosols, with typical VOC levels of 88–92%, a drop to 80% can be accomplished with total water addition of about 15%, which causes an increase in solution viscosity and a heavier, wetter spray. In aerosols with typical VOC level of 93–98%, a much bigger decrease is needed. The propellant amount is fixed, requiring an even greater percent of the concentrate to be water, about 23%. This in turn causes difficulties in spraying and may require the use of dimethyl ether as a component of the propellant system. Significant discontinuities occur when 55% VOC levels are attempted. Polymer suppliers are working to develop new materials which may be adaptable to these high water content formulas with the overall goal of approaching the performance standards of anhydrous fixative sprays.

Coloring Preparations

Hair coloring preparations have been in use since the ancient Egyptians, and recorded recipes exist in many cultures. These followed the traditional application of plant extracts or metallic dyes, both of which still are used. In the latter part of the nineteenth century, synthetic organic compounds were discovered which eventually led to modern hair coloring.

Among the desired properties of a good hair dye, toxicological safety is of primary importance. Coloration should be achievable in 10–30 min at ambient temperature and from a limited dyebath. These requirements mean that only relatively small molecules can penetrate into hair keratin and for this reason oxidative and nitro dyes have received considerable attention. In addition, the dye should impart a natural appearance to the hair under a variety of lighting conditions. It should produce a minimum of scalp staining and be convenient to use. Sunlight fastness must be good, and though fading depends on the type of product, the various components of the dye mixture must fade uniformly. Though hair tends to be nonuniform in diameter, shade, and history of abuse, the color should be fairly uniform or level from root to tip.

Modern hair colorants can be divided into temporary, semipermanent, and permanent systems. These categories are characterized by the durability of the color imparted to the hair, the type of dye employed, and the method of application (see DYES AND DYE INTERMEDIATES).

Temporary Hair Colorants. Temporary hair colorants give a color that is easily removed from hair. This is done by using large dye molecules that deposit on the surface of the hair without penetrating the cuticle. Dyes used for this class of hair color are shown in Table 5 (see COLORANTS FOR FOOD, DRUGS, COSMETICS AND MEDICAL DEVICES). Products that use these dyes are applied to the hair, usually after shampooing, and left there without any after-treatment. Under normal conditions color that is applied in this way is very fugitive and the dye can be completely removed by one shampoo. The retail products in this category are

Table 5. Temporary Dye Colors

FDA Designation	CAS Registry Number	CI name	Type
Ext. D&C Violet No. 2	[4430-18-6]	Acid Violet 43	anthraquinone
D&C Red No. 33	[3567-66-6]	Acid Red 33	azo
D&C Brown No. 1	[1320-07-6]	Acid Orange 24	disazo
D&C Green No. 5	[4403-90-1]	Acid Green 25	anthraquinone
Ext. D&C Yellow No. 7	[846-70-8]	Acid Yellow 1	nitro
D&C Red No. 22	[7372-87-1]	Acid Red 87	xanthene
FD&C Blue No. 1	[2650-18-2]	Acid Blue 9	triphenylmethane
FD&C Green No. 3	[2353-45-9]	Food Green 3	triphenylmethane

called temporary rinses. The transitory nature of these dyes provides both an advantage and a disadvantage to consumers. The products can be considered foolproof in that the color can be easily removed if the result is not satisfactory. However, there is a limitation in the amount of color that can be applied, and it is not possible to obtain very dark shades. Complexing the anionic dyes with quaternary amines enables more color to be deposited and a composition with these ingredients with benzyl alcohol and a nonionic resin has been patented (26).

Under certain conditions, so-called temporary dyes can be made to last longer than one shampoo. In salons, these dyes may be applied under a bonnet style hair dryer. At the temperature of the hair dryer ($\sim$50°C), the dyes can diffuse through the outer layers of the cuticle of the hair, and the color will last for many shampoos.

Temporary hair dye products usually are formulated at a neutral or slightly acidic pH. Besides the dyes, the formulations may contain a small amount of a quaternary amine to neutralize the negative charge on the dyes, a fragrance, a small amount of a solvent or surfactant to solubilize the fragrance, and a preservative (Table 6).

Semipermanent Hair Colorants. The term semipermanent defines hair color products that give a coloration lasting through 5–6 shampoos. This system uses so-called direct dyes which penetrate into the cortex but slowly diffuse out again when the hair is washed. The depth of coverage is limited and no lightening

Table 6. Formulation of Temporary Hair Dye Product

Substance	CAS Registry Number	Wt %
dyes		0.500
quaternized amine		0.200
fragrance		0.100
ethoxydiglycol	[111-90-0]	2.000
methylparaben	[99-76-3]	0.100
nonoxynol-6	[27177-01-1]	0.500
water	[7732-18-5]	96.600

of color can take place. Nitro and anthraquinone dyes are used mainly, and azo-benzenes less frequently. The color produced by a particular commercial product may result from the combination of many individual dyes, eg, three yellows, two reds, two violets, and one blue. A blend is necessary to achieve the desired color and to obtain a match between the roots and the more permeable ends. The nitro compounds, mainly nitrophenylenediamines and nitroaminophenols, provide an excellent range of yellow, orange, red, and violet hues but are deficient in blue; to provide blue tones, aminoanthraquinones are included. These are known as disperse dyes because of their insolubility in water. A partial list of semipermanent dyes includes HC Yellow No. 2 [4926-55-0], HC Yellow No. 4 [52551-67-4], Disperse Blue 3 [2475-46-9], Disperse Violet 1 [128-95-0], HC Red No. 1 [2784-89-6], and HC Orange No. 1 [54381-08-7]. More extensive data are available (27,28).

There is a wide variety of dyes unique to the field of hair coloring. Successive N-alkylation of the nitrophenylenediamines has an additive bathochromic effect on the visible absorption to the extent that violet-blue dyes can be formed. Since the simple N-alkyl derivatives do not have good dyeing properties, patent activity has concentrated on the superior N-hydroxyalkyl derivatives of nitrophenylene-diamines (29,30), some of which have commercial use (31). Other substituents have been used (32). A series of patents also have been issued on substituted water-soluble azo and anthraquinone dyes bearing quaternary ammonium groups (33).

Semipermanent hair color products are formulated at an alkaline pH, usually between 8.5 and 10. At this pH the cuticle of the hair lifts away from the hair a little, allowing for easier penetration of dye. An alkyl amine buffered with an organic acid normally is used to obtain the desired pH. The formulations contain a mixture of solvents and surfactants to solubilize the dyes and a thickening agent is added so that the product stays on the hair without running or dripping. A 20–30 min application time is normal for this type of product. A representative formula for a semipermanent dye product is given in Table 7.

Semipermanent hair color products have the advantage of being removable; if a consumer is not satisfied with the result, the color is gradually washed out of the hair. The products are perceived as very gentle. The ease of removal of these

Table 7. Formulation of Semipermanent Hair Dye Product

Substance	CAS Registry Number	Wt %
dyes		1.500
triethanolamine	[102-71-6]	1.200
oleic acid	[112-80-1]	0.600
propylene glycol	[57-55-6]	3.000
ethoxydiglycol	[111-90-0]	2.000
sodium lauryl sulfate	[151-21-3]	0.600
nonionic surfactant		2.500
hydroxyethylcellulose	[9004-62-0]	1.800
fragrance		0.300
water	[7732-18-5]	86.500

products is also a disadvantage because a consumer needs to reapply the color after every 6–8 shampoos to maintain the color.

The naturally derived dyes used by the Egyptians and other ancient civilizations are actually examples of semipermanent dyes. The best known dye of this kind comes from the Henna plant and is still in use after thousands of years. The extract of the Henna plant contains lawsone [83-72-7] (2-hydroxy-1,4-naphthoquinone). This dye produces a reddish color on hair, which is best used to produce a warming effect on brown hair. The dye is complicated to use in that it must be made into paste by mixing with hot water. The paste is then left on the hair for 30–45 minutes. Henna can enhance the color of brown hair by adding warm tones but the color on gray hair is an unattractive orange shade. Other dyes extracted from natural materials include chamomile, indigo, logwood, and walnut. Although all these materials can be used to dye hair, only henna is permitted in the United States as a hair colorant (ca 1993). Products using natural dyes other than henna are occasionally seen in the marketplace, but because of the lengthy application times and weak color results they are never very successful.

Permanent Hair Colorants.   Permanent colorants produce hair coloration that lasts until the hair grows out. Color is formed inside the hair by hydrogen peroxide-induced coupling reactions of colorless dye precursors. A full range of shades can be obtained with this system and the permanent or oxidative hair colorants are considered to be the most important class of hair dyes.

Oxidation Hair Colorant.   Color-forming reactions are accomplished by primary intermediates, secondary intermediates, and oxidants. Primary intermediates include the so-called para dyes, *p*-phenylenediamine, *p*-toluenediamine, *p*-aminodiphenylamine, and *p*-aminophenol, which form a quinone monoimine or diimine upon oxidation. The secondary intermediates, also known as couplers or modifiers, couple with the quinone imines to produce dyes. Secondary intermediates include *m*-diamines, *m*-aminophenols, polyhydroxyphenols, and naphthols. Some of the more important oxidation dye colors are given in Figure 1. An extensive listing is available (24,28).

The mechanism of oxidative dyeing involves a complex system of consecutive, competing, and autocatalytic reactions in which the final color depends on

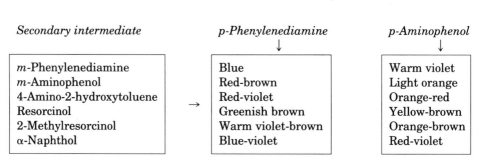

Primary intermediates

Secondary intermediate	*p*-Phenylenediamine ↓	*p*-Aminophenol ↓
m-Phenylenediamine	Blue	Warm violet
m-Aminophenol	Red-brown	Light orange
4-Amino-2-hydroxytoluene	Red-violet	Orange-red
Resorcinol →	Greenish brown	Yellow-brown
2-Methylresorcinol	Warm violet-brown	Orange-brown
α-Naphthol	Blue-violet	Red-violet

Fig. 1.   Quinine imines from primary intermediates couple with secondary intermediates to form various colors.

the efficiency with which the various couplers compete with one another for the available diimine. In addition, hydrolysis, oxidation, or polymerization of diimine may take place. Therefore, the color of a mixture cannot readily be predicted and involves trial and error. Though oxidation dyes produce fast colors, some off-shade fading does occur, particularly the development of a red tinge by the slow transformation of the blue indamine dye to a red phenazine dye.

Also present but not essential in permanent hair colorants are nitro dyes which dye hair without oxidation. These dyes, nitro derivatives of aminophenols and benzenediamines, impart yellow, orange, or red tones. Although they have good tinctorial value, they are not as colorfast as the oxidative dyes. They also are used in semipermanent hair colorants.

Attempts to broaden the range of materials available as dye precursors have been made (34,35). Oxidative dyes based on pyridine derivatives produce less sensitization than those based on benzene derivatives (36); however, they lack tinctorial power, lightfastness, and availability. Derivatives of tetraaminopyrimidine are claimed to act as primary intermediates to give intense shades with good fastness and excellent toxicological properties (37).

Oxidative hair dye products usually are formulated at a pH of 9.5–10.5. Ammonia, buffered with oleic acid, is the most commonly used alkalizing agent. Formulations contain a mixture of solvents and surfactants to solubilize the dyes and help in wetting the hair. A reducing agent is included in the formula to prevent premature oxidation of the dyes during storage. A typical formulation for a permanent hair color product is given in Table 8.

The colorant formulation is mixed just before application with an oxidizing agent, ie, developer. Hydrogen peroxide is the preferred developer usually at a concentration of 6%. An important function of hydrogen peroxide apart from inducing the color formation is the bleaching of the melanin pigment which constitutes the natural hair color. This process allows the development of light shades

Table 8. Permanent Hair Colorant Formulation for a Medium Brown Shade

Ingredients	CAS Registry Number	Wt %	Function
oleic acid	[112-80-1]	15	surfactant
sodium lauryl sulfate	[151-21-3]	1	surfactant
oleyl alcohol	[143-28-2]	10	thickener
ammonia	[7664-41-7]	2	alkali
isopropyl alcohol	[67-63-0]	10	solvent
propylene glycol	[57-55-6]	5	solvent
sodium sulfite	[7757-83-7]	0.2	antioxidant
EDTA	[60-00-4]	0.05	sequesterant
p-phenylenediamine	[106-50-3]	0.7	reactant
resorcinol	[108-46-3]	0.4	reactant
1-naphthol	[90-15-3]	0.1	reactant
p-aminophenol	[123-30-8]	0.05	reactant
m-aminophenol	[591-27-5]	0.1	reactant
water	[7732-18-5]	55.4	

which are not available either with semipermanent or temporary dyes. Although the amount of lightening is limited, it is possible to dye hair one to two levels lighter. Thus dark brown hair can be lightened to light brown and light brown to medium blonde.

If lighter colors than these are desired it is necessary to decolorize all the melanin in the hair in a preliminary step, and then add color back to the desired depth in a second treatment. This is known as a double-process treatment. The decolorization step consists of treating the hair with an alkaline mixture of per-sulfate salts and peroxide. The persulfate is added to the peroxide as a dry powder immediately before applying to the hair. Although the persulfate salts alone do not have any bleaching effect, the persulfate–peroxide mixture can remove all the melanin in the hair. Dark brown or darker hair can be lightened to a light blonde shade in about an hour.

After the hair is bleached it has an unnatural straw-like color and is then dyed to the desired tone with either a semipermanent or a permanent hair color product. The dye products designed to color bleached hair to a natural looking blonde shade are called toners.

The presence of ammonia and hydrogen peroxide in permanent hair color products is a disadvantage. Both are considered by consumers to be harsh chemicals. The odor of ammonia is unpleasant for a personal care product. Mono-ethanolamine has been used as a substitute for ammonia in some commercial permanent hair color products. It is not as effective as ammonia in allowing the hair to be lightened but it does not have as strong an odor.

Longer Lasting Semipermanent Hair Dyes. Several products have appeared on the market that are positioned as being more gentle than the usual permanent hair color products. They use the same dyes as other permanent hair color and also are mixed with hydrogen peroxide immediately before use. They differ from other permanent hair dye products in that they employ an alkalizing agent other than ammonia to obtain a pH about one unit lower than conventional products. Sometimes they use lower concentrations of peroxide. Typically the color results obtained using these products are not as durable as the products formulated at a higher pH. These products are positioned as longer lasting semipermanent hair color products.

Melanin Drying. One development (ca 1993) in hair coloring involves the formation of pigments within the hair that are very similar to natural melanin. Thus either catalytic or air oxidation of 5,6-dihydroxyindole [*3131-52-0*] can be effectively used to permanently dye hair within a short time (38). The formed color can, if required, be further modulated with dilute H_2O_2 or can be even totally removed from hair by this oxidant.

Metallic Dyes. Metallic dyes are among the older hair color materials known. Commercial products are based on a 1% solution of lead acetate in an aqueous, slightly acidic, alcoholic medium. Precipitated sulfur appears to be essential. The convenience aspect is stressed by the leave-in application method. Actually, the color development is so slow, taking about a week to ten days, that there is no alternative to this technique. Daily application is needed at first.

It has been shown that keratin [*9008-18-8*], and not cellulose-type fibers, are dyed. It is speculated that a lead–sulfur–keratin complex is formed. The color penetrates the hair fiber to a limited extent, forming a ring around the outside edge and imparting a lifeless appearance. Once developed, the color cannot be removed. The shades are limited yellows or light browns. Appealing mainly to men, the products are often called color restorers because of the gradual build-up of color.

Although lead acetate [301-04-2] is the only metallic dye used in the early 1990s, salts or silver, copper, nickel, cobalt, bismuth, and iron have been utilized in the past. A patent (39) refers to the use of bismuth citrate in a solution made alkaline with triisopropanolamine.

Hair Bleaching. Hair owes its natural color to the internal deposits of melanin pigment, distributed in the form of granules (0.3 by 1 μm) within the hair cortex. The density of the pigment distribution together with the chemical nature of the pigment, ie, eumelanin, phaeomelanin, or both, determine the actual hair color. The purpose of bleaching is to lighten or altogether decolorize the natural pigment with minimal damage to the hair itself.

When hair is bleached, the physicochemical character of melanin undergoes profound alterations, resulting in a predictable change of hair color. Black or brown hair progressively changes to reddish brown, auburn, reddish blonde, and finally pale blonde. The process may be stopped at any point or continue to the lightest shade. The latter provides a good background for a variety of blonde shades that can be obtained in a subsequent coloring step. This bleaching and coloring combination is known as double-process blonding.

An ammoniacal solution is added just before use to activate the hydrogen peroxide. Ammonia is preferred over sodium carbonate (40) or ethanolamines for maximum bleaching. The alkaline solution can be formulated into a shampoo vehicle with oleate soaps or ethoxylated fatty alcohols. When the bleach is applied to areas such as new hair growth, a viscous cream or paste may be preferred, formulated with fatty alcohols, alkanolamides, or other thickeners.

Bleaches of the simple ammoniacal peroxide type give limited lightening, which can be increased with bleach accelerators or boosters, including one or more per salts such as ammonium, potassium, or sodium persulfate or their combinations. These salts, which are susceptible to decomposition in aqueous solution, are packaged as dry powders and added just before use. In the absence of hydrogen peroxide, however, persulfates do not have any bleaching effect (41).

The amount of oxygen evolved is not related to the degree of bleaching (40). Oxidative decoloring is caused by hydrogen peroxide or by the HO_2^- ions present in alkaline solution. Hydrogen peroxide is also an effective solvent for melanin (41).

In the reaction of alkaline hydrogen peroxide with keratin (41,42), the disulfide cross-links of cystine are oxidized to sulfonic acid groups. This causes the hair to lose a portion of its tensile properties during bleaching. Despite these changes, the hair fibers retain their integrity and the results are usually pleasing to the user. However, care must be taken during repetitive applications to limit the bleach mixture to the regrowth, especially when persulfates are used. To this end it is important to employ high viscosity mixtures with minimum tendency to spread onto previously bleached hair.

Hair Coloring Regulation Issues. In the United States the classification of color additives is complex. Under the Federal Food, Drug and Cosmetic Act, all cosmetic colors must be the subject of an approved color additive petition to the

Food and Drug Administration; there is an exception for coal-tar colorants used to color hair. Based on the composition of these colorants, FDA can require a certification on each manufactured batch of colorant to assure conformance with the approved specifications. In the early 1990s FDA has required certification only for synthetically derived coal-tar type colors. Many of the approved color additives, both certified and noncertified, are restricted in their potential use. These restrictions can be found in the color additive regulations in the Code of Federal Regulations at 21 CFR 73 and 74.

Any of these approved color additives may also be used to color the hair as long as they are not specifically restricted from this use. However, as noted before, there is an exception to the use of coal-tar colorants to color hair. This exception, which further complicates the regulation of colorants, is due to the presence of the so-called hair dye exemption in the Federal Food, Drug and Cosmetics Act. When this Act was passed in 1938 and again when the Color Additive Amendments of 1960 were enacted, U.S. Congress recognized the sensitizing potential of some hair colors and specifically separated the use of coal-tar colors for coloring hair from the adulteration and color additive provisions of the law. However, in order to use these coal-tar colors, Congress did provide for specific warnings and conditions for their use. Nowhere in the Act, however, did Congress define coal-tar colors. FDA has interpreted this term to mean colors synthetically derived from aromatic chemical sources. They have recognized that the term coal-tar in the Federal Food, Drug and Cosmetics Act is a generic term that describes the original synthetic sources of these dyestuffs and is not meant to be restrictive as to the specific organic feedstock. Most coal-tar colors used in 1993 are derived from petrochemicals.

Based on the lack of a definition for coal-tar colors in the Act, and on a liberal interpretation of Section 601(e) (43), many materials have been used to color hair under this exemption. The use of coal-tar color for coloring hair was clarified by a ruling of the United States Court of Appeals for the second circuit in Toilet Goods Association v. Finch in 1969 and by a subsequent notice in the *Federal Register* (44). Under this ruling the Court stated that Congress did not intend to exempt noncoal-tar colors from Section 601(e). The notice further states that metallic salts and vegetable substances are not coal-tar derivatives and their use for coloring hair, without an approved color additive petition, would be considered adulteration. Following this Notice, petitions for henna, bismuth citrate, and lead acetate were approved by FDA for use in coloring hair. These materials are the only vegetable materials and metallic salts that can be used (ca 1993) to color the hair and are restricted for use.

Lead acetate can be used only for coloring scalp hair at a level not to exceed 0.6%, as lead, weight/volume of the product. The regulations provide specific restrictions (including label specifications) that lead acetate must not be used to color mustaches, eyelashes, eyebrows, or hair on parts of the body other than the scalp.

Bismuth citrate can be used only for coloring scalp hair such that the amount of bismuth citrate does not exceed 0.5% weight/volume of the product. Specific restrictions prohibit the use of bismuth citrate for coloring eyelashes, eyebrows, or hair on parts of the body other than the scalp; they also indicate label specifications.

Henna can be used only to color hair with the exception of eyelashes, eyebrows, or generally in the area of the eye. The label for products containing henna must caution the consumer not to use in the area of the eye or on cut or abraded scalp.

The use of nonapproved coal-tar colors for coloring hair does require special warnings and precautions; a cosmetic is deemed adulterated if it bears or contains any poisonous or deleterious substance which may render it injurious to users under the conditions of use (43). Coal-tar hair dye is exempt if the label conspicuously contains a warning that the product contains ingredients which may cause skin irritation on certain individuals, that a preliminary test according to accompanying directions should first be made, and that the product must not be used for dyeing the eyelashes or eyebrows; to do so may cause blindness. In addition, the labeling must include directions for doing the preliminary patch test. It should be noted, however, that the hair dye exemption for coal-tar dyes relates only to coal-tar dyes not listed as approved color additives. The certified colors approved by FDA for use in cosmetics are also coal-tar colors. The use of those specific coal-tar dyes for coloring hair does not require the warnings and patch test instructions (43).

Permanent Waving Preparations

The interest in and appreciation of the beauty of waved hair were shared already by ancient Assyrians who, as depicted on numerous base reliefs, wore cascades of curls falling over their shoulders. Little is known of the technology used at that time to produce their curls but it may not have been different from the waving techniques employed by Egyptians who curled their hair with mud and then dried it in the hot sun. Since then, improvements in waving techniques have been associated with the progress of technology; these historical developments are well documented (45,46). Basic precepts of modern permanent waving are due to pioneering work in the 1930s on the chemical and physical properties of wool (47,48). Over the years, these principles have been creatively utilized and adapted by cosmetic chemists to yield safe and efficacious products enjoyed by millions of consumers every year.

Chemistry of Hair Waving. A particular geometry innate to each individual hair is the result of processes of keratinization and follicular extrusion that transform viscous mixtures of polypeptide chains into strong, resilient, and rigid keratin fiber. Waving entails softening keratin, molding it to a desired shape, and finally annealing the newly imparted configuration. The underlying mechanism of waving is thus essentially molecular and involves a manipulation of physicochemical interactions that stabilize the keratin structure.

In native keratin the conformation stability is derived primarily from covalent cross-linking by cystine and from an extensive network of hydrogen bonds. Some contribution also comes from the electrostatic interaction of basic and acidic side chains, ie, salt linkages, as well as from the hydrophobic bonding on nonpolar residues present in keratin; however, the contribution of the latter is small, and in the intact fiber the covalent and polar interactions greatly overshadow the nonpolar ones. The hydrogen bonds, although weaker than either the disulfide cross-links or salt bridges, are much more numerous, ie, an average of 10 hydrogen bonds per each disulfide or salt link, and are largely responsible for the dry strength and rigidity of hair. Water readily breaks hydrogen bonds and this plasticizing action is utilized in imparting a temporary wave (set) to hair. The process is simple, being accomplished by wrapping wet hair around a roller or rod and drying it in the curled state. On drying, hydrogen bonds are reformed in a new configuration which is additionally stabilized by the recovery of fiber stiffness. Exposure to moisture quickly reverses the setting process and the hair returns to its natural shape.

Attempts to obtain permanent waving effects by manipulation of secondary bonds alone, ie, hydrogen bonds, salt-linkages, or nonpolar interactions, have proved unsuccessful. The cleavage of covalent disulfides in addition to secondary bond rupture is essential for imparting a durable wave stable to repeated shampooing (48). The cystine cross-links play the dominant role in controlling the recovery process in fiber. Severance of at least some of these cross-links is necessary to allow for molecular rearrangement, and some degree of molecular flow, that must take place during the molding step to attain relaxation of imposed stress and effective rebuilding of stabilizing bonds. Reductive fission of hair disulfides by mercaptans has become the preferred technique.

$$\text{keratin—S—S—keratin} \ \rightleftarrows \ 2 \ \text{keratin—SH}$$

Under relatively mild conditions, adequate cleavage can be achieved and the ensuing formation of cystine residues provides a welcome opportunity to reform the severed linkages by simple oxidative treatment to complete the process cycle. The reduction usually is carried out under alkaline conditions which favor the generation of the attacking nucleophile or mercaptide ion; potentiate the swelling of keratin; and promote the molecular rearrangement via sulfhydryl–disulfide interchange.

A viable alternative to reduction fission is the sulfitolysis of hair. The action of sulfite on the combined cystine in keratin yields one cystine residue and a Bunte salt. The reaction is highly reversible, particularly at pH 8. This feature of reversibility associated with limited fission of disulfide cross-links renders the waving with alkali sulfite simple and less aggressive than that with thioglycolic acid [68-11-1]. The attained curl is more of a body wave type. Higher cleavage levels, similar to those obtained with mercaptans, can be generated at neutrality but this downward shift in pH impairs the molecular rearrangement process and manifests itself in a softer and less durable wave.

The final step of the permanent waving reforms the disulfide bonds; it is usually called neutralization. To rebuild a disulfide cross-link in hair from two residues of combined cysteine [52-90-4] requires a close proximity of the reacting side chains. Interfering factors are numerous, yet the rebuilding reaction is very effective. Attesting to this is evidence of the recovery of the mechanical strength of the fiber and direct analysis of combined cystine [56-89-3] prior to and after neutralization. Data suggest that between 80–90% of the cysteine residues formed during the reduction step are converted back into cystine upon neutralization (49,50). It is not known what happens with the remaining 10–20% of the residues. Excluding dithiodiglycolic acid, cysteic acid residues seem to be the only identified by-products of the neutralization reaction. The latter are formed, however, in quantities too small to account for the balance of lost cysteine.

Hair Waving Process.   In the typical waving procedure, freshly shampooed and still damp hair is divided into 40–60 tresses. Each tress is wetted with waving lotion and wound onto plastic curlers with the help of porous end papers or sponges. The size of the curler decides the nature of the resultant wave. The smaller the curler, the tighter the wave. Hair is then left to process, rinsed thoroughly after 10–20 min, and neutralized while still on rods. After neutralization, hair is unwound, rinsed again, and either freely dried or set in the desired style.

Depending on the type of waving product used, there may be several variations to the procedure outlined above. Thus, instead of wrapping with lotion, the hair is wound wet and the lotion applied to curled hair. Some instructions also suggest a creep stage for better tightness and durability. This is simply a 30 min wait between rinsing off the lotion and application of the neutralizer.

Waving Lotions.   The reagent most frequently used for the reduction of hair is thioglycolic acid [68-11-1]. Although a variety of other mercaptans have been screened (51), none has been able to match the unique combination of efficacy, safety, and low cost that is a hallmark of thioglycolic acid.

Conventional waving lotions contain 0.5–0.8 M thioglycolic acid adjusted to and maintained at pH 9.1–9.5. The neutralizing base is ammonia [7664-41-7], alkanolamines, or both. Ammonia appears more effective than sodium hydroxide [1310-73-2] in facilitating diffusion of the thioglycolate through the hair. It is also preferred over nonvolatile amines because it escapes during processing. The resultant drop in pH reduces the activity of the lotion with time and thus minimizes the danger of over processing. Under practical waving conditions, ammonium thioglycolate [5421-46-5] fully penetrates the hair in 15–20 min (49,50,52), although longer time may be required for very coarse hair (1). The extent of disulfide bond cleavage that typically occurs during waving varies between 20–40% (53), the lower figure being a representative value for previously untreated hair.

Alkali sulfites have gained a stronghold in the hair waving market, focusing on the soft wave and casual styles for which the expectations of the waving performance are less rigorous. Reduced danger of over processing and lack of odor are clear benefits. Both ammonium sulfite [10196-04-06] and sodium sulfite [7757-83-7] are used in concentrations ranging from 0.5–1.0 M at pH 6.5–10.2 (54,55).

Waving products have appeared on the market formulated in the neutral pH range, ie, so-called acid waves. They are based either on the thioglycolic acid or its glyceryl esters. The waving performance of these products is mediocre. The resulting wave lacks the crispness and durability of the conventional alkaline wave although this is somewhat compensated by lower hair damage. Often heat is used to improve the result.

In the Orient, particularly Japan, the use of cysteine as a waving agent is widespread. This amino acid is claimed to provide a natural and nonodorous alternative to thioglycollic acid and to wave the hair without the damage. As a waving agent cysteine is a poor performer and, in most formulations, thioglycolic acid is added to improve waving efficacy.

Waving lotions frequently are formulated with a number of additives with the intention of enhancing the efficacy and the aesthetics of the process. Thus surfactants of the nonionic type are used to improve the wetting of hair and penetration, a hydrogen bond breaking agent such as urea [57-13-6] is added to intensify the swelling of hair, ammonium sulfate [7783-20-2] is used to decrease swelling, and latex emulsions and polyacrylates are employed as opacifiers. Conditioning materials used include mineral oil [8012-95-1], lanolin [8020-84-6], and hydrolyzed protein; the addition of cationic polymers has been patented (56,57). Perfuming of the thioglycolate lotions, although essential and desirable, is very difficult as the odor of the mercaptan is augmented by the unpleasant smell of the reduced hair. The latter is particularly evident in the case of sulfite lotions which are usually odorless.

Neutralizing Lotion. The principal active ingredient of cold wave neutralizers is usually an oxidizing agent. The most popular is hydrogen peroxide [7722-84-1], employed at a concentration of 1–2%; it continues to find widespread use. Aqueous solutions of sodium bromate [7789-38-0] at a concentration of 10–20% occasionally are used and are technically preferred over the peroxide formulations because of excellent stability and absence of hair bleaching. Neutralizing powders appear to be on the decline but formulations still in use consist of sodium perborate [7632-04-4] combined with hexametaphosphates to improve solubility in hard water.

Wetting and foaming additive agents occasionally are used to improve spreading and retention of the neutralizer in the hair. Acids such as citric acid [77-92-9] and tartaric acid [526-83-0] are suggested for the deswelling of hair and thus improvement in its overall condition. Conditioning agents such as stearalkonium chloride [122-19-0] are frequently employed to assure smooth texture, easy combing, and control of flyaway. Conditioning additives based on polymeric silicones have been patented (58).

Evaluation. Two parallel approaches are used in the industry to assess the efficacy of waving formulations. These are full- and half-head tests against established products, and laboratory evaluation of hair tresses processed according to waving instructions. The latter consists of a battery of tests related to the waving performance of the product, eg, degree of curl, durability of curl to repeated shampooing, etc; incidental hair damage, ie, measurements of swelling and tensile strength; and the final hair condition, ie, assessment of luster, dry and wet combability, and flyaway. Descriptions of laboratory techniques used in context of these tests have been published (59–62).

Health and Safety. The dermal toxicology of alkaline solutions of thioglycolic acid has been reviewed extensively (63–65). The reagent has been found harmless to normal skin when used under conditions adopted for cold waving. Some irritation is observed on abraded skin but this appears to be associated with the alkaline component of the waving solution (65). Hand protection is recommended for the professional hairdressers who routinely handle these products.

Manufacturing. The highly reactive nature of the active components of the permanent waving products requires rigorous control at every production stage. Cleanliness must be exemplary. Metal contaminants must be avoided and all tanks, valves, hoses, and bottling-machine parts in contact with the chemicals must be inert or lined with glass or nonreactive plastic (Teflon, polyethylene). Only high purity materials should be used and must be adequately checked prior to manufacture.

Hair Straightening Preparations

There are functionally different types of hair straightening preparations, ie, those which produce temporary straightening and those which are designed to accomplish permanent effects.

Temporary Hair Straightening. The most frequently used technique in this category is hot combing. An oily substance, ie, pressing oil, is applied to hair which is then combed under tension with a heated comb. The function of the pressing oil is to act simultaneously as a protective heat-transfer agent between the comb and the hair and as a lubricant to reduce the drag of the comb. The straightening effect is immediate but is lost quickly on exposure of hair to moisture. Pressing oils are usually based on petrolatum [8027-32-5] and mineral oil mixed with some wax and a perfume. Frequent hot combing dulls and damages the hair and leads ultimately to hair breakage.

Permanent Hair Straightening. The basic technical premise underlying permanent hair straightening is similar to that adhered to in waving. Hair is softened; maintained straight under tension for a period of time, usually accomplished by means of the high viscosity of the product and repeated combing; rinsed; and rehardened by application of the neutralizer. It thus is not surprising that many hair straightening compositions are just thickened versions of permanent waving products. Alkaline thioglycolate (6–8%) is formulated into a thick oil-in-water (o/w) emulsion or cream using generous concentrations of cetyl alcohol [124-29-8] and stearyl alcohol [112-92-5] and high molecular weight polyethylene glycol, together with a fatty alcohol sulfate as emulsifier. The emulsifier provides an added advantage of ready rinsability. Processing time may be anywhere from 30 min to 2 h, depending on the initial curliness of the hair. Conventional oxidizing neutralizers, eg, H_2O_2, bromates, and perborates, are used in the final step of the process.

Hair straightening compositions based on mixtures of ammonium bisulfite [10192-30-0] and urea [57-13-6] have been introduced and have found some application in the Caucasian hair market. The reformulation of the cystine crosslinks in bisulfite-reduced hair is best accomplished by a rinse, pH 8–10, rather than by the use of oxidizing agents (66).

An important class of permanent straighteners in use is that based on alkali as an active ingredient. Sodium hydroxide [1310-73-2], potassium hydroxide [1310-58-3], or a sodium carbonate [496-19-8] combination with guanidine [593-85-1] are used at concentrations of 1.5–3% in a heavy cream base. Even though the recommended treatment time is only between 5 and 20 min, the straightening effects, in general, surpass those obtained with either thioglycolates or bisulfites. This is due to the very different chemistry of the process as well as greater aggressiveness of the alkaline relaxers. A 15 min treatment irreversibly decreases the cystine content of hair to two-thirds of its initial value (2). The damaging action of strong alkali on hair is not restricted to the disulfide bonds alone. Apart from the potential of the main chain scission, ie, peptide bond hydrolysis, the very nature of the high pH base leads to a build-up of negative charges in hair which results in increased swelling, the latter being intensified by concurrent breakdown of the disulfide bonds. Great care must be exercised in the use of the alkaline relaxers as even a short contact with skin can cause blistering.

Professional Use Products

Many products in the hair care and hair color categories are distributed solely for professional use by cosmetologists, beauticians, and hairdressers in their places of business.

The Fair Packaging and Labeling Act does not apply to products used in professional establishments. Specifically, this means that these products are not required to have an identity statement or a list of ingredients. This exception, however, is limited only to those products actually intended for professional use. Products sold by the professional establishments to their customers for personal use are considered retail products and must be fully labeled as such.

Although these professional-use-only products do not require ingredient labeling, the cosmetics industry has developed a program to voluntarily list the components of professional products. However, under this voluntary program, the ingredients are listed in alphabetical order rather than descending order. This has been done to make it easier for the professional hairdresser to locate a specific compound that may be of interest.

Economic Aspects

Retail sales of hair preparations have more than doubled from 2×10^9 in 1978 to ca 4.2×10^9 in 1991 (67). While price increases over this 13-year period were clearly a factor, a variety of novel and functional products have been introduced into this market. A large rise in the shampoo sales, from 777×10^6 in 1978 to 1510×10^6 in 1991, was helped by the new and highly effective category of conditioning shampoos. However, conditioning shampoos had little adverse effect on the conditioner market which exploded from sales of 121×10^6 in 1978 to well over 800×10^6 in 1992. Diversification and new product lines in the hair styling category, particularly styling mousses, have pushed sales from 327×10^6 in 1978 to 870×10^6 in 1991. Permanent waves generated renewed appeal to consumers by growing from 39×10^6 in 1978 to well over 120×10^6 in 1992. There has been some growth in hair coloring, helped primarily by introduction of products for men, with total sales in 1992 of 571×10^6 as compared to 300×10^6 in 1978.

Regulations

Definitions.   Cosmetic products in the United States are regulated by FDA under the authority of two different laws, ie, the Federal Food, Drug and Cosmetics Act and the Fair Packaging and Labeling Act. Each of these Acts imposes slightly different conditions and labeling requirements for the products under their jurisdiction.

The Food, Drug and Cosmetics Act defines a cosmetic as a substance intended to be rubbed, poured, sprinkled, or sprayed on, introduced into, or otherwise applied to the human body or any part thereof for cleansing, beautifying, promoting attractiveness, or altering the appearance. A drug is defined as an article intended for use in the diagnosis, cure, mitigation, treatment, or prevention of disease in humans or other animals, and articles intended to affect the structure or any function of the body of humans or other animals.

The Fair Packaging and Labeling Act, which uses the same definitions for drugs and cosmetics as the Food, Drug and Cosmetic Act, only has jurisdiction over retail products sold to the consumer for use at home. This condition exempts free samples and professional use products not sold to a consumer for personal use.

Labeling Regulations. The Food, Drug and Cosmetics Act requires that the cosmetic product be safe under conditions of use and that labeling is not false or misleading. Under this Act, the labeling of a cosmetic product must contain the name and address of the manufacturer, packer, or distributor; the net contents; and any appropriate warnings. This information must appear on the label of the product, both inner and outer containers.

The Fair Packaging and Labeling Act is designed to provide the consumer with information to help them make value comparisons in the marketplace. It requires the name and address of the manufacturer, packer, or distributor; the net contents; an identity statement of the function of the product; and for cosmetics, a listing of the ingredients in descending order of concentration. The above information must appear only on the outer label; the identity statement and the ingredient labeling need not appear on inside packaging, if there is any. These items are only mandatory on the outside package where the consumer can see them at point of purchase. Since both acts require the name and address of the manufacturer, packer, or distributor, and the net contents, these items must appear on all labeling copy.

Labeling compliance with these regulations is complex, and further labeling information is available (67,68).

Drug Products. Although most hair care products are cosmetics and are regulated as such, some products also can be drugs. The regulatory status of these products is determined by the intention or claims made for the product. If the product is intended simply to beautify, cleanse, or alter the appearance, the product is a cosmetic. If, however, the product is intended to treat or prevent a disease condition or to affect the structure or function of the body, the product is a drug. Therefore, any product that makes a representation that it can control dandruff, treat psoriasis or seborrheic dermatitis, grow hair, prevent baldness, or other similar claims, is considered a drug product. Products that make both drug and cosmetic claims are considered drugs and must be in compliance with both the drug and cosmetic regulations.

Drug products must meet the requirements established by the Center for Drug Evaluation and Research at FDA. They must comply with the appropriate OTC Drug Review Final Rule published in the *Federal Register* or must be the subject of an approved New Drug Application (NDA) filed with FDA. Products not meeting these requirements are considered by FDA to be New Drugs without an NDA and subject to regulatory action.

FDA has published two final rules for hair products as of this writing (69). Any over-the-counter (OTC) drug product labeled or promoted for external use as a hair grower or for hair loss prevention is regarded as a new drug and must be the subject of an approved new drug application (NDA). Products making these claims without an NDA are considered to be in violation of the Federal Food, Drug and Cosmetic Act and are also mislabeled.

The conditions whereby dandruff, seborrheic dermatitis, and psoriasis drug products are generally recognized as safe and effective and are not misbranded is available (70). Specific active ingredients that can be used as well as the statement of identity, indications for use, and required warnings, are identified. Products that do not meet all of these requirements are considered new drugs and must have an approved NDA for the nonmonograph conditions.

There are a variety of other regulations and restrictions that affect all cosmetics including hair products. At the federal level, there are required warnings and restrictions for aerosol products and for products where safety has not been adequately substantiated (71). Other issues have been generally initiated by various state governments. California Proposition 65 contains warnings for products that expose consumers to specified levels of chemicals that cause cancer or reproductive toxicity. Many states, including California, New York, Rhode Island, and Vermont, have enacted legislation that restricts product environmental claims including degradability, recycling issues, and environmental safety. The Federal Trade Commission has issued guidelines for environmental claims but these may not be sufficient to meet the requirements of all the states. Several states, notably California and New York, have enacted legislation limiting the amount of volatile organic compounds (VOCs) in various product types, including hair sprays and other hair care products. For hair mousses the level of VOCs must not exceed 16% by January 1, 1994, and for hair styling gels that level must not exceed 6% by January 1, 1997.

BIBLIOGRAPHY

"Shampoos and Hair Preparations" in *ECT* 1st ed., Vol. 12, pp. 221–243, by F. E. Wall; "Tints, Hair Dyes and Bleaches" in *ECT* 1st ed., Vol. 14, pp. 166–189, by F. E. Wall; "Hair Preparations" in *ECT* 2nd ed., Vol 10, pp. 768–808, by W. R. Markland, Chesebrough-Pond's Inc.; in *ECT* 3rd ed., Vol. 12, pp. 80–114, by R. Feinland, F. E. Platko, L. White, R. DeMarco, J. J. Varco, and L. J. Wolfram, Clairol, Inc.

1. J. Menkart, L. J. Wolfram, and I. Mao, *J. Soc. Cosmet. Chem.* **17**, 769 (1966).
2. L. J. Wolfram and R. Yare, in J. Orfanos, ed., *Proceedings of 1st International Conference on Hair Research*, Springer-Verlag, Berlin, 1981.
3. W. R. Markland, *Am. Perfum.* **67**, 57 (1957).
4. A. Lanteri, *Soap Cosmet. Chem. Spec.* **54**, 31 (1978).
5. G. S. Kass, *Cosmet. Perfum.* **90**, 105 (1975).
6. K. R. Dutton and W. B. Reinisch, *Manuf. Chem.* **34**(1), 4 (1963).
7. J. R. Hart and E. F. Levy, *Soap Cosmet. Chem. Spec.* **53**, 31 (1977).
8. R. R. Riso, *Soap Chem. Spec.* **39**, 82 (1963).
9. G. Barker, M. Barabash, and P. Sosis, *Soap Cosmet. Chem. Spec.* **54**, 38 (1978).
10. A. K. Reng, *Cosmet. Toiletries* **93**, 95 (1978).
11. V. Kinglake, *Soap Perfum. Cosmet.* **51**(5), 206 (1978).
12. G. Barker, *Cosmet. Perfum.* **90**, 70 (1975).
13. U.S. Pat. 2,950,255 (Aug. 23, 1960) S. R. Goff (to the Gillette Co.).
14. J. I. Yablonski and C. L. Goldman, *Cosmet. Perfum.* **90**, 45 (1975).
15. U.S. Pat. 2,999,069 (Sept. 5, 1961), J. N. Masci and N. A. Poirier (to Johnson and Johnson Co.).

16. U.S. Pat. 3,055,836, (Sept. 25, 1962), J. N. Masci and N. A. Poirier (to Johnson and Johnson Co.).
17. H. S. Manheimer, *Am. Perfum.* **76**, 36 (1961).
18. J. J. Leyden and A. M. Kligman, *Cosmet. Toiletries* **99**, 23 (1979).
19. U.S. Pat. 4,788-006 (Nov. 29, 1988), R. E. Bolich, J. (to Procter and Gamble Co.).
20. *CTFA International Cosmetic Ingredient Dictionary*, 4th Ed., CTFA, Washington, D.C., 1991.
21. R. Lockhead, *HAPPI*, 60 (Apr. 1990).
22. *Merck Index*, 11th ed., Merck & Co., Rahway, N.J., 1989.
23. *Fed. Reg.* **42**, 24,536 (May 13, 1977).
24. *Code of Federal Regulations 21, Parts 701,* and *740,* U.S. Government Printing Office, Washington, D.C., Apr. 1, 1992.
25. *California Code of Regulations,* Title 13, Sec. 94,500–94,517.
26. U.S. Pat. 3,653,797 (Appr. 4, 1972) C. R. Reis, A. V. Forbriger, and K. I. Patel.
27. J. F. Corbett, in K. Ventakamaran, ed., *Chemistry of Synthetic Dyes*, Vol. 5, Academic Press, Inc., New York, 1971, p. 475.
28. G. S. Kass, in M. G. deNavarre, ed., *Chemistry and Manufacture of Cosmetics*, 2nd ed., Vol. 4, Continental Press, Orlando, Fl., 1975, p. 841.
29. U.S. Pat. 3,088,878 (May 7, 1963) W. H. Brunner and A. Halasz (to Clairol, Inc.).
30. Brit. Pat. 1,061,515 (Mar. 15, 1967); 1,104,970 (Mar. 6, 1968) G. Kalopissis and A. Bugant (to L'Oreal).
31. U.S. Pat. 5,041,143 (Aug. 20, 1991) G. Lang and A. Junino (to Clairol, Inc.).
32. U.S. Pat. 5,024,673 (Jun. 18, 1991) Y. Pan and L. Hochman (to Clairol, Inc.).
33. U.S. Pat. 3,100,739 (Aug. 13, 1963) W. Kaiser and P. Berth (to Therachemie Chemische Therapeutische Geselschaft).
34. Brit. Pat. 1,025,916 (Apr. 14, 1966), R. Charles, G. Kalopissis, and J. Gascon (to L'Oreal).
35. U.S. Pat. 3,884,627 (May 20, 1975) F. Brody and S. Pohl (to Clairol, Inc.).
36. F. Lange, *Am. Perfum. Cosmet.* **80**, 33 (1965).
37. U.S. Pat. 4,003,699 (Jan. 18, 1977) D. Rose, F. Saygin, and E. Weinrich (to Henkel).
38. K. Brown and co-workers, *J. Soc. Cosmet. Chem.* **40**, 65 (1989).
39. U.S. Pat. 3,954,393 (May 4, 1976), H. Lapidus (to Combe).
40. V. Bollert and L. Eckert, *J. Soc. Cosmet. Chem.* **19**, 275 (1968).
41. L. Wolfram, K. Hall, and I. Hui, *J. Soc. Cosmet. Chem.* **21**, 875 (1970).
42. C. Robbins, *J. Soc. Cosmet. Chem.* **22**, 339 (1971).
43. *Federal Food, Drug and Cosmetic Act, Section 601(e)*, U.S. Government Printing Office, Washington, D.C.; *Ibid., Section 601(a)*.
44. *Fed. Reg.* **38**, 2996 (Jan. 31, 1973).
45. M. J. Sutter, *J. Soc. Cosmet. Chem.* **1**, 103 (1948).
46. *The Basic Science of Hair Treatments*, Nestle-LeMur Inc., New York, 1935.
47. W. T. Astbury and H. J. Woods, *Philos. Trans. R. Soc. London, Ser. A* **232**, 338 (1933).
48. J. B. Speakman, *J. Soc. Dyers Color.* **52**, 335 (1936).
49. H. Zahn, T. Gerthsen, and M. Kehren, *J. Soc. Cosmet. Chem.* **14**, 529 (1963).
50. J. G. Gumprecht, K. Patel, and R. P. Bono, *J. Soc. Cosmet. Chem.* **28**, 717 (1977).
51. J. W. Hoefele and R. W. Broge, *Proc. Sci. Sect. Toilet Goods Assoc.* **36**, 31, (1961).
52. R. E. Reed, M. DenBeste, and F. L. Humoller, *J. Soc. Cosmet Chem.* **1**, 109 (1948).
53. H. Freytag, *Fette Seifen Anstrichm.* **58**, 245 (1956).
54. M. S. Balsam and E. Sagarin, eds., *Cosmetics, Science and Technology*, Wiley-Interscience, New York, 1972, p. 214.
55. U.S. Pat. 3,864,476 (Feb. 4, 1975) F. J. Altieri.
56. U.S. Pat. 3,912,808 (Oct. 14, 1975) P. E. Sokol (to Gillette Co.).

57. U.S. Pat. 4,416,297 (Nov. 22, 1983), L. J. Wolfram, D. Cohen, and N. Tehrani (to Clairol, Inc.).
58. U.S. Pat. 4,770,873 (sept. 13, 1988) L. J. Wolfram and D. Cohen (to Clairol, Inc.).
59. E. J. Stavrakas, M. M. Platt, and W. J. Hamburger, *Proc. Sci. Sect. Toilet Goods Assoc.* **31**, 36 (1959).
60. E. J. Valko and G. Barnett, *J. Soc. Cosmet. Chem.* **3**, 108 (1951).
61. M. Garcia and J. Diaz, *J. Soc. Cosmet. Chem.* **27**, 379 (1976).
62. A. C. Lunn and R. E. Evans, *J. Soc. Cosmet. Chem.* **28**, 549 (1977).
63. R. Whitman and M. G. Brookins, *Proc. Sci. Sect. Toilet Goods Assoc.* **25**, 42 (1956).
64. J. H. Draize, E. Alvarex, and M. F. Whitesell, *Proc. Sci. Sect. Toilet Goods Assoc.* **7**, 29 (1947).
65. J. A. Norris, *Food Cosmet. Toxicol.* **3**, 43 (1965).
66. L. J. Wolfram and D. L. Underwood, *Text. Res. J.* **36**, 947 (1966).
67. *Cosmetics, Toiletry and Fragrance Association (CTFA) Labeling Manual*, 5th ed., CTFA, Washington, D.C., 1990.
68. *Code of Federal Regulations, Title 21, Part 701*, U.S. Government Printing Office, Washington, D.C., Apr. 1, 1993.
69. *Fed. Reg.* **54**, 28,772 (July 7, 1989).
70. *Fed. Reg.* **56**, 63,554 (Dec. 4, 1991); Ref. 24, pp. 358.701 to 358.750.
71. *Code of Federal Regulations, 21 Parts* 2.125, 740.11, and 740.10, U.S. Government Printing Office, Washington, D.C., Apr. 1, 1993.

STANLEY POHL
JOSEPH VARCO
PAUL WALLACE
LESZEK J. WOLFRAM
Clairol, Inc.

HALOGENATED FIRE RETARDANTS. See FLAME RETARDANTS, HALOGENATED.

HANSA RED. See AZO DYES.

HARDNESS

Hardness is a measure of a material's resistance to deformation. In this article hardness is taken to be the measure of a material's resistance to indentation by a tool or indenter harder than itself. This seems a relatively simple concept until mathematical analysis is attempted; the elastic, plastic, and elastic recovery properties of a material are involved, making the relationship quite complex. Further complications are introduced by variations in elastic modulus and frictional coefficients.

As a consequence, although the precise analysis of the indentation process continues, numerous practical applications of indentation hardness are in use and others are being developed. The impetus to this development is that whatever the numerical value of indentation hardness, it is clearly related to many other material properties of greater interest to engineers such as strength, wear resistance, and machinability. The relationship to the strength properties of materials is the most important. The indentation hardness test provides at once a simple, rapid, and essentially nondestructive means of testing a material and discovering its strength.

A hardness indentation causes both elastic and plastic deformations which activate certain strengthening mechanisms in metals. Dislocations created by the deformation result in strain hardening of metals. Thus the indentation hardness test, which is a measure of resistance to deformation, is affected by the rate of strain hardening.

Anisotropy in metals and composite materials is common as a result of manufacturing history. Anisotropic materials often display significantly different results when tested along different planes. This applies to indentation hardness tests as well as any other test.

Many types of hardness tests have been devised. The most common in use are the static indentation tests, eg, Brinell, Rockwell, and Vickers. Dynamic hardness tests involve the elastic response or rebound of a dropped indenter, eg, Scleroscope (Table 1). The approximate relationships among the various hardness tests are given in Table 2.

Although indentation hardness tests are usually classified as nondestructive they do in fact leave a permanent indentation on the surface of the workpiece. Thus the nondestructiveness of indentation hardness testing depends on the criticality of the tested surface and the location of the indentations.

Indentation Tests

Brinell. The first reliable indentation hardness test was developed by Brinell in 1900 and used ball bearings to make indentations in steel (1). The technique has remained reliable and essentially unchanged for nearly 100 years. The test, described by ASTM Standard E10 (2), is still in use.

The principle of the Brinell hardness test is that the spherical surface area of a recovered indentation made with a standard hardened steel ball under specific load is directly related to the property called hardness. In the following, HBN =

Table 1. Hardness Tests Described by ASTM Standards

Common name	Title	ASTM number[a]
Brinell	Brinell Hardness of Metallic Materials	E10
Rockwell	Rockwell Hardness and Rockwell Superficial Hardness of Metallic Materials	E18
Vickers DPH	Test Method for Vickers Hardness of Metallic Materials	E92
Knoop/DPH	Test Method Microhardness of Materials	E384
Scleroscope	Recommended Practice for Scleroscopic Hardness Testing of Metallic Materials	E448
International Rubber	Test Method for Rubber Property International Hardness	D1415
Durometer	Test Method for Rubber Property Durometer Hardness	D2240
Barcol	Test Method for Indentation Hardness of Rigid Plastics via Barcol Impresser	D2583
Portable	Test Method for Indentation Hardness of Metal using Portable Hardness Testers	E110
Webster	Webster Hardness Gauge	B647

[a]Ref. 2.

Table 2. Approximate Relation Between Hardness Scales[a]

Vickers	Brinell	Rockwell		Superficial		Knoop	Scleroscope
		B	C	15 N	30 N		
900			67	92.9	83.6	895	95
800	722		64	91.8	81.1	822	88
700	656		60.1	90.3	77.6	735	81
600	564		55.2	88.0	73.2	636	74
500	471		49.1	85.0	67.7	528	66
400	379		40.8	80.8	60.2	412	55
300	284		29.8	74.9	50.2	309	42
250	238	99.5	22.2	70.6	43.4	262	36
200	190	91.5				216	29
150	143	78.7				164	22
100	95	56.2				112	

[a]This table shows the relationship between hardness testing scales, but should not be used for hardness converson. See ASTM E140 (2) for specific materials conversions.

Brinell hardness number, P = load in kgf, D = diameter of the ball in mm, and d = diameter of the impression in mm. (For load expressed in Newtons, the denominator must be $9.807\pi D$ ... to obtain the same HBN).

$$\text{HBN} = \frac{2P}{\pi D(D - \sqrt{D^2 - d^2})}$$

In commercial practice a 10-mm steel ball is considered standard, although other diameters may be used, and a 29.4 kN (3000 kgf) load is most common. Lesser loads are used for materials softer than steel such as aluminum and copper.

Because of the geometric limitations of the indenting ball the relationship between indentation area and computed hardness number deviates from linearity when the recovered indentation diameter of a 10-mm ball is less than 2.5 mm or greater than 6.0 mm.

In practice it is still necessary to read the diameter of the Brinell impressions with a calibrated microscope; however, the computations to derive the Brinell hardness number are unnecessary for standard loads and indentors. Table 1 of ASTM E10 (2) contains the tabulated relation between indentation diameter and hardness number.

Test pieces for Brinell testing must have two parallel sides and be reasonably smooth for proper support on the anvil of the test machine. Minimum sample thickness must be 10 times indentation depth. Successive indentations must not be closer than three indentation diameters to one another or to the edge of the test piece.

Thus the Brinell test in its original manifestation is a laboratory test in which cut pieces are brought to it for testing. The lack of portability spawned several modifications to achieve that property.

The simplest of the portable modifications is a lightweight version of the original machine in which the hydraulic loading system is replaced by a spring. This machine still requires a sample be cut from large pieces for testing.

The pin Brinell tester takes the form of a large C clamp with the ball indenter on the end of the screw. Load is controlled by a built-in shear pin. A modification of this device employs impact loading by a hammer to achieve similar results.

There are also strap-on type Brinell testers in which the anvil is supplanted by a chain or other clamping device and the indenter is spring-loaded. These have the advantage of being able to test directly very large objects without the need for cutting samples.

The hand-held comparative Brinell tester is the most portable device. With this device a hammer blow is substituted for the static load which is transmitted first through a standard bar of known hardness and then through the indenter into the workpiece. The indentations in both the standard bar and the workpiece are measured and from the ratio of the diameters the HBN is derived. The loss in accuracy is made up for by the excellent portability.

The latest portable Brinell testers are spring-loaded, hand-held, and digitized to read directly in Brinell hardness units. Their accuracy is questionable because of extreme surface sensitivity and they are not in fact Brinell testers but Rockwell indenters calibrated to read in Brinell numbers.

The Brinell test range is limited, by the capability of the hardened steel ball indenters used, to HBN 444. This range can be extended upward to HBN 500 by using special cold work-hardened steel balls and to as high as HBN 627 by using special tungsten carbide balls.

Standard practice for Brinell testing is to measure the diameter of each indentation twice and average the measurement before entering the tables to determine HBN. The same averaging principle is applied on nonflat (curved) surfaces which yield an elliptical, not a round, indentation.

Rockwell. The invention of the Rockwell hardness tester in 1919 was an advance over previous indentation tests requiring accurate indentation measurement and tabular reduction to derive a hardness number. In the Rockwell test the hardness number is read directly from the instrument dial (1,3).

The principle of the Rockwell hardness test is that the depth of the indentation between a minor and a major load applied through an indenter is inversely proportional to the hardness number. Using a minor load to set the indenter helps to reduce backlash in the measuring system.

In the Rockwell test a spheroconical diamond (Brale) indenter or a hardened steel ball is used with various load ranges to achieve a series of scales identified by a suffix letter (Table 3). The suffix letter defines both load and indenter. The most popular scales used are "C" for hard materials and "B" for soft materials. A Rockwell hardness number is meaningless without the letter suffix, eg, HRC 54 or HRB 95.

The Rockwell testing machine is thus a framework permitting stable support of the workpiece on one side and means to impress the indenter under specified load on the other. A dial indicator attached to the indenter spindle is used to read directly the depth of indentation in hardness numbers.

The relationship between depth of penetration and the Rockwell hardness number is

$$HRC = C - (d/0.002)$$

where HRC = Rockwell C hardness number, C = indenter constant (100), and d = depth indentation in mm. Similarly the relationship for Rockwell B hardness is

$$HRB = B - (d/0.002)$$

where HRB = Rockwell B hardness number, B = ball constant 130, and d = depth indentation in mm. This relationship is most often used in computing min-

Table 3. Rockwell Hardness Testing Scale Designations

	Major[b] load, N[c]		
Indenter[a]	588	98	1471
Brale[d]	A	D	C
1.6 mm ball	F	B	G
3.2 mm ball	H	E	K
6.4 mm ball	L	M	P
12.7 mm ball	R	S	V

[a]Hardened steel balls of various diameters are used for materials significantly softer than the ball itself. Brale[d] penetrators are used for harder materials.
[b]Minor load is 98 N = 10 kgf for all scales.
[c]To convert N to kgf, divide by 9.807.
[d]A spheroconical diamond indenter having a 120° included cone angle.

imum sample thicknesses which for Rockwell tests is a minimum of 10 times indentation depth.

The A, D, and C Rockwell scales used primarily for steel and hard materials yield hardness numbers from 20 to about 85. Hardness numbers lower than HRC 20 are invalid; the Rockwell B, G, or F scales should be used. Hardness conversions from one scale to another are available for some common materials. (see Table 4, ASTM E140)

The Rockwell superficial test was developed to accommodate smaller and thinner samples than the standard Rockwell test. The test principle is identical to the standard Rockwell test but the major and minor loads are substantially smaller. The test machine is also similar but modified to accommodate the smaller loads. Dual purpose machines can handle both standard and superficial tests.

The indenter/load combinations used for superficial Rockwell testing are listed in Table 5. As with the standard Rockwell test it is necessary to include the

Table 4 ASTM Standards Related to Hardness Testing[a]

ASTM number	ASTM standard
A370	Brinell Tests of Steel Products
A833	Comparison Hardness Tester Practice
B294	Rockwell Test on Cemented Carbides
B347	Rockwell Test for Sintered Materials
B578	Knoop Test for Electrodeposited Coatings
B647	Webster Hardness Gauge
B648	Barcol Test of Aluminum Alloys
B724	Newage Portable Hardness Tests for Aluminum Alloys
C569	Indentation Test for Thermal Insulation
C661	Durometer Test for Elastomeric Sealants
C730	Knoop Test for Glass
C748	Rockwell Test for Graphitic Materials
C849	Knoop Test on Ceramic Whitewear
C886	Scleroscope of Carbon and Graphite
D1414	Hardness Tests on Rubber O-Rings
D1415	Test Method for Rubber Property–International Hardness
D2240	Test Method for Rubber Property–Durometer Hardness
D2583	Test Method for Indentation Hardness of Rigid Plastics by Means of Barcol Impressor
D617	Rockwell Test for Phenolic Laminated Sheet
D785	Rockwell Test on Electrical Insulating Materials
E92	Test Method for Vickers Hardness of Metallic Materials
E103	Rapid Indentation Hardness Tests
E384	Test Method for Microhardness of Materials
E1077	Test for Decarburization of Steel
E140	Hardness Conversion Tables for Metals
E448	Recommended Practice for Scleroscope Hardness Testing Metallic Materials
F451	Rockwell Hardness of Bone Cements
F500	Hardness of Neurosurgical Acrylic Resin

[a] Ref. 2.

Table 5. Rockwell Hardness Testing Scale Designations

Indenter[a]	Major[b] load, N[c]		
	147	294	441
Brale[d]	15N	30N	45N
1.6 mm ball	15T	30T	45T
3.2 mm ball	15W	30W	45W
6.4 mm ball	15X	30X	45X
12.7 mm ball	15Y	30Y	45Y

[a]Hardened steel balls of various diameters are used for materials significantly softer than the ball itself. Brale[d] penetrators are used for harder materials.
[b]Minor load for all scales is 29.4 N = 3 kgf.
[c]To convert N to kgf, divide by 9.807.
[d]A spheroconical diamond indenter having a 120° included cone angle.

superficial load/indenter combination used for the hardness number to be meaningful, eg, HR30N 65 or HR30T 65.

Most laboratory and shop-use Rockwell hardness testers are nonportable, lever operated, deadweight machines. Newer versions have digital readouts rather than the traditional analogue dial. Some designs of Rockwell testers employ a spring-loading system instead of deadweights.

Portable hand-held direct reading Rockwell testers have been developed and are in use, as are numerous C clamp configurations, all intended for field or shop use. The newest computerized digital readout Rockwell tester provides the ultimate in portability at the cost of some loss of sensitivity due to the very light loads used.

The standard Rockwell test requires a relatively smooth surface (120 grit or better) for reproducibility. Superficial Rockwell test samples must be ground to 600 grit or better for accuracy and reproducibility.

Although the Rockwell test is intended to be used on flat parallel-sided specimens, its use can be extended to rounded surfaces by using a curvature correction factor. Compound surfaces such as gear teeth can be tested but the results must be corrected for curvature.

As with all tests, frequent calibration of the test equipment using standard hardness blocks is a prerequisite for reliable hardness testing (see ASTM E18). Standard hardness blocks are available through commercial sources in the United States but do not have traceability to internationally accepted standards as in Europe.

Rockwell hardness testing has been extended to both low and high temperature regimes usually by enclosing the sample and part of the machine in an environmental chamber and using extensions for the anvil and indenter.

Recommended procedures for use with Rockwell and superficial Rockwell tests are detailed in ASTM E18 (2).

Vickers Hardness. The Vickers or diamond pyramid hardness (DPH) developed in 1924 was an improvement over the Brinell test. The Vickers test used a pyramidal diamond as the indenter. This permitted the hardness testing of much harder materials, and the constant 136° angle of the indenter eliminated the problem of variable indentation shape encountered using spherical indenters (1).

Vickers hardness numbers are calculated from measurement of the indentation diagonals as follows, where HV = Vickers hardness, P = applied load in N, D = indentation diagonal in mm, and θ = 136°. For load in kgf, omit 9.807 in the denominator.

$$\text{HV} = \frac{2P\sin(\theta/2)}{9.807\,D^2} \quad \text{or} \quad = \frac{1.8544\,P}{9.807\,D^2}$$

The Vickers hardness test is a macrohardness test in which loads are commonly varied from 9.8 to 1180 N (1 to 120 kgf). Vickers hardness numbers are invariant with load within the stated limits.

The Vickers hardness test is commonly made on a flat specimen on which the indenter is hydraulically loaded. When the desired number of indentations have been made, the specimen is removed and both diagonals of the indentations, measured using a calibrated microscope, are then averaged. The Vickers hardness number may be calculated, or for standard loads taken from a precalculated table of indentation size vs VHN. The preferred procedures are described in ASTM E92 (2).

The Vickers hardness test, developed in the United Kingdom, is more popular there than in the United States. VHN (Vickers hardness number) and DPH (diamond pyramid hardness) are synonymous terms.

Surface finish requirements for the Vickers test vary with the test load. Heavy load tests can be made on a 120 grit ground surface. At low loads increasingly finer surface preparation is required, approaching that for metallographic specimens, to permit accurate diamond indentation measurements.

Minimum thickness requirements are $1\frac{1}{2}$ times the indentation diagonal measurement, and there should be no visible marking or bulge visible on the side opposite the indentation.

The Vickers test is based on a plane surface; however, correction tables are available for both convex and concave surfaces (see ASTM E92) (2).

Conversion to other hardness scales from Vickers is approximated for specific materials listed in ASTM E140 (2). Conversions outside the stated areas should be avoided unless supported by test data.

Microhardness. Given a sufficiently accurately ground Vickers diamond indenter, and the load insensitivity of the Vickers test, a natural extension of this technology has been to reduce loads to less than 9.8 N (1 kgf). The load limit defines the area called microhardness testing (ASTM E384). To accomplish this test, machines capable of accurately applying loads as low as 9.8 mN (1 gf) have been developed. All use the Vickers 136 diamond indenter. This development permits the determination of the hardness of discrete metallurgical or mineralogical constituents not previously possible with macrohardness techniques (4).

In 1939 the Rhomb-shaped diamond indenter, having a length-to-breadth ratio of 7:1, was introduced to microhardness testing (5). The advantages of the

Knoop indenter are that it requires measurement of only the long diagonal of the indentation and it is said to be more sensitive and accurate when used on highly recoverable materials such as glass.

In time most commercially available microhardness testers accepted both Vickers and Knoop indenters. The Vickers remained almost universally used in Europe but shared acceptance with the Knoop in the United States.

The Knoop hardness number is computed from the measured long diagonal by the following formula where HK = Knoop hardness, P = load in Newtons, and d = long diagonal, mm.

$$HK = \frac{14229\ P}{d^2}$$

All microhardness testing machines are touchy to operate owing to the very small loads involved and the danger of inertial effects when loading. In addition it has been found that the duration of load application affects results. Commercial testers, therefore, have automatic load/unload cycles and interval timers. More modern versions have autodata recording and in some cases programmable sequences of indentation locations. Microhardness testing due to the careful sample preparation required (metallographic polish) and the delicacy of the test apparatus, is a laboratory test only.

Applications of microhardness testing greatly extend the conventional indentation hardness test to glass and ceramics, metallographic constituents, and to thin coatings or other surface treatments not otherwise testable.

The shortcomings of microhardness tests include numerous sources of errors not found in macrohardness tests such as friction, vibration, inertia, windage, and the skill of the test operator.

Ultrasonic Microhardness. A new microhardness test using ultrasonic vibrations has been developed and offers some advantages over conventional microhardness tests that rely on physical measurement of the remaining indentation size (6). The ultrasonic method uses the DPH diamond indenter under a constant load of 7.8 N (800 gf) or less. The hardness number is derived from a comparison of the natural frequency of the diamond indenter when free or loaded. Knowledge of the modulus of elasticity of the material under test and a smooth surface finish is required. The technique is fast and direct-reading, making it useful for production testing of similarly shaped parts.

Scratch Hardness. *Mohs'.* An early (1822) hardness comparison test involved assigning a relative number to all known materials (usually minerals and pure metals) by virtue of their relative ability to scratch one another. The results of this classification are not relatable to other properties of materials or to other measures of hardness. As a result of this limited usefulness, the Mohs' hardness test is primarily used for mineral identification. Some examples of the Mohs' hardness scale, which ranks materials from 1 to 10, are listed in Table 6.

Scratch Test. The scratch microhardness test is a refinement of the Mohs' test. The corner of a cubic diamond is drawn across the surface of a metallographically polished sample under a constant load, usually 29.4 N (3 kgf). The width of the resultant Vee groove scratch varies inversely with the hardness of the ma-

Table 6. Mohs' Hardness Numbers for Some Materials

Mohs' hardness	Material
1	talc/graphite
2	cadmium/anthracite
3	calcite/boric acid
4	bell metal/fluorite
5	manganese/asbestos
6	feldspar/pumice
7	flint/quartz
8	topaz/beryl
9	corundum/chromium
10	carbon/diamond

terial displaced where H = scratch hardness number and λ = groove width in micrometers.

$$H = \frac{10000}{\lambda^2}$$

This test finds application on finely polished and/or etched metallographic samples and mineralogical samples. It is useful for distinguishing variations in hardness between adjacent microconstituents. The test is extremely delicate and therefore is little used commercially. Use is largely restricted to research institutions. There is no established means of converting scratch hardness data to other hardness scales.

Sceleroscope Rebound Tests. The Scleroscope is a rebound-type hardness tester invented in 1907 (7). The principle involves dropping a diamond-tipped hammer from a specified height onto the test piece and measuring the height of rebound. In practice the hammer or drop weight is enclosed in a calibrated glass tube, raised by air pressure, then dropped. Rebound height is read from a scale on the glass tube. Later models of the Scleroscope contain a friction clutch that stops the hammer at the maximum point of rebound for easier reading. Preferred test procedures are described in ASTM E448 (2).

The Scleroscope scale ranges from 0 to 140; the calibration point of 100 is the hardness of fully quenched but untempered steel. Standard test blocks embodying this condition are used for calibration.

Portability, simplicity, and high speed are the main advantages of this portable hardness tester. It uses a single numerical scale encompassing the hardness of all metals.

Friction due to lack of vertical positioning of the tube is a source of error, as is sensitivity to the surface condition of the test piece. Samples of small mass cannot be tested except when supported on a heavy anvil.

Scleroscope hardness numbers are convertible to other hardness scales (see ASTM E140) (2).

Special-Purpose Testers

Barcol Indenter.   The Barcol hardness tester is a hand-held, spring-loaded instrument with a steel indenter developed for use on hard plastics and soft metals (ASTM D2583) (2). In use the indenter is forced into the sample surface and a hardness number is read directly off the integral dial indicator calibrated on a 0 to 100 scale. Barcol hardness numbers do not relate to nor can they be converted to other hardness scales. The Barcol instrument is calibrated at each use by indenting an aluminum alloy standard disk supplied with it. The Barcol test is relatively insensitive to surface condition but may be affected by test sample size and thickness.

Durometer.   The Durometer hardness test was developed for and is used for determining the hardness of elastomers. The Durometer is a hand-held, spring-loaded instrument which when pressed against the sample forces a conical steel indenter into the surface. Durometer hardness numbers range from 0 to 100 and are read directly from the attached dial indicator. Several load scales are available, but the A scale (8 N = 822 gf) and the D scale (44.5 N = 4.54 kgf) are most common. Specifics of the test procedure are discussed in ASTM D2240 (2). Lighter load scales and larger diameter indenters are available for very soft materials such as foam.

Operator skill and experience are necessary to obtain consistent results using a Durometer. Speed of load application, dwell time, and sample thickness can affect reproducibility of results. Durometer calibration prior to each test series is done using a test block provided with the instrument. When large numbers of tests are required, improved consistency of results are obtained if the Durometer is used with the accessory vertical stand rather than hand held.

International Rubber Hardness.   The International rubber hardness test (ASTM D1415) (2) for elastomers is similar to the Rockwell test in that the measured property is the difference in penetration of a standard steel ball between minor and major loads. The viscoelastic properties of elastomers require that a load application time, usually 30 seconds, be a part of the test procedure. The hardness number is read directly on a scale of 0 to 100 upon return to the minor load. International rubber hardness numbers are often considered equivalent to Durometer hardness numbers but differences in indenters, loads, and test time preclude such a relationship.

Webster Gauge.   The Webster hardness gauge (ASTM B647) (2) looks like a large pair of pliers or a paper punch. The spring-loaded head contains a conical steel indenter which is forced against the sample surface by squeezing the handles. A direct reading gauge indicates relative sample hardness on a scale of 0 to 20. This test is less precise than the Rockwell or Brinell but has the advantage of greater speed and portability. The Webster gauge was developed specifically for determination of the hardness of sheet aluminum products and its use remains largely in that industry.

Hardness Conversions

Despite variations in hardness test procedures and the variations in physical properties of the materials tested, hardness conversions from one test to another are possible (see ASTM E140 and Table 2). This approximate relationship is only consistent within a single-material system, eg, iron, steel, or aluminum.

Conversion of hardness data to some measure of strength is also possible and has been done for several common materials (Fig. 1). Rules of thumb have also been developed relating tensile strength to hardness for steel, eg,

$$\text{tensile strength} = \frac{\text{HBN}}{2} \times 1000$$

Caution should be exercised when using materials strength data obtained from hardness conversion tables.

Although Vickers and DPH microhardness tests should yield the same numerical results on a given material, such is not always the case. Much of the observed variance may be a function of differences in the volume of sample material displaced by the macro and micro indentations.

Many special applications of indentation hardness testing techniques to unusual materials or conditions have been developed, some of which are listed in Table 4.

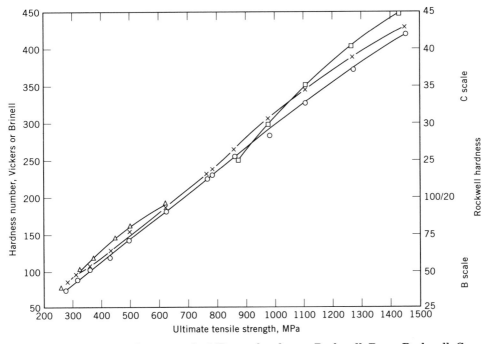

Fig. 1. Ultimate tensile strength, MPa vs hardness. Rockwell B, △; Rockwell C, □; Brinell, ○; and Vickers, ×. To convert MPa to psi, multiply by 145.

BIBLIOGRAPHY

"Hardness" in *ECT* 1st ed., Vol. 7, pp. 362–369, by R. Madden, Johns Hopkins University; in *ECT* 2nd ed., Vol. 10, pp. 808–819, by C. J. McMahon, Jr., University of Pennsylvania; in *ECT* 3rd ed., Vol. 12, pp. 118–128, by G. Langford, Drexel University.

1. L. Small, *Hardness Theory and Practice*, Service Diamond Tool Co., 1960.
2. *ASTM Standards*, American Society for Testing and Materials, Philadelphia, Pa.
3. U.S. Pat. 1,516,207 (1919), S. P. Rockwell.
4. L. Small, *Hardness Testing*, American Society for Metals International, 1987.
5. F. Knoop, "A Sensitive Pyramidal Diamond Tool for Indentation Measurement," NBS res. paper RP 1220, *J. Res. Natl. Bur. Stand.*, (July 1939).
6. C. Kleesattel, "The Ultrasonic Contact Impedance Testing Method," *The ECHO*, Vol. 27, Krautkramer GmbH.
7. A. R. Fee and co-workers, *Mechanical Testing*, Vol. 8, American Society for Metals International, 1985, pp. 71–113.

General References

H. Scott and co-workers, *Metals Handbook*, American Society for Metals International, pp. 93–104.
V. E. Lysaght and A. DeBellis, *Hardness Testing Handbook*, American Chain and Cable Co., 1969.

RONALD D. CROOKS
Consultant

HAZARD ANALYSIS AND RISK ASSESSMENT

The hazards associated with any facility which produces or uses chemicals can be quite numerous, perhaps in the hundreds or thousands for larger facilities. These hazards are the result of the physical properties of the materials, the operating conditions, the procedures, or the design, to name a few. Most of the hazards are continually present in a facility.

Without proper control of hazards, a sequence of events (scenario) occurs which results in an accident. A hazard is defined as anything which could result in an accident, ie, an unplanned sequence of events which results in injury or loss of life, damage to the environment, loss of capital equipment, or loss of production or inventory.

Risk consists of two components: the probability of the accident and the consequence. It is not possible to completely characterize risk without both of these components. Thus, a hazard could have low probability of accident but high consequence or vice versa. The result for both cases is moderate risk.

The purpose of hazard analysis and risk assessment in the chemical process industry is to (*1*) characterize the hazards associated with a chemical facility; (*2*) determine how these hazards can result in an accident, and (*3*) determine the risk, ie, the probability and the consequence of these hazards. The complete procedure is shown in Figure 1 (see also INDUSTRIAL HYGIENE; PLANT SAFETY).

Most of the techniques for determining risk or identifying hazards that are discussed herein require analysis by committee. The committee must be formed from individuals having specific and relevent experience to the chemical process under consideration. Furthermore, the management of this committee is paramount to the success of the project. Members must focus on the problem at hand and continue to make satisfactory progress.

The first step is to have a complete and detailed description of the system, process, or procedure under consideration. This must include physical properties of the materials, operating temperatures and pressures, detailed flow sheets, instrument diagrams of the process, materials of construction, other detailed design

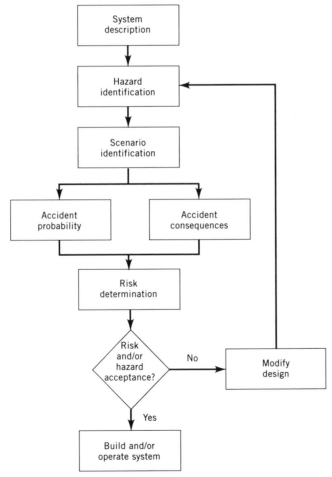

Fig. 1. Flow chart representing the complete hazard identification and risk assessment procedure.

specifications, and so forth. The more detailed and up-to-date this information is, the better the result of the analysis.

The next step is to identify the hazards. This is done using a number of established procedures. It is not unusual for several hundred hazards to be identified for a reasonably complex process.

The subsequent step is to identify the various scenarios which could cause loss of control of the hazard and result in an accident. This is perhaps the most difficult step in the procedure. Many accidents have been the result of improper characterization of the accident scenarios. For a reasonably complex chemical process, there might exist dozens, or even hundreds, of scenarios for each hazard. The essential part of the analysis is to select the scenarios which are deemed credible and worst case.

The next part of the procedure involves risk assessment. This includes a determination of the accident probability and the consequence of the accident and is done for each of the scenarios identified in the previous step. The probability is determined using a number of statistical models generally used to represent failures. The consequence is determined using mostly fundamentally based models, called source models, to describe how material is ejected from process equipment. These source models are coupled with a suitable dispersion model and/or an explosion model to estimate the area affected and predict the damage. The consequence is thus determined.

The final part of the procedure is to decide if the risk is acceptable. If it is not, then a change must be made and the entire procedure restarted. If the risk and/or hazard are acceptable, then the process and/or procedure are implemented.

The hazard analysis and risk assessment procedure can be applied at any stage in the lifetime of a process or procedure including research and development (qv), initial conceptual design (see PLANT LAYOUT; PLANT LOCATION), pilot-plant operation (see PILOT PLANTS AND MICROPLANTS), construction and start-up, operation, maintenance (qv), plant expansion, and final plant decommissioning. For economic reasons it is best to begin this procedure during the very initial stages when changes are easier and less costly.

There are a large number of standard methods suitable for each stage in the hazard analysis and risk assessment procedure. The selection of the proper method depends on several factors. Some of these are the type of process, the stage in the lifetime of the process, the experience and capabilities of the participants, and the step in the procedure that is being examined. Information regarding the selection of the proper procedure is available in an excellent and comprehensive reference (1).

Hazard analysis does have limitations. First, there can never be a guarantee that the method has identified all of the hazards, accident scenarios, and consequences. Second, the method is very sensitive to the assumptions made by the analysts prior to beginning the procedure. A different set of analysts might well lead to a different result. Third, the procedure is sensitive to the experience of the participants. Finally, the results are sometimes difficult to interpret and manage.

For chemical facilities in the United States, hazard analysis is not an option if inventories of hazardous chemicals are maintained in amounts greater than the threshold quantities specified by the Occupational Safety and Health Administration (OSHA) regulation 1910.119. Many facilities are finding that hazard analysis has many benefits. The process or procedure often works better, the quality of the product is improved, the process experiences less down time, and the employees feel more comfortable in the work environment after a hazard analysis has been completed.

Hazard Identification Procedures

Methods for performing hazard analysis and risk assessment include safety review, checklists, Dow Fire and Explosion Index, what-if analysis, hazard and operability analysis (HAZOP), failure modes and effects analysis (FMEA), fault tree analysis, and event tree analysis. Other methods are also available, but those given are used most often.

Safety Review. The safety review was perhaps the very first hazard analysis procedure developed. The procedure begins by the preparation of a detailed safety review report. The purpose of this report is to provide the relevant safety information regarding the process or operation. This report is generally prepared by the process engineer. A typical outline for this report follows.

   I. Introduction
      A. Process summary
      B. Reactions and stoichiometry
      C. Engineering data
  II. Raw materials and products (refers to hazards and special handling requirements)
 III. Equipment set-up
      A. Equipment description
      B. Equipment specifications
  IV. Procedures
      A. Normal operating procedures
      B. Safety procedures
            1. Emergency shutdown
            2. Fail-safe procedures
            3. Primary release procedures
      C. Waste disposal procedures
      D. Clean-up procedures
   V. Safety checklists
  VI. Chemical hazard sheets (MSDS)

The next step in the procedure is to form a committee comprised of people with expertise specific to the process and chemistry involved. The committee could also include an industrial hygienist, an environmentalist, the process operators, a consultant, and others. The committee should not contain more than a dozen individuals.

The safety review report is distributed to the committee which meets to work its way through the report, section by section, discussing safety concerns and potential improvements to the process or procedure. An individual must be designated to take minutes at the meeting and record suggested modifications. If the review concerns an existing process, the committee should perform a site visit to examine the actual equipment.

At the completion of the review of the report, an action plan is formulated and changes agreed upon by the committee are implemented. A final check must be made by management to ensure that these changes are actually completed.

The safety review technique is also useful for small laboratory operations and small changes in existing processes. In these cases, the committee often consists of two or three people and any changes are often less formally recommended.

Checklists. A checklist is simply a detailed list of safety considerations. The purpose of this list is to provide a reminder to safety issues such as chemical reactivity, fire and explosion hazards, toxicity, and so forth. This type of checklist is used to determine hazards, and differs from a procedure checklist which is used to ensure that the correct procedure is followed.

The hazards checklist usually has three columns next to each item on the list. Items can number in the hundreds or even the thousands. The first check is marked if the issue has been considered and complete. The second check is marked if additional consideration or work is required, and the last check is marked if the item does not apply. An example of a detailed checklist can be found in the literature (2).

Dow Fire and Explosion Index. The Dow Fire and Explosion Index (3) is a procedure useful for determining the relative degree of hazard related to flammable and explosive materials. This Index form works essentially the same way as an income tax form. Penalties are provided for inventory, extended temperatures and pressures, reactivity, etc, and credits are applied for fire protection systems, process control (qv), and material isolation. The complete procedure is capable of estimating a dollar amount for the maximum probable property damage and the business interruption loss based on an empirical correlation provided with the Index.

The procedure begins by using a material factor that is a function only of the physical properties of the chemical in use. The more hazardous the material, the higher the material factor. A table containing factors for common materials is provided with the Index. Additionally, a procedure is detailed for determining the material factor for unlisted materials.

The next step is to apply penalties for general process hazards such as exothermic or endothermic reactions, material handling and transfer, enclosed or indoor units, access, drainage, and for special process hazards, eg, toxic materials, low or high pressure, flammable dusts, low or high temperature, leakage, rotating equipment, quantity of material. Correlations are provided to assist in determining reasonable penalties for these items.

Finally, the penalties are factored into the original material factor to result in a fire and explosion index value. The higher this value, the higher the degree of hazard.

The next step is to apply a number of loss control credit factors such as process control (emergency power, cooling, explosion control, emergency shutdown, computer control, inert gas, operating procedures, reactive chemical reviews), material isolation (remote control valves, blowdown, drainage, interlocks) and fire protection (leak detection, buried tanks, fire water supply, sprinkler systems, water curtains, foam, cable protection). The credit factors are combined and applied to the fire and explosion index value to result in a net index.

The net index is used with correlations provided to determine the maximum probable property damage and business interruption loss in the event of an accident.

The Dow Fire and Explosion Index is a useful method for obtaining an estimate of the relative fire and explosion hazards associated with flammable and combustible chemicals. However, the technique is very procedure oriented, and there is the danger of the user becoming more involved with the procedure than the intent.

What-If Analysis.   The what-if analysis is simply a brainstorming technique that asks a variety of questions related to situations that can occur. For instance, in regards to a pump, the question What if the pump stops running? might be asked. An analysis of this situation then follows. The answer should provide a description of the resulting consequence. Recommendations then follow, if required, on the measures taken to prevent an accident.

A what-if form, consisting of columns assigned to identify the item under consideration, lists the question, describes the potential consequence/hazard, and lists the recommendations. Additionally, columns can be employed to assign work and to indicate completion.

The what-if analysis approach is useful throughout the entire lifetime of a process and is frequently used in conjunction with the checklist approach. However, the approach is very unstructured and depends heavily on the experience of the analysts to ask the correct questions.

Hazard and Operability Analysis.   The hazard and operability analysis (HAZOP) procedure is quite popular because of its ease of use, the ability to organize and structure the information, minimal dependence on the experience of the analysts, and the high level of results. Furthermore, the approach is capable of finding hazards associated with the operation of a facility, hence the incorporation of the word operability in the name.

The HAZOP procedure, performed by committee, is mostly an organizational one. There is little technology associated with the process. The HAZOP approach is capable of identifying hundreds of items for a reasonably complex process. This information must be organized and managed properly.

The HAZOP committee must be composed of people with specific experience related to the process at hand. The chair, or facilitator, responsible for managing the committee should be highly familiar with the HAZOP procedure and should have excellent committee management skills. This person must ensure that the discussion is focused and productive, and then oversee the paperwork and progress of the work.

The first step in the procedure is to define the purpose, objectives, and scope of the study. The more precisely this is done, the more focused and relevant the committee discussions can be. The next step is to collect all relevant information on the process under consideration. This includes flow diagrams, process equipment specifications, nominal flows, etc. The procedure is highly dependent on the reliability of this information. Efforts expended here are worthwhile. Many committees use the flow sheet as the central structure to organize their discussions.

After the first two steps are completed, the committee conducts the review. The facilitator divides the flow sheet into a number of sections containing one principal equipment piece and auxiliaries. A section is chosen and the following procedural steps performed (4): (1) a study node, ie, vessel, line, operating instruction is chosen; (2) the node's design intention, ie, flow, cooling, etc, is described; (3) a process parameter such as temperature, pressure, pH, component, viscosity, etc, is chosen; (4) a guide word (Table 1) to determine a possible deviation is applied; (5) if the deviation is applicable, the possible causes should be determined and any protective systems noted; (6) the consequences of the deviation should be evaluated; (7) specific action should be recommended when spelling out what, when, and by whom; and (8) all information should be recorded on HAZOP forms. Steps 4 through 8 should be repeated until all guide words have been applied to the chosen process parameter. Steps 3 through 9 should be repeated until all applicable process parameters have been considered for the given study node. Finally, steps 1 through 10 should be repeated until all study nodes have been completed in a given section. Then the next section should be examined. The guide words provided in Table 1 represent a standard set. Most companies customize

Table 1. List of Guide Words for HAZOP Procedure[a]

Guide word	Meaning	Comments	Example
no, not, none	the complete negation of	no part of the design intention is achieved, but nothing else happens	no flow
more, higher, greater	quantitative increase	applies to quantities such as flow rate and temperature as well as activities like heat and reaction	more flow
less, lower	quantitative decrease	same as above	less flow
as well as	qualitative increase	all the design and operating intentions are achieved along with some additional activity, such as contamination of process streams	something else with the flow
part of	qualitative decrease	only some of the design intentions are achieved, some are not	partial flow
reverse	the logical opposite of	most applicable to activities such as flow or chemical reaction; also applicable to substances	reverse flow
other than	complete substitution	no part of the original intention is achieved; the original intention is replaced by something else	something else flows
sooner than	too early or in wrong order	applies to process steps or actions	flow started early
later than	too late or in wrong order	applies to process steps or actions	flow started late
where else	in additional locations	applies to process locations or locations in operating procedures	flow goes some other place

[a]Ref. 4.

their sets of guide words and many companies use different sets based on the type of unit operation being examined.

The committee must carefully regulate its time to ensure that the participants do not experience HAZOP burnout. Many meetings might be required over a period of months to complete a particularly large process, but meetings should be limited to not more than three hours every other day.

A reactor system is shown in Figure 2 to which the HAZOP procedure can be applied. This reaction is exothermic, and a cooling system is provided to remove the excess energy of reaction. If the cooling flow is interrupted, the reactor temperature increases, leading to an increase in the reaction rate and the heat generation rate. The result could be a runaway reaction with a subsequent increase in the vessel pressure possibly leading to a rupture of the vessel.

Performing a HAZOP on this process with the assigned task of considering runaway reaction episodes would lead to a completed form such as that shown in Figure 3. The process is already small enough to be considered a single section. Four study nodes are cooling water line, stirring motor, monomer feed line, and reactor vessel. Figure 3 shows the HAZOP form completed for the cooling water and stirring motor study nodes.

The HAZOP analysis would reveal the following potential process modifications: (1) installation of a cooling water flow meter and low flow alarm to provide an immediate indication of cooling loss; (2) installation of a high temperature alarm to alert the operator in the event of cooling function loss; (3) installation of a check valve in the cooling line to prevent reverse flow of cooling water; (4) periodic inspections and maintainance of the cooling coil; and (5) evaluation of the cooling water source to consider any possible interruption and contamination of the supply. Once the recommendations have been completed, it is the job of management to rate the recommendations with respect to importance and then to ensure that the recommendations are implemented.

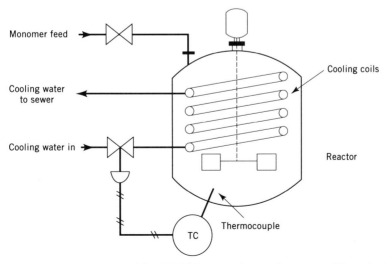

Fig. 2. Reactor systems used for HAZOP example (5). Courtesy of Prentice Hall.

Project name: Example 1						Date: 1/1/93	Page 1 of 2		Completed:
Process: Reactor shown in Figure 2									No action:
Section: Reactor shown in Figure 2						Reference drawing: Figure 2			Reply date:
Item	Study node	Process parameters	Deviations (Guide words)	Possible causes	Possible consequences	Action required	Assigned to:		
1A	Cooling water	Flow	No	1. Control valve fails closed	1. Loss of cooling, possible runaway	1. Select valve to fail open	DAC 1/93		
				2. Plugged cooling coils	2. Same	2. Install filter with maintenance procedure	DAC 1/93		
						Install cooling water flow meter and low flow alarm	DAC 2/93		
						Install high temperature alarm to alert operator	DAC 2/93		
				3. Cooling water service failure	3. Same	3. Check and monitor reliability of water service	DAC 2/93		
				4. Controller fails and closes valve	4. Same	4. Place controller on critical instrumentation list	DAC 1/93		
				5. Air pressure fails, closing valve	5. Same	5. See 1A.1			
1B			High	1. Control valve fails open	1. Reactor cools, reactant conc builds, possible runaway on heating	1. Instruct operators and update procedures	JFL 1/93		
				2. Controller fails and opens valve	2. Same	2. See 1A.4 above			
1C			Low	1. Partially plugged cooling line	1. Diminished cooling, possible runaway	1. See 1A.2 above			
				2. Partial water source failure	2. Same	2. See 1A.2 above			
				3. Control valve fails to respond	3. Same	3. Place valve on critical instrumentation list	JFL 1/93		
1D			As well as	1. Contamination of water supply	1. Not possible here	4. None		X	
1E			part of	1. Covered under 1C				X	
1F			reverse	1. Failure of water source resulting in backflow	1. Loss of cooling, possible runaway	1. See 1A.2			
				2. Backflow due to high backpressure	2. Same	2. Install check valve	JFL 2/93		
1G			Other than	1. Not considered possible				X	
1H			Sooner than	1. Cooling normally started early	1. None			X	
1I			Later than	2. Operator error	1. Temperature rises, possible runaway	1. Interlock between cooling flow and reactor feed	JEH 1/93		
1J			Where else	1. Not considered possible				X	

Fig. 3. Hazards and Operability (HAZOP) analysis example.

Failure Modes and Effects Analysis. Failure modes and effects analysis (FMEA) is applied only to equipment. It is used to determine how equipment could fail, the effect of the failure, and the likelihood of failure. There are three steps in an FMEA (4): (*1*) define the purpose, objectives, and scope. Large processes are broken down into smaller systems such as feed or cooling. At first, the failures are only considered to affect the system. In a more general study, the effects on a plant-wide basis can be considered. (*2*) Define the problem and boundary conditions. This includes identifying the system to be studied, establishing the physical boundaries, and labeling the equipment with a unique identifier for use in the FMEA procedure. (*3*) Complete an FMEA table (4) by beginning at the system boundary and evaluating the equipment items in the order these appear in the process.

An FMEA table contains a series of columns for the equipment reference number, the name of the piece of equipment, a description of the equipment type, configuration, service characteristics, etc, which may impact the failure modes and/or effects, and a list of the failure modes. Table 2 provides a list of representative failure modes for valves, pumps, and heat exchangers. The last column of the FMEA table is reserved for a description of the immediate and ultimate effects of each of the failure modes on other equipment and the system.

Fault Tree Analysis. Fault trees represent a deductive approach to determining the causes contributing to a designated failure. The approach begins with the definition of a top or undesired event, and branches backward through intermediate events until the top event is defined in terms of basic events. A basic event is an event for which further development would not be useful for the purpose at hand. For example, for a quantitative fault tree, if a frequency or probability for a failure can be determined without further development of the failure logic, then there is no point to further development, and the event is regarded as basic.

Table 2. Failure Modes for Process Equipment[a]

Equipment type	Failure modes
valve, normally open	fails to open (or fails to close when required)
	closes unexpectedly
	leaks to external environment
	valve body rupture
pump, normally operating	fails on (fails to stop when required)
	stops unexpectedly
	seal leak/rupture
	pump casing leak/rupture
heat exchanger, high pressure on tube side	leak/rupture, tube side to shell side
	leak/rupture, shell side to external environment
	tube side plugged
	shell side plugged

[a]Ref. 4.

Figure 4 shows a fault tree for a flat tire on an automobile. The top event, the flat tire, is broken down into two immediate contributing events, road debris and tire failure. The contributing event, road debris, is a basic event. This event, which cannot be broken down into other events unless additional information is provided, is enclosed in a circle to denote it as a basic event. The other event, tire failure, is enclosed in a rectangle to denote it as an intermediate event.

These two events are related to each other through an OR gate, ie, the top event can occur if either road debris or tire failure occurs. Another type of gate is the AND gate, where the output occurs if and only if both inputs occur. OR gates are much more common in fault trees than AND gates, ie, most failures are related in OR gate fashion.

The next step is to define the intermediate event, tire failure. There are two events which could contribute: a worn tire resulting from much usage or a tire that is defective owing to a manufacturing problem. These are both basic events because additional information is needed for any further definition.

An important part of fault tree analysis is the initial problem definition. Failure to adequately define the problem can produce unclear results. The top event must be precisely defined. Events such as FIRE IN PLANT, or EXPLOSION OF EXTRACTOR, are too vague and general. Likewise, top events such as LEAK IN VALVE V24 are too specific. Appropriate events would include RUNAWAY REACTION IN REACTOR R1, HIGH PRESSURE IN VESSEL V1, HIGH LEVEL IN VESSEL V2, and so forth. The analysis boundary conditions, ie, all of the equipment under consideration, and the state of this equipment must also be defined; the open valves, the material flowing, etc, must be designated. Then the level of resolution must be defined; eg, the valve itself or the positioner on the valve must be designated. Additionally any unallowed events such as wiring failures, lighting, etc, should be defined as should any assumptions made in the analysis.

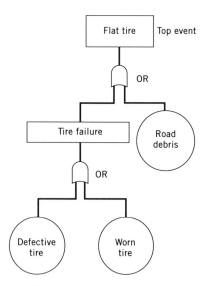

Fig. 4. Fault tree analysis for a flat tire (6). Courtesy of Prentice Hall.

Other considerations for fault tree construction are (1) assume that faults propagate through normally operating equipment. Never assume that a fault is stopped by the miraculous failure of another piece of equipment. (2) Gates are connected through labeled fault events. The output from one gate is never connected directly into another.

It is important in fault tree analysis to consider only the nearest contributing event. There is always a tendency to jump immediately to the details, skipping all of the intermediate events. Some practice is required to gain experience in this technique.

The principal problem in using fault trees is that for reasonably complicated processes the analysis is most likely to produce a huge fault tree. Fault trees involving hundreds or even thousands of intermediate events are not uncommon. The effort involved in fault tree development can also be substantial, requiring several years.

Another problem for fault trees is the uniqueness of the result. Fault trees produced by two different teams of analysts most often show a different structure. However, this problem is reduced as the detail in the problem definition increases.

A pumped storage facility having two tanks and three pumps is shown schematically in Figure 5. Any one tank can be connected to any of the pumps to provide raw material. The first step in the fault tree analysis procedure is to define the problem. If the top event is defined as the failure to pump raw material from pumped storage, then the analysis boundary conditions and equipment state: the equipment is configured as shown in Figure 5; both tanks contain the same raw material; any one pump can be connected to either of the two tanks to provide raw material. The level of resolution is the equipment configuration shown in Figure 5. Unallowed events include wiring failures, electrical failures, lighting, tornadoes, etc.

The resulting fault tree is shown in Figure 6, in which the top event is defined in terms of two intermediate events: failure of the tank system or failure of the pumping system. Failure in either system would contribute to the overall system failure. The intermediate events are then further defined in terms of basic events. All of the basic events are related by AND gates because the overall system failure requires the failure of all of the individual components. Failures of the tanks and pumps are basic events because, without additional information, these events cannot be resolved any further.

Event Trees. Event trees use an inductive logic approach to consider the effects of safety systems on an initiating event. The initiating event is propagated

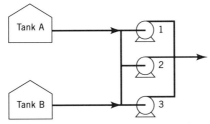

Fig. 5. Schematic of a pumped storage facility.

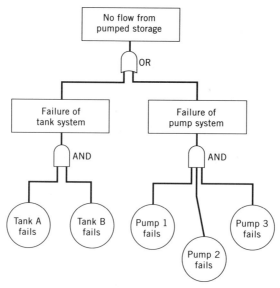

Fig. 6. A fault tree for the pumped storage example of Figure 5. For a real system the tank and pump failures would be more precisely defined, or set as intermediate events having further definition by subsequent basic events and more detailed failure modes.

through the various safety functions. Branching is dependent upon the success or failure of the safety function.

Consider again, for example, the case of the flat tire on an automobile. The initiating event in this case is the flat tire. There are two safety functions which can be defined: a spare tire and an emergency road patrol. Other safety functions might be included depending on the particular situation.

The event tree is drawn by first identifying the initiating event, on the left-hand side of the drawing sheet, as shown in Figure 7. The two safety functions are identified on the top of the sheet. A line is drawn from the initiating event to a position immediately below the first safety function, in this case the spare tire. At this point the line branches, the upper branch representing the success of the safety function and the lower branch representing the failure of this safety function. The lines are continued in this fashion so that branching occurs below each safety function.

In some cases the safety function is meaningless. For the example provided, if the spare tire is successfully mounted, then the safety function for the emergency road patrol is meaningless. In this case the line is drawn directly through the safety function.

The branching is continued until all of the safety functions are considered. At this point a conclusion is reached about the result. For the flat tire example, only two results are possible: the driver is either stranded or back on the road. The circle used to terminate the stranded result is given an X to denote it as an unfavorable outcome.

The initiating event is given a unique letter designation. In Figure 7 it is assigned the letter A. Each safety function is also assigned a unique letter des-

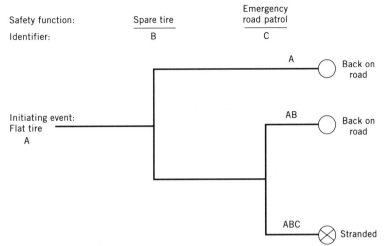

Fig. 7. Event tree for a flat tire.

ignation, different from the letter used for the initiating event. These letters are used to identify each line on the event tree. Thus, letter sequence AB identifies initiating event A, followed by the failure of safety function B.

It is not coincidental that the top event of the fault tree is the initiating event for the event tree. The fault tree shows how an event is decomposed into basic events whereas an event tree demonstrates the effect of the various safety functions. The disadvantage of event trees is that the outcomes are difficult to predict. Thus the outcome of interest might not arise from the analysis.

Scenario Identification

An important part of hazard analysis and risk assessment is the identification of the scenario, or design basis by which hazards result in accidents. Hazards are constantly present in any chemical facility. It is the scenario, or sequence of initiating and propagating events, which makes the hazard result in an accident. Many accidents have been the result of an improper identification of the scenario.

It is not practicable to perform detailed studies on all possible scenarios; thus many studies focus on identifying the worst practicable scenario and the worst potential scenario. The worst practicable scenario considers scenarios which have a reasonable chance for occurrence. This includes pipe ruptures, holes in storage tanks, ground spills, and so forth. The worst potential scenario is a scenario leading to the largest catastrophe. This includes complete spillage of tank contents, rupture of large bore piping, explosive rupture of reactors, and so forth. Examples maybe found in the literature (9).

Most hazard identification procedures have the capability of providing information related to the scenario. This includes the safety review, what-if analysis, hazard and operability studies (HAZOP), failure modes and effects analysis (FMEA), and fault tree analysis. Using these procedures is the best approach to identifying these scenarios.

Source Modeling and Consequence Modeling

Once the scenario has been identified, a source model is used to determine the quantitative effect of an accident. This includes either the release rate of material, if it is a continuous release, or the total amount of material released, if it is an instantaneous release. For instance, if the scenario is the rupture of a 10-cm pipe, the source model would describe the rate of flow of material from the broken pipe.

Once the source modeling is complete, the quantitative result is used in a consequence analysis to determine the impact of the release. This typically includes dispersion modeling to describe the movement of materials through the air, or a fire and explosion model to describe the consequences of a fire or explosion. Other consequence models are available to describe the spread of material through rivers and lakes, groundwater, and other media.

The dispersion model is typically used to determine the downwind concentrations of released materials and the total area affected. Two models are available: the plume and the puff. The plume describes continuous releases; the puff describes instantaneous releases.

An explosion model is used to predict the overpressure resulting from the explosion of a given mass of material. The overpressure is the pressure wave emanating from a explosion. The pressure wave creates most of the damage. The overpressure is calculated using a TNT equivalency technique. The result is dependent on the mass of material and the distance away from the explosion. Suitable correlations are available (2). A detailed discussion of source and consequence models may be found in References 2, 8, and 9.

Probability

In order to complete an assessment of risk, a probability must be determined. The easiest method for representing failure probability of a device is an exponential distribution (2).

$$R(t) = e^{-\mu t} \tag{1}$$

where $R(t)$ is the reliability, μ is the failure rate in faults per time, and t is the time.

There are other distributions available to represent equipment failures (10), but these require more detailed information on the device and a more detailed analysis. For most situations the exponential distribution suffices.

Once the reliability is defined, the failure probability, $P(t)$, follows.

$$P(t) ~ 1 - R(t) = 1 - e^{-\mu t} \tag{2}$$

Figure 8 compares the failure probability and reliability functions for an exponential distribution. Whereas the reliability of the device is initially unity, it falls off exponentially with time and asymptotically approaches zero. The failure probability, on the other hand, does the reverse. Thus new devices start life with high reliability and end with a high failure probability.

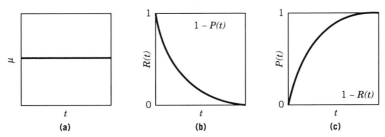

Fig. 8. (a) Failure rate, (b) reliability, and (c) failure probability.

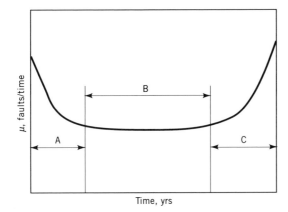

Fig. 9. Failure rate curve for real components. A, infant mortality; B, period of approximately constant μ; and C, old age.

A considerable assumption in the exponential distribution is the assumption of a constant failure rate. Real devices demonstrate a failure rate curve more like that shown in Figure 9. For a new device, the failure rate is initially high owing to manufacturing defects, material defects, etc. This period is called infant mortality. Following this is a period of relatively constant failure rate. This is the period during which the exponential distribution is most applicable. Finally, as the device ages, the failure rate eventually increases.

Table 3 lists typical failure rate data for a variety of types of process equipment. Large variations between these numbers and specific equipment can be expected. However, this table demonstrates a very fundamental principle: the more complicated the device, the higher the failure rate. Thus switches and thermocouples have low failure rates; gas–liquid chromatographs have high failure rates.

The next step is to develop a method to determine the overall reliability and failure probability for systems constructed of a variety of individual components. This requires an understanding of how components are linked. Components are linked either in series or in parallel. For series linkages, overall failure results from the failure of any of the components. For parallel linkages, all of the components must fail. An example of a series linkage is an automobile. The car is disabled if a flat occurs in any one of the four tires. This situation is linked in

Table 3. Failure Rate Data for Process Hardware[a]

Instrument	Failure rate, faults/yr
controller	0.29
control valve	0.60
flow measurement	
fluids	1.14
solids	3.75
flow switch	1.12
gas–liquid chromatograph	30.6
hand valve	0.13
indicator lamp	0.044
level measurement	
liquids	1.70
solids	6.86
oxygen analyzer	5.65
pH meter	5.88
pressure measurement	1.41
pressure relief valve	0.022
pressure switch	0.14
solenoid valve	0.42
stepper motor	0.044
strip chart recorder	0.22
thermocouple temperature measurement	0.52
thermometer temperature measurement	0.027
valve positioner	0.44

[a]Ref. 9.

parallel to the spare tire. The car is completely disabled only if a flat occurs and the spare tire is flat.

The computational technique for the two linkages is shown in Figure 10. For series linkages (Fig. 10**a**), the reliabilities of the individual components are multiplied together. For parallel linkages (Fig. 10**b**) the failure probabilities are multiplied together. This method for combining the distributions assumes that the failures of the individual devices are independent of each other, and that the failure of one device does not strain an adjacent device causing it, too, to fail. It also assumes that devices fail hard, that is, the device is obviously failed and not in a partially failed state.

Another problem with this approach is common mode failures. A common mode failure is a single event which could lead to the simultaneous failure of several components at the same time. An excellent example of this is a power failure, which could lead to many simultaneous failures. Frequently, the common mode failure has a higher probability than the failure of the individual components, and can drastically decrease the resulting reliability.

The numbers computed using this approach are only as good as the failure rate data for the specific equipment. Frequently, failure rate data are difficult to acquire. For this case, the numbers computed only have relative value, that is, they are useful for determining which configuration shows increased reliability.

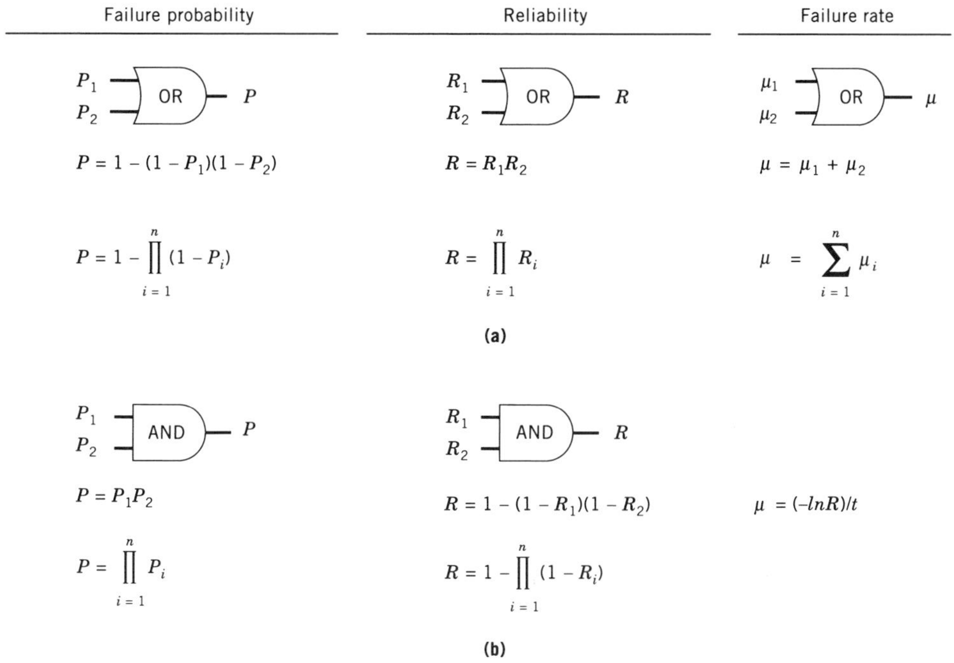

Failure probability
Reliability
Failure rate
P_1
P_2
OR
P
$P = 1 - (1 - P_1)(1 - P_2)$
$P = 1 - \prod_{i=1}^{n} (1 - P_i)$
R_1
R_2
OR
R
$R = R_1 R_2$
$R = \prod_{i=1}^{n} R_i$
μ_1
μ_2
OR
μ
$\mu = \mu_1 + \mu_2$
$\mu = \sum_{i=1}^{n} \mu_i$
(a)
P_1
P_2
AND
P
$P = P_1 P_2$
$P = \prod_{i=1}^{n} P_i$
R_1
R_2
AND
R
$R = 1 - (1 - R_1)(1 - R_2)$
$R = 1 - \prod_{i=1}^{n} (1 - R_i)$
$\mu = (-lnR)/t$
(b)

Fig. 10. Reliability and failure probability computations for components in (**a**) series linkage where the failure of either component adds to the total system failure, and (**b**) parallel linkages where failure of the system requires the failure of both components. There is no convenient way to combine the failure rate (11). Courtesy of Prentice Hall.

Figure 11 shows a system for controlling the water flow to a chemical reactor. The flow is measured by a differential pressure (DP) device. The controller decides on an appropriate control strategy and the control valve manipulates the flow of coolant. The procedure to determine the overall failure rate, the failure probability, and the reliability of the system, assuming a one-year operating period, is outlined herein.

These process components are related in series, thus if any one of the components fails, the entire system fails. The failure rates for the various components are given in Table 3. The reliability and failure probability are computed for each individual component using equations 1 and 2 and assuming a one-year period of operation. The results are shown in Table 4.

The overall reliability for components in series is computed using the appropriate equation in Figure 10. The result is

$$R = \prod_{i=1}^{3} R_i = (0.55)\,(0.75)(0.24) = 0.10$$

The failure probability is computed from equation 2.

$$P = 1 - R = 1 - 0.10 = 0.90/\text{yr}$$

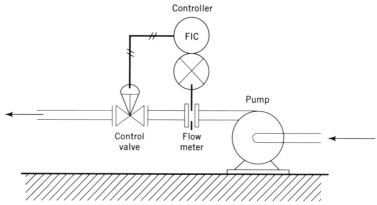

Fig. 11. Flow control system (11). FIC = flow indicator controller. Courtesy of Prentice Hall.

Table 4. Risk Assessment of Flow Control System[a]

Component	Failure rate, μ, faults/yr	Reliability, $R = e^{-\mu t}$	Failure probability, $P = 1 - R$
control valve	0.60	0.55	0.45
controller	0.29	0.75	0.25
DP cell	1.41	0.24	0.76

[a]Fig. 11; $t = 1$ yr.

The overall failure rate is computed using the definition of the reliability, equation 1.

$$\mu = -ln(0.10) = 2.30 \text{ failures/yr}$$

Hazard Acceptance and Inherent Safety

The remaining step in the hazard identification and risk assessment procedure shown in Figure 1 is to decide on risk acceptance. For this step, few resources are available and analysts are left basically by themselves. Some companies have formal risk acceptance criteria. Most companies, however, use the results on a relative basis. That is, the results are compared to another process or processes where hazards and risks are well-characterized.

If the hazards and/or risk are unacceptable, then something must be done to change them. The process can be modified, the raw materials changed, and/or the process relocated, for example. In extreme cases, the process might be abandoned as too hazardous.

A more recent concept which could have significant impact on future designs is that of inherent safety (12). This basic principle states that what is not there cannot be blown up or leak into the environment. Thus, the idea is to avoid the hazard in the first place.

Hazard avoidance is performed by three techniques. First, there is substitution. This means substituting a less hazardous material for the material in use and asking whether that flammable solvent is really necessary. Or is that toxic chemical the only possible reaction pathway? The second method for inherent safety is attenuation, ie, operating the process at lower temperatures and pressures. The last inherent safety technique is intensification. This means using much smaller inventories of hazardous raw and intermediate materials, and reducing process hold-up and inventories. These inventories are readily reducible if the management practices associated with the resource are improved. Details on the technical management requirements for a successful hazards analysis and risk assessment program are provided elsewhere (13). These techniques are still under development and substantial changes can be expected.

BIBLIOGRAPHY

1. *Guidelines for Hazards Evaluation Procedures: Second Edition with Worked Examples*, American Institute of Chemical Engineers, Center for Chemical Process Safety, New York, 1992.
2. D. A. Crowl and J. F. Louvar, *Chemical Process Safety: Fundamentals with Applications*, Prentice Hall, Englewood Cliffs, N.J., 1990.
3. *Dow's Fire and Explosion Index Hazard Classification Guide*, 6th ed., American Institute of Chemical Engineers, New York, 1985.
4. G. A. Page, *Hazard Evaluation Manual*, American Cyanamid, Wayne, N.J., 1990.
5. Ref. 2, p. 324.
6. Ref. 2, p. 356.
7. S. R. Hanna and P. J. Drivas, *Guidelines for Use of Vapor Cloud Dispersion Models*, American Institute of Chemical Engineers, Center for Chemical Process Safety, New York, 1987.
8. *Guidelines for Chemical Process Quantitative Risk Analysis*, American Institute of Chemical Engineers, Center for Chemical Process Safety, New York, 1989.
9. F. P. Lees, *Loss Prevention in the Process Industries*, Butterworths, London, 1986.
10. K. C. Kapur and L. R. Lamberson, *Reliability in Engineering Design*, John Wiley & Sons, Inc., New York, 1977.
11. Ref. 2, p. 343.
12. T. Kletz, *Plant Design for Safety, A User Friendly Approach*, Hemisphere Publishing, New York, 1991.
13. *Guidelines for Technical Management of Chemical Process Safety*, American Institute of Chemical Engineers, Center for Chemical Process Safety, New York, 1989.

DANIEL A. CROWL
Michigan Technological University

HAZARDOUS WASTE TREATMENT.   See WASTE TREATMENT, HAZARDOUS.

HEAT-EXCHANGE TECHNOLOGY

HEAT TRANSFER

The selection of a heat exchanger is not a trivial task. In order to select a proper heat exchanger for a given application, various factors such as pressure, temperature, size, fouling factor, and the use of toxic or corrosive fluids must be considered. Among these, pressure and temperature requirements mainly dictate the type of heat exchanger selected. Heat exchangers that have to be operated at extreme pressures and temperatures can be very large and expensive. On the other hand, heat exchangers to be operated at moderate pressures and temperatures can be small and inexpensive. In general, for high pressures and temperatures, tubular heat exchangers that conform to safety regulations and manufacturing codes are used. For moderate pressures, small but very efficient plate heat exchangers can be employed.

There are three heat-transfer modes, ie, conduction, convection, and radiation, each of which may play a role in the selection of a heat exchanger for a particular application. The basic design principles of heat exchangers are also important, as are the analysis methods employed to determine the right size heat exchanger.

Heat-Transfer Theory

Heat transfer is a science which tries to predict the energy transfer between materials of different temperatures. The three modes of heat transfer are conduction, convection, and radiation. A heat exchanger is designed and built based on heat-transfer principles; thus, some understanding of these basic principles is essential to design or select a heat exchanger. Efficiency and economics may depend directly on how effectively fundamental heat-transfer principles are applied in the design of the heat exchanger.

Conduction Heat Transfer. When there is a temperature difference in a body, there is an energy transfer from the high temperature region to the low temperature region, a phenomenon called an energy transfer by conduction. Although conduction occurs in liquids and gases, the contribution to heat transfer is relatively small as compared to convection or radiation for these cases. In a solid such as a metal tube wall or a flat wall made of multicomponent materials, however, conduction is the dominant heat-transfer mode. In most conventional heat exchangers, heat transfer occurs between two fluids separated by solid walls, which are either a tube wall in tubular heat exchangers or a plane wall in plate heat exchangers.

Fourier's Law of Heat Conduction.   The heat-transfer rate, Q, per unit area, A, in units of W/m^2 ($Btu/(ft^2 \cdot h)$) transferred by conduction is directly proportional to the normal temperature gradient:

$$\frac{Q}{A} \sim \frac{dT}{dx} \tag{1}$$

or in equation form the heat-transfer rate Q becomes,

$$Q = -kA\frac{dT}{dx} \tag{2}$$

where the proportionality constant, k, is called the thermal conductivity of the material. The minus sign, required in equation 2 to ensure that the direction of the heat transfer is positive when the temperature gradient is negative, is necessary because thermal energy flows in the direction of decreasing temperature.

Values of thermal conductivity are temperature-dependent and vary widely for different materials. Table 1 summarizes the thermal conductivity values of a few materials relevant to heat-exchanger analysis (1,2).

The Plane Wall.   To calculate the heat-transfer rate through a plane wall, Fourier's law can be applied directly.

$$Q_x = -\frac{kA}{\Delta x}(T_2 - T_1) = \frac{T_1 - T_2}{R_{th}} \tag{3}$$

where Δx and A are the thickness and surface area of the plane wall, respectively, and k is the thermal conductivity of the plane wall. R_{th} is the thermal resistance to heat-transfer, which is equal to $\dfrac{\Delta x}{kA}$ (see Fig. 1a).

Table 1. Values of Thermal Conductivity and Specific Heat for Various Materials[a]

Material	Thermal conductivity, $W/(m \cdot {}^\circ C)$[b]			Specific heat at 20°C, $J/(kg \cdot {}^\circ C)$[c]
	at 20°C	at 100°C	at 300°C	
aluminum[d]	204	206	228	896
carbon steel, 1.0% C	43	43	40	473
copper[d]	386	379	369	381
brass, Cu85%, Zn30%	111	128	147	385
stainless steel, AISI304	14.9	17	19	477
water	0.613	0.683		4239
air	0.0261	0.0331	0.0456	1014

[a]Refs. 1 and 2.
[b]To convert $W/(m \cdot {}^\circ C)$ to $Btu/(h \cdot ft \cdot {}^\circ F)$, multiply by 0.578.
[c]To convert $J/(kg \cdot {}^\circ C)$ to $Btu/(lbm \cdot {}^\circ F)$, multiply by 0.00023885.
[d]Pure material.

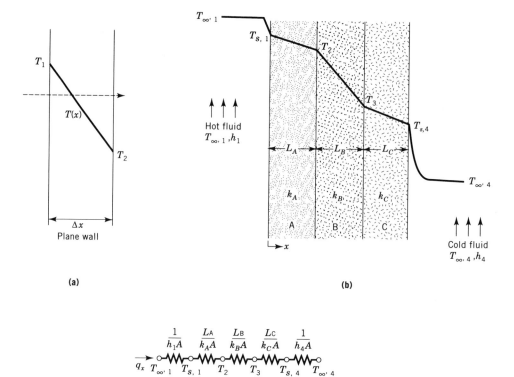

Fig. 1. Sketch of heat flow through (**a**) a plane wall where the arrow indicates the direction of heat flow and (**b**) a series of composite walls. (**c**) The equivalent thermal circuit for the series. Terms are defined in text.

When the plane wall is made of more than one material (Fig. 1**b**), the heat-transfer rate Q_x is given as

$$Q_x = \frac{\Delta T_{\text{overall}}}{\Sigma R_{th}}$$

$$= \frac{T_{\infty,1} - T_{\infty,4}}{[1/(h_1 A) + L_A/(k_A \cdot A) + L_B/(k_B \cdot A) + L_C/(k_C \cdot A) + 1/(h_4 \cdot A)]}$$

(4)

where L_J and k_J are the thickness and thermal conductivity of panel J, respectively (1). The convection resistances resulting from fluids $1/(h_1 \cdot A)$ and $1/(h_4 \cdot A)$ are included in the denominator, which represents the sum of all the thermal resistances; h is the convective heat-transfer coefficient.

The Overall Heat-Transfer Coefficient. For the design and sizing of heat exchangers, the heat-transfer rate Q is often expressed as

$$Q = UA\Delta T$$

(5)

where ΔT is a local temperature difference between two fluids and the overall heat-transfer coefficient, U, is defined as

$$U = \frac{1}{A\Sigma R_{th}} \qquad (6)$$

The Tube Wall. Tubular heat exchangers are built using a number of circular (or noncircular) tubes; thus, the heat-transfer rate across tubular walls, following Fourier's law of heat conduction, becomes

$$Q = \frac{(T_i - T_o)}{\ln(r_o/r_i)/2\pi Lk} \qquad (7)$$

where T_i and T_o are the temperatures at the inner and outer walls, r_i and r_o are the radii of the inner and outer diameters of the tube, and L is the length of the tube. The expression in the denominator is the thermal resistance of a tube wall.

Convection Heat Transfer. Convective heat transfer occurs when heat is transferred from a solid surface to a moving fluid owing to the temperature difference between the solid and fluid. Convective heat transfer depends on several factors, such as temperature difference between solid and fluid, fluid velocity, fluid thermal conductivity, turbulence level of the moving fluid, surface roughness of the solid surface, etc. Owing to the complex nature of convective heat transfer, experimental tests are often needed to determine the convective heat-transfer performance of a given system. Such experimental data are often presented in the form of dimensionless correlations.

Convective heat transfer is classified as forced convection and natural (or free) convection. The former results from the forced flow of fluid caused by an external means such as a pump, fan, blower, agitator, mixer, etc. In the natural convection, flow is caused by density difference resulting from a temperature gradient within the fluid. An example of the principle of natural convection is illustrated by a heated vertical plate in quiescent air.

Newton's Cooling Law of Heat Convection. The heat-transfer rate per unit area by convection is directly proportional to the temperature difference between the solid and the fluid which, using a proportionality constant called the heat-transfer coefficient, h, becomes

$$Q = hA(T_{\text{fluid}} - T_{\text{solid}}) \qquad (8)$$

From empirical studies, values of h are known for many common heat-transfer fluids. Typical values for air and water are given in Table 2 (1–3).

In the forced convection heat transfer, the heat-transfer coefficient, h, mainly depends on the fluid velocity because the contribution from natural convection is negligibly small. The dependence of the heat-transfer coefficient, h, on fluid velocity, V, which has been observed empirically (1–3), for laminar flow inside tubes, is $h \sim V^{1/3}$; for turbulent flow inside tubes, $h \sim V^{3/4}$; and for flow outside tubes, $h \sim V^{2/3}$. Flow may be classified as laminar or turbulent. Laminar flow is generally characterized by low velocities and turbulent flow by high velocities. It is

Table 2. Values of the Convective Heat-Transfer Coefficient[a]

Material	Convection process	Heat-transfer coefficient, h	
		W/(m²·°C)	Btu/(h·ft²·°F)
air	free	2–25	0.4–5
air	forced	25–250	5–50
water	free	50–1,000	10–200
water	forced	50–20,000	10–4,000
water	boiling	2,500–100,000	500–20,000
water vapor	condensing	5,000–100,000	1,000–20,000

[a] Refs. 1–3.

customary to use the Reynolds number, Re, to identify whether a flow is laminar or turbulent.

Dimensionless Numbers used in Convection Heat-Transfer Analysis. Reynolds Number. The Reynolds number, Re, is named after Osborne Reynolds, who studied the flow of fluids, and in particular the transition from laminar to turbulent flow conditions. This transition was found to depend on flow velocity, viscosity, density, tube diameter, and tube length. Using a nondimensional group, defined as $\rho V D_i/\mu$, the transition from laminar to turbulent flow for any internal flow takes place at a value of approximately 2100. Hence, the dimensionless Reynolds number is commonly used to describe whether a flow is laminar or turbulent. Thus

$$Re = \frac{\rho V D_i}{\mu} \qquad (9)$$

where ρ = fluid density, V = average flow velocity, D_i = tube inside diameter, and μ = fluid (dynamic) viscosity.

Noncircular tubes are often used in various compact heat exchangers and the Reynolds number in these tubes is of interest. For noncircular tubes such as square, rectangular, elliptic, and triangular tubes, the so-called hydraulic diameter, D_H, defined as

$$D_H = \frac{4A}{P_m} \qquad (10)$$

is used in the definition of the Reynolds number where A and P_m are the cross-sectional area and the wetted perimeter of tube, respectively. The noncircular tubes are used in compact heat exchangers mainly because of their geometry, ie, a greater surface area per unit volume of exchanger is possible. Secondary flow at the corners of square and rectangular ducts are of relatively minor importance except for non-Newtonian fluids.

Nusselt Number. Empirical correlations can be obtained for a particular size of tube diameter and particular flow conditions. To generalize such results and to apply the correlations to different sizes of equipment and different flow conditions, the heat-transfer coefficient, h, is traditionally nondimensionalized by the use of the Nusselt number, Nu, named after Wilhelm Nusselt,

$$Nu = \frac{hL}{k_f} \qquad (11)$$

where L is a characteristic length for the flow geometry under consideration, eg, for flow inside a tube, $L = D_i$ the tube internal diameter; and k_f is the thermal conductivity of the fluid.

    Prandtl Number.   The Prandtl number, Pr, is the ratio of the kinematic viscosity, v, to the thermal diffusivity, α.

$$Pr = \frac{v}{\alpha} = \frac{c_p \mu}{k_f} \qquad (12)$$

Unlike the Reynolds and Nusselt numbers, which depend on flow conditions, the Prandtl number is independent of flow conditions and represents the thermo-physical property of a fluid. Values of the Prandtl number of air and water at room temperature are approximately 0.7 and 7.0, respectively. The Prandtl number of air remains almost constant with increasing temperature, whereas that of water decreases significantly, ie, for water $Pr = 1.90$ at 93°C. Some well-known heat-transfer fluids have relatively high Prandtl numbers. For example, the Prandtl numbers of glycerin, ethylene glycol, and engine oil are 12.5, 204, and 10,400, respectively (1,2).

    Friction Coefficient.   In the design of a heat exchanger, the pumping requirement is an important consideration. For a fully developed laminar flow, the pressure drop inside a tube is inversely proportional to the fourth power of the inside tube diameter. For a turbulent flow, the pressure drop is inversely proportional to D_i^n where n lies between 4.8 and 5. In general, the internal tube diameter, D_i, plays the most important role in the determination of the pumping requirement. It can be calculated using the Darcy friction coefficient, f_D, defined as

$$f_D = \frac{(-dp/dx)D_i}{\rho V^2/2} \qquad (13)$$

dp/dx represents the pressure gradient along the axial direction of flow, ρ is the density, and V the average flow velocity. The negative sign in front of the pressure gradient is necessary to make the Darcy friction coefficient positive because dp/dx is always negative. This definition of the friction coefficient is valid for both laminar and turbulent flows.

    Correlations for Convective Heat Transfer.   In the design or sizing of a heat exchanger, the heat-transfer coefficients on the inner and outer walls of the tube and the friction coefficient in the tube must be calculated. Summaries of the various correlations for convective heat-transfer coefficients for internal and external flows are given in Tables 3 and 4, respectively, in terms of the Nusselt number. In addition, the friction coefficient is given for the determination of the pumping requirement.

Table 3. Correlations for Convective Heat-Transfer and Friction Coefficients for Circular Tube Flow[a]

Flow type[b]	f_D	Nu_d	Pr
laminar	$64/Re_d$	$4.36,$[c] 3.66[d]	
turbulent, smooth tube			
$\quad Re_d \geq 10{,}000$[e]		$0.027Re_d^{0.8}Pr^{1/3}(\mu/\mu_s)^{0.14}$	$0.7 \leq Pr < 16{,}700$
$\quad Re_d \leq 20{,}000$	$0.316Re_d^{-0.25}$	$0.023Re_d^{0.8}Pr^{nf}$	$0.6 \leq Pr < 160$
$\quad Re_d \geq 20{,}000$	$0.184Re_d^{-0.20}$		$0.6 \leq Pr < 160$

[a]Refs. 1 and 2.
[b]Fully developed flow.
[c]Constant wall heat flux.
[d]Constant wall temperature.
[e]For temperature-dependent viscosity, μ_s = viscosity at wall temperature.
[f]Where $n = 0.4$ for heating, 0.3 for cooling.

Table 4. Correlations for Convective Heat-Transfer and Friction Coefficients for External Flow[a]

Laminar flow surface	Nusselt number	Re_d	C	m
flat plate where $0.6 \leq Pr \leq 50$	$0.664Re_x^{0.5}Pr^{1/3}$ [b]			
cylinder where $0.4 \leq Re_d \leq 400{,}000$	$CRe_d^m Pr^{1/3}$ [c]	$1-40$	0.75	0.4
		$40-1000$	0.51	0.5
		$1 \times 10^3 - 2 \times 10^5$	0.26	0.6
		$2 \times 10^5 - 10^6$	0.076	0.7

[a]Refs. 1 and 2.
[b]$Nu_{x,avg}$.
[c]Nu_d.

The convective heat-transfer coefficient and friction factor for laminar flow in noncircular ducts can be calculated from empirically or analytically determined Nusselt numbers, as given in Table 5. For turbulent flow, the circular duct data with the use of the hydraulic diameter, defined in equation 10, may be used.

Basic Thermal Design Methods for Heat Exchangers

The basic heat-transfer principles of sizing and rating heat exchangers are important to design. Sizing refers to determining the amount of heat-transfer surface area required to transfer a specified quantity of thermal energy from one fluid to another for given fluid conditions and thus usually applies to the design of a new heat exchanger. Rating refers to determining the rate of heat transfer for given fluid-inlet conditions and given heat-exchanger geometry and thus applies to the performance of an existing heat exchanger. However, the sizing and rating calculation methods can be used interchangeably to obtain either piece of heat exchanger information.

Table 5. Correlations for Heat-Transfer and Darcy Friction Coefficients for Noncircular Laminar Duct Flow[a]

Duct design	$Nu_{H2}{}^b$	$Nu_T{}^c$	$f_D Re$
square	3.091	2.976	56.9
rectangular, aspect ratio			
0.5	3.017	3.391	62.2
0.25	4.35	3.66	74.8
triangular, isosceles	1.892	2.47	53.3

[a] Refs. 1 and 2.
[b] Nu_{H2} is the Nusselt number for uniform heat flux boundary condition along the flow direction and periphery.
[c] Nu_T is the Nusselt number for uniform wall temperature boundary condition.

The heat-transfer surface area determined by the basic sizing or rating method described herein is considered the minimum required area. There are also additional surface area requirements in the final sizing of a heat exchanger.

The economics of equipment fabrication usually call for a minimum heat-transfer surface area in an overall compact arrangement. However, this may result in excessive pressure drop, which would increase the operating cost of fluid-moving devices such as pumps and fans. In this sense, proper design of a heat-exchanger requires a balanced approach between thermal sizing and pressure drop. Therefore, a general method of calculating pressure drops in heat exchangers is needed. The heat exchangers considered in this section are those in which the two fluids are separated by partition walls or cylindrical tubes.

Basic Heat-Transfer Equations. Consider a simple, single-pass, parallel-flow heat exchanger in which both hot (heating) and cold (heated) fluids are flowing in the same direction. The temperature profiles of the fluid streams in such a heat exchanger are shown in Figure 2**a**.

An energy balance over a differential length, dx, yields

$$dQ = -\dot{m}_h c_{ph} dT_h = \dot{m}_c c_{pc} dT_c = U(P_m dx)(T_h - T_c) \tag{14}$$

where $\dot{m}_h$ and $\dot{m}_c$ are the hot and cold fluid flow rates, respectively; c_{ph} and c_{pc} are constant-pressure specific heat for the hot and cold fluids; P_m is the perimeter of the tube; T_c and T_h are the temperatures of the hot and cold fluids; and U is the overall heat-transfer coefficient which may be expressed in the form:

$$U = \cfrac{1}{\cfrac{A}{h_i \eta_{t1} A_i} + r_{fi}\cfrac{A}{A_i} + R_w + r_{fo}\cfrac{A}{A_o} + \cfrac{A}{h_o \eta_{to} A_o}} \tag{15}$$

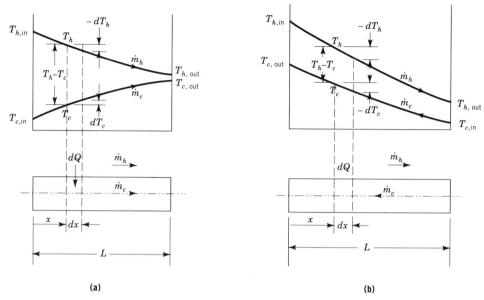

Fig. 2. Fluid temperature profiles in (**a**) a parallel flow heat exchanger and (**b**) a counterflow heat exchanger. Terms are defined in text.

where the thermal resistance of the tube wall, R_w, for a cylindrical wall becomes

$$R_w = \frac{A\ln(D_o/D_i)}{2\pi k_w L} \qquad (16)$$

and for a plane wall of thickness, t

$$R_w = \frac{t}{k_w} \qquad (17)$$

The subscripts i and o correspond to inner and outer surfaces of tube, respectively. In these equations, A is a reference area $P_m L$, for which U is defined, and η_t is the total efficiency of a finned heat-transfer surface and is related to the fin efficiency, η_f, by

$$\eta_t = 1 - \frac{A_f}{A_t}(1 - \eta_f) \qquad (18)$$

where A_f is the area of fins only and A_t is the total area of finned surface, ie, finned area plus base nonfinned area. When there is no fin, η_t becomes unity and the associated area becomes that of bare surface.

Equation 14 contains two independent equations and three unknown variables (Q, T_h, T_c). An additional relation needed to solve this equation may be obtained by writing an equation of energy balance from the inlet to an arbitrary location, x:

$$\dot{m}_h c_{ph}(T_{h,\text{in}} - T_h) = \dot{m}_c c_{pc}(T_c - T_{c,\text{in}}) \tag{19}$$

Assuming that U, c_p, and P_m are invariant with respect to temperature and space, one can integrate equation 14 subject to equation 19, and obtain, after rearrangement, a basic heat-transfer equation for a parallel-flow heat exchanger (4).

$$Q = UA\frac{(T_{h,\text{out}} - T_{c,\text{out}}) - (T_{h,\text{in}} - T_{c,\text{in}})}{\ln\dfrac{(T_{h,\text{out}} - T_{c,\text{out}})}{(T_{h,\text{in}} - T_{c,\text{in}})}} \tag{20a}$$

$$= UA\frac{\Delta T_a - \Delta T_b}{\ln\left(\dfrac{\Delta T_a}{\Delta T_b}\right)} = UA(\text{LMTD}) \tag{20b}$$

where $\Delta T_a = T_{h,\text{out}} - T_{c,\text{out}}$; $\Delta T_b = T_{h,\text{in}} - T_{c,\text{in}}$; and

$$\text{LMTD} = \frac{\Delta T_a - \Delta T_b}{\ln\left(\dfrac{\Delta T_a}{\Delta T_b}\right)} \tag{21}$$

The LMTD, ie, logarithmic mean temperature difference, is an effective overall temperature difference between the two fluids for heat transfer and is a function of the terminal temperature differences at both ends of the heat exchanger.

A similar derivation can be made for a single-pass counterflow heat exchanger in which the hot and cold fluids are flowing in the opposite direction (see Fig. 2b). The resulting heat-transfer equation is still

$$Q = UA(\text{LMTD}) \tag{22}$$

where $\text{LMTD} = (\Delta T_a - \Delta T_b)/\ln(\Delta T_a/\Delta T_b)$, but the terms ΔT_a and ΔT_b for counterflow are

$$\Delta T_a = T_{h,\text{in}} - T_{c,\text{out}} \tag{23}$$

$$\Delta T_b = T_{h,\text{out}} - T_{c,\text{in}} \tag{24}$$

The equations for counterflow are identical to equations for parallel flow except for the definitions of the terminal temperature differences. Counterflow heat exchangers are much more efficient, ie, these require less area, than the parallel flow heat exchangers. Thus the counterflow heat exchangers are always preferred in practice.

For heat exchangers other than the parallel and counterflow types, the basic heat-transfer equations, and particularly the effective fluid-to-fluid temperature differences, become very complex (5). For simplicity, however, the basic heat-transfer equation for general flow arrangement may be written as

$$Q = UA\Delta T_m \tag{25}$$

where ΔT_m is the mean temperature difference (MTD) and is expressed in terms of the counterflow heat-exchanger LMTD (as defined by eqs. 21, 23, and 24):

$$\Delta T_m = F \cdot (\text{LMTD}) \tag{26}$$

In equation 26, F is the MTD correction factor and, in general, is a function of the flow configuration and the two temperature factors defined:

$$P = \frac{(T_{t,\text{out}} - T_{t,\text{in}})}{(T_{s,\text{in}} - T_{t,\text{in}})} \tag{27}$$

$$R = \frac{(T_{s,\text{in}} - T_{s,\text{out}})}{(T_{t,\text{out}} - T_{t,\text{in}})} = \frac{\dot{m}_t c_{p_t}}{\dot{m}_s c_{p_s}} \tag{28}$$

where subscripts s and t refer to shellside and tubeside fluids, respectively. Examples of F-charts may be found in the literature (5–7). F is less than unity, indicating that all heat exchangers are less efficient than a pure counterflow heat exchanger. Thus, from the MTD correction factor, F, the degree of the inefficiency relative to the pure counterflow heat exchanger can be determined.

Equations 21, 23, 24, 25, and 26 form the basic heat-transfer equations for all heat exchangers, except for parallel-flow exchangers for which equations 20 and 21 apply. However, the following are special cases: (1) if either one or both fluids are at constant temperature (as is the case with boiling or condensation), then $F = 1$; (2) if ΔT_a is equal to ΔT_b, then LMTD $= \Delta T_a = \Delta T_b$; (3) if physical properties and the overall heat-transfer coefficient, U, vary, or if the fluid temperature profile(s) is not smooth along the tube length, then the entire length should be divided into a number of small heat-exchange elements, and basic heat-transfer equations should be summed up as follows:

$$Q = \sum_i Q_i = \sum_i U_i A_i \Delta \bar{T}_i \tag{29}$$

or

$$Q = \sum_i Q_i = \sum_i U_i A_i (F_i \cdot \text{LMTD}_i) \tag{30}$$

where i is an index referring to the ith element and $\Delta \bar{T}_i$ is the difference in the average fluid temperature in the ith element.

Thermal Sizing Method. Thermal sizing of a heat exchanger means determining the required heat-transfer surface area for a given set of thermal duty and fluid stream inlet–outlet conditions. The calculation procedure for the thermal sizing utilizes the basic heat-transfer equations. For example, given the thermal duty, Q, flow rates, and inlet and outlet temperatures of both hot and cold streams, the following steps are used. (1) Choose the heat-exchanger type or configuration. (2) Compute the heat-transfer coefficients based on the flow geometry and thermal–hydraulic conditions of the fluids. (3) Compute the overall heat-transfer coefficient, U, using equations 15 and 16 or 17. (4) Compute the LMTD using given inlet and outlet fluid temperatures, from equations 21, 23, and 24. (5) Compute the temperature parameters P and R from equations 27 and 28. (6) Read the MTD correction factor F using an appropriate F-chart corresponding to a chosen heat exchanger configuration (5). A value of F greater than 0.9 is virtually always used for design. This is because the F values tend to fall off rapidly below about 0.8–0.9, making it inefficient to operate a heat exchanger below this range. In other words, because the heat-exchanger efficiency drops off dramatically in this area, small errors in calculations may have dire consequences. The use of this criterion, ie, $F > 0.9$, may require the choice of a different exchanger configuration, ie, more tube and shell passes, in order to obtain an $F > 0.9$. (7) Using equation 26, compute $\Delta T_m = F \cdot \text{LMTD}$. (8) Compute the required surface area from equation 25:

$$A = \frac{Q}{U \Delta T_m} \tag{31}$$

(9) If U varies along the tube length or the stream temperature profile is not a smooth curve, then divide the entire tube length into a number of small heat-exchange elements, apply steps (2) through (8) to each element, and sum up the resulting area requirements as follows:

$$A = \sum_i A_i = \sum_i \frac{Q_i}{U_i (\Delta T_m)_i} \tag{32}$$

(10) Translate the heat-transfer area determined in steps (8) or (9) into corresponding tube bundle dimensions (ie, number of tubes, diameter, and tube length).

In actuality, tube bundle dimensions are assumed prior to step (2) in order to compute convective or radiative heat-transfer coefficients. Steps (2) through (10) are repeated until a close agreement between assumed and calculated values is reached. If the final heat-transfer surface arrangement so determined is not satisfactory, the heat-exchanger configuration can be changed (step 1) and the entire computation repeated. The calculation steps can be readily computer-programmed or calculated using Lotus or Excel software.

Heat-Exchanger Effectiveness Method. The method of heat-exchanger effectiveness is useful in determining or rating the performance of a given heat exchanger. This method can also be used in sizing a new heat exchanger.

The heat-exchanger effectiveness, ϵ, is defined as the ratio of the actual rate of heat transfer in a given heat exchanger to the maximum, ie, thermodynamically possible, heat-transfer rate. The latter quantity would be realized in a counterflow heat exchanger having infinite heat-transfer surface area. To determine this maximum possible rate of heat exchange, consider two extreme bounding cases as depicted in Figure 3. In the first case (Fig. 3**a**), the thermal capacity rate, C, which is the mass-flow rate times specific heat, of the hot stream is much greater than that of the cold stream, ie, $C_h >> C_c$. Then, the maximum outlet temperature attainable by the cold stream would be the inlet temperature of the hot stream, and therefore:

$$Q_{\max} = \dot{m}_c c_{pc}(T_{c,\text{out}} - T_{c,\text{in}}) = \dot{m}_h c_{ph}(T_{h,\text{in}} - T_{h,\text{out}}) \tag{33}$$

where $C_c = \dot{m}_c c_{pc}$.

In the second bounding case (Fig. 3**b**), the thermal capacity rate of the cold stream is much greater than that of the hot stream. Then, the minimum outlet temperature attainable by the hot stream would be the inlet temperature of the cold stream, and therefore:

$$Q_{\max} = \dot{m}_h c_{ph}(T_{h,\text{in}} - T_{h,\text{out}}) = C_h(T_{h,\text{in}} - T_{c,\text{in}}) \tag{34}$$

where $C_h = \dot{m}_h c_{ph}$. Denoting $C_{\min}$ and $C_{\max}$ as the smaller and the larger of C_h and C_c, one can combine equations 33 and 34 into a single expression:

$$Q_{\max} = C_{\min}(T_{h,\text{in}} - T_{c,\text{in}}) \tag{35}$$

Hence, the heat-exchanger effectiveness can be written as follows:

$$\epsilon = \frac{Q}{Q_{\max}} = \frac{C_h(T_{h,\text{in}} - T_{h,\text{out}})}{C_{\min}(T_{h,\text{in}} - T_{c,\text{in}})} = \frac{C_c(T_{c,\text{out}} - T_{c,\text{in}})}{C_{\min}(T_{h,\text{in}} - T_{c,\text{in}})} \tag{36}$$

These equations can be rearranged to obtain:

$$Q = \epsilon C_{\min}(T_{h,\text{in}} - T_{c,\text{in}}) \tag{37}$$

$$T_{h,\text{out}} = T_{h,\text{in}} - \epsilon \frac{C_{\min}}{C_h}(T_{h,\text{in}} - T_{c,\text{in}}) \tag{38}$$

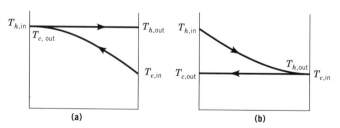

Fig. 3. Extreme cases of counterflow heat exchangers having infinite surface area (**a**) $C_h >> C_c$ and (**b**) $C_h << C_c$. Terms are defined in text.

$$T_{c,\text{out}} = T_{c,\text{in}} + \epsilon \frac{C_{\min}}{C_c}(T_{h,\text{in}} - T_{c,\text{in}}) \tag{39}$$

The importance of equations 37–39 is that once the heat-exchanger effectiveness, ϵ, is known for a given heat exchanger, one can compute the actual heat-transfer rate and outlet stream temperatures from specified inlet conditions. This process is known as rating a given heat exchanger.

It should be noted from equation 36 that

$$\epsilon = \frac{(T_{h,\text{in}} - T_{h,\text{out}})}{(T_{h,\text{in}} - T_{c,\text{in}})} \quad \text{if } C_h = C_{\min} \tag{40}$$

$$\epsilon = \frac{(T_{c,\text{out}} - T_{c,\text{in}})}{(T_{h,\text{in}} - T_{c,\text{in}})} \quad \text{if } C_c = C_{\min} \tag{41}$$

For cases in which one fluid is at constant temperature, ϵ is often defined as equation 40 if T_c is constant, eg, boiling, and as equation 41 if T_h is constant, eg, condensation.

The expressions for ϵ can be obtained using basic governing heat-transfer equations, such as equations 14 and 19, with proper substitutions of equation 36. For simple, single-pass, parallel- and counterflow heat exchangers, the following expressions result:

parallel flow
$$\epsilon = \frac{1 - e^{-N_{\text{tu}}(1 + C_R)}}{1 + C_R} \tag{42}$$

counterflow
$$\epsilon = \frac{1 - e^{-N_{\text{tu}}(1 - C_R)}}{1 - C_R e^{-N_{\text{tu}}(1 - C_R)}} \tag{43}$$

where

$$N_{\text{tu}} = \text{number of (heat-) transfer units} = \frac{UA}{C_{\min}} \tag{44}$$

and

$$C_R = \frac{C_{\min}}{C_{\max}} \tag{45}$$

As indicated in equations 42 and 43, ϵ is a function of N_{tu} and C_R, and furthermore, it is also a function of flow configuration. Examples of ϵ-charts for different flow configurations are shown in Figure 4 (8). Note that the effectiveness is limited, even for infinite surface area, for parallel and parallel–counterflow heat exchangers. For the case of $C_R = 1.0$, the limiting value is 0.5 for a pure parallel-flow exchanger (Fig. 4**a**) and 0.586 for a combined parallel–counterflow exchanger (Fig. 4**c**), but ϵ can be as high as unity for a pure counterflow exchanger (Fig. 4**b**)

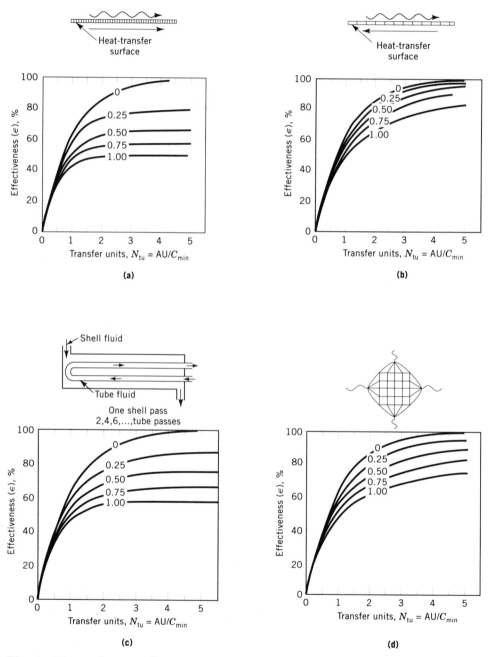

Fig. 4. Heat-exchanger effectiveness where numbers on the curves represent the ratio C_{min}/C_{max} (8). (**a**) Parallel flow; (**b**) counterflow; (**c**) parallel counterflow; and (**d**) crossflow, both fluids unmixed.

and cross-flow exchanger with both fluids unmixed (Fig. **4d**). Other flow configurations are available (8).

Optimum heat-transfer results when the thermal capacity rates of the two fluid streams are balanced, ie, when $C_h = C_c$ or $C_R = 1.0$. Heat-exchanger effectiveness is useful in rating existing heat exchangers. This concept can also be used in sizing new heat exchangers. The calculation procedures for thermal sizing using the $\epsilon - N_{tu}$ method are as follows: given the conditions of thermal duty, Q, flow rates, and inlet and outlet temperatures of both hot and cold streams, the following step-by-step procedures are used. (*1*) Choose the heat-exchanger configuration. (*2*) Assume tube bundle geometry and dimensions. (*3*) Compute the heat-transfer coefficients based on flow geometry and thermal–hydraulic conditions of the fluids. (*4*) Compute the overall heat-transfer coefficient, U, using equations 15 and 16 or 17. (*5*) Compute C_h and C_c, and denote the smaller and the larger of the two as C_{min} and C_{max}, respectively. Also compute $C_R = C_{min}/C_{max}$. (*6*) Compute heat-exchanger effectiveness, ϵ, using given inlet and outlet fluid temperatures, from equation 36. (*7*) Read N_{tu} based on the values of ϵ and C_R from an appropriate ϵ-chart corresponding to chosen heat-exchanger configuration. (*8*) Compute required surface area from the definition of N_{tu}, equation 44:

$$ A = N_{tu} \frac{C_{min}}{U} $$

(*9*) Translate the heat-transfer area determined above into corresponding tube bundle dimensions. If different from those assumed in step (*2*), repeat steps (*2*) through (*8*) until satisfactory agreement is reached. The ϵ-N_{tu} method cannot be applied to cases in which U varies along the tube length or the stream temperature profile is not smooth, ie, boiling or condensation is included.

More recently, a set of charts has been presented enabling both the design and performance calculations to be done using a single chart (9,10). These charts have the N_{tu} parameter curves superimposed on the standard MTD correction factor curves. Thus the four pertinent groups, P, R, F, and N_{tu}, are displayed together on the same chart. The design calculation is thus reduced to finding F and N_{tu}, given P and R; the performance calculation is reduced to finding F and P, given R and N_{tu}.

Design Margins. The heat-transfer surface area determined using the sizing or rating methods is considered the minimum required area. There are additional surface-area requirements for design margins in the final sizing of a heat exchanger.

Effect of Uncertainties in Thermal Design Parameters. The parameters that are used in the basic sizing calculations of a heat exchanger include heat-transfer coefficients; tube dimensions, eg, tube diameter and wall thickness; and physical properties, eg, thermal conductivity, density, viscosity, and specific heat. Nominal or mean values of these parameters are used in the basic sizing calculations. In reality, there are uncertainties in these nominal values. For example, heat-transfer correlations from which one computes convective heat-transfer coefficients have data spreads around the mean values. Because heat-transfer tubes cannot be produced in precise dimensions, tube wall thickness varies over a range of the mean value. In addition, the thermal conductivity of tube wall material cannot be measured exactly, adding to the uncertainty in the design and performance calculations.

If a heat exchanger is sized using the mean values of the design parameters, then the probability, or the confidence level, of the exchanger to meet its design thermal duty is only 50%. Therefore, in order to increase the confidence level of the design, a proper uncertainty analysis must be performed for all principal design parameters.

The degree of data spread around the mean value may be quantified using the concept of standard deviation, σ. If the distribution of data points for a certain parameter has a Gaussian or normal distribution, the probability of normally distributed data that is within $\pm\sigma$ of the mean value becomes 0.6826 or 68.26%. There is a 68.26% probability of getting a certain parameter within $\overline{X} \pm \sigma$, where $\overline{X}$ is the mean value. In other words, the standard deviation, σ, represents a distance from the mean value, in both positive and negative directions, so that the number of data points between $\overline{X} - \sigma$ and $\overline{X} + \sigma$ is 68.26% of the total data points. Detailed descriptions on the statistical analysis using the Gaussian distribution can be found in standard statistics reference books (11).

The values of σs are experimentally determined for all uncertain parameters. The larger the value of σ, the larger the data spread, and the greater the level of uncertainty. This effect of data spread must be incorporated into the design of a heat exchanger. For example, consider the convective heat-transfer coefficient, h, where the probability of the true value of h falling below the mean value $\overline{h}$ is of concern. Or consider the effect of tube wall thickness, t, where a value of t greater than the mean value $\overline{t}$ is of concern.

The effects of data spread should be examined for all individual parameters. These individual effects usually take place simultaneously, and the combined effect is assessed using the root–sum–square (RSS) method. The total additional surface area required to obtain a certain level of design confidence is calculated from

$$\Delta A = \sqrt{\sum_i (\Delta A_i)^2} \qquad (46)$$

where ΔA is the excess heat-transfer area over the nominal area associated with each of the uncertainties. ΔA_i is determined from the thermal performance calculations of the exchanger based on the uncertainty in each individual parameter and nominal values of the remaining parameters. A detailed method of assuming the data uncertainties for the design of heat exchangers may be found in Reference 11.

Shellside flow maldistribution effect is normally included in the h-correction. Tubeside flow maldistribution effect is relatively small in most designs because the principal resistances are tube friction and entrance/exit pressure drop.

Other Effects. Bypass Flow Effects. There are several bypass flows, particularly on the shellside of a heat exchanger, and these include a bypass flow between the tube bundle and the shell, bypass flow between the baffle plate and the shell, and bypass flow between the shell and the bundle outer shroud. Some high temperature nuclear heat exchangers have shrouds inside the shell to protect the shell from thermal transient effects. The effect of bypass flow is the degradation of the exchanger thermal performance. Therefore additional heat-transfer surface area must be provided to compensate for this performance degradation.

Entrance and Exit Span Areas. The thermal design methods presented assume that the temperature of the shellside fluid at the entrance end of all tubes is uniform and the same as the inlet temperature, except for cross-flow heat exchangers. This phenomenon results from the one-dimensional analysis method used in the development of the design equations. In reality, the temperature of the shellside fluid away from the bundle entrance is different from the inlet temperature because heat transfer takes place between the shellside and tubeside fluids, as the shellside fluid flows over the tubes to reach the region away from the bundle entrance in the entrance span of the tube bundle. A similar effect takes place in the exit span of the tube bundle (12).

This implies that the LMTD or MTD as computed in equations 20 through 26 may not be a representative temperature difference between the two heat-transferring fluids for all tubes. The effective LMTD or MTD would be smaller than the value calculated, and consequently would require additional heat-transfer area. The true value of the effective MTD may be determined by two- or three-dimensional thermal–hydraulic analysis of the tube bundle.

Baffle–Tube Support Plate Area. The portion of a heat-transfer tube that passes through the flow baffle–tube support plates is usually considered inactive from a heat-transfer standpoint. However, this inactive area must be included in the determination of the total length of the heat-transfer tube.

Plugged Tube Allowance. Heat-transfer tubes may need to be plugged during fabrication or operation for a variety of reasons. An additional tube allowance should be made at the initial design stage to compensate for possible loss of some tubes. It is customary that the percentage of tubes permitted for plugging during the lifetime of the exchanger is specified in the equipment specification.

Miscellaneous Effects. Depending on individual design characteristics, there are other miscellaneous effects to consider in the determination of the final sizing of a heat exchanger. These include effects of flow maldistribution of both the shellside and tubeside fluids, stagnant or inactive regions in the tube bundle, and inactive length of the tube in tubesheets. These effects should be individually assessed and appropriate additional areas should be provided.

In summary, the final heat-exchanger sizing should be based on the basic theoretical value plus all other additional allowances as described.

$$A_{\text{total}} = A_{\text{min}} + \Delta A \text{ (from equation 46)} + \Delta A \text{ (bypass flow)}$$
$$+ \Delta A \text{ (entrance/exit effects)} + \Delta A \text{ (baffle plates)} \qquad (47)$$
$$+ \Delta A \text{ (plugging)} + \Delta A \text{ (miscellaneous)}$$

Pressure Drop Calculations. There are two principal costs to consider in sizing a heat exchanger: manufacturing costs and operating costs. From a manufacturing standpoint, in general, the less the heat-transfer surface area, the lower the manufacturing cost. The operating cost of a heat exchanger results primarily from the cost of the power to run fluid-moving devices such as pumps and fans, and this power consumption is directly proportional to fluid stream pressure drop. Therefore, an optimum design of a heat exchanger requires a proper balance between thermal sizing and pressure drop. Thus a general method of calculating pressure drops is needed.

The pressure difference between the inlet and outlet nozzles on either the shellside or tubeside of a heat exchanger may be written in the form

$$p_{\text{in}} + \left(\frac{1}{2}\rho V^2\right)_{\text{in}} + (\rho z)_{\text{in}} = p_{\text{out}} + \left(\frac{1}{2}\rho V^2\right)_{\text{out}} + (\rho z)_{\text{out}} + (\Delta p)_{\text{loss}} \quad (48\text{a})$$

or

$$p_{\text{in}} - p_{\text{out}} = \left[\left(\frac{1}{2}\rho V^2\right)_{\text{out}} - \left(\frac{1}{2}\rho V^2\right)_{\text{in}}\right] + [(\rho z)_{\text{out}} - (\rho z)_{\text{in}}] + (\Delta p)_{\text{loss}}$$
$$= (\Delta p)_{\text{kinetic energy}} + (\Delta p)_{\text{elevation}} + (\Delta p)_{\text{loss}} \quad (48\text{b})$$

The first bracketed term represents the pressure difference between the inlet and outlet nozzles resulting from the difference in fluid velocity, acceleration or momentum, and the second bracketed term represents the pressure difference between the inlet and outlet nozzles from the difference in elevation. Both of these pressure difference terms are not a lost energy because the energy is recovered in other parts of the fluid flow circuit if the circuit is a closed one. In an open flow circuit, the circulating pump or fan must work against these pressure differences or drops.

The term pressure drop usually refers to the pressure loss that is not recoverable in the circuit, and it is lost energy that is dissipated into the fluid stream in the form of heat energy. The pressure drop in a flow circuit is associated with various forms of energy dissipation owing to friction, change in flow area, flow turning, and others:

$$(\Delta p)_{\text{loss}} = (\Delta p)_{\text{friction}} + (\Delta p)_{\text{area change}} + (\Delta p)_{\text{turning}} + (\Delta p)_{\text{misc}} \quad (49)$$

$(\Delta p)_{\text{misc}}$ represents miscellaneous pressure drops associated with valves, fittings, etc, in the flow circuit, and these devices are not usually present inside a heat exchanger.

Frictional Pressure Drop.   The frictional pressure drop inside a heat exchanger results when fluid particles move at different velocities because of the presence of structural walls such as tubes, shell, channels, etc. It is calculated from a well-known expression of

$$(\Delta p)_{\text{friction}} = f_D \frac{L}{D_H}\left(\frac{1}{2}\rho V^2\right) \quad (50)$$

where f_D is the Darcy friction factor $= 4f$ and f is the Fanning friction factor; D_H is the hydraulic or equivalent diameter $= 4(\text{flow area})/(\text{flow perimeter})$; and

$\frac{1}{2}\rho V^2$ is the dynamic pressure. The Darcy friction factor, f_D, can be derived from a theoretical analysis of fluid flow in a closed conduit; for laminar flow

$$f_D = \frac{64}{Re} \tag{51}$$

and for turbulent flow

$$f_D = \frac{0.184}{Re^{0.2}} \tag{52}$$

(see also Table 3). Equation 52 is applicable to smooth-walled conduits for 10,000 $< Re <$ 120,000. In general, f_D is a function of the Reynolds number, Re, and the relative roughness of conduit surface, ϵ_w/D_H, as shown in the Moody diagram (Fig. 5) (13). For laminar flow, equation 50 is used with Table 5. Equations 50 to 52, the Moody diagram, and Table 5 may be used to calculate frictional pressure drops inside heat-transfer tubes and for unbaffled or longitudinal shellside flow.

Pressure Drop from Area Change.   Pressure drop from area change occurs as a result of energy dissipation associated with eddies formed when a flow area is suddenly expanded or contracted. It is expressed in the following form:

$$(\Delta p)_{\text{area change}} = K\left(\frac{1}{2}\rho V^2_{\text{max}}\right) \tag{53}$$

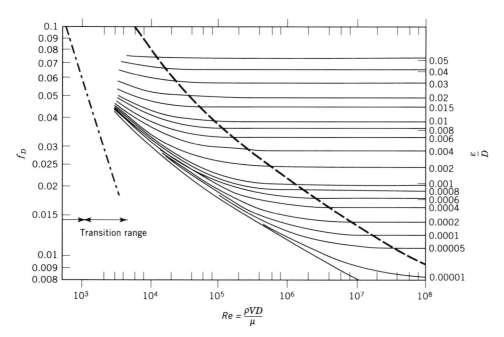

Fig. 5.   Moody diagram for Darcy friction factor (13): (——), smooth flow; (————), wholly turbulent flow; (—·—·—), laminar flow.

where K is the pressure loss coefficient and V_{max} is the flow velocity based on smaller or minimum flow area. K is a function of flow–area ratio and is given in Figure 6 for both sudden contraction and expansion cases. Equation 53 may be used to calculate pressure drops from inlet nozzles to shell or channel, from shell or channel to outlet nozzles, and tubeside inlet and exit losses. When there is a gradual change in tube diameter, the corresponding pressure–loss coefficient can be determined using the curve given in Figure 6c.

Pressure Drop Owing to Flow Turning. When a fluid turns along a curved surface or mitered bend, a secondary flow is formed as a result of centrifugal force acting on fluid particles. An energy dissipation follows, and the pressure decreases. The pressure drop associated with flow turning is expressed as

$$(\Delta p)_{turning} = K\left(\frac{1}{2}\rho V^2\right) = f_D\left(\frac{L}{D_H}\right)_{equivalent} \left(\frac{1}{2}\rho V^2\right) \tag{54}$$

where K is the turning-loss coefficient (see Fig. 7), V is the flow velocity calculated based on the upstream unaffected flow area, and L/D_H is an equivalent length-to-diameter ratio. The equivalent length-to-diameter ratio concept can also be used in conjunction with a familiar friction pressure-drop formula of equation 50. The pressure drops in a U-bend section of U-tubes in a heat exchanger can be calculated using a similar approach as given in equation 54.

Shellside Tube Bundle Pressure Drop. The frictional pressure drop for unbaffled or longitudinal shellside flow can be calculated from equation 50 using the hydraulic or equivalent diameter concept described. However, most heat exchangers are baffled on the shellside with disk-donut or segmental types. The resulting flow field is very complex, and the unrecoverable pressure drop is a combination of pressure losses owing to friction, area changes, and flow turning.

There are no general, reliable pressure–drop formulas for baffled shellside flow available in the open literature. Therefore, the Donohue correlation for a pair of disk-donut or segmental baffles is presented herein because of its simplicity (15):

$$\Delta p = (\Delta p)_{window\ flow} + (\Delta p)_{cross\ flow} \tag{55a}$$

$$(\Delta p)_{window\ flow} = \frac{1.087\ (G_w)^2}{(\text{specific gravity})}, \Delta_p \text{ in kPa} \tag{55b}$$

$$(\Delta p)_{cross\ flow} = \frac{60N}{\left(\frac{s_t}{D_o} - 1\right)(Re_{max})}\left(\frac{1}{2}\rho V^2_{max}\right) \quad \text{for } Re_{max} < 100 \tag{55c}$$

$$= \frac{3N}{\left(\frac{s_t}{D_o} - 1\right)^{0.2} (Re_{max})^{0.2}}\left(\frac{1}{2}\rho V^2_{max}\right) \text{ for } 500 < Re_{max} < 30{,}000 \tag{55d}$$

where G_w is the mass flow at the window in units of kg/(m²·s); V_{max}, the maximum intertube cross-flow velocity at bundle center; and N, the number of tube rows

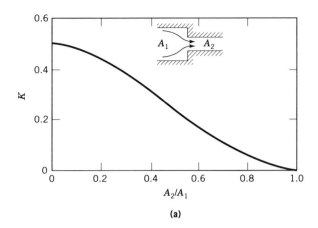

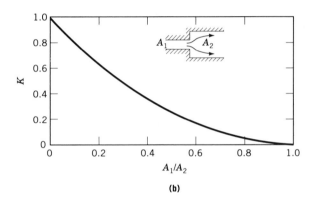

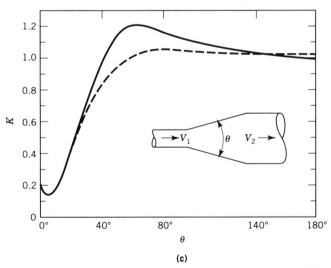

Fig. 6. Pressure–loss coefficient, K (**a**) for sudden contraction, $h_L = K(V_2^2/2)$; (**b**) for sudden expansion, $h_L = K(V_1^2/2)$; and (**c**) versus θ in a gradually expanding section, $h_L = K\dfrac{(V_1 - V_2)^2}{2}$, where (——), $D_2/D_1 = 1.5$; (— — —), $D_2/D_1 = 3$ (13).

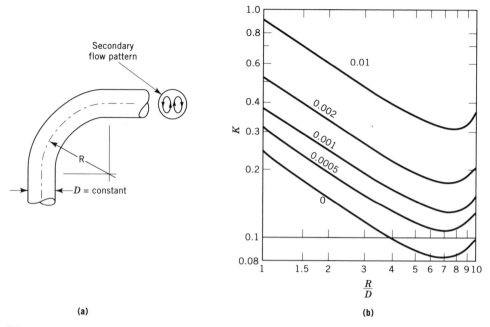

(a)   (b)

Fig. 7. (**a**) Configuration for flow turning. The frictional resistance resulting from the bend length must be added; (**b**) pressure–loss coefficient, K, for 90° flow turning where the numbers on the lines refer to ϵ/D values (14).

traversed by fluid in cross flow in going from one baffle window to the next. The total shellside bundle pressure drop is obtained by multiplying equation 55a by the appropriate number of pairs of window and cross flows in the heat exchanger.

Types of Heat Exchangers.

The shell-and-tube exchanger is the workhorse of power, chemical, refining, and other industries (Fig. 8). One fluid flows on the inside of the tubes whereas the other fluid is flowing through the shell and over the outside of the tubes. Baffles are used to ensure that the shellside fluid flows across the tubes, thus inducing high heat transfer.

Plate heat exchangers, which are used as an alternative to shell-and-tube heat exchangers in relatively low temperature and pressure applications involving liquids and two-phase flows, have some important advantages over shell-and-tube exchangers. Plate heat exchangers include plate–frame heat exchangers (Fig. 9**a**), plate–fin heat exchangers (Fig. 9**b**), and plate–coil heat exchangers (Fig. 10). The advantages of using plate heat exchangers are (16) less surface for heat transfer is required, resulting in weight, volume, and cost advantage over shell-and-tube and other noncompact heat exchangers; thermal rating of plate heat exchangers can readily be increased or decreased by varying the number of plates which is important if substantial changes in load occur; and the increased effect-

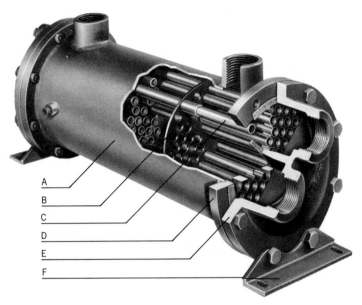

Fig. 8. Shell-and-tube heat exchanger: A, shell of high strength; B, tube sheet; C, tubes (normally small diameter tubes are seamless, but large diameter tubes (> 1 in.) are welded tubes); D, bonnets; E, baffles to assure more efficient circulation by providing minimum clearance between tubes and tube holes as well as baffles and shells; and F, mounting brackets. Courtesy of Basco.

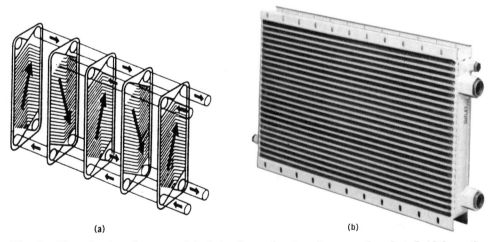

(a)   (b)

Fig. 9. Plate heat exchangers: (**a**) plate–frame heat exchanger when hot fluid from the heat source enters the heat exchanger through connections in the stationary frame plate and is channeled over one side of each plate. Cold fluid enters through different frame plate connections and flows on the other side of each plate in a direction opposite to the hot fluid direction. Courtesy of Bell & Gosset. (**b**) Limco model 6502 plate–fin heat exchanger having compact plate–fin aluminum brazed liquid-to-air construction. The brazed plate–fin construction provides the most efficient heat-exchanger system in terms of size, weight, and performance. Courtesy of Limco.

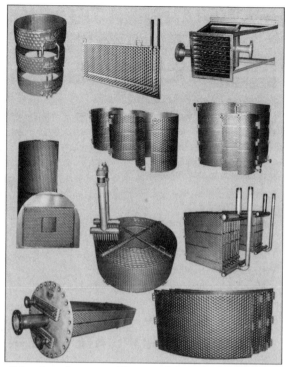

Fig. 10. Plate–coil heat exchangers. Plate–coil is a very efficient and versatile prime surface-type heat exchanger. Courtesy of Karbate Vicarb Inc.

iveness of plate heat exchangers reduces the required cooling flow rate, resulting in savings relative to piping, pumps, valves, and operating cost. In spite of these advantages, however, the plate heat exchangers are rarely used, even in relatively low temperature and pressure applications. This may be because of the widespread familiarity with shell-and-tube exchangers and the large number of manufacturers of shell-and-tube exchangers (3). Plate heat exchangers are not normally used in nuclear applications for safety considerations. Note that ASME Boiler and Pressure Vessel codes do not recognize the plate heat exchangers.

The principal disadvantage of plate–frame heat exchangers is the large number of surfaces that must be sealed by gaskets. The maximum temperature and pressure at which the plate heat exchangers can be used are limited to 2.7 MPa (400 psi) and 145°C (300°F). Imperfect sealing of the gasket at higher temperature ranges has been a problem using plate heat exchangers (17–19).

Cold-plate heat exchangers for electronics cooling applications (Fig. 11) operate at heat-flux levels typically on the order of 2–10 W/cm^2. The electronics industry has targeted heat-flux capacities of up to 25 W/cm^2 for the next generation cold-plate for advanced applications. Achieving this level of cooling requires development of methods of providing high surface-density cooling within coolant passages in the cold plate. Potential scenarios that might provide high heat flux cooling may include high density ribbed or finned surfaces or impinging jet cooling

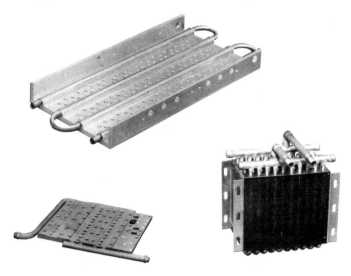

Fig. 11. Liquid-cooled cold plates or heat sinks have been developed as thermal management solutions to cool components for liquid-cooled computer systems and other electronic systems where heat removal becomes one of the important design criteria. Courtesy of EG & G.

of the cold-plate primary surface (see ELECTRONIC MATERIALS; ELECTRONICS COATINGS).

Use of Heat Exchangers

Heat exchangers are used whenever energy has to be transferred, and the proper design and use of heat exchangers are vitally important for efficient operation of an industrial system, for energy conservation, and ultimately for the protection of the environment (3,17–20). The heat-exchanger industry had an estimated gross revenue of nearly one-quarter of a trillion dollars in 1991 (20). Despite decades of continuous research and development, there are numerous design and operating problems originating from a lack of understanding of basic flows and heat-transfer phenomena such as flow distribution in manifolds, flow-induced vibration in two-phase flows, heat-transfer enhancement, fouling, etc.

In order to help companies and organizations overcome problems associated with heat exchangers, Heat Transfer Research Inc. (HTRI) was established in the United States in 1962, and the Heat Transfer and Fluid Flow Service (HTFS) was established in the United Kingdom in 1967. These organizations provide results of heat-transfer and fluid-flow research, design methods, supporting computer programs, and proprietary equipment testing (17,18). In addition, the American Society of Mechanical Engineers (ASME), American Society for Testing and Materials (ASTM), and Tubular Exchanger Manufacturers Associations Inc. (TEMA) provide various safety and design codes and technical services. More recent reference books come with software disks, and analyses or design calculations of various energy systems can be conducted (21).

Fundamental issues involved in the use of various heat exchangers have been summarized in a thermal science workshop sponsored by the National Science Foundation (20). There are a number of areas that require different types of heat exchangers. Some of the emerging technologies where heat exchangers are expected to play a critical role are electronic cooling, micro and macro gravity applications, ozone (qv) depletion, global warming and other environmental issues, biotechnology (qv), high temperature superconductors (see SUPERCONDUCTING MATERIALS), and ultrahigh temperature waste-heat recovery. For example, in future designs, heat-exchanger passages may be incorporated into silicon substrates for the purpose of cooling microelectronic chips. The passage dimensions could be made as small (<1 μm) as those of the chip features. Similar microscale heat exchangers may be necessary for the biotechnology industry which processes various macromolecules at controlled temperatures.

Shell-and-Tube Exchangers. The single most important problem unique to shell-and-tube exchangers is the existence of various flow-leakage and bypass streams on the shellside. Proprietary empirically based computer programs are available for the analysis of shellside single-phase and multiphase flows from HTRI and HTFS. To further improve shell-and-tube exchangers, innovative shellside designs must be used to maintain high heat transfer at reduced pressure drop.

Plate Heat Exchangers. The plate heat exchanger technology has advanced considerably since the early 1970s. Some basic issues still need to be addressed for further advancement of technology and utilization of plate heat exchangers in order to compete against shell-and-tube exchangers. The pressure profiles in inlet and outlet manifolds (headers) have a very significant effect on the flow rate through the plate heat exchangers and on flow distribution through the plates, thereby affecting heat-transfer performance.

High Temperature and Waste Heat Recovery Exchangers. Heat exchangers in industrial process heating applications are commonly used to improve process efficiency by preheating the combustion air using waste heat from the flue gases. Advances in these heat exchangers should have significant impact on energy savings in primary metal, glass (qv), and ceramic firing industries (see CERAMICS; METALLURGY).

The primary need in the development of heat exchangers for recuperation in high temperature industrial processes does not relate mainly to heat-transfer issues but rather to material durability. Heat exchangers need to withstand high temperatures and high corrosiveness of gases such as those containing sodium silicates and chlorine and potassium salts. As of this writing, no low cost ceramic materials are available for heat exchangers for these conditions. In addition, because these ceramic components must be mated to metallic components, methods to obtain ceramic-to-metallic seals must be obtained to simplify manufacturing and installation of recuperation equipment in these applications. Moreover the auto regenerative burners are expensive; thus, new burners that produce low NO_x and are reasonable in price should be developed. There is good potential for the application of low cost, easily retrofitted passive heat-transfer enhancement techniques to high temperature waste heat recovery as well as to high temperature process heat exchangers/reactors such as fired heaters, steam reformers, and other process vessels used in the petrochemical industries. A better understanding of the radiative properties of exhaust gas streams would allow better design of such heat exchangers (see also PROCESS ENERGY CONSERVATION).

Low Temperature Difference Heat Exchangers. Many applications exist in which the extraction of thermal energy from low temperature differences between the source and sink is discussed. Some examples are ventilation application, ocean thermal energy conversion (OTEC) power plants, and atmosphere thermal energy conversion (ATEC) power plants to recover thermal energy owing to a temperature difference between a mountaintop and the valley. In the latter two examples, abundant free energy can be harnessed if the appropriate inexpensive durable heat exchangers are available. These low temperature difference heat exchangers would require very large areas. This is the principal disadvantage and limitation (see THERMAL ENERGY CONVERSION).

The success of low temperature difference applications depends on the development of special inexpensive materials, heat-exchanger constructions, and surfaces having high heat-transfer performance at a very low pressure drop, ie, low pumping power requirements. Whereas low cost paper and plastic exchangers are commercially available for ventilation applications, specialized materials and exchanger constructions are needed to provide cost-effective exchangers with durability and good performance. OTEC power plants have not materialized because heat-exchanger capital and operating costs are too high, and the life is too short.

Direct Contact Heat Exchangers. In a direct contact exchanger, two fluid streams come into direct contact, exchange heat and maybe also mass, and then separate. Very high heat-transfer rates, practically no fouling, lower capital costs, and lower approach temperatures are the principal advantages.

Cooling towers are excellent examples of direct contact heat exchangers. Sparging, ie, a process of forcing gas or water vapor through liquid to remove undesirable gases using low pressure stream (qv) is a common way to heat up vats of material in the food industry (see FOOD PROCESSING). When chemical reactions must be stopped very quickly, direct quenching downstream of the reactor is often employed. Beds of solids are often heated by direct contact with fluidizing gas, eg, in fluidized-bed dryers (see DRYING).

Direct contact heat exchangers have, however, received limited use in conventional power and process applications. These exchangers appear to offer advantages in some geothermal power applications and have been proposed for use in ocean thermal energy conversion-system designs (see GEOTHERMAL ENERGY; RENEWABLE ENERGY RESOURCES). Direct contact heat transfer also plays an important role in some nuclear accident scenarios in which vapor produced in an accident is condensed by bringing it into direct contact with colder water (see NUCLEAR REACTORS, SAFETY IN NUCLEAR FACILITIES). Applications of direct contact exchanger technology could come in the biotechnology area, where direct contact heat-transfer/mass-transfer/reactor units may be used in the production and processing operations in genetic engineering (qv) technology. Whereas the uses of direct contact exchanger technology are not completely understood, the potential impact could be great, particularly in aerospace power and heat-exchanger designs.

Heat Exchangers Using Non-Newtonian Fluids. Most fluids used in the chemical, pharmaceutical, food, and biomedical industries can be classified as non-Newtonian, ie, the viscosity varies with shear rate at a given temperature.

In contrast, Newtonian fluids such as water, air, and glycerin have constant viscosities at a given temperature. Examples of non-Newtonian fluids include molten polymer, aqueous polymer solutions, slurries, coal–water mixture, tomato ketchup, soup, mayonnaise, purees, suspension of small particles, blood, etc. Because non-Newtonian fluids are nonlinear in nature, these are seldom amenable to analysis by classical mathematical techniques.

The optimum design of process equipment which handles non-Newtonian fluids could be significantly improved once predictive capability were increased. However, the basic understanding of the fluid mechanical and heat-transfer behavior of non-Newtonian, ie, viscous and viscoelastic, fluids is limited (22). A better understanding of pressure drop and heat-transfer behavior of non-Newtonian flows applicable to typical heat-exchanger geometries should lead to the design and development of more energy-efficient processes and to better quality control of the final products. In general, the viscosity of a non-Newtonian fluid can be significantly larger than that of water. Therefore, the selection of a pump size to provide enough flow rate and subsequently to ensure adequate heat removal or supply is necessary.

A significant heat-transfer enhancement can be obtained when a noncircular tube is used together with a non-Newtonian fluid. This heat-transfer enhancement is attributed to both the secondary flow at the corner of the noncircular tube (23,24) and to the temperature-dependent non-Newtonian viscosity (25). Using an aqueous solution of polyacrylamide the laminar heat transfer can be increased by about 300% in a rectangular duct over the value of water (23).

A knowledge of the viscous and thermal properties of non-Newtonian fluids is essential before the results of the analyses can be used for practical design purposes. Because of the nonlinear nature, the prediction of these properties from kinetic theories is as of this writing in its infancy. For the purpose of design and performance calculations, physical properties of non-Newtonian fluids must be measured.

Micro-Heat Exchangers. A better understanding of transport phenomena in microchannel heat exchangers appears to be vital to the development of some advanced microelectronic devices. In future designs, heat-exchanger passages are expected to be incorporated into silicon substrates for the purpose of cooling substrate-mounted microelectronic chips. The passage dimensions could be made as small (<1 μm) as those of the chip features, in which case the passage size may be comparable to the mean free path of air molecules pumped through the passages. The spacing between two molecules of gas is on the order of 1 μm, whereas that of liquid is on the order of 0.1 μm (13).

Further research on convective transport under low Reynolds number, quasicontinuum conditions is needed before the optimal design of such a micro heat exchanger is possible. The cooling heat exchanger is usually thermally linked to a relatively massive substrate. The effects of this linkage need to be explored and accurate methods of predicting the heat-transfer and pressure-drop performance need to be developed.

Electrohydrodynamic-Based Heat Exchangers. Electrohydrodynamics refers to the coupling of an electric field and a velocity field in a dielectric fluid continuum. Electric-field effects on heat transfer in polar gases generally take place via a modification of the gas velocity and temperature boundary layers.

Electric fields in complex flows act to change the character of flow stability. Applications of electrohydrodynamics in convective heat transfer are diverse such as in heating ventilation or air conditioning (HVAC) cooling of electronic equipment applications, space power applications, micromachines, ultrasmall high duty heat exchangers, and noninvasive flow control techniques.

Characterization and influence of electrohydrodynamic secondary flows on convective flows of polar gases is lacking for most simple as well as complex flow geometries. Such investigations should lead to an understanding of flow control, manipulation of separating, and accurate computation of local heat-transfer coefficients in confined, complex geometries. The typical Reynolds number of the bulk flow does not exceed 5000.

Flow-Induced Vibrations.   One of the critical limitations of the increased performance of shell-and-tube heat exchangers is the onset of flow-induced vibrations at high shellside fluid flows that result in a loud acoustic (noise) vibration of more than 150 dB, or the vibration of tubes to the extent that the tube walls are worn through. Whereas a great deal of research has been done to understand vibration excitation mechanisms and to develop design guides, much is unknown about predicting flow-induced vibration occurrence, the location and type of damage, and the rates of wear. There is a substantial dependence on experience with the hardware in service, and this severely limits the development of the next generation of heat exchangers. Elimination or substantial minimization of flow-induced vibrations would have a significant impact in power, process, petroleum (qv), and other industries that use shell-and-tube exchangers.

Flow Maldistribution.   One of the principal reasons for heat exchangers failing to achieve the expected thermal performance is that the fluid flow does not follow the idealized anticipated paths from elementary considerations. This is referred as a flow maldistribution problem. As much as 50% of the fluid can behave differently from what is expected based on a simplistic model (18), resulting in a significant reduction in heat-transfer performance, especially at high N_{tu} or a significant increase in pressure drop. Flow maldistribution is the main culprit for reduced performance of many heat exchangers.

In addition to the reduction in performance, flow maldistribution may result in increased corrosion, erosion, wear, fouling, fatigue, and material failure, particularly for liquid flows. This problem is even more pronounced for multiphase or phase change flows as compared to single-phase flows. Flow distribution problems exist for almost all types of exchangers and can have a significant impact on energy, environment, material, and cost in most industries.

For gross flow maldistribution in heat exchangers, modeling is available for heat-transfer performance prediction, but no modeling is available for pressure-drop prediction. This is because, in most of the cases, the static pressure distribution is not uniform at the exchanger inlet and outlet faces, and no modeling or computational fluid dynamic analysis is possible without the boundary conditions. Gross flow maldistribution significantly increases pressure drop. In addition, because there are an infinite number of gross flow maldistributions possible, the only approach is to analyze the problem numerically for idealized uniform pressure boundary conditions.

No systematic study is reported to quantify the effect of manifold induced-flow maldistribution on a single-phase pressure drop and heat transfer in a heat exchanger. Such flow maldistribution is common in gas-to-gas and liquid-to-gas exchangers with manifolds, and in a plate heat exchanger in which many parallel passages are connected by inlet and outlet pipe manifolds created by plate ports. For two-phase flow distribution, however, no practical methods exist for ensuring the adequate distribution of the vapor and liquid phases among many parallel-flow channels. The result in the cryogenic gas processing area is, for example, that phases are separated and introduced into separate heat exchangers for further vaporization or condensation at a significant penalty in overall thermodynamic optimization of the system. Viscosity-induced flow maldistribution has been hardly analyzed to quantify the influence on heat transfer and pressure drop. Very meager information is available in the literature on natural convection-induced flow maldistribution and its effect on the exchanger heat transfer and pressure drop. A combination of hot- and cold-fluid maldistributions, both tubeside and shellside, can create a more serious problem than the individual maldistributions alone. Heat exchangers involving multiphase flow appear to have the highest likelihood of flow maldistribution and the resulting thermal and mechanical performance loss and flow instability. This is especially critical where multiphases exist at inlet.

Header Design

Headers, ie, manifolds and tanks, are the chambers or transition ducts at each end of the heat-exchanger core on each fluid side for distributing fluid to the core at the inlet and collecting fluid at the exit. These may be classified broadly as normal, turning, and oblique flow headers. Poor design of headers reduces heat-transfer performance significantly and may also increase pressure drop substantially owing to flow maldistribution, flow separation, and jet effects. Thus header design is an important problem for all heat exchangers where fluid from the inlet pipe is distributed to the exchanger core via manifolds and tanks. If novel heat-exchanger applications are contemplated, the header volume must be a very small fraction of the total exchanger volume, particularly for highly compact heat-exchanger applications.

No design theory and modeling is available to obtain uniform flow for normal headers, ie, diffusers having downstream flow resistance and turning headers, with or without vanes. Only very limited design information is available for oblique flow headers. Manifolds in a heat exchanger can be further classified into four types: dividing, combining, parallel, and reverse flow manifolds. Parallel and reverse flow manifolds are those which combine dividing and combining flow manifolds. In a parallel flow manifold, the flow directions in dividing and combining flow headers are the same; in a reverse flow manifold, the flow directions are opposite. The objective of the manifold design is to obtain a uniform flow distribution in the heat-exchanger core, with the manifold occupying the smallest fraction of volume of the total heat exchanger.

Most experimental studies on manifolds are limited to turbulent flow in circular pipes. The flow characteristics of branch points in manifolds (26) and the effect of the Reynolds number and branch pipe resistance on the flow distribution in dividing and combining flow manifolds (27) have been studied using water. In the latter case, the branch pipe resistance was varied by using different size orifice plates. The flow distribution, which was independent of the Reynolds number in a range of $Re = 30,000 - 100,000$, was found to improve with the increase of branch pipe resistance. Reverse and parallel flow manifolds have been studied both analytically and experimentally using hair (28,29). A manifold having a smaller area ratio ($AR = 1.41$) had a better flow distribution than the one having a larger area ratio ($AR = 2.81$), regardless of the resistance in branch pipes. The area ratio, AR, is the ratio of the total channel cross-sectional area to the dividing flow header cross-sectional area.

Several investigators have conducted analytical and numerical studies on dividing and combining flow manifolds. Friction was shown always to increase the flow imbalance in a combining flow manifold and friction might either increase or decrease the flow imbalance in a dividing flow manifold depending on the area ratio (30). The larger the cross-sectional area of dividing and combining flow manifolds, the better the flow distribution has been reported to be (31). A mathematical model having a one-dimensional elliptic solution procedure has been proposed (32) in which a uniform flow distribution could be achieved at relatively low values of $(AR)C_D C_T^{0.5}$. Analytical and numerical solutions for uniform flow distributions in perforated conduits have been obtained by using variable cross sections of circular and rectangular shapes (33).

Numerical results from the study of parallel and reverse flow manifolds (34,35) showed that the area ratio had a more significant effect on the flow distribution than the header friction. The variation in the flow distribution increased as AR increased. Manifold header shapes for uniform flow-through resistances have been determined analytically (36). A numerical study on the flow distribution in a manifold-shaped direct cooling system of varying geometrical characteristics has been conducted (37). It was found that a manifold having inclined channels had a better flow distribution than the one having channels positioned normal to the dividing flow header. A summary of studies on the flow distribution in manifolds is summarized in Table 6.

It has become quite popular to optimize the manifold design using computational fluid dynamic codes, ie, FIDAP, Phoenix, Fluent, etc, which solve the full Navier-Stokes equations for Newtonian fluids. The effect of the area ratio, AR, on the flow distribution has been studied numerically and the flow distribution was reported to improve with decreasing AR.

A numerical study of the effect of area ratio on the flow distribution in parallel flow manifolds used in a liquid cooling module for electronic packaging demonstrate the usefulness of such a computational fluid dynamic code. The manifolds have rectangular headers and channels divided with thin baffles, as shown in Figure 12. Because the flow is laminar in small heat exchangers designed for electronic packaging or biochemical process, the inlet Reynolds numbers of 5, 50, and 250 were used for three different area ratio cases, ie, $AR = 4, 8$, and 16.

The qualitative flow distribution in a manifold can be estimated by examining a streamline plot. Figure 13 shows the streamline plot for the manifold having $AR = 4$. Note that the same amount of fluid flows between two consecutive streamlines. The area ratio is an important parameter affecting the flow distribution in a manifold, as shown in Figure 14a, which shows the percent flow rate in each channel for three cases. As the area ratio increases, the percent flow rate

Table 6. Studies on Flow Distribution in Manifolds

Solution	Type	Parameters	Fluid	Assumptions[a]	Reference
numerical	parallel	AR	water laminar		38
numerical	parallel reverse				35
analytical	parallel reverse	$m = m(AR, D_c/D_d)$		1-D flow, no friction in headers	39
numerical analytical	parallel reverse		air turbulent		30
numerical	parallel reverse	flow rate heat load	turbulent	1-D flow	40
numerical	parallel reverse	AR, friction	air turbulent	1-D flow	34
analytical experiment	parallel reverse	AR, office resistance	air turbulent		29
analytical experiment	parallel reverse	header geometry D_c/D_d	air turbulent		36
analytical	dividing combing	friction	turbulent	1-D flow	31
numerical	dividing combing	$(AR)\, C_D C_T^{1/2},\ \dfrac{4\, fLD_d}{nd^2}$	turbulent		33
analytical	dividing combing		turbulent	no friction	41
experiment	dividing combing	branch pipe resistance Re, $(D_d/d)^2$	water turbulent		28
numerical	dividing combing	$AR,\ \dfrac{4\, fLD_d}{1.58\, nd^2}$			42
numerical experiment	dividing combing	friction, momentum, number of branch pipes	air turbulent	1-D flow	32
experiment	dividing combing		water		26
step by step	dividing	$AR, L/D_d$			43
numerical	box-shaped	branch inclination variable cross section		1-D flow	37
analytical numerical	perforated conduit	variable cross section	laminar turbulent		33

[a] 1-D = unidimensional.

982

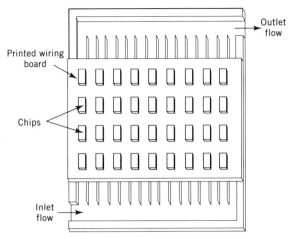

Fig. 12. Schematic diagram of a liquid-cooling module manifold for electronic packaging.

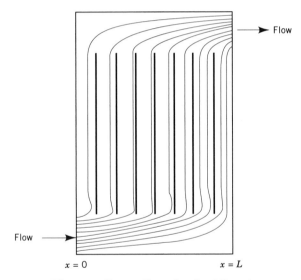

Fig. 13. Streamline plot for $AR = 4$.

increases in channels no. 1 and no. 8, whereas the percent flow rate decreases in the middle channels.

The flow distribution is a direct consequence of the static pressure difference between dividing and combining flow headers (38). There are two factors controlling the pressure variations in manifold headers: friction and momentum. In a combing flow header, these two factors lower the pressure along the header in the flow direction. However, in a dividing flow header, these two factors work in opposite directions. The friction effect lowers the pressure along the header, whereas

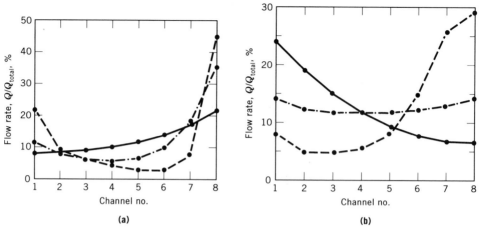

Fig. 14. Flow rates for (**a**) area ratios of (——) 4, (–·–·–) 8, and (- - -) 16, each having the same flow rate and $Re = 50$; and (**b**) Reynolds numbers of (——) 5, (–·–·–) 50, and (- - -) 250 for a liquid-cooling module having $AR = 8$ and $D_c/D_d = 4.0$.

the momentum effect increases the pressure. The flow velocity decreases in the flow direction owing to fluid loss into channels, creating momentum deficiency along the dividing flow header and thus increasing pressure. Furthermore, the pressure increases near the end of the dividing flow header due to the conversion of kinetic energy to stagnation pressure.

The flow distribution in a manifold is highly dependent on the Reynolds number. Figure 14**b** shows the flow distribution curves for different Reynolds number cases in a manifold. When the Reynolds number is increased, the flow rates in the channels near the entrance, ie, channel no. 1–4, decrease. Those near the end of the dividing header, ie, channel no. 6–8, increase. This is because high inlet velocity tends to drive fluid toward the end of the dividing header, ie, inertia effect.

Figure 15 shows the effect of the width ratio D_c/D_d, the ratio of the combining header width to the dividing header width, on the flow distribution in manifolds for Reynolds number of 50. By increasing D_c/D_d, the flow distribution in the manifold was significantly improved. The ratio of the maximum channel flow rate to the minimum channel flow rate is 1.2 for the case of $D_c/D_d = 4.0$, whereas the ratio is 49.4 for the case of $D_c/D_d = 0.5$.

The improvement of the flow distribution by increasing the value of D_c/D_d results from the decrease of momentum gain in the combining header. In a combining header, the fluid velocity increases in the flow direction because fluid is added from the channels. The momentum gain through this fluid addition results in a pressure decrease in the flow direction. For a given flow rate and a fixed D_d, the flow velocity in the combining header decreases with increasing value of D_c/D_d. Therefore, increasing the value of D_c/D_d results in a reduced pressure drop in the combining header in the flow direction.

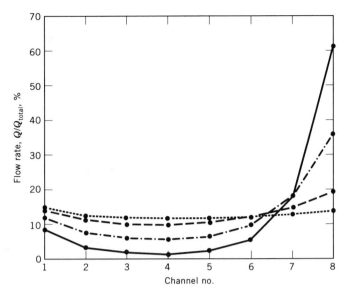

Fig. 15. Flow rates for D_c/D_d of (——) 0.5, (—·—·—) 1.0, (———) 2.0, and (···) 4.0 in a liquid-cooling module having $D_d = 0.492$ cm, $AR = 8$, and $Re = 50$.

Performance Enhancement in Heat Exchangers

Static Mixer. Heat exchangers are commonly built using conventional straight tubes. The laminar layer near tube walls reduces the heat exchange between fluid and tube wall, even in a turbulent flow condition. To enhance the performance of conventional shell-and-tube heat exchangers, one can use static mixer elements inside tubes as shown in Figure 16. Process fluid is continuously mixed, thus producing performance enhancement. The pressure drop using a static mixer can be calculated by multiplying the calculated pressure drop in the empty tube by a factor K_{mixer}:

$$\Delta p_{\text{with mixer}} = K_{mixer}\,\Delta p \qquad (56)$$

where K_{mixer} can be in the range of 5–70, indicating that there is a substantial increase in the pressure drop.

(a)   (b)

Fig. 16. Static mixers which provide a continuous mixing and processing unit with a non-moving part. These static mixers can be easily installed in new and existing pipelines. (**a**) Courtesy of Ross; (**b**) courtesy of Chemineer.

Advanced Heat-Transfer Fluid. A conventional heat-exchanger system requires a high volumetric flow rate, resulting in the consumption of a large amount of pumping power. Using an advanced heat-transfer fluid has been proposed to increase the convective heat-transfer coefficient by increasing the effective thermal capacity of working fluids, a technique that would permit the use of a smaller volumetric flow rate and smaller heat exchangers.

In order to validate this concept, an experiment was performed using an ice-water slurry and it was found that a 25% ice slurry had a two-to-four-times higher thermal capacity than chilled water (44). As the concentration of ice particles in the ice-slurry mixture increased up to 30%, no significant change of pressure drop was reported compared to pure water.

A slurry using phase-change materials has been tested for high temperature applications. When applying liquid–solid phase-change materials to a heat exchanger, the principal challenge consists in how to continuously circulate the phase-change material through the heat-transfer flow loop. Use of a microencapsulated phase-change-material (PCM) slurry has been suggested as a working fluid (45). The microencapsulation technique uses hollow spheres smaller than 1 mm in diameter to encapsulate a phase-change material. The investigators (45–48) reported that the convective heat-transfer coefficient as well as the thermal capacity would increase because of two mechanisms affecting the effective thermal conductivity, k_{eff}, and the effective thermal capacity, C_{eff}.

The effective thermal conductivity of a liquid–solid suspension has been reported to be (46) larger than that of a pure liquid. The phenomenon was attributed to the microconvection around solid particles, resulting in an increased convective heat-transfer coefficient. For example, a 30-fold increase in the effective

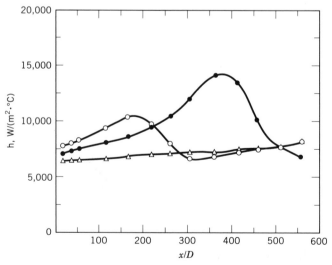

Fig. 17. Heat-transfer coefficient comparisons for the same volumetric flow rates for ($\triangle$) water, 6.29 kW, and a phase-change-material slurry; ($\circ$), 10% mixture, 12.30 kW; and ($\bullet$), 10% mixture, 6.21 kW. The Reynolds number was 13,225 to 17,493 for the case of water.

thermal conductivity and a 10-fold increase in the heat-transfer coefficient were predicted for a 30% suspension of 1-mm particles in a 10-mm diameter pipe at an average velocity of 10 m/s (45).

A manufacturing technology to produce very small encapsulated phase-change materials has been developed (47). These encapsulated phase-change materials were applied in a convective heat-transfer test section, and a 50–100% higher heat-transfer coefficient was reported.

A numerical simulation of laminar flow using a phase-change slurry has been attempted (48). The slurry temperature at the inlet of the heating test section was assumed to be the melting temperature of the particles. The local Nusselt number was maximum at the inlet of the heating test section and decreased along the flow direction, which was the same trend as for single-phase flow. However, the Nusselt number of the phase-change-slurry flow was 2–4 times higher than those of a single-phase flow.

As an alternative technology to produce phase-change-material particles smaller than 0.1 mm in diameter, a method using an emulsifier has been developed (49–51). A 10% suspension of solid hexadecane particles in water had approximately a 13% higher pressure drop than a 10% emulsion of liquid hexadecane in water. The local convective heat-transfer coefficient for a phase-change-material slurry was found to be significantly greater than the value of water, as shown in Figure 17. However, it varied significantly along the heating test section, a phenomenon that made the LMTD method or the effective thermal capacity method difficult to use for the analysis of convective heat transfer with a two-phase liquid–solid mixture.

NOMENCLATURE

Symbol	Parameter	Units
A	heat-transfer surface area, flow area	m^2
AR	area ratio	
C	thermal capacity rate	W/K
C_D	coefficient of discharge for channels	
C_R	ratio of C_{min} to C_{max}	
c_p	specific heat for constant pressure	$J/(kg{\cdot}K)$
C_T	turning loss coefficient	
D	diameter	m
D_c, D_d	width of combining and dividing flow headers	m
D_H	hydraulic or equivalent diameter (eq. 10)	m
F	MTD correction factor	
f	Fanning friction factor	
f_D	Darcy friction factor	
G	mass flux	$kg/(m^2{\cdot}s)$
h	heat-transfer coefficient	$W/(m^2{\cdot}K)$
K	pressure loss coefficient	
k	thermal conductivity	$W/(m{\cdot}K)$
L	length or manifold length	m

LMTD	logarithmic mean temperature difference (eq. 21)	K
$\dot{m}$	mass flow rate	kg/s
N	number of tube rows traversed by fluid	
Nu	Nusselt number	
N_{tu}	number of transfer units	
P	temperature parameter (eq. 27)	
p	pressure	N/m^2
P_m	perimeter or circumference	m
PCM	phase-change material	
Pr	Prandtl number	
Q	heat-transfer rate	W
R	temperature parameter (eq. 28)	
Re	Reynolds number	
r_f	fouling resistance	$m^2 \cdot K/W$
R_{th}	thermal resistance to heat transfer	$m^2 \cdot K/W$
R_w	thermal resistance of wall	$m^2 \cdot K/W$
s_t	transverse tube pitch	m
T	temperature	K
t	thickness	m
U	overall heat-transfer coefficient	$W/(m^2 \cdot K)$
V	flow velocity	m/s
x	length	m
z	elevation	m
ΔA	additional surface area	m^2
Δp	pressure difference or drop	N/m^2
ΔT	temperature difference	K
ΔT_m	mean temperature difference	K
α	thermal diffusivity	m^2/s
ϵ	heat-exchanger effectiveness	
ϵ_w	surface roughness	m
η_f	fin efficiency	
η_t	total efficiency of surface with fins	
μ	viscosity	$N \cdot s/m^2$
ν	kinematic viscosity	m^2/s
ρ	density	kg/m^3
σ	standard deviation	

Subscripts

c	cold or heated fluid, crossflow
eff	effective
f	fin, fouling
h	hot or heating fluid
i	inside
in	inlet
max	maximum
min	minimum
o	outside
out	outlet
s	shellside
t	tubeside
w	wall, window

BIBLIOGRAPHY

"Heat Transfer" in *ECT* 1st ed., Vol. 7, pp. 370–389, by G. T. Skaperdas, The M. W. Kellogg Co.; in *ECT* 2nd ed., Vol. 10, pp. 819–846, by G. T. Skaperdas, The M. W. Kellogg Co., Division of Pullman Inc.; "Heat Transfer" under "Heat Exchange Technology" in *ECT* 3rd ed., Vol. 12, pp. 129–170, by J. P. Fanaritis, J. A. Kwas, and A. T. Chase, Struthers Wells Corp.

1. F. P. Incropera and D. P. DeWitt, *Fundamentals of Heat and Mass Transfer*, 3rd ed., John Wiley & Sons, Inc., New York, 1990.
2. J. P. Holman, *Heat Transfer*, 7th ed., McGraw-Hill Book Co., Inc., New York, 1990.
3. G. Walker, *Industrial Heat Exchangers: A Basic Guide*, 2nd ed., Hemisphere Publishing Corp., New York, 1990.
4. F. Kreith, *Principles of Heat Transfer*, International Educational Publishers, New York, 1973.
5. R. A. Bowman, A. C. Mueller, and W. M. Nagle, *Trans. Am. Soc. Mech. Eng.* **62**, 263, (1940).
6. *Standards of Tubular Exchanger Manufacturers Association*, TEMA, Inc., New York, 1974.
7. A. C. Mueller, in W. M. Rohsenow and J. P. Hartnett, eds., *Handbook of Heat Transfer*, McGraw-Hill Book Co., Inc., New York, 1973, Sec. 18.
8. W. M. Kays and A. L. London, *Compact Heat Exchangers*, McGraw-Hill Book Co., Inc., New York, 1964.
9. J. Bowman and R. Turton, *Chem. Eng.*, 92–99 (July 1990).
10. R. Turton, D. Ferguson, and O. Levenspiel, *Chem. Eng.*, 81–88 (Aug. 1986).
11. S. M. Cho, *J. Heat Trans. Eng.* **8**(2), 63–74 (1987).
12. M. M. AbuRomia, A. W. Chu, and S. M. Cho, ASME Paper No. 76-WA/HT-73, 1976.
13. B. R. Munson, D. F. Young, T. H. Okiishi, *Fundamentals of Fluid Mechanics*, John Wiley & Sons., Inc., New York, 1994.
14. F. White, *Fluid Mechanics*, 2nd ed., McGraw-Hill Book Co., Inc., New York, 1986.
15. D. A. Donohue, *Industrial and Engineering Chemistry,* **41**, 2499 (1949).
16. Ref. 3, p. 96.
17. H. Martin, *Heat Exchangers*, Hemisphere Publishing Corp., New York, 1992.
18. E. A. D. Saunders, *Heat Exchangers: Selection, Design, and Construction*, John Wiley & Sons, Inc., New York, 1988.
19. A. P. Fraas, *Heat Exchanger Design*, 2nd ed., John Wiley & Sons., Inc., New York, 1989.
20. H. R. Jacobs, J. P. Hartnett, *Thermal Sciences: Emerging Technologies and Critical Phenomena*, NSF Thermal Science Workshop, Chicago, April 18–21, 1991.
21. B. K. Hodge, *Analysis and Design of Energy Systems*, 2nd ed., Prentice-Hall, Inc., Englewood Cliffs, N.J., 1990.
22. R. B. Bird, R. C. Armstrong, and O. Hassager, *Dynamics of Polymeric Liquids*, John Wiley & Sons, Inc., New York, 1987.
23. J. P. Hartnett and M. Kostic, *Int. J. Heat Mass Trans.* **28**, 1147–1155 (1985).
24. C. Xie and J. P. Harnett, *Int. J. Heat Mass Trans.* **35** 641–648 (1992).
25. S. Shin, W. K. Gingrich, Y. I. Cho, and W. Shyy, *Int. J. Heat Mass Trans.* **36**, 4365–4373 (1993).
26. J. S. McNown, *Am. Soc. Chem. Eng. Trans.* **119**, 1103 (1954).
27. T. Kubo and T. Ueda, *Bull. of Jpn. Soc. Mech. Eng.* 12, 802, 1969.
28. R. A. Bajura and E. H. Jones, Jr., *J. Fluids Eng.* **98**, 654 (1976).
29. R. L. Pigford, M. Ashraf, and Y. D. Miron, *Ind. Eng. Chem. Fundam.* **22**, 463 (1983).
30. P. I. Shen, *J. Fluids Eng.* **114**, 121 (1992).
31. A. Acrivos, B. D. Babcock, and R. L. Pigford, *Chem. Eng. Sci.* **10**, 112 (1959).

32. A. K. Majumdar. *Appl. Mathemat. Modeling* **4**, 424 (1980).
33. J. Berlamont and A. Van der Beken, *ASCE J. Hydraulics Div.*, 1531 (1973).
34. A. B. Datta and A. K. Majumdar, *Int. J. Heat Fluid Flow* **2**, 253 (1980).
35. A. K. Singhal, L. W. Keeton, A. K. Majundar, and T. Mukerjee, *An Improved Mathematical Formulation for the Computations of Flow Distributions in Manifolds for Compact Heat Exchangers*, paper presented at The ASME Winter Annual Meeting, Anaheim, Calif., 1986, p. 105.
36. M. Permutter, *ASME J. Basic Eng.* **83**, 361 (1961).
37. C. Pisoni and L. Tagliafico, *Warme Stoffubertragung* **24**, 97 (1989).
38. S. H. Choi, S. Shin, and Y. I. Cho, *Int. Comm. Heat Mass Trans.* **20**, 221–234 (1993).
39. M. K. Bassiouny and H. Martin, *Chem. Eng. Sci.* **39**, 693 (1984).
40. A. B. Datta and A. K. Majumdar, *Int. J. Heat Mass Trans.* **26**, 1321 (1983).
41. R. A. Bajura, *ASME J. Eng. Power* **93**, 7 (1971).
42. F. A. Zenz, *Hydrocarbon Process. Pet. Refiner* **41**, 125 (1962).
43. J. D. Keller, *ASME J. Appl. Mech.*, 71, 77 (1949).
44. C. Cleary and co-workers, *Hydraulic characteristics of ice slurry and chilled water flow*. Advanced Energy Transmission Fluids–Final Report of Research, IEA District Heating, 1990.
45. K. E. Kasza and M. M. Chen, *J. Solar Energy Eng.* **107** 229–236 (1985).
46. C. W. Sohn and M. M. Chen, *J. Heat Trans.* **106**, 539–542 (1984).
47. D. P. Colvin, J. C. Mulligan, and Y. G. Bryant, *Enhanced heat transfer in environmental systems using microencapsulated phase change materials*, 22nd International Conference on Environmental Systems, 1992. U.S. Pat. 4,807,696, D. P. Colvin and J. C. Mulligan (1989).
48. P. Charunyakorn, S. Sengupta, and S. K. Roy, *Int. J. Heat Mass Trans.* **34**, 819–833 (1991).
49. E. Choi, Y. I. Cho, and H. G. Lorsch, *Int. J. Heat Mass Trans.* **37,** 207–215 (1994).
50. E. Choi, Y. I. Cho, and H. G. Lorsch, *Int. Comm. Heat Mass Trans.* **19**, 1–15 (1992).
51. E. Choi, Y. I. Cho, and H. G. Lorsch, "Two-Phase Flow in Energy Exchange Systems," ASME Publication. HTD, Vol. 220, 1992, pp. 45–50.

YOUNG I. CHO
Drexel University

S. M. CHO
Foster Wheeler Energy Corporation

HEAT-TRANSFER MEDIA OTHER THAN WATER

Water, in any of its phases, is an excellent heat-transfer medium. High rates of heat transfer and high latent heat of vaporization make water attractive when condensing steam (qv) for heating or boiling water when cooling. Liquid water affords high rates of heat transfer and resultant economical pumping costs. Steam is also used to remove or to add sensible heat when the steam is superheated. Water is relatively inexpensive, thermally stable, nontoxic, and nonflammable, and is frequently the heat-transfer medium of choice. There are, however, definite limitations. For example, water freezes at 0°C and the critical temperature, T_c, is 374.1°C. Additionally, the vapor pressure is relatively high and the saturation pressure at T_c is 22.10 MPa (218.2 atm). At temperatures above 100°C, the saturation pressure of water may require the use of expensive equipment to contain the heat-transfer medium. The economics of using water as a heat-transfer medium must be evaluated for each installation. At temperatures below 0°C or above 200°C, heat-transfer media other than water often are the more optimum choice.

Cooling high temperature streams using typical cooling waters is not economically attractive. In addition to the obvious disadvantage of the loss of the high level heat, there is also the possible corrosion and fouling of hot surfaces by cooling waters unless expensive water-treatment methods are employed. Ambient air frequently is used as a coolant to avoid corrosion problems when high level energy cannot be recovered. Warm streams cannot be cooled below the summer design temperatures of cooling water systems.

Glycol/water, typically ethylene glycol or propylene glycol solutions, are widely used for liquid-phase secondary cooling and heating applications (see GLYCOLS). Using appropriate inhibitors, the glycol-based fluids can be used over a temperature range of -50 to 175°C. When mixed with water, glycols are not flammable in concentrations up to 80 vol % glycol. Undiluted glycols have flash points not far removed from 100°C, and it is therefore possible to ignite the pure glycols if glycol concentration is increased above 80 vol %. Glycol/water solutions intended for heat-transfer systems are manufactured by Union Carbide Corp. and Dow Chemical (1,2).

High Level Heat-Transfer Media

The ideal high level heat-transfer medium would have excellent heat-transfer capability over a wide temperature range, be low in cost, noncorrosive to common materials of construction, nonflammable, ecologically safe, and thermally stable. It also would remain liquid at winter ambient temperatures and afford high rates of heat transfer. In practice, the value of a heat-transfer medium depends on several factors: its physical properties in relation to system efficiency; its thermal stability at the service temperature; its adaptability to various systems; and certain of its physical properties.

Physical properties of heat-transfer media that are important, in addition to the thermophysical properties include vapor pressure, low temperature pumpability, freezing or pour point, flash point, fire point, and autoignition temperature. The heat-transfer medium must exhibit sufficient thermal stability at the service temperature and high enough flash and fire points to permit safe operation. Most high level heat-transfer fluids are used at temperatures above the flash and fire points under proper protection from flames and arcs, but are not used above their autoignition temperatures. Other factors that must be considered include ease of reprocessing fluid that experiences thermal degradation and ease of monitoring the system to detect the presence of decomposition products or contaminants. Compatibility with process fluids and the ability to resist damage from nuclear radiation also may be important.

The highest economical service temperature of a fluid may depend on whether the fluid is being used for heating or cooling. Thermal stability at the service temperature must permit a long operating life so that system failures and costly equipment shutdowns are limited. A heating fluid normally reaches its highest temperature in a fired or electric heater or vaporizer. The important criterion is the maximum film (or wall) temperature to which the fluid can be economically subjected. A coolant fluid reaches its maximum temperature in a heat exchanger or reactor, both of which frequently result in more moderate heat fluxes than do fired heaters. High heat fluxes are, however, possible when cooling. When the coolant does not undergo a phase change, conditions associated with control at reduced rates of heat removal may establish the maximum film temperature to which the fluid is subjected. This may result in a bulk fluid temperature for the application that is lower than that which is practical when the fluid is used as a heating medium. Vapor-phase systems are sometimes preferred in order to avoid this condition.

Several generalizations can be made concerning thermal stability and degradation of organic heat-transfer media. (1) Aromatic materials exhibit thermal stabilities that generally are superior to aliphatic compounds. (2) The recommended maximum operating temperature for commercially available products is a rough measure of relative thermal stability. (3) Polymer formation is detrimental. Polymers increase fluid viscosity and promote carbonization which leads to fouling. However, none of the heat-transfer fluids noted herein exhibit exothermic polymerization unless contaminated with oxygen, organic material, or a polymerization catalyst. (4) Fluid degradation should produce a minimum of volatile materials. Volatile components increase operating losses and may present a safety hazard. The formation of volatile materials can be handled by proper design of the expansion-tank venting system. (5) Degradation should not produce reactive or corrosive materials. (6) Oxidation stability may be an important factor if air is present at high temperatures. Low insoluble sludge formation is an advantage.

Vapor-Phase and Liquid-Phase Operation

When establishing whether liquid-phase or vapor-phase systems are better, it is necessary to consider the overall process and economics, the thermal tolerance of the process, and the required equipment. In many cases, the costs for the two systems do not differ significantly. In vapor-phase systems, heat is transferred at the saturation temperature of the vapor, which affords uniform and precisely controlled temperatures. In liquid-phase systems, the temperature of the fluid necessarily changes as heat is transferred, therefore, temperatures are not uniform even if large circulation rates are employed for the heat-transfer fluid. In systems having multiple heat users, a combination of both vapor and liquid phase may be preferred. For small, compact systems, natural-convection vapor-phase systems generally are preferred. Electrically heated packaged units are commonly used for small liquid-phase systems. For large systems, heat losses from fluid piping may be greater for vapor-phase systems. Larger vapor-phase systems frequently require forced circulation condensate return when there are several users at different temperature levels.

Advantages of liquid-phase systems over vapor-phase systems are (1) no condensate return system is required, an important factor when there are multiple users operating at widely different temperatures; (2) simple and more easily operable systems when heating and cooling must be alternated; (3) there is no temperature gradient as a result of pressure drop in the supply piping; (4) liquid systems afford a positive flow through equipment and minimize problems associated with improper venting and natural-convection vaporizer tube burnout; (5) liquid phase eliminates the problems associated with condensate removal from complex geometries; and (6) liquid-phase systems generally have less mechanical leakage.

Advantages of vapor-phase systems over liquid-phase systems are (1) vapor-phase systems provide much more heat per unit mass of heat-transfer fluid; (2) condensing or boiling affords more uniform heat removal or addition and more precise temperature control; (3) vapor-phase heat transfer has an advantage when using equipment that does not permit easy control of liquid flow pattern and velocity; (4) natural circulation systems can be employed, thereby obviating pumps; (5) vapor systems require lower working inventories of the heat-transfer fluid; and (6) vapor systems frequently permit higher rates of heat transfer.

Heat-Transfer Fluids

Petroleum Oils. The most widely used heat-transfer medium at temperature levels above that obtained with moderate pressure steam is a high boiling petroleum fraction. Several oils are used. In general these are safe, essentially nontoxic, relatively low cost, noncorrosive fluids that have been refined to standard physical property specifications for heat-transfer service. These oils are flammable and a compromise must be established between flash point and viscosity. Usually, it is not practical to select an oil having a flash point above the maximum operating temperature of the system. To obtain a high flash point entails the use of a heavier stock having a resultant decrease in heat-transfer efficiency. The poorer thermophysical properties of the fluid reduce the rate at which heat can be transferred. If oils having high flash points leak from the system, insulation fires can occur.

Petroleum oils are subject to two kinds of deterioration: oxidation and thermal cracking. Petroleum oil at temperatures above 200°C and in contact with air or oxygen is oxidized at relatively high rates. Oxidation results in the buildup of organic acids and the formation of insoluble materials or sludge. Sludge causes the fluid viscosity to increase. Further, the sludge is deposited on the heat-transfer surfaces, thereby reducing the rate of heat transfer. Petroleum oils generally produce more oxidation sludge than do synthetic fluids. Many heat-transfer oils are compounded with an inhibitor to reduce oxidative deterioration (see ANTIOXIDANTS). However, closed systems in which oxygen is excluded from contact with the hot oil are recommended. Thermal cracking occurs when petroleum oils are exposed to high temperatures. Thermal cracking results in the formation of new materials, some of which are light, relatively volatile products. The more volatile products lower the flash point of the oil; the heavier and more viscous products

Table 1. Commercially Available Heat-Transfer Fluids

Fluid	Chemical composition	Temperature range, °C Min	Temperature range, °C Max	Viscosity,[a] mPa·s(= cP)	Vapor pressure,[b] kPA[c]	Pour point, °C	Flash point, °C	Fire point, °C	AIT,[d] °C
Petroleum oils									
Mobiltherm 603	paraffinic oil	40	290	20	21	-7	170		350
Caloria HT 43	paraffinic oil	40	315	5	80	-9	204		354
Thermia Oil C	paraffinic oil	40	290	230	5	-12	235		
Calflo FG	paraffinic oil	40	260	15	9	-21	194	204	285
Calflo AF	paraffinic oil	40	290	25	10	-18	204	225	343
Calflo HTF	paraffinic oil	40	325	35	15	-16	212	239	355
Multitherm PG-1	mineral oil	65	315	7	32	-40	171	196	366
Multitherm IG-2	paraffinic oil	65	315	10	7	-18	227	260	371
Multitherm 503	1-decene dimer	-50	260	400	20	-60	160	174	324
Paratherm NF	mineral oil	65	315	7	32	-40	174	196	366
Paratherm HE	paraffinic oil	65	315	10	6	-15	227	260	371
Therminol HFP	paraffinic oil	4	300	300	5	-15	227	263	385
Therminol XP	mineral oil	-15	315	1200	32	-29	182	196	324
Synthetic fluids									
Tetralin	hydronaphthalene	40	310	5	675	-32.5	77	90	384
UCON HTF-500	polyalkylene glycol	40	260	50	0	-37	244	316	415
Dowtherm A	diphenyl/diphenyl oxide	40	400	2.5	1050	12[e]	116	135	621
Dowtherm G	aryl ethers	0	370	75	365	4	141	146	584
Dowtherm LF	alkylated aromatic	-40	340	400	410	-40	115	125	467
Dowtherm J	alkylated aromatic	-70	315	10	1205	-73	56	68	420
Dowtherm Q	alkylated aromatic	0	330	10	330	-40	120	124	411
Dowtherm HT	hydrogenated terphenyls	10	340	250	110	0	179	191	350
Therminol 55	alkylated aromatic	-20	290	1150	26	-40	177	210	357
Therminol 59	alkylated aromatic	-45	315	1000	161	-61	146	154	404
Therminol 60	polyaromatic mixture	-50	315	2100	165	-68	154	160	446
Therminol 66	hydrogenated terphenyls	0	340	500	100	-26	177	193	374

Therminol 75	alkyl polyphenyls	160	400	2	130	70	199	227	538
Therminol LT	alkylated aromatic	−70	315	10	1510	−75	57	66	429
Therminol D-12	synthetic hydrocarbon	−45	260	10	395	−95	59	71	277
Therminol VP-1	diphenyl/diphenyl oxide	40	400	2.5	1065	12[e]	124	127	621
Marlotherm S	dibenzylbenzenes	−15	350	1000	315	−35	190	235	500
Marlotherm L	benzyl toluenes	−50	350	200	4200	−70	120	145	500
Thermalane L	synthetic paraffin	−45	260	100	90	−84	165	183	332
Thermalane 600	synthetic paraffin	−15	300	500	7	−65	240	271	377
Thermalane 800	synthetic paraffin	−15	325	250	30	−73	229	257	377
Syltherm 800	dimethylsiloxane polymer	−40	400	50	1360	−60[e]	177	193	385
Syltherm XLT	polydimethylsiloxane	−70	260	12	525	−105	47	54	350
Hitec Salt	nitrates and nitrites	150	540	20	535	145[e]			

[a]At minimum temperature.
[b]At maximum operating temperature.
[c]To convert kPa to psi, multiply by 0.145.
[d]AIT = autoignition temperature.
[e]Melting point.

reduce the rate at which heat is transferred. The less volatile products also increase fouling of heat-transfer surfaces.

Heat-transfer oils that are offered by most suppliers resist thermal cracking and chemical oxidation. Generally, these are intended for use in closed systems, and, because vapor pressures are low, are limited mostly to liquid-phase systems having forced circulation. Most have viscosities that permit pumping readily at both starting and operating temperatures. When applied according to the recommended guidelines, these fluids provide long service life and afford good heat-transfer performance in a wide range of applications. Properly designed systems which are well maintained and operated permit maximum film (or wall) temperatures approximately 20°C higher than the recommended maximum operating temperature for the fluids.

Several companies offer oils for heat-transfer service. Physical characteristics are summarized in Table 1. The oils discussed herein are widely used. Product brochures on the fluids are available from the manufacturers (3–12).

Mobiltherm 603. Mobiltherm 603, manufactured by Mobil Oil Corp., is a high paraffinic oil suitable for systems in which combined heating and cooling cycles are used. It functions efficiently at both low and high temperatures and withstands repeated thermal cycling.

Caloria HT 43. Caloria HT 43, manufactured by Exxon Co., is a paraffinic-based oil that is compounded with an oxidation inhibitor.

Thermia Oil C. Thermia Oil C, Shell Oil Co., is a selected mineral-oil fraction containing appropriate antioxidants.

Petro-Canada Oils. Petro-Canada manufactures three oils specially refined for use as heat-transfer fluids. Calflo FG is a semisynthetic, paraffinic heat-transfer fluid specifically developed for use in systems where incidental contact with food may result. Calflo AF is a saturated paraffinic oil containing inhibitors to minimize oxidation. Calflo HTF is a saturated paraffinic oil inhibited to minimize oxidation.

Monsanto Chemical Company Oils. Monsanto Chemical Co. manufactures two oils for heat-transfer applications. Therminol HFP is a solvent refined paraffinic oil; Therminol XP is a clear white mineral oil essentially identical to Multitherm PG-1 and Paratherm NF.

Multitherm Corporation. Multitherm Corp. manufactures two oils for heat-transfer applications: Multitherm PG-1, a clear white mineral oil essentially identical to Therminol XP and Paratherm NF, and Multitherm IG-2, a solvent refined paraffinic oil essentially identical to Paratherm HE.

Paratherm Corporation. Paratherm Corp. manufactures two oils for heat-transfer applications: Paratherm NF, a clear white mineral oil essentially identical to Therminol XP and Multitherm PG-1, and Paratherm HE, a solvent refined paraffinic oil essentially identical to Multitherm IG-2.

Synthetic Fluids. Petroleum oils are products of nature and this is reflected in their cost. There are some serious limitations imposed by the chemical characteristics of natural products. These have led to the development of synthetic heat-transfer fluids that supplement natural products and permit more efficient operation at both lower and higher operating temperatures than can generally be obtained using natural products. These synthetic fluids also are subject to thermal cracking and chemical oxidation, and systems should be designed so that oxygen and other contaminants do not contact the fluids. Maximum fluid film and bulk temperatures should not exceed those recommended by the manufacturer in order to achieve long operating life. Properly designed systems which are well maintained and operated permit film (or wall) temperatures approximately 20°C greater than the recommended maximum operating temperature for the fluids.

Synthetic fluids are safe, noncorrosive, essentially nontoxic, and thermally stable when operated under conditions recommended by the manufacturers. Generally, these fluids are more expensive than petroleum oils, but the synthetics can usually be reprocessed to remove degradation products. There are several classes of chemicals offered permitting a wide temperature range of application. Any heat-transfer fluid in use should be examined periodically to monitor degradation or contamination.

The manufacturers of synthetic fluids offer technical service and consultation, and fluid reprocessing service can be arranged between the supplier and the user. Complete physical properties and detailed information concerning synthetic fluids are reported in the manufacturers' product literature (13–36). The physical characteristics of the synthetic fluids can be found in Table 1.

Tetralin.   Tetralin is a trade name of Du Pont for 1,2,3,4-tetrahydronaphthalene [*119-64-2*], $C_{10}H_{12}$. Tetralin, a derivative of naphthalene, is made by hydrogenating one ring completely and leaving the other unchanged. Tetralin is produced by several manufacturers and is one of the oldest heat-transfer fluids. Tetralin can be used both in liquid- and vapor-phase systems. The normal boiling point is 207°C.

Dowtherm Heat-Transfer Fluids.   Dow Chemical Co. manufactures a family of heat-transfer fluids to meet differing applications. Dow Chemical also markets the Syltherm fluids produced by Dow-Corning Corp. Design and operating guidelines are offered in many of the company publications describing the Dowtherm fluids.

Dowtherm A, an eutectic mixture of 73.5% diphenyl oxide [*101-84-8*], $C_{12}H_{10}O$, and 26.5% diphenyl [*92-52-4*], $C_{12}H_{10}$, is one of the oldest and most widely used synthetic heat-transfer fluids (see BIPHENYL AND TERPHENYLS). It has a characteristic aromatic odor even at low concentrations. Dowtherm A is used in both liquid- and vapor-phase systems. Because of its high freezing point (12°C), it often requires protection against freezing. The normal boiling point is 257°C. The recommended maximum service temperature is 400°C at a pressure of 16 mPa (150 psia); however, higher operating temperatures can be used where the higher decomposition rate can be justified economically. Dowtherm A has been used in applications to 430°C in properly designed systems. Dowtherm A and Therminol VP-1 are essentially identical fluids.

Dowtherm G is a mixture of di- and triaryl compounds and has good flow characteristics at low temperatures. Dowtherm G is highly stable, and the products of decomposition consist of high molecular weight materials which remain in solution in the liquid. Dowtherm G is intended for use in liquid-phase systems. The fluid has a striking odor even at extremely low concentrations.

Dowtherm LF is a mixture of diphenyl oxide and methylated biphenyl for use in liquid-phase systems. The low crystal point and low viscosity obviate protection from freezing at temperatures down to −30°C.

Dowtherm J is a mixture of isomers of an alkylated aromatic that contains only carbon and hydrogen. Dowtherm J can be used in liquid-phase systems at temperatures as low as −73°C and in vapor-phase systems at temperatures from 185 to 315°C. Dowtherm Q is a mixture of diphenylethane and alkylated aromatics intended for liquid-phase systems. It can be used at temperatures as low as −34°C. Dowtherm HT is a mixture of hydrogenated terphenyls intended for liquid-phase systems. Dowtherm HT and Therminol 66 are essentially identical.

Therminol Heat-Transfer Fluids.   Monsanto Chemical Co. manufactures a series of synthetic heat-transfer fluids that offer a wide operating temperature range. All of these fluids, except for Therminol VP-1 and Therminol LT, are intended for liquid-phase operation only. Monsanto also offers the two natural petroleum oil-based heat-transfer fluids discussed.

Therminol 55 is an alkylated aromatic available for use at moderate temperatures. Therminol 59 is an alkyl-substituted aromatic available for both heating and cooling applications. Therminol 60 is a mixture of polyaromatic compounds available for both heating and cooling applications. Therminol 66 is a modified terphenyl intended for liquid-phase systems. Therminol 66 and Dowtherm HT are essentially identical. Therminol 75 is a mixture of terphenyls and quaterphenyls. Therminol 75 was developed for systems operating at temperatures from 370–400°C. At ambient conditions, Therminol 75 is a soft solid material having a melting range of about 40–70°C. Therminol LT is an alkyl substituted aromatic intended for both liquid- and vapor-phase operation. Its normal boiling point is 181°C. Therminol D-12 is a synthetic hydrocarbon intended for liquid-phase systems operating over a wide temperature range developed for systems in which incidental contact with food products may result. Lastly, Therminol VP-1 is essentially identical to Dowtherm A.

Marlotherm Heat-Transfer Fluids.   Two heat-transfer fluids are manufactured by Hüls America: Marlotherm S is a mixture of isomeric dibenzylbenzenes intended for liquid-phase systems, and Marlotherm L is a mixture of benzyl toluenes that are suitable for both liquid- and vapor-phase applications. Marlotherm L can be pumped readily at temperatures as low as −50°C and can be used in vapor-phase systems at temperatures from 290–350°C. The low temperature characteristics of Marlotherm enable it to be used in processes involving both heating and cooling.

Thermalane Heat-Transfer Fluids.   Coastal Chemical Co. manufactures three heat-transfer fluids intended for liquid-phase systems. Thermalane L is a synthetic paraffin intended for low temperature applications. Thermalane 600 and Thermalane 800 are synthetic paraffins.

Multitherm 503.   Multitherm 503 is manufactured by Multitherm Corp. and is a synthetic hydrocarbon intended for liquid-phase systems in which both heating and cooling are required. It was developed for systems in which incidental contact with food products may result.

Ucon HTF-500.   Union Carbide Corp. manufactures Ucon HTF-500, a polyalkylene glycol suitable for liquid-phase heat transfer. The fluid exhibits good thermal stability in the recommended temperature range and is inhibited against oxidation. The products of decomposition are soluble and viscosity increases as decomposition proceeds. The vapor pressure of the fluid is negligible and it is not feasible to recover the used fluid by distillation. Also, because the degradation products are soluble in the fluid, it is not possible to remove them by filtration; any spent fluid usually must be burned as fuel or discarded. The fluid is soluble in water.

Syltherm Heat-Transfer Fluids. Dow Corning Corp. manufactures two heat-transfer fluids which are silicone polymers rather than organic fluids and are less susceptible to fouling as a result of fluid degradation (see SILICON COMPOUNDS, SILICONES). These fluids are marketed by Dow Chemical Co. All lines containing Syltherm 800 volatiles must be maintained at temperatures above 65°C because one of the breakdown components of the volatiles freezes at 61°C. This requirement generally requires heat tracing of lines to safety relief devices, instrumentation, and vent systems.

Syltherm XLT is a polydimethylsiloxane intended for liquid-phase systems which operate at low temperatures. Syltherm 800 is a modified dimethylsiloxane polymer intended for liquid-phase systems. The recommended maximum fluid temperature is greater than the autoignition temperature.

Hitec Heat-Transfer Salt. Hitec heat-transfer salt, manufactured by Coastal Chemical Co., is an eutectic mixture of water-soluble inorganic salts: potassium nitrate (53%), sodium nitrite (40%), and sodium nitrate (7%). It is suitable for liquid-phase heat transfer at temperatures of 150–540°C. The melting point of fresh Hitec is 142°C. Hitec heat-transfer salt is very stable but does undergo a slow endothermic breakdown of the nitrite to nitrate, alkali metal oxide, and nitrogen. The nitrogen must be vented from the system. The nitrite also slowly oxidizes to nitrate in the presence of oxygen. Any carbon dioxide present is absorbed to form carbonates which may precipitate. Water vapor also is absorbed to form alkali metal hydroxides. All of these reactions occur accompanied by a rise in the freezing point of the salt. When all of the nitrite changes to nitrate, the freezing point rises to 220°C. Further, the formation of only 0.5 mol % sodium carbonate causes the freezing point to rise to 250°C.

Although Hitec is nonflammable, it is a strong oxidizer and supports the combustion of other materials. Consequently, combustible materials must be excluded from contact with the molten salt. Hitec is compatible with carbon steel at temperatures up to 450°C. At higher temperatures, low alloy or austenitic stainless steel is recommended. Adding water to Hitec does not appreciably alter its corrosion behavior.

Because Hitec is an excellent heat-transfer fluid which does not foul heat-transfer surfaces, American Hydrotherm Corp. developed a proprietary salt dilution technique to counter the disadvantage of high freezing point and to extend the range of application of the molten salt. Controlled dilution of the salt with water permits reduction of its freezing point to any desired level down to ambient temperatures. This extends the range of applicability and enables the fluid to be used in operations where temperature requirements vary widely. The water that is added for dilution gradually is evaporated and removed as the system is heated. As the system is cooled, water is gradually added to prevent freezing of the salt. It is essential that all of the water be removed before heating the mixture to high temperatures in order to avoid steam explosions which result from rapid liberation of steam. If water dilution is not employed, Hitec systems are generally limited to bath types owing to the large volumetric contraction upon freezing.

Other Synthetic Fluids. Other synthetic fluids are used as heat-transfer fluids although most of them are not sold specifically for this purpose. Fluids that sometimes are used include diethylene glycol, triethylene glycol, propylene glycol, butyl carbitol, *para*-cymene, several silanes, several silicone fluids, some silicate fluids, other polyalkylene glycols, other organic ethers, and other molten salts. Fluidized solids also are used as heat-transfer media.

Gases. The common permanent gases can be used as heat-transfer media and are the only substances capable of spanning the entire range of temperatures required in industrial applications. These gases exhibit excellent thermal stability and are relatively easy to handle. Unfortunately, they also exhibit relatively poor heat-transfer characteristics because of thermophysical properties. Gas systems for heat transfer are characterized by low rates of heat transfer, large volumetric flow rates, and high pumping costs. However, most of these disadvantages can be offset by operating at moderate pressures, eg, 2 MPa (20 atm), and by using extended (finned) surfaces for heat transfer. The choice of a particular gaseous medium is frequently a compromise between the use of inexpensive gases, eg, air and flue gases, or the use of commercially prepared gases, eg, nitrogen (qv) and carbon dioxide (qv). Commonly used gases include air, flue gases, nitrogen, carbon dioxide, hydrogen, helium, and argon. Superheated steam is also frequently used as a heat-transfer fluid.

Liquid Metals. Liquid metals (37,38) are used as heat-transfer media at temperature levels as high as can be contained by suitable materials of construction. High rates of heat transfer are achieved with liquid metals, thus they are suitable for operation requiring high heat flux or low temperature differences. Liquid metals offer a broad operating temperature range, ie, the difference between their melting and boiling points ranges from 500–1000°C. Liquid metals also have low vapor pressures and exhibit high thermal stability.

Liquid metals, however, present several disadvantages. Their weights must be considered with regard to equipment design. Additionally, liquid metals are difficult to contain and special pumps must be used for system safety. Alkali metals react violently with water and burn in air. Liquid metals also may become radioactive when used for cooling nuclear reactors (qv).

The most commonly used liquid metal is sodium–potassium eutectic. Sodium, potassium, bismuth, lithium, and other sodium–potassium alloys also are used. Mercury, lead, and lead–bismuth eutectic have also been used; however, these are all highly toxic and application has thus been restricted.

Comparison of Heat-Transfer Fluids. A large number of heat-transfer fluids are available for use at moderately high (100–300°C) temperatures. Several are utilized at temperatures up to, and sometimes exceeding, 400°C. A dozen fluids may fulfill the operating requirements of a specific application. Final fluid selection should be based on safety of the fluid during service, heat-transfer rate, operating pressure drop, and system cost. Table 2 offers a comparison of liquid-phase heat-transfer fluids on the basis of pumping rate, heat-transfer coefficient, frictional pressure drop, and minimum velocity required for turbulent flow. Table 2 shows the relative performance of each fluid as physical properties change with temperature.

The minimum velocity required to maintain fully developed turbulent flow, assumed to occur at Reynolds number (Re) of 8000, is inside a 16-mm inner diameter tube. The physical property contribution to the heat-transfer coefficient inside and outside the tubes are based on the following correlations (39):

inside tubes $h/cG = 0.023\ Re^{-0.2}Pr^{-2/3}$

outside tubes $h/cG = 0.33\ Re^{-0.4}Pr^{-2/3}$

where h is the heat-transfer coefficient, c the specific heat, G the mass flow rate per unit area, and Pr the Prandtl number. These expressions are used for the physical property terms, yielding the following equations:

inside tubes $c^{1/3}k^{2/3}\rho^{0.8}\mu^{-0.467} = (h_i D_i^{0.2})/(0.023 v_i^{0.8})$

outside tubes $c^{1/3}k^{2/3}\rho^{0.6}\mu^{-0.267} = (h_o D_o^{0.4})/(0.33 v_o^{0.6})$

where k is the fluid thermal conductivity; ρ, fluid density; μ, fluid viscosity; D the tube diameter; and v linear velocity. Each value is normalized so that the lowest value of the physical property terms is unity.

Some physical properties, such as heat capacity and thermal conductivity, are difficult to measure accurately at higher temperatures and error as great as 20% are common. For critical applications, consult the heat-transfer fluid manufacturer concerning methods that were employed for these measurements.

The relative pressure drop, ΔP, expression is based on the Fanning pressure-drop equation:

$$\Delta P = (4fL/D)\ (\rho v^2/2)$$

If the friction factor f is assumed to be $\alpha Re^{-0.2}$ where α is a constant, and the Fanning equation is solved for physical property terms, the result is

$$\mu^{0.2}\rho^{0.8} = (\Delta P D^{1.2})/(2L\alpha v^{1.8})$$

For pressure drop inside tubes, α is 0.046 and L is the fluid-flow path length. Across tubes banks, α is 0.75 and L is the product of the number of tube rows and the number of fluid passes across the tube bank. The physical property term is again tabulated after being normalized so that the lowest value is approximately unity.

When the heat duty requirement, Q, is specified and the fluid temperature change, ΔT, is fixed, as a result of operating or equipment limitations, the required volumetric pumping rate from the heat balance is

$$V = (Q/\Delta T)\ (1/c\rho)$$

Table 2. Performance Comparison of Heat-Transfer Fluids

Fluid	Temperature, °C	Minimum velocity, m/s	Pumping rate factor	Pressure drop factor	Heat-transfer factor Outside tubes	Heat-transfer factor Inside tubes
Mobiltherm 603	100	2.18	1.31	2.59	1.29	1.42
	150	1.04	1.26	2.15	1.55	1.98
	200	0.67	1.20	1.90	1.73	2.42
	250	0.46	1.18	1.68	1.88	2.82
	300	0.34	1.16	1.50	2.00	3.22
Caloria HT 43	100	3.56	1.34	2.82	1.00	1.00
	150	1.37	1.28	2.22	1.25	1.52
	200	0.76	1.24	1.89	1.42	1.93
	250	0.52	1.22	1.65	1.51	2.22
	300	0.39	1.21	1.48	1.54	2.40
Thermia Oil C	100	3.55	1.28	2.91	1.11	1.11
	150	1.42	1.23	2.31	1.41	1.66
	200	0.81	1.17	1.98	1.64	2.18
	250	0.50	1.16	1.73	1.82	2.70
	300	0.37	1.16	1.58	1.95	3.07
Calflo FG	100	1.84	1.32	2.48	1.35	1.55
	150	0.98	1.26	2.11	1.60	2.07
	200	0.63	1.21	1.87	1.79	2.54
	250	0.44	1.18	1.69	1.97	3.00
Calflo AF	100	2.84	1.35	2.70	1.20	1.25
	150	1.32	1.26	2.25	1.51	1.80
	200	0.77	1.22	1.96	1.71	2.32
	250	0.51	1.16	1.74	1.91	2.82
	300	0.36	1.11	1.57	2.08	3.29
Calflo HTF	100	3.13	1.28	2.78	1.19	1.22
	150	1.45	1.23	2.31	1.46	1.75
	200	0.84	1.19	2.00	1.69	2.25
	250	0.55	1.15	1.78	1.89	2.73
	300	0.39	1.10	1.62	2.07	3.22
Multitherm PG-1	100	2.14	1.27	2.64	1.16	1.28
	150	0.95	1.18	2.16	1.43	1.86
	200	0.55	1.14	1.85	1.61	2.34
	250	0.39	1.11	1.63	1.72	2.68
	300	0.28	1.11	1.43	1.79	2.98
Multitherm IG-2	100	3.45	1.20	2.83	1.12	1.13
	150	1.44	1.16	2.28	1.39	1.67
	200	0.82	1.12	1.95	1.57	2.11
	250	0.55	1.11	1.72	1.68	2.45
	300	0.45	1.11	1.57	1.69	2.55
Multitherm 503	100	0.88	1.24	1.97	1.71	2.25
	150	0.51	1.21	1.69	1.94	2.86
	200	0.34	1.20	1.48	2.11	3.38
	250	0.25	1.18	1.32	2.16	3.69
Paratherm NF	100	2.14	1.27	2.64	1.16	1.28
	150	0.95	1.18	2.16	1.43	1.86

Table 2. (*Continued*)

Fluid	Temperature, °C	Minimum velocity, m/s	Pumping rate factor	Pressure drop factor	Heat-transfer factor	
					Outside tubes	Inside tubes
	200	0.55	1.14	1.85	1.61	2.34
	250	0.39	1.11	1.63	1.72	2.68
	300	0.28	1.11	1.43	1.79	2.98
Paratherm HE	100	3.45	1.20	2.83	1.12	1.13
	150	1.44	1.16	2.28	1.39	1.67
	200	0.82	1.12	1.95	1.57	2.11
	250	0.55	1.11	1.72	1.68	2.45
	300	0.45	1.11	1.57	1.69	2.55
Tetralin	100	0.41	1.21	2.12	2.02	3.12
	150	0.29	1.21	1.87	2.12	3.51
	200	0.19	1.21	1.61	2.29	4.13
	250	0.14	1.20	1.42	2.41	4.60
	300	0.11	1.17	1.26	2.46	4.95
UCON HTF-500	100	5.74	1.09	3.76	1.29	1.17
	150	2.39	1.05	3.03	1.57	1.69
	200	1.44	1.04	2.59	1.69	2.02
	250	0.89	1.00	2.29	1.77	2.34
Dowtherm A	100	3.75	1.28	2.37	1.88	2.78
	150	1.23	1.24	2.07	2.08	3.38
	200	0.65	1.21	1.84	2.21	3.94
	250	0.42	1.26	1.59	2.20	3.98
	300	0.30	1.20	1.50	2.25	4.28
	350	0.24	1.21	1.36	2.16	4.26
	400	0.20	1.25	1.22	2.05	4.30
Dowtherm G	100	1.12	1.24	2.90	1.56	1.96
	150	0.62	1.21	2.47	1.80	2.55
	200	0.39	1.18	2.16	1.97	3.08
	250	0.27	1.23	1.92	2.07	3.47
	300	0.19	1.16	1.72	2.20	3.93
	350	0.15	1.16	1.55	2.24	4.23
Dowtherm LF	100	0.57	1.23	2.32	1.86	2.69
	150	0.36	1.21	2.03	2.04	3.23
	200	0.23	1.16	1.81	2.13	3.63
	250	0.20	1.11	1.75	2.18	3.88
	300	0.17	1.09	1.51	2.15	3.97
Dowtherm J	100	0.25	1.35	1.66	2.21	3.73
	150	0.20	1.31	1.49	2.33	4.10
	200	0.17	1.29	1.36	2.37	4.34
	250	0.16	1.29	1.23	2.38	4.52
	300	0.15	1.31	1.10	2.35	4.45
Dowtherm Q	100	0.56	1.19	2.21	1.79	2.59
	150	0.33	1.31	1.90	1.96	3.25
	200	0.22	1.23	1.67	2.06	3.57
	250	0.17	1.20	1.54	2.13	3.91
	300	0.14	1.20	1.39	2.12	4.03

Table 2. (*Continued*)

Fluid	Temperature, °C	Minimum velocity, m/s	Pumping rate factor	Pressure drop factor	Heat-transfer factor Outside tubes	Heat-transfer factor Inside tubes
Dowtherm HT	100	1.88	1.33	2.92	1.25	1.42
	150	0.84	1.25	2.39	1.54	2.06
	200	0.50	1.18	2.07	1.79	2.65
	250	0.34	1.13	1.84	1.98	3.12
	300	0.26	1.09	1.67	2.11	3.56
	350	0.22	1.07	1.57	2.18	3.80
Therminol 55	100	1.88	1.24	2.56	1.29	1.46
	150	0.86	1.20	2.11	1.55	2.06
	200	0.52	1.17	1.82	1.73	2.54
	250	0.35	1.17	1.60	1.84	2.91
	300	0.26	1.18	1.47	1.88	3.13
Therminol 59	100	0.73	1.18	2.35	1.59	2.18
	150	0.43	1.22	2.02	1.80	2.75
	200	0.29	1.19	1.79	1.95	3.22
	250	0.22	1.16	1.60	2.04	3.57
	300	0.17	1.15	1.44	2.08	3.83
Therminol 60	100	0.82	1.26	2.59	1.62	2.17
	150	0.48	1.21	2.13	1.85	2.76
	200	0.32	1.17	1.89	2.00	3.22
	250	0.25	1.14	1.71	2.07	3.54
	300	0.20	1.13	1.56	2.08	3.73
Therminol 66	100	1.73	1.28	2.90	1.22	1.46
	150	0.80	1.21	2.39	1.56	2.08
	200	0.48	1.16	2.08	1.74	2.59
	250	0.34	1.12	1.86	1.87	2.98
	300	0.27	1.09	1.69	1.93	3.23
	350	0.23	1.08	1.56	1.95	2.36
Therminol 75	100	3.75	1.23	3.73	1.14	1.13
	150	1.23	1.18	2.88	1.53	1.90
	200	0.65	1.14	2.44	1.80	2.54
	250	0.42	1.12	2.15	1.99	3.05
	300	0.30	1.10	1.93	2.11	3.45
	350	0.24	1.09	1.76	2.15	3.64
	400	0.20	1.09	1.61	2.14	3.81
Therminol LT	100	0.25	1.36	1.64	2.03	3.47
	150	0.18	1.32	1.45	2.07	3.78
	200	0.14	1.31	1.30	2.07	3.95
	250	0.12	1.32	1.15	2.00	3.98
	300	0.10	1.35	1.00	1.87	3.82
Therminol D-12	100	0.33	1.31	1.53	1.74	2.81
	150	0.22	1.29	1.34	1.80	3.14
	200	0.17	1.29	1.17	1.79	3.30
	250	0.14	1.32	1.03	1.67	3.19
Therminol HFP	100	3.45	1.20	2.83	1.18	1.13
	150	1.44	1.16	2.28	1.39	1.67
	200	0.82	1.12	1.95	1.57	2.11
	250	0.54	1.11	1.72	1.68	2.45
	300	0.45	1.11	1.57	1.69	2.55

Table 2. (*Continued*)

Fluid	Temperature, °C	Minimum velocity, m/s	Pumping rate factor	Pressure drop factor	Heat-transfer factor	
					Outside tubes	Inside tubes
Therminol XP	100	2.14	1.27	2.64	1.16	1.28
	150	0.95	1.18	2.16	1.43	1.86
	200	0.55	1.14	1.85	1.61	2.34
	250	0.39	1.11	1.63	1.72	2.68
	300	0.28	1.11	1.43	1.70	2.98
Therminol VP-1	100	3.75	1.28	2.37	1.88	2.78
	150	1.23	1.24	2.07	2.08	3.39
	200	0.65	1.21	1.84	2.21	3.94
	250	0.42	1.26	1.59	2.20	3.98
	300	0.30	1.20	1.50	2.25	4.28
	350	0.24	1.21	1.36	2.16	4.26
	400	0.20	1.24	1.22	2.05	4.30
Marlotherm S	100	1.51	1.27	2.88	1.37	1.63
	150	0.81	1.20	2.45	1.60	2.15
	200	0.49	1.13	2.13	1.82	2.71
	250	0.34	1.09	1.91	1.96	3.13
	300	0.26	1.06	1.73	2.06	3.60
	350	0.21	1.04	1.59	3.11	3.71
Marlotherm L	100	0.55	1.30	2.24	1.76	2.56
	150	0.36	1.24	1.97	1.94	3.16
	200	0.26	1.20	1.76	2.07	3.49
	250	0.21	1.17	1.62	2.10	3.68
	300	0.18	1.15	1.49	2.11	3.85
	350	0.16	1.13	1.39	2.09	3.88
Thermalane L	100	0.82	1.30	1.94	1.71	2.30
	150	0.48	1.29	1.66	1.96	2.91
	200	0.31	1.20	1.44	2.15	3.49
	250	0.24	1.18	1.29	2.26	3.85
Thermalane 600	100	0.75	1.19	2.45	1.72	2.35
	150	0.42	1.19	2.08	1.92	2.94
	200	0.25	1.20	1.79	2.09	3.54
	250	0.19	1.20	1.59	2.15	3.87
	300	0.16	1.21	1.45	2.11	3.93
Thermalane 800	100	1.94	1.20	2.37	1.44	1.63
	150	0.96	1.18	1.97	1.71	2.22
	200	0.60	1.17	1.70	1.89	2.70
	250	0.46	1.16	1.55	1.98	2.98
	300	0.35	1.16	1.39	2.06	3.28
Syltherm XLT	100	0.37	1.59	1.72	1.63	2.55
	150	0.26	1.60	1.49	1.74	2.94
	200	0.21	1.65	1.29	1.80	3.18
	250	0.18	1.75	1.12	1.80	3.27
Syltherm 800	100	1.76	1.50	2.64	1.22	1.41
	150	1.05	1.51	2.26	1.33	1.69
	200	0.69	1.53	1.96	1.39	1.94
	250	0.48	1.56	1.71	1.43	2.13
	300	0.36	1.62	1.49	1.43	2.26
	350	0.28	1.70	1.29	1.38	2.30
	400	0.24	1.82	1.13	1.30	2.23

Table 2. (*Continued*)

Fluid	Temperature, °C	Minimum velocity, m/s	Pumping rate factor	Pressure drop factor	Heat-transfer factor	
					Outside tubes	Inside tubes
Hitec Salt	150	4.39	0.69	7.21	3.66	3.51
	200	1.97	0.70	6.05	4.51	5.08
	250	1.25	0.71	5.44	5.06	6.24
	300	0.90	0.72	4.98	5.49	7.23
	350	0.64	0.76	4.55	6.42	8.34
	400	0.48	0.80	4.21	6.30	9.00
	450	0.41	0.82	3.98	6.30	9.67
	500	0.39	0.83	3.39	6.33	9.88
water	100	0.15	0.56	1.80	10.38	19.44
	150	0.10	0.57	1.58	11.53	23.82
	200	0.08	0.57	1.43	12.36	26.73
	250	0.07	0.59	1.28	11.66	26.14
	300	0.07	0.54	1.15	11.28	24.36

Table 2 tabulates the physical property contribution of the required volumetric pumping rate ($1/c\rho$) normalized to assign the value of unity to the lowest data point.

The relative energy requirement is the product of pressure drop and volumetric flow rate when comparing fluids at the same velocity. Fluid performance comparisons on the basis of constant pumping energy or constant power require the application of a multiplier. Consider two fluids, X and Y, having relative heat-transfer coefficients h_x and h_y and relative pressure drops ΔP_x and ΔP_y. If the fluids are operating under the same energy requirement, the coefficient ratio for fluids X and Y (h_x/h_y) should be multiplied by the factor $(\Delta P_y/\Delta P_x)^\beta$. Using the correlations previously presented, β should be 0.8/2.8 for fluids inside tubes and 0.6/2.8 for fluids outside tubes.

The equations presented herein do not include any viscosity correction to reflect the difference between the viscosity at the wall temperature and the bulk fluid temperature. This effect is generally negligible, except at low temperatures for organic fluids having viscosities that are strongly temperature dependent. For such conditions, the values tabulated in Table 2 should be appropriately modified.

Low Level Heat-Transfer Media

Refrigeration is required to cool to temperatures lower than those attainable using cooling water or ambient air. Low temperature processing also requires a suitable low temperature fluid. Several fluids are used for both of these types of service. There are several types of refrigeration systems, each of which requires a suitable working fluid or refrigerant. Refrigerants absorb heat not wanted or needed and reject it elsewhere (see REFRIGERATION AND REFRIGERANTS). Some refrigeration systems use only gases, eg, air. In others, heat is removed from the system by evaporation of a liquid refrigerant and is rejected by condensation of the refrigerant vapor. This evaporation–condensation process occurs in absorption refrigeration systems as well as in mechanical compression and steam-jet refrigeration systems.

Gas-Cycle Systems. In principle, any permanent gas can be used for the closed gas-cycle refrigeration system; however, the prevailing gas that is used is air. In the gas-cycle system operating on the Brayton cycle, all of the heat-transfer operations involve only sensible heat of the gas. Efficiencies are low because of the large volume of gas that must be handled for a relatively small refrigeration effect. The advantage of air is that it is safe and inexpensive.

Steam-Jet Systems. Low pressure water vapor can be compressed by high pressure steam in a steam jet. In this way, a vacuum can be created over water with resultant evaporation and cooling; water, therefore, serves as a refrigerant. This method frequently is used where moderate cooling (down to 2°C) is needed. The process is inefficient and usually is economically justified only when waste steam is available for the motive fluid in the steam jet.

Absorption Systems. Absorption refrigeration cycles employ a secondary fluid, the absorbent, to absorb the primary fluid, refrigerant vapor, which has been vaporized in the evaporator. The two materials that serve as the refrigerant–absorbent pair must meet a number of requirements; however, only two have found extensive commercial use: ammonia–water and water–lithium bromide.

Water–lithium bromide systems cannot be used for low temperature refrigeration because the refrigerant turns to ice at 0°C. Lithium bromide crystallizes at moderate concentrations and therefore usually is limited to applications in which the absorber is cooled with cooling water. Other disadvantages are associated with the low pressure required and with the high viscosities of the lithium bromide solution. The pair does offer the advantages of safety and stability and affords a high latent heat of vaporization. The system has wide application in air-conditioning (qv).

Ammonia–water systems are more complex than water–lithium bromide systems, but can be used at temperatures down to −40°C. Ammonia–water systems operate under moderate pressures and care must be taken to avoid leaks of the irritating and toxic ammonia (qv). Sometimes a third material with a widely different density, eg, hydrogen, is added to the cycle in order to eliminate the need for mechanical pumping.

Mechanical Compression Systems. The equipment in a refrigeration system serves only to provide the refrigerant in the liquid state at the place where cooling is desired. Because evaporation of the liquid is the only step in the refrigeration cycle that produces cooling, the properties of the refrigerant should permit high rates of heat transfer and minimize the volume of vapor to be compressed. High rates of heat transfer also should be afforded at the condenser, where heat must be rejected. Generally, the selection of a refrigerant is a compromise between conflicting requirements and often is influenced by properties not directly related to its ability to transfer heat. For example, flammability, toxicity, environmental effects, density, molecular weight, availability, cost, corrosion, electrical characteristics, freezing point, and the critical properties are often important factors in selection of a refrigerant.

Several types of fluids are used as refrigerants in mechanical compression systems: ammonia, halocarbon compounds, hydrocarbons, carbon dioxide, sulfur dioxide, and cryogenic fluids. A wide temperature range therefore is afforded. These fluids boil and condense isothermally. The optimum temperature or pressure at which each can be used can be determined from the economics of the system. The optimum refrigerant can be determined only for the specific refrigeration requirements needed. A discussion of refrigerants and a compilation of their physical properties has been reported (40).

Secondary Coolants.   In many refrigeration applications, heat is transferred to a secondary coolant which is in turn cooled by the refrigerant. The secondary fluid may be any liquid that transfers heat without a change in its state. Secondary coolants, often known as brines (40), are frequently mixtures of water and an appropriate material that can form eutectic mixtures. Secondary coolants may be used where building codes restrict the use of certain refrigerants, in order to minimize the frequency and cost of leakage, where the users of refrigeration cannot be grouped close to the central refrigeration system, and where several users must be controlled at different temperature levels. The use of secondary coolants reduces the efficiency of the refrigeration system and often increases the investment and operating costs.

Secondary coolants frequently are called brines because such fluids originally were mixtures of salts and water. Common refrigeration brines are water solutions of calcium chloride or sodium chloride. These brines must be inhibited against corrosion.

Organic fluids also are mixed with water to serve as secondary coolants. The most commonly used fluid is ethylene glycol. Others include propylene glycol, methanol (qv), ethanol, glycerol (qv), and 2-propanol (see PROPYL ALCOHOLS, ISOPROPYL ALCOHOL). These solutions must also be inhibited against corrosion. Some of these, particularly methanol, may form flammable vapor concentrations at high temperatures.

Water mixtures of salts or organic fluids offer a range of operating temperature levels down to $-60°C$. Heat-transfer characteristics vary with concentration and temperature. The viscosities of all of the brines are relatively high at the temperatures frequently encountered in refrigeration systems. Because of the relatively high viscosities of brines, many of the common refrigerants are sometimes employed as secondary coolants. Economic system design usually requires viscosities less than 10 mPa·s(=cP) for the secondary coolants. Halocarbons are the most commonly used because of their nonflammability, but ozone depletion is an increasing concern with these fluids. Some of the synthetic heat-transfer fluids such as Dowtherm J, Therminol LF, and Syltherm XLT, are often used as the secondary fluid of choice because these offer excellent heat-transfer properties and viscosity remains low (ca 10 mPa·s(=cP)) at $-70°C$.

Heat Pumps.   Heat pumps involve the application of external power to pump heat from a lower temperature to a higher temperature. Heat pumps are frequently used for space heating and are simply refrigeration cycles operated in reverse. The heat rejected in the condenser becomes the primary objective of operation. Consequently, refrigerants used for mechanical compression refrigeration have similar application in heat pumps.

Heat pumps also have application in processing plants. Frequently, a refrigerant is evaporated in the condenser of a distillation column and the refrigerant vapor is compressed to a level at which the vapor condenses in the reboiler of the same column, thereby providing the heat needed to operate the distillation process. Sometimes the vapor from the column serves as the refrigerant if the vapor properties are similar to those of the optimum refrigerant. Heat pumps applied to distillation (qv) processes have been economical for low temperature processing. They have been less frequently applied when processing at moderate temperatures. Changing energy economics require that this approach be examined periodically.

Thermal Engine Cycles. Thermal engine cycles operating with organic refrigerants are employed to recover energy from waste heat streams at temperatures below 150°C. Recovery of such heat is justified only when recovery cannot be effected through process-oriented heat utilization. Typical systems employ the Rankine cycle to produce electrical or shaft power. Thermal efficiencies of Rankine cycles are low, but thermal engines are applied when large quantities of waste heat are available. The most frequently used refrigerants are halocarbons and hydrocarbons. Mixed refrigerants that evaporate and condense over a selected temperature range are used when the available heat source undergoes a wide temperature change as heat is removed. Cascaded systems, ie, thermal engine cycles in series operating at different temperatures, sometimes are employed to increase thermal efficiencies. In such systems, each thermal engine cycle may employ a different refrigerant.

NOMENCLATURE

	Parameter	Units
c	specific heat	J/(kg·K)
D	tube diameter	m
f	friction factor	
G	mass flow rate per unit area	$(v\rho)$, kg/(m^2·s)
h	heat-transfer coefficient	W/(m^2·K)
k	fluid thermal conductivity	W/(m·K)
L	length	m
Pr	Prandtl number	$c\mu/k$
Q	heat-transfer rate	W
Re	Reynolds number	DG/μ
V	volumetric flow rate	m^3/s
v	linear velocity	m/s
X	arbitrary fluid	
Y	arbitrary fluid	
α	constant in friction factor equation	
β	exponent for correction factor multiplier	
ΔP	fluid pressure drop	
ΔT	fluid temperature change	°C
ρ	fluid density	kg/m^3
μ	fluid viscosity	Pa·s

Subscripts

i	inside
o	outside
x	fluid X
y	fluid Y

BIBLIOGRAPHY

"Heat-Transfer Media Other than Water" in *ECT* 1st ed., Vol. 7, pp. 390–397, by W. J. Danziger, M. W. Kellogg Co.; in *ECT* 2nd ed., Vol. 10, pp. 846–862, by W. J. Danziger, M. W. Kellogg Co., Div. of Pullman Inc.; under "Heat Exchange Technology" in *ECT* 3rd ed., Vol. 12, pp. 171–191, by P. E. Minton and C. A. Plants, Union Carbide Corp.

1. *Glycols*, Union Carbide Corp., Danbury, Conn.
2. *UCAR Thermofluids*, Union Carbide Corp., Danbury, Conn.
3. *Heating With Mobiltherm*, Mobil Oil Corp., New York.
4. *Mobil Product Data Sheet, Mobiltherm*, Mobil Oil Corp., New York.
5. *Caloria HT 43*, Exxon Co., Houston, Tex.
6. *Thermia Oils Technical Bulletin*, Shell Oil Co., Houston, Tex.
7. *Calflo Heat Transfer Fluids*, Petro-Canada Products, Atlanta, Ga.
8. *Multitherm Heat Transfer Fluids*, MultiTherm Corp., Colwyn, Pa.
9. *Paratherm Heat Transfer Fluids*, Paratherm Corp., Conshohocken, Pa.
10. *Therminol XP*, Monsanto Chemical Co., St. Louis, Mo.
11. *Therminol HFP*, Monsanto Chemical Co., St. Louis, Mo.
12. P. L. Gehringer, *Handbook of Heat Transfer Media*, Reinhold Publishing Co., New York, 1962.
13. *Dowtherm A Heat Transfer Fluid*, Dow Chemical Co., Midland, Mich.
14. *Dowtherm G Heat Transfer Fluid*, Dow Chemical Co., Midland, Mich.
15. *Dowtherm HT Heat Transfer Fluid*, Dow Chemical Co., Midland, Mich.
16. *Dowtherm J Heat Transfer Fluid*, Dow Chemical Co., Midland, Mich.
17. *Dowtherm LF Heat Transfer Fluid*, Dow Chemical Co., Midland, Mich.
18. *Dowtherm Q Heat Transfer Fluid*, Dow Chemical Co., Midland, Mich.
19. *Equipment for Systems Using Dowtherm Heat Transfer Media*, Dow Chemical Co., Midland, Mich.
20. *Health, Environmental, and Safety Considerations for High Temperature Organic Heat Transfer Systems*, Dow Chemical Co., Midland, Mich.
21. *Design and Operational Considerations for High Temperature Organic Heat Transfer Systems*, Dow Chemical Co., Midland, Mich.
22. *Therminol Heat Transfer Fluids*, Monsanto Chemical Co., St. Louis, Mo.
23. *Therminol 55 Heat Transfer Fluid*, Monsanto Chemical Co., St. Louis, Mo.
24. *Therminol 59 Heat Transfer Fluid*, Monsanto Chemical Co., St. Louis, Mo.
25. *Therminol 60 Heat Transfer Fluid*, Monsanto Chemical Co., St. Louis, Mo.
26. *Therminol 66 Heat Transfer Fluid*, Monsanto Chemical Co., St. Louis, Mo.
27. *Therminol 75 Heat Transfer Fluid*, Monsanto Chemical Co., St. Louis, Mo.
28. *Therminol D-12 Heat Transfer Fluid*, Monsanto Chemical Co., St. Louis, Mo.
29. *Therminol LT Heat Transfer Fluid*, Monsanto Chemical Co., St. Louis, Mo.
30. *Therminol VP-1 Heat Transfer Fluid*, Monsanto Chemical Co., St. Louis, Mo.
31. *Marlotherm Heat Transfer Media For a Wide Range of Temperature*, Huls America, Inc., Piscataway, N.J.
32. *Thermalane Heat Transfer Fluids*, Coastal Chemical Co., Houston, Tex.

33. *UCON Heat Transfer Fluid 500*, Union Carbide Corp., Danbury, Conn.
34. *Syltherm Heat Transfer Liquids*, Dow Corning Corp., Midland, Mich.
35. *HITEC Heat Transfer Salt*, Coastal Chemical Co., Houston, Tex.
36. *Hydrotherm Molten Salt Heat Transfer System*, American Hydrotherm Corp., New York.
37. R. N. Lyon, ed., *Liquid Metals Handbook*, Atomic Energy Commission and Department of the Navy, Washington, D.C., 1952.
38. C. B. Jackson, ed., *Liquid Metals Handbook, NaK Supplement*, Atomic Energy Commission and Department of the Navy, Washington, D.C., 1955.
39. P. E. Minton, in *Encyclopedia of Chemical Processing and Design*, Vol. 25, Marcel Dekker, Inc., New York, pp. 190–299.
40. *ASHRAE Handbook of Fundamentals*, American Society of Heating Refrigerating, and Air-Conditioning Engineers, New York.
41. R. L. Green, A. H. Larsen, and A. C. Pauls, "Get Fluent About Heat Transfer Fluids," *Chem. Eng.* (Feb. 1989).

PAUL E. MINTON
Union Carbide Corporation

HEAT PIPES

Heat pipes are used to perform several important heat-transfer roles in the chemical and closely allied industries. Examples include heat recovery, the isothermalizing of processes, and spot cooling in the molding of plastics. In its simplest form the heat pipe possesses the property of extremely high thermal conductance, often several hundred times that of metals. As a result, the heat pipe can produce nearly isothermal conditions making an almost ideal heat-transfer element. In another form the heat pipe can provide positive, rapid, and precise control of temperature under conditions that vary with respect to time.

The heat pipe is self-contained, has no mechanical moving parts, and requires no external power other than the heat that flows through it. The heat pipe, which has been called a thermal superconductor, was described initially in 1944 (1) but commercial use did not follow. The same basic structure was again described in 1963 in conjunction with the space nuclear power program (2).

Principles of Operation

The heat pipe achieves its high performance through the process of vapor state heat transfer. A volatile liquid employed as the heat-transfer medium absorbs its latent heat of vaporization in the evaporator (input) area. The vapor thus formed moves to the heat output area, where condensation takes place. Energy is stored in the vapor at the input and released at the condenser. The liquid is selected to have a substantial vapor pressure, generally greater than 2.7 kPa (20 mm Hg), at the minimum desired operating temperature. The highest possible latent heat of vaporization is desirable to achieve maximum heat transfer and temperature uniformity with minimum vapor mass flow.

When an atom or molecule receives sufficient thermal energy to escape from a liquid surface, it carries with it the heat of vaporization at the temperature at which evaporation took place. Condensation (return to the liquid state accompanied by the release of the latent heat of vaporization) occurs upon contact with any surface that is at a temperature below the evaporation temperature. Condensation occurs preferentially at all points that are at temperatures below that of the evaporator, and the temperatures of the condenser areas increase until they approach the evaporator temperature. There is a tendency for isothermal operation and a high effective thermal conductance. The steam-heating system for a building is an example of this widely employed process.

The unique aspect of the heat pipe lies in the means of returning the condensed working fluid from the heat output area, or condenser, to the heat input end, or evaporator. Condensate return is accomplished by means of a specially designed wick. The surface tension of the liquid is the active force that produces wick pumping, which is a familiar process in lamp wicks and sponges. Using proper design, a substantial flow rate can be sustained against the pressure head of the counterflowing vapor or even against a slight gravitational head. In those applications where the heat source is below the heat sink, the condensate returns by gravity, ie, without the wick (see also EVAPORATION).

The heat pipe consists, then, of the following components: a closed, evacuated chamber (evacuation is required to establish a contaminant-free system and to prevent air or other gases from interfering with the desired vapor flow), a wick structure of appropriate design, and a thermodynamic working fluid having a substantial vapor pressure at the desired operating temperature. A schematic drawing of an elemental heat pipe is shown in Figure 1. The following basic condition must be satisfied for proper operation:

$$\Delta P_c > \Delta P_l + \Delta P_v + \Delta P_g \qquad (1)$$

that is, for liquid return, the pressure difference owing to capillarity, ΔP_c, must exceed the sum of the opposing evaporator-to-condenser pressure differential in the vapor, ΔP_v, plus the pressure differential in the liquid caused by gravity, ΔP_g, plus that caused by frictional losses, ΔP_l. Under this condition, there is liquid flow toward the evaporator, and heat can be transferred. The pressure difference in the vapor is a direct function of the mass flow rate and an inverse function of the

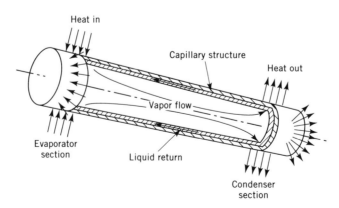

Fig. 1. Cutaway view of a heat pipe.

cross-sectional area of the vapor space. The mass flow rate is related directly to the transferred power and inversely to the latent heat of vaporization. The gravitational head is the elevation of the evaporator with respect to the condenser. It can be either positive or negative, depending on whether it aids or opposes the desired flow in the wick.

Design Features

The heat pipe has properties of interest to equipment designers. One is the tendency to assume a nearly isothermal condition while carrying useful quantities of thermal power. A typical heat pipe may require as little as one thousandth the temperature differential needed by a copper rod to transfer a given amount of power between two points. For example, when a heat pipe and a copper rod of the same diameter and length are heated to the same input temperature (ca 750°C) and allowed to dissipate the power in the air by radiation and natural convection, the temperature differential along the rod is 27°C and the power flow is 75 W. The heat pipe temperature differential was less than 1°C; the power was 300 W. That is, the ratio of effective thermal conductance is ca 1200:1.

A second property, closely related to the first, is the ability of the heat pipe to effect heat-flux transformation. As long as the total heat flow is in equilibrium, the fluid streams connecting the evaporating and condensing regions essentially are unaffected by the local power densities in these two regions. Thus the heat pipe can accommodate a high evaporative power density coupled with a low condensing power density, or vice versa. It is common in heat-transfer applications for the intrinsic power densities of heat sources and heat sinks to be unequal. This condition may force undesired performance compromises on the equipment or process in question. The heat pipe can be used to accomplish the desired matching of power densities by simply adjusting the input and output areas in accordance with the requirements. Heat flux transformation ratios exceeding 12:1 have been demonstrated in both directions, ie, concentration and dispersion of power density. It is not uncommon in chemical applications for flame heat sources to be employed to establish desired reaction temperatures and rates. The natural power density from the flame can be appreciably greater than that desired locally within the reaction vessel. A heat pipe can collect the power at high density from the flame and distribute it at low density over large areas within the vessel.

The third characteristic of interest grows directly from the first, ie, the high thermal conductance of the heat pipe can make possible the physical separation of the heat source and the heat consumer (heat sink). Heat pipes >100 m in length have been constructed and shown to behave predictably (3). Separation of source and sink is especially important in those applications in which chemical incompatibilities exist. For example, it may be necessary to inject heat into a reaction vessel. The lowest cost source of heat may be combustion of hydrocarbon fuels. However, contact with an open flame or with the combustion products might jeopardize the desired reaction process. In such a case it might be feasible to carry heat from the flame through the wall of the reaction vessel by use of a heat pipe.

The fourth characteristic, temperature flattening, makes use of all three of the preceding properties. The evaporation region of a heat pipe can be regarded as consisting of many subelements, each receiving heat and an influx of liquid working fluid and each evaporating this fluid at a rate proportional to its power input. Within the limitations discussed in the following sections, each incremental unit of evaporation area operates independently of the others, except that all are fed to a common vapor stream at a nearly common temperature and pressure. The temperature of the elements is, therefore, nearly uniform. It can be seen that the power input to a given incremental area can differ widely from that received by other such areas. Under other circumstances, a nonuniform power profile would produce a nonuniform temperature profile. In the case of the heat pipe, however, uniformity of temperature is preserved; only the local evaporation rate changes. In this fashion the heat pipe can flatten the very nonuniform power input profile from a flame, delivering heat to the sink with the same degree of uniformity as if the heat source were uniform. Another example is the use of a heat pipe to cool simultaneously, and to nearly the same temperature, a number of electronic components operating at different power levels.

A modified version of the basic heat pipe (4) has a series of unique properties of considerable value in the regulation of temperature and heat flow. This device, known as the gas-controlled or variable-conductance heat pipe, operates so that its access to the heat sink varies in proportion to changes in power input, while preserving its operating temperature at a very nearly constant value. Changes of power input by a factor exceeding 30:1 have been recorded with a change in temperature of less than 1°C. This extremely precise temperature regulation is accomplished through simple principles and without resort to external sensing and control mechanisms. The operation is as follows: the heat pipe vessel is extended to include a volume of inert gas at a predetermined pressure (Fig. 2). The effect upon the heat pipe of this gas pressure is similar to the effect on the boiling point of water of the ambient air pressure, ie, the operating temperature is established as the point on the fluid vapor pressure–temperature curve where the vapor pressure equals the gas pressure. During heat pipe operation, the kinetic energy of the highly directional vapor flow sweeps the gas to the condenser of the heat pipe. The gas and vapor remain highly segregated as long as the mean free path of a vapor molecule in the gas is short, corresponding to a pressure of ca $\geq$20 kPa ($\geq$0.2 atm). Under these conditions, the gas–vapor interface is extremely sharp and heat pipe action ceases beyond this point. The location of the interface is indicated by an abrupt drop in temperature.

As the heat input to such a gas-controlled, constant-temperature heat pipe is increased, the operating temperature tends to remain constant because the location of the interface moves so as to expose to the vapor an increased access area to the heat sink. The degree of temperature control is determined by the

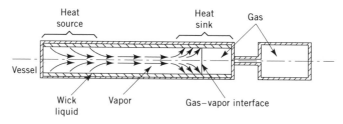

Fig. 2. Schematic diagram of a gas-controlled heat pipe.

ratio of the total gas volume to the displaced gas volume. This volume need not be large to effect precise temperature control because the temperature is a very slow function of the fluid's vapor pressure. A device of this type provides similar regulation under conditions where the heat-sink properties vary with time. It also starts quickly and smoothly from a cold, frozen condition under which a conventional heat pipe might stall. The control point of a gas-controlled heat pipe can be varied with the pressure of the gas. Devices of this type have been used for measuring vapor pressures, regulating the temperature of semiconductors (qv), and establishing thermal control of orbiting spacecraft.

Operational Limits

The wick has a finite pumping capacity for returning the condensed working fluid from condenser to evaporator against a frictional or gravitational head. The total thermal power transfer capability of the heat pipe is the product of the latent heat of vaporization and maximum mass flow rate of the fluid that can be sustained by the wick. Operation at greater power produces complete evaporation of the returning fluid before it reaches the end of the heat pipe. The resulting dryness can lead to an uncontrolled rise in temperature in the uncooled section of the evaporator, and ultimate failure. The effect of gravity is similar. If the desired operation requires that the liquid flow be upward against gravity or another accelerating force, operation is affected adversely to the degree that it is a function of the lift height, liquid density, and the mass flow rate. Under the zero gravity conditions of space flight, heat pipe operation is unimpaired.

In many applications, especially in the chemical and semiconductor fields, the closest possible approach to isothermal operation may be desired. Under these conditions, the effects of vapor velocity must be considered if the velocity of the vapor exceeds about Mach 0.1, when a noticeable temperature differential shows itself in the heat pipe. If near isothermal operation is desired, designers restrict the vapor velocity to lower levels.

An absolute upper limit on operating temperature exists for any given fluid and vessel combination. This limit is determined by the creep or rupture strength of the vessel, ie, the ability of the vessel to contain the increasing vapor pressure of the working fluid.

Although there are several limits which apply to heat pipe operation, these generally lend themselves to specific design solutions or occur at sufficiently high levels of performance to permit a wide latitude of practical applications. The envelope of these limits is shown generically in Figure 3.

Selection of Materials

Working Fluid. Qualitatively, for high power throughput under typical operating conditions, it is advantageous to have a high latent heat of vaporization, high surface tension, high liquid and vapor densities, and low liquid and vapor viscosities (5). The fluid vapor pressure, a contributor to the vapor density, is one of the fastest changing functions of temperature. The melting point of the fluid is not only important in determining the minimum operating temperature, but also

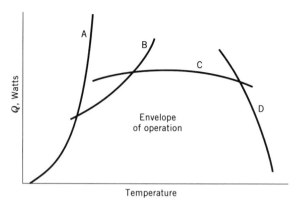

Fig. 3. The throughput, Q, vs operating temperature showing the envelope of heat pipe operating limits, where curve A represents limits associated with vapor flow, ie, either insufficient working fluid vapor pressure is available to transport vapor along the length of the heat pipe (viscous limit) or the vapor flow has reached the sonic velocity (sonic limit). Curve B represents the entrainment limit which occurs when friction with the outgoing vapor prevents the returning liquid from reaching the evaporator. This is sometimes referred to as a flooding limit. Curve C describes a wicking limit and occurs when the capillary pressure developed in the wick can no longer support the total pressure drop in the fluid flow path (includes liquid, vapor, and gravitational effects). Curve D is the boiling limit which occurs when vapor is generated within the capillary structure in an uncontrolled manner, much the same as film boiling.

determination of the start-up and storage characteristics. If solidification of the fluid is expected, care must be taken to avoid stresses, caused by density changes, which may distort the vessel or wick. The relationship of the liquid density to the surface tension is used to determine the lifting height of the fluid in a given wick structure and the extent to which operation can be expected in opposition to an accelerative force such as gravity.

Operating Lifetime. The operating lifetime of a given heat pipe is usually determined by corrosion mechanisms (see CORROSION AND CORROSION CONTROL). The repetitive distillation or refluxing of the working fluid can cause rapid mass transport of dissolved material unless careful attention is paid to compatibility. The result can be solution of the wick or vessel in the condenser region and clogging of the wick in the evaporator. Small quantities of impurities can accelerate the corrosive action in some systems. A number of pairs of materials have long (thousands of hours) undegraded life, when properly processed (Table 1).

Wick. The selection of a suitable wick for a given application involves consideration of its form factor or geometry. The basic material generally is chosen on the basis of the wetting angle and compatibility considerations. The wick and vessel materials are generally the same to minimize electrochemical effects. The factors that must be considered in wick design are often conflicting so that each specific heat pipe requires a separate study to determine the optimum structure. Compromises are made between the small capillary pore size desired for maximum pumping pressure and the large pore size required for minimum viscous drag, especially for very long heat pipes or for those that must pump against gravity.

Table 1. Pairs of Materials Having Long, Undegraded Life

Fluid	Wick-vessel material	Temperature range, K
liquid nitrogen	aluminum	60–100
	304 stainless steel	
Freon 21	aluminum	230–300
	iron	
ammonia	aluminum	200–300
	stainless steel	
	carbon steel	
	nickel	
acetone	aluminum	325–400
	stainless steel	
	copper	
	brass	
	glass	
water	copper	325–550
mercury	304 stainless steel	500–900
rubidium	304 stainless steel	600–1100
	molybdenum	
	niobium	
potassium	304 stainless steel	800–1200
	nickel	
	molybdenum	
	niobium	
sodium	304 stainless steel	900–1500
	316 stainless steel	
	Inconel	
	molybdenum	
	tungsten	
	niobium	
	Hastelloy X	
lithium	molybdenum	1300–1900
	tungsten	
	niobium	
	tungsten–rhenium	
	tantalum	
silver	tungsten	1700–2200
	tantalum	
	tungsten–rhenium	

Several wick structures are in common use. First is a fine-pore (0.14–0.25 mm (100–60 mesh) wire spacing) woven screen which is rolled into an annular structure consisting of one or more wraps inserted into the heat pipe bore. The mesh wick is a satisfactory compromise, in many cases, between cost and performance. Where high heat transfer in a given diameter is of paramount importance, a fine-pore screen is placed over longitudinal slots in the vessel wall. Such a composite structure provides low viscous drag for liquid flow in the channels and a small pore size in the screen for maximum pumping pressure.

Where complex geometries are desired, the wick can be formed by powder metallurgy techniques, ie, a dry powder is sintered in place, often around a central mandrel, which is then removed (see METALLURGY, POWDER). Such wicks can be made with extremely small pore sizes, providing good pumping pressures, but tend to have high viscous drag properties. The drag may be offset by longitudinal liquid passages formed in the metal powder. For heat pipes of considerable length where minimum viscous drag is required, an arterial wick geometry has been employed, ie, a tubular artery, often formed of screen, is attached to the wick which lines the inside walls of the heat pipe. The inside of the artery provides a low drag passage for liquid flow.

The cross-sectional area of the wick is determined by the required liquid flow rate and the specific properties of capillary pressure and viscous drag. The mass flow rate is equal to the desired heat-transfer rate divided by the latent heat of vaporization of the fluid. Thus the transfer of 2260 W requires a liquid (H_2O) flow of 1 cm^3/s at 100°C. Because of porous character, wicks are relatively poor thermal conductors. Radial heat flow through the wick is often the dominant source of temperature loss in a heat pipe; therefore, the wick thickness tends to be constrained and rarely exceeds 3 mm.

Vessel.   The vessel in which a heat pipe is enclosed must be impermeable to assure against loss of the working fluid or leakage into the heat pipe of air combustion gases or other undesired materials from the external environment. In the quiescent, cold state, the heat pipe is evacuated except for the working fluid and is generally under an external atmospheric pressure. As operation is initiated, the vapor pressure of the working fluid rises and offsets the external pressure. Frequently, the heat pipe operates at a vapor pressure exceeding the external pressure. Under these conditions, a heat pipe may be designed to conform with established pressure vessel codes, considering both rupture and creep strengths.

The vessel, as well as the wick, must be compatible with the working fluid. Where possible, the wick and vessel are made of the same material to avoid the formation of galvanic corrosion cells in which the working fluid can serve as the electrolyte. In addition to its role within the heat pipe, the vessel also serves as the interface with the heat source and the heat sink.

Applications in the Chemical Industry

Heat/Solvent Recovery.   The primary application of heat pipes in the chemical industry is for combustion air preheat on various types of process furnaces which simultaneously increases furnace efficiency and throughput and conserves fuel. Advantages include modular design, isothermal tube temperature eliminating cold corner corrosion, high thermal effectiveness, high reliability and options for removable tubes, alternative materials and arrangements, and replacement or add-on sections for increased performance (see FURNACES, FUEL-FIRED).

The principal competing technology is provided by the rotary regenerative heat exchanger. The main advantages of the heat pipe exchanger lie in its ease of cleaning and its ability to sustain a high pressure differential between the air and gas streams without leakage. When compared to rotary regenerative and tubular exchangers, heat pipe units are generally smaller in size than tubular exchangers and somewhat larger than rotary units. Cold spots within the exchanger, and resulting corrosion, are generally less of a problem than with either alternative design. Payback can be attractive. In one case study (6), installation of a heat pipe preheater in a 30-MW utility-scale boiler resulted in an estimated fuel saving of roughly $250,000 and a payback on the equipment cost of slightly more than eight months at a fuel cost of $1.50/GJ (1 GJ $\approx$ 10^6 Btu).

Typically, an array of finned heat pipes is placed so that heat is transferred from a hot exhaust gas stream to an incoming cold air stream. Heat pipe heat exchangers range in size from a few hundred watts for electronic control cabinets to several hundred thousand watts for large-scale preheating of combustion air. A typical exchanger is shown in Figures **4a** and **4b**. Solvent recovery is accomplished when outgoing exhaust gases from chemical processes are cooled below their dew point, forcing condensation of solvents they may contain.

Mold Coolers for Plastic Injection Molding.   Heat pipes are used for local temperature control in the injection molding of plastics (see POLYMER PROCESSING). A heat pipe is often used to force local cooling within a mold to speed

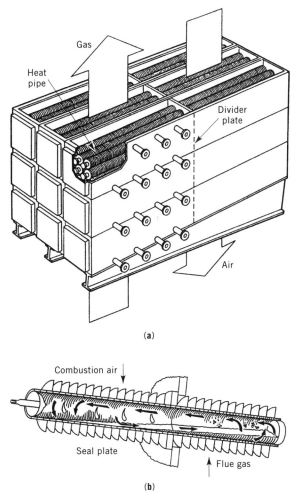

(a)

(b)

Fig. 4.   (**a**) Air preheater using heat pipes; (**b**) typical heat pipe used in air preheater. Courtesy ABB Air Preheater, Inc.

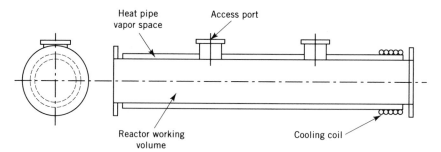

Fig. 5. Isothermal tubeflow reactor. Courtesy Dynatherm Corp.

operation, control viscosity, retention of material in a difficult mold area, or to reduce thermal stresses on cooling.

Chemical Reaction Vessels and Isothermal Furnaces. High temperature heat pipes using liquid metals as working fluids are used to provide a uniform environment for a variety of chemical processes. Isothermal furnace liners, used as inserts in conventional tubular furnaces, are annular heat pipes which receive their heat input along the outer diameter and provide an extremely uniform, ie, near isothermal, temperature environment to the work zone on the inside. Units of this type are used for the growth of semiconductor crystals, vapor deposition, and other processes requiring uniform temperatures. Isothermal chambers ranging up to 1.2 m in diameter and 2.4 m in height have been made, providing temperature uniformly of ± 1°C or better throughout this volume at temperatures in the range of 500–1000°C. Figure 5 is a line drawing of an isothermal reactor.

BIBLIOGRAPHY

"Heat Pipe" in *ECT* 2nd ed., Suppl. Vol, pp. 488–499, by G. Y. Eastman, Radio Corp. of America; under "Heat Transfer Technology" in *ECT* 3rd ed., Vol. 12, pp. 191–202, by G. Y. Eastman and D. M. Ernst, Thermacore, Inc.

1. U.S. Pat. 2,350,348 (June 6, 1944), R. S. Gaugler (to General Motors).
2. G. M. Grover and co-workers, *J. Appl. Phys.* **35**, 1990 (1964).
3. E. D. Waters and co-workers, "The Application of Heat Pipes for the Trans-Alaska Pipeline," *Proceedings of the 10th Intersociety Energy Conversion Engineering Conference,* Newark, Del., 1975.
4. U.S. Pat. 3,613,773 (Oct. 19, 1971), W. B. Hall and F. G. Block (to RCA).
5. G. Y. Eastman and D. M. Ernst, *The Heat Pipe, A Unique and Versatile Device for Heat Transfer Applications,* RCA, Lancaster, Pa., 1966.
6. Technical data, ABB Air Preheater, Wellsville, N.Y.

General References

Reference 2.
T. P. Cotter, *Theory of Heat Pipes, LA-3246-MS,* Los Alamos Scientific Laboratory, University of California, Los Alamos, N.M., 1965.
T. P. Cotter, *Heat Pipe Startup Dynamics, LA-DC-9026,* Los Alamos Scientific Laboratory, University of California, Los Alamos, N.M., 1969.
J. E. Kemme, *Heat Pipe Design Considerations, L-4221-MS,* Los Alamos Scientific Laboratory, University of California, Los Alamos, N.M., 1969.

H. Cheung, *A Critical Review of Heat Pipe Theory and Applications, UCRL-50453*, Lawrence Radiation Laboratory, University of California, Livermore, Calif., 1968.

R. C. Turner, *Feasibility Investigation of the Vapor Chamber Fin, AD839-469*, RCA Electronic Components, Lancaster, Pa., 1968.

G. Y. Eastman, *Sci. Am.* **218**(5), 38 (1968).

G. Y. Eastman, *The Heat Pipe—A Progress Report, ST-4048*, RCA Electronics Components, Lancaster, Pa., 1969.

R. A. Freggens, *Experimental Determination of Wick Properties for Heat Pipe Applications, ST-4086*, RCA Electronics Components, Lancaster, Pa., 1969.

W. L. Haskin, *Cryogenic Heat Pipe, Report AFFDL-TR-025*, Flight Dynamics Lab., Wright Patterson Air Force Base, Ohio, 1967.

P. D. Dunn and D. A. Reay, *Heat Pipes*, Pergamon Press, Inc., New York, 1976.

Heat Pipe Design Handbook, Dynatherm Corp., Cockeysville, Md., 1972.

P. Vinz and C. A. Busse, "Axial Heat Transfer Limits of Cylindrical Sodium Heat Pipes Between 25 W/cm^2 and 15.5 kW/cm^2," *International Heat Pipe Conference*, Stuttgart, Germany, 1973.

C. A. Busse, *Int. Heat Mass Transfer* **16**, 169 (1973).

J. E. Kemme, *Vapor Flow Considerations in Conventional and Gravity-Assist Heat Pipes, LA-UR-75-2308*, Los Alamos Scientific Laboratory, University of California, N.M., 1975.

WALTER B. BIENERT
Dynatherm Corporation

DONALD M. ERNST
Thermacore, Inc.

G. YALE EASTMAN
DTX Corporation

NETWORK SYNTHESIS

The discipline of process synthesis has been used since the 1970s to assist in the systematic design of chemical process plants (see SEPARATION SYSTEMS SYNTHESIS; SIMULATION AND PROCESS DESIGN). Synthesis of heat-exchange (energy) networks has shown large economic rewards and the most promise for future universal use. Figure 1 is a representation of the material, energy, and economic value flows for a process plant identifying the four principal highly interdependent equipment subsystems.

Process design begins with potential product slates, the selection of appropriate raw materials, and a comparison of their relative economic values. A process plant can be viewed as an engine for transformation of the raw materials into products and powered by the flow of energy through it. The required supply and removal of the energy to the process has long been recognized as an important factor in the economic viability of the process. Process methodology, eg, catalytic vs high pressure, can radically alter energy and capital needs. Traditionally, external energy requirements were controlled almost exclusively by altering the process, and heat recovery was included as an addition to the process design that

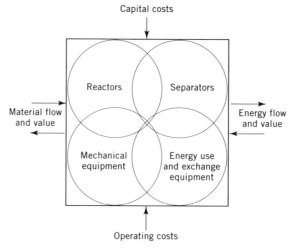

Fig. 1. Subsystems and flows associated with a process plant.

saved heat, ie, energy use and exchange equipment was not viewed as a system that could be optimized.

In every process plant, heat exchangers are used for process–process-stream heat transfer, and heaters or coolers are used for process-utility heat transfer. More sophisticated process design led to more focus on optimizing heat exchanger size as part of the process energy flow. Initial optimizations usually involved only a process–process heat exchanger and a cooler. Economic credit was taken for reduction in operating costs but it was expected that capital equipment cost could only increase. Later it was realized that greater saving could be achieved if whole networks of exchangers could be optimized. Graphical techniques were subsequently developed to visualize the flow of energy through the process. However, most attempts involved the mathematical optimization of equation sets centered on assumed cost functions plus appropriate energy balance and exchanger design equations. Early in the 1970s, a firm thermodynamic foundation was laid for the minimum required energy flow in a process plant. At the same time it was demonstrated that costs for operating and capital savings might both be expected if the pairing and sequencing of process streams for heat exchange was carefully considered. Increased heat utilization does not always mean a trade-off, and many studies have shown a reduction in energy consumption as well as capital cost which in itself is a remarkable recommendation for a systematic approach to network design.

The technique of establishing a lower bound for external utility requirements also reveals significant information about how energy is or can be used in the process. Hence, once the process has been chosen and conditions outlined, the designer can take a focused view of the energy subsystem in the plant without all of the intermediate design work previously required. More recently, these concepts have been extended to include energy flows associated with mechanical heat engines (ie, letdown turbines) present in most plants (see POWER GENERATION).

The high cost of energy mandates carefully planned heat-exchange networks for economically and environmentally viable plants (see HEAT-EXCHANGE TECHNOLOGY, HEAT TRANSFER). The process of developing a good heat-exchange or energy network is most easily viewed as a multiple-tier optimization problem. Minimum lifetime plant cost is the objective. The complexity of the design process leads to multiple optimizations of the energy subsystem. The process includes minimizing capital and operating costs plus an emphasis on developing a robust design that includes the necessary flexibility and operability characteristics.

Design Scheme

The process design engineer must understand and have a plan for using the principles, design rules, and techniques for energy, capital, and operability trade-offs. Figure 2 is a schematic diagram of such a multitiered plan. Iteration with the systematic design of other plant subsystems and between the tiers is used where appropriate; however, any significant change in results from a tier usually requires a completely different solution from the following tiers.

Once the network has been synthesized, traditional design techniques are used to set stream flows through parallel units and analyze individual heat exchangers, heaters, and coolers. Optimization of these individual units is also considered at this point. The optimization of heat exchangers involves such things as baffle and tube sheet layout as well as pressure drop. This multiple tier approach to the problem is necessitated by the complexity of the design process which must span the gulf from conceptual network development to detailed mechanical rating of heat exchangers.

Efficient network synthesis principles and techniques relieve the process engineer of the burden of accepting designs based on art which cannot be shown to be superior. The opportunities for process improvement are best before the structure of the network is determined. Methods for rating individual heat exchangers are well developed and can be used to design new exchangers or simulate the performance of existing units. The overall approach described herein is not limited to new plants but can be used for modification of existing plants as well.

Simulation tools are available for sizing and analyzing plants. However, these tools do not replace the designer as the architect of the plant because selection of process and the sequencing of units are the designers' choices. The same is true for heat-exchanger networks. Most of the commercial process simulator companies market computer modules that perform some of the tedious steps in the process but none is able to remove the designer from the process.

Problem Specification. The problem of heat-exchange network synthesis can be described as follows. A set of cold streams, $i = 1$ to M, initially at supply temperature T_{si} is to be heated to target temperature T_{ti}. Simultaneously, a set of hot streams, $j = 1$ to N, initially at supply temperature T_{sj} is to be cooled to target temperature T_{tj}. Variation in these supply or target temperatures may be permitted in a particular problem if the underlying process conditions are flexible and changes result in significant savings. In addition, some streams may require special alloy heat exchangers or be prohibited from matching with certain other streams. Several hot (heat source) and cold (heat sink) utilities are available for use. The enthalpy vs temperature relationship is known for all these streams. The appropriate physical properties for determining heat-transfer characteristics are also given. The best network of heat exchangers, heaters, and coolers to accomplish the required temperature changes is desired. Appropriate placement of let-

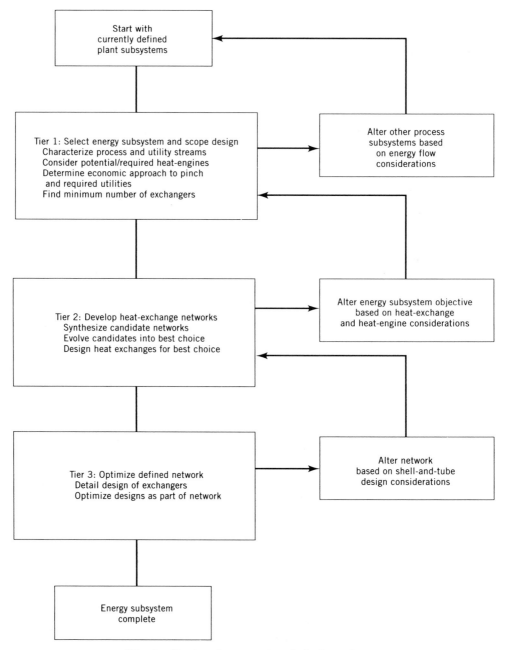

Fig. 2. Heat-exchange network design scheme.

down turbines and similar energy conversion devices is also needed. Best usually means most economic for the capital cost and utility costs available.

Representation. The impetus for heat-exchange network design was the development of an adequate means to represent the problem. The temperature and heat (enthalpy) relationship for any process stream can be represented on a temperature–enthalpy diagram. Figure 3 shows four different streams on such a diagram. It is possible to slide any of the streams horizontally without changing the amount of energy (enthalpy) required for heating or cooling. Vertical changes are not permitted, however, because these alter process temperatures. This temperature–enthalpy domain representation of the network problem dates back to what was called a heat picture (1). This style of representation was expanded and is discussed at length in Reference 2.

An important feature of the temperature–enthalpy representation is that individual streams can be lumped together in an ensemble of single hot and single cold composite streams. The composite representation helps structure the solution to a given heat-exchange network problem. Figure 4a shows two hot streams and two cold streams that are to be considered for a heat-exchange network. A typical composite representation of these streams is shown in Figure 4b. If one composite curve is slid toward the other, the curves usually approach each other at a single pinch point. This is the point where the curves would first touch when slid horizontally. The pinch is analogous to the minimum temperature of approach in the design of an individual heat exchanger. The importance of the temperature pinch point to network synthesis cannot be overstated.

Alternative representations of stream temperature and energy have been proposed. Perhaps the best known is the heat-content diagram, which represents each stream as an area on a graph (3) where the vertical scale is temperature, and the horizontal is heat capacity times flow rate. Sometimes this latter quantity is called capacity rate. The stream area, ie, capacity rate times temperature change, represents the enthalpy change of the stream.

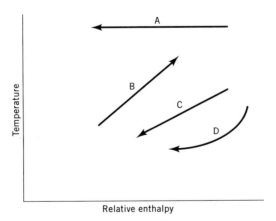

Fig. 3. Temperature–enthalpy representation of stream where A represents a pure component that is condensing, eg, steam; B and C represent streams having constant heat capacity, C_p, that are to be heated or cooled, respectively; and D represents a multicomponent mixture that changes phase as it is cooled.

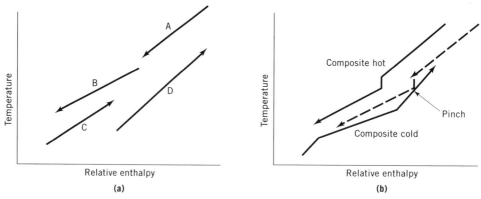

Fig. 4. (**a**) Process streams in heat-exchange network where A and B represent hot streams, C and D cold streams. (**b**) The solid lines represent superstreams, composites constructed from the process streams of (**a**). The dashed line represents the hot stream horizontally repositioned to generate a pinch point. See text.

In addition to a thermodynamic representation of streams, a means of representing the matching of streams for heat transfer is needed. The flowsheet representation has traditionally been used. This diagram shows stream flows as both horizontal and vertical lines intersecting at exchangers or nodes represented as circles (Fig. 5**a**). This topological representation can also be shown by temperature–energy diagrams such as Figures 3 and 4; however, both are somewhat awkward to use. What is needed is a way to represent an evolving network. A most useful representation (4) is a grid method that draws streams as horizontal lines with arrows to indicate heat sources (right-pointing arrow) or cold sinks (left-pointing). Heat exchange between two streams, a match or node, is shown as two circles connected by a vertical line. This diagram gives the essential process topology while permitting the viewer to concentrate on the pairing and sequencing of streams. Figure 5**b** shows the grid representation for a system of five streams with the pinch temperature indicated.

Idealization. The final rating or design of a heat exchanger requires detailed knowledge of the fluids to be heated and cooled as well as information on the specific type of exchanger. The overall heat-transfer coefficient U usually includes a film coefficient for fluid on the outside of the tubes, a film coefficient for fluid inside the tubes, resistance of the tube wall to heat transfer, and a fouling resistance both inside and outside the tube. These coefficients, correlated with fluid properties, streamflow velocities, and exchanger type, are unobtainable until a specific network is totally defined. By then, of course, the network synthesis is complete and the designer must work with a specific network configuration.

For the purpose of network synthesis, the overall heat-transfer coefficient is usually idealized as a constant value. This independence of the heat-transfer coefficient makes possible the iterations necessary to solve the network problem. Usually, the overall heat-transfer coefficient for each exchanger (match) is defined as

$$U = 1/(1/h_i + 1/h_j)$$

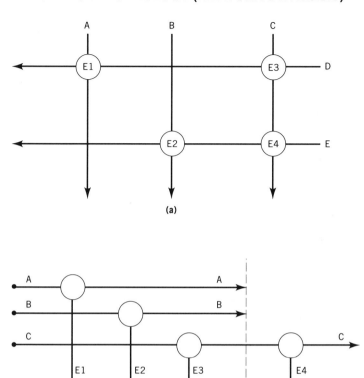

Fig. 5. Heat-exchange network: (**a**) flow representation, and (**b**) grid representation. See text.

where h_i is the coefficient for the cold stream, and h_j is the coefficient for the hot stream. This simple formulation permits quick calculation of the heat-transfer coefficient for any pair of streams. It does not include the practical consideration of exchanger geometry. In fact, exchangers are often considered to be of the double-pipe variety where the overall exchanger area is given simply by

$$A = Q/(U\Delta T_{LM})$$

where A is total heat-transfer area required, Q is the heat transferred in the exchanger, U is the overall heat-transfer coefficient, and ΔT_{LM} is the log–mean–temperature driving force. In industrial practice, more complex exchanger geometries are used. The above formula is usually modified to permit these complex geometries by including a factor, F, to correct ΔT_{LM}:

$$A = Q/(UF\Delta T_{LM})$$

Frequently, the difference in exchanger type does not influence the desired topology to any significant extent. For industrial problems, however, it is necessary to consider individual heat-exchanger shells rather than just the match that is called the heat exchanger. If a high level of heat recovery is desired, the effect of the F factor can be important. This problem has been solved but is beyond the scope of this article.

Limiting Network Conditions

Heat-exchange network design has been made easier by the development of various limiting conditions. The technique of limiting conditions is common in many areas of chemical engineering and can provide great insight into a problem. A notable example is the concept of minimum reflux and minimum stages in distillation. Actual distillations must operate between these limits, and economic designs are frequently based on a reflux that is a fixed percentage above the minimum. A detailed understanding of any model requires a clear view of the model at limiting conditions. A variety of limits has been developed and is exploited in heat-exchange network design.

Feasibility (Heat-Transfer Pinch). A network must be in heat balance, but much more can be said about required heating and cooling utilities. The concept of composite streams permits quick determination of utility requirements. If one composite stream is slid horizontally, it will approach the other (Fig. 6). The limiting condition of maximum energy recovery (MER_+) occurs when the composite streams just touch (zero-temperature driving force) or when the need for either a hot or cold utility is eliminated. If the streams touch (Fig. 6**a**), any possible network having limiting utility rates requires infinite heat-exchange area. On the other hand, if only a heating or cooling utility is required, the limiting case (Fig. 6**b**) might represent a practical solution.

Furthermore, the limiting utility rates for any given temperature of approach at the pinch, ΔT_p, can be found from the temperature–enthalpy diagram. Figure 6**c** shows the composite streams for a problem in which a temperature of approach was specified. The composite streams are slid horizontally until the desired ΔT_p is obtained. The maximum energy recovery for this approach temperature would be $\text{MER}_{\Delta T}$. Minimum heating and cooling requirements are shown. A problem table that is used for feasibility as well as network synthesis may be found in Reference 4. Generalized procedures can take into account restrictions on stream matching and are based on formulating the problem as a classic transportation problem.

A low temperature of approach for the network reduces utilities but raises heat-transfer area requirements. Research has shown that for most of the published problems, utility costs are normally more important than annualized capital costs. For this reason, ΔT_p is chosen early in the network design as part of the first tier of the solution. The temperature of approach, ΔT_p, for the network is not necessarily the same as the minimum temperature of approach, ΔT_e, that should be used for individual exchangers. This difference is significant for industrial problems in which multiple shells may be necessary to exchange the heat required for a given match (5). The economic choice for ΔT depends on whether

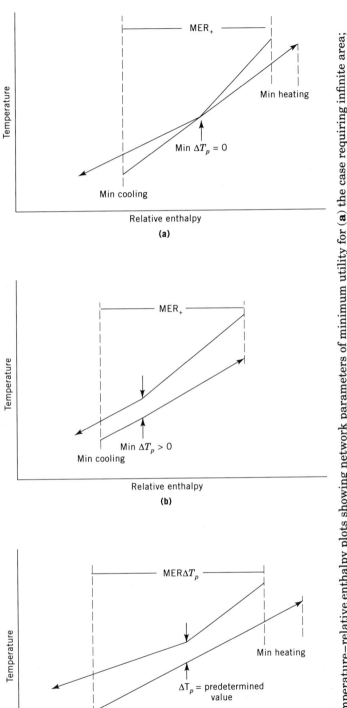

Fig. 6. Temperature–relative enthalpy plots showing network parameters of minimum utility for (**a**) the case requiring infinite area; (**b**) where heat balance gives minimum utility; and (**c**) showing the requirement at ΔT_p. Terms are defined in text.

the process environment is heater- or refrigeration-dependent and on the shape of the composite curves, ie, whether approximately parallel or severely pinched. In crude-oil units, the range of ΔT_p is usually 10–20°C. By definition, $\Delta T_e \geq \Delta T_p$. The best relative value of these temperature differences depends on the particular problem under study.

The concept of a temperature pinch has additional ramifications. If heat passes across the pinch, additional utilities are required (4). This observation can be proven by examining Figure **6a**. Suppose heat from the hot composite stream above the pinch was to be used to heat the cold composite stream below the pinch. This would displace an equal amount of heating from the hot stream below the pinch. The displaced heat would be available at a temperature suitable for heating only below the pinch, and thus additional cooling would be required to dispose of the displaced heat. Additionally, the heat from above the pinch would have to be replaced by a utility. An important consequence of the total energy balance around a network is that any change in the hot-utility rate is exactly reflected in the cold-utility rate. Careless design can easily cause a double penalty in utilities. Insightful design can give a double savings in utilities. This savings obviously reduces operating costs but may reduce capital costs as well, because for every two units of utility heat saved, only one additional unit of process–process heat transfer is needed. The unique feature of this analysis is that the thermodynamic feasibility of constructing a network with any given set of process streams and utilities can be determined without resorting to a detailed examination of possible heat-exchange networks.

Minimum Area.   The limit of minimum network area is presented in References 2 and 3. If idealized double-pipe exchangers are used, a heat-exchange network having minimum area can quickly be developed for any ΔT_p. In the limiting case, where all heat-transfer coefficients are assumed to be equal, the area for this network can easily be obtained from the composite streams by

$$A_{\Delta T_p} = \frac{1}{U} \int_{T_{\text{lowest}}}^{T_{\text{highest}}} \frac{\partial Q}{T_j - T_i}$$

where A is the minimum possible network area at the given ΔT_p, U is the heat-transfer coefficient, and Q is the heat transferred. The vertical distance between the two composite stream curves, $T_j - T_i$, is the driving force at each point along the composite streams. This integration is similar to that used for the design of individual heat exchangers. The minimum-area value is not of great interest in itself; however, it does provide a limiting value much like the Carnot cycle for heat engines. Figure 7 shows the feasible solution space for a system of streams (2). The minimum area required for any network is plotted as a function of utility rates. The temperature of approach ΔT_p is shown at selected points.

Methods have been developed that are able to generate a minimum-area network for any given energy recovery, ΔT_p. The minimum-area networks developed generally employ many heat exchangers and are not an economic solution. These may be used to start an evolutionary process to develop better networks. Rather than minimum total area, the concern should be with the way in which

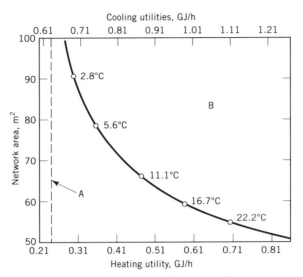

Fig. 7. Heat-exchange network solution space (2), where line A represents the minimum utility for feasibility, ie, infinite area required. Region B is the feasible solution space. The temperatures correspond to ΔT_p values.

the area is distributed among the exchangers. The installed cost of exchangers can best be characterized as

$$\text{cost} = a \cdot A^b + c$$

where a, b, and c are constants that depend on economic conditions and materials of construction. The constant b ranges from 0.5 to 1.0, and c is frequently neglected. Recall that

$$[A_1^b + A_2^b + \cdots + A_E^b] \geq [A_1 + A_2 + \cdots + A_E]^b$$

for $0 \leq b \leq 1$. Therefore, the distribution of area among heat exchangers is most important, and minimum area alone is not of paramount importance.

Minimum Number of Exchangers.   The fewest number of matches or exchangers that are required in a network can be developed as a limit. The number needed, E_{min}, is generally one less than the total number of streams, S (process and utility), involved in the network:

$$E_{min} = S_{process} + S_{utilities} - 1$$

When this equation holds, each stream match (exchanger) must provide that one of the two streams involved reaches its target temperature. Such a network is called acyclic. In an acyclic network, it is not possible to trace a closed path along stream lines from exchanger to exchanger and return to the starting point without retracing some of the path.

There are exceptions to this simple equation that occur infrequently but nevertheless must be considered. A more complete relationship for the number of exchangers, E, in a network is obtained by applying Euler's network relation from graph theory (6):

$$E = S_{\text{process}} + S_{\text{utilities}} - P + L$$

where P is the number of independent heat loads that can be identified. This means P is the number of possible subproblems, each of which are in heat balance. In practice this is almost always one, but if it exceeds one, the minimum number of exchangers is reduced. Cyclic paths in a network are called heat-load loops. Figure 8 is an illustration of a cyclic network ($E1$, $E5$, $E6$, and $E2$). The number of loops or cycles, L, that exist in a network add to the required number of exchangers. If $P = 1$ (single problem) and $L = 0$ (acyclic), the generalized equation reduces to the equation given for E_{min}. It is not always possible to construct an acyclic network, and thus the minimum number of exchangers may be increased above the simple minimum.

Unfortunately, the minimum number of exchangers is not the same as the number of shells required if conventional shell-and-tube heat exchangers are used. If ΔT_p is large, the difference is not significant. However, as more energy recovery is obtained, ΔT_p must be reduced and exchangers near the pinch may require several shells in series to maintain an acceptable F factor. As for the limit of minimum area, the number of exchangers is not of paramount concern. Rather, the number of exchanger shells and the area distribution among the shells are of primary importance. The cost equation given before should be modified as

$$\text{cost} = a \cdot A_{\text{shell}}^b + c$$

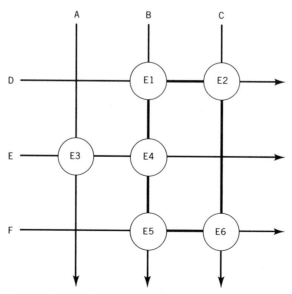

Fig. 8. Heat-exchange network showing heat-load loop.

to emphasize that the area of each shell must be considered when calculating costs. The two temperatures of approach, ΔT_p and ΔT_e, provide a means of including this effect in the multiple tiered optimization process.

Synthesis Algorithms

The synthesis of heat-exchange networks takes place after the feasibility of developing a network has been established for the selected ΔT_p and other constraints. The primary objective is to obtain the lowest cost structure for the utility rates selected. Like many engineering optimization problems, there is not a single distinct solution but rather a set of nearly equal cost solutions. This set of solutions has been called juxtaoptimum (2). The selection of the best solution for a given application can be highly dependent on the difficult-to-quantify factors: operability, controllability, and flexibility. Ideally, synthesis methods would lead to all members of the juxtaoptimum set without wasting time developing poor networks. The designer could then pick a design based on these secondary factors. Differences in costs of a few percent are below the accuracy of the models used and certainly in the range where nonquantifiable network characteristics would be used to make a selection.

Synthesis methods can be classified in many different ways. The categories chosen herein are combinatorial, heuristic, inventive, and evolutionary.

Combinatorial. Combinatorial methods express the synthesis problem as a traditional optimization problem which can only be solved using powerful techniques that have been known for some time. These may use total network cost directly as an objective function but do not exploit the special characteristics of heat-exchange networks in obtaining a solution. Much of the early work in heat-exchange network synthesis was based on exhaustive search or combinatorial development of networks. This work has not proven useful because for only a typical ten-process-stream example problem the alternative sets of feasible matches are ca 1.55×10^{25} without stream splitting.

Heuristic. The heuristic approach is another early synthesis technique based on the application of sets of rules to lead to a specific objective such as network cost. Unfortunately, heuristics cannot guarantee that the objective will be reached although they generally offer quick solutions.

Inventive. Methods that exploit characteristics of heat exchange among groups of hot and cold streams are inventive. Strategies to develop networks that can take advantage of some special characteristic of the problem have been the most successful. One set of characteristics can lead to a minimum network area for a given ΔT_p. Other inventive methods have been able to guarantee networks that recover maximum energy and use the minimum number of heat exchangers. These strategies are far more efficient than any previous methods but do not directly utilize network cost as an objective.

Networks having minimum total area can be quickly and easily developed using the technique of stream splitting to avoid temperature contention (2,3). The difficulty with reaching this subobjective is that the networks usually contain a large number of small heat exchangers. In fact, the number of exchangers is an effective maximum.

A thermodynamic–combinatorial (TC) method follows exactly the opposite tactic with regard to stream splitting (7). The TC method permits construction of all possible networks without stream splitting and using the minimum number of heat exchangers. A prescribed degree of energy recovery, $MER_{\Delta T}$, is specified. The objective to construct all of the networks meeting these criteria without tedious enumeration of many unwanted networks is indeed tantalizing. The method provides for the possibility that no such networks exist for any given problem. The small five-stream problem known as 5SP1 serves as an example of the TC method. The basic stream information for this problem is shown in Figure 9 using grid representation. The heat exchanger shown connecting streams 1 and 4 is compulsory in all networks that meet the stated synthesis criteria. Because a single heating utility is permitted, the minimum number of heat exchangers is five. Figure 10 is a table used to enumerate the possible stream matches (exchangers) for this problem. The potential exchangers and heaters are shown on the grid in the upper part of the figure. The twenty-four possible networks are listed as rows in the table with five Xs indicating the particular matches in each network. The right-hand columns show an N for those matches that would violate any one of three different constraints. The eight selections remaining are indicated with a check mark. When these were examined, three were found to violate the ΔT_p constraint that was applied to each exchanger. The five other networks are shown in Figure 11. Equivalent annual cost was used to select Figure 11**b** as the optimum network, although all five had costs within a 1% band.

The temperature interval (TI) method is an alternative to the TC approach of enumerating all networks having maximum energy recovery and minimum number of heat exchangers without stream splitting (4). The TI method was the first to pursue this network goal and represents a significant advance in the development of networks. The key to the method is the analysis of a number of subnetworks as synthesis tasks. The original problem is partitioned into subnetworks based on ordered temperature levels. Each level is examined individually for potential matches. Heat is not passed from a higher level to a lower one until provision has been made for all of the cold streams in the higher level. In this way, the maximum variety of subnetwork designs is available. The process en-

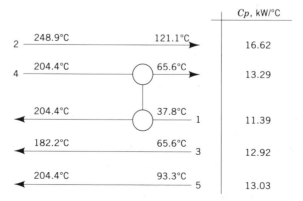

Fig. 9. Streams for problem 5SP1 (7). See text.

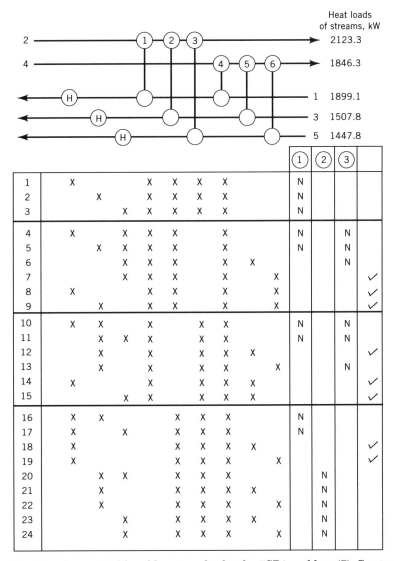

Fig. 10. Development of feasible networks for the 5SP1 problem (7). See text.

gineer selects among these designs to arrive at the final network. Subsequently, an evolutionary technique (ED) can be used to improve the networks generated with the TI method.

The pinch method for network synthesis gives an important insight (8). Unlike the other methods discussed, the pinch method starts the synthesis at the network pinch. The pinch is used to split the synthesis problem into two separate subproblems which are solved starting at the pinch and working toward the opposite temperature end of each subproblem. Starting the design at the pinch has the distinct advantage that it is easy to identify whether or not a stream split is required to meet the chosen energy constraint, ΔT_p, for synthesis of a network.

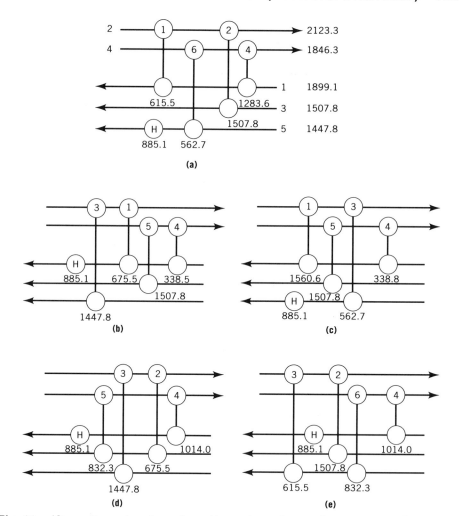

Fig. 11. Alternative network configurations where the numbers not in circles represent heat loads of streams in kW (7). See text.

Although a design-method summary is presented, the detailed implementation is left up to the engineer. The method cannot be classified as algorithmic. A manual design of the network is advocated. In this way, the engineer can probably guide the design to the best network. A very readable account of the whole network development process using the pinch method for network synthesis is given in Reference 9; however, no specific evolutionary rules for use with the pinch method are given.

Evolutionary. The evolutionary approach to network development starts with a good network developed by some alternative means and proceeds to change it into an optimal network from the standpoint of cost. The success of any evolutionary method depends greatly on the starting network. Evolutionary methods are specific to starting network characteristics and are usually matched with the method used to develop the initial network, eg, TI followed by ED. There are two extremes for possible starting networks.

The first is a network that has minimum area but a maximum number of exchangers as proposed by the algorithmic–evolutionary approach (10). The algorithmic part of this method is the development of a minimum area network. The evolutionary part employs a set of rules to modify systematically the initial network. The three rules presented are heuristic in nature and seek to combine exchangers and stream splits to reduce network cost. The problem of reducing stream splits appears difficult to researchers.

An alternative starting network is one without stream splits. The networks from the TI method maximize energy recovery and may introduce heat-load loops. Stream splits are not made in the initial steps of network invention. The ED method is proposed to be one in which heuristic rules and strategies would be used to improve the networks developed by the TI method. The importance of a thermodynamic base for evolutionary rules is stressed in this proposal, but there is no explicit guidance for the evolutionary process.

Some modifications to the TI method and the introduction of a sound loop-breaking strategy that is algorithmic in nature are shown in Reference 11. Also shown is the concept of level of loop to indicate the number of source and sink streams involved in any heat-load loop. Loops can range from first level (one source and one sink stream) to Nth level. The smaller of the number of source or number of sink streams is N. The evolutionary procedure searches for loops from the lowest level first and tries to break (eliminate) them to reduce the number of exchangers; the procedure also introduces stream splits (11). Reference 5 includes the concept of using ΔT_p to find utility rates and thereafter using ΔT_e for network synthesis. This indirectly allows minimization of the number of heat-exchanger shells rather than just the number of exchangers. A case study analysis is necessary to implement this method, which is called the double temperature of approach (DTA).

All these evolutionary methods follow an initial synthesis method. The objective is to develop the best network design prior to the start of detailed and costly calculation of individual network components. In practice, an approximate cost is usually determined for the several candidates before one or more are selected for optimization.

Network Optimization

Process calculations for traditional unit-operations equipment can be divided into two types: design and performance. Sometimes the performance calculation is called a simulation (see SIMULATION AND PROCESS DESIGN). The design calculation is used to roughly size or specify the equipment. Following the design guideline, a particular piece of equipment is chosen. It is then necessary to calculate the performance of the selected item in the service of interest. The problem of choosing the pipe size required to deliver a fixed amount of fluid illustrates this sequence. The design equation gives a pipe diameter for fixed fluid properties, flow rate, and pressure drop. It is not very likely that the diameter determined is one of the standard sizes in which pipe is made. Hence, a standard size having a somewhat larger diameter must be chosen. To determine how much fluid flows through this new pipe, a performance calculation is required. If a pump provides the pressure to drive the fluid through the pipe, an optimization problem exists. Pump-operating costs plus capital cost for the pipe and pump are minimized by proper selection of pipe diameter (see PIPING SYSTEMS; PUMPS).

A corresponding problem occurs in the third tier of heat-exchange network design (Fig. 2). Limits on utilities and number of heat exchangers were set as a first-tier objective. Candidate networks were proposed with the help of synthesis algorithms in the second tier. Where network topology is fixed, the detailed design calculations for individual heat exchangers can proceed. Heat exchangers can be designed as single- or multiple-shell units. A low ΔT for a given exchanger may make it necessary to use several shells in series to improve the F factor. In some cases, large flow rates or area requirements may make it necessary to have several shells in parallel. In any case, the design of the various units in the network can be completed. Subsequently, individual equipment is specified, and a performance calculation is needed for the network. Variation in exchanger sizes and stream flows provides an opportunity for further optimization of the network.

The detailed network calculations are time-consuming, and computer-aided methods are virtually required to evaluate the alternatives in the optimization process. The number of variables to consider is large. Parameters that can be varied include exchanger areas, pressure drops, and detailed mechanical aspects of the individual exchangers. These include such items as baffle cut and spacing, tube diameter, gauge, and pitch. If the network under consideration is in an existing plant, installed exchangers may be selected for use. Network operation (start-up, turndown, and shutdown) should also be of concern at this point.

A flow-sheeting program is needed to effectively handle all of these detailed design calculations. Moreover, a program capable of optimizing the selected variables is desirable. Such programs have been developed and are available as adjuncts to the principal commercial process simulation packages. The objective function can be any convenient accounting criterion, eg, payout.

Design Practice

The economic incentive to recover process energy has been discussed at great length in the literature (see PROCESS ENERGY CONSERVATION). Escalation of energy costs during the 1970s was a significant factor as was the concern about environmental impact (qv) in the late 1980s. There are certainly many instances where additional capital for energy-recovery equipment can reduce operating expense. The trade-offs involved in adding insulation to pipes are a classic problem. On the other hand, the double savings (heat and cooling) that result from careful use of energy can lead to both a capital and operating cost savings. Not using the energy at all is a sure way to reduce environmental impact. Whereas belief in the inevitability of the capital vs operating cost trade-off exists because of the assumption that there are no basic faults in the engineering designs, this assumption is not warranted if there has been a reckless energy transfer across the pinch in the heat-exchange network design. As an example, 12 energy-saving projects at Imperial Chemical Industries were analyzed using network techniques (13). Energy savings were 6–60% of the original design, and capital savings were as high as 30% of the original design. At least half of the projects resulted in capital savings.

Many of the calculations necessary for heat-exchange network design can be done by hand and need to be understood by every process engineer. In fact, a clear understanding of the methodology involved can be most easily accomplished using a sheet of graph paper with development of some of the curves and networks discussed. Emphasis needs to be placed on developing an overall understanding of process-energy use rather than simply adding heat exchangers to recover energy.

Computational resources make it attractive for the average process designer to use a computer program to determine and evaluate alternative network configurations. Commercial heat-exchange network programs are available and these programs typically take the raw stream data available to the process engineer for the plant and calculate the limiting network conditions. Alternative networks based on synthesis algorithms are then determined. Some of the programs even do a more detailed design of individual exchangers and subsequent optimization of the network. Following the accurate determination of flows and other stream conditions, several of the available simulators prepare the input necessary for traditional process simulation programs. However, none of these programs relieves the process engineer of the need to be the architect of the energy system.

The earliest use of heat-exchange network synthesis was in the analysis of crude distillation (qv) units (1). The crude stream entering a distillation unit is a convenient single stream to heat while the various side draws from the column are candidate streams to be cooled in a network. So-called pumparounds present additional opportunities for heating the crude. The successful synthesis of crude distillation units was accomplished long before the development of modern network-synthesis techniques. However, the techniques now available ensure rapid and accurate development of good crude unit heat-exchange networks.

The heat-recovery opportunities in a 20,700 m³ (130,000 bbl) per stream-day crude-oil column and some of the practical aspects of optimizing heat recovery in a crude unit have been discussed (14) (see PETROLEUM). Most crude units process a variety of crudes and produce a broad range of product slates requiring flexibility in the energy network. A crude-oil feed stream and five hot streams leaving the crude column are shown in Figure 12**a**. Side stream strippers and other items are not shown on the column to simplify the drawing. Four of the hot streams are available for heat exchange with the crude: residuum, gas oil product, kerosene pumparound, and overhead reflux. The overhead product is not exchanged with the crude because of possible contamination of the naphtha product if an exchanger leaks. The remaining hot stream kerosene product is not cooled so that it can be used directly in another unit. The crude-oil feed is heated with these various streams, and finally, a fired heater is used to increase the temperature to that required in the column flash zone. The composite enthalpy curve for the hot streams and the crude-stream enthalpy curve are shown in Figure 12**b**. Alternative design cases are shown in Figure 13. Multiple shells are required for several of the heat exchangers. The sensitivity of the various cases to utility and capital cost changes for several different crudes has been investigated (14).

The complexity of problems that can be handled by commercially available programs is illustrated in Reference 15. A 15,900 m³ (100,000 barrels) per day

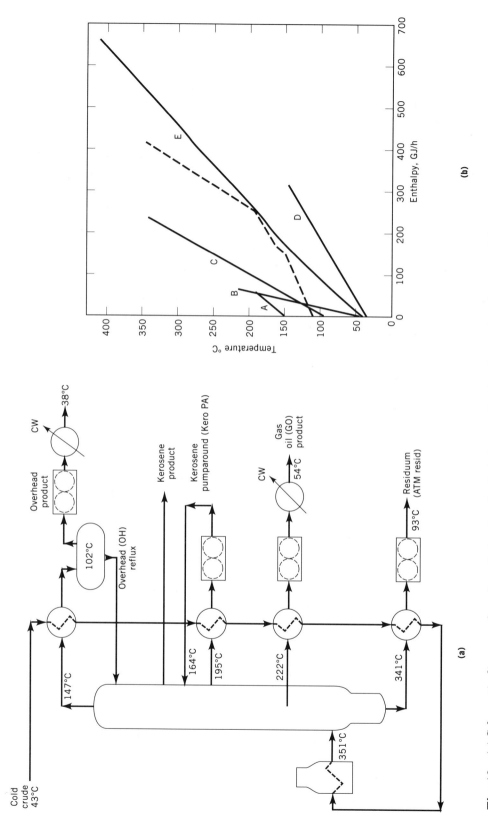

Fig. 12. (a) Schematic diagram of crude unit showing five hot streams leaving the crude column, and (b) temperature enthalpy diagram for the streams in (a) where A represents the kerosene pumparound; B, the gas oil product; C, the residuum; D, the overhead reflux; and E, the crude. The dashed line is the total available. To convert GJ/h to Btu/h, multiply by 9.48×10^6 (14).

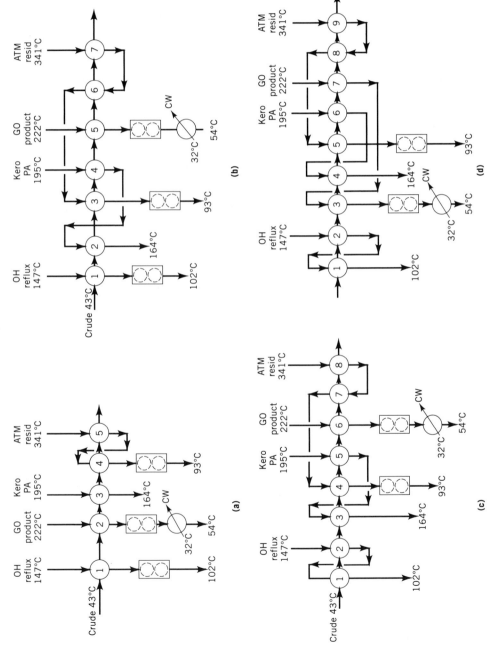

Fig. 13. Alternative network configurations for the system shown in Fig. 12 (14).

crude unit consisting of an atmospheric and vacuum crude column, two heaters, and a total of 21 heat exchangers were analyzed.

Future Developments

Heat-exchanger network design was first explored in the industrial environment and subsequently became a topic of academic research. Substantial progress has been made in developing techniques that can be used successfully for industrial problems. An excellent review of several classes of process-synthesis problems including heat-exchange network design is given in Reference 16. In a realistic process-design environment, the various synthesis techniques must be used simultaneously.

Energy Networks.   Heat-exchange networks are concerned with just a portion of the total energy in a process plant. The essential considerations when turbine-compressor systems and heat-exchange networks are combined, outlined in Reference 17, is now a mature area of process synthesis. The temperature–enthalpy representation can still be used to conceptualize the addition or removal of energy from a process. Energy (work) simply enters or leaves via an idealized heat engine that operates over a specified temperature range. The usual operation of a Carnot heat engine is assumed, but the concept of the pinch for a heat-exchange network can be expanded to an energy pinch for the total system. Introduction of a heat engine (eg, letdown turbine) in such a way as to cause energy flow across the pinch wastes process energy. On the other hand, if the engine is introduced properly, free energy can be obtained for the capital cost of a machine. Heat-pump placement can be analyzed in a similar manner. Proper placement reduces total energy demand, whereas improper placement may increase it.

Ideally, a process plant should be examined for its total energy consumption (see ENERGY MANAGEMENT). Other plant energy systems are under consideration (18) and should eventually be included in this type of analysis. This would include not only process thermal energy and shaft energy, but pumping requirements and electrical power as well.

Separation Networks.   The area of separation networks is important in chemical-process plants. Many separations are conducted by distillation, and there has been considerable interest in the process of designing sequences of distillation columns. Columns are separation-process-run by energy, and there is an obvious tie-in with heat exchange. Some progress has been made in the analysis of energy input to and withdrawal from a distillation column. The techniques for separation-process synthesis must be integrated with those for energy network synthesis. Several limits have been developed for separation processes similar to those for heat-exchange networks (see SEPARATION SYSTEMS SYNTHESIS). This is the first step toward synthesis of entire flow sheets.

NOMENCLATURE

A	heat-transfer area
a	constant in the heat-exchanger cost equation
AGO	atmospheric gas oil
b	constant in the heat-exchanger cost equation
C_p	heat capacity

c	constant in the heat-exchanger cost equation
CW	cooling water
DTA	double temperature of approach
E	number of heat exchangers in network
$E_{\min}$	minimum number of heat exchangers in network
ED	evolutionary technique
F	correction factor for ΔT_{LM} in heat exchangers
GO	gas oil
H	relative enthalpy
h_i	heat-transfer coefficient for stream h_i
h_j	heat-transfer coefficient for stream h_j
HVGO	high vacuum gas oil
Kero	kerosene
L	number of heat-load loops in a heat-exchange network
LVGO	low vacuum gas oil
M	number of cold process streams in network
MER_+	maximum energy recovery for given set of process streams
$\text{MER}_{\Delta T}$	maximum energy recovery with network approach of ΔT
N	number of hot process streams in network
NAP	naphtha
NAP REB	naphtha reboiled
OH	overhead
P	number of independent stream subsets in a network
PA	pumparound
Q	heat transferred in exchanger
$Sproc$	number of process streams in network
$Sutil$	number of utility streams in network
SP	process-stream problem
Stm	steam
T_{si}	supply temperature of cold process streams i
T_{sj}	supply temperature of hot process streams j
T_{ti}	target temperature of cold process stream i
T_{tj}	target temperature of hot process stream j
ΔT_e	minimum temperature of approach for any heat exchanger
ΔT_p	minimum temperature of approach at the network pinch
ΔT_{LM}	log–mean temperature difference for a given heat exchange
TC	thermodynamic–combinatorial
TI	temperature internal
U	overall heat-transfer coefficient

BIBLIOGRAPHY

"Heat Exchange Technology, Network Synthesis" in *ECT* 3rd ed., Suppl. Vol., pp. 521–545, by E. Hohmann, California State Polytechnic University.

1. A. M. Whistler, *Pet. Ref.* **27**(1), 57 (1948).
2. E. C. Hohmann, *Optimum Networks for Heat Exchange*, Ph.D. dissertation, University of Southern California, Los Angeles, 1971.
3. N. Nishida, S. Kobayashi, and A. Ichikawa, *Chem. Eng. Sci.* **26**, 1841 (1971).
4. B. Linnhoff and J. R. Flower, *AIChE J.* **24**, 633 (1978).
5. R. W. Colbert, *CEP* **78**(7), 47 (1982).
6. B. Linnhoff, D. R. Mason, and I. Wardle, *Comp. Chem. Eng.* **3**, 295 (1979).
7. J. R. Flower and B. Linnhoff, *AICHE J.* **26**, 1 (1980).
8. B. Linnhoff and E. Hindmarsh, *Chem. Eng. Sci.* **38**, 745 (1983).

9. B. Linnhoff and J. A. Turner, *Chem. Eng.* **88**(22), 56 (1981).
10. N. Nishida, Y. A. Lir, and L. Lapid, *AICHE J.* **22**, 539 (1976).
11. J. L. Su, *A Loop-Breaking Evolutionary Method for Synthesis of Heat Exchanger Networks*, Master's thesis, Washington University, St. Louis, Mo., 1979.
12. T. B. Chilland and S. F. Yang, *Oil Gas J.*, 73 (Feb. 6, 1978).
13. B. Linnhoff and co-workers, *A User Guide on Process Integration for Efficient Use of Energy*, Institute of Chemical Engineers, Rughy, U.K., 1982.
14. F. Huang and R. Elshout, *Chem. Eng. Prog.* **72**(7), 68 (1976).
15. F. Kleinschrodt and G. Hammer, *Chem. Eng. Prog.* **79**(7), 33 (1983).
16. N. Nishida, G. Stephanopoulos, and A. W. Westerberg, *AICHE J.* **27**, 3 (1981).
17. B. Linnhoff and B. W. Townsend, *CEP* **78**(7), 72 (1982).
18. B. Linnhoff, *Trans IChemE* **71**, 503 (1993).

General References

D. Boland and B. Linnhoff, *Chem. Eng.*, 22 (Apr. 1979).
V. Cena, C. Mustacchi, and F. Natali, *Chem. Eng. Sci.* **32**, 1227 (1977).
J. Cerda and A. W. Westerberg, *Minimum Utility Usage in Heat Exchanger Network Synthesis-A Transportation Problem*, DRC Report No. 06-16-80, Carnegie-Mellon University, Pittsburgh, Pa., 1980.
H. Coben, G. F. C. Rogers, and H. I. H. Saravanamutto, *Gas Turbine Theory*, Longman, Inc., New York, 1972.
P. R. Cooper, *Royal Society Esso Energy Award*, press release, Royal Society, London, Aug. 1981.
H. A. Dunford and B. Linnhoff, *Cost Saving in Distillation Symposium*, paper no. 10, Institution of Chemical Engineers, Leeds, U.K., 1981.
R. V. Elshout and E. C. Hohmann, *Chem. Eng. Prog.*, 72 (Mar. 1979).
R. A. Greenkorn, L. B. Koppel, and S. Raghavan, "Heat Exchanger Network Synthesis— A Thermodynamic Approach," *71st AlChE Meeting*, Miami, Fla., 1978.
L. E. Grimes, *The Synthesis and Evolution of Networks of Heat Exchange That Feature the Minimum Number of Units*, Master's thesis, Carnegie-Mellon University, Pittsburgh, Pa., 1980.
L. E. Grossman and R. W. H. Sargent, *Comp. Chem. Eng.* **2**, 1 (1978).
P. W. Haywood, *Analysis of Engineering Cycles*, 3rd ed., Pergamon Press, New York, 1980.
P. Hinchley, *Chem. Eng. Prog.* **73**, 90 (1977); *Chem. Ing. Tech.* **49**, 553 (1977).
K. C. Hohmann and F. J. Lockhart, "Optimum Heat Exchanger Network Synthesis," *AICHE 82nd National Meeting*, Atlantic City, N.J., 1976.
E. C. Hohmann and D. B. Nash, "A Simplified Approach to Heat Exchanger Network Analysis," *85th National AlChE Meeting*, Philadelphia, Pa., June 1978.
S. Kobayashi, T. Umeda, and A. Ichikawa, *Chem. Eng. Sci.* **26**, 3167 (1971).
B. Linnhoff, *Thermodynamic Analysis in the Design of Process Networks*, Ph.D. dissertation, University of Leeds, Leeds, U.K., 1979.
B. Linnhoff, *Design of Heat Exchanger Networks—A Short Course*, University of Manchester Institute of Science and Technology, Manchester, U.K., 1982.
B. Linnhoff and K. J. Carpenter, "Energy Conservation by Energy Analysis—The Quick and Simple Way," *Second World Congress of Chemical Engineering*, Montreal, Canada, 1981.
B. Linnhoff and J. A. Turner, *Chem. Eng.*, 742 (Dec. 1980).
B. Linnhoff and J. A. Turner, *Chem. Eng.*, 56 (Nov. 2, 1981).
R. L. McGalliard and A. W. Westerberg, *Chem. Eng. J.* **4**, 127 (1972).
A. H. Masso and D. F. Rudd, *AICHE J.* **15**, 10 (1969).
R. L. Motard and A. W. Westerberg, *Process Synthesis*, AIChE advanced seminar lecture notes, New York, 1988.
D. B. Nash and co-workers, "A Simplified Approach to Heat Exchanger Network Analysis," *AIChE 85th National Meeting*, Philadelphia, Pa., June 1978.

T. K. Pho and L. Lapidus, *AICHE J.* **19**, 1182 (1973).

R. N. S. Rathore and G. J. Powers, *Ind. Eng. Chem. Process Des. Dev.* **14**, 175 (1975).

J. V. Shah and A. W. Westerberg, "Evolutionary Synthesis of Heat Exchanger Networks," *AIChE Annual Meeting,* Los Angeles, Calif., Nov. 1975.

J. J. Siirola, "Status of Heat Exchanger Network Synthesis," *AIChE National Meeting,* Tulsa, Okla., 1974.

R. A. Smith, "Economic Velocity in Heat Exchangers," *ASME/AIChE 20th National Heat Transfer Conference,* Milwaukee, Wis., 1981.

G. Stephanopoulos, *Synthesis of Networks of Heat Exchangers—A Self-Study Block Module,* Project PROCEED, Massachusetts Institute of Technology, Cambridge, 1977.

J. Taborek, in E. U. Schlunder and co-workers, eds., *Heat Exchanger Desiln Handbook,* Hemisphere Publishing Corp., New York, 1982, Chapts. 1–5.

D. W. Townsend and B. Linnhoff, *AIChE J.,* (1982).

D. W. Townsend and B. Linnhoff, *Chem. Eng.,* 91 (Mar. 1982).

T. Umeda, T. Harada, and K. Shiroko, "A Thermodynamic Approach to the Synthesis of Heat Integration Systems in Chemical Processes," *Proceedings of the 12th Symposium on Computer Applications in Chemical Engineering,* Montreaux, Switzerland, 1979, p. 487.

L. Wells and M. G. Hodgkinson, *Process Eng.,* 59 (1977).

EDWARD HOHMANN
California State Polytechnic University

HEAT PUMPS. See POWER GENERATION; THERMOELECTRIC ENERGY CONVERSION.

HEAT-RESISTANT POLYMERS

The search for heat-resistant polymers began about 1960 and there is continued growth in the development of these materials that can perform long-term service at elevated temperatures. The intensive interest in these polymers results from new technological processes requiring higher use temperatures for development. Thermally stable or high performance polymers dictate high melting (softening) temperatures, resistance to oxidative degradation at elevated temperatures, resistance to other (nonoxidative) thermolytic processes, and stability to radiation and chemical reagents.

The definition of polymer thermal stability is not simple owing to the number of measurement techniques, desired properties, and factors that affect each (time, heating rate, atmosphere, etc). The easiest evaluation of thermal stability is by the temperature at which a certain weight loss occurs as observed by thermogravimetric analysis (tga). Early work assigned a 7% loss as the point of stability; more recently a 10% value or the extrapolated break in the tga curve has been used. A more realistic view is to compare weight loss vs time at constant temperature, and better yet is to evaluate property retention time at temperature: one set of criteria has been 177°C for 30,000 h, or 240°C for 1000 h, or 538°C for 1 h, or 816°C for 5 min (1).

A number of thermally stable polymers have been synthesized, but in general the types of structures that impart thermal resistance also result in poor processing characteristics. Attempts to overcome this problem have largely been concentrated on the incorporation of flexible groups into the backbone or the attachment of stable pendent groups. Among the class of polymers claimed to be thermally stable only a few have achieved technological importance, some of which are polyamides, polyimides, polyquinoxalines, polyquinolines, and polybenzimidazoles. Of these, polyimides have been the most widely explored.

Polyimides

Polyimides (PI) were among the earliest candidates in the field of thermally stable polymers. In addition to high temperature property retention, these materials also exhibit chemical resistance and relative ease of synthesis and use. This has led to numerous innovations in the chemistry of synthesis and cure mechanisms, structure variations, and ultimately products and applications. Polyimides (qv) are available as films, fibers, enamels or varnishes, adhesives, matrix resins for composites, and molding powders. They are used in numerous commercial and military aircraft as structural composites, eg, over a ton of polyimide film is presently used on the NASA shuttle orbiter. Work continues on these materials, including the more recent electronic applications.

Synthesis and Properties. Several methods have been suggested to synthesize polyimides. The predominant one involves a two-step condensation reaction between aromatic diamines and aromatic dianhydrides in polar aprotic solvents (2,3). In the first step, a soluble, linear poly(amic acid) results, which in the second step undergoes cyclodehydration, leading to an insoluble and infusible PI. Overall yields are generally only 70–80%.

dianhydride   diamine   soluble poly(amic acid)

polyimide

A viscous solution of poly(amic acid) can be processed into films, fibers, and coatings, and the final product undergoes thermal cyclodehydration.

Numerous diamines and aromatic dianhydrides have been investigated. Wholly aromatic PIs have been structurally modified by incorporating various functional groups, such as ether, carbonyl, sulfide, sulfone, methylene, isopropylidene, perfluoroisopropylidene, bipyridyls, siloxane, methyl phosphine oxide, or various combinations of these, into the polymer backbone to achieve improved properties. The chemistry and applications of PIs have been described in several review articles (4).

PIs commonly have been synthesized from reactions of pyromellitic dianhydride [26265-89-4] (PMDA) or 3,3′,4,4′-benzophenone tetracarboxylic dianhydride [2421-28-5] (BTDA) with a number of diamines like 2,2-bis(4-aminophenyl) propane, 2,2-bis(4-amino-3-methylphenyl)propane, 1,1-bis(4-aminophenyl)-1-phenylethane, and 1,1-bis(4-amino-3-methylphenyl)-1-phenylethane (5). The PMDA-based PIs were thermally more stable than the corresponding PIs obtained from BTDA.

A successful synthesis of novel, soluble aromatic PIs involving 3,4-bis-(4-aminophenyl)-2,5-diphenylfuran by polymerization with aromatic tetracarboxylic dianhydrides through the conventional two-step method has been reported (6) (Fig. 1).

The polymers are stable up to 550°C (10% weight loss by tga) in N_2 atmosphere with the glass-transition temperatures, T_g, ranging from 281–344°C. The inherent viscosities of the polymers in H_2SO_4 are up to 0.32 dL/g. Also prepared were aromatic PIs containing triphenylamine units (7). These polymers show a 10% weight loss at 520°C in air, with T_gs in the range 287–331°C. Synthesis of some bismaleimides has been reported from epoxy resins which show thermal stability up to 370°C (8). Thermotropic poly(ester–imides) have been obtained from trimellitic acid with phosphonate or phosphate groups in the main chain (9). These polymers are stable in the range 410–425°C (5% weight loss) in air.

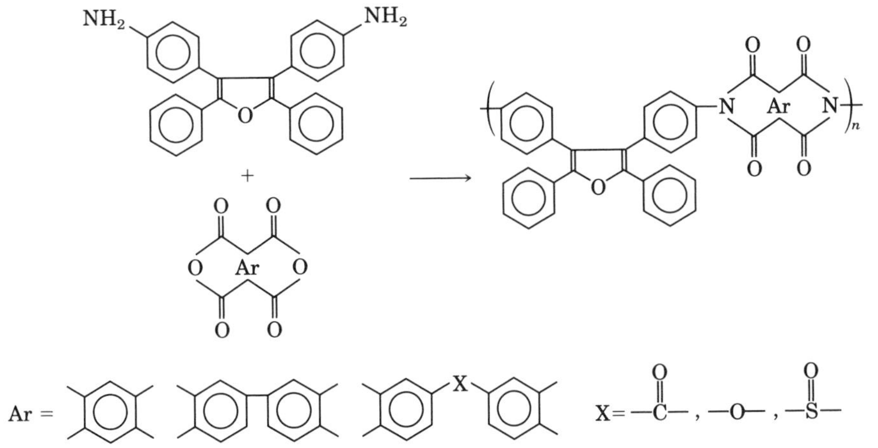

Fig. 1. Wholly aromatic PIs made by conventional methods.

Structurally different PIs have been synthesized with alternating rigid (pyromellitimide) and semiflexible (polymethylene) units along the chain backbone (10). A change in T_g of the polymers obtained with different spacer groups follows the expected trend; that is, as the aliphatic segment increases in length (C-8 to C-12) the T_g decreases (383 to 305°C). Further mixing of aliphatic-type linkages drives the T_g even lower. Polyesterimides are well known for their thermal stability coupled with processing ease. A moderately stable polyesterimide has been synthesized from trimellitic anhydride [552-30-7], bisphenol A [80-05-7], and 4,4'-diaminodiphenyl sulfone [80-08-0] by three different routes (11).

These polymers are soluble in polar solvents and stable up to 290–325°C.

Copoly(amide–imides) comprise an important class of copolyimides that have been developed into a commercial product. Incorporating the amide linkage into the PI makes the polymer more tractable than simple PIs, but involves a loss in thermal stability. However, copoly(amide–imides) still possess quite good thermal stabilities, intermediate between those of polyamides and PIs (12). They are relatively inexpensive to synthesize.

A series of cross-linkable copoly(amide–imides) is known to be possible from aromatic diamines and substituted isophthaloyl chlorides containing unsaturated imide rings as a pendent function (13).

These polymers possess enhanced solubility compared to the aromatic polyamides with no deterioration in thermal stability. Their T_gs vary from 264–311°C depending on the nature of the X group. An additional feature of these polymers is that their tensile strengths increase after heat treatment due to cross-linking.

Aliphatic–aromatic poly(amide–imides) based on N,N'-bis(carboxyalkyl)-benzophenone-3,3',4,4'-tetracarboxylic diimides have shown a 10% weight loss at 400°C (14).

where $m = 1,2,3,5,10$

The T_gs of the polymers were in the range 122–250°C, which showed a predictable increase with a decrease in length of the flexible aliphatic chains.

Aromatic copoly(amide–imide)s with s-triazine rings in the repeating unit of the backbone are also possible from a diacyl chloride reacting with preformed imide groups and diamines containing s-triazine rings (15).

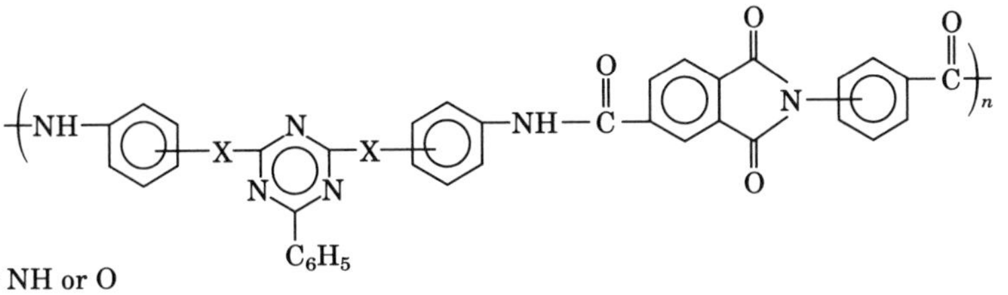

where X = NH or O

The polymers have been obtained by conventional solution condensation at low temperature using polar organic solvents in N_2 atmosphere and are stable up to 400°C in air (5% weight loss). Their inherent viscosities range from 0.40 to 1.17 dL/g. This method overcomes the disadvantages of PI preparation, ie: (1) a two-step process, and (2) the instability of the intermediate (poly(amic acid)) in the presence of moisture. In another study (16), novel phosphorylated bismaleimides and nonphosphorylated tetramaleimides containing substituted s-triazine rings (chain extended by imide, amide, or urea groups) were prepared and polymerized. These polymers were stable up to 312–370°C in air or N_2.

Poly(phenylquinoxaline–amide–imides) are thermally stable up to 430°C and are soluble in polar organic solvents (17). Transparent films of these materials exhibit electrical insulating properties. Quinoxaline–imide copolymer films prepared by polycondensation of 6,6'-methylene bis(2-methyl-3,1-benzoxazine-4-one) and 3,3',4,4'-benzophenone tetracarboxylic dianhydride and 4,4'-oxydianiline exhibit good chemical etching properties (18). The polymers are soluble, but stable only up to 200–300°C.

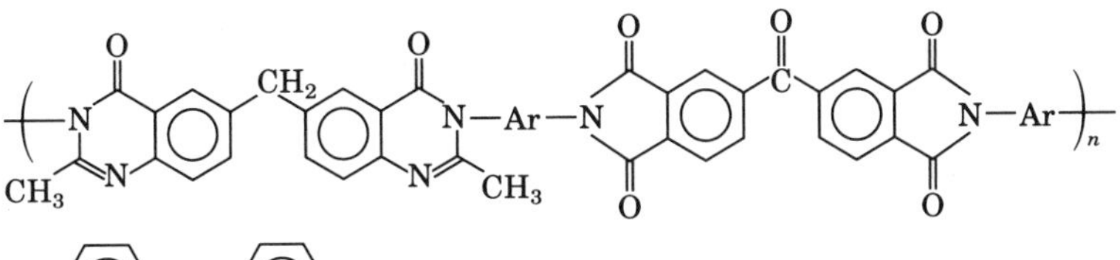

where Ar =

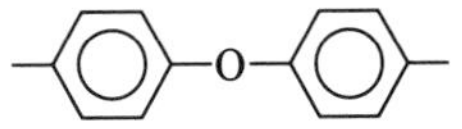

Polyetherimide synthesis has been achieved by reaction of a dianhydride containing an ether linkage with a diamine, reaction of a diamine containing an ether linkage with a dianhydride, or nucleophilic displacement of halo or nitro groups of a bisimide by bisphenol dianion (19,20). Such PIs exhibit good thermal stability and melt processability.

A large variety of bisimides and polymers containing maleimide and citraconimide end groups have also been reported (21–26). Thus polymers based on bisimidobenzoxazoles from the reaction of maleic anhydride and citraconic anhydride with 5-amino-2-(p-aminophenyl)benzoxazole and 5-amino-2(m-aminophenyl)benzoxazole are found to be thermally stable up to 500°C in nitrogen.

A cross-linked and crystalline copoly(ester–imide) containing an alkene function was made by reaction of an unsaturated diacid chloride containing a cyclic imido group with ethylene glycol at low temperature (27).

$$+\left(C-\underset{O}{\underset{\|}{}}\bigcirc N-\bigcirc-CH=CH-\underset{O}{\overset{O}{\underset{\|}{C}}}-O-CH_2-CH_2-O\right)_n$$

These polymers lost 5% of their weight at 325°C in a nitrogen atmosphere and the T_g of the polymers varied from 230–262°C. The inherent viscosity of the polymers is low (0.13–0.17 dL/g in DMF at 30°C).

Terpoly(amide–imide–urethanes) have been synthesized in yields up to 50–75% by the reaction of 4-carboxy-N-(p-hydroxyphenyl)phthalimide with diisocyanates in N-methyl-2-pyrrolidinone containing 5% lithium chloride (28).

$$+\left(O-\bigcirc-N\bigcirc\overset{O}{\underset{O}{}}C-NH-Ar-NHC\right)_n$$

where Ar = $-\bigcirc-$, $-\bigcirc-CH_2-\bigcirc-$, $-\bigcirc-O-\bigcirc-$

These polymers show thermal stability in the range 300–340°C (10% weight loss) and inherent viscosities of 0.15–0.45 dL/g.

Polypyromellitimide films based on cyclotriphosphazene and bisaspartimide-derived diamines have shown thermal stabilities up to 800°C and char yields of 56–68% in N_2 and 24% in air (29). High strength fire- and heat-resistant imide resins containing cyclotriphosphazene and hexafluoroisopropylidene groups have also been synthesized by the thermally induced melt polymerization of maleimido–phenoxy cyclotriphosphazenes linked by hexafluoroisopropylidenediphthalimide groups (30). The polymers are fire resistant with char yields of 78–80% at 800°C in N_2 and 60–68% in air at 700°C. The laminates have a flexural strength of 490 MPa (71,000 psi) and a modulus of 56,600 MPa (8.2 × 10^6 psi), demonstrating better properties than those of many other presently used resins.

Some novel copolyimides containing metal phthalocyanines are possible by treating copper, cobalt, nickel, and zinc phthalocyaninotetramines with PMDA and BTDA (31–35). They are self-extinguishing, do not burn when exposed to flame, and oxidize in air at 500–600°C, leaving metal oxides. In nitrogen, degradation occurs at 600–800°C leaving a 75–85% char yield.

Silicon-containing PIs, useful as insulation and protective materials, demonstrate adhesion to fibers, fabrics, glass, quartz, and carbon (36). The synthetic method used is the reaction of the silicon-containing dianhydride with diamines.

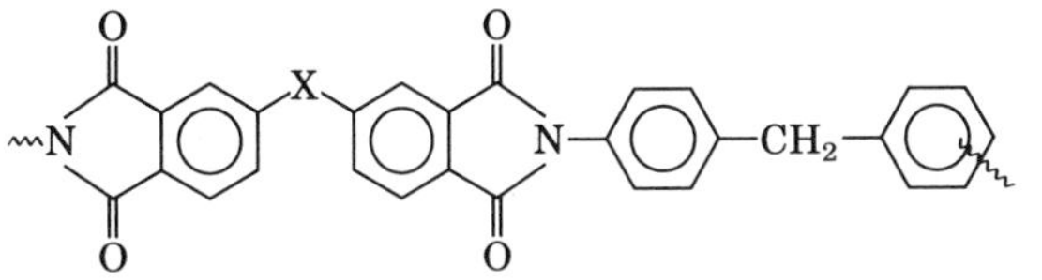

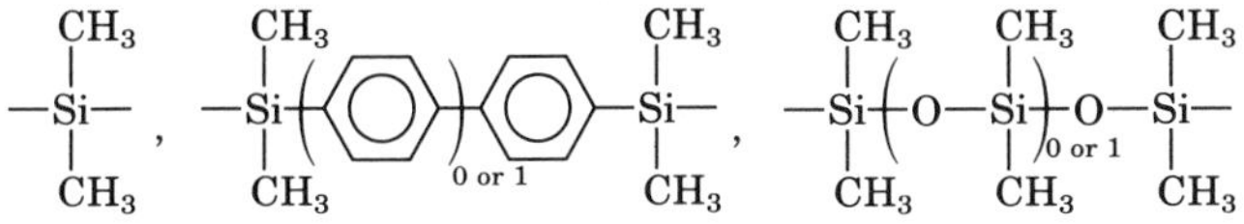

Three important linear aromatic PIs, namely LARC-TPI, LARC-160, and LARC-13 were developed by researchers at NASA-Langley Research Center (37–39) (Fig. 2).

LARC-TPI is a linear thermoplastic PI which can be processed in the imide form to produce large-area, void-free adhesive bonds. Mitsui Toatsu Chemicals,

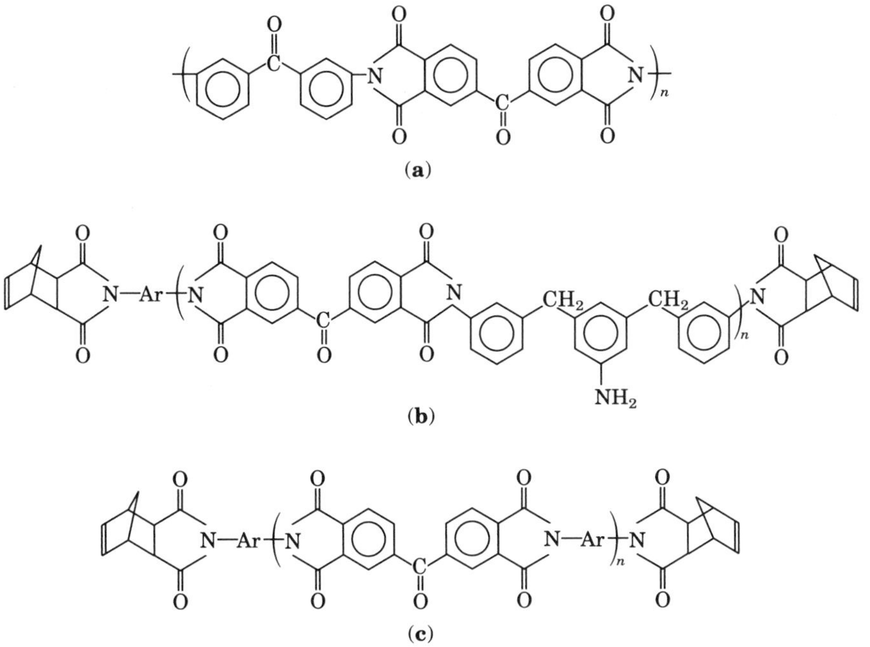

(a)

(b)

(c)

Fig. 2 LARC polyimides where LARC stands for Langley Research Center. (a), LARC-TPI from 3,3′-diaminobenzophenone and BTDA; (b), LARC-160, where Ar = $-C_6H_4CH_2(C_6H_3NH_2CH_2)_nC_6H_4-$; (c), LARC-13, where Ar = 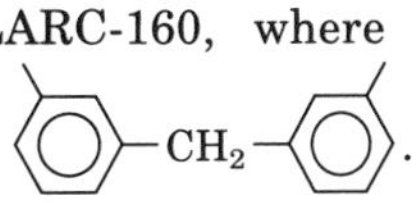 .

Inc., has obtained license to produce this product commercially for applications such as adhesives, films, molding compounds, etc. These are thermooxidatively stable and show essentially no loss in weight at 300°C in air. Weight loss does not exceed 2–3% after isothermal aging in air at 300°C for 550 h.

LARC-13 is a nadimide-terminated addition-curing adhesive. Because of its high degree of flow during cure, it is autoclave processible. It has been successfully used to bond a high temperature composite to a ceramic for missile applications which achieve several seconds of performance at 595°C. A significant use of this adhesive is in the bonding of honeycomb sandwich structures. PMR-15 polyimides, based on mixtures of reactive components, have also been investigated (40–42) which exhibited thermooxidative stability up to 316°C. PMR is a designation of NASA-Lewis and stands for polymerization of monomeric reactants *in situ*.

Since 1975, numerous polyimide backbones containing hexafluoroacetone or hexafluoroisopropoxybenzene groups have been investigated (43,44). These polymers show greatly enhanced solubility (up to 20% in amide solvents) and significant promise in gas separation research and technology.

Applications. The excellent chemical, mechanical, and electrical properties of PIs have led to their widespread use and make them attractive as high performance materials. Since the 1980s, there has been an accelerated interest in the use of PIs in a variety of applications ranging from aerospace to microelectronics to medicine. Addition-type PIs have also emerged as state-of-the-art matrix resins for high strength, high temperature-resistant composites for use as structural materials in aerospace applications (45). These low molar mass compounds end capped with reactive imide groups (eg, maleimide or nadimide) can be thermally polymerized to give cross-linked PIs. Their chemical, thermal, and mechanical properties are amenable to a variety of modifications by way of structural changes of the diamine precursor or the reactive imide groups. One of the most recent uses of PIs is as reverse osmosis membranes. The polymers substituted with polar groups such as methoxy show excellent stability toward hydrolytic and bacterial attack (46).

Polyimide films such as Du Pont's Kapton or H-film are used as wire and cable wrap, motor-slot liners, and in transformers and capacitors. These films are excellent choices in applications where high dielectric strength must be retained under thermal stress. Although not yet of commercial significance, several types of PIs have been spun into fibers and used in various high technology applications, eg, fire-resistant fabric, thermally stable composites, etc. In silicon dioxide or nitride layers in integrated circuit technology, the PI films are excellent alternative candidates due to their toughness, low conductivity, and ease and reliability of processing (47,48).

In the electronics industry, PIs find wide applications as a dielectric material for semiconductors due to thermal stability (up to 400°C) and low dielectric constant. PIs are being considered for use in bearings, gears, seals, and prosthetic human joints. The intended part can be machined or molded from the PI, or a film of PI can be applied to a metallic part. Because of their superior adhesion, dielectric integrity, processing compatibility, and lack of biological system impact, PIs have been used in many biological applications with particular success as body implants.

Polyoxadiazoles

Poly(1,3,4-oxadiazole) (POD) is a widely used isomer of the oxadiazole family of thermally stable polymers. The general structure of POD is

where R is often an aromatic ring, eg, m- or p-phenylene. The popularity of this particular isomer is attributed to the superior thermal stability it adds to a polymer structure as compared to the 1,2,4- or 1,2,5-isomers. Detailed studies began on this class of polymers in the late 1950s; since then, a large number of PODs have been developed. The literature up to 1979 on polyoxadiazoles has been reviewed (49).

Synthesis and Properties. Polyoxadiazoles containing aromatic moieties with aliphatic linkages/groups have been widely explored in the literature. The aromatic moieties increase the rigidity of the polymer; the presence of aliphatic groups makes the chain more flexible and processible.

One series of POD has been prepared from the corresponding dicarboxylic acid/acid chlorides and hydrazine sulfate in polyphosphoric acid (PPA) (50,51), one of the most common techniques for this type of backbone.

where X = OH or Cl, R = C_6H_5, $(CH_2)_4$,$(CH_2)_8$, *cis* CH=CH

Thermal stability of the polymers ranges from 328 to 390°C in N_2 atmosphere (10% weight loss).

Aromatic PODs containing amide and imide groups have been synthesized by the solution polycondensation method (52).

Decomposition temperatures are in the range of 360–375°C and inherent viscosities range from 0.62 to 0.90 dL/g in conc H_2SO_4. The polymers are insoluble in DMAC.

Carbazole-containing PODs have been obtained (53) by cyclodehydration (in the presence of $POCl_3$) of polyhydrazides prepared by polycondensation of N-ethyl-3,6-carbazoledicarbonyl chloride with dihydrazides of the corresponding dicarboxylic acids. Thermal decomposition of the polymers containing aliphatic units occurs at 365–380°C, compared to 400–405°C for polymers containing aromatic units.

where R = $(CH_2)_4$, $(CH_2)_8$, m-C_6H_4, p-C_6H_4, 9-ethylcarbazole-3,6-diyl

Novel 1,3,4-oxadiazole-containing polyazomethines have been synthesized by the polycondensation of diamines, 2,5-bis(m-aminophenyl)-1,3,4-oxadiazole and 2,5-bis(p-aminophenyl)-1,3,4-oxadiazole with aromatic dialdehydes, iso-phthaldehyde, and terephthalaldehyde (example follows), in m-cresol at 20°C (54).

These polymers have reduced viscosities up to 1.13 dL/g and electric conductivity as high as 10^{-11}–10^{-12} S/cm. All the polymers are insoluble in common organic solvents but soluble in conc H_2SO_4. Thermal degradation begins around 400°C in air and nitrogen according to tga.

Poly(1,3,4-oxadiazole-2,5-diyl-vinylene) and poly(1,3,4-oxadiazole-2,5-diyl-ethynylene) were synthesized by polycondensation of fumaramide or acetylene-dicarboxamide with hydrazine sulfate in PPA to study the effect of the two repeating units on polymer electronic and thermal properties (55).

Both the polymers are dark in color and exhibit semiconductivity and paramagnetism. The electric conductivity measurements are performed on pellets and on thin films in sandwich and surface cells.

Research activities in the area of PODs containing aromatic groups have been centered around the production of highly processible, soluble, and thermally stable polymers. In this particular class of PODs, the imide- and phenylene-containing backbones have been widely explored.

Fully aromatic, thermally (up to 250°C) and hydrolytically resistant films of PODs have been realized from polyhydrazides (56). Films of these polymers are useful as seawater desalination membranes.

where R = H, OCH$_3$; R' = H, OCH$_3$, CN

Polyoxadiazole-imides containing hexafluoroisopropylidene (HFIP) groups are soluble in common solvents and still retain good mechanical and thermal properties (57). The HFIP-based POD has the best solubility in common solvents of this series of polymers, and its tough, transparent films are the lightest in color. All the PODs show nearly identical thermal stabilities (10% weight loss) at 415–430°C in air and argon by tga. These films exhibit an initial modulus of 296 MPa (42,900 psi), an ultimate strength of 95.6 MPa (13,900 psi), and an ultimate elongation of 32.4%.

Thermally stable POD films containing pyridine rings have potential application as reverse osmosis membranes (58).

A general method for the preparation of copolyoxadiazoles, ie, poly(aryl ether oxadiazoles), has been developed where the generation of an aryl ether linkage is the polymer-forming reaction (59) (Fig. 3). Synthetic methods are based on either an oxadiazole-activated or an hydrazide-activated halo-displacement with phenoxides. The hydrazide is subsequently thermally dehydrated to the oxadiazole heterocycle. An appropriately substituted diarylfluorooxadiazole is prepared and polymerized with various bisphenols in NMP and N-cyclohexyl-2-pyrrolidinone (CHP) mixture in the presence of K$_2$CO$_3$ to give the copolyoxadiazole. The inherent viscosities range from 0.44 to 0.76 dL/g in NMP at 25°C and the T_gs from 190 to 210°C. The excellent thermal stability of these materials (polymer decomposition temperatures in excess of 450°C) allow them to be melt processed in spite of their high T_g.

The synthesis of phenoxaphosphine-containing PODs by the cyclodehydration of polyhydrazides obtained from 2,8-dichloroformyl-10-phenylphenoxaphosphine-10-oxide and aliphatic and aromatic dihydrazides has been described (60). All polymers are soluble in formic acid, *m*-cresol, and conc H$_2$SO$_4$, but insoluble or partially soluble in benzene, chloroform, and hexamethylphosphoric triamide.

R = (CH$_2$)$_4$, (CH$_2$)$_8$, *p*-C$_6$H$_4$, *m*-C$_6$H$_4$

The PODs obtained from aromatic dihydrazides are partly soluble in dimethylacetamide, DMSO, etc; the others dissolve in these solvents. The thermal stabil-

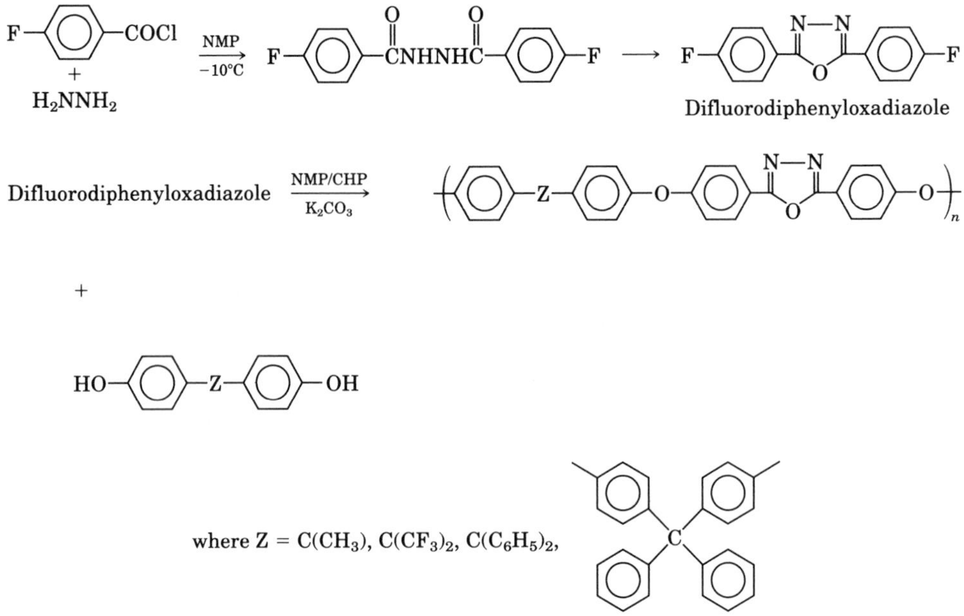

Fig. 3. Preparation of copolyoxadiazoles.

ities (up to 464°C in air and 476°C in nitrogen (10% weight loss)) of the polymers were determined by tga.

Applications. The uses of PODs have been extensively investigated as evidenced by the large number of patents, but not widely developed into commercial products. The polymers are hydrolytically stable and partially crystalline, and they can be oriented to a high degree of crystallinity by drawing. By far the most widely known uses of this type of polymer are in fibers and films; the films are often transparent yellow or brown with a percent elongation-to-break reported as high as 140, but more commonly in the range of 25 to 75%. Tensile strengths are near 118 MPa (1200 kgf/cm^2), and upon weathering or heat aging this value often drops to around 78 MPa (800 kgf/cm^2). These polymers are frequently processed from sulfuric acid solution.

The fibers of aromatic PODs are known to have a combination of good properties, such as strength and stiffness, fatigue resistance, and relatively low density, in the range of 1.2 to 1.4 g/cm^3. PODs have been used to improve the heat resistance of many synthetic fibers. This is usually done by dissolving the POD in sulfuric acid and then treating the fibers with this solution. Poly(p-phenylene-1,3,4-oxadiazole) is the most commonly used commercial polyoxadiazole, and the fiber spun from this polymer is called Oksalon.

Conducting polymers having good thermal resistance are often prepared by heat treating POD alone. The heat-treated POD can be obtained in the form of a strong, flexible film composed of highly ordered graphite crystallites. POD is also used in liquid crystal display cells in which electrode plates are coated with the polymer. A patent has been filed for battery electrodes and batteries fabricated with PODs and other heterocyclic ring system polymers, which reportedly exhibit an open-circuit voltage of 0.75 V after charging from a 2.7 V dry cell for a few minutes.

In other areas, POD has been used to improve the wear resistance of a rubber latex binder by incorporation of 25% of Oksalon fibers. Heat-resistant laminate films, made by coating a polyester film with POD, have been used as electrical insulators and show good resistance to abrasion and are capable of 126% elongation. In some instances, thin sheets of PODs have been used as mold release agents. For this application a resin is placed between the two sheets of POD, which is then pressed in a mold, and the sheets simply peel off from the object and mold after the resin has cured. POD-based membranes exhibit salt rejection properties and hence find potential as reverse osmosis membranes in the purification of seawater. PODs have also been used in the manufacturing of electrophotographic plates as binders between the toner and plate. These improved binders produce sharper images than were possible before.

Polyquinoxalines

Polyquinoxalines (PQ) have proven to be one of the better heat-resistant polymers with regard to both stability and potential application. The aromatic backbones are derived from the condensation of a tetramine with a bis-glyoxal, reactions first done in 1964 (61,62). In 1967, a soluble, phenylated version of this polymer was produced (63). The chemistry and technology of polyquinoxalines has been reviewed (64).

Polyphenylquinoxalines (PPQ) are easier to make than the polyquinoxalines and offer superior solubility, processibility, and thermooxidative stability (65). The PPQs exhibit excellent high temperature adhesive, composite, and film properties. However, to increase the use temperature of PPQs, acetylene groups have been placed on the backbone and subsequently thermally cured (66). In order to raise the T_g of the polymer, attempts were made to introduce the pendent ethynyl and phenylethynyl groups along the PPQ backbone followed by thermal curing (67) and their mechanical and thermal properties have been studied (68,69). The cured, acetylene-terminated PPQs are less thermooxidatively stable than the parent polymers. The thermooxidative stability of cured PPQ containing pendent ethynyl or phenylethynyl moieties is lower than that of PPQ without ethynyl moieties. However, introduction of phenylethynyl groups improves the mechanical properties at 232°C as demonstrated on adhesives (70).

Synthesis and Properties.   A number of monomers have been used to prepare PQs and PPQs, including aromatic bis(o-diamines) and tetramines, aromatic bis(α-dicarbonyl) monomers (bisglyoxals), bis(phenyl-α-diketones) and α-ketones, bis(phenyl-α-diketones) containing amide, imide, and ester groups between the α-diketones. Significant problems encountered are that the tetraamines are carcinogenic, difficult to purify, and have poor stability, and the bisglyoxals require an arduous synthesis.

Polyquinoxalines are prepared by the solution polymerization of aromatic bis(o-diamines) such as 3,3',4,4'-tetraminobiphenyl and aromatic bis(glyoxal hydrates) such as 4,4'-oxybis(phenylglyoxal hydrate):

The aromatic bis(o-diamine) is added as a fine powder or slurry to a stirred slurry of the bis(glyoxal hydrate) to form PQ or bis(phenyl-α-diketone) to form PPQ in m-cresol at ambient temperature. The reaction temperature is maintained below 40°C because during the early stages of polymerization higher temperatures lead to branching. After the exotherm subsides, the mixture is stirred at ambient temperatures for several hours to give a viscous solution of the polymer. Solutions of high molar mass PPQs can be readily prepared this way without gel formation. High molar mass, linear PQs cannot be prepared by melt polycondensation because the bisglyoxals decompose upon heating. The linear high molar mass PPQs are generally soluble in phenolic and chlorinated solvents but insoluble in polar solvents.

Some representative backbone structures of PQs and PPQs and their T_g data are given in Table 1. As in other amorphous polymers, the T_gs of PQs and PPQs are controlled essentially by the chemical structure, molecular weight, and thermal history. Several synthetic routes have been investigated to increase the T_g and also to improve the processibility of PPQ (71). Some properties of PPQ based on 2,3-di(3,4-diaminophenyl)quinoxaline and those of 1,1-dichloro-2,2-bis(3,4-diaminophenyl)ethylene are summarized in Table 2.

Acetylene-terminated phenylquinoxaline oligomers have been reported for possible adhesives in the aerospace industry (72–74). Several PQs containing 4-substituted phenyl groups on the quinazolene ring have been prepared and their thermal properties studied (75). However, in an attempt to improve the thermal stability of PQs, poly(phenylquinoxaline-co-naphthoylene) benzimidazoles have been synthesized through catalytic copolycondensation of bis(o-phenylenediamines) with bis(α-diketones) and bis(naphthalic anhydrides), a number of these having acenaphthylene groups (76).

A rigid-rod polyimide derived from biphenyldianhydride and p-phenylenediamine was modified by the incorporation of diamines containing phenylquinoxaline and aryl ether linkages and the morphology and mechanical properties of the resulting imide–aryl ether–phenylquinoxaline polymers were investigated (77). The films displayed T_gs in the 300°C range, and their thermal stabilities were comparable to that of the parent PI. Quinoxaline-activated poly(aryl ether) synthesis has been demonstrated to be an efficient route for the preparation of poly(aryl ether–phenylquinoxalines) (78,79).

New heat-resistant polymers containing p-nitrophenyl-substituted quinoxaline units and imide rings as well as flexible amide groups have been synthesized by polycondensation reaction of a diaminoquinoxaline derivative with diacid dichlorides (80). These polymers are easily soluble in polar aprotic solvents with inherent viscosities in the range of 0.3–0.9 dL/g in NMP at 20°C. All polymers begin to decompose above 370°C.

Table 1. Glass-Transition Temperatures of PQ and PPQ

X	Ar	$T_g°, °C^a$
	Polyquinoxalines	
—		393
—		354
	Polyphenylquinoxalines	
—O—		298
—O—		279
—O—		257
$\overset{O}{\underset{\parallel}{-C-}}$		325
$\overset{O}{\underset{\parallel}{-C-}}$		269
$\overset{O}{\underset{\underset{O}{\parallel}}{-S-}}$		346
$\overset{O}{\underset{\underset{O}{\parallel}}{-S-}}$		293

$^a T_g$ is determined by dsc at a heating rate of 20°C/min.

Applications. Films of PQs and PPQs can be readily prepared by melt pressing at temperatures above the T_g. Thermally and chemically stable polyquinoxalines find potential applications as films, coatings, adhesives, ultrafiltering materials, and composite matrices that demand stability in harsh environments. PPQs exhibit good stability toward strong acids and bases. Small

Table 2. Some Properties of PPQs

Ar[a]	$[\eta]$, dL/g	T_g°, C
	1.64	330
	0.65	280
	0.40	325
	0.57	290

[a]All show a 10% weight loss at 500°C.

composite nozzles of PPQ have shown excellent stability upon deep submergence in geothermal energy wells (superheated steam, brine, and sulfur compounds.)

PPQs possess a stepladder structure that combines good thermal stability, electrical insulation, and chemical resistance with good processing characteristics (81). These properties allow unique applications in the aerospace and electronics industries (82,83). PPQ can be made conductive by the use of an electrochemical oxidation method (84). The conductivities of these films vary from 10^{-7} to 10^{-12} S/cm depending on the dopant anions, thus finding applications in electronics industry. Similarly, some thermally stable PQs with low dielectric constants have been produced for microelectronic applications (85). Thin films of PQs have been used in nonlinear optical applications (86,87).

Polyquinolines

Polyquinolines are some of the most versatile thermally stable polymers; they were developed during the 1970s in response to increasing demand for high temperature resistant materials and are undergoing commercial development (Maxdem, Inc., San Dimas, California). Evidence of their stability is manifested by weight losses in nitrogen of only 15–30% when heated to 800°C and by demonstration of useful lifetimes in air at 300°C. Polyquinolines are characterized by repeating quinoline units, which display catenation patterns of 2,6-, 2,4-, or 3,6-.

Synthesis and Properties. Polyquinolines are formed by the step-growth polymerization of o-aminophenyl (aryl) ketone monomers and ketone monomers with alpha hydrogens (mostly acetophenone derivatives). Both AA–BB and AB-type polyquinolines are known as well as a number of copolymers. Polyquinolines have often been prepared by the Friedlander reaction (88), which involves either an acid- or a base-catalyzed condensation of an o-amino aromatic aldehyde or ketone with a ketomethylene compound, producing quinoline. Surveys of monomers and their syntheses and properties have been published (89–91).

The wide variety of ketomethylene and amino ketone monomers that could be synthesized, and the ability of the quinoline-forming reaction to generate high molar mass polymers under relatively mild conditions, allow the synthesis of a series of polyquinolines with a wide structural variety. Thus polyquinolines with a range of chain stiffness from a semirigid chain to rod-like macromolecules have been synthesized. Polyquinolines are most often prepared by solution polymerization of bis(o-amino aryl ketone) and bis(ketomethylene) monomers, where R = H or C_6H_5, in m-cresol with di-m-cresyl phosphate at 135–140°C for a period of 24–48 h (92).

Polyquinolines have also been obtained by a post-polymerization thermal treatment of poly(enamino nitriles) (93). The resulting polymers show excellent thermal stability, with initial weight losses occurring between 500 and 600°C in air (tga); under nitrogen, initial weight loss occurs at about 600°C and there is a 20% weight loss up to 800°C.

In an effort to increase the processibility of polyquinolines, fluoromethylene groups have been successfully incorporated into the chain in place of Ar in the bis(ketomethylene) moiety (94). In fact, a small percentage of perfluorobutylene groups in the polyquinoline chain was sufficient to decrease the T_g significantly while still retaining other desired mechanical and thermal properties. Another approach was to prepare a series of oligomeric polyquinolines containing pendent biphenylenes and capped with either phenyl or biphenylene moieties (95). A representative list of different polyquinoline backbones and their properties appears in Table 3.

The glass-transition temperatures of the polyquinolines are influenced greatly by the catenation pattern, the position of phenyl substitution, and the type of linkage connecting quinoline units in the polymer chain. Although many of the polyquinolines display melt-transition temperatures, x-ray analysis has shown that they are largely amorphous, typically with less than 20% crystallinity; however, rigid polyquinolines can be annealed above the T_g to a high (up to 60%) degree of crystallinity. The thermal stabilities of polyquinolines are excellent both in air and nitrogen. In general, polyquinolines with ether linkages are less stable than their rigid counterparts.

Table 3. Representative Polyquinolines and their Properties

Polymers	R	T_g,°C	Tga data, °C[a] Air	N$_2$
Flexible chains				
	H	266	545	565
	C$_6$H$_5$	305	530	555
Rigid chains				
	H	340	570	580
	C$_6$H$_5$	360	570	595
Semirigid chains				
	H	308	530	555
	C$_6$H$_5$	351	520	570

[a]Temperature of 10% weight loss.

All the flexible polyquinolines are readily soluble in chlorinated hydrocarbons such as methylene chloride and chloroform. Semirigid polyquinolines are soluble in tetrachloroethane or *m*-cresol, but rigid polyquinolines are soluble only in strong acids like sulfuric and trifluoromethane sulfonic acid. Dilute solution properties of polyquinolines have been investigated by techniques such as membrane osmometry, light scattering, viscometry, and gel-permeation chromatography (96,97).

Applications. Most of the stilbene-based polyquinolines display photoresponsive (98) and photomechanical effects as manifested by a contraction in polymer film samples upon irradiation.

Polyquinolines have been used as polymer supports for transition-metal catalyzed reactions. The coordinating ability of polyquinoline ligands for specific transition metals has allowed their use as catalysts in hydroformylation reactions (99) and for the electrochemical oxidation of primary alcohols (100).

Polyquinolines are good electrical insulators as indicated by conductivity values in the order of 10^{-15}–10^{-12} S/cm (101). However, by virtue of their extended conjugation, doped, wholly aromatic polyquinolines offer potential for high conductivity. Rigid polyquinolines display highest values of conductivity, generally on the order of 8–11 S/cm.

Some biphenylene end-capped polyquinolines have been used to make carbon-fiber reinforced composites (102). However, properties of these composites dropped off significantly when oxidatively aged for 50–100 h at 316°C.

Hexafluoroisopropylidene (HFIP)-Containing Polymers

Much attention has been paid to the synthesis of fluorine-containing condensation polymers because of their unique properties (43) and different classes of polymers including polyethers, polyesters, polycarbonates, polyamides, polyurethanes, polyimides, polybenzimidazoles, and epoxy prepolymers containing pendent or backbone-incorporated bis-trifluoromethyl groups have been developed. These polymers exhibit promise as film formers, gas separation membranes, seals, soluble polymers, coatings, adhesives, and in other high temperature applications (103,104). Such polymers show increased solubility, glass-transition temperature, flame resistance, thermal stability, oxidation and environmental stability, decreased color, crystallinity, dielectric constant, and water absorption.

Synthesis and Properties. In 1972, Du Pont marketed a series of linear aromatic polyimides called NR-150 (105) based on 2,2-bis(3,4-dicarboxyphenyl) hexafluoropropane dianhydride (6FDA) and diaminobenzene or 1,5-diaminonaphthalene. These polymers displayed excellent high temperature adhesive properties but were difficult to process because of the presence of condensation volatiles and high boiling solvent residues. Thermoplastic PIs for use as heatseals have been developed by Du Pont by reaction of 6FDA with a variety of diamines (106,107). Similarly, Hoechst-Celanese Corp. (108) has synthesized a class of fully imidized, soluble PIs based on hexafluoro-2,2-bis(aminophenyl)propanes (4,4'-6FDA and 3,3'-6FDA). These polymers form transparent, colorless, flexible films with good resistance to uv radiation and humidity at elevated temperature. Additionally, they exhibit higher long-term thermooxidative stability and better electrical insulating properties than the conventional PIs.

Polyimides of 6FDA and aliphatic diamines with good low temperature processing and low moisture swelling are known to be useful as hot-melt adhesives (109). Aluminum strips bonded by this polymer (177°C/172 kPa (25 psi) for 15 min) exhibited a lap-shear strength of 53 MPa (7690 psi) at room temperature and 35 MPa (5090 psi) at 100°C. The heat- and moisture-resistant 6F-containing PIs useful in electronic devices are prepared from 1,3-bis[4-(4-aminophenoxy)-α,α-bis-(trifluoromethyl)benzyl]benzene [89444-72-4] and PMDA (110). The T_g of the films (after being annealed at 300°C) is approximately 350°C.

In another study (111), several PIs were synthesized from 6FDA with different diamines. These polymers form tough, clear films with enhanced solubility in amide solvents. Inherent viscosities range from 0.28 to 1.10 dL/g with the number-average molar mass, M_n, from 22,100 to 170,000. Thermal stabilities (temperature of 10% weight loss by tga) are 480 to 530°C. The permeability of hydrogen for these films is roughly four times higher than in cellulose acetate or polysulfone.

A number of HFIP-derived polyethers are known which exhibit good mechanical, thermal, and electrical properties (112,113). Aromatic polyethers have been synthesized from bisphenol A (R = H) or AF (R = F) and fluorinated aromatics (Ar = perfluorophenyl, perfluorobiphenyl, or 2,4-difluorophenyl) (114–

116). Polymerization of bisphenol with 1,2,4,5-tetrafluorobenzene was not observed, and with hexafluorobenzene proceeded only if the ratio of potassium carbonate to bisphenol was carefully controlled. The polymer derived from deca-fluorobiphenyl and bisphenol AF is produced in 77% yield and has an inherent viscosity of 1.07 dL/g. These polymers are highly soluble and thermally stable and exhibit low water uptake (0.3%) and a very low dielectric constant (2.21).

$$\left(O-\!\!\!\bigcirc\!\!\!- \underset{\underset{CR_3}{|}}{\overset{\overset{CR_3}{|}}{C}} -\!\!\!\bigcirc\!\!\!- O-Ar \right)_n$$

Some amorphous copoly(ether–sulfone) films have been prepared (117) with T_gs around 130°C with no loss in weight up to 400°C in air or N_2. Other backbones investigated in this class of polymers are copoly(ether–amides) (118) and co-poly(ether–ketones) (119). These polymers show good mechanical properties, flow characteristics, and abrasion resistance.

A polyester backbone with two HFIP groups (12F aromatic polyester of 12F-APE) was derived by the polycondensation of the diacid chloride of 6FDCA with bisphenol AF or bisphenol A under phase-transfer conditions (120). These poly-mers show complete solubility in THF, chloroform, benzene, DMAC, DMF, and NMP, and form clear, colorless, tough films; the inherent viscosity in chloroform at 25°C is 0.8 dL/g. A thermal stability of 501°C (10% weight loss in N_2) was observed.

The first HFIP-based polycarbonate was synthesized from bisphenol AF with a nonfluorinated aromatic diol (bisphenol A) and phosgene (121,122). Incorpora-tion of about 2–6% of bisphenol AF and bisphenol A polycarbonate improved the dimensional stability and heat-distortion properties over bisphenol A homopoly-carbonate. Later developments in this area concern the flame-retardant proper-ties of these polymers (123,124).

$$\left(O-\!\!\!\bigcirc\!\!\!- \underset{\underset{CF_3}{|}}{\overset{\overset{CF_3}{|}}{C}} -\!\!\!\bigcirc\!\!\!- O-\overset{\overset{O}{\|}}{C}-Ar-\overset{\overset{O}{\|}}{C} \right)_n \qquad \left(O-\!\!\!\bigcirc\!\!\!- \underset{\underset{CF_3}{|}}{\overset{\overset{CF_3}{|}}{C}} -\!\!\!\bigcirc\!\!\!- O-\overset{\overset{O}{\|}}{C} \right)_n$$

polyester   polycarbonate

High molar mass polyurethanes were obtained from condensation of 4,4'-(hexafluoroisopropylidene)bis(phenylchloroformate) with various diamines (125). These polymers could be cast into transparent, flexible, colorless films or spun into fibers which showed promise as crease-resistant fabrics. Other polyurethanes discovered are good candidates for naval and aerospace applications (126).

$$\left(\bigcirc\!\!\!- \underset{\underset{CF_3}{|}}{\overset{\overset{CF_3}{|}}{C}} -\!\!\!\bigcirc\!\!\!- \overset{\overset{H}{|}}{N}-\overset{\overset{O}{\|}}{C}O-R-O\overset{\overset{O}{\|}}{C}-\overset{\overset{H}{|}}{N} \right)_n$$

Applications. The applications sought for these polymers include compos-ites, structural plastics, electronics/circuit boards, aircraft/spacecraft coatings, seals, dental and medical prosthetics, and laser window adhesives. However, other than the early commercialization by Du Pont of the NR-150 B material, little development has occurred. These polymers are quite expensive ($110 to $2200 per kg for monomers alone).

Hexafluoroisopropoxy (HFIP-O) Group-Containing Polymers

Several classes of polymers containing the HFIP-O group have been reported. These polymers show promise as film formers, gas separation membranes, coatings, seals, and other high temperature applications due to the properties imparted by this function, similar in many ways to the HFIP group.

Synthesis and Properties. Several polymers containing HFIP-O groups have been investigated, the most common beeing epoxies and polyurethanes. The development of fluorinated epoxy resins and the basic understanding of their chemistry has been reviewed (127).

Numerous avenues to produce these materials have been explored (128–138). The synthesis of two new fluorinated bicyclic monomers and the use of these monomers to prepare fluorinated epoxies with improved physical properties and a reduced surface energy have been reported (139,140). The monomers have been polymerized with the diglycidyl ether of bisphenol A, and the thermal and mechanical properties of the resin have been characterized. The resulting polymer was stable up to 380°C (10% weight loss by tga).

One of the first attempts to produce polyurethane was from the reaction of an intermediate polyol of 1,3- and 1,4-bis(hydroxyhexafluoroisopropyl)benzene (m- and p-12F-diols) by reaction with epichlorohydrin. This polyol was subsequently allowed to react with a commercial triisocyanate, resulting in a tough, cross-linked polyurethane (129,135,139). ASTM and military specification tests on these polyurethanes for weather resistance, corrosion prevention, blister resistance, and ease of cleaning showed them to compare quite favorably with standard resin formulations.

The next approach to incorporate the 12F-diol into a polyurethane matrix was reaction of the m-12F-diol with aliphatic diacid chlorides (where $x = 3$ or 4) to give low molar mass polyesters (141):

$$\text{HO-C}\underset{\overset{|}{CF_3}}{\overset{\overset{CF_3}{|}}{\bigcirc}}\text{C-OH} + \text{ClC}\text{-(CH}_2\text{)}_x\text{CCl} \xrightarrow{130\,°C,\,48\,h} \text{A}\!\left(\!\text{O-C}\underset{\overset{|}{CF_3}}{\overset{\overset{CF_3}{|}}{\bigcirc}}\text{C-O-C-(CH}_2\text{)}_x\text{C}\!\right)_{\!n}\!\text{B}$$

where A = H or aliphatic acid
B = OH or 12F-diol

For OH terminals chain extension leads to polyurethanes, polyesters, etc. For acid terminals, chain extension leads to polyesters, polyamides, etc. Interestingly, a stoichiometric balance between the acid and diol did not yield a high molar mass polyester, and further heating to 150°C resulted in degradation of the product. However, by reaction of a slight excess of the diol or diacid chloride, either an oligomeric (M_n = 2100–2700) diol or diacid, respectively, was achieved, either of which could then be chain extended to high molecular weights.

Following this work, the m-12F-diol was used for the direct reaction with hexamethylene-1,6-diisocyanate in the presence of dibutyltin dilaurate to produce a cross-linked elastomer or a reactive prepolymer which was terminated with either isocyanate or hydroxyl groups, depending on which reactant was in excess (142,143).

Acrylate or methacrylate resins (R = H or CH_3) have been obtained from 1,3-bis(hexafluoroisopropanolyl)-4-fluoroalkylbenzene (144), where R_f = $-C_nF_{2n-1}$ and n = 1–18, and X is as shown.

$$\text{XO-C}\underset{\overset{|}{CF_3}}{\overset{\overset{CF_3}{|}}{\bigcirc}}\text{C-OX;}\quad \text{X} = -\overset{\overset{O}{\|}}{C}-CR{=}CH_2{-}CH_2{-}\overset{\overset{OH}{|}}{C}H{-}CH_2{-}O{-}\overset{\overset{O}{\|}}{C}{-}CR{=}CH_2$$

These monomers were mixed with nonfluorinated acrylates and cured conventionally, such as by free-radical mechanism. Similar monomers and their polymers have also been reported (144,145). Monomer was synthesized by the condensation of acryloyl chloride with a 5-fluoroalkyl-substituted m-12F-diol in a chlorofluoroalkane solvent with triethylamine acid acceptor.

Applications. These polymers have been proposed or evaluated for a number of applications such as structural plastics, special aircraft coatings, aircraft windshield coatings, lubricant barrier films, coatings for ships and ice breakers, and as fillings for teeth or the molding of false teeth. More recent applications of these polymers include laser window adhesives and optical cement. The largest application, however, has been as a chemically resistant coating for storage tanks.

HFIP-O-containing fluoroacrylate polymers have been suggested earlier for biomedical applications such as artificial human organs and dental materials (146). This is largely due to the ability to polymerize monomers containing a suspension of polytetrafluoroethylene. The result is a tough product with low water absorption and a low friction surface. These polymers are also useful as textile impregnants (147). Epoxy resins have potential for use in computer composite circuit boards. The resin with branched ether suggests possible application in blood oxygenators or, eventually, artificial lungs (148).

BIBLIOGRAPHY

"Heat-Resistant Polymers" in *ECT* 3rd ed., Vol. 12, pp. 203–225, by J. Preston, Monsanto Triangle Park Development Center, Inc.

1. H. H. Levine, *Ind. Eng. Chem.* **54**, 22 (1962).
2. P. E. Cassidy and N. C. Fawcett, *Encycl. Chem. Technol.* **18**, 704 (1982).
3. P. E. Cassidy, *Thermally Stable Polymers–Synthesis and Properties*, Marcel Dekker, Inc., New York, 1980.
4. J. W. Verbicky, Jr., in J. I. Kroschwitz, ed., *Encyclopedia of Polymer Science and Engineering*, 2nd ed., Vol. 12, Wiley-Intersicence, New York, 1988, p. 364.
5. N. D. Ghatge, B. M. Shinde, and U. P. Mulik, *J. Macromol. Sci. Chem.* **A22**, 1109 (1985).
6. H. J. Jeong and co-workers, *J. Polym. Sci., Polym. Chem. Ed.* **29**, 39 (1991).
7. Y. Oishi and co-workers, *J. Polym. Sci., Polym. Chem. Ed.* **30**, 1027 (1992).
8. J. O. Park and S. H. Jang, *J. Polym. Sci., Polym. Chem. Ed.* **30**, 723 (1992).
9. H. R. Kricheldorf and R. Huner, *J. Polym. Sci., Polym. Chem. Ed.* **30**, 337 (1992).
10. J. R. Evans, R. A. Orwall, and S. S. Tang, *J. Polym. Sci., Polym. Chem. Ed.* **22**, 3559 (1984).
11. S. K. Dolui, D. Pal, and S. Maiti, *J. Appl. Polym. Sci.* **30**, 3867 (1985).
12. J. F. Dezern, *J. Polym. Sci., Polym. Chem. Ed.* **26**, 2157 (1988).
13. F. J. Serna, J. D. Abajo, and J. G. De la Campa, *J. Appl. Polym. Sci.* **30**, 61 (1985).
14. S. H. Hsiao and C. P. Yang, *J. Polym. Sci., Polym. Chem. Ed.* **29**, 447 (1991).
15. E. Butuc and G. Gherasim, *J. Polym. Sci., Polym. Chem. Ed.* **22**, 503 (1984).
16. J. A. Mikroyannidis and A. P. Melissaris, *J. Polym. Sci., Polym. Chem. Ed.* **26**, 1405 (1988).
17. V. V. Korshak and co-workers, *Acta Polym.* **39**, 8 (1988).
18. S. Kubota and T. Ando, *J. Appl. Polym. Sci.* **35**, 695 (1988).
19. R. O. Johnson and H. S. Burlhis, *J. Polym. Sci., Polym. Symp.* **70**, 129 (1983).
20. B. K. Mandal and S. Maiti, *J. Polym. Sci., Polym. Lett. Ed.* **23**, 317 (1985).
21. I. K. Varma, G. M. Fohlen, and J. A. Parker, *J. Polym. Sci., Polym. Chem. Ed.* **20**, 283 (1982).
22. A. V. Galanti and D. A. Scola, *J. Polym. Sci., Polym. Chem. Ed.* **19**, 451 (1981).
23. C. P. R. Nair and co-workers, *J. Polym. Sci., Polym. Chem. Ed.* **24**, 1109 (1986).
24. B. S. Rao, *J. Polym. Sci., Polym. Letts.* **26**, 3 (1988).
25. I. K. Varma, S. P. Gupta, and D. S. Varma, *Die Angew. Makromol. Chem.* **153**, 15 (1987).
26. K. N. Ninan and co-workers, *J. Appl. Polym. Sci.* **37**, 127 (1989).
27. S. Maiti and A. Ray, *Makromol. Chem.* **183**, 2949 (1982).
28. K. Kurita and H. Murakoshi, *Polym. Commun.* **26**, 179 (1985).
29. D. Kumar, *J. Polym. Sci., Polym. Chem. Ed.* **22**, 3439 (1984).
30. D. Kumar, G. M. Fohlen, and J. A. Parker, *J. Polym. Sci., Polym. Chem. Ed.* **22**, 927 (1984).
31. B. N. Achar, G. M. Fohlen, and J. A. Parker, *J. Polym. Sci., Polym. Chem. Ed.* **20**, 773 (1982).
32. B. N. Achar, G. M. Fohlen, and J. A. Parker, *J. Polym. Sci., Polym. Chem. Ed.* **20**, 2781 (1982).
33. B. N. Achar, G. M. Fohlen, and J. A. Parker, *J. Polym. Sci., Polym. Chem. Ed.* **21**, 3063 (1983).
34. B. N. Achar, G. M. Fohlen, and J. A. Parker, *J. Polym. Sci., Polym. Chem. Ed.* **22**, 319 (1984).
35. B. N. Achar, G. M. Fohlen, and J. A. Parker, *J. Polym. Sci., Polym. Chem. Ed.* **23**, 1677 (1985).

36. G. N. Babu, in K. L. Mittal, ed., *Polyimides, Syntheses, Characterization and Applications*, Vol. 1, Plenum Press, New York, 1984, p. 51.
37. A. K. St. Clair and T. L. St. Clair, in Ref. 36, Vol. 2, p. 977.
38. H. D. Burks and T. L. St. Clair, in Ref. 36, p. 117.
39. H. D. Burks and T. L. St. Clair, *J. Appl. Polym. Sci.* **30**, 2401 (1985).
40. G. D. Roberts and R. W. Lauver, *J. Appl. Sci.* **33**, 2893 (1987).
41. D. Wilson, *Brit. Polym. J.* **20**, 405 (1988).
42. D. Garcia and T. T. Sarafini, *J. Polym. Sci., Polym. Phys. Ed.* **25**, 2275 (1987).
43. P. E. Cassidy, T. M. Aminabhavi, and J. M. Farley, *J. Macromol. Sci., Rev. Macromol. Chem. Phys.* **C29**, 365 (1989).
44. P. E. Cassidy and co-workers, *European Polymer J.* (in press).
45. J. Malinge, J. Garapon, and B. Sillion, *Brit. Polym. J.* **20**, 431 (1988).
46. K. Taguchi, *J. Polym. Sci., Polym. Lett. Ed.* **18**, 525 (1980).
47. G. A. Brown, *Org. Coat. Plast. Prepr.* **43**, 476 (1980).
48. Y. K. Lee and J. D. Craig, *Org. Coat. Plast. Prepr.* **42**, 451 (1980).
49. P. E. Cassidy and N. C. Fawcett, *J. Macromol. Sci., Rev. Macromol. Chem.* **C17**, 209 (1979).
50. I. K. Varma and C. K. Geetha, *Indian J. Chem.* **16A**, 352 (1978).
51. I. K. Varma and C. K. Geetha, *J. Appl. Poly. Sci.* **22**, 411 (1978).
52. S. U. Ahmed and S. I. Ahmed, *Polym. Mater. Sci. Eng.* **59**, 994 (1988).
53. N. D. Negodyaev and T. P. Sokolova, Deposited Doc., VINITI 3390-75, (1975); *Chem. Abstr.* **88**, 74564.
54. Y. Saegusa, T. Koshikawa, and S. Nakamura, *J. Polym. Sci., Polym. Chem. Ed.* **30**, 1369 (1992).
55. I. Schopov and M. Vodenicharova, *Makromol. Chem.* **179**, 63 (1978).
56. A. Klimmek and J. Krieger, *Angew. Makromol. Chem.* **109/110**, 165 (1982).
57. C. J. Thaemlitz, W. J. Weikel, and P. E. Cassidy, *Polym. Prepr.* **32(2)**, 260 (1991).
58. E. Oikawa and H. Nozawa, *Polym. Bull.* **13**, 481 (1985).
59. J. L. Hedrick, *Polym. Bull.* **25**, 543 (1991).
60. M. Sato and M. Yokoyama, *J. Polym. Sci., Polym. Chem. Ed.* **18**, 275 (1980).
61. G. P. de Gaudemaris and B. J. Sillion, *J. Polym. Sci., Polym. Phys. Ed.* **2**, 2203 (1964).
62. J. K. Stille and J. R. Williamson, *J. Polym. Sci., Polym. Chem. Ed.* **2**, 3867 (1964); and *Polym. Phys. Ed.* **2**, 209 (1964).
63. P. M. Hergenrother and H. H. Levine, *J. Polym. Sci., Poly. Chem. Ed.* **5**, 1453 (1967).
64. P. M. Hergenrother, in J. I. Kroschwitz, ed, *Encyclopedia of Polymer Science and Engineering*, 2nd ed., Vol. 13, Wiley-Interscience, New York, 1988, p. 55.
65. P. M. Hergenrother, *J. Marcromol. Sci. Rev. Macromol. Chem.* **6**, 1 (1971).
66. F. L. Hedberg and F. E. Arnold, *J. Polym. Sci., Polym. Chem. Ed.* **14**, 2607 (1976).
67. P. M. Hergenrother, *Macromolecules* **14**, 8918 (1981).
68. P. M. Hergenrother, *Polym. Eng. Sci.* **21**, 1072 (1981).
69. R. F. Kovar, G. F. L. Ehlers, and F. E. Arnold, *J. Polym. Sci., Polym. Chem. Ed.* **15**, 1081 (1977).
70. P. M. Hergenrother, *J. Appl. Polym. Sci.* **28**, 355 (1983).
71. N. M. Belomoina and co-workers, in M. J. M. Abadie and B. Silliori, eds., *Polyimides and Other High Temperature Polymers*, Elsevier Science Publishers, Amsterdam, 1991, p. 143.
72. S. Lin and C. S. Marvel, *J. Polym. Sci., Polym. Chem. Ed.* **22**, 1939 (1984).
73. R. F. Kovar, G. F. L. Ehlers, and F. E. Arnold, *J. Polym. Sci., Polym. Chem. Ed.* **15**, 1081 (1977).
74. F. L. Hedberg and F. E. Arnold, *J. Appl. Sci.* **24**, 763 (1979).
75. S. Kubota and co-workers, *J. Polym. Sci., Polym. Chem. Ed.* **24**, 2047 (1986).
76. V. V. Korshak and co-workers, *Acta Polym.* **39**, 455 (1988).
77. J. L. Hedrick and J. W. Labadie, *J. Polym. Sci., Polym. Chem. Ed.* **30**, 105 (1992).

78. J. L. Hedrick and J. W. Labadie, *Macromolecules* **21**, 1883 (1989).
79. J. W. Labadie, J. L. Hedrick, and S. K. Boyer, *J. Polym. Sci., Polym. Chem. Ed.* **30**, 519 (1992).
80. M. Bruma and co-workers, *Angew. Makromol. Chem.* **193**, 113 (1991).
81. L. Fengcai, W. Baigeng, and C. Jinbao, *Polym. Sci. Technol.* **26**, 261 (1984).
82. L. Y. Chiang and co-workers, *Polym. Mater. Sci. Eng.* **64**, 216 (1991).
83. N. H. Hendricks and co-workers, *Int. SAMPE Electron.* [*Conf.*] **4**, 544 (1990).
84. Z. Chi, L. Zhugan, and L. Fengcai, *Polymer* **32**, 3075 (1991).
85. N. H. Hendricks and co-workers, *Int. SAMPE Electron.* [*Conf.*] **5**, 365, (1991).
86. A. K. Agarwal and S. A. Jenekhe, *Chem. Mater.* **4**, 95 (1992).
87. A. K. Agarwal and co-workers, *Polym. Prepr.* **32**(3), 124 (1991).
88. P. Friedlander, *Chem. Ber.* **15**, 2572 (1882).
89. D. M. Sutherlin, in J. I. Kroschwitz ed., *Encyclopedia of Polymer Science and Engineering,* 2nd ed., Vol. Index, Wiley-Interscience, New York, 1990, p. 279.
90. J. F. Wolfe and J. K. Stille, *Macromolecules* **9**, 489 (1976).
91. J. K. Stille, *Contemp. Top. Polym. Sci.* **5**, 209 (1984).
92. P. D. Sybert and J. K. Stille, *Macromol. Synth.* **9**, 49 (1985).
93. J. A. Moore and D. R. Robello, *Macromolecules* **22**, 1084 (1989).
94. D. T. Clark and co-workers, *Macromolecules* **17**, 1871 (1984).
95. D. M. Sutherlin and J. K. Stille, *Macromolecules* **19**, 251 (1986).
96. P. M. Cotts and G. C. Berry, *Polym. Prepr. Am. Chem. Soc. Div. Polym. Chem.* **24**, 328 (1983).
97. P. M. Cotts, *J. Polym. Sci., Polym. Phys. Ed.* **24**, 1493 (1986).
98. E. K. Zimmermann and J. K. Stille, *Macromolecules* **18**, 321 (1985).
99. M. Ding and J. K. Stille, *Macromolecules* **16**, 839 (1983).
100. S. J. Stoessel, C. M. Elliott, and J. K. Stille, *Chem. Mat.* **1**, 259 (1989).
101. S. E. Tunney, J. Suenaga, and J. K. Stille, *Macromolecules* **20**, 258 (1987).
102. J. P. Droske, J. K. Stille, and W. B. Alston, *Macromolecules* **17**, 14 (1984).
103. W. J. Koros and D. R. B. Walker, *Polym. J.* **23**, 481 (1991).
104. C. J. Thaemlitz, W. J. Weikel, and P. E. Cassidy, *Polymer* **33**, 3278 (1992).
105. H. H. Gibbs, *Proc. 17th Natl. SAMPE Symp.* **17**, 1 (1972).
106. H. H. Gibbs and C. V. Breder, *Polym. Prepr.* **15**(1), 775 (1974).
107. H. H. Gibbs and C. V. Breder in N. A. J. Platzer, ed., *Copolymers, Polyblends, and Composites, Adv. Chem. Ser., No. 142*, Am. Chem. Soc., Washington, D.C., 1975, p. 442.
108. W. H. Mueller and R. Vora, *Product Data Sheet*, American Hoechst Corp., SPG-Central Research, Rhode Island, 1987.
109. U.S. Pat. 4,569,988 (Feb. 11, 1986), D. A. Scola and R. H. Pater (to United Technologies Corp.).
110. Jpn. Kokai Tokkyo Koho, JP. 58,180,531 (Oct. 22, 1983), (to Hitachi Chem. Co., Ltd.).
111. G. R. Husk, P. E. Cassidy, and K. L. Gebert, *Macromolecules* **21**, 1234 (1988).
112. Fr. Pat. 1,394,897 (Apr. 9, 1965), G. S. Stamatoff and J. W. Wittaman, (to E. I. du Pont de Nemours & Co., Inc.).
113. U.S. Pat. 3,332,909 (July 25, 1967), A. G. Farnham and R. N. Johnson (to Union Carbide Corp.).
114. J. Irvin and co-workers, *J. Polym. Sci., Polym. Chem. Ed.* **30**, 1675 (1992).
115. F. W. Mercer and co-workers, *J. Polym. Sci., Polym. Chem. Ed.* **30**, 1767 (1992).
116. U.S. Pat. 5,115,082 (May 19, 1992), F. W. Mercer and R. C. Sovish (to Raychem Corp.).
117. M. Shimizu, M. Kakimoto, and Y. Imai, *J. Polym. Sci., Polym. Chem. Ed.* **25**, 2385 (1987).
118. Jpn. Kokai Tokkyo Koho JP, 58,149,944 (Sept. 6, 1983) (to Hitachi Chem. Co., Ltd.).
119. P. E. Cassidy, G. L. Tullos, and A. K. St. Clair, *Macromolecules* **24**, 6059 (1991).
120. L. S. Wells, P. E. Cassidy, and K. M. Kane, *High Perform. Polym.* **3**, 191 (1991).

121. Eur. Pat. Appl. 29,111 (May 27, 1981), S. Krishnan and A. L. Baron (to Mobay Chemical Corp.).
122. U.S. Pat. 4,346,211 (Aug. 24, 1982), S. Krishnan and A. L. Baron (to Mobay Chemical Corp.).
123. PCT Int. Appl. WO 82 02,402 (July 22, 1982), V. Mark and C. V. Hedges (to General Electric Co.).
124. Jpn. Kokai Tokkyo Koho, JP., 62,141,061 (June 24, 1987), T. Tokuda and K. Furukawa (to Teijin Chemicals Ltd.).
125. U.S. Pat. 3,373,139 (Mar. 12, 1968), P. W. Morgan (to E. I. du Pont de Nemours & Co., Inc.).
126. T. M. Keller, *J. Polym. Sci., Polym. Chem. Ed.* **23**, 2557 (1985).
127. J. R. Griffith, J. G. O'Rear, and S. A. Reines, *Chemtech*, 311 (May 1972).
128. J. R. Griffith and D. E. Field, *NRL Progr. Rep.*, (June 1973).
129. J. G. O'Rear and J. R. Griffith, *Am. Chem. Soc. Div. Org. Coat. Plast. Chem. Prepr.* **33**(1), 657 (1973).
130. U. S. Pat. 3,879,430 (Apr. 22, 1975), J. G. O'Rear and J. R. Griffith (to U. S. Navy).
131. U.S. Pat. 3,852,222 (Dec. 31, 1974), E. F. Donald, C. V. Falls, and J. R. Griffith (to U. S. Navy).
132. J. R. Griffith, D. E. Field, and J. G. O'Rear, *Am. Chem. Soc. Div. Org. Coat. Plast. Chem. Prep.* **34**(1), 709 (1974).
133. Can. Pat. 956,398 (Oct. 15, 1974), J. R. Griffith (to U. S. Navy).
134. D. E. Field and J. R. Griffith, *Ind. Eng. Chem. Prod. Res. Dev.* **14**, 52 (1975).
135. J. R. Griffith, J. G. O'Rear, and J. P. Reardon, *Polym. Sci., Technol. (Adhes. Sci. Technol.)* **9A**, 429 (1975).
136. U.S. Pat. Appl. 329,229 (Mar. 27, 1989), J. R. Griffith (to U. S. Navy).
137. J. R. Griffith and J. G. O'Rear, *Am. Chem. Soc. Div. Org. Coat. Plast. Chem. Prep.* **40**, 781 (1979).
138. U.S. Pat. 4,045,408 (Aug. 30, 1977), J. R. Griffith and J. G. O'Rear (to U. S. Navy).
139. R. F. Brady and A. M. Sikes, *Macromolecules* **24**, 688 (1991).
140. A. M. Sikes and R. F. Brady, *Polym. Prep.* **31**(1), 358 (1990).
141. T. M. Keller, *J. Polym. Sci., Polym. Chem. Ed.* **22**, 2719 (1984).
142. T. M. Keller, *J. Polym. Sci., Polym. Chem. Ed.* **23**, 2557 (1985).
143. B. S. Holmes and T. M. Keller, *Am. Chem. Soc., Div. Polym. Chem. Prepr.* **25**(1), 338 (1984).
144. U.S. Pat. Appl. 237,838 (Aug. 14, 1981), J. R. Griffith and J. G. O'Rear (to U. S. Navy).
145. J. R. Griffith, *Am. Chem. Soc., Div. Polym. Mater. Sci., Eng. Prepr.* **50**, 304 (1984).
146. U.S. Pat. 4,356,296 (Oct. 26, 1982), J. R. Griffith and J. G. O'Rear (to U.S. Navy).
147. U.S. Pat. 3,544,535 (Dec. 1, 1970), E. E. Gilbert (to Allied Chemical Corp.).
148. J. R. Griffith and J. G. O'Rear, *Polym. Mater. Sci. Eng.* **53**, 766 (1985).

PATRICK E. CASSIDY
TEJRAJ M. AMINABHAVI
V. SREENIVASULU REDDY
Southwest Texas State University

HEAT STABILIZERS

Heat stabilizers protect polymers from the chemical degrading effects of heat or uv irradiation. These additives include a wide variety of chemical substances, ranging from purely organic chemicals to metallic soaps to complex organometallic compounds. By far the most common polymer requiring the use of heat stabilizers is poly(vinyl chloride) (PVC). However, copolymers of PVC, chlorinated poly(vinyl chloride) (CPVC), poly(vinylidene chloride) (PVDC), and chlorinated polyethylene (CPE), also benefit from this technology. Without the use of heat stabilizers, PVC could not be the widely used polymer that it is, with worldwide production of nearly 16 million metric tons in 1991 alone (see VINYL POLYMERS).

The discussion centers on heat stabilizers for PVC because this polymer is the most important class of halogenated polymers requiring these chemical additives. PVC of ideal chemical structure (1) should be a relatively stable compound as predicted from model studies using 2,4,6-trichloroheptane [13049-21-3] (2) (1).

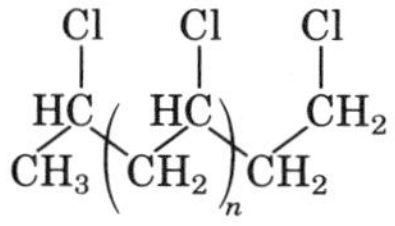

(1) 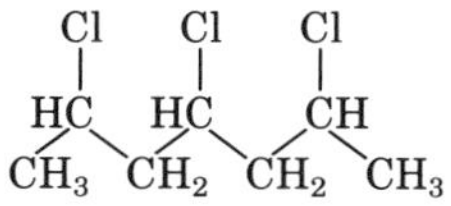   (2)

During the polymerization process the normal head-to-tail free-radical reaction of vinyl chloride deviates from the normal path and results in sites of lower chemical stability or defect sites along some of the polymer chains. These defect sites are small in number and are formed by autoxidation, chain termination, or chain-branching reactions. Heat stabilizer technology has grown from efforts to either chemically prevent or repair these defect sites. Partial structures (3–6) are typical of the defect sites found in PVC homopolymers (2–5).

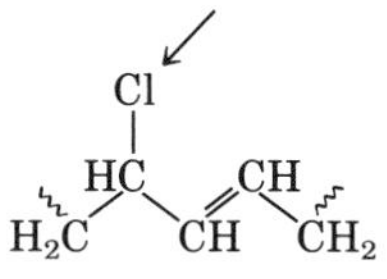

(3) 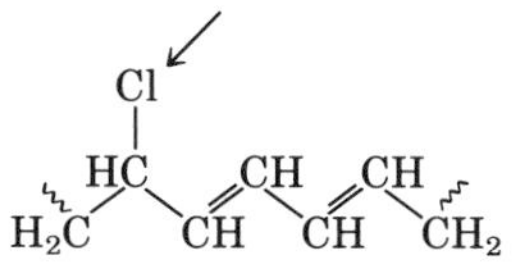   (4)   (6)

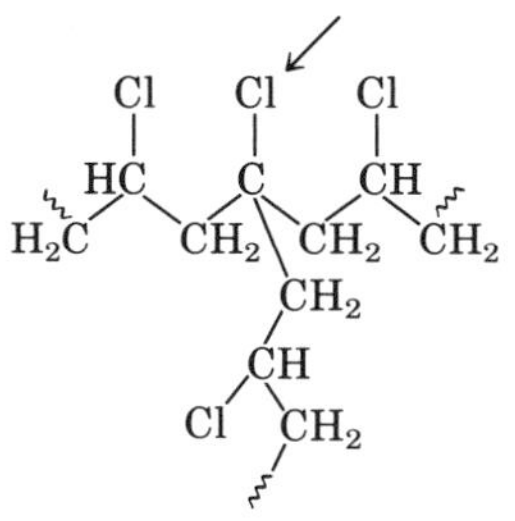

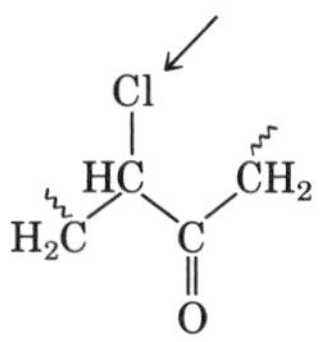

(5)

The dissociation energies for the highlighted (by pointing arrows) carbon–chlorine bonds are significantly lower than that of a normal secondary C–Cl bond and can lead to thermal dehydrochlorination of the polymer backbone. In addition, the released HCl acts to catalyze further dehydrochlorination, indicating that both homolytic and ionic processes are involved. As the conjugated polyene sequences grow in length, further weakening of the carbon–chlorine bond occurs leading eventually to rapid, catastrophic dehydrochlorination, cross-linking, and chain scission resulting in loss of mechanical, electrical, and rheological properties in the final articles. A good indication of the onset of thermal degradation is the appearance of color in the polymer. When the conjugated polyenes reach a length of about seven double bonds, this chromophore begins to absorb visible light and appears yellow in color. Further degradation leads to brown and eventually black-colored products.

Some work has been conducted on the *in situ* or preventative stabilization of PVC, but these efforts have been largely unsuccessful and costly and have primarily centered on copolymerizing vinyl chloride with other vinyl monomers to block the conjugative ordering of the chlorine atoms in the polymer (6). The commercially important heat stabilizers are arrestive in nature. They chemically repair the defect sites or in some way reduce the deleterious nature of these sites during the processing and use of PVC articles. Just as important, most of the active heat stabilizers are also good HCl scavengers and reduce the catalytic effects of this deleterious by-product. PVC is so widely used because its properties are so easily manipulated by an inexhaustible variety of added components. Applications can range from rugged PVC pipes to crystal clear drinking water bottles to colorful toys to supple artificial leather depending on the choice of additives in the formulation. The needs and choices of the heat stabilizer are dependent upon the final desired physical properties and the processes used to form the articles. Figure 1 shows the complexity of PVC formulation alternatives. Only the heat stabilizers and, normally, lubricants are essential additives to process the polymer. All of the other classes of additives are entirely discretionary.

In normal operations, PVC resin is intimately mixed with the desired ingredients under high intensity shear mixing conditions to result in a homogeneous dry powder compound. The heat stabilizers can be either liquids or powders and

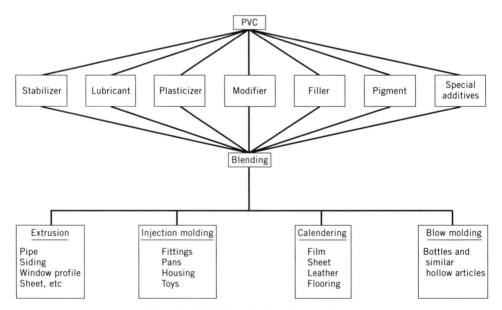

Fig. 1. PVC formulation alternatives.

are added early in the blending cycle to afford stabilizing action during this operation. Preheating the resin to about the glass-transition temperature facilitates the adsorption of the liquid additives giving the final compound better powder flow properties and decreasing the bulk density. Post-compounding operations, eg, extrusion pelletizing, can increase the overall heat history of the polymer, thus necessitating slightly higher levels of heat stabilizers to compensate for this.

History of Stabilizer Development

Although PVC was discovered in the late nineteenth century in Germany, it was not until the discovery that certain lead soaps improved the thermal stability of the polymer in the early 1930s that commercialization of PVC began. Lead-based stabilizers have continually been used in certain applications, such as wire and cable coatings; however, because of toxicity and ecotoxicity concerns, considerable efforts are being made to eliminate all uses of lead-based heat stabilizers. In the mid-1930s, workers in the United States found that organotin carboxylates also provided good heat stability. The high activity, nonstaining nondusting characteristics, and complete compatibility, allowing total transparency, made the organotin stabilizers a good choice for rigid, flexible, and even plastisol end uses. During the 1940s, alkali and alkaline-earth metal soaps, especially those of cadmium, barium, zinc, and calcium, were discovered and commercialized as PVC heat stabilizers (see DRIERS AND METALLIC SOAPS). By the 1950s combinations of these metal soaps with organic costabilizers, such as phosphites and epoxides, demonstrated great processing latitude and cost effectiveness, particularly in plasticized PVC. Some of these formulations remain state-of the-art technology for certain end uses. Just as in the case for lead, considerable pressure has been exerted on the plastics industry to discontinue uses of cadmium for both stabilizers and pigments because of its high toxicity. Organotin mercaptide chemistry was introduced as a new class of PVC heat stabilizers during the 1950s and played a significant role in the development of many rigid PVC applications. The efficiency per unit dose remains unsurpassed for the organotin mercaptide-stabilized formulations for food packaging (qv), water pipes, weatherable construction panels, and molded parts. Antimony mercaptides were also discovered in the 1950s and did compete with the organotin mercaptides in rigid PVC applications for several years, but again, toxicity concerns have greatly diminished the use of this class of stabilizer.

Function of Stabilizers

Although great progress has been made since the 1960s to sort out the many reactions ongoing during the thermal degradation of PVC, much of the formulating and use of stabilizers remains an art in commercial practice. PVC degradation proceeds by both free-radical and ionic reactions, although the latter appears to be the more important route. Lewis acid catalysts, such as zinc chloride or hydrogen chloride, can greatly accelerate the rate of dehydrochlorination of the polymer. Heat stabilizers serve several distinct functions during PVC processing: absorption of hydrogen chloride, replacement of labile chlorines, prevention of autoxidation, and disruption of polyunsaturated sequences. An ancillary function of many heat stabilizers is provision of uv stability, leading to good weathering properties for the final articles. Good uv stability depends on having adequate heat stability because partially degraded PVC is susceptible to uv degradation regardless of the choice of uv absorber additives used.

Absorption of Hydrogen Chloride. Effective heat stabilizers have the ability to bind hydrogen chloride. Most stabilizer systems contain one or more metallic soaps or salts which readily undergo a simple acid–base reaction with the by-product hydrogen chloride as the PVC degrades:

$$M(X)_n + HCl \rightarrow M(X)_{n-1}Cl + HX$$

$$M(X)_{n-1}Cl + HCl \rightarrow M(X)_{n-2}Cl_2 + HX, \text{ etc}$$

where M is a metal, usually Pb, Zn, Ca, Ba, or Cd, and X is a carboxylic acid radical, usually a weakly acidic fatty acid ligand. Typical examples of effective metal soaps and salts used as PVC stabilizers include lead stearate, dibasic lead phthalate, tribasic lead sulfate, zinc octanoate [557-09-5], barium tallate, cadmium 2-ethylhexanoate, calcium stearate, calcium nonylphenate [30977-64-1].

A new type of inorganic metal complex, called hydrotalcite has appeared (ca 1990). These synthetic minerals, functionally akin to zeolites, have layered structures of Al and Mg and function to trap hydrogen chloride between these layers (7). The hydrotalcite minerals are generally used with other stabilizers as part of a stabilizer system. A typical hydrotalcite may be represented by a formula such as $Mg_4Al_2(OH)_{12}CO_3 \cdot 3H_2O$. Many modifications can be made by changing the Al to Mg ratio and by including other metal salts, such as zinc oxide. Unlike most metallic salts, the hydrotalcites are compatible with PVC and can provide completely transparent PVC articles.

Many PVC stabilizer formulations also contain one or more organic costabilizers that can also absorb hydrogen chloride. Typical of these additives are epoxidized fatty acid esters and organophosphites:

$$R\!-\!\overset{O}{\overset{/\,\backslash}{CH\!-\!CH}}\!-\!R' + HCl \rightarrow R\!-\!\overset{OH}{\underset{|}{CH}}\!-\!\overset{Cl}{\underset{|}{CH}}\!-\!R'$$

$$P(OR)_3 + HCl \rightarrow H\!-\!\overset{O}{\overset{\|}{P}}(OR)_2 + R\!-\!Cl$$

Organotin mercaptides can also absorb hydrogen chloride.

$$R\!-\!Sn(SR')_3 + HCl \rightarrow R\!-\!Sn(SR')_2Cl + HSR'$$

$$R\!-\!Sn(SR')_2Cl + HCl \rightarrow R\!-\!Sn(SR')\,Cl_2 + HSR', \text{ etc}$$

Although some of these stabilizers are added specifically to react with evolved hydrogen chloride, when the primary function of the stabilizer is to repair defect sites or disrupt autoxidation reactions, the degree that these stabilizers react with hydrogen chloride can actually detract from their primary function necessitating the use of higher levels of stabilizers in the PVC formulation.

Replacement of Labile Chlorines. When PVC is manufactured, competing reactions to the normal head-to-tail free-radical polymerization can sometimes take place. These side reactions are few in number yet their presence in the finished resin can be devastating. These abnormal structures have weakened carbon–chlorine bonds and are more susceptible to certain displacement reactions than are the normal PVC carbon–chlorine bonds. Carboxylate and mercaptide salts of certain metals, particularly organotin, zinc, cadmium, and antimony, attack these labile chlorine sites and replace them with a more thermally stable C–O or C–S bound ligand. These electrophilic metal centers can readily coordinate with the electronegative polarized chlorine atoms found at sites similar to structures (**3–6**).

In the early 1960s, two different ^{14}C and one ^{113}Sn radio-labeled di-n-butyltin bis(isooctylthioglycolate) stabilizers were synthesized. Heat stability studies with these organotin compounds demonstrated the important replacement reactions leading to the following proposed mechanism (8). It is believed that the salts of zinc, cadmium (9), and lead (10) also undergo these reactions resulting in their respective ligands substituting on the PVC chains.

This mechanism not only accounts for the substitution of the more labile chlorine atom on the polymer chain, it also results in the elimination of a new potential initiation site by moving the double bond out of conjugation with any adjacent chlorine atoms. The newly formed C–O or C–S bonds, with $\Delta H > 484$ kJ/mol (100 kcal/mol), are significantly more thermally stable than even the normal C–Cl bonds in PVC at about 411 kJ/mol (85 kcal/mol) (11).

Ultimately, as the stabilization reactions continue, the metallic salts or soaps are depleted and the by-product metal chlorides result. These metal chlorides are potential Lewis acid catalysts and can greatly accelerate the undesired dehydrochlorination of PVC. Both zinc chloride and cadmium chloride are particularly strong Lewis acids compared to the weakly acidic organotin chlorides and lead chlorides. This significant complication is effectively dealt with in commercial practice by the co-addition of alkaline-earth soaps or salts, such as calcium stearate or barium stearate, ie, by the use of mixed metal stabilizers.

Displacement of activated chlorine atoms also proceeds with certain types of organic compounds, but only in the presence of Lewis acid catalysts. Particular examples include epoxides, polyhydric alcohols, trialkylphosphites (12), and β-aminocrotonates (13). These additives are commonly used in conjunction with metallic stabilizers to provide complete, high performance, commercial stabilizer packages.

Prevention of Autoxidation.   The observation that PVC thermally degrades more rapidly in air than in a nitrogen atmosphere leads to the conclusion that the prevention of oxidative reactions can improve the thermal stability of PVC (14). When phenolic antioxidants (qv) are included in the formulation, the rate of hydrogen chloride evolution, at 180°C in air, is noticeably retarded. Therefore, good stabilizers must also provide antioxidant protection. Many of the additives previously discussed, which provide HCl scavenging and labile chlorine displacements, are also fairly good antioxidants as evidenced by their activity in other nonhalogenated polymers, particularly trialkyl- and triarylphosphites, β-aminocrotonates, and organotin mercaptides. The organotin mercaptides are particularly efficient in their reduction of hydroperoxides (15):

$$R_2Sn(SR')_2 \; + \; -CH_2\!-\!\underset{\underset{\textstyle OOH}{|}}{CH}\!-\!CH_2\!- \; \longrightarrow \; [R_2SnO]_n \; + \; -CH_2\!-\!\underset{\underset{\textstyle OH}{|}}{CH}\!-\!CH_2\!- \; + \; R'S\!-\!SR'$$

Disruption of Polyunsaturated Sequences.   Sequential lengthening of the conjugated double bonds along a PVC molecule leads to the gradual development of color resulting from absorption of visible light (see COLOR). This lengthening conjugation further weakens the already weak allylic C–Cl bonds and leads to further loss of hydrogen chloride which, in turn, catalyzes further degradation. Left unchecked, the complete catastrophic degradation of the polymer is imminent.

Disruption of the conjugation of these long polyene groups can lead to color bleaching of the partially degraded PVC. Simple mercaptans effectively add to these highly reactive double bonds (16). Organotin mercaptides are also known to undergo this reaction, although it is likely that the actual reagent is the free mercaptan resulting from reaction with evolved HCl (17):

$$R_2Sn(SR')_2 \; + \; HCl \; \longrightarrow \; R_2\underset{\underset{\textstyle Cl}{|}}{Sn}SR' \; + \; HSR'$$

The organotin maleate and maleate half-ester derivatives also exhibit this bleaching effect reportedly by a Diels-Alder addition reaction (18). The reaction is specific to the organotin maleates; other organotin carboxylates containing normal dieneophiles fail to produce similar results (19).

Stabilizer Test Methods

Heat stabilizers are tested in a variety of ways to simulate their performance during PVC processing, ie, calendering, extrusion, pressing, heat curing, and molding. Because of the wide number of these applications, it is impossible to provide one or two definitive laboratory tests for all stabilizer products. In general, heat stabilizers are tested as a component in a complete formulation where each ingredient has a measured effect on the overall performance. Stabilizer performance is generally evaluated by visually inspecting the color of the test pieces as a function of heating and processing time. Static oven aging, dynamic two-roll milling, and torque rheometry are three of the most common tests used to evaluate heat stabilizers. A whole host of tests are conducted on the final products of the various processes to make judgments as to the effectiveness of any particular PVC formulation. Standard tests for stabilizer evaluation include ASTM D2115, Oven Stability of PVC Compounds; ASTM D2538, Fusion of PVC Compounds Using a Torque Rheometer; ASTM D1499, Stability of PVC to Light Exposure; and DIN 53-381F, Heat Stability of PVC Compounds by Metrastat Oven.

Classes of Heat Stabilizers

Organotin Compounds. Organotin-based heat stabilizers are the most efficient and universally used PVC stabilizers. Nearly 40% of the 64,000 metric tons of stabilizers used in the United States during 1992 were organotin-based products. These are all derivatives of tetravalent tin, and all have either one or two alkyl groups covalently bonded directly to the tin atom. The commercially important alkyltins are the methyltin, n-butyltin, and n-octyltin species. For about 10 years between 1980 and 1990, estertin or β-carboalkoxyethyltin derivatives were produced commercially. Aryltin and branched alkyltin derivatives are relatively poor heat stabilizers (20). The class of tri-n-alkyltin compounds are known to be toxic, and because they are very poor stabilizers, they are completely avoided in PVC heat stabilizer applications. The anionic ligands can be chosen from a wide variety of groupings but are normally chosen from maleate, alkylmaleate, esters of thioglycolic acid or esters of mercaptoethanol, depending on the end use application and the processing conditions for the PVC compound. Almost all organotin stabilizer products are formulated with mixtures of monoalkyltin and dialkyltin species in a ratio to maximize the stability of the resin together with all of the other microingredients in a given PVC formulation.

Alkyltin Intermediates. For the most part, organotin stabilizers are produced commercially from the respective alkyltin chloride intermediates. There are several processes used to manufacture these intermediates. The desired ratio of monoalkyltin trichloride to dialkyltin dichloride is generally achieved by a redistribution reaction involving a second-step reaction with stannic chloride (tin(IV) chloride). By far, the most easily synthesized alkyltin chloride intermediates are the methyltin chlorides because methyl chloride reacts directly with tin metal in the presence of a catalyst to form dimethyltin dichloride cleanly in high yields (21). Coaddition of stannic chloride to the reactor leads directly to almost any desired mixture of mono- and dimethyltin chloride intermediates:

$$Sn^0 + 2\ Cl_2 \rightarrow SnCl_4$$

$$Sn^0 + 2\ CH_3Cl \rightarrow (CH_3)_2SnCl_2$$

$$(y + x)\ (CH_3)_2SnCl_2 + x\ SnCl_4 \rightarrow y\ (CH_3)_2SnCl_2 + 2x\ CH_3SnCl_3$$

The direct reaction of other alkyl chlorides, such as butyl chloride, results in unacceptably low overall product yields along with the by-product butene resulting from dehydrochlorination. All alkyl halides having a hydrogen atom in a β- position to the chlorine atom are subject to this complication.

The other important direct alkylation processes involve reaction of electron-rich olefinic compounds with either tin metal or stannous chloride (tin(II) chloride) in the presence of stoichiometric amounts of hydrogen chloride (22). Butyl acrylate ($R = C_4H_9$) was used commercially in this process to prepare the estertin or β-carboalkoxyethyltin chlorides as illustrated in the following.

$$Sn^0 + 2\ HCl + 2\ H_2C{=}CHCOOR \rightarrow (ROOCCH_2CH_2)_2SnCl_2$$

$$SnCl_2 + HCl + H_2C{=}CHCOOR \rightarrow ROOCCH_2CH_2SnCl_3$$

A number of activated olefinic compounds react very well in this scheme including methacrylates, crotonates, acrylonitrile, and vinyl ketones. These reactions are typically run in an etherial solvent and can be run without the complications of undesirable side reactions leading to trialkylated tin species.

The other commercially important routes to alkyltin chloride intermediates utilize an indirect method having a tetraalkyltin intermediate. Tetraalkyltins are made by transmetallation of stannic chloride with a metal alkyl where the metal is typically magnesium or aluminum. Subsequent redistribution reactions with additional stannic chloride yield the desired mixture of monoalkyltin trichloride and dialkyltin dichloride. Both *n*-butyltin and *n*-octyltin intermediates are manufactured by one of these schemes.

$$2\ R_2Mg + SnCl_4 \rightarrow R_4Sn + 2\ MgCl_2$$

$$4\ R_3Al + 3\ SnCl_4 \rightarrow 3\ R_4Sn + 4\ AlCl_3$$

$$R_4Sn + 2\ SnCl_4 \rightarrow 2\ RSnCl_3 + R_2SnCl_2$$

Stabilizer Synthesis.   The selected alkyltin chloride intermediate reacts with either a carboxylic acid or a mercaptan in the presence of an appropriate base, such as sodium hydroxide, to yield the alkyltin carboxylate or alkyltin mercaptide heat stabilizer. Alternatively, the alkyltin chloride can react with the base to yield the alkyltin oxide, which may or may not be isolated, for subsequent condensation with the selected carboxylic acid or mercaptan.

$$R_2SnCl_2 + 2\ R'SH + 2\ NaOH \rightarrow R_2Sn(SR')_2 + 2\ NaCl + H_2O$$

$$R_2SnCl_2 + 2\ NaOH \rightarrow (R_2SnO)_n + 2\ NaCl + H_2O$$

$$(R_2SnO)_n + 2\ R'COOH \rightarrow R_2Sn(OOCR')_2 + H_2O$$

Typically, alkyltin carboxylates are prepared from isolated alkyltin oxides because this route leads to higher efficiency and purity for these products. In many of the modern alkyltin mercaptide stabilizers, sulfide sulfur ligands are also used in combination with mercaptide ligands. Usually the sulfide groups are introduced as a metal sulfide and added along with the mercaptan and base to the alkyltin chloride intermediate. The resulting stabilizer is a mixture of products having both sulfide and mercaptide groups bound to each of the alkyltin compounds.

Costabilizers. In most cases the alkyltin stabilizers are particularly efficient heat stabilizers for PVC without the addition of costabilizers. Many of the traditional coadditives, such as antioxidants, epoxy compounds, and phosphites, used with the mixed metal stabilizer systems, afford only minimal benefits when used with the alkyltin mercaptides. Mercaptans are quite effective costabilizers for some of the alkyltin mercaptides, particularly those based on mercaptoethyl ester technology (23). Combinations of mercaptan and alkyltin mercaptide are currently the most efficient stabilizers for PVC extrusion processes. The level of tin metal in the stabilizer composition can be reduced by up to 50% while maintaining equivalent performance. Figure 2 shows the two-roll mill performance of some methyltin stabilizers in a PVC pipe formulation as a function of the tin content

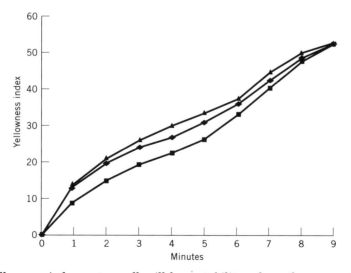

Fig. 2. Yellowness index vs two-roll mill heat stability, where the mercaptide ligands, $-SCH_2COOR$ (▲), $-SCH_2CH_2COOR'$ (♦), and $-SCH_2CH_2COOR'$ + $HSCH_2CH_2COOR'$ (■), are 17, 9, and 6% Sn, respectively. $R = C_8H_{17}$; $R' = C_{18}H_{33}$.

and the mercaptide groups at 200°C. The test formulation contains 100 parts of PVC (Fikentscher K = 65), 1.2 parts of paraffin wax, 0.6 parts of calcium stearate, and 0.4 parts of methyltin-based stabilizers.

The various lubricants formulated into PVC to improve the processing can also enhance the performance of the stabilizer. In pigmented applications, calcium soaps, eg, calcium stearate, are commonly used as internal lubricants to promote PVC fusion and reduce melt viscosity. This additive is also a powerful costabilizer for the alkyltin mercaptide stabilizers at use levels of 0.2 to 0.7 phr. Calcium stearate can significantly improve the early color and increase the long-term stability at low levels; however, as the concentration increases, significant yellowing begins to occur.

Commercial Stabilizers. A wide variety of alkyltin stabilizers have been used commercially since the 1960s because no particular compound universally satisfies every requirement of PVC processing. In general, the alkyltin mercaptides exhibit the highest overall heat stability together with imparting excellent rheological properties to the polymer. The alkyltin carboxylates, on the other hand, are unsurpassed for imparting excellent weathering properties but generally give poor rheological characteristics. Table 1 lists the commercially important alkyltin stabilizer compounds. These compounds typically are formulated with several different adjuvants, both active and inert, to tailor their performance to given commercial applications. Table 2 lists the stabilizer manufacturers and the trade names associated with each.

Economics. The pricing of stabilizers is generally based on the PVC processing application, the type of PVC used, and the other microingredients present in the formulation. In facile extrusion operations, such as the manufacture of PVC pipes, stabilizer formulations usually contain relatively low levels of the alkyltin compounds; about 5–10% tin is usual. More difficult extrusion applications, such as window lineals, extruded sheets, and house siding, require significantly higher levels of more efficient stabilizers. Injection molding of pipe fittings and accessories also requires more efficient stabilizers. Blow molding and calendered sheet applications, particularly where glass-like clarity is required, are the most de-

Table 1. Commercially Important Alkyltin Compounds

Name	CAS Registry Number	Structure
poly(dibutyltin maleate)	[32076-99-6]	$\text{--}(C_4H_9)_2SnOOCCH\!\!=\!\!CHCOO\text{--}_n$
poly(dioctyltin maleate)	[32077-00-2]	$\text{--}(C_8H_{17})_2SnOOCCH\!\!=\!\!CHCOO\text{--}_n$
dibutyltin bis(butyl maleate)	[17209-76-6]	$(C_4H_9)_2Sn(OOCCH\!\!=\!\!CHCOOC_4H_9)_2$
dimethyltin bis(2-ethylhexyl thioglycolate)	[26636-01-1]	$(CH_3)_2Sn(SCH_2COOC_8H_{17})_2$
dibutyltin bis(2-ethylhexyl thioglycolate)	[25168-24-5]	$(C_4H_9)_2Sn(SCH_2COOC_8H_{17})_2$
dioctyltin bis(2-ethylhexyl thioglycolate)	[26401-97-8]	$(C_8H_{17})_2Sn(SCH_2COOC_8H_{17})_2$
dibutyltin sulfide	[4253-22-9]	$[(C_4H_9)_2SnS]_3$
methyltin tris(2-ethylhexyl thioglycolate)	[54849-38-6]	$CH_3Sn(SCH_2COOC_8H_{17})_3$
butyltin tris(2-ethylhexyl thioglycolate)	[25852-70-4]	$C_4H_9Sn(SCH_2COOC_8H_{17})_3$
octyltin tris(2-ethylhexyl thioglycolate)	[26401-86-8]	$C_8H_{17}Sn(SCH_2COOC_8H_{17})_3$
methyltin tris(2-mercaptoethyl oleate)	[59118-79-5]	$CH_3Sn(SCH_2CH_2OCOC_{17}H_{33})_3$
methyltin (2-mercaptoethyl oleate)sulfide	[68442-12-6]	$CH_3Sn(SCH_2CH_2OCOC_{17}H_{33})(S)$

Table 2. U.S. Producers and Trade Names of Alkyltin Stabilizers and Mixed Metal Stabilizers

Producers	Alkyltin trade names	Mixed metal trade names
Akcros Chemicals Inc., Dobbs Ferry, N.Y.	Stanclear	Interstab
Atochem North America, Philadelphia, Pa.	Thermolite	
Cardinal Stabilizers, Columbia, S.C.	Cardinal Clear	
Morton International, Cincinnati, Ohio	Advastab	
Polymer Additives Group, Witco Inc., New York, N.Y.	Mark	Mark
Ferro Corp., Bedford, Ohio		Therm-Chek
M-R-S Chemicals, Inc., Maryland Heights, Mo.		MiRaStab
Synthetic Products Co., Cleveland, Ohio		Synpro/Synpron

manding processing conditions. Stabilizers for these applications are typically the highest efficiency formulations in the North American marketplace. Table 3 summarizes the typical applications of the tin stabilizers and their price range.

Health and Safety Aspects. Many of the alkyltin stabilizers are considered safe to use in almost every conceivable end use for PVC (24). Particularly, the U.S. FDA, German BGA, and Japanese JHPA have sanctioned the use of mixtures of dimethyltin and monomethyltin isooctyl thioglycolate (25), mixtures of di-*n*-octyltin and mono-*n*-octyltin isooctyl thioglycolate (26), and poly(di-*n*-octyltin ma-

Table 3. Typical Uses And Prices For Alkyltin Stabilizers

Application	Stabilizer type	Use level[a]	Tin, %	Price,[b] $/kg
	Extrusion			
PVC pipe	mercaptide/mercaptan	0.3–0.6	5–10	2.75–4.50
window profile/siding	mercaptide	1.3–1.7	14–19	7.00–10.00
siding substrate	mercaptide/mercaptan	0.7–1.2	9–14	4.00–8.50
weatherable clear	carboxylate	2.5–3.5	14–20	9.00–14.00
	Molding			
pipe fittings	mercaptide	1.3–2.0	16–19	8.00–10.00
	mercaptide/mercaptan	1.3–2.0	11–16	4.50–8.50
food bottles	methyl/octyltin mercaptides	0.8–1.4	16–19	8.00–11.00
	Calendered sheet			
food packaging film	methyl/octyl tin mercaptides	0.8–1.4	16–19	8.00–11.00
general-purpose film	mercaptides	1.0–2.0	16–19	8.00–10.00

[a]Parts per hundred parts of PVC resin.
[b]1992.

leate) as the primary heat stabilizers for PVC used for food packaging purposes. These same n-octyltin products are also approved for use in pharmaceutical applications such as pill containers and PVC tubings (27). Since the migration into water of the alkyltin mercaptide-based products from PVC is found to be extremely low, most of these stabilizers are suitable for pipes carrying drinking water according to NSF International (28), a private industry supported regulatory agency complying with all current U.S. EPA guidelines. Tin-stabilized formulations for drinking water pipes have been in use since the late 1960s throughout North America. A risk analysis on the use of these stabilizers in PVC has been compiled (29) which emphasizes the low toxicity and ecotoxicity of this class of PVC stabilizers. The key to the safety and low toxicity for the alkyltin stabilizers lies in the fact that modern manufacturing practices eliminate the production of any significant amounts of the more toxic trialkyltin species from all stabilizers.

Several studies have chosen to focus on the volatility of the alkyltin stabilizers and their by-products of PVC stabilization, alkyltin chlorides, during the calendering operation because this process presents a worst case scenerio for PVC processing: relatively high stabilizer levels, very high exposed surface area of hot PVC melt, and high processing temperatures. In two of these studies conducted by the NATEC Institute in Germany, extremely low levels of volatilized alkyltin compounds were observed (30). Similar studies conducted by Morton International confirmed these results (31). All of these studies demonstrate that the level of volatile tin compounds in the air during PVC processing operations are significantly below the TLV of 0.1 mg/m^3 for tin compounds established for the United States workplace.

Mixed Metal Stabilizers.   The second most widely used class of stabilizers, with nearly 24,000 metric tons sold in the United States in 1992, are the mixed metal combinations. These products predominate in the flexible PVC applications in the United States; however, they find competition from the lead-based products in Europe. The only noteworthy flexible PVC application where the mixed metal products do not dominate is for electrical wire and cable coatings where the lead products are preferred, although alternative mixed metal stabilizers are continually being sought to replace the leads in this application, as well. In Europe, mixed metal stabilizers are preferred for the extruded rigid building profiles because they provide good weathering and physical properties to the PVC in this use.

The commercially important alkali and alkaline-earth metals used in these stabilizer systems are based on the salts and soaps of calcium, zinc, magnesium, barium, and cadmium. These metal salts and soaps are combined to make a stabilizer system; there is synergy between these compounds during PVC processing. Because the chloride salts of both zinc and cadmium are easily formed during PVC processing and are strong Lewis acids, it is not surprising that short stability times and catastrophic degradation are observed when either zinc or cadmium soaps are used alone. In fact, zinc carboxylates, by themselves, are worthless as PVC stabilizers. When zinc and cadmium salts are combined with other compounds, which prevent or delay the formation of the respective Lewis acids, good PVC stability can be obtained. Particularly, the salts of calcium and barium serve this purpose. Other organic compounds, such as phosphites, epoxides, polyols, and β-diketones, can also be added to enhance the performance further.

The most popular commercial products are combinations including calcium–zinc, barium–calcium–zinc, barium–zinc, and barium–cadmium. Barium–cadmium combinations were, at one time, the most widely used mixtures, but as of the early 1990s their use has decreased considerably due to toxicity and ecotoxicity concerns surrounding cadmium compounds. Modern calcium–zinc, barium–zinc, and barium–calcium–zinc mixtures are touted as effective replacements for many of these barium–cadmium formulations. The safety of these newer products are unquestioned and certain calcium–zinc mixtures are widely used to stabilize PVC food packaging, mineral water bottles, and pharmaceutical containers throughout the world. In many applications, particularly in plasticized PVC, the mixed metal products effectively offer the right combination of processibility, heat and light stability, low odor, and nonsulfur staining characteristics to be the best choice of stabilizer.

Stabilization Mechanism. Zinc and cadmium salts react with defect sites on PVC to displace the labile chloride atoms (32). This reaction ultimately leads to the formation of the respective chloride salts which can be very damaging to the polymer. The role of the calcium and/or barium carboxylate is to react with the newly formed zinc–chlorine or cadmium–chlorine bonds by exchanging ligands (33). In effect, this regenerates the active zinc or cadmium stabilizer and delays the formation of significant concentrations of strong Lewis acids.

Reaction with defect site

$$\text{Zn(OOCR)}_2 + 2\,\text{CH}=\text{CH}-\underset{\underset{\text{Cl}}{|}}{\text{CH}} \longrightarrow \text{ZnCl}_2 + 2\,\text{CH}-\underset{\underset{\text{OOCR}}{|}}{\text{CH}}=\text{CH}\sim\sim$$

Regeneration of stabilizer

$$\text{ZnCl}_2 + \text{Ca(OOCR')}_2 \longrightarrow \text{Zn(OOCR')}_2 + \text{CaCl}_2$$

The chloride salts of calcium and barium are weak Lewis acids and do not tend to promote PVC degradation. By carefully choosing the ratio of zinc or cadmium salt to calcium and/or barium soap, the overall stabilizing effects can be tuned to an optimum level for a given application or process. The typical mixed metal products usually contain between a 2:1 to 1:2 ratio of the metal salts. Ultimately, though, these delaying tactics are spent and the zinc or cadmium chlorides form resulting in rapid hydrogen chloride evolution and cross-linking reactions leading to a brittle, black, crumbling product of no value. Although it is not completely understood, the maximum level of stability for these mixtures is reached at a concentration of stabilizer of about 4–5% of the polymer. Adding higher amounts of stabilizer has little effect on the overall stability of the PVC; this same phenomenon is also observed for the alkyltin stabilizers.

Mixed Metal Stabilizer Synthesis. The mixed metal salts and soaps are generally prepared by reaction of commercially available metal oxides or hydroxides with the desired C_8–C_{18} carboxylic acids. The liquid stabilizer products sometimes employ metal alkylphenates and over-based metal alkylphenates, particularly calcium or barium alkylphenates, in place of the metal carboxylates. The desired ratio of metal salts can be achieved by coprecipitation from the appropriate ratio of metal oxides or, more often, from isolated metal salts by blending to give the correct ratios. During the blending process, a variety of other coadditives or secondary stabilizers, such as phosphites, polyols, epoxides, β-diketones, or antioxidants, can be added to complete the stabilizer package. Modern stabilizers are provided as either liquids or nondusting powders which are easily handled in totally automated compounding operations.

Commercial Stabilizers. There is a great variety of commercial formulations utilizing the mixture of the alkali and alkaline-earth metal salts and soaps. In many cases, products are custom formulated to meet the needs of a particular application or customer. The acidic ligands used in these products vary widely and have dramatic effects on the physical properties of the PVC formulations. The choice of ligands can affect the heat stability, rheology, lubricity, plate-out tendency, clarity, heat sealability, and electrical and mechanical properties of the final products. No single representative formulation can cover the variety of PVC applications where these stabilizers are used.

Typically, solid stabilizers utilize natural saturated fatty acid ligands with chain lengths of C_8–C_{18}. Zinc stearate [557-05-1], zinc neodecanoate [27253-29-8], calcium stearate [1592-23-0], barium stearate [6865-35-6], and cadmium laurate [2605-44-9] are some examples. To complete the package, the solid products also contain other solid additives such as polyols, antioxidants, and lubricants. Liquid stabilizers can make use of metal soaps of oleic acid, tall oil acids, 2-ethylhexanoic acid, octylphenol, and nonylphenol. Barium bis(nonylphenate) [41157-58-8], zinc 2-ethylhexanoate [136-53-8], cadmium 2-ethylhexanoate [2420-98-6], and overbased barium tallate [68855-79-8] are normally used in the liquid formulations along with solubilizers such as plasticizers, phosphites, and/or epoxidized oils. The majority of the liquid barium–cadmium formulations rely on barium nonylphenate as the source of that metal. There are even some mixed metal stabilizers supplied as pastes. The U.S. FDA approved calcium–zinc stabilizers are good examples because they contain a mixture of calcium stearate and zinc stearate suspended in epoxidized soya oil. Table 4 shows examples of typical mixed metal stabilizers.

Costabilizers. The variety of known costabilizers for the mixed metal stabilizers is a very long listing. There are, however, a relatively small number of commercially used costabilizers. Some of these additives can also be added by the PVC compounder or processor in addition to the stabilizer package to further enhance the desired performance characteristics. The epoxy compounds and phe-

Table 4. Formulations of Mixed Metal Stabilizers, %

Liquid formula	Solid formula	Paste formula
barium tallate overbase, 30	barium stearate, 25	zinc stearate, 15
barium bis(nonylphenate), 20	cadmium laurate, 50	calcium stearate, 15
zinc 2-ethylhexanoate, 15	bisphenol A, 5	tris(nonylphenyl) phosphite, 30
diphenyl decylphosphite, 30	pentaerythritol, 20	
dibenzoylmethane, 5		epoxidized soya oil, 40

nolic antioxidants are among the most commonly used costabilizers with the mixed metal stabilizers.

Epoxy Compounds.   Epoxidized soya oil (ESO) is the most widely used epoxy-type additive and is found in most mixed metal stabilized PVC formulations at 1.0–3.0 phr due to its versatility and cost effectiveness. Other useful epoxy compounds are epoxidized glycerol monooleate, epoxidized linseed oil, and alkyl esters of epoxidized tall oil fatty acid.

Antioxidants.   Phenolic antioxidants, added at about 0.1–0.5 phr, are usually chosen from among butylated hydroxytoluene [*128-37-0*] (BHT), and *p*-nonylphenol [*104-40-5*] for liquid stabilizer formulations and bisphenol A [*80-05-7*] (2,2-bis-(*p*-hydroxyphenyl)propane) for the solid systems. Low melting thioesters, dilauryl thiodipropionate [*123-28-4*] (DLTDP) or distearyl thiodipropionate [*693-36-7*] (DSTDP) are commonly added along with the phenolics to enhance their antioxidant performance. Usually a 3:1 ratio of thiodipropionate to phenolic antioxidant provides the desired protection. Most mixed metal stabilizer products contain the antioxidant ingredient.

Polyols.   Polyols, such as pentaerythritol [*115-77-5*], dipentaerythritol [*126-58-9*], and sorbitol [*50-70-4*], most likely chelate the active metal centers to reduce their activity toward the undesired dehydrochlorination reaction. These additives are generally included in the stabilizer formulation, used in the range of 0.2 to 0.7 phr.

Phosphites.   Tertiary phosphites are also commonly used and are particularly effective in most mixed metal stabilizers at a use level of 0.25–1.0 phr. They can take part in a number of different reactions during PVC processing: they can react with HCl, displace activated chlorine atoms on the polymer, provide antioxidant functionality, and coordinate with the metals to alter the Lewis acidity of the chloride salts. Typical examples of phosphites are triphenyl phosphite [*101-02-0*], diphenyl decyl phosphite [*3287-06-7*], tridecyl phosphite [*2929-86-4*], and polyphosphites made by reaction of PCl_3 with polyols and capping alcohols. The phosphites are often included in commercial stabilizer packages.

β-*Diketones.*   A new class of costabilizer has emerged that is effective with the mixed metal systems. The β-diketones can significantly enhance the performance of the calcium–zinc and barium–calcium–zinc systems when used at 0.1 to about 0.7 phr. Although relatively expensive, the β-diketones greatly improve early color stability and rheological performance while benefitting the weatherability of the final PVC articles. Typical of these additives are dibenzoylmethane [*2929-86-4*] and stearoyl benzoyl methane [*58446-52-9*]. These additives are generally formulated as part of the mixed metal stabilizer package.

Specialty Amines.   Some substituted nitrogenous compounds can provide similar benefits. Esters of 2-aminocrotonate and bis-2-aminocrotonate, and appropriately substituted dihydropyridines, eg, 3,5-bis-lauryloxycarboxy-2,6-dimethyl-1,4-dihydropyridine [*37044-66-7*] and 3,5-bis-ethoxycarboxy-2,6-dimethyl-1,4-dihydropyridine [*1149-23-1*], are examples of these costabilizers. These relatively expensive costabilizers are used at 0.1–0.7 phr and are particularly effective when added to the calcium–zinc stabilizers.

Hydrotalcite.   Synthetic hydrotalcite minerals are gaining commercial acceptance for their ability to costabilize PVC in the presence of other primary stabilizers (see Table 2). The performance of the mixed metal stabilizers are particularly boosted when an equal part level, about 2–3 phr, of hydrotalcite is added to the PVC formulation. These minerals function by trapping HCl within the layered lattice arrangement of atoms. The formula, $Mg_4Al_2(OH)_{12}CO_3 \cdot 3H_2O$, is commonly written; however, these minerals are generally nonstoichiometric by nature and can include some amounts of alternative elements in their compositions. They function similarly to the zeolites but exist in layered structures and have a different trapping mechanism. In addition to their performance enhancement, the hydrotalcite minerals are compatible with PVC and can be used effectively in clear PVC applications as well as the pigmented formulations.

Economics. As with the alkyltin stabilizers, the market pricing of the mixed metal stabilizers tend to be directed by the particular application. The calcium–zinc and barium–cadmium packages are typically used at 2.0–4.0 parts per hundred of PVC resin (phr) in the formulation. These completely formulated products are sold for $2.50–$4.40/kg for the liquid products and $3.20–$6.50/kg for the solids and pastes. The higher efficiency products aimed at rigid applications tend toward the higher end of the cost range.

The basic metal salts and soaps tend to be less costly than the alkyltin stabilizers; for example, in the United States, the market price in 1993 for calcium stearate was about $1.30–$1.60, zinc stearate was $1.70–$2.00, and barium stearate was $2.40–$2.80/kg. Not all of the coadditives are necessary in every PVC compound. Typically, commercial mixed metal stabilizers contain most of the necessary coadditives and usually an epoxy compound and a phosphite are the only additional products that may be added by the processor. The required costabilizers, however, significantly add to the stabilization costs. Typical phosphites, used in most flexible PVC formulations, are sold for $4.00–$7.50/kg. Typical antioxidants are bisphenol A, selling at $2.00/kg; *p*-nonylphenol at $1.25/kg; and BHT at $3.50/kg, respectively. Pricing for ESO is about $2.00–$2.50/kg. Polyols, such as pentaerythritol, used with the barium–cadmium systems, sells at $2.00, whereas the derivative dipentaerythritol costs over three times as much. The β-diketones and specialized dihydropyridines, which are powerful costabilizers for calcium–zinc and barium–zinc systems, are very costly. These additives are $10.00 and $20.00/kg, respectively, contributing significantly to the overall stabilizer costs. Hydrotalcites are sold for about $5.00–$7.00/kg.

Health and Safety Aspects. Overall, the mixed metal stabilizer industry is undergoing significant change during the early 1990s due to the increasing restrictions on cadmium compounds. Most of the research effort has focused on new products to replace the traditional barium–cadmium formulations with technical and cost-effective products. In some regions, cadmium is allowed only in applications where there are no effective replacement technologies. The replacement products generally contain salts of barium, zinc, calcium, and/or potassium; all these compositions are considered safe in the many flexible PVC end uses.

Calcium–zinc soaps are used in many PVC food container applications because these heat stabilizers are universally accepted as safe by the U.S. FDA, German BGA, Japanese JHPA, and other government regulatory groups.

Lead Stabilizers. In use since the 1940s, the lead-based stabilizers have played an extremely important role in the development of PVC as a high performance polymer. The myriad of toxicological and ecotoxicological problems surrounding the use of any lead chemicals has restricted lead stabilizers to uses in flexible

PVC wire and cable coatings in the United States with consumption in 1992 estimated at 15,000 metric tons. In Europe and Asia, the lead stabilizers predominate for wire and cable uses and are also widely used to stabilize PVC pipe and weatherable building profiles. These are solid products and are supplied as powders, flakes, or strands, usually in special packaging to control dusting.

The commonly used commercial lead-based PVC stabilizers rely on one or more lead(II) oxide groups bound to the primary bivalent lead salt. These overbased lead compounds have higher levels of lead and are more basic, thus reacting more readily with evolved hydrogen chloride during PVC processing. It is typical to find combinations of mono-, di-, and tribasic lead compounds as the primary heat stabilizers because these often work together to provide a good balance of both early and long-term stability. The choice of anion also effects the performance by reducing the reactivity of the lead oxide toward other ingredients in the formulation as well as the polymer. The stabilization by-product, lead dichloride, is nearly inert, ie, it is white in color, nonionic, insoluble in water, and has low Lewis acidity. It is these properties that give lead-stabilized PVC such a low conductivity in electrical applications. Commercially, most lead stabilizers are combinations containing lead stearates which also provide good lubrication to PVC compounds. Additional lubricants often are not necessary for thermally processing these PVC formulations.

Their high toxicity has greatly limited the applications for lead stabilizers in North America and is now spreading around the world. Environmental agencies such as The World Health Organization (WHO) and the U.S. EPA are continually lowering recommended human exposures to lead compounds. Another limitation is that lead products have a high refractive index and as a result can only be used in opaque applications. Overbased lead salts have a high degree of reactivity and tend to interact, many times disfavorably, with other ingredients in the formulation. Also, they have a very high specific gravity compared to other stabilizers, resulting in higher density PVC products. Lastly, the lead stabilizers can react with almost any source of sulfur to form black lead sulfide, the so-called lead stain phenomenon. Despite these drawbacks, they remain highly effective PVC heat stabilizers.

Stabilization Mechanism. Traditionally, lead salts were thought to perform only as acid scavengers during PVC stabilization; it is likely that this activity leads to good long-term stability. Bivalent lead compounds can readily form complexes, and recently workers have proposed that these products also displace labile chlorines on the polymer in a fashion similar to the mixed metal stabilizers. A free-radical mechanism is proposed for this displacement which improves the early color hold of the PVC (10). Because the lead chloride is such a weak Lewis acid, costabilizers used in the mixed metal systems are generally ineffective and unnecessary. Increasing stabilizer concentration in the polymer generally leads to increased stability times.

Lead Stabilizer Synthesis. Most commercial stabilizers are produced by reaction of a water slurry of lead oxide with the appropriate acid while heating, yielding a solid product with a particle size of about 1 μm. This condensation proceeds leaving the desired level of overbasing in the final product. Most often, the lead product is treated with a coating agent to reduce dusting and improve dispersability in the PVC. The stabilizer is then filtered, dried, and packaged.

During the drying and coating step, other coadditives such as pigments, lubricants, and fillers can be blended into the mixture to make a total package formulation.

Commercial Stabilizers.　There are six lead salts and soaps that typically are used in the commercial PVC stabilizers. The lead stearate soaps are often combined with the lead salts to provide lubrication and added stabilizer activity. The key to the high activity of these stabilizers is the very high lead content. Table 5 describes six commonly used lead stabilizers.

By far the most common lead salt used for PVC stabilization is tribasic lead sulfate. It can be found either alone or combined with another lead salt in almost every lead-stabilized PVC formulation. Many of the combinations are actually coprecipitated hybrid products, ie, basic lead sulfophthalates. Dibasic lead stearate and lead stearate are generally used as costabilizers combined with other primary lead salts, particularly in rigid PVC formulations where they contribute lubrication properties; dibasic lead stearate provides internal lubrication and lead stearate is a good external lubricant. Basic lead carbonate is slowly being replaced by tribasic lead sulfate in most applications due the relatively low heat stability of the carbonate salt which releases CO_2 at about 180°C during PVC processing.

Flexible Applications.　The mainstay of the lead stabilizers in the United States is in flexible wire and cable coating applications. The nonconductive nature of lead stabilizers is unsurpassed by other classes of stabilizers. Rather high levels of stabilizers are necessary for these uses because of the required heat aging specifications for most insulating materials. Typically 5–8 phr of lead stabilizer is needed in most insulation compounds. Careful consideration must be given to the choice of lead stabilizer due to the high reactivity of the basic lead oxide groups with other ingredients in the formulation. This is particularly true of the plasticizer choice. The more demanding, high temperature applications require stabilizers rich in dibasic lead phthalate to provide high levels of heat aging stability; the less demanding applications, jacketing and low temperature insulation, usually rely on tribasic lead sulfate as the primary stabilizer. Under high temperature aging conditions, tribasic lead sulfate tends to react with the ester-type plasticizers to increase volatility and reduce the resilience of the PVC insulation.

Rigid Applications.　The use of the lead stabilizers is very limited in the United States; but, they are still used in several rigid PVC applications in Europe and Asia. The highest use of lead stabilizers in rigid PVC is for pipe and conduit applications. Tribasic lead sulfate is the primary heat stabilizer with lead stea-

Table 5. Principal Lead Stabilizers

Stabilizer	CAS Registry Number	Formula	PbO, %	Specific gravity
tribasic lead sulfate	[12202-17-4]	$PbSO_4 \cdot 3PbO \cdot H_2O$	89	6.9
dibasic lead phosphite	[12141-20-7]	$PbHPO_3 \cdot 2PbO \cdot \frac{1}{2}H_2O$	90	6.1
dibasic lead phthalate	[17976-43-1]	$C_4H_4(COO)_2Pb \cdot 2PbO$	80	4.2
basic lead carbonate	[1319-46-6]	$2PbCO_3 \cdot Pb(OH)_2$	87	6.7
dibasic lead stearate	[56189-09-4]	$Pb(OOCC_{17}H_{35})_2 \cdot 2PbO$	55	2.0
lead stearate	[1092-35-7]	$Pb(OOCC_{17}H_{35})_2$	29	1.4

rates included to provide lubrication. The lead products are typically fully formulated, usually including lubricants and pigments for pipe extrusion applications. These lead one-packs, when used at about 1.8–2.5 phr, provide all of the stabilizer and lubrication needed to process the polymer. A lead one-pack contains tribasic lead sulfate, dibasic lead stearate, calcium stearate, polyethylene wax, paraffin wax, ester wax, and pigments.

Dibasic lead phosphite is used in rigid building profile applications because the PVC weathering properties are found to be very good. Normally, the titanium dioxide pigment loading is increased to about 5 phr in these formulations. In less critical uses, ie, interior profiles, tribasic lead sulfate remains the standard. There are three U.S. producers of note of lead stabilizers: Anzon (Philadelphia), a division of Synthetic Products Co. (Cleveland), with trade names Tribase, Dythal Dyphos, Lectro, and Leadstar; Eagle-Picher Ind. (Cincinnati) markets Epistatic; and Hammond Lead Co. (Hammond, Indiana) sells Halstab, Halbase, Halphal, and Halphos.

Economics. The lead-based stabilizers tend to be priced relatively low, with 1992 prices ranging from $1.40 to $3.60/kg in the United States. The lead phthalates tend toward the higher end of this range, whereas the pipe one-pack products fall into the low end.

Health and Safety Aspects. Worldwide, there is continuing pressure by environmental and human toxicologists to reduce the use of heavy metals such as lead in every application. In the United States, the EPA and Occupational Safety and Health Administration (OSHA) provide regulations over the producers and users of lead stabilizers. The permissible exposure level (PEL) of workers is regulated at 50 $\mu g/m^3$ of airborne lead per 8-h workday. Further, workers are not allowed to be exposed to greater than 30 $\mu g/m^3$ for more than 30 days per year (34). The state of California increased the safety factor further in 1990 by regulating exposure to less than 5 $\mu g/m^3$ for any exposure. Warning labels describing the toxic effects of lead compounds must be applied to all lead stabilizer packaging. In some states, ie, New Jersey and California, any product containing more than 0.1% lead must be labeled as containing lead.

Lead stabilizers have not been used to manufacture drinking water pipes in the United States since 1970 due to the migration levels of lead found in the water from the stabilizers. In 1991 the U.S. EPA further reduced the allowable level of lead in drinking water to zero (action level of only 15 parts per billion (ppb) lead). In 1993, The World Health Organization provided new guidelines targeting the level of lead in drinking water to less than 10 ppb. All uses of lead in PVC pipes are under scrutiny and the stabilizer industry is responding with nontoxic tin-based and calcium–zinc-based technologies. Significant efforts are underway to reduce the uses of lead stabilizers in flexible wire and cable applications as well. New mixed metal formulations are now being touted as providing the needed high levels of both heat stability and nonconducting electrical properties.

Antimony Mercaptide Stabilizers. In the mid-1950s antimony mercaptides were first proposed as PVC heat stabilizers (35). Antimony tris(laurylmercaptide) [6939-83-9] and antimony tris(isooctyl thioglycolate) [27288-44-4] are typical of this class of heat stabilizers. These compounds were used mainly in rigid PVC applications, particularly pipes, competing with the alkyltin mercaptides. Their use has greatly diminished during the late 1980s and early 1990s for a variety of reasons. Particularly, questions raising doubts about the toxicological safety of antimony compounds have arisen. The performance of the antimony products has also reduced their uses in many processes. For example, they are less compatible with PVC which leads to a cloudy appearance in clear applications; they react with sulfur sources to form Sb_2S_3, an orange-colored by-product; and they detract from the weatherability of PVC formulations. Further, because the cost of tin metal decreased significantly during the 1980s, antimony pricing has continued to rise, making these products less economically attractive than they once had been.

Commercial Stabilizers. The performance of the antimony stabilizers is significantly enhanced by adding polyhydroxybenzene compounds, eg, catechol, to the PVC (36). In commercial practice, about 5–10% catechol is formulated with the antimony mercaptide stabilizer products. The antimony mercaptides are normally prepared by heating antimony oxide with the appropriate mercaptan, normally isooctyl thioglycolate, under conditions to remove water.

$$Sb_2O_3 + 6\ HSCH_2COOR' \rightarrow 2\ Sb(SCH_2COOR')_3 + 3\ H_2O$$

The liquid antimony tris(isooctyl thioglycolate) is then treated with the required additives to prepare the commercial products.

Current manufacturers of these products are the Polymer Additives Group of Witco Corp. (New York), with the trade name Mark, and Synthetic Products Co. (Cleveland), with the trade name Synpron. The antimony-based stabilizers are typically used for rigid PVC extrusion applications at about 0.4 to about 0.8 phr, priced at about \$3.80–\$4.50/kg.

Health and Safety Aspects. The U.S. EPA has significantly reduced the allowed levels of antimony compounds in drinking water causing a toxicity cloud over the viability of this class of stabilizers. Presently, antimony products are no longer allowed for use as potable water pipe stabilizers pending completion of NSF International's review (28). For these reasons, the future of this stabilizer technology appears limited.

BIBLIOGRAPHY

"Heat Stabilizers" in *ECT* 3rd ed., Vol. 12, pp. 225–249, by L. I. Nass, Consultant.

1. G. Ayrey, R. C. Poller, and I. H. Siddiqui, *J. Polym. Sci.*, **B-8**, 1 (1970).
2. D. Braun, in G. Geuskens, ed., *Degradation And Stabilization Of Polymers*, John Wiley & Sons, Inc., New York, 1975, pp. 23–41.
3. W. H. Starnes, Jr., *Am. Chem. Soc., Div. Polym. Chem. Polym. Prepr.* **18**, 493 (1977).
4. A. Guyot, M. Bert, P. Burille, M. F. Llauro, A. Michel, *J. Pure Appl. Chem.* **53**, 401 (1981).
5. G. Ayrey, B. C. Head, and R. C. Poller, *J. Polym. Sci. Macromol. Rev.* **8**, 1 (1974).
6. M. J. R. Cantow, C. W. Cline, C. A. Heiberger, D. Th. A. Huibers, and R. Phillips, *Mod. Plast.* **46**(6), 126 (1969).
7. U.S. Pat. 4,963,608 (Oct. 16, 1990), M. Kunieda and H. Takida (to Kyowa Kagaku Kogin Kabushiki Kaisha).
8. A. H. Frye, R. W. Horst, and M. A. Paliobagis, *J. Polm. Sci.*, A-2, 1765, 1785, 1801 (1964).
9. A. H. Frye and R. W. Horst, *J. Polm. Sci.* **40**, 419 (1959); **45**, 1 (1960).

10. E. W. Michell, *J. Vinyl Technol.* **8**, 55 (1986).
11. R. C. Weast, ed., *CRC Handbook of Chemistry and Physics*, 72nd ed., CRC Press, Boca Raton, Fla., 1991.
12. E. D. Owen, in E. D. Owen, ed., *Degradation And Stabilization Of PVC*, Elsevier, London, 1984, pp. 223–236.
13. A. Michel, T. V. Hoang, and A. Guyot, *J. Macromol. Sci., A* **12**, 411 (1978).
14. D. Dresedow and C. F. Gibbs, *Nat. Bur. Stand. Circ.* **525**, 69 (1953).
15. D. E. Winkler, *J. Polym. Sci.* **35**, 3 (1959).
16. W. H. Starnes and co-workers, *Polym. Prep., Am. Chem. Soc., Div. Polym. Chem.* **19**, 623 (1978).
17. H. O. Wirth and H. Andreas, *Pure Appl. Chem.* **49**, 627 (1977).
18. L. S. Troitskaya and B. B. Troitski, *Plast. Massy.*, 12 (1968).
19. E. Parker, *Kunstoffe* **47**, 443 (1957).
20. G. Ayrey and R. C. Poller, in G. Scott, ed., *Developments In Polymer Stabilization-2*, Applied Science, London, 1980, p. 1.
21. U.S. Pat. 3,857,868 (Dec. 31, 1974), R. C. Witman and T. G. Kugele; U.S. Pat. 3,862,198 (Jan. 21, 1975), T. G. Kugele and D. H. Parker (to Cincinnati Milacron, Inc.).
22. R. E. Hutton, J. W. Burley, and V. Oakes, *J. Organometal. Chem.* **156**, 369 (1978).
23. U.S. Pat. 4,701,486 (Oct. 20, 1987), R. E. Bresser and K. R. Wursthorn (to Morton International, Inc.).
24. K. Figge, *Pack. Technol. Sci.* **3**, 27, 41 (1990).
25. *U.S. Code Of Federal Regulations, Title 21: Food and Drugs*, 21 CFR 178.2010, Washington, D.C., 1992
26. *Ibid.*, 21 CFR 178.2650.
27. *Ibid.*, 21 CFR 314.420.
28. *Drinking Water System Components—Health Effects*, ANSI/NSF Standard 61, NSF International, Ann Arbor, Mich., 1988.
29. K. A. Mesch and T. G. Kugele, *J. Vinyl Tech.* **14**, 131 (1992).
30. D. Van Battum, *Report 06236/76 and 07822A.75*, CIVO, Zeist, the Netherlands, 1976; A.-M. Dommröse, *Report 88 9787*, Natec Institut, Hamburg, Germany, 1988.
31. T. G. Kugele, *China Plast. Rubber J.* **20** 40 (1989).
32. A. Guyot and A. Michel, in Ref. 20, p. 89.
33. P. P. Klemchuk, *Adv. Chem. Ser.* **85** (1968).
34. OSHA Standard For Occupational Exposure To Lead (29 CFR 1910-1025) Washington, D.C.; *Fed. Reg.* **43** 52952 (Nov. 14, 1978).
35. U.S. Pat. 2,680,726 (June 8, 1954), E. L. Weinberg and co-workers (to M&T Chemicals); U.S. Pat. 2,684,956 (July 27, 1954), E. L. Weinberg and co-workers (to M&T Chemicals).
36. U.S. Pat. 4,029,618 (June 14, 1977), D. Dieckmann (to Synthetic Products Co.).

General References

L. I. Nass, in L. I. Nass, ed., *Encyclopedia of PVC*, Vol. 1, Marcel Dekker, New York, 1976.
E. D. Owen, in E. D. Owen, ed., *Degradation and Stabilization of PVC*, Elsevier, London, 1984, Chapt. 5.
J. Edenbaum, ed., *Plastics Additives and Modifiers Handbook*, Van Nostrand Reinhold, New York, 1992, Sect. II.
H. Andreas, in R. Gächter and H. Müller, eds., *Plastics Additives Handbook*, Hanser, Munich, 1983, Chapt. 4.
Proceedings SPE Vinyl RETEC, New Brunswick, N.J., Sept. 29–Oct. 1, 1992, Society of Plastics Engineers, Stamford, Conn., pp. 33–82.

KEITH A. MESCH
Morton International, Inc.